THE LIBRARY
ST. MARY'S COLLEGE OF MARYLAND
ST. MARY'S CITY, MARYLAND 20686

SMITHSONIAN INSTITUTION

WASHINGTON, D.C.
1973

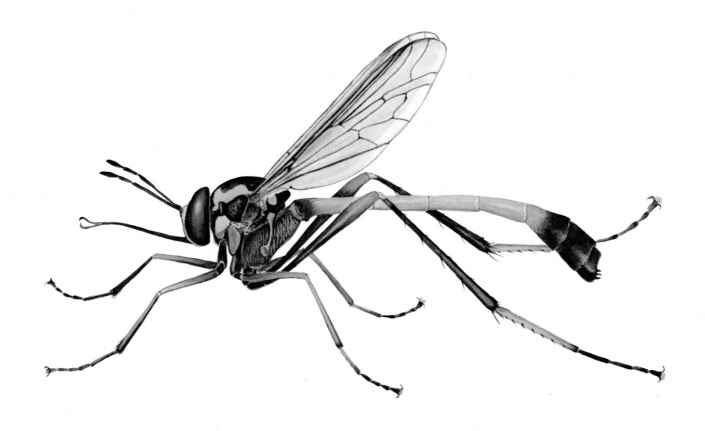

Bee Flies of the World
The Genera of the Family Bombyliidae

By FRANK M. HULL
Research Associate, Smithsonian Institution

Smithsonian Institution Press   City of Washington   1973

United States National Museum Bulletin 286
SI Press Number 4903

Library of Congress Cataloging in Publication Data
Hull, Frank Montgomery, 1901-
Bee Flies of the World.
Bibliography: p.
1. Bombyliidae. 2. Parasites—Insects. I. Title.
QL537.B75H85   595.7'71   73-1581

ISBN 0-87474-131-9

Two species of the remarkable wasp mimics of the Genus *Systropus* Wiedemann: *S. arizonensis* Banks (top) and *S. elegans* Hull. ARTIST: Arthur A. Cushman.

Contents

	Page
Dedication	ix
Acknowledgments	xi
Introduction	3
History of Bombyliidae Studies	4
Generalizations of the Host Relationships of Bee Flies During Immature Stages	6
Review of Published Information on Host Relationships	8
Ethology and Interrelationships of the Adult	33
Morphology of the Immature Stages	36
Morphology of the Adult Bee Fly	47
Terminology Used in Structures of the Male Genitalia of Bee Flies	52
Chromosomes	56
Fossil Bombyliidae	56
Zoogeography of the Bombyliidae	59
Synopsis and Distribution	61
Phylogeny and Subdivision of the Family Bombyliidae	63
Homeophthalmae	68
Bombyliinae	68
Bombyliini	69
Dischistini	69
Heterostylini	69
Cythereini	69
Conophorini	69
Corsomyzini	70
Mariobezziini	70
Acrophthalmydini	70
Paratoxophorini	70
Eclimini	70
Crocidiini	70
Phthiriinae	194
Gerontinae	202
Usiinae	212
Heterotropinae	222
Toxophorinae	231
Toxophorini	232
Lepidophorini	232
Systropinae	242
Dolichomyiini	243
Systropini	243

	Page
Xenoprosopinae	251
Platypyginae	253
Cyrtosiini	254
Psiloderini	254
Platypygini	254
Mythicomyiinae	264
Mythicomyiini	265
Empidideicini	265
Tomophthalmae	278
Cylleniinae	278
Cylleniini	279
Henicini	279
Tomomyzini	279
Peringueyimyiini	279
Neosardini	279
Lomatiinae	302
Lomatiini	302
Myonematini	303
Antoniini	303
Aphoebantini	303
Prorostomatini	303
Xeramoebini	303
Comptosiini	303
Exoprosopinae	364
Villini	364
Villoestrini	364
Exoprosopini	364
Anthracinae	435
Anthracini	435
Walkeromyini	445
Supplemental Text-Figures	447
Illustrations of Bombyliidae (figures 1–1030)	461
Bibliography	573
Index	657

Text-Figures

		Page
1.	Eggs, larvae, and pupae of *Anthrax limatulus* Say	37
2.	First-stage larvae	38
3.	Mouthparts of bee fly larvae	39
4–7.	Pupae of bee flies	40–44
8.	Armament of pupae of three subfamilies	46
9.	Mouthparts of adult bee flies of *Exoprosopa* Macquart	48
10.	Thorax of a bombyliid	49
11.	Wing of a generalized Recent bee fly	51
12–14.	Male genitalia of *Poecilanthrax californicus* Cole	53–55
15–16.	Fossil bee flies	57, 58
17.	Concentration areas of bee flies of the world	60
18.	Provisional phylogenetic chart of bee flies	65
19.	Distribution pattern of the Bombyliinae	68
20.	Habitus, *Bombylius major* Linné	77
21.	Habitus, *Germinaria canalis* Coquillett	115
22.	Habitus, *Corsomyza simplex* Wiedemann	164
23.	*Eclimus gracilis* Loew, after Austen	178
24.	Distribution pattern of Phthiriinae	194
25.	Habitus and lateral aspect of *Phthiria* species	196
26.	Distribution pattern of Gerontinae	202
27.	Habitus, *Geron senilis* Fabricius	204
28.	Distribution pattern of Usiinae	213
29.	Habitus, *Usia aenea* Rossi	214
30.	Distribution pattern of Heterotropinae and Xenoprosopinae	222
31.	Habitus, *Heterotropus aegyptiacus* Paramonov	224
32.	Distribution pattern of Toxophorinae	231
33.	Habitus, *Toxophora leucopyga* Wiedemann	233
34.	Distribution pattern of Systropinae	242
35.	Habitus, *Systropus (Cephenius) maccus* Enderlein	244
36.	Distribution pattern of Platypyginae	253
37.	Habitus, *Cyrtosia nitens* Loew	255
38.	Habitus, *Psiloderoides mansfieldi* Hesse	258
39.	Distribution pattern of Mythicomyiinae	264
40.	Habitus, *Mythicomyia monacha* Melander	266
41.	Distribution pattern of Cylleniinae	278
42.	Habitus, *Cyllenia maculata* Latreille	281

	Page
43. Distribution pattern of Lomatiinae	302
44. Habitus, *Lomatia lateralis* Meigen	308
45. Distribution pattern of Exoprosopinae	364
Seven maps showing distribution patterns of species of *Poecilanthrax*	393
46. Habitus, *Exoprosopa rhea* Osten Sacken	410
47. Distribution pattern of Anthracinae	432
48. Habitus, *Anthrax tigrinus* deGeer	434

Supplemental Text-Figures

1–7.	Bee fly pupae	449
8–15.	Heads of pupae	450
16–20.	Bee fly larval head structure	451
21.	Female genital structures	452
22.	Wings of fossil bee flies	453
23.	Map of collections in 1962	454
24.	Habitus, *Zinnomyia karooensis*, after Hesse	455
25.	Habitus, *Dischistus pulchellus* Austen, from Austen	455
26.	Habitus, *Cytherea barbara* Sack, from Austen	456
27.	Habitus, *Callastoma fascipennis* Macquart, from Austen	456
28.	Habitus, *Legnotomyia trichorhoea* Loew, from Austen	457
29.	Habitus, *Heterotropus nigrithorax*, new species, after François	457
30.	Habitus, *Caenotus minutus* Cole, from Cole	458
31.	Habitus, *Dolichomyia gracilis* Williston, from Cole	458
32.	Habitus, *Amictus minor* Austen, from Austen	459
33.	Habitus, *Lomatia lepida* Austen, from Austen	459

Dedication

To my late and beloved wife, Mary Marguerite Chappell Hull, I owe an immeasurable debt. She gave of her time patiently and untiringly to the typing and checking of thousands of pages of manuscript in its several revisions. To the compilation and checking of nearly three thousand bibliographic references she gave diligent and painstaking attention. At the time of her death she had also compiled a bibliography of the Syrphidae, comprising perhaps two thousand titles. Not the least of my debt to her lies in her encouragement and her faith in the worth and completion of these two world monographs on important groups of insects. Neither this present work, nor the earlier work, *Robber Flies of the World*, would have been possible without her devotion and help. All of these tasks were cheerfully undertaken in addition to caring for a family and a home. Her noble and saintly life is done. In God's home she rests.

Acknowledgments

This work was begun in 1952 and was based upon a study of the extensive collections of the United States National Museum (now the National Museum of Natural History), the American Museum of Natural History, the Museum of Comparative Zoology at Harvard University, and the British Museum (Natural History) in London. The National Museum of Natural History is rich in Bombyliidae. It includes the extensive collections made by D. W. Coquillett in the western United States and much other material added in recent years. This also includes the acquisition of the Melander Collection. The British Museum (Natural History) afforded material drawn from many quarters of the World. The collections in England were visited in 1952 and again in 1953 and 1954.

The National Science Foundation gave three grants which made possible the study of museum collections in the United States, England, Europe, and Australia. To this Foundation I should like to express appreciation both for the support of this monograph and for the helpfulness of its officials. The University of Mississippi provided a temporary Research Professorship which permitted travel time and absence from teaching. I wish especially to thank the late Honorable Walter Sillers of Rosedale, Mississippi, and the Honorable Robert D. Morrow of Brandon, Mississippi, for encouraging state support for this work. I am deeply indebted to Dr. M. B. Huneycutt, Chairman of the Department of Biology, University of Mississippi, for continuous help, support, and encouragement through the long task. Dr. Lewis Nobles, recent Dean of the University of Mississippi Graduate School, and his successor, Dr. Joseph Sam, have been most helpful.

I wish to express my indebtedness particularly to the many persons in numerous countries who have generously helped me in this work on the bee flies. The staff members at the United States National Museum, both of the Department of Agriculture and of the National Museum staffs, were most helpful. I want to thank Dr. Willis Wirth and Dr. Curtis Sabrosky for their interest, suggestions, and guidance concerning problems of nomenclature. Mr. George Steyskal was most helpful in recommending preferred nomenclature for the morphology of the male genitalia and for his remarkable knowledge of appropriate gender in specific names. I shall always have a deep feeling of gratitude toward the present Chairman of the Department of Entomology, Dr. Karl V. Krombein, and his predecessor, Dr. J. F. Gates Clarke. I wish to thank the staff members of the Smithsonian Institution Press for their patience and skill in getting the manuscript and illustrations into print. My thanks also to Dr. L. V. Knutson.

Many other persons have been most helpful. I am deeply indebted to Dr. A. J. Hesse of the South African Museum in Capetown for furnishing material which, owing to the illness of my late wife, I would not have been able to see except by travel and visitation. The late Dr. Reginald H. Painter of Kansas State University lent several genera I had not found elsewhere. I wish to thank Dr. Frank Morton Carpenter of the Biological Laboratories, Harvard University, Dr. J. Bequaert, also others of the Museum of Comparative Zoology; also the late Dr. Howard Curran of the American Museum of Natural History; Dr. Edward Ross and P. H. Arnaud of the California Academy of Sciences; Dr. Frank Cole, Dean E. Gorton Linsley, Dr. P. H. Hurd, and Mr. A. T. McClay of the University of California. I especially appreciate the helpfulness of Mr. Jack Hall of the Riverside branch of this University, also Carl Rettenmeyer and F. L. Truxen.

I wish to thank Dr. W. G. Downs, then of the Rockefeller Institute staff in Mexico City, for having been instrumental in 1951 in interesting me in both Bombyliidae and Asilidae during my visit to Mexico City and to Cuernavaca. Professor Candide Bolivar of the Instituto Politecnico was most helpful then and on later visits. Professor H. F. Strohecker, University of Miami, sent a few bee flies from Florida. Dr. Z. Kazab, Director General of the Hungarian Natural History Museum, very kindly sent some bee flies from Mongolian expeditions; no new genera were found.

I owe a debt of gratitude to the authorities at the British Museum (Natural History), especially to Mr. N. Riley, Dr. W. E. China, Mr. Harold Oldroyd, the late Mr. R. L. Coe, Mr. Clark, and the late Dr. Fritz van Emden. All of these gentlemen were extremely helpful. The late Professor J. Timon-David of Marseille, Dr. Willi Hennig and Dr. Erwin Lindner, and the late Professor Fred Keiser have all been most helpful. I acknowledge also help from Mr. B. R. Stuckenberg of the Natal Museum and of Mr. John Bowden of the University of Ibadan, Nigeria.

To the many in Australia who gave generously of time and helped in numerous ways, I owe an unforgettable debt. These people include Dr. A. J. Nicholson, Director (when I was there) of the Entomology Division of the CSIRO, and his efficient staff as follows: the late Dr. S. J. Paramonov, Mr. Edgar Riek, Mr. K. R. Norris, Mr. Tom Campbell, Dr. M. J. D. White, Mr. Frank Gay, Mr. Paul Wilkerson, Mr. M. M. H. Wallace, Mr. John Callaby, Mr. Keith Taylor, and Mr. Don Wilson.

Others in Australia who gave assistance of great value were Mr. R. T. M. Pescott and Mr. Alexander Burns of the National Museum in Melbourne; Dr. and Mrs. (Dr.) I. M. Mackerras of the Queensland Institute of Medical Research; Professor F. A. Perkins and Dr. Elizabeth Marks of the University of Queensland; Mr. George Mack of the Queensland Museum; Mr. Anthony Musgrave of the Australian Museum in Sydney; Dr. P. D. F. Murray, Dr. A. R. Woodhill, and Mr. J. R. Henry of the University of Sydney; Mr. D. J. Lee of Public Health and Tropical Medicine at the University in Sydney; the late Mr. H. Womersley of the South Australian Museum; Mr. Harry Lower of the Waite Agricultural Research Institute, Adelaide; Mr. G. H. Hardy of Katoomba, New South Wales; Mr. Athol Douglas of the Perth Museum; and Mr. E. F. H. Jenkins, the Government Entomologist at Perth.

The fine and valuable drawings prepared by Mr. Arthur Smith deserve special mention. He is responsible for most of the faces of the bee flies—identified by the initials A.S. His exceptional, fine drawings have added greatly to the project. Many of the remaining drawings and all of the drawings of the genitalia were done by Mr. Kenneth Weisman—bearing the initials K.E.W. I am greatly indebted to him for his fine and skillful work. Mr. Arthur Cushman prepared the frontispiece and the text-figures identified with initials A.D.C. Most other drawings were done by the author, a few were made by my graduate students, Clyde Sartor and William Martin.

I wish to thank the many librarians at all institutions who have used their special talents to help locate obscure references. The staffs at the Smithsonian Institution Library and at the University of Mississippi Library have been particularly helpful. I wish to especially thank Miss Mahala Saville. I owe a deep debt of gratitude to my principal typist, Mrs. Emily H. Peacock. My daughter, Cecil Hull, also was most helpful in typing manuscript. The index has been prepared by my youngest son, Clovis Hull.

Among others who helped with the numerous minor tasks, assisted with manuscript and illustrations, I wish to thank my eldest son, Frank M. Hull and his wife; the late Professor Ralph B. Brundrett, Jr., for checking translations of German descriptions; and Rex Paul, a diligent graduate student in my classes.

I especially wish to express my appreciation to my former graduate student, William C. Martin, a skilled insect illustrator and photographer, for his assistance in preparing several drawings and charts, and also figures 1-9, 15, 16. Maps were taken from the thesis of my graduate student, Miss Dee Sellers.

It is with great happiness that I thank my dear present wife, Mrs. Laura Felton Hull, for her help in the heavy task of reading manuscript and of bookkeeping of the numerous descriptions and illustrations; also, I add my thanks to her for her patience and skill.

Since the beginning of these monographs in 1952, the following great dipterists who collaborated in these works are now deceased: Dr. Reginald H. Painter, Dr. Charles Howard Curran, Dr. S. J. Paramonov, Dr. Stanley Bromley, Dr. Fritz van Emden, Dr. Richard Frey, Prof. J. Timon David, Prof. Fred Keiser, Mr. Hassan C. Efflatoun, Mr. R. L. Coe, Mr. H. Womersley. Possibly there are others. I also acknowledge with deep gratitude the assistance and encouragement of eight great dipterists and entomologists of former years, now deceased: Dr. Raymond C. Osburn, Dr. J. M. Aldrich, Raymond Shannon, Prof. James Stewart Hine, Dr. P. P. Calvert, Prof. Robey Wentworth Harned, Mario Bezzi, and A. L. Melander.

The author wishes it had been possible to incorporate all of the suggestions made by friends who read the final manuscript. Unfortunately, this has not been possible, because the three generous grants given for preparation were exhausted; also because of my regular teaching duties and of several years of illness in my family. What changes were possible have been made. I make these statements lest I be deemed unappreciative. I bespeak the tolerance of my readers in my efforts to produce a work that will be useful and provide a comparative, worldwide view of a family of remarkable beauty and usefulness in man's increasing battle against destructive insects.

Bee Flies of the World
The Genera of the Family Bombyliidae

Introduction

The bee flies, or Bombyliidae, constitute at the present time one of the large and diverse families of the order Diptera. This family is worldwide in distribution; nearly 4,000 species are known, distributed among 194 genera and 24 subgenera. These flies are significantly more abundant in arid and semiarid regions, but even in such areas they are more numerous at well-watered spots, arroyos, or canyons, which support the rank vegetation necessary for the population of insects that serve as hosts for these flies.

In numbers of present-day species, this family is smaller than the Asilidae, yet we know 46 species of fossil bee flies found in 31 genera, 20 of which are extinct genera. Of the 39 species of fossil asilids, only 3 fall into extinct genera. So far all fossil bee flies are from the Oligocene and the Miocene, with none known from the Eocene period.

Wherever the habits of the larval stages of bee flies have become known, they have proved to be destroyers of locust eggs or parasitic or hyperparasitic upon a wide variety of other insects. The net effect of the many species which attack and destroy egg pods of hungry locusts is certainly of great benefit at times in controlling the numbers of these insects. Other species parasitize the extensive group of noctuid moths, whose larvae form the destructive cutworms and sod worms. Adult bee flies of many species visit flowers in search of nectar and pollen.

Perhaps only three groups of these flies affect man's economy adversely: (1) one species of the small genus *Heterostylum* Macquart is known to destroy the bees that pollinate forage crops; (2) the group of secondary parasites, which by destroying the scoliid enemies of white grubs, such as *Anomala* and *Phyllophaga*, operates to maintain scarab populations, instead of diminishing them. This second group includes *Ligyra* Newman and some species of *Exoprosopa* Macquart; (3) a third group is hurtful, like some species of *Hemipenthes* Loew, because the flies destroy the tachinid parasites of caterpillars. Probably the useful species of *Hemipenthes* outnumber the hurtful species.

Thirty-two genera of bombyliids in eleven subfamilies have been reared from the hosts, and as time goes on, and more rearing data are accumulated, other genera and species will undoubtedly be added to the list.

The hosts of bee flies are found within six orders of insects: Lepidoptera, Orthoptera, Hymenoptera, Diptera, Coleoptera, and Neuroptera. In any case it is clear that the Bombyliidae exert a considerable though annually variable effect upon the populations of insects. Occasionally bee fly populations are extensive and in such instances may have a key role in maintaining the natural balance of the components of faunas. Even where the percentage of parasitism remains at a relatively low level, perhaps 5 percent, bee flies are still a factor in insect control; nor do we know that low-level bee fly parasitism is not somehow due to competition by high-level tachinid or Hymenoptera parasitism.

While the Bombyliidae stand fifth in number of species in the order, they really fall into what might be called the third level of speciation, for bee flies are exceeded in number of species only by the approximately equal Tipulidae and Tachinidae, which two families are of nearly the same size, and the Asilidae and Syrphidae, which are also of nearly the same size. The latter two families have fewer species than either the Tipulidae or the Tachinidae but are considerably larger in number of species than the Bombyliidae. This leaves Bombyliidae in third place.

The fossil record suggests that the Bombyliidae, like the smaller family Nemestrinidae, are rather ancient, and further that today they meet with considerable competition from both the parasitic Tachinidae and the parasitic Hymenoptera.

Bee flies generally are large and often beautiful and graceful insects with attractively patterned wings and brightly colored bodies; it is not surprising that many persons have given much time and study to them. Few other Diptera match these flies in powers of flight, and to watch a giant species of *Exoprosopa* such as *ingens* Cresson—at times common in the Rio Grande area—with its long, delicately slender legs, as it comes to lightly rest upon a flower head is an attractive sight to students of insects. Austen (1937) states: "the charm of spring in Palestine is enhanced by visits to low-growing flowers of dainty little Bombyliinae like *Anastoechus stramineus*." Certainly, this is true over a wide part of the globe. It has been 42 years since I had the opportunity to see large numbers of early bee flies in the New Mexico desert coming to verbenaceous blooms, but it is one of the sights I have long remembered.

Many dipterists have contributed to our knowledge of the family Bombyliidae, and today a great deal of attention is being directed to these insects by a younger generation of entomologists. It is the hope of the author that this work will acquaint persons interested in the family with a knowledge of these insects in other regions and will generally stimulate interest in a most attractive family of Diptera. This work is designed to integrate all previous studies in order that we may have an effective world concept of this group. The phylogeny, distribution, and life histories have been treated, in addition to the morphology and taxonomy of the group, and a world checklist of species is included.

The accompanying bibliography covers a span of years and contains the work of more than 500 authors and nearly 1,600 titles; it ends with the year 1964, but a few later papers have been added. Complete bibliographies in all groups of insects are much needed.

The author takes full responsibility for the arrangement of genera and subgenera in this work. With many ardent workers in different parts of the world it is hardly to be expected that all will find themselves in full agreement with the disposition of genera made here. Nevertheless the present treatment may serve as a basis toward further studies of these flies.

History of Bombyliidae Studies

Linné (Linneaus), in 1758, in his tenth edition of "Systema Naturae" erected the genus *Bombylius*, in which he included three species (*Bombylius major*, *B. medius*, and *B. minor*) all of which are valid today. At the same time (1758) in his genus *Musca* he described three species of bombyliids (*Musca morio*, *M. maurus*, and *M. hottentotus*). In his twelfth edition Linné, besides adding *Bombylius capensis*, described the species *Musca denigratus*. This last species later proved to be the same as his earlier *Musca morio*. All four species of *Musca*, three of them valid, were soon taken out of *Musca* and placed in the new genus *Anthrax*, which had been created by Scopoli (1763).

Scopoli (1763) based his genus *Anthrax* on a single species, at first believed to be *Musca morio* Linné (1758), but it was later shown by Bezzi (1902, 1908) and Aldrich (1926) that Scopoli's type-species actually consisted of a 1746 unnamed species of Linné's, and oddly a species which Linné himself still confused with his own *Musca morio* in his 1761 paper.

This unique type-species of *Anthrax* Scopoli, erroneously believed by him to be *Musca morio* Linné, proved to be the species that was described by Schrank (1781) as *Musca anthrax* Schrank. The critical character of separation of the genus consisted of a tuft or pencil of hairs at the apex of the antennal style, which Scopoli had penetratingly observed in setting forth the characters of his genus.

All of these several species of bee flies left by Linné in his genus *Musca*, together with *Musca anthrax* Schrank and others added to this group of flies by later authors, came to be regarded as a separate family of flies known as the Anthracidae, or more commonly published as Anthracides. One of the Linnean species, *Musca morio* (1758), was indeed a mixture of three species—*Musca morio* Linné, *Musca denigratus* Linné (1767, 12th edition), and *Musca anthrax* Schrank (1781). It was not until many years later that Meigen and, still later, Macquart astutely recognized that these *Musca* species were indeed diverse members of the same family as *Bombylius* Linné, and were united with it. The student of bee flies is referred to Aldrich (1926).

Linné's species, *Bombylius major*, stands as the type of the genus and may be reckoned relatively generalized as far as the other members of the family are concerned. Moreover, *Bombylius major* Linné and *Musca morio* Linné (now *Hemipenthes morio*) are Bombyliidae which are holarctic in distribution and therefore occur in both Europe and America. It is interesting that they each belong to widely separated major subdivisions of the family.

Out of the many species placed by later authors in the second division, typified by Linné's species of *Musca* and in early days called Anthracides, the genus *Anthrax* itself is restricted to species having the tuft of hairs at the apex of the antenna. Later dipterists were able to discover important characters supplementary to the pencil of hairs and the group gradually became known as the Bombyliidae Tomophthalmae, while the group *Bombylius* and allied forms were grouped under what has been called the Bombyliidae Homeophthalmae. This nomenclature is discussed later in this work. The genus *Anthrax* Scopoli is restricted to those forms that do have the pencil of hairs at the apex of the antenna; all others in this group fall into one or another of many later genera.

Shortly after Linné, Fabricius (1775-1805) in several publications described 63 species of bee flies, and 2 genera which are now the basis of tribes or subfamilies. Mikan (1796) must have been an ardent enthusiast of bee flies, publishing his monograph of the Bombyliidae of Bohemia with 4 color plates. Still later Wiedemann (1817-1830) described 113 species and erected 4 important genera, one of which is the basis of a subfamily. Meigen, in the same period, besides describing many species, erected 4 genera. Subfamilies are now based on three of these genera. About the middle of the nineteenth century, a little before or after, many dipterists made significant contributions to our knowledge of the family. Among these were Loew, Macquart, Walker, Rondani, and Bigot. Macquart described 14 valid genera. Loew erected 18 genera that are valid today. In "Die Dipteren-Fauna Südafrika" Loew gave us our first comprehensive study of the Diptera of this vast region. Most of Loew's work was on the Palaearctic and Nearctic fauna. Other dipterists of this century who made contributions were Schiner, Philippi, Jaennicke, Roeder, Williston, Coquillett, and Osten Sacken.

The works of the last three were particularly important in supplying us with knowledge of the fauna of the United States and Mexico. Williston erected 6 genera, Coquillett 8, and Osten Sacken 12 new genera that are now in use.

The early years of the present century ushered in a period of greatly intensified work on bee flies. While both Becker and Bezzi began publishing on bee flies prior to 1900, almost all of their many publications fall in the period of 1900-1926. The untimely and tragic death of Bezzi ended a career of brilliant work in dipterology, much of which had gone to the Bombyliidae. Bezzi's 45 publications on bee flies of the Palaearctic and Ethiopian faunas stand as a monument to his astuteness and penetration, as well as to his industry. Throughout his life Bombyliidae were favorite objects of study.

Becker's fine series of publications totaled more than 30 and is particularly important in relation to the fauna of eastern Europe and Asia Minor. Beginning in 1924, Paramonov made numerous, exhaustive studies of the bee flies, especially of the region that is now Soviet Republics. His publications total more than 60 on this family. Two additional monumental studies of the bee flies of the Palaearctic region have appeared: The fine, illustrated volume by Engel (1932-1937) as a part of Lindner's *Die Fliegen der palaearktischen Region*, and the great work by the late Efflatoun Bey (1945), "A monograph of Egyptian Diptera. Part 6. Family Bombyliidae," which represents only the first half of his treatment of the Bombyliidae. The second half lies written but unpublished because of his untimely death. His fine color plates remain unpublished for lack of a sponsor. Efflatoun comments that as a result of his studies and those of his students, the number of known Egyptian Bombyliidae has quadrupled since 1919. Verrall, Austen, Edwards, and Oldroyd have also worked with bee flies and each has given us several publications.

In the United States in recent years several students have undertaken a serious study of this family. The late Dr. R. H. Painter became interested in this family in 1920. Since 1925 he has published 20 papers on bee flies. The untimely death of this able and enthusiastic worker, which occurred suddenly in Mexico City on December 23, 1968, cut short a brilliant career at its zenith, and it deprived me of a valued friendship which began when we were tent-mates Japanese beetle scouting in New Jersey in 1920, each turning to Diptera for life's work. We were also roommates two years at Ohio State University. In recent years he was ably assisted by his wife, Mrs. Elizabeth M. Painter.

Other recent workers include the late A. L. Melander, C. H. Curran, J. C. Hall, D. E. Johnson, and R. B. Priddy. In South America, Messias Carrera has given some attention to this family. In South Africa, A. J. Hesse has devoted more than a quarter of a century to intensive study of South African Bombyliidae, resulting in three astounding and comprehensive volumes, illustrated with more than 800 figures. Recently John Bowden has completed a fine volume on the Bombyliidae of Ghana. In Australia, Roberts made significant studies on bee flies of that continent, and G. H. Hardy also interested himself in this group.

This work attempts to supply the minimum of adequate illustrations for each genus. Past publications of Diptera having to do with bee flies have been sparse in illustrations. The most notably illustrated works are those of Engel, with 239 text-figures and 15 plates with 198 figures, and of Efflatoun Bey, with 551 figures, all in black and white. Bezzi (1924) used 46 figures; Austen (1937) illustrated with 72 figures by Terzi and 1 color plate. Hesse's South African work with 823 figures has already been mentioned. There is also a color plate in the initial volume of *Die Fliegen der palaearktischen Region*. Painter and Hall (1960) provided color photographs of the species of *Poecilanthrax* Osten Sacken in their monograph of this genus. Efflatoun Bey, in a communication to this author, stated that he was unable to publish any of his beautiful color plates of Bombyliidae. This is a great loss to students of this family. Séguy and Becker have illustrated some of their publications.

Every student of insects stands deeply indebted to the work of others before him. I wish to pay especial tribute to the fine work not only of Engel, but also to Bezzi, who assisted me in my early studies 40 years ago, and to Dr. Paramonov with whom I have had pleasant excursions, and to the very generous assistance of Dr. A. J. Hesse of the South African Museum at Capetown. Certainly, the student of the Bombyliidae is deeply indebted to the authors of the more than 200 papers that deal with the fascinating and little known life histories of this family. These are summarized here with the special interest and hope of drawing much increased interest to this important aspect of this family.

The species list in this work, which has been included with the genera and subgenera, is offered as a practical device for the student, new and old, to check on the contents of each genus. Where a species was described in a different genus from which it now stands, this is indicated by the earliest generic assignment enclosed in parentheses. The dates accompanying each species as listed will readily give the bibliographic clue to the place of publication. In the preparation of the species list I have followed Kertész (1909) in his world catalog of Diptera, except for the Palaearctic region where I have turned to Engel and Paramonov for the disposition of these species. Of course Hesse (1938, 1956) has been followed for the assignment of South African species. Also the assignment of North American species reflects the thinking of recent students in this area and many of them have been checked by the author. Wherever I have been privileged to see types of species at various museums, I have used the information to reassign them where necessary, according to the concepts of this study. Many of the exotic species of older authors stand in need of additional study and especially of illustration. Kertész (1909) listed 84 genera and 1,696 species; the number of both genera and species has more than doubled since that time. This checklist ends with the year 1965.

Generalizations of the Host Relationships of Bee Flies During Immature Stages

The intricate relationships of bee flies with other insects, upon which they are parasitic, or more infrequently hyperparasitic, provide deeply interesting and fascinating material for study by the future generations of entomologists. Hosts are known for comparatively few genera, and adequate life-history studies have been made for only a few species. The account by Shelford (1913a) of *Anthrax analis* Say (as *Spogostylum*), the work of Clausen (1928) in India on the hyperparasite *Ligyra oenomaus* Rondani (as *Hyperalonia*), and the work of Bohart, Stephen, and Eppley (1960) on *Heterostylum robustum* Osten Sacken, as well as Krombein (1967) on trap nesting stand as excellent and highly satisfactory examples of able life-history studies. Such painstaking studies require much time and effort.

Some generalizations do emerge. We know hosts of a few of the genera of eleven subfamilies out of fourteen. Of five subfamilies we know the host of only one genus. We know that these hosts are derived from six orders of insects, but much more commonly from Lepidoptera, Orthoptera, and Hymenoptera. The first group of bee fly genera—no less than nineteen, and probably the vast majority—seeks soil-hidden hosts; females of many genera have acanthophorites. These include *Sparnopolius* Loew, which attacks white grubs in the soil, and at least one species of *Villa* Lioy, which attacks the tenebrionid *Meracantha* Kirby. Genera, like *Poecilanthrax* Osten Sacken, that seek sod-hidden cutworms also fall into this group. Four genera attack mud cells of wasps or stem-mining wasps, and three genera presumably attack exposed caterpillars. No less than thirteen genera in seven tribes or subfamilies specialize in the consumption of egg pods of locusts; probably many other genera do likewise. We know some of the hosts for thirty-two genera.

A second group of bee flies attacks subterranean insects which are miners, for example *Heterostylum* Macquart parasitizing the mining bees *Nomia* Latreille and *Anthrax analis* Say, a parasite of the mining beetles of the genus *Cicindela* Linné; both of these flies flip their eggs directly into or near the mine or tunnel. Presumably, *Dipalta* Osten Sacken, which has been reared from ant lion larvae, deposits its eggs the same way, as ant lion pits are conspicuous objects. From this group it is perhaps but a small extension to the two genera *Anthrax* Scopoli and *Toxophora* Meigen, of separate subfamilies, which attack the earthen nests of mud daubers, eumenids, or the cliff-bank nests of *Anthophora* Latreille and stem tunnels of *Xylocopa* Latreille. In all of these cases, the egg is flipped by a quick, sudden, forward jerk of the abdomen in the direction of the desired objective, and with surprising accuracy.

Finally, a third group of genera seeks exposed hosts, such as lepidopterous larvae. Here belong the Systropinae, seeking larvae or pupae of the family Eucleidae. The Geroninae, which parasitize psychids and pyralids, are also properly placed in this group.

Bee fly populations are supported by some families of Lepidoptera: in this order six families—Noctuidae, Psychidae, Eucleidae, Cossidae, Pyralidae, and Tortricidae—of widely divergent habits are known to be parasitized by three subfamilies of bee flies. None of the diurnal Lepidoptera are included.

1. The family Noctuidae: cutworms, sod worms, and army worms. These species, the largely nocturnal larvae of the owlet moths, are exceptionally abundant in many parts of the world. They belong to the family Noctuidae of worldwide distribution and contain some 500 genera and more than 20,000 species (Essig, 1942). Maximum development appears in the Nearctic and Neotropical faunas (3,500 Nearctic species against 1,800 Palaearctic fauna (Imms, 1948). Walkden (1950), who reared several species of *Poecilanthrax* Osten Sacken from noctuid larva, obtained nearly 300 species of noctuids from light traps in Kansas. I have been astonished at the enormous numbers of these moths attracted to lights at wayside restaurants by night in August on Kansas highways.

Walkden also reared large numbers of some 53 species of noctuids, and he suggests that powerful repressive factors evidently work against them since so few species are major pests. He found the egg count for some to range to more than 2,000, and many species with egg production of 600 to 800. Walkden classifies these into larval habit as: (a) subterranean, (b) tunnel makers, (c) surface feeders, (d) climbers, and (e) boring types. Of the species he reared, some in limited numbers, six were attacked by bombyliids; these bee flies consisted of three species of *Poecilanthrax* Osten Sacken and two unidentified bee flies. Moreover, all of the above ecological groups of cutworms were attacked by bee flies, except the boring types.

Some interesting results appear from these studies. Of 53 species of noctuids he reared, many of them in considerable numbers, 27 species were attacked by hymenopterous parasites. Where these were identified, 18 species were attacked by an average of 3.5 species of Hymenoptera; 10 species were attacked by 2.7 species of tachinids; only 6 species of larvae were attacked by 1.3 species of bombyliids; 5 species of larvae were parasitized by 1 species of bee fly. From 1 species, *Chorizagrotis auxiliaris* Grote, 3 species of bee flies were reared, which is perhaps not surprising, as this is a surface-feeding larva. The other genera attacked by bee flies consisted of 3 climbing species, *Euxoa scandens* Riley, *Euxoa messoria* Harris, and *Peridroma margaritosa* Haworth; the tunnel maker, *Agrotis venerabilis* Walker, and the subterranean *Agrotis orthogonia* Morrison. The above data suggest that tachinids are more efficient parasites and Hymenoptera still more efficient than are the bombyliids, as far as the cutworm group of noctuids is considered.

2. The family Psychidae: caseworms and bagworms. Mik (1896) has recorded the parasitization of a Palaearctic species of *Fumea* Stephens by bee flies of the genus *Geron* Meigen. *Fumea* Stephens is rather exceptional since in this species the female leaves the case during mating. The family Psychidae is large and distributed throughout temperate and tropical parts of the world, the larvae feeding on many kinds of trees and shrubs. There are about 150 Palaearctic species. The species of *Geron* Meigen are numerous in the Nearctic region and so far scanty in the Old World and related genera extend down into South Africa. They are also present in Australia. One species of fossil bee fly from the Miocene has been tentatively assigned to *Geron* Meigen, and it is of interest perhaps that both the psychid moths and the eucleid moths, which are parasitized by the almost worldwide Systropinae, are placed in the same superfamily, the Zygaenoidea of Gravenhorst (Psychoidea of Imms). Both these bee flies have reduced venation. Until recently we did not know the host for any Nearctic species of *Geron* Meigen; Donahue (1968) notes that *Solenobia walshella* Clemens is a host of *Geron calvus* Loew.

3. The family Tortricidae. The bee fly *Geron argentifrons* Brunetti has also been recorded by Maxwell-Lefroy (1909) as attacking *Laspeyresia* Hübner.

4. The family Pyralidae. Mik (1896) recorded the genus *Geron* Meigen as a parasite of the European pyralid *Nephopteryx* Hübner.

5. The family Cossidae: carpenter moths or goat moths. A widely distributed, small, and apparently ancient family of wood-boring moths; both larvae and moths are usually of large or very large size. One genus and species of bee fly is known to live at the expense of these moths. Oldroyd (1951) has recorded, described, and illustrated a very large bee fly, *Oestranthrax goliath* Oldroyd, from Malaya, which was reared from a cossid pupa. Many interesting questions might be raised as to how the fly reaches the larvae. He adds that the general appearance of the adult strongly suggests a nocturnal or crepuscular habit.

6. The family Eucleidae (Limacodidae of authors): slug caterpillars. These makers of egglike cocoons represent a family of some 850 or more species of widespread distribution, chiefly tropical, but found also in temperate areas of moderate rainfall. The bee flies of the genus *Systropus* Wiedemann, likewise found all over the world, are exceptionally beautiful insects and quite remarkable for their mimicry of the long, slender wasps of the genus *Ammophila* Kirby (*Sphex* Linné), and they have been taken from the genera *Parasa* Moore and *Sibine* Herrich-Schaeffer.

It is a curious fact that the adults of *Systropus* Wiedemann, among the most beautiful of all bombyliids, seem to be rare. In 40 years of collecting at Oxford, Mississippi, I have taken but a single specimen of one species. In Mexico in 1962, however, with three pairs of eyes searching—my wife, my son, and myself—we found a single specimen at one locality, but before leaving Mexico, at milepost 1258 on the Mexican Pacific Highway, we found a large population, and each of us collected many mating pairs of two associated species. This was from shortly before noon to about 2:00 P.M.; Dr. R. H. Painter (1962) in the same summer also found two associated species in eastern Mexico, in smaller numbers.

Schaffner (1959) in extensive rearing experiments with temperate Lepidoptera only obtained a single specimen of *Systropus* Wiedemann.

Since Cockerell points to the marked scarcity of muscoid flies as fossils in the Colorado Miocene, it seems likely that tachinids have been steadily replacing bombyliids. Finally, I point to the total number of Diptera and Hymenoptera reared by Walkden from 19 species of cutworms. Of 10,108 larvae of these 19 species, mostly reared over a period of 12 to 15 years or more, approximately 523 were Diptera emergents, which were not broken down into respective number of tachinids and bombyliids, and approximately 788 were Hymenoptera of the several species. One species was attacked by as many as 6 species of tachinids; 1 was attacked by 13 species of Hymenoptera, and the maximum number of bee flies from 1 species was 3.

Four families of Coleoptera have been known to be attacked by bombyliids. There is a single record by Hyslop (1915) of a bee fly, *Villa alternata* Say, which he obtained from a pupa of the tenebrionid *Meracantha contracta* Beauvois. Numerous larvae of this beetle had been obtained in rotting wood at the base of a stump at the top of South Mountain, Maryland; some of the larvae transformed to pupae but considerably more of them had been parasitized by a tachinid.

Recently, Jack Hall recorded the parasitism of a cerambycid beetle by the bee flies of the genus *Thevenemyia* Bigot.

Shelford (1913a) showed that tiger beetles of the genus *Cicindela* Linné serve as host for at least one species of bee fly, *Anthrax analis* Say.

The soil-living scarabaeid population, however, supports a greater number of bee flies. Forbes (1907) called attention to the larvae of *Sparnopolius lherminierii* Macquart acting as an ectoparasite upon white grubs. Davis (1919) in an extensive study of enemies of white grubs (*Phyllophaga* spp.) noted that the scoliid wasp *Tiphia* Fabricius, abundant parasites of grubs of the genera *Phyllophaga* Harris, *Ligyrus* Burmeister, *Cyclocephala* Latreille are in turn parasitized by the bee flies *Exoprosopa fascipennis* Say and *Exoprosopa pueblensis* Jaennicke, as well as by *Rhynchanthrax parvicornis* Loew. Swezey (1915) found *Chrysanthrax cypris* Meigen attacking the scoliid wasp *Elis* Fabricius in Illinois. Clausen (1928) in a remarkable study made in India found the closely related genus *Ligyra* Newman (as *Hyperalonia* Rondani), species *oenamaus* Rondani, parasitizing *Tiphia* Fabricius, *Scolia* Fabricius, and *Campsomeris* Lepeletier cocoons.

Riley (1877, 1878) was the first to find bombyliids living in egg cases of locusts. He published a series of seven papers on this subject from 1877 to 1881. In Russia, however, Stepanov at very nearly the same time discovered the relation of these flies to locusts and published his first paper in 1881 and two others in 1882. Also, Saunders, Waterhouse, and Fitch reported bee flies feeding on locust eggs in the Troad, specimens having been sent to London by Calvert. In the years immediately following, several other European workers gave some attention to acridophile bombyliids, in particular Kunckel in three works (1889, 1893-1905, 1894). After the turn of the century and recently still more attention has been paid to this group, because of the important effect of devastative plagues of locusts on man's food crops. These findings will be reviewed later. No less than 13 genera in 5 subfamilies have the habit of consuming locust eggs. While some workers refer to this habit as predation, I here treat it as parasitism extended to the egg stage, inasmuch as only a single egg pod is consumed by a single fly; after the bee fly larvae has reached the second or third instar, it is too portly to go searching for a second meal ticket; furthermore, bee flies have remarkable powers of maturing on small victims, as evidenced by the wide range in size of adults, a matter discussed elsewhere.

The Diptera infrequently furnish hosts to this family. Austen (1914) noted that *Thyridanthrax* Osten Sacken extends its host behavior to include tsetse flies, *Glossina* Wiedemann, in Africa, and because of its important significance in disease this relationship has been followed up with other studies.

Several tachinids and even calliphorids furnish host relationship to bee flies, which thus become hyperparasites. Examples are: *Hemipenthes morio* Linné parasitizing pupae of *Parasetigena* Brauer and Bergenstamm and *Ernestia rudis* (Fallén) (Baer, 1920), and according to Vassiliew (1905) *Hemipenthes morio* Linné and *Villa velutina* Meigen (as *Anthrax*) parasitizing *Masicera silvatica* (Fallén), a tachinid that attacks Lepidoptera.

One Nearctic genus, *Dipalta* Osten Sacken, has been reared from the pupal cases of ant lions (Myrmeleonidae).

Four subfamilies and six genera of bee flies are known to depend upon Hymenoptera for hosts. These bombyliids attack a wide variety of solitary types of bees and wasps. Twenty-two hymenopterous genera have been noted to produce bee flies in rearing experiments. It should be further noticed that *Ligyra* Newman and *Exoprosopa* Macquart act as secondary or hyperparasites by attacking scoliids of the genera *Tiphia* Fabricius, *Scolia* Fabricius, *Campsomeris* Lepeletier, the sphecid *Sceliphron* Klug, and the psammocharid *Pseudogenia*. Also *Chrysanthrax* Osten Sacken has been reared from *Elis* Fabricius. It is also most interesting that Finlayson and Finlayson (1958) reared *Hemipenthes morio* Linné from cocoons of the European pine sawfly in Ontario, but again the bee flies, as parasites, were greatly outnumbered by tachinids and both of these dipterous families by Hymenoptera. The primary parasites in this group attack nests of mining bees or wasps in earth or cliff sides and also isolated nests of mud daubers and potter wasps.

Hyperparasitism has developed, as far as is now known, only in the rather dominant subfamily Exoprosopinae and to a lesser extent in the Anthracinae, which I believe should represent a tribe within the Exoprosopinae. Flies of the genera *Exoprosopa* Macquart and *Ligyra* Newman attack scoliid cocoons of *Tiphia* Fabricius and other genera. *Hemipenthes* Loew is a hyperparasite of *Ophion* Fabricius and *Banchus femoralis* Thomson which themselves have attacked nocturnal Lepidoptera. *Anthrax* Scopoli and *Hemipenthes* Loew are hyperparasites of tachinid flies.

Review of Published Information on Host Relationships

(Abbreviations used for area: AU, Australian; ET, Ethiopian; HO, Holarctic; NE, Nearctic; NEOT, Neotropical; OR, Oriental; PA, Palaearctic.)

BOMBYLIINAE: *Bombylius* Linné

Species	Hosts	Authors
boghariensis (PA)	Not known	Lucas (1852)
canescens (PA)	*Odynerus spinipes* Linne	Chapman (1878)
	Halictus Latreille spp.	Yerbury (1902)
major (HO)	Pupa extruded from earth	Imhoff (1834)
	*Andrena	Dufour (1858)
	Andrena labialis (Kirby)	Chapman (1878)
minor (as *pumilus* Meigen) (PA)	*Colletes daviesana* Smith	Neilsen (1903)
	Epeolus productus Thomson (hyperparasite)	
vulpinus (as *fugax*) (PA)	*Panurgus dentipes* Latreille	Séguy & Baudot (1922)
sp.	*Colletes fodiens* (Fourcroy)	Schmidt-Goebel (1876)
sp.	*Halictus farinosus* F. Smith	Bohart (1960) in Bohart, Stephen, & Eppley

* Junior synonym of *A. ovina* Klug.

Bombylius boghariensis Lucas. A pupa of this species was obtained in North Africa by Lucas (1852), who gave a fairly good, though diagrammatic figure of the stage, as well as the adult. He found these pupae emerging from the soil, but he was unable to learn the host.

Bombylius canescens Mikan. Chapman (1878) observed a species, believed to be *B. canescens*, ovipositing freely in nests of *Odynerus spinipes* Linné upon a bank. Yerbury (1902) noted many individuals ovipositing in nests of several species of *Halictus* Latreille. This was done by sudden swoops or jerks of the abdomen; the egg was clearly seen as it left the insect but was never found.

Bombylius major Linné. Though Imhoff (1834) described a pupa that he found extruded from the earth, Dufour (1858) was the first person to give much information concerning the life history of this species, the type-species of the family. Dufour found the species attacking *Andrena nitidiventris* Dufour. He described and figured both larvae and pupae. The nearly mature larva (third stage) still had thoracic bristles.

Chapman (1878), during some excavation of earth in England, found a colony of *Andrena labialis* Kirby, among which he found a few overwintering pupae of this bee fly. The pupa was described but not illustrated. Chapman adds: "our present subject (*B. major* pupa) has to force a passage through the clay filling placed by the bee in its burrows, and to comb through 6 to 10 inches of these burrows. . . . though many pupae have to force their way through obstructions, this is the only one I know of actually provided with a mattock and shovel with which to do its own navigating."

In view of the abundance of this widespread, Holarctic species, which is extraordinarily common in Mississippi and other parts of the United States, as well as in Europe, it is surprising that so little accurate information is known concerning the life history of *Bombylius major* Linné.

Bombylius mexicanus Wiedemann. On May 28, 1964, the author found a considerable population of this large species ovipositing in dry soil about mesquite trees on the outer limits of Laredo, Texas. A sunken depression with a half-filled pond and old, undisturbed mesquite surrounding it had produced a wealth of both vegetation and insects. This fly was found ovipositing both by jerky thrusts of the abdomen toward the soil near the base of mesquite and by distinctly inserting the abdomen into the somewhat friable and powdery soil.

Bombylius minor Linné. J. C. Nielson (1903) gives a good account of the biology of this species as *Bombylius pumilus* Meigen. Austen (1937) in a footnote on p. 14 states: "*Apud* Engel, the subject of this paper was really *B. minor* L.; *B. pumilus* does not occur in Denmark, where Nielsen's observations were made." Neilsen's report, with the exception of Riley's work on *Aphoebantus* Loew, was perhaps the best of its kind up until this date. He gives detailed figures from the larvae and pupae of the species, which are only slightly diagrammatic. These flies were found in the nest of *Colletes daviesana* Smith in Denmark. Neilsen also found them attacking *Epeolus* larvae, which had parasitized *Colletes*, thus acting in this case as a secondary parasite. For a time the first-stage larvae feed upon the pollen stored up by the bee for its own brood, switching over later to a parasitic existence upon the bee grub.

Bombylius vulpinus Wiedemann. Under the name of *Bombylius fugax* Wiedemann, Séguy and Baudot (1922) describe the egg, larva, and pupa of this species, which they found parasitizing *Panurgus dentipes* Latreille. They give excellent figures of all of the stages.

Bombylius sp. Schmidt-Goebel (1876) found a species of *Bombylius* as a parasite of *Colletes fodiens* Fourcroy.

Bombylius sp. Bohart (1960) in Bohart, Stephen, and Eppley notes that many two-year old larvae of *Bombylius* sp. have been observed parasitizing *Halictus farinosus* F. Smith.

Allen and Underhill (1876) relate having found the young larvae of *Apalus* (as *Sitaris*) clinging to "humble bee flies," by which they assume the bee fly was in a definite way associated with nests of *Anthophora* Latreille; the species of bee fly was unfortunately not given, but the term used above suggests *Bombylius* Linné, but it may have been *Anthrax* Scopoli.

BOMBYLIINAE: *Systoechus* Loew

Species	Hosts	Authors
	Egg pods:	
acridophagus (as *albidus*) (ETH)	*Locustana pardalina* Walker	Potgieter (1929) Hesse (1938)
aurifacies (ETH)	*Schistocerca gregaria* Forskål	Greathead (1958)
autumnalis (PA)	*Dociostaurus* (*Stauronotus*) *maroccanus* Thunberg	Stepanov (1881)
		Ingenitsky (1898)
	Several locust	Troitsky (1914)
	Stauroderus scalaris (F.-W.)	
	Gomphocerus sibiricus (L.)	
	Pararcyptera microptera (F.-W.)	Vorontsovsky (1926)
	Dociostaurus kraussi (Ingen.)	
	Dociostaurus albicornis (Ev.)	
gradatus (PA)	*Stauronotum vastator* Stev.	Stepanov (1882b)
	Calliptamus italicus Linne	Lindeman (1902)
marshalli (ETH)	*Acrotylus deustus* Thunberg	Paramonov (1931)
		Hesse (1938)
oreas (NE)	*Camnula pellucida* (Scudder)	Riley (1877, 1880)
	Melanoplus spretus (Walsh)	
	Melanoplus mexicanus Sauss., *femur-rubrum* De Geer, & *differentialis* Thomas	Parker & Wakeland (1957)
	Camnula pellucida (Scudder) & other grasshoppers	Treherne & Buckell (1924)
socius (OR)	*Colemania sphenarioides* I. Bolivar	Fletcher (1916)
somali (ETH)	*Schistocerca gregaria* Forskål	Hynes (1947) Greathead (1958)
sulphureus (PA)	*Dociostaurus maroccanus* Thunberg, *Gomphocerus sibiricus* L.	
	Gomphocerus sibiricus (L.) *Stauroderus scalaris* (F.-W.)	Bezrukov (1922)
	Stauronotas vastator Stev.	Stepanov (1882b)
	Dociostaurus maroccanus Thunberg	Paoli (1937)

Species	Hosts	Authors
vulgaris (NE)	*Melanoplus mexicanus* Saussure	Parker & Wakeland (1957)
	Melanoplus femur-rubrum DeGeer	
	Melanoplus differentialis Thomas	
	Melanoplus differentialis Thomas	Painter (1962)
	Melanoplus bivittatus Say	
	Melanoplus bilituratus Walker	
	Syrbula admirabilis Uhler	
	Melanoplus mexicanus Saussure	Gilbertson & Horsfall (1940)
xerophilus (ETH)	*Locustana pardalina* Walker	Potgieter (1929) Hesse (1938)

Research on *Systoechus* Loew and other genera that attack grasshopper egg pods have centered around ethological and quantitative studies by some workers, and taxonomic and morphological studies by others.

This important aspect of bee fly behavior is usually listed as predatism, but I should prefer to think of it as ectoparasitic ovoparasitism, following Austen (1937). Some egg-destroying species of bee flies in other genera are apparently able to vary their host selection from locust eggs to caterpillars or to wasps.

Systoechus acridophagus Hesse. This is one of the two species reared by Potgieter (1929), its biology is quite similar and its host species the same as that of *Systoechus xerophilus* Hesse. Hesse (1938) illustrated and described the pupa. Some of the many South African species of *Systoechus* Loew are extremely large, with 35 to 40 mm. wing spread; their biology should be very interesting.

Systoechus aurifacies Greathead. Greathead (1958) records this new species as attacking egg pods of *Schistocerca gregaria* Forskål in Eritrea. *Systoechus aurifacies* was found in a limited area of about 200 by 300 square yards. Of the 80 locust pods found, 22 were infested with a small number of larvae, the yield was 25 larvae of what proved to be this species. Greathead notes that the biology of *S. aurifacies*, which is apparently scarce, is very similar to that of *Systoechus somali* Oldroyd.

Systoechus autumnalis Pallas. Stepanov (1881) bred this species from egg pods of *Dociostaurus* (*Stauronotus*) *maroccanus* Thunberg in Russia. Ingenitsky (1898) bred it from several locusts in eastern Siberia; Troitsky (1914) records it from egg pods of *Stauroderus scalaris* (F.-W.) and *Gomphocerus sibiricus* (L.) in Siberia; Vorontsovsky (1926) records this species from *Paracyptera microptera* (F.-W.), *Dociostaurus kraussi* (Ingen.), *D. albicornis* (Ev.) from western Siberia.

Systoechus gradatus Wiedemann. Stepanov (1882) records this species under the name *Systoechus leucophaeus* Wiedemann with host egg pods from *Stauronotum vastator* Stev. and in 1881 from *Dociostaurus maroccanus* Thunberg in the Crimea. Lindeman (1902) reared it from *Calliptamus italicus* Linné in Russia.

Systoechus marshalli Paramonov. This species was recorded from egg packets of *Acrotylus deustus* Thunberg by Paramonov in 1931.

Systoechus oreas Osten Sacken. Riley (1877, 1880) was the first worker who discovered flies of this genus in their role of locust-egg destroyers. In his first annual report of the U.S. Entomological Commission he notes that numerous larvae, which he tentatively assigned to the Ichneumonidae but which later proved to be *Systoechus* Loew, were feeding within the egg pods. Riley states: "It exhausts the eggs, and leaves nothing but the shrunken and discolored shells." Riley quotes a communication from S. D. Payne in Minnesota: "as I was strolling through the fields, I stopped to examine some eggs (locust). I found the ground in spots quite full of white grubs, worms or maggots.... many of them were in the egg-pods, busy at work." In 1880 Riley had realized the true identity of the maggots he had found in the locust-egg pods, and in an interesting report in the *American Entomologist* he gives the first published figure of the larva of *Systoechus oreas* Osten Sacken with details of head capsule, the pupa, and the imago. Further, he summarizes knowledge of bee fly histories at that time. In 1881 he gives a good description of the larva and pupa and more adequate figures of both this species and of *Aphoebantus mus* Osten Sacken (as *Triodites*). See also Parker and Wakeland (1957) under *Systoechus vulgaris* Loew.

Systoechus somali Oldroyd. Considerable information was given on the bionomics of this species by Hynes (1947), a staff member of the Anti-Locust campaign in Italian Somaliland. In May and June of 1945 it was noticed that egg pods of locusts were being destroyed in many places by bee fly larvae, which proved to be this species. The host locust was *Schistocerca gregaria* Forskål. Egg packets of this species contain about 70 eggs, and the number of larvae in each one ranged from 0 to 30, but averaged 10. Hynes states that they were all third-stage larvae and that no first- or second-stage larvae were found. The larvae in the pods at the time of collection ranged from 2 to 14 mm., which he was inclined to believe were not all of the same age. Growth was rapid; 10 or 11 days after the laying, a large proportion were full grown. Most of the larvae, he states, were found with their mouthparts applied to the middle of a locust egg with nearby eggs shrunken. He estimated that 5 or 6 larvae were necessary to destroy one locust-egg pod. He did not learn the fate of the larvae when excessive numbers were present in a single pod. The final size variation amounted to 6 to 16 mm.

In a few days the translucent white of the larvae became opaque yellow; larvae burrowed a short distance from the egg packet, and came to rest in small ovoid cavities, similar to those described by Potgieter (1929) for *Systoechus acridophaga* Hesse (as *albidus* Loew);

they were found about 5 to 10 cm. below the soil level. A week after the locust had hatched, the soil was becoming dry and caked and it remained so for months; the usual interval between rains is 4 to 5 months. An examination 2 months later showed the larvae slightly shrunken, active if disturbed, and resting in diapause. Although the irregular areas laid over by locusts amounted roughly to some 300 or 400 miles in each direction, the *Systoechus* were abundant only in a compact area about 120 by 160 miles, a situation maintained through two seasons. Hynes says that a staff member reported about 40 percent of the locust eggs were destroyed in soft soils and nearly 100 percent destroyed in hard soils. Hynes found that in very soft, sandy soils bee fly larvae were nearly absent. In late May the locusts had laid heavily, and in the harder soils he found 96 percent of the egg packets infested with an average of 4 larvae, but only 32 percent infestation in sandy soil, and these packets had only 1.6 larvae. In one place actually 100 percent of the eggs were infested with 18 larvae per packet.

Hynes (1947) carried out laboratory experimentation which indicated the importance of rain as a factor concerned in the emergence of adults. Some 1,400 larvae in diapause were buried in dry soil within shallow metal pans with drainage. One pan was immediately watered heavily, and a few days later adults started to emerge. Another pan was watered lightly on the surface; another pan was kept air dry. The first pan was kept saturated. Emergences were recorded daily for over a year. Only 41 flies emerged, but at the end of the period many larvae were alive and in diapause; after 15 months 40 percent of the larvae were still alive in the dry cage, but no flies had emerged from it. Hynes concluded that light rain was insufficient to break diapause. Oddly, Hynes, working with *Systoechus somali* Oldroyd, and Potgieter (1929), with *S. acridophaga* Hesse (as *albidus* Loew), and De Lepiney and Mimeur (1930), with *Cytherea infuscata* Meigen (as *Glossista*), observe that even when conditions are favorable for emergence, some larvae remain in diapause. Larvae in diapause in dry soil, if wetted, swelled to original size, appeared to have a slightly reduced fat body, and if refusing to pupate, they could remain normal for a long time, move actively, and again form new soil cavities. Apparently emergences from pupae occurred only in the day time. Pupal period was short, but the larvae would not pupate in soil-filled tubes. Under laboratory conditions the pupal period was 9 to 15 days; for *Systoechus acridophaga* Hesse (as *albidus* Loew), Potgieter found the pupal period to be 7 to 9 days in the laboratory, and 14 to 23 days outdoors in the summer. De Lepiney and Mimeur record the pupal period of *Cytherea infuscata* (as *Glossista*) to be 10 days.

Hynes quotes unpublished notes of Dr. D. L. Gunn on oviposition. Dr. Gunn states that the fly hovers over the hole in the sand left by a locust at a height of 1 to 3 cm.; an egg can be seen ejected from the extended abdomen, which is curved forward, and a few grains of sand can be seen to move at the hole entrance; no eggs or first-stage larvae were recovered. The number of jerks of the bee fly ranged from 10 to 20 on an average, and to 40 in one case.

Greathead (1958) adds considerable additional data on *Systoechus somali* Oldroyd beyond that supplied by Hynes (1947). He notes that Hynes' larvae represented both second and third stages, and calls attention to the patchiness or irregularity of the distribution of the species. Greathead gives a table of the percentage parasitism for this as well as other parasites from 13 localities. He also gives a good figure of the larva of this species and a very valuable series of comparative figures of details on 4 species of *Systoechus* Loew larvae and pupae. Greathead notes that larval growth is quite rapid, in a few cases completed in about 4 days. Moreover, he finds that there are usually not enough larvae in a pod to completely destroy all eggs; however, he quotes a personal communication from Ellis and Ashall who commonly found 20 to 40 larvae in an egg pod and who noted as many as 66 larvae in one egg pod. Greathead found that the larvae puncture an egg and suck out the contents, leaving the chorion dry and collapsed. This method does not result in the general putrefaction such as accompanies feeding of *Stomorhina* Rondani. Greathead states that *Systoechus somali* larvae sent to laboratories in England remained unchanged for as long as 3 years at temperatures of 25° to 30°C.

Systoechus sulphureus Mikan. Recorded by Canizo (1944) from egg cases of *Dociostaurus maroccanus* Thunberg, *Gomphocerus sibiricus* (L), and *Stauroderus scalaris* (F.-W.). He gives a figure of the mature larva, and pupa with details of its armament. Paoli (1937) gives an excellent, illustrated account of the morphology of this species.

Systoechus vulgaris Loew. Parker and Wakeland (1957) collected the data made in 1938, 1939, and 1940 from grasshopper-egg surveys. These studies were made in 16 study areas in the United States in the states of Arizona, California, Kansas, Minnesota, Montana, North Dakota, and South Dakota. This work assessed the average annual destruction of locust-egg pods by bee flies, as well as blister beetles and ground beetles. The bee flies included *Systoechus oreas* Osten Sacken, and *S. vulgaris* Loew, *Aphoebantus hirsutus* Coquillett, and *A. barbatus* Osten Sacken. Of these *S. vulgaris* was not only the most common, but also the most widely distributed species occurring in Michigan, Minnesota, Montana, Nebraska, North and South Dakota. *Aphoebantus hirsutus* was next most frequently found, but limited to California and Oregon.

These authors found that the average annual destruction of egg pods amounted to 17.87 percent, of which bee flies accounted for 6.18 percent. The highest egg-pod destruction for a single year reached 77.52 percent in North Dakota. However, average egg-pod destruction for many counties reached 50 percent or more, with bee flies taking a proportionately larger total. Criddle

(1933) found that the combined efforts of *Systoechus vulgaris* Loew and other predators accounted for over 20 percent egg-pod destruction in the whole province of Manitoba in 1932. Shotwell (1939) in an account of grasshopper eggs destroyed in 6,277 fields in 11 states noted that bee flies per square foot of soil amounted to 0.31, which was higher than either blister beetles or carabids. Gilbertson and Horsfall (1940) in a study in South Dakota noted that of 59.3 percent destruction of *Melanoplus mexicanus* Saussure, bee flies accounted for 35.6 percent, again a greater proportion than meloid larvae. Parker and Wakeland (1957) discuss methods of taking data and results of findings by each state.

Systoechus xerophilus Hesse. This was one of two species of *Systoechus* Loew reared by Potgieter (1929) from *Locustana pardalina* Walker, the brown or Tek locust of South Africa. Its larva is figured by Hesse (1938).

BOMBYLIINAE: *Anastoechus* Osten Sacken

Species	Hosts	Authors
baigakumensis (PA)	*Locusta migratoria* Linné	Zakhvatkin (1931)
barbatus (NE)	*Melanoplus mexicanus* Saussure	Parker & Wakeland (1957)
melanohalteralis (NE)	*Dissosteira longipennis* Thomas	Painter (1962)
nitidulus (PA)	*Locusta migratoria* Linné	Sacharov (1913)
	Locusta migratoria-gallica Remaudiere	Roerich (1951)
	Dociostaurus maroccanus Thunberg	Stepanov (1882) Chinkevitsch (1884) De Lepiney & Mimeu (1930) Seabra (1901)
	Gomphocerus sibiricus (L.) *Stauroderus scalaris* (F.-W.)	Troitsky (1914)
	Pararcyptera microptera (F.-W.)	Moritz (1915)
	Calliptamus italicus Linné	Bezrukov (1922)

Anastoechus baigakumensis Paramonov. Zakhvatkin (1931) notes that he reared some individuals many years earlier from *Locusta migratoria* Linné.

Anastoechus barbatus Osten Sacken. Parker and Wakeland (1957) give an extensive report covering the results of grasshopper-egg destruction from 7 states over a 3-year period. Their data include three other species of bee flies, besides meloids and carabids, without any separation of hosts attacked by the several predators. They list 13 species of grasshoppers, which were the dominant or second dominant species in the areas studied. They show very interesting maps of from 15 to 30 states for each year, giving the percentage of egg pods destroyed by all 3 major groups of predators. The greatest efficiency of destruction tended to lie in a north-south belt including Minnesota to Texas. *Anastoechus barbatus* was collected by the survey only in Montana. *Melanophus mexicanus* Saussure was the dominant locust in all Montana areas. I have, however, found adults of *A. barbatus* exceptionally abundant in northern Arizona and in New Mexico west of Socorro. Painter (1962) found *A. barbatus* common in Wyoming; he gives a map of the distribution of this wide-ranging species. Parker and Wakeland conclude: "Without destruction of grasshopper egg pods by larvae of bee flies, blister beetles and ground beetles, and by other natural factors, the frequency of major outbreaks over extensive areas would be greatly increased."

Anastoechus melanohalteralis Tucker. Painter (1962) records having reared this species from egg pods of *Dissosteria longipennis* Thomas.

Anastoechus nitidulus Fabricius. In a paper on the parasites and predators of *Locusta migratoria* Linné in Gascony, Roerich (1951) gives a brief discussion of this species with photographs of the larvae amid an egg case and of mature larva and pupa. He states that the larva consumes about 6 eggs. Stepanov (1882) and Chinkewitsch (1884) both record this species as preying on the eggs of *Dociostaurus maroccanus* Thunberg in Transcaucasia. Zakhvatkin (1931) says that the species *A. nitidulus* is of little importance in Turkestan. Seabra (1901) found this bee fly attacking the same locust in Portugal, while Sacharov (1913) reared these bee flies from *Locusta migratoria* Linné in Astrachan. Troitsky (1914) reared *A. nitidulus* from *Gomphocerus sibiricus* (L.) and *Stauroderus scalaris* (F.-W.) in Siberia. Likewise in Siberia, Moritz (1915) found it attacking eggs of *Pararcyptera microptera* (F.-W.). Finally, Bezrukov (1922) reared this species from the eggs of *Calliptamus italicus* Linné in Siberia. De Lepiney and Mimeur (1930) also recorded this species from *Dociostaurus maroccanus* Thunberg in the region of Guerrouaou and Djebel Guilliz (Rift). Thus, it may be clearly seen that *A. nitidulus* is a species of wide distribution, which feeds in the egg pods of at least 6 acridid hosts.

BOMBYLIINAE: *Sparnopolius* Loew

Species	Hosts	Authors
lherminierii (NE)	*Phyllophaga* Harris	Forbes (1907) Davis (1919) Malloch (1915)

Sparnopolius lherminierii Macquart. Forbes (1907) found a larva of this bee fly attached to the back of a *Phyllophaga* grub. Davis (1919) believed that *S. lherminierii* probably deposited its eggs in cracks within the ground on the chance of finding a scarabaeid grub. Although Davis found *Sparnopolius* Loew far from abundant, I have seen this species in profusion on a few occasions. One autumn I colleced about 150 flies in an hour upon a patch of *Helianthus* of about one-sixth of an acre in extent. Malloch (1915) described the pupa in detail.

BOMBYLIINAE: *Heterostylum* Macquart

Species	Hosts	Authors
robustum (NE)	*Nomia melanderi* Cockerell	Bohart, Stephen, & Eppley (1960)
	Nomia bakeri Cockerell	
	Nomia triangulifera Vachal	
	Nomadopsis anthidia Fowler	
	Nomadopsis scutellaris Fowler	
	Halictus rubicundus (Christ.)	
	Nomia melanderi Cockerell	Frick (1962)

Heterostylum robustum Osten Sacken. An excellent account of the biology of this species was published by Bohart, Stephen, and Eppley (1960). This bombyliid is a major enemy of the alkali bee *Nomia melanderi* Cockerell over a large part of its range, and the bee *Nomia* Latreille is considered an extremely valuable pollinator of alfalfa and in consequence responsible for the high yields of seed. They found that the percentages of host prepupae destroyed varied from 90 percent in some areas in Utah to as low as 5 percent in parts of Washington. In Oregon a nesting site with half a million bees had a parasite incidence of 91 percent in 1956. Before 1946 these authors state that several nesting sites contained over a hundred thousand nests; they quote Todd from personal communication, who described the ground as covered by thousands of empty bombyliid pupal skins. Two factors operate to maintain reduced bee populations; some of them nest away from congested areas; other late-emerging bees continue to lay eggs after most flies are gone. Other hosts included *Nomia bakeri* Cockerell, *N. triangulifera* Vachal; *Nomadopsis anthidius* Fowler, *N. scutellaris* Fowler, and *Halictus rubicundus* (Christ.).

The first bee flies appear 10 days earlier than the bees, but their emergence period overlaps the bee-emergence period by several weeks. Bee flies begin emerging at 8 A.M. and continue until 1 P.M., with the peak occurring between 9 and 10 A.M. After emergence the adults seek nectar before mating, and ovipositing then begins; by 2 P.M. most of the adults have sought shelter in surrounding vegetation in the deepest shade. Bohart, Stephen, and Eppley point out that dispersal must be great, for though several hundred flies may emerge from 2 or 3 square yards of soil, generally only 1 or 2 adults can be seen hovering over this area. Of 162 flies marked at emergence, only 2 were seen later. Caged flies were kept alive over a week. Blocks of soil with bees in them were moved a mile away from all nesting sites, the emerging bee flies killed, but new ones came in as soon as bee activity began.

In oviposition, the female perceives a hole or crack in the ground, dips downward, gives the abdomen a downward flick, touches hind tarsi to the ground, hovers again for a second or two, releases another egg; after this has been repeated 5 or 6 times, *H. robustum* is apt to rise up 5 or 6 inches and pursue her way, searching for more holes; the bee fly never alights during oviposition. One female was observed to make 210 dips in 20 minutes, laying an estimated 1,000 eggs during the day, even with occasional inactive periods. Twenty-five females dissected had from 44 to 424 eggs in the lower ovaries and oviducts, and a greater number undeveloped in the upper ovaries.

Out of a series of vials, blackened and buried in the soil, 50 or more eggs were collected in a single day in each vial, these eggs were 1.2 mm. long, 0.7 mm. wide, oval, and tapered equally at each end; they had an adhesive coating. In the laboratory, incubation takes 8 to 11 days. First-stage larvae are planidial, 1.6 mm. long, nearly white, very active, and besides 3 pairs of thoracic bristles, they have a long, caudal pair also. Head and neck have 8 pairs of smaller bristles.

These first-stage larvae progress much like those of *Anthrax* Scopoli, according to these authors, grasping the surface with their mouth hooks, making contact with posterior pseudopods, then pushing forward. Little is known of their early history, for they never made contact with the host until in its final instar, and out of many cells examined, the few which contained planidia prior to such time showed that these minute larvae were either inside the pollen ball or on the wall of the cell against the ball. They surmise that the planidia can enter the cells through the entrance plug after it is sealed, and that this probably accounts for the occasional appearance of *H. robustum* Osten Sacken larvae in early summer feeding on overwintered prepupae. Likewise, they suggest that early in the season the flies may flip eggs into reopened burrows of the previous season, the planidia attacking hosts that are late in breaking diapause. They suggest that the planidia are able to survive for a considerable period and point to the existence of moderate parasitism in the bee *Nomia triangulifera* Vachal, which provisions its nests after the bee fly has virtually disappeared. These authors note that recently hatched bee fly larvae are not attracted to host larvae of any age, and confined with prepupae at room temperatures, they wander about and die in 24 hours, but live longer at lower temperatures.

The question of what kind of food, if any, these planidia take in their long wait seems not to have been clearly settled. Bohart, Stephen, and Eppley noted that it was common to find 2 or 3 first-stage larvae on a single host larva, but only one made significant growth and moulted to the next stage; the fate of the other planidia is unknown, but there was no evidence of combat as in clerids or meloids. As in other bee flies, the first-instar larvae do not attach themselves firmly and often change position before starting to feed.

Third-stage bee fly larvae of *Heterostylum robustum* Osten Sacken have the mouth tightly appressed against the host; this is believed to assist in the suction necessary to draw out the fluids of the host; the wound must be indeed minute. Fabre, with lens of limited powers, was unable to observe any mouth hooks and came to believe that suction alone was sufficient to withdraw the host fluids. He even tried to inflate the empty skins of host larvae, and, while he observed no leakage of air, we know now that the larvae do possess minute, knifelike mouth hooks as described under the morphology of larvae. It is a curious fact noted by both Fabre and Bohart, Stephen, and Eppley that the host larvae not only remain alive but maintain a fresh and healthy hue during the period of engorgement of the parasite. Eventually the host becomes shrunken and its final remains are a collapsed shell. For further comments see *Anthrax trifasciatus* Meigen.

Bohart and his associates note that the fourth-instar larva will readily release its hold upon the host, if disturbed, but will soon again attach itself with the maxillae until the mouth hooks have sunk into the host's integument; they observed that the disturbed larva will just as readily attack a new host and will accept a wasp or hive bee larva with no hesitation. Still more interesting is the fact that several larvae placed together become cannibalistic, due doubtless to a very strong urge to feed. These fourth-stage larvae feed 3 or 4 days and almost double their length. It is peculiar and very interesting that in the Utah area the fourth-stage larva consumes all of one host and about half of another. As it leaves the first cell it plugs it tightly with dirt leaving host skin against the wall. It then burrows through the soil until it finds another cell with a prepupal bee; it requires about 2 days to consume half of this bee's contents. In the Oregon area it appears that only one host larva was sufficient to bring the fly parasite to maturity; as the parasite grub weighed 0.19 to 0.22 mg., and the host bee 0.126 to 0.198 mg., the margin is close. In overwintering, Bohart, Stephen, and Eppley found that the full-grown larva burrows laterally or upward from the brood chamber, which may have lain buried some 5 to 10 inches, until it lies 2 or 3 inches below the soil surface; here it forms a large, oval, overwintering cell.

This active behavior of the last instar of *Heterostylum robustum* Osten Sacken is very much in contrast to so many bee fly larvae, particularly the group attacking the Hymenoptera, which I have observed to be totally helpless and which never leave tight, tough cocoons of host. Bohart, Stephen, and Eppley note that a mature larva placed in plaster of Paris will hollow out a pupal cell; the *Heterostylum* pupa is very active and restless, alternately extending and telescoping its abdominal segments. These authors found that the pupa will gyrate the abdomen, packing soil behind it, then gyrate the anterior end to loosen more soil in front of it, repeating the process until the surface is reached. Breaking the crust, the head is thrust free, and the abdominal segments are rotated rhythmically until both head and thorax are free. Breaking the pupal skin, it crawls out as the split extends backward; for a few minutes the abdomen and rear legs remain in the case; struggling free with its hind legs, it crawls to the nearest lump of dirt or other object and remains until wings expand and dry, releasing fluid from the anus. They state that in 10 minutes it is ready to fly. Bohart, Stephen, and Eppley have provided the interested student with 16 magnificent photographs depicting stages in the life history of *Heterostylum robustum* Osten Sacken.

Frick (1962) made an extraordinarily fine study of ecological factors of *Nomia melanderi* Cockerell and *Heterostylum robustum* Osten Sacken in Washington State. Populations were sampled and percentage of parasitism of many lots was determined, the average percent being 12.1 in that area. Much seasonal emergence data were collected and a great deal of data accumulated concerning the effect of temperature. High temperatures proved rather lethal to stages beyond the prepupa, and bee fly larvae had a threshold of development about 5°F. higher than bee prepupae. Soil moisture in the spring had a marked effect on both beginning date and rate of emergence. Efforts have been made by Bohart (1958) to reduce the number of bee flies with the use of methoxychlor. While bee flies average 169 mature ova, the alkali bee has only 25, but Frick (1962) observes that the shorter life span of the bee fly works in favor of the bee. Moreover, the bee fly uses only a small part of the day for oviposition, and indiscriminate oviposition cancels out many bee flies, as Frick observes that they will even cast eggs at eyelets of shoes.

Nye and Bohart (1959) present a remarkable photographic record of the emergence stages of *Heterostylum robustum* Osten Sacken.

BOMBYLIINAE: *Callostoma* Macquart

Species	Hosts	Authors
desertorum (PA)	*Dociostaurus maroccanus* Thunberg *Calliptamus turanicus* Tarb. *Dociostaurus kraussi* (Ingen.) *Dociostaurus crucigerus-tartarus* Stschelk. *Dociostaurus albicornis turkemenus* Uv., *D. maroccanus* Thunberg *Schistocerca gregaria* Forskål *Ramburiella turcmana* F. W.	Portschinsky (1894) Zakhvatkin (1931)
fascipennes (PA)	*Dociostaurus maroccanus* Thunberg	La Baume (1918) Saunders, Waterhouse, & Fitch (1881) Austen (1937)

Callostoma desertorum Loew. This species as reported by Portschinsky (1894) feeds on egg pods of locusts, probably *Dociostaurus maroccanus* Thunberg, in Turkestan. Zakhvatkin (1931) adds six species to the list of hosts. Zakhvatkin says of this species: "... this large, brightly colored fly is one of the most valuable insect enemies of *D. maroccanus* Thunberg throughout Turkestan." He notes that it is a polyphagous species, feeding freely on eggs of several important xerophilus locusts.

Egg laying occurs in the places where the locusts swarm; most of the eggs are laid in holes or fissures in the earth, with as many as 80 or 100 deposited simultaneously, others are placed in pulverized soil by the abdomen of the female, with its excavating spines placed directly in the soil. Potential egg production ranges from 1,600 to 2,000. Primary larvae 2.5 to 2.8 mm. emerge in about a week. They hunt energetically through the soil in different directions, and on reaching an egg pod, they enter on the lower part. After 2 moults full growth ensues, and the larva leaves the egg pod for a narrow, elliptical hibernation cell. Pupation is in May, lasting 3 weeks. Average destruction of *Dociostaurus maroccanus* eggs is 20 percent, often much higher, and sometimes accomplishing almost complete destruction.

Callostoma fascipennes Macquart. This fly has been known as an effective destroyer of locust eggs since 1881, when Saunders, Waterhouse, and Fitch recorded the species destroying locust eggs in great numbers in Cyprus. They indulge in some interesting, if perhaps estimated, statistics on the abundance of the species: "Biddulph's dispatch informs us that from 5 to 8 percent of the locust eggs are this year devoured by these larvae. Since 800,000 okes of locust egg-cases have been destroyed in Cyprus this season to the end of October, it follows, from the lowest computation, that about eighty millions of our powerful natural allies—the bee flies—were associated with them, and must have been sacrificed if the destruction of the egg-cases took place before the larvae of the fly had left the cases." In their December 1881 statement they say: "Of the egg-cases received 1 oz., avoirdupois, contains 48 white-earth cases or 38 red-earth cases; say 45 cases average. An oke being 2¾ lbs. English (44 oz. avoirdupois), there would be thus about 2000 egg-cases to the oke, and if 5 percent were affected, one oke of these cases would contain about 100 Bombyliid larvae." These authors give a good drawing of the imago, and for the times a good figure of the larva, head capsule, and pupa.

La Baume (1918) discusses the biology of this species in Syria, and figures imago, larva, and pupa. The host for his material was *Dociostaurus* (as *Stauronotus*) *maroccanus* Thunberg. Austen (1937) gives a fine figure of *Callostoma fascipennes* done by Terzi. He quotes Calvert (1881):

> A grey pupa I was holding in my hand suddenly burst its envelope, and in half a minute on its legs stood a fly: thus identifying the perfect insect. . . . I found the fly, now identified, sucking the nectar of flowers, especially of the pink scabious and thistle, plants common in the Troad. (Later on I counted as many as sixteen flies on a thistle-head.) The number of flies rapidly increased daily until the 13th, when the ground appeared pitted all over with small holes from whence the parasite had issued. A few pupae were then still to be found—a larva the rare exception. The pupal state thus appears to be of short duration. It was very interesting to watch the flies appearing above ground: first the head was pushed out; then with repeated efforts the body followed; the whole operation was over in two or three minutes: the wings were expanded, but the colours did not brighten until some time after. Occasionally a pupa could not cast off its envelope and came wriggling out of the ground, when it was immediately captured by ants. Unfortunate flies that could not detach the covering membrane, adhering to the abdomen, also succumbed, as indeed many of the flies that could not get on their legs in time.

BOMBYLIINAE: *Cytherea* Fabricius

Species	Hosts	Authors
holosericea (PA)	*Panurgus canescens* Latreille (bees)	Séguy (1930)
	Egg pods:	
infuscata (as a *Glossista*) (PA)	*Dociostaurus maroccanus* Thunberg	De Lepiney & Mimeur (1930)
obscura (PA)	*Dociostaurus maroccanus* Thunberg	Stepanov (1881) Stefani-Perez (1913) Paoli (1937)
transcaspia (as *setosa*) (PA)	*Calliptamus italicus* Linné *Calliptamus turanicus* Tarbinsky	Zakhvatkin (1931)

Cytherea holosericea Fabricius. As *Chalcochiton holosericea*, Séguy (1930) illustrates a pupa of this species taken from the nest of *Panurgus canescens* Latreille. This pupa is noteworthy for the fringe of exceptionally long hairs lying transversely along the margins of the abdominal segments.

Cytherea infuscata Meigen. De Lepiney and Mimeur (1930) give a short account of the larval and pupal stages of this species, which they found destroying the egg pods of *Dociostaurus maroccanus* Thunberg in Guerouaou and Djebel Guilliz. They recorded this species as *Glossista infuscata* Meigen.

Cytherea obscura Fabricius. Stepanov (1881) reared this species from locust eggs of *Dociostaurus maroccanus* in the Crimea. Stefani-Perez (1913) likewise discusses the development of this species in Sicily and gives figures of the stages. Paoli (1937) in a fine series of three papers gives an account, not only of the biology of this species including all stages, but a very good study of the morphology of the larvae, its mouthparts, head capsule, sectional studies, and fine illustrations. Paoli's third paper concerns the morphology of the female abdomen.

Cytherea transcaspia Becker. As the synonym of this species (*setosa* Paramonov), Zakhvatkin (1931) states that this species is an important parasite on egg pods of *Calliptamus italicus* Linné in the Zeravshan Valley and likewise of *Calliptamus turanicus* Tarbinsky, in nearby deserts. He finds that this bee fly destroys an average of 13 percent of the egg pods of both species and reaches a maximum at times of 40 percent destruction. He lists no less than four species of bee flies competing for the eggs of *C. turanicus* Tarbinsky in addition to a species of *Mylabris* Muller.

BOMBYLIINAE: *Thevenemyia* Bigot

Species	Hosts	Authors
sp.	*Phymatodes* Mulsant	unpublished

In a personal communication, Mr. Jack Hall, Senior Museum Scientist, at the Riverside Banch of the University of California, tells me that he has an undescribed species of *Thevenemyia* Bigot that was reared from the cerambycid *Phymatodes* Mulsant from a maple tree. This is a very important discovery. I have before me several pupae taken from the deadwood of *Umbellularia californica* Nuttall, at Los Gatos, California, from which *Thevenemyia luctifer* Osten Sacken (as *auripilus* Bigot) was reared. These were supplied by the courtesy of Dr. Krombein of the National Museum of Natural History and Dr. Wirth of the U.S. Department of Agriculture.

Hall (1954) obtained individuals of *Thevenemyia* Bigot which were reared from logs that contained anobiid beetle larvae, *Ptilinus acuminatus* Casey. Exact host was undetermined, but very possibly consisted of these beetle larvae. Hall (1969) notes that they have been reared on several occasions from wood, usually deadwood. He names also *Ceanothus thyrsiflorus* Eschscholtz and *Pinus contorta* Douglas, and also from "dry firewood" in Australia. Several persons, including myself, have noticed a propensity of the adults for fire-blackened trees, especially around Mt. Rainier.

TOXOPHORINAE: *Lepidophora* Westwood

Species	Hosts	Authors
lepidocera (NE)	*Euodynerus foraminatus apopkensis* Robertson	Krombein (1967)
	Stenodynerus saecularis rufulus Bohart	
	Trypargilum tridentatum archboldi Krombein	
	Trypoxylon politum Say	Spears, Sartor and Hull (in litt.)
	Podium rufipes Fabricius	
	Unidentified spiders and cockroaches in wasp nests	Krombein (1967)

Lepidophora lepidocera Wiedemann. Krombein (1967) was able to rear four individuals of this remarkably interesting bee fly in the Lake Placid and Highlands Ridge area of Florida. The host wasps proved to be *Euodynerus foraminatus apopkensis* Robertson and also *Stenodynerus saecularis rufulus* Bohart. The trap-nests were tied to the side of the dead trunk of a small tree, one placed on a dead stump, and others suspended beneath the limb of live scrub hickory and oak trees. The borings for these nests were 4.8 mm. and also 6.4 mm. Krombein found this genus to be unique because the bee fly larva fed only upon the olethreutid caterpillars stored for the host wasp; it was also unique in that the content of several cells was necessary for the growth and maturation of the parasite. He noted that the adult fly eclosed a few minutes after one of the pupae wriggled free from the trap nest. He found a fifth nest, occupied by *Trypargilum tridentatum archboldi* Krombein which was probably attacked by *Lepidophora* sp., but he was unable to rear the larva as far as the pupa; this larva had sucked dry all of the spiders that this wasp had stored in two cells.

Krombein also reported (p. 255) the species *lepidocera* Wiedemann, as *appendiculata* Macquart, parasitizing *Podium rufipes* Fabricius at Lake Placid, Florida, and feeding upon the cockroaches stored by this wasp.

Sartor, in Spears, Sartor, and Hull (in litt.), reports that he found a strange and curious larva in an unplugged organ pipe mud nest of *Trypoxylon politum* Say in Lafayette County, Mississippi; this nest was filled with caterpillars and was discovered beneath a bridge in a small valley known to have a high population of *Lepidophora lepidocera* Wiedemann (more than fifty adults have been taken in this restricted area). Without realizing the nature of his find and without illustrating its cephalic morphology he allowed it to pupate, and a very short time later a perfect adult eclosed from the pupa.

TOXOPHORINAE: *Toxophora* Meigen

Species	Hosts	Authors
amphitea (NE)	*Pachodynerus erynnis* Lepeletier	Krombein (1967)
	Stenodynerus lineatifrons	
	S. beameri Bohart	
	S. saecularis rufulus Bohart	
	Euodynerus megaera Lepeletier	
	E. schwartzi Krombein	
	Ancistrocerus campestris Saussure	
	Eumenes fraternus Say	Osten Sacken (1877)
leucopyga (NE)	*Eumenes fraternus* Say	Osten Sacken (1877)
maculata (PA)	*Eumenes, Pelopaeus* (=Sceliphron), *Odynerus*	Séguy (1926)
	Eumenes pomiformis Fabricius	
pellucida (NE)	*Dianthidium dubium dilectum* Timberlake	
virgata (NE)	*Odynerus* sp.	Townsend (1893a)
	Stenodynerus toltecus Saussure	Krombein (1967)

Toxophora amphitea Walker. Krombein (1967) was able to rear more than twenty individuals of this species from vespid nests both in the Plummer's Island, Maryland, area and the Lake Placid, Florida, area. Most of his nests were suspended from limbs of living hickory in the sand-scrub areas of Florida; others on the side of standing dead tree trunks. The host wasps for these interesting flies proved to be: *Pachodynerus erynnis*

Lepeletier, *Stenodynerus lineatifrons* Bohart, *S. beameri* Bohart, *S. saecularis rufulus* Bohart, *Euodynerus megaera* Lepeletier, *E. schwartzi* Krombein, and *Ancistrocerus campestris* Saussure. He estimated a life cycle of 31 to 35 days in Maryland. He obtained no information on duration of egg stage or elapsed time until the planidiiform larvae attack the host. His data suggest that *amphitea* males emerge 4 to 7 days earlier than the females. The planidia generally attached themselves transversely on the dorsum of the host prepupa on a segment near the head; they reached larval maturity in less than a week. The parasites were in random distribution within the cells of the host wasps. Dr. Krombein notes that Osten Sacken (1877) reported Glover had reared *amphitea* from a jug nest of *Eumenes fraternus* Say feeding either on the caterpillars or the wasp larva.

Toxophora leucopyga Wiedemann. Under the name *T. fulva* Gray, this species was reported by Osten Sacken as bred from a cell of *Eumenes fraterna* Say.

Toxophora maculata Rossi. Cros (1932) gives a very detailed description without illustration of the pupa of this species, which he reared from a larva taken in the nest of *Eumenes pomiformis* F. Séguy (1926) illustrates the immature stages.

Toxophora pellucida Coquillett. Hurd and Linsley (1950) record rearing this species from a number of nests of *Dianthidium dubium dilectum* Timberlake, a megachilid bee. This bombyliid was one of five parasites, and the material was collected in Santa Clara County, California.

Toxophora virgata Osten Sacken. Townsend (1893) describes the pupal skin of this species in great detail; it was bred from a nest of *Odynerus* Latreille at Fort Collins, Colorado.

Krombein (1967) records rearing this species from the nest of the vespid *Stenodynerus toltecus* Saussure in the Portal, Arizona, area, probably others of the same species and also several others from unidentified vespids in the same area. Nests were placed beneath a cedar limb, or on fence posts, or by bridges in this desert area.

USIINAE: *Usia* Latreille

Species	Hosts	Authors
atrata (PA)	*Cataglyphis cursor* (Fonscolombe)	Xambeu (1898)

Usia atrata Fabricius. The only record of this species concerns a larva and pupa which Xambeu (1898) said he had found in the vicinity of the nest of the ant *Cataglyphis cursor* (Fons.) upon which he believed the *Usia* larva had been feeding. He gives a brief description of them. Some species of *Usia* are confined to the warmer parts of the Palaearctic region. It is difficult to understand how they could attack the ants themselves, but they might have a parasitic or scavenger role in connection with ant larvae as do *Microdon* Meigen species in the Syrphidae.

P. du Merle (1971) reared *Usia atrata* Fabricius and *U. ?aenea* Rossi from a tenebrionid beetle.

CYRTOSIINAE: *Psiloderoides* Hesse

Species	Hosts	Authors
mansfeldi (ET)	*Locustana pardalina* Walker	Hesse (1967)

Psiloderoides mansfeldi Hesse. Hesse (1967) states that this very remarkable fly was bred from the egg packets of the brown trek locust, *Locustana pardalina* Walker. These flies were obtained in the Kenhardt district of the North-Western Cape, South Africa.

CYRTOSIINAE: *Cyrtomorpha* White

Species	Hosts	Authors
flaviscutellaris (AU)	Egg pods: *Austroicetes cruciata* (Saussure)	Fuller (1938b)

Cyrtomorpha flaviscutellaris Roberts. Fuller (1938) described the mature larvae and pupae of this species but was unable to add any information on the younger stages or behavior of the adults. The larvae were found among empty egg pods of *Austroicetes cruciata* (Saussure) in a rather dense egg bed of this species in West Australia. In the laboratory she found that they remained as prepupa several months and sometimes for more than a year. She gives good illustrations of the larval mouthparts and the armature of the pupa, as well as the exceptionally minute spiracles of the larva. The larva and pupa are unusually strongly arched, a shape perhaps correlated with the unusually compact form of the adults.

MYTHICOMYIINAE: *Mythicomyia* Coquillett

Species	Hosts	Authors
sp.	*Anthophora edwardsii* Cresson	unpublished

In a recent communication Jack Hall, Senior Museum Scientist of the University of California, Riverside, notes that these little gnatlike flies have been reared from mud cells of the bee *Anthophora edwardsii*, and that numerous examples were recovered from a cell. This is a very important discovery. It is not difficult to see that such a large host could well provide food for a considerable number of these small flies.

According to Hall the recovery of many individuals from a single bee cell suggests a gregarious condition of the larvae, a multiple parasitism. Such might result from these flies alighting directly upon the mud cells, or very close to them, and depositing a number of ova, resulting in many planidia.

GERONTINAE: *Geron* Meigen

Species	Hosts	Authors
argentifrons (OR)	Laspeyresia jacculatrix Meyer (Tortricidae)	Maxwell-Lefroy (1909)
gibbosus (PA)	Nephopteryx sublineatella Strg. (Psychidae)	Mik (1896)
	Fumaria crassiorella Bruand (Psychidae)	Mik (1896)
calvus (NE)	Solenobia walshella Clemens (Psychidae)	Donahue (1968)

Geron argentifrons Brunetti. Maxwell-Lefroy (1909) reports the rearing of this species from the tortricid *Laspeyresia jacculatrix* Meyer in India.

Geron gibbosus Olivier. Mik (1896) relates that O. Werner reared this species from caterpillars of the pyralid *Nephopteryx* in Dalmatia, and from *Fumea crassiorella* Brd., a psychid, collected in Vienna.

Geron calvus Loew. Donahue (1968) reared this species from a psychid, *Solenobia walshella*, in Michigan. W. C. Martin (1968) reared thousands of bag worms from North Mississippi without finding any *Geron* species.

PHTHIRIINAE: *Phthiria* Meigen

Species	Hosts	Authors
sulphurea (NE)	? Romalea Serville	unpublished

Data taken from flies reared from sand at Crescent City, Florida, from which *Romalea* Serville were emerging. From pupae supplied by Dr. Krombein of the National Museum of Natural History and Dr. Wirth of the U.S. Department of Agriculture.

SYSTROPINAE: *Systropus* Wiedemann

Species	Hosts	Authors
barnardi (ET)	? Parathosea sp.	Hesse (1938)
bicuspis (ET)	Stenomutilla beroe Peringuey	Bezzi (1924)
conopoides (NEOT)	Sibine bonaerensis Berg	Kunckel (1904)
crudelis (ET)	Eucleid caterpillar	Westwood (1876b)
	Coenobasis amoena Felder	Hesse (1938)
fumipennis (NEOT)	Miresa clarissa (Stoll)	Goncalves (1946)
macer (NE)	Prolimacodes scapha Harris	Brooks (1952)
	Lithocodes fasciola H. S.	Schaffner (1959)
	Limacodes sp.	Walsh (1864)
	Adoneta spinuloides H. S.	Lugger (1899)
marshalli (ET)	Parasa urda Druce	Seydel (1934)
nitidus (NEOT)	Sibine fusca Stoll	Dyar (1900)
sp. (OR)	undetermined eucleid on cherry in India	Clausen (1928)

Systropus barnardi Hesse. This author illustrates the pupa of this species, which was reared from cocoons of a eucleid, presumed to be *Parathosea* sp. Hesse describes and compares the pupa with that of *Systropus crudelis* Westwood; he notes that the cephalic ridge is much reduced and abdominal spines are longer, stronger, and reduced in number.

Systropus bicuspis Bezzi. In his description of this Central African species Bezzi (1924) notes that the specimens are labeled: "Bred from cocoons of *Stenomutilla beroe* Peringuey."

Systropus conopoides Künckel. This is a South American species from Argentina, which was made the subject of a short essay by Künckel (1904). The flies were reared from the eucleid (limacodid) *Sibine bonaerensis* Berg.

Systropus crudelis Westwood. Westwood (1876), in his monograph of the eleven species of this genus known to him, describes the above species from reared material from Natal. While the host species, a caterpillar feeding on mimosa, was not named, the cocoons were of the eucleid type (limacodid), capsular in shape, and are illustrated by him, together with several views of the *Systropus* pupa. "No one," says Westwood, looking at the robust pupa, "would have supposed that it could have produced such an elongated, slender imago as the *Systropus*." Within each of the cocoons, a *Systropus* pupa was found, which was quite unlike that of other bombyliid pupae. A strong conical projection was on its head, with which Westwood surmised the pupa was able to dislodge the operculum of the cocoon. On the underside of the head was a long appendage extending as far as the first ventral segment with the basal half grooved and with the remaining distal part jointed in the middle. Westwood suggested that the paired, basal parts were likely antennal cases and that the remainder represented the proboscis and sheath. On the robust, convex abdomen the first 7 segments bore a strong, short, curved bristle laterally, and the dorsal surface was provided with a transverse row of very short, fine spines. Westwood speculates as to the type of parasitism involved. No trace of the larval skin could be found in the cocoon:

> Was its larva an internal parasite like the larvae of Tachinae or was it external, like the larva of *Scolia*. . . . the latter seems to imply difficulties in the formation of a compact, oval cocoon, like that before us, by a caterpillar infested by an external parasite, unless we suppose that it was not until the cocoon had been formed, that the egg of the parasite hatched, so as to enable the parasitic larva to feed without hindrance upon its prey within the enclosed cell of the cocoon.

Hesse (1938) records material reared from the cocoons of the eucleid *Coenobasis amoena* Felder feeding on *Acacia*.

Systropus fumipennis Westwood. Goncalves (1946) recorded this South American species bred from *Miresa clarissa* (Stoll).

Systropus macer Loew. This species has been reared on several occasions. Walsh (1864) reared this species, which he mistook for a conopid, from undetermined eucleids found on oak. Lugger (1899) also reared the species. Schaffner (1959) in rearing many thousands of Microlepidoptera only chanced upon a single individual of this species, which came from the host *Lithacodes fasciata* H.-S. in Connecticut. He believed that there were at least two generations per year. He felt that the hibernation was probably in the puparium.

Brooks (1952) reared one individual in Ontario from *Prolimacodes scapha* Harris, which emerged in April 1944, but this date probably bears no relation to the natural date of emergence. Brooks illustrates the anterior and lateral aspects of the head of the pupa and the last segment of the abdomen; these figures show it to be very different from the African species such as *Systropus crudelis* Westwood.

Systropus marshalli Bezzi. Seydel (1934) records a long series of 65 individuals of this species reared from cocoons of *Parasa urda* Druce. These he compared with *Belonogaster griseus* Fabricius as a probable mimetic model.

LOMATIINAE: *Aphoebantus* Loew

Species	Hosts	Authors
hirsutus (NE)	*Camnula pellucida* Scudder	Wilson (1936)
mus (NE)	*Camnula pellucida* Scudder	Riley (1880, 1881, 1882) Lemmon (1879)
sp. (OR)	*Tiphia* grubs (secondary parasite)	Clausen (1928)

Aphoebantus hirsutus Coquillett. A remarkable study of this species was made by Wilson (1936). He recorded results of control measures applied against an outbreak of *Camnula pellucida* Scudder in northern California in 1927; by using poisoned bran plus the natural enemies, the breeding ground in the Tule Lake area was reduced from 5,000 acres of dense egg beds to local and sparse beds in 1929. Wilson describes his method of sampling the *Aphoebantus* Loew population in some detail. The method of mating is especially interesting; he states that the male inserts its abdomen into the soil beside that of the ovipositing female—as many as 10 males may attend one female. The congregation of mating flies helped locate egg beds. A count of 104 locust capsules showed 2,886 eggs, an average of 27.75. Heavier populations of both were encountered in moist peat or sandy loam soils with sparse vegetation. A tendency of migration by the locusts to and from selected oviposition beds was noted. Oviposition beds, once established, become permanent. The *Aphoebantus* Loew attacked from 0.7 to 62.0 percent of the grasshopper-egg capsules in the egg beds of the Tule Lake district and constituted an important control factor.

Aphoebantus mus Osten Sacken. Riley was the first to discover that these small bee flies live in the egg pods of locusts during their larval life. In 1881 he describes the large, fat, white larva that J. G. Lemmon sent to him from the Pacific Coast. He quotes Lemmon (1879): "We don't know certainly what this larva becomes, but at a venture he is hailed with great joy. The ground that was first filled with locust eggs by the end of September looked as if scattered with loose shells, so thorough was the work of destruction." Riley (1881) gives a description of the stages and illustrates details of the species, which used the locust *Camnula pellucida* Scudder for its host. Lemmon (1879) recorded his experiences with the locust plague and its predators in the Sacramento, California, *Weekly Record-Union*, for November 29: "the grubs ate out and destroyed thousands of eggs last fall, . . . having lain dormant all winter, and being now found still among the eggs, which are fast hatching out (June)."

Aphoebantus sp. Clausen (1928) records for the first time an undetermined species of this genus attacking *Tiphia* Fabricius in Assam, India, in the role of a secondary parasite.

LOMATIINAE: *Lomatia* Meigen

Species	Hosts	Authors
hamifera (PA)	*Schistocerca gregaria* Forskål	Séguy (1932) quoting Régnier et al.

Régnier, Lespes, and Rungs (1931) record the obtaining of *Lomatia hamifera* Becker from egg pods of *Schistocerca gregaria* Forsk. (From Séguy 1932).

ANTHRACINAE: *Anthrax* Scopoli

Species	Hosts	Authors
albofasciatus (NE)	*Tachysphex terminatus* Smith	Marston (1964)
analis (NE)	*Cicindela scutellaris* Say subsp. *lecontei* Hald.	Shelford (1913a)
anthrax (PA)	*Hoplomerus spinipes* Linné	Verhoeff (1891)
	Osmia rufa Linné	du Buysson (1888)
	Eumenes unguiculus Villeneuve	Bugnion (1886)
	Megachile sp.	Giraud. See Brauer (1883)
	Odynerus sp., and *Osmia* sp.	Giraud. See Brauer (1883)
	Chalicodoma muraria Fabricius	Jacquelin du Val (1851)
	Odynerus spinipes Linné	Laboulbéne (1858a)
	Chalicodoma muraria Fabricius	Lampert (1886)
	Ophion sp., and *Banchus compressus* (Fabricius)	Lassman (1912)
argyropyga (NE)	*Trypargilum tridentatum archboldi* Krombein	Krombein (1967)
	Trypargilum c. rubrocinctum Packard	
	Trypargilum striatum Provancher	
	Ancistrocerus c. catskill Saussure	
	Stenodynerus f. fulvipes Saussure	
	S. pulvinatus surrufus Krombein	

Species	Hosts	Authors
aterrimus (NE)	S. saecularis rufulus Bohart S. beameri Bohart Pachodynerus erynnis Lepeletier Trypargilum striatum Provancher T. clavatum Say T. collinum rubrocinctum Packard Isodontia auripes Fernald Euodynerus megaera Lepeletier Monobia quadridens Linné Ancistrocerus spinolae Saussure	Krombein (1967)
	Bembix sp.	Rau, 1946; Spears, Sartor, and Hull (in litt.)
	Trypoxylon politum Say Sceliphron caementarium Drury	
atriplex (NE)	Megachile gentilis Cresson	Krombein (1967)
binotatus (PA)	Chalicodoma muraria Fabricius	Schaffer (1764) Frauenfeld (1861)
caffer (ET)	Ceratina nasalis Fr.	Hesse (1956)
cintalapa (NE)	Megachile concinna Smith	Butler and Ritchie (1965)
diffusus (ET)	Megachile sp.	Hesse (1956)
distigmus (PA)	Megachile nipponica Cockerell Odynerus mikado Karsch Rhychium mandarineum Saussure Sceliphron tubifex Latreille Trypoxylon obsonator Smith Solenius spp. (7 unnamed spp.) Megachile sculpturalis Smith	Iwata (1933)
fur (NE)	Sceliphron caementarium Drury	Marston (1964)
	Chalybion californicum Saussure	Rau (1940)
	Rygchium sp., Osmia sp.	Osten Sacken (1887)
irroratus (NE)	Megachile gentilis Cresson Megachile mendica Cresson Dianthidium heterulkei fraternum Timberlake Ashmeadiella bucconis denticulata Cresson Hylaeus asininus Cockerell and Casad	Krombein (1967)
isis (PA)	Egg pods of Dociostaurus maroccanus Thunberg	La Baume (1918)
jazykovi (PA)	Prepupal larvae of Epicauta erythrocephala Pallas feeding on Calliptamus italicus Linné	Zakhvatkin (1931)
limatulus s. str. (NE)	Anthophora abrupta Say	Frison (1922)
	Trypoxylon politum Say	Spears, Sartor, and Hull (in litt.)
limatulus artemesia (NE)	Sceliphron caementarium Drury Chalybion californicum Saussure Anthophora sp. Dianthidium curvatum sayi Cockerell	Custer (1928)
	Sceliphron caementarium Drury Chalybion californicum Saussure	
	Trypoxylon texense Saussure	Marston (1964)
	Anthophora occidentalis Cresson also reared from cells of Anthophora parasitized by Melecta californica miranda Fox	Marston (1964)
limatulus columbiensis (NE)	Anthophora sp.	Marston (1964)
limatulus larrea (NE)	Anthophora sp.	Marston (1964)
	Anthophora flexipes Cresson	Torchio and Youssef (1968)
limatulus vallicola (NE)	Anthophora linsleyi Timberlake	Marston (1964)
monachus (PA)	Mylabris scabiosae scabiosae in egg pods of Ramburiella turcomana F. W.	Zakhvatkin (1931)
nidicola (NE)	Diadasia consociata Timberlake	Linsley, Mac Swain, and Smith (1952, 1957)
	Diadasia enavata Cresson	Cole (1952)
oophagus (PA)	Dociostaurus maroccanus Thunberg Dociostaurus kraussi Ingenetisky Dociostaurus crucigerus tartarus Stechelk Calliptamus turanicus Tarbinskii Ramburiella turcomana Fischer von Waldheim	Zakhvatkin (1931)
tigrinus (NE)	Xylocopa virginica virginica Linné	Angus (1868); Rau (1926)
	Xylocopa californica arizonensis Cresson	Hurd (1959)
	Xylocopa tabaniformis orpifex Smith	Davidson (1893) Nininger (1916)
	Xylocopa augustii Lepeletier	Hurd (1959)
trifasciatus (PA)	Chalicoderma sp.	Fabre (1879)
trifasciatus leucogaster (PA)	Cemonus sp.	Frauenfeld (1864)
trimaculatus (NE)	Diadasia sp.	Marston (1964)
zonabriphagus (PA)	In locust egg pods feeding on other bombyliids and on Mylabris sp.	Portschinsky (1895) Troitsky (1914)

Anthrax albofasciatus Macquart. Marston (1964) figures a pupae of this species sent to him from New England, which had been taken from cells of *Tachysphex terminatus* Smith; females were seen ovipositing in the depressed areas resulting from the closure of cells; and pupae, as well as shed pupal skins, were found on the sand of the same spot in the spring.

Anthrax analis Say. This species selects a host quite unlike that of all other known members of the genus. Shelford (1913) has given a fine account of the biology of the species and its relation to tiger beetles of the genus *Cicindela* Linné. Oviposition flight is described as taking an irregular zigzag flight about 2 inches above the sand until the fly passes above a hole in the sand; as this happens the fly suddenly halts, moves backward and downward so as to touch the sand 5-10 mm. from the hole; some sand moves into the burrow; such thrusts are repeated a number of times. Shelford notes that the host larvae may appear at the surface during this time, and on two occasions the fly stopped ovipositing when it did so. By using gentle pressure on a female bee fly Shelford was able to squeeze out a large number of eggs; they were light brown ellipsoids, 0.28 by 0.12 mm. in size and not adhesive. Young larvae were most commonly found singly on the ventral side of the third instar of the host larva, where they clung between the legs. He found no second-instar larvae with parasites; here they cannot be reached by the host and do not easily come into contact with the burrow. He noted that not infrequently hosts had more than one parasite between the legs, or on other parts of the body.

Shelford (1913) was unable to learn how the bee fly larvae reach their host; he did note that host larvae, dug from a spot where a fly was seen depositing eggs on July 16, had parasites of first and second instars when removed from the burrows on September 23. He noted that an average of about 7 percent of the host larvae had parasites; some collections showed a parasitization as high as 16 percent; in pine areas only one out of several hundred larvae had a bee fly parasite upon it. The head segment of the larvae bears long, curved mandibles according to Shelford; the smallest larvae found were 0.5 to 0.6 mm. in length and were taken in late summer and autumn, occasionally in spring. Most of these larvae moulted in the fall, passing the winter attached to the host; he was not sure when the second moult took place, but believed the host had fed about a month in early June in the Chicago area. A third moult took place about the time the host stopped feeding, and in the observed cases before the pupal cell was constructed. Host pupation, Shelford found, was delayed for about a month after the cell was constructed. The parasite did not grow rapidly until the host had been in its cell about 3 weeks; by this time the old organs of the host had broken down and the internal parts were in a semifluid condition. At this time the host larva is torpid and helpless; the parasite shifts its position to the middle of the ventral side of the host; tapping food here, it grows very rapidly, increasing in length from 4.5 mm. to 10 mm. in 48 hours. Parasite growth is completed in 144 hours, its length becoming 1.8 cm., all this later growth is without further moults; it remains 6 or 7 days in a quiescent state before pupating.

Shelford (1913) describes the pupa and gives an excellent figure; he describes the digging motion of the larvae that he confined to a wide glass tube filled with sand. The average rate of progression was 1 cm. per hour; the path traveled may take the pupa an extensive distance as it works its way from the pupal chamber in a devious path to the surface. Shelford found the parasitism of this species almost confined to *Cicindela scutellaris lecontei* Haldeman.

Anthrax anthrax Schrank. This European species, the type-species of *Anthrax* Scopoli, was reared from a nest of *Eumenes unguiculus* Villeneuve, by Bugnion (1886), who reported it as *Anthrax sinuata* Fallén (lapsus for Meigen). Percheron, as early as 1835, commented briefly upon a pupa of this species, also under the name *A. sinuata*. About the same time de Buysson (1888) reported that he found the same species (again reported as *A. sinuata* Fallén) in the nest of bees of *Osmia rufa* Linné. Von Roser (1840) obtained the species from nests of *Anthophora* Latreille. Laboulbéne (1858) obtained this species from nests of *Odynerus spinipes* Linné. In 1857 he describes a pupa and reviews knowledge of bee fly life histories up to this date, pointing out that several dipterists(Zetterstedt, 1842; Walker, 1851) considered them as living in roots of plants. Lampert (1886), in a review of parasites of the wall bee *Chalicoderma muraria* Fabricius, states that he has reared many *Anthrax anthrax* (as *A. sinuata* Meigen) from the larvae of this bee. Lampert also reared no less than nine parasites from these wall bee nests, which shows that the bombyliid has considerable competition.

Verhoeff (1891) found the young larvae of this species.

Anthrax argyropyga Wiedemann. Krombein (1967) in a unique study of trap nesting of bees and wasps carried out over a period of twelve years reports finding *Anthrax argyropyga* Wiedemann and three other species of *Anthrax* Scopoli as parasites of several species of wasps and bees. The species *argyropyga* Wiedemann was obtained from the following species of wasps: *Trypargilum collinum rubrocinctum* Packard, *T. striatum* Provancher, *T. collinum* Smith, *T. tridentatum archboldi* Krombein, *Stenodynerus fulvipes* Saussure (parasitism ranging as high as 40 percent of available cells), *S. pulvinatus surrufus* Krombein, *S. saecularis rufulus* Bohart, *S. beameri* Bohart, *Pachodynerus erynnis* Lepeletier, and *Ancistrocerus collinum catskill* Saussure.

Krombein found that about 28 days were required between attachment of the first-instar larva and the emergence of the adult fly. He believed that about 2 weeks represented the elapsed time between hatching of the egg and attachment of the young larva. He notes that it is entirely possible for the first-stage larvae to remain in this instar as long as 2 months if a suitable host is not available. This species attacks the host prepupa or pupa, and 8 days were required to exhaust a host. This species occurred in the company of *Anthrax aterrimus* Bigot in one instance. They were likely to occupy any of the cells in series in a nest. His research supplies considerable additional seasonal data.

Schmidt and Hull obtained this bee fly from a trap nest containing an unidentified wasp in Mississippi.

Anthrax aterrimus Bigot. Krombein (1967) in his remarkable studies of trap-nested bees and wasps found this species living in the nests of the following wasps: *Trypargilum striatum* Provancher, *T. clavatum* Say, *T. collinum rubrocinctum* Packard, *Euodynerus megaera* Lepeletier, *Ancistrocerus spinolae* Saussure, and *Monobia quadridens* Linné, and *Isodontia auripes* Fernald. They were all from shaded settings in open woods. In many of the nests the bombyliids overwintered as diapausing larvae, which coincides with the findings of Spears, Sartor, and Hull (in litt.) in Mississippi. Krombein supplies much additional data on the ethology of this species and notes that an example has been also reared from *Isodontia philadelphica* Lepeletier.

Rau (1946) records this species as *Anthrax slossonae* Johnson, as a parasite of *Trypoxylon politum* Say.

Anthrax atriplex Marston. Krombein (1967) reports that he reared this species from the cocoons of the bee *Megachile gentilis* Cresson from Arizona. The bee fly larva fed on the bee prepupa in the cocoon. In another case it fed upon the resting bee larva. The trap nests were placed beneath dead limbs of mesquite in open desert.

Anthrax binotatus Wiedemann. Under the names *Argyramoeba subnotata* and *A. binotata* this species has been bred by several persons from larvae of the European wall bee *Chalicodoma muraria* Fabricius, of which the earliest were Schaffer (1764) and Frauenfeld (1861).

Anthrax caffer Hesse. This species was bred from the nests of the bee *Ceratina nasalis* Fr. and reported by Hesse (1956), who figured the anterior horns of the pupa.

Anthrax cintalapa Cole. Butler and Ritchie (1965) found that the bee fly *Anthrax cintalapa* Cole parasitized the bee *Megachile concinna* Smith in both Arizona and California.

Anthrax diffusus Wiedemann. Hesse (1956) figures the pupa and caudal spines of this species, the material being bred from clay nests of a *Megachile* sp. nesting in sand.

Anthrax distigmus Wiedemann. Iwata (1933) provided nesting tubes on a shelf in his house in Japan; he states that *Anthrax distigmus* visited the shelf every clear day, especially around noon in the summer of 1931. The bombyliid alighted on the nest tube, touched the entrance floor with the tip of the abdomen, and flew away, but soon returned to repeat the same procedure. Although Iwata could not find eggs, his nests proved to be heavily parasitized. He states that the active, hatched larva, finding a host larval cell enters it, attaches itself onto the larva, and remains "inoffensive" until the larva finishes the spinning of its cocoon and becomes dormant; the bombyliid then begins to suck in the body fluid of the host larva.

Iwata (1933) noted that if the host was small, the bee fly larva consumed all, leaving only the skin; if large, as in the case of *Megachile sculpturalis* Smith, it reached maturity, leaving half of the prey unconsumed. When *Anthrax distigmus* attacked *M. sculpturalis* there was only one brood a year, but if it attacked wasps and bees with two or three generations, then the bee fly likewise went through two or three generations. It is especially interesting that Iwata found the bee fly larva overwintered generally in the mature larval state, but not infrequently overwintered in the first larval instar. He noted that the pupa uses its strong spiny cephalic and caudal armature to pierce the partitions of resin or mud and works its way even from as many as nine cells to the entrance. The bands of hooks and setae on the abdomen support the pupa securely at the entrance as it emerges. Iwata found that the larva does not excrete in the larval period, the meconia is discharged for the first time after emergence, the first is white, the second black. Iwata lists other hosts from which he reared *Anthrax distigmus*.

Anthrax fur Osten Sacken. Marston (1964) found that *Anthrax fur* overwintered mostly as full-grown larvae; he found that in some cases the first-stage larvae remained as such through the winter; he found two such larvae on *Rygchium* larvae on April 19, and another on a *Sceliphron* Klug larva. Marston introduces the interesting speculation that the host larvae produces some substance in its body that is voided about the time it deposits its fecal pellet, and that such substance is necessary to stimulate the moulting of the parasite larva to the second and actively feeding stage; it could also operate as a retained substance in the host larvae, which no longer effects retarded feeding, once it is eliminated.

Marston (1964) found that about 7 days before pupation, the bee fly larvae elongates and begins to "arch" itself in the characteristic shape of the pupa; pupal spines can be seen through the hyaline integument of the larva, and the pupa, at first transparent, as it emerges by contractions of the body, quickly becomes opaque yellow. This yellow color changes and darkens some 6 or 8 days prior to emergence or even sooner, depending on temperature. Its final preemergence color is quite black. On emergence, the wings expand in 2 minutes. Marston found that it required about 2 hours before it was ready to fly. It is of particular interest that the pupa must be anchored for successful emergence; this accounts for the fact that pupal exuviae are so frequently seen attached to and extending halfway out of nest entrances with the thoracic and anterior abdominal segments free.

Under the name *Argyramoeba fur*, Rau (1940) reports a bee fly found heavily infesting nests of Mexican mud daubers in Mexico, 40 km. south of Victoria. He also reports that he found these same bee flies in mud dauber nests from Franklin, Texas. In 1964 Hull reared this species from a mud dauber nest beneath a bridge north of Laredo, Texas.

Anthrax irroratus Say. Rau (1940) records having taken a dead bee fly from a mud nest of a wasp in February from Canyon de Galeana, Mexico. He called this bee fly *Spogostylum* sp. near *oedipus* Fabricius. Townsend (1893) gives a very full description of the pupa of *Anthrax irroratus* under the name of *Argyramoeba oedipus* Fabricius. This species had been bred from a nest of *Odynerus* sp. Brooks (1952) reared individuals from *Megachile nivalis* Fries in Saskatchewan. He describes the pupa and illustrates details of the armament.

Krombein (1967) provides considerable data on the ethology and hosts of this widespread species. At Scottsdale, Arizona, he reared it from the following hosts: the megachilids, *Megachile gentilis* Cresson, *Dianthidium heterulkei fraternum* Timberlake, *Ashmeadiella bucconis denticulata* Cresson; the colletid bee *Hylaeus asininus* Cockerell and Casad and an unidentified vespid wasp also in Arizona. He gives data on the pupal period and notes that this bee fly may overwinter either as an egg or an unfed planidial larva in the nests of *Hylaeus* and *Ashmeadiella* or as a diapausing larva in the *Dianthidium* nests.

Other records for this species consist of *Megachile nivalis* Friese listed by Brooks (1952) in Saskatchewan, Baker (1895) for an odynerid type wasp in Colorado, and Cooper (1954) reared this species from *Euodynerus foraminatus* Saussure (under the name *Rygchium rugosum* Saussure).

Anthrax isis Meigen. In 1918 La Baume recorded this species as destroying the eggs of *Dociostaurus maroccanus* Thunberg in Syria, in the vicinity of the Euphrates in Wilajet Aleppo. La Baume further states that two examples of another, undetermined, species of *Anthrax* were also taken from similar egg pods.

Anthrax jazykovi Paramonov. Zakhvatkin (1931) says that the larvae of this species inhabit the egg pods of *Calliptamus italicus* Linné and are in these egg pods parasitic upon prepupal larvae of *Epicauta erythrocephala* Pallas. Second-stage larvae were found feeding upon them in May, the full-grown larvae not uncommon in June and July. Pupation occurred in July and emergence in August in the Zaravashan Valley. Between 800 and 1,000 eggs were laid. Hibernation is presumed to be as first-stage larvae.

Anthrax limatulus Say. Marston (1964) has given an extended account of the biology of *Anthrax limatulus* Say and its several subspecies. This author has reduced *Anthrax fur* Osten Sacken to subspecies status of *Anthrax limatulus* Say, but in spite of the fact that the pupal spines are alike, I am not satisfied with this status. I therefore leave *Anthrax fur* Osten Sacken as a distinct species characterized by its definite range, unique pattern and coloration, and other particulars. The author has observed that *Anthrax fur* extends farther west, and rearings show that quite distinct species of *Anthrax* do overlap and do compete for available wasp's nests for parasitism, in which numerous complex factors doubtless operate. Marston states that his reared individuals of *Anthrax fur* presumably parasitized *Sceliphron caementarium* Drury and *Chalybion californicum* Saussure (Specidae). He also reared this species from the vespid *Rygchium* sp. and the megachilid bee *Osmia* sp.

Females of *Anthrax limatulus* were observed to oviposit August 18 and 19; one of these began ovipositing in an old abandoned nest. This bee fly, close to the bridge ceiling, kept the body at an angle and when about 3 inches away from a site flipped the egg with a quick jerk of the abdomen. Marston points out that the egg has an adhesive coating that adheres to whatever it touches. Many of the oviposition sites were small holes in the concrete. Marston found that the egg-laying stimulus seems to be black or dark colored spots and noted that the bee fly even pelted eggs against the black spots on the wing tip of a resting moth. Egg-laying periods lasted from 1 to 4 minutes, the fly going away to rest nearby between layings. Marston collected some 400 eggs from six sites by means of an aspirator; 100 were taken on September 7.

Marston (1964) points to the evident high mortality of the young larvae, which are seemingly unable to use the 3 pairs of thoracic hairs with the complete efficiency of legs; he introduced as many as 5 young planidial bee fly larvae onto a single bee grub. Only 1 larva survived and Marston conjectures that this may be due to cannibalism. Incubation period of the eggs ranged upward to 18 days or more; eclosion is by means of mouth hooks from the end of the egg, the hole enlarging as the larva forces its way through, dragging the shell, until, because of wedging, it is forced off. First-instar locomotion was found by Marston to be caterpillar-like, an alternate extension of thoracic segments, attachment with mandibles, drawing up the remainder of the body in an undulatory fashion, and taking hold with proleg-like pseudopods. The first-instar larva remains within the cell for about 4 days, then attaches itself between first, second, or third thoracic segments, feeding irregularly; with restless periods it moves about over the hosts back; it moulted to the second instar about 9 days after the host larvae began its cocoon, or some 20 days after the host began to feed. In order to moult, the bee fly larva attaches to its host, leaves its cast skin holding on by the mandibles. In the second instar the bombyliid larva loses its thoracic and anal setae and abdominal pseudopods. It at once becomes more sluggish, feeding rapidly and growing about 0.5 mm. a day; about 7 days after the first moult, the second occurs and the larva still feeding rapidly grows about 2.5 mm. a day, Marston found. Host punctures apparently close up without exudate; the parasite frequently shifts its position and feeds from new spots.

Spears, Sartor, and Hull (in litt.) made extensive studies of *Anthrax limatulus* Say in northern Mississippi. The work of Spears dealt with the attack of this species and also of *Anthrax aterrimus* Bigot on mud dauber nests, of which she collected 2,638 nests. Of these nests 1,070 were *Trypoxylon politum* Say and the remainder were all *Sceliphron caementarium* Drury. Her nests came from 14 sites in 6 counties. From the nests of *Trypoxylon politum* Say she obtained 165 males and 180 females of *Anthrax limatulus* Say, but of *Anthrax aterrimus* Bigot, only 28 males and 22 females. From the 1,568 nests of *Sceliphron caementarium* Drury she obtained 19 males and 27 females of *limatulus* Say and no males and a single female of *aterrimus* Bigot. The percentage parasitism was found to be:

Host	Parasite	Parasitism by nest	Parasitism by individuals
T. politum	*A. limatulus* Say	32.24	5.27
	A. aterrimus Bigot	4.67	0.76
S. caementarium	*A. limatulus* Say	2.93	6.97
	A. aterrimus Bigot	0.06	0.15

It is very interesting that although Rau (1916, 1946) recorded no less than six species of bee flies as parasites of *Sceliphron caementarium* Drury, Spears secured only two species in northern Mississippi.

The abundance of mud dauber nests of species of *Sceliphron* Klug, and of the organ pipe mud wasp *Trypoxylon texense* Saussure is truly remarkable at times. Beneath the bridge across the Tallahatchie River in Lafayette County, Mississippi, a bridge four-tenths of a mile long, I found the mud nests approximated one hundred thousand. Many cells contained parasitized grubs.

Sartor's work concerned entirely the attack by *Anthrax limatulus* Say upon the nests of the cliff bee *Anthophora abrupta* Say. This was the only species of bee fly which he found attacking this bee. He found that under laboratory conditions no more than one larva ever developed from a host grub; when more than one was placed upon a host grub, all but one bee fly larva would leave the brood cell within 3 days. He was unable to decide positively whether cannibalism takes place under conditions of crowding, but he noted that on three occasions dead larvae were found in the same brood cells with actively feeding larvae. He noted that feeding occurs irregularly in the first instar; at each feeding it appears to make a different puncture at new sites on the dorsal side of one of the thoracic segments. The first-instar larvae moulted to the second instar 12 to 18 days after hatching. Seven days later it moulted to the third stage.

Sartor found that an *Anthrax limatulus* Say larva would consume fully the host tissues of a grub within 24 days. Pupation took place around the first of May. He comments on the wear and tear consequent to the pupal armature. Oviposition may begin as early as 8:00 A.M. on clear days and may continue for 4 hours, the active periods lasting 5 to 10 minutes followed by rest periods of 5 to 20 minutes. Oviposition was observed as early as June 10 and as late as September 15, but it is most intense in July and August. Twenty-one nesting sites in 4 counties in northern Mississippi were studied. His work extended over 1964 and 1965 and involved a study of about 6,000 brood cells and revealed a bee fly parasitism ranging from 1.6 percent to 35.4 percent.

Clifford Osborn, a former student of the author, in some uncompleted studies of bee fly parasites reared *Anthrax limatulus* Say from cocoons of a cuckoo wasp, *Chrysis* sp., where it was acting in the role of a hyperparasite.

Anthrax limatulus artemesia Marston. Marston reared this subspecies from nests of several hosts as *Sceliphron caementarium* Drury, *Chalybion californicum* Saussure, and *Trypoxylon texense* Saussure. Custer (1928) recorded this subspecies, *A. limatulus artemesia*, from nests of the leaf-cutting bee *Dianthidium curvatum sayi* Cockerell. Linsley and MacSwain (1942) pointed out that the parasitic anthophorid *Melecta californica miranca* Fox is subject to secondary parasitism by a species of *Anthrax*, which Marston identified as *A. limatulus artemesia*. Not only does the subspecies *Anthrax limatulus artemesia* attack the *Melecta*, but it also attacks the bee *Anthophora occidentalis* Cresson, which is host of the *Melecta*.

Anthrax limatulus columbiensis Marston. The only three known individuals of this subspecies were reared from nests of *Anthophora* Latreille.

Anthrax limatulus larrea Marston. One individual of this subspecies was reared from a cell of a species of *Anthophora* Latreille.

Anthrax limatulus vallicola Marston. This subspecies was first reported by Linsley and MacSwain (1942) as *Anthrax* sp. near *fur* Osten Sacken and attacking *Anthophora linsleyi* Timberlake.

Anthrax monachus Sack. Zahkvatskin (1931) found these larvae in the egg pods of *Ramburiella turcomana* F. W. In these egg pods they were hyperparasitic on the pseudopupae of *Mylabris scabiosae scabiosae* Olivier.

Anthrax nidicola Cole. In a remarkable study of the biology of *Diadasia consociata* Timberlake, an emphorine bee, which nests in colonies upon the ground, Linsley, MacSwain, and Smith (1952) discovered two bee fly parasites, which belonged to undescribed species in different genera. They found that fungi at times destroy 50 percent or more of the cells in those ground-nesting, turret-making species, but they were able to secure a good assortment of parasites by covering square-yard areas with screens placed in position in early morning; at the end of the nesting season each burrow was excavated and contents recorded; field samples of parasites were collected and were reared in depressions upon a coat of paraffin, overlaying damp (sterilized) sand in petrie dishes. The bombyliids remained as larvae in the cells until May or June, appearing then as pupae, emerging as adult flies in late May until early July. These authors state that oviposition occurs by the female hovering over the nesting area, throwing the eggs into cracks, crevices, or burrows by a rapid movement of the abdomen.

They observed that when a first-instar bombyliid was introduced into a bee burrow of the current season, it moved to an open cell by using the six long thoracic setae, which seemed to serve as legs. It remained inactive within this cell until the host larva had completed feeding and then oriented itself along one of the intersegmental membranes; here it started feeding through small, integumental punctures and continued until the host was reduced to a mere shell. Linsley, MacSwain, and Smith further noted that feeding might be delayed until the next spring. When both bombyliid and rhipiphorid larvae were introduced into the same cell the bee fly larva was the one that survived. In one case a rhipiphorid larva had almost entirely consumed the bee larva and was in consequence able to destroy the bee fly larva. These authors also concluded that bee fly parasitism was materially aided in areas where many bee turrets had been injured or destroyed. In seven samples containing 1,181 cells, the percentage of bee flies ranged from 1.2 percent to 29.3 percent. These data apply both to the *Anthrax nidicola* Cole and the *Paravilla apicola* Cole. Linsley and MacSwain (1957) also record *Anthrax nidicola* from nests of *Diadasia enavata* Cresson.

Anthrax oophagus Paramonov. Zakhvatkin (1931) found this species to live both as a parasite and as a hyperparasite within the egg pods of several species of locusts in Turkestan, such as *Dociostaurus maroccanus* Thunberg, *D. crucigerus tartarus* Stechelk, *Ramburiella turcomana* F.-W., and *Calliptamus turanicus* Tarbinsky. In this bee fly the larvae were also living hyperparasitically, both on the bee fly *Callostoma* Macquart, and the meloid, *Mylabris atrata* (Gebler). Percentage infestation of pods never rose above 6 to 9 percent. Zakhvatkin stated further that the cycle includes a partial second generation. Flies of the first generation emerge in early June and oviposit in the same places as *Callostoma* Macquart and the Moroccan locust. Eggs are placed in the powderlike, thin, upper layer of soil or thrown into fissures. After incubation of 5 to 9 days the young larvae emerge, and, while much like those of *Callostoma* Macquart, are only half as large; after undergoing two moults, they rapidly mature; some remain in the pod without change until the next spring, then pupate for 16 to 34 days; others pupate in July, the flies emerge in August and September and oviposit in the same places.

Anthrax tigrinus De Geer. This large and conspicuous Nearctic and Neotropical species of wide distribution has been observed by a number of entomologists, and the best studies so far were made by Nininger (1916) and Hurd (1959). Four species of *Xylocopa* Latreille in three subgenera of this bee have been found to harbor this bee fly. Hurd (1959) points out that many of the remaining 150 species of American carpenter bees will be found to be hosts. Interested students are especially referred to Hurd's excellent summary of this bee fly, and to his monograph on carpenter bees, for the distribution of these prospective hosts.

Synonymy of *Anthrax tigrinus* De Geer is listed under the description of the genus *Anthrax* Scopoli, but in various studies on *Xylocopa* Latreille parasites the bombyliid species has been referred to as *Anthrax simson* Fabricius, and as *Argyramoeba simson* Fabricius, or *Argyramoeba tigrina* De Geer.

Angus (1868) was the first person to find the pupa of *Anthrax* associated with *Xylocopa* Latreille. In this and in another brief paper (1868) Angus refers to this fly as *Anthrax sinuosa* Wiedemann, and while we suspect that he actually had *Anthrax tigrinus* De Geer, we cannot be certain. At any rate he observed the fly ovipositing and casting eggs toward the openings of *Xylocopa* Latreille nests "in the same manner as a bot fly depositing its eggs on the horse." The eggs he found "quite numerously around the openings of the cells of the insects, and also to extend some distance from them." In pricking some of these freshly deposited eggs, he noted that small maggots made their appearance. Angus sent his specimens of larvae to Parkard, who acknowledged them (1868) and later described them (1897).

Balduf (1962) in his comments merely repeats the notes of Angus. Davidson (1893) comments on *Anthrax tigrinus* (as *Argyramoeba simson*) and says: "it was interesting to observe this pupa with its rings of hooked hairs on its body, preventing it from going backward as it gradually wriggled itself through the partitions to the external opening leaving its case hanging to the edge of the opening." Out of 6 cells, Davidson found 3 with bee fly pupae, 1 with a chalcid pupa, and 2 live bees.

Nininger (1916) studied behavior of *Anthrax* Scopoli in *Xylocopa* cells during approximately a year in the San Dimas Mts., California. He points out that the bee fly first appeared upon stored pollen as a very minute, but extremely active, larva that "restlessly creeps about over food-mass, egg and larva, feeding promiscuously, then finally settles down and fastening itself by means of its hooked beak to the sixth or seventh segment of the *Xylocopa* larva." The first-instar larva appeared before the bee egg had hatched and was 4 weeks old or more when it attached itself, at which time its length was 3-5 mm. Growth was very slow, increasing in 2 weeks to only 4 or 5 mm., after which growth increased explosively, doubling the size of the parasite in 24 hours. From this point on, the bee grub rapidly shriveled. Nininger comments upon the advantage of retarded growth rate, which enabled the bee larva to become large enough to serve for its complete development. The full-grown fly larva remains motionless for 10 to 12 days, becomes active, moults 2 days later, and pupates; it remains as a pupa for 15 to 20 days before emergence. He found parasitism amounted to about 10 percent, and was well distributed, with usually only one parasitized larva in the same brood.

Hurd (1959) collected *Xylocopa* Latreille nests from Yucca stalks near Rodeo, New Mexico, in September. He speculated upon an apparent lack of synchrony with respect to bee fly emergence at this late season and available pollen filled bee nests; had the pupae he found in the cells been left there, it is possible that they would have remained in this state until the following spring. I certainly agree that the adult bee fly can hardly survive a winter of hibernation, although I have known emerged *Anthrax* Scopoli species to live in a bottle 6 or 7 days without food at room temperatures. Much more likely, it seems that either eggs or planidia overwinter, instead of adults.

Anthrax trifasciatus Meigen. Fabre (1908), in a truly delightful style mixed with curious teleological interpretations, recounts his observations upon the larvae of this species, which were plentiful in the nests of the European wall bee, *Chalicodorma muraria* Fabricius. It is a meticulous account by a masterly observer.

Anthrax trifasciatus leucogaster Wiedemann. Engel (1937) makes *Anthrax leucogaster* a subspecies of *trifasciatus*. Under the name *Argyramoeba leucogaster*, Frauenfeld (1864) reared the species from the wasp *Cemonus* sp. and described the pupa.

Anthrax trimaculatus Macquart. Marston (1964) illustrates a pupa of this species obtained in southern Brazil. Numerous individuals were ovipositing there in the openings of *Diadasia* nests in a forest path.

Anthrax zonabriphagus Portschinsky. Both Portschinsky (1895) and Troitsky (1914) record this species as found within the egg pods of grasshoppers, but feeding upon other bombyliid larvae, and upon *Mylabris* sp. larvae instead of the locust eggs.

The genus *Anthrax* Scopoli. Present information shows that different species of *Anthrax* have been reared from no fewer than 24 genera and 47 species of bees, wasps, locust egg pods, and beetles as parasites and sometimes as hyperparasites. According to Clausen (1928) a species has been reared from *Schistocera*. Another species of *Anthrax* attacks beetles of the genus *Cicindela* Linné. Many species attack mud nests of wasps or bees. These are easily found and brought into laboratories for observation.

Krombein (1967) in his very remarkable work on trap-nesting wasps and bees records three species of *Anthrax* as parasites of no less than twenty species and subspecies of bees and wasps; this does not take into account the host species discovered by other authors.

ANTHRACINAE: *Walkeromyia* Paramonov

Species	Hosts	Authors
lurida Walker (NEOT)	Xylocopa submordax Cockerell	unpublished

Data taken from flies reared from pupae supplied by Dr. Krombein of the National Museum of Natural History and Dr. Wirth of the U.S. Department of Agriculture.

These pupae were quite different from those of other members of the Anthracinae. They were almost identical with the pupae of *Anthrax tigrinus* Fabricius. From these studies of pupae I conclude that these very large anthracine flies should fall within a separate tribe, the Walkeromyini.

EXOPROSOPINAE: *Villa* Lioy

Species	Hosts	Authors
alternata (NE) (also as scrobiculata Loew)	Cut worms: Cut worm larvae	Riley and Howard (1890)
	Agrotis orthogonia Morrison	Brooks (1952)
	Euxoa flaviocollis Smith	
	Euxoa ochrogaster Guenee	
	Euxoa tesselata Harris	
	Feltia ducens Walker	
fulviana (NE)	Euxoa sp. pupae	Brooks (1952)
handfordi (NE)	Agrotis vetusta Walker	Brooks (1952)
hottentota (also as flava Meigen) (PA)	Lycophotia porphyrea (Denis & Schiff)	De Geer (1776) Walker (1851) Westmaas (1861) Ritsema (1868)
	Euxoa segetum (Denis & Schiff)	
	Euxoa forcipula (Denis & Schiff)	Rogenhofer in Brauer (1883)
	Dichronia aprilina (Linné) pupa	Mulsant (1852)
	Barathra brassicae (Linné)	Wahlberg (1838)
	Panolis piniperda pupa	Vassiliew (1905)
hypomelaena (NE)	Feltia herilis (Grote)	Riley and Howard (1890)
molitor (NE)	Similar to Taeniocampa rutula Grote	Riley and Howard (1890)
	Agrotis orthogonia Morrison	Brooks (1952)
paniscus (PA)	Lepidopterous pupa	Yerbury (1900)
quinquefasciata (PA)	Pine processionary moth pupa	Biliotti, Demolin, and Du Merle (1965)
sexfasciata (ET)	Spodoptera exempta (Walker)	Hesse (1956)
vitripennis (ET)	Noctuidae pupa	Hesse (1956)
alternata (NE)	Beetles: Meracantha contracta Beauvois	Hyslop (1915)
pygarga (PA)	Podonta nigrita Fabricius	Portschinsky (1915)

Villa alternata Say. Hyslop (1915) recorded this species as bred from one or two of the numerous larvae of the tenebrionid, *Meracantha contracta* Beauvois, which he found buried a few inches about the base of an old tree stump. The single bombyliid pupa emerged from the pupa of the beetle and was figured by Hyslop. The *Meracantha* Kirby larvae were quite numerous and the bee fly suffered competition from tachinid parasites of the genus *Neopales*. Brooks (1952), in a fine research report, records this species from several caterpillars and pupae of the genus *Agrotis*. He gives a key to 12 species of bombyliid pupae and to 10 species of larvae and illustrates many details; this is a very important contribution to the morphology of this family.

Villa fulviana Say. Brooks (1952) reports *Euxoa* sp. as host to *V. fulviana*, and host pupae were collected from June 30 to July 12.

Villa handfordi Curran. This species is reported by Brooks (1952) to parasitize *Agrotis vetusta* Walker.

Villa hottentota Linné. Under the name *Anthrax flava* Meigen, Wahlberg (1838) recorded rearing this species from caterpillars of the noctuid, *Barathra brassicae* (Linné). This species has been bred from a wide variety of hosts. Mulsant (1882) reported it from *Dichronia aprilina* (Linné) pupae; Vassiliew (1905) gives *Panolis piniperda* pupa as host. No less than five authors have reported *Agrotis* as host to *hottentota*.

According to Biliotti, Demolin, and Du Merle (1965), d'Androic (1956) has recorded this species as a parasite of *Thaumatopoea* (as *Cnethocampa*) *pityocampa* Schiff. I have not seen d'Androic's paper.

Villa hypomelaena Macquart. Riley and Howard (1890) record having received four individuals bred in Indiana by F. M. Webster from cutworm pupae of *Feltia herilis* (Grote).

Villa molitor Loew. Riley and Howard (1890) record a pupa resembling *Taeniocampa rufula* Grote sent to them. They give a few details of the pupal armature. The lepidopterous pupa was consumed, the bee fly issuing from a hole in one end, presumably in the larval state. Brooks (1952) lists *Agrotis orthogonia* Morr. as host of this bee fly.

Villa paniscus Rossi. Yerbury (1900) reported this species bred from a lepidopterous pupa found in sand at St. Helen's, Isle of Wight, which pupa was found July 7 and the fly emerged July 12.

Villa pygarga Loew. This species attacks tenebrionid beetle larvae, according to Portschinsky (1915); the host was *Podonta nigrita* Fabricius. According to this author, parasitism reached as high as 50 percent in some localities.

Villa quinquefasciata Wiedemann. Biliotti, Demolin, and du Merle (1965) made a very important study of the ethology of this species. They found it to be an important parasite of the pine processionary moth *Thaumetopoea pityocampa* Schiff. Egg-laying behavior was divided into two phases: In the first phase the females alight on ground composed of very fine sand where it fills up the perivaginal pouch with such sand. In phase two, it flies away, and the egg now covered with adhesive substance is released into the perivaginal pouch, there it becomes covered completely with a fine earthen film. It is then ejected into situations that avoid direct exposure to the sun, such as at the base of stones or cracks in the earth. It continues to do so until all the sand in the perivaginal pouch is used up, whereupon it refills the pouch for a new egg-laying sequence. The authors note the importance of soil types in meeting satisfactory conditions for this species.

Villa sexfasciata Wiedemann. Hesse (1956) reared this species from the noctuid *Spodoptera exempta* (Walker), both in Pretoria and Southern Rhodesia. The pupa is described, but not illustrated.

Villa vitripennis Loew. Hesse (1965) notes that a large individual of this species was reared from a moth pupa of the family Noctuidae in 1941. Hesse figures parts of the pupal skin.

EXOPROSOPINAE: *Poecilanthrax* Osten Sacken

Species	Hosts	Authors
lucifer (NE)	Wasps: Hyperparasite of *Elis haemorrhoidalis* Fabricius	Box (1925)
alcyon (NE)	Cut worms: *Chorizagrotis thanatologia* Dyar	
	Euxoa ochrogaster Guenee	
	Euxoa flavicollis Smith	
	Pseudaletia unipuncta Haw.	Brooks (1952)
	Peridroma margaritosa Haw.	Walkden (1950)
fasciatus (NE)	*Chorizagrotis auxiliaris* Grote	Walkden (1950)
flaviceps fuliginosus (NE)	*Agrotis subterranea* Fabricius	Hall (Painter and Hall, 1960)
lucifer (NE)	*Laphygma frugiperda* A. & S.	Allen (1921)
sackeni monticola (NE)	*Crymodes devastator* Brace	
	Agrotis orthogonia Morrison	Brooks (1952)
tegminipennis (NE)	*Euxoa flavicollis* Smith	
	Agrotis c-nigrum L.	Brooks (1952)
willistoni (NE)	*Agroperina dubitans* Walker	
	Auxoa ochrogaster Guenee	
	Euxoa flavicollis Smith	
	Euxoa tesselata Harris	
	Feltia ducens Walker	
	Crymodes devastator Brace	
	Chorizagrotis thanatologia Dyar	Brooks (1952)
	Euxoa scandens Riley	Walkden (1950)
	Chorizagrotis auxiliaris Grote	Painter and Hall (1960)

All except one of the known hosts of *Poecilanthrax* Osten Sacken are noctuids. Box (1925) reported the rearing of *P. lucifer* Fabricius from *Elis haemorrhoidalis* Fabricius in Puerto Rico. Seven species of this Nearctic genus have been recorded as endoparasites of noctuids of the sod worm or cutworm type. Very little information is available on the earlier phases of these life histories, and we are left to conjecture with respect to certain details. Brooks (1952) thought the flies overwintered as eggs in soil or vegetation. Painter and Hall (1960), while admitting that this may be true in the north, surmise that in southern areas the eggs may hatch soon after oviposition and the planidia search around for any small caterpillars lurking among grass or vegetation roots. Another factor, the number of generations, plays a part here.

In Mississippi the present writer found *P. lucifer* Fabricius apparently bivoltine. On May 1, 1963, extraordinary numbers of this species were present on a small plot beside the University of Mississippi campus highway, over a hundred were collected in a little more than 2 hours; about 60 were collected near 10:00 A.M. from abundant flowers of *Coreopsis*, until scarcely any were left. On return an hour later, almost as many more were collected. With *P. lucifer* were a few *Parabombylius* sp. In Mississippi *P. lucifer* almost disappears in late May and June, reappearing in late July or August. They again become quite abundant in September and October. H. W. Allen (1921) reared numerous individuals of *P. lucifer* from the southern grass worm *Laphygma frugiperda* A. & S. Parasitized larvae pupated, but shortly thereafter the bee fly larvae matured and pupated within the pupa of the grass worm, from which it emerged by twisting and wriggling until free. Harned (1921) also comments on the abundance of this parasite.

Painter and Hall (1960) suggest that the planidial life in this genus is short, perhaps no more than a week, and that the remaining larval instars are of brief duration. The flies emerge soon after the host pupates. Emergence from the host pupal skin is in the usual way, the parasite cutting its way out by means of its cephalic spines, resting a short time, then wriggling its way to the surface; the pupal case projects partly from the soil.

The extent of parasitism by *Poecilanthrax* Osten Sacken in cut worm populations varies widely. Walkden (1950) found it relatively low in Kansas. The several species known to Brooks (1952) gave parasitism of 2 to 5 percent. *P. flaviceps fuliginosus* had a low incidence, only 1.3 percent. This species was reared by Hall (Painter and Hall, 1960) from *Agrotis subterraneum* Fabricius in Calexico. Clausen (1940) gave 18 to 25 percent for the range of parasitism in *P. lucifer* Fabricius. Almost certainly the parasitism of this species may rise to 50 percent in some areas at certain times. Most species of this genus was univoltine, but Painter and Hall (1960) note that several species appear to be bivoltine, at least at times, these are: *Poecilanthrax flaviceps flaviceps* Loew, *P. flaviceps fuliginosus* Loew, *P. effrenus* Coquillett, *P. lucifer* Fabricius, and more rarely *P. poecilogaster poecilogaster* Osten Sacken.

EXOPROSOPINAE: *Paravilla* Painter

Species	Hosts	Authors
apicola (NE)	Hymenoptera: *Diadasia consociata* Timberlake, *D. bituberculata* Cresson	Linsley & MacSwain (1952)
gorgon (NEOT)	*Elis* sp.	Wolcott (1923-1924)
perplexa (NE)	*Diadasia diminuta* Cresson	Linsley & MacSwain (1957)
tricellula (NE)	*Diadasia bituberculata* Cresson	Linsley & MacSwain (1952)
sp. (near *P. flavipilosa*) (NE)	*Diadasia vallicola* Timberlake	Linsley & MacSwain (1957)

Paravilla apicola Cole. This was one of two species of bee flies found parasitizing *Diadasia consociata* Timberlake, in San Joaquin County, California. The discussion under *Anthrax nidicola* Cole applies to this species as well. *P. apicola* also parasitizes *Diadasia bituberculata* Cresson.

Paravilla gorgon Fabricius was reared by Wolcott (1923-1924) from cocoons of *Elis haemorrhoidalis* Fabricius in Puerto Rico, behaving in this case as a secondary parasite.

Paravilla perplexa Coquillett. According to Linsley and MacSwain (1957), this species was seen commonly ovipositing in nesting sites of *Diadasia diminuta* Cresson near Carlsbad, New Mexico. They state, "The female hovers over a turret and drops lower to oviposit, repeating this activity several times at each burrow. After ovipositing in several burrows, the bee fly rests on the ground a few feet from the nesting site for several minutes. Later she resumes oviposition and frequently includes some of the same burrows in the second visit." Still another species of *Paravilla*, near *P. flavipilosa* Cole, was found ovipositing within the nests of *D. vallicola* Timberlake. While they suggest that certain species of bee flies may be regularly associated with *Diadasia*, they note that some species at least are not host specific, and host size is not a limiting factor. *P. apicola* was reared from both the small cells of *Diadasia consociata* and the large cells of *D. bituberculata*.

Paravilla tricellula Cole. Linsley and MacSwain (1952) obtained this species from ground-nesting bees of *Diadasia bituberculata* Cresson in San Diego County, California. This fly, one of five parasites of this bee, was obtained from 102 cells, or a total percentage of 27.1. Linsley (1958) evaluates this and other species of bombyliids that attack ground-nesting bees. Writing of bees, he states, "potentially, the most effective insect parasites of gregarious, ground-nesting species are probably bombyliid flies, which in a *Nomia* Latreille colony are capable of infesting almost 100 per cent of the burrows, which are open at a given time."

EXOPROSOPINAE: *Rhynchanthrax* Painter

Species	Hosts	Authors
parvicornis (NE)	*Tiphia* sp. (on *Phyllophaga* spp.)	Davis (1919)

Rhynchanthrax parvicornis Loew. Davis (1919) obtained this species from *Tiphia* Fabricius cocoons; the exit hole in the cocoon, always made at the end, is circular and sharp, as if made with a sharp knife. *Tiphia* emergence holes are ragged and made a little to one side of the tip.

EXOPROSOPINAE: *Chrysanthrax* Osten Sacken

Species	Hosts	Authors
cypris (NE)	*Myzinum ephippium* (Fabricius)	Swezey (1915) Davis (1919)
edititia (NE)	*Anthophora montana* Cresson	Davidson (1900)

Chrysanthrax cypris Meigen. Swezey (1915) reared this species from cocoons of *Myzinum ephippium* (Fabricius), at Urbana, Illinois.

Chrysanthrax edititia Say. Davidson (1900) relates the curiously long-drawn-out larval history of 10 individuals of bee fly larvae he took, which were parasitizing *Anthophora montana* Cresson. He got them in July 1895; one bee fly hatched out in the same season; 2 years later another pupated, but died; in July 1899, after a short pupation period, four others succeeded in emerging as adult flies.

EXOPROSOPINAE: *Oestranthrax* Bezzi

Species	Hosts	Authors
goliath (OR)	Cossid moth: Pupa of a cossid moth (indet.)	Oldroyd (1951)

Oestranthrax goliath Oldroyd. In 1951 Oldroyd recorded, described, and illustrated this very large bee fly from Malaya, which was reared from a cossid pupa. Many interesting questions might be raised as to how the fly reaches the larvae. Oldroyd says that the general appearance of the adult strongly suggests a nocturnal or crepuscular habit.

EXOPROSOPINAE: *Dipalta* Osten Sacken

Species	Hosts	Authors
serpentina (NE)	*Myrmeleon immaculatus* De Geer	Smith, Roger C. (1934)

Dipalta serpentina Osten Sacken. Roger C. Smith (1934) relates that he obtained individuals of this species, which he reared from cocoons of the above ant lion collected in Kansas. This record extends the hosts of bee flies into the order Neuroptera.

EXOPROSOPINAE: *Thyridanthrax* Osten Sacken

Species	Hosts	Authors
fenestratus (PA)	Egg pods: *Pararcyptera microptera* (Fischer-Waldheim)	Portschinsky (1895) Bezrukov (1922)
	Dociostaurus maroccanus Thunberg	Séguy (1930)
	Ocneridia volxemii (I. Bolivar)	Kunckel (1893, 1894)
pallidipennis (PA)	*Dociostaurus maroccanus* Thunberg	Zakhvatkin (1931)
perspicillaris (PA)	*Paracyptera microptera* (F.-W.)	Troitsky (1914)
abruptus; argentifrons; alliopterus; beneficus; brevifacies; burtii; lloydi; lugens; salutaris; transiens (All ET)	Tsetse flies: *Glossina morsitan* Westwood, *brevipalpis* Newstead, *austeni* Newstead, *pallidipes* Austen, *tachinoides* Westwood	Austen (1914, 1922, 1929) Lloyd (1916) Hegh (1929) Lamborn (1915) Fiedler and Kluge (1954)
leucoproctus (ET) lugens (ET)	Cutworms: *Loxostege frustalis* (Zeller) (Pyralidae)	Hesse (1956)
atratus (NE)	Wasps: *Bembix occidentalis beutenmulleri* Fox	Bohart and MacSwain (1939)
lugens (ET)	Muscid fly: Muscid fly found in a nest of *Odontotermes badius* (Haviland)	Hesse (1956)
velutinus (PA)	*Thaumatopoea* (as *Cnethocampa*) *pityocampa* Schiff	Biliotti, Demolin, and du Merle (1965)

Some of the Palaearctic species of *Thyridanthrax* Osten Sacken attack the egg cases of locusts. A Nearctic species attacks bembecids, while South African species utilize tsetse flies (*Glossina*) as hosts. One South African species has been reared from a muscid fly. Still others act as both primary parasites, and as secondary parasites of braconids, in noctuid caterpillars of the army worm and cut worm type.

Thyridanthrax fenestratus Fallén. Bezrukov (1922) found this species living in the egg pods of the locust *Pararcyptera microptera* (F.-W.) in Siberia. Kunckel (1894) records it from eggs of *Ocheridia* (*Ocneridia*) at Biska, and *Stauronotus* at R'hiras. Séguy (1930) lists the larva of this species as found in the egg pods of *Dociostaurus maroccanus* Thunberg, *Ocneridia volxemii* (I. Bolivar), and *Pararcyptera microptera* (F.-W.)

Thyridanthrax pallidipennis Paramonov also destroys locust egg pods.

Thyridanthrax perspicillaris Loew. Troitsky (1914) records this species as feeding in the egg pods of *Pararcyptera microptera* (F.-W.) in Siberia.

Thyridanthrax lloydi Austen. This species was described by Austen (1914); he noted it to be the first species of Diptera recorded as a parasite of tsetse flies, the genus *Glossina* Wiedemann. The original discovery was made by Lloyd, who comments upon the relationship in 1916. He states that the pupa of the bee fly bursts forth from the *Glossina* pupa anteriorly and works its way to the surface of the ground by means of the long fringe of hairs which are alternately raised and depressed, with the peglike legs acting as a fulcrum. On the surface of the ground it becomes quiescent, elongates, splits along the dorsal surface, and the time from emergence to flight occupies only 2 or 3 minutes.

Austen (1914, 1929) and Hesse (1956) have shown that 10 species of *Thyridanthrax* Osten Sacken are parasites of 5 species of *Glossina* Wiedemann.

The additional species of *Thyridanthrax* known to use *Glossina* Wiedemann as hosts are: *T. abruptus* Loew, *argentifrons* Austin, *alliopterus* Hesse, *beneficus* Austen, *brevifacies* Hesse, *burtii* Hesse, *lugens* Loew, *salutaris* Austen, *transiens* Bezzi. Records available to the author show the following: (1) Of these species, 5 have been bred from only a single species of *Glossina*. From *Glossina morsitans* were bred the 4 species of *T. beneficus*, *lloydi*, *salutaris*, and *transiens*, and from *Glossina austeni* was bred the species *T. burtii*. (2) Two species of bombyliids were bred from 2 species of *Glossina*. *T. argentifrons* was bred from *Glossina morsitans* and *G. tachinoides*. The species *T. lugens* was reared in small numbers from both *Glossina morsitans* and *G. austeni*. (3) However, Hesse (1956) points out that the remaining 3 species of *Thyridanthrax* were each reared from 3 species of *Glossina*. Twenty individuals of *T. alliopterus* were secured from *Glossina austeni*, *brevipalpis*, and *pallidipes*. And 98 specimens of *T. brevifacies* were obtained from these same 3 species of *Glossina*. No less than 371 individuals of *T. abruptus* were reared from *Glossina morsitans*, *brevipalpis*, and *pallidipes*. No less than 359 of the *T. abruptus* species were bred from *G. morsitans* alone. Thus, it will be seen that *Glossina morsitans* is parasitized by no less than 7 species of *Thyridanthrax* and is the most heavily parasitized species. Hesse (1956) notes that the extension of several of these species of *Thyridanthrax*, namely *T. abruptus*, *brevifacies*, *lugens*, and *transiens*, into areas where tsetse flies do not occur indicates still other hosts for these species of *Thyridanthrax*.

Thyridanthrax leucoproctus Loew. Hesse (1956) calls attention to the fact that this species was bred by S. J. S. Marias from caterpillars of the Karoo army worm *Loxostege frustalis* (Zeller). He notes further that Marias was able to show that this bee fly was actually a secondary parasite of the army worm, since the bred individuals emerged from pupae lodging within 2 braconid cocoons of the genus *Macrocentrus*. *Thyridanthrax leucoproctus* then becomes a harmful species of bombyliid.

Thyridanthrax lugens Loew was also bred from the Karoo army worm, the noctuid *Loxostege frustalis* (Zeller).

Thyridanthrax atratus Coquillett. This large, black, American species was found by Bohart and MacSwain (1939) to be parasitizing *Bembix occidentalis beutenmulleri* Fox associated with *Exoprosopa eremita* Osten Sacken as a competitor. The area studied was Contra Costa County, California. See comments under *Exoprosopa* applying to both species.

Thyridanthrax lugens Loew. Hesse (1956) reports this species as parasitizing a muscoid fly found in a nest of *Odontotermes badius* (Haviland).

Thyridanthrax velutinus Meigen. Biliotti, Demolin, and du Merle (1965) point out that d'Androic (1956) records *T. velutinus* as a parasite of *Thaumatopoea* (as *Cnethocampa*) *pityocampa* Schiff. I have not seen d'Androic's paper.

EXOPROSOPINAE: *Hemipenthes* Loew

Species	Hosts	Authors
catulina (NE)	*Bessa harveyi* Townsend, parasite of the sawfly *Pristiphora* sp.	Brooks (1952)
maura (PA)	Cocoons of *Ophion* sp. and *Banchus* sp. as hyperparasite of *Panolis piniperda* Panzer	Portschinsky (1895)
morio (PA)	*Ophion* sp., *Banchus compressus* F.	Lassmann (1912)
	Nemoraea sp. on *Panolis piniperda* Panzer	Sack (1899)
	Masicera silvatica (Fallén)	Vassiliew (1905)
	Parasetigena sp., and *Ernestia rudis* (Fallén)	Baer (1920)
	Dendrolinus sp.	Séguy (1930)
	Gonia spp. and *Bonnetia comta* Fallén out of noctuid pupae	Brooks (1952)
sinuosa (NE)	*Neodiprion sertifer* Geoff.	Finlayson and Finlayson (1958)
	Neodiprion sertifer Geoff.	Griffiths (1959)
velutina (PA)	*Masicera silvatica* (Fallén) as hyperparasite of *Dendrolinus pini* L.	Vassiliew (1905)
sp. (NE)	*Winthemia* sp.	Clausen (1928)
sp. (NE)	*Paraphyto opaca* Coquillett	Brooks (1952)

Hemipenthes catulina Coquillett. This species acts as a hyperparasite. Brooks (1952) obtained flies from the tachinid *Bessa harveyi* Townsend, which is a parasite of the sawfly *Pristiphora* sp. in British Columbia.

Hemipenthes maura Linné. In 1895, Portschinsky obtained this species from the cocoons of *Ophion* sp. and *Banchus* sp., where it assumed the role of a hyperparasite of *Panolis piniperda* Panzer.

Hemipenthes morio Linné. This species, recorded as *Anthrax morio* Linné, was reported by Lassmann (1912) as a hyperparasite, that is a secondary parasite, from a species of the genus *Ophion* Fabricius and also from *Banchus compressus* F. In 1899, Sack recorded this bee fly from the tachinid *Nemoraea*, which had been parasitizing *Panolis piniperda* Panzer.

Baer (1920) reared this species from the tachinids *Parasetina* sp., and *Ernestia rudis* (Fallén) where *Hemipenthes morio* was acting as a secondary parasite to them. Vassiliew (1905) had previously reared this bee fly from *Masicera silvatica* (Fallén), a tachinid parasite of other Lepidoptera. It seems likely that its chief role, perhaps its only role, is that of a hyperparasite, in which case *H. morio* must be classed as a harmful species in man's economy. Baer raises some interesting questions. Does the *Hemipenthes* larva enter the tachinid larva before or after it leaves the body of the caterpillar? Does it penetrate the puparium, or does it reach the tachinid host while the latter is crawling about the ground seeking a place to pupate? Baer felt the latter possibility was more probable. If true, bee fly eggs must be laid before pupation of the tachinid, and after hatching, the bee fly larvae must await the maturing of its host, which might take place after attachment.

There seems to be no evidence, as far as *Hemipenthes* Loew is concerned, whether the bee fly is endoparasitic, or feeds externally upon the host within its puparium, which it could certainly do. It is believed that those bee flies which parasitize noctuid pupae do lead an endoparasitic existence.

Brooks (1952) recorded the rearing of this species (as *H. moriodes* Say) from the tachinids *Gonia* spp. and *Bonnetia comta* Fallén out of noctuid pupae in Canada.

Hemipenthes sinuosa Wiedemann. Finlayson and Finlayson (1958) reared this species from large numbers of cocoons of the European pine sawfly *Neodiprion sertifer* Geoffroy in Ontario during 5 of the years between 1941-1949. The percentage of parasitism by *Hemipenthes* Loew was given as 0.3 in 1941 on 626 cocoons; 0.1 in 1943 on 2,308 cocoons; 4.5 in 1946 on 1,892 cocoons; 0.4 in 1947 on 3,225 cocoons; and 12.0 in 1949 on 275 cocoons.

Their study of the parasites of this sawfly recorded 23 species; 13 species in 9 genera were Ichneumonidae; 6 species in 5 genera were chalcids; and 4 species in 4 genera were Diptera. The other flies included tachinids *Neophorocera hamata* A. & W. and *Spathimeigenia erecta* Ald. They give a description of the last instar larva of the bee fly, a figure of the buccopharyngeal apparatus, and an arched, dorsal view of a pupal exuvium, which did not show the details of the coronal armament. The exit hole on the sawfly cocoon is directly on the end of the cocoon with the cap attached. Brooks (1952) also reared this bee fly from *Neodiprion* sp. in British Columbia.

Hemipenthes velutina Meigen. Vassiliew (1905) reared this species from pupae of *Masicera silvatica* (Fallén), where it was acting as a hyperparasite of *Dendrolinus pini* L.

EXOPROSOPINAE: *Exoprosopa* Macquart

Species	Hosts	Authors
apicalis (ET)	*Tachypompilus ignitus* (Smith)	Hesse (1956)
caliptera (NE)	Parasitic Hymenoptera	Brooks (1952)
eremita (NE)	*Bembix occidentalis beutenmulleri* Fox	Bohart and MacSwain (1919)
fasciata (NE)	*Phyllophaga* spp. (pupae)	Richter and Fluke (1935)
fascipennis (NE)	*Tiphia* Fabricius on *Phyllophaga* Kirby	Davis (1919), Richter and Fluke (1935)
pueblensis (NE)	*Tiphia* Fabricius on *Phyllophaga* Kirby	Davis (1919)

Exoprosopa apicalis Wiedemann. Hesse (1956) noted that an example of this species had been reared from the pupal cocoons of the large pompilid wasp *Tachypompilus ignitus* (Smith), which in turn preys upon the large theraphosid spider *Harpactira*.

Exoprosopa caliptera Say. Brooks (1952) reared an individual from an unidentified parasitic hymenopteron in Alberta. He describes the pupa and illustrates the cephalic spines.

Exoprosopa eremita Osten Sacken. This species was given considerable study by Bohart and MacSwain (1939) in a sand-dune area in Contra Costa County, California. It was accompanied by *Thyridanthrax atrata* Coquillett, which behaved quite similarly. Both species were parasitizing *Bembix occidentalis beutenmulleri* Fox. Flies became active only after dunes were warmed by the late morning sun, and they became inactive as early as 3:00 P.M. In ovipositing, the flies hovered a few inches from a *Bembix* nest hole, flipping the tip of the abdomen forward; as this is done several times at each burrow, it seems clear that several eggs are cast toward each tunnel. After such behavior the female rests on sand a few inches from the tunnel before searching for another burrow. Perhaps this rest is needed to allow additional eggs to move downward to the exit of the ovipositors.

Bohart and MacSwain found that only prepupae were attacked; the first moult left the larvae maggot-like, without its bristles, and with head region retracted. The bee fly overwinters as a mature larva; pupation takes place in the summer within the *Bembix* cocoon. The combined percentage of parasitism of *Bembix* Fabricius by the two species amounted to about 1 percent. These authors surmise that the habit of this species of *Bembix* of keeping its entrance covered while hunting and after complete provisioning accounts for the low degree of parasitism, inasmuch as the bombyliids will only oviposit at an opened burrow while the wasp is inside. Bohart and MacSwain figure the adults and larvae in situ in cocoons.

Exoprosopa fasciata Macquart. Richter and Fluke (1935), in a brief but remarkable study, record finding 27 bombyliid larvae of this species attacking exclusively the pupae of *Phyllophaga* sp. in 4 localities of Wisconsin pasture land in 1933 and 1934. Most of them were found at an average depth of 15.5 inches, while 237 unparasitized pupae and prepupae were arranged at an average depth of 16.2 inches. Parasitism was slightly below 10 percent at the principal locality. These authors further state that the small larvae of the bee fly feed on the ventral surface of the white grub pupae from the middle of July and increase rapidly in size, leaving nothing but shriveled grub pupal skins; these mature larvae then remain in old pupal grub cells during the following winter. They note further that 3 full-grown bee fly larvae collected in April were still larvae in December of the same year. One larva taken in 1933 became full grown that season, was kept over winter at room temperature, pupated on September 30 of the following year, and emerged in November.

Exoprosopa fascipennis Say. This species was first reared from *Tiphia* Fabricius cocoons by Forbes (1907). Davis (1919), in his excellent report on the natural enemies of *Phyllophaga*, notes that *E. fasci-*

pennis had been reared from such cocoons in several parts of Indiana and Illinois; he did not see any adults ovipositing and surmised that the flies might lay their eggs on flowers or directly on *Tiphia* Fabricius.

Exoprosopa pueblensis Jaennicke. Davis (1919) figures a pupa and adult of this species, which had been reared from a *Tiphia* cocoon collected in Kansas.

EXOPROSOPINAE: *Litorrhynchus* Macquart

Species	Hosts	Authors
dilatatum (ET)	*Sceliphron spirifex* (Linné)	Bezzi (1921) Hesse (1956)

Litorrhynchus dilatatum Bezzi. In 1921, Bezzi relates that a specimen of *Litorrhynchus tollini* Loew was bred by Dr. Peringuey from the mud nest of the ubiquitous South African sphecid *Sceliphron quartinae* (Gribodo) together with a large ichneumonid, *Osprynchotus capensis* Spinola, and a mutillid, *Dolichomutilla sycorax* (Smith). However, Hesse (1956) states that the bee fly is properly the species *Litorrhynchus dilatatum* Bezzi, and the wasp properly *Sceliphron spirifex* (Linné). Actual host of the fly within the nest was not determined.

EXOPROSOPINAE: *Ligyra* Newman

Species	Hosts	Authors
morio (as erythrocephalus) (NEOT)	*Monedula* sp. *Pompilus* Fabricius	Ruiz Pereira (1929, 1930) Copello (1933) Sorensen (1884)
oenomaus (OR)	*Campsomeris* Lepeletier *Scolia* Fabricius *Tiphia* Fabricius	Clausen (1928)

Ligyra morio Fabricius. In 1929 and 1930, Flaminio Ruiz Pereira published some notes on this species under the name *Exoprosopa erythrocephala* Fabricius. These bee flies were found in the nest of *Bembix* spp. in the Chilean province of Atacama and were utilizing the wasps as hosts. Copello (1933) recorded the same species as a parasite of *Monedula surinamensis* De Geer from the barancas of San Isidro. He described egg and larval stages and gave rather crudely executed figures of several stages; these suffice to show that the planidium is of the usual type; the spines as shown upon the pupa are unusually stout and heavy.

Ligyra oenomaus Rondani. Under the old genus name *Hyperalonia* Rondani, Clausen (1928) presented a remarkable ethological and statistical study of the relationship of this species of bee fly to its hosts, *Tiphia* sp. This *Tiphia* sp. was attacking the grubs of *Anomala dimidiata* Hope.

Field collections of cocoons made by a large species of *Tiphia* at Shillong, India, in the summer of 1925 showed parasitism of more than 60 percent by this bee fly, consequently causing a sharp reduction in the effective parasitism of *Anomala dimidiata* by scoliids.

Clausen's data showed that *Tiphia* adults began to emerge May 25 and ended July 20. *Tiphia* larvae that were feeding were found beginning June 16 and ended August 17. *Tiphia* larvae in the cocoons were first found June 26 and continued beyond September; whereas, *Ligyra* adults first emerged July 8 and ended August 29. Parasitism of the *Tiphia* cocoons by *Ligyra* Newman began August 1, mounting slowly at first and then sharply after September 2, to a peak of 55 percent on September 26. It is noteworthy that the adult population of *Ligyra* began as much as 5 to 6 weeks after that of its host and the greater part of the *Tiphia* larvae had spun cocoons before any adult *Ligyra* appeared. In investigations extended over two summers, 1925 and 1926, more than 1,400 cocoons of *Tiphia* were collected. In the first summer the parasitism ranged from 57.6 to 65.3 percent. In 1926, beginning with zero parasitism until July 15, the percentage gradually rose from 1.9 on August 2 to 56.3 on September 30.

No eggs or egg laying by *Ligyra* Newman was observed, and only two first-instar larvae were found, these in September. A dissected, gravid female showed 537 mature, slightly brownish, ellipsoidal eggs, each 0.5 mm. long. As the life of the female fly is long, the author pointed out that the ultimate total number of eggs per female may be much greater. Clausen (1928) concluded that the period of the first instar must be a very short one, but that the time spent by them in search of cocoons was considerable. The *Tiphia* cocoons lay in sandy, light soil at a depth of 1 to 3 inches. Clausen surmised that the planidial larvae must be obliged to seek out the *Tiphia* cocoons. He felt that it was more likely that the planidium would be able to penetrate the cocoon than that the adult *Ligyra* would be able to detect the presence of the *Tiphia* cocoons beneath the soil, but this must remain an unproved point.

Clausen also concluded that the planidium must penetrate the cocoons of *Tiphia* Fabricius, since such a large part of the *Tiphia* population is so encased by the time the bee flies are about. The planidium is at first 0.9 mm. long and reaches 2.0 mm. before moulting. Three pairs of long, thoracic setae and a pair of caudal setae are lost at the first moult. Clausen states that feeding begins almost immediately when they enter the cocoon and is largely in the thoracic region. The second-instar larva measures from 2.7 to 3.5 mm., is capable of free movement, and the host larva begins to show signs of numerous feeding punctures in its integument on both thorax and abdomen. In the third and final instar the thoracic segments are yellowish, the abdomen is milky white, due to the numerous fat bodies beneath. Winter and spring are passed in this state and pupation is late in June. Emergence is by means of a circular cut at the head end of the cocoon, in contrast to an irregular hole made at one side of the anterior end by *Tiphia* Fabricius. This author found that the *Ligyra* Newman normally attacks the larva of the host and rather uncommonly the pupa. A few individuals of this bee fly were reared from cocoons of *Scolia* sp. and *Campsomeris* sp.

Clausen (1928) reports that *Ligyra* sp. has been reared from the psammocharid *Pseudagenia* Kohl, but I have not seen the reference.

Ethology and Interrelationships of the Adult

Bee flies are characteristically sun-loving insects. They are expert in flight, and in most species the wings are large in proportion to the body, rendering them highly adapted to hovering and poising, whether in search of hosts or in the process of taking nectar from flowers. This they do as beautifully as nemestrinids or syrphid flies. Since the rather long and slender legs are weak, particularly in the tarsi, these flies generally confine themselves to brief rest stops upon the ground, stones, or tops of flowers, and sometimes on leaves. From such perches they often spring upward in rapid, powerful flight, usually coming back down within a short distance. The larger species of *Exoprosopa* Macquart may not alight within fifty or a hundred feet, after being disturbed. Some species of *Geron* Meigen and of *Thevenemyia* Bigot have a way of bobbing up and down above flower heads before alighting. Bee flies are more active during the late morning and early afternoon hours. During a bee fly hunting trip to Mexico and the southwestern United States in 1962, it was noted that these flies seldom appeared before 8 o'clock in the morning; by 10 o'clock peak populations could be expected in any locality with activity falling off toward the middle of the afternoon. No nocturnal activity has been reported for any species, but the comment of Oldroyd (1951) on a possible crepuscular species is interesting.

Males of *Bombylius* Linné may spend several days hovering during much of the daylight hours beneath trees in forest glades; poising at one spot with very little change in position, but so sensitive are they to motion or disturbance that I was once obliged to make three tries with a net before I secured a male of *Bombylius major* Linné. Hardy (1920) comments on the extreme rapidity with which *Systoechus crassus* Walker darts away from its hover position 8 or 10 feet above the ground.

Holmes (1913) made some interesting observations on numerous individuals of a species of *Bombylius* Linné that he found in the hills east of Berkeley, California. All of the hovering individuals had their heads turned away from the sun, and all were hovering in the sunlight; when a shadow was thrown upon a fly it immediately darted elsewhere; moreover, he noted the habit of darting toward objects that approached, for example, several times after honey bees and twice after yellow jackets, a reaction doubtless associated with mating; he further noted that when a fly met a member of its own species of the same sex, they would spin around in a whirl.

We can therefore, distinguish three separate occasions when bee flies hover or poise: in a mating or premating exhibition; hovering in search of hosts or the nesting sites of hosts together with egg deposition or egg thrusting; and hovering before flowers in seeking nectar.

In this connection I should like to take note of the marked attention that several of the larger species of *Anthrax* Scopoli pay to people. On many occasions they have come within inches of my bare arm or neck and often refuse to be intimidated, and I have had them light upon my arm for brief periods. They will also hover about automobiles regardless of their color.

I have noted some evidence that the large *Anthrax tigrina* Fabricius and similar species seem to have a route over which they hover and flutter in search of oviposition sites, although to positively demonstrate this would require marked individuals. Some bee flies evince a territorial distaste for other species. In September recently, about the few wide clumps of desert blossoms along the highway southwest of Riverside, California, I watched a series of *Villa crocina* Coquillett chase away individuals of a species of *Lepidanthrax* Osten Sacken of about half the size of the *Villa*.

Washburn (1906) personally claims to have been "quite severely bitten" by bombyliids. He quotes Lugger (1898, p. 43) as follows: "This proboscis can be used for other purposes besides sipping nectar, as the writer found out to his sorrow, when he attempted to catch some of them in his hand and succeeded; violent pain, a swollen finger and added knowledge were other results of the catch not bargained for." Both authors apparently refer to *Systoechus oreas* Osten Sacken as the vicious one.

While bee flies do exhibit a brief, artificially stimulated catalepsy or death-feigning, it is finished within seconds.

The longevity of bee flies varies. Species that seldom or never feed, such as those in the genus *Anthrax* Scopoli, have a more limited life. However, I have kept captured individuals of both sexes of *Anthrax limatulus* Say alive up to 6 days in a jar with food and water. I have induced them to accept a mixture of water and honey from the end of a pipette. While I have never seen the flies of this genus feeding from flowers, they have occasionally been seen to feed according to the records of other observers. Robertson (1928) recorded two small species of *Anthrax* Scopoli on 11 flowers, and feeding on 6 species of flowers. With the species of genera that have long mouthparts adapted for probing flowers, the longevity may be greatly extended and may in large measure depend upon the length of the blooming season of their favorite food plants. Some bee flies are bivoltine, especially in the southern ranges, but certainly the great majority of all members of the family are univoltine.

Painter and Hall (1960) list 4 out of the 33 known species of *Poecilanthrax* Osten Sacken that are known to be bivoltine in some areas; among these 4 is *Poecilanthrax lucifer* Fabricius, very common in Mississippi. One brood emerges in late April, but the principal brood emerges in midsummer and lingers until frost.

Many bee flies, especially those of arid regions with little rainfall, are able to resist adverse conditions and remain in diapause for long periods. Davidson (1900) records how several bee fly larvae of *Thyridanthrax edititia* Say, taken from cells of the turret-making ground bee *Anthophora montana* Cresson, survived in the larval state for four years before pupating and emerging as adults; one of these appeared to have been in the ground at least a year earlier. Other writers have occasionally noted a larval residence extending into three years. The author has maintained numerous larvae of *Anthrax limatulus* Say in enforced diapause up to three years by keeping them on damp sand in petri dishes in refrigerators. These were all from mud dauber wasp nests; at the time of writing I still have many unopened refrigerated cells of the cliff-dwelling bee *Anthophora abrupta* Cresson that contain bee fly larvae.

Bee flies appear to produce two types of sounds, although both kinds may possibly have the same basic origin. Some of the larger species of *Exoprosopa* when frightened suddenly from a flower head will rise up in the air with a loud whir of wings. Some species produce a distinct humming sound at different pitches. The shrillest bee fly with which I am acquainted is *Heterostylum robustum* Osten Sacken; in Mississippi, at least, this species has a loud, shrill hum pitched almost as high as nemestrinids and audible at some distance. Those I collected in western Kansas were scarcely noticeable. It is possible there is a sex difference in the hum. On the other hand the lowest notes I have encountered come from a species of *Geron* Meigen which are only audible when the fly is placed within a tube.

The hum of *Anastoechus barbatus* Osten Sacken is pitched quite low; several years ago Marguerite Hull, my late wife, and I found a population of over a hundred in a very small patch of blooming *Helianthus* in the New Mexico mountains; their combined sound was very audible indeed.

Mimicry is highly developed within the subfamily Systropinae, where it is indeed remarkable; the flies in this worldwide subfamily resemble wasps of the slender, thread-waisted type. Many American species mimic the species of *Sphex* (*Ammophila*). Brues (1939) points out that many of the Oriental species of *Systropus* mimic vespoid types of wasps which have conspicuous yellow spotting on the thorax and abdomen. He notes that this is not universally true and finds that the ammophiloid pattern is more widespread and therefore apparently older.

Bezzi (1924) calls attention to the beautiful bee flies of the genus *Antonia* Loew; these he considers to be mimics of vespids and crabronids and to resemble strongly such syrphid flies as *Sphaerophoria* Lepeletier and Serville and *Xanthogramma* Schiner; he notes that Kneucker captured individuals of *Antonia suavissima* Loew upon flowers of *Zygophyllum coccineum* L., which strongly matched the coloration of the fly, a cryptic coloration situation.

I am able to add another interesting example of mimicry. The very large robust, stout bodied species *Bryodemina valida* Wiedemann closely resembles a large species of *Bombus* found associated in the same area and same time sequence in the area of Guadalajara, Mexico; this species is abundant at least as far south as Cuernavaca.

Cryptic coloration also occurs rather frequently among those species which rest upon sand of different colors; some species of *Paravilla* Painter have become conspicuously whitish, a close match of unusually whitish desert situations.

Very few observations have been published concerning the courtship and mating of bee flies. In south Texas, south of San Antonio, June of 1964, I had an opportunity to witness the nuptial activity of a pair of small individuals of *Exoprosopa fascipennis* Say; they pursued each other in small circles among the low vegetation 5 or 6 inches above the ground, facing each other; this continued several minutes; mating was unsuccessful, perhaps the flies were frightened away. In July of 1964 in Minnows Valley, Lafayette County, Mississippi, at 2 P.M., I found a mated pair of *Toxophora amphitea* Walker, attached end to end and resting upon leaves of low-growing sneezeweed, *Helenium amarum* (Raf.) Rock. After watching them some time I tried to catch them, missed them; they flew away, but when I returned to almost the same spot 15 minutes later I found them still mated. Linsley and MacSwain (1942) comment on the mating of *Anthrax limatulus vallicola* Marston (reported as *Anthrax* sp., near *fur* Osten Sacken) which took place during the warmest part of the day. The pairs remained in copulation for a considerable period of time and when disturbed flew away without separating. Sartor (in litt.), in extended studies of *Anthrax limatulus* and its relationship to *Anthophora abrupta* Cresson, observed mated pairs only twice, one of these at midmorning. He noted that the male in resting on a cliff was oriented upward, the female downward. The author has found paired *Lepidophora lutea* Painter resting on sneezeweed flowers in July, and has seen others attempting to mate.

Du Merle (1966) has recently published an exceptional study of the mating and related behavior of *Villa quinquefasciata* Wiedemann. He was able to contrive a remarkably successful breeding cage and was able to induce the species to mate easily in captivity. He found that the individuals refuse to feed and their longevity averaged 4 days with a maximum of 6 days, which compares closely with our own findings for *Anthrax* Scopoli species. The females fill up their perivaginal pouches with sand provided for them in sand boxes within the cage and then eject their eggs toward the base of smooth white stones resting on mesh wire. Using this technique he was, in 1965, able to secure 42,000 eggs from 503 adults of this species. Eggs were collected with a vacuum cleaner. See figures of cages in this paper.

Generally speaking bee flies oviposit by flicking the eggs toward the oviposition site or in many species by inserting the abdomen within the soil, as do certain species of asilid flies. In May 1964, on the outskirts of San Antonio, I watched a large, orange-red species of *Bombylius* Linné use both methods of oviposition in dense, old-mesquite forest. It is of interest that the color of this species is very impermanent, the bright red fading to a dirty brownish yellow.

The population of bee flies will vary and will depend in any area on the extent to which the host species are available and also upon the abundance of flowers to which the species is attracted. At times they are extraordinarily abundant. I have found two hundred or more *Sparnopolius lherminierii* Macquart concentrated in less than an acre of *Helianthus* sp., and even greater swarms of western species about clumps of flowers in the highlands of New Mexico. In the genus *Poecilanthrax* Osten Sacken, which attacks sod worms and cutworms, a few species are quite abundant, others distinctly rare. Van Duzee (1931) comments on swarms of a species of *Conophorus* Meigen along the Salinas River of California. Probably no instance of greater concentration of bee flies has ever been noticed than that recorded by Calvert (1881) having to do with the abundance of *Callostoma fasciatum* Fabricius, a locust-egg predator. In eastern New Mexico, in 1965, my wife and I found a species of *Phthiria* Meigen so abundant that as many as 25 individuals crowded in upon each *Helianthus* flower head; this occurred over a wide area.

One of the factors concerned in bee fly populations must certainly be competition for available hosts. Walkden (1950), in studying life histories and in rearing large numbers of cutworms and sodworms from western prairies, shows that besides being attacked by bee flies these worms have many tachinid and hymenopterous enemies as well. Most areas have meloid competitors. Many bee fly eggs must be lost by indiscriminate scattering.

Concentration of bee flies in terms of genera and species is especially interesting when considered in terms of small areas or specific localities. Such species concentration depends largely upon the complexity of the local environment. It also varies from month to month. Jack Hall (personal communication) informed me that a year of continuous intensive collecting in a single long canyon in the Colorado desert yielded nearly 50 species of bee flies.

No such richness of bee fly fauna can be found in the eastern United States. I have collected 28 of the 38 species of bee flies so far known from Mississippi in one single small valley, lying between wooded hills and cleared for the development of more than a hundred small minnow-rearing ponds. This could hardly be matched by any other spot in northern Mississippi. However, the whole fauna of Mississippi is depauperate with respect to this family. Numerous trips have been made by the author, accompanied by his late wife, into the southwestern United States, southern Texas, and one extensive trip into Mexico for the purpose of collecting and studying these flies. In this extensive trip into Mexico, in August of 1962, about 75 collecting sites were sampled by the three collectors on the trip. The total bee fly fauna for any one spot did not exceed 22 species and only one such spot yielded this many bee flies; this was approximately 40 miles east of Guadalajara. Similarly in south Texas in May of 1964, only one spot, a growth of virgin mesquite together with dense undergrowth and thickets, yielded 23 species. It would require year-round collecting to learn what the total fauna of these two bee fly-rich spots might be.

The enemies of bee flies are numerous, and there are at least 11 kinds. For soil-living bee flies there is probably no greater enemy than mold, which in hot, wet seasons undoubtedly destroys both larvae and pupae.

Next in importance in all probability are the birds. Frick (1962) obtained some interesting data; he found that birds were the most important enemy of emerging bee fly pupae of *Heterostylum robustum* Osten Sacken. Blackbirds, larks, sparrows, and magpies all fed upon them at the nesting sites; while these birds destroyed bees also, they appeared to remove a higher percentage of the flies because the wriggling of the pupae at or near the surface tended to attract attention. In 1956 Frick found 29 bees and 48 bee flies in the stomachs of 24 birds; in 1959 he found 35 birds contained 119 bees and 44 bee flies; none of these were blackbirds. Almost certainly the flycatchers and the vireos catch some bee flies; these types of birds become more prominent in the tropics. Hardy (1920) noted an insectivorous bird in the act of catching a species of *Systoechus* Loew.

Krombein (1967) comments upon the prevalence of eulophid parasites in trap nests placed by him for the attraction of bees and wasps. He noted one species of the bee fly was attacked by these chalcidoid parasites.

Shrews and mice may be expected to consume some larvae and pupae, especially of species like some *Villa* Lioy which live in a forested cover of leaf mold. Frick (1962) found that mice destroyed some bee fly prepupae which were removed, eaten, and the skeletons dropped in runways; he also noted the skunks destroyed large numbers of pupae, and he believed that they ate the larvae as well since none could be found where they had dug.

Robber flies have been recorded as including bee flies in their prey. Fattig (1945) and Linsley (1960) both record such instances. Frick (1962) found only one instance of a robber fly having seized an individual of *Heterostylum robustum* Osten Sacken, but he noted that tiger beetles and the ant *Formica fusca* Latreille were minor predators. I have captured a number of asilids that were holding bee flies on which they were feeding.

I have found numerous instances of where crab spiders (*Misumena*) and the green lynx spider *Peucetia viridans* Walckenaer, which hide in flower heads, have captured bee flies. These lynx spiders, family Oxyopidae, occur from coast to coast, but the group is more common southward. This is also true of the ambush bugs of the family Phymatidae; they catch bee flies.

Another enemy is represented by certain species of bembicid wasps which favor dipterous insects as a source of prey.

Presumably some lizards, more especially the anolids, capture some bee flies, although I have seen no records of such captures. *Anolis carolinensis* Voigt is not often seen but is likely more abundant than realized. Anolids increase in number in the warmer latitudes. Studies by the author of stomach contents of *Sceleporus* species and of *Eumeces* species show that their food consists almost totally of spiders, at least in Mississippi.

I have seen no record of either mantids or reduviids capturing bee flies, yet in all probability they do sometimes clutch these flies when such predators are lurking in flower heads to which bee flies come.

The relationship of bee flies to flowers is a matter of considerable interest and one to which collectors and students are just beginning to pay attention. While I have listed most recorded flower preferences under each genus, some generalizations are pertinent here.

Robertson (1928) made an extended study of insect visitors to flowers in the vicinity of Carlinsville, Illinois. Out of 396 species of flowers in 62 plant families, a total of 169 flower species in 41 plant families were visited by bee flies. A total of 29 species of bee flies were found in the area. There were 8 species of Bombyliinae in 5 genera and 13 species of Exoprosopinae in 3 genera; the remainder consisted of 1 species of Lomatiinae, 2 species of Phthiriinae, 2 of Toxophorinae, 1 of Systropinae, and 2 of Geroninae. Thirteen species of bee flies came to *Rudbeckia triloba* Linnaeus, the most attractive flower. Of the flowers, 86 species were visited by only 1 species of bee fly, 53 of them by *Bombylius* spp. only. The Bombyliinae visited 109 species of flowers in 11 families, of which 41 were composites, and the Exoprosopinae visited 89 flower species in 19 families, of which many were composites. Seasonal factors were important: there were only 3 records of the genus *Bombylius* having visited a member of the Compositae, all the other Bombyliinae were of the genus *Systoechus* Loew or *Sparnopolius* Loew. The genus *Toxophora* Meigen visited 18 species of flowers in 7 families; *Lepidophora* Westwood appeared on 3 species of Compositae, and *Systropus* Wiedemann visited 10 species of flowers in 3 families. See these genera in text.

Langhoffer (1902) discusses for a few European species of bee flies their color preferences and the length of time spent on each floret. Knuth (1898-1909) gives some interesting but scattered comments on the relation of Bombyliidae to flowers in his extensive handbook. He notes that the species of *Bombylius* Linné can bore into succulent tissues as well as extract nectar, and he had seen them probing into nectarless flowers. Mueller (1881) in his work "Alpenblumen" has a few comments on the bee flies that visit flowers.

Morphology of the Immature Stages

EGGS

Bee fly eggs are produced in large numbers. Apart from watching various bee flies thrust their eggs toward the oviposition site, few writers have handled or observed the eggs. Eggs of known species vary from a length of 0.28 mm. and a width of 0.12 mm. in the case of *Anthrax analis* Say (Shelford, 1913) to 1.2 mm. in length and 0.7 mm. in width in the eggs of *Heterostylum robustum* Osten Sacken collected by Bohart et al. (1960). It is clear that the eggs of many of the smaller species of bee flies such as *Phthiria* Meigen and *Aphoebantus* Loew, and the still smaller *Mythicomyia* Coquillett must be quite minute. *Heterostylum* eggs have a smooth, pearly white chorion with a mucilaginous coating. Marston (1964) found the eggs of *Anthrax limatulus* Say to be 0.5 mm. in length and 0.29 mm. in width. Sartor (in litt.) collected 850 eggs of the same species but stated that the eggs averaged 0.2 mm. in width; they were oval, deposited in two-days time by two ovipositing females. This bee fly was attacking cliff nests of *Anthophora abrupta* Say and as many as 97 eggs were thrust into a single vial embedded in the bank. They were densely coated with mucilaginous material, obscuring the iridescence, and completely covered with minute, soil particles, leaving the color varying from white to tawny. According to Shelford (1913) the eggs of *Anthrax analis* Say are brown and not adhesive. Du Merle (1966) describes the methods by which he was able to obtain many thousands of eggs of *Villa quinquefasciata* Wiedemann; these eggs measured 0.7 mm. in length, 0.5 mm. in width.

FIRST-INSTAR LARVA

These larvae are planidial. Sartor (in litt.) found this larva in *Anthrax limatulus* Say to be about 1 mm. to 1.2 mm. in length. It is slender, elongate, cylindrical. The coloration is white, the internal tissues revealed by the transparent integument. He found 3 thoracic and 9 abdominal segments; each segment was approximately equal in length and similar in shape except the terminal segment which was conical and smaller in size. There is a pair of long stiff bristles on each thoracic segment, which Clausen calls heavy spines, and an additional pair of caudal setae extending backward from the terminal segment. Bohart et al. (1960) found that the planidium of *Heterostylum robustum* Osten Sacken has large mouthparts, with a median pair of tonglike hooks, flanked by long, slender, maxillae that bear club-shaped palpi tipped with long bristles; moreover he found that the head and neck have 8 pairs of small bristles. Sartor found a pair of pseudopods located on the anterior margin of abdominal segments 2 to 6 and on the posterior margin of the eighth segment. Bohart described these on the eighth segment as double, and he considered the respiratory system as peripneustic, although the

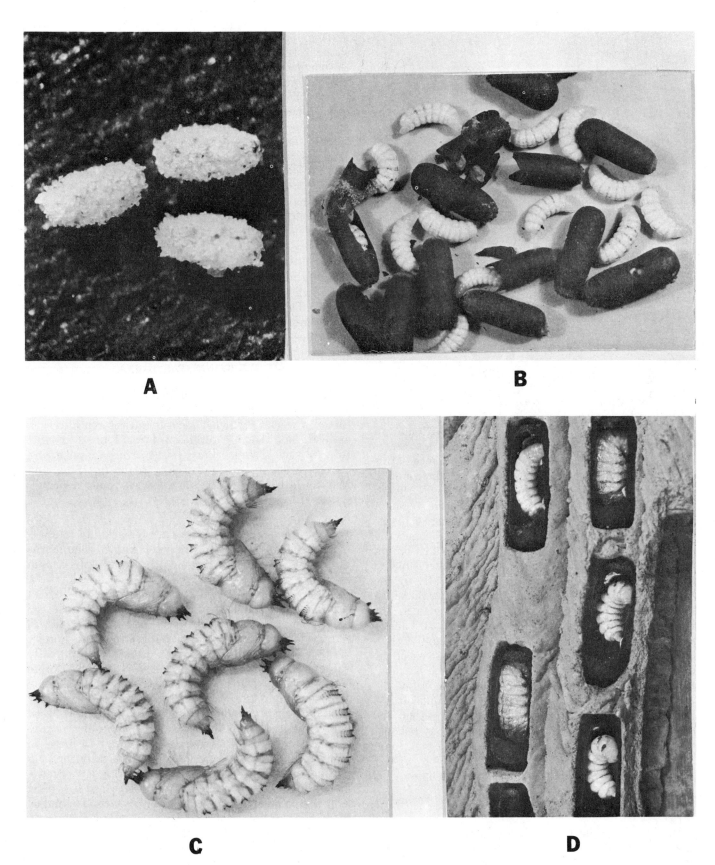

TEXT-FIGURE 1.—A–C: Eggs, larvae, and pupae of *Anthrax limatulus* Say; D, larvae, "in situ," within wasp nest.

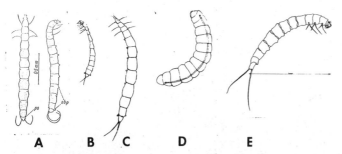

TEXT-FIGURE 2.—Planidial or first-stage larvae of bee flies: A, *Systoechus* species, after Brooks; B, *B. pumilus* Meigen; C, *B. vulpinus* Wiedemann, after Engel; D, late-stage larva, *B. vulpinus* Wiedemann; E, *Heterostylum robustum* Osten Sacken, after Bohart.

abdominal spiracles anterior to the penultimate ones were poorly developed and possibly vestigial. Sartor described the first-instar larva as highly active and having a humped appearance in the thoracic region when it is moving forward.

Berg (1940) presented a remarkable morphological study of the larval stages of *Systoechus vulgaris* Loew; see figures reproduced here.

SECOND-INSTAR LARVA

Working with *Anthrax limatulus* Say, Sartor found that these larvae were 3 mm. long after the first moult, but Marston (1964) found them only 1.5 to 2.0 mm. in length at the time of this first moult. Sartor also states that the milky white tissues are much in evidence in this stage, and that 12 body segments are easily counted. The larva is compressed dorsoventrally, with broad pleural protuberances on the third thoracic segment and on the first to fifth abdominal segments; each segment is well marked by segmental lines. He found the head shallowly invaginated with only the dark brown mouth hooks discernible. The spiracles were visible on the eighth segment and partly hidden dorsally by the overhanging seventh segment, but project over the ninth segment. The prothoracic spiracle according to Sartor is crescentic and opens posteriorly with 11 barlike sclerotic spots. The caudal spiracle is round with 8 barlike schlerotic spots. Berg (1940) gives an excellent description of this larval stage as well as the first and third stage in his work on *Systoechus vulgaris* Loew.

Both Sartor and Bohart comment on the difficulty of finding second-instar larvae. Sartor states that the second-instar larvae of *Anthrax limatulus* Say reached the third stage in 20 hours and was then 5 mm. long. Bohart considered that this stage in *Heterostylum robustum* Osten Sacken lasted only about 12 hours.

THIRD-INSTAR LARVA

In this stage the bee fly larva increases greatly in bulk and length and also becomes characteristically crescentic or C-shaped. There are no bristles or tubercles for locomotion; it is now a grub with very little movement. The tight application of the concave portion about the mouth with doubtless some vacuum developed helps to hold it attached to the host. Perhaps the arched shape of the larva doubly accommodates the parasite to the surface of the host and the surface of the cell. It is milky white due to a transparent integument and white internal tissue; numerous oval, opaque white fat bodies make their appearance in the abdomen. This instar is amphineustic and cylindrical, with the bend in the body located in the anterior half. Segmental lines are emarginate, says Sartor, and are visible on segments 2 to 5; pleural protuberances are present on the first through the seventh abdominal segments. The eighth abdominal segment is small and forms a ledge between the seventh and ninth segments upon which the posterior spiracle is located. This spiracle is circular with usually 8 to 9 sections. The anterior spiracle is crescentic, open posteriorly and has 11 sections. See accompanying figures of the third-instar larva.

Not all bee fly larvae become C-shaped in the final stages. Those mature larvae of *Lepidophora* Westwood which we have obtained from caterpillar-filled nests of species of wasps were straight and linear in form.

FOURTH-INSTAR LARVA

Bohart et al. (1960) describes a fourth-instar larva for *Heterostylum robustum* Osten Sacken, but this is the only work in which I have found reference to such an instar and the only authors as far as I am aware who refer to such a stage. Berg (1940) refers to none, and Sartor, working with *Anthrax limatulus* Say, and Marston (1964), with the same species, found none.

PUPAE

The pupal stages of bee flies are remarkably interesting. Pupae of this family are better known than larvae of bee flies, because so often only the pupal skins remain in chance rearing discoveries.

These pupae are free and mobile and possess distinct sheaths which encase the mouthparts, the antennae, the wings, and the legs. When the fourth-instar larvae changes over to a prepupa, or pupa, it is at first milky white and gradually turns pale yellow; it gradually darkens to brown and may become almost black before emergence and tends to become very active in the late stages of pupal life, wriggling about and extending or telescoping the abdominal segments. We have kept many hundreds of both larvae and pupae of *Anthrax aterrimus* Bigot and *Anthrax limatulus* Say alive on damp sand. They may be kept in either state for long periods on sand in a refrigerator. As this is written I have plump, white larvae that have remained in a petri dish within a refrigerator for several years. The pupae of bee flies resemble those of asilids in many respects.

Asilid pupae bear thornlike spines upon the dorsum of the abdominal segments which are sharp or blunt, but usually sharp, and are vertically erect even if curled backward. Bee fly pupae may in rare instances as in

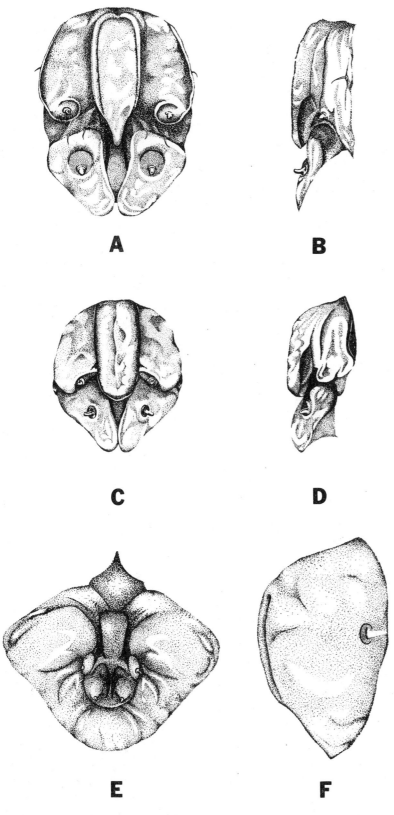

TEXT-FIGURE 3.—Mouthparts of bee fly larvae of the two principal divisions of the family: A and B, front and lateral aspects of larva of *Systoechus vulgaris* Loew; C and D, front and lateral aspects of larva of *Anastoechus barbatus* Osten Sacken; E and F, front and lateral aspects of *Anthrax limatulus* Say. Figures A-D redrawn from Painter (1962); Figures E-F drawn by William Martin.

Phthiria spp. show similar, sharply pointed erect thorns, but in such cases they arise from the anterior part of what might possibly be called the anlage of the thorns, whereas in asilids they arise from the posterior edge.

Bee fly pupae rather characteristically show a transverse band of close-set chitinous rods attached across many of the abdominal segments, from which at either posterior or anterior end or from both ends, the apices

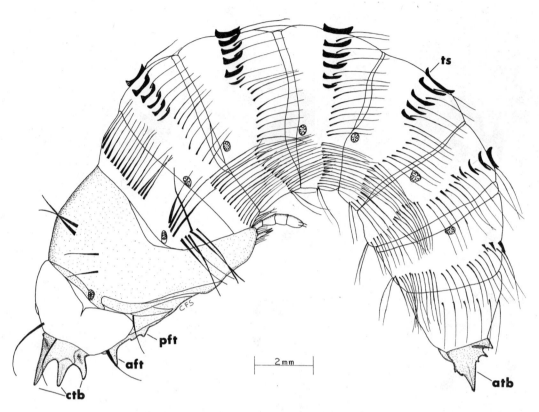

Text-Figure 4.—Pupa of a bee fly, *Anthrax limatulus* Say.

of the rod are apt to be sharply erected into a sharp hooklike spine or thorn. From a study of pupae of 8 subfamilies, 18 genera, and 26 species that are before me, this distinction holds good. Occasionally, as in *Aphoebantus* (*Triodites*) *mus* Osten Sacken, these longitudinally placed chitinous rods are so short that the posteriorly erect spine is suggestive of the spine armor of an asilid. Malloch (1915) states that long, slender, curled hairs alternate with the chitinous rods and spines; while such hairs may be irregularly present there are none in *Systropus* Wiedemann.

A second character of importance in distinguishing bee fly pupae from other families lies in the presence and character of strong, chitinized thorns upon the head capsule. In asilids this cephalic armament as noted by Malloch (1917) consists of a unified, fused group of 3 very strong spines present on each antennal sheath and lying below the anterior, apical, cephalic thorn. He notes that such are absent from the pupae of *Leptogaster* Meigen. In the opinion of this author this is indeed not an argument for separate family status of these obviously asilid flies. There are bee fly genera that lack the usual cephalic spines, as in the case of *Systropus* Wiedemann and *Toxophora* Meigen, yet these are in every respect bombyliids; it does not make sense to create separate families for these two bee flies any more than it does to put the leptogastrine asilids into a separate family. Characteristically asilid pupae have the postcephalic thorn group therefore consisting of 3 well-developed clustered thorns; whereas, there are never more than 2 in this position in the bee flies, and these two are often separated at the base. It should be here noted that other bee fly pupae vary from the above characterization, such as the curious *Psiloderoides* in the Cyrtosiinae and *Glabellula* Bezzi (as *Pachyneres* Greene), and for which figures are here reproduced.

All bee fly pupae and pupal exuviae tend to show some curvature. The four most important areas from the standpoint of generic and subfamily characters lie in the head armament, that of the dorsum of the principal abdominal segments, and the apical, terminal processes of the pupa together in some instances with the armament of the abdominal pleurites.

The head armament usually consists of four groups of strongly developed heavily chitinized thorns, each paired, so that there are 8 thorns. The posteriormost pair may be very much reduced, as in *Thevenemyia auripila* Osten Sacken. In *Neodiplocampta* Curran the outer spines of the lateral pairs are much reduced, but there is an additional small pair below the lowermost of the principal spine groups, making 10 in all. In *Hemipenthes* Loew all the spines are reduced except the anterior pair, which are approximated, semitruncate, and shovellike. In *Triodites* Osten Sacken there are 2 additional unpaired spines, one on front and one behind the lower pair, and moreover the anterior spine

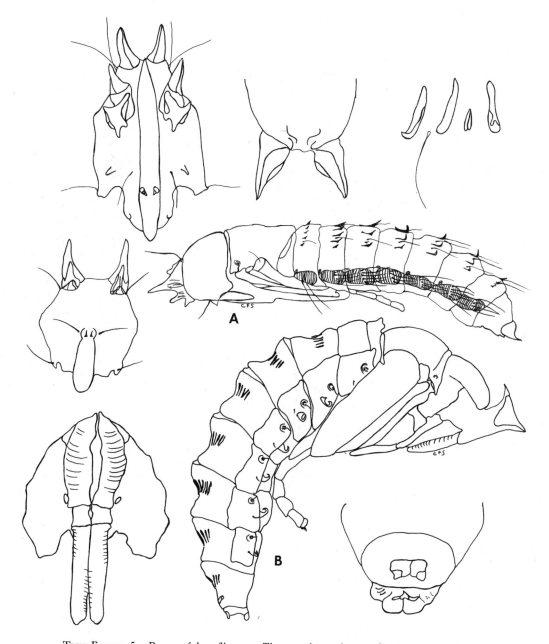

TEXT-FIGURE 5.—Pupae of bee flies: A, *Thevenemyia* species; B. *Systropus macer* Loew.

and both members of the lateral pair are fused into groups of 3. In Anthracinae each group of 3 becomes fused into a hemicircular arch of 6 spines. The base of the wing sheath in *Heterostylum* Macquart bears 2 spines on each side. The dorsal abdominal armament varies from genera in which neither end of the longitudinal chitinous rods is raised into erect thorns or spines, as in *Hemipenthes* Loew; here each end is about equally well developed. In *Exoprosopa* Macquart usually, but with exceptions, each end is raised into remarkably stout, erect, sharply thinned but bluntly rounded, shovellike plates.

In *Walkeromyia* Paramonov the dorsal rods end in sharper spikes and give way laterally to curious, conspicuous, very wide, quite long, flattened, closely appressed swordlike outgrowths. In *Phthiria* Meigen the few central rods end in long, strong, strongly curved, erect, sharp spikes or spines, which arise from the posterior ends of the rods. Terminally the bee fly pupae vary from the Anthracinae where the paired processes are heavily chitinized, elongate, narrowly separated, bifid or fin-shaped and accompanied above and below on each side by short, small, tubercular spines, and another, paired or unpaired, placed medially upon the preceding segment, to the majority of genera in which these terminal processes are small, slender, consisting of one slender, sharply pointed protuberance with or without an additional shorter one below. In *Walkeromyia*

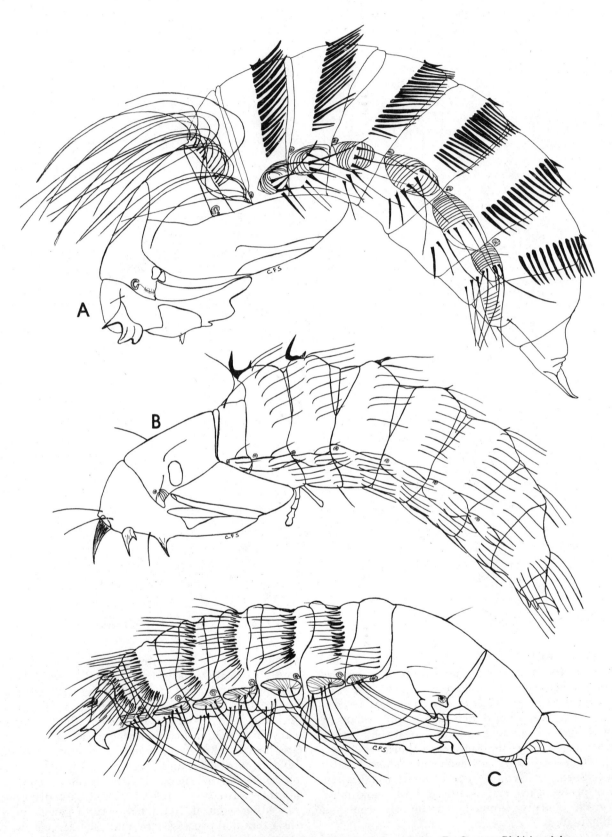

TEXT-FIGURE 6.—Pupae of genera of three subfamilies of bee flies: A. *Anthrax tigrinus* De Geer; B. *Phthiria sulphurea* Loew; C. *Toxophora amphitea* Walker.

Paramonov and related forms the pupa ends in a curious, long, widely separated, spine tipped, fleshy protuberance. Finally in some genera there are important differences on the abdominal pleurites. They may have a tuft of 3 very long, very stiff, uniformly thick hairs, as in *Toxophora* Meigen, to as many as 9 very long, flattened curved, spikelike, or swordlike outgrowths, or as in most genera only fine, long hairs. In most genera the anterior, circular, flattened spiracle is composed of corrugated ridges which exactly resemble a coiled millipede. But there may be curious specific differences in the number of lobes; they are reduced to 5 in *Anthrax irroratus* Say.

I have prepared a list of the subfamily characteristics of all those subfamilies for which I have pupae. To aid others in the fascinating study of the pupae I have illustrated all those for which I have material. To further help students of these flies I have included Malloch's key (1917) to 16 species, 9 of which are not among the 26 species before me.

In bee flies the last larval stage is usually followed by a prepupa (pronymph of authors): in this stage the head thorns are partially developed and the longitudinal plates of chitin are already laid down upon the dorsum of the abdomen. Séguy (1926) who illustrates the pupae of 11 species, all figures rather small, also illustrates the pronymph of several species.

The presence of these stout spines is associated with the difficulties the pupa meets in emerging to the surface from mud nests of wasps, which do not usually offer much obstruction, to the deeper lying layers of clay and earth under which many subterranean forms find themselves.

If the comparative morphology of the pupae is valid, as I believe it is, we must conclude that the nearest relative to the subfamily Systropinae is the genus *Toxophora* Meigen, and that these two subfamilies are relatively closely related. Also it is at once apparent that *Lepidophora* Westwood does not find its relationship with *Toxophora* Meigen but rather is related to the Bombyliinae; while I suspect it should be regarded as a tribe within the Bombyliinae, I place it with some reluctance in the Toxophorinae. Also it is clear from these studies that *Anthrax tigrinus* Fabricius should be placed in a genus entirely separate from *Anthrax* Scopoli, *Argyramoeba* Schiner, or *Spogostylum* Macquart; it should be placed within the genus *Walkeromyia* Paramonov or a new genus should be erected for it. The former course is preferable.

KEY TO THE SUBFAMILIES OF BEE FLIES AS KNOWN FROM PUPAE

1. Antennal sheaths present and prominent, distinct and fleshy, and arising from and proceeding backward from anterocephalic thorns which may be almost completely fused, side by side. Dorsum of pupae with 7 rows of close-set, longitudinal, slightly curved, posteriorly distinct but also appressed spines which are not raised above the surface at apices. Cephalic thorns range from 1 fused pair to 2 distinct pairs 2
   No distinct antennal sheaths present. Usually 4 pairs of cephalic thorns present; some of which may be reduced, semifused or vestigial. Dorsum almost always with chitinous longitudinal ridges, one or both ends of which are raised into sharp, erect spines 3

2. Cephalic thorns reduced to a single, almost completely fused pair, with however a minute stublike tubercle adjacent at the level of the second antennal segment. Apex of pupa with 1 or 2 pairs of mammillate nodules, the more posterior pair semifused. Dorsal rows of curved spines indistinct anteriorly. Lower lateral edge of first 6 abdominal pleurites with curved, attenuate, sharp-tipped, fleshy protrusions. Some species characters afforded by the flutings on the thorax. Proboscis sheath fused. Based on different species of *Systropus* Wiedemann . . . . . SYSTROPINAE
   Second pair of cephalic thorns small, borne on the middle of the divergent antennal sheaths. Anterior thorns fused except the divergent apices. Proboscis sheath divided longitudinally. Posterior apex of pupa with 2 pairs of widely separated, short, triangular thorns; lower pair curved. Upper and lower pair as far apart as thorns of each set are separated. First 6 abdominal pleurites each with 3 extraordinarily long, uniformly stout, finely pointed bristles. Based on different species of *Toxophora* Meigen . . . . TOXOPHORINAE

3. Chitinized spinous longitudinal ridges restricted to first two abdominal segments and these two segments have only 3 ridges, each raised anteriorly into strong, erect, long, backwardly curved spines. Posterior apex of pupa with a pair of widely separated, thin, posteriorly attenuate and pointed, bladelike protrusion. Thorn bearing part of cephalic capsule circular, with 3 pairs of strong, sharp thorns, the anterior pair especially sharp, widely separated . . . . PHTHIRIINAE
   Chitinized ridges present on second to eighth segments and usually with the posterior part ending as an erect, vertical, sharp or blunt spine, sometimes both ends with erect spines. Sometimes first segment also with a few spines. 4

4. Each anterior spine and each lateral pair gathered together into a unit of 3. Units of each side may be separated or fused together in front 5
   All of the cephalic spines are separate and distinct even if some of them are small 7

5. Both units fused together at base transversely so that all 6 spines though distinct, stout and long, are gathered together into a hemicircular unit remote from the posterior pair of spines; all 6 arise from an elevated common base. Apex of pupa with a large, chitinous, elongate, black, bifid or finlike process, narrowly but distinctly separated. Also the base of the tenth segment has a pair of smaller, black spines above and below; also a paired or unpaired small spine dorsally in the middle of the ninth segment. Based on many species of *Anthrax* Scopoli.
   ANTHRACINAE
   Each anterior unit of three spines on the head capsule separate 6

6. The spines of the head capsule are large, blunt, or rounded or spadelike. Chitinous ridges of the dorsum with sharp erect spines both anteriorly and posteriorly at least on the middle segments. The spines give way laterally to long, thick, appressed, flattened, swordlike hairs and on the sides of the first segment to many, long, curled hairs. Pleurites with from 4 to 8 or more long, flattened, conspicuous, swordlike hairs or outgrowths. Apex of pupa

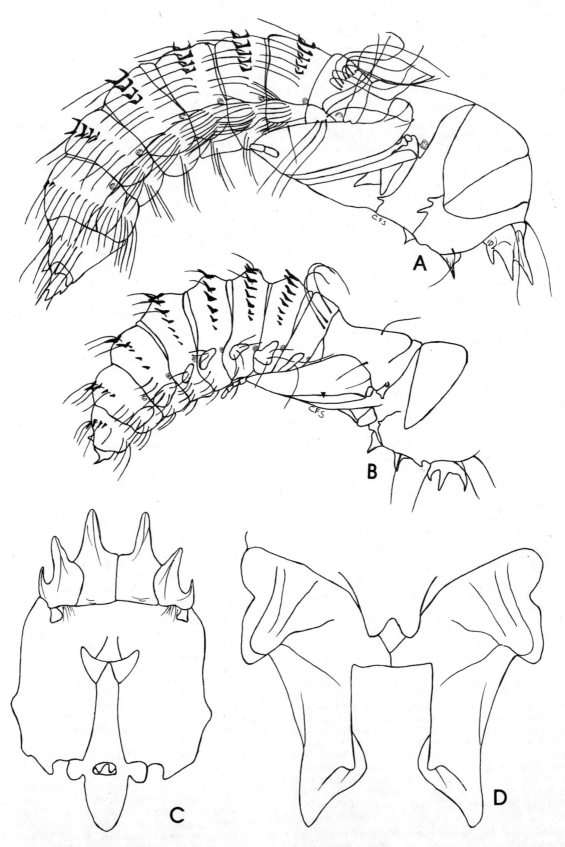

Text-Figure 7.—Pupae of genera of two subfamilies of bee flies together with dorsal views of anterior and posterior armament of *Villa lateralis* Say.

with a pair of widely separated, fleshy, attenuated processes ending in a spine. Very large flies. Tribe Walkeromyini . . . . ANTHRACINAE
The spines of the head capsule are long and sharp. Posterior pair of spines strong with an unpaired additional spine in front and behind. Apex of pupa with a single pair of very small spines, widely separated, arising from a tubercular swelling. Longitudinal ridges quite short ending in an erect spine only posteriorly. No spectacular hairs, swordlike outgrowths, etc. *Aphoebantus mus* Osten Sacken . . . . LOMATIINAE

7. Posterior pair of spines of the head capsule weak or much reduced 8
   Posterior pair of spines well developed 9

8. Apex of pupa with a pair of small, widely separated, attenuate spines borne on a basal protuberant swelling. *Thevenemyia auripilus* Osten Sacken, Tribe Eclimini.
   BOMBYLIINAE
   Apex of pupa with a large, triangular, heavily chitinized more or less triangular, widely separated plate on each side. *Lepidophora lepidocera* Wiedemann. Tribe Lepidophorini . . . . BOMBYLIINAE

9. Chitinous ridges of dorsum with spines in front as well as behind 11
   Chitinous ridges of dorsum spined only behind or not at all. 10

10. Chitinous ridges without spines either in front or behind. All the cephalic spines much reduced, except the anterior pair which are blunt, obliquely truncate, narrowly separated. Spines at apex minute, slender, widely separated arising from a vertical ridge which bears below two other even smaller, tubercular spines. *Hemipenthes* Loew.
    EXOPROSOPINAE
    Chitinous ridges with a row of erect spines on their posterior ends. Villini; *Paravilla* Painter, *Neodiplocampta* Curran, etc. *Exoprosopa* Macquart, spp.
    EXOPROSOPINAE

11. Apex of pupa with stout, heavily chitinized, upward thrust spines. Chitinous ridges of dorsum very strong, the anterior and posterior erect spines sharp, long and conspicuous. *Exoprosopa* Macquart . . . . . . EXOPROSOPINAE
    Apex of pupa with a pair of slender, weak, upward thrust spines on each side 12

12. Apex of pupa with a pair of additional, stout, shovellike plates or spines in the middle, dorsally, semifused, above the slender apical spines. Base of wing with a pair of small spines on each side. Tribe Heterostylini.
    BOMBYLIINAE
    Apex of pupa with only a pair of weak, slender, upward turned, slender, lateral spines. No spines at base of wing. *Bombylius* Linné, *Systoechus* Loew, etc. . . . BOMBYLIINAE

KEY TO PUPAE (FROM MALLOCH, 1917)

"1. Upper central pair of cephalic processes thorn-like, widely separated from their entire length; lateral cephalic process or processes thorn-like, but little if any shorter than the central pair 2
– Upper central pair of cephalic processes stout, not thorn-like, contiguous for the greater portion of their length; lateral cephalic processes tubercle-like, much shorter than the central pair 13

2. Apical abdominal segment terminating in a pair of long, tapering, backwardly directed thorns; first abdominal segment with the postspiracular hairs as long as head and thorax combined (*Spogostylum*) 3
– Apical abdominal segment usually more or less truncated and with an upwardly and backwardly directed upper process and one or two smaller protuberances below it; first abdominal segment with the postspiracular hairs much shorter than head and thorax combined 5

3. Head with 4 long thorns on upper anterior margin, the lower one on each side with a small protuberance at base on under side; the pair of thorns on lower portion of central line of face large, their bases contiguous; hairs on head and thorax very long; laterad of the short thorns the transverse armature of dorsal abdominal segments 2-6 consists of 2-3 long, widely placed rounded hairs . . . . *Spogostylum anale*.
– Head with 6 short, stout thorns on upper anterior margin 4

4. The pair of thorns on lower portion of central line of face small, their bases subcontiguous; hairs of abdomen, except those of verse armature of dorsal abdominal segments 2-6 consists of 12-20 long, closely placed, flattened hairs . . . . *Spogostylum simson* (p. 393).
– The pair of thorns on lower portion of central line of face large, their bases, subcontiguous; hairs of abdomen, except those of basal segment, normal . . . . *Spogostylum albofasciatum* (p. 395).

5. Antero-lateral margins of head each with 1 strong thorn, upper anterior margin with 2 such thorns, making 4 in all; labrum with a bifid thorn . *Chrysanthrax fulvohirta*.
– Antero-lateral margins of head each with 2 strong thorns, upper anterior margin with 2 such thorns, making 6 in all 6

6. Labrum with a strong bifid thorn; wing with a median subcostal protuberance . . . . *Aphoebantus mus*.
– Labrum unarmed 7

7. The stout thorns on dorsal abdominal segments turned up at bases and apices 8
– The stout thorns on dorsal abdominal segments turned up at apices only 11

8. Lower lateral cephalic thorn with a palp-like organ projecting on its under surface at base, the apex of which is armed with several hairs . . . . *Bombylius*.
– Lower lateral cephalic thorn without a palp-like organ on under surface 9

9. Apical 3 segments without the dorsal transverse series of short thorns, armed only with slender hairs; no slender hairs interspersed between the short thorns of median portion of series on other segments . . . . *Argyramoeba oedipus*.
– At least the penultimate and antepenultimate segments with dorsal transverse series of short thorns; slender hairs interspersed between the short thorns on all segments . 10

10. Wings extending to apex of third abdominal segment, their color pale . . . . *Systoechus oreas*.
– Wings extending short of apex of second abdominal segment, fuscous apically . . . . . *Exoprosopa fasciata*?

11. Transverse armature of first abdominal dorsal segment consisting of a series of short, stout thorns on middle portion, and a number of long, slender, closely placed hairs on each side . . . . *Exoprosopa fascipennis*.
– Transverse armature of first abdominal dorsal segment consisting of a few widely placed hairs, the middle portion either entirely bare or with very slight indications of small tubercles which do not appear as distinct thorns . 12

12. Lower one of the pair of lateral cephalic thorns simple apically, but with a small wart-like protuberance at base on lower surface, the small wart bearing 2 distinct hairs; wings without discal protuberances . . . . *Sparnopolius fulvus*.
– Lower one of the pair of lateral cephalic thorns with a short subapical protuberance, the apex of thorn turned upward, base simple; wings each with a pair of protuberances, one about one fourth from base and the other near middle . . . . *Anastoechus nitidulus*.

13. No well-developed pair of thorns on lower median portion of face . . . . *Toxophora virgata*.

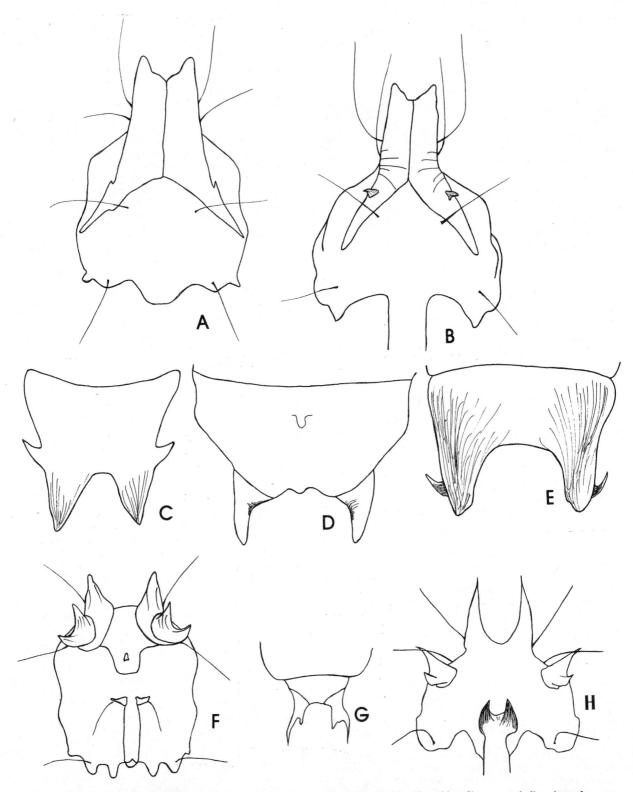

TEXT-FIGURE 8.—Anterior and posterior armament of the pupae of three subfamilies of bee flies: A, cephalic spines of *Toxophora virgata* Osten Sacken; B, cephalic spines of *T. amphitea* Walker; C, caudal spines of *T. virgata* Osten Sacken; D, caudal spines of *Aphoebantus mus* Osten Sacken; E, caudal spines of *Toxophora amphitea* Walker; F, cephalic spines of *Aphoebantus mus* Osten Sacken; G, caudal spines of *Phthiria sulphurea* Loew; H, cephalic spines of *P. sulphurea* Loew.

- A well-developed pair of thorns on lower median portion of face 14
14. Eighth ventral abdominal segment without hairs on disc . . . . *Hyalanthrax hypomelas.*
- Eighth ventral abdominal segment with hairs on disc . 15
15. Eighth ventral abdominal segment with 2 hairs on each side of disc; distance from the pair of thorns on lower central portion of head to apex of basal portion of sheath of mouth-parts about 4 times as great as distance from the latter to apex of proboscis . . *Hyalanthrax lateralis.*
- Eighth ventral abdominal segment with 10-12 long hairs on disc; distance from the pair of thorns on lower central portion of head to apex of basal portion of sheath of mouth-parts about twice as great as distance from the latter to apex of proboscis . . . *Hyalanthrax alternata.*"

Morphology of the Adult Bee Fly

Bee flies range in size from the large and conspicuous species, such as members of the genus *Exoprosopa* Macquart which may reach a wing span of 64 mm., down to very minute, deserticolous floriphiles, such as *Mythicomyia* Coquillett, which have much reduced wing venation; many of these are not more than 1.2 mm. in length, a few range down to 0.9 mm.

Most species have a compact form, both in abdomen and thorax, the former being broad and short. In the remarkable subfamily Systropinae and to a lesser extent in a few other genera, the form is exceptionally long and slender and wasplike; these flies are rather bare in appearance, the microsetae scarcely noticeable. Most bee flies have fine, rather long vestiture, sometimes with coarser elements intermixed, sometimes composed entirely of dense plushlike pile. Some bee flies such as the genus *Toxophora* Meigen have well-developed macrochaetae. A few species have closely adpressed scales, either fine and narrow or wide. A few bombyliids resemble syrphids, therevids, or empidids. While the ground color of the integument is more often black or dark brown, in many groups it is pale yellow, whitish, or red, and still other species show a pronounced brassy or bluish metallic sheen. Wings of bee flies show a highly variable venation and may be hyaline or may show very beautiful patterns of spots; iridescence varies.

The head is usually approximately hemiglobular, but in the Tomophthalmae it may be almost completely globular due to the extensive cuplike development of the occiput. The vertex and frontal area are nearly plane with the eye; the eye is only rarely raised a trifle; the frons may be slightly higher than the adjacent eye margins. The head is usually of nearly the same width as the thorax but may be more narrow and not infrequently is much wider. The frons may show a transverse depression or fovea immediately in front of the antenna; it is rarely inflated or tumid.

The eyes are always large and extensive, almost always occupying much the greater part of the head; they are convex in front, with the upper facets enlarged in the male, and in the subfamily Heterotropinae a distinct dividing line separates an area of larger facets above, smaller ones below; this is true in males of *Geminaria* Coquillett. While generally bare, Bowden (1964) has pointed out that a few bee flies (Exoprosopinae) do show fine, scattered ocular pile. In the male the eyes are usually holoptic or at least nearly touching, but in many groups they are dichoptic, or there are sometimes exceptions within the same genus (*Eurycarenus* Loew). In the genus *Systropus* Wiedemann the eyes are widely holoptic in both sexes. The postocular margin is variable and important second only to the character of the occiput in separating the two main divisions of the family. It may be nearly or quite plane, throughout its vertical length in the more generalized bee flies, or sinuous, slightly concave, and in the higher forms it is characteristically indented to a varying extent and with or without a short anteriorly directed bisecting line. Three ocelli are present and they are very rarely reduced in size.

The occiput varies from flattened to convex, and in the higher members of the family is characteristically tumid, greatly expanded posteriorly, with a dorsal, vertical, bilobate, foveate depression and a deep central sunken cavity. Many Usiinae have ventral lateral bullae upon the occiput. The character of the occiput is of primary importance in separating the major divisions of the family and is discussed further under the phylogeny and evolution of these flies.

The oral opening or buccal cavity is always large but does vary some in size and still more in position in respect to the principal axis of the head; it is generally deeper below and at the point of emergence of the proboscis and the opening may slope almost or quite to the base of the antenna, eliminating what would properly constitute the face as separated from the clypeal space. I restrict the face to the nonsunken part of the front aspect of the head, below the antenna, and before the clypeal area begins; below is the clypeal space which I call the oragenal cavity, or oral cup; its sides are usually sharp, or as Hesse (1938) says, carinate, and with a vertical wall. This whole area of the bee fly head from proboscis to antenna is most variable and important. The face as here defined may be almost entirely absent, or restricted to a small triangle extending laterally below the front, ending with the gena, or again the face be extensive and even conical, or also tumidly swollen.

At the apex of the labium are to be found the paired labellae, which are separable by the bee fly. Dimmock notes that on the inner side of each labellum there are 3 longitudinal grooves or channels held open by semirings of chitin at right angles to their axes and together on each side; they have been termed pseudotracheae but are not likely comparable to muscoid pseudotracheae.

Curiously, in a few bee flies the mouthparts are greatly reduced or even totally absent, with a mere slit remaining in the oral space. The gena I define as the lateral-ventral space outside the oragenal cup or clypeal space and only rarely extending below the eyes; posteriorly it would merge with the lower occiput; genae

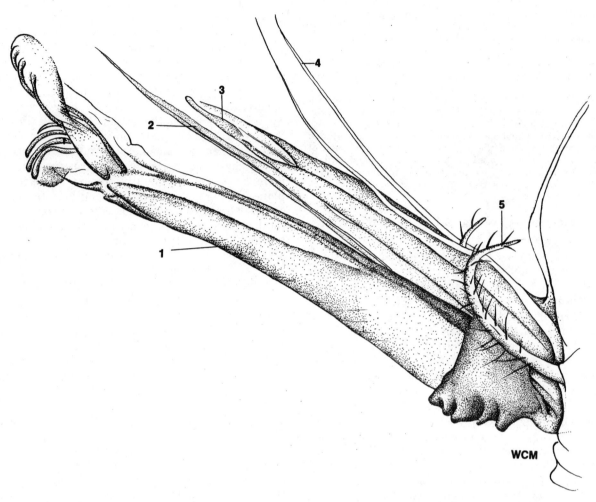

Text-Figure 9.—Mouthparts of adult bee flies of *Exoprosopa* sp. Explanation: 1, labrum, with labellum at apex; 2, mandible; 3, labrum-epipharynx; 4, maxilla; 5, maxillary palpus.

may be absent; it is bare, pubescent, or scaled.

The tentorial fissure lies outside the oragenal cup and is often knife-thin, or it may constitute a more conspicuous fissure. Medially it encloses the clypeal space.

The proboscis of the bee fly varies from stout and short, as in the Anthracinae and Lomatiini, or even absent (Villoestrini) to very long and slender as in *Bombylius* Linné and *Phthiria* Meigen; it is sometimes 4 times as long as the head. It is entirely absent in a few bee flies. According to Dimmock (1881) and Peterson (1916), it is composed of 5 parts. At rest the hypopharynx lies within a groove on the dorsal aspect of the labium and is covered by the labrum (labrum-epipharynx) Beneath and at either side are to be found the maxillae, which are very fine and slender, shorter than the other parts; all of these may be disclosed if the mouthparts are teased apart with a very fine-pointed needle or insect pin, the maxillae being the most troublesome to separate. The maxillary palpi lie at the base on each side and consist usually of 2 segments; there may, occasionally however be 3 segments, sometimes only 1, and the palpus is fused and semivestigial in *Comptosia* Macquart; the apical segment may be somewhat dilated and clavate, one or both segments may be covered with prominent hairs or scales; the trend is toward reduction in number of segments.

Besides the long, slender type of proboscis seen in Phthiriinae and most Bombyliinae there is the shorter, stouter type with the more or less expanded, almost muscoid-like labellum seen in Anthracinae which flies, when they feed at all, probably utilize pollen; Bezzi (1924) speculates upon the possible use of the curious facial brush in *Corsomyza* Wiedemann in flower pollination. Finally, in 4 genera the mouthparts range from vestigial to a mere slit, as in *Villoestrus* Paramonov; this trend in reduction was compared by Bezzi (1924) to similar phenomena in other Diptera, such as the oestrids and the tabanid *Adersia* Austen. In the bee fly *Euanthobates* Hesse there are curious coecalike processes below the head (Hesse, 1965), which may be involved in pollination.

The antennae of bee flies are usually set close to one another but may be widely separated in the Cythereni and a few genera of other groups. In the Bombyliinae

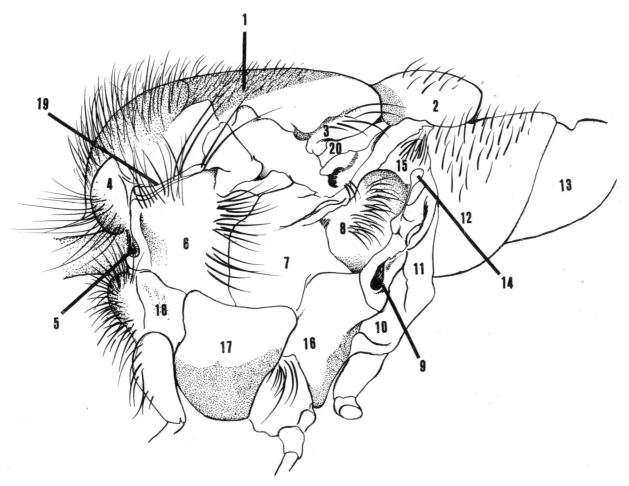

Text-Figure 10.—Lateral aspect of the thorax of a bombyliid, *Antonia suavissima* Loew. Explanation: 1, mesonotum; 2, scutellum; 3, postalar callosity; 4, humerus or humeral callosity; 5, anterior spiracle; 6, mesopleuron; 7, pteropleuron; 8, metapleuron; 9, posterior spiracle; 10, metasternum, 11, posterior metasternum; 12, first tergite; 13, second tergite; 14, halter; 15, metanotum; 16, hypopleuron; 17, mesosternum or sternopleuron; 18, prosternum; 19, notopleuron; 20, anterior metapleuron.

the antenna tends to be rather long and slender with the first and third segment of nearly equal length. The second segment is always small and beadlike in nearly every genus; *Neosardus* Roberts is an exception. The third segment may end without a style, or have a conspicuous one as in *Othniomyia* Hesse. Moreover, as a marked exception, *Prorachthes* Loew has the third segment rather widely dilated. The third segment frequently has one or two small microsegments at the apex, and sometimes an apical, or subapical spine set within a cup. The third segment is occasionally more or less plumose. The style has a whorl of hairs at the apex in the Anthracinae.

The thorax in this family is usually quadrate, and widest posteriorly, but may be a trifle longer or shorter. Except in the Toxophorinae, where it forms a conspicuous collar, the pronotum is lower than the level of the prominent mesonotum which abuts against the upper occiput. The mesonotum itself is usually only slightly convex, but may be strongly arched anteriorly, and in some genera it is conspicuously humped; in such species the pleuron tends to be compacted and high vertically. Surface of mesonotum with usually only simple pile, often long and dense but looser, more scattered and thinned when scales are also present as in Anthracinae and various other genera. Bristles often present on prealar, supraalar, and postalar areas, and in the Toxophorinae there are large, long, conspicuous, arched bristles. There may be also acrostical, dorsocentral, and anteromarginal bristles. The scutellum is large and semicircular, subtriangular, either flattened or slightly tumid, the margin entire or rarely bilobate or emarginate.

The mesopleuron is large, and frequently bears a dense tuft of upturned stiffened hair with sometimes additional bristles intermixed. The metapleuron may be quite bare in some forms but usually has a distinct tuft of stiffened or bristly hairs which may take the form of a vertical fanlike row. The hypopleuron is bare and the pteropleuron usually bare. The metanotum is variable, usually well developed and lying beneath the scutellum. The propleuron usually bears conspicuous pile and sometimes bristles. In Systropinae the strigula is prominent; as Painter and Painter (1963) point out, this structure has been variously named the membranous tubercle by Verrall (1909), the scutellar callosity by Bezzi (1924), and the foliate scutellar callosity by Hesse (1938). The term strigula was adopted by Williston (1901) and reused by Carrera and D'Andretta (1950), and by Painter. It has the double advantage of priority and brevity.

The prosternum varies from widely dissociated from the propleuron, as in *Comptosia* Macquart and many other genera, to barely touching on the mesal aspect of the latter, as in *Toxophora* Meigen. There is also much variation in the form of the propleuron, and in many aspects of the posterior pleuron.

The metasternum is sometimes very prominent and strongly developed so much as to compact the anterior sternum and coxae and crowd them forward; this is the situation in the Systropinae. On the metapleuron two divisions can be recognized corresponding to the episternum (or mesopleuron) and the epimeron (or combined pteropleuron, hypopleuron, etc.) of the middle pair of legs; in this work these two divisions are merely called the anterior and posterior metapleuron; the line that divides them reaches the top of the metacoxa.

The legs of bee flies with few exceptions are weak; they are adapted for alighting and resting in the often brief rest stops these insects make upon flowers, leaves, soil, rocks, logs, etc. In most genera the hind legs are much longer than the others and in some bee flies the anterior pair are greatly reduced. Middle and posterior coxae are short, except in those forms with a highly arched mesonotum; in Toxophorinae they are all three of equal length. In some bee flies there are moderately strong bristles beneath the femora but they are usually restricted to the hind pair and even there they may be lacking.

The tibiae bear distinct rows of spicules or straight spinelike bristles of varying degrees of thickness, but here again there is often much variation and they may be absent even upon the anterior tibia; their presence or absence upon the front tibia has been used in classification, but it is a weak character; in a tibia which otherwise appears to be devoid of bristles very close inspection will usually reveal minute, slender, bristly hairs erected obliquely above the basal coat of microsetae. Most tibiae have a well-developed apical circlet of spines as is again seen in asilids. In males of *Walkeromyia* Paramonov and in both sexes of *Exoprosopa* Macquart, subgenus *Pterobates*, the hind tibiae are feathered with long scales. The lower surface of the front tarsi may have the typical, oblique, sharply pointed bristles modified into a fine, erect, fuzzy pile, and in a few genera the tarsal spiculae are thickened and rather blunt. The claws vary from almost straight, slightly curved to sickle-shaped. Sometimes the apical tarsal segment is broadened. The claws may be of about the same size throughout or the anterior pair may be quite minute; in some Exoprosopinae the claws have a long, sharp, basal tooth, and in the Bombyliinae the genus *Zinnomyia* Hesse is so characterized. Pulvilli are usually well developed, wide or narrow, sometimes reduced or vestigial or absent; some genera have a trace of an empodial or more probably an aroliar structure. The halteres vary and more extended study might be profitable; usually with slender stalks the knob varies from oval to various shapes, apically truncate or excavated.

The wing is well developed in most subfamilies of bee flies. It is large and extensive, as would be expected in insects which have such well-developed powers of flight. Posterior cells range in number from 5 in rare instances, as for example in three subgenera of *Exoprosopa* Macquart to 4, the usual number, down to 3 in the subfamilies Usiinae, Phthiriinae and Gerontinae. The remigial area, or radial field of these flies, often shows complicated venation reminiscent of ancient and archaic types but yet greatly reduced in the subfamily Cyrtosiinae, still more in the Mythicomyiinae. The genus *Empidideicus* Becker, as Melander (1946b) has remarked, shows the greatest reduction of neuration in the entire family; here the auxiliary vein is vestigial, the second vein fused with the first, the third vein unbranched, the discal cell open on account of the loss of the posterior crossvein, and finally the ambient vein is absent. In certain African elements in the family, such as *Tomomyza pictipennis* Bezzi, and related species, and also in *Bombylius namaquensis* Hesse, the radial field is unstable and, though normally with 3 submarginal cells, may show 4 or more such cells because of the presence of stump veins and supernumerary crossveins which may be remnants of a much more primitive venation; Hennig (1954) calls these accessory veins. Some species of *Comptosia* Macquart from Australia show an even more erratic radial field. Erratic venation often appears in the form of spur veins below the discal cell.

The position of the anterior (radiomedial) crossvein is quite variable; it may lie near the radial sector or base of third vein, at the middle of the discal cell, or quite distally placed near the outer third or fourth or sixth of this cell. The first posterior cell while commonly open may be closed and stalked; the same is true of the anal cell. The alula may be broad and lobate, reduced, almost vestigial, or even absent and scales if present upon the thoracic squama are likely to border the alula as well. The ambient vein is commonly present but may extend only part way around the wing or be wholly absent.

The wing of bee flies often shows beautiful patterns, which are especially characteristic within the subfamily Exoprosopinae. In many genera the wing is characteristically hyaline or even vitreous hyaline with at most the costal or costal and auxiliary cells tinged; there are numerous genera and species where the basal part of the

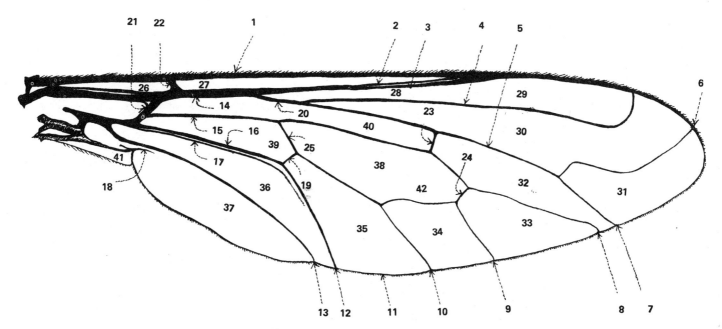

Text-Figure 11.—The wing of a generalized Recent bee fly, *Dischistus mystax* Wiedemann. Explanation: 1, costa; 2, subcosta or auxiliary vein; 3, first branch of the radius, or first longitudinal vien (R_1); 4, second longitudinal vein (R_2 and R_3); 5, third longitudinal vein (R_4 and R_5); 6, anterior branch of third vein (R_4); 7, posterior branch of third vein (R_5); 8, first branch of medius or fourth vein (M_1); 9, second branch of fourth vein (M_2); 10, third branch of fourth vein, or posterior intercalary vein (M_3), with anterior branch of the cubitus; 11, ambient vein; 12, second branch of fifth vein (Cu_2); 13, second anal vein; 14, main stem of the radius (R); 15, main stem of the medius (M); 16, main stem of the cubitus (Cu); 17, first anal vein; 18, remnant of third anal vein; 19, anterior branch of cubitus; 20, radial sector; 21, arculus; 22, humeral crossvein; 23, anterior, small, or middle crossvein; 24, anterior intercalary vein; 25, lower branch of the medius; 26, first or basal costal cell; 27, costal cell; 28, subcostal or mediastinal cell; 29, marginal cell (R_1); 30, first submarginal cell (R_2 and R_3); 31, second submarginal cell; 32, first posterior cell; 33, second posterior cell; 34, third posterior cell; 35, fourth posterior cell; 36, third basal, anal or lower basal cell; 37, axillary cell, anal angle; 38, discal cell (1st M_2); 39, second basal cell; 40, first basal cell; 41, alula, axillary or posterior lobe; 42, posterior, or medial crossvein.

wing is tinged with gray or brown or yellow; many species in various genera have the wings spotted, banded, or fenestrate with odd windowlike spots, or are mottled and infuscated. The apex of the wing in certain species of *Comptosia* Macquart may be milky white.

The costal border at the base may or may not have a widened margin, may be with or without a bristly comb, and may or may not be preceded by a curious hook, called variously the costal hook, prealar hook, basal hook. Efflatoun (1945) points out that in Exoprosopinae there is often a kind of fan formed of long scales on the base of the wing, actually arising from the mesonotum and which may conceal the hook; he notes that this was called the patagium erroneously by Lundbeck (1908) and erroneously the wing scale by Becker (1916); the term "wing scale" he reserves for a minute hook at the base of the alula, resting on the alar squama. Efflatoun calls the frenulum—again an apparent misuse of terms—a narrow skinlike band joining the inner base of the squama with the lateral base of the scutellum, and he further states that this so-called frenulum may bear on its margins the plumula composed of a small tuft of conspicuous and colored hairs or scales, especially in Anthracinae and some *Exoprosopa* Macquart species; this may be the structure from which, by modification, the strigula is developed in the Systropinae.

Near the upper border distally within the second basal cell there is very often a small, pale, unpigmented spot, called the prediscoidal spot; it is much in evidence in the Villini. In a few genera there are scales upon the wings.

In this family the abdomen is generally obtuse, stout, robust, scarcely as long as wide, or even more or less globular in genera in which the whole body is thick dorsoventrally, as in *Platypygus* Loew, *Cyrtomorpha* White, and a few other genera. In many genera, however, the abdomen is moderately lengthened, short to long-conical or tubular; in the subfamily Systropinae the abdomen is remarkable for the long, slender, petiolate, wasplike form with the last segments a trifle widened. The vestiture of the abdomen varies from dense, erect, very fine, plushlike pile concealing the ground color, to modifications in which the lateral margins bear appressed, extended, fanlike, radiating extensions of the pile or scales. Also the surface may be nearly bare, except for dense, minute, appressed micropubescence or microscalation, or there is very commonly a mixture of loose, long hairs together with micropubescence or scales. Not infrequently the posterior margin bears a row of conspicuous, stout bristles, as in *Cytherea* Fabricius and many other genera. *Lepidanthrax* Osten Sacken is characterized by patches of bright scales on lateral margins and terminal segments, and it is sometimes more glittering in the male sex; it is also true of some Anthracinae that the apex of the male abdomen is covered with brilliant silvery scales, which perhaps are important in courtship.

As Hesse (1938) points out, the abdomen shows a minimum of 6 to 9 segments, males with usually one less visible segment and even in females the last segment may be recessed beneath the one above, as in *Heterostylum* Macquart and many others. The last segments in the male form a hypopygium sometimes small, sometimes as in *Usia* Fabricius or *Phthiria* Meigen large and terminal, or again in many genera the hypopygium, which is composed of the associated parts of the male genitalia, may be reflexed ventrally, inconspicuous, almost hidden, twisted, or turned to the right side (if viewed dorsally), as in the case of *Exoprosopa* Macquart. Young (1921) in his study of the dipterous thorax found 7 abdominal spiracles, all lying in membrane. He noted the prominent first tergum and sternum inflated, large, and jammed up against the metathorax beneath the scutellum. I have not adopted his terminology for the thorax as I think a simpler one is better, so I have adopted most of Williston (1908) and Curran (1934).

The female genital system often involves a series of lateral spines placed and attached just beyond the lateral valves of the seventh sternite. These are shown in the diagram here reproduced from the remarkable studies of Biliotti, Demolin, and du Merle (1965). These spines, characteristic of species of *Villa* Lioy and *Exoprosopa* Macquart and other genera, are often utilized in deposition of eggs, although the eggs are sometimes ejected directly from hovering bee flies instead of being placed directly by the bee fly alight. I have seen the same species of *Bombylius* Linné in southern Texas utilize both methods of oviposition. Many other genera of bee flies have the female terminalia enclosed apically by an extraordinarily dense cloak of long, fine hairs; du Merle notes that the females of the species of *Villa* Lioy have a special perivaginal cavity.

Hesse (1938) was the first to point out how truly indispensable are the male genitalia for the final elucidation of species. In his remarkable and incomparable works (1938, 1956), he illustrated dissections of 519 species in many genera of South African bee flies and has done the world a great service. Hesse devised his own terminology, overlooking the excellent work of Maughan (1935) and of Cole (1927) who dissected and illustrated and labeled the genitalia of 5 bee fly genera. Bowden (1964) has brought the terminology of male genitalic parts in line with those used by Snodgrass (1957) for the Tabanidae; Snodgrass did not, as far as I am aware, ever make any study of the Bombyliidae. I prefer the terminology of Cole and of Maughan, and I have used it on the accompanying series of text figures showing four levels of bee fly male genitalia. Following is a list of terms used herein compared to my understanding of equivalent terms as used by several researchers.

TERMINOLOGY USED IN STRUCTURES OF THE MALE GENITALIA OF BEE FLIES

Terms used in this work	Terms of other authors believed equivalent
1. Basistylus: Bowden, Snodgrass. This is the large, basal, ventral structure, to which the dististyli are attached.	Penis sheath: Metcalf. Tenth segment: Cole. Basal parts: Hesse. Basimere: Snodgrass. Gonocoxite.
2. Dististylus: Cole.	Clasper: Metcalf. Telomere: Bowden, Snodgress. Beaked apical joint: Hesse.
3. Epandrium: Cole. This is the large, dorsal structure (tergum 9) to which the cercus is attached.	Epandrium: Van Emden, Hennig. Ninth (IX) segment. Tergum IX.
4. Axial system: Metcalf.	Aedeagal complex: Hesse. Aedeagal sheath: Bowden.
5. Basal part of aedeagus: This is the ejaculatory sac.	Middle part of aedeagal complex: Hesse. Chitinous box: Metcalf.
6. Ejaculatory duct and process. This is the apical part of aedeagus.	Aedeagus of authors.
7. Lateral and basal (ventral) apodemes. This, together with ventral, or basal apodeme, forms the sperm pump.	Lateral ejaculatory apodeme. Basal and lateral keels of authors. Basal and lateral struts of Hesse. Paraphyses: Bowden. Sustentacular apodemes of authors.
8. Ramus.	Basal margin of basimere (gonocoxite). Ramus: Hesse. Dorsal rami: Bowden.
9. Epiphallus.	Anterior, or basal margin of ninth sternum; apex may be split, or may be medially sutured.
10. Sternum IX. Segment lying below at base of basistylus.	Hypandrium.
11. Cercus.	Cercus of authors.

Any male of a species of *Bombylius* Linné makes a good starting place for the study of the male genitalia. The hypopygium consists of two, symmetrical, capsular, shell-like structures, which are not independently movable. They are usually more or less recessed within the last visible tergum and sternum, but there are exceptions; they may be visible only from below, and more or less reflexed backward, as in *Exoprosopa* Macquart, but in *Bombylius* Linné, although directed downward, are visible apically.

The double structure comprising the hypopygium consists of the large ninth and much reduced following somites. The morphologically dorsal part is the ninth tergum or epandrium, which bears the remains of the following somites, ending in the cerci and the anal

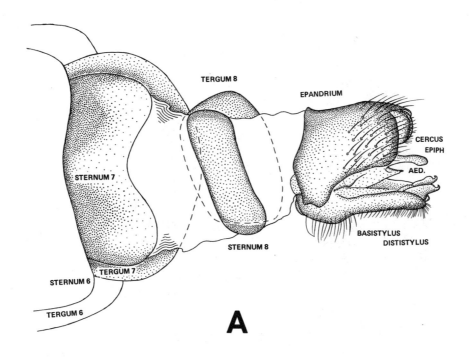
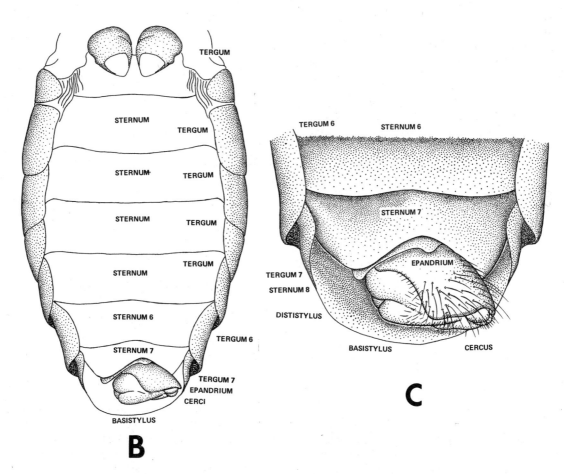

Text-Figure 12.—Genitalia of a male bee fly, *Poecilanthrax californicus* Cole. Explanation: A, sternum of abdomen; B, parts "in situ"; C, lateral aspect of male genitalia exserted.

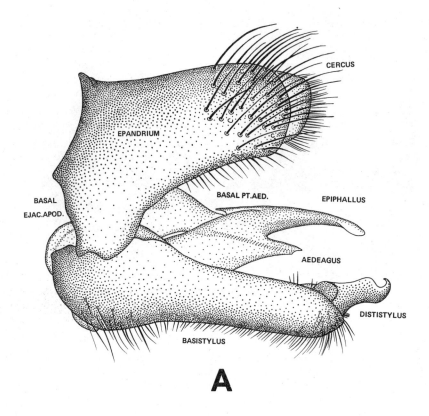

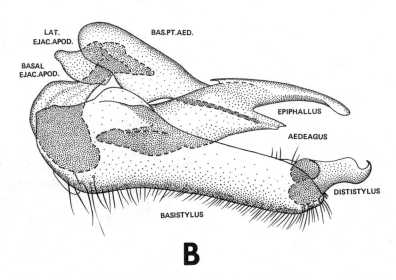

Text-Figure 13.—Male genitalia of *Poecilanthrax californicus* Cole: A, surface levels; B, second level, with the epandrium removed.

orifice. The ventral part most in evidence consists of a pair of basistyli between which ventrobasally may sometimes be seen a more or less rudimentary ninth sternum.

In most bombyliids there is a movable structure attached at the apex of the basistylus, which is the telomere of Snodgrass, the beaked apical part of Hesse, and the dististylus of Maughan and Cole; the latter term is adopted here. It may be noted that the base of the basistylus may itself be produced into a lobe or process.

In the Gerontinae this dististylus is more or less immovable and lobelike. The dististylus takes on almost an endless variety of complex shapes and forms, short and wide, or oval, elongate, etc., and as Hesse points out, flattened, hollowed, or convex.

Found within the pair of basistyli is the aedeagal apparatus or aedeagal complex of Hesse; it is attached to the basistylar shell by membranes and also by the ramus, the ramal element from each side generally is

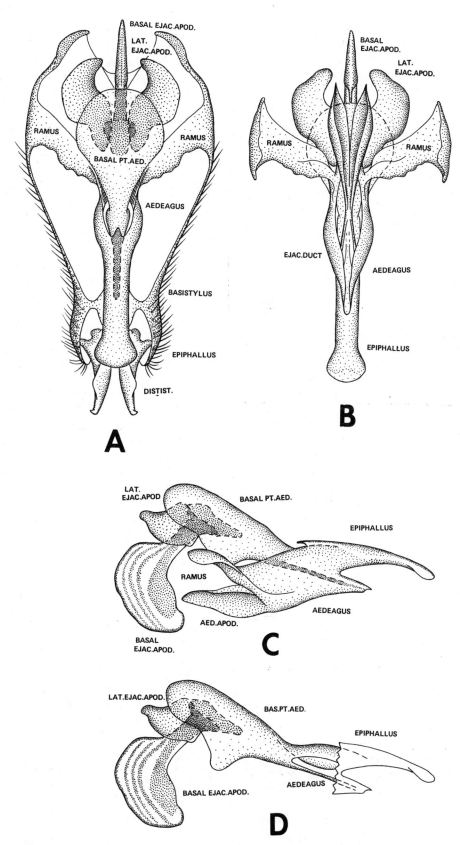

Text-Figure 14.—Male genitalia of *Poecilanthrax californicus* Cole. Third level: A and B, dorsal aspects; C, the basistylus removed; D, Fourth level, the axial system.

fused or coalesced to form an outer rim from which proceeds the aedeagus. The aedeagus is very variable: short or long, slender, straight or curved, or even tubular. Bowden (1964) considers that the ventral aedeagal processes of Hesse are not part of the aedeagus but belong entirely to the last abdominal segment and really form an aedeagal sheath which is connected by strong lateral processes, the dorsal rami of Hesse. He considers that these connect the whole to the base of the basistyle. According to Bowden, the aedeagus is easily dissectable from the sheath and consists of the aedeagus, and in addition a basal strut (Hesse, 1938) or apodeme, and lateral struts or paraphyses. We consider the aedeagus to be overlaid and often enclosed in a dorsal, wider sheath which we call the epiphallus. Hesse observes that the aedeagus is the intromittent organ of the bee fly, serving as a guide for the penis itself and the seminal duct within. All of these structures are especially complex within the Systropodinae and Gerontinae, with additional spines, prongs, hooks, etc.

For the study of the male genitalia, it was possible with the aid of the National Science Foundation grant to secure the services of Mr. Kenneth Weisman as a technician and delineator.

In preparing the male genitalia for the purpose of illustrating, for the most part, the technique of Metcalf (1921) was adopted.

The dried specimens were placed in a relaxing jar containing a solution of approximately 3 percent carbolic acid for 12 to 16 hours, after which the genitalia were removed. They were then cleared from 24 to 36 hours in a 10 percent solution of potassium hydroxide. This step of the procedure necessitates frequent observation in order to make sure the genitalia do not become overcleared. After clearing for the appropriate time the genitalia were removed to a neutralizing solution of 5 percent acetic acid. While in the neutralizing solution adhering muscle and other tissues that still remained were removed, after which the structures were deposited in glycerine-filled microvials and subsequently fixed on the pin that held the dried specimen.

Study and drawings were made while the genitalia were supported by glycerine in a spot dish.

The primary illustration for each genus consists of the complete genitalia in lateral aspect. After this figure was completed the epandrium was removed and the dorsal aspect of the basistylus (penis sheath) and its accompanying structures were illustrated. The final illustration, that of the lateral aspect of the axial system, necessitated this structure to be dissected from the basistylus. This view, called the "axial system," has been prepared for the more important genera. After completion of the illustrations, all the genital structures were again placed in the individual microvials and affixed to the pin carrying the dissected example of the genus.

Chromosomes

Chromosome studies have shed much light upon the interrelationships of the groups in the family Syrphidae. Similar studies in the Bombyliidae should be very fruitful, but according to personal communication from J. W. Boyes there are very few records in the literature, possibly not more than five or six. Metz (1916), in a very interesting paper entitled "Chromosome studies on the Diptera: II. Referring to *Villa lateralis* Say (as *Anthrax*)," states: "no more conspicuous cases of chromosome pairing have come to my attention than those exhibited by this and other species of the Bombyliidae." For this species he shows 5 large pairs of chromosomes and 1 small pair (Figures 129-133); the diploid number is 12. He was not able to identify the sex chromosomes, but believed them to be the very small pair that was present. For *Hemipenthes sinuosa* Wiedemann (as *Anthrax*), he found 9 pairs of chromosomes, all of different sizes, with the x and y chromosomes identified and the diploid number is 18 (Figures 134-137). For *Anthrax tigrinus* De Geer (as *Spogostylum simson* Fabricius) he finds the diploid number is 12, with a less decided variation in size (Figures 141-142).

Fossil Bombyliidae

Forty species of fossil bee flies have been given names. These species are distributed within 30 genera, of which 20 are extinct genera. This compares with the Asilidae which at present have the same number of known fossil species but only 18 genera, 3 of which are extinct. There are 3 genera of asilids known from an Eocene formation, 2 of which are extinct genera; no Eocene bombyliids are known. Four entomologists have concerned themselves extensively with fossil bee flies. Cockerell in a series of papers in the early part of the present century described numerous genera and species but with perhaps an inadequate background knowledge of the family. Many of his determinative decisions were based upon fragmentary portions of the bee fly wing, as illustrated in his work of 1914. Paramonov, with more exhaustive knowledge of the family, published a good summary of known species of bee fly fossils in 1939. Melander (1949) added several species which he described from material in the American Museum of Natural History. Hennig (1966) has provided a fine illustrated study of the bee flies known from the Baltic amber; I am able to add another from drawings I made in 1936, at which time I searched the collection at Königsberg for Syrphid fossils.

I present a list of the bee flies that are known from fossil forms with comment where I am able to make such. I have not been able to examine the type material of many species.

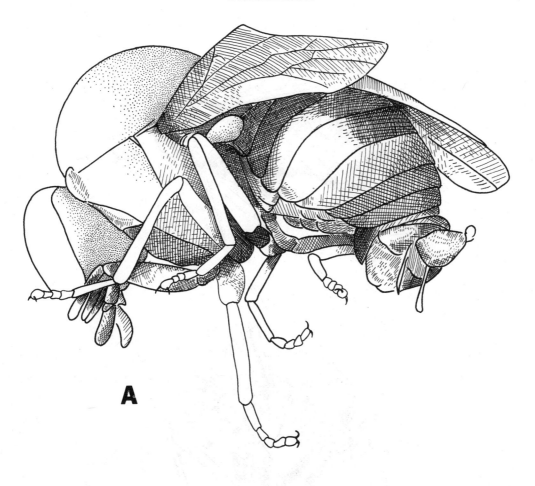
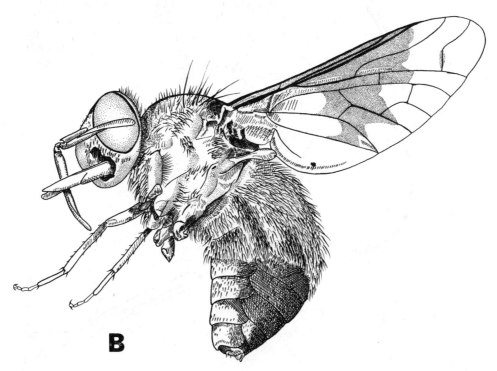

Text-Figure 15.—Fossil bee flies: A, *Proglabellula electrica* Hennig; B, *Paracorsomyza crassirostris* Loew. Redrawn after Hennig.

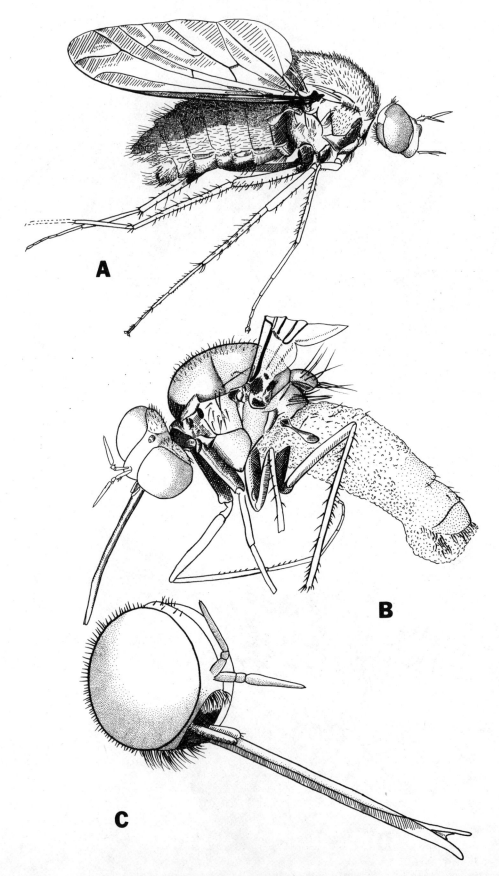

TEXT-FIGURE 16.—Fossil bee flies: A, *Amictites regiomontana* Hennig; B, *Glaesamictus hafniensis*; C, *Glaesamictus hafniensis* Hennig, face. Redrawn after Hennig.

FOSSIL BOMBYLIIDAE ACCORDING TO SUBFAMILIES

BOMBYLIINAE
Fossil Genera
  Praecythera sardii Theobald, 1937b, Oligocene (Aix-en-Provence)
  Paracorsomyza crassirostris Loew, 1850, Meunier, 1910, 1915, 1916, Oligocene (Baltic Amber)
Recent Genera
  Bombylius depereti Meunier, 1915, Oligocene (Aix-en-Provence)
  Bombylius sp. Berendt, 1830, Lower Oligocene (Amber of Samland, Poland)
  Bombylius sp. Schlotheim, 1820, Upper Miocene (Oeningen Shales, Baden)

TOXOPHORINAE
Fossil Genera
  Alepidophora pealei Cockerell, 1909, Miocene (Florissant)
  Alepidophora minor Melander, 1949, Miocene (Florissant)
  Alepidophora cockerelli Melander, 1948, Miocene (Florissant)

PHTHIRIINAE
Fossil Genera
  Protophthiria palpalis Cockerell, 1914, Miocene (Florissant)
  Protophthiria atra Melander, 1949, Miocene (Florissant)
  Acreotrichites scopulicornis Cockerell, 1916, Miocene (Florissant)
Recent Genera
  Phthiria oligocenica Timon-David, 1944, Lower Oligocene (Camoins)
  Apolysis magister Melander, 1946a, Miocene (Florissant)

GERONTINAE
Fossil Genera
  Palaeogeron vetustus Meunier, 1914, Oligocene (Aix-en-Provence)
  Geronites stigmalis Cockerell, 1914, Miocene (Florissant)
Recent Genera
  Geron? platysome Cockerell, 1914, Miocene (Florissant)

USIINAE
Fossil Genera
  Lithocosmus coquilletti Cockerell, 1909, Miocene (Florissant)
Recent Genera
  Usia atra Statz, 1940, Upper Oligocene (Upper Oligocene, from Rott)

SYSTROPINAE
Fossil Genera
  Melanderella glossalis Cockerell, 1909, Miocene (Florissant)
  Melanderella testea Melander, 1949, Miocene (Florissant)
  Pachysystropus rohweri Cockerell, 1909, Miocene (Florissant)
  Pachysystropus condemnatus Cockerell, 1910, Miocene (Florissant)
Recent Genera
  Systropus rottensis Meunier, 1917, Oligocene (l'Aquitanien de Rott, Siebengebirge, Rhineland)
  Systropus acourti Cockerell, 1921, Upper Oligocene (Isle of Wight)
  Dolichomyia tertiaria Cockerell, 1917, Miocene (Florissant)

MYTHICOMYIINAE
Fossil Genera
  Proglabellula electrica Hennig, 1966, Lower Oligocene (Baltic Amber)

LOMATIINAE
Fossil Genera
  Protolomatia antiqua Cockerell, 1914, Miocene (Florissant)
  Protolomatia recurrens Cockerell, 1916, Miocene (Florissant)
  Alomatia fusca Cockerell, 1914, Miocene (Florissant)

Recent Genera
  Xeramoeba gracilis Giebel, 1862, Lower Oligocene (Copal)

CYLLENIINAE
Fossil Genera
  Megacosmus mirandus Cockerell, 1909, Miocene (Florissant)
  Megacosmus secundus Cockerell, 1911, Miocene (Florissant)
  Palaeoamictus spinosus Meunier, 1916, Upper Oligocene (Baltic Amber)
  Protepacmus setosus Cockerell, 1916, Miocene (Florissant)
  Verrallites cladurus Cockerell, 1913, Miocene (Florissant)
  Amictites regiomontana Hennig, 1966, Upper Oligocene (Baltic Amber)
  Glaesamictus hafniensis Hennig, 1966, Upper Oligocene (Baltic Amber)
Recent Genera
  Amphicosmus delicatulus Melander, 1949, Miocene (Florissant)

ANTHRACINAE (Tribe ANTHRACINI)
Fossil Genera
  Anthracida xylotona Germar, 1849, Upper Oligocene (Orsberg bei Rott im Siebengebirge, Rheinlande)

EXOPROSOPINAE (Tribe VILLINI)
Recent Genera
  Hemipenthes s. l. sp. Goldfuss, 1831, Upper Oligocene (Rheinlande [Aquitain?])
  Hemipenthes s. l. sp. Burmeister, 1832, Lower Oligocene (Baltic Amber)
  Hemipenthes s. l. sp. Keferstein, 1834, Upper Miocene (Oeningen, Baden)
  Hemipenthes s. l. provincialis Handlirsch, 1908, Lower Oligocene (Aix-en-Provence)
  Hemipenthes s. l. tertiarius Handlirsch, 1908, Upper Miocene (Gabbro, Italy)
  Hemipenthes s. l. gabbroensis Handlirsch, 1908, Upper Miocene (Gabbro, Italy)

Zoogeography of the Bombyliidae

Bee flies, while worldwide in distribution, are far more abundant in arid lands. They are not absent from rainy temperate zones or tropical rain forest but in the latter they may be quite scarce. In several weeks of collecting in the Panama rain forest in August 1938 1 saw only one bombyliid. In a study of bombyliids in 1962 over a five-week period, in Mexico, ranging from arid eastern country to the upper edges of the rain forest near Tamazunchale, to the western desert part of the country, the number of bombyliids varied significantly according to the rainfall and area visited. Four or five areas were sampled each day by three collectors spending about an hour at each locality. Generally it was found that an hour of intensive collecting by three persons was sufficient to obtain practically all the Bombyliidae that were present. More time given to the same place seldom resulted in additional species. These areas and the number of bombyliids found at each are shown on the accompanying map. While the total bombyliid fauna must be significantly greater for each of these localities, if accumulated on a year around basis, it is nevertheless possible to gauge the relative richness of the fauna from this sampling.* Of course, there is an additional factor which is important too; the amount of rainfall at any one time has a distinct bearing on the

---
* Hall found 50 species in one canyon in California on a year basis.

bee fly fauna as indeed upon all of the insect populations of an area. When bee flies are mapped to the point of location from which the type was described, it is at once apparent that the annual rainfall has some peculiar relationship to bee fly populations. Desert rains often "trigger" the emergence of bee fly populations which appear seven days or a week or more later.

The peculiarities in the distribution of bee flies throughout the world are most interesting. Since, as far as is known, all bee flies are parasites or hyperparasites, it is not to be expected that species density centers would be the same as they are for the syrphid flies and the asilid flies, in both of which families the Neotropical region has much the greater number followed by the Palaearctic region. Even in the Syrphidae the special conditions affect these densities; for example, the wet, New World rain forests, so extensive, lend themselves to the great development of the flies in the Syrphidae, which often are aphidophagous and coccidophagous. In the Bombyliidae the development of the several subfamilies must be tied in very closely with the development and abundance of species and genera of the two great sources of hosts, the fossorial bees and wasps and the egg masses and packets of the locustids. A recent review of the Orthoptera of South Africa shows a total of about 2,000 species of locustids; it is not surprising then to find such an immense development of the genera *Systoechus* Loew and allies and of *Lomatia* Meigen.

As of the end of 1965 the world count of bee flies stood at about 3,924 species, distributed through 194 genera and 27 subgenera. The total species and genera for the Homeophthalmae division, 1,884 in 109 genera, is not greatly different from the 2,012 species in 85 genera within the Tomophthalmae; there are more species in the latter division but fewer genera. Only genera are considered in these totals, the subgenera are omitted.

Now if we look at the separate world regions we find 625 species in the first division from the Palaearctic, and 559 species from the Ethiopian area, against 494 Palaearctic species and 641 Ethiopian species in the second division. Note that these two world regions lead very considerably; almost 60 percent of the known bee flies are described from these two regions. We must not lose sight of the fact, however, that many arid lands have not been adequately explored for bee flies; such certainly are the southern areas of South America, still parts of the western United States, and still more parts of Australia and Central Asia. The bee fly fauna of Asia, apart from Asia Minor and the Transcaucasus, is depauperate indeed upon the basis of present-day records, with only 169 species listed from there; even more species—263—are known from Australia.

Only 9 species have been described from what is usually called Oceania, or the South Pacific Islands. There is a single unique genus from New Zealand.

After an inspection of the accompanying tabulation, if we look more closely at some of the subfamilies, we will note that the subfamilies Bombyliinae, and the Lomatiinae, Anthracinae, and Exoprosopinae as well

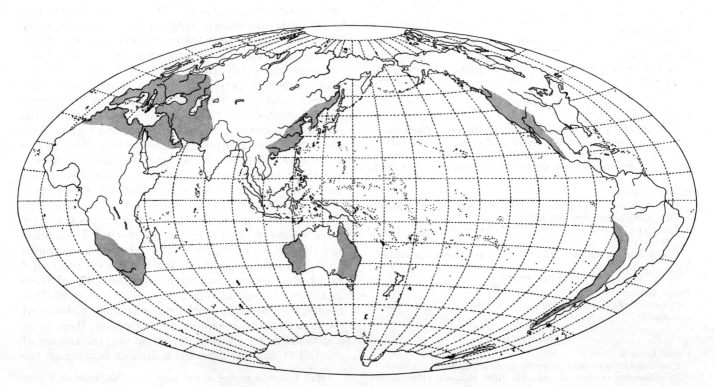

TEXT-FIGURE 17.—Concentration areas of the bee flies of the world.

as the small subfamily Cylleniinae are all dominant in what may be called the Afro-Eurasian landmass—the more easterly parts of Asia omitted. The home of the Heterotropinae is unmistakably in southern Europe and Asia Minor; this is largely true of the Cyrtosiinae. Other peculiarities of distribution will be noted; such is the great speciation of a single genus of the Mythicomyiinae, *Mythicomyia* Coquillett, in the western United States; the confinement of the moderately large genus *Usia* Latreille to the southern and western Palaearctic region. The smaller subfamilies tend to be absent or much reduced in one or more world regions, as the Cyrtosiinae in the Neotropical area, the Heterotropinae from both Australia and the Neotropical area.

There are odd phenomena with respect to genera. Species of *Anastoechus* Osten Sacken are more numerous in the Northern Hemisphere, and species of *Systoechus* Loew are more abundant in the southern half of the world. Both appear to be locust-egg pod consumers. One of the outstanding features of the zoogeography of this family is the large number of endemic and restrictive or peculiar genera of certain regions. For example, of the 50 genera endemic to South Africa, 20 are monotypic. Only 16 genera are endemic in the Palaearctic region and 16 in the Neotropical region. There are 18 genera peculiar to the Nearctic region. Some of these, like *Lordotus* Loew from the first division of the family and *Poecilanthrax* Osten Sacken from the second division, are well developed and have prominent features. Not surprising are the 12 genera peculiar to Australia, including unique genera such as *Neosardus* Roberts, *Myonema* Roberts, and *Comptosia* Macquart.

The virtual restriction of the relict genera of the Comptosini to Australia and the Chilean subregion will be noted with interest by everyone concerned with these flies.

Synopsis and Distribution

DIVISION HOMEOPHTHALMAE

Bombyliinae, Latreille, 1802
Bombyliini:

	NE	NEOT	PA	ET	OR	AU	OC	Total
Acanthogeron			14	1				15
Systoechus	5	1	22	82	2	8		120
Lissomerus			1					1
Anastoechus	3		50	21		2		76
Bombylius	28	23	139	137	15	45	8	395
Bombylodes			3					3
Isocnemus				1				1
Brychosoma				1				1
Zinnomyia				2				2
Parabombylius	9							9
Sisyromyia				3				3
Nectaropota			1					1
Eusurbus						2		2
Sisyrophanes				7				7
Eucharimyia				1				1
Staurostichus						1		1

Acrophthalmydini, new:

	NE	NEOT	PA	ET	OR	AU	OC	Total
Acrophthalmyda				1				1

Eclimini, new:

	NE	NEOT	PA	ET	OR	AU	OC	Total
Eclimus			2					2
Thevenemyia	28	2	3			3		36
Tillyardomyia						1		1

Paratoxophorini, new:

	NE	NEOT	PA	ET	OR	AU	OC	Total
Paratoxophora				1				1

Dischistini, new:

	NE	NEOT	PA	ET	OR	AU	OC	Total
Dischistus		2	20	27	1	3		53
Bombylosoma			1					1
Chasmoneura				8				8
Doliogethes				15				15
Gonarthrus				28				28
Legnotomyia				9				9
Lordotus	23							23
Geminaria	2							2
Othniomyia				1				1
Lepidochlanus				1				1
Prorachthes			7	1				8
Adelidea				7				7
Sosiomyia				1				1
Paramonovella, new						2		2
Pilosia, new						1		1
Sparnopolius	5	5	1			1		12
Cacoplox		3						3
Sericusia		1						1
Platamomyia		1						1
Bromoglycis						1		1
Cryomyia						1		1
Hallidia	1							1
Neodischistus	2							2
Conophorina				1				1
Euprepina	3							3

Conophorini, Becker, 1912

	NE	NEOT	PA	ET	OR	AU	OC	Total
Conophorus	19		42					61
Aldrichia	2							2

Corsomyzini, new:

	NE	NEOT	PA	ET	OR	AU	OC	Total
Corsomyza				30				30
Callynthrophora				4				4
Gnumyia				2				2
Hyperusia				6				6
Megapalpus				3				3
Zyxmyia				1				1

Mariobezziini, Becker, 1912

	NE	NEOT	PA	ET	OR	AU	OC	Total
Mariobezzi				6				6

Cythereini, Becker, 1912

	NE	NEOT	PA	ET	OR	AU	OC	Total
Cytherea			6	52				58
Callostoma			3					3
Pantarbes	3							3
Sericosoma			6					6
Gyrocraspedum				1				1
Oniromyia			2					2

Heterostylini, new:

	NE	NEOT	PA	ET	OR	AU	OC	Total
Heterostylum	6	8						14
Eurycarenus		1		13				14
Triploechus	2	5						7
Efflatounia			1					1
Karakumia			1					1

Crocidiini, new:

	NE	NEOT	PA	ET	OR	AU	OC	Total
Crocidium			2	20				22
Mallophthiria		1						1
Desmatomyia	2							2
Adelogenys				3				3
Apatomyza				1				1
Semiramis			1					1
Tamerlania			1					1

	NE	NEOT	PA	ET	OR	AU	OC	Total
PHTHIRIINAE, Becker, 1912								
Acreotrichus						3		3
Phthiria	29	21	26	13	1	6		96
GERONTINAE, Hesse, 1938								
Amictogeron				21				21
Geron	21	10	10	27	2	6	1	77
Pseudoamictus				3				3
USIINAE, Becker, 1912								
Apolysis	8		3	8				19
Oligodranes	53		7	3				63
Dagestania			2					2
Usia			50	2				54
HETEROTROPINAE, Becker, 1912								
Apystomyia	1							1
Caenotus	4							4
Heterotropus	1		25	1	2			29
Prorates	1	2						3
TOXOPHORINAE, Schiner, 1868								
Toxophorini:								
Toxophora	7	11	12	16	1	1		48
Lepidophorini, new:								
Lepidophora	2	5						7
Cyrtomyia		2						2
Marmasoma						1		1
Palintonus				1				1
SYSTROPINAE, Brauer, 1880								
Dolichomyiini, new:								
Dolichomyia	1	2				3		6
Systropodini:								
Systropus	4	33		31	39	17	2	126
XENOPROSOPINAE, Hesse, 1956								
Xenoprosopa				1				1
PLATYPYGINAE (Cyrtosiinae, of authors), Verrall, 1909								
Cyrtosiini, Becker, 1912								
Cyrtosia			31	5		1		37
Ceratolaemus				4				4
Onchopelma				3				3
Psiloderoidini, new:								
Psiloderoides				1				1
Cyrtomorpha						2		2
Platypygini:								
Platypygus	1		15	1				17
Cyrtisiopsis				2				2
MYTHICOMYIINAE, Melander, 1902								
Mythicomyiini:								
Mythicomyia	130	1						131
Aetheoptilus				1				1
Doliopteryx				2				2
Glabellula	6		6	2		1		15
Pseudoglabellula				1				1
Empidideicini, new:								
Empidideicus	4		9	3				16
Anomaloptilus			1	4				5
Euanthobates				2				2
Leylaiya				1				1

DIVISION TOMOPHTHALMAE

	NE	NEOT	PA	ET	OR	AU	OC	Total
CYLLENIINAE, Becker, 1912								
Cylleniini:								
Cyllenia			1	7				8
Sphenoidoptera	1							1
Amictus			1	26				27
Sinaia				2				2
Henicini, new:								
Henica				1				1
Nomalonia				6				6
Tomomyzini, new:								
Tomomyza				11				11
Pantostomus				9				9
Amphicosmus	4							4
Metacosmus	3							3
Paracosmus	6							6
Peringueyimyini, new:								
Peringueyimyia				1				1
Neosardini, new								
Neosardus						4		4
LOMATIINAE, Schiner, 1868								
Lomatiini:								
Lomatia		1	34	97		4		136
Anisotamia				1				1
Ogcodocera	1	1						2
Bryodemina		3						3
Edmundiella				1				1
Myonematini, new:								
Myonema						1		1
Docidomyia						5		5
Antoniini, new:								
Antonia			6	7		4		17
Aphoebantini, Becker, 1912								
Aphoebantus	52	6	21			1		80
Pteraulax				8				8
Epacmus	13							13
Prorostomatini, new:								
Prorostoma				1				1
Coryprosopa				1				1
Epacmoides				4				4
Stomylomyia		1						1
Plesiocera		6		7				13
Conomyza				1				1
Exepacmus	1							1
Eucessia	1							1
Xeramoebini, new:								
Xeramoeba				2				2
Chionamoeba			6	2				8
Petrorossia			8	17	1			26
Desmatneura	1		1					2
Chiasmella			1					1
Comptosiini, new:								
Comptosia					1	46		47
Lyophlaeba		24						24
Ylasoia		2						2
Doddosia						1		1
Ulosoma	1							1
Oncodosia						3		3
EXOPROSOPINAE, Becker, 1912								

	NE	NEOT	PA	ET	OR	AU	OC	Total
Villini, new:								
Villa	50	152	80	53	34	35	29	433
Astrophanes	1	1						2
Chrysanthrax	22							22
Cyananthrax		1						1
Deusopora		1						1
Dipalta	2							2
Diplocampta		3						3
Hemipenthes	24	1	16					41
Lepidanthrax	9	3			2	2		16
Mancia	1							1
Neodiplocampta		3						3
Paravilla	26							26
Poecilanthrax	33							33
Prothaplocnemis				1				1
Pseudopenthes						1		1
Stonyx	1	3						4
Synthesia				1				1
Thyridanthrax	9	3	64	44				120
Rhynchanthrax	3							3
Oestrimyiini, new:								
Oestranthrax	1		5	4				10
Oestrimyza		1						1
Marleyimyia				1				1
Villoestrus			1	1				2
Exoprosopini:								
Atrichochira				1				1
Diatropomma				2				2
Exoprosopa	41	39	74	236	40	13		443
Heteralonia				2				2
Isotamia				1				1
Ligyra	2	13	11	16	17	22		81
Litorhynchus				29	1			30
Micomitra								
Colossoptera					1			1

ANTHRACINAE, Latreille, 1804

Anthracini:

	NE	NEOT	PA	ET	OR	AU	OC	Total
Anthrax	29	25	121	69	27	7		278
Dicranoclista	2			1				3

Walkeromyiini, new:

	NE	NEOT	PA	ET	OR	AU	OC	Total
Walkeromyia		2						2
Coniomastix					1			1

DIVISION HOMEOPHTHALMAE

110 Genera

	NE	NEOT	PA	ET	OR	AU	OC	Total
BOMBYLIINAE	14	18	23	30	4	16	1	106
PHTHIRIINAE	1	1	1	1	1	2		7
GERONTINAE	1	1	1	3	1	1	1	9
USIINAE	2		5	2	1			10
HETEROTROPINAE	4		2	1	1			8
TOXOPHORINAE	2	3	1	2	1	2		11
SYSTROPINAE	2	2	1	1	1	2		9
XENOPROSOPINAE				1				1
CYRTOSIINAE	1		2	6		2		11
MYTHICOMYIINAE	3	1	4	8		1		17

DIVISION TOMOPHTHALMAE

81 Genera

	NE	NEOT	PA	ET	OR	AU	OC	Total
CYLLENIINAE	5	2	3	5		1		16
LOMATIINAE	6	7	11	14	2	8		48
EXOPROSOPINAE	15	14	8	15	6	5		63
ANTHRACINAE	2	2	1	2	2	1		10
Totals from each region	58	51	63	91	20	41	2	326

DIVISION HOMEOPHTHALMAE

	NE	NEOT	PA	ET	OR	AU	OC	Total
No. of Species								
BOMBYLIINAE	137	66	381	421	19	75	8	1107
PHTHIRIINAE	29	21	26	13	1	9		99
GERONTIINAE	21	10	10	51	2	6	1	101
USIINAE	61		62	11	2			136
HETEROTROPINAE	7		27	1	2			37
TOXOPHORINAE	9	18	12	17	1	2		59
SYSTROPINAE	5	35	31	39	17	5		132
XENOPROSOPINAE				1				1
CYRTOSIINAE	1		46	16		3		66
MYTHICOMYIINAE	140	1	17	15		1		174

DIVISION TOMOPHTHALMAE

	NE	NEOT	PA	ET	OR	AU	OC	Total
No. of Species								
CYLLENIINAE	14	2	35	28		4		83
LOMATIINAE	69	38	86	148	2	65		408
EXOPROSOPINAE	225	224	252	395	95	73	29	1293
ANTHRACINAE	31	27	121	70	28	7		284
Totals from each region	749	442	1106	1226	169	250	38	3980

Phylogeny and Subdivision of the Family Bombyliidae

One of the principal attractions of this ancient and fascinating family lies in the remarkable diversity of the many suprageneric basitypes which it contains. I define the word basitype as being a uniquely different genus within a higher category, such as a subfamily, which sets it at once apart; such genera have often been used as an excuse for the creation of new taxons within a family; sometimes such treatment is justified, more often not. An example would be the rather unique genus *Antonia* Loew; I have made this genus the basis of a new tribe, and a full dissection of its genitalia seems to warrant and support such treatment.

Universally, previous students of this family have expressed dissatisfaction and uncertainty with regard to the relationships of several groups. The taxonomy of the family has been overburdened and today is top-heavy with subfamilies, of which no less than 23 have been proposed by various authors; certainly these are not all of equal value. In my classification of the family I have found it necessary to rectify this situation by reducing many of these to a tribal status, which I believe expresses more accurately the basic relationships within the family. I have named no new subfamilies, although I have added new tribes and suggested other new tribes.

Distinguished students have even suggested separate subfamily status for several more "groups," such as *Tomomyza* Wiedemann, *Pantostomus* Bezzi, *Plesiocera* Macquart, and *Antonia* Loew. These groups do deserve recognition. Yet it is my firm conviction that all values are better preserved if they are placed within tribes of the most appropriate subfamily. This I have done. There are twelve subfamilies of scarcely disputable validity; eight are in the Homeophthalmae—Bombyliinae, Phthiriinae, Toxophorinae, Systropinae, Gerontinae, Usiinae, Platypyginae, Mythicomyinae—and four are in the Tomophthalmae—Cylleniinae, Lomatiinae, Exoprosopinae, and Anthracinae. It is likely that the Heterotropinae should be left as a tribe within the Bombyliinae. While I believe that Xenoprosopinae should be placed as a tribe within the Bombyliinae (even as *Oestranthrax* Bezzi is placed in Exoprosopinae as an aberrant specialized offshoot), I have refrained from doing so until such time as *Xenoprosopa* Hesse is better known. The classification in this work recognizes 11 tribes within the Bombyliinae, 7 within the Lomatiinae, 5 within the Cylleniinae, and 3 within the Exoprosopinae.

The evolution of the Bombyliidae has proceeded along two main lines of development. On the one hand, the division or section (or series) Homeophthalmae has the posterior eye margin almost always entire, except for the small group of genera centered around *Heterostylum* Macquart, and has the occiput flat, or even slightly concave, not expanding posteriorly and not bilobate above; this is the older group; Cockerell (1914) shows it to comprise the most ancient known forms (in his study of the fossil bee flies) and Cockerell also considered this division to be of Old World origin. Moreover, in this first division the cervix of the thorax is short, the head less freely movable, its mobility still further reduced by a general occurrence of a long, slender proboscis, more hovering and poising genera, and perhaps a greater dependence upon nectar. This division in large measure tends toward seasonal difference and dependence upon different kinds of flowers, but of course with exceptions. This division, the Homoeophthalmae, contains the Bombyliinae itself with 11 tribes, and also 9 additional, highly modified, subfamilies, which are specialized and frequently show a much reduced venation. Included within this division there are 110 genera of which the greater number are Palaearctic; only 15 are Holarctic; only 6 are worldwide. All subgenera are omitted, the division contains approximately 1,885 species. Known hosts seem to favor the solitary wasps and bees but also egg pods of locusts; three subfamilies have turned to families of Lepidoptera for hosts. Hosts are known for 11 genera.

On the other hand we have those flies in which the occiput becomes more and more extensively developed, thickened, and with a bilobed aspect dorsally, and hollowed parts with extensive deep central cup; also they have an increasingly conspicuous indentation in the posterior eye margin which may or may not have an accompanying, horizontal, linear, bisecting line. There is a tendency in this division for both the antenna and the proboscis to become shorter, and in fact rather conspicuously shorter, with a fleshy, muscoidlike labellum or even for the complete reduction of the parts. The cervix becomes longer, the head more freely movable, a trend toward different kinds of flowers, especially the clustered, late season composites. This division, the Tomophthalmae, contains especially the Exoprosopinae, the Anthracinae, the Lomatiinae, and the Cylleniinae, together having a total of 17 tribes.

But the characterization of the family is not quite this simple. There are still other characters which trend in certain directions but are quite variable. In a sense all of these subdivisions fail to be clear-cut and all of them right down to genera have to be characterized collectively with an ensemble of characters because of overlapping features. For this reason I have tried to select what seems to be the most important single feature for each group to which I add additional distinguishing characteristics. Each of the several divisions that have been proposed in the past, regardless of the rank assigned them, together with new ones here recognized, seem to me to constitute what might be called so many basitypes. In the Tomophthalmae we find 81 genera. The Palaearctic region contains 27 genera; there are 9 Holarctic and 4 worldwide genera. There is a total of 2,068 species from the world. Hosts are known for 16 genera.

The host preference is known for comparatively few genera but many species seek out subterranean wasp grubs, another tribe seeks almost exclusively mud-nest-building wasps, still another seeks out nocturnal Lepidoptera; others parasitize *Glossina* or *Calliphora* pupae or myrmeleonids. Some attack cossid larvae, beetle larvae, or egg pods of locusts; as at present known their larval host range is much more diverse than in the first division of the family. There are numerous examples of hyperparasitism.

My studies convince me that of still greater importance in the segregation of second subdivisions of the family is the deeply hollowed and conspicuous cuplike bowl of the central occiput; for actually this deep hollowed cup is present in all four of the subfamilies in the second division, whereas, in many of the tribes or genera of the Cylleniinae there is scarcely any backward overgrowth or overdevelopment of the occiput, or forward recession of the eye, and no indentation or bisecting line such as seen in the Lomatiinae and Exoprosopinae. The Cylleniinae falls into this second division because they have the hollowed, cuplike occiput; they are the most primitive or generalized members of this division and while the extent of the occiput varies from little in *Paracosmus* Osten Sacken, to much in *Pantostomus* Bezzi, there is at most only a faint indication of the bilobed indentation. They also show the characteristic modification of the antenna, which is a feature of this division. In order to clarify the morphology of the posterior occiput I have prepared a series of figures illustrating the conditions of the occiput in each tribe or major subdivision; I refer to Figures 745, 748, 750, 752, and 775.

While Bezzi (1924) took note of the hollowed occiput, he placed first emphasis upon the thickened occiput and the bisected and indented posterior eye margin. Phylogenetically I believe there is no question but that the hollowing of the occiput is the primary divisionary character. It began in the Cylleniinae, and hence I place these flies as the first subfamily within the Tomophthalmae (section [2]). With it and accompanying it there follows a profound anterior recession of the eye (as begun in *Tomomyza* Wiedemann and ending in *Oestranthrax* Bezzi, etc.), the consequence of which is the greatly thickened occiput. The end result is a short, occipital fringe around the rim of the cup, a long cervix, complete freedom of head movement, generally shortened proboscis, more and more muscoidlike labellum, change of flower preference, and perhaps a greater consumption of pollen.

The more time and study I expend upon this family the more apparent it seems to me that many characters represent a situation of what must be thought of as duplication or as parallel development or convergence, with the reappearance of certain characteristics independently in different parts of the family. These I shall indicate below.

Cellular content of wings: from a condition where there are commonly as many as 4, and sometimes even

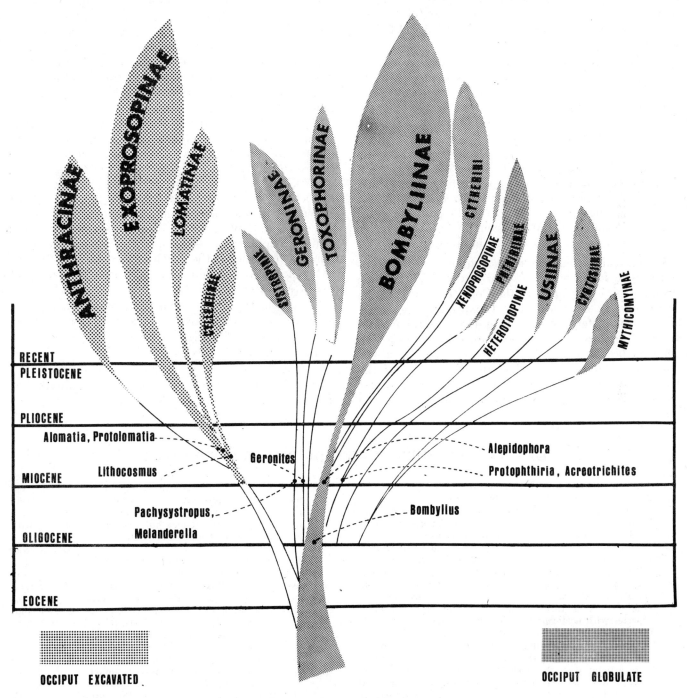

TEXT-FIGURE 18.—A provisional phylogenetic chart of bee flies and the relationship of subfamilies and tribes.

5, submarginal cells (some of which result from supernumerary crossveins), and as many as 5 posterior cells in occasional instances, the trend has been toward a gradual reduction in the number of veins and cells, to such genera as *Empidideicus* Becker, which has a simple, unbranched third vein, open discal cell with loss of the posterior crossvein, the second vein fused with the first vein, the auxiliary vein vestigial and has no ambient vein. As far as present time and presently known bee flies, we find that *Empidideicus* Becker shows the greatest reduction and the greatest simplification. The most unstable field in the bee fly wing is first the radial field, and second the posterior field as centered around the discal cell and the third and fourth posterior cells. We find:

a. The radial field with from 5 submarginal cells, as in *Bombylius namaquensis* Hesse, to one, as in *Cyrtosia* Perris.
b. From a 3-branched radial sector in many genera (4 in *Adelidea* Macquart) to 1 branch in *Euanthobates* Hesse.
c. From a free second vein, as in most genera, to one ending in the costa, as in *Glabellula* Bezzi.
d. From a complete auxiliary vein, the usual state, to a vestigial one, as in *Mythicomyia* Coquillett.
e. In the postremigial field, from 4 posterior cells, the usual state, and occasionally 5, to 3 posterior cells in no less than five subfamilies.

It is notable that the most complex venation appears to be found within *Ligyra* Newman in the tribe Exoprosopini and again in the Comptosini within the subfamily Lomatiinae. The Comptosini are today almost completely restricted to the Australian and Chilean regions. Another example of survival of a complex bee fly is the Australian *Neosardus* Roberts, with its 4 submarginal cells. In other parts of the world the flies within these subfamilies have rather simplified venation, and, except for the close relation of *Ligyra* Newman to *Exoprosopa* Macquart, they seem to be rather distantly related to the other members of their respective groups. The Comptosini and Neosardini seem indeed to be relict genera in the sense that they tend to have an extreme southern distribution much as *Trichophthalma* Schiner in the family Nemestrinidae. Other peculiarities of venation are noted under subfamilies and genera.

Among the bee flies the subfamily Bombyliinae with some 74 genera and subgenera and approximately 1,107 species is quite large, and we believe these to be the most generalized. The next largest subfamily is the Exoprosopinae, which has 33 genera (all subgenera omitted) and has approximately 1,293 species. Though much more specialized in some respects this subfamily has several very large and seemingly active and evolving or differentiating groups. Such are *Villa* Lioy, used in the wide sense, with about 433 species. *Exoprosopa* Macquart has about 443 species.

The widespread distribution of such semiphylogeront groups as *Toxophora* Meigen and *Systropus* Wiedemann seems to indicate an ancient origin. Since both of these groups parasitize highly successful types of host insects, they seem to be in no way threatened with extinction. The genus *Lomatia* Meigen appears to have arisen in the Palaearctic but has become remarkably successful in South Africa. Ideal conditions and isolation have apparently been responsible for the enormous number of species of the gnatlike *Mythicomyia* Coquillett which have developed in southern California and to a lesser extent in nearby states.

Gigantism appears within a few genera. Some species of *Exoprosopa* Macquart attain a wing spread of 64 mm.; perhaps the bulkiest bee fly is *Bryodemina valida* Wiedemann, from Mexico. Great size reduction or dwarfing (nanism) has appeared in several genera and subfamilies but none more so than in the minute flies of the genus *Mythicomyia* Coquillett.

KEY TO THE SUBFAMILIES OF THE BOMBYLIIDAE

1. Third longitudinal vein with two branches 3
   Third longitudinal vein with only one branch 2
2. Second longitudinal vein vestigial or absent, or if present, and complete, it ends in the first vein and not in the costa and results in a short, triangular marginal cell.
   . . . . MYTHICOMYINAE
   Second longitudinal vein normal and well formed and always ending separately and independently in the costa, beyond the end of the first vein. Second basal cell generally shorter than the first . . . . PLATYPYGINAE
3. Wing with 4 or rarely with 5 posterior cells. Includes species with posterior veins partly atrophied 6
   Wing with only 3 posterior cells 4
4. Abdomen long and slender, many times longer than wide; laterally compressed throughout its length, or more or less attenuate on the basal half or more. Metasternum oblique and often greatly and remarkably drawn out and extended posteriorly . . . . SYSTROPINAE
   Abdomen robust, oval or subspherical, or abdomen short conate, attenuate apically but never extensively lengthened. Metasternum not drawn out extensively posteriorly; never more than normal in development . 5
5. Third antennal segment long but gradually attenuate, with a short, still more narrowed microsegment or style.
   . . . . GERONTINAE
   Third antennal segment blunt and obtuse at apex, with a minute, subapical, subdorsal spine . . . . . . USIINAE
6. Flies without a large, conspicuous cavity in the occiput. Occiput usually not extensive, or if thickened, the outer part slopes gradually away from the eye margin. Occiput not bilobate above, with deep, vertical, postocellar fissure. Middle of posterior eye margin almost always smooth, entire and continuous, without indentation; an indentation is very rarely present in this group 9
   Flies with a large, central, deeply sunken cavity in the posterior occiput of the head. Posterior occiput extensive and tumid posteriorly beyond the lateral dorsal eye margin. Occiput always bilobate above and with resultant

deep, vertical postocellar fissure. Middle of lateral posterior eye margin with or without, more often with, an indentation, which occasions even greater exposure of the occiput. Indentation, when present, generally conspicuous, more rarely the eye margin is recessed forward on the entire upper half 7

7. Lateral eye margin neither indented nor recessed forward above . . . . CYLLENIINAE
Lateral eye margin with distinct indentation or at least the upper half recessed forward 8

8. Second longitudinal vein arising before the middle crossvein and at an acute angle, rarely with a bluntly rounded angle . . . . LOMATIINAE
Second vein arising opposite to middle crossvein or nearly so, and always at right angle . . . . . . EXOPROSOPINAE

9. Prothorax prominent, bearing remarkably long, stout, straight or curved macrochaetae on lateral, dorsal or both portions. The anterior margin of the mesonotum and postmargin of the scutellum bear long, slender or stout, macrochaetae or bristles. Femora and tibiae with appressed pile and scales. The tibiae often have numerous, long, spiculate spines or bristles. Thorax gibbous and often strongly and conspicuously arched. Abdomen arched and decumbent, or straight, cylindroid and elongate. I have treated *Lepidophora* Westwood as a separate tribe under this subfamily, although my later studies convince me it should be regarded as a tribe under the Bombyliinae.
TOXOPHORINAE
Prothorax short and normal. Macrochaetae if present never exaggerated. Thorax and abdomen not exceptionally arched, gibbous, and decumbent 10

10. Small, compact flies of 2 to 5 millimeters length, in which the third antennal segment is blunt and obtuse at the apex, and bears a short, thornlike spine, concealed or exposed, lying in a recess which may be either lateral and subapical, or dorsal and subapical, or apical and situated between two short protuberances above and below the spine. Anterior margin of oral opening always knife-sharp, the inner walls vertical and the oral cavity deep. Proboscis long and slender and always longer than the head; palpus likewise slender, usually long. Wings broad at the base and upon the axillary portion. Never more than 2 marginal cells present. Discal cell always much wider apically and first posterior cell widest at the wing margin. Males holoptic and head of male flattened across the top of eyes. Head of female with the front flattened and bearing a transverse groove or fovea over the depressed middle portion. Tibiae and femora with fine pile only; the hairs sometimes stiffened; bristles and spicules absent . . . . PHTHIRIINAE
Flies not with these characters collectively 11

11. "Facial and buccal part of head remarkably transformed and aberrant, depressed or excavated; mouth parts very aberrant, represented by a slight, central, boss-like elevation from the lower part of which projects a short, central, blunt, downwardly directed, spline-like process (medial anterior part of buccal rim), bounded on each side below by an oval, inflated lobe; antenna somewhat close together, with hairs on all the segments, segment 1 produced below into a large, conspicuous, densely haired, bladder-like or lobe-like extension, and segment three stoutish, bluntly tapering; front remarkably broad; occiput not deeply excavated behind, only slightly concave, very broadly and shallowly depressed groove-like behind and below ocellar tubercle, bounded on each side by a tumid, lobe-like prominence; ocellar tubercle remarkably broad, slightly elevated boss-like, its posterior ocelli wide apart and reniform and its anterior one much reduced; hind margin of eyes only feebly sinuous, not indented and not bisected; third posterior cell markedly narrowed and converging apically; legs more or less shortened, without spines on femora below; spicules and spurs on tibiae and tarsi feebly developed; vestiture in form of fine hairs and scaling not densely developed, without stiff, bristly hairs or bristles on any part of body, and hairs on abdomen markedly short and poorly developed."
XENOPROSOPINAE
Mouthparts functional and otherwise not as above.
BOMBYLIINAE

12. Discal cell (discoidal cell) broadened at the end (before the ultimate narrowing) and much of the widening due to the shortening of the anterior crossvein. Discal cell broader than the second posterior cell. Eyes of male separated into two distinct size groups of facets, the small ones below, the large ones above, but without dividing line. Inner margin of eye often indented near antennal base. Relatively bare flies . . . . HETEROTROPINAE
Discal cell not broadened at the end and not broader than the second posterior cell. Anterior crossvein long. Eyes of male not divided into two sizes of facets.
PHTHIRIINAE

Division HOMEOPHTHALMAE
Subfamily Bombyliinae Latreille, 1802

This is a large and important subfamily. The adults are primarily nectar feeders and the proboscis as a rule is long and slender, yet there are exceptions. The hosts as far as known are usually solitary bees, but *Systoechus* Loew destroys locust egg pods and two genera attack beetle larvae. For a great many genera, however, the hosts are unknown.

The following characters will serve to define this subfamily. A discal cell is always present. There are never less than 2 submarginal cells, and 3 submarginal cells are frequently seen in the Corsomyzini, Conophorini and Cythereini, as well as in some other genera. The posterior cells are never less than 4. The posterior eye margin is very rarely indented; the only exception is the Heterostylini. The occiput is not prominent and bilobed, but not concave either. The antennae are always subadjacent, except in the Cythereini. I do not consider the interesting group of genera around *Cytherea* Fabricius comparable to the other subfamilies within the Bombyliidae, and because I regard the family classification as top-heavy with subfamilies, I reduce these to tribal status; I have done the same with the *Eclimus* Loew group of genera. The antennae are generally long and slender. The wings tend to be either hyaline or at most smoky and basally tinged with brown, but there are exceptions. This is in strong contrast to the frequency of distinct and characteristic patterns in the Exoprosopinae and Lomatiinae of the second great division of the family. I have treated *Lepidophora* Westwood as a separate tribe under Toxophorinae, although my later studies convince me it should be regarded as a tribe under the Bombyliinae.

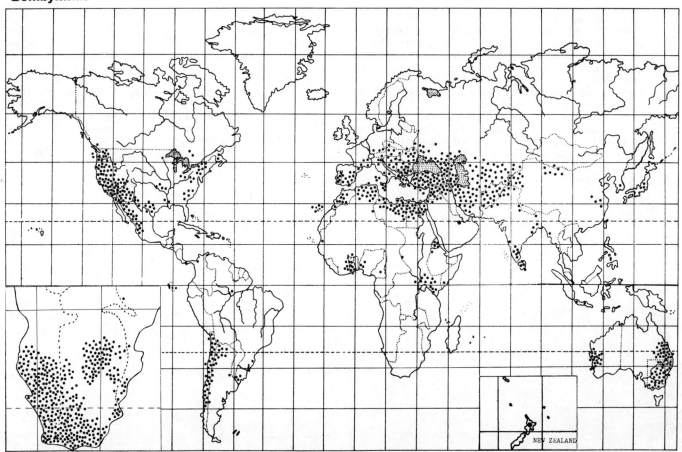

TEXT-FIGURE 19.—Pattern of the approximate world distribution of the species of the subfamily Bombyliinae.

Tribe Bombyliini

In addition to the genus *Bombylius* Linné, on which the family is founded, those other genera in which the first posterior cell is closed and stalked fall here; such are *Systoechus* Loew, *Anastoechus* Loew, *Sisyrophanes* Karsch, *Acanthogeron* Bezzi, etc.

Tribe Dischistini

These flies have the first posterior cell open and usually widely open. They may be regarded as the most generalized of the Bombyliidae and contain much the greater number of genera within the subfamily. While it is true that this cell at the wing margin narrows in some genera and this division is probably an unnatural one; it is nevertheless highly useful. Probably there is a slow trend in certain genera in the direction of the closure of this cell.

Tribe Heterostylini

Here I place the 4 genera that have a distinct indentation near the middle of the posterior eye margin. It includes, besides the Nearctic *Heterostylum* Macquart, the genus *Triploechus* Edwards, *Eurycarenus* Loew, an Ethiopian genus, and *Efflatounia* Bezzi and *Karakumia* Paramonov. Again in this indentation of the posterior eye margin, we may possibly see a development parallel to that so commonly found in the Tomophthalmae.

Tribe Cythereini

Because of the great variability in the width of the face and still greater variability in the degree of separation of the antennae, I can see little justification for placing *Cytherea* Fabricius and *Callostoma* Macquart in a separate subfamily. It would be just as reasonable to place the odd *Corsomyza* Wiedemann and the *Eclimus* Loew group in separate subfamilies. *Cytherea* Fabricius and allies are separable on only three particulars; in *Cytherea* the division or separation of the antennae reaches its greatest extent. It is somewhat less in *Callostoma* Macquart. I note much variation even within the species of a genus; *Pantarbes willistoni* Osten Sacken has the antennae but little separated, but the distance is twice as great in certain other species. I have followed Becker (1913) in placing the rather striking *Gyrocraspedum* Becker in this tribal group. I have done so in spite of the nearness of the antennae, because the base of the second longitudinal vein arises from the third longitudinal unusually close to the small crossvein and also abruptly and is in consequence suggestive of the second great division of the Bombyliidae, the so-called Tomophthalmae. This character is also variable; it is not present for instance in *Pantarbes* Osten Sacken where this vein arises normally even though the face is very wide. *Sericosoma* Macquart has the antennae very widely separated but like *Pantarbes* Osten Sacken the second longitudinal vein arises normally. I cannot escape the conclusion that we are here likely dealing with a case of parallel development, or convergence of characters, in this tendency of the point of origin in the Cythereini to edge outward distally as is present in so many Tomophthalmae genera.

The genus *Oniromyia* Bezzi from South Africa, an odd, aberrant bee fly, belongs rather doubtfully to this tribe. It has the antennae quite close together, the head not especially wide, and the second vein arises quite normally. Perhaps the only character that suggests a relationship to the Cythereini lies in the widely bossed or inflated frontovertical area. Its other characteristics are considered below. I reduce *Glossista* Rondani and *Chalcochiton* Loew to the status of subgenera where they still serve a useful purpose. Thus, there are 8 genera and subgenera within the tribe as represented here.

Tribe Conophorini

Flies with an exceptionally tumid, swollen occiput, in which the posterior expansion and swelling is quite unlike the Tomophthalmae division; in *Conophorus* Meigen the central part of the occiput is most prominent, rounding gradually off toward the eye margin, increasing or contributing toward a subglobular effect; coincidentally the first two segments of the antennae tend to be swollen, often much swollen, reminding one of the condition often seen in therevids or in *Symphoromyia* Frauenfeld; finally the venation, especially in *Conophorina* Hesse, is remarkably generalized with all cells widely open; in others there are 3 submarginal cells with a tendency to minor expansion of the apical remigial portion of the wing.

I place in this tribe *Conophorus* Meigen, *Conophorina* Hesse, *Aldrichia* Coquillett, and the more specialized *Prorachthes* Loew and *Legnotomyia* Bezzi, and following Becker I include *Codionus* Rondani, with subgeneric status only; *Codionus* has only 2 submarginal cells, and the apical venation is rather different in all those species that have only 2 submarginal cells, as these cells are broadly rounded instead of being blocked out rectangularly. Several European species would fall in *Codionus*, as I have shown in the discussion of *Conophorus;* also *Calopelta fallax* Greene.

Tribes Corsomyzini and Mariobezziini

These are rather unique flies, but I cannot see any justification for placing them in a separate subfamily when they are much better left in a tribe where they are obviously related. Probably these two tribes must be united. I have not had males of *Mariobezzia* Becker to dissect, though Dr. A. J. Hesse has generously furnished abundant material of *Corsomyza* Wiedemann. The facial, antennal, and venational characters are unique.

Tribe Acrophthalmydini

Because this Chilean genus has such a peculiar pygidium with its peculiar sleevelike collar, very different from the condition which obtains in *Bombylius* Linné where the genitalia are small, recessed, and turned downward and sideways beneath the apex of the abdomen, I believe these flies deserve to be placed in a separate tribe.

Tribe Paratoxophorini

Here I place the peculiar single genus *Paratoxophora* Engel; it bears no relationship whatever to *Toxophora* Meigen, and the name is an unhappy one. It may later prove to be related to *Amictus* Wiedemann since its genitalia are similar in one respect. There is only one species of this genus known and while there are several species of *Amictus* Wiedemann, I have so far been able to dissect the male of only one species. At present I do not believe these genera are related.

Tribe Eclimini

Here I place 2 unusually interesting genera. *Eclimus* Loew, which as Jack Hall (1969) has shown is restricted to southern Europe, and the more numerous species of *Thevenemyia* Bigot, the center for which lies in the southwestern United States. There is no justification for placing these flies in a separate subfamily. If every unique bee fly is placed in a separate subfamily there would be no less than 40 subfamilies; the true relationships of the bee flies and the phylogeny of the family is better served by placing these numerous peculiar groups in tribes within one of the giant subfamilies.

Tribe Crocidiini

I place here a group of bee flies that clearly fall within the definition of the Bombyliinae but which are notable for several characteristics which preclude them being placed in other tribes. *Crocidium* Loew is found in Egypt and more abundantly in South Africa. *Mallophthiria* Edwards is from southern South America. *Desmatomyia* Williston with its peculiar antennae is known only from two species from the southwestern United States. *Apatomyza* Wiedemann is a unique genus from South Africa and so is *Adelogenys* Hesse. The remaining genera are from the Transcaucasus region. All the flies in this tribe have 4 posterior cells and only 2 submarginal cells. They can be recognized by the straight condition of the anterior branch of the third vein which reaches the margin of the wing at or very close indeed to the apex of the wing. The antennae show some peculiarities in some genera.

The male terminalia in the Bombyliinae are variable in some respects but comparatively uniform in a number of features. The dististylus is almost always stout basally, narrowing apically into a hornlike point of variable sharpness, usually convergent at the apex. It may be also directed outward. Only in rare instances such as *Heterostylum* Macquart is the dististylus supplied with an additional or accessory lobe or prong. Nevertheless, it would be a very rash move to segregate *Heterostylum* Macquart and its related forms such as *Triploechus* Edwards and *Eurycarenus* Loew into a separate subfamily. Rather I have placed this group within a tribe within the Bombyliinae. Several of these have the aedeagal complex, which includes the epiphallus and the ejaculatory process, divided by a sinuslike space, resulting in a clasper dorsally. In general the entire structure is much shorter and more compact than the majority of genitalia such as in the Lomatiinae or the Anthracinae. The aedeagal complex is shorter and less spoutlike.

KEY TO GENERA OF BOMBYLIINAE

1. Middle of lateral posterior eye margin distinctly indented, but central occiput normal and without enlarged cavity. 2
   Lateral posterior eye margin continuous and entire, without indentation. Central occiput likewise normal, with at most a small central depression 6
2. Third antennal segment with a pencil of hairs at the apex. **Karakumia** Paramonov
   Apex of third antennal segment, bare, without a cluster of hairs 3
3. First posterior cell open or closed in the margin. Pulvilli quite minute and vestigial. Male claw toothed. Alula

almost or quite absent . . . . ***Efflatounia* Bezzi**
First posterior cell closed and petiolate; alula well developed 4

4. Three submarginal cells. End of marginal cell recurrent but plane, leaving the adjacent submarginal cell very broad against the costa, vertical in position and greatly narrowed below. Anterior crossvein at middle of discal cell. Abdomen with scanty appressed pile upon the tergites, the dorsum with very scanty long, fine, erect, tiny hairs, but the side margin with rather dense tufts of similar pile . . . . ***Triploechus* Edwards**
Two submarginal cells only. End of marginal cell gently curved 5

5. End of first posterior cell attenuate and acuminate, the stalk comparatively short. Upper anterior intercalary vein present though short. Anterior crossvein lies before the middle of the discal cell. Sides of mesonotum and posterior margins of abdominal tergites with strong bristles. Eyes in male usually holoptic or approximate, more rarely dichoptic. Male terminalia at apex with strong toothlike comb. Genae wide below, opposite the proboscis, and continuously densely pilose in contrast to the related genera . . . . ***Eurycarenus* Loew**
End of first posterior cell not acuminate, the anterior branch of the fourth vein ends close to the fork of the third vein. Upper anterior intercalary vein usually quite absent, the medial crossvein and fourth vein cross at this point. Anterior crossvein lies at or beyond the middle of the discal cell. Bristles on mesonotum weak, wanting on the abdominal tergites. Males holoptic. The bare oragenal cup margin unusually wide, obliterating the gena in some species . . . . ***Heterostylum* Macquart**

6. Antennae widely separated at the base. Second vein usually arising more or less abruptly close to the anterior crossvein (Cythereini) 7
Antennae closely adjacent, rarely if ever separated by more than the thickness of one segment 13

7. Posterior branch of the third vein drops downward to meet the fourth vein; it closes the first posterior cell and because of its shortness it simulates a crossvein. Second vein arises abruptly, before and quite near the anterior crossvein. First vein turned upward beside the auxiliary vein, both ending closely together. Proboscis 2 or more times as long as the head. Face bluntly conical, the face and front with short, fine, loose pile. Large flies with elongate wings and short, stout, reddish-brown wings . . . . ***Gyrocraspedum* Becker**
First posterior cell open or closed; if closed it ends rather acutely. Face and front often widely covered with dense, appressed scales which, however, may be restricted to the eye margins. Groups with dense facial and frontal scales have scanty pile; if the pile is dense the scales are more restricted 8

8. Second vein arises at or before the base of the discal cell and more or less acutely. No bristles anywhere upon the abdominal tergites; sides of first three tergies with a few stiff hairs. Pulvilli well developed 9
Second vein arises well beyond the base of the discal cell and more or less abruptly. Face with scanty pile but widely and densely covered with appressed scales. Posterior tergomargins of the abdomen beset with numerous long, stiff, more or less erect, conspicuous bristles. Pulvilli usually rudimentary 11

9. First posterior cell closed with a stalk 10
First posterior cell open maximally. Antennae widely separated. First antennal segment not swollen ventromedially. Metapleuron in front of haltere bare; metapleuron in front of spiracle with a tuft of long pile.
***Sericosoma* Macquart**

10. Stalk of first posterior cell quite short. Wing with 2 submarginal cells; slender and long, much narrowed and reduced in size. Antennae separated at base by a little less than twice the thickness of one segment. First antennal segment swollen and expanded ventromedially somewhat as in the Lomatiinae. Antennae situated low upon the head, the extensive front covered by loose, very long, erect hairs and scattered appressed scales between the pile. Abdomen moderately broad at base, elongate, and tapered toward apex. Surface covered with small, slender, appressed scales; no bristles present along the posterior margins of abdominal terga . ***Oniromyia* Bezzi**
Stalk of first posterior cell long. Wing with 3 submarginal cells. Wings of normal width. Antennae widely separated. First antennal segment simple, the third segment long, slender on the basal half, widened beyond. Face with dense, long pile and dense scales restricted to the eye margins. Abdomen with dense, long, erect, fine pile only; similar pile on the thorax. Metapleuron in front of haltere with dense tufts of pile; metapleuron in front of haltere bare, overshadowed by the multiple fringes of the squama . . . . ***Pantarbes* Osten Sacken**

11. First posterior cell closed, usually with a rather long stalk, or rarely closed in the margin. Rather large species with elongate, regularly tapered and banded abdomen with tergomarginal bristles posteriorly. Wings often whitish in tint, generally with a broad band of pale brown across the middle of the wing. Third antennal segment narrowed near base then tapered to the apex.
***Callostoma* Macquart**
First posterior cell open. Somewhat more robust flies, the abdomen tapered but shorter, not or inconspicuously banded. Wings with basal half or two-thirds tinged or marked or clouded with brown. Tergomarginal bristles present also 12

12. Face extended forward only bluntly. Whole face, front, and genae densely covered with pale, or often silvery scales, wings with basal half or two-thirds strongly tinged with brown. Male eyes very widely separated. Proboscis long and slender, usually quite long, always extended beyond the oral cup. Usually with 3 submarginal cells (*Cytherea*, sensu stricto); if only 2 submarginal cells (subgenus *Glossista* Rondani) . . . . ***Cytherea* Fabricius**
Small flies. Pulvilli absent. Face and front with sparse, fine, erect hair in place of appressed scales. Oragenal cup and face extended conically forward, separated by a sharp, distinct crease. Wing clear hyaline, faintly tinged basally with brown. Male eyes less widely separated; 2 submarginal cells . . . . ***Cytherea (Chalcochiton)* Loew**

13. First posterior cell open, even if narrowly open 14
First posterior cell always closed and usually with a long stalk, rarely closed in the margin 63

14. First antennal segment cylindrical and extraordinarily stout and wide, 2 or 3 times as wide as the second segment, and 4 or 5 times as wide as the third segment. First antennal segment as long or nearly as long as the outer two combined. Occiput very strongly developed from both dorsal and lateral aspect; marginal cell strongly widened apically, the ending rounded or plane, or even slightly recurrent. Metapleuron pilose, or base; 3 submarginal cells 15
First antennal segment not extensively widened and at the same time elongate 16

15. Three submarginal cells 79
Two submarginal cells . ***Conophorus (Codionus)* Rondani**

16. Second antennal segment as long as the first or longer, and both cylindrical, stout, much thickened, and bristly. Marginal cell not usually widened apically. Third segment flat but much widened in the middle, the apex with some long, stiff hairs. Three submarginal cells.
***Aldrichia* Coquillett**

Second segment of antenna shorter than the first, usually much shorter; neither segment extensively thickened. 17

17. Both marginal and first submarginal cells greatly and exceptionally widened in the outer half. Second submarginal cell as wide as long or wider. Three submarginal cells present 18
Marginal and submarginal cells usually not widened; only the marginal cell widened or the first submarginal cell, or neither, but not both. Second submarginal cell long or short, but longer than wide; usually with 2 submarginal cells; rarely with 3 submarginal cells 20

18. Head as wide as high and the face wide; third antennal segment without a long style 19
Head higher than wide, the face narrow; third antennal segment with a long style . . . . . . **Othniomyia** Hesse

19. Anterior crossvein enters discal cell at the outer fifth. Eyes with or without a complete dividing line from middle of posterior margin to middle of anterior margin. Abdomen long, conical and tapered, the pile mostly quite short; face with scanty pile and a few bristles. Small flies . . . . **Geminaria** Coquillett
Anterior crossvein enters discal cell near the outer fourth. Large flies with dense, fine, conspicuous, long, erect pile, plushlike on thorax and abdomen and also on the face. Males holoptic; females with some appressed pile.
Lordotus Loew

20. Face very broad, with dense, erect, but rather short pile in a hemicircular band around the shallow oragenal cup. Head, thorax, and abdomen clothed primarily with dense, usually dark, flat-appressed, glittering scales. All antennal segments relatively short, the second especially so; third segment flattened laterally, greatly expanded dorsoventrally and bearing a tuft of spines dorsally. Wings dark colored and characteristically bear a mottled, marbled, or irrorate pattern . . . . . **Prorachthes** Loew
Not such flies 21

21. First basal cell distinctly longer than the second basal cell, the anterior crossvein situated close to the middle of the discal cell or more generally much beyond the middle of the discal cell 22
Anterior crossvein situated near the base of the discal cell and at least before the middle; consequently the first basal cell is approximately of the same length as the second basal cell, or at most very slightly longer . . . 60

22. Second antennal segment more than half as long as the first segment, and 2 or 3 times as long as wide and swollen through the middle; first segment also a little swollen and both bear numerous, long, coarse, bristly hairs, especially below. Rather small flies with semidecumbent abdomen with scattered, very long, erect, bristly hairs along the postmargins of the tergites.
Conophorina Hesse
Not such flies 23

23. Head very broad and sometimes remarkably broadened; as broad or broader than the thorax. Eyes in male always separated at least as much as width of ocellar tubercle and often very much more widely separated. Both front and facial region broad. Face not only broad but extensive downward, genae likewise very wide; face may be much inflated and tumid, sometimes with characteristic dense brush of long, erect hair; third antennal segment elongate but thickened or thinned basally only to become clavate apically and apex sometimes excavated, and microsegments reduced or absent. Apex of last male sternite notched medially; mesonotum low and not humped; thorax and especially abdomen with dense, long, fine erect, conspicuous pile. Wings with 2 or 3 submarginal cells. Dististyli generally compressed, claw-shaped or hook-shaped. Wing short, only twice as long as wide, wide basally, usually widest at base. Tribes Mariobezziini and Corsomyzini 24
Not as above 29

24. Face extraordinarily broad and tumid or swollen; front also tumid and swollen. First antennal segment thickened, very short, more or less barrel-shaped. Oral cavity situated very low on head due to great extension of face vertically downward 25
Face not conspicuously tumid and swollen or inflated; genae less widened, swollen and conspicuous. Front likewise less swollen and tumid. First antennal segment not thickened, barrel-shaped, usually much longer. Proboscis always longer, often much longer, always projecting well beyond oral cavity, or antenna. Oral cavity always much more frontal in position, the face above therefore shortened. Dististyli distinctly clawlike and compressed laterally 27

25. Abdomen elongate, and tapered, and with fine scattered, short hairs. Face continued down to bottom of eyes and even below them so that the oral cavity though oblique is ventral in position. Face generally with loose, rather scattered, fine pile. End of second antennal segment with distinct spur dorsally. Genae also very wide. Wings widest at base. Anal cell closed and stalked; always 3 submarginal cells. Proboscis moderately elongate and not confined to buccal cavity . . . . . **Mariobezzia** Becker
Abdomen shorter and not conspicuously or distinctly elongate and tapered. Pile of abdomen longer and erect and dense. Proboscis short and confined to oral cavity, or at least not extending beyond antenna. Oral cavity also very low upon head due to vertical extension of face downward. Face and front have a conspicuous, circular brush of long, erect pile. Dististyli less compressed . 26

26. First antennal segment short and much swollen; third antennal segment elongate and strongly narrowed on the basal half; circular brush of pile around face and gena remarkably long and dense. A small metapleural tuft is present; wings with either 2 or 3 submarginal cells present. No dense, feathery pile on tibiae, at least not on hind ones; last hind tarsal segment with no conspicuous long hairs. Antennae attached right above oragenal recess rim; second segment with dense spinule-like pubescence; third segment clavate and broadened apically. Proboscis longer, more slender, labellum slender.
Callynthrophora Schiner
First antennal segment not or scarcely wider than second; widely separated at base; base of third segment not greatly narrowed. Circular brush of facial pile less developed, the pile not so long. The head has the face, sides of face, and gena remarkably broad and tumid and inflated, therefore, whole face below antennae more swollen and inflated. Antennae attached high up, much nearer to anterior ocellus. Second segment without fine spinulose pubescence; third segment more rodlike, thus not conspicuously dilated or clavate apically. Proboscis quite short and stout, nearly confined to oral cavity, rather muscoidlike, with broad, fleshy labellum. No metapleural tuft present. Wings with 3 submarginal cells. Legs with dense pile and dense, feathery pile on hind tibia and one or more long hairs apically and dorsally on the last hind tarsal segment. Males with dististyli broader on the basal half, and broadly dorsoventrally compressed, pilose, culminating in a slender, nonclawlike beak . . . . **Gnumyia** Bezzi

27. Head in front remarkably broad, facial region very broad. Sides of face and genae both more tumid; head below much broader than the vertex, even in the female; inner margins of the eyes clearly and strongly diverge down the sides of the face in both sexes. Antennae attached rather high upon the head, half or nearly half the distance from front ocellus to edge of oragenal recess. Pile of the body

dense and more conspicuous, particularly in males; pile of the facial region forms a curious, dense, characteristic, distinct, circular brush especially in males, but usually even in females. Empodium and pulvilli longer in these flies . . . . **Corsomyza** Wiedemann

Frontal aspect of head not especially widened or broad. Also face not widened, and lateral face and gena not tumid. Inner margins of eyes along sides of face, not much widened; if broader in some males, there is no circular brush of dense facial hair. Antennae inserted considerably lower, either just above oragenal rim or much less than half the distance from oragenal rim to anterior ocellus. Pile of thorax and abdomen much more sparse, often with a rather bare appearance. Empodium and pulvillus shorter 28

28. Eyes in both sexes quite broadly separated, the interocular space considerably greater than width of the tubercle; hence, inner margins of eyes parallel or nearly so; oragenal recess rim above, below the face, prominently protruding, thin, generally rectangular, as the margin of a cup. Antennae attached rather high, close to middle between oral recess and ocelli. Proboscis longer and the palpi long and slender. A prominent, stout bristle on the notopleuron. Pleuron almost completely shining and bare; no metapleural tuft present. Wings rather longer, alula more reduced and vestigial, axillary lobe more reduced; legs with only sparse pile; no long bristly hair above apical tarsal segment. Anal cell either closed and stalked or very widely open . . . **Megapalpus** Macquart

Eyes in male more narrowly separated above, by only the width of the tubercle or very little more; hence much more narrow than female. Oragenal rim not protruding (and spoutlike) in the two sexes; antennae attached almost immediately above oragenal rim; proboscis generally *much* shorter, the palpi shorter, wider; mesonotum without distinct notopleural bristle. Pleuron with sparse pile, but more pile, a metapleural tuft generally present. Wings longer, the alula and axillary lobe wider, even subtriangular. Legs with considerably more pile, even in females; last tarsal segment with one or more curious long dorsal hairs . . . . **Hyperusia** Bezzi

29. Small, stout flies with short, obtuse, decumbent abdomen. First antennal segment short, the second segment even shorter, beadlike, not as long as wide. Third segment elongate and cylindrical, with two quite stout microsegments of about the same thickness as the basal part of segment; first of these microsegments about as long as the second antennal segment, the terminal microsegment about 2 or 3 times as long as the preceding one; at apex it is blunt and bears an apicolateral cavity with spine. Second submarginal cell long and quite slender; anterior branch of third vein arises very gently, nearly plane, and ends barely above wing apex; second vein ends quite straight and gradually. Shining brown flies with little pile . . . . **Desmatomyia** Williston

Entirely different flies. Third antennal segment not with such stout peculiar microsegments 30

30. Head loosely and not closely attached to thorax. Occiput bimammillate, with very long, loose pile; medially grooved but not bilobed. First antennal segment long and slender, about 8 times as long as the short, beadlike second segment, and with a dense, narrow band of quite long, stiff, bristly, distinctly appressed pile on the underneath surface, especially near the apex. Face rather long and pollinose, with fringe of long, fine pile along the sides. Proboscis long and slender. Males holoptic and the hypopygium developed into a long beaklike process composed of the elongate, diverging, terminal tergum and sternum. Abdomen cylindroid, rather long and slender. The aedeagus forms a long curled filament.

Paratoxophora Engel

Entirely different flies. Hypopygium of male never drawn out into a large, long, beaklike structure with upper and lower valves, and flies otherwise not the same 31

31. End of submarginal cell widened and expanded; wing in known species with some brown at base or middle; second vein meeting the costa at a right angle. Wing narrowed at base, alula present but reduced. Anal cell widely open and the ambient vein unusually strong. Abdomen short, broad, much wider than thorax, subcircular, more or less flattened; tergites dark with narrow, opaque yellow, linear posterior margins; tergal pile fine, erect, and rather sparse. Third antennal segment arched below and plane above, or wide basally and gradually attenuate toward the apex. Characteristically the third antennal segment bears a tuft of long, bristly hairs above the middle and below and again at the apex; these hairs rarely reduced in length or number but always present. Oragenal cup circular, deep, with vertical walls and the outer rim distinct, extending only a short distance beyond the eye . . . . **Legnotomyia** Bezzi

Third antennal segment not with tufts of bristles in middle or apex. Abdomen not wide, subcircular with yellow bands and short loose pile. Entirely different flies . . 32

32. The anterior branch of the third vein, whether gently or strongly curved or nearly straight, never turned upward at wing apex. Usually rather small bare flies 55

The anterior branch of third vein always distinctly turned upward at wing apex, no matter whether curved, sinuous, sigmoid, or nearly straight 33

33. Metapleuron with a distinct tuft of pile in front of the haltere 34

Metapleuron clearly without a distinct tuft of pile in front of the haltere; dense, long, squamal pile may sometimes be matted overlying the metapleuron, not arising from it. 43

34. Whole face, thorax and pleuron, abdomen, and femora plastered with dense scales in addition to straggly, coarse, erect, pile. Male eyes separated by at least the distance between the posterior ocelli; discal cell very wide apically.
Lepidochlanus Hesse

Scales confined to legs 35

35. Large, very robust flies with wide short abdomen, and dense pleural and thoracic pile. Abdominal pile likewise very dense, with widely flared, matted, flattened tufts along the side of the abdomen. Proboscis elongate but labellum much enlarged and palpus with 2 segments; anterior crossvein at outer third. Anal cell open; hind femur spiny. Australian . . . . **Eusurbus** Roberts

Not such flies 36

36. Oragenal cup polished and bare and prominently protruded. Femoral spines absent; abdomen moderately elongate.
Euprepina Hull

Oragenal recess with outer walls either pollinose or pilose. 37

37. Second antennal segment 3 or 4 times as long as wide, and densely covered with long, stiff or bristly hairs on all sides. Genofacial space extraordinarily wide, the genae particularly wide and with deep, rounded, sloping tentorial fissure; face and genal slopes with dense, long pile and pollen. Quite large flies . . . . . . **Cacoplox** Hull

Second antennal segment beadlike, quite short. Genofacial space less extensive 38

38. Anal cell closed and stalked. Flies with long, quite dense, erect pile, and abdomen not quite 1½ times as long as wide, somewhat narrowed apically. Proboscis elongate. labellum quite slender, palpus exceptionally long and filiform, of one segment. Anterior crossvein lies at the outer sixth of discal cell. Intercalary vein of moderate length but first posterior cell open maximally.
Pilosia Hull, new genus

Anal cell open, usually widely open 39
39. Anal cell narrowly open; anterior crossvein at middle of discal cell. Intercalary vein long. First two antennal segments with short pile. Upper occiput with very long erect bushy pile. Small gray or black flies with whitish pollen and long, dense pile . . . . . **Sericusia** Edwards
Not such flies 40
40. Small flies with short, obtuse abdomen; whole thorax and abdomen with dense, long, erect pile. Whole sides of face and gena continuously covered with fine, very long pile. Males widely holoptic. Female front without a transverse depression. First antennal segment slender and long; third long, slender, attenuate. Anterior crossvein at the middle of the discal cell . . . . **Bombylisoma** Rondani
Not such flies 41
41. Rather large flies, with stout, short, robust abdomen, the terga with rows of prominent bristles on posterior borders. Face rather prominently extended forward and genae wider than usual. Hind femur with stout, spiny bristles and the tibiae stout . . **Bombylodes** Paramonov
Face recessive. Small flies, at most medium sized. Gena usually quite narrow and hind femur with only a few slender spines below. Anterior crossvein at middle of discal cell, or before 42
42. Small to medium-sized flies. Male eyes narrowly separated, approximate, or broadly separated. Frontal depression in female situated so as to lie in the middle of the front. Segment one of antenna usually longer, segment three usually shorter. Anterior crossvein always much before the middle, sometimes even a trifle before the basal third; intercalary vein always long. Legs always with a few spines on middle femur below as well as hind femur. Last male sternite with posterolateral angles rounded, not angularly acute; hypopygium with only membrane, or apicolateral process below. Flies with much bright red, the color of iron rust, on face, pleuron, and abdomen in both sexes . . . . **Doliogethes** Hesse
Male eyes always meeting for at least a short distance above; female eye less widely separated, usually less than 3 times width of tubercle. First antennal segment short, about 2 to 2½ times as long as the second; third antennal segment longer, rodlike, straight. Basal comb smaller. Pile everywhere more sparse, even more so in females. Intercalary vein characteristically much more narrowed or shortened. Flies almost always blackish, rarely with red on pleuron or abdomen; usually without a silvery tuft on each side of antenna in females; usually with silvery white scaliform pile on abdomen in both sexes, particularly males. Spines below only on hind femur. Last male sternite with posterolateral angle acute or even angularly produced. A distinct process on each side of aedeagus below ending in a curved or recurved hook or prong; process below aedeagus with spine or hook . . . . **Chasmoneura** Hesse
43. Three submarginal cells 44
Two submarginal cells 46
44. Third antennal segment simple or thin.
Adelidea Macquart
Third antennal segment broad at base, attenuate, or with stiff bristles. Males with eyes separated by nearly the width of the ocellar tubercle 45
45. Third antennal segment with stout bristles dorsally, weak bristles apically . . . . **Sosiomyia** Bezzi
Third antennal segment stout basally becoming attenuate near apex and without peculiar bristles. Second basal cell sometimes with extra crossvein . **Adelidea** Macquart
46. Anterior branch of third vein unusually strongly turned upward toward costa. Marginal cell widened apically, ending in a plane, rectangular vein simulating a crossvein; first posterior cell a little narrowed, the anterior crossvein at outer fourth of discal cell or beyond. Proboscis stout, little longer than the head, the labellum leaflike. Mesonotum humped anteriorly; haltere with a tuft of pile upon it at base. Third antennal segment rodlike, scarcely narrowed, notched at apex.
Bromoglycis Hull
Wing venation not so displayed. Third antennal segment not long, stout throughout and notched apically . . . 47
47. Large flies with very short, stout or robust, somewhat convex abdomen, considerably wider than long; much dense pile present. First posterior cell strongly narrowed; anterior crossvein at basal third. Third antennal segment quite long, slender, more slender on basal half and ending in spine and 3 or 4 long, stiff hairs or weak bristles.
Sisyromyia White
No bristly hairs at apex of antennae; venation different. Abdomen longer 48
48. Wing long and slender, alula absent, axillary lobe narrow. Anal cell very widely open and anterior crossvein at the outer tenth of the discal cell; first posterior cell a little widened. Anterior margin of wing sepia, with sepia spots along middle of wing at crossveins and at base and middle of anterior branch of third vein. Antennae long and slender, the second segment about 3 times as long as wide and nearly as long as the first segment; third segment long, widest on base and middle, a little narrowed distally. Proboscis long, labellum slender. The male hypopygium extraordinarily large, short, subglobose, capsulelike and terminal . . . . **Acrophthalmyda** Bigot
Not such flies 49
49. Small flies with conically tapered abdomen, black and densely covered with close-lying brilliant silvery scales and scaliform hair. First posterior cell open maximally and flared toward apex. Eyes of the holotypic male with a transverse impressed horizontal line. Hind femur with spines below; all the femora and tibiae also densely covered with silvery scales. Pulvilli quite long but also quite slender. Antennae elongate and slender. Proboscis moderately lengthened, the labellum narrowly leaflike, with longitudinal stria . . . . . . **Cryomyia** Hull, new genus
Without silver scales over whole abdomen and posterior thorax. Eyes of male not with a horizontal impressed brown line 50
50. Head loosely attached to thorax. Occiput prominent, bimammillate, with shallow, barely indicated medial groove, not bilobed. Occiput on upper slopes with very dense, long, stiff, white or shining yellow pile, sometimes black hairs intermixed. First antennal segment elongate, a little thickened toward the base and 4 or 5 times as long as wide and about 4 times as long as the short second segment; second segment longer than wide; third segment elongate and gradually but only slightly narrowing to the apex. The first segment bears short, stiff, bristly hair above, conspicuously longer hair below. Sides of face and gena with rather dense fringe of long, shining pile. Abdomen tends to be slightly cylindrical and a little elongated but not conspicuously. Alula reduced; mesonotum a little humped, and basal comb wanting.
Gonarthrus Bezzi
Occiput poorly developed; flies quite different from above.
51
51. Second vein has a kink before expanded bulbiform apex of marginal cell; anterior branch of third vein very strongly sigmoid and thrust straight upward to costa beyond the end of the marginal cell. Anal cell open. Small black flies with gray white pollen, long, erect white pile; abdomen conical, a little elongate, and distinctly compressed. Third antennal segment elongate and attenuate and ending in an apical spine . . . **Hallidia** Hull
Second vein without a kink and anterior branch of third vein not exceptionally sigmoid, or ending by a strong upward thrust 52
52. Oragenal cup prominent, extended forward and shining, polished and bare, or pollinose with a thin fringe of

ocularmarginal hairs. Antennae elongate but the first antennal segment somewhat stout and thickened; sometimes much thickened; third segment stout or much thickened in the middle and spindle-shaped. Anterior crossvein at basal third or middle of discal cell. Abdomen of male with dense, fine, erect pile; of females with dense fine, curled, flat-appressed, glittering hair and sparse erect hair. Labellum filiform . . . **Sparnopolius** Loew
Oragenal cup not prominent, shining and bare, or if prominent abdomen elongate, narrow, and cylindrical. Pile generally much more abundant on sides of face, or below antennae 53

53. Gena very narrow and linear. Labellum large and leaflike. Pile relatively fine and stiff . . . . **Laurella** Hull
Gena very wide, or if narrow labellum narrow or filiform and the oragenal and genofacial pile scanty or absent. 54

54. Short obtuse flies with stout robust abdomen and much dense, fine, erect pile. Whole of upper occiput, the ocellar tubercle of male, the very wide front of female, all sides of the elongate first antennal segment and the whole sides of face and gena densely covered with long, stiff, erect, bristly pile. Gena quite wide. Third antennal segment rather slender but distinctly spindle-shaped. Labellum slender and elongate as well as proboscis.
Dischistus Loew
Longer, cylindrical flies with short, dense pile, a different type of antenna, or small largely bare, strongly tapered flies with much pale color on head, thorax, and abdomen. 75

55. First antennal segment long and stout. Abdomen especially elongate; *Thereva*-like flies; base of third antennal segment swollen . . . . **Apatomyza** Wiedemann
First antennal segment not long and also stout, the third segment not swollen at the base 56

56. Head more spherical, the occiput less concave; genae absent or nearly so; abdomen more slender and elongate, the axillary lobe of wings rather narrow and the wing quite long in proportion to thorax and abdomen.
Adelogenys Hesse
Head less spherical; genae present and well developed. Abdomen broader at base, more conical, the sides a little compressed. Wings distinctly broad at the base and the axillary lobe in consequence well developed . . . 57

57. Proboscis extremely short and stout or robust, much shorter than the length of the head. Wing like *Crocidium* Loew. Apex of antenna with a *Phthiria*-like, minute spine above instead of apical . . . . **Mallophthiria** Edwards
Proboscis long and slender, at least longer than the head; often much longer. Last section of fourth vein distinctly arched or bowed forward 58

58. Anterior crossvein at the middle of the discal cell. Anal cell broadly open. Branches of the third vein proportionately shorter than in *Semiramis* Becker.
Tamerlania Paramonov
Anterior crossvein distinctly placed well beyond the middle of the discal cell; at outer fourth or fifth. Anal cell usually closed with a long stalk 59

59. Fourth vein strongly arched; gena and occiput not very extensive below the eye. Discal cell longer and narrower apically and wing not greatly widened at base.
Semiramis Becker
Fourth vein barely arched forward. Discal cell shorter and quite wide apically. Occipito-genal area much more extensive below the eye. Wing usually strongly widened toward base; anal cell closed and stalked or open. Hind femur without spicules or bristles . . **Crocidium** Loew

60. Quite small, compact flies with short, more or less decumbent abdomen. Abdominal pile on tergites entirely of glittering, appressed scales. First two antennal segments short and slender; third segment expanded basally and then attenuate and narrowed to the blunt apex which bears a small, preapical microsegment and spine, set within an elongate recess. Wings hyaline.
Neodischistus Painter
Small to large size flies. No scales on abdomen. Abdomen short and obtuse with long pile present on the terga. All three segments of antenna slender; third segment attenuate at apex with apical bristles or spine, or microsegment, or both 61

61. Third antennal segment with a tuft of hairs at apex. First posterior cell strongly narrowed near the wing margin; discal cell also greatly narrowed apically. Large flies with wide face and front, and body and densely covered with fine, long pile everywhere . . . **Sisyromyia** White
Third antennal segment with only an apical spine or none at apex. These cells not or but little narrowed near the distal ends 62

62. Oragenal cup, walls, and face polished and bare, pile absent or very scanty. Males widely holoptic; genae very narrow and linear, without pile except below.
Sparnopolius Loew
Oragenal cup and genae both with long pile. Usually, small reddish flies, the abdomen at least shining. Pile long, fine, erect, plushlike. Scales and scaliform hairs absent.
Doliogethes Hesse

63. First posterior cell closed in the margin, or with a very short stalk. The anterior crossvein enters the discal cell at the basal third, therefore the first basal cell is clearly but not greatly longer than the second basal cell. Both males and females usually with a patch of pale, or silvery hairs or scales on the eye margin opposite the antenna . . . . **Parabombylius** Williston
First posterior cell closed with a long stalk. The anterior crossvein enters the discal cell at the base, the middle or beyond 64

64. Three submarginal cells. Terga with conspicuous bristles on the margins. Form decumbent . **Nectaropota** Philippi
Two submarginal cells; if 3 submarginal cells present, no rows of strong bristles on posterior margins of terga . 65

65. First basal and second basal cells of nearly equal length, the first basal cell at most only slightly longer than the second 66
First basal cell clearly much longer than the second . . 69

66. Claws with a basal tooth; pulvilli entirely absent . . . 67
Claws without a basal tooth; pulvilli present, usually well developed 68

67. Tibiae without spicules. Males holoptic. Anterior crossvein beyond the middle of the discal cell and first posterior cell rather pointed apically and short petiolate; marginal cell widest apically, ending obtusely and the second vein at a right angle. Abdomen shining, oval, attenuate at apex, quite flattened and depressed, reddish with a pattern of black bands and medial spots. Proboscis 3 or 4 times as long as the head. Third antennal segment quite narrow and linear . . . . **Isocnemus** Bezzi
All tibiae with short spicules. Males rather widely dichoptic. Anterior crossvein not farther than middle of discal cell and first posterior cell wider and long petiolate; marginal cell equally wide over outer third and second vein ending obliquely. Abdomen short and subcircular, shorter than wide, gently convex, shining black, with dense, short, erect, black pile and conspicuous postmarginal fringes of appressed white pile. Proboscis at least 6 times as long as head; third antennal segment much widened . . . . **Zinnomyia** Hesse

68. Face and oragenal cup well hidden by dense, long pile. Upper anterior intercalary vein long, as long or longer than the anterior crossvein. Anterior branch of third vein often with a basal spur. Posterior margin of terga usually if not always with rows of long, erect bristles. Pulvilli reduced in length. Head as broad or broader than mesonotum. No transverse groove or depression in front of female . . . . **Anastoechus** Osten Sacken

Face and oragenal cup with long, but loose, scattered pile. Upper anterior intercalary vein much reduced, shorter than the anterior crossvein. Abdominal pile dense, fine, long, erect, plushlike. Anterior branch of third vein without spur. Pulvilli well developed. Head narrower than mesonotum. Female front often with transverse depression. Labellum filiform. If labellum wide and leaflike, subgenus *Parasystoechus* . . . **Systoechus** Loew

69. Front very wide, nearly as wide as genofacial space. Front, upper middle face, and oragenal cup polished, shining black; upper half of front on each side with broad, dense tuft of long, erect, coarse, black pile; lateral face and genae opaque and pollinose, together with a dense, encircling brush of coarse, erect, opaque, white pile. Males widely holoptic. Hind femur without bristles below.
Sisyrophanus Karsch
Not as above. Face and front if wide are pollinose, not shining. Hind femur rarely without extensive bristles below 70

70. First antennal segment stout and robust, 5 or 6 times as long as the second and with long, dense, bristly pile above and especially below 71
First antennal segment slender, the bristly hairs long but scattered. Males holoptic. Hind femur with conspicuous bristles below 72

71. No spines below on hind femur; eyes of male holoptic.
Lissomerus Austen
Spines present below on hind femur; eyes of male distinctly separated . . . . **Acanthogeron** Bezzi

72. Contact of discal cell and second posterior cells punctiform; intercalary vein absent 73
Intercalary vein short or long, but distinct 74

73. Third antennal segment, very slender, strongly attenuate at apex, ending in a spiney bristle.
Staurostichus Hull, new genus
Third antennal segment thick throughout to the apex; apex notched, with the spine-bristle and microsegment lying closed some distance from the apex.
Brychosoma Hull, new genus

74. Three submarginal cells . . **Bombylius (Triplasius)** Loew
Two submarginal cells . . . . **Bombylius** Linné

75. Large or small flies. Antenna relatively elongate and distinctly slender, but of varying length, second segment short, first and third of approximately equal length, first segment usually with long, subappressed, shaggy hair, third segment attenuate, ending in a minute, cylindrical microsegment with apical spine. Oragenal cup usually produced and sometimes quite long, rarely short; always with knife-thin edge and vertically sided, deep interior recess, the outer wall usually polished and shining. Palpus long and slender, composed of 2 or 3 segments, the last segment sometimes shorter and considerably more narrow than the preceding segment, rarely dilated and swollen; third segment may be fused with second segment. Abdomen long, rather slender, cylindrical or slightly compressed laterally, even narrowed in the middle. Wings long, slender, the anterior branch of third vein strongly arched and sigmoid. Medial crossvein arises below, near the middle of the discal cell. Male costa often spinose. Anterior branch of the third vein strongly arched at the base, remainder sigmoid, usually ending at wing apex. Anterior crossvein usually distinctly shortened. Occiput very gently and shallowly bimammillate; separated by a weak furrow in the middle, scarcely greater than *Bombylius* Linné; not bilobed and not with central cavity. Tribe Eclimini 76
Not such flies 78

76. Apical part of anterior branch of third vein very strongly thrust upward. Three submarginal cells. Anterior crossvein not notably shortened. Wing narrow at base, axillary lobe quite narrow, alula absent. First submarginal cell subcuboidal . . . . **Tillyardomyia** Tonnoir

Apical part of anterior branch of third vein curved but ending at apex of wing. Anterior crossvein distinctly contracted, pinching inward the vein on each side above and below. Two or 3 submarginal cells 77

77. Three submarginal cells. Axillary lobe wider than anal cell. First submarginal cell as wide and as long and slender as the cell beneath it . . **Thevenemyia (Arthroneura)** Hull
Face long, slender and oragenal recess extended forward and downward as a shining, bare cup; without pile. Occiput generally much swollen laterally. Body thinly pilose, surface punctate or rugose, abdominal segments with deep, transverse depressions. Wings slender at base, much narrowed. Alula absent, anal cell widely open.
Eclimus Loew
Face shorter, though usually extensive, with long, shaggy pile, pollinose or bare of pollen. Body densely pilose, surface smooth, abdominal segments with small lateral dents or none. Base of wing as wide as middle of wing. Occiput flat, or evenly swollen. Mesonotum sometimes with nodules (subgenus *Epibates* Osten Sacken); male costa sometimes spinose or nodose. Alula partly developed, anal cell narrowly open . . . **Thevenemyia** Bigot

78. Proboscis reduced to a mere stub but palpus large, long, curled upward and prominent. Abdomen elongate, strongly tapered, nearly bare and shining. Ambient vein absent. Anal cell recessive, closed with a long stalk. Blackish flies. Females. (Males with lower eye facets much smaller). See Heterotropinae . . **Caenotus** Cole
Proboscis much longer; never reduced to a mere stub. Abdomen elongate and tapered in both sexes, especially males. Small flies usually largely pale in coloration and some blackish, paired, regular markings on the abdomen as a rule. Anterior crossvein placed always at or very close to the middle of the short, apically widened discal cell. Anal cell closed and stalked. Females with face peaked forward abruptly, even rectangularly in some species, shortly beneath the antenna; gena beneath the eye usually prominent. Wing veins often colorless and whole wing sometimes whitish. Females. (Males with lower facets very much smaller.) See Heterotropinae.
Heterotropus Loew

79. Apex of marginal cell plane, scarcely extending beyond the apex of the auxiliary cell . . . **Conophorus** Meigen
Apex of marginal cell bulbouslike, protruding beyond the apex of the second branch of the radius where it joins the thickened costa. Auxiliary cell quite narrow, ending opposite the anterior crossvein. Costa thickened beyond. Base of both outer submarginal cells equally wide and each with a nearly plane simulated crossvein at the base. Eyes of male widely separated by more than the width of the large, abrupt, high, ocellar tubercle. Male front triangular with a mat of dense, long, shining, yellowish, completely appressed pile. Abdomen triangular, as wide as long. Thorax, scutellum, and abdomen entirely opaque yellowish-brown pollinose with pile, partly curled, partly upright; metapleuron, metanotum, and pteropleuron bare.
Platamomyia Brèthes

Genus *Bombylius* Linné

FIGURES 1, 4, 7, 244, 245, 247, 432, 433, 458, 487, 520, 786, 787, 788

Bombylius Linné, Systema Naturae, Ed. 10, p. 606, 1758. Type of genus: *Bombylius major* Linné, 1758, by designation of Latreille, p. 443, 1810.
Choristus Walker, Insecta Saundersiana, Diptera, part 3, p. 197, 1852. Type of genus: *Choristus bifrons* Walker, 1852, by monotypy.

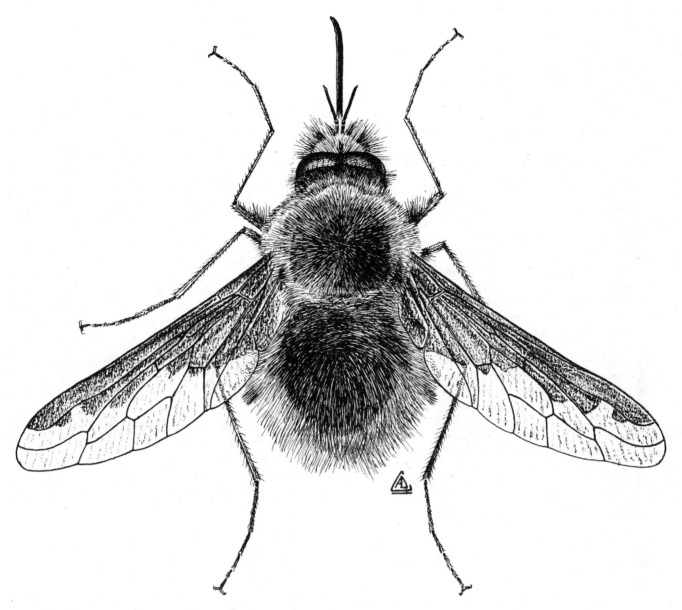

Text-Figure 20.—Habitus, *Bombylius major* Linné.

Parisus Walker, Insecta Saundersiana, Diptera, part 3, p. 196, 1852. Type of genus: *Parisus paterculus* Walker, 1852, by monotypy.

Triplasius Loew, Neue Beitrage III, p. 7, 1855. Type of subgenus: *Bombylius bivittatus* Loew, 1855, by monotypy.

Johnson (1907), pp. 95-100, key to 10 New England species.
Séguy (1926), pp. 244-247, key to 28 species found in France.
Bezzi (1924), pp. 32-58, key to 68 Ethiopian species.
Engel (1934), pp. 199-209, key to 75 Palaearctic species.
Maughan (1935), pp. 58-60, key to 8 species from Utah and western states.
Austen (1937), pp. 15-18, key to 19 ssp. from Palestine and nearby area.
Hesse (1938), pp. 44-111, key to males of 104 species and 91 females.
Painter (1939), pp. 268-270, key to 17 species of the United States.
Paramonov (1940), pp. 101-122, key to males of 90 and females of 83 species and subspecies.
Efflatoun (1945), pp. 303-306, key to 13 species from Egypt.
Paramonov (1955), pp. 159-161, key to 20 African species.
Bowden (1964), p. 15, key to 8 species from Ghana.

Small to large flies with widely varying appearance but always with rather short, broad abdomen densely covered with pile which varies from erect, long and "pufflike" to straggly, sparse vestiture exaggerated in tufts on the side of the abdomen. In still other species, as in the great *Bombus*-like species *punctatus* Fabricius, the abdominal vestiture while dense is short and the facial pile is reduced.

The antennae in this genus are elongate; the first segment is from 2 to 3 times as long as wide, usually slender, sometimes a little thickened, its pile either long and conspicuous and completely encircling the segment, as in *fuscus* Fabricius, or shorter and scantier above and even reduced somewhat below; second segment usually beadlike, at most twice as long as wide; the third segment slender and nearly the same thickness throughout or sometimes considerably dilated toward the base; there are two minute microsegments, the last one with a spine.

Proboscis elongate. The pile of the face and genae is usually very long indeed, dense and fine, obscuring the surface.

The anterior crossvein is typically at the middle of the discal cell and the first posterior cell is closed and bears a long stalk. Wings often spotted, sometimes clear hyaline or tinged with brown, especially basally. Species with the anterior crossvein far toward the apex probably require segregation, as do any with the first posterior cell narrowly open or closed at the margin.

Because of the large size of this genus there are many subgroups in different parts of the world; Hesse (1938) placed the 94 South African species known to him in 3 main groups, the first group he subdivided into 5 sections. Many of the species are difficult to distinguish except on a complex of characters. An inspection of Hesse's valuable drawings of genitalia shows that several types are found in the genus. American species and Palaearctic forms can also be divided into subgroups. The relationships of *Bombylius* Linné, on which the family was founded, are discussed in the section on phylogeny and will not be repeated here; as in other such old genera it has served as a dumping ground by early dipterists; there are doubtless quite a few names still placed in the genus which require removal. The genus is worldwide in distribution.

Length: 5 to 15 mm.; most species 6 to 10 mm.; wing about the same length.

Head, lateral aspect: The head is more or less triangular, the face angularly produced forward to a varying degree. The occiput is rather prominent and swollen behind, but gradually sloping outward to the eye margin. The oragenal cup is either oblique and strongly recessive in those species where the head is more triangular, or in those forms where the face is shorter and less extended, the oragenal cup is apt to be more extended below, and vertical in profile. In the typespecies there is much, dense, long pile along the sides of the widened oragenal cup. The face bears long pile completely across the entire width beneath the antennae, as in the type-species; or it is quite bare widely beneath the antennae and the pile restricted to the outlying portion of the face, as in *Bombylius punctatus* Fabricius. The pile continues long and erect down the sides of the gena, or in some species it may be relatively short and include scaliform hairs. Occiput with short, fine pile close to the eye margin or sometimes the upper third with a fringe of bristly hairs close to the eye margin, followed by dense, erect, coarse pile behind. In still other species the pile is long, coarse, and quite loose and scanty.

The eye varies from hemispherical with posterior margin entire to species like *Bombylius fuscus* Fabricius where there is a very extensive long but quite shallow concavity on the posterior margin. The proboscis is long and slender; the labella are likewise long and slender. Palpus rather short, the basal segment shorter than the apical one, the apical segment uniformly slender or a little thickened and spindle-shaped, sometimes clavate and grooved below and usually with several long bristles either at the apex or on the outer margin. The front in males is large and slightly convex with a shallow median groove and in length rather more than half the total length from antennae to the posterior ocelli. The pile of the front in males varies from tufts of short, bristly hairs to a clump of long, slender, erect bristles with or without underlying shorter hair. Sometimes these hair tufts are strongly divided in the middle, parted, and turned outward.

The antennae are attached at the upper fifth of the head and are elongate and slender, approximately as long as the head. In some species, as in *Bombylius fuscus* Fabricius, the first segment is robust and approaches *Acanthogeron* Bezzi. This segment may be anywhere from twice as long as wide to three or more times as long as wide. Second segment always short, about as long as wide. Third segment generally quite slender, narrower than the second segment but occasionally much widened on the basal half as in *Bombylius punctatus* Fabricius and gradually tapered to the narrow apex. Apex with 1 or 2 very small microsegments and a spine at the apex. The first segment characteristically bears much long, shaggy, bristly pile, sometimes reaching half the length of the first segment. The third segment occasionally bears a few hairs dorsally.

Head, anterior aspect: The head is small and much narrower than the thorax, the eyes are in contact in the male for a distance a little greater than the length of the ocellar triangle and are sometimes separated by a distance half or more than the width of the ocellar triangle as in *Bombylius fuscus* Fabricius. Female front only moderately wide, in some instances nearly as wide as the face, the pile in the female front is apt to be restricted to the sides and in *Bombylius cruciatus* Fabricius a transverse depression is across the female front. Antennae narrowly separated, the ocellar tubercle is a little raised, small, with the ocellar triangle equilateral or obtuse, the ocelli in males are adjacent to the eye margin and a clump of long, slender, bristly hairs arises between the ocelli. The oral opening is large, short oval, or sometimes somewhat triangular, its outer slopes may be wide and extensive as in the typespecies, its inner wall in all species is strongly widened and inflated medially on sometimes more than the lower half. The gena itself is always narrow, the tentorial fissure narrow but deep and the oral opening is deep. From above, the occiput is shallowly bilobate with a narrow crease behind the vertex.

Thorax: The thorax is short and broad, somewhat convex, both head and abdomen slightly decumbent in some species, less so in others. The color is generally black with gray or black pollen, and covered very densely in most species with a matted, appressed, furry pile, shining and sometimes more erect medially. On the anterior half it is generally so dense that it is impossible to see the ground color. There are 4 or 5 slender bristles on the notopleuron; the humerus may be completely obscured by dense pile or bare as in *Bombylius fuscus* Fabricius. Scutellum generally covered with long, fine, erect pile that is loose or dense. Mesopleuron and metapleuron pilose with the hypopleuron and pteropleuron wholly bare, and sternopleural pile confined to a few hairs dorsally, or in some species the pile is extraordinarily dense, including pteropleuron and sternopleuron and only the hypopleuron is bare. Squamal fringe long and fine and fanlike.

Legs: The legs are rather long and slender, the anterior tibia with posterodorsal and posteroventral rows of minute, slender spicules, 10 in each row; middle and hind tibiae with 4 rows, slender but longer, and with 15 to 20 in each row. The femora all tend to be a little thickened and dilated toward the base and the hind femur characteristically bears a row of long, prominent spinous bristles, which may contain from 10 to 15 in a row. Pile of the legs consist of minute, flattened scales in some, larger, appressed scales in others; claws very slender, sharp, curved; the pulvilli long.

Wings: The wings either slender throughout including the base or sometimes broad basally as in *Bombylius punctatus* Fabricius. Wings hyaline as in *Bombylius niveus* Meigen with the anterior half of the wing dark brown and rather sharply delimited, irregularly behind, as in the type-species and in *Bombylius fimbriatus* Meigen, or more or less lightly tinged with brown, wholly tinged and dark brown as in *Bombylius fuscus* Fabricius, or the wing often spotted. Two submarginal cells only, except three in the subgenus *Triplasius* Loew. Marginal cell either narrow or expanded apically, intercalary vein usually long, sometimes almost wanting. The first posterior cell always closed with a long stalk, the anal cell always open, the alula always large, the ambient vein always complete. The anterior crossvein is quite variable in position, being found either just before the middle, at, or beyond the middle of the discal cell. In any case the first basal cell is always longer than the second. A well-developed costal comb is present.

Abdomen: The abdomen is short and broad, wider than the thorax, nearly circular as viewed from above, or sometimes a little narrowed apically. Generally black in ground color, nearly opaque as in *Bombylius cruciatus* Fabricius or quite shining as in *Bombylius fuscus* Meigen. Like the pile of the thorax, the pile is quite variable in color and varies from snow white forms to jet black species and others where it is a drab brown or yellow. The pile may be entirely of one kind, long, fine, erect hairs, or short, matted-appressed, scaliform pile in posteriorly arranged bands on the tergites with additional postmarginal fringes of bristles and bristly hairs. Male terminalia small and recessed below at the apex of the abdomen.

Because of the large number of species that Dr. A. J. Hesse (1938, p. 44) was able to study and summarize, I quote his work on the hypopygium of the males:

Hypopygium of males extremely variable in shape, with the beaked apical joints very variable, never with a subapical lobe or with the outer part very angularly or lobularly prominent, often elongate and sometimes broad and almost leaf-shaped, with the apical part of the aedeagus never very broad and spout-like or very slender and arcuately curved upwards, with or without a ventral aedeagal process below, with the dorsal part sometimes produced basally into a strap-like process projecting basally on each side, with the basal strut assuming various shapes. A comparison of the numerous figures with those of other, and sometimes related, genera will give a much better conception of the type of hypopygium found in this genus. Though there is some considerable uniformity in the structure of the hypopygium in *Bombylius* there are marked structural differences as well.

The male terminalia from dorsal aspect, with epandrium removed, are comparatively broad with a moderately large, membranous area. The dististyli are small but convergent with sharply pointed curved apices. It is somewhat larger laterally, and turned upward. The epandrium is short, but high with large cerci. The basistylus moderately stout, narrowing somewhat apically. The basal apodeme is large.

Material studied: I have before me a number of variable and characteristic European species, a long series of American species, and also considerable Australian material. I have seen the African material in the British Museum (Natural History).

Immature stages: Not well known, but those species that are known prove to be parasites of mining bees. See further under life histories.

Ecology and behavior of adults: The adults are readily attracted to flowers of many kinds which they probe with their long proboscis. Males are rapid fliers and especially wary. Holmes (1913) comments upon the orientation of these flies to light. Painter (1939) records *medorae* Painter from blossoms of *Asclepias tuberosa* Linné; he took *major* Linné from both lilac and plum. *Bombylius major* Linné is very abundant in Mississippi on plum and on *Cardamine bulbosa* (Schreb.) BSP.

Distribution: Nearctic: *Bombylius albicapillus* Loew, 1872 [=*atricapillus* Wulp, 1881, lapsus], *albicapillus diegoensis* Painter, 1933; *atriceps* Loew, 1863, *atriceps fulvibasoides* Painter, 1962; *aurifer* Osten Sacken, 1877; *aurifer pendens* Cole, 1919; *austini* Painter, 1933; *cachinnans* Osten Sacken, 1877; *canadensis* Curran, 1933; *comanche* Painter, 1962; *duncani* Painter, 1940; *eboreus* Painter, 1940; *facialis* Cresson, 1919; *fraudulentus* Johnson, 1907; *incanus* Johnson, 1907 [=*philadelphicus* Johnson, not Macquart, 1899]; *lancifer* Osten Sacken, 1887; *lassenensis* Johnson and Johnson, 1965

[=*pallescens* Johnson and Maughan, 1953]; *maculifer* Walker, 1852 [=*azaleae* Shannon, 1916]; *major* Linné, 1758 (see Palaearctic, also); *medorae* Painter, 1940; *metopium* Osten Sacken, 1877; *mexicanus* Wiedemann, 1821 [=*fulvibasis* Macquart, 1855, =*philadelphicus* Macquart, 1840]; *pulchellus* Loew, 1863; *pygmaeus* Fabricius, 1781, *pygmaeus canadensis* Curran, 1933; *ravus* Loew, 1863; *silvus* Cole, 1919; *subvarius* Johnson, 1907; *texanus* Painter, 1933; *validus* Loew, 1863; *varius* Fabricius, 1805; *cinerivus* Painter, n.n. for *cinereus* Bigot not Olivier, 1789.

Neotropical: *Bombylius abdominalis* Wiedemann, 1828 [=*mesomelas* Wiedemann, 1830]; *aurovittatus* Macquart, 1849; *bellus* Philippi, 1865; *bicolor* Loew, 1861; *clio* Williston, 1901; *flavipilosus* Cole, 1923; *frontatus* Philippi, 1865; *goyaz* Macquart, 1840; *haywardi* Edwards, 1837; *helvus* Wiedemann, 1821; *hyalinus* Fabricius, 1805; *hyalipennis* Macquart, 1846; *io* Williston, 1901; *landbecki* Philippi, 1865; *melampogon* Philippi, 1865; *nigricornis* Philippi, 1865; *ochraceus* Bigot, 1892; *plumipes* Drury, 1773; *ravus* Loew, 1863; *rufoanalis* Macquart, 1849; *ruizi* Edwards, 1937; *semirufus* Loew, 1872; *valdivianus* Rondani, 1863 [=*valdivianus* Philippi, 1865].

Palaearctic: *Bombylius agilis* Olivier, 1789; *albaminis* Séguy, 1949; *albissimus* Zaitsev, 1964; *ambustus* Pallas in Wiedemann, 1818 [=*dispar* Meigen, 1820]; *altaicus* Paramonov, 1940; *androgynus* Loew, 1855; *angulatus* Macquart, 1826; *antennipilosus* Paramonov, 1926; *appendiculatus* Bezzi, 1906; *apulus* Cyrillo, 1787; *arenosus* Paramonov, 1940; *argentarius* Séguy, 1934; *argentifacies* Austen, 1937; *argentifrons* Loew, 1873 [=*similis* Loew, 1873]; *armeniacus* Paramonov, 1925; *armeniacus* Paramonov, 1926; *atlanticus* Abreu, 1926; *aurulentus* Paramonov, 1926; *axillaris* Meigen, 1830; *barbula* Pallas in Wiedemann, 1818; *bedouinus* Efflatoun, 1945; *boghariensis* Lucas, 1852 [=*alveolus* Becker, 1906]; *brevirostratus* Austen, 1937; *brevirostris* Olivier, 1789; *callopterus* Loew, 1855, *callopterus umbripennis* Paramonov, 1926; *candidifrons* Austen, 1937; *canescens* Mikan, 1796 [=*cinereus* Meigen, 1804, =*ctenopterus* Walker, not Mikan, 1849, =*favillaceus* Meigen, 1820, =*minor* Meigen, not Linné, 1820, =*minor* Curtis, not Linné, 1836, =*pumilus* Meigen, 1820, =*pusillus* Meigen, 1830, =*variabilis* Loew, 1855]; *capillatus* J. Palm, 1876; *catheriniensis* Efflatoun, 1945; *chinensis* Paramonov, 1929; *chinensis* Paramonov, 1931; *cinerarius* Pallas in Wiedemann, 1818, *cinerarius eversmanni* Paramonov, 1926, *cinerarius karelini* Paramonov, 1926, *cinerarius pallasi* Paramonov, 1926; *cinerascens* Mikan, 1796 [=*favillaceus* Meigen, 1820, =*floralis* Meigen, 1820, =*minor* Meigen, not Linné, 1820]; *cinereus* Olivier, 1789; *citrinus* Loew, 1855; *collaris* Becker, 1906; *crassitarsis* Paramonov, 1940; *cruciatus* Fabricius, 1798 [=*analis* Olivier, 1789, =*leucopogon* Meigen, 1804, =*posticus* Fabricius, not Meigen, 1805], *cruciatus leucopygus* Macquart, 1849; *debilis* Loew, 1855; *deserticola* Paramonov, 1940; *desertivagus* Paramonov, 1940 [=*dorsalis* Loew, not Olivier, 1873]; *discoideus* Fabricius, 1794 [=*analis* Fabricius, 1794, =*analis* Thunberg, 1827, =*charopus* (Lichtenstein), 1796 (as *Tabanus*), =*nigrita* Cyrillo, 1787, =*suffusus* Walker, 1849, =*thoracicus* Fabricius, 1805], *discoideus* (as *analis*) *waterbergensis* Hesse, 1938; *discolor* Mikan, 1796 [=*concolor* Zeller, not Mikan, 1840, =*medius* Scopoli, not Linné, 1763], *discolor shelkovnikovi* Paramonov, 1927; *dorsalis* Olivier, not Loew, 1789; *elongatus* Rossi, 1794; *eploceus* Séguy, 1949; *exiguus* Walker, 1871; *fallax* Austen, 1937; *fimbriatus* Meigen, 1820 [=*dimidiatus* Wiedemann in Meigen, 1820], *fimbriatus expletus* Loew, 1855, *fimbriatus ventralis* Loew, 1855; *firjuzanus* Paramonov, 1929; *flavescens* J. Palm, 1876; *flavipes* Wiedemann, 1828; *floccosus* Loew, 1857; *fuliginosus* Wiedemann in Meigen, 1820 [=*brevirostris* Meigen, 1830], *fuliginosus polypogon* Loew, 1855, *fuliginosus rhodius* Loew, 1855, *fuliginosus tavrizi* Paramonov, 1926; *fulvescens* Wiedemann in Meigen, 1820 [=*apicalis* Meigen, 1820, =*longirostris* Meigen, 1820], *fulvescens candidus* Loew, 1855; *fulvibasis* Macquart, 1955; *fulvipes* Villers, 1789; *fumosus* Duffour, 1852; *fuscus* Fabricius, 1781; *gracilipes* Becker, 1906; *iranicus* Paramonov, 1940; *koreanus* Paramonov, 1926; *kozlovi* Paramonov, 1926; *kutshurganicus* Paramonov, 1926; *lejostomus* Loew, 1855; *lugubris* Loew, 1857; *lusitanicus* Meigen, 1830; *maculatus* Fabricius, 1755; *maculithorax* Paramonov, 1926; *maculipennis* Macquart, 1849; *maculipennis melanopus* Timon-David, 1952; *major* Linné, 1758 [=*aequalis* Fabricius, 1781, =*albipectus* Macquart, 1855, =*antenoreus* Lioy, 1864, =*fratellus* Wiedemann, 1828, =*lanigerus* Fourcroy, 1785, =*sinuatus* Mikan, 1786, =*variegatus* DeGeer, 1776, =*vicinus* Macquart, 1840], *major australis* Loew, 1855, *major basilinea* Loew, 1855, *major consanguineus* Macquart, 1840; *marginatus* Cyrillo, 1787; *maurus* Olivier, 1789; *medius* Linné, 1758 [=*confrater* Loew, 1855, =*concolor* Mikan, 1796, =*discolor* Meigen, not Mikan, 1804, =*intermedius* Walker, 1849 =*major* Samouelle, 1819, =*medii* Strobl, 1898, =*punctatus* De Geer, 1776]; *medius albomicans* Loew, 1855, *medius caucasicus* Paramonov, 1940, *medius dalmatinus* Strobl, 1893, *medius pallipes* Loew, 1855, *medius pictipennis* Loew, 1855, *medius punctipennis* Loew, 1855, *medius seminiger* Becker, 1906; *megacephalus* Portschinsky, 1887; *melanopygus* Bigot, 1860; *mendax* Austen, 1937; *minimus* Scopoli, 1772; *minimus* Macquart, 1834; *minor* Linné, 1758 [=*albibarbis* Zetterstedt, 1842, =*cinereus* Meigen, 1820, =*dilutus* Wiedemann in Meigen, 1820, =*pumilus* Zetterstedt, not Meigen, 1842, =*subcinctus* Wiedemann in Meigen, 1820, =*venosus* Meigen, 1804], *minor ochraceus* Paramonov, 1926; *minusculus* Efflatoun, 1945; *mobilis* Loew, 1873, *mobilis obscuripennis* Paramonov, 1929; *modestus* Loew, 1873, *modestus alexandri* Paramonov, 1940, *modestus phaeopterus* Bezzi, 1924; *moldavanicus* Paramonov, 1926; *morio* Olivier, 1789; *moussayensis* Efflatoun, 1945; *mus* Bigot, 1862; *nanus* Meigen, 1838 [=*pygmaeus* Macquart, 1834]; *nigrifrons* Becker,

1913; *nigrilobus* Collart, 1940; *nigripes* Macquart, 1834; *nigropenicillatus* Bigot, 1892; *nivifrons* Walker, 1871 [=*niveifrons* Engel]; *niveus* Meigen, 1804, *niveus hololeucus* Loew, 1873; *nubilus* Mikan, 1796; *nubilus algericus* Villeneuve, 1932, *nubilus monticola* Paramonov, 1926; *nudus* Villers, 1789; *numida* Macquart, 1849; *obliquus* Brulle, 1832; *oceanus* Becker, 1908; *olivierii* Macquart, 1840; *olsufjevi* Paramonov, 1940; *pallens* Wiedemann in Meigen, 1820; *pallidicris* Brulle, 1832; *pericaustus* Loew, 1873; *persicus* Paramonov, 1926; *pictus* Panzer, 1794 [=*planicornis* Fabricius, 1798]; *pilirostris* Loew, 1855; *podagricus* Paramonov, 1940; *pseudargentatus* Paramonov, 1929; *pumilus* Meigen, 1820 [=*pusillus* Meigen, 1830]; *punctatus* Fabricius, 1794 [=*sticticus* Boisduval, 1835, *subluna* Walker, 1849]; *pusio* Meigen, 1830; *quadrifarius* Loew, 1855; *repeteki* Paramonov, 1940, *repeteki submodestus* Paramonov, 1940; *semifuscus* Meigen, 1820 [=*cincinnatus* Becker, 1891, =*nigripes* Strobl, 1898, =*senilis* Jaennicke, 1867]; *senex* Rondani, 1863, *senex violaceipes* Strobl, 1909; *shah* Paramonov, 1940; *shibakawae* Matsumura, 1916; *simulans* Austen, 1937; *striatifrons* Becker, 1906 [=*striatus* Becker, 1915, =*turanicus* Paramonov, 1926]; *sytshuanensis* Paramonov, 1926; *tephroleucus* Loew, 1855; *testaceiventris* Paramonov, 1925, *testaceiventris bergi* Paramonov, 1926; *torquatus* Loew, 1855 [=*undatus* Wiedemann in Meigen, 1820]; *trichurus* Pallas in Wiedemann, 1818; *turcmenicus* Paramonov, 1926; *undatus* Mikan, 1796, *undatus diagonalis* Wiedemann in Meigen, 1820; *ushinskii* Paramonov, 1940; *uzbekorum* Paramonov, 1926; *vagans* Meigen, 1830; *venosus* Mikan, 1796 [=*cinereus* Meigen, 1820, =*favillaceus* Meigen, 1820, =*holosericeus* Wiedemann in Meigen, 1820, =*minor* Meigen, not Linné, 1820, =*pusillus* Meigen, 1830]; *vertebralis* Dufour, 1833; *versicolor* Fabricius, 1787; *vlasovi* Paramonov, 1940; *vulpinus* Wiedemann in Meigen, 1820 [=*fugax* Wiedemann in Meigen, 1820, =*micans* Meigen, 1804, not Fabricius, 1798, =*posticus* Meigen, 1820], *vulpinus desertorum* Paramonov, 1926, *vulpinus palaestinus* Paramonov, 1926; *wadensis* Efflatoun, 1945; *zarudnyi* Paramonov, 1940.

Ethiopian: *Bombylius acroleucus* Bezzi, 1921; *actus* Bezzi, 1924: *aemulus* Hesse, 1938; *aequisexus* Paramonov, 1956; *albiventris* Macquart, 1840; *ammophilus* Hesse, 1938; *anastoechoides* Hesse, 1938; *angulosus* Bezzi, 1921; *annuliventris* Hesse, 1938; *anomalus* Hesse, 1938; *argentatus* Fabricius, 1805; *argentifer* Walker, 1849; *argyrolomus* Bowden, 1964; *arnoldi* Hesse, 1938; *ater* Scopoli, 1763; *atronotatus* Hesse, 1938; *aurantiacus* Macquart, 1840; *auricomus* Bezzi, 1924; *auriferus* Hesse, 1938, *auriferus melas* Hesse, 1938; *auriferus nigripes* Hesse, 1938; *aurimystax* Hesse, 1938; *basifumatus* Speiser, 1914; *bezzii* Hesse, 1938; *bicoloratus* Bezzi, 1924; *bifidus* Bezzi, 1924; *bivittatus* Loew, 1855 (as *Triplasius*); *bombiformis* Bezzi, 1921 (see 1924); *brachyrhynchus* Bezzi, 1921; *braunsi* Bezzi, 1921; *brunnipennis* Loew, 1852; *calviniensis* Hesse, 1938; *capensis* Linné, 1767; *claripennis* Macquart, 1840; *cockerelli* Hesse, 1938; *damarensis* Hesse, 1938; *darlingi* Hesse, 1938; *delicatus* Wiedemann, 1830; *disjunctus* Bezzi, 1924; *elegans* Wiedemann, 1828; *erythrocerus* Bezzi, 1901; *eurhinatus* Bezzi, 1921, *eurhinatus bechuanus* Hesse, 1936; *extraneus* Hesse, 1938; *femoralis* Bezzi, 1924; *fenestralis* Hesse, 1938; *flagrans* Bezzi, 1924; *flaviceps* Macquart, 1840; *flavus* Macquart, 1840; *fucatus* Bezzi, 1921; *fulvonotatus* Wiedemann, 1818; *fulvosetosus* Hesse, 1956; *furiosus* Walker, 1860; *fuscilobus* Bezzi, 1924; *globulus* Bezzi, 1921; *haemorhoidalis* Bezzi, 1921, *haemorhoidalis waterbergensis* Hesse, 1938; *hirtus* Loew, 1860; *horni* Paramonov, 1931; *hottentotus* Hesse, 1938; *hypoleucus* Wiedemann, 1821; *hypoxanthus* Loew, 1863 [=*plagiatus* Bezzi, 1921]; *icteroglaenus* Hesse, 1938; *imitator* Hesse, 1938; *impurus* Loew, 1863; *inermis* Hesse, 1938; *kaokoensis* Hesse, 1938; *karasanus* Hesse, 1938; *karooensis* Hesse, 1938; *kilimandjaricus* Speiser, 1910; *lateralis* Fabricius, 1805; *latipectus* Hesse, 1938; *leucolasiua* Hesse, 1938; *lugens* Bezzi, 1924; *luteipennis* Bezzi, 1924; *marginellus* Bezzi, 1921; *mauritanus* Olivier, 1789; *megaspilus* Bezzi, 1921; *melanolomus* Hesse, 1938; *melanopus* Bezzi, 1924, *melanopus maculipennis* Timon-David, 1952; *melanurus* Loew, 1860; *meltoni* Hesse, 1938; *micans* Fabricius, 1798; *minusculus* Hesse, 1938, *minusculus pallidiventris* Hesse, 1938; *miscens* Walker, 1871; *molitor* Wiedemann, 1830; *mollihirtus* Hesse, 1938; *mollis* Bezzi, 1921; *monticolus* Francois, 1955; *montium* Francois, 1955; *montivagus* Hesse, 1938; *mundus* Loew, 1863; *muscoides* Hesse, 1938; *mutillatus* Bezzi, 1921; *namaquensis* Hesse, 1938; *neithokris* Jaennicke, 1867; *nieuwveldensis* Hesse, 1938; *nigrilobus* Bezzi, 1924; *nigripecten* Bezzi, 1921, *nigripecten cinctutus* Hesse, 1938; *niveus* Macquart, 1855; *obesus* Bezzi, 1921; *obtusus* Bezzi, 1924; *okahandjanus* Hesse, 1938; *ornatus* Wiedemann, 1828, *ornatus pleuralis* Bezzi, 1924; *pallescens* Hesse, 1938; *pallidulus* Walker, 1849; *parallelus* Bezzi, 1924; *parvus* Bowden, 1964; *paterculus* (Walker, 1852 (as *Parisus*); *pentaspilus* Bezzi, 1921; *peringueyi* Bezzi, 1921; *permixtus* Hesse, 1938; *plorans* Bezzi, 1924; *pruinosulus* Hesse, 1938; *pseudoargentatus* Paramonov, 1929 [=*argentatus* Fabricius, 1805]; *pseudopsis* Hesse, 1938; *punctatelloides* Hesse, 1938; *punctatillus* Bezzi, 1921; *punctifer* Bezzi, 1921; *purpureus* Bezzi, 1921; *rhomboidalis* Hesse, 1938; *rufescens* Hesse, 1938; *ruficeps* Macquart, 1840; *rufiventris* Macquart, 1846; *rufoantennatus* Becker, 1909; *rufus* Macquart, 1840; *senegalensis* Macquart, 1840; *servillei* Macquart, 1840; *simplicipennis* Bezzi, 1924; *simulans* Hesse, 1938; *spinibarbus* Bezzi, 1921; *subacutus* Hesse, 1938; *terminatus* Becker, 1909; *tinctipennis* Hesse, 1938, *tinctipennis thornei* Hesse, 1938; *transitus* Hesse, 1938; *tripudians* Bezzi, 1924; *tuckeri* Hesse, 1938; *turneri* Hesse, 1938; *uniformis* Paramonov, 1955; *vansoni* Hesse, 1936; *volucer* Hesse, 1938; *xanthocerus* Bezzi, 1921; *zoutpansbergianus* Hesse, 1938, *zoutpansbergianus occidentalis* Hesse, 1938.

Oriental: *Bombylius albosparsus* Bigot, 1892; *ardens* Walker, 1894; *brunetti* Senior-White, 1922; *comastes* Brunetti, 1909; *erectus* Brunetti, 1909; *fulvipes* Bigot, 1892; *morosus* de Meijere, 1914; *orientalis* Macquart, 1840; *propinquus* Brunetti, 1909; *pseudoterminalis* Senior-White, 1923 [=*vicinus* Brunetti, not Macquart, 1840]; *pulchellus* Wulp, 1880 [=*dives* (Bigot), 1888 (as *Eucharimyia*)]; *scintillans* Brunetti, 1909; *socius* Walker, 1852; *terminalis* Brunetti, 1909; *tricolor* Guérin-Méneville, 1835.

Australian: *Bombylius albavitta* Macquart, 1849; *albiceps* Macquart, 1848; *albicinctus* Macquart, 1847; *alienus* Hardy, 1942 [=*pulchellus* Roberts, 1928]; *altus* Walker, 1849; *antecedens* Walker, 1849; *areolatus* Walker, 1857; *australianus* Bigot, 1892; *australis* Guérin-Menéville, 1838; *bifrons* (Walker), 1852, (as *Choristus*); *brevirostris* Macquart, 1849; *chrysendetus* White, 1917; *consobrinus* Macquart, 1847; *crassirostris* Macquart, 1849; *crassus* Walker, 1849; *distinctus* Walker, 1852; *dulcis* Roberts, 1928; *fuscanus* Macquart, 1849; *hilaris* Walker, 1849; *immutatus* Walker, 1849; *lobalis* Thomson, 1869; *loewii* Jaennicke, 1867; *matutinus* Walker, 1849; *nanus* Walker, 1849; *notatipennis* Macquart, 1855; *palliolatus* White, 1917; *penicillatus* Macquart, 1849; *pictipennis* Macquart, 1850; *pinguis* Walker, 1849; *platyrus* Walker, 1849; *primogenitus* Walker, 1849; *probellus* Hardy, 1942 [=*bellus* Roberts, 1928]; *proprius* Roberts, 1928; *punctipennis* Thomson, 1869; *pycnorrhynchus* Thomson, 1869; *robertsi* Paramonov, 1934; *rubriventris* Bigot, 1892; *rutilus* Walker, 1849; *sericans* Macquart, 1849; *spinipes* Thomson, 1869; *succandidus* Roberts, 1928; *tenuicornis* Macquart, 1846; *tenuirostris* Roberts, 1928; *vetustus* Walker, 1849; *viduus* Walker, 1852.

Oceanica: *Bombylius albifacies* Macquart, 1855; *bicoloricornis* Macquart, 1855; *fulviceps* Macquart, 1855; *guttatipennis* Macquart, 1855; *minimus* Macquart, 1855; *niveus* Macquart, 1855; *rufilabris* Macquart, 1855; *watanabei* Matsumura, 1916.

Patria Ignota: *Bombylius albifrons* Gmelin, 1790; *cephalotes* Walker, 1849; *dimidiatus* Macquart, 1840; *fuscus* Thunberg, 1827; *inornatus* Walker, 1849; *limbipennis* Macquart, 1840; *maculifer* Walker, 1852; *maculipennis* Eversmann, 1834; *pulchellus* Eversmann, 1834; *recedens* Walker, 1852; *signifer* Walker, 1852; *tibialis* Walker, 1849; *tinctus* Walker, 1849; *tripunctatus* Macquart, 1840; *vagabundus* Meigen, 1830.

Genus *Systoechus* Loew

Figures 6, 10, 236, 442, 444, 789, 790

Systoechus Loew, Neue Beitrage III, p. 34, 1855. Type of genus: *Bombylius sulphureus* Mikan, 1796, designated by Coquillett, 1910, p. 611, the sixth species, of 14 (as 15) species.

Becker (1916), pp. 60-62, key to 13 Palaearctic species.
White (1924), p. 83, key to 3 species from Australia.
Bezzi (1924), pp. 62-64, key to 17 Ethiopian species.
Engel (1934), pp. 279-280, key to 10 Palaearctic species and many subspecies.
Maughan (1935), pp. 62-63, key to 2 Utah species.
Hesse (1938), pp. 293-347, key to 54 species *Systoechus* and 17 species *Anastoechus* and many subspecies.
Efflatoun (1945), pp. 269-270, key to 2 Egyptian species.
Painter (1962), p. 256, key to 5 Nearctic species.
Bowden (1964), pp. 22-23, key to 9 species from Ghana.

Flies ranging in size from small to medium size and in rare instances large. In general appearance the ground color is opaque black with pollenlike bloom but so densely covered with long, erect, yellow to whitish pile as to obscure the ground color, which gives these flies a yellowish appearance. Legs generally pale yellow with darkened tarsi. These flies and the close relative *Anastoechus* Osten Sacken are in general similar to *Bombylius* Linné, and like that genus the first posterior cell is closed and stalked; they differ from *Bombylius* in the position of the anterior crossvein, which is basal, lying well before the middle of the discal cell, so that the first basal cell is approximately the same length as the second basal cell. From *Acanthogeron* Bezzi they are separated by the same distinctions as in *Bombylius* Linné, and in addition the first antennal segment is not thickened or swollen, and in *Systoechus* Loew the eyes of the male are dichoptic. In *Systoechus* the basal comb is strongly and conspicuously developed, and the hind femur has spines below; front tibia with weak spines. *Systoechus* Loew is distinguished from *Anastoechus* Osten Sacken by the more moderate amount of hair on the face. The facial hair forms a complete circumoral fringe in *Systoechus*. In *Anastoechus* the whole anterior face and genae are so extremely, densely hairy that all surface details are obscured until the hair is removed; also its intercalary vein is longer.

Systoechus Loew is a large genus found in all major world regions.

Length: 6 mm. to 17 mm., excluding proboscis; wings 6 mm. to 18.5 mm.; most species are about 6 to 9 mm. The largest species is the South African *Systoechus goliath* Bezzi, 17 mm. long, proboscis 12 to 13 mm. additional, wing spread about 45 mm.

Head, lateral aspect: The head is subtriangular, the face rather extensively produced. The occiput rather prominent medially, sloping gradually out to the eye margin. The oragenal cup is prominent above, reduced in length below. The pile of the face is comparatively dense, extending broadly across beneath the antennae, erect and long and coarse and among this pile there are usually some shorter, appressed, glittering hairs. In some species the facial pile is more scanty and scattered; it continues on down the sides of the oragenal cup, some of it arising from within the tentorial fissure. Occiput with quite dense pile, longer medially, the more lateral hairs shorter, flat-appressed and ending at the eye margin, and the pile is continued long and abundant beneath the head. The eye is approximately hemicircular but flattened above in males. The posterior margin is entire and almost plane. Proboscis quite elongate and slender, the labellum likewise long and slender. The whole proboscis about 3 times as long as the head. Palpus relatively short and slender with a few apical hairs, the basal segment much shorter. Front is swollen in the males, depressed a short distance in front of the ocellar tubercle and with a shallow, medial groove. Females with the front similar but with the flattened, depressed area occupying the whole upper half of the front. The raised areas are set off posteromedially by a circular line. Male front with scattered, fine, erect, bristly hairs, pollen, and some appressed, scaliform pile in the middle or along the eye margins in some species. Female front with similar pile.

Antennae placed at the upper fifth of the head, shorter than the head in length; the slender first segment is 3 times as long as wide. The short second segment is beadlike, a little widened apically. The third segment is as long as the first two segments together or barely longer, knobbed at the base and widest on the basal half. It ranges from narrower than the second segment to a little wider, strongly attenuated and narrowed on the apical half with a minute microsegment and spine. The first segment bears much long, stiff, bristly pile or weak bristles above and below extending beyond the second segment and on the upper side sometimes reduced to a few short hairs.

Head, anterior aspect: The head is distinctly narrower than the thorax. The front is large and triangular in males, wide in females but not quite as wide as the face. Eyes in males separated by approximately the distance between the posterior ocelli. Antennae separated by the thickness of one segment. Ocellar tubercle small, protuberant, the large ocelli almost touch the eye in the male. Anterior ocellus much wider than the others. Ocellar tubercle with a few long, bristly hairs. Oral opening large and short oval, deep behind. In the middle the inner walls of the oragenal are extraordinarily inflated, more so than in *Bombylius* Linné, the outer sloping wall of the oragenal cup is extensive and extensively pilose. The gena is very narrow, the tentorial fissure deep and rather conspicuous. From above the occiput is distinctly bilobate with a vertical groove behind the column of the vertex.

Thorax: The thorax is short, broad, rather convex anteriorly, the head semidecumbent, the abdomen less so. The mesonotum is black with opaque brown or gray pollen, very densely furry, especially anteriorly in front of the wings, the pile not exceptionally long but very dense. Notopleuron at most with a few bristly hairs. Humerus with long, dense pile and the scutellum likewise. Pleuron everywhere pilose including pteropleuron, hypopleuron, and metapleuron. Squama with a long fringe.

Legs: The legs long and slender, anterior tibia with 2 rows of minute spicules, 5 or 6 in each row; the middle tibia is similar with about 8 spicules in each of 4 rows. Hind tibia with slender but longer spicules in 4 rows, 6 to 8 in each row, except ventrally, where there are only 3 or 4. Ventral margin of the short, hind femur with a row of 5 to 8 long, slender bristles. Pile of the legs consists of scanty, scattered, long, slender, flat-appressed scales on the femora; they are much more minute on the tibiae. Claws long, slender, sharp, the pulvilli long but slender.

Wings: The wings are comparatively large, broad at the base, hyaline, sometimes tinted at the immediate base with 2 submarginal cells. The first posterior cell closed with a long stalk, the intercalary vein long, although the discal cell is narrowed apically both above and below. The anal cell is widely open, characteristically the anterior crossvein lies very near the base of the discal cell, so that the first basal cell is either no longer than the second, or only very slightly longer. Axillary lobe wide, alula large, ambient vein complete, costal comb strongly developed.

Abdomen: The abdomen is elongate oval, narrowed a little apically, black in ground color, feebly shining through the faint pale gray or brown pollen and covered with extraordinarily dense, plushlike pile of even length continued in the same length over the sides of the abdomen and posteriorly, so that the abdomen appears to be one mass of plush. I find 7 tergites in the male.

The male terminalia are large, are extruded but ventral in position. Sternites also with long, dense pile. The pile of the species known to me ranges from snowy white to bright, golden yellow. Also, beneath the scutellum there is a fringe of very short, dense hair on the first abdominal segment.

According to Hesse:

Systoechus Loew differs from both *Bombylius* and *Anastoechus* by the entirely different type of aedeagus or structures associated with the aedeagus in the hypopygium of the males. These differences are very characteristic and very constant in the genus. The aedeagus is either sickle-shaped and with a characteristic flattened keel below or the aedeagus is normal but has on each side a styletlike, rod-like, clavate or even racket-shaped process projecting apically from a girdle-like or bridge-like basal part which is continuous on each side with the lateral ramus to each basal part. The beaked apical joints are usually elongate, narrowish, somewhat laterally compressed and not foveately depressed above.

The male terminalia from dorsal aspect, epandrium removed, show extensive membranous areas, slender and elongated dististyli from both dorsal and lateral aspects. There is a curious, transverse, comparatively slender, circular, bridgelike loop proceeding distally from the ramus of each side, and crossing the epiphallus from each side of the epiphallus and arising from this bridge there is an elongate, knife-shaped, apically expanded, long, slender process of about the same width as the bridge itself. This process is apt to be double-pronged, dorsoventrally and separated by a deep incision. There is a very characteristic difference from the genus *Anastoechus* Osten Sacken.

Material available for study: The type-species and several American and some Ethiopian and Australian species.

Immature stages: The larvae are important consumers of locust eggs. See further under life histories.

Ecology and behavior of adults: Efflatoun (1945) says of *gradatus* Wiedemann that it is an extremely rapid flier and expert at hovering, and that if not caught at the first stroke of the net one can not expect them to return as do the species of *Bombylius* Linné; thus they are easily frightened.

Distribution: Nearctic: *Systoechus candidulus* Loew, 1863; *fumipennis* Painter, 1962; *oreas* Osten Sacken, 1877; *solitus* (Walker), 1849 (as *Bombylius*); *vulgaris* Loew, 1863.

Neotropical: *Systoechus seniculus* (Philippi), 1865 (as *Bombylius*).

Palaearctic: *Systoechus albicans* (Macquart), 1849 (as *Bombylius*); *arcticus* Strobl, 1910; *autumnalis* (Pallas in Wiedemann), 1818 (as *Bombylius*), *autumnalis albibarbis* Engel in Lindner, 1935; *glabellula* Strobl, 1910; *gomez menori* Rubio, 1959; *gradatus* (Wiedemann in Meigen), 1820 (as *Bombylius*) [=*cinereus* (Meigen), 1820 (as *Bombylius*), =*leucophaeus* Wiedemann in Meigen, 1820, =*quadratus* Loew lapsus, 1855], *gradatus gallicus* Villeneuve, 1904, *gradatus lucidus* Loew, 1855, *gradatus tesquorum* Becker, 1916, *gradatus validus* Bezzi, 1925; *grandis* Paramonov, 1940; *laevifrons* Loew, 1855; *longirostris* Becker, 1916; *microcephalus* Loew, 1855; *nivalis* Brunetti, 1912; *pallidipilosus* Austen, 1937; *pallidulus* Greathead, 1958; *pumilio* Becker, 1915; *sericeus* (Meigen), 1820 (as *Bombylius*), [=*exalbidus* (Meg. in litt. in Meigen), 1820 (as *Bombylius*), =*nubilis* (Meigen), 1804 (as *Bombylius*)]; *sibiricus* Becker, 1926 (Olchon Isl.); *sinaiticus* Efflatoun, 1945; *solitus* (Walker), 1849 (as *Bombylius*); *somali* Greathead, 1958; *sulphureus* (Mikan), 1796 (as *Bombylius*) [=*ctenopterus* (Mikan), 1796 (as *Bombylius*), =*fulvus* (Meigen), 1820, (as *Bombylius*), =*minimus* (Fabricius), 1787 (as *Bombylius*)], *sulphureus aurulentus*, (Wiedemann in Meigen), 1820 (as *Bombylius*), *sulphureus convergens* Loew, 1855; *sulphureus dalmatinus* Loew, 1855; *sulphureus orientalis*, Zakhvatkin, 1955; *unicolor* Becker, 1916; *vulpinus* Becker, 1907 (Engel gives date as 1910).

Ethiopian: *Systoechus aberrans* Hesse, 1938; *acridophagus* Hesse, 1938; *affinis* Hesse, 1938, *affinis discrepans* Hesse, 1938; *albidus* Loew, 1860, *albidus auripilus* Hesse, 1938; *albipectus* Hesse, 1938; *altivolans* Hesse, 1938; *anthopilus* Hesse, 1938; *argyroleucus* Hesse, 1938; *argyropogonus* Hesse, 1938; *atriceps* Bowden, 1964; *aureus* Hesse, 1938; *auricomatus* Bowden, 1959; *aurifacies* Greathead, 1958; *austeni* Bezzi, 1924; *badipennis* Hesse, 1938; *badius* Hesse, 1938; *bechuanus* Hesse, 1936; *bombycinus* Hesse, 1938, *bombycinus bedfordi* Hesse, 1938, *bombycinus pallidispinis* Hesse, 1938; *brunnibasis* Hesse, 1938; *candidus* Hesse, 1938; *canescens* Hesse, 1938; *canicapillis* Bowden, 1959; *canipectus* Hesse, 1938; *cellularis* Bowden, 1959; *cervinus* Loew, 1860; *chlamydicterus* Hesse, 1938; *chrystallinus* Bezzi, 1924; *damarensis* Hesse, 1938; *deceptus* Hesse, 1938; *eremophilus* Hesse, 1936; *exiguus* Hesse, 1938; *exilipes* Bezzi, 1923; *faustus* Hesse, 1938; *ferrugineus* Macquart, 1834; *flavicapillis* Bowden, 1959; *fuligineus* Loew, 1863; *fumitinctus* Hesse, 1938; *fusciventris* Hesse, 1938; *goliath* Bezzi, 1922; *heteropogon* Bowden, 1959; *inordinatus* Hesse, 1938; *kalaharicus* Hesse, 1936; *lacus* Bowden, 1964; *leoninus* Bowden, 1964; *leucostictus* Hesse, 1938; *lightfooti* Hesse, 1938; *litoralis* Bowden, 1964; *marshalli* Paramonov, 1931; *melampogon* Bezzi, 1912; *mentiens* Bezzi, 1924; *mixtus* (Wiedemann), 1821 (as *Bombylius*) [=*scutellaris* (Wiedemann), 1828 (as *Bombylius*), =*scutellatus* (Macquart), 1840 (as *Bombylius*), =*stylicornis* (Macquart), 1834 (as *Bombylius*)]; *montanus* Hesse, 1938; *monticolanus* Hesse, 1938; *namaquensis* Hesse, 1938; *neglectus* Hesse, 1938; *nigribarbus* (Loew), 1852 (as *Bombylius*), *nigribarbus falsus* Hesse, 1938; *nigripes* Loew, 1863, *nigripes nomteleensis* Hesse, 1938, *nigripes plebeius* Hesse, 1938; *niveicomatus* Bowden, 1959; *phaeopterus* Bezzi, 1924; *polioleucus* Hesse, 1938; *poweri* Hesse, 1938; *rhodesianus* Hesse, 1938; *robustus* Bezzi, 1912; *rubricosus* (Wiedemann), 1821 (as *Bombylius*); *rudebecki* Hesse, 1956; *rufiarticularis* Hesse, 1938; *salticolus* Hesse, 1938; *scabrirostris* Bezzi, 1921; *segetus* Bowden, 1964; *silvaticus* Hesse, 1938, *silvaticus turneri* Hesse, 1938; *simplex* Loew, 1860; *somali* Oldroyd, 1947; *spinithorax* Bezzi, 1921; *stevensoni* Hesse, 1938; *subcontiguus* Hesse, 1938; *subulinus* Bowden, 1964; *transvaalensis* Hesse, 1938; *tumidifrons* Bezzi, 1921; *ventricosus* Bezzi, 1921; *waltoni* Hesse, 1938; *xanthoplocamus* François, 1964; *xerophilus* Hesse, 1938.

Oriental: *Systoechus eupogonatus* Bigot, 1892; *flavospinosus* Brunetti, 1920, *socius* Walker, 1852.

Australian: *Systoechus albohirtus* Roberts, 1928; *callynthrophorus* Schiner, 1868; *cinctiventris* Roberts, 1928; *flavovillosus* Roberts, 1928; *leucopygus* Wulp, 1885; *pallidus* Roberts, 1928; *pausarius* Jaennicke, 1867; *rubidus* Roberts, 1928.

Patria Ignota: *Systoechus canus* (Macquart), 1840 (as *Bombylius*); *fasciculatus* (Macquart), 1840 (as *Bombylius*).

Genus *Anastoechus* Osten Sacken

FIGURES 8, 239, 240, 447, 465, 778, 780, 791, 792

Anastoechus Osten Sacken, Bull. U. S. Geol. Survey, vol. 3, p. 251, 1877. Type of genus: *Anastoechus barbatus* Osten Sacken, 1877, by monotypy. [Not *Bombylius nitidulus* Fabricius, 1794; see Coquillett, p. 506, 1910.]

Becker (1916), pp. 51-55, key to 20 Palaearctic species.
Bezzi (1924), pp. 73-74, key to 11 South African species.
Paramonov (1930), pp. 424-437, key to many Palaearctic species.
Engel (1934), pp. 288-290, key to 14 Palaearctic species.
Engel (1934), pp. 290-296, key (taken from Paramonov) to 41 species and many subspecies.
Maughan (1935), pp. 61-62, key to 2 Utah species.
Austen (1937), pp. 40-41, key to 4 species from Palestine area.

Hesse (1938), pp. 293-347, key to 17 species of *Anastoechus*; 54 species of *Systoechus* and many subspecies.
Paramonov (1940), pp. 256-267, key to 37 males, 44 females of Palaearctic species.
Efflatoun (1945), pp. 278-279, key to 6 Egyptian species.
Painter (1962), p. 266, key to 3 Nearctic species.

Broad, stout flies of short abdomen and compact form. They are densely furry or pilose in character and rather like *Bombylius* Linné in venation, with closed first posterior cell, and long proboscis. In both *Anastoechus* Osten Sacken and in *Systoechus* Loew the first and second basal cells, however, are of equal length, the anterior crossvein enters the discal cell quite close to the base, and the discal cell is wide apically. It is separated from *Systoechus* Loew in several particulars. In shape the discal cell is much wider apically with its sides nearly parallel; the pilosity of the face and front is much more extensive and conspicuous. The front in the female is convex and lacks a transverse depression; the male is without a central, furrowlike depression. Moreover, on the postmargins of the tergites *Anastoechus* Osten Sacken has rows of erect, stout bristles.

All of the species agree, except 2 South African species, in having the anterior crossvein quite basal in position. Hesse described 2 transitional species, *Bombylius bezzi* Hesse and *Bombylius anastoechoides* Hesse from South Africa, which have the crossvein placed as it is in *Bombylius* Linné, but otherwise agree with *Anastoechus* Osten Sacken and *Bombylius* Linné as based on the 20 or more South African species of the former genus. I do not see how the concept of *Anastoechus* Osten Sacken can be modified in respect to its characteristic venation to include the two species named above, which I therefore refer to a new subgenus of *Bombylius* Linné.

Anastoechus Osten Sacken is a large and important genus of wide distribution. Two-thirds of the species are from the Palaearctic, but at least 20 are from South Africa; 2 are from Australia; 3 from the Nearctic. None are known from South America.

Length 6 mm. to 17 mm.

Head, lateral aspect: The head is hemispherical with the face only moderately produced but everywhere covered by long, fine, erect, and generally bristly pile with, in addition, matted, ocular tufts of subappressed, somewhat flattened tomentum which extends inward from the eye margin. Antenna situated rather near the top of the head so that the front is relatively short. Occiput prominent, sloping inward from the eye and densely covered with very long, erect, coarse pile which becomes bristly on the upper half of the vertex. Also, the outer margin of the occiput adjacent to the eye is rather extensively covered with matted tomentum which curls outward and is appressed against the eye margin. Posterior eye margin plane with no indentation. Proboscis 2 to 4 times as long as the head. Palpus filiform but small with some long hairs apically and also laterally toward the base. Antenna elongate and slender, the first segment may be 2 to 6 times as long as the second segment, which is quite small. Third segment either spindle-formed or bulbous at the base; in any case it is drawn out and greatly attenuate apically.

Head, anterior aspect: Males with the eyes separated by about the width of the ocellar triangle. Females with eye separation nearly 3 times that of the males. The head is narrower than the mesonotum. The ocellarium is rather low with numerous, coarse hairs and a larger, dense tuft of longer hairs arising behind the ocelli. Front everywhere covered with conspicuous pile; in some species long, erect, bristly pile predominates; in others matted tufts of flattened tomentum are more in evidence. Oral opening large and short oval. Eyes bare. The attachment of the antennae is close; the antennae are slightly divergent.

Thorax: At the wings the thorax is approximately as wide as long. Like *Systoechus* Loew the pile is extraordinarily dense and erect and for the most part fine but with an enclosed tuft of 10 or more bristles arising on the notopleuron. Weak bristles arise from the upper mesopleuron and long, conspicuous bristles arise from the postalar callosity. Pleuron almost everywhere densely matted pilose. Squama and plumula with long, dense fringe of fine pile. Anterior hypopleuron largely bare.

Legs: The legs are slender, all the tibiae with bristles, the anterior and middle tibiae with 3 or 4 rows of stout, short to moderately long bristles. Hind basitarsus only half as long as the tibia. Hind femur with a ventrolateral row of 10 long, comparatively stout bristles, besides several bristles dorsally on each side near the apex. Hind basitarsus half as long as the tibia. Claws quite long and slender, the pulvilli long.

Wings: The wings are large, venation similar to that of *Systoechus* Loew, except that the discal cell is much wider apically. Like *Bombylius* Linné and *Systoechus* Loew, the first posterior cell is closed with a long stalk. Wings hyaline or more usually tinted with brown on the basal half, darkest at the base. There are 4 posterior cells and the anal cell is quite widely open. Alula moderately large; ambient vein complete. Second vein takes origin at the base of the discal cell. The costa has a very conspicuous bristle-comb at the base of the wing, often with additional pile or flattened tomentum. Basal spur or hook present. The wing is densely microvillose.

In one species from the Gobi Desert sent me by the Hungarian Museum, the basal comb is extraordinarily extensive and prominent, extended way forward from the wing border.

Abdomen: The abdomen is moderately long oval and distinctly convex. The vestiture consists of rather dense, extraordinarily long, erect, fine pile with a conspicuous row of stout, erect bristles of the same length, or even longer, which is located along the margin of each tergite and with similar rows mixed with long pile on the sternites. Female terminalia enclosed by a dense band of reddish bristly hairs. Male terminalia with dististyli generally quite broad on the basal two-thirds, more or less leaf-shaped, with the dorsal part depressed or even excavated and scooped out.

Dr. A. J. Hesse, in his work on South African bee flies, describes the male hypopygium as follows: "Hypopygium of males with the aedeagus never falcate or sickle-shaped, its lower part not produced into a sharp keel and never with a stylet-like or rod-like process on each side below."

The male terminalia from dorsal aspect, with epandrium removed, show in *barbatus* Osten Sacken a relatively narrow body with short, swollen, stout dististyli which bear a divergent hooklike process apically. The epiphallus is single and not divided. It is long, slender and attenuate. There is a complete absence of the girdle-like circular band across the epiphallus. There is in *barbatus* Osten Sacken a medial extension of the basistylus on each side of the ejaculatory process. The basal apodeme is small, the entire structure is much simpler than in *Systoechus* Loew.

Material available for study: I have before me a long series of the Nearctic species *barbatus* Osten Sacken, and I have had the privilege of studying all of the species of this genus in the British Museum (Natural History). I also have considerable material from the Expedition of the Hungarian Museum to the Gobi Desert.

Immature stages: *Anastoechus* Osten Sacken is an important consumer of locust egg pods throughout the Holarctic region. See further discussion under the section on life histories.

Ecology and behavior of adults: Austen (1937) comments on the charm lent to the spring season in Palestine by the visits of dainty little bee flies like *Anastoechus stramineus* Meigen visiting low-growing flowers. I have found species of this genus visiting late summer flowers in great multitudes in such states as New Mexico and Arizona; a single clump of blossoms may have fifty or more of these bee flies hovering about it. Efflatoun (1945) records taking *trisignatus* Portschinsky from flowers of *Acantholimon* sp. He comments on the extraordinary abundance of the genus in Egypt. Austen (1945) gives a color plate of *stramineus* Meigen.

Distribution: Nearctic: *Anastoechus barbatus* Osten Sacken, 1877; *hessei* Hall, 1958 [=*deserticola* Hall, 1956, not Hesse]; *melanohalteralis* Tucker, 1907, *melanohalteralis fulvipennis* Tucker, 1907.

Palaearctic: *Anastoechus acuticornis* (Macquart), 1840 (as *Bombylius*); *aegyptiacus* Paramonov, 1930; *andalusiacus* Paramonov, 1930; *angustifrons* Paramonov, 1930; *anomalus* Paramonov, 1940; *araxis* Paramonov, 1926, *araxis nigrisetosa* Paramonov, 1926; *asiaticus* Becker, 1916 [=*villosus* Paramonov, 1930], *asiaticus albulus* Paramonov, 1940; *aurifrons* Efflatoun, 1945; *bahirae* Becker, 1915 [=*fascipennis* Becker, 1916, =*fuscipennis* Becker, 1915], *bahirae pyramidum* Paramonov, 1940; *baigakumensis* Paramonov, 1926; *bitinctus* Becker, 1916; *caucasicus* Paramonov, 1930; *chinensis* Paramonov, 1930; *elegans* Paramonov, 1930; *exalbidus* Wiedemann in Meigen, 1820; *firjuzanus* Paramonov, 1930; *flaveolus* Becker, 1916; *fulvescens* Becker, 1913; *fuscus* Paramonov, 1926; *hyrcanus* (Pallas in Wiedemann), 1818 (as *Bombylius*) [=*hircanus* (Wiedemann), 1828 (as *Bombylius*), =*stramineus* Becker, =*viduus* Becker, 1916], *hyrcanus aralicus* Paramonov, 1940; *intermedius* Becker, 1916; *latifrons* (Macquart), 1838 (as *Bombylius*); *longirostris* Wulp, 1885; *mongolicus* Paramonov, 1930; *monticola* Paramonov, 1930; *montium* Becker, 1916; *nigricirratus* Becker, 1913; *nitidulus* (Fabricius), 1794 (as *Bombylius*) [=*caudatus* (Meigen), 1804 (as *Bombylius*), =*cephalotes* (Walker), 1849 (as *Bombylius*), =*diadema* (Meigen), 1804 (as *Bombylius*)], *nitidulus aberrans* Paramonov, 1940, *nitidulus hummeli* Paramonov, 1934; *niveicollis* Enderlein, 1933; *niveus* Hermann, 1909 [=*albopectinatus* Becker, 1930]; *nomas* Paramonov, 1930; *olivaceus* Paramonov, 1930, *olivaceus corsikana* Paramonov, 1930; *pulcher* Paramonov, 1930; *retardatus* Becker, 1913; *rubriventris* Paramonov, 1930; *setosus* (Loew), 1855 (as *Systoechus*); *sibiricus* (Becker), 1916 (as *Anastoechus*); *smirnovi* Paramonov, 1926, *smirnova nigrifemorata* Paramonov, 1926; *stackelbergi* Paramonov, 1926; *stramineus* (Wiedemann in Meigen), 1820 (as *Bombylius*); *suzukii* Matsumura, 1916; *syrdarpensis* Paramonov, 1926; *trisignatus* Portschinsky), 1881 (as *Systoechus*), *trisignatus retrogradus* (Becker), 1902 (as *Systoechus*), *trisignatus werneri* Paramonov, 1930, male only; *turanicus* Paramonov, 1926; *turkestanicus* Paramonov, 1926; *turkmenorum* Paramonov, 1930; *vlasovi* Paramonov, 1930, *vlasovi albicans* Paramonov, 1930; *werneri* Paramonov, 1930, female only; *zimini* Paramonov, 1940.

Ethiopian: *Anastoechus argyrocomus* Hesse, 1938; *deserticolus* Hesse, 1938, *deserticolus coloratus* Hesse, 1938; *dolosus* Hesse, 1938; *erinaceus* Bezzi, 1921; *eurystephus* Hesse, 1938; *flavosericatus* Hesse, 1938; *fuscianulatus* Hesse, 1938; *innocuus* Bezzi, 1921; *latifrons* (Macquart), 1839; *leucochroicus* Hesse, 1938; *leucosoma* Bezzi, 1921; *macrophthalmus* Bezzi, 1922; *macrorrhynchus* Bezzi, 1924; *meridionalis* Bezzi, 1912; *nitens* Hesse, 1938; *phaleratus* Hesse, 1938, *phaleratus albicerus* Hesse, 1938; *pruinosus* Hesse, 1938; *rubicundus* Bezzi, 1924; *sericatus* Hesse, 1938; *sericophorus* Hesse, 1938, *sericophorus congruens* Hesse, 1938; *varipecten* Bezzi, 1921.

Australian: *Anastoechus annexus* Roberts, 1928; *perspicuus* Roberts, 1928.

Genus *Sisyrophanus* Karsch

FIGURES 241, 243, 434, 446, 466, 477

Sisyrophanus Karsch, Ent. Nachrichten, vol. 12, p. 53, 1886.
  Type of genus: *Sisyrophanus homeyeri* Karsch, 1886, by monotypy.

Bezzi (1924), p. 84, key to 6 Ethiopian species.
Hesse (1938), pp. 524-525, key to 2 South African species.

Small to large flies, generally quite black, but the legs sometimes reddish or yellowish, with long, slender proboscis and slender antennae rather like *Bombylius* Linné, but varying in the character of the apex of the third segment which may end in a minute terminal style or microsegment, or may have the apex concavely excavated, with a dorsal hook-shaped spine; in other species only the dorsal hook is present.

The acute, closed, and stalked third posterior cell suggests the venation of *Eurycarenus* Loew, but here the resemblance ends as the posterior contour of the eye is simple.

The front, the dorsally extensive oragenal cup, and the cheeks are all shining vitreous black; the abdomen is apt to be shining, the thorax often dull or nearly opaque. Thorax and abdomen rather densely covered with gray to white pile, erect and fine, and the post margins of the tergites generally with rows of slender, black bristly hairs. The wide space on each side of the oragenal cup has a conspicuous pretty fringe of dense long, erect, white pile, both above and below, and the palpi tend to have odd radiating tufts of similar white pile. Hind femur without bristles below but tibial spicules well developed on all tibiae. Front separated from face by a distinct, shallow, transverse furrowlike depression.

The type of wing venation together with the bare and shining face, conically produced above to variable extent, sometimes quite prominent as in *pyrrhocerus* Bezzi, is sufficient to distinguish this genus. It has been variously related to *Dischistus* Loew, *Doliogethes* Hesse, and *Sparnopolius* Loew. From all of the *Dischistus* Loew complex it is distinguished by the venation and the smooth shining face, except the smooth-faced section of *Chasmoneura* Hesse, and here again the venation sets it apart as it also does separate it quickly from *Sparnopolius* Loew. The genus is most closely related to *Dischistus* Loew, sensu stricto, and bears no relation to the Pthiriinae as indicated by Bezzi (1924).

In this genus *Sisyrophanus* Karsch, the hind femur is without bristles, the metapleuron lacks pile, the wings are quite short, the basal comb absent, the anal cell open. The abdomen varies from broadly oval to elongate conical but bears dense, erect, fine, mostly pale pile, with or without rows of bristles on the hind margins. Also it must be noted that while the oragenal cup and more or less extended face is bare, there is a marginal or frontogenal fringe of pile, varying from long to short which may hide the face in the lateral aspect. Eyes holoptic in males, widely separated in females.

This genus is largely confined to central Africa, only two species ranging as far south as Rhodesia and Bechuanaland.

Length: 5 to 13 mm.; wing smaller, 5 to 10 mm.

Head, lateral aspect: The head is hemispherical in profile with the face completely replaced in the middle by the oragenal cup, which is large, oblique, strongly retreating below but is extensive above reaching well beyond the eye margins. The face is restricted to the dorsal part of the oragenal cone and to a lateral area just below the antennae and continues below the antennae and continues below with the quite wide gena. The face and oragenal cone, both long or short, are totally bare and polished as in *Sparnopolius* Loew. Along the front genal ocular margin is a dense, circular fringe or band of long, conspicuous, erect, coarse, white pile. Below, this pile is somewhat appressed and bends inward, as a thick brush. Face and front likewise continuous although just above the antenna a wide, somewhat crescentic, shallow sulcus begins, which is fluted or grooved on each side, and which continues transversely across the front. The front is depressed in the middle and is nearly as wide as the face. The occiput is moderately well developed although laterally it is shallow. It becomes more extensive medially, and from a dorsal aspect it appears more tumid because the upper eye corners turn forward. There is no conspicuous central cavity. The occipital pile is quite dense, long and opaque white below, still longer, coarse, and shining above, but along the sides near the eye margins it becomes much shorter, opaque, rather blunt, and does not extend beyond the eye margin itself. Palpi short and rather stout, conspicuously clothed with long, radiating, coarse, white pile along ventral, lateral, and medial surfaces. The proboscis is slender and is 2 or 3 times the head length. Antenna attached near the upper fourth of the head, elongate and slender, the first segment is cylindrical, somewhat more stout than the second segment and at least 3 times as long as the second segment. It is conspicuously pilose with a dorsal tuft or clump of pile at the base, a few shorter dorsal setae beyond and the whole ventral surface rather conspicuously and densely covered with opaque, white pile nearly as long as that found on the face. The second segment is as long or longer than the first two segments together, rather strongly though regularly tapered from the base to the apex or a little widened through the middle and in some species narrow both at base and apex, and with or without a minute microsegment. There is a minute spine at the tip. Some species have the apex hollowly excavated, with a dorsal hooklike spine; others have only a dorsoapical hook.

Head, anterior aspect: The head is very much wider than the thorax and is wider below since the eyes are narrowed inwardly above. Genofacial space very wide and from a third to two-fifths of the total width. Oral cavity narrow and long; the sides of the oragenal cup are inflated and swollen, rounded laterally, and the pile of the lower portion of the head and occiput behind the proboscis is not only dense but remarkably long, white, and bushy. The eyes in the male are touching for a considerable distance in spite of the great width of the front in the female. The antennae are narrowly separated at the base by the width of one segment. The front has conspicuous, erect, and often rather long, coarse pile on the upper half. It may extend on to the vertex. The ocellar tubercle is rather high but gently raised or swollen and nowhere abrupt. Anterior ocellus quite enlarged, the three forming an obtuse triangle.

Thorax: The thorax is somewhat longer than wide, flattened and depressed posteriorly, and only shallowly convex anteriorly. There is a rather deep, wide, sulcate, shallow depression separating the posterior mesonotum from the prominent postalar callosity. The mesonotum is densely covered with long, erect, coarse, shining pile, bristles absent, except that upon the margin of the scutellum there are 4 or 5 pairs of long, slender bristles or bristly hairs, and the whole scutellum is densely covered with erect, long, fine pile. Pleuron remarkably hairy although there is no pile on the metapleuron and the hypopleuron has only a narrow, vertical band of short hairs on the posterior section adjacent to the metasternum and behind the spiracle. The mesopleuron and entire upper half of the sternopleuron are very densely covered with bushy, rather shining, long, coarse pile which tends to radiate out from the upper corner of the mesopleuron, part of it directed forward and more of it, and still longer hairs appressed and directed backward and some of it downward. Squama moderately long, and with dense tufts of long pile on the dorsal surface, somewhat shorter pile on the ventral surface and its actual posterior margin.

Legs: The legs are moderately stout and rather long, the pile consisting primarily of many exceptionally minute, flat-appressed setae but with in addition some coarse, long, shaggy somewhat appressed pile and a few scaliform hairs on the ventral surfaces of the femora. Hind femur without bristles or spines. All of the tibiae with appressed spicules in several rows. They are rather long and strong on the hind tibia.

Wings: The wings are rather short, they are hyaline or with the base and costal cells tinted yellow or infuscated. There are 2 submarginal cells, the first posterior cell narrowed and closed a short distance before the margin. Third posterior cell quite rhomboidal in shape, possibly even wider than long. Anal cell widely open. The long, anterior crossvein enters the discal cell a little beyond the middle; the radial sector is short; the third vein arises proximal to the base of the discal cell. Costal comb absent. Alula narrow, without fringe, the ambient vein complete; the whole wing is somewhat narrowed toward the base.

Abdomen: The abdomen is much wider than the thorax, either broadly oval or elongate oval, narrowed and pointed at the apex, and densely covered with long, erect, fine, mostly pale pile over the rather depressed and flattened tergites which are only gently convex. Postmargins of the tergites with from 1 to 2 rows of long, slender bristles.

Hesse (1938) notes that the dististyli in the male resemble *Dischistus* Loew, sensu stricto, and are elongate, slightly curved and directed outward, more or less flattened or depressed above, the dorsal edges being slightly carinate, the apex acute. The aedeagus, he states, is well developed and elongate, without any ventral aedeagal process below. It also has a "prominent, posterior aedeagal strut on each side of the feebly developed lateral and basal struts."

Material available for study: Two species in the British Museum (Natural History).

Immature stages: Unknown.

Ecology and behavior of adults: Not on record.

Distribution: Ethiopian: *Sisyrophanus abdominalis* Bezzi, 1924; *homeyeri* Karsch, 1886; *leptocerus* Bezzi, 1912; *minor* Bezzi, 1924; *neavei* Bezzi, 1924; *ogilviei* Hesse, 1938; *pyrrhocerus* Bezzi, 1912.

Genus *Acanthogeron* Bezzi

Figures 2, 237, 467, 784

Acanthogeron Bezzi, Bull. Soc. Roy. Ent. Egypte, p. 164, 1925.
   Type of genus: *Bombylius senex* Wiedemann in Meigen, 1820, by original designation.

Bezzi (1925c), pp. 164-165, key to 5 Palaearctic species.
Séguy (1930b), pp. 90-107, key to species.
Engel (1934), pp. 272-273, key to 11 Palaearctic species.
Efflatoun (1945), pp. 254-255, key to 4 Egyptian species.

Flies of medium to large size, but unlike *Bombylius* Linné in general appearance and covered with extraordinarily dense, fine, long, erect, soft pile, which gives these flies a furry appearance. They are separated from *Bombylius* Linné by the slightly longer abdomen, by the separated eyes in the male, the separation being narrow but distinct and usually approximating the distance between the posterior ocelli. Also separated from *Bombylius* Linné by strongly thickened, first antennal segment and by the position of the anterior crossvein, which enters the discal cell well beyond the middle of the discal cell.

Hesse (1938) notes that Bezzi was unaware of the close relationship of his genus *Acanthogeron* to *Dischistus* Loew and that it differs from the true members of the genus *Dischistus* Loew, a genus which has been in much confusion because of misunderstood type designation. *Acanthogeron* Bezzi is properly separated from *Dischistus* Loew through the open first posterior cell of *Dischistus* Loew and the closed first posterior cell in *Acanthogeron* Bezzi; the latter is therefore much more like *Bombylius* Linné. The species *Acanthogeron grandis* Efflatoun would accordingly go into the genus *Dischistus* Loew. Hesse found the male terminalia of the South African *Acanthogeron hirticeps* Bezzi to be similar to *Dischistus* Loew and treated it in that genus. Because of the general fact that the open or closed first posterior cell is an important distinction in separating major groups of bee flies, I have included this species under *Acanthogeron* Bezzi. A genitalic study of all of the Palaearctic species of *Acanthogeron* and comparison with the species of *Dischistus* Loew is much needed.

Efflatoun (1945) points out that Bezzi established this genus entirely on the presence of spines on the ovipositor, not realizing that spines are present also on *Bombylius* Linné and *Systoechus* Loew. However, Efflatoun notes other well-defined differences as shown above. From *Lissomerus* Austen, which this genus resembles very strongly, both in the form of the marginal cell and the stout, thickened shape of the first antennal segment, it may be distinguished by the separated eyes in the male, the somewhat greater separation of the antennae, the longer face, the very dense oral beard and the relatively larger wings. The shape of the discal cell is acuminate at the apex like that of *Lissomerus* Austen. In *Acanthogeron* Bezzi the ocelli are flattened, the front quite wide anteriorly, bare in the middle, but with dense, long, fine, appressed mats of pile laterally along the eye margins. Moreover, in *Lissomerus* Austen there is no costal comb even in the male, and the hind femur has only sparse, fine hairs restricted more or less to the basal half, and its femur also lacks spines. It should be noted that spines may be also more or less lacking in some species of *Acanthogeron* Bezzi.

This is a Palaearctic genus with one representative in South Africa.

Length: 9 mm. to 11 mm.

Head, lateral aspect: The head is hemispherical. The face is well developed with the oragenal cup prominent. The whole of both face and genae bear a dense, erect fringe of long, rather coarse pile which completely hides the genal cup and extends laterally to the eye margin and above to the antenna, mixing with the copious, long pile of the latter and extending around the lower cheeks and the whole of the occiput. The pile becomes shorter on the occiput next to the eye margin but is everywhere erect. No scales or flattened pile present. The antenna is located near the top of the head with the front relatively short. Posterior eye margins plane, the occiput prominent, sloping inward from the eye. Proboscis 5 or 6 times as long as the head. Palpus prominent, slender, curved inward close to the base and with dense, long, coarse pile laterally and at the apex. The antennae are relatively a little shorter than in *Bombylius*. The first segment is about as long as the second and third combined and rather stout as in *Conophorus* Meigen; the thickness varies to some extent. The first segment bears conspicuous, long, dense, coarse or bristly pile dorsally and especially laterally and below. Third segment is no wider than the short, beadlike second segment, but with a rather conspicuous annular constriction at the immediate base and rather strongly tapered and narrowed on the outer half; third segment with a few setae dorsally.

Head, anterior aspect: Eyes bare and always separated in the male to a variable extent but usually approximately the distance between the posterior ocelli. The front is quite wide anteriorly and bare in the middle but bears dense, long, fine, more or less appressed mats of pile laterally along the eye margin. Oral opening long, oval and rather narrow but large. The ocelli are flattened. Ocellar tubercle low and bearing a tuft of long, coarse hairs.

Thorax: The thorax is somewhat longer than wide, and slightly convex, especially on the scutellum. Whole thorax everywhere covered with dense, long, soft pile. On the upper mesopleuron, intermixed with the pile, there are a number of coarse hairs; on the notopleuron there are 2 or 3 that might be called very slender bristles, and 10 or 12 similar bristly hairs on the postalar callosity. Efflatoun (1945) says of this genus, "the metapleuron is sometimes bare." I find in *Acanthogeron senex* Meigen both metapleuron and upper hypopleuron bare. Squama and plumula with very long, dense pile.

Legs: The legs are slender, the anterior tibia with 3 rows of short, stout spines with 6 to 12 in each row. Hind femur ventrolaterally with 5 rather longer, more prominent spines on the outer half and some bristly hairs toward the base; also there are several spines dorsally on each side near the apex; pile of hind femur is coarse and flat-appressed. Spicules of hind tibia rather longer, stouter and more prominent, situated in 4 rows; the ventrolateral and ventromedial rows contain about 10, the dorsolateral row may contain as many as 25 appressed spicules. Hind basitarsus with many short spines which are less than half as long as the hind tibia. Pulvilli present.

Wings: The wings are relatively somewhat smaller than in *Bombylius* Linné but not as small as in *Lissomerus* Austen. Marginal cell rather wide apically, the second vein recurrent to meet the costa at a right angle. First posterior cell closed and stalked. Upper anterior intercalary vein short and of about the same length as *Lissomerus* Austen and much shorter than *Dischistus* Loew. The anterior crossvein enters the discal cell well beyond the middle. In both wings in the *Acanthogeron senex* Meigen material before me there is a spur vein back from the anterior branch of the third vein arising a short distance from the base of this branch. Anal cell widely open. Costa with a prominent, matted comb at base. Two submarginal and 4 posterior cells present. Ambient vein complete.

Abdomen: The abdomen is broad and short oval and somewhat depressed; it is everywhere covered with dense, erect, fine, long pile which continues down over the sides of the abdomen and the pile is long but somewhat less dense on the sternites. Male genitalia small, the slightly protuberant, adjacent lobes well concealed by a dense fringe of long, fine pile.

Material available for study: The series representing this genus in the collections of the British Museum (Natural History).

Immature stages: Unknown.

Ecology and behavior of adults: Efflatoun (1945) writes of the species *separatus* Becker that it is abundant in some of the Egyptian deserts from February 25 to the middle of April, and that the males when poised apparently motionless in midair glitter like a ball of silver; he states that they can easily be detected by their shadow or their shrill hum, and easily dodge any ordinary stroke of the net. He took females hovering in front of low-lying blossoms and captured some upon mauve flowered sweet stock, *Matthiola humilis.*

Distribution: Palaearctic: *Acanthogeron albatus* Seguy, 1934; *auripilus* Seguy, 1930; *biroi* (Becker), 1906 (as *Bombylius*); *blanchei* Efflatoun, 1945; *dayas* Seguy, 1934; *efflatounbeyi* François, 1961 [=*auripilus* Efflatoun, 1945, not Séguy]; *grandis* Efflatoun, 1945; *maroccanus* Seguy, 1930; *senex* (Wiedemann in Meigen), 1820 (as *Bombylius*), *senex deses* (Meigen), 1838 (as *Bombylius*), *senex violaceipes* (Strobl), 1909 (as *Bombylius*); *separatus* Becker, 1906 (as *Bombylius*) [=*perniveus* Bezzi, 1925, =*talboti* Séguy, 1930]; *syriacus* Villeneuve, 1912 (as *Bombylius*); *tripolitanus* Paramonov, 1940; *versicolor* (Macquart), 1840, not Fabricius, 1840 (as *Bombylius*) [=*mittrei* Séguy, 1930].

Ethiopian: *Acanthogeron hirticeps* Bezzi, 1921.

Genus *Lissomerus* Austen

FIGURES 13, 238, 435, 448, 781

Lissomerus Austen, The Bombyliidae of Palestine, p. 33, 1937. Type of genus: *Lissomerus niveicomatus* Austen, 1937, by original designation.

Medium size, blackish flies but with dense, fine, long, erect, pale pile. Very closely related to both *Acanthogeron* Bezzi, and to *Bombylius* Linné. Paramonov (1941) considered it nearest to *Bombylius* Linné, of which perhaps it is only a subgenus. From *Bombylius* Linné it is readily distinguished by the complete lack of bristles beneath the hind femur; *Acanthogeron* Bezzi is intermediate in this respect having bristles, usually slender, confined to an area near the apex of the femur. In *Lissomerus* Austen the antennae are nearly adjacent; in *Acanthogeron* Bezzi there is a distinct and appreciable gap between them at the base.

There are some minor venational distinctions. The anterior crossvein is placed a little beyond the middle of the discal cell; the upper part of the anterior intercalary vein is very greatly shortened; the veins closing the first posterior cell are plane, not curved. Most of these venational characteristics are included in the very wide range of venational differences within the genus *Bombylius* Linné.

Only the type-species known.

Length: 9 to 10 mm.

Head, lateral aspect: The head is comparatively short, the face only slightly produced beyond the eye margin. While the occiput is rather prominent and thick, the posterior eye margin is entire and straight, the eye considerably narrowed below with the posteroventral angle subrectangular. The face is slightly produced above and recedes gradually to the cheeks, which do not extend above the eye. The male front is quite small and short, obtusely triangular, bare and pollinose only in the middle with a band of dense, subappressed, short, fine, black pile along the eye margins. Outer portion of upper face with a curving band containing scattered, quite fine and quite long, erect, black, bristly hairs. Inner portion of face adjacent to the oragenal cup, and bordered by a dense fringe of fine, long, erect, white hairs, which are continued below and widened to cover the entire cheeks and ventral portion of the occiput. The occiput itself bears abundant, rather long, coarse, white hairs centrally which change to finer and shorter hairs along the eye margins. There is a shallow crease with narrow fissure in the middle of the occiput dorsally behind the ocellar tubercle. The ocelli are large, occupying an equilateral triangle; the ocellar tubercle is low.

The proboscis is elongate and slender, 4 or 5 times the length of the head. What may be palpi may be represented by tiny, flat, obliquely truncate protuberances on each side at the base of the proboscis and which may also be a portion of the basal sheath. The palpi are not described by Austen (1937). The antennae are situated at the upper fourth of the head and elongate, generally similar to *Acanthogeron* Bezzi. The first segment is 5 or 6 times as long as wide and characterized by rather dense, long, bristly pile dorsally, laterally and below, only a little less conspicuous than in *Acanthogeron*. The second segment is small, short and beadlike, the third segment is as long as the first and more narrow, capitate or annulate at the base, a little widened on the basal half dorsoventrally, and gently tapered on the outer half and bears at the apex a minute, conical spine.

Head, anterior aspect: The head is a little wider than the thorax, from frontal aspect rather transverse below and rather arched above. The genofacial opening is quite wide, not quite half the head width and distinctly divergent below. Oral opening broadly oval. The oragenal cup is rather short but more prominent shortly below the antennae, from which point it extends obliquely downward and is produced a distance about equal to that of the face; it recedes below and is quite short opposite the proboscis. Front and face separated by a line but no depression. Upper eye facets are slightly enlarged, the eyes meet in front of the anterior ocellus and are quite narrowly separated for the upper half of what properly is the front. Ocellar tubercle bears 4 or 5 slender bristles and as many bristly hairs.

Thorax: The thorax is densely covered with fine, delicate, erect, shining white pile, especially long along the anterior margin and upon the sides of the mesonotum, the whole disc and the margin of the scutellum; the marginal scutellar hairs are slightly more coarse. Mesopleuron everywhere with a dense, more or less matted and depressed and more or less radiating tuft of long, fine, shining, white hairs. Sternopleuron with similar shorter hairs. Posterior hypopleuron with a tuft of long, fine, white pile. Metapleuron and remainder of the pleuron bare. Squama well developed with a long, dense fringe of fine pile.

Legs: The legs are slender and rather long, the middle and anterior femora at least twice as wide at base as apex, hind femur quite slender, all of them with some tufts of dense, extremely fine, long, white pile, partly appressed, partly erect. All the pile of the hind femur appressed, scaliform outwardly but with a fringe of long, fine hairs which are erect from the base to the middle and situated on the ventral surface. Femora without bristles or spicules below. All the tibiae with conspicuous spicules in 3 or 4 rows.

Wings: The wings are hyaline, marginal cell widened and gently widened apically; 2 submarginal cells; first posterior cell closed with a long stalk; anterior crossvein rectangular, which enters at the outer third of the discal cell. Medial crossvein is 5 or 6 times as long as the upper anterior intercalary vein. Anal cell widely open, alula quite large, without fringe. Ambient vein complete and the wings minutely villose. The male costa lacks a setiferous comb.

Abdomen: The abdomen is elongate, subtriangular, but little convex anteriorly, more compressed and cylindrical posteriorly. The tergites and sternites very densely covered with long, erect, fine, silky, white, conspicuous pile. There are tufts of very short, dense pile on each side underneath the scutellum.

Material available for study: The unique type collected at Jerusalem by P. A. Buxton and now in the British Museum (Natural History).

Immature stages: Unknown.

Ecology and behavior of adults: Not on record.

Distribution: Palaearctic: *Lissomerus niveicomatus* Austen, 1937.

Genus *Bombylodes* Paramonov

FIGURES 5, 262, 438

Bombylodes Paramonov, Faune de l'U.S.S.R., no. 25, pp. 65, 339, 1940b. Type of genus: *Dischistus multisetosus* Loew, 1857, by original designation.

Paramonov (1924), pp. 65-66, key to 4 Palaearctic species.

Large, stout flies with dense, long, furry or plushlike pile upon thorax and abdomen much as in many species of *Acanthogeron* Bezzi and *Bombylius* Linné. Body form stout and "plump"; abdomen wider than thorax. Hind femur with numerous distinct bristles. Paramonov (1941) states that *Bombylodes* Paramonov differs also from *Dischistus* Loew in the position of the anterior crossvein, which vein is placed at the end of the second third of the discal cell in *Bombylodes* Paramonov and is situated in *Dischistus* Loew before or after the middle of the discal cell. However, this does not hold good for all Palaearctic species formerly placed in *Dischistus* Loew, as for example *Dischistus trigonus* Bezzi, as figured by Paramonov.

Moreover Paramonov's 1941 conception of the genus *Dischistus* Loew must give way to the real *Dischistus* Loew, which Hesse has made very clear is based upon the South African species *mystax* Wiedemann. It is clear that the wing of *Bombylodes* Paramonov is very much like the true *Dischistus* Loew; both have the anterior crossvein placed far beyond the middle of the discal cell; both have the discal cell itself narrowed and the first posterior cell narrowed and spoutlike. Both have the cheeks wide, widest in *Bombylodes* Paramonov; this genus appears distinct from *Dischistus* Loew in its spiny hind femur and simple antennae; the femur of *Dischistus* Loew in the strict sense is unspined, the first antennal segment incrassate.

Apparently Efflatoun was unaware of Paramonov's genus *Bombylodes*, or of Becker's designation of *mystax* Wiedemann as type-species of *Dischistus*, for he included this common species *multisetosus* Loew of the Egyptian coast in *Dischistus* Loew.

Paramonov placed 3 other species besides the type-species in his genus. The species *giganteus* Villeneuve and *eximius* Becker he declared unknown to him; *asiaticus* Becker he stated might prove a variety of *multisetosus* Loew; he gave a very fine whole figure of *asiaticus* Becker, a more slender species. Engel (1934) placed both *eximius* Becker and *giganteus* Villeneuve in synonymy under *multisetosus* Loew, and placed *asiaticus* Becker in *Sparnopolius* where it certainly does not belong.

My own observations have been taken from material of *multisetosus* Loew and *eximius* Becker in the British Museum (Natural History).

The range of these species is in the southern Palaearctic; the type-species is from North Africa; *asiaticus* Becker is from Persia.

Length: 10 to 13 mm.; wing, 10 to 13 mm.

Head, lateral aspect: The head is hemispherical with both face and cheeks prominent. The occiput is extended backward through the middle of the head almost as far as the face extends forward; it slopes rapidly upward as it approaches the vertex but is less narrowed below. Occipital pile very fine and dense but not longer than the first antennal segment. The antennae are set near the upper fifth of the head and in consequence the face is extensive vertically below the antennae but bluntly rounded and only subtriangular; whereas many of the Palaearctic species formerly placed in *Dischistus* Loew have the face sharply and triangularly produced and oblique. Oragenal cup confined to the lower half of the head. Pile of face long, fine, dense, and no where longer than the third antennal segment. The eye is large and hemicircular, its posterior margin not quite plane. The proboscis is very slender, very long, nearly if not quite 4 times as long as the head. Palpus slender and elongate, with a few fine setate hairs apically. Front flattened with pile similar to that of the occiput and face. The antennae are relatively small, with the first segment stout but not conspicuously enlarged as in

Acanthogeron Bezzi; it is about 4 times as long as the second segment which is not as long as wide. The first segment bears numerous, fine, long bristles especially above and below. Third segment only a little longer than combined length of the first two segments; the basal two-thirds is swollen, at its middle wider than the second segment; the outer third or more is tapered to a thick, bristle-tipped style; there is no microsegment. Second segment at most with setae.

Head, anterior aspect: Head a little wider than thorax. The eyes in the male are holoptic. The antennae are almost adjacent and only very narrowly separated. Ocellar tubercle distinct; ocelli large. The oral recess is twice as long as wide, confined to the lower half of the head and its width slightly greater below; it is rather shallow. The genal area is very wide, as wide or wider on each side as the width of the oral recess.

Thorax: Subquadrate, gently convex and densely covered with fine, erect, soft pile including scutellum and nearly all of pleuron. The thoracic pile is not or scarcely longer than the pile on the back of the head. Bristles absent. Metapleuron pilose.

Legs: The legs are comparatively slender. The hind femur ventrally has a considerable series of stiff spinous bristles in two or more rows. Claws and pulvilli both well developed. Anterior tibiae without spines.

Wings: The wings are hyaline, long and narrow and the base and anterior border tinged with deep brown. The discal cell is comparatively narrow and distinctly narrowed at its apex due to the marked shortening of the upper anterior intercalary vein; medial crossvein very long, nearly straight, almost paralleling the wing margin. The first posterior cell is narrowed characteristically in the same fashion that *Efflatounia aegyptiaca* Bezzi is narrowed; actually the wing of this species and *multisetosus* Loew are remarkably similar, the basic difference between the two genera lying in the excavation or indentation in the post eye margin of *Efflatounia* and its allies. In *Bombylodes multisetosus* Loew the anal cell is narrowed rather abruptly near the wing margin, but the cell is open by nearly the width of the costal cell. The alula is well developed, and there is a narrow basal costal comb without hook. Near the base of the anterior branch of the third vein there is according to Efflatoun (1945) an occasional spur vein directed backward.

Abdomen: The abdomen is elongate-conical with soft furry pile of long, erect character in the male; the female has dense pile which in the type-species is of two lengths and two colors, the whitish pile forming transverse stripes is much shorter than the longer, coarse, bristlelike black hairs which are really erect, slender, black bristles. Efflatoun says the female ovipositor apparently lacks spines; he figures the male genitalia in situ, the dististyli symmetrical, placed vertically side by side, directed upward, large in size.

Material available for study: The series of species in the British Museum (Natural History).

Immature stages: Unknown.

Ecology and behavior of adults: I have seen no comments on the behavior of these flies which presumably are similar to *Bombylius* Linné or *Systoechus* Loew.

Distribution: Palaearctic: *Bombylodes asiaticus* (Becker), 1913 (as *Dischistus*); *eximius* (Becker), 1906 (as *Dischistus*); *multisetosus* (Loew), 1857 (as *Dischistus*).

Genus *Isocnemus* Bezzi

FIGURES 246, 441, 445

Isocnemus Bezzi, The Bombyliidae of the Ethiopian Region, p. 101, 1924. Type of genus: *Isocnemus nemestrinus* Bezzi, 1924, by original designation.

A fly of medium size which Bezzi (1924) described as having a superficial resemblance to a nemestrinid in general aspect and coloration. This peculiar fly was described by Bezzi (1924) from a unique holoptic male from Abyssinia; he placed it within the Usiinae where it obviously does not belong and failed to include the genus in his key to Ethiopian genera. I have redescribed and illustrated the unique type in the British Museum (Natural History).

The venation of *Isocnemus* Bezzi is almost identical with any one of several genera, as *Lissomerus* Austen, *Acanthogeron* Bezzi, and still more like *Sisyrophanus* Karsch. The profiles of the head and face are similar to both *Acanthogeron* Bezzi and *Sisyrophanus* Karsch. It is undoubtedly most closely related to the latter genus, and it differs from both of them primarily in the flattened occiput, which is very short in profile and especially reduced toward the vertex; it may be considered to be a very aberrant relative of *Sisyrophanus* Karsch. From *Acanthogeron* Bezzi it is also distinguished by the absence of the long, slender style attached to the long, slender third antennal segment. The stout basal tooth of the claws and minute pulvilli are distinctive.

Other characters of *Isocnemus* Bezzi, besides those mentioned, consist of the rather flattened, stout, robust, short oval, somewhat inflated, mostly reddish abdomen with blackish base and black basal tergal bands. There are no bristles on abdomen, virtually none upon the rather short, stout legs. Basal comb of wing virtually indistinguishable. The slender proboscis is 3 or more times as long as the head; the slender hairy palpus not as long as the antenna; third antennal segment with short, small, pointed microsegment. Legs comparatively short and stout, red with black coxae, and contrary to Bezzi's statement, with weak bristles below.

The face is very short with the oragenal opening extended to the base of the antenna. The occiput not prominent, comparatively flat centrally, not completely undeveloped, as Bezzi asserts.

The shining abdomen is flattened, oval, a little narrowed at apex and has for a bee fly an odd pattern of red and black; first segment black, second to fifth segments red with black anterior border dilated in the middle to form a subquadrate spot; sides of tergites also black; sternites yellow. The pile of both abdomen and mesonotum consists of rather sparse, comparatively long, yellowish-gray hairs, denser along hind borders and sides of the abdominal segments.

The genus bears no relationship to the Usiinae, as I have shown elsewhere that the spiculate character of the tibiae can be variable and deceptive. It is very closely related to *Sisyrophanus* Karsch as an inspection of the figures shows. Indeed, the only two distinctions I find concern the character of the antenna, *Isocnemus* Bezzi having a microsegment and lacking the curious hook at the apex of the third segment, whereas, the reverse is true in *Sisyrophanus* Karsch; the occiput and the pile of the head and antenna are all much more extensively developed in *Sisyrophanus* Karsh. The shape of the eye differs. Both genera have very long proboscis.

This genus is monotypic and based upon a single male from Abyssinia.

Length: 9.5 mm.; wing expanse 21 mm.

Head, lateral aspect: Face short, extended rather abruptly for a short distance forward from the base of the antennae and deeply creased rectangularly along the sides leaving the lateral margins of the face plane with the antennae. The eye is large, strongly convex anteriorly, longest above the middle and with a tendency to be flattened somewhat obliquely on the dorsal half. Male eye holoptic for nearly twice the length of the front. The occiput is thick ventrally and in the middle disappearing upon the upper fifth. Pile of occiput short, dense, and fine. Proboscis exceptionally long and slender, directed forward and upward and 4 times the head length; the blades at the apex are thin and laterally compressed. Palpus quite long and slender, cylindrical, very slightly clavate apically, with a narrow ventral fringe of numerous stiff bristly hairs; the basal segment is short and slightly more robust; the subapex has 2 or 3 short bristles. Antennae attached at the upper third of the head, the first segment is 3 times as long as the second; the third segment is as long as the first two combined, as wide on the basal third as on the second segment, strongly attenuate apically with a short, stout attenuate microsegment carrying an apical spine. The first segment has numerous moderately long, slender, bristly hairs laterally and ventrolaterally. Both second segments, both above and below, have short setae. Posterior margin of the eye without indentation.

Head, anterior aspect: The lower portion of the head anteriorly is broad and nearly plane; the cheeks are slightly produced below, less wide and broadly rounded dorsally. The subepistomal area is long, narrow and deeply concave, oblique and bare. Face pubescent laterally with a long, subocular fringe of coarse, pale pile, longest immediately below the antenna; this pile grows shorter below and is continued to the bottom of the eyes, but not quite to the posterior corners. The front is small, forming an equilateral triangle; it is pubescent laterally with a tuft of short, stiff hair in front of each antenna. Vertical triangle nearly equilateral, rising above the eye and with a tuft of moderately long, fine, erect pile. The anterior eye facets are slightly enlarged and convex.

Thorax: Pile of mesonotum dense, fine and erect, somewhat longer laterally. The postalar callosity has numerous, long, rather matted, bristly hairs which are more isolated posteriorly. Humerus with long pile; scutellum large, wide, more or less flattened across the disc, swollen toward the base; the disc is densely long pilose, with equally numerous, bristly hairs on the margin. Pleuron with long dense pile and with especially long, matted tufts directed backward along the posterior mesopleuron and sternopleuron. There is a wide, vertical band of long, matted pile on the metapleuron. Squamae with long, matted pile. Lateral metasternum pilose. Ventral metasternum narrow with a long, posterior, matted fringe of pile nearly overlapping the first sternite.

Legs: Slender, the posterior femora and posterior tibiae with some long appressed pile dorsally and laterally. Hind tibiae with 2 rows of rather long, slender, subappressed bristles medially, also a row dorsolaterally, 3 or 4 ventrolaterally, located chiefly on the basal half, and 6 ventrally. Middle tibiae with a posterior, and a posteroventral row, an anterodorsal, and an anterior row, and an anteroventral row of 5 bristles. Anterior tibiae with the anterior bristles, as well as the anterodorsal and posterodorsal rows of bristles, quite short. It has a posterior row and 7 or 8 longer posteroventrals. Tarsi with the basitarsus quite long and slender, half as long as the tibiae. Claws slender, sharp, curved from the base with a basal tooth and with minute pulvilli and a still smaller triangular medial lappet.

Wings: Ambient vein complete, base of the costa simple, the comb very small.

The wings are rather slender and short. Marginal cell progressively widened to apex where it is rounded, the second vein meeting the costa at nearly a right angle or recurrent. The anterior branch of the third vein arises rather far out toward apex, arises abruptly, then curves gently outward, then sharply upward. First posterior cell closed and stalked.

Abdomen: Broad, moderately swollen. There are 7 tergites in the male; the last is one-fourth as long as the sixth, and there is a short rounded lobe of the eighth tergite laterally. The pile of the abdomen is short, shining, rather fine, curled and appressed and comparatively scanty with additional scanty, fine, long, erect, delicate pile intermixed which becomes considerably more abundant on the first and second tergites. The first tergite laterally has long, dense, erect, coarse pile. The lateral margins have longer pile which tends to form distinct, somewhat matted tufts in the posterior corners, but these are not conspicous as such. The first sternite is apilose.

Material studied: I studied the unique type at the British Museum (Natural History). I believe this is the only known individual.

Immature stages: Unknown.

Ecology and behavior of adults: Not on record.

Distribution: Ethiopian: *Isocnemus nemestrinus* Bezzi, 1924.

Genus *Zinnomyia* Hesse

FIGURES 3, 242, 449, 463

Zinnomyia Hesse, In Hanström, Brinck, and Rudebeck: South African Animal Life, vol. 2, p. 387, 1955. Type of genus: *Zinnomyia karooensis* Hesse, 1955, by original designation.

Peculiar flies of considerable size which resemble some species of *Bombylius* Linné in several respects. The venation is similar, the anterior mesonotum is humped, and the small head is much narrower than the thorax, snugly attached to it, and sets low upon the thorax. The abdomen is considerably wider than the thorax, convex, short, with only 5 tergites visible from above; the fifth is scarcely noticeable because it is turned downward. The abdomen is very broad, shining black, with dense, quite short, erect, black, setate pile upon the second, third, and fourth tergites. The hind margins of these tergites each bear a long, dense flat-appressed fringe of opaque, white hair, overlapping the next tergite.

The principal distinguishing characteristics of this genus, apart from the unusual appearance, consist of the distinct, long, slender, accessory claws on each tarsus. They are especially long on the hind tarsi. Also, the pulvilli are entirely lacking.

The pile of the face and gena is short and quite similar to that of *Bombylius punctatus* Fabricius. The third segment of the antenna is leaf-shaped, pointed at only the immediate apex, but typically the apex is drawn out into a long style composing a third or more of the total length. The length of the third segment is not quite equal the combined length of the first two segments; the base of the third antennal segment is sharply annulate.

Hesse notes the general similarity between this species and *Bombylius analis* Fabricius.

The proboscis is quite long and slender.

Hesse calls attention to the remarkable resemblance of this fly to some species of the tabanid genera *Diatomineura* Rondani or *Pangonia* Latreille. It also resembles some species of nemestrinids such as *Fallenia* Meigen.

This unique fly is found only in the Kaokoveld District and the Koup Karoo area of South Africa.

Length: 10.5 to 14.5 mm.; wing 10.5 to 15 mm.

Head, lateral aspect: Head short and hemicircular; the face extends forward beneath the antenna a distance equal the distance of the first antennal segment. The front is covered with a dense tuft of curled, flat-appressed, white hair extending to below the antenna. Still longer white hair extends over the whole face, curling over the buccal cavity in an overlapping tuft and continuing with shorter hair along the sides of the gena to the bottom of the eye. The oragenal cavity is almost vertically truncate. The occiput is comparatively short, hidden, and obscured by dense, opaque white pile; this pile is very short adjacent to the eye margin; it is a little longer behind. The posterior pile of the occiput, however, is a little longer than the anterior pronotal pile, and leaves a crease behind it. The palpus is minute, very slender, with only a few, fine hairs; it is of uniform thickness, has a minute apical segment and barely extends beyond the oral cup. The proboscis is quite long and slender and more than 5 times as long as the head and thrust straight forward. The black antennae are attached at the upper fifth of the head and are held obliquely upward; the first segment is more than twice as long as the second. The second segment is of the same width as the first and slightly longer than wide. The third segment is sharply annulate. It is widened through the middle, much flattened laterally. Ventrally the third segment is convex on the basal half, but it is also convex throughout the entire dorsal length, and typically the third segment ends in a slender, nonarticulate, spine-tipped style, its length a third, or sometimes more, of the entire length of the segment.

Head, anterior aspect: Viewed anteriorly the head has a slight rectangular appearance. The oragenal space is about one-third the total head width. The eyes are separated in the male by the width of the ocellar tubercle, which is quite far to the rear, so that the posterior ocelli actually set behind the eye itself. Ocelli tubercle low and set in an obtuse triangle and all of the ocelli small. The front is depressed. The head is much narrower than the thorax. The oral opening has parallel sides but above the proboscis all of the lower wall is much widened and inflated and bears fine, oblique, grooves or scoring. The antennae are narrowly separated at the base.

Thorax: The thorax is short and robust, distinctly elevated anteriorly. The mesonotum is shining black, completely obscured by the dense pile anteriorly, but the ground color is shining through across the middle and behind. There are no bristles on the notopleuron or at the base of the wing, and at most there are 2 or 3 pairs of slender bristles and some additional bristly hairs, which are rather numerous on the postalar callosity. The scutellum is large and rather convex posteriorly, hemicircular, with the base also narrowly circular, and with a rather deep crease behind the thorax. Its pile is very dense, and erect over the entire scutellum, brownish-orange in *karooënsis* Hesse. There are no bristles on the scutellar margin. The mesopleuron and propleuron are very densely covered with opaque, subappressed, and partly tufted, white or brownish-white pile merging into the pile of the occiput. The pteropleuron is also densely pilose and there is a conspicuous, vertical fringe of opaque, yellowish or whitish pile on the metapleuron in front of the haltere but, however, with a narrow visible bare space between. The metanotum appears to be bare, but the abdomen is so tightly joined to the thorax that I cannot verify this without detaching it. The hypopleuron has a dense, conspicuous tuft of pile beneath the spiracle. The squama bears a conspicuous fringe of pile. The alary squama is large and blackish, much as is the base of the wing excluding the alula. The sternopleuron also has bushy pile.

Legs: The legs are slender, and comparatively long, especially the hind pair, orange and yellowish, with the knees narrowly blackish; the femora may be blackish basally. The middle femur tends to be a little thickened toward the base. This is also true of the anterior femur, rather more so in males, and the front femur has a few short bristles in *brincki* Hesse. Hind femur with 10 to 14 short, oblique bristles or spiny setae in *karooënsis* Hesse. The hind tibia has about the same number of short bristles ventrolaterally and 2 rows of still shorter bristles dorsally. Anterior tibia with 4 rows of fine, short, opaque bristles. The pile of the legs, in addition to the bristles, consists of flat-appressed, rather dense pale, flattened hairs. The tarsi are elongated and slender with stout, short, oblique bristles placed ventrally. All of the claws bear an additional conspicuous basal tooth longer on the hind tarsi, but conspicuous on all of the legs, and this tooth is somewhat flattened. The pulvilli are absent. There appears to be a minute, stublike trace of arolium between the claws.

Wings: Wings are hyaline, long and quite slender, at least 3½ times as long as wide. The base of the wing before the basal cells is dark sepia, and there is a diffused small cloud of brown where the third vein arises, and this vein arises with a wide, knotlike swelling; radial sector quite short. The anterior crossvein enters the discal cell at the middle. The anterior margin of the discal cell is convex, the vein toward the apex pulled backward, so that the intercalary vein is short. The first posterior cell is closed with a quite long stalk. The anal cell is widely open. Although the wing is narrow at the base, the axillary lobe is well developed and the alula is wide. The basal comb is wide but bears soft, appressed hairs, and the preceding hook is triangular and has a minute apical spur.

Abdomen: The abdomen is extraordinarily broad and quite short, with a bare appearance due to the shiny black ground color, which shines through the dense, short, erect, black pile; it is only obscured by the postmarginal bands of dense, opaque, white, flat-appressed pile, extending over onto the succeeding tergites behind the second, third, and fourth tergites. The pile of the first tergite is entirely erect, coarse, very dense, short and yellow beneath the scutellum, longer and brownish white laterally, but the character of the pile remains the same. All sternites have dense, rather long, white pile. From the dorsal aspect, only 5 tergites are visible, the second, third, and fourth are equally long; the fifth is smaller, narrower, and from above it is only barely visible apically in the middle of the abdomen.

Hesse notes that the male hypopygium resembles that of the species *Bombylius* Linné.

The dististyle, or beaked-apical joint, is shaped very much as in many species of *Bombylius* Linné, with the aedeagal apparatus and the aedeagus itself well developed. He notes further that the inner apical and basal parts of the shell-like clasper are produced. The aedeagus is projecting, is long, and armed with a minute, recurved, hook-like process near the apex; there is no ventral aedeagal process. . . . that the posterior, projecting, flattened structure of the middle part of the aedeagal apparatus is well developed.

The male terminalia are so well concealed and recessed below in the curved-under tergites of the abdomen that I temporarily mistook my material for the female sex.

Material available for study: I have studied a male paratype of *karooënsis* Hesse sent me for study and dissection through the generosity of Dr. A. J. Hesse of the South African Museum at Capetown.

Immature stages: Unknown.

Ecology and behavior of adults: Not on record.

Distribution: Ethiopian: *Zinnomyia brincki* Hesse, 1955; *karooënsis* Hesse, 1955.

Genus *Parabombylius* Williston

FIGURES 9, 272, 456, 732, 793, 794

Parabombylius Williston, Journ. New York Ent. Soc., vol. 15, p. 1, 1907. Type of genus: *Thlipsogaster coquilletti* Williston, 1899, as *ater* Coquillett, 1894, preoccupied in *Bombylius* by Scopoli, 1763.

Painter (1926a), pp. 74-75, key to 8 North American species.
Curran (1930), p. 7, key to 5 North American species.

Small, obtuse flies greatly resembling a small species of *Bombylius* Linné to which they are closely related. In some species the abdominal pile is furry and long and dense; in others it is sparse and straggly. Some species are golden, or golden red pilose; others black pilose. Many species have patches of silver scaliform pile on thorax and abdomen. They are readily recognized in all forms by the small patch of silvery pile in both sexes; this patch varies from 2 or 3 hairs to an extensive patch; moreover, the greatly narrowed first posterior cell is closed in the margin; the anterior crossvein lies before the middle of the discal cell and the anal cell is open, the alula large, and the proboscis long. The whole face and at least upper half of gena bear long, dense, fine pile. The moderately lengthened antenna has the third segment rather conspicuously dilated in the middle; apex with a small bristle-tipped microsegment.

Parabombylius Williston is restricted to the southwestern part of the Nearctic, primarily in desert country, except that 1 or 2 species range into high rainfall areas in Arkansas, Louisiana, and Mississippi. A long series of *Parabombylius coquilletti* Williston was collected by the author in Oktibbeha County, Mississippi, and it also occurs in Lafayette County, Mississippi. Other species range into northern Mexico.

Length: 5 mm. to 8 mm. antennae included; wing 5 mm. to 7 mm.

Head, lateral aspect: The head is as long as high, the face moderately extended, in height equal the length of the first antennal segment. The oragenal cup retreats and in consequence is shorter below. The pile of the face is long, fine, dense, and erect and continuous from one side to the other, the underlying surface is pollinose. In both males and females, along the eye margin opposite the antennae where front and face meet, there is a patch of bright, silvery scales which is sometimes small, sometimes broad and extensive. The long, fine pile continues on down the sides of the oragenal cup. The occiput is shallow, sloping gradually out to the eye margin and densely covered with short, coarse pile forming a fringe that extends only a short distance above the eyes. Sometimes this pile is replaced by stiff bristles, and in some species in the females there may be a supplementary patch of dense, appressed, silvery scales on the eye at the middle of the occiput. Curiously the pollinose ventral occiput between the eyes is almost lacking in pile. The eye is hemicircular with the posterior margin forming a very broad, shallow curve over much of the middle portion.

Proboscis slender about twice as long as the head, the palpus short and slender with a few long setae. The front in males is comparatively small but wide and the eye and the front are sunken where they meet. The medial facets of males are considerably enlarged. Male front with numerous, long, fine, erect hairs; female front transversely depressed in front of the antennae and this portion has only appressed scaliform pile. The upper, outer corners of the front sometimes with some erect pile. The antennae are placed at the upper fifth of the head; they are nearly or quite as long as the head. The first segment is 3 or 4 times as long as wide, the second segment is a little longer than wide and widest apically. The third segment is slender, spindle-shaped, widest in the middle, narrowed to a varying extent on the basal half, but more strongly narrowed from just beyond the middle to the apex. It bears 1 or 2 microsegments, and has a thick, apical spine. In the type-species the antenna, while of the same pattern, is somewhat shorter and is distinctly shorter than the head; the first antennal segment bears much long, bristly pile above and below. In some species this pile is longer below. In *Parabombylius albopenicillatus* Bigot it is longer and consists partly of bristles. In this species the third segment has several long, fine hairs dorsally, others below at the apex.

Head, anterior aspect: The head is oval, about $1\frac{2}{3}$ times as wide as high; the head is not as wide as the thorax, the eyes in the male meet for a distance longer than the front, the ocellar tubercle is a little protuberant and bears a tuft of long, fine, bristly hairs centrally and also a row behind. The large ocelli lie on the eye margin and are set in an equilateral triangle. The oral opening is large, rectangular but rounded at each corner and extends far back under the eye. Its outer wall is comparatively short, the inner wall characteristically thickened much as in *Bombylius* Linné and the oral recess is deep. The gena is reduced to a mere line along the eyes, the tentorial fissure narrow but deep.

Thorax: The thorax is comparatively elongate, shallowly convex, the head and abdomen slightly decumbent, opaque black in ground color, in males covered with very dense, rather long, plushlike, shining pile, golden red or yellowish white. The pile is dense on the humerus, postalar callosity, somewhat less so on the posterior half of the mesonotum and the disc of the scutellum. Both these latter areas sometimes contain additional fine, appressed, glittering pile, nonscaliform. Also, in some species over the middle of the mesonotum in front of the wing there are 2 patches of white, flat-appressed scales, another medially in front of the scutellum. Bristles are absent. Whole mesopleuron with dense, long, coarse, appressed pile. In other species the mesopleural pile is restricted to a few black bristly hairs, and a border of white pile above. In every case the metapleuron has a fanlike band of long hair, and behind the spiracle there is a tuft of hairs which in some species forms a second fanlike band. Pteropleuron with or without some erect hair. Hypopleuron always bare. Squama with a long fringe.

Legs: The legs are comparatively long and slender. Anterior femur with a row of 8 minute spicules posterodorsally, and 4 or 5 below. Middle tibia spicules equally minute. Hind tibia with 4 rows of somewhat longer, slender spicules, 8 in a row, and with 5 stouter spicules anteroventrally. Hind femur with 4 stout or 7 or 8 more slender bristles below. Claws small, fine, sharp. The pulvilli are slender, about two-thirds as long as the claws.

Wings: The wings are rather large, hyaline, or slightly tinged with brown at the base. Two submarginal cells, the first posterior cell is closed in the margin, or has a very short stalk. Anterior crossvein lies at or just beyond the basal third of the discal cell. Intercalary vein short and anal cell open. Axillary lobe moderately large, the alula quite large. Ambient vein complete. Costal comb present but weak.

Abdomen: The abdomen short and broad, wider than the thorax, black and more or less shining, usually with some bluish or greenish reflections, in some species quite opaque and the last 4 or 5 tergites tend to be somewhat compressed triangularly. Pile usually of 2 or 3 types and often several colors. In the type-species it is almost entirely dense, erect, fine and reddish golden in the males. In females the flat-appressed, reddish-golden hairs of the mesonotum together with scattered silver spots are both continued onto the abdomen. In males of other species the dense, erect hair is pale yellow or black and there is sometimes intermixed rather dense, flat-appressed, curled golden hair on the posterior tergites. There may be a band of short, dense, opaque white pile along the anterior margin of the second segment and as many as 6 small, semiconfluent, medial spots of dense, appressed, white pile on the remaining tergites. Posterior corners of tergites sometimes with patches of dense, white pile. Male terminalia large, often twisted to one side and extruded and visible below.

The male terminalia from dorsal aspect, with epandrium removed, show a body not greatly unlike *Bombylius major* Linné. The membranous area is apparently lacking. The dististyli are larger, stouter but also convergent. In lateral aspect the epandrium is subrectangular with a spoutlike posteroventral process, short cerci, and in other respects quite similar to *Bombylius* Linné.

Material studied: Five species in my collection including the type-species.

Immature stages: Unknown.

Ecology and behavior of adults: These flies can be found hovering about over sandy ground or upon flowers which in desert areas are usually of the low-growing varieties. I have taken them upon *Coreopsis* sp. in May near the University of Mississippi campus.

Distribution: Nearctic: *Parabombylius coquilletti* (Williston), 1899 (as *Bombylius*) [=*ater* (Coquillett), 1894 (as *Thlipsogaster*; preocc. in *Bombylius* by Scopoli, 1763)]; *maculosus* Painter, 1926; *nigrofemoratus* Painter, 1939 [=*vittatus* Curran not Painter, 1926]; *pulcher* Painter, 1926; *subflavus* Painter, 1926; *syndesmus* (Coquillett), 1894 (as *Thlipsogaster*); *vittatus* Painter, 1926.

Neotropical: *Parabombylius albopenicillatus* (Bigot), 1892 (as *Bombylius*); *dolorosus* (Williston), 1901 (as *Bombylius*).

Genus *Sisyromyia* White

FIGURES 11, 252, 436, 443

Sisyromyia White, Papers Proc. Roy. Soc. Tasmania, p. 197, 1917. Type of genus: *Bombylius auratus* Walker, 1849, given by Bezzi, p. 7, 1924, the first of 2 species.

White in Hardy (1924), pp. 84-85, key to 7 Australian species. Roberts (1928), pp. 426-427, key to 6 Australian species.

Large flies, densely long pilose in which there is much matted, appressed glistening, somewhat scaliform pile especially upon thorax and abdomen. From *Bombylius* Linné or *Systoechus* Loew it is at once separated by the open though narrowed first posterior cell; from *Dischistus* Loew it is separated by the presence of at least some spinelike bristles beneath the hind femur and the presence of 2 tufts of pile on the metapleuron. In some species the femoral spines are short, oblique, and more or less hidden by the femoral pile; also there is a variable amount of pile beneath the hind femur, and it may become a conspicuous fringe of long, downturned hairs, the full length of the femur, or it may be more or less restricted to the basal part of the femur. The apex of the long, slender, third antennal segment has a distinctive and curious tuft of 4 or 5 slender bristles.

Characteristically there is a medial row of spots or continuous vitta of contrasting, bright colored, matted pile upon the abdomen. Wings spotted, or uniformly tinged with brown or hyaline with irregular, sepia, foreborder.

This is a distinctive Australian genus with 5 or 6 species. I can see no relationship with the genus *Sparnopolius* Loew as asserted by Roberts. It is closer to *Dischistus* Loew.

Length: 8 to 12 mm.; wing sometimes a little shorter, or in *decorata* Walker distinctly longer, and slender.

Head, lateral aspect: The head is more or less triangular since the face is prominently extended forward and the oragenal cup slants or recedes backward. The occiput is quite shallow in both males and females; the oragenal recess is prominent especially above, but less so below. The face bears dense, quite fine, erect, moderately long pile. This pile is a little curled downward laterally and across the middle. The face bears even more glittering, golden or coppery, slightly scaliform pile, erect or only subappressed, extending laterally over the eye margin itself and anteriorly where it is even longer over the margin of the face and the oragenal cup. All of this pile extends down the sides of the oragenal cup and the gena. The pile of the occiput is moderately bushy and dense, especially toward the center of the occiput where it tends to curl outward and be subappressed and a little shorter toward the eye margin, which it partly overlaps. Surface of occiput pale pollinose. Below the occiput has extremely long, dense, fine pile extending downward beginning at the bottom of the eye and extending across the very broad ventral part of the head. The eye is large, flattened above, the medial facets somewhat enlarged. The postmargin of the eye is entire, the proboscis is not quite twice as long as the head, robust and stout with the labella large, broad, short, and the palpus long and slender, bent upward from near the base at what appears to be the fusion of segments and its entire pile is microscopically short and covers all of the outer segment.

Front wide and very slightly swollen on each side in the female but small and triangular in the male due to the position of the antennae. The pile in the female consists of bushy tufts of very fine, scattered, erect tufts of rather long hairs and a shorter, denser mat of semi-appressed, golden or coppery, scaliform pile. In the male the front has only pile of this latter character and perhaps 2 or 3 short, bristly hairs along the eye margin. Antennae attached almost at the top of the head; they are elongate and quite slender, the first 2 segments are a little wider than the third. The first segment is flattened medially, and is 1½ to 2½ times as long as the second. The second segment barrellike, longer medially than laterally, barely longer than wide. The third segment is very slender, more than twice as long as the 2 basal segments and is knobbed at the base, very slightly widened through the middle of the outer half, bearing at the narrowed apex a spine and a tuft of 4 or 5 bristly hairs which are 3 or 4 times as long as the spine. Second segment with 2 or 3 dorsal setae. The first segment bears dorsally, laterally, and ventrally some scattered, long, quite fine, bristly, black hairs.

Head, anterior aspect: The head is very wide, as wide or wider than the thorax and more than twice as wide as high. The eyes in the male are in contact for a very short distance, the more so as they are rounded on their anteromedial corners. The antennae are narrowly separated at the base, the ocellar tubercle is moderately large and prominent, smaller in males, the ocelli large, in males not quite reaching the eye margin, and the ocellar tubercle more obtuse in females. Pile of the tubercle long, erect, and fine. The oral opening is large and wide and oval, moderately deep behind, the walls much thickened outwardly but nearly vertical interiorly. The genae are also moderately wide because of the great width of the head and below the genae shelve inward from the eye to the prominent tentorial fissure. Occiput very shallowly bilobate behind and with no indication of medial groove behind the vertex.

Thorax: The thorax is somewhat longer than wide, strongly convex from side to side, the head a little decumbent, the abdomen also a little decumbent. Mesonotum black and opaque gray pollinose, densely covered over the whole with glittering, flat-appressed, brassy or coppery pile and also some short, erect, brassy pile, which is also fairly dense especially on the whole mesonotum in front of the wings. Notopleuron with 5 bristles, postalar callosity with only coarse hairs, its posterior wall bare. The scutellum either light in color or black with very fine, erect, long, black pile and short, curled metallic hair close to the surface. Mesopleuron pilose over its whole surface, the pile somewhat semi-appressed. Sternopleuron and metapleuron also extensively pilose. Pteropleuron pollinose only. Squamal fringe moderately long and thin.

Legs: The legs are comparatively long and slender, the anterior tibia has extremely minute, pale red spicules. The middle and hind tibiae have slightly longer, slender, red spicules in 4 rows containing 10 or more spicules. The anterior and middle femora are swollen on the basal half, especially on the first pair. Posterior femur with a row of 11 ventrolateral red bristles, the pile minute, scaliform, flat-appressed and brassy or coppery. The bases of all of the femora with some fine, long, erect, brassy hairs, especially shaggy and tufted on the first four. Claws fine, curved, sharp, the pulvilli long and spatulate.

Wings: The wings are large, tinged with brown throughout, or subhyaline with pale micropubescence and the anterior half irregularly sepia. Wings moderately broad basally, a little pointed apically, 2 submarginal cells, the marginal cell gradually widened apically and the second vein making a rounded turn forward to the costa. First posterior cell open a little less than the length of the anterior crossvein, this crossvein lies at or barely beyond the basal third of the discal cell. Discal cell a little narrowed apically both from front and behind. Anal cell widely open, axillary lobe sometimes narrow, sometimes wide basally; the alula elongate and broad, or relatively narrow. There may be a spur vein backward from the anterior branch of the third vein not far from the base. Ambient vein complete.

Abdomen: The abdomen is broad, as wide or wider than the thorax. In the males it is very short and convex, in the females subtriangular. The abdomen is generally blackish in ground color with opaque pollen and pile of several kinds. There is a band of rather short, but dense, erect pile on the sides of the first tergite, the sides of the remaining tergites have extensive, curled, glittering, scaliform pile, either short and completely appressed, or longer and subappressed.

In all species before me this pile is diminished toward the middle of the tergites but leaves a dense, medial vittae of similar pile which stands out conspicuously, in addition there is some rather dense, erect pile along the anterior half of the second segment, the whole lateral margins are covered with bushy, long, erect pile, either coppery or golden, sometimes tufted and over the remainder of the dorsum bears much extraordinary long, fine, erect, black pile. I find 9 tergites in males, the last 2 small and short, but only 7 in the females. Sternites with long, shaggy, pale yellow or brassy pile. Male terminalia minute, turned to one side and exposed from below.

Material available for study: Males and females of *Sisyromyia decorata* Walker and *S. aurata* Walker.

Immature stages: Unknown.

Ecology and behavior of adults: Unknown.

Distribution: Australian: *Sisyromyia aurata* (Walker), 1849 (as *Bombylius*); *brevirostris* (Macquart), 1849 (as *Bombylius*) [=*eulabiatus* (Bigot), 1892 (as *Systoechus*)]; *decorata* (Walker), 1849 (as *Bombylius*) [=*scutellaris* (Thomson), 1869 (as *Bombylius*)].

Genus *Nectaropota* Philippi

FIGURES 20, 253, 1027, 1028, 1029

Nectaropota Philippi, Verh. zool.-bot. Ges. Wien, vol. 15, p. 679, 1865. Type of genus: *Nectaropota setigera* Philippi, 1865, by monotypy.

Small flies of no great width with moderately decumbent abdomen and head characterized by the bare aspect; the vestiture consists of fine, opaque pollen on a blackish background. The mesonotum has an obscure bluish or grayish-white vitta and sutural fascia. The abdomen is robust, rather high dorsally from a lateral aspect; 6 tergites are visible dorsally and all but the first has a postmarginal border of 10 or 12 long, slender, erect, conspicuous, black bristles or bristly hairs. The occiput is shallow; the eye almost triangular from above and narrowly separated in the male. The antennae are relatively short with the third segment short oval and a little narrowed at the apex and flattened medially. The proboscis is long and slender with slender labellum. The wing is characteristic; it is slender, with 3 submarginal cells, closed and stalked first posterior cells, a series of 10 or more diffused blackish spots on crossveins or furcations, and moreover the anal cell is very widely open.

This fly is from Santiago Province, Chile.

Length: 6 mm. excluding proboscis; length with proboscis 7.5 mm.; wing 5.5 mm.

Head, lateral aspect: The head is subglobular with the occipital component shallow. The occipital pile is bristly and sparse. There is a clump of rather small, reddish bristles on each side dorsally, strong, erect, and rather long. These are restricted to the upper fourth of the head and in the middle of the occiput there are only coarse, flattened, scalelike and highly or entirely appressed bristles which reach the edge of the eye margin. Similar bristles actually extend in front all the way to the vertex. Ventrally the occiput has a few fine hairs of no great length. While the upper occiput is shallowly bilobate, and the eye is excavated forward on the upper part resulting in a greater exposure of occiput from the dorsal aspect, there is, however, no central cavity of any kind. The posterior eye margin is approximately plane, except for being slanted or excavated forward near the top. The eyes are distinctly flattened with much enlarged facets over a small area along the middle of the front. The lower eye margin is much rounded at the bottom of the head.

The oragenal cup from the lateral aspect is distinct, extending forward obliquely for a distance about equal to the thickness of the first antennal segment. It bears a sparse fringe of long, shining-white, stiffened hairs arising along the sides of the face and extending down the narrow gena about halfway to the bottom of the head. The narrow tentorial fissure is long and distinct. The face proper beneath the antennae is of about the same extent as the oragenal cup, perhaps a little longer; it is pale yellow and pollinose; sides of oragenal cup polished and without pollen, its inner walls thickened in the middle, and the oral recess fairly deep. The entire lateral aspect of face and oral cup is plane and oblique. All of the face is pale yellow and yellow brown in ground color; all of the occiput and upper half of the front and vertex blackish. The proboscis is rather long and slender, nearly twice the length of the head with small, short, slender labellum. The palpus is small, quite slender, pale yellow and not extending beyond the oral cavity. It appears to have a short basal segment.

The antennae are attached a little above the middle of the head and are relatively short. The first segment is about 2½ times as long as wide; it is brownish yellow with minute, whitish micropubescence and 3 or 4 pale yellow bristles above. The second segment is not as long as wide. It is more or less beadlike and pale brown; the pubescence is extremely minute. There are 1 or 2 bristles below and 3 or 4 dorsolateral. The third segment is short oval and on the dorsal aspect a little convex on the basal half and plane beyond. Ventrally it is gently rounded below and to the apex. This segment is rather thick, reddish brown, in length about equal to the first two segments and at the apex it bears a dorsal pit containing a minute yellowish spine.

Head, anterior aspect: The head is wider than the thorax and the eyes relatively large. The eyes are not in contact and for a short distance are separated only by the width of the anterior ocellus. The antennae are narrowly separated but touching at the end of the first segment. The ocellar tubercle is moderately prominent and bears a clump of quite stiff blackish bristles. The ocelli lie in an equilateral triangle. The front is small and forms an equilateral triangle covered with matlike appressed pale, flattened pile. It is rather bare immediately in front of the antennae. The oral opening is short-oval and rather transverse anteriorly.

Thorax: The mesonotum is opaque and densely pollinose with an olive-brown color and an indistinct, medial, grayish stripe or vitta, a stripe lying along the transverse suture and still another one curving across the anterior mesonotum behind the light brownish-yellow humerus. The humerus bears several long, stiff, reddish bristles. There are 2 or 3 quite stout, long, reddish, notopleural bristles arising from black tubercles. There are at least 3 others on the postalar callosity equally stout and 2 or 3 somewhat shorter reddish bristles above the wings. In addition there is an acrostical row and a sublateral band of reddish bristles rather long, quite erect, almost as long as those which are placed laterally. In addition there are numerous small, very coarse, short, suberect, reddish, almost scalelike bristles arising sparsely over the surface of the mesonotum; they also arise from black spots. Scutellum olive brown pollinose with similar, sparse, subappressed, coarse, reddish, scalelike bristles, and along the margin with 3 or 4 pairs of conspicuous long, reddish bristles and also a submarginal and a discal row of similar long reddish bristles. The pleuron is mostly brownish yellow to grayish brown but densely overlaid by very fine, pale pollen. The upper border of the mesopleuron has some long yellowish bristly hairs, scanty in number, and some tufts of opaque, appressed, coarse, whitish hairs directed backward on the posterior half. The pteropleuron, metapleuron, metanotum, and hypopleuron are all bare. There are a few hairs on the upper sternopleuron.

Legs: The femora are unusually stout without, however, being conspicuously swollen. The anterior and middle pair are twice as wide basally as apically and bear dense, flat-appressed, pale, yellowish, scalelike hairs. The thickened hind femora are similar. These bear 2 or 3 fine blackish bristles at the apex dorsally, but only minute hairs ventrally. All of the rather short tibiae are similar; very minute, pale suberect setae are on the slender, short, anterior tibiae, but only 5 or 6 in each row. The setae on the hind tibiae are also quite fine, pale in color and not much larger than those on the anterior tibiae. The legs are mostly pale brownish yellow, the femora becoming a little more darker brown apically, and the tarsi which are rather short, dark brown above and in place of the scalelike hairs, they have mostly fine, appressed, bristly pile. The apices of the tarsi ventrally are excavated. The claws are rather stout, black, sharply curved at the apex and the slender pulvilli are quite long.

Wings: The wings are quite slender and also somewhat narrowed at the base with a long, narrow, axillary lobe and quite narrow alula. The ambient vein is strong and complete. The wings are rugosely wrinkled and iridescent, rather grayish hyaline due to the villi and have a rather conspicuous series of light brown more or less diffusely margined series of brown spots, especially about the crossveins, near the apex of the veins and one in the middle of the second vein. There are 12 such spots. There are 3 submarginal cells, the anterior branch of the third vein makes a conspicuous bend forward, the marginal cell is a little widened apically and obtuse. It is connected to the lower submarginal cells by a distinct, plane, crossveinlike branch representing the third branch of the radius. The anterior branch of the third vein arises obliquely and is not quite plane. The first posterior cell is closed with a long stalk. The anterior crossvein enters the discal cell barely beyond the middle. Anal cell very widely open. Fringe along the posterior margin, long, conspicuous, and fine. There is a shallow bend in the second vein nearly opposite the anterior crossvein.

Abdomen: The abdomen is semidecumbent. It is very stout leaving the abdomen subcylindrical. In length it is about equal to the mesonotum if the scutellum is included. The pygidium is very conspicuous with a long, ventral, shining, thin, pale colored, brown margined, bristle-fringed, flatlike extension of the sternite on each side. Although I have not had a specimen for dissection, these flaps may constitute an inverted epandrium. The general color of the abdomen is opaque, brownish gray, with conspicuous, diffused, brownish-yellow postmargin. The pollen is extremely thin. The pile sparse, brassy yellow, curled, quite flat-appressed, except on the base of the first segment where there is a rather conspicuous band of coarse, erect, pale, bristly pile. The posterior margin of each segment bears extremely conspicuous, erect, long, stiff, black bristles confined to a single row; they are reddish in color on the first segment and restricted to the sides. I find 7 tergites visible on one side and 8 upon the other. The pygidium is quite large with the dististyle sharply falcate.

Material available for study: A male of the unique type-species kindly lent me by Jack Hall of the University of California at Riverside. It was collected in Santiago Province, Chile, at Quebrada de la Plata by Evert Schlinger and M. Irwin on October 27 at an elevation of 510 meters.

Immature stages: Unknown.

Ecology and behavior of adults: Not on record.

Distribution: Neotropical: *Nectaropota setigera* Philippi, 1865.

Genus *Eusurbus* Roberts

FIGURES 22, 257, 439, 451

Eusurbus Roberts, Proc. Linnean Soc. New South Wales, vol. 54, p. 580, 1929. Type of genus: *Bombylius crassilabris* Macquart, 1855, by original designation.

Large, remarkably robust and stout flies with short broad abdomen, appearing wider because of the dense, flattened tufts of hair along the sides of the abdomen. The head and antennae resemble *Bombylius* Linné with minor differences. The occiput is poorly developed above but is nevertheless densely pilose. The male eyes are holoptic but the front and vertex are both small. There is dense, matted tomentum on face, occiput, anterior half of mesonotum, pleuron, sides of the abdomen, and the dorsum of abdomen, except where denuded; posteriorly there are numerous extremely fine, longer, bristly hairs on the abdomen. Notopleuron with a tuft of 10 or more bristles. Proboscis short, the bristly palpus quite short and concealed in the oral recess. There are 2 submarginal cells and all posterior cells are open, the first being narrowed apically. Shortly after it arises there is a backward spur from the anterior branch of the third vein. The anterior crossvein lies just a little beyond the middle of the wide but apically narrowed discal cell. By the open first posterior cell this genus shows some transitional relationship to *Dischistus* Loew.

This fly belongs to the Bombyliinae and not to the Usiinae where Roberts (1929) placed it. Apparently there are no representatives of the Usiinae in Australia.

Known only from the Australian region.

Length: 10 to 19 mm.

Head, lateral aspect: Head exceptionally short, the occiput of moderate extent, and extraordinarily densely covered with short, fine, plushlike pile, curled and tufted and extending plushlike into the soft pile on the anterior margin of the mesonotum. The extreme anterior margin of the occiput along the eye margin is pollinose at the bottom of the head but throughout all the middle portion has a fringe of minute, quite short, posteriorly directed, pale gray hairs, changing toward the vertex to the longer pile which is abundant everywhere. Also, on the upper third of the occiput there are numerous, fine, slender, black bristles or bristly hairs intermixed with the pile and also of the same length. Front in the male small, flattened, forming an equilateral triangle and with some fine semiappressed hair directed forward. The face is moderately protuberant, obtusely triangular and extends forward only about half the length of the first antennal segment. It is rather obscured by dense, long, fine, bristly pile and with some additional short, curled, more or less matted pile along the eye margin. The long pile continues completely across the face mingling with the long, slender, black bristles of the antenna. The pile of the face, both kinds, continues on down the sides of the gena. The eye is relatively short, of considerable depth. It is at least

2½ times as high as long and somewhat flattened above. The upper facets in the male are only slightly enlarged. Traces of a greenish reflection remain in dried material.

The proboscis is elongate but not more than twice as long as the head. It is stout but a little flattened; the labellum is elongate oval and strongly flattened. Palpus is slender and elongate but does not extend without the oral cavity. However, it bears toward the base 5 long, quite stout, finely pointed, very long bristles. The antenna are attached at the upper fourth of the head. They are comparatively large and elongate. The first segment is stout and somewhat swollen, about 4 times as long as wide and with very numerous, long, slender, black bristles on all sides, except medially, and directed obliquely forward. The second segment is attached obliquely, about twice as long as wide, slightly widened apically, and bears some long setae, especially dorsally. Third segment strongly knobbed at the base, the knob not quite as wide as the second segment, suddenly narrowed and then slightly widened on the basal sixth or less and very gradually drawn out apically and bearing a distinct spine at the tip. The second segment is as long as the first two together.

Head, anterior aspect: The head is much narrower than the thorax and almost hemispherical from the anterior aspect. The gena and lower occiput extend or bulge downward slightly below the eye margins so that the head, although very wide at the bottom, is not quite plane. Also the head is a little flattened at the top of the eyes and vertex. The eyes in the male are holoptic for about half the distance from the vertex to the antenna and the upper facets are only slightly enlarged. The ocellar triangle is low, the tubercle extending back behind the eye margin as far as it does in front of the eye margin, but the large ocelli are placed in an obtuse triangle, the posterior pair resting on the eye margin at the posterior eye corners. Frontal triangle equilateral. Antenna separated at the base by the thickness of one segment and then convergent so that they lie side by side. The oral cup occupies about one-third of the width of the face, a little less anteriorly. It is deep; the walls are rather thick and obscured by dense hair curling within. There is a rather steep, vertical tentorial fissure of considerable length shelving inward from the eye margin, so the gena, while wide, tends to slope inward until the bottom of the fissure is reached. Whole sides of face, gena, and bottom of the occiput rather densely covered with comparatively long, soft pile, and laterally with some fine, long, bristly hairs.

Thorax: Slightly convex, especially anteriorly. Both head and abdomen somewhat drooping. The thorax is about as long as wide, covered with extraordinarily dense, fine, matted pile, which is plushlike, completely obscuring the opaque brownish-black ground color, except where denuded. The disc of the scutellum and the area immediately in front of it and on the postalar callosity are apt to be covered with a somewhat more sparse vestiture of very fine, erect, stiff hairs, a little more bristlelike on the margin of the scutellum. Notopleuron with a tuft of 8 or 10 slender bristles extending a little above the pile. The whole pleuron everywhere appears to be densely covered with soft, rather long, fine, dense, matted pile, appressed backward. It covers the whole sternopleural, prosternal area and continues as a dense mat anteriorly around the thorax, completely merged with the pile of the occiput. Squamal fringe extraordinarily long, forming a wide, radiating, fanlike, relatively thin, umbrellalike structure. The arrangement of the pile on the thorax of this species is reminiscent of many species of nemestrinids.

Legs: Stout; anterior and middle femora a little swollen basally. Hind femur and tibia and basitarsus stout without being swollen. The anterior pair of legs is the most greatly reduced of any bombyliid known to me. The anterior femur is not more than two-thirds as long as the middle femur and scarcely more than one-third as long as the hind femur. Anterior tibia reduced and smooth. Middle tibia with 7 fine, short bristles anterodorsally, 8 or 9 slightly longer posterodorsally. Anterior middle femur with some bushy hair on the basal half below. Hind femur with 10 moderately long, slender anteroventral bristles. Bristles of hind tibia unusually slender and small. There are 8 anteroventral, 8 or 10 posteroventral, 8 or 9 posterodorsal and an anterodorsal fringe of many appressed setae. Claws long, bent at the apex. Pulvilli long and spatulate.

Wings: Rather large, broad at the base, somewhat pointed apically, hyaline, except that costal and subcostal cells are apt to be brownish yellow. There are 2 submarginal cells and apt to be a spur from the base of the anterior branch of the third vein and the marginal cell is apically widened backward and broadly rounded. First posterior cell strongly narrowed but not closed. Anterior intercalary vein also very strongly narrowed, almost eliminated. Anal cell very narrowly open, axillary lobe broad, alula large, the distal margin plane. Ambient vein complete, costal comb a little thickened at base and axillary sclerites with a considerable tuft of fine hair.

Abdomen: Very broad, somewhat wider than the thorax, comparatively convex across the dorsum, ground color feebly shining black. There are 7 tergites visible from above. The pile of the abdomen in the middles of the tergites consists of moderately abundant, very fine, slender, stiff hairs, curled backward, becoming longer and more conspicuous along the posterior margins. In addition there is a light coating of fine, short, curled, pale, appressed hair. In addition the lateral margins of the second to the fifth tergites have conspicuous, dense, matted tufts of very long, stiff black hairs which extend outward and backward from the tergal margins; these tufts accentuate the width of the abdomen considerably, in some species this hair is black, but with a small amount of similar white hair at each anterior corner and a large tuft of dense, fine, white hair on the outer fourth of the first and second tergites. This white hair mixed in with the black hair gives some species a banded appearance. Sixth and seventh tergites with some depressed, matted tufts of long, fine hair directed backward. The male terminalia symmetrical, forming an exposed, long, downwardly directed, conical hypopygium, bearing a tuft of bristly hairs laterally on each side just beyond the middle.

Material available for study: A representative of the type-species supplied through the kindness of Dr. S. J. Paramonov and the Commonwealth Scientific Industrial Research Organization. I also have several undescribed species.

Immature stages: Unknown.

Ecology and behavior of adults: I have seen no comments upon the adults or the flowers they visit. Roberts (1929) notes attraction to flowers.

Distribution: Australian: *Eusurbus crassilabris* (Macquart), 1855 (as *Bombylius*); *nigricinctus* Roberts, 1929.

Genus *Eucharimyia* Bigot

Eucharimyia Bigot, Bull. Soc. Ent. France, ser. 6, vol. 8, p. cxl, 1888b. Type of genus: *Bombylius pulchellus* Wulp, 1880, as *Eucharimyia dives* Bigot, 1888, by monotypy.
Euchariomyia Wulp, Cat. Diptera South Asia, p. 73, 1896, lapsus.

I have not seen this species. I quote Bigot:

Male. Body narrow, distinctly elongate; head hemisphere, not at all wider than the thorax; antennae of head bearing three long, basally contiguous segments; first two segments equally long and long pilose, third segment very short, that is bare, cylindrical, the apex abruptly truncate, with minute bristles, simple, filiform and obtuse. Proboscis rigid, as long as head and thorax; palpus hidden. The eyes are holoptic, in contact, bare. The face is prominent, setae and hairs few, arranged as hairs (barba) or mystax. Thorax muricate. Abdomen narrow and depressed, the genitalia hidden. Legs slender, distinctly elongate, the posterior femur below with short spines; anterior tibia and middle tibia bearing much abbreviated and shortened spinules, pulvilli minute; (alis, abdomine parum longioribus cellulis marginalibus et submarginalibus daubus tantum, posticis quatour). First longitudinal cell (first posterior cell) closed at the margin, at the base unequal. Anal cell open, the third longitudinal vein arising one-fourth the distance (longe ante) before the transverse external vein, the base oblique.

All black, face pubescent below the antennae and on the front above, ornamented with four silver spots. The humerus and pleura on both sides with similar silver spots. Orbits (occiput) and cervix (collo) covered with yellowish-gold pile. Thorax before and behind on both sides covered sparsely with yellowish-gold pollen or pruinescence. Abdomen at the base and apex of the segment short, yellowish, golden tomentose, the apical fourth above entirely silver; wings and halteres black on all parts.

This fly was described from Ceylon.

Distribution: Oriental: *Eucharimyia pulchella* (Wulp), 1880 (as *Bombylius*) [=*dives* Bigot, 1888, as *Eucharimyia*].

Brychosoma, new genus

Figure 14

Type of genus: *Brychosoma pulchra*, new species.

Flies of about the same size as a small species of *Bombylius* Linné. They are distinguished by the curious stublike style emitted at the base of the microsegment of the third antennal segment with its U-shaped pocket between style and apex of antenna. Other interesting characteristics consist of the long, continuous band of long, shaggy pile beneath the hind femur, obscuring a row of rather stout spinous bristles of the same length. The lower veins of the first posterior and discal cells cross at a point much as in *Heterostylum* Macquart, but the posterior margin of the eye is plane. The face and the antenna, with the exception of the apical sulcus, are very similar to *Acrophthalmyda* Bigot; that genus is probably its nearest relative, from which it differs in the venation.

Found so far only in southern Queensland, Australia.

Length: 10 mm., including antenna; antenna 2 mm.; wing 11 mm.

Head, lateral aspect: The head is semiglobular from the lateral aspect because of the rather prominent, extended face and gena and the very prominent occiput, especially produced backward near the middle of the head. The occiput slopes gradually to the eye margin at the upper eye corner and below at the lower eye corner. The whole head is densely gray pollinose, except for a large, sepia-brown spot, which encloses the ocellar tubercle and extends medially down the occiput. There is also a sepia brown crossband of pollen placed transversely across from the eye margin to the base of the antennae, which narrowly borders the antennae above. This band is also densely covered with short, appressed, shiny, red-brown pollen, and overlaps the more ventrally placed, long, brownish-black hairs. The face is loosely but conspicuously covered with a mixture of more or less curled, nearly erect, light brownish red, stiffened hairs; it also has additional similar hairs that are brownish black. The light red hairs tend to form an inner encircling fringe about the oragenal cavity. The dark hairs tend to be peripheral. This combined brush of light red and dark brown hairs continues on down to the upper margin of the deep, pocketlike tentorial fissure. The pile of the occiput is loose and scattered, more dense and longer medially becoming short and equally sparse toward the eye margin. Ventral pile beneath the occiput long and white and similar in quantity.

The posterior eye margin is plane or nearly so on the upper two-thirds and only slightly rounded on the lower third. It is somewhat angularly produced backward below the middle so that the shape of the eye is a little peculiar. The palpus is slender, comparatively short, cylindrical, composed of one segment which has short, fine, stiff, bristly hairs. The proboscis is long and slender, at least 2½ times as long as the head. The labellum is short, oval, flattened and leaflike with microsetae below. The antennae are attached at the upper fifth of the head and the head is slightly decumbent. The antennae are as long as the maximum length of the head, and quite slender. The first segment is about 5 times as long as wide with scattered, long, erect, slender, bristly hairs dorsally and laterally, and a few shorter, similar hairs ventrally near the apex. The second segment is short and beadlike with a few fine setae above and below. The third segment is more than twice as long as the first segment; viewed laterally it is of equal thickness throughout; viewed dorsally it is very slightly narrowed and ends in a curious, oblique, stublike, conical microsegment which bears the same, dense, appressed microsetae as the remainder of the segment and has a thick, bristle-tipped, rather short style emitted dorsally at the base of the microsegment so that there is a curious U-shaped space between style and apex of the antennae.

Head, anterior aspect: The head is quite wide, much wider than the anterior part of the thorax and as wide or slightly wider than the posterior part of the thorax. Viewed from in front the head is nearly 3 times as wide as high with the eyes more convex on the upper half. The front is quite wide, at least 3 times as wide as the large, obtuse tubercle, and the front at the antennae is nearly 5 times as wide as the tubercle; the genofacial space continues to widen so that at the bottom of the head it is at least three-fifths of the total head width. The genae are prominent, sloping inward medially, pollinose only and ending in a deep, tentorial pocket below. Walls of the oragenal cup are wide and prominent, nearly vertically interiorly. The oral opening is oval and the antennae at base are separated by little less the thickness of one segment.

Thorax: The mesonotum is oval, quite strongly narrowed anteriorly, widest across the postalar callosity; transverse across the base of the scutellum, which has a shallow declivity. The scutellum is large, gently convex, quite wide, with gently rounded posterior margin. The disc has a few fine, scattered hairs, and on the margin on each side there are about 10 long, stiff, bristlelike hairs. The squama is short with a dense fringe of long, white hairs in one row. The thoracic squamae are more distinctly tufted. There appears to be a stiff, fanlike fringe of plumula extending downward in front of the haltere. The disc of the mesonotum is opaque, brownish black, due to dense pollen. Anteriorly the pollen is rather fine, dense, erect, and reddish, but of no great length. Sides of the mesonotum as far back as the anterior margin of the postalar callosity are covered with light gray pollen and moderately dense, erect, shaggy, white pile. There are 4 light red bristles on the notopleuron, and 4 or 5 long, quite slender, black, bristly hairs on the anterior part of the postalar callosity. Mesopleuron with dense, coarse, white pile appressed downward and backward, but not upward. The pteropleuron and metapleuron are bare. There is a little scattered pile above the hind coxae. The haltere is short and knoblike.

Legs: The legs are of no great length, the hind pair longer. All tibiae and tarsi quite slender, but the femora distinctly thickened toward the base. The hind femur is of nearly the same width throughout and more than twice as thick as the tibia. The first 4 femora bear some tufts of appressed, shaggy hairs ventrally and anteriorly. It forms a considerable fringe on the middle femur. The hind femur has a curious, dense fringe of subappressed, black, shaggy hairs on the medial and ventral medial surfaces. Laterally on the outer half there are 5 oblique short bristles on the hind femur no longer than the shaggy pile. Anterior tibia with a dorsal lateral row of 10 or 12 short, oblique spicules, and a few shorter ones below. Middle tibia with a double dorsal row of 8 or 10 similar, slightly larger spicules, and 6 placed ventral laterally. On the hind tibia the spicules are slightly longer and a little slender. There are 5 lateral, 8 dorsal lateral, 9 dorsal medial, and 7 ventral bristles. The apex has 2 or 3 slender spicules ventrally, 2 or 3 others dorsally. Claws slender, curved on the apical half, the pulvilli long and slender; the medial stub is black.

Wings: The wings are long and slender, narrowed both basally and apically, although the alula is prominent with plane posterior margin slanting base and rounding apex. The wing is longer than the entire insect with a dark sepia anterior border extending irregularly backward leaving the posterior margin of the wing hyaline. Marginal cell brown with a large, clear spot in the middle; first submarginal cell clear apically, except immediately behind the costa, and with the apex of the anterior branch of the third vein brown on each side. Base of second submarginal cell dark brown but the base of the anterior branch of the third vein narrowly margined and diffusely margined by a nearly hyaline area. The hyaline postmargin of the wing extends through the middle of the first posterior cell, almost to the third vein. Basal half of the discal cell and first posterior cell brown with color slightly lighter along the crossveins. Basal half of anal cell and extreme anterior base of axillary cell also dark brown. There is a moderately large, clear spot placed anteriorly which lies in the second basal cell opposite the base of the discal cell. The first posterior cell is closed with a long stalk; the anterior crossvein enters the discal cell beyond the middle and the intercalary vein is completely absent. The lower veins of the first posterior cells and discal cells cross at a point very much like *Heterostylum* Macquart. The costa is scarcely expanded at the base and there is a very tiny, short spine-tipped hook at the base.

Abdomen: The abdomen is very short and wide, wider than the thorax. It is black, obscured by opaque, brownish-black pollen on the second and third tergites, light brown on the fourth and grayish on the fifth, sixth, and seventh tergites. It is also pale-grayish brown on the first tergite. First tergite covered densely with erect, fine, shining, yellowish-white pile, longer laterally, shorter but conspicuous, and completely continuous beneath the scutellum. Posterior border of this tergite with tufts of pile appressed inwardly on each side near the middle. The second tergite is recessed under the shelflike first tergite laterally, and has fine, scattered, erect hairs. This tergite is as long as the next three. The sixth and seventh tergites are all short.

Remainder of abdomen with fine, scattered, stiffened, loose and sparse hairs. The pile is more dense and tufted laterally on the sides of the fourth and fifth segments and from the sides of the more or less concealed eighth tergite. The last four tergites are more or less pinched in or compressed from each side. The female terminalia are completely concealed by a dense fringe of fine pile.

Material available for study: One female of the unique type.

Immature stages: Unknown.

Ecology and behavior of adults: Not on record.

Distribution: Australian: *Brychosoma pulchrum*, new species.

Staurostichus Hull, new genus

Type of genus: *Staurostichus chiastoneura*, new species.

Bombylius-like flies in which the venation is like *Heterostylum* Macquart, but the posterior margin of the eye is quite entire and plane. First posterior cell closed and with a long stalk. These flies while related to *Bombylius* Linné show several rather interesting differences. Like many of the Australian species the labella of the short proboscis is very widely oval and leaflike. The pile of the face is very fine, erect, turned forward and scanty. There appears to be a short basal segment with a very slender palpus. At this point it is bent upward.

These flies are known from Queensland, Australia. Male and female of *Staurostichus chiastoneura*, new species, from Queensland.

Length: 10 to 12 mm. Wing: 10 mm.

Head, lateral aspect: Hemispherical, the occiput is quite shallow and flattened, the face is rather prominent, extended triangularly forward at least the length of the first antennal segment; it is then obliquely and almost vertically truncate below. The pile of the face and upper part of the gena is quite fine, erect, and rather scanty. There is similar pile on the broad, rather flattened front of the female; the very minute front of the male has pollen only. Female vertex with a conspicuous tuft of long, fine, erect, black hairs. The ocellar tubercle is obtuse. The proboscis is about twice as long as the head, stout without being conspicuously so; the labellum is conspicuously and ovally lamellate, like a broad leaf. The palpus is quite slender with a long stiff, apical bristle, and shorter bristles. It has a widened and flattened basal segment, the apical segment slightly longer. The antenna is attached very close to the top of the head. It is elongate and quite slender, the first segment is 4 or 5 times as long as wide, with long, fine, scattered, stiffened hairs on all sides except medially, and the antennae are separated by the width of a segment at the base; second segment beadlike; third segment annulate, almost equally slender throughout, considerably longer than the first two, narrowed only at the apex and with a short, attenuate microsegment which bears an apical spine.

Head, anterior aspect: Head as wide as mesonotum. The eyes in the female are widely separated, not much wider toward the lower plane of the base. In the males the eyes are holoptic by 4 or 5 times the length of the minute front, and about 3 times the length of the ocellar tubercle. There is a small area of enlarged facets where the eyes meet. The oragenal cup is large, oval, quite deep, especially below, and does not have thickened walls at its outer edge. The tentorial fissure is conspicuous but narrow.

Thorax: The mesonotum is distinctly longer than wide, rather strongly convex, so that both head and abdomen are somewhat decumbent, as much as in *Sparnopolius* Loew. Mesonotum opaque, brownish black, brownish gray along the side on the postalar callosity and narrowly above the scutellum. Scutellar crease narrow. The mesonotal pile is dense, fine, erect, becoming scanty on the posterior half and pale brownish yellow; it has a few, equally fine, blackish hairs intermixed with 3 or 4 not very prominent reddish notopleural bristles. The 7 or 8 postalar bristles are fine, and black. Scutellum reddish and opaque with fine, long black bristles posteriorly in a wide irregular band. Also, some short, curled, glittering yellow pile is on the lateral margin of the mesonotum, on its posterior margin, and also on the base and posterior margin of the scutellum. Pleuron obscurely opaque, brownish yellow, becoming grayish above the middle and posterior coxa. The pile is abundant, but rather strongly appressed, downward and backward, and is pale yellow. The squamal pile forms a conspicuous radiating fringe. Metapleuron and metanotum bare.

Legs: Slender throughout. The anterior and middle femora are little widened at the base. Middle femur with one or two bristles. Hind femur with a ventral fringe of 8 conspicuous reddish bristles. Anterior tibia with 10 minute, slender, dorsal bristles, 8 similar posterior bristles. Bristles of the remaining tibia are a little longer and thicker. Claws slender, pulvilli well developed.

Wings: Wings are slender, a little narrowed basally but are very large, prominent, elongate, wide basally and with apically rounded alula. Marginal cell slightly and progressively widened, but with apex broadly rounded below. Two submarginal cells present. Basal origin of anterior branch of third vein broadly rounded, first posterior cell closed with a long petiole. The anterior crossvein enters the discal cell barely beyond the middle; contact of discal and second posterior cells punctiform as in *Heterostylum* Macquart. Anal cell widely open. Wing faintly tinged with brown due to reddish-brown villi. Anterior border of the wing to just beyond the middle, diffusely and obliquely tinged with reddish brown covering the whole base of the wing and alula.

Abdomen: Oval, about the same length of the thorax, or slightly shorter, beginning to narrow at the end of the second tergite so that the remaining five tergites form an equilateral triangle. The abdomen is black, faintly shining, mostly covered with opaque, dark reddish brown pollen and rather dense, appressed and partly curled, glittering brownish-golden pile, especially in male; the similar pile of females being a little more scattered, less appressed. Posterior margin of all tergites with numerous, erect, rather long, quite fine, black, stiffened hairs. Male terminalia reflexed below; rather large.

Material available for study: A pair collected by the author at Acacia Ridge, Queensland, in September to October of 1953.

Immature stages: Unknown.

Ecology and behavior of adults: These flies come to the flowers of *Leptospermum* species.

Distribution: Australian: *Staurostichus chiastoneura*, new species.

Genus *Dischistus* Loew

FIGURES 25, 29, 275, 453, 980, 982, 984

Dischistus Loew, Neue Beiträge III, p. 45, 1855. Type of genus: *Bombylius mystax*, Wiedemann, 1818. Designated by Becker, p. 494, 1913a.

Bezzi (1924), pp. 88-98, key to 18 species, sensu stricto, and related groups.
White in Hardy (1924), p. 83, key to 3 Australian species (*Dischistus* of authors).
Paramonov (1926a), pp. 159-161, key to 13 species (*Dischistus* of authors).
Engel (1933), pp. 182-184, key to 17 Palaearctic species (*Dischistus* of authors).
Hesse (1938), pp. 531-534, key to 4 South African species, sensu stricto.
Paramonov (1940b), pp. 70-73, key to 16 species (*Dischistus* of authors).
Efflatoun (1945), pp. 241-242, key to 3 Egyptian species (*Dischistus* of authors).

This genus early became a depository for a number of disparate elements in the family, many of which have been removed by later dipterists, especially Hesse (1938), who has pointed out correctly that the type-species necessarily is *Bombylius mystax* Wiedemann, a South African species designated by Becker (1913a), antedating the designation of *Bombylius minimus* Schrank by Brunetti (1920). There is reason to believe that this action by Becker limits this genus to South Africa, but leaves it related to the chiefly Palaearctic genus *Acanthogeron* Bezzi.

As based upon *Bombylius mystax* Wiedemann, the genus *Dischistus* Loew is distinguished by the open first posterior cell, the narrow wing, the much reduced alula, the depauperate basal comb, and the much thickened first two antennal segments. Especially characteristic is the strongly spindle-shaped first antennal segment. The body is covered with a coarse, dense pubescence, affording it what Hesse calls a "pufflike" appearance, not greatly unlike that seen in well-preserved individuals of *Systoechus* Loew, *Acanthogeron* Bezzi, etc. The male eyes are narrowly separated, or as much as the width of the antennal tubercle. The sides of the face and upper part of genae are characteristically covered with long dense pile.

This is a South African genus, as noted above. However, as I have been unable to check each Palaearctic species personally, I leave all these in the genus for the time being, only too well aware that they will require critical study at some future date. The Neotropical species listed for *Dischistus* Loew almost certainly do not belong there.

Length: 7 to 13 mm., usually 9 to 10; wing a little shorter.

Head, lateral aspect: The head is comparatively short and is prominent upon the somewhat swollen occiput behind the eye. The oragenal cup is sometimes twice as long as in the type-species; the proboscis may be extraordinarily long and slender, as in *plumipalpus* Bezzi, where it is 8 times as long as the head. In front the face and oragenal cavity are steeply oblique. From the lateral aspect, very little of the face shows and the quite wide gena do not extend beyond the eye margin on the lower half of the head, except beneath the eye. In front the eye is somewhat flattened and almost plane. The posterior eye margin is strongly rounded below but almost plane on the upper two-thirds. It is a little rounded near the vertex, and a little flattened from vertex to antennae. The face is extensive below the antennae but obscured by very long, stiff, black, almost bristly pile extending along the lateral side from antennae to the bottom of the head, among which there are a very few, long, and very few, shorter, curved, golden hairs. Face in the middle without pile, except for some golden hairs proceeding downward from the first antennal segment.

The occiput while swollen is strongly sloping from the center to the eye margin, so that it is tumid anteriorly. Laterally the outer slopes of the occiput have numerous, conspicuous, long, black, bristly hairs, both above and below, but also with some shorter, curled, subappressed, glittering, golden hairs. Also, the front, especially on the female, has a long, dense tuft of black bristly hairs on each side; ground color of the front shining and black. There are a few golden hairs on the front. The palpi are slender, comparatively long, although not conspicuously projecting from out of the cavity. They have dense long hairs on the sides of the basal segments. Proboscis long and slender, thrust forward, and slightly more than twice as long as the head. It extends well beyond the antennae. The antennae are attached at the upper fifth of the head, and are almost as long as the head. The first segment is unusually long, stout and thickened but much less than in *Conopohrus* Meigen. It bears on all sides numerous quite long, stiff, slender, black bristles, changing largely to stiff, golden hairs below. The second antennal segment is minute, short, cylindrical, and slightly narrow at the base. This segment is also distinctly narrower than the first segment. The third segment is as long as the first but much more slender, strongly annulate at the base, conspicuously spindle-shaped and especially attenuate apically. The apex is drawn out to a rather fine point. The annulate base is quite narrow.

Head, anterior aspect: Head much wider than the thorax. In the male the eyes are separated by the width of the ocellar tubercle and in females by nearly 3 times the width of the very broad, obtuse ocellar tubercle. The ocellar tubercle is conspicuous with sloping side forming an obtuse triangle, with large ocelli, and numerous coarse, long bristly hairs. The area surrounding the tubercle is largely bare. Viewed from in front the sides of the face and gena are nearly parallel and the whole face and gena very wide, constituting at least half the width of the head. The width of the oragenal cup is a little more than half the width of the oragenal space; its inner walls are steeply sloping, outer walls vertical with a distinct tentorial crease and the gena rounded and sloping medially. The proboscis arises from the middle of the oragenal cup. The antennae are separated by the width of one segment.

Thorax: The thorax is barely longer than wide, scutellum excluded. It is dull, brownish-black, nearly opaque, with a thin, obscured brownish pollen, but is conspicuously covered with dense, stiff, erect pile, mostly brownish yellowish, especially dense anteriorly, but also very dense along the sides, and upon the postalar callosity. Among this yellow pile there are a few, fine, long, erect, black hairs, more numerous laterally. The pile of the mesonotum in *plumipalpus* Bezzi is especially dense and collarlike anteriorly, but is long, erect, and dense over the entire mesonotum, and there is no black hair intermixed with the yellowish-white pile. The scutellum is relatively small, semicircular, convex on the disc with sharp anterior depression and the disc and margin covered with similar, stiff, erect, mostly yellow pile, similar to the disc of the mesonotum. The mesopleuron is completely covered with dense, mat of long, stiff, yellow pile, extending upward and even longer backward. Pteropleuron bare. The metapleuron and hypopleuron are also bare, except for 3 or 4 fine hairs behind and beneath the haltere. The halteres are small with short, compact club turned ventrally downward a short distance. The squamae have a dense fringe of long, articulate hairs in several rows.

Legs: The legs are slender and comparatively long but all of the femora, beginning at the middle, are thickened basally especially upon the anterior and middle pairs; hind femur less conspicuously thickened. The anterior femur has a scanty fringe of long, fine, black hairs, with some yellow hairs intermixed and dorsally some flat-appressed, slightly flattened, scanty yellowish pile. The hind femur has a few, long, subappressed, slender, black, bristly hairs from near the middle to the apex, and also a few, shorter, appressed bristly hairs, all of them slender. There are no true bristles on the hind femur. The anterior tibia has a single row, posteroventral in position, of 5 long, erect, sharp, slender, black bristles, and also an apical circlet of 6 shorter, black bristles. In *plumipalpus* Bezzi the anterior tibia has 3 rows of fine, sharp, black spicules or bristles, each row with 10 to 12 elements and 3 or 4 fine, ventral lateral bristles. The ventrals on the hind tibia are also more numerous than in the type-species. The very slender hind tibia has a dorsal row of 4 long, slender, black bristles. The first two elements are doubled and also a lateral row of bristles; none of these are longer than the bristles lying along the front tibia. Also, ventromedially there are 4 or 5 slender, shorter, black bristles. The tibiae are slender, but not especially long. Hind tarsus not especially long, hind basitarsus with a lateral fringe of 5 moderately long, oblique, black bristles. The short ventral setae are yellow. The claws are thick but sharp at apex, the pulvilli long but rather slender.

Wings: The wings are hyaline, slender, approximately as long as thorax and abdomen combined. The marginal cell is quite blunt and slightly widened at the apex. The anterior branch of the third vein is sigmoid but it ends well above the wing apex. The anterior crossvein is rectangular, entering the discal cell at the outer fourth. The intercalary vein is short, the first posterior cell is open but distinctly narrowed. The anal cell is widely open, the axillary lobe is narrow. The alula is quite narrow, and the basal comb absent. The wings are strongly iridescent and slightly wrinkled or fluted.

Abdomen: Short, and broad, rather flattened, the width distinctly greater that the thorax. The length is scarcely more than the length of the mesonotum, scutellum excepted. Posterior half of abdomen rather triangular and the abdomen widest at the end of the third tergite. I find 7 tergites visible, the last 2 turned downward; third to the fifth of approximate length. The abdomen is black, faintly shiny and faintly pollinose. It is covered with sparse but conspicuous, very long, stiff, bristly, black hairs, which are erect and form irregular fringes on the posterior margins of the tergites. There is about an equal amount of scattered, curled, partly erect, partly appressed, much shorter, golden-yellow hair. This golden-yellow hair tends to become longer, and denser on the outer third of the tergite, and especially dense and conspicuous laterally along the curled-over margins of the tergites. There is a fringe of short, erect, yellow pile on the base of the first tergite extending under the scutellum except in the middle. The abdomen of *plumipalpus* Bezzi has only erect stiff yellow and black hairs. Female terminalia enclosed by a dense, mat of yellow pile. Hesse describes the male hypopygium as follows:

The dorsal and sides of the basal part is distinctly rugulose or striate; the lower inner apical neck region is prominent. The beaked apical segments usually directed outward, and with a peculiar twisted structure or shape. The aedeagus is prominent, stoutish, broad, tubular and spout-like and with or without the ventral aedeagel process below. The lateral struts are broad and short, and the basal struts incised dorsally.

Material available for study: I have a pair of the type-species sent to me for study and dissection through the generosity of Dr. A. J. Hesse of the South African Museum, Cape Town, who also sent a male of *Dischistus plumipalpus* Bezzi.

Immature stages: Not on record.

Ecology and behavior of adults: Not on record.

Distribution: Neotropical: *Dischistus amabilis* Wulp, 1881; *transatlanticus* (Philippi), 1865 (as *Bombylius*).

Palaearctic: *Dischistus algirus* (Macquart), 1840 (as *Bombylius*); *barbula* Loew, 1855; *breviusculus* Loew, 1855; *croaticus* Kertesz, 1901; *flavibarbus* Loew, 1855 [=*lutescens* Loew, 1855, female]; *giganteus* Villeneuve, 1920; *imitator* Loew, 1855; *lutescens* Loew, 1855, male; *melanocephalus* (Fabricius), 1794 (as *Bombylius*) [=*argyropygus* Macquart, 1849]; *nigricephalus* Séguy, 1926; *nigriceps* Loew, 1862; *notatus* Engel, 1933; *pulchellus* Austen, 1937; *simulator* Loew, 1855; *sinaiticus* Efflatoun, 1945; *singularis* (Macquart), 1849 (as *Bombylius*); *transcaspicus* Paramonov, 1924; *trigonus* Bezzi, 1925; *turkmenicus* Paramonov, 1926, *turkmenicus flavisetis* Paramonov, 1940; *unicolor* Loew, 1855.

Ethiopian: *Dischistus aurifluus* Bezzi, 1924; *capito* Loew, 1860; *capito longirostris* Hesse, 1938; *coracinus* Loew, 1863; *dimidiatus* Bezzi, 1912; *farinosus* Bezzi, 1924; *frontalis* Loew, 1863; *gemmeus* Bezzi, 1924; *gibbicornis* Bezzi, 1923; *heterocerus* (Macquart), 1840 (as *Bombylius*); *heteropterus* (Macquart), 1840 (as *Bombylius*); *hirticeps* (Bezzi), 1921 (as *Bombylius*), *hirticeps karooensis* Hesse, 1938; *hirtus* Bezzi, 1912; *iris* Szilady, 1942; *lepidus* Loew, 1860; *melanurus* Bigot, 1892; *mystax* (Wiedemann), 1818 (as *Bombylius*); *niveus* (Macquart), 1840 (as *Bombylius*); *nucalis* Bezzi, 1924; *ovatus* Bezzi, 1921; *plumipalpis* Bezzi, 1921; *pusio* (Wiedemann), 1828 (as *Bombylius*); *rubicundus* Bezzi, 1921; *rufirostris* Bezzi, 1924; *variegatus* (Macquart), 1840 (as *Bombylius*); *vitripennis* Loew, 1855; *vittipes* Bezzi, 1921.

Oriental: *Dischistus resplendens* Brunetti, 1909.

Australian: *Dischistus formosus* Roberts, 1928; *pallidoventer* Roberts, 1928; *perparvus* Roberts, 1928.

Genus *Bombylisoma* Rondani

FIGURES 27, 273, 473, 513

Bombylisoma Rondani, Dipterol. Ital. Prodr., vol. 1, p. 164, 1856. Type of genus: *Bombylius minimus* Schrank, 1781, as *Bombylius sulphureus* Fabricius, 1805, by original designation.
Bombylosoma Marshall, Nomencl. Zool., p. 323, 1873.
Bombyliosoma Verrall in Scudder, Nomencl. Zool., p. 47, 1882.

Hesse (1938) has made it abundantly clear that the genus *Dischistus* Loew must be based upon the South African species *mystax* Wiedemann and other closely related species from the same area; I have been able to study and illustrate some of these through the kindness of Dr. Hesse. *Dischistus* Loew proves to be quite different from the species from the Palaearctic area currently placed under that name, and for these I take Rondani's name *Bombylisoma*, of which I have the type-species before me. These flies are related to *Bombylodes* Paramonov, from which they differ in several respects. In *Bombylodes* Paramonov the anterior crossvein is at the outer third of the discal cell, there are numerous distinct bristles on the underside of the hind femur, the apex of the tibia is thickened, and the cheeks are wide and broad, the head is strongly oblique, etc. In *Bombylisoma* Rondani (*Dischistus* of authors) there are few bristles below the hind femur, the tibiae are very slender, the anterior crossvein lies at or before the middle of the discal cell; the genae may be relatively wide or narrowed, but the head, according to Paramonov, is only moderately oblique (quer). From this genus the true *Dischistus* Loew is readily separated by the strongly spindle-shaped third antennal segment, the basal segments with spikelike bristles, the very broad genae, the position of the anterior crossvein which lies at the outer fifth of the discal cell; the first posterior cell is narrowed, the intercalary vein is short, the anterior branch of the third vein sigmoid, the end of the marginal obtuse and widened. The femoral armament consists of long, stiff hairs, the tibiae slender. Face short but with dense, long bristles.

The Palaearctic species formerly going under this name, of which Paramonov was familiar with many, are still a disparate group and will require further subdivision when restudied in detail, including genitalic studies. It seems likely that *Dischistus* Loew is restricted to South Africa.

Length: 7 to 13 mm.; of wing, 6.5 to 13 mm.

Head, lateral aspect: The head is hemispherical. The occiput is only shallowly developed but bears loosely scattered long hairs curled upward on the upper half and outward below. The head seems to be rather tightly apposed to the thorax. The front is very small, flat and sunken and appears to be pollinose only. The face, however, is rather strongly produced but quite rounded; it is obtuse or domelike and extends forward at least half the length of the long first antennal segment. The pile of the face is extraordinarily dense, quite fine and very long, and extends forward as a rounded brush and reaches half the length of the long antennal segment. The face is extensive below the antenna, but recedes rapidly down to the level of the eye on the lower third of the head; the long pile extends more thinly down to the bottom of the head. The eye is hemicircular, very slightly and broadly excavated a little below the middle so that the posterior margin is not quite plane.

The proboscis is long and slender, about twice as long as the head; palpus very slender and a little more than half as long as the first antennal segment. The long, slender antennae are attached at or a little below the upper fourth of the head. The slender first segment is two-thirds as long as the slender third segment, and there are rather dense, quite long, fine yet stiff hairs which extend obliquely outward on all sides. The second segment is quite small and short and beadlike. The third segment is of almost uniform thickness; it is slightly bent downward beyond the middle and the immediate apex is slightly narrowed and bears a very minute, bristle-tipped microsegment.

Head, anterior aspect: The head is considerably wider than the thorax and is at least 1½ times as wide as high; it is strongly oval in appearance; the eyes are tightly holoptic for a distance distinctly greater than the length of the equilateral vertical triangle. The ocellar tubercle comprises the whole of the vertical triangle; it is equilateral, and the small ocelli set upon each corner, and with a tuft of rather long, fine, erect hairs which arise in the center. The antennae are attached adjacent to one another. The oral opening is elongate, obliquely ventral in position, slightly narrowed anteriorly; the cavity is deep, but the lateral wall is slightly inflated, and it becomes shallow just below the face. The proboscis has a sheath at the base. The genae are quite wide when viewed from below. The tentorial fissure is narrow.

Thorax: The thorax is relatively short and opaque black on both mesonotum and scutellum, and slightly convex in front of the scutellum. The pile is dense, fine, and erect especially on the anterior half and moderately long. The scutellar pile, however, is rather sparse and consists of a few short, fine hairs and rather more scattered, quite long, slender, erect, stiff black hairs. The mesonotal pile in the type-species is pale brownish yellow, except laterally and on the postalar callosity. There are 3 very weak notopleural bristles obscured by pile and several longer ones on the postalar callosity. The pleural pile is dense, but the hypopleuron is bare, the metapleuron has dense abundant pile.

Legs: The femora are stout without being conspicuously swollen. Only the first 4 femora are a little swollen basally, and these have a fringe of ventral hair. The hind femur ventrally has a single row of 5 or 6 rather long, comparatively stout, oblique bristles. Anterior tibia with 10 or 12 minute bristly setae and barely longer slender setae posterodorsally; middle tibia similar; posterior tibia with 4 or 5 long, slender bristles anterodorsally, 10 minute ones dorsomedially and a few longer ones posteroventrally and anteroventrally. Claws small, slender, and sharp; pulvilli long and slender.

Wings: Nearly hyaline. In the type-species they are faintly tinged with brown near the base. The anterior crossvein enters the discal cell at or very near the middle. The first posterior cell is widely open and the anal cell is open; base of the costa without a comb; ambient vein complete; alula large, elongate, apically rounded.

Abdomen: Short oval, a little narrowed apically. At the base it is as wide as the thorax. Seven tergites are visible in the male. Male terminalia quite large, symmetrical, and extended straight downward. The pile of the opaque black abdomen is quite loose and rather scattered and pale yellow in the type-species, but in addition on the posterior margin of each tergite there is a row of remarkably long, erect, fine, black, bristly hairs which are twice as long as the yellow hairs. Middles of posterior tergites also with some scattered, appressed, slightly flattened, curled, golden hair.

Material available for study: Representative of the type-species presented to me through the courtesy of Dr. Joseph Bequaert in 1949.

Immature stages: Unknown.

Ecology and behavior of adults: Not on record. Probably very similar to *Bombylius* Linné or *Systoechus* Loew.

Distribution: Palaearctic: *Bombylisoma minimus* Schrank, 1781 (as *Bombylius*) [=*flavus* (Meigen), 1804 (as *Bombylius*), =*nigrifrons* (Becker), 1913 (as *Bombylius*), =*sulphureus* (Fabricius), not Mikan, 1805 (as *Bombylius*)].

Genus *Chasmoneura* Hesse

FIGURES 41, 42, 280, 562

Chasmoneura Hesse, Ann. South African Mus., vol. 34, p. 586, 1938. Type of genus: *Bombylius argyropygus* Wiedemann, 1821, by original designation.

Hesse (1938), pp. 589-593, key to 8 South African species.

This genus of small, *Bombylius*-like flies has been separated from *Dischistus* Loew by Hesse (1938) by a lengthy ensemble of characters. Perhaps the most dependable, quickly visible character is the presence of a distinct, rather conspicuous tuft of pile on the metapleuron, lacking in *Gonarthrus* Bezzi and *Dischistus* Loew, sensu stricto. The eyes are holoptic in males. The body is always black, never with red upon the scutellum. There are always conspicuous mats of flattened scalelike pile on some part, either the sides of thorax, sides of scutellum, on the sides of the female abdomen, sometimes in the male, and sides of the face in female; such pile is more likely to be present upon the abdomen. There are no long brushlike fringes of hairs beneath any of the femora and spines present only beneath the hind femur. The wings with basal comb small, the alula always well developed, in contrast to *Gonarthrus* Bezzi where it is reduced. Wings hyaline, tinged with brown or rarely with infuscated crossveins.

Chasmoneura Hesse is a South African genus with 8 species.

Length: 4 to 8.5 mm., wing a little longer.

Head, lateral aspect: The head is somewhat triangular due to the rather flattened front and upper part of eyes and the short, rounded projection of the face in front. The occiput is short, especially above, gradually sloping out to the eye margin. The pile of the occiput is quite dense, coarse and bristly, and rather short, much shorter than in *Gonarthrus* Bezzi. It scarcely rises above the posterior eye margin. The total projection of the face is a little less than the length of the first antennal segment. The oragenal cup is quite low and short in profile. The face bears numerous, slender bristles and bristly hairs both laterally and beneath the antennae and the surface has graying micropubescence or pollen. The pile of the occiput is comparatively dense, short and sparse below, becoming more abundant laterally but not extending beyond the eye margins, although suberect medially. Dorsally this coarse, almost bristly pile of the occiput makes a rather wide, dense band which scarcely rises above the ocellar tubercle. Eyes in the male depressed or sunken where they join above. The upper facets a little enlarged, the posterior margin of the eye entire but not quite plane.

Proboscis slender, elongate, and about 1½ times as long as the head. The palpus quite short, recessed, and more or less concealed. It has 2 segments and may or may not project beyond the oral cavity. The front has a distinct, transverse, furrowlike depression in females. The antennae are attached at the upper fourth of the head. They are slender, and moderately long, nearly as long as the head. The first segment is nearly straight but a little stout, a little swollen laterally, 2 to 3 times as long as wide, about twice as long as the second segment and bears numerous long, stout bristles over the whole of the dorsal and lateral surfaces and still longer bristles below. Second segment not quite twice as long as wide, with stout, long, black setae dorsally and laterally. Third segment is knobbed at the base, slender, slightly flattened laterally, more than 1½ times as long as the first 2 segments together and bearing at the apex small spine-tipped microsegments, one or two.

Head, anterior aspect: The head is slightly more narrow than the width of the thorax across the wings. The eyes in the male are holoptic and meet for a little more than the length of the ocellar tubercle. The antennae are distinctly separated at the base by about half the width of the first segment. The ocellar tubercle is an equilateral triangle, comparatively low, with large ocelli resting on the eye margin in the male. The oral opening is large and in width more than half the interocular space. The oral recess is quite deep with the side walls not quite vertical, and a little thickened. The gena is narrow, about as wide, however, as the oragenal wall and sloping into an unusually deep, wide, tentorial fissure.

Thorax: In all the species of *Chasmoneura* Hesse the thorax is black with sometimes a little red on the pleuron but never on the scutellum. The pile on the thorax is usually dense and usually rather conspicuously scaliform, either flattened and scaliform or fine and hairlike, with similar pile continued on to the scutellum. Upper mesopleuron with long, coarse, erect pile, pteropleuron and anterior hypopleuron bare. There are a few hairs posteriorly on the hypopleuron. There is a conspicuous, definitive cluster of long, coarse, flattened hairs standing erect on the upper two-thirds of the convex metapleuron. This character immediately separates these flies from *Gonarthrus* Bezzi.

Legs: The legs are comparatively long and slender. The femora are without any long, fringing or brushlike hairs below and the front and middle pair have no ventral spines. However, the quite slender, short hind femur always has several conspicuous, stout, rather long, generally black spines ventrally which are appressed against the femur. Hind tibia quite long and slender with strongly appressed, rather stout, black, spiculate bristles or spines; there are 6 ventromedially, 5 dorsomedially, all short, and 3 long, stout ones dorsolaterally. Pulvilli long, extending to the apex and the claws sharply curved downward apically.

Wings: The wings are generally nearly hyaline but often with the base or costal areas slightly yellowish or brown or gray. In *Chasmoneura argyropyga* Wiedemann the wing is much darker and is rather heavily tinged with brown, except near the apex on the posterior border. There are 2 submarginal cells and the anterior branch of the third vein is rather strongly curved and may have a trace of a spur near the base. The marginal cell is very little widened apically, but more so than in *Gonarthrus* Bezzi. The first posterior cell is slightly narrowed apically, the anterior intercalary vein is short, the anterior crossvein enters the discal cell from the middle, and anal cell is widely open, the axillary lobe is a little wider near the base and the alula is well developed. The ambient vein is complete. There is a basal comb developed to a varying extent in contrast to *Gonarthrus* Bezzi and a well developed hook lying before the base of the comb.

Abdomen: The abdomen is short and wide, rapidly narrowed apically. The abdomen is generally black, rarely reddish brown along the sides, thinly pollinose and with dense pile, often scalelike and flat-appressed or fine and hairlike and with some additional erect pile. There is a prominent wide band of unusually long, coarse, erect pile on the first abdominal segment, extending under the scutellum where it is shorter. I find 7 tergites present in the male. The scaling on the sides of the abdomen is often conspicuous and silvery and extends in matted masses around the curled-over edges of the abdomen.

Because of the large number of species that Dr. A. J. Hesse was able to study and summarize, I quote his work on the hypopygium of the males:

Hypopygium of males with the beaked apical joints elongate or bird-head shaped; lateral ramus, on each side from basal parts, coalescing under aedeagus and forming a ventral aedeagal process, which is sometimes complex and which always ends in an apical hook, spine or prong-like, more or less slender and curved, process on each side; aedeagus itself either long or short and more or less slender and tubular.

Material available for study: Several species supplied through the kindness of Mr. Harold Oldroyd and Mr. John Bowden. Dr. A. J. Hesse also kindly sent two species for study and dissection.

Immature stages: Unknown.

Ecology and behavior of adults: Not on record.

Distribution: Ethiopian: *Chasmoneura argyropygus* (Wiedemann), 1821 (as *Bombylius*) [=*argyropus* Loew, 1860, lapsus]; *cinereitincta* Hesse, 1938; *diademata* Bezzi, 1912; *flavipes* Hesse, 1938; *horni* Hesse, 1938; *kaokoensis* Hesse, 1938; *loewi* Hesse, 1938; *rhodesiana* Hesse, 1938.

Genus *Doliogethes* Hesse

FIGURES 37, 260, 468, 471, 744, 772

Doliogethes Hesse, Ann. South African Mus., vol. 34, p. 545, 1938. Type of genus: *Bombylius seriatus* Wiedemann, 1821, by original designation.

Hesse (1938), pp. 547-555, key to 15 species from South Africa.

Small flies, generally black, especially on thorax and abdomen, but vary greatly in the appearance of the species. The legs are reddish, blackish, or part yellow and the antennae sometimes of lighter color; lower part of body and face usually with much red. The abdomen is usually more or less elongate oval. Related to *Dischistus* Loew, it is a variable group containing many South African species. While the pile may be long and fine in some species of *Doliogethes* Hesse, it is generally less dense and shaggy than in *Dischistus* Loew, and it is especially shorter on the head. Eyes in male either very narrowly separated or divided by as much as the width of ocellar tubercle; female front and vertex broad. The antennae tend to be shorter than in *Dischistus* Loew and, particularly, more slender and less robust; third segment never spindle-shaped. Palpus of 2 segments but very short and concealed in the oragenal cavity. Wings frequently brown at base, the crossveins occasionally clouded but generally the remainder of wing hyaline. First posterior cell always open; a well differentiated costal comb present. Alula quite well developed but reduced in *Dischistus* Loew. Legs usually with some spicules or spinous bristles below femora, but sometimes restricted to the last pair; never with the long, dense, bristly hairs found in *Dischistus* Loew. Tibial spines unusually long and pale in color on all legs.

This is a South African genus of about 14 species.

Length: 4 to 11 mm.; usually 5 to 7 mm.

Head, lateral aspect: The head is rather short, the eye only about half as long as high, the face, however, is prominent and extensive, rather like *Bombylius* Linné. The oragenal cup is strongly and obliquely retreating so that it is scarcely extended beyond the eye opposite the proboscis; laterally its walls are inflated and rounded with posterior, oblique grooves and ridges. The face is everywhere covered with rather dense, long, erect, coarse, silvery white pile growing shorter below and beside the eye margin to which it extends. Face and front continuous, the lower half of the front with pile equally long and similar to that of the face; the upper and narrowed portion has appressed, scaliform pile. The occiput is only moderately swollen and developed behind the eye, without deep central cavity; it bears dense, long, erect, fine pile becoming much shorter near the eye margin where it is appressed and partly flattened and more or less curled and matted. Across the upper part of the occiput but not encroaching upon the marginal band of short, matted pile there is a conspicuous band of long, erect, white, bristly hairs. The posterior eye margin is entire but plane only through the middle half of the eye, which is more or less rounded both above and below. The eye is also distinctly wider on the upper part.

Palpus small, short, filiform, composed of 2 segments; the basal segment is a little flattened and widened through the middle with long, slender bristles; the apical segment has 2 or 3 minute setae at the apex. Antenna attached at the top of the head and almost on a plane with the upper eye margin. It is quite slender, not quite as long as the head; the first segment is 4 or 5 times as long as the bead-shaped second segment. Third segment more narrow than the basal two segments; its length is about equal the combined basal segments or slightly longer. It is gradually tapered from the middle so that the apex is narrower and it ends in a minute, spine-tipped microsegment.

Head, anterior aspect: The genofacial space is distinctly more than one-third the head width with the sides nearly parallel. The genae are wide with the tentorial fissure also wide and restricted to the lower half of the oral space; oral cavity short oval. The eyes of the male are widely separated by more than the width of the ocellar triangle. Antennae attached quite close together. Ocellar tubercle low with large, protruding ocelli in an equilateral triangle.

Thorax: The thorax is short, about as long as broad and not quite as wide as the head. The mesonotum is rather flattened and only slightly convex anteriorly and laterally and rather sharply raised above the base of the scutellum. It bears dense, long, tufted or matted, erect or semierect, coarse hairs, either dark or pale and rather variable according to species, and especially dense anteriorly, where it is also likely to be in matted tufts. The scutellum is quite large, with pile similar to the mesonotum. On the notopleuron and postalar callosity there are several quite slender, long, pale bristles. Pleuron rather hairy. The mesopleuron is widely covered with long, dense pile, some tufts of shorter matted pile and below some still finer, shorter, appressed pile. The sternopleuron is pilose; the hypopleuron has a conspicuous clump of moderately long, fine hairs behind and a vertical band of similar pile in front; metapleuron with a large clump of long, bristly hairs. Squama large, with a long fringe.

Legs: The legs are slender with the femora about twice as thick as the tibia and the anterior femur a little widened toward the base. The pile consists of loose but more or less appressed, long, flattened or scaliform hairs. Hind femur with 5 unusually long, conspicuous, ventral bristles. All of the tibiae have prominent spicules in several rows. Claws long and slender;

the pulvilli are half as long as the claw, the arolium short.

Wings: The wings are hyaline outwardly with the base and basal crossvein often tinted brown or yellow; basal half of wing broad, the apex narrowed. There are 2 submarginal cells, the 4 posterior cells are widely open, the anal cell likewise. The anterior crossvein enters the basal third of the discal cell. Base of wing with a strong flared and expanded, matted costal comb. Ambient vein complete, the alula large and bearing a short fringe.

Abdomen: The abdomen is broad and short, rather strongly narrowed but blunt at apex. The last several segments are more or less laterally compressed, and therefore with considerable depth. The abdomen is covered with remarkably long, erect, fine pile, silvery white in some species, highly varying in other species. Along the sides it becomes somewhat appressed. Sternites with similar pile but not as long.

Beaked apical segments of hypopygium of no great length. The aedeagus may have a ventrally directed apical process on each side, or an upright membranous type of process.

Material studied: Four species kindly furnished by Dr. A. J. Hesse of the South African Museum. Others seen at the British Museum (Natural History).

Immature stages: Unknown.

Ecology and behavior of adults: Not on record.

Distribution: Ethiopian: *Doliogethes aridicolus* Hesse, 1938; *chionoleucus* Hesse, 1938; *consobrinus*, Hesse, 1938; *imbutatus* Hesse, 1938; *luridus* Hesse, 1938; *melanops* Hesse, 1938; *meridionalis* Hesse, 1938; *ovatus* Bezzi, 1921; *pallidulus* Hesse, 1938; *psammocharus* Hesse, 1938; *rubicundus* Bezzi, 1921; *seriatus* (Wiedemann), 1821 (as *Bombylius*), *seriatus pullatus* Hesse, 1938; *seriatus puniceus* Hesse, 1938, *seriatus vagens* Hesse, 1938; *tripunctatus* Macquart, 1921; *trivirgatus* Hesse, 1938; *vittipes* Bezzi, 1921.

Genus *Gonarthrus* Bezzi

Figures 24, 263, 561, 1015, 1016, 1017

Gonarthrus Bezzi, Bombyliidae of Ethiopian Region, p. 109, 1924.
  Type of genus: *Dischistus cylindricus* Bezzi, 1906, by original designation.

Bezzi (1924), p. 111, key to 6 species from Ethiopia.
Hesse (1938), pp. 622-633, key to 26 species from South Africa.

Small, densely long pilose flies with clear, hyaline wings and a rather characteristic form to the anterior branch of the third vein, which is nearly straight. First posterior cell open, alula greatly reduced or even vestigial; basal comb absent or much reduced. The anterior crossvein is placed only at the outer third of the discal cell, but the latter cell is relatively broad, not triangular or anywhere angulate, and its apex is greatly narrowed. The male eyes are holoptic, the female front with a transverse depression in front of the slender, contiguous antennae. The hind femur has spines below, separating it from *Dischistus* sensu stricto; as in that genus the metapleural tuft of hairs is absent. The pile is dense and long below, on the first two antennal segments, also on sides of face, its middle, however, bare, and the pile is long on the gena and occiput. On the occiput the pile often forms a very dense, long, conspicuous, wide encircling band of considerable depth, merging with the dense, long pile of mesonotum and pleuron. Abdomen with dense, long pile in which weak, long bristles or bristly hairs are intermixed.

This characteristic South African genus contains some 26 species, many of which Hesse notes are not easily distinguished from others within the genus.

Length: 4 to 10 mm., usually 6 to 8; wing of approximately the same length.

Head, lateral aspect: The head in lateral aspect is curiously high and oval, scarcely extended beyond the eye but the posterior occiput rather remarkably protuberant and swollen backward in the middle of the head but sloping gradually down to the eye margin in all directions above, laterally, and below. Pile of the occiput is indeed extraordinary and consists of numerous, very long, slender, bristly hairs standing upright and not quite reaching the eye margin but extending to the upper eye corners vertically. The hairs are more or less directed backward in the middle of the occiput and downward below the eye, and nowhere obscuring the dense pollen on the surface of the occiput. The front has a broad, transverse, concave depression in the female, with the pile almost flat-appressed, long and bristly, arising on the more posterior slope and extending forward. Opposite the antennae, there are tufts of extraordinarily long, pale, fringing, bristly pile, which curves forward and extends slightly downward and continues down as a wide band along the entire length of the gena and the sides of the face. In some species there is a tuft of long, sometimes black, erect pile on each side of the vertex near the ocellar tubercle. In males the front is very small, triangular, pollinose only, and sunken. The eye is wider below, in fact may be almost rectangular at the bottom. The posterior margin is entire but not plane, being slightly concave. The proboscis is long, rather stout, with rather large labella, more narrow in some species. The whole proboscis is about 1½ to 2 times as long as the head.

The palpus is long and slender with the second segment short and bearing a few short, apical hairs. Antennae are attached at the upper fourth of the head. They are rather long and slender, the slender first segment straight and not incrassate as in *Dischistus* Loew. It is 4 or 5 times as long as wide and bears some long, coarse pile dorsally and laterally, near the base, but on more than the outer half it bears conspicuous, very long, coarse bristly pile. This segment is generally pale pollinose. The second segment is nearly twice as long as wide with a few, fine, short, appressed, setate hairs. Third segment knobbed at the base, long and slender, scarcely widened in the middle and attenuate near the apex. The apex of the third segment bears a small, spine-tipped microsegment. The surface pile of the third segment is of a very minute, dense, erect type.

Head, anterior aspect: The head is a little wider than the thorax. The eyes are holoptic in the male and in full contact. In extent the line of contact is equal more than half the distance from the ocellus to the antennae. Actually it appears to be much greater than this but there is a very tiny, narrow triangle above the frontal triangle which is itself quite small and sunken and pollinose only. Female front wide, at least a third of the head width, narrowing a little toward the vertex and distinctly concave transversely in the middle. The antennae are very narrowly separated at the base, the ocellar tubercle high but quite small and bears in the middle a longitudinal row of many long, slender bristles. The ocelli rest on the eye margin. The oral opening is large, shallow and about one-third the interocellar space. The side margins of the oragenal cup are inflated into a large, rounded, tubelike roll on each side. The oral recess on each side, except at the bottom, likewise has an almost equally large, vertical, tubelike expansion on each side, leaving a narrow, deep crease on the upper half. Gena as wide as the wall of the oragenal cup and with a long, very deep tentorial fissure, the gena sloping down medially into the fissure. Whole surface of the head beneath the antennae, except immediately above the proboscis, covered with fine, dense, pale-colored pollen.

Thorax: The thorax is shallowly convex, a little more so along the anterior border. The mesonotum is thinly pollinose, opaque, and quite densely covered with long, erect, coarse, bristly pile over the whole surface, including the humerus and scutellum. The notopleuron has 2 stout generally blackish bristles. The whole of the mesopleuron, upper sternopleuron, anterior propleuron covered with long, conspicuous, rather dense, coarse pile. Pteropleuron and metapleuron bare. The posterior hypopleuron has a tuft of shorter hairs beneath the haltere and behind the spiracle. Squama short, with a comparatively long fringe of hairs. Halteres small with short, oval, closed knob. There is no scaliform pile upon the thorax.

Legs: The legs are of average length, the anterior femur shorter than usual and rather slender, only slightly thickened toward the base. All of the femora have dense, matted, long, scaliform pile, often snow white, becoming brownish or blackish apically. In addition the anterior femur has tufts of scattered, quite long, somewhat flattened, blunt-tipped hairs at the base below, growing shorter toward the middle of this femur. Middle femur with a tuft of such hairs just beyond the middle. Hind femur with a few similar hairs on the basal half and with 3 stout, long spines laterally on the outer half, not ventral in position. Anterior tibia with an anterior and posterior dorsal row of small spicules, each row containing 3, sometimes 4 bristles. Middle tibia with much more conspicuous bristles but equally few in number. There are 3 anterodorsal, 4 posterodorsal, and 3 posteroventral bristles. Hind tibia with similar rows and bristles of about the same length and thickness. Pulvilli long and rather wide, the claws strongly curved at the apex.

Wings: The wings are hyaline. There are 2 submarginal cells with the anterior branch of the third vein nearly straight and scarcely curved at the base. Marginal cell not widened apically. First posterior cell opened maximally. Anterior intercalary vein short, the anterior crossvein enters the discal cell a little before the outer third. Anal cell narrowly open. Axillary lobe narrow, the alula distinctly narrow, the ambient vein complete. Wing uniformly villose, the villi quite small.

Abdomen: The abdomen is a little elongate, slightly longer than the mesonotum. I find 7 tergites. The surface is largely bare, more or less shining with a thin dark pollen and no scales present. However, there is much comparatively dense, long, coarse, bristly pile across the tergites and a very dense, conspicuous fringe along the curled-over side margins and apex of the whole abdomen, generally pale yellowish in color. The pile on the abdomen between the curled-over margins is erect. The posterior margins of the tergites has a fringing row of equally long, slender, black bristles. Sternites obscured by the dense pile from the lower edges of the tergites.

Hesse states that the hypopygium of males have the beaked apical joints elongate, subcylindrical or cylindrical and provided with a subapical lobe or spinelike process and a tuft or clump of spinelike bristles on dorsum, with the aedeagus long and sometimes very slender.

Material available for study: Several species of this distinctive South African genus. Mr. John Bowden kindly sent some material, and I have seen the material in the British Museum (Natural History).

Immature stages: Unknown.

Ecology and behavior of adults: I have seen no comments upon the behavior of the adults or of the flowers they visit.

Distribution: Ethiopian: *Gonarthrus chioleucus* Hesse, 1938; *chioneus* Bezzi, 1921; *chloroxanthus* Hesse, 1938; *citrinus* Hesse, 1938; *clavirostris* Hesse, 1938; *culiciformis* Hesse, 1936; *cygnus* Bigot, 1892; *cylindricus* (Bezzi), 1906 (as *Dischistus*) female; *irvingi* Hesse, 1938; *kalaharicus* Hesse, 1938, *kalaharicus venustus* Hesse, 1936; *labiosus* Hesse, 1938; *leucophys* (sic) Bigot, 1892; *mimus* Hesse, 1938; *monticolus* Hesse, 1938; *mutabilis* Bowden, 1964; *namaënsis* Hesse, 1938; *natalensis* Hesse, 1938; *nivalis* Hesse, 1938; *phileremus* Hesse, 1938; *rhodesiënsis* Hesse, 1938; *subtropicalis* Hesse, 1938 [=*cylindricus*, Bezzi, in part, 1906]; *tenuirostris* Hesse, 1938; *turneri* Hesse, 1938, *turneri melalophus* Hesse, 1938; *versfeldi* Hesse, 1938; *vumbuensis* Hesse, 1938; *willowmorensis* Hesse, 1938; *xanthinus* Bezzi, 1921; *xantholeucus* Bowden, 1964.

Genus *Lordotus* Loew

Figures 45, 46, 47, 48, 290, 575, 690, 807, 808, 809

Lordotus Loew, Berliner Ent. Zeitschr., vol. 7, p. 303, 1863. Type of genus: *Lordotus gibbus* Loew, 1863, by monotypy.

Painter (1939), p. 283, key to 13 Nearctic species.
Hall (1954b), pp. 6-8, key to Nearctic species; 15 species and subspecies.
Johnson and Johnson (1959), pp. 9-26, key to numerous western species.

Small to large flies easily recognized by the peculiar venation coupled with the dense, fine, erect pile. Many species are bright golden yellow or reddish pilose; others have almost entirely black pile, or the pile of legs and face may be black while that of thorax and abdomen is white. On the wing the first posterior cell is open. There are 3 submarginal cells. The marginal cell is greatly widened and expanded distally, ending obtusely since the anterior branch of the second vein is placed nearly vertically, ending in costa nearly or quite at a right angle. The posterior branch simulates a crossvein and joins the anterior branch of the third vein relatively close to its base, so that its short basal section is not greatly longer; the remainder of this anterior branch of the third vein continues, after union with the posterior branch of second vein, upward to end before the apex of the wing. Wings largely or quite hyaline; costa sometimes denticulate.

In the genus *Lordotus* Loew there are species with a broad, robust, obtuse abdomen and others with the abdomen slender and conically elongate; the eyes may be barely contiguous or extensively holoptic. The oval cup is wide and long pile is continued down the whole lateral margin of face and gena. The antennae are slender and elongate; the third segment is annulate at base, rarely with a short microsegment or style.

Two other genera, *Geminaria* Coquillett and *Othniomyia* Hesse, have almost exactly similar venation. *Geminaria* Coquillett contains small flies which in appearance resemble an attenuate *Aphoebantus* Loew with bilobed scutellum; however, its head and occiput clearly place it within the first division of the Bombyliidae; moreover, its holoptic eyes are divided by a horizontal line, as in *Caenotus* Cole and it further differs by the long palpus which is as long as the antenna. *Othniomyia* Hesse is more peculiar, it too has a sulcate scutellum; it has a much shorter proboscis, a shorter and different type of antenna and the male is dichoptic; the whole face is quite narrow besides other peculiarities dealt with under that genus.

There may be a distant relationship with *Conophorus* Meigen; the venation is somewhat similar and some species of both genera have a denticulate costa, but the swollen first antennal segment of *Conophorus* Meigen is unique; one cannot rely heavily upon the character of the venation alone, which is somewhat similar in several other genera.

Johnson and Johnson (1959) distinguish three species groups: those with an antennal microsegment as in *miscellus* Coquillett; those with male costa denticulate and sexes often dimorphic, the *gibbus* Loew group; those with simple male costa and sexes usually alike, the *apicula* Coquillett group. Hall (1954) found the genitalia quite uniform with little or no species value. This suggests that the genus is a recently evolving group. Seven subspecies have been described. Hall (1954) found the antennal segmental ratios undependable and notes that the principal specific characters are coloration and pile, etc.

Lordotus Loew is restricted to the western United States including Mexico; the species are concentrated in the southwestern United States.

Length: 4 mm. to 16 mm. including antenna, but excluding proboscis; the wing 3.5 to 11 mm., thus rather reduced. Some species are extraordinarily variable in size of the sexes.

Head, lateral aspect: The head in profile forms more than a hemicircle, since the occiput is comparatively prominent; the face is not prominent and is scarcely visible in profile, although it may extend beneath the antennae as much as the length of the first antennal segment and sometimes has a crimped margin above the oragenal cup. The pile of the face varies from scattered, coarse, erect hairs of moderate length to still longer, dense, somewhat scalelike hairs, which continue all down the entire extent of the gena. The occiput is distinctly swollen medially but slopes off gently to the outer eye margin, its pile is very dense, long, shining, and scaliform in the pale pilose species and likewise covered with a close-lying mat of appressed, short, scaliform pile. In species like the type-species this mat of pile may be black punctate where the long bristles emerge. The occiput is less prominent than in some of the other species. The eye is large and plane behind, scarcely convex in front but strongly rounded above and below.

The proboscis is elongate and slender, about twice as long as the head, or even longer in some species. The labellum is slender, the palpus slender with a row of moderately long, coarse hairs below in some species but with a similar dosal row in the type-species, together with a basal ventral tuft. Basal segment short. The front in males is quite small, flat, appressed micropubescent, without medial groove. In females it bears considerable coarse, erect pile on each side. Antennae attached at the upper third of the head, only moderately elongate, the first segment is about twice as long as the second, or shorter. The second antennal segment varies from only as long as wide to other species where it is two or more times as long as wide. Both of these segments are slender and bear some long, bristly hair, which is rather fine in some species, longer and conspicuous below in others, but always short pilose above. The third segment is spindle-shaped, knoblike at the base, longer and more attenuate apically and at the apex has a minute spine; a few species have a short microsegment.

Head, anterior aspect: The head is small and distinctly narrower than the thorax, the front in the female is a little more narrow than the face. The eyes in the male vary from almost contact to species where they are distinctly holoptic. The antennae are adjacent, the ocellar tubercle is small in the female and minute in the male. Ocelli small, ocellar triangle obtuse, the pile long. The oral opening is narrow, moderately deep, and its inner wall very narrow or sometimes thickened. The gena is quite wide, the tentorial fissure a mere line. From above, in males and females, there is a shallow groove beneath the medial posterior collar of the vertex.

Thorax: The thorax is distinctly though only moderately humped anteriorly, the extent varying and somewhat greater in the type-species than in other species. The abdomen is also slightly decumbent in those species that are humped. Thorax black is ground color, opaque, with or without gray pollen, with or without appressed, shining curled pile, always with rather dense, erect, fine or coarse pile, denser and more prominent on the anterior half where it appears as if it were brushed backward. Scutellum with long, erect, abundant pile. Bristles absent. Pteropleuron and hypopleuron bare, mesopleuron with dense, long pile, metapleuron with a conspicuous tuft of long pile immediately in front of the halteres and an additional tuft varying in extent behind the spiracle. Squama with a wide band of long conspicuous pile.

Legs: The legs are comparatively short and stout and very spiny. There are 3 rows of stout spicules on the anterior tibia, each containing about 7, there are 4 rows on the middle tibia, each containing 10 to 14 stronger, spiny bristles and the hind tibia are similar. Femora without bristles; the middle and hind pair have a fringe of coarse, ventral hairs; other pile flattened, dense scale-like and quite appressed. Pulvilli well developed, claws simple, and rather stout.

Wings: The wings are relatively small, hyaline except that in a few species they are tinged with brown at base and centrally in front of the discal cell; in one species there are 6 to 8 small brownish spots. The venation is especially interesting; there are 3 submarginal cells; all posterior cells and the anal cell widely open. The discal cell is wide across the middle, angular below; the anterior crossvein lies at the outer third or even the outer fourth of the discal cell. The most unique feature of the wing lies in the remarkable expansion of the marginal cell on its outer half. It bends rather strongly backward at or just beyond its midpoint so that it is very wide apically and the anterior branch of the second vein, although curved, turns sharply backward up to the costa. Axillary lobe moderately developed; the alula is small, the ambient vein complete. There is a brush of pile at the base of the costa but no costal hook. At least 7 species have the male costa denticulate or spinose.

Abdomen: The abdomen is as wide as the thorax but tends to be elongate. It is at least twice as long as wide, some species are more slender than others. The ground color varies from shining black to subopaque. A few species have the apical tergites reddish or the tergal postmargins yellow. One group of species has varying amounts of completely appressed, shining, curly pile on the tergites; all species have comparatively abundant, loose, long, erect pile covering the whole of the tergites, the sides of the abdomen and the sternites. *Lordotus* Loew species vary from those which are wholly white pilose, partly opaque, partly shining to others which are brown, black, or reddish black pilose, and in many species the pile is a beautiful golden yellow. Both male and female terminalia small, inconspicuous, somewhat recessed, the dististyli of the male visible from below.

The male terminalia from dorsal aspect, epandrium removed, show stout dististyli, curved outward at apex with a sharp point. The epiphallus is large, becoming progressively wider toward the base. It extends well beyond the base of the dististyli. From the lateral aspect the basistylus resembles that of *Anastoechus* Osten Sacken and so does the epiphallus. The dististylus is more slender, and quite attenuate apically. The epandrium has a prominent, ventral, apical lobe.

Material studied: Extensive material of the type-species and six other species in my collection; also others seen in the National Museum of Natural History and the University of California at Berkeley.

Immature stages: Hall (1954) notes that a pupal skin is attached to a specimen of *L. gibbus gibbus* Loew in the Canadian National Collection, source unknown. He also quotes an observation by J. W. McSwain, who at Antioch, California, in September, watched a female of *gibbus striatus* Painter investigating burrows of sphecids and bembecids, flying about a foot away, bury at least the posterior half of the abdomen in loose sand in apparent pulsating oviposition, the abdomen gradually withdrawn.

Ecology and behavior of adults: Adults are readily collected on flowers, but I only remember seeing them on composites—*Helianthus* in New Mexico and *Solidago* at Mesa Verde. They are also seen resting on desert sand. They are more abundant in late summer or autumn, and range up to altitudes of 5,000 or even as much as 9,000 feet.

Distribution: Nearctic: *Lordotus abdominalis*, Johnson and Johnson, 1959; *albidus* Hall, 1954; *apiculus* Coquillett, 1887; *arizonensis* Johnson and Johnson, 1959; *arnaudi* Johnson and Johnson, 1959; *bipartitus* Painter, 1939; *bucerus* Coquillett, 1894; *cingulatus* Johnson and Johnson, 1959, *cingulatus lineatus* Johnson and Johnson, 1959; *diversus* Coquillett, 1891, *diversus diplasus* Hall, 1954; *divisus* Cresson, 1919 [=*niger* Cresson, 1923]; *ermae* Hall, 1952; *gibbus* Loew, 1863 [=*flavus* (Jeannicke), 1867 (as *Adelidea*)], *gibbus striatus*, Painter, 1939 (1940); *hurdi* Hall, 1957; *junceus* Coquillett, 1891; *lutescens* Johnson and Johnson, 1959;

miscellus Coquillett, 1887, *miscellus melanosus* Johnson and Johnson, 1959; *perplexus* Johnson and Johnson, 1959; *planus* Osten Sacken, 1877; *puellus* Williston, 1893; *pulchrissimus* Williston, 1893 [=*carus* Cresson, 1923, =*pulcherrimus* Aldrich, 1905, emend., *pulchrissimus luteolus*, Hall, 1954]; *rufotibialis* Johnson and Johnson, 1959; *sororculus* Williston, 1893, *sororculus nigriventris* Johnson and Johnson, 1959; *zonus* Coquillett, 1887.

Genus *Geminaria* Coquillett

Figures 35, 281, 461, 571

Geminaria Coquillett, Trans. American Ent. Soc. Philadelphia, vol. 21, p. 109, 1894a. Type of genus: *Lordotus canalis* Coquillett, 1887, by original designation, the first of 2 species.

Small, quaint, pollinose flies with elongate, slender, and attenuate abdomen and quite long slender proboscis at least 4 times as long as the head; the hairy palpus is also long. Like some *Aphoebantus* Loew, to which it is not related and like the South African *Othniomyia* Hesse, it has a sulcate or bilobed scutellum. The venation is typically like that of *Lordotus* Loew, yet the first submarginal cell bulges outwardly. Males holoptic, and with a distinct horizontal line in both sexes. Costa microspiculate or tuberculate. The wings may be hyaline or may be spotted.

The few species are Nearctic, from the southwestern United States. Ground frequenting desert species, attracted to low-growing flowers.

Length: 6 to 8 mm., excluding proboscis; proboscis 5 mm.; wing 5.5 mm.

Head, lateral aspect: The head is more or less subglobular, the occiput only moderately extensive and the face very little produced forward. The oragenal cup still less extensive and scarcely visible in profile. The pile of the face consists of a double row of rather strong bristles laterally beginning close to the antennae and continuing on down nearly to the middle of the

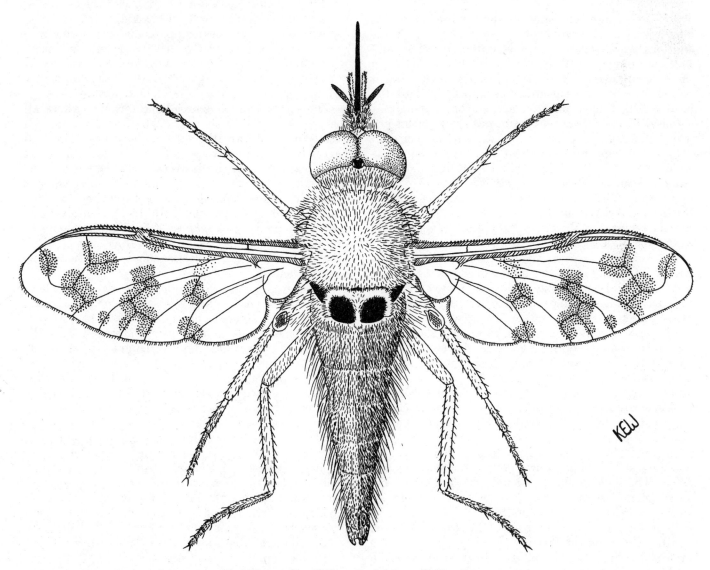

Text-Figure 21.—Habitus, *Geminaria canalis* Coquillett.

oragenal cup, where they are replaced by equally long, white, coarse, bristly hairs. Middle of the face beneath the antennae with dense, appressed micropubescence only, which continues over the whole face and gena. The occiput is pollinose with rather dense, extremely coarse pile around the middle extending to the eye margin where it is appressed and longer medially, where it is more or less erect and becoming very dense and much longer at the bottom of the longitudinally extended head. Vertically the occipital pile consists of similar, short, subappressed, bristly hairs and a band of loose, scattered, long, slender bristles visible either from the sides or from the front. The eye is quite large and proportionately relatively long, that is extended forward, strongly flattened above, a little flattened ventrally and broadly rounded in front below. Also, there is a blackish, narrow band dividing the facets on the lower part of the eye but not separating them into groups of different sizes. However, the dorsal facets on the flattened area above are much larger. Posterior margin of the eye entire.

Proboscis quite long and slender, about 3 times as long as the head, palpi quite prominent, rather thick and long with long, shaggy pile, especially below, and shorter, appressed pile dorsally. Front of the male small, triangular with some appressed, light colored scales. Antennae attached at the upper third of the head, a little longer than the eye, not quite as long as the head. The first segment is stout, $2\frac{1}{2}$ times as long as wide, the second segment quite stout, a little longer than wide, third segment knoblike at the base, not quite as wide as the middle of the second segment, somewhat narrowed on the apical third and ending bluntly. The first segment of the antennae has numerous, slender bristles and bristly hairs dorsally, laterally, and below. Second segment with long setae apically on all sides, except the middle.

Head, anterior aspect: The head is not quite twice as wide as high, and almost or quite as wide as the thorax. The eyes are in contact extensively, the vertical triangle reduced to a very small equilateral triangle, a little raised and swollen. The rather large ocelli lie adjacent to the eye margin. The antennae are very narrowly separated, the oral opening is large, short oval, wider beneath the head and extends far back beneath the eyes. It is also quite deep both anteriorly and posteriorly. Anteriorly the walls are a little thickened on the inside, the genae are wide, the tentorial fissure absent or closed. From above, the head is shallow, bilobate, the shallow occiput sloping gradually to the eye margin.

Thorax: The thorax is a little longer than wide, narrow anteriorly, opaque brown pollen over black ground color. The mesonotum is covered with flat-appressed, short, pale scales, rather dense anteriorly and along the sides and in front of the scutellum. There are similar scales narrowly along the base of the scutellum, vertically on each side and narrow, vertical band down the middle dividing the bisulcate scutellum into 2 rounded, jet black, polished, mammate protuberances. The high, dorsally plane, and vertical posterior boundary of the narrow, postalar callosity is jet black. On the mesonotum there are also some scattered, rather short, curled, reddish, bristly hairs turned backward. Moreover, there are a few longer erect red bristles arising from the black swollen sides of the scutellum. Pleuron blackish and gray pollinose with some small brown and yellowish areas. The mesopleuron and propleuron have considerable long, partly erect, partly appressed, coarse, white, scaliform pile and some dorsal red bristles on the mesopleuron. Notopleuron with a patch of 10 or 12 slender, red bristles. There are conspicuous tufts of coarse, white, scaliform pile on the metapleuron in front of the haltere and likewise behind the spiracle. Squamal fringe composed of the same pile, and sternopleuron with much similar pile. Metapleuron flattened.

Legs: The legs are comparatively short and relatively slender. Anterior tibia with a posteroventral row of 5 very small spicules, and with other still more microscopic setae. Hind tibia with 4 rows of more conspicuous black spicules containing 6 to 8, with the exception of the ventrolateral row which contains only 3 or 4. Femora without bristles; hind femur with a row of 9 long, coarse hairs ventrally. All the femora and tibiae with numerous but largely isolated long, wide, flat-appressed, cream-colored scales. Claws slender, sharp, pulvilli long.

Wings: The wings are broad and relatively short, all the posterior cells widely open, the first posterior cell greatly flared toward the edge of the wings. Anal cell narrowly but distinctly open. Anterior crossvein lies at the outer seventh of the discal cell. Axillary lobe wide, the alula moderately wide. Ambient vein complete, costa with minute spinules, the marginal cell is flared and widened apically, somewhat like *Lordotus* Loew but even more like *Conophorus* Meigen and *Othniomyia* Hesse. The wing may be either hyaline or spotted on all crossveins and other places along veins. There are 3 submarginal cells.

Abdomen: The abdomen elongate conical, as wide as the thorax only on the first segment, beyond this gently and gradually tapered until quite narrow at the apex. The abdomen tends to be a little compressed laterally, the tergites in fact are almost triangular in cross section. The color is opaque black overlaid by opaque, reddish sepia pollen but on the turned-over sides the tergites are brownish cream colored. Pile on the tergites loose, curled scaliform, and flat-appressed. It is white on the first segment and along the posterior margins of the tergites, except in the middle, along the lateral margins, but golden brown on all the remainder of the tergites. First segment laterally and anteriorly with a dense brush of long, erect, shining, coarse, or flattened pile. Male terminalia partly enclosed by a posterior lobelike projection from the eighth tergite. The dististyli are large, adjacent, and seen only from below.

Material available for study: I have material in my own collection of both known species.

Immature stages: Unknown.

Ecology and behavior of adults: Adults are found resting on desert sand or coming to low-growing flowers.

Distribution: Nearctic: *Geminaria canalis* Coquillett, 1887; *pellucida* Coquillett, 1894.

Genus *Othniomyia* Hesse

FIGURES 264, 470, 475

Othniomyia Hesse, Ann. South African Mus., vol. 34, p. 707, 1938.
 Type of genus: *Othniomyia tylopelta* Hesse, 1938, by original designation.

This small fly is obviously closely related to the American genera *Lordotus* Loew and *Geminaria* Coquillett, since the peculiar venation is almost identical. However, there are significant differences. From *Geminaria* Coquillett it is separated by the much shorter, rather fleshy type of proboscis and the shorter antennae which has a long but thick, apically attenuate style or microsegment, bearing at its apex a spine; intermediate microsegments not separately distinguishable according to Hesse (1938). The principal, first, microsegment is longer than the short, stout third segment. In *Geminaria* Coquillett, the proboscis is quite long and slender, several times as long as the head, and the eyes are divided by a distinct dark, horizontal line in some species. In this genus and also in *Othniomyia* Hesse the scutellum is sulcate or fissured medially.

In *Lordotus* Loew while the proboscis is not as long as in *Geminaria* Coquillett, it is still much longer and more slender than in *Othniomyia* Hesse; also *Lordotus* Loew has a longer, more slender antenna, and third antennal segment which has a minute, tiny, microsegment at the apex.

These are exceptionally compact, humpbacked flies which tend to be a little compressed laterally. The pile is short, scanty, and appressed. The face is unusually narrow for a bee fly and the male eyes are separated by nearly half the width of the face. In profile the face is quite short, the oragenal cup demarcated above, also laterally, by a well-marked groovelike crease. There are a few long bristles along the sides of the face and the 3 or 4 which extend along the sides of the upper half of the oragenal cup are stouter.

The three above-named genera all have the anal cell open, but it is more widely open in *Othniomyia* Hesse than any bee fly I have seen; in these genera the only wing area modified from what must have been a rather generalized condition is the anteroapical area above the third vein where the cells have been expanded backward. It is rather distantly related to the above genera.

Othniomyia Hesse is known from a single unique species from the southern Karoo. Hesse described the female in 1956. The eyes are separated by twice the width of the posterior ocelli.

Length: 5.5 mm.; wing 5.5 mm.

Head, lateral aspect: The face is very short, its length approximately as long as the antennae, the style excepted, and on the ventral portion it is narrowly extended forward as a sort of a flared lip above the oral opening, the latter continued obliquely down to the cheeks and becoming short below. The cheeks are both short and narrow. The eye is strongly convex anteriorly without lateral indentation but extending back farther on the ventral portion so that it is gently recessive on the upper half, and this portion is also plane. The occiput is moderately thick, especially on the upper part, obscured by the close attachment to the thorax and the thick anterior pile of the prothorax. The proboscis is short and slender, directed forward, the terminal lamellae somewhat spatulate and a little narrowed apically and this portion approximately as long as the basal part of the proboscis. The lateral margins of the face have 10 or 12 long, slender bristles on each side in a vertical row. The antennae are attached at the upper third of the head, the first two segments short with the first segment twice as long as the second, and bearing 2 long and 3 or 4 shorter slender bristles ventrally. Second segment with only minute setae, placed everywhere apically, except medially. The third segment is quite short, pyriform, the upper neck narrow, short and bearing a thick, rather long attenuate style ending in a quite short bristle; the whole style is as long as the third segment.

Head, anterior aspect: Front with yellowish scales and some long, coarse, pale pile. Ocellarium low, at the posterior eye corners, obtuse, filling the entire width of the vertex. The front and vertex are narrow, very slightly convex through the middle.

Thorax: The mesonotum is rather strongly raised and arched anteriorly with scattered, long, fine, bristly hairs across the anterior margin between the humeri and the wing and intermixed with curled, coarse, whitish, scale-like hairs which are again especially prominent along the posterior margin of the mesonotum and on the sides and middle of the apex of the scutellum. Scutellum with 2 or 3 long, slender bristles laterally and other bristly hairs, large, convex with a rather deep, wide, medial sulcus. Pleuron with slender, bristly hair widely over the mesopleuron and some of the white scalelike hair. Metapleuron with a dense tuft of long, blackish, bristly hair; squamae with a white fringe of flattened hair.

Legs: The femora are moderately stout with a few, short, dark scales continued onto the tibiae, the color reddish brown for the scales. All bristles rather slender; on the hind femora there is a scattered, ventrolateral row and some lateral subapical elements. The hind tibiae with the usual 4 rows, most of them rather short and containing 6 to 8 elements each. Middle tibiae similar, the anterior tibiae and femora short. The length of the anterior femora 4 mm., of their tibiae 4 mm., but for the middle tibiae and femora 7.4 mm. Anterior tibiae with minute anterodorsal bristles and 8 somewhat longer posteroventral bristles and some 6 or 7 short posterodorsals. Claws small, slender, sharp;

the pulvilli are well developed and with only a minute, medial lappet.

Wings: Hyaline, except that most of the costal cell, basal half of first basal cell, and extreme base of and axillary cells are dark blackish brown. Venation very similar to *Lordotus* Loew. There are 3 submarginal cells and 4 posterior cells. The apex of the marginal cell is greatly expanded, its apex obtuse, gently rounded.

Abdomen: Obtuse, rather densely covered with short, yellowish and brownish scales, and with some fine, bristly hairs along the posterior margins. Sides of the first tergite without conspicuous bristles; the scalose hairs are a little longer sublaterally and there is similar scaly pile on the sternites.

The dististyli, according to Hesse (1938) are "broad and elongate, hollowed out below, convex above, not depressed; aedeagus without any ventral processes, its main body at the middle produced into a basally directed process on each side, and lying dorsal to the medial aedeagal apparatus."

Material studied: The unique type in the British Museum (Natural History).

Immature stages: Not known.

Ecology and behavior of adults: Not on record.

Distribution: Ethiopian: *Othniomyia tylopelta* Hesse, 1938.

Genus *Lepidochlanus* Hesse

FIGURES 43, 270, 464, 476, 773

Lepidochlanus Hesse, Ann. South African Mus., vol. 34, p. 613, 1938. Type of genus: *Lepidochlanus fimbriatus* Hesse, 1938, by original designation.

Small flies of dark coloration and dense, erect, long, fine, white pile and hyaline wings. They are further characterized by the separation of the eyes in the male, the small front, the white scales on the front and upper face. There are flattened, quite appressed hairs on the thorax and abdomen, sometimes white, sometimes reddish, or both, which lie between the very long, erect, slender hairs and delicate bristles. Another character is the condition of the anterior crossvein, which enters the discal cell near the basal fourth. The first two segments of the antennae are rather stout, the third segment is very like *Chasmoneura* Hesse but has the apical spine largely concealed. It also differs from *Chasmoneura* Hesse in the dichoptic male eyes, in vestigial pulvilli, broad, short discal cell and absence of frontofacial depression. The occiput is much less prominent than in *Gonartharus* Bezzi. From *Dischistus* Loew it differs in the presence of dense, scaliform pile above and below; spines are present below the hind femur; also in the well-developed alula and hypopygial characters; it is distinguished from *Doliogethes* Hesse by the shape of the discoidal cell, the dense scaliform pile and the absence of a frontofacial depression.

Hesse (1938) emphasizes the blunt aspect of the bristly hairs and their fimbriate character; from a study of a pair determined by Dr. Hesse and most kindly sent me for dissection and study, I find all the stiff hairs blunt but none with split ends. Except for bluntness the bristly pile is much like *Chasmoneura* Hesse and *Doliogethes* Hesse, and like them rather sparse.

The only known species is from South Africa.

Length: 2.5 to 6.5 mm.

Head, lateral aspect: The head is hemispherical although a little flattened anteriorly. The face is moderately produced but rounded, the oragenal cup is confined to the area opposite the proboscis where there is a deep, widely flared, tentorial pocket; the gena is moderately wide on each side. The face and upper part of cheeks bear rather dense, long, erect, and partly appressed, shining, white pile. The entire front and upper half of the face and all the anterior eye margins are densely covered with long, appressed, truncate, white scales; side margins of front with several long, suberect, bristly white hairs. The antennae are placed at the top of the head, almost on the same plane as the upper eye margin. The occiput is rather poorly developed, smooth in the middle above and without fissure. It bears dense, long, fine, white pile behind forming weak bristles anteriorly, but near the eye margin there is only appressed, white, scaliform pile; posterior eye margin quite entire. The palpus is short, the apical segment is about the same length as the basal one. The proboscis is long and slender, 3 or 4 times as long as the head. The first two segments of the antenna are stout; the first is nearly 3 times as long as the bead-shaped second and bears conspicuous, long, bristly hairs dorsally, laterally, and below. The third segment is not quite twice as long as the combined length of the first two; it is considerably more slender with annulate base and parallel sides for more than the basal half. Distally it is rather strongly narrowed and tapered and ending obliquely with a very short, more or less concealed spine-tipped microsegment.

Head, anterior aspect: The head is about as wide as the thorax across the wings, possibly a little wider. The genofacial space is more than one-third the head width. The eyes of the male are widely separated by the distance between the outer edges of the posterior ocelli. The ocellar tubercle is distinct but not high; the posterior ocelli are a little farther apart than from the anterior ocellus. The ocellar tubercle bears 7 or 8 long, coarse hairs; the upper eye facets are very slightly enlarged. The oral cavity is oval, and of no great length.

Thorax: The thorax is broad, moderately convex anteriorly; the mesonotum and scutellum have rather dense slightly flattened, flat-appressed pile; they are also rather densely covered with coarse, long, silvery white, slightly flattened hairs, with distinct, long, slender bristles on the notopleuron and still longer, more conspicuous bristles on the margin of the scutellum. The pleuron has a dense, appressed tuft of long pile radiating upward and backward from the center in which there is much appressed, shorter, scaliform pile. Metapleuron densely long pilose; the hypopleuron also pilose; squama large with a long fringe.

Legs: The legs are quite slender and have fine, flat-appressed scales. The posterior femur has at least 3 long, rather stout, oblique, white bristles beneath. The anterior tibia has only 2 minute spicules and these are situated posteriorly; posterior tibia with 3 rather longer, slender, anteroventral spicules and 2 minute, dorsal spicules. The tarsi are remarkably slender, especially the terminal segments which are rather less than half as wide as the slender tibia; claws unusually long and quite slender and slightly curved; the pulvilli are about half as long as the claws; the arolium is reduced to a minute stub.

Wings: The wings are hyaline; there are 2 submarginal cells, and the 4 posterior cells are widely open; the discal cell is unusually short and broad, the anterior crossvein enters near or just beyond the basal fourth, the alula is well developed, the ambient vein complete, the anal cell is widely open. There is a moderately well-developed basic, costal comb and sharp beaklike process at the base of the male wing. The costa is thickened from the junction of the first vein nearly to the apex of the wing.

Abdomen: The abdomen is short, obtuse, but rather strongly narrowed at the apex, broad at the base and rather convex. It has a rather abundant but not dense covering of quite long, erect, coarse hairs which on the last 2 or 3 tergites become either dorsally or laterally increased in length and are remarkably long. In addition the surface of the tergites is rather densely covered with completely flat-appressed, short, scaliform pile which may be either white or reddish.

Dr. A. J. Hesse in his work on South African bee flies describes the male hypopygium as follows: "Hypopygium of male much like that of some species of *Bombylius*, with the beaked apical joints elongate and somewhat laterally compressed; aedeagus straight and without any ventral aedeagal process."

Material studied: A pair of the type-species kindly furnished by Dr. Hesse.

Immature stages: Unknown.

Ecology and behavior of adults: Hesse (1938) states that the adults settle on sand in the hottest part of the day, or may be swept from flowers of *Mesembryanthemum*. Hesse further notes an apparent procryptic, sympathetic coloration, the forms on reddish sand displaying a beautiful cinnabar scaling, those upon paler materials being colored accordingly.

Distribution: Ethiopian: *Lepidochlanus fimbriatus* Hesse, 1938 [=*niveus* Bezzi not Macquart, 1921 (as *Dichistus*)].

Genus *Adelidea* Macquart

FIGURES 34, 256, 266, 478, 516, 739, 1006, 1007, 1008

Adelidea Macquart, Diptères exotiques, vol. 2, pt. 1, p. 84, 1840. Type of genus: *Bombylius anomalus* Wiedemann, 1821, by monotypy, as *Adelidea fuscipennis* Macquart, 1840.

Sobarus Loew, Neue Beiträge III, p. 39, 1855. Type of genus: *Bombylius anomalus* Wiedemann, 1821, by monotypy.

Hesse (1938), pp. 681-686, key to 7 South African species.

Small flies, dark colored and pollinose; all species have fine, yet stiff, sparse, scattered, rather long, and erect hairs on the abdomen, which are more concentrated along the sides, on the mesonotum, on the pleuron, and on the sides of the extended face and oragenal cup. Thus, it will be seen that these flies are relatively bare in appearance. The male eyes are dichoptic. There are 3 submarginal cells and the widely open first posterior cell immediately separates it from *Triplasius* Loew. Wings in the known species are always infuscated; spots are often found on the crossveins. The anterior crossvein lies at middle of the discal cell; sometimes a supernumerary crossvein lies nearer the base of that cell; anal cell open. Basal comb and hook weak. Beneath the scattered longer hairs there is a dense coat, sometimes worn thin, of brassy yellow, slightly flattened, adpressed hairs, which do not obscure the underlying pollen. The abdomen is short, oval, and broad, or rather flattened.

Antennae elongate; third segment knoblike at base, widest near the base and attenuate beyond, with short, small, spine-tipped microsegment; third segment frequently with dense layer of spinules below. Legs weak and slender; hind femur with 2 or 3 weak bristles below; anterior tibia with minute short bristles; hind tibia with 3 or 4 longer weak bristles in rows.

This genus is superficially like *Sparnopolius* Loew to which it bears no relation. It may be near the Chilean genus *Platamomyia* Brethes, which has 3 submarginal cells and flattened abdomen.

Adelidea is a characteristic element of the South African fauna. I place it in the Dischistini on the basis of its *Bombylius*-like facies and open posterior cell. *Sosiomyia* Bezzi is only a highly aberrant related form.

Length: Species vary from 5 to 12 mm., usually 6 to 7 mm.; wing slightly longer.

Head, lateral aspect: The head is triangular due to the strong forward triangular projection of the face, carrying the oragenal cup with it below the face. Occiput is prominent throughout but especially so in the middle near where it joins the thorax; posterior eye margins entire. The pile of the occiput is abundant, erect, appressed near the margin, moderately long and bristly inwardly; it becomes still longer beneath the eye. The pile of the face is rather similar to that upon the occiput, consisting of scattered, long, slender but stiff or wiry hairs of long curled, silky hairs, all pale, or partly black, that extend entirely from the lower gena all around the sides and middle of face and extend loosely in a fringe over the edge of the oral opening. In the related *Sosiomyia* Bezzi the circular fringe is more bushy and longer. The front is flat, longitudinally depressed in males, with a tuft of bristly pile near the eye and just above the antennae and in the middle a few appressed, bristly setae. The front may have dense, wiry, flattened pile, but in species like *pterosticta* Hesse the front is nearly bare, but the lateral tuft is present. Face, front, and occiput everywhere are densely pollinose obscuring the ground color. The antennae carry the same pollen. The eye is quite regular and high, the posterior margin is entire but lacks much of being plane. In fact it is almost as much rounded as the anterior half of the eye.

Palpus of 2 segments, conspicuous, slender and rather long; it bears a few appressed, stiff hairs along the sides and a few at apex; the apical segment often very elongate. The proboscis is long and slender, the labellum likewise; the proboscis is about 2 to 4 times as long as the head. The antennae are attached from the upper third even as high as the upper fifth of the head or almost in the same plane as the vertex. The antennae are elongate, relatively slender but not large. They are not as long as the head but are a little longer than the width of the eye. The first segment is 2 to 3 times as long as wide, nearly 4 times as long as the second segment, a little wider at the apex with 2 or 3 setae at the apex below and several longer, bristly hairs dorsally in the middle; second segment distinctly short and beadlike and rounded with a few minute setae around its middle. The third segment has an annulate base, is narrower than the second segment and then a little widened on the basal half both above and below; this segment bears a few, scattered, short, suberect, bristly hairs on the expanded portion and 2 quite small microsegments and the last one has a strong bristle or short, oblique spine at the tip.

Some of the species of *Adelidea* Macquart, but by no means all of them, bear an erect band of spinulose pubescence on the ventral side of the third antennal segment; on the upper surface of the base there are 2 tiny setae; this pubescence is lacking in *braunsii* Bezzi, *immaculatus* Bezzi, *nigrifemoris* Hesse, and in *pterosticta* Hesse; it is, however, present in *anomolus* Wiedemann and *maculata* Hesse.

Head, anterior aspect: The head is not quite as wide as the thorax, the eyes in the male are widely separated by a distance slightly less than the width of the ocellar triangle. The head is considerably wider than high. The antennae are narrowly separated, the ocellar tubercle is only moderaly high, not sharply separated from the vertex in front but a litle more distinctly demarcated behind and laterally. It is in fact broadly and gently rounded, pollinose with some 20 long, slender bristles or bristly hairs rising from it. The upper eye facets are scarcely enlarged. The ocelli are large, the ocellar triangle not quite equilateral, the ocelli even in males are widely separated from the eye margin and the posterior pair lie well behind the posterior eye corners. The posterior occiput has a most shallow, medial, vertical depression. The oral opening is deep, rather large, its width certainly more than half the interocular space. The oragenal cup is a little thickened laterally; it is thin above and below, the upper part of the recess is pollinose and smooth and has no pile; gena narrow, except at the lower corner of the eye; tentorial fissure deep.

Thorax: The thorax is quite shallowly convex, the mesonotum blackish or brown, often yellowish or lighter colored on the humerus and side margins, whole surface thinly opaque pollinose. The pile of the mesonotum consists of fine, curved, rather short but scattered and scanty bristly hairs and approximately an equal amount both of flat-appressed, brassy, slightly scaliform pile and also, especially anteriorly and laterally somewhat longer pile of fine, yellowish, erect hairs; females have the erect pile partly black. The notopleuron has 3 or 4 distinct long bristles, not very stout; scutellum clothed like the mesonotum. Nearly the whole of the mesopluron is pilose, the posterior border has tufts of long, stiff pile apt to be brassy and a few bristly hairs turned upward along the upper border. The whole pleuron has very fine pollen, pteropleuron and metapleuron lack pile; the hypopleuron, however, has a tuft posteriorly behind the spiracle and beneath the haltere; squama large with a rather short, scanty fringe. Propleuron with a tuft of pile above the front coxa. The haltere is rather large, though the knob is thin and flattened.

Legs: The first 4 legs are relatively short; all the femora are relatively slender, the hind legs are somewhat longer. The anterior tibia has weak, rather minute spicules, 7 or 8 dorsally, and 3 or 4 posteroventrally; spicules of middle tibia better developed; there are 5 dorsally, a like number anteroventrally and posteroventrally, and likewise posterodorsally. The posterior femur bears dense, appressed, rather long, coarse, scaliform pile; there is no long pile anywhere, but ventrally along the middle it has 3 or 4 weak, short bristles or in *anomalus* Wiedemann with 5 to 7 strong, pale bristles; its tibia has about the same number of spicules in each row as on the middle pair. The pulvilli are long, the claws quite slender and sharp and a little curved at the apex; arolium short and stublike.

Wings: Elongate and slender and in all of the known species of *Adelidea* Macquart the wings are brownish infuscated and at most hyaline in only part; diffuse, darker spots are often found on crossveins in some species; darker species may have clear sinuses in cells. There are 3 submarginal cells; the marginal cell itself is rather strongly expanded apically, the end of the cell rounded, the second vein at first pulled backward, then turning to meet the costa at a right angle. The anterior branch of the third vein arises rather sharply, is joined rectangularly by the branch from the second vein, then continues to make a rounded bend to join the costa some distance from the apex of the wing. The first posterior cell is widely open, the anterior crossvein joins the discal cell at the middle, or a little beyond, or even at the outer fourth, and sometimes as in *pterosticta* Hesse it has an additional crossvein nearer the base of the discal cell. The anal cell is widely open, axillary lobe narrow, alula moderately well developed, with an inconspicuous fringe, ambient vein complete. The base of the costa has a weak comb.

Abdomen: The abdomen is short, rather broad and flattened, or at most shallowly convex, scarcely longer than wide, shining but with thin pollen. The pile may consist of thinly scattered hairs, a few very long, wiry hairs over the middle of the tergites which are more numerous, however, along the sides of the second and third tergites and they are mixed with some long, stiff pile; also there is a scanty amount of shorter, flat-appressed, wiry, slightly flattened pile present on the tergites tending to overlap on the posterior margin onto the next segment. First segment with a band of erect pile which is rather loose. In some species, such as *anomalus* Wiedemann, the tergites are covered with dense, flat-appressed, curled, glittering pile, and the postmargins have a row of long, erect, wiry hairs, yellow in males, black in females.

In his summary of the male terminalia, Hesse (1938) found *Adelidea* Macquart to be characterized by the shape of the dististyli (beaked apical segments):

These are more or less dorsoventrally compressed, scarcely depressed and more planiform above, more hollowed out below, with the outer, apical angle almost always prominently subangular or even subacute and with the dorsum provided with backwardly directed, stiff, bristly hairs; aedeagus more or less hidden from below by the ventral aedeagal process which forms either a single, forwardly projected process, or a more complex form, more or less divided apically into two, broadened, thin, lamellate, flaplike lobes curled downward and inward to form a hood or cowl.

Material studied: The series of flies in this genus at the British Museum (Natural History) and material of *anomala* Wiedemann kindly sent by Dr. A. J. Hesse for study and dissection. I also have material of *pterosticta* Hesse given to me by the British Museum.

Immature stages: Unknown.

Ecology and behavior of adults: Hesse (1938) says that several species like to settle in wet environments along mossy banks, damp soil, rocks along streams or even along dried-up water courses.

Distribution: Ethiopian: *Adelidea anomala* (Wiedemann), 1821 (as *Bombylius anomalus*) [=*fuscipennis* Macquart, 1840, =*longirostris* (Wiedemann), 1828 (as *Cyllenia*)], *anomala fuligineipennis* Hesse, 1938; *braunsi* Bezzi, 1921; *immaculata* Bezzi, 1922; *maculata* Hesse, 1938; *nigrifemoris* Hesse, 1938; *pterosticta* Hesse, 1938; *ruficornis* Bezzi, 1921.

Genus *Sosiomyia* Bezzi

FIGURES 32, 279, 452, 810, 811

Sosiomyia Bezzi, Ann. South African Mus., vol. 18, p. 67, 1921a. Type of genus: *Sosiomyia carnata* Bezzi, p. 69, 1921a, by original designation. Also, as *Sosiomyia comata* Bezzi, p. 470, 1921b, lapsus.

Medium-sized flies, with sparse, shaggy, bristly pile and with short, obtuse abdomen. The whole fly is black, except for the yellow tibiae and tarsi and yellowish brown delicately mottled wings. The pile on the dorsal half is likewise yellowish to golden red; the ventral pile is everywhere grayish white. The ground color is opaque covered with blue-gray pollen or bloom. The genus is distinguished by the peculiar third antennal segment together with the 3 submarginal cells in the wing. This third antennal segment is broad and moreover is laterally compressed, especially in the female, and I have illustrated the antenna of both sexes; this segment bears 2 or 3 long, stoutish bristles dorsally near the base and short, stiff pubescence along the medial surface ending in 2 or 3 short bristles. All of the bristles and pile are exaggerated in the females. Male eyes widely dichoptic, separated by nearly the width of the ocellar tubercle. Another peculiarity lies in the unusually bristly character of the covering on face, front, occiput, mesonotum, pleuron, scutellum, postmargins of the tergites and sides of the tergites where it is even more outstanding.

The alula is prominent and all the cells are widely open, the first posterior cell only slightly narrowed.

This fly seems to be most closely related to *Adelidea* Macquart, differing in the characters given above. It is entirely South African; a single species known.

Length: 5 to 9 mm.; wing 6 to 11 mm.

Head, lateral aspect: The head somewhat triangular, the face extended forward beneath the antennae. The eye itself is quite triangular in males. The oragenal cup is much as in *Systoechus* Loew; it is more extensive above, retreating below. The occiput rather tumid, especially below, although sloping gradually to the eye margin. The face bears long, bristly pile, loose and scattered in character, golden above and extended down as equally long, coarse, silvery hairs along the sides of the oragenal cup, but the genal pile is shorter, subappressed, shining, silvery white. Whole surface of face densely grayish white pollinose, but brown pollinose on each side of the antennae. Pile of occiput rather sparse and straggly but with much coarse hair and slender bristles concentrated more or less medially and deeply on the upper part of the occiput. Whole occiput gray pollinose. Lower pile of the occiput, however, short, fine, and white but extraordinarily long and dense and shining white beneath the head. The eye has the upper facets a little enlarged, the posterior margin entire and the upper part of the eye a little flattened in males. Proboscis slender, about twice as long as the head, the labellum quite slender. Palpi short and very slender. Front in the male flattened with tufts of bushy, golden pile, somewhat scaliform, somewhat appressed and directed obliquely outward.

In females the front is also flattened, somewhat depressed and covered with some scanty patches of short, curled, appressed, reddish golden, scaliform pile, and on each side near the eye with a small cluster of long, reddish golden bristles curled forward. Antennae attached in the upper fifth of the head considerably shorter than the head in length. The first segment is at most 3 times as long as wide and rather robust. Second segment quite short, not as long as high; third segment in the male barely longer than the first two, strongly knobbed at the base, compressed laterally, rather broad and stout basally and then beginning near the middle narrowed to the apex where it bears 2 small microsegments and an apical spine. In the female this segment is a little shorter but much widened and broadened and greatly wider than the second segment; it is vase-shaped, narrowed at the immediate base and near the apex strongly narrowed. In both sexes the third segment has several long, stout bristles dorsally and shorter ones at the apex. In both sexes the first segment has long, scattered bristles laterally and below, some shorter ones above.

Head, anterior aspect: The head is as wide as thorax, twice as wide as high, male face widening below. The upper part of the female face is as wide as the front and both very wide. The face is only slightly wider below. The eyes are widely separated in the male by a distance equal to the total width of the ocellar tubercle. The antennae are narrowly separated at the base, the ocellar tubercle is a little swollen and only moderately protuberant. The large ocelli are formed in an obtuse triangle and are separated from the eye margin by less than their width. In the center is a tuft of moderately long, bristly hairs. The oral opening is of average size, placed near the bottom of the head, deep posteriorly with the inner wall only slightly thickened; the outer wall opposite the proboscis bears some long, coarse pile. The gena is wide, pollinose, and bears short pile. The tentorial fissure is deep and rather conspicuous, except as hidden by pile. From the dorsal aspect the occiput is very shallowly bilobate, the groove behind the vertical pollen disappearing below.

Thorax: The thorax is short and broad, slightly narrowed anteriorly, black, with opaque gray pollen, more or less vittate and covered with fine, flat, short, appressed, curled, reddish golden pile, rather loosely and also scattered, erect, not very long, shining reddish, bristly pile. Notopleuron with a tuft of 5 long, slender red bristles and others along the lateral margin. Several long bristles on the postalar callosity. The scutellum is thin, gray pollinose, with short, fine, appressed pile. The scutellum has a double, irregular row of some 20 long, slender, red bristles on each half of the posterior border. Vertical postalar border with a little pile above only. Mesopleuron in the upper posterior corner with a tuft of red bristles and over most of its surface is fine, white, semiappressed pile. Pteropleuron anteriorly with a scattered band of pile. Hypopleuron bare, metapleuron with a large tuft of pile only behind the posterior spiracle. Squamal fringe, long, thin and yellow.

Legs: The legs are rather long, tibiae and tarsi quite slender, the femora distinctly thickened. Anterior tibia with 3 rows of conspicuous bristles; the anterobasal row has 5 long and 2 short bristles; ventral row with 3, the posterior row with 7. Middle tibia similar. Apical bristles of the midtibia quite long. The hind tibia has unusually long, stout bristles in 4 rows; dorsal row with 5, lateral row with 5, the ventrolateral row with 3, and the medial row with 5. Middle and anterior femora widened basally, hind femur with a ventral row of 7 long, stout, spinous bristles and an additional even stouter bristle laterally on the apical fifth. Pile consists of long, coarse, dense, flat-appressed scaliform pile, mostly of a silvery color. Claws long, sharp, the long pulvilli are slender and spatulate.

Wings: There are 3 submarginal cells, the second vein makes a rectangular bend forward into the costa. All posterior cells and the anal cell widely open. Anterior crossvein lies at the outer third of the discal cell, intercalary vein long, axillary lobe of no great width, but the alula large and evenly rounded. The base of the wing, including the alula, is light reddish brown. There are brown spots on the crossvein and vein forks. Costal and subcostal cells yellow and yellow spots in the middle of the first basal cell followed by brown. Remainder of wing tinged with reddish golden villi. Ambient vein complete.

Abdomen: Abdomen black with dense, gray pollen, rather strongly flattened, and a little wider than the thorax. The abdomen is only about as long as wide, slightly narrowed beyond the third segment. First tergite with a band of short, yellow pile beneath the scutellum, becoming longer laterally. All other tergites with scattered, appressed, curled, scaliform, golden hairs, and some fine, scanty, erect, long hairs and likewise with a row of quite long, slender, reddish golden bristles along the posterior margins of the tergites. However, the lateral margins of the tergites have some bushy tufts of yellow and whitish yellow pile, and a considerable number quite long, backwardly turned, slender, black bristles. Sternites with white pile both appressed and erect. Male terminalia recessed at the apex of the abdomen and turned to one side.

The male terminalia from dorsal aspect, epandrium removed, show a comparatively slender figure. The epiphallus is rather strongly expanded toward the base. There is a considerable membranous area and a slender, convergent, conspicuous, backward prong from the posterior part of the basistylus and an unusually large, conspicuous, short, swollen dististylus. Apically the dististylus bears a small, divergent, rather short process, which is not very sharply pointed. In the lateral aspect, however, the dististylus is long, slender, attenuate and curved upward. From the epiphallus the ejaculatory process is free and ventral. The basal apodeme is short, small, and inconspicuous, the epandrium large and quadrate.

Material available for study: A pair of the type-species sent me for dissection and study through the kindness of Dr. A. J. Hesse.

Immature stages: Unknown.

Ecology and behavior of adults: I have found no comment in the literature.

Distribution: Ethiopian: *Sosiomyia carnata* Bezzi, 1921 [=*comata* Bezzi, 1921, lapsus].

Genus *Sparnopolius* Loew

FIGURES 204, 420, 692, 805, 806

Sparnopolius Loew, Neue Beiträge III, p. 43, 1855. Type of genus: *Bombylius l'herminierii* Macquart, Diptères exotiques, vol. 2, pt. 1, p. 103, 1840, as *Bombylius fulvus* Wiedemann, 1821, by original designation; preoccupied by *Bombylius fulvus* Meigen, 1820. Next available species. Not *Bombylius breviorostris* Macquart, 1840, as designated by Coquillett, p. 606, 1910a; preoccupied by *Bombylius breviorostris* Olivier, 1789.

Painter (1939), p. 280, key to 6 Nearctic, Neotropical species.

Small to medium-sized flies. The venation of the hyaline wings is of the *Dischistus* Loew type in that the first posterior cell is open maximally, and the anterior crossvein is at the basal third of the widened discal cell; moreover, the first two antennal segments are thickened though not as much as in *Dischistus* Loew. The eyes of the male are extensively holoptic, depressed where they join, and the male front is tiny and minute. Two other characteristics stand out in this genus. The oragenal cup which is distinctly protruded and obliquely truncate is, together with the face as well, bare of pile and shining. There are a few, fine, short hairs on the lower part of the very narrow gena wall and outer edge of oragenal cup very little thickened. Proboscis twice the head length; palpus long and slender. Secondly, the whole thorax and abdomen are densely covered with erect, fine pile; bristles absent even on notopleuron. Females with some appressed nonscaliform pile on the abdomen among the longer erect hairs. Hind femur with appressed pile and a row of moderately long stiff hairs below; no bristles. Anterior tibia with short spicules only ventrally; hind tibia with weak, scattered bristles, more numerous below.

Sparnopolius Loew is a Holarctic genus; a few of the species range into Mexico. I have not seen the 3 species supposed to come from Brazil, nor the *Sparnopolius limbatus* Bigot from Australia, but I suspect they belong elsewhere.

Length: 6 to 10 mm., including antenna; wing 4 to 6 mm.

Head, lateral aspect: The head is almost hemispherical with flattened front in the females. In the males the front is flat and the eyes flattened and also depressed at contact. The occiput is only moderately developed behind the eye. The face below the antennae projects forward angularly somewhat less than the length of the first antennal segment. The oragenal cup is large, retreating, and quite plane across the front. In other species the face is quite short, constituting a mere rim beneath the antennae. It is shining and bare, except for a pollinose spot along the eye margin. The occiput is a little more pronounced near the middle of the eye; it slopes gradually outward to the eye margin, and bears dense, coarse, rather long, bristly pile on its whole surface, which becomes shorter and somewhat appressed close to the eye margin. The eye is hemispherical, except as it is flattened above in males, and there is a very broad, quite shallow concavity along the posterior margin. Pile absent. Proboscis 1½ times as long as the head, with rather stout, long, lanceolate labellae. Palpus of 2 segments, rather long and slender, scarcely extending beyond the oral cavity, and with scattered, weak hairs laterally and at the apex. The basal segment is very short. The front is extremely small in males, no larger than the ocellar triangle and bears only appressed micropubescence.

Antenna attached just above the middle of the head. The first segment is always stout and swollen; it is 2½ times as long as the short, swollen second segment; it shows a relationship to *Conophorus* Meigen. Both of these segments have rather numerous, long bristles below and somewhat shorter bristles above and laterally. Third segment more elongate, and very little longer than the first segment but more slender. The base is knoblike. The third segment is widest in the middle but never as wide as the second segment and is attenuate on the outer half. It bears a small spine-tipped microsegment.

Head, anterior aspect: The head is circular on the upper half, much less so below and is a little wider than the thorax. The female front is almost as wide as the face and in males the eyes are holoptic and touch for more than twice the length of the small front. The antennae are narrowly separated, the ocellar tubercle is prominent but small and nearly equilateral. Ocelli large, the oral opening is widest below, the lateral wall of the oragenal cup is slightly thickened, especially on its inner surface and is knife-thin posteriorly, shallow anteriorly and deep behind. The gena is linear and quite narrow; the tentorial groove deepest across the middle of the oragenal opening. The gena bears a row of scattered, rather long, fine hairs on the lower half. Vertically there is a shallow crease dorsomedially on the occiput below the vertex.

Thorax: The thorax is about 1½ times as long as wide and very densely covered with moderately long, erect, shining pile, which becomes more sparse posteriorly. Scutellum similarly covered, the pile longer and finer in males, more coarse, flattened and appressed in females. Bristles absent, even on the notopleuron. Pteropleuron, hypopleuron and all of metapleuron bare, except for a small tuft of hairs behind the spiracle. Squama with a dense, brushlike fringe of rather short, coarse hair.

Legs: The legs are comparatively short and slender. Anterior tibia with a row of 5 or 6 fine, short bristles ventrally, middle tibia with 3 or 4 similar anteroventral bristles, a like number of anterodorsal bristles and posterodorsal bristles, but with a fringe of nearly 20 posteroventral bristles in the type-species only. Hind tibia with several rows of a few bristles and with some 12 or 15 minute bristles ventromedially and ventrodorsally.

Femora without bristles below but with a fringe of scattered, coarse hairs beneath the hind femur. Pile of the legs coarse, slightly flattened, shining, and flat-appressed. Pulvilli slender and well developed, claws fine and slender.

Wings: The wings are hyaline, the venation generalized; all cells open; two submarginal cells; first posterior cell opened maximally, discal cell rather short and comparatively broad, angulate below, the anterior crossvein lies near the basal fourth of the discal cell. The first basal cell is a little longer than the second. Basal comb absent though there is a little coarse pile at the base of the costa. No costal hook present. The axillary lobe is well developed. The alula is moderately wide and ambient vein complete. The marginal cell is somewhat expanded apically.

Abdomen: The abdomen is elongate oval, as wide or a little wider than the thorax and narrowing slightly at the apex. It is slightly flattened over the middle and bears long, scattered, rather fine, erect hairs on the dorsum of all the tergites in the male. This pile becomes much more conspicuous, denser, and brushlike on the rolled over side margins of the tergites. In females the pile is not as long and is largely flat-appressed, somewhat flattened, more brightly shining and scale-like and is more conspicuous among the posterior margins of the tergites. Apical tergites sometimes with a very few long, fine, bristly hairs of contrasting color. In females I find 7 segments, the terminalia enclosed ventrally by overlapping, dense fringes of hair. The last sternite is oval. Males with 9 segments visible, the genitalia recessed and lying side by side beneath the last segment.

The male terminalia from dorsal aspect, epandrium removed, show a curious broad, triangular aspect of the general figure. The epiphallus is broad and wide basally ending at the base of the long, slender, attenuate dististyli which have no hooks or process. The aedeagus appears to be largely merged with the epiphallus. From the lateral aspect the large epandrium is subtriangular and the cerci are large. Except for the absence of the epandrium lobe, it is very much like *Lordotus* Loew in general appearance.

Material available for study: I have a large series of the type-species which is an abundant species in Mississippi, and I also have a large series of the smaller, western species that my wife and I collected in the Western States in September 1964.

Immature stages: The larvae of the type-species attack white grubs in the soil. Life history of smaller species unknown.

Ecology and behavior of adults: These flies are readily taken at composites such as sneezeweed and sunflowers in Mississippi during the month of September.

Distribution: Nearctic: *Sparnopolius anomalus* Painter, 1939 (1940); *brevicornis* Loew, 1872; *coloradensis* Grote, 1867; *cumatilis* Grote, 1867; *lherminierii* (Macquart), 1840 (as *Bombylius* [=*brevirostis* (Macquart), 1840 (as *Bombylius*), =*fulvus* (Wiedemann), 1821 (as *Bombylius*), =*fuscipes* (Bigot), 1892 (as *Dischistus*)].

Neotropical: *Sparnopolius apertus* (Macquart), 1847 (as *Bombylius*); *bicinctus* (Wiedemann), 1830 (as *Bombylius*); *caminarius* Wiedemann, 1830; *confusus* (Wiedemann), 1824 (as *Bombylius*) [=*senilis* (Wiedemann), not Fabricius, 1821]; *diversus* Williston, 1901; *nigriventris* (Philippi), (as *Sparnopolius*).

Palaearctic: *Sparnopolius asiaticus* Becker, 1913.

Australian: *Sparnopolius limbatus* Bigot, 1892.

Genus *Sericusia* Edwards

FIGURES 30, 50, 267, 491, 517

Sericusia Edwards, Rev. Chilena Hist. Nat., p. 36, vol. 40, 1936. Type of genus: *Sericusia lanata* Edwards, 1936, by original designation.

Small, black flies appearing opaque because of the fine, gray-white pollen of head, thorax, scutellum, and abdomen, and the abundant, long, erect, silvery white pile. It appears to be related to *Sparnopolius* Loew. The antenna is similar and the venation of the hyaline wing is similar, except that the anterior crossvein lies slightly beyond the middle of the discal cell. An important difference lies in the presence of the long, conspicuous pile upon the face and oragenal cup, which are bare in *Sparnopolius* Loew; also the metapleuron is densely hairy, but bare in *Sparnopolius* Loew. It also appears to be related, perhaps even closer, to the African *Lepidochlanus* Hesse, from which it differs in the absence of bristles beneath the hind femur, and the position of the anterior crossvein; in *Lepidochlanus* Hesse this crossvein is basal. In *Sericusia* Edwards there are 2 submarginal cells and the posterior cells are widely open; the anal cell is closed in the margin or may be very narrowly open; the alula is well developed, with a long fringe.

This small genus is restricted to the Chilean section of the Neotropical region; the type-species was described from Argentina.

Length: 6 to 7 mm.; wing 5 mm.

Head, lateral aspect: The head is hemispherical in profile but the face is rather short, the oragenal cup retreating. There is a conspicuous, circular fringe of long, dense, erect, fine, shining white pile on the whole of the face which is curled downward over the oral cavity and continued on down the gena, where it becomes much shorter opposite the proboscis. The pile of the front, however, while moderately long is contrastingly appressed and lies very close to the surface. The entire head is everywhere grayish white pollinose over a black background. The sides of the vertex close to the eye have a tuft of some 10 long, slender, more or less erect, white bristles. The ocellar tubercle is low but very gradually raised and without demarcation. The occiput is quite prominent but only centrally as it gradually slopes down to the eye margin outwardly. There is no conspicuous central cavity, and it is rather densely covered with long, slender bristles medially and dorsally, which curve backward, and give way to shorter pile anteriorly which becomes quite short near the eye margin; pile of the lower occiput moderately long and quite fine. The palpus is slender, and short especially on the second segment which is about as long as the first; second segment with a row of long, fine, bristly hairs laterally and 1 or 2 setae at the apex.

Proboscis quite long and slender, the labellum exceptionally small, short, and even more slender. The proboscis is at least 5 or 6 times as long as the head. The antennae are attached closely adjacent and they are attached near the top of the head on a plane nearly equal that of the upper eye margin. They are slender and elongate, the first segment twice as long as the second and with rather long, fine hairs laterally and below and somewhat shorter hairs above. The second segment is bead-shaped, a little narrowed at the base. The third segment is narrowed basally and annulate at the base; it is slightly widened on the basal third then gently tapered to the narrow apex which lacks a microsegment; it has a minute spine at the apex.

Head, anterior aspect: The head is wide and considerably wider than the thorax with the genofacial space occupying distinctly more than one-third of the head width; oral cavity short oval, sides of the orgenal cup wide but sunken and somewhat concave. The genae are also rather wide. The face and front are continuous, the female front almost as wide as the face. Antennae quite narrowly separated, almost adjacent.

Thorax: The thorax is a little longer than wide and more or less gibbous and rather strongly convex, especially on the anterior half. It is gray pollinose and opaque and everywhere conspicuously and densely covered with long, erect, coarse, silvery white pile which is almost bristly on the notopleuron and above the wing. The scutellum is similarly covered with long pile, rather triangular in shape and its marginal pile is somewhat more coarse and bristly; squama large, with a long, fine fringe on the margin. The pleuron is pollinose and densely pilose; the metapleuron is without pile, though mesopleural pile extends backward over it; the hypopleuron has a tuft of dense, appressed pile only along its posterior border but which changes to erect hairs beneath the halteres.

Legs: The legs are comparatively short, stout, especially on the femur, the anterior femur unusually shortened, the tarsi and to some extent the tibiae are longer and a little more slender. The pile is dense, flat-appressed, silvery white and scaliform and rather long in character. The hind femur lacks any long pile beneath although the anterior and middle femora, especially the middle one has some rather long, shaggy, more or less erect pile below. Bristles are quite absent on the femora; anterior tibia with 2 or 3 minute spicules posteriorly in 2 rows, only 2 in each row. Both middle and hind tibiae however, especially the latter, have somewhat longer and more conspicuous spicules but none of them large. The claws are quite slender, bent at the immediate apex, the pulvilli long, the arolium minute and stublike.

Wings: The wings are quite hyaline, the veins pale. The base of the costa has some fine, long hairs but no comb. The subcosta and the first vein end quite close together and are really fused and a little expanded where they enter the costa. The fourth branch of the radius turns up distinctly though not conspicuously. There are 2 submarginal cells, and all 4 posterior cells are widely open; the anterior crossvein enters the discal cell distinctly beyond the middle. The anal cell is closed in the margin or narrowly open; alula twice as long as wide, somewhat produced apically and with a rather long fringe.

Abdomen: The abdomen is elongate, subconical, and distinctly though moderately decumbent. At the base it is not quite as wide as at the thorax and begins to be somewhat compressed laterally, even near the base. It is covered loosely both with long, fine, erect, white pile and some shorter, equally fine, long, flat-appressed, white pile. Postmargins of second to fifth tergites with 1 or 2 rows of long and more coarse and somewhat more bristly hairs. First tergite laterally with dense, long, erect, coarse pile.

Material studied: The type-species in the British Museum (Natural History), from which my description and illustrations are taken.

Immature stages: Unknown.

Ecology and behavior of adults: Not on record.

Distribution: Neotropical (Argentina): *Sericusia lanata* Edwards, 1936.

Genus *Platamomyia* Brèthes

Figures 21, 286, 486, 1009, 1010, 1011

Platamodes Loew, Neue Beiträge III, p. 40, 1855. Type of genus: *Platamodes depressus* Loew, 1855, by monotypy. Preoccupied by Ménetries, Coleoptera, 1849. No. 79 in Loew's collection as *Bombylius depressus*.
Platamomyia Brèthes, Rev. Chileña Hist. Nat., vol. 28, p. 105, 1925. Change of name.

Medium-sized flies of unusually compact robust form. The abdomen is moderately flattened, subtriangular, though rounded laterally on the basal half of each side and much wider than the thorax. Thorax, abdomen and scutellum are opaque with comparatively dense pile, especially dense on the abdomen, and where it is matted, curled, bristly and glittering brassy yellow with additional erect hairs. Mesonotum obscurely vittate. The head is wide from the anterior aspect with very wide, opaque gena which bears a fringe of numerous, long, erect, brassy hairs. The true face, beneath the antennae and above the oral recess, is extensive. On each side of the gena is a bare, shining, black, vitreous, oral recess which is short-oval and wider below. The proboscis is like that of *Bombylius* Linné or *Conophorus* Meigen; it is long and slender. The antenna resembles *Conophorus* Meigen. The first segment is stout, swollen, and robust. The third segment has numerous, moderately long, slender hairs, placed dorsally and also below near the apex. The eyes are well separated in the male by a little more than the width of the large, high, conspicuous ocellarium; in fact from the anterior aspect the eyes are sunken on each side of the ocellar tubercle. The triangular front bears a dense mat of pale, brassy yellow pile, the very long, matted hairs are extended downward on each side of the antennae, and throughout the full width of the front.

This genus is further characterized by the venation; the wing is relatively slender, the marginal cell strongly recurrent at the apex; there are 3 submarginal cells. The anterior branch of the third vein ends well before the apex of the wing; first posterior cell a little narrowed apically; anal cell open; ambient vein unusually stout and complete, alula about half as wide as long and circular. The anterior crossvein enters the discal cell distinctly beyond the middle, but not at the outer third. Male terminalia enclosed by a cuplike collar, short and downturned but otherwise somewhat similar to *Acrophthalmyda* Bigot.

This fly is known from Chile.

Length: 8 mm. excluding proboscis; 10 mm. including proboscis; wing 7½ mm.

Head, lateral aspect: The head is at least as long as high. The posterior eye margin is plane, but the occiput is extensive both above and below, rounding off near the top and bottom. While there is a narrow fissure dorsally behind the ocelli, the cavity in the center of the occiput is small as in *Bombylius* Linné. The occipital pile is very dense and subopaque, at most moderately shining in part; it is shorter near the ocular margin. Whole occipital surface pollinose and punctate. The face is moderately extended beyond the eye and is extensive beneath the antennae. There is a broad, wide gena extending downward, also pollinose and punctate, with long, brassy yellow, erect pile continuing below the eyes and below the occiput. The gena is rectangularly separated from the shining black, bare, rather long, oragenal cup. The proboscis is long and slender with a short, slender labellum. It is about twice as long as the head. The palpus is large and apparently composed of 3 segments. The apical segment, short, distinct, clavate, turned upward, and all of the segments with long, fine hairs. The front is widely separated in the male by a little more than the width of the large, relatively high ocellar tubercle, the sides of which are vertical. Remains of front widely triangular, with a dense mat of pale, shining yellow pile. The hair is quite long and flat-appressed, the mat extending down on each side of the antennae and to the base of the antennae in the middle. The ocellar tubercle has several long, slender hairs curved forward and upward.

The antennae are rather conspicuous, with the first segment swollen and robust as in *Conophorus* Meigen and about 2½ times as long as wide. This segment bears much long, pale yellow, bristly pile curved out in a matlike fringe, with fewer hairs arising dorsally; the pile of this segment continues as a loose fringe around the bottom of the segment. These segments are separated at the base by almost the thickness of the second segment, but because they are swollen also toward the middle, they are more narrowly separated along the greater part of their length. Second segment short and as long as wide, not as stout as the first segment. Third segment about as long as the first segment and much less stout; it is attenuate apically and bears numerous brassy yellow hairs along its entire dorsal length and a few other hairs ventrally near the apex.

Head, anterior aspect: The head although wide is more narrow than the thorax. The upper eye facets are only a little enlarged. The ocellar tubercle is prominent and the anterior ocellus large. The oral recess is large, slightly narrowed anteriorly, and is comparatively deep, but with vertical sides and the walls thin. The tentorial fissure is present but narrow; the genal area is prominent and wide and thinly pollinose and also punctiform.

Thorax: The mesonotum is posteriorly as wide as long, it is slightly narrowed anteriorly, the transverse suture is rather short, its medial ends very widely separated. The scutellum is wide, at least twice as wide as long, and together with the mesonotum entirely opaque and covered with olive brown punctate pollen over a black background. There are a pair of distinct, slightly darker, narrowly separated, medial vittae which end just beyond the base of the wings. Laterally there is a much more obscure vitta. Mesonotal and scutellar pile mostly fine and erect, rather scanty, but with perhaps more abundant erect or suberect, short, fine, brassy yellow pile which tends to be curled or curved. Humerus with much long, erect pile; the anterior collar of the mesonotum merges with the occipital pile. There are at most a few very slender yellow bristles among the dense pile of the notopleuron and postalar callosity. The pleuron is black with dense reddish-brown pollen and much light yellow, matted, tufted, long, pale pile on the mesopleuron. Some of its extends upward; much of it matted and turned backward. Pteropleuron, metapleuron, hypopleuron pollinose only. The abdomen is so compacted that it is difficult to inspect the short metanotum. It appears to be without pile. Hyperpleuron above the middle coxae bare, with a tuft of fine hairs behind the spiracle. The haltere is triangularly oval, swollen and rounded from the dorsal aspect but with strong cuplike recess ventrally. Squama moderately well developed with fine, long fringe and with an ambient veinlike margin.

Legs: The legs are weak; all the femora are quite slender and they are everywhere light brown. The pile is appressed brassy yellow rather scanty on the femora; the anterior femur bears a fringe of loose long hairs ventrally and behind. The middle femur is similar. Hind femur with 10 or 12 very slender, erect, long, stiff hairs ventrally but with a more conspicuous fringe of equally long, erect hairs on the entire medial surface from top to bottom. It also has scattered, appressed, brassy pile. The tibia has long slender bristles in several rows in addition to short, appressed, glittering, brassy, subscaliform pile. The tarsi are large and slender, as long or longer than the tibiae; the slender claws are sharp, curved on the outer half, the pulvilli long and spatulate; I find no empodium or other medial structure.

Wings: Wings comparatively slender, curved down anteriorly toward the apex so that the posterior margin seems to be largely plane near the apex. The subhyaline wing is tinted with brown due to the dense villi. There are 3 submarginal cells, the marginal cell is strongly dilated or swollen backward near the apex and also the second vein is recurrent. First posterior cell slightly narrowed at the apex. Anterior crossvein much thickened in the middle and placed a little beyond the middle of the discal cell. Anal cell open. Ambient vein very stout, the fringe minute, fine, and dense. The alula is about twice as long as wide. The subcosta and the first branch of the radius are close together and the wing margin much thickened beyond this point. Costal setae small and fine.

Abdomen: The abdomen is much broader than the thorax, wide, short, subtriangular, rounded on each side anteriorly and in the male with the hypopygial apex blunt. The whole surface of the moderately flat abdomen from which the name *depressus* derives is completely opaque with yellowish to olive-colored pollen over a black background and extremely dense, short, curled, much appressed, glittering, brassy yellow pile. In addition to this appressed brassy pile, there are numerus fine, erect hairs of the same color but of no great length. These form a fairly dense fringe along the more or less curved-over lateral margins. I find 8 tergites visible dorsally, the last 2 quite short, and much narrowed, compared to the wide anterior tergite. The male hypopygium is enclosed by a short, downturned, sleevelike or cuplike receptacle formed apparently of the epandrium, and perhaps similar to the even more exaggerated cuplike wrapper of *Acrophthalmyda* Bigot. The basistyli are large and lie side by side within this cup.

Material available for study: A male of the type-species from Santiago Province, Chile, lent me by Jack Hall for the study of this species and its genitalia; collected October 3, 1956, near Maipu by E. I. Schlinger.

Immature stages: Unknown.

Ecology and behavior of adults: No information on record.

Distribution: Neotropical: *Platamomyia depressus* (Loew), 1855 (as *Bombylius*, later as *Platamodes*).

Genus *Neodischistus* Painter

FIGURES 26, 261, 472, 518

Neodischistus Painter, American Mus. Nov., no. 642, p. 1, 1933b. Type of genus: *Neodischistus currani* Painter, 1933, by original designation.

Quite small, stout, compact flies of bare aspect, with gently arched thorax and with decumbent abdomen. The front, mesonotum, and scutellum, and abdomen are more or less plastered with dense, dark, flat-appressed scales. Face and oragenal cup both well developed but with only a few fine, short hairs. Basal segments of antenna slender; third both rounded and dilated at base, then tapering until at apex it is no wider than the second segment. Wings hyaline with 2 submarginal cells; the 4 posterior cells are widely open and the anal cell unusually widely open; anterior crossvein enters the discal cell at or before the basal fifth; alula prominent and ambient vein complete. The third segment of the antenna has, close to the apex and dorsally placed, a small spine, lying erect in an apical cuplike extension of the segment; a more or less similar arrangement is seen in *Phthiria* Meigen and *Oligodranes* Loew. These small flies of *Neodischistus* Painter show no close relationship to *Dischistus* Loew, the type-species of which is before me. *Neodischistus* has a tiny spot of silvery scales on each side of the face below the antennae; these suggest *Parabombylius* Williston, in which the similar scalose silvery spots originate a little higher; however, they are sharply separated from that genus by their venation, facial pile, and antennal characteristics, as well as the surface covering of scales. *Neodischistus* Painter appears to be most closely related to *Sparnopolius* Loew, but differs in the character of the pile and the antennae.

The genus consists of 2 Neotropical species, which range from Panama to Argentina.

Length: 4.5 mm. to 5 mm.

Head, lateral aspect: The head is hemispherical, the face only slightly projected but somewhat more prominent above, extended outwardly at least the length of the first 2 antennal segments. The oragenal cup is plane and truncate, rather large and makes a right angle with the face; it is almost vertical. The face is slightly concave on the upper half, slightly convex on the lower half; considered together they are very much like *Sparnopolius* Loew and moreover are almost bare. On each side of the margin of the oral cup, near its rim, there are about 15 scattered, erect, black setae and a few others on the face. The face is separated from front by a somewhat arcuate, shallow depression, bearing on each side, by the eye margin, a patch of broad, silvery scales. The posterior eye margin is entire, though perhaps not quite plane. The occiput is not very prominent, perhaps as thick as the width of the third antennal segment, and dorsally the postocciput is smooth, and without fissure. Occipital pile dense and coarse, more conspicuous above and shorter below. The palpi are quite slender and filiform and contrary to the author's description, I believe it has 2 segments, each of which bears a few short hairs laterally. Proboscis long and slender, about 3 times as long as the head. The antennae are attached at the top of the head upon almost the same plane as the eye at the vertex.

Head, anterior aspects: The head is a little wider than the thorax; the genofacial space is a little less than one-third the total head width and slightly narrowed below;

oral space long oval. The eyes of the male are holoptic, the upper facets somewhat enlarged. Front of female wide with parallel sides, divided by a medial groove and bearing dense, appressed, metallic scales. Ocellar tubercle low, with the ocelli relatively far apart. Posteriorly there is a band of stiff setae behind the ocelli.

Antenna moderately elongate, of about the same length as the head itself, or slightly less. The first 2 segments are cylindrical; first segment twice as long as broad and both with minute setae. Third segment widened toward the base, widest at the basal fourth where it is somewhat less than twice the width of the second segment. It is then narrowed to a blunt, rounded apex which is only half as wide as the second segment. Near the apex dorsally there is a short, spine-tipped style lying exposed and erect in a cuplike receptacle of the apex.

Thorax: The thorax is quite short and broad and somewhat arched and convex, though only to a moderate extent. The mesonotum bears quite coarse, short bristly, erect setae or short bristles laterally, continued from the humerus above the wing over the postalar callosity and giving way on the margin of the scutellum to somewhat long but slender bristles. There are bristly hairs of the same length placed transversely in front of the scutellum. The anterior margin of the mesonotum, the sides and the prescutellar area have numerous, glittering, wide, metallic scales; there are more scattered scales on the scutellum and as these are easily rubbed off, it is likely that they are a continuous color on mesonotum and scutellum. Pleuron with a band of coarse, short, erect, bristly pile along the upper border and with longer, slender bristles, rather dense in the upper posterior corner and with a tuft of metallic scales below; metapleuron bare. The squama is large and bears a long, bristly fringe.

Legs: The legs are slender but short, hind femur a little more than twice as thick as tibia, the pile is dense, appressed and scaliform. Ventrally near the apex the hind femur has 2 or 3 moderately long, rather stout bristles. Anterior tibia with 2 or 3 rows of weak, pale spicules, 4 in each row. Hind tibia with more prominent, somewhat longer pale spicules present, also few in number. Pulvilli present, nearly as long as claws, arolium minute.

Wings: The wings are hyaline. There are 2 submarginal cells, the second vein and anterior branch of the third vein only gently curved at apex. There are 4 posterior cells all widely open; the anal cell is unusually widely open. Alula large with a fringe of long hairs. Ambient vein complete, base of costa with coarse setae. The anterior crossvein enters the discal cell at the basal fifth; the radial sector is short. The upper anterior intercalary vein is at most only one-fourth as long as the medial crossvein. First and second basal cells are of nearly equal length.

Abdomen: The abdomen is about as long as broad, or slightly longer, distinctly wider than the thorax and slightly convex. The whole surface is covered with dense, broad, flat, glittering scales. There are some slender bristly hairs laterally, especially along the margins of the tergites. At the base of the abdomen on each side there is a very dense, extensive lateral band of coarse, bristly pile, partly slender, partly flattened and truncate, and all of it erect. It is situated on the sides of the first segment.

Material studied: Type-species material in the British Museum (Natural History).

Immature stages: Unknown.

Ecology and behavior of adults: Dr. C. H. Curran stated that the females he collected in Panama were in a patch of high grass in a wide trail.

Distribution: Neotropical: *Neodischistus collaris* Painter, 1933; *currani* Painter, 1933.

Genus *Bromoglycis* Hull

FIGURES 38, 271

Bromoglycis Hull, Journ. Georgia Ent. Soc., vol. 6, no. 1, pp. 5-7, 1971. Type of genus: *Bromoglycis robustus* Hull.

Large, unusually stout, black flies with grayish pollen and moderately dense, suberect, fine, bristly black pile on the unusually short, broad abdomen. These flies belong to the Dischistini but are quite different from the South African type-species.

They are further characterized by the short, stout proboscis with the large flattened, short, oval labellum, short, slender palpus, and the broad, apically tapered and apically sulcate third antennal segment. The metapleuron in front of haltere is bare and the lower surface of the hind femur has only a few, quite fine, erect hairs. From *Laurella* Hull and *Pilosia* Hull, this new genus is quickly separated by the notably expanded marginal cell. These flies appear to be nearest related to *Bombylodes* Paramonov but differ in venation, character of pile, and antenna, etc. The anterior half of mesonotum is quite humped resulting in a drooping aspect somewhat like *Laurella* but more pronounced. Males holoptic.

These flies are from Eastern Australia.

Length: 14 mm., including antenna; wing 11 mm.

Head, lateral aspect: The head is much smaller than the bulky, convex, wide thorax, and has a prominent occiput in both sexes. The occiput is more than usually extensive on the upper part of the head; however, it is evenly rounded and sloping backward and downward as is characteristic in Bombyliinae. There is no noticeable vertical groove or fissure behind the faintly impressed posterior column of the ocellar tubercle. The pile of the occiput is dense and stiff and not very long; it barely rises above the eye margin when viewed in profile. The occipital pile is much shorter close to the eye margin, and behind on the more central portion of the occiput it forms distinct weak bristles. The face is prominent. The oragenal cup is rather strongly extended forward a distance nearly equal to the long first antennal segment; in the lateral aspect it is obliquely truncate due to the much shorter extension of the oral cup at the bottom of the head. The pile is scanty, curved, and wiry; these weak, pale red bristles make a fringing border extending outward along the upper and lateral margins of the oral cup and the sides of the gena. The entire head is black with pale brownish-gray pollen. The eye is broadly rounded below and more or less obliquely flattened on its upper border in both sexes but more so in males; the posterior and anterior borders of the eye are parallel and both nearly plane; the upper eye facets of the male are scarcely enlarged; they are barely larger than the ventral facets.

The proboscis is stout, relatively short, cylindrical, scarcely if at all longer than the head and bears a large, short-ovate, muscoidlike labellum. The relatively small palpus is cylindrical and black with a distinct short apical segment carrying several long bristly hairs; the long basal segment has some very long, slender, stiff hairs on the outside with some shorter hairs dorsally and medially. The frons is a small triangle and bears gray pollen and a few fine, appressed, setate golden hairs attached along the eye margin and directed forward. The antennae are rather stout and elongate; they are as long as the head from post eye margin to the nodular and tuberclelike apex of the oragenal cup; all the segments are cylindrical and black, except the extreme apex of second and base of the third, which are brown. The first segment is curiously granulate with minute, black setae dorsally and with a few, oblique, stiff hairs ventrally of no great length. Second segment has only minute setae above, slightly longer setae below; the second segment is as long as wide and forms a very short cylinder. The third segment is strongly annulate at the base; it is somewhat more narrow than the second segment and is slightly narrowed on the outer half and ends with an apical notch. The third antennal segment is $1\frac{1}{3}$ times longer than the first two antennal segments.

Head, anterior aspect: The head in males is much more narrow than is the broad, wide, short thorax; in females it is proportionately a little wider. The antennae are narrowly separated at the base; the oragenal space occupies a little more than one-third the total head width at the bottom of the eye, and most of this space is taken up by the thick wall of the oragenal cup. The gena is narrow but the tentorial fissure is deep. The ocellar tubercle in males is small but prominent, forming an equilateral triangle; it bears a few long, thin, erect, bristly hairs centrally.

Thorax: The thorax is quadrate; the distance across the base of the wing at least in males being perhaps even greater than the length of the mesonotum if the scutellum is excluded. It is strongly convex on the anterior half and somewhat convex or inflated in front of the scutellum. The head is drooping or decumbent. Except for the reddish-brown humerus and the postalar callosity, the thorax and scutellum are black. The mesonotum and scutellum are nearly opaque and bear thin brownish pollen; the pile is rather short and dense on the anterior half, thinner posteriorly, very fine, tangled and curly, and with shorter, curled, wiry pile posteriorly. All this pile is pale yellow or pale brassy in color or even almost yellowish white. There are some fine, stiff, wiry hairs on the notopleuron, on the humerus, also above the wing, and on the postalar callosity; these form at most weak bristles. Pleuron black with very pale brownish-white or brownish-yellow pollen. The immediate area in front of the haltere appears to be without pile; it is obscured with a strong tuft of pile reaching downward from the thoracic squama and other tufts curling upward from below the spiracle. The haltere is quite small. The metanotum appears to be vertical and very closely jammed against the thorax and also bare.

Legs: Legs small but stout, the anterior pair is especially reduced; the femora are slightly widened basally but nowhere swollen; they bare rather loose, completely appressed, long, somewhat flattened whitish hairs. All the legs are brownish orange; the apical tarsal segments become dark brown. There is a scanty fringe of very fine, short, white hairs on the lower surface of all the femora. Bristles of the tibiae very fine, sharp, oblique and black; anterior tibia with 10 dorsal microbristles and 3 longer bristles posteroventrally. Middle tibiae with the 10 anterodorsal microbristles and like number of more minute setae posterodorsally and about the same number of longer, more distinct, slender bristles posteroventrally. Posterior tibiae with the bristles shorter and more poorly developed; however, there are 3 or 4 slender ventral bristles. Pulvilli well developed and broad.

Wings: Not widened basally; the ambient vein is complete; the axillary lobe is moderately wide; the alula is crescentic and not very wide. All of the veins are heavy and dark brown; the anal cell is widely open, the anterior crossvein is oblique and enters the discal cell near the outer fifth, the intercalary vein is only half as long as the anterior crossvein. The first posterior cell is open but narrowed; its space on the posterior margin is a little less than the length of the anterior crossvein. The marginal cell is much widened discally and ends by a rather long plane vein which enters the costa at a right angle; this vein is about twice as long as the anterior crossvein and bears a spur at the outer corner and back from the corner there is another short, anterior spur which is probably variable. The anterior branch of the third vein turns strongly upward and bears a spur at its basal angle.

Abdomen: The abdomen is quite short, thick, and stout and distinctly wider than the mesonotum. I find 7 tergites visible from above in the male; the abdomen viewed from above is roughly in the shape of an equilateral triangle with the sides a little rounded and with the apex blunt. The small male terminalia are posterior in position and turned to the right side. The abdomen is black with faint greenish reflections and a very faint, dark brown pollen. The dorsal pile is chiefly very fine, suberect, black, and wiry and mixed with it in a male is some shorter, appressed, curled, glittering brassy hairs. Along the lateral margin the pile forms thick, bushy tufts of partly brassy and partly opaque brownish-white pile together with a few of the very fine, black hairs; these tufts are more or less directed backward.

What is presumably the female of this species, taken at the same time, has matching venation and differs principally in the light brownish-red scutellum and also in the laterally compressed third antennal segment which is stouter and wider but has the same type apex; the first antennal segment is relatively a little shorter, lighter in color, with longer dorsal setae. It certainly belongs to *Bromoglycis* but may be a different species.

Material available for study: Several individuals of both sexes.

Immature stages: Unknown.

Ecology and behavior of adults: Collected while hovering and feeding upon bush-type species of *Leptospermum* located within open forests and rank undergrowth some miles north of Brisbane.

Distribution: Australian: *Bromoglycis robustus* Hull.

Genus *Cacoplox* Hull

FIGURES 44, 276, 489, 515

Cacoplox Hull, Journ. Georgia Ent. Soc., vol 5, no. 3, pp. 163-165, 1970. Type of genus: *Cacoplox griseata* Hull.

Large, blackish flies, with very wide, brown or gray pollinose face and much long, dense pile from the prominent, bullose sides of the face and equally prominent pile from the wide gena below. The tentorial fissure is wide, flared, deep, and conspicuous. The proboscis is long and slender and has long, slender labellum. The antennae are elongate, the third segment slender, or slightly flattened, scarcely narrowed at the base or apex and bears a minute microsegment followed by a short oblique spine; the second segment has more conspicuous bristly pile than the first segment; this second segment is nearly as long as the first segment and has a complete circlet or band of radiating stiff, long hairs. The wings are mostly hyaline with a gray or brown tinge basally and anteriorly. The first posterior cell is slightly narrowed and the anterior crossvein lies at the outer fifth or outer sixth of the discal cell. The abdomen is somewhat longer than in *Dischistus* Loew or *Sparnopolius* Loew and bears extraordinarily dense, long, matted pile especially laterally and terminally; this pile becomes erect and long and dense anteriorly upon the abdomen. The metapleural pile in front of the haltere is especially dense, long, and conspicuous. This genus is related to *Euprepina* Hull, differing in the wider face, bearing the dense pile. Also in these flies in contrast to *Euprepina* the tibiae and tarsi, especially on the hind legs, bear dense, short, thick, minute scales rather different from those of many bombyliids.

These flies are found in the southern part of the Neotropical region, particularly Argentina and Chile.

Length: 10 to 16 mm., excluding proboscis; of wing 10 to 15 mm.

Head, lateral aspect: The head is very broad, wider than the thorax, but from the lateral aspect it is rather short on the ventral half and above, but with extraordinary swollen face. The occiput is short and flattened with dense short pile along the middle and long, stiff, bristly pile above. Ventrally the pile of the occiput is very fine and plushlike, very long dense and white and mixed in the with the equally long, dense, white pile beneath the eye. Pollen of the occiput pale brownish yellow or grayish yellow. The eye is very slightly curved backward and plane only on the upper three-fourths. It is turned inward somewhat on the posterior margin, and the eye contour is broad and obtusely rounded below with greenish reflections. The front is extraordinarily broad, much wider than either eye and occupies nearly half of the very wide head. Its pollen is chocolate brown, and there is a deep, conspicuous, trenchlike fossa or depression extending halfway between the antenna and the ocellar triangle. The front has a patch of pile laterally on each side of the flattened slope constituting the posterior part of the broad depression. Also there is a row of stiff bristly hair arising from the lower edge of this posterior slope and this hair is appressed forward with 2 isolated tufts in the midline.

Ocellar tubercle large, tumid, set between the eye corners with the ocelli occupying an obtuse triangle on the anterior half of the tubercle. The tubercle bears numerous, long, erect fine, black hairs and bristly hairs, some of which are directed forward. The face is extraordinarily swollen laterally, rounded, tumid, and produced forward rather farther than the length of the first antennal segment. This swollen area, brownish or grayish yellow pollinose, is rolled or convexly curved inward to form a deep crease beside the anterior part of the moderately long, wide oral cup and this crease or fissure becomes deeper ventrally below to end in a large, deep pit or pocket. Below the pocket the gena becomes somewhat tumid or swollen in front of the lower fourth of the eye. The middle of the face beneath the antenna, constituting the anterior part of the oral cup, has pollen only and sometimes several transverse ridges. In addition there is an extraordinarily long fringe of pile arising from the deep inner slope of the face and genofacial area.

Proboscis and labellum very slender and elongate and more than 3 times as long as the head. The palpus is large, elongate, cylindrical with a small clavate apical segment and the whole bearing conspicuous, dense, long pile dorsally and laterally and ventrolaterally. The antennae are attached at the upper fourth of the head. They are elongate and slender and distinctly separated at the base by the thickness of one segment. The first segment is 3½ times as long as wide, covered dorsally, laterally, and below with some very long, loose, scattered bristly hairs. There are also a few such hairs placed medially, some black, some yellow. The second segment is four-fifths as long as the first segment, slightly narrowed along the middle, barely widened apically, and bears numerous long, oblique, black bristles, mostly reddish in some species. Third segment as long as the first two, more slender, the base knobbed, slightly wider at the base of the third, but continuing to be wide nearly to the apex. It has 2 small microsegments at the apex, the last one bearing a yellow spine.

Head, anterior aspect: Head exceptionally wide and wider than the thorax. The vertex is as wide as either eye, the front expands rapidly, and face is even a little wider below across the genal area. Antennae distinctly separated at the base. The oral recess is broad, large and more or less triangular. In some species it is very shallow, the sharp lateral edge scarcely sloping inward. In others it may be a little deeper with the edge thickly inflated and rounded. In any case there is a raised medial, longitudinal ridge down the middle of the oral recess. Tentorial fissure deep and conspicuous, the pit even deeper.

Thorax: Short, broad, and slightly convex. The scutellum is large, thick, and convex posteriorly and a deep groove separates it from the mesonotum. It is opaque, black, with dark brown or gray pollen and the mesonotal pile is extraordinarily dense, fine and plushlike and may obscure the ground color. In one species the mesonotal pile is light reddish brown. In another it is pale brownish or yellowish white. Scutellum pollinose and densely closed like the mesonotum. Scutellum margin with some long, stiff hairs. There are several stout, short, black bristles on the notopleuron. The pleuron is very densely pilose everywhere, except on the pteropleuron and the anterior hypopleuron; posterior pleuron with a conspicuous vertical band of pile; squamae with long slender bristles.

Legs: The femora stout, the first four a little swollen toward the base and bearing flat-appressed, scaliform pile above the long shaggy hairs ventrally, longest basally, decreasing in length toward the apex. Hind femur with similar appressed scaliform hair, long hair basally on the ventral half, and with 7 oblique, sharp, rather stout black bristles on the outer half. All tibia spiny or bristly. The hind pair has 4 rows, 8 to 14 anteroventrally, 10 or 12 dorsolaterally, 6 to 10 dorsomedially, and 9 to 15 ventromedially. These bristles are rather stout. Hind tibia either with long slender scales, or dense, minute, broad, short, flat-appressed scales. Claws large, sharp, strongly curved, with long, large pulvilli and short arolium.

Wings: Large elongate, pointed apically, hyaline or slightly tinged with brown, especially anteriorly and toward the base. There are 2 submarginal cells. The oblique anterior crossvein enters the discal cell at the outer sixth or seventh. The first posterior cell is slightly narrowed. The anal cell is widely open; alula large; ambient vein complete. The costal comb is present and thickened with fine setae at the base; basal hook absent.

Abdomen: Broad basally, wider than the thorax, more or less triangular, sometimes a little produced apically. The color is opaque, black or brown or clay-colored pollen, or greenish pollen. Base of the first tergite with a curious, erect fringe of short, dense pile extending well under the scutellum. The outer lateral pile long and direct, equally long and dense on the sides of the second tergite, but remaining pile, extraordinarily long and matted and flat-appressed over the back of the abdomen, forming brushlike mats, and leaving the last two tergites largely free of pile except for a few fine, short hairs. Female terminalia with a fringe of spines on each side.

Material available for study: The unique types.
Immature stages: Unknown.
Ecology and behavior of adults: Unknown.
Distribution: Neotropical: *Cacoplox griseata* Hull.

Cryomyia, new genus

FIGURES 39, 265

Type of genus: *Cryomyia argyropila*, new species, by present designation.

Small flies with a stout abdomen, especially thick basally, the abdomen conical from the lateral aspect and tapered from the dorsal view. The pile is long, erect, extraordinarily delicate and fine, dense, and shining white in color, and the whole surface of the abdomen is densely plastered with appressed silver pile of a flattened, scaliform character. On the mesonotum the erect pile is similar to that of the abdomen, but the opaque, blackish disc, shining laterally, has scattered, minute, appressed, curled whitish hairs. Legs quite slender, clothed in snow-white appressed, scaliform pile and hind femur below with 4 or 5 minute, fine, short, stiff, pale hairs; bristles absent. Antenna quite slender and elongate, third segment slightly narrowed and this segment a little longer than the first two. The proboscis is slender but not very long and the labellum is flattened but narrowly open. This fly belongs to the Dischistini but is not close to any other genus. The water-clear hyaline wing has the first posterior cell flared to the apex, the anal cell open.

This genus can be readily separated by the distinctly impressed horizontal line running across the eyes and immediately below the antenna.

Flies are found in Western Australia.

Length: 7 mm.; wing 6 mm.

Head, lateral aspect: Head hemicircular with a comparatively shallow occiput vertically behind the prominent equilateral ocellar tubercle. It becomes gradually and distinctly produced backward and inward on the middle and lower part of the head. There is no occipital cup; the foramen is small and reduced and these flies belong to the Homeophthalmae and to the subfamily Bombyliinae. The face is barely projected forward as a blunt, low, triangular elevation which extends a short distance outward from the base of the antennae.

There is a circular fringe of long, white pile around the gena and upward across the upper margin of the oragenal cup. The area immediately below the antennae bears pollen only; the pollen is white on a black background. The pile on the occiput consists of wide, appressed, white scales, especially laterally, that reach to the eye margin and rather dense, shining white, long, erect, coarse hairs that become shorter near the eye margin and longer and denser toward the central part of the occiput. The front forms an equilateral triangle covering about half the distance from the antennae to the posterior eye margin; it is densely covered with shining, silvery white pollen and a lateral tuft of long, appressed, flattened, silvery white hairs which are directed forward. The eyes of the male are tightly apposed, and holoptic on the upper part with the upper facets rather strongly enlarged and the eye near the middle divided by a complete, oblique, yet more or less horizontal, distinctly impressed line from posterior eye margin to anterior eye margin. This is not a line of demarcation of the eye facets.

The ocellar tubercle forms a somewhat obtuse triangle which extends backward from the posterior eye corners so that the posterior ocelli overlooks the occiput. There is a tuft of fine, erect, brownish hairs in the middle of the tubercle. The proboscis is rather strongly compressed laterally, not slender, and the elongate oval, much compressed labellum has longitudinal grooves and is microsetate but not spinulate. The base of the proboscis in recessed within the head. The long slender brownish-yellow palpus is likewise recessed and bears a short terminal segment. The proboscis is nearly twice as long as the head. The antenna is attached above the middle of the head and is comparatively elongate and slender; the first and second segments are cylindrical, with the first segment 2½ times as long as the second and annulate at the apex; the second segment is a little longer than wide. The third segment at the base is annulate and nearly as wide as the second segment; its length is little greater than the first two segments and it grows slightly narrower on the apical half; it is a little compressed laterally and ends bluntly with a minute blunt spine.

Head, anterior aspect: Short oval with the oragenal space slightly more than one-third the head width. The oragenal recess is deep and oval with nearly vertical walls. The tentorial fissure is shallow. The exterior of the oragenal cup and gena is black; the latter has a long fringe of shiny white, flattened hairs growing shorter below and with a basal mat of flattened, silver scales that extend below around the bottom of the eye to the occiput. Interior of oragenal cup creamy white, the lateral part of the walls scarcely thickened.

Thorax: Shining, black, with faint brown pollen over the middle part changing to a rather distinct, erect, coarse, dense, grayish-white micropubescence along the sides of the mesonotum. Dorsum of the mesonotum densely covered with long, silvery white, shaggy, slightly flattened pile which is erect, almost bushy, and merges completely into similar, long, more opaque pile on the mesopleuron. Also on the mesonotum there are numerous, short, silvery-white scales which are appressed and lie loosely scattered over the surface. They are continued densely on the disc of the scutellum, over its whole posterior border, and likewise densely over the vertical surface of the postalar callosity. The pleuron, like the mesonotum and scutellum, is black and it is covered with a thin, grayish-white pollen and a dense mat of silvery-white scales like hairs and tufts of rather long, silvery-white pile. Only the metapleuron in front of the haltere appears to be without pile. Base of haltere with a tuft of flattened hairs; the knob is large and thick and subglobose, cream colored with a brown base. There is a plumula of long, radiating white hairs.

Legs: Black; the first pair of legs are comparatively small; the first four femora are stout, basally widened but not swollen, and densely matted with appressed, silvery-white scales. Both pairs, especially the middle pair, have a posteroventral fringe of long, erect, fine, white hairs. Hind femur twice as wide as hind tibiae; it is a little flattened, of uniform width, completely covered with matted, appressed, white scales and 5 or 6 short, fine, white bristlelike hairs below. Hind tibia as long as the femur; it is slender, with rather dense white scales laterally somewhat more loose and scattered scales medially. There are 7 or 8 short, fine suberect, white bristles placed anteroventrally and 5 to 8 similar bristles on both sides dorsally. In addition there is a posteroventral row of 8 similar bristles. Hind basitarsus equal to the length of remaining segments. Pulvilli long and slender; claws slender, sharp, bent apically, and yellow on the basal half.

Wings: Quite hyaline but with the distal part of the auxiliary cell yellow. The wing is strongly iridescent. All the veins are brown, except the auxiliary which is yellow, and the ambient vein on the wide and prominent alula is white. The first posterior cell is widely open and flared. The second posterior cell is flared. The third posterior is as wide on the edge of its border as it is on its base. The long intercalary vein is rectangular and plane. The anterior cross vein is rectangular and lies near but just before the outer third of the discal cell. Anal cell widely open. Marginal cell widened and ending quite obtusely with a rectangular bend from the second vein. The anterior branch of the third vein is sinuous and gently sigmoid. Base of costa narrowly widened and bearing silver scales and a pale tuberclelike lobe. The fringe on the axillary lobe and alula are long and shining white.

Abdomen: Elongate conical and at the base, at least as wide as, perhaps a little wider than, the mesonotum. It rapidly narrows from the end of the third tergite. I find 8 tergites readily visible from the above; they are all black, only the terminal and the postmargin of the last tergite and sternite brownish yellow. The entire surface of the abdomen is very densely covered with elongate, narrow, silvery scales and flattened hairs completely obscuring the ground color, and also with dense, very fine, quite long and erect, faintly shining, white pile, which is apt to be a little denuded posteriorly and tends to form wide transverse bands. The terminalia are small, posterior in position, and recessed.

Material available for study: The unique type, a female, from Western Australia, near Perth. Collected by the author in early January 1954.

Immature stages: Unknown.

Ecology and behavior of adults: Collected hovering and feeding upon low-growing small plants of *Leptospermum* sp.

Distribution: Australian: *Cryomyia argyropila*, new species.

Genus *Euprepina* Hull

FIGURES 49, 419, 455, 815, 816

Euprepina Hull, Proc. Ent. Soc. Washington, vol. 73, no. 2, pp. 181-183, 1971. Type of genus: *Euprepina nuda* Hull by original designation.

Densely pilose flies with the abdomen a little more than usually elongate and with a slender proboscis and slender, small labellum. Related to *Sparnopolius* Loew which they resemble in the bare, black and shining, prominent and extended oragenal cup. They differ from that genus in the position of the anterior crossvein, which is placed at or beyond the outer third of the discal cell; in *Sparnopolius* Loew this crossvein lies at the basal third or always before the middle of the cell. Also in *Euprepina* there is a dense, long tuft of pile in front of the haltere, but this space is bare in *Sparnopolius* Loew, which appears to be a Nearctic genus.

Euprepina is found from Rio de Janeiro, Brazil, southward to Uruguay.

Length: 10 to 13 mm., excluding proboscis; wing 8 to 10 mm.

Head, lateral aspect: Head slightly triangular in appearance because of the flattened eyes of the male, flat front in the female, and more especially the triangularly produced and obliquely truncate, or plane oral cup, its sides convex above, sharp and polished like glass without pile but with pollen on a short area beneath the antenna separated from the oral cup by a crease. The occiput is rather thick inwardly, that is to say, it is considerably produced in profile, but it is of the flat *Bombylius* type with small central foramen. The pile of the occiput is dense and shaggy dorsally, extending only a short distance above the eye. The hair is rather blunt, shorter along the middle, but becoming long and fine ventrally. Posterior margin of the eye is not quite plane, very shallowly concave, and the upper facets are strongly enlarged in the males. The front is reduced to a minute triangle in front of the antenna in the male, but it is slightly raised on the upper half and on the vertex in the female, and there is a shallow transverse depression in front of the antenna. The area on which the antenna rests protrudes not more than the thickness of the second antennal segment in females, much less so in the males. The proboscis and labellum are very long and slender, 4 or 5 times as long as the head. Palpus cylindrical with short clavate terminal segment, both bearing numerous, long, coarse, bristly hairs laterally and ventrolaterally.

Head, anterior aspect: Wider than the thorax, from the anterior view at least twice as wide as high. In the male the upper half of the eye is broadly curved, holoptic for two-thirds the distance from antenna to ocellus. The male ocellar tubercle is equilateral, raised above the eye by the depth of the ocellus or a little more and with a tuft of fine, long bristly hairs and the ocellar tubercle is placed opposite the sharply rectangular posterior eye corners. The ocellar tubercle is much larger in females and scarcely at all raised. Pile of the front bristly and appressed forward. The black ground color is covered with brown pollen. The small male front has pollen only. The antennae are narrowly separated, convergent and then divergent on the outer segments. The oral cup is almost circular, large, extending to the eye margins, except for a very narrow inwardly depressed genal strip which widens below. The anterior half of the oral cup has the walls greatly widened and inflated, curled over and inward from the outside, leaving only the central part deep and with a narrow medial ridge anteriorly. The antennae are attached at the upper fifth of the head and are rather large, elongate, and slender. First segment slightly longer than the second, and bearing numerous, rather long, coarse bristles and bristly hairs on all sides, longer laterally and below. The second segment, a little widened apically, has shorter but similar bristles on all sides. Third segment slender, spindle-shaped, the base knobbed, the outer half attenuate, with 2 small microsegments and a bristle at the tip.

Thorax: Thorax longer than wide, rather convex, the head is a little drooping. The mesonotum is opaque black with dense, brown pollen and dense, fine, erect pile of no great length. Notopleuron with 4 slender bristles, postalar callosity with many long, fine, slender bristly hairs. Scutellum clothed like the mesonotum with fine, long hairs on the margin. Pleuron densely pilose including the propleuron, whole of the mesopleuron and metapleuron. There are some fine, long hairs on the posterior hypopleuron above the hind coxa, but the pteropleuron and anterior hypopleuron are bare.

Legs: Slender on all pairs. All femora with fine fringe of very slender, long hairs ventrally. Bristles absent, though in some species the hairs on the hind femur are a little thicker. Femora also covered with flat-appressd, long, slender scales or scaliform pile. Tibial bristles extremely weak, oblique, and sharp. The posterior tibia has 7 or 8 such bristles ventrolaterally, about the same number dorsolaterally and on the other rows. Claws small, slender, the pulvilli long.

Wings: Long and slender, subhyaline, tinged with pale brown anteriorly and basally. There are 2 submarginal cells, first posterior cell very slightly narrowed. Anal cell widely open, the anterior crossvein enters the discal cell near the outer fourth. Hence the first basal cell is much longer than the second. Alula wide, widest apically. Ambient vein complete. Costal comb absent and hook absent or merely stublike.

Abdomen: Rather elongate distinctly narrowed, tapered and compressed posteriorly. It is more than 1½ times as long as the mesonotum, the scutellum excepted. The seventh tergite and corresponding sternite form a long cone in the female, about as long as wide and obtuse apically. Color of the abdomen black, opaque in females with dark to light brown pollen and with appressed, glittering, somewhat flattened hair over the middle of the abdomen and in addition some erect, moderately long, loose, scattered, yellow hairs basally, changing to fine, erect, black hairs along the posterior margins of the more apical tergites. The curled-over side margins of the first four or five tergites with long, conspicuous, dense, erect, yellow or sometimes reddish or whitish pile in the females, or in some species with the pile on the first two tergites shining white in color. Females in some species have part of the first tergite and basal half of the second tergite opaque white. Females of other species may lack this completely. The male abdomen is similar in form but rather different in pile. In some species the male has the first two tergites opaque white across the middle and opaque across all of the middle of the fifth tergite. Some have only the fifth tergite pale. The pile of the male is dense, fine and erect and tending to form bands across the middle on posterior margins of the tergites. Also on the sixth and seventh tergites the males have a dense, expanded tuft of hair which is quite conspicuous; curled over lateral margins with very dense hair. Male terminalia asymmetrical and recessed.

The male terminalia from dorsal aspect, epandrium removed, show a stout, broad figure. The basistyli have a long, backward process, convergent and becoming stout at the base. The epiphallus is particularly stout and swollen in the middle and toward the base. The dististyli are short and stout with divergent apical hook. From the lateral aspect the basal apodeme is quite large and extensive and the epandrium looks somewhat like *Lordotus* Loew because of the ventral apical lobe.

Material available for study: A series of both males and females found among material lent by the National Museum of Natural History. Also a series purchased from a South American collector.

Immature stages: Unknown.

Ecology and behavior of adults: Not on record; unknown.

Distribution: Neotropical: *Euprepina nuda* Hull.

Genus *Hallidia* Hull

FIGURES 36, 278, 519, 812, 813, 814

Hallidia Hull, Journ. Georgia Ent. Soc., vol. 5, no. 3, pp. 165-166, 1970. Type of genus: *Hallidia plumipilosa* Hull.

A small fly with rather strongly humped mesonotum and with decumbent head and abdomen. It is densely covered with pale pinkish or ochre-colored pollen over a black background with quite narrow, laterally widening, brownish-yellow margins on the tergites of the abdomen. The head is hemispherical in lateral aspect but it is nearly twice as wide as high. The proboscis is long and slender; the antennae are elongate and slender. The marginal cell is strongly bulbous or swollen downward at the apex; the anterior branch of the third vein thrust strongly upward before the wing apex. First posterior cell maximally open. The anterior crossvein is at the outer fourth of the discal cell and is short. The veins on each side of the cross vein are drawn close together. The face and front and occiput and gena have long, erect, rather dense, in places somewhat curled, silvery-white hairs. There is similar pile on the thorax, scutellum, and abdomen. It is easily denuded, and is apt to be sparse and straggly. There are 2 submarginal cells and 4 posterior cells. The anal cell is distinctly open. The abdomen is short, with long, loose, straggly pile; it is strongly compressed laterally. Males holoptic and the male terminalia small, terminal, and somewhat recessed.

These small, curious flies which very superficially resemble *Phthiria* Meigen are from Chile. They appear to be rather aberrant. Several of these individuals were received among miscellaneous bombyliids lent by the California Academy of Sciences.

This genus was named in honor of Jack Hall.

Length: 6 mm.; wing, 5 mm.

Head, lateral aspect: The head is nearly hemispherical from a lateral view. The occiput is comparatively thick but it is not cupulate, not bilobed. The pile of the occiput is comparatively sparse, but it is long, coarse, erect, especially so on the upper half. The occiput is much thinner at the bottom of the eye and upper and lower eye corners are broadly rounded. The face extends downward, a little obliquely for the length of the first antennal segment. The large, oval, oragenal cavity is nearly vertical. The eye is large, the posterior margin quite plane and in males the upper eye facets are scarcely enlarged, and there is no horizontal line across the eye. The gena is rather wide, wider than the thickness of the antenna and bears a continuous band of long, white pile like the somewhat longer pile curling downward transversely across the face. The genal pile extends to the bottom of the eye mixing with the lower occipital pile. The proboscis is long and slender, not more than 2½ times as long as the head. The small, slender palpus is not as long as the first antennal segment. It appears to consist of one segment only. The front viewed from above is about one-third the head width in females and forms only a small triangle anteriorly which is covered with fine, grayish-white pollen without pile in the male, but with abundant dense, long, erect or curved, stiff or bristly whitish hairs in the females. The bristly pile in the female is confined to a large triangle resting along each eye margin, and in front of this bristly pile there is a patch of appressed shorter, bristly pile. The middle of the female front constitutes a nonpilose, pollinose area widening toward the antennae.

The antennae are attached at the upper tenth of the head. They are at least as long as the eye, slender, the first segment 3 or 4 times as long as wide. The second segment is short, barrellike or beadlike, the third segment with strongly annulate base is slightly narrowed, then a little widened for a short distance, then tapered to the apex which bears a minute spine.

Head, anterior aspect: The head is as wide as the thorax; the eyes in the male are separated by half the width of an ocellar triangle and set quite low. The ocelli are set in an equilateral triangle. The antennae are separated by less than the thickness of one segment. The oral opening is large, wide, and short oval, and comparatively shallow. The gena is at least as wide as the first antennal segment with a well-developed tentorial fissure.

Thorax: The thorax is slightly longer than wide, strongly humped, black and ground color, with dense yellowish-brown pollen with a pinkish cast, and with coarse moderately pale pile and without distinct bristles. The scutellum is pilose and concolorous with abundant, long, erect, yellowish-white bristles and bristly pile. The metanotum is bare, the metapleuron is also bare, and the pteropleuron likewise. There is a tuft of pile beneath the haltere and the squama pile is extraordinarily long and fine. Plumula absent.

Legs: The anterior femora are a little thickened toward the base; the hind femur is almost uniformly thick, as wide as the base of the anterior femora. All of them have comparatively dense, flat-appressed, flattened or scalelike pile and the hind femur has 5 distinct, sharp, black, ventral, oblique spines, becoming larger toward the apex. The anterior tibia has a posterodorsal and a ventral row of small, sharp, black, oblique spicules or spines. Middle tibia with 4 rows of similar more conspicuous sharp, short bristles. Each row has about 7 elements. Hind tibia likewise with 4 rows of slightly longer oblique black bristles, 7 or 8 in each row. All of the legs are light yellow, the last tarsal segment, however, brownish black. Pulvilli long.

Wings: Wings are rather slender, narrowing slightly at the base, the alula twice as long as wide, the ambient vein complete, and with a long fringe. The marginal cell is slightly curved backward before its middle then curves forward again and then more strongly backward toward the apex, joining the costa at a right angle. The anterior branch of the third vein is strongly sigmoid and ends obliquely, well above the apex of the wing. First posterior cell ending quite widely open. Anterior crossvein rectangular and quite short, much as in *Eclimus* Loew. It enters the discal cell at the outer fourth; intercalary vein long; four posterior cells present; anal cell very widely open. There is no basal comb, and no spur or hook. The wing is rather densely brown villose with a small diffuse brown spot at the end of the second basal cell. There is a brown spot at the base of the third posterior cell, likewise the second and third posterior cells and again cutting across the middle of the marginal cells, together with a long, anterior, brown streak in the middle of the discal cell.

Abdomen: The abdomen is tapered. At the base it is less wide than the mesonotum, very little wider than the scutellum. It is distinctly compressed laterally. There are 7 tergites visible in the male, and the same number in the female. Ground color black, except for narrow, brownish-yellow postmargins on tergites and the whole abdomen overlaid with pale, somewhat pinkish-brown pollen. The pile is comparatively dense, quite fine, quite erect, long and whitish. Male terminalia small and recessed, but directed posteriorwardly.

The male terminalia from dorsal aspect, epandrium removed, show that the epiphallus is conical, and drawn out to a fine point apically, far beyond the apex of the dististyli. The dististyli are leaflike with blunt apex. From the lateral aspect the epandrium is strongly quadrate, a little concave below with prominent knoblike cercus, the basal half of the basistylus is also quadrate, rather high, with a process both at the base above and a process in the middle distally. The bristles are very stout and long on the lower border.

Material available for study: Males and females from Chili, both Olmue and Valparaiso.

Immature stages: Unknown.

Ecology and behavior of adults: Not on record.

Distribution: Neotropical: *Hallidia plumipilosa* Hull.

Genus *Laurella* Hull

FIGURES 23, 274, 514

Laurella Hull, Journ. Georgia Ent. Soc., vol. 6, p. 3, 1971. Type of genus: *Laurella auripila* Hull, by original designation.

Robust, opaque black flies of considerable size which belong to the Dischistini. The broad, rather flattened and relatively short, posteriorly tapered abdomen, together with the whole of mesonotum, is covered with minute, curled and appressed, glittering, brassy hairs and the whole of the tergites are also covered, as well as the posterior mesonotum, with numerous, very fine, long, erect, black hairs. Anterior third of the anteriorly convex, more or less humped mesonotum and the mesopleuron, sternopleuron, the occiput, sides of face, and sides of the abdomen with long, stiff, greenish brassy pile.

These flies appear to be very different from either the true *Dischistus* Loew, or the European *Bombylosoma* Rondani, or the American *Sparnopolius* Loew. They are perhaps nearest to *Euprepina* Hull from South America which differs markedly in the type of face; they are also related to the Australian genus *Pilosia* Hull, which has a densely hairy metapleuron, a long, filiform labellum and an entirely different type of pilosity.

This genus was named for my wife, Laura Hull.

These flies are known from Eastern Australia in the Brisbane area.

Length: 11 to 12 mm., wing 10 mm.

Head, lateral aspect: Hemiglobular with occipital component shallow and virtually absent above. It is slightly more prominent in females. The pile of the occiput is quite dense and becomes shorter near the eye margin; it is very fine, erect, only slightly extending above and beyond the eye margin. This pile is rather light, shiny straw yellow in color. Posterior eye margin vertically plane but with the faintest suggestion of a long concave portion below the middle. The face extends forward obliquely a distance almost or quite equal to the length of the first antennal segment, together with a minute, triangular front. The front is yellowish brown with pollen of the same color and with fine, subappressed, reddish brown to blackish hairs of no great length which curl and curve forward. This pile changes over to brassy yellow hairs on the sides of the oragenal cavity; some of these hairs acutely curve inward on the interior thickened wall of the shallow oragenal cup. This oragenal cup is divided by a slender fissure dorsally. The eye is large and hemicircular and extensively holoptic with the upper facets a little enlarged and with reddish to greenish reflections. The frons forms a quite small, equilateral triangle with brownish-yellow pollen only. Proboscis elongate and slender, slightly compressed, but the sides rounded rather than flat and the labellum though much compressed is large and short oval with fine setae below. The palpus is long, slender, and cylindrical. The second segment is almost as long as the first. It has only a few fine, short hairs.

The antennae are attached just above the middle of the head. They are slender and moderately elongate. The first segment is 1½ times as long as the second segment and pollinose. This pollen possibly consists of extremely fine micropubescence. The first segment has scanty long, stiff, slender, wiry black hairs on the side and below and shorter ones above. The second segment is narrower basally and nearly twice as long as wide from the dorsal aspect but more strongly widened from the lateral aspect; it bears moderately long, fine, stiff, black hairs above and below. The third segment is strongly annulate at the base and distinctly longer than the first two segments. From the dorsal aspect it is compressed basally and of a nearly uniform thickness beyond. From the lateral aspect it is distinctly widened over the basal half and narrowed to a blunt apex. At apex it bears a rather long, spinous bristle; laterally this segment is distinctly spindle-shaped or shaped like a ninepin. Antennae entirely black, except for the brown spine.

Head, anterior aspect: Head distinctly wider than the mesonotum. The antennae are adjacent. The small ocellar tubercle is rather prominent and equilateral in shape and between the large ocellar there is a tuft of numerous, quite fine, erect black hairs some of them curved forward. The oragenal space in males forms a large triangle; the gena is narrow, except below, and there is a deep tentorial fissure. Walls of the oragenal cup rather thick; pile along the sides of the gena rather long and brassy yellow, except above, and curved inward.

Thorax: The thorax is distinctly longer than wide and rather abruptly convex on the anterior third of the mesonotum so that the head appears to be decumbent. Pile in males dense, fine, erect, and brownish yellow on the anterior third and apparently giving way to fine, much more scanty black hairs over most of the mesonotum and scutellum among which there is much short, flat-appressed, curved, glittering, brassy yellow pile. Ground cover of mesonotum opaque black, becoming olive along the lateral margin and on the postalar callosity. Notopleural with 6 to 9 distinct, short, only moderately stout black bristles. Pleuron black, except along the sutures on the lower sclerites, and with the whole surface very densely pinkish to yellow-white pollinose. Mesopleuron very densely covered with long, stiff or slightly flattened, appressed, brassy yellow hairs; it is divided in the middle into a group curving upward and a group curving downward. The hypopleuron and the lower metapleuron are without pile, but there is an extensive tuft of long, coarse pile immediately below the triangular spiracle. There is no pile on the metapleuron in front of the haltere, but the metanotum is heavily pilose; it is flared outward basilaterally and deeply recessed behind the haltere. Haltere with a tuft of pile at the base.

Legs: Legs distinctly stout, the anterior pair are well developed. The first four femora are stout and a little widened basally and all the femora are medium brown. The tibiae are more blackish and all the femora bear shaggy, appressed, long, yellowish, flattened hairs and they all bear a scanty fringe of fine, erect, pale yellow, down-turned hairs. Middle femur with 2 or 3 fine black bristles below hind femur with 3 or 4 more stout, conspicuously longer, stout, black bristles. Anterior tibiae with 3 or 4 minute, black bristles dorsally and 7 or 8 longer, oblique bristles below. Hind tibiae with four rows of slender black bristles, 8 or 9 bristles in each row. The second and remaining segment of the anterior tarsi have a curious, erect, fine, fuzzlike pubescent pile in female. Pulvilli large, claws stout, curved apically.

Wings: Narrow basally as much as apically. They are uniformly tinted with pale brown due to villi and hence are not quite hyaline. The veins are heavy and dark brown and ambient vein complete. The alula is distinct though narrow and about four times as long as wide. The anal cell is widely open. The discal cell is much widened apically, the rectangular intercalary vein is as long as the anterior crossvein; the latter enters the discal cell at the outer fourth. First posterior cell maximally wide on the wing margin; the marginal cell, while a little widened apically, ends quite narrow and the second vein curves only quite shallowly to the apex. Basicosta a little widened, the costal comb composed of short, fine setae; there is a blunt lobe at the base but no hook.

Abdomen: As long as thorax with the scutellum included; basally it is a little wider than long and rather rapidly and regularly narrowed, and in males it may be only as long as mesonotum proper, scutellum excluded. Abdomen opaque black, densely covered in both sexes with short, appressed, curled, glittering brassy yellow pile and over the middle of the tergite many extremely fine, erect, slender black hairs which are longer in the males merging into similar, denser, erect, brassy yellow pile along the lateral margins of the tergite and at the base of the abdomen. Sternites very densely covered with matted, appressed, shining yellow hair. I find 7 tergites visible in both sexes with a trace of an eighth which is recessed in the males. Male terminalia posterior and slightly recessed, brownish orange, and exposed only from below.

Material available for study: Several flies including both sexes collected by the author in Queensland while on a collecting trip with Dr. I. Mackerras of Brisbane.

Immature stages: Unknown.

Ecology and behavior of adults: Collected while hovering and feeding upon flowers of a bush-type *Leptospermum* sp. in gum forest.

Distribution: Australian: *Laurella auripila* Hull.

Pilosia, new genus

FIGURES 28, 282, 512

Type of genus: *Pilosia flavopilosa*, new species.

Flies of stout robust form, with rather convex, tapered abdomen that is barely longer than the thorax. They are opaque black with dense, fine, erect, long, brownish-yellow pile upon the whole dorsal and lateral surface of the abdomen, whole thorax and head. The wing venation is somewhat similar to *Dischistus* Loew, to which tribe these flies belong; however, the anterior branch of the third vein is nearly straight and plane, and the face and antenna are quite different from *Dischistus* Loew.

Its nearest relative is *Laurella* Hull, from which it differs in the presence of dense pile upon the metapleuron, the closed and stalked anal cell, and the filiform labellum. In *Laurella* Hull the labellum is large, ovate, flattened, and conspicuous and the anal cell widely open besides a very different type of vestiture.

These flies are from Western Australia.

Length: 12 mm., wing 10 mm.

Head, lateral aspect: Head short and hemicircular with the occiput prominent, except above where it is shallow and the lobes are separated by a distinct shallow crease. The pile is fine, dense, and erect; it scarcely extends above the eye margin but grows longer toward the deeper portions of the occiput and shorter behind the lower, rounded, ventral eye margin and then much longer at the bottom of the head. This long pile is continued densely without break up the sides of the very wide gena and nearly to the base of the antenna. This leaves the face beneath the antenna bare, except for pollen, and leaves the upper part of the oragenal cup bare. The face proper is quite short but present beneath the antenna, but it is demarcated from the upper part of the oragenal cup by a creaselike continuation of the rather deep shelving tentorial fissure. The eye is prominent and large but distinctly flattened above. The anterior and posterior margins are almost parallel and both of them nearly plane so that the shape of the eye is very much as a sausage-shaped ellipsoid. The lower eye facets on the ventral third are much finer and are rather sharply and transversely demarcated from the extensive area of enlarged facets, although there is no crease or impressed line. Only large facets may be seen from the dorsal aspect. The postmargin of the eye is plane. The front is quite small and has brownish yellow pollen only. The proboscis is long and slender with a slender microspinulate labellum.

The proboscis is about $2\frac{1}{2}$ times as long as the head. The long, very slender palpus is composed of a single segment nearly half the length of the proboscis; it extends as far as the slender antenna extends, and it has a few fine, lateral, setate hairs. The antennae are attached at the upper fourth of the head; they are slender and moderately elongate and all three segments are cylindrical. The first segment is $2\frac{1}{2}$ times as long as the second; the second is at least $1\frac{1}{2}$ times as long as it is wide from the dorsal aspect but viewed laterally it is wider apically. The third segment is annulate at the base and at least $1\frac{1}{2}$ times as long as the combined length of the first two segments, or twice as long as the first segment. The third segment is a little narrow near the apex and a little curved downward apically; it bears a short, thick microsegment followed by a somewhat less thick, much longer, spine-tipped microsegment. First segment with numerous, very fine, rather long, blackish hairs above and below but which are longer below; they do not extend any further than the long pile of the face.

Head, anterior aspect: The head is fully as wide, perhaps a little wider, than the mesonotum; the antennae are adjacent. The male eyes touch and are holoptic but nevertheless they are rounded at the point of junction. The ocellar tubercle is prominent and high and forms an equilateral triangle; centrally it bears a tuft of fine, erect, black hairs of no great length. The oragenal space is very wide and much flared and ventrally occupies at least three-fifths of the head width, perhaps as much as three-fourths; consequently the wide gena grows wider below. The oragenal recess is short oval, quite deep with vertical sides; it is not much wider than the gena.

Thorax: Scarcely, if any, longer than wide and gently convex; the head is not decumbent. The whole thorax is black, dully shining bluish green with faint brown pollen and quite dense, fine, erect, brownish pile over the whole surface front to back, including the scutellum. The vertical sides of the postalar callosity have a rather dense tuft of long, fine pile; the plumula appears absent. The dense long pile of the mesonotum is continued over the mesopleuron and the whole of the pteropleuron and there is pile on the metapleuron in front of the haltere, abundant pile laterally on the metanotum. There is a tuft of pile below the spiracle, also a tuft on the base of the haltere; the haltere is light yellow with dorsal part of the knob conspicuously black.

Legs: Anterior legs rather short and stout, the femur distinctly widened toward the base; the tibiae have minute bristles ventrally. The femur has a multiple fringe of pile on the posteroventral surface; the middle femur is similar in every respect; the ventral fringe wider and more conspicuous. Hind femur of uniform width throughout and with 4 or 5 quite short ventral bristles on the distal half and with only a few long hairs toward the base. All the femora are black and narrowly brown at the apex; all bear coarse and rather long, flat-appressed, yellow or brownish yellow scaliform pile. Hind tibiae with 4 rows of slender oblique bristles of no great size; the two dorsal rows contain 12 or more, the ventral medial row contains 8 slightly longer, more slender bristles and the anteroventral row contains about the same number. The tarsi are short, the hind basitarsi are about as long as the remaining segments; the pulvilli are long. Tibiae and tarsi rather light brown.

Wings: Much wider at the base; the axillary lobe is prominent and the alula is rounded with a rather long, brownish-yellow fringe similar in color to the long squamal fringe. The ambient vein is complete; all the veins are heavy and dark brown with the third and fifth a little more yellowish; the anterior crossvein enters the apically narrowed discal cell at or beyond the outer eighth. The intercalary vein is plane and not quite as long as the anterior crossvein; the anal cell is closed with a distinct stalk. Third posterior cell a little narrower on the margin; the fourth wider. The first posterior cell is open maximally. The anterior branch of the third vein is almost straight from the base and is barely curved. It ends well above the wing apex. Although the marginal cell is widened apically the second vein ends obliquely in a low, gentle curve. The costal and subcostal cells are dark brown and somewhat yellowish toward the base; basicosta not much widened bearing an inconspicuous border of stiff hairs somewhat more appressed and setate against the costal cells. There is no basal hook but a curious, curled, horizontal tuft of pile reaches out over the base of the wings which is not part of the mesonotal pile.

Abdomen: About the same length as the mesonotum the scutellum included, or slightly longer; the base of the abdomen is at least as wide as the mesonotum, perhaps very slightly wider. It is progressively narrowed. I find 8 tergites visible from above; the last one is quite short and together with the terminalia it is twisted to the left. The terminalia and the hoodlike conical and triangular epandrium is placed in a posterior position and twisted to the left. The abdomen is somewhat convex, especially laterally, and almost opaque black; it is feebly shining and with a very dense pile over the whole. This pile consists of quite fine, erect, pale brownish-yellow hairs which extend over the sides of the abdomen and also there are a few, fine, short, scattered, curled, brassy yellow hairs more or less appressed on the surface of the tergites; they are partially erect and inconspicuous and likewise quite fine. Sternites with a few, long, fine hairs but mostly with subappressed, fine, curled hair curling close to the segments.

Material available for study: The unique, male type of the genus.

Immature stages: Unknown.

Ecology and behavior of adults: Collected while hovering and feeding upon flowers of low growing, small plant types of *Leptospermum* sp., in Western Australia.

Distribution: Australian: *Pilosia flavopilosa*, new species.

Genus *Heterostylum* Macquart

FIGURES 54, 55, 283, 504, 559, 834, 835, 836

Heterostylum Macquart, Diptères exotiques, suppl. 3, p. 35, 1848.
 Type of genus: *Heterostylum flavum* Macquart, by monotypy.
Comastes Osten Sacken, Bull. U.S. Geol. Surv., vol. 3, p. 256, 1877. Type of genus: *Comastes robustum* Osten Sacken, 1877, by monotypy.

Painter (1930a), pp. 1-2, key to 10 North American species.

All the members of the *Heterostylum* Macquart group, which includes *Triploechus* Edwards, *Efflatounia* Bezzi, and *Eurycarenus* Loew, are alike in having a rounded excavation or notch at the middle of the posterior border of the eye. Likewise, they all have notopleural macrochaetae.

Medium-sized flies of unusually robust appearance due to the comparatively short abdomen, which is covered rather densely with erect or semierect pile of moderate length. The whole thorax, with the exception of the dull, bare, propleuron and hypopleuron is covered with extraordinarily dense, recumbent pile brushed backward in mats, which obscure the ground color. The pile upon the mesonotum is not very long; it is longer on the pleuron, so much so as to make the head seem distinctly narrower than the thorax. The head is only slightly narrower in the males; in females it is approximately of the same width as the thorax, except for the pile. Oragenal cup very shallow, the walls extraordinarily widened, and flattened, eliminating the gena at the sides. Eyes holoptic in males, widely separated in females.

Heterostylum Macquart is unique in its venation. The first branch of the medius joins the fifth branch of the radius exactly at or sometimes only a little distance from the point of origin of the fourth branch of the radius. Also, the discal cell ends in a point, as the second branch of the medius crosses the first, the intercalary vein being eliminated.

The flies of this genus are principally distributed throughout the western part of the United States; one species extends into Mississippi and Georgia, one is

West Indian, and one or two species are found in northern South America.

Length: 10 to 15 mm., without proboscis; wing 10 to 16 mm.

Head, lateral aspect: The head is subtriangular in profile due to the flattened front and subconical extension of the face. The occipital part of the head is more extensive in the middle; this is only partly due to the distinct and characteristic indentation on the posterior margin of the eye just below the middle of the head. The face is obliquely sloping. It forms a subtriangular extension which reaches very little farther than the end of the first antennal segment but its depth or length is perhaps equal to that of both the first and second antennal segments. The oragenal cup is rather longer than the face and extends obliquely backward. The face is densely covered with fine pile and has a conspicuous fringe of rather denser shining pile along the lower margin; its surface is pollinose. The gena is quite narrow below, due perhaps to the curiously and greatly widened walls of the oragenal cup which appear to have numerous, oblique riblike ridges. This whole outer thicker portion is opaque and micropubescent. The eye is narrow in profile, at least twice as high as long, and without pile. The proboscis is quite long and slender, and about 3 times as long as the head. The labella likewise is quite long and slender, often divergent and curled backward in museum specimens. Palpus short but slender, the apical segment no more than one-third the length of the first segment.

Head, anterior aspect: The head becomes somewhat narrowed from side to side above. The shape of the eye is more or less rectangular in females, naturally rounded above and below, the front being almost as wide as the face. The sides of the eye, except above and below, are rather flattened. The head is more narrow than the thorax. The eyes of the male are in contact a distance equal the length of the first antennal segment. The antennae are narrowly separated at the base. The frontal groove runs between them. The vertex is sunken on each side opposite the eye, conspicuously sunken in some species. In both males and females the ocellar tubercle is high, prominent, and conspicuous, and there is a rather dense band or brush of pile transversely in front of the posterior ocelli lying upon the obtuse, ocellar triangle; ocelli large. The oral opening is rather long or high, a little wider in the middle, fringed with pile from face which curls over it from the upper half. It is rather shallow in depth, and the walls below on the lower half or more are expanded into a curious, wide, inflated, opaque, micropubescent, ribbed structure. Gena quite narrow on each side of this inflated oragenal border, the tentorial fissure is deep at this point. From above the occiput is gently bilobed to the depth of the vertex behind in males. In females a shallow groove runs from the posterior corner of the eye of the much wider vertex down to the occipital foramen. Pile on the occiput quite long and dense above but not rising beyond the ocellar pile. Elsewhere the pile on the occiput is dense but shorter.

Front and vertex pollinose. The front in females is flat or more or less sunken, with or without longitudinal grooves. In males the upper part of the eye and the anterior part of the front is flattened, the greater lower part of the front is divided by a medial groove, each outer half a little raised. The pile consists of long tufts in the males but scattered, erect hairs and appressed tufts in the females. Antenna slender and elongate; the first segment is 3 to 4 times as long as wide, wider apically; the second segment is barely longer than wide and wider apically. The third segment is long, not wider than the second segment, slightly narrowed at the base and gradually narrowed on the apical half; it is quite slender in some species, less so in others. At its apex it bears a minute, spine-tipped microsegment. The antennae are attached at the upper third of the head, the first segment bears a number of long, moderately coarse hairs above and below.

Thorax: The thorax is convex but not humped; it is densely covered with relatively short pile obscuring the opaque blackish ground color. In the species known to me the scutellum at least and sometimes the prothoracic sclerites, the hypopleuron, and the conically tuberculate humerus are paler in color. Bristles are restricted to 3 notopleural bristles in front of the wing of varying stoutness, 6 or 8 long, slender bristles on the postalar callosity and fine, rather long, bristly hairs on the posterior margin of the scutellum. The scutellum is large, thick, and convex, the pteropleuron and hypopleuron are bare, the former completely overlaid by the dense, appressed brush of long pile extending posteriorly from the mesopleuron. Metapleural pile is restricted to a convex lobe behind and below the spiracle that lies above the posterior coxae. Squamal fringe moderately long and dense.

Legs: The legs are quite slender and rather long. Anterior tibiae with only very minute, inconspicuous, spiny bristles. They are slightly longer on the middle tibia, longer and more conspicuous on the hind tibia where there are 4 rows containing 7 to 10 bristles. Middle femur with 4 or 5 bristles, hind femur with 5 to 10 bristles below and sometimes short bristles dorsally at the apex. The pile is short and appressed, slightly flattened. Pulvilli vary from about half as long as the very sharp, slender, curved claw, to other species where they are fully as long as claws and spatulate.

Wings: The wings are hyaline or brownish at the base, rather long, pointed apically and with a characteristic venation. There are 2 marginal cells. The first posterior cell is closed by a vein, which in some species has receded to a point directly across from the base of the anterior fork of the third vein, as in *Heterostylum robustum* Osten Sacken. Also, the medial crossvein or intercalary vein has disappeared so that the posterior branch of the medius crosses at the end of the discal cell. Anal cell widely open, axillary lobe rather wide and the alula quite extensive and long. Ambient vein complete.

Abdomen: The abdomen is short and broad, wider than the thorax. The pile forms dense and appressed bands which arise at the base of the tergites and sometimes along the posterior borders, and above this flat-lying pile there is a rather dense covering of long, fine, erect or semierect pile, sometimes bicolored, and sometimes uniform. Male hypopygium recessed. Last segment in females elongate and compressed together sidewise, with dense, fine, appressed, flattened pile.

The male terminalia from dorsal aspect, epandrium removed, show a broad, compact body with very little membrane. The dististyli are large, stout, bifid with the two apices rounded and club-shaped, with extra lobes and the whole turned backward; this retrorse state may be unnatural. The epiphallus bears a hook at the apex. From the lateral aspect the epandrium is long and the basistylus is long, stout and arched. Basal apodeme small.

Material studied: I have a long series of the type-species collected by myself and material of *rufum* Olivier and *croceum* Painter.

Immature stages: The larvae of these flies are important destroyers of mining bees involved in the pollination of crops. See further discussion under life histories.

Ecology and behavior of Adults: Adults use the long proboscis to probe many flowers such as thistles in western Kansas and button bush, *Cephalanthus occidentalis*, in Mississippi. They often emit a very shrill, high-pitched hum when hovering, of nearly the same pitch and intensity as nemestrinids.

Distribution: Nearctic: *Heterostylum croceum* Painter, 1930; *deani* Painter, 1930; *engelhardti* Painter, 1930; *laticeps* (Bigot), 1892 (as *Bombylius*); *robustum* (Osten Sacken), 1877 (as *Comaste*); *sackeni* (Williston, 1893 (as *Comastes*).

Neotropical: *Heterostylum bicolor* Wulp, 1888; *ferrugineum* Fabricius, 1805; *flavum* Macquart, 1848; *haemorrhoicum* (Loew), 1863 (as *Bombylius*); *pallipes* Bigot, 1892; *rufum* Olivier, 1789 [=*basilare* Wiedemann, 1819, =*deustum* Thunberg, 1827, =*histrio* Walker, 1849]; *stigmatias* Knab, 1913; *xanthobase* Curran, 1929.

Genus *Triploechus* Edwards

FIGURES 53, 304, 524, 551, 843, 844, 845

Triploechus Edwards, Rev. Chilena Hist. Nat., vol. 40, p. 31, 1936. Type of genus: *Bombylius heteroneurus* Macquart, 1849, by original designation.

Medium-size flies of broad, short, rounded abdomen, which strongly resemble *Bombylius* Linné in general appearance. They are readily recognized through the presence of the rounded notch at the middle of the postmargin of the eye and through the characteristic venation. In the type-species the pile on the surface of the abdomen is fine, rather dense, not very long and is completely appressed or matted to the surface, yet with a few erect, long, wiry hairs scattered over the surface; in addition, along the sides of the abdomen there are rather dense tufts of long, outwardly directed pile, black in the middle and at apex, yellowish in between and at the base; whole first segment with dense, erect, yellowish pile. On the wing the basicostal comb is weak, the hook absent, the apical half of costa microspinose or tuberculate. The marginal cell continuously widens at the apex, is recurrent and plane, simulating a crossvein, leaving the adjacent submarginal cell broadly based against the costa, vertical in position and greatly narrowed below. There are 3 submarginal cells. The anterior crossvein lies just beyond the middle of the discal cell. The anal cell is open; alula prominent, and first posterior cell closed with a long stalk. The antenna resembles that of *Bombylius* Linné. The face is prominently projected with a projecting fringe of fine hairs; oragenal cup deep, its walls narrow. Pulvilli reduced to one-third length of claw.

Notwithstanding the marked difference in venation, I place this genus in the *Heterostylum* Macquart group. The presence of 3 submarginal cells is similar to *Triplasius* Loew, subgenus of *Bombylius* Linné, and it is possible that the notched eye margin has been acquired independently.

The type-species of this small genus is from Chile; one species is found in the southwestern United States.

Length: 9 to 11 mm., without proboscis.

Head, lateral aspect: The head is long and more or less triangular. The occiput is very slightly developed beyond the eyes on the upper part, but is more extensive below. The oragenal cup and face are quite extensive, in length at least two-thirds the length of the eye, or even more. Front large, slightly swollen on each side with a medial groove in the male. Both face and front are opaque pollinose and bear considerable, very fine, long, stiff, erect hairs. The occipital pile is dense, fine, and erect and of about equal length above and below and extends deeply inward. The eye has the upper facets very slightly enlarged and the posterior margin is deeply indented below the middle. The proboscis is long and slender with a small, very slender labellum. It is about 2½ times as long as the head, and is generally thrust forward or a little downward. Palpus is long and slender, extending well beyond the oragenal cup. It is composed of 2 segments; the much longer basal segment bears a few long setae laterally near the apex. The blunt second segment has a few minute, short hairs at the apex. The antenna is attached at the upper fourth of the head; it is slender and elongate but not large. The first segment is about 2½ times as long as wide with long coarse hairs above and below; the dorsal elements are especially long and bristly. The second segment is equally wide but short and beadlike. Third segment less wide even at the base and gradually narrowed to the apex where it bears a short microsegment tipped by a stout microsegment and followed by a much smaller one bearing a short apical spine.

Head, anterior aspect: The head is not quite as wide as the thorax. The eyes in the male are in contact a distance somewhat less than the length of the ocellar triangle and the eye margin is depressed along the point of junction. The antennae are separated at the base by a distance equal to more than half the width of the first segment. The ocellar tubercle is small, moderately high, the large ocelli rest close to the eye margin in the male and they form an equilateral triangle with a tuft of numerous long, fine, stiff, erect hairs in the middle of the triangle. The oral opening is large, wide, a little narrow below and deeper below. It is quite shallow anteriorly but the pollinose interior is divided by a groove in the middle and a rounded, inflated, longitudinal ridge on each side. Posteriorly the walls of the oragenal cup are much thickened and inflated with oblique grooves. The gena is narrow below, except at the bottom of the eye, and the tentorial fissure quite deep. The sides of the oragenal cup bear on the upper half much extensive, long, coarse pile and the whole surface is covered with minute, appressed, scaliform pile.

Thorax: The mesonotum is a little longer than wide, a little humped and convex on the anterior half, more or less flattened posteriorly, sloping gradually back to the large, wide, thick, marginal convex scutellum. The mesonotum bears abundant but not dense, quite fine, erect, stiff pile and in addition some appressed scaliform pile especially on the anterior half, beginning at the transverse suture. Notopleuron with 4 long, stout, black bristles. Humerus with coarse, long pile especially below. Postalar callosity with a few long, slender bristles. Disc of the scutellum with some long, scattered, erect, stiff, bristly hairs, longer but not thicker on the margin. Mesopleuron with much long, coarse pile especially on the upper half. Sternopleuron with some long, fine pile anteriorly. Pteropleuron and metapleuron without pile. Hypopleuron with some long pile only in the upper posterior corner beneath the spiracle. Squama quite short but the marginal band of the pile dense, long and conspicuous. Halteres small, the cup hollowed apically.

Legs: The legs are slender and rather long. First four femora slightly thickened basally, hind pair long and slender. The anterior tibia with 7 or 8 short spicules anteriorly, with about 15 dorsally and 7 or 8 somewhat longer posteriorly. Middle tibia with longer spicules. Hind femur with coarse, long, appressed, scaliform pile, more dense dorsally, very scanty elsewhere and with a ventrolateral row of a few long, slender hairs and in addition on the outer half 3 or 4 distinct, stout, reddish spines. Hind tibia long and slender with 10 spicules dorsomedially, 7 or 8 longer ones dorsolaterally and 8 or 10 still stouter and longer spinuous bristles ventrolaterally. Pulvilli about half as long as the claw, narrowed apically, the large claws are only very slightly curved.

Wings: The wings are often diffusely tinged with brown basally. The marginal cell is greatly widened apically. There are 3 submarginal cells, uniformly in the type-species, 2 in certain American species. The anterior branch of the third vein is turned sharply forward to end before the apex of the wing. The type-species has the male costa, at least, spinous tuberculate. The first posterior cell is closed with a long stalk, the anterior crossvein enters the discal cell a little beyond the middle. The anterior intercalary vein is not quite as long as the long anterior crossvein. The anal cell is widely open, the alula large and broad. Ambient vein complete. There is a moderately developed comb and thickened costa.

Abdomen: The abdomen is short, fully as wide as the thorax or a little wider and certainly not longer than the thorax from the base of the first segment to the end of the abdomen. It is more or less shining with very thin pollen and covered with rather abundant, very fine, erect, stiff black hairs and in addition there is some flat-appressed, fine, somewhat scaliform yellowish pile throughout, but especially on the fourth and fifth tergites at the end of the abdomen. I find only 7 tergites visible in the male, the last one being rather longer than the preceding one. In addition there are characteristic and conspicuous, very dense tufts of pile along the side margin of the tergites. They are long and striking, some of the tufts black, some of them pale. First segment with long, erect dense pile extending under the scutellum. Male terminalia large, turned to the left and recessed beneath the apex of the abdomen.

The male terminalia from dorsal aspect, with epandrium removed, are broad and short. The dististyli are short, and broad with curved or curled, divergent, pointed apices. Membranous pockets large. The lateral aspect shows the epiphallus to have 2 arms fused distally with dorsal hooklike process. The epandrium is quite short.

Material available for study: The type-species and *Triploechus novus* Williston from New Mexico.

Immature stages: Unknown.

Ecology and behavior of adults: These flies are readily attracted to the blossoms of low-growing desert flowers.

Distribution: Nearctic: *Triploechus novus* (Williston), 1893 (as *Triplasius*) [=*recurvus* (Coquillett), 1902 (as *Bombylius*)]; *vierecki* Cresson, 1919.

Neotropical: *Triploechus angustipennis* Edwards, 1937; *heteroneurus* (Macquart), 1849 (as *Bombylius*); *minor* Edwards, 1937; *ornatus* (Rondani), 1863 (as *Triplasius*); *pallipes* Edwards, 1937.

Genus *Eurycarenus* Loew

FIGURES 51, 52, 299, 525, 533, 831, 832, 833

Eurycarenus Loew, Ofvers. Kongl. Vet. Akad. Forhandl., vol. 17, p. 83, 1860. Type of genus: *Bombylius laticeps* Loew, 1852, by monotypy.

Bezzi (1924), p. 70, key to 6 Ethiopian species.
Hesse (1938), pp. 508-512, key to 7 South African species.

Bowden (1964), pp. 32-33, key to 8 species or species groups of West Africa.

Medium-size flies of rather characteristic appearance. The abdomen is broad, short-oval, only moderately convex. The pile of the abdomen is short, quite appressed, either fine, wiry and curled, or flattened and scaliform, in varying amounts, with similar flattened, shining pile on the mesonotum. Moreover, the abdomen frequently has a pattern of bands or spots because of contrasting pile and because of prominent, postmarginal rows of long, upright, black bristles and pile on the tergites. The posterior eye margin shows a deep, rounded indentation at the middle. Of the 5 members of the *Heterostylum* group, *Eurycarenus* Loew appears to be nearer to *Efflatounia* Bezzi though the alula is absent and the pulvilli is reduced. Of the nonmembers of this group, there is a general similarity between *Eurycarenus* Loew venation and that of *Acanthogeron auripilus* Efflatoun; the former has the anterior crossvein lying before the middle of the discal cell, the latter has it placed beyond the middle. Those who have the material before them will be astonished by the curious, yet superficial, resemblance between the American *Heterostylum robustum* Osten Sacken and *Eurycarenus laticeps* Loew, which is restricted to the color, size, form, and pattern.

Still other salient characters of the genus consist of the rather extraordinarily stout bristles along the sides of the female front, the presence of 2 or 3 microsegments on the apex of the antenna, and the nature of the first posterior cell, which, because both upper and lower veins closing the cell are plane and not curved, is acutely pointed. The stout bristles or macrochaetae in front of the wing, varying from 2 to 10, are found in all 4 gena of this group.

This genus is confined entirely to South Africa.

Length: 9.5 mm. to 12.5 mm.; wing of about the same length.

Head, lateral aspect: The head is comparatively short, somewhat peculiar in several respects. The front and face are continuous, both prominent, and well extended beyond the eye, shallowly convex throughout and the antennae, which arise from close to the middle represent the only point of division between front and face. Both front and face rather densely covered with fine, erect pile, either pale or dark, together in some species with considerable matted and flat-appressed pile. The surface of both front and face is pollinose. Because the face is not greatly extended, the oragenal cup is only moderately oblique in profile; its walls are only a little thickened and the oragenal fissure is rather long and moderately wide. The genae also are rather wide and quite densely covered with conspicuous, shining, sometimes flattened hairs of considerable length, matted and turned downward, and extended quite to the eye margin. Occiput only moderately developed, a little more swollen centrally, but without any conspicuous cavity and its pile is long, erect, coarse, and bristly dorsally behind the upper eye corners, still longer laterally and below; this pile is finer and still longer beneath the head. The posterior eye margin has a characteristic, deep indentation in the middle that is quite unusual in members of this subfamily.

Palpus rather large and conspicuous, of 2 segments, the second leaf-shaped but slender and with a few minute setae. Proboscis rather stout and about twice as long as the head with the labellum consisting of at least one-third the total length. It is bare throughout, except for quite microscopic setae scarcely visible. Antennae attached above the upper fourth of the head; they are slender and moderately long. First antennal segment from $1\frac{1}{2}$ to 2 times as long as the second segment, with long setae above and sometimes with 2 or 3 still longer bristles below. The second segment is a little longer than wide, the third is about $1\frac{1}{2}$ times the combined length of the first two segments. The third segment is gently tapered from the base, becoming much more narrow at the apex and sometimes has 3 or 4 oblique setae dorsally. Apex with 2 microsegments, both short, the second a little longer and spine-tipped.

Head, anterior aspect: The genofacial space is at least one-third the head width or somewhat more. Eyes of the male narrowly touching and approximated, and rounded medially, or separated by as little as the width of the anterior ocellus and sometimes the eyes are angulate above and separated by as much as the distance between the outer edges of the posterior ocelli. Ocellar tubercle only slightly raised, the ocelli large and situated in an obtuse triangle. Separation at base of antennae equal to the thickness of one segment.

Thorax: The thorax is a little wider than the head and scarcely longer than broad, but with a very large scutellum, which is slightly convex and raised until it is on a level with at least the posterior mesonotum. Whole surface of the mesonotum densely covered with fine, suberect, or partly appressed, shining pile of no great length but which becomes longer along the lateral margins. There are strong, characteristic bristles present, which are long, conspicuous, and rather stout. On the notopleuron there is in the posterior corner 1 long bristle and before it 4 to 10 additional bristles equally stout but much shorter, which may extend as far as the humerus. Postalar callosity with 4 quite stout, long bristles, the scutellar margin with 6 pairs of similar, long, stout bristles, the discal pile like that of the mesonotum. Lateral margins of the postalar callosity with an especially dense tuft of long, fine, opaque pile. Squama large and elongate with a quite long marginal fringe. Pleuron densely pilose almost everywhere with 2 long bristles on the upper corner of the mesopleuron; both the metapleuron and the entire posterior half of the hypopleuron are extensively covered with long, dense, fine, erect pile. The anterior pile of the pleuron has the appearance of being subappressed and brushed backward.

Legs: The legs are relatively short and stout, especially upon both tibiae and femora. The hind tibia is more than half as wide as its femur, the anterior femur enlarged toward the base and with an anterior and posterior row of short setae or spicules. Middle femur with 3 spines anteriorly, hind femur with conspicuous spines in 2 irregular rows, 7 of them long, 8 others short. Pile of the femora consists of long, broad, conspicuous scales. Tibiae with similar scales which are as wide but shorter. Ventral surface of anterior and middle femora with some shaggy hairs, especially toward the base. These hairs are erect or semierect. Tibial spicules very strongly developed on all tibiae in 3 or 4 rows with 8 to 12 spicules in each row. Claws curved from the base; the pulvilli long.

Wings: The wings are hyaline or sometimes with the base a little tinted, at least above the alula. Base of costa with a pronounced and highly developed costal comb and sharp, proximal, beaked spur. There are 2 submarginal cells; the first posterior cell is closed and stalked, and it appears to be the fourth vein which reaches the wing margin and receives the lower branch of the third. Anal cell widely open, its apex slightly widened, the anterior crossvein enters the discal cell at or close to the basal fourth, hence the first basal cell tends to be very slightly longer and sometimes more distinctly longer than the second basal cell. Discal cell rounded or convex near the apex above and below, but especially above, and the medial crossvein moderately reduced in length. Alula unusually large with a dense but not very long fringe.

Abdomen: The abdomen is short and robust, as wide as the thorax, gently convex, and distinguished by the transverse bands of pile upon the tergites, which are rather dense, often of 2 or more colors. These bands generally contain long, coarse, and suberect pile on the basal margin followed by partly hidden, rather dense, appressed, scaliform pile. Finally, the entire postmargin of the tergites characteristically has a single row of long, erect, strong bristles, which are usually black and form spikelike palisades. Male genitalia recessed. Female genitalia guarded by a fringe of hairs among which are, on each side, 2 or 3 very stout, curved, hooklike spines.

Since Dr. A. J. Hesse was able to base his conclusions upon a greater number of species, I quote his remarks upon the male hypopygium:

> The hypopygium of the males, in members of this genus, is also peculiar in having the dorsal apical part of basal parts produced forward and upward or obliquely upward into a transverse flattened lobe or process the upper edge of which is usually black, curled over slightly backwards and armed with ctenate spines. The outer lower margin of the basal parts in neck region has a very distinct tuft of long, stiff bristles and the dorsum of neck region is usually much flattened or even depressed. The dististyles (beaked apical joints) have the base above hollowly depressed, its apical part also flattened below and ending in a short claw-like spine.

Although the male genitalia are recessed, one can sometimes observe that the dorsoapical portion of the basal part is directed obliquely upward as a flattened lobelike process bearing comblike spines on its outer edge or margin. Lower margin of the basal part outwardly with a tuft of stiff, rather long bristles. The dististyli of the hypopygium usually have the base above hollowly depressed; the apical part is flattened below and ends in a short spine.

From my study I find the male terminalia from dorsal aspect, epandrium removed, show a rather broad, short body with large membranous areas. Dististyli with a narrow back-turned, medial, hooklike process with a characteristic ctenate band of stout spines. From the lateral aspect the basistylus is pear-shaped, the epandrium large and quadrate, the epiphallus has a long slender process. Basal apodeme small.

Material available for study: The material studied consists of several species in the British Museum (Natural History) and in the collection of the author. John Bowden also very kindly sent me some material.

Immature stages: Unknown.

Ecology and behavior of adults: Unknown.

Distribution: Neotropical: *Eurycarenus chilensis* Paramonov, 1934.

Ethiopian: *Eurycarenus albicans* Bezzi, 1924; *argentifrons* Bowden, 1964; *cingulatus* Hesse, 1936; *dichopticus* Bezzi, 1924; *flavicans* Bowden, 1964; *lanatus* Bowden, 1964; *laticeps* (Loew), 1852 (as *Bombylius*) [=*latifrons* Loew, 1863, lapsus]; *loewi* Hesse, 1938; *melanurus* Bezzi, 1924; *minimus* Bezzi, 1921; *propinquus* Hesse, 1938; *sessilis* Bezzi, 1921; *subater* Paramonov, 1960.

Genus *Efflatounia* Bezzi

FIGURES 300, 529, 557

Efflatounia Bezzi, Bull. Soc. Roy. Ent. Egypte, vol. 8, p. 171. 1925c. Type of genus: *Efflatounia aegyptiaca* Bezzi, 1925, by original designation.

Medium size, robust flies of rather similar appearance to *Bombylius* Linné, or *Systoechus* Loew, and like them with dense, long, plushlike pile.

Efflatounia Bezzi possesses in common with the five genera in the subfamily Bombyliinae a rounded, notchlike excavation or indentation near the middle of the posterior margin of the eye. It differs from the other three by having the first posterior cell open, usually sharply narrowed, or sometimes closed in the margin. It is further unique in the very short, or even completely absent alula, which is prominent in the other three genera. The anterior cross vein lies beyond the middle of the discal cell. The male claw is toothed. Efflatoun (1945) contrasts this genus with *Dischistus* Loew and *Acanthogeron* Bezzi, Egyptian genera, noting that in *Efflatounia* Bezzi the face is more narrow, more produced forward, and bears sparser and shorter pubescence. The pulvilli are minute and vestigial, and there is no pile or vertical fan of hairs on the metapleuron.

Described from Egypt; I have also seen material from southern Algeria.

Length: Without proboscis, 9 mm. to 13 mm.

Head, lateral aspect: The head is strongly and more or less conically produced, extending below as far as the end of the first antennal segment, obliquely truncate below by the large, wide, oral opening. The length of the face above the opening is as long as the first 2 antennal segments, and this portion of the face also extends obliquely forward below the antenna. The face is exceptionally long on the ventral portion. The eye is rather narrow or high, with a deep posterior indentation below the middle. Occiput prominent and well developed. Pile of the occiput dense, rather short and fine, and erect, becoming coarse on the upper half. The proboscis is long and very slender, the terminal piece equally slender and quite long, occupying about one-fourth the length of the proboscis. Palpus long and conspicuous, slender and compressed basally, a little attenuate distally, and composed of one segment.

Head, anterior aspect: The face shows a deep narrowed crease, quite close to the eye margin, thereby reducing the middle of the gena. The oral opening begins quite a long distance from the occiput. Sides of the face and the whole of the front rather densely long, fine, dark brown pilose, the face and front with thick pollen. Sides of the front swollen, but the greatly narrowed portion leading to the ocellar triangle sunken; this upper half of the front is very greatly and abruptly narrowed, vertex swollen and raised above the eye margin, the whole of it given over to the ocelli, and with 12 to 15 long, fine bristly hairs directed more or less forward between the ocelli and 4 or 5 pairs behind the ocelli. Antenna attached at the upper fourth of the head, the first segment 3 or 4 times as long as the second and bearing abundant, long, bristly hairs ventrally and laterally. Second segment quite short and beadlike. Third segment elongate, slender, the basal half usually broader than the rather attenuate apical half.

Thorax: The thorax is pollinose, opaque, and densely covered with long, erect, coarse, shining pile. There are 3 or 4 notopleural bristles, which are rather stout, 7 or 8 postalar bristles, which are long and slender. Margin of the scutellum with 3 or 4 pairs of very long, bristly hairs, the disc thickly long, pale pilose; the scutellum is quite large, thick, and convex apically. Ventral margin of the postalar callosity with a long brush of very coarse, long pile. The callosity is convex and bears a rather conspicuous crease or fossa medially. Propleuron protuberant and convex above the coxae; together with the mesopleuron and the humerus, it bears very dense, long, coarse, pale pile. On the mesopleuron this tufted pile extends outward dorsally and posteriorly. Sternopleuron with some scattered pile, the metapleuron and the posterior metasternum each with a tuft of long pile. Squamae with especially dense, long, fine pile.

Legs: The legs are gently swollen toward the base with coarse, appressed pile continued onto the tibiae, where it is shorter and flattened. The pile is also more or less flattened on the femora, but not scalelike. Ventral surface of the hind femur with 4 long, reddish bristles on the outer half, the basal ventral portion with long, black, bristly hairs. Hind tibiae with 10 dorsomedial bristles, 7 or 8 ventromedial, 8 to 10 dorsolateral, and a like number of ventrolateral bristles. Claws rather long, but quite slender, only gently curved, and toothed in the male. The pulvilli are minute and narrow; the moderately long, slender, medial empodium is not bristlelike and is obtuse apically.

Wings: Hyaline but with costal and subcostal cells brownish. The venation is similar to *Dischistus* Loew but the first posterior cell which is usually narrowly open, with spoutlike shape apically, may also be closed at the margin. The anterior crossvein is situated well beyond the middle of the discal cell. The anal cell is rather widely open and the alula is much reduced or absent.

Abdomen: The abdomen is conical to short oval, a little wider than the thorax, opaque, pollinose with thick, long, very fine, black, bristly pile and with almost equally long, pale yellow pile, which is more or less tufted laterally and tends to be concentrated in transverse bands. While the pile in the male is dense and furry, pale colored, in the female the pile is shorter and with the long, erect, black hairs much more abundant. Sternites with long, fine pile. Genitalia concealed in both male and female, the latter without spines.

Material studied: The series of individuals in the British Museum (Natural History).

Immature stages: Unknown.

Ecology and behavior of adults: Efflatoun (1945) comments on the remarkable ability of the males to hover and the difficulty of capture; he collected a series of 30 specimens of this rare fly; the genus is not, however, endemic to Egypt.

Distribution: Palaearctic: *Efflatounia aegyptiaca* Bezzi, 1925.

Genus *Karakumia* Paramonov

FIGURE 307

Karakumia Paramonov, Trav. Mus. Zool., Kieff, no. 1, p. 77, 1926. Type of genus: *Karakumia nigra* Paramonov, 1926, by original designation.

With the genus *Efflatounia* Bezzi very closely related, distinguished from that genus, as well as other genera of that subfamily, through presence of a pencil of hairs at the apex of the third antennal segment, it stands near the genus *Bombylodes* Paramonov and is distinguished from *Acanthogeron* Bezzi by the presence of a distinct indentation on the middle of the hind margin of the eyes, a characteristic found only in and with a few exotic genera.

The habitus is very similar to *Bombylius* Linné. The width of the head is equal to or even exceeds the width of the thorax. The front has a deep transverse groove; it is divided into two parts. The upper part of the front in profile is woolly, as are also the lower parts. The middle of the face forms a large protuberance. Proboscis and antenna as in *Bombylius* Linné. Venation as in *Efflatounia* Bezzi. First posterior cell broad, at the apex narrowed, yet, however, open. The anterior crossvein is placed at the end of the second third of the discal cell. Legs as in *Bombylius* Linné. The underside of the hind femur has distinct bristles. Claws toothed (indented), the pulvilli entirely rudimentary. Pile of the body thick, but also relatively long.

Material available for study: None. I have quoted Paramonov's description from his 1940 work. He gave a full figure of this fly.

Immature stages: Unknown.

Ecology and behavior of adults: Not on record.

Distribution: Palaearctic: *Karakumia nigra* Paramonov, 1926.

Genus *Cytherea* Fabricius

FIGURES 73, 75, 296, 298, 459, 503, 521, 569, 825, 826, 827

Cytherea Fabricius, Entomologica Systematica, vol. 4, p. 413, 1794. Type of genus: *Cytherea obscura* Fabricius, 1794 [designated by Becker, p. 455, 1913a].

Mulio Latreille, Précis des caractères génériques des Insectes, p. 155, 1796. Type of genus: *Cytherea obscura* Fabricius, 1794. "This genus proposed for *Cytherea* under the false impression that this last name was preoccupied by *Cythere*, Müller 1785, Crust" [from Bezzi footnote p. 2, 1924].

Chalcochiton Loew, Stettiner Ent. Zeitung, vol. 5, p. 157, 1844. Type of subgenus: *Chalcochiton speciosus* Loew, 1844, by monotypy.

Glossista Rondani, Dipterol. Ital. Prodr., vol. I, p. 163, 1856. Type of subgenus: *Mulio infuscatus*, Meigen, 1820, by original description.

Logcocerius Rondani, Arch. Zool. Anat. Fisiol., vol. 3, (sep.) p. 61, 1863. Type of genus: *Anthrax holosericeus* Fabricius, 1794, by original designation; 1 species, as *Mulio holosericeus* Loew, lapsus.

Loncocerius, emendation; in Kertesz, p. 70, 1909.

Lonchocerius Bezzi, p. 4, 1924, emendation.

Becker (1903), pp. 23-28, keys to 26 Palaearctic species (as *Mulio* Latreille).
Paramonov (1930), p. 7, keys to many Palaearctic species.
Engel (1934), pp. 321-324, keys to 24 Palaearctic species.
Austen (1937), pp. 46-47, key to 8 Palestine species.
Efflatoun (1945), pp. 368-369, key to 9 Egyptian species.

Bristly flies of small to large size, which are quickly recognized through the widely separated antennae, inflated front, and the point of origin of the second longitudinal vein. The species have a rather conically elongate abdomen that is only slightly and gradually narrowed apically, the surface of which is covered with moderately dense, completely appressed, narrowly flattened hairs of varying color, and the postmargins of the tergites have conspicuous rows of stiff, upstanding, abundant bristles. Also, there is some erect, whitish pile beginning on the sides of the several tergites, encroaching more to the middle on the basal ones, but extending rather conspicuously down upon the curled-over side margins. The wings are generally tinted at middle and base with pallid brown. The second longitudinal vein arises from the third longitudinal vein near the small crossvein and rather abruptly. The scutellum, postalar callosity, and sides of mesonotum in front of the wings bear a number of long, slender bristles. The wide front is distinctly bossed or inflated and bears thick mats of wide, silvery scales, which extend down upon sides and middle of the face. The face in profile is inextensive, but little produced; the oragenal cup is shallow, its margin knife-thin. Pulvilli very greatly reduced.

Cytherea Fabricius is an important element of the Palaearctic bee fly fauna. It is confined to this region. I believe Neotropical species belong elsewhere.

Length: 4.5 mm. to 15.0 mm., excluding proboscis. A majority of the species are 6 mm. to 10 mm. The largest known species is *Cytherea barbara* Sack with 31 mm. wing spread.

Paramonov (1930) called attention to the difficulty in drawing a sharp dividing line between the groups of the genus, such as those with 2 and those with 3 submarginal cells and the presence, or absence, of pulvilli, and length of proboscis. While I would not have erected genera on these differences, because genera have, in fact, already been erected to represent these differences, I believe that they do serve a useful place in evaluating the phylogenetic trends within the family. I point to the further fact that these same distinctions are used not only elsewhere in this family, but in other families as basis for genera. For this reason, since they are already in the literature, I recognize the following subgenera: *Glossista* Rondani (1856) and *Chalcochiton* Loew (1844).

The species of *Cytherea* Fabricius are numerous and largely distinguished by differences in the faint wing pattern, the coloration of the legs and likewise of the scalation and appressed pile; this accounts for numerous synonyms. Efflatoun, 1945, states that he has been obligated to drop 4 or 5 species, and adds that of 3 other species he is unable to show sharp distinctions.

Two areas of the wing show some instability, the submarginal area perhaps less than the discal cell area, where the basal part of the second branch of the medius sometimes shows a basally directed spur vein.

Species of *Cytherea* Fabricius in the strict sense have 3 submarginal cells, very minute pulvilli, and proboscis nearly twice the head length.

The subgenus *Glossista* Rondani, comprising most species, has 2 submarginal cells.

The subgenus *Chalcochiton* Loew has 2 submarginal cells, fully developed pulvilli, and short proboscis. See below for full description.

Head, lateral aspect: The head is short and high. The occiput is shallowly convex, not very extensive behind the eye and about equally prominent above and below. From the middle at the point where it joins the thorax it slopes gently toward the eye margins, the bottom of the head and the vertex. The occipital pile is of two types. There is a rather dense mat of broad, flat-appressed, moderately long scales, white or yellowish on the outer borders of the occiput, extending to the eye margin. In addition deep within the occiput there is an area adjacent to the thorax with dense, upright pile. In addition in the upper corners of the occiput on each side of the ocellar tubercle there is a patch of stiff, bristly pile directed upward and still deeper on the interior of the occiput there is a row of 4 or 5 stout bristles curved outward on each side. The face is rather prominent and extensive in height as well as horizontally. The height of the face is as much as or more than the length of the antennae, and it is produced from the eye margin as much as or more than the length of the antenna.

The front is almost plane with the eye and densely covered with matted, silvery white, broad scales. Laterally above each antenna there is a patch of semierect, stiff, bristly pile which is a continuation of the broad band of dense, erect, often blackish pile that stretches in front of the ocelli across the entire interocular space and over the whole of the ocellar triangle.

The pile of the face consists of very dense, large, broad, silvery white, flat-appressed scales extending over the edge of the oral opening and in addition numerous, long, slender, down-curved, bristly white hairs. These scales and pile continue on down the genae and the sides of the oragenal cup but are completely wanting beneath the eye where there are only a few hairs and the surface is pollinose. The eye is large, plane for about the middle third then broadly rounded above and below. The posterior margin is entire. The proboscis is slender and elongate but only 1½ times as long as the head. The labellum is slender. The palpus is long and slender, the first segment has numerous bristly hairs laterally. The second segment is quite short, clublike with a few setae. The antennae are attached a little above the middle of the head and are small but slender. The first segment is about as long as wide, has a tuft of broad scales dorsally, some small scales medially, and short setae at the apex. The second segment is very small and beadlike with apical setae above and minute setae below. Third segment knobbed at the base, even more narrow than the second segment, very slightly widened just beyond the base, narrowed and attenuate beyond the middle, and the apex bears a small spine.

Head, anterior aspect: The head is quite broad, a very little wider than the mesonotum, the eyes in the male are widely separated, the interocular distance being as much as or a little more than the antennal distance. The antennae are separated by more than their length, the ocellar tubercle is quite low, the large ocelli set in an obtuse triangle far from the eye margin. In addition to the bristly hairs in the ocellar triangle, there are some slender bristles. The oral opening is relatively small, elongate, narrowed anteriorly and only moderately deep. The lateral ridge in the recess is low and barely indicated. The genae are wide as is to be expected with such a wide head. They extend, however, very little beyond the eye and slant down into the deep tentorial fissure on each side of the rather prominent oragenal cup. The outer walls of the oragenal cup bear the same scales and pile as the face and genae. The recess within has no pile.

Thorax: The thorax is unusually flattened and very shallowly convex, a little more convex anteriorly. It is opaque, pollinose and black. The pile is not very dense and consists of curled, wiry, flat-appressed, and somewhat flattened hairs over the disc of the mesonotum, and a considerable amount of suberect, stiff pile of no great length, some of it very fine and blackish and some of it much more coarse and white, especially along the anterior border of the mesonotum. Along the anterior border of the mesonotum the pile is a little more erect, stiff, and whitish.

The humerus, the whole notopleuron, and the area immediately above the wing have much fine, curled, matted, opaque white pile. Also, the anterior border of the mesonotum, and the lower part of the humerus especially, has many moderately stout, curved, red bristles; the whole notopleuron has a cluster of 20 or more bristles, several are quite long and stout. There are 4 laterally above the wing, 4 long, stout bristles on the postalar callosity, and 2 or more irregular rows of long, stout, red bristles on the outer half of the scutellum. The scutellum is polished, shining black. The disc of the scutellum has much curious, curled hair basally and especially laterally and also some flat-appressed scales. The whole mesopleuron is pilose, the pile coarse, long, and whitish, some of it directed backward, and that on the upper anterior portion directed upward with a cluster of red bristles. Pteropleuron with some tufts of flattened, curled pile. Metapleuron with a large clump of dense, long, erect pile. The greater part of the hypopleuron is bare. However, there is a tuft of pile along its entire upper border beneath the spiracle divided into 3 portions. Squama large with a long, dense fringe of pile. Haltere with long and spatulate knob, the apex cup-shaped outwardly.

Legs: The legs are relatively short, all of the femora are a little stout although not swollen toward the base and together with the tibiae they are covered with numerous, short, broad, whitish to brownish scales which are flat-appressed. Anterior tibia with well-developed spicules, though slender; there are 8 to 10 in the anterodorsal row, a like number posterodorsally and 5 or 6 posteroventrally. The anterior femur has a few oblique hairs below, and a few slender bristles anteriorly. Middle femur with 3 rather strong bristles anteriorly on the outer half, the tibia similar to the anterior pair. Posterior femur with a ventrolateral fringe of a few, scattered, stiff hairs and on the outer half 5 bristles which are slender, 2 dorsally at the apex and 1 medially. The hind tibial bristles or spicules are very similar to those on the other legs. Pulvilli minute and vestigial. The slender sharp claws are scarcely curved at all.

Wings: The wings are wide and are broad basally, more pointed apically. All the species known to me have the basal half or two-thirds yellowish brown with some lighter spots along the crossveins or vein furcations, or have a brownish band across the middle of the wing leaving the apical third of the wing more or less hyaline. In the type-species and in *Cytherea* sensu stricto there are 3 submarginal cells. The first posterior cell is a little narrowed apically, the second vein arises abruptly but not rectangularly about midway between the anterior crossvein and the base of the discal cell. The anterior crossvein enters the discal cell at the middle. The medial crossvein is long and biangulate, often with a trace of a spur at the first angle. Anal cell quite widely open, axillary lobe very little wider at the base than apically, the alula more than twice as long as wide. There is a very strongly developed basal comb with long setae and some scales. The ambient vein is complete.

Abdomen: The abdomen is elongate oval, narrowing apically. I find 6 tergites in the male with 2 others shortened and concealed beneath the sixth. The abdomen is as wide as the mesonotum at the base. The tergites are feebly shining black, rather densely covered with appressed and subappressed long, slender scales, often of several colors but generally pale. In addition along the basal half of each tergite, especially on the outer third of the second to the fifth tergite, there are numerous, long, erect, coarse hairs, generally white. On the first segment this pile is even more extensive and more bristly in character; laterally it reaches as far as the halteral knob and touches it, medially it becomes much shorter beneath the scutellum. The white pile on the sides of the tergites become more abundant laterally and more conspicuous. Finally, each tergite including the first has a complete row postmarginally, or double row, of long, stout, erect, spikelike bristles, generally reddish. End of abdomen obliquely truncate leaving the male terminalia recessed within the opening. Male genitalia slightly protrusive, triangular, the outer valves bear a medial patch of stiff, sharp, fine bristly hairs which are appressed upward.

The male terminalia from dorsal aspect, epandrium removed, show a broad base, the remaining structure narrowed distally. The epiphallus has a flat, expanded, shovellike apex, the whole structure is elongate, but expanded toward the base, and laterally it has a snakelike downward curve with the apex attenuate and turned upward in another curve. The dististylus is large, elongate, the apex divergent, sharply pointed dorsally but quite blunt laterally. The basal apodeme is large, the ejaculatory process is separated below from the epiphallus. The male terminalia of *Cytherea* Fabricius do not appear to differ in any remarkable extent from that of such genera as *Legnotomyia* Bezzi, or *Sosiomyia* Bezzi. Its most unique features are the spadelike apex of the epiphallus, its curious curved figure from the lateral aspect and the peculiar shape of the dististyli.

Materal studied: The type-species and other species material in the British Museum (Natural History) and in my own collection.

Immature stages: The larvae of this genus are important destroyers of locust eggs in the Palaearctic region. See section under life histories for further details.

Ecology and behavior of adults: Becker (1903) states that he collected two species upon flowery sand dunes by the Mediterranean shores. Efflatoun (1945) speaks of *aurea* Fabricius as a very difficult species to catch, hovering in barley fields in April, and matching the background. Several individuals of *barbara* Sack were captured at Sinai at altitudes of 1,800 meters.

Subgenus *Chalcochiton* Loew

Chalcochiton Loew has been treated as a synonym of *Cytherea* Fabricius by Paramonov (1930), and Austen (1937), but it was nevertheless retained as a subgenus by Engel (1934), and I have adopted the same procedure. Although Paramonov (1930) appropriately pointed out the variability of the pulvilli in the quite numerous species of *Cytherea*, the subgenus *Chalcochiton* Loew still has a useful place, based on the short proboscis, not longer than the oral cavity, the quite fully developed pulvilli, and 2 submarginal cells. The subgenus *Chalcochiton* is as worthy to retain as the many subgenera of *Exoprosopa* Macquart. I have before me representatives of *Chalcochiton pallasii* Loew, which I have compared with *Cytherea obscura* Fabricius, the type-species of *Cytherea*.

Length: 7 to 11 mm.

Head, lateral aspect: The head is hemispherical with the face and front extremely wide, the face prominent and bluntly rounded. Face laterally and in the middle and the genae likewise densely covered with moderately long, erect, fine, black pile changing to bristles in the middle above the oral margin. Occiput not at all prominent, not or scarcely extending beyond the eye and without any central, conspicuous cavity. The posterior eye margin is entire and the occipital pile dense but short, except below, where it is a little longer. Antenna with the first 2 segments short and the first segment twice as long and considerably wider than the second segment. Third segment annulate at the base, a little more than twice as long as the combined length of the first 2 segments, attenuate and narrowed from the base to the slender apex, which has a minute, spine-tipped microsegment. Proboscis short, not longer than the oral opening. Palpus long but with the second segment quite short, somewhat enlarged and short oval. The first segment bears a few long hairs near the base.

Head, anterior aspect: In the middle of each side of the face there is a creaselike extension of the tentorial fissure, which below is represented by a deep pit, and this lateral creaselike depression runs to the base of the antenna. The oral cavity anteriorly is remarkably shallow and not very deep posteriorly. The antennae, as characteristic of *Cytherea* Fabricius, are extremely widely separated, perhaps even more so. Across the area between the antennae on the lower half of the front and the upper border of the face there are numerous, subappressed, silvery-white scales and scaliform hairs. Outer half and upper half of the front with conspicuous, dense, erect, coarse, black pile of the same length and character as that found on the outer portion of the face. It also continues in the same character and length over the large, low, ocellar tubercle and the entire upper postvertex. Eyes of male broadly separated.

Thorax: The thorax is shallowly convex, opaque black, and rather densely covered with erect, black hairs, some fine, and some coarse and bristly. Notopleuron with 3 long and 4 or 5 shorter, distinct, black bristles. At least 1 long, black bristle over the wing, 3 upon the postalar callosity, and 3 or 4 pairs on the scutellar margin; the scutellar disc has a few fine, erect, long, black hairs. Pleuron with dense, erect, black pile dorsally and posteriorly on the mesopleuron, and also on the metapleuron. Hypopleuron without pile, except for a tuft of fine, short hairs immediately below the halteres and behind the spiracle. The pleuron is dully shining black. Squama brownish black with a dense, long fringe of dark, sepia hairs.

Legs: The legs are long, the femora well developed without being stout, the tibiae and tarsi also well developed. Middle and hind femora with conspicuous bristles, middle femur with 5 long, anteroventral bristles, and with 4 or 5 dorsolateral bristles at the apex. Pile of legs consists of dense, broad, flat-appressed, brownish white and black scales intermixed. Middle femur with a conspicuous, posterior fringe of long, black hairs. Anterior femur with less abundant, similar hairs. Anterior tibia with long, conspicuous spicules in 2 rows, the dorsal row with 9, posterior row with 6 spicules, or bristles. Middle tibia with 4 rows and hind tibia also with 4 rows, all quite long and conspicuous. Claws long and sharp, the pulvilli well developed and more than half as long as the claw.

Wings: See Figure 298. Basal half of wing obliquely brownish black, outer half hyaline.

Abdomen: The abdomen is opaque black, as wide as the thorax at the base, gently tapered and narrowed, the end obtuse and the length about 1½ times as long as the basal width. The tergites are covered with quite long, erect, slender hairs, which are quite dense laterally and especially dense and conspicuous on the lateral margin, and which are likely to be denuded across the middles of the tergites. This pile is white, except on the first tergite where it is entirely black or brownish black, long laterally but short and also dense beneath the scutellum. There are 7 tergites visible, the last one somewhat shortened and the sixth a little shorter than the fifth. The outer portion of the posterior margin of the fifth to seventh tergites each bear quite long, conspicuous, comparatively stout, black bristles, which are directed backward.

Distribution: Neotropical: *Cytherea ?cinerea* (Lynch Arribalzaga), 1878 (as *Mulio*); *costata* (Bigot), 1892 (as *Glossista*); *?dubia* (Macquart), 1846 (as *Mulio*); *lateralis* (Rondani), 1868 (as *Mulio*); *marginalis* (Rondani), 1868 (as *Mulio*); *multicolor* (Bigot), 1892 (as *Glossista*).

Palaearctic: *Cytherea adumbrata* Paramonov, 1930; *albifrons* (Loew), 1873 (as *Mulio*); *albolineata* (Bezzi), 1925 (as *Glossista*); *alexandrina* (Becker), 1902 (as *Mulio*); *angusta* Paramonov, 1930; *araxana* Paramonov, 1930; *arenicola* Paramonov, 1930; *argentifrons* (Macquart), 1849 (as *Anthrax*); *argyrocephala* (Macquart), 1840 (as *Anthrax*); *aurea* Fabricius, 1794 [=*?punctipennis* (Macquart), 1840, (as *Mulio*)]; *bisalbifrons* (Bezzi), 1922 (as *Chalcochiton*); *brevirostris* (Olivier), 1811 (as *Mulio*); *bucharensis* Paramonov, 1930; *carmelitensis* (Becker), 1903 (as *Mulio*); *cinerea* Fabricius, 1805; *cinerea* Wiedemann in Meigen, not Fabricius, 1820 (as *Mulio*); *claripennis* (Becker), 1903 (as *Mulio*); *cyrenaica* (Bezzi), 1926 (as *Glossista*); *delicata* (Becker), 1906 (as *Mulio*); *deserticola* Paramonov, 1930, *deserticola albifrons*, Paramonov, 1930; *dichroma* Paramonov, 1930; *dispar* (Loew), 1873 (as *Mulio*); *disparoides* Paramonov, 1930; *elegans* Paramonov, 1930; *farinosa* (Loew), 1873 (as *Mulio*); *fenestrata* (Loew), 1873 (as *Mulio*), *fenestrata barbara* Sack, 1906 (as *Mulio*); *fenestrulata* (Loew), 1873 (as *Mulio*), [=*armeniaca* Paramonov, 1930]; *fratellus* (Becker), 1903 (as *Mulio*); *holosericea* (Fabricius), 1794 (as *Anthrax*) [=*aberrans* (Walker), 1849, (as *Cyllenia*), =*semiargenta* (Macquart), 1840 (as *Anthrax*)]; *infuscata* (Meigen), 1820 (as *Mulio*); *latifrons* Paramonov, 1930; *lugubris* (Loew), 1873 (as *Mulio*), [=*?aberrans* (Walker), 1853 (as *Cyllenia*)]; *maroccana* (Becker), 1903 (as *Mulio*), [=*discepes* Becker, 1915, =*nitidapex* (Bezzi), 1925 (as *Glossista*)]; *melaleuca* (Loew), 1873 (as *Mulio*) [=*melanoleuca* Becker, 1903]; *mervensis* Paramonov, 1926; *nucleorum* (Becker), 1902 (as *Mulio*); *obscura* Fabricius, 1794 [=*taurica* (Becker), 1903 (as *Mulio*)], *obscura beckeri* Paramonov, 1930; *pallasii* (Loew), 1856 (as *Mulio*) [=*albivilla* (Pallas in Wiedemann), 1818 (as *Nemotelus*), =*bipunctata* (Pallas in Wiedemann), 1818 (as *Nemotelus*), =*holosericea* (Wiedemann), 1818 (as *Mulio*), sine descr., =*schineri* (Nowicki), 1867 (as *Chalcochiton*), =*striata* (Pallas in Wiedemann), 1818 (as *Nemotelus*)]; *pamirensis* Paramonov, 1930; *persicana* (Becker), 1903 (as *Mulio*); *rungsi* Timon-David, 1952; *semiargyrea* (Strobl), 1906 (as *Mulio*); *speciosa* (Loew), 1844 (as *Chalcochiton*); *syriaca* (Loew), 1868 (as *Mulio*) [=*pallasii* (Nowicki, not Loew), 1867 (as *Chalcochiton*)]; *thyridophora* (Bezzi), 1925 (as *Glossista*); *transcaspia* (Becker), 1903 (as *Mulio*) [=*setosa* Paramonov, 1930]; *trifaria* (Becker), 1906 (as *Mulio*); *turanica* Paramonov, 1930; *turkestanica* Paramonov, 1930; *turkmenica* Paramonov, 1930; *wadensis* Efflatoun, 1945.

Australian: *Cytherea lipposa* (Bigot), 1892 (as *Glossista*).

Genus *Callostoma* Macquart

FIGURES 74, 76, 306, 511, 531

Callostoma Macquart, Diptères exotiques, vol. 2, pt. 1, p. 77, 1840. Type of genus: *Callostoma fascipenne* Macquart, 1840, by monotypy.
Callistoma Kertész, vol. 5, p. 107, 1909. Emendation.

Paramonov (1929), pp. 38, 99-100, a consideration of Palaearctic species.
Engel (1934), p. 351, key to Palaearctic species.

A genus of medium-sized to large flies readily recognized and separated from its closely related ally, *Cytherea* Fabricius, by the more elongate and slender, yet tapered, abdomen, the distinct cross band or middle spot upon the wing, and the closed and stalked first posterior cell. Moreover, the third antennal segment is different; it tends to be swollen and moderately enlarged basally and narrowed and attenuate beyond this basal part. The face and front are much as in *Cytherea* Fabricius, the latter swollen above but with a curious, distinct fovea beginning just in front of the ocelli and extending nearly to and flattening out before the antennae. Antennae widely separated. Scalation upon sides of face, sides of mesonotum and pleuron, and sides of abdomen dense, whitish, and conspicuous; the thoracic and abdominal scales are more wide, the face scales less wide than *Cytherea* Fabricius.

All tibiae with numerous spiculate bristles, all femora with a few. Pulvilli vestigial. Proboscis 2 or 3 times as long as head. Face moderately produced, the oral recess deep.

This Palaearctic genus has few species.

Length: 11 to 18 mm.; most species large. The species *imperator* Nurse from Quetta, India, is the largest known—18 mm. long.

Head, lateral aspect: The head is relatively short, the face not being extensive, except below the middle of the head, where it is somewhat bluntly and angularly produced. The pile of the face consists of very dense, appressed, broad, white scales which tend to be rubbed off in the middle of the face, but which extend as an unbroken mat to the eye margin and on down the sides of the gena and oragenal cup. In addition to the scales there are a few coarse, scattered, shining white hairs over the whole area. The occiput is shallowly convex and not extensive; its pile dorsally consists of erect, dense band of coarse, not very long, white pile along the inner margin of the concavity and more scattered, similar hairs across the upper part of the occiput. Similar, erect, coarse hairs ventrally on the occiput but with along the middle portion, adjacent to the eye, a dense band of flat-appressed, white scales in several rows. The eye is narrow and high, nearly twice as high as wide, evenly hemicircular anteriorly and the posterior half of the eye is also rounded, though less extensively than the anterior part. The proboscis is quite long and slender, more than twice as long as the head. The labellum also slender. Palpus short but slender, not extending beyond the oral cavity. Front a little swollen laterally and below the vertex but distinctly depressed in the middle as a broad, shallow concavity widening out before the antennae. Antennae attached at the upper fourth of the head, short and small, the first segment about 1½ times as long as wide, and bearing a few short hairs, especially laterally. The second segment is only half as long as high, the third segment consists of a small bulbous basal portion and a quite long, attenuate stylelike part with spine at the tip.

Head, anterior aspect: The head is about 1½ times as wide as the eye, in the females the front is nearly as wide as the face, and the head is wider than the thorax. The eyes in the male are separated. The antennae are very widely separated, lying closer to the eye margin than to each other. The ocellar tubercle is quite large, swollen, the large ocelli lie in an equilateral triangle. The pile on the front consists chiefly of broad, white, flat-appressed scales on the anterior half and of coarse, loose, scattered, erect hairs on the upper half of the front. There is a dense tuft of black hairs between the ocelli in some species, not longer than the remaining pile, and behind the vertex there are numerous, long, coarse, pale hairs. The oral opening is elongate oval, quite deep anteriorly as well as posteriorly. The gena is wide, more or less rectangularly separated from the extensive oragenal cup as it recedes backward. Tentorial fissure narrow and hidden by dense scales; the occiput is deeply hollowed out centrally.

Thorax: The thorax is somewhat elongate and rather convex, shining black, covered with numerous, slender, white scales and some short, suberect, shining white hairs. Notopleuron with 2 long, slender yellowish bristles and several shorter ones. The scutellum is thick, convex, with pale, narrow scales and along the margin several pairs of long, slender, bristly hairs. Mesopleuron and sternopleuron both densely plastered with narrow, white scales and a little coarse, long, pale pile dorsally on the mesopleuron. Remainder of pleuron bare, except for an anterior vertical fringe on the propleuron, a patch of scales vertically in front of the haltere and another patch behind the spiracle.

Legs: The legs are relatively short and stout, anterior tibia spiculate in three rows of 6 or 8 each; all of the bristles slender. Middle and hind tibiae with similar, somewhat longer, equally slender spicules, 6 to 8 in a row and 4 rows present. Hind femur with 7 or 8 minute, fine, bristly setae ventrally, the pile consisting of broad, short, flat-appressed, truncate scales, similar on all the tibiae. Claws long, slender, but scarcely curved at all apically. Pulvilli reduced to mere minute stubs.

Wings: The wings are quite large, often with a brown band across the middle, brown at the base above the alula and often with a whitish sheen on each side of the brown band. First posterior cell closed with a long stalk. Anal cell widely open. The anterior crossvein lies at or near the middle of the discal cell. Vein on the anterior margin of the third posterior cell strongly sigmoid. Axillary lobe quite narrow. Alula also narrow.

Abdomen: The abdomen is elongate, at least 1½ times as long as the thorax, at the base as wide as the thorax and gradually narrowing. Abdomen shining black, tergites with a wide band of flat-appressed, narrow white, or brownish white scales on the anterior of the second, third, and fourth tergites and most of the remaining tergites. In addition, the postmargins of all of the tergites have a row of depressed, coarse, slender bristles or bristly hairs, pale yellow in the type-species.

Material available for study: The type-species and other material in the British Museum (Natural History). Also, a male of *fascipenne* Macquart presented to me some years back from the Museum of Comparative Zoology.

Immature stages: The larvae of these flies are important consumers of egg pods of locusts. See further discussion under life histories.

Ecology and behavior of adults: Austen (1937) related how these flies are found in abundance in the Mediterranean on thistle heads and in blossoms of pink scabious as well. He notes that many freshly emerged adults fall a quick prey to ants.

Distribution: Palaearctic: *Callostoma desertorum* Loew, 1873; *fascipenne* Macquart, 1840, *fascipenne palaestinae* Paramonov, 1931; *imperator* Nurse, 1922; *persicum* Paramonov, 1929; *soror* Loew, 1873.

Genus *Gyrocraspedum* Becker

Figures 56, 303, 501, 510

Gyrocraspedum Becker, Ann. Mus. Zool. Acad. Imp. Sci. St. Pétersbourg, vol. 17, p. 456, 1913a. Type of genus: *Gyrocraspedum pleskei* Becker, 1913, by original designation.

Quite large, dark colored, stout bodied, robust flies with large and extensively dark colored wings and decumbent, relatively short, subcylindrical abdomen, which is a little tapered toward the wide, cylindrical apex. Abdomen at base as wide or wider than thorax. The well developed, rather long wings and the type of antennae and general robust form of thorax and abdomen remind one of some nemestrinids.

The second vein arises abruptly quite close to the anterior crossvein, which together with the widely separated antennae brings this genus within the scope of the Cytherini, where both Becker and Bezzi have placed it. However, the postocciput has a rather widely separated, deep crease on each side, and although it is not distinctively bilobate as in the Cylleniinae, still this characteristic is additional reason for regarding the members of this tribe as somewhat transitional to the Tomophthalmae.

The single known species is from Persia.

Length: 16 mm.; wing 19 mm.

Head, lateral aspect: Head with the posterior margin of the eye straight, without indentation. Occiput moderately protuberant above, less so below. Eyes bare. Face obtusely conical and moderately produced. Proboscis elongate and slender and about 3 times as long as the head in length. Palpus relatively short but slender with long hairs above; it is composed of 2 segments, the basal one quite short. The first antennal segment is about as long as wide, wider apically, the second segment less than half as long as the first segment; both segments have minute, short, bristly hairs; third segment flattened laterally, elongate conical in lateral aspect and nearly twice as long as the first two segments combined. The basal half of the third segment is slightly dilated but not wider or higher than the first segment, and it is rather strongly attenuate, beginning at the middle of the segment; its apex is simple. Pile of the occiput fine, short, and except near the eye margin, directed backward. Along the eye margin there are numerous, flattened hairs or scales, dark in color, plastered close to the occiput and directed forward.

Head, anterior aspect: Face with the lateral grooves shallow, except for a deep recession opposite the middle of the epistoma. These grooves are shallowly continued nearly to the base of the antenna. The antennae are set very much apart. Vertex wide, strongly swollen, the ocellarium not conspicuous but the ocelli are large. There is a depression, broad in extent, covering the lower part of the front. Ocellarium and the area behind with dense, coarse, erect pile. Sides of front with a patch of rather stout, subappressed, reddish bristles above each antenna. Eye margins of the face with appressed, scaliform pile. There is some similar pile, more erect, scattered over the face among which are a few, short, coarse hairs. Postocciput above with a widely separated, deep crease on each side, partly obscured by pile.

Thorax: The thorax is large, wide, and stout. The mesonotum is covered with minute, appressed setae, which are rather scattered and scanty for the most part and are replaced by appressed scales in front of the scutellum, densely above the wing, and to a more scattered extent behind the humeri and between the humeri. Humerus with a dense tuft of reddish-brown bristles anteriorly. Similar but shorter bristles continue over the notopleuron, and the posterior corners of the notopleuron bear 2 or 3 rather long, stout bristles. Postalar callosity with 2 such bristles, scutellum large, thick, convex, the postmargin especially thick and bearing 2 or 3 pairs of comparatively short, stout bristles, and some other short, bristly pile basally on the sides but scattered, appressed scales basally on the disc. Pleuron mostly bare and shining. Upper mesopleuron, however, with 20 or 30 long, erect, quite stout, reddish-brown bristles arising from tubercles. Those on the notopleuron also rise from tubercles. Posterior mesopleuron and upper sternopleuron with scattered scales. Metapleuron and hypopleuron with a few, fine, long hairs. Ventral metasternum membranous.

Legs: The legs are rather long and stout but not swollen. Hind femur and tibia each with a few short, subappressed, black bristles. Hind tibia anteroventrally with 8 such bristles, dorsolaterally with 12, and posterodorsally with 4 or 5 bristles; apex with a circlet of 4 stout and 4 or 5 fine bristles. Anterior tibia with 9 small, short bristles anteriorly, and with 7 bristles at the apex. All of the femora and tibiae with dark, short scales. Whole fly dark, reddish brown, almost black.

Wings: The wings are deeply tinged with dark, reddish brown, the middle of the cells scarcely lighter. The second vein arises more or less steeply and is oblique quite close to the anterior crossvein. The first posterior cell is closed with a long stalk; the upper part of the posterior branch of the third vein and the second vein each describe a strong, wide, regular curve or recurrent turn backward to join the costa obliquely. The marginal cell is left especially wide but uniformly wide. The first vein and the auxiliary vein or subcosta end very close together, the distance apart less than the length of the anterior crossvein; this crossvein is situated barely before the middle of the discal cell. Because of its prominent, apical width the wing resembles those of *Comptosia* Macquart species to some extent. Anal cell narrowly open. Alula moderately large. Basal comb and hook absent.

Abdomen: The abdomen is wide and robust, and at the base as wide or wider than the thorax, and moderately tapered to the wide, cylindrical apex. Posterior margins of the segments each with a fringe of rather stout, numerous, appressed, reddish-brown bristles. Remaining surface of the tergites with scattered, appressed, dark colored scales. Whole of the first abdominal segment, however, with very dense, erect bristles long and conspicuous on the outer third of the tergite, shorter beneath the scutellum and less stout, but equally numerous. On the posterior margin of the segments, the bristles are directed backward. Sternites with coarse, scattered, bristly hairs. Abdomen decumbent and arched into the figure of "C." Male terminalia recessed and forming 2 vertical, flattened, adjacent plates. Material for dissection not available.

Material available for study: Consists of a male very kindly loaned to me by Dr. R. H. Painter.

Immature stages: Unknown.

Ecology and behavior of adults: Not on record.

Distribution: Palaearctic: *Gyrocraspedum pleskei* Becker, 1913.

Genus *Pantarbes* Osten Sacken

Figures 82, 83, 305, 499, 507, 828, 829, 830

Pantarbes Osten Sacken, Bull. U.S. Geol. Survey, vol. 3, p. 254, 1877. Type of genus: *Pantarbes capito* Osten Sacken, 1877, by monotypy.

Painter (1939), p. 295, key to 3 species.

Medium sized, robust, densely long, woolly pilose, nonbristly and nonscaly flies which are very peculiar in several ways. The wings, hyaline or nearly so, have 3 submarginal cells, closed and stalked first posterior cell, and the discal cell is angularly widened below. The robust abdomen is short and obtusely conical or even rounded at the apex. The face and front are somewhat inflated but only moderately and with a rather dull background. The antennae vary greatly in the amount of separation but the eyes are extremely far apart indeed, especially in females. Eyes of male widely separated but less so than in females. The long, slender, third antennal segment is slightly dilated beyond the middle, with knob or disclike base and minute, apical microsegment represents a characteristic feature. The oral cavity is small, moderately deep, the proboscis long and slender. A notable feature is the extraordinarily wide gena due to the extensively wide face and widely separated eyes. Some males have to a marked extent a coronal crown or circlet of dense, upstanding, long pile around the upper sides of the face meeting in the middle of the front. All tibiae spiculate, the hind femur with a few dorsoapical bristles. Femora densely scaled, the tibiae less so; a few scales on lower sides of face.

The second vein arises normally, the prefurca short. For this reason one must conclude that these flies do not perhaps really belong in the Cytherini. The curious type of face and facial pile and the form of the antenna show a resemblance to the South African *Corsomyza* Wiedemann and *Gnumyia* Bezzi. The genus is restricted to the Nearctic and Mexican regions in the Southwest.

Length: 6 to 10 mm., excluding proboscis. Wing 5 to 9 mm.

Head, lateral aspect: The head from the side is more or less hemispherical, if only the posterior eye margin is considered. The occiput is only moderately developed in either sex but is more prominent medially extending gently out to the eye margin. The face from the side is very short and inconspicuous but vertically is rather long. The pile of the entire front and vertex, face and gena is extraordinarily dense and fine and long in both sexes, especially in males where it forms a curious brush over the whole front extending down the sides along the eye much as in some Corsomyzini. In addition there are scales along the sides of the greatly widened face, which extend on down along the gena. Occipital pile dense but curiously short, especially along the middle of the occiput and also below. The eye is large, almost hemispherical, a little narrower on the ventral half, the posterior margin entire and plane and no pile present. In males the proboscis is quite slender and elongate with short labella and about 1½ times as long as the head. In the females it is much shorter and stouter and the labella also larger though slender. The palpus is quite short and inconspicuous with a terminal tuft of hairs. The front is swollen and raised in the male and also in the female. The antennae are attached at the upper third of the head, though elongate, they are rather small and quite slender. The first segment in males is well hidden by the pile, is about twice as long as wide, or a little longer; the second segment is minute, beadlike; the third segment even more slender, a little dilated on the outer half, narrowed at the immediate apex and bears 2 small, bristle-tipped microsegments. The pile of first segments consists of a few fine, short, bristly hairs.

Head, anterior aspect: The head is at least twice as wide as high in both males and females and much wider than the thorax. The male eyes are widely separated by almost the length of the third antennal segment, the vertex depressed at the medial eye margin but then rising to the swollen, obtuse, ocellar triangle; the ocelli rather large. The pile in the male on the front forms such a conspicuous brush that viewed anteriorly one can scarcely see the vertex. In the female the front and face are both excessively wide and almost equally wide, and in fact even wider than *Corsomyza* Wiedemann. The antennae, however, are widely separated at the base in strong contrast to *Corsomyza* Wiedemann; they are adjacent and the face in the present genus is much wider than any other Cytherini. The front and face are continuous and the face is extensive, as long or longer than the slender, third antennal segment with the oragenal cup located quite ventrally in position, and in the male of the type-species quite obscured from any view by the density of the pile. In *Pantarbes willistoni* Osten Sacken the pile, while abundant, is much finer, looser and does not obscure the shining, somewhat pollinose ground color. There are fewer scales along the sides of the eye, than in other species. The gena is very extensive, as wide as the third antennal segment is long, and the tentorial fissure is found at the medial edge of a broad, circular, medially depressed genal border. There is a marked, wide depression vertically behind the ocellar tubercle in the males, much more shallow in females.

Thorax: The thorax is a little longer than wide, gently convex, opaque or subopaque and covered rather densely with moderately long, fine, erect pile nowhere obscuring the ground color. There are very weak bristly hairs on the notopleuron, still more weak on the postalar callosity and the scutellar margin. Scutellum thin, opaque, pollinose. Mesopleuron, however, with long, coarse pile. The pteropleuron is bare. There is a conspicuous fanlike tuft of coarse pile on the upper part of the hypopleuron immediately in front of the spiracle. Likewise, there is a tuft of somewhat shorter pile behind the spiracle, and there is a tuft of pile immediate above the posterior coxa on the hypopleuron. Squama with a dense, wide brush of pile.

Legs: The legs relatively short and slender, the femora a little thickened. Anterior tibia with a posterodorsal and posteroventral row of black, fine spicules, 10 to 15 in each row. Middle tibia with equally fine, longer spicules in 4 rows. There are about 15 in the dorsal row, fewer in the others. Hind tibia with 4 rows of spicules of the same character, 10 or 12 in the dorsal row, 7 to 10 below. Femora without bristles but with a few long, coarse hairs below and copiously plastered with unusually large and conspicuous loose scales: brownish pink in *Pantabarbes willistoni* Osten Sacken, snow-white in other species. The pulvilli are oval, as long as the claw, claws fine and sharp.

Wings: The wings are hyaline with the costal cell and the base of the marginal and submarginal cells, and the end of the basal cell sometimes tinted lightly with brown. There are 3 submarginal cells, the marginal cell is widened apically, extends beyond the forking of its lower branch of the second vein. The anterior branch of the third vein, after meeting the second vein, continues first posteriorward and then makes a strong, rounded bend anteriorly to end near the apex of the wing. The end of the anterior branch of the second vein may be strongly recurrent in some species. First posterior cell closed with a long stalk, discal relatively short and broad, the medial crossvein quite long and the anterior crossvein lies at or just before the middle of the discal cell. Anal cell widely open, axillary lobe moderately well developed, the alula also of moderate or average development. There is a well-developed costal comb at the base but no costal hook. Ambient vein complete.

Abdomen: The abdomen is subtriangular, or short and broadly oval, wider than the thorax, shining black or dull and subopaque with a faint pollen. These flies have a very furry abdomen, the pile, while not dense, is quite abundant, extraordinarily fine, long, and erect, not obscuring the ground color. In *Pantarbes willistoni* Osten Sacken it is almost bristly. I find 7 segments in both sexes. The terminalia of the male are recessed into a rather large opening visible below. Female terminalia hidden by a fringe of dense, pale bristly hair apposed from each side.

The male terminalia from dorsal aspect, epandrium removed, show a figure with a broad, wide base, made more slender distally by the long, peculiar dististylus which is unusually long and swollen distally and also basally, so that there is a long, narrow neck in between. At the apex of the dististylus, is a short, divergent, small hook-shape process. In lateral aspect this apical, hooklike process on the dististylus is more conspicuous. The entire structure is unusually short. The epandrium is subrectangular, posteriorly with a short, pointed lobe above and another below. The epiphallus is quite large, short, and stout from both aspects, and from the dorsal aspect with a crescentic arch leaving an attenuate process at the base on each side.

Material available for study: I have material of the type-species and also of *Pantarbes pusio* Osten Sacken kindly furnished me by Frank Cole and also of *Pantarbes willistoni* Osten Sacken which I found abundant in spring at verbenaceous flowers of the New Mexico desert, and I have also undescribed species.

Immature stages: Unknown.

Ecology and behavior of adults: These flies are readily attracted to low-growing desert plants in early spring where they may be found hovering about the blossoms.

Distribution: Nearctic: *Pantarbes capito* Osten Sacken, 1877; *pusio* Osten Sacken, 1887; *willistoni* Osten Sacken, 1887.

Genus *Sericosoma* Macquart

FIGURES 40, 302, 502, 508, 1000, 1001, 1002

Sericosoma Macquart, Diptères exotiques, suppl. 4, p. 115, 1849. Type of genus: *Sericosoma fascifrons* Macquart, 1849, by monotypy.

Paramonov (1947b), p. 361, key to 6 species from Chile and Argentina.

Small to medium-sized flies characterized by the dense, fine, erect pile and tapered, conically elongate abdomen and the hyaline wing with maximally open first posterior cell. There are 2 submarginal cells and the anterior crossvein lies beyond the middle of the widened discal cell. The antennae are set widely apart and the males are widely dichoptic. The slender proboscis is at least twice as long as the head and the palpal pile is so long and dense as to fill and obscure the oral recess.

This genus is certainly related to *Pantarbes* Osten Sacken, which it resembles in the character of the pile about the face and the arrangement of the erect fringe across the front of males. Like *Pantarbes* the palpus is very densely pilose. It differs from *Pantarbes* in the presence of only 2 submarginal cells and the widely open first posterior cell; also in *Sericosoma* Macquart the metapleuron in front of the haltere is bare, in front of the spiracle tufted with pile; the reverse is true in *Pantarbes* Osten Sacken. The antenna differs in that the third segment is shorter and is attenuated to about half its basal thickness on the outer apical half of this segment.

Sericosoma Macquart is a Neotropical genus that is known only from the Chilean and Argentinian subregion.

Length: 10 to 12 mm., excluding proboscis; wing 8 to 9 mm.

Head, lateral aspect: The head viewed laterally is more or less restricted to the very large oval eye which is slightly wider above than below, almost as much rounded and convex on the postmargin as it is anteriorly. There is very little occiput visible laterally and it is more extensive in the middle of the head where the head joins the thorax, sloping above, below, and laterally to the eye margin. The pile of the occiput is short and very much in contrast to the extraordinary dense, long pile of vertex, front and face, and below the eyes. This occipital pile consists of a fringe of long, spatulate or lanceolate scales laid down along the lateral portion of the occiput and extending to the margin of the eye, among which there are a few more slender hairs, although these striking scales predominate. Further inward there is more dense, coarse pile, especially on the upper fourth of the occiput which extends to the vertex but does not reach above it.

Surface of occiput pollinose. The front is nearly flat transversely, it is pollinose, and except on the lower third, very densely covered with long, erect, rather stiff pile, which is almost bristly and obscures the ground color. This pile is generally black or partly pale. The lower third of the front is without pile except for 4 or 5 scaly hairs on each side. Sometimes there is a large matted tuft of appressed, silvery, scaly hairs. The dense, complete extensive band of long hairs encircles each antennal segment and meets the shelflike, dense brush of pile on the upper half of the front. Whole vertex and ocellar triangle likewise densely covered with pile of the same character.

The face in lateral aspect is very short and the genae and the oragenal area likewise, but the pile on all this facial and genal area is extraordinary. It forms a fluffy mass of long, shining white, dense, fine, silvery hairs extending outward in matted, curled tufts, which largely obscure the ground color. This bushlike pile continues below and throughout under the bottom of the eye. The postmargin of the eye is entire. The proboscis is quite slender and relatively long but is about 1½ times as long as the head. The labellum is quite slender. It is surrounded at the base by masses of pile. The palpi are completely hidden. The antennae are attached at the upper fourth of the head, they are comparatively slender although the first segment is distinctly swollen and robust, more especially convex laterally and ventrally, but the whole antenna is not as long as the eye. The first segment is a little more than twice as long as its middle width and it is pale pollinose and on all sides bears a dense band from base to apex of long, comparatively coarse radiating shining white pile, which gives the appearance of rosettes when viewed from the front. The second segment is quite small, short and beadlike, more narrow than the first segment, the third segment at the base is a little more narrow than the second segment and is about as long as the first two segments together; strongly attenuate on the outer half it bears a small, short microsegment.

Head, anterior aspect: the head is much wider than the thorax, the eyes are widely separated in the male by almost the length of the antennae. The antennae are very widely separated indeed by nearly the same amount of separation as the interocular space across the front ocellus. The ocellar tubercle is low; the very large ocelli lie in an obtuse triangle quite removed from the eye margin. The oral opening is ventral in position, rather small, short oval, a little less than one-third the interocular space. It is rather deep, especially behind, the outer wall is quite low, in fact posteriorly it does not extend beyond the genal surface which is flattened posteriorly. However, the gena above the proboscis and below the face shelve deeply and convexly down into a deep, triangular recess just above the junction with the thorax.

Thorax: The thorax is opaque black covered rather densely with erect, coarse, almost bristly pile, pale over the middle of the mesonotum, brassy brownish laterally.

It becomes almost black on the upper mesopleuron where, including all the anterior pile on the mesonotum it is very dense, directed upward and collarlike. Humeral pile equally dense and long; notopleuron without bristles, except for the very dense, bristly pile over the whole notopleuron. Scutellum similar to the mesonotum in color and pile. Some species, however, instead of a few basilateral bristles have very dense tufts of coarse, bristly pile basilaterally, becoming less abundant over the disc of the scutellum, but also with lateral patches of long, appressed scales. The posterior half of the mesopleuron is densely covered with pile of a rather finer quality and directed backward. Pteropleuron bare. Metapleuron with an extensive cluster of long, fine, erect pile. Hypopleuron bare, except for a tuft of long, backwardly directed hairs immediately behind the spiracle. Squama short with a dense fringe of long pile. Halteres small with short, oval, closed knob.

Legs: The legs are rather long, especially the hind pair. All of the femora are slender. The anterior femur has a conspicuous brush of rather dense, long, fine ventral hairs, longer at the base, shorter at the apex and partly continued around on the posterior surface. The hind femur has a very scanty fringe of long hairs anteriorly rather than ventrally, which contains some 20 or more hairs in 2 irregular rows. Several of these hairs on the outer half are slightly bristly. All of the femora are covered with matted, large, broad scales, pale in color. The anterior tibia has 5 or 6 minute setae posterodorsally and a like number posteriorly. Middle tibia with some considerably longer, slender, black, oblique spicules, 4 or 5 in each row. Hind tibia quite slender with 12 moderately long, black spicules anterodorsally, a like number posterodorsally and 6 anteroventrally, and 12 ventromedially. Pulvilli as long as the claw and spatulate. The claws gently curved on the apex.

Wings: The wings are broad, shining vitreous hyaline; the minute somewhat reddish villi present everywhere do not affect the color. There are 2 submarginal cells, the anterior branch of the third vein is very little curved, bends only a short distance from its base and sometimes has a trace of a spur. Marginal cell not swollen apically. The first posterior cell is widest at the wing margin, the discal cell widest on the basal half, the anterior crossvein enters the discal cell near the outer third, the anterior intercalary vein is quite long, the anal cell is open, the extent varying. Axillary lobe widest in the middle. Alula width less than half its length. Ambient vein complete and strong. Basal comb very weak, but there are strong matted scales beneath the costa basal to the humeral crossvein.

Abdomen: The abdomen is elongate, strongly tapered from the base, rather convex. The base of the abdomen is considerably wider than the large scutellum but not as wide as the mesonotum. The tergites bear coarse, extremely long, erect, pale pile which becomes thin and scanty down the middle third of the abdomen and outwardly becomes increasingly dense until it forms conspicuous, wide band or fringe of lateral pile, which curls over the sides of the abdomen and which extends straight outward and at the apex backward. Most of this pile is yellowish or whitish but on the underneath side of the curled-over tergites it becomes brownish black. This long, erect, pale pile extends further inward on the first segment, becoming shorter under the scutellum. The middles of the tergites including the whole surface of the outer margin are thickly plastered with matted, broad, long, pointed, whitish scales directed backward and somewhat toward the medial line. In some species small triangles of blackish scales are left in the middle of the tergites. I find 7 segments in the male. The male terminalia are small and enclosed in a small circular ventral opening surrounded by pile.

Material available for study: Two species from Chile.

Immature stages: Unknown.

Ecology and behavior of adults: Not on record.

Distribution: Neotropical: *Sericosoma argentinae* Paramonov, 1947; *bigotianum* Edwards, 1937; *fascifrons* Macquart, 1849; *furvum* (*fulva* in err.) Edwards, 1937; *pubipes* Edwards, 1937; *squamiventre* Edwards, 1937.

Genus *Oniromyia* Bezzi

Figures 80, 297, 498, 509, 822, 823, 824

Oniromyia Bezzi, Ann. South African Mus., vol. 18, p. 71, 1921a. Type of genus: *Eurycarenus pachyceratus* Bigot, 1892, by original designation.

A curious, unique fly which because of the reduced wings is almost phylogeront in character. The elongate abdomen is only slightly or moderately tapered and bears only short, appressed, adherent, sparse, brownish-yellow pile on the tergal surfaces, except that laterally on the first three segments there are small tufts of fine, bristly hairs at the margin.

The head is wider than the thorax, the front is wider than the face, and the small, relatively short antennae are but little separated and sit oddly low upon the face profile; the first segment is bulging or extended below at the apex which reminds one of the Lomatiinae; however, the head, eye contour, and occiput are entirely of the Bombyliinae. The front and vertex are swollen or bossed and bear appressed, narrowly scalelike pile intermixed with rather dense, erect, fine pile. Face very short, oral recess deep, the walls of the oragenal cup paper-thin. The pile on the lower half of the first antennal segment is dense and very long and the lateral facial pile is long. Only the posterior ocelli are present; these are small, sunken, obscured, and far apart. Proboscis with scales on the basal half above.

The feeble wings are quite slender as well as small. All tibiae with prominent spiculate bristles.

The genus, of which only 2 species are known, is confined to the region south of the Sahara.

Length: 9 to 11 mm.; wings 7 to 8 mm., and 4 times as long as wide.

Head, lateral aspect: The head is hemispherical, the occiput is not very extensive, rather plane or flattened, extending very little beyond the eye margin, either above or below. The oragenal cup is short and thin with a very sharp-edged wall and extends almost to the antennae. Pile is absent on the face on this small area immediately beneath the antennae but the long, shaggy, coarse, appressed pile from the lower front curls over and downward on the lateral portion of the front. The front in female is extraordinarily wide and broad and conspicuously swollen, except upon the lower fifth in front of the antennae where there is a medial, more or less triangular area, flattened and without pile. The whole front is conspicuously covered across the swollen portion with many long, fine, erect, stiff black hairs beneath which are numerous, subappressed, long slender scales, whitish in front, brownish yellow on the upper half. Both hair and scales are loose, not obscuring the polished, shining black ground color. Anteriorly toward the antennae on each side the scales become more dense, shining white in color and there is considerable shaggy, suberect, stiff, white pile bent forward and completely obscuring the ground color. These tufts reach downward below the antennae. The pile of the occiput consists of rather short, erect, stiff, somewhat curled hairs at the upper eye corners and behind the vertex. The greater part of the occiput on the outer third consists of yellowish brown scales on the upper half and longer whitish scales on the lower half which extended to the eye margin, or even lap over it, and some intermixed, quite short, erect hairs medially. At the bottom of the occiput and beneath the eyes there is much long, erect, coarse, shining, black pile. The posterior margin of the eye is entire, all the facets small and of a uniform size.

The proboscis is long and quite slender and at least 2½ times as long as the head with small, slender labellum and it is thrust directly forward. The palpus is extremely slender, scarcely extended beyond the oragenal cup. I find only one palpal segment, widened a little apically, bearing a few apical hairs. The antennae are attached at the middle of the head; they are comparatively small; the first segment is, however, remarkable. It is about twice as long as wide, bears very long, coarse, shining white pile ventrolaterally on the outer half, still longer pile ventrally at the apex. This first segment has a curious, very large, bullalike or tuberclelike, swollen extension ventroapically, which is demarcated by an oblique crease and which bears a dense tuft of very long, coarse, shining, yellowish-white pile. The second segment is small and short but is about 1½ times as long as wide, wider apically with a few short setae medially at the base and others along its dorsal surface. Third segment is as long as the first two together, but small and slender; on the basal half it is swollen above and below and therefore quite long oval or spindle-shaped and the outer half is uniformly slender, less than one-third as wide as the widest part of the base, blunt at the apex and bears on this narrow portion 7 or 8 lateral, bristly hairs, somewhat appressed. Apex of third segment with a short, terminal bristle.

Head, anterior aspect: The head is distinctly wider than the thorax, the female front is not quite half the total head width, has parallel sides, and the genofacial space is very nearly as wide as the front. In males the eyes are maximally divided, the front with parallel sides to the vertex, the space at the vertex is distinctly greater than that of either eye. The ocellar tubercle is low and forms an equilateral triangle that has a longitudinal, wide, shallow, sulcate trough. The posterior ocelli are quite small but the anterior ocellus is wanting entirely. The width of the oral opening is a little more than half the total width between the eyes. The oragenal cup has exceptionally thin walls and is as extensive below and posteriorly as it is along the sides. It is quite deep, elongate oval, slightly narrowed anteriorly. The genae are wide with dense tufts of long, white, appressed scales. The tentorial fissure is shallow. The whole area beneath the head is densely covered with short, broad scales, besides the long, erect pile.

Thorax: The mesonotum is nearly quadrate, scarcely convex, though abruptly slanting to the scutellum posteriorly and somewhat more convex anteriorly. It is opaque black covered with loose, scattered, brownish-yellow scales and fine, scattered, stiff, erect hairs which become longer and more numerous anteriorly and which extend also to the postalar callosity, onto the disc and especially onto the margin of the scutellum. Disc of the scutellum also covered with loose scales. The humerus bears a dense tuft of long, subappressed, whitish scales. Notopleuron with 4 long, slender, red bristles. The postalar callosity with 8 or more slender reddish bristles. The upper mesopleuron bears a conspicuous fringe of long, upturned, curled, yellow, bristly hairs. The lower half and the sternopleuron bears some matted white pile and tufts of long white scaliform hairs. The pteropleuron bears a tuft of scaliform pile attached to its anterior border. The metapleuron is extensively long, coarsely pilose. Posteriorly the hypopleuron above the hind coxae has an odd tuft of dense, scaliform pile. Squama short with a very short fringe of hair. Halteres especially small.

Legs: The legs are rather short. The anterior and middle femora are somewhat swollen toward the base. The posterior femur is thickened throughout without being strongly swollen and all 3 femora covered densely with large, broad, flat-appressed, white scales changing to reddish brown apically on the hind femur. Anterior and middle tibiae rather stout. The anterior tibia has a dorsal row of 8 appressed setae, a posterior row of 12 longer, oblique setae; middle tibia similar but with the setae longer and stouter, a little more numerous. Apex of both of these tibiae posteriorly and ventrally with 4 quite long, reddish, spinous bristles. The hind tibia is extraordinarily stout, of uniform width throughout with a row of 7 or 8 long, reddish-brown bristles dorsomedially, a row of similar bristles ventromedially on the basal half, 5 equally long, stout bristles dorsolaterally and 5 ventrolaterally. Apex with 4 long, quite stout, spinous bristles below, 2 others on each side laterally and medially and the whole of the hind tibia with broad, large, appressed, reddish-brown scales. All basitarsal segments thickened, the remaining segment rapidly becoming slender. Pulvilli long and slender, claws minute and slender, scarcely curved at the apex.

Wings: The wings are much reduced in size and quite slender. The anterior branch of the third vein recedes considerably backward from the apex of the wing. The marginal cell is a little widened apically, the second vein ends at right angle with the costa. First posterior cell long and narrow and closed with a short stalk. Intercalary vein short, anal cell widely open, axillary lobe quite long and narrow, the alula absent, the ambient vein complete, the whole wing covered with fine, reddish-brown villi. There is no basal comb or hook.

Abdomen: The abdomen is elongate but broad basally, tapered toward the apex and only shallowly convex especially anteriorly. I find 8 segments with the last being quite short and bearing laterally a tuft of long, slender, stiff, black hairs. Across the middle the tergites are covered very loosely with small, yellowish brown scales, flat-appressed and with the scales on the posterior margin extending over the edge. Ground color shining black. Posterior corners of each tergite with a dense tuft of appressed, white scales preceded by a few black, bristly hairs on the second and third segments and on the first segment preceded by a dense tuft of erect, white hairs which extends inward under the scutellum.

The male terminalia from dorsal aspect, epandrium removed, show a stout, rather wide comparatively short figure. The dististyli are prominent and viewed dorsally turn outward at an obtuse rounded lobelike structure with stout spines. The epiphallus is wide and elongate and seen laterally is suggestively like *Marmasoma* White in its long, snakelike, S-shaped figure, with apex turned downward, and to which it bears no relationship. Laterally the basistylus is long on the outer half becoming a narrow extensive lobe with many stubby retrorse spines. What appears to be perhaps a dorsomedial outgrowth of the basistylus forms in the lateral aspect a very wide, blanketing or flanking structure on each side of the prominent epiphallus with enclosed aedeagus, and this structure is beset with very numerous minute sensory hairs. Epandrium quite long, narrowing gradually apically and truncate at the apex.

Material available for study: A pair of the type-species received for dissection and study through the courtesy of Dr. A. J. Hesse of the South African Museum, Capetown.

Immature stages: Unknown.

Ecology and behavior of adults: Unrecorded.

Distribution: Ethiopian: *Oniromyia caffrariae* Hesse, 1960; *pachyceratus* (Bigot), 1892 (as *Eurycarenus*).

Genus *Conophorus* Meigen

FIGURES 66, 69, 70, 71, 72, 284, 484, 490, 820, 821

Conophorus Meigen, Illiger's Magazin fur Insektenkunde, vol. 2, p. 268, 1803. Type of genus: *Bombylius virescens* Fabricius, 1787, as *Bombylius maurus* Mikan, 1796. By monotypy.
Ploas Latreille, Dictionnaire d'histoire naturelle, vol. 24, p. 190, 1804. Type of genus: *Bombylius virescens* Fabricius, 1787, as *Ploas hirticornis* Latreille, 1804. By monotypy.
Codionus Rondani, Ann. Mus. Genova, vol. 4, p. 299, 1873. Type of genus: *Codionus chlorizans* Rondani, 1873. By monotypy.
Calopelta Greene, Proc. Ent. Soc. Wash., vol. 23, p. 23, 1921. Type of genus: *Calopelta fallax* Greene, 1921, by original designation.

Coquillett (1894b), pp. 101-102, key to 9 Nearctic species (as *Ploas* Latreille).
Séguy (1926), p. 224, key to 7 species from France.
Paramonov (1929), p. 157, key to Palaearctic species.
Engel (1933), pp. 29-31, key to 32 Palaearctic species.
Paramonov (1940), pp. 8-13, key to 37 Palaearctic species.
Priddy (1958), pp. 5-7, key to 16 Nearctic species.

Small, or nearly medium-sized flies which are easily recognized by the extraordinarily swollen, inflated, and elongated first antennal segment, together with the swollen occiput and weak but conspicuously bristly vestiture of the whole fly. The abdomen is broad, flattened, short oval, wider than the thorax and densely covered with fine, curled hairs, which in the type-species are appressed to the tergites, but in other species are more or less erect. In the type-species and most species the wing has 3 submarginal cells; the costal comb is weak, the base has a protruding lappet instead of a hook; the costal margin is sometimes microtuberculate; eyes of male shortly holoptic or approximated. The type-species tends to be decumbent and slightly humped. The marginal and submarginal cells tend to be somewhat expanded distantly but much less so than in the *Lordotus* Loew group.

Conophorus Meigen was made the basis of a subfamily by Becker. The family Bombyliidae appears to me to be overrich in subfamilies, many of unequal value. I believe a truer picture of the family is obtained by subordinating many of the so-called subfamilies to tribal rank and this I have done.

Conophorus Meigen is a Holarctic genus.

Length: Species in this genus vary in length from about 3.5 mm. to 12 mm., including antenna.

Conophorus Meigen has one subgenus, *Codionus* Rondani (1873), containing those species with 2 submarginal cells. *Calopelta* Greene (1921) is a synonym.

Head, lateral aspect: Head subglobular, the occipital component is extensive and medially slopes gradually outward to the eye margin. The face is greatly reduced, its height or extent less than the length of the second antennal segment. The oragenal cup is not visible in profile and is described below. There are some quite long, scattered, fine hairs on the sides of the face opposite the antennae where the face emerges from the gena, and this long, loose pile continues all the way down the gena to the bottom of the head. Middle of face with no pile. The occiput is tumid in the middle behind and bears quite long, fine, somewhat bristly pile, which is rather loose and scattered. It is more prominent in females. The eye is large and in females may be somewhat rectangular. In other females and in males it may be approximately hemispherical and bare. Postmargin of eye plane. The proboscis is short, stout, compressed laterally, with short, stout labellum. It is very little longer than the head or even shorter. Palpus slender but nearly as long as the proboscis. It bears fine, scattered hairs. The front is small in males, flat, pollinose, and without pile. In females, however, it bears numerous, long, fine, scattered, erect, bristly hairs and is slightly inflated or raised.

The antennae are attached just a little above the middle of the head and are characteristic. The first segment is elongate to a varying extent, as long or even longer than the next 2 segments and is extraordinarily swollen, stout and robust. It is cylindrical in form with abundant, long, fine, bristly hairs, more abundant below and lacking only on the medial surface. The second segment is barely longer than wide, much more narrow and likewise bears similar long bristles, more abundant below. The third segment in the type-species is much reduced, knob-shaped at the base, very slender, still more so on the outer half, slightly spindle-shaped, expanded at the tip and bears a tiny, spine-tipped microsegment. In other species it is wider, but similar. In some species at least laterally it is nearly as wide as the second segment, narrowed only near the apex.

Head, anterior aspect: The head is not quite as wide as the thorax, the eyes are rounded medially and are in contact for only a short distance. The antennae are adjacent in some species, but very narrowly separated in the type-species. The ocellar tubercle is minute, the ocelli quite tiny lying in an equilateral triangle. The tubercle bears a few long, fine, erect hairs. The oral opening is large and wide, extending almost to the antennae in some species. The genae are also quite wide and the oragenal cup, while comparatively deep in the middle and behind, extends outwardly only as a narrow, low, creased rim, usually not showing from the side. Genal pile loose but long and extensive. The tentorial fissure is a mere line. From the dorsal aspect the occiput meets medially in a shallow groove below the vertical column.

Thorax: The thorax is short and broad, slightly humped, rather more so in the type-species, and subopaque black. The type-species is quite opaque. Whole surface covered with two kinds of pile, short, glittering, curled, subappressed pile in many species, especially the type-species, in others with very fine, pale, more or less erect hairs, and all species in addition have a considerable amount of fine, long, erect, bristly hairs. Bristles absent. Scutellum rather thin, the vestiture like that of the mesonotum. Mesopleuron with moderately dense, long, pale pile and sometimes with additional, fine, black, bristly hairs. Pteropleuron and hypopleuron bare, metapleuron with a conspicuous vertical brush of long, coarse pile in front of the halteres. There is also a tuft of hairs behind the spiracle. Metapleural pile may be present or absent. Squama with a wide fringe of long pile.

Legs: The legs are rather short and relatively stout. Anterior tibia with minute, fine, bristly spicules in 2 rows posteriorly. Middle tibia with much longer, spiny bristles in 4 rows with 4 or 5 in each row. Hind tibia similar with 10 to 12 bristles in each row. There are no bristles or spines beneath the femora but there are a few, scattered, long, fine hairs ventrally beneath all 3 pairs of femora. Pile of legs flat-appressed, rather shaggy in Nearctic species, more scalelike and more appressed in the type-species. Pulvilli long and spatulate, claws small and fine and sharp.

Wings: The wings are rather slender in most species, the costa sometimes with spiny tubercles in the males. There are 3 submarginal cells, all posterior cells and anal cell widely open, the discal cell is rather long, the anterior crossvein lies near or just beyond the basal third of this cell. The wings are hyaline or sometimes show a mottled brown pattern. Marginal cell gradually widened apically; at its end the vein turns up abruptly and is either rounded or plane, but in species with 2 submarginal cells it is bulbous at apex. Axillary lobe narrow, the alula likewise. Ambient vein complete. The base of costa has a weakly developed comb. Costal hook not present.

Abdomen: The abdomen is unusually broad and flattened, varying from short oval to elongate oval. It is much wider than the thorax, in most species it has abundant, short, curled, glittering, appressed pile and in addition over flat surface of the abdomen numerous long, scattered, erect hairs, which are very slender and fine but stiffer in the type-species. While usually the abdomen is opaque or nearly so, some species have the abdomen very shiny without the appressed pile, and without lateral tufts. Many species, including the type-species, have odd, dense tufts of pile along the sides of the abdomen, long at the base of a tergite and shorter toward the end. Terminalia small, recessed, the dististyli visible from below in the male. Sternites with long, dense pile.

The male terminalia from dorsal aspect, epandrium removed, show a short, compact, broad body. The epiphallus is short, and quite wide basally with slender, triangular or conical apex. The dististyli are long, elongated, and pointed apically. From the lateral aspect they are hooked upward and backward; they are also convergent. The epandrium is large and subquadrangular.

Material studied: I have before me both sexes of the type-species and four Nearctic species from the Western States, and I have seen additional species in the collections of the National Museum of Natural History.

Immature stages: The hosts are unknown but Priddy

(1939, 1958) recalls finding a pupal skin that had emerged from soil. See comments under life histories.

Ecology and behavior of adults: Adults are attracted to spring flowers where I have taken them in the New Mexico desert. Van Duzee (1931) records an extraordinary abundance of a species of *Conophorus* Meigen on flowers along the Salinas River, California.

Distribution: Nearctic: *Conophorus amabilis* (Osten Sacken), 1877 (as *Ploas*); *atratulus* (Loew), 1872 (as *Ploas*); *auratus* Priddy, 1954; *chinooki* Priddy, 1954; *collini* Priddy, 1958; *columbiensis* Priddy, 1954; *cristatus* Painter, 1940; *fallax* (Greene), 1921 (as *Calopelta*); *fenestratus* (Osten Sacken), 1877 (as *Plaos*); *hiltoni* Priddy, 1958; *limbatus* (Loew), 1869 (as *Ploas*); *melanoceratus* Bigot, 1892; *nigripennis* (Loew), 1872 (as *Ploas*); *obesulus* (Loew), 1872 (as *Ploas*); *painteri* Priddy, 1958; *pictipennis* (Macquart), 1840 (as *Ploas*); *rufulus* (Osten Sacken), 1877 (as *Ploas*); *sackenii* Johnson and Maughan, 1953; *serratus* (Coquillett), 1894 (as *Ploas*).

Palaearctic: *Conophorus aduncus* (Loew), 1870 (as *Ploas*), male only; *aegyptiacus* Bezzi, 1925; *alpicola* (Villeneuve), 1904 (as *Ploas*); *antennatus* Paramonov, 1940; *asiaticus* Paramonov, 1929; *bellus* (Becker), 1906 (as *Ploas*); *bivittatus* (Loew), 1862 (as *Ploas*); *bombyliiformis* (Loew), 1873 (as *Ploas*) [=*aduncus* (Loew), 1870 (as *Ploas*), female only]; *caucasicus*, Zaitsev, 1960; *chinensis* Paramonov, 1929; *chlorizans* (Rondani), 1873 (as *Codionus*); *decipiens* (Loew), 1873 (as *Ploas*); *engeli* Paramonov, 1947 [=*decipiens* Engel, 1934, not Loew]; *flavescens* (Meigen), 1820 (as *Ploas*); [=*hamilkar* Paramonov, 1929]; *fuliginosus* (Wiedemann in Meigen), 1820 (as *Ploas*) [=*atratus* Meigen, 1830]; *fuminervis* (Dufour), 1852 (as *Ploas*); *fuscipennis* (Macquart), 1840 (as *Ploas*); [=*fumipennis* Schiner, 1862]; *glaucescens* (Loew), 1863 (as *Ploas*); *greeni* Austen, 1936; *griseus* (Fabricius), 1787 (as *Bombylius*) [=*ater* (Lamarck), 1816 (as *Ploas*), =*atratus* (Meigen), 1820 (as *Ploas*), =*hannibal* Paramonov, 1929]; *heteropilosus* Timon-David, 1952; *hindlei* Paramonov, 1931; *kozlovi* Paramonov, 1940; *loewi* Paramonov, 1929; *luctuosus* (Loew), 1870 (as *Ploas*); *lusitanicus* (Guérin-Méneville), 1835 (as *Ploas*); *mauritanicus* Bigot, 1892; *monticola* Paramonov, 1929; *nobilis* (Loew), 1873 (as *Ploas*), *nobilis iranicus* Paramonov, 1940; *paraduncus* Paramonov, 1929; *pictipennis* (Macquart), 1840 (as *Ploas*) [=*macroglossa* (Dufour), 1852, (as *Ploas*)]; *pseudaduncus* Paramonov, 1929; *pusillus* (Loew), 1869 (as *Ploas*); *rjabovi* Paramonov, 1929; *rossicus* Paramonov, 1929; *simplex* Loew, 1869 [=*similis* Kertész, 1909, lapsus]; *syriacus* Paramonov, 1929; *talyshensis* Zaitsev, 1960; *turkestanicus* Paramonov, 1929; *ussuriensis* Paramonov, 1940; *validus* (Loew), 1869 (as *Ploas*); *virescens* (Fabricius), 1787 (as *Bombylius*); [=*hirticornis* Latreille, 1804, =*latus* (?Dufour in Vernall), 1909 (as *Ploas*), =*luridus* (Wiedemann in Meigen), 1820 (as *Ploas*), =*maurus* (Mikan), 1796 (as *Bombylius*)].

Genus *Conophorina* Becker

FIGURES 33, 269, 481

Conophorina Becker, Ent. Mitt., vol. 9, p. 181, 1920. Type of genus: *Conophorina bicellaris* Becker, 1920, by monotypy.

Small flies with the general form of *Conophorus* Meigen, but with a different type of antenna and with only 2 submarginal cells. The face and gena are more prominent, likewise more hairy. The scanty, scattered, long, erect hairs on the postmargins of the tergites, upon the mesonotum, occiput above, vertex, sides of face and basal antennal segments are very characteristic. The pile is even longer and more abundant in males. Also the pile in part in females is differently colored from males, the hair of the face and antenna being golden rather than brownish black.

The eyes in the male are extensively holoptic; the front is reduced and small although equilateral and it bears a few, long, golden, flattened hairs on each side. Genitalia protruded, but relatively small and inconspicuous, surrounded by only a few, sparse, bristly hairs.

The venation is quite simple and generalized. The first two antennal segments are thickened through their middles, but much less so than in *Conophorus* Meigen; both segments are elongate and hairy, and segment two is almost as long as the first. The palpus is distinctly 3 segmented, which is unusual.

This genus is based upon a single South African species.

Length: 4 to 5 mm.

Head, lateral aspect: Head hemispherical, the occiput short with pile which is long and erect, abundant but not dense. Just below the middle of the eye the bristly pile becomes shorter, appressed, and directed outward toward the eye margin. The front is broad, rather flattened, but with the ocellar tubercle exceptionally large with the area in front of it swollen, leaving the middle of the front below, sunken. The ocellar triangle obtuse. Pile at vertex between the ocelli consists of 4 or 5 very long, slender, bristly hairs. There is a tuft of other equally long hairs at each eye corner and curled forward. Sides of front and middle of front above antenna with numerous appressed or subappressed, golden, scaliform hairs. The front in the male is very small, triangular, pollinose, without pile, and the small but steep ocellar tubercle has an extensive tuft of long, erect, bristly hairs. The face is moderately produced forward beneath the antenna, a distance certainly not more than half the length of the first antennal segment, and vertically it is rather short, much encroached upon by the large oral recess. The pile of the face consists of a conspicuous, sublateral fringe of numerous, coarse, golden, bristly hairs. On the upper part this fringe mingles with the equally long, golden bristly hairs of the first two antennal segments. The fringe continues shorter and a little more sparse, down the sides of the gena to the bottom of the head. The eye is large, slightly wider below, the upper facets much enlarged in the male.

Proboscis elongate, rather stout, about twice the length of the head, extending a little beyond the very long antenna. The palpi are rather large, long, flattened and pointed, a little more than half as long as the proboscis and with a few oblique setae laterally and apically. Palpus with 2 segments. The antenna is attached a little above the middle of the head, the first segment is stout, a little swollen, 3 or 4 times as long as wide, with dense, long bristly pile on all sides. The second segment is equally stout, elongate barrel-shaped, about 2½ times as long as wide and also with considerable long bristly pile. The pile of the head and antenna in the male is chiefly, almost entirely black. In the female, almost entirely golden. The third antennal segment is more slender, a little longer than the first segment, with a distinct knob at the base, widened beyond, strongly flattened, and narrowed beyond the middle. It has a small, short, attenuate microsegment, bearing a spinous bristle apically.

Head, anterior aspect: From the front the head is wider than the thorax, not quite twice as wide as high. The eyes of the male are in contact for a long distance, leaving the male front a small equilateral triangle. Also, the ocellar triangle is small but high, and the anterior ocellus rests upon the eye margin. In the female the eyes are widely separated by almost a third the width of the head. The ocellar tubercle is much larger, rather abruptly elevated, and there is a sunken depression in the middle of the front. The antennae are narrowly separated. The oral cup is large, short oval, comparatively deep posteriorly, slanting outward anteriorly with narrowly separated pubescent ridges anteriorly. Margin of cup especially on the posterior half lineally thin and extending only a short distance forward from the very wide gena, between which there is a linear crease.

Thorax: Short and broad and rather strongly convex both in front and behind with a large, wide, thick, posteriorly convex scutellum. The mesonotum is largely opaque black with traces of shining vittae anteriorly. The postalar callosity and scutellum are polished and shining. The pile is comparatively scanty, semi-appressed or flat-appressed and consisting of golden scaliform pile in both sexes. It is apt to be denuded over the central part of the mesonotum, but a little better preserved and a little more abundant on the notopleuron, and postalar callosity lying above the wing, in front of the scutellum, and on the margin of the scutellum. In addition to this appressed, golden scalation there are scattered, very long, erect, slender, blackish hairs in length equal to those found upon the postvertex or upon the abdomen. They are distributed loosely and scantily over the mesonotum and scutellum, except where they are denuded. Mesopleuron with a large tuft of long scaliform golden pile. Pteropleuron, metapleuron, and hypopleuron bare. Fringe of squamae long and forming a comparatively thick brush. Halteres large, and the large knob rather flattened, and the whole dark sepia.

Legs: Moderately stout, but the femora are not swollen. All 3 pairs have a loose fringe of long, stiff hairs ventrally, perhaps bristly hairs. In addition they are covered with long, flat-appressed, golden scales or scale-like hairs. Anterior tibiae and tarsi with numerous, oblique, short, stout spicules or spinous setae below, and 4 or 5 tiny, slender bristles above on the tibia. Middle tibia with 3 or 4 slender, oblique bristles below, the same number a little more stout above. Hind tibia with 7 slightly longer, slender, oblique, curved bristles anterodorsally and 12 or 15 shorter ones posterodorsally, but with only 2 or 3 ventrally. Claws stout, short, rather blunt, with long, slender, spatulate pulvilli.

Wings: Slender, nearly hyaline, tinged by comparatively dense, brown villi. Only 2 submarginal cells, the second submarginal cell is long and relatively narrow. The anterior crossvein enters the discal cell at the outer third. The anterior intercalary vein is rectangular, plane, and not quite as long as the anterior crossvein. The first posterior cell is of uniform width throughout. The anal cell is rather widely open. The second anal vein curved distally. Axillary lobe narrow. Alula absent. Ambient vein complete and the marginal fringe long. The third vein arises from the second vein rather than the converse, and it does so well before the end of the second basal cell. Costal comb absent.

Abdomen: About 1½ times as long as the thorax. At the base as wide as the thorax and then rather strongly tapered, conical, and pointed apically in the female. The seventh tergite forms an equilateral triangle, but it is shorter and transverse in the male. The abdomen is largely opaque black with small, polished, shining areas laterally. The pile forms a tangled mat of partly flat-appressed and partly erect, golden, scaliform pile, in which, however, there are a number of scattered, very long, erect, quite slender, black, bristly hairs, which are very readily denuded. Females with the last sternite yawning backward and downward, but the terminalia within recessed and enclosed by a fringe of hair. Male terminalia protrusive, the hypopygium turned sidewise. Hesse (1938) states that the basimeres are not narrowed in the neck region and that the beaked distimeres are elongate and narrow, the aedeagus with a ventral process below and the lateral struts short.

Material available for study: A male of the type-species and other material in the British Museum (Natural History). Also, material kindly sent by Dr. A. J. Hesse for dissection and study.

Immature stages: Unknown.

Ecology and behavior of adults: Not on record.

Distribution: Ethiopian: *Conophorina bicellaris* Becker, 1920.

Genus *Aldrichia* Coquillett

FIGURES 65, 285, 482, 495

Aldrichia Coquillett, Trans. American Ent. Soc. Philadelphia, vol. 21, p. 93, 1894a. Type of genus: *Aldrichia ehrmanii* Coquillett, 1894, by monotypy.

These Nearctic flies are approximately the same size range as *Conophorus* Meigen but are more slender. The pile is relatively fine and erect and true bristles are lacking. There are 3 submarginal cells but the venation differs in several particulars. The marginal cell is of uniform width meeting the costa at a right angle with a rounded apex; the first submarginal cell likewise is not expanded, is relatively long, and is rectangular at the base where the posterior branch of the third vein forms a crossvein. The wing is likewise uniquely different in the form origin of the second vein which arises rectangularly, sometimes with a backward spur, close to the anterior crossvein.

Aldrichia Coquillett differs from *Conophorus* Meigen in the form and character of all three antennal segments. The first segment is shorter and in fact not quite as long as the second segment, both quite hairy above and below, and the short third segment is flattened and oval with 3 small microsegments and bristly apical spine.

Length: 9 to 11 cm.

Head, lateral aspect: The head is globular, the occiput prominent in females, less so in males, the face is scarcely visible in the lateral aspect and the oragenal cup not at all. The vertical depth of the face is less than the thickness of the second antennal segment at its apex. The face is pollinose and bare. The pile of the occiput consists of numerous, though not dense, bristly hairs which are much longer in the female and in both sexes are much longer medially than laterally, except for 2 or 3 hairs near the eye margin. The eye is large, hemispherical, the posterior margin entire. Proboscis short, very stout, with large, flared, four-lobed labellum, fleshy and muscoidlike. The whole length of the proboscis no longer than the eyes. Palpus consists of 1 very long, rather stout, outwardly curved, apically rounded segment almost as long as the eye, which bears numerous long, fine hairs on all sides, except medially. The front in females is a little inflated and bears a number of quite long, slender, bristly, erect hairs, as long as those of the vertex, among which are a few short, glittering, curled, subappressed hairs. Male front minute, flat, pollinose, without pile, forming an equilateral triangle and without a groove. Antennae attached at the head; they are quite large and conspicuous, actually a little longer than the head, the first 2 segments both strongly swollen and robust though not as much swollen as in *Conophorus* Meigen. These first 2 segments are approximately equal in length and each contain much moderately long, shaggy, or somewhat bristly pile on all sides except medially and medial pile is present apically on the second segment. Third segment large, flattened, quite wide, except at the immediate apex, including the small, spine-tipped microsegment. The third segment nearly or quite as long as the second segment.

Head, anterior aspect: The head is not quite as wide as the thorax and not quite twice as wide as high. The front widens slightly toward the antennae, and there is a transverse groove shortly in front of the antennae. In the male the eyes are in contact for a distance of nearly 3 times the length of the front, and eyes bulge upward from the point of contact. The antennae are narrowly separated, the ocellar tubercle small, prominent, the ocelli forming an acute triangle in the male, and equilateral triangle in the female and the tubercle bears 10 or more long, slender, bristly hairs. The oral opening is large, though shallow, its walls narrow and thin. The genae are moderately wide, recessed inward from the eye margin down to a linear tentorial fissure. Viewed from above both males and females are shallowly bilobate.

Thorax: The thorax is elongate, slightly convex, opaque black and covered loosely with fine, erect, rather short pile, and about the same amount of fine, short, appressed, curled, glittering pile. The pile is much more abundant in the female than in male. Humerus with a tuft of comparatively dense, coarse, rather long, shining pile. Scutellum shining with curled, glittering pile on the disc, and disc and margin with numerous, fine, long, erect, stiff hairs, more numerous in the female, and the curled hair seemingly absent in the male. Notopleuron with 3 or 4 long, slender, black bristles. Pleuron sparsely pollinose, largely bare, the posterior border and the upper border of the mesopleuron with long, loose, abundant, coarse pile. Remainder of pleuron bare, except for a tuft of hairs behind the spiracle. Squamal fringe wide but rather short.

Legs: The legs are comparatively short and the femora a trifle thickened. Anterior tibia with 6 short, fine bristles or spicules posterodorsally. Middle and hind tibiae with 4 rows of longer, but slender, bristles, 5 or 6 in each row. Femora ventrally with a few scattered short hairs, the pile appressed and somewhat scaliform though not conspicuously so. Claws short, sharp, strongly curved, pulvilli long and broad.

Wings: The wings are slender, tinged with brown, all cells widely open, 3 submarginal cells, first submarginal cell curved forward to meet the costa just before the tip of the wing. Marginal cell very little widened apically. The medial crossvein or intercalary vein long and quite rectangular. Anterior crossvein lies a little before the middle of the discal cell. The second vein arises curiously and characteristically as a rectangular crossvein just before the anterior crossvein and it has a short backward spur. Axillary lobe quite narrow, the alula also quite narrow. Costal comb weak.

Abdomen: The abdomen is comparatively elongate, at the base as wide or a little wider than the thorax and loosely covered with flattened, curly, appressed, glittering, brassy pile, which is more abundant in the female; and in addition, sparse, erect, black, bristly hairs along the postmargin of the tergites. Sides of the first and the curled undermargin of all of the tergites with numerous, long, erect, brassy-white pile. Females with 7 tergites, males with 9 tergites. Last sternite in the female more or less triangular forming a down-turned flap and with a tuft of bristly hairs at the apex and concealing on each side a row of dense, brassy pile, which in turn conceals the female terminalia. In the male the dististyli lie side by side, only visible from below.

Material available for study: I have before me a pair of the type-species presented to me by Dr. R. H. Painter, and I have seen other material in museums.

Immature stages: Unknown.

Ecology and behavior of adults: The adults are taken on low-growing flowers in the North Central States of America, south of the Great Lakes. Painter (1939, 1940) records them from flowers of *Amorpha* sp., from *Thaspium barbinode*, and also from dewberry.

Distribution: Nearctic: *Aldrichia auripuncta* Painter, 1939; *ehrmanii* Coquillett, 1894.

Genus *Prorachthes* Loew

FIGURES 115A, 259, 479, 494, 1012, 1013, 1014

Prorachthes Loew, Berliner Ent. Zeitschr., vol. 12, p. 381, 1868. Type of genus: *Prorachthes ledereri* Loew, 1868, by monotypy.
Dumontiella Séguy, Encycl. Ent., ser. B, II, vol. 6, p. 162, 1932. Type of genus: *Dumontiella vespertilio* Séguy, 1932, by monotypy.
Cheilohadrus Hesse, Ann. South African Mus., vol. 34, p. 674, 1938. Type of genus: *Cheilohadrus conspersipennis* Hesse, 1938, by original designation. Subgenus of *Prorachthes* Loew.

Paramonov (1931a), p. 104, key to 6 Palaearctic species.
Engel (1932), pp. 49-50, key to 6 species, after Paramonov.

Small blackish flies that are peculiar because of the diffuse, curiously marbled or mottled patterns of whitish spots on the wings which characterize all of the known species. There are 2 submarginal cells; the first posterior cell is open maximally and discal cell is triangular with the anterior crossvein lying beyond the middle. Marginal cell expanded apically with the end of the cell broadly rounded. The pattern of spots varies in the different species; the wing presents an effect superficially like a species of *Haematopota* Meigen. The alula is virtually absent.

The face is extraordinarily tumid and swollen, the occiput rather swollen also. The antennae are stout, and within Palaearctic species the third segment is conspicuously and characteristically flattened, dilated, and widened and nearly as broad as long. The occiput, thorax, and abdomen bear rather dense, flattened, shining scales.

The genus is not closely related to others, but with the expanded occiput, shortened proboscis, and bristly, inflated antennal segments, I believe it shows some relationship to the Conophorini.

This genus is restricted to the Palaearctic, especially the Transcaucasus area from which or near which most species are known. One subgenus, *Cheilohadrus* Hesse, is known from South Africa, differing markedly in the form of the antennae, which is only slenderly spindle-shaped with dorsal spinules.

Length: 4.5 mm. to 7 mm.; wing the same length.

Head, lateral aspect: The head is subglobular but largely because of the tumid face and swollen occiput. Posterior eye margin both plane and entire. The lower half of the front is separated from the upper, comparatively flattened portion by a transverse, shallow depression. The lower part of the front bearing the antenna is almost if not quite as produced and tumid as the face. The face is produced, but recedes only below and opposite the proboscis. The oragenal cup is in evidence only medial to the deep, rounded, punctate depression opposite the proboscis. The face bears dense, erect, moderately long, black bristles. Pile of the front similar but more sparse and shorter. The occiput is tumid and bears scattered, shining, metallic scales along the sides and a little hair only ventrally between the lower eye corner with some additional hairs near the upper eye margin. Palpus quite short, the base very slender, the apical portion gradually and clavately widened with the apex truncate. Proboscis short, little greater in length than the head but extending as far as the antenna. It is markedly flattened laterally and quite narrow from the dorsoventral aspect. The antenna is large and conspicuous, the first segment is robust, twice as wide at apex and bears numerous long setae, stout and more conspicuous along the apical margin. The second segment is only about one-fourth or one-fifth as long as the first segment and much reduced in width. The third segment in Palaearctic species is greatly widened and dilated dorsoventrally and sharply and rather abruptly reduced in size at the apex.

Head, anterior aspect: The head is very little wider than high but with the genofacial space very extensive and occupying more than half the total width. The genae occupy two-thirds of this space. Oral opening short oval. Eyes of the male usually holoptic for some distance, the upper facets a little enlarged. In some species, the eyes of the male are narrowly separated. The eyes of the female are widely separated. The antennae lie adjacent to one another. The ocelli are large, set on a low tubercle and are equilaterally placed.

Thorax: The thorax is about as long as wide and slightly narrowed anteriorly. The head is wider than the anterior thorax but more narrow than the posterior thorax. Mesonotum dull black, or dark colored with scattered, shining, metallic, and rather broad scales over the surface, together with a few minute, appressed setae which are more in evidence on the sides of the notopleuron. There are especially dense patches of scales on each side in front of the scutellum, and above the wing, and likewise on the sides of the scutellum. The disc of the scutellum bears a number of not very long, erect, black setae or bristly hairs. Pleuron largely bare, except for a few scales and setae in the upper corner of the mesopleuron. Metapleuron with a tuft of rather dense, erect, bristly, black pile and posterior hypopleuron with a similar shorter tuft.

Legs: The legs are comparatively short; all the femora are slightly thickened through the middle and bear minute, flat-appressed, scaliform pile. The anterior femur below has a few quite short, fine, erect setae. Claws fine, long, and sharp with short pulvilli about half as long as the claws. Arolium minute, short, and conical.

Wings: The wings are rather large and elongate with an opaque whitish pattern greatly modified by slaty or sooty blackish color in a characteristic mottled or marbled pattern for each species. The wing is slightly narrowed basally, the axillary cell narrowed toward the base and the alula, except in *conspersipennis* Hesse, quite narrow but with a fringe of a few long hairs. There are 2 submarginal cells; the marginal cell is greatly widened apically, the 4 posterior cells are widely open, the anal cell also widely open. Discal cell rather short, much widened distally, the medial crossvein but little longer than the upper intercalary vein, and the anterior crossvein enters the discal cell at the outer third or even beyond. Ambient vein complete.

Abdomen: The abdomen is short oval but broad at the base and a little wider than the thorax; it is rather shining dark brown or blackish and bears rather numerous, short, wide, appressed, metallic scales. Posterior margins of tergites with a few fine, erect, coarse, or bristly hairs. Basal portions of the tergites sometimes tend to have foveate depressions. For *conspersipennis* Hesse, that author notes that the "basimere of the male has bristly hairs, the distimeres (beaked apical joints) are elongate, taper to a point, provided with dense, bristly, short hairs, not depressed above, but hollowed out below; aedeagus without a ventral process, its dorsum not produced on either side; lateral struts broad and leaf-like."

Material available for study: The series of this genus in the British Museum (Natural History).

Immature stages: Unknown.

Ecology and behavior of adults: Engel (1933) records taking *beckeri* Paramonov upon flowers of a *Euphorbia* species. Paramonov found his exampes on the border of a woods.

Distribution: Palaearctic: *Prorachthes beckeri* Paramonov, 1926; *ledereri* Loew, 1868; *longirostris* Bezzi, 1925 [=*crassipalpis* Villeneuve, 1930, =*vespertilio* (Séguy), 1932 (as *Dumontiella*)]; *pleskei* Paramonov, 1926; *portschinskyi* Paramonov, 1926; *stackelbergi* Paramonov, 1926; *vespertilio* (Séguy), 1932 (as *Dumontiella*).

Ethiopian: *Prorachthes conspersipennis* (Hesse), 1938 (as *Cheilohadrus*), *conspersipennis xerophilus* (Hesse), 1938 (as *Cheilohadrus*).

Genus *Legnotomyia* Bezzi

FIGURES 31, 254, 255, 469, 474, 800, 801, 802

Legnotus Loew, Neue Beiträge III, p. 41, 1855. Type of genus: *Legnotus trichorhoeus* Loew, 1855, by original designation, as no. 80 in Loew's collection as *Bombylius trichorhoeus* Loew. Preoccupied by Schiødte, Hemiptera, 1848.
Legnotomyia Bezzi, Zeitschr. syst. Hymen. und Dipt. Teschendorf, vol. 2, p. 191, 1902. Change of name.
Legonotus Bischof, Wiener Ent. Zeitung, vol. 22, p. 42, 1903.
Psiatholasius Becker, Zeitschr. syst. Hymen. und Dipt., vol. 6, p. 145, 1906. Type of genus: *Psiatholasius bombyliiformis* Becker, 1906, by original designation.

Engel (1932), pp. 58-59, key to 4 Palaearctic species.

Odd, medium-size flies; the head, legs, and thorax shining black, the abdomen dull, feebly shining black. The abdomen is very broad, short, and rounded, considerably wider than the thorax and but little convex. The postmargins of the tergites are opaque, creamy yellow, the dorsal pile quite sparse, fine, long, and erect; pile on the sides and apex of abdomen more abundant, and stiffer, and all of the pile golden yellow; sides of tergites turned under laterally. The wing may be tinged with brown in the middle, brown on the basal half, or with a middle cross band; it is wide apically, but the axillary lobe is much narrowed and the alula is narrow. Marginal cell bulbous and sometimes greatly expanded backward at its apex. First posterior cell open maximally; anal cell widely open. Anterior crossvein quite variable; it is situated at the middle, or just beyond the middle of the discal cell. In *Legnotomyia cineracea* Austen it lies before the basal third and the discal cell is short, wide, angulate below. In addition to the above peculiarities the antennae are unique. The first segment is short, about twice as long as wide; the second segment is very short; the third segment is elongate, angularly expanded below in the middle or swollen over the whole basal half and narrowed apically. At the apex it bears 4 or 5 long, distinctive, yellow bristles. Face with a few long hairs; gena and oral cup bare. Proboscis long; occiput swollen, especially below. *Legnotomyia* Bezzi has 4 posterior cells, 2 submarginal cells. It is a somewhat aberrant member of the Bombyliinae. I cannot see the slightest basis for placing *Legnotomyia* Bezzi in the Usiinae, which subfamily has 3 posterior cells, closed and stalked anal cell, eliminated face, and other distinctions.

Legnotomyia Bezzi is restricted to the Palaearctic with the exception of one species from South Africa. *Psiatholasius* Becker is Palaearctic.

Length: 5 mm. to 9 mm.

Head, lateral aspect: Head globular, the face and oragenal cup rather extensive and well developed. The occiput prominent, even in males, but gradually sloping to the eye margins. Face densely micropubescent and also with about 20 long, shining, coarse or flattened hairs extended forward on each side of the face. Pile of occiput rather scanty laterally and shorter, becoming longer, coarse and somewhat more abundant medially.

The eye is hemicircular, the posterior margin entire, but also the posterior border of the eye extending or bulging outward from the occiput. Proboscis very long and slender, at least 4 times as long as the head. The labellum is slender, short, and cylindrical. Palpus relatively short, slender with 2 or 3 relatively long, apical, bristly hairs, and the basal segment quite short. Front in the male flattened and pollinose with a few, long, scaliform hairs appressed forward. Antennae attached at the upper sixth of the head; they are relatively short; the first segment is a little over twice as wide, more than twice as long as the very short second segment, which is only half as long as high. Both first and second segments have a pronounced dorsal, medial spur. Third segment rather wide in the middle and somewhat flattened, a little narrowed at base and apex, which segment bears a small, bicolored microsegment at the apex and this segment characteristically bears 8 or 10 short bristles dorsally and laterally in the middle, 4 or 5 longer bristles below in the middle, and 5 equally long at the apex. The first segment bears some moderately long, stiff hairs dorsolaterally, and a few below in the middle.

Head, anterior aspect: The head is about 1½ times wider than high, distinctly oval, the front in the male comparatively large and triangular, the eyes in the male almost but not quite touching for a short distance, the upper facets enlarged. The ocellar tubercle is comparatively large and protuberant, the large ocelli lie adjacent to the eye in a short isosceles triangle, the oral opening is quite large, short rectangular and quite deep in the middle and behind; its walls are thin, the gena is relatively sharp and narrow and more or less sharply and rectangularly separated from the oragenal cup, the outer walls of which are polished and bare; the tentorial fissure is narrow, but deep opposite the proboscis. From above, the postvertex is not at all bilobate and the central cavity of the occiput is small.

Thorax: The thorax is short, wide, gently convex, shining black, covered with loose, scattered, erect, fine hairs of moderate length and little denser anteriorly. There is a tuft of similar hair in the upper posterior corner of the mesopleuron which extends outward, a fringe along its upper border which extends upward. The scutellum is short, covered with the same, fine, erect pile. The pleuron is shining black. The metapleuron, hypopleuron, and pteropleuron entirely quite bristly. Squamal fringe sparsely pilose and comparatively short. Halteres with a large knob and an elongate patch of setae at the base of the knob.

Legs: The legs are comparatively short, the tibiae, however, rather long and slender. All of the tibiae rather densely appressed, brassy setate and without bristles, although the hind pair has a ventrolateral row of some short, fine bristly hairs. Femora without bristles below, all the femora slender throughout, the hind pair with a ventrolateral row of some 10 or 12 long, coarse hairs that could not well be called bristles. The pile is setate, appressed, slightly flattened but no scales present. Claws long, slender, sharp. Pulvilli long and spatulate.

Wings: The wings are comparatively broad, narrowed at the base, broad at the apex. There are 2 submarginal cells; the first posterior cell is open maximally, the anal cell very widely open, axillary lobe and alula both narrow. The anterior crossvein lies just beyond the middle of the discal cell. The marginal cell is greatly expanded apically, broadly rounded at the apex below, meeting the costa at nearly a right angle. All of the veins are unusually stout, especially the costa, the radial trunk, and first vein. The radial trunk bears setae. There is an extra cell in both wings of my material formed by a crossvein near the end of the radial sector. Ambient vein unusually stout and complex. The wings in most species have a diffuse brownish band across the middle and sometimes the base is brownish.

Abdomen: The abdomen is remarkably broad and short and gently convex. It is much wider than the thorax, the tergites are subopaque black, a little shining in the middle and with conspicuous, though narrow opaque, cream-colored margins posteriorly. The pile consists of rather abundant, erect, moderately long, shining, fine, yellowish hair spread uniformly over the tergites and perhaps a little accentuated posteriorly and a little laterally. I find 8 tergites in the male, and the eighth being extremely short and tucked under the short seventh. Male terminalia comparatively large, thrust downward from the apex of the abdomen.

The male terminalia from dorsal aspect, epandrium removed, shows a body longer than wide, scarcely narrowed apically. Dististyli short, quite stout basally, almost triangular, convergent at apex and bearing numerous, bristly hairs. From the lateral aspect, it is longer, slender, turned upward, attenuate, pointed at the apex. The epiphallus becomes strongly swollen progressively toward the base. Epandrium, cerci, and basal apodeme all quite large.

Material available for study: I have a male of the type-species, and I also have seen the material in the British Museum (Natural History). I also have a female of *fascipennis* Bezzi sent me through the courtesy of the Cairo University, Cairo, Egypt.

Immature stages: Unknown.

Ecology and behavior of adults: I have seen no comments on adult behavior or flowers visited.

Distribution: Palaearctic: *Legnotomyia bombyliiformis* (Becker), 1906 (as *Psiatholasius*); *cineracea* Austen, 1937; *erivanensis* Paramonov, 1925; *fascipennis* Bezzi, 1925; *leyladea* Efflatoun, 1945; *palestinae* Paramonov, 1947; *persica* Paramonov, 1933; *trichorhoea* (Loew), 1855 (as *Legnotus*).

Ethiopian: *Legnotomyia striata* (Bischof), 1903 (as *Legnotus*).

Genus *Corsomyza* Wiedemann

Figures 63, 64, 68, 287, 462, 497, 729, 817, 818, 819

Corsomyza Wiedemann, Nova Dipterorum Genera, p. 13, 1820. Type of genus: *Corsomyza simplex* Wiedemann, 1820, as given by Bezzi, p. 3, 1924.
Lasioprosopa Macquart, Diptères exotiques, suppl. 5, p. 82, 1855. Type of genus: *Corsomyza nigripes* Wiedemann, 1820, as *Lasioprosopa bigotii* Macquart, 1855, by monotypy.
Pusilla Paramonov, Proc. Roy. Ent. Soc. London, ser. B, vol. 23, p. 27, 1954b. Type of genus: *Pusilla longirostris* Paramonov, 1954, by original designation.
Denamyza, new subgenus. Type of subgenus: *Corsomyza ochrostoma* Hesse, 1938.

Bezzi (1924), pp. 105-106, key to 9 South African species.
Hesse (1938), pp. 716-728, key to 13 male species and 15 female species and subspecies.

Small to medium-sized flies that are unique and peculiar in the wide or sometimes very wide head in both sexes, often wider than the thorax and the remarkably wide front in both sexes, the males being strongly dichoptic. The separation of the eyes in males is at least as wide as the wide, swollen ocellar tubercle. The front is somewhat inflated and the broad, wide face is apt to be inflated laterally. There is often a broad, wide, circular, fanlike brush of long, erect, dense pile reaching in full width from the lower gena up past the face and up around to the base of antenna, and above it. This brush of pile is particularly likely to be developed in this way in males. The gena is as much expanded as the face leaving the moderately deep, rather small, oragenal cup hidden by pile. The antennae are characteristic; they are long and slender; the short first two segments have a circular band of moderately long pile all around them. The slender, elongate third segment is knob-shaped at the base; it is then quite slender beyond, but increases in width apically and becomes clavate and ends in an apical clubbed part that is grooved or hollowed out, and has a tiny apical microsegment. Male eyes sharply divided into large facets above.

Wings chiefly hyaline with basal comb and alula reduced, usually much reduced; first posterior cell open maximally; the anterior crossvein lies just beyond the apical third of discal cell; this cell is short, angularly extended below, the intercalary vein unusually long. There may be either 2 or 3 submarginal cells. The antennae lie closely adjacent.

Corsomyza Wiedemann may be taken as the major representative of a group of characteristically South African genera not found elsewhere. The expanded genal area carrying into the face and front above is unique.

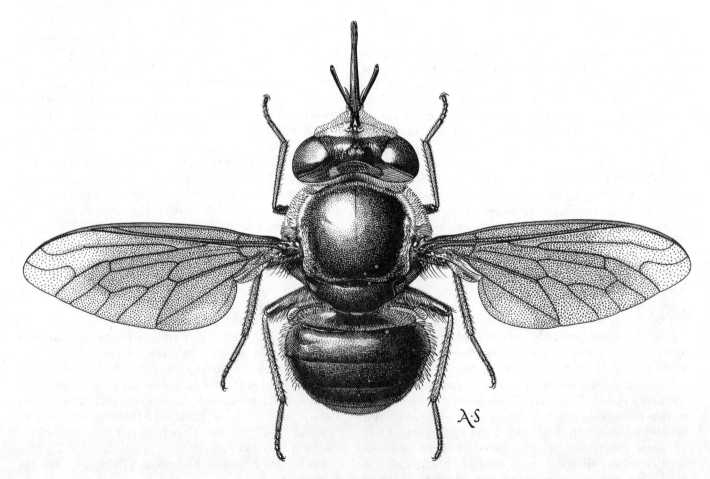

Text-Figure 22.—Habitus, *Corsomyza simplex* Wiedemann.

It seems to be that the species *Corsomyza ochrostoma* Hesse (1938) clearly deserves subgeneric assignment, and I place it in *Denamyza*, new subgenus, derived from Greek: denaios, old, and myzo, suck.

Hesse (1938) gives the distinguishing differences thus: costal cell scarcely broadened, distinctly widened anal cell, more reduced axillary cell, distinctly raised and elevated ocellar tubercle, shorter first antennal segments, shorter labellum, distinctly longer apical spines on tibiae, several bristly hairs across last tarsal segment, much shorter than in other species, and less strongly curved claws. Found so far only in Namaqualand.

Length: 4.5 to 11 mm.; usually 6 to 7 mm. Wing a little shorter, or same length.

Head, lateral aspect: Head more or less hemispherical, the occiput short, strongly sloping inward when viewed from above to a point not greatly distant from the inner eye margin, and from this point the occiput slopes inward and forward so that it is slightly concave, but there is no large deep extensive cup as in the second division of the family. The pile of the occiput is scanty and distinctly flat-appressed, short and bristly over all of the middle of the occiput, becoming longer and shaggy at the bottom of the head. At the top of the occiput behind the vertex a row of long, conspicuous, erect, stiff hairs radiates out like a fan. The front is convex and tumid and swollen and bears conspicuously long, fine pile all across the middle portion. Beginning on the lower third it is a little reduced medially and there is a peculiar, very dense, conspicuous fringe of pile, extending obliquely forward and encircling the face and gena below, even below and around the proboscis. This odd, circular brush encloses centrally the antenna above with its long radiating circlet of pile and encloses also laterally and below a dense, wide, tufted outgrowth of hair of equal length, perhaps of softer texture, and often of a different color.

The postmargin of the eye is nearly, but not quite, plane above. The eye is rounded forward and reduced in size below and with the lower third of the proboscis very much smaller and rather clearly demarcated. The proboscis is quite slender and rather long and about twice the head length with very slender, pointed, curved, microspinulate labellum. Palpus long and slender, distinctly projecting beyond the facial brush and apparently unsegmented. First antennal segment moderately long and slightly thickened and from 4 to 6 times as long as wide and always with a very conspicuous, dense, radiating circlet of pile confined to the dorsal, lateral, and ventral segments of each antenna. This pile does not extend beyond the facial brush, and this segment is at most slightly thickened apically. Second segment short, robust, somewhat barrel-shaped with short setae dorsally, slightly longer setae medially and laterally. The third segment is characteristic in shape, long and slender on the basal half or sometimes more than half, either gently widened apically or rather abruptly clavate on the outer fourth or fifth and sometimes grooved on the inner margin, and sometimes with spinulose pubescence apically, or with an apical notch. The third segment is at least 1½ times the combined length of the first two segments and sometimes longer.

Head, anterior aspect: Distinctly wider than the thorax. Approximately twice as wide as high. Eyes in the male widely separated by at least the length of the first two antennal segments. Ocellar tubercle quite large, broad, the ocellar triangle extraordinarily obtuse, the tubercle high, but gently sloping. Eye of female much more widely separated. Front convex and inflated, shining blackish, a tuft of long pile lies between the ocelli. From the anterior aspect, the facial brush, both the outer fringe and the central part are dense and plushlike, but with the facial space immediately below the antenna exposed and bare. Oral recess rather narrow and deep posteriorly. Gena extraordinarily wide.

Thorax: The thorax is quite short and broad at least as wide as long if not wider, generally opaque blackish with very dense, erect, fine plushlike pile of great length over the entire mesonotum and the whole of the short, very broad scutellum. Mesopleural pile equally dense, long, tufted and sometimes matted. Pteropleuron, hypopleuron, and metapleuron without pile.

Legs: Quite small and weak. All of the short femora bear dense fringes of long, fine, erect pile directed downward, obliquely forward, and obliquely toward the rear middle. Dorsal pile minute. Anterior tibia comparatively smooth dorsally; remaining tibia with a few, long, oblique, stiff hairs dorsally. Claws quite slender on the apical half, stout at base, sharply curved apically. Pulvilli long and wide.

Wings: Hyaline, or sometimes tinged with yellowish brown, the costal cell and the base more yellowish. Wings sometimes with an opaque, milky-white appearance. There are usually 2, sometimes 3, submarginal cells. Anterior crossvein always enters the discal cell well beyond the middle and usually near the outer third. The first posterior cell is open maximally. The anal cell is rarely open, more usually closed at the margin or with a short stalk. Alula reduced, generally narrow, even vestigial. Basal comb absent, ambient vein complete. The second vein arises abruptly at the end of the first basal cell.

Abdomen: Approximately as long as thorax, somewhat conical, narrowed and bluntly pointed apically, at the base somewhat less wide than the thorax. There are 7 tergites visible from above, the color varies from reddish brown to black, feebly shining, the pile generally extraordinarily long, fine, and quite erect over the whole abdomen, but forming a dense, long conspicuous band or fringe along the side margins. Pile varies from deep brownish yellow to white. The male terminalia are small, surrounded by long hair. Hesse (1938) states that the basimere always has an extra lobe or flap at the base and that the dististyli are clawlike, laterally compressed, and the apical part curved downward and pointed; he states that the apical lobelike part of the aedeagus is curved upward, very uniform in the species, and without a ventral process at the base.

The male terminalia from dorsal aspect, epandrium removed, show a long, rather slender body, basally widened with extensive membranous area. The large long dististyli are convergent, rather strongly attenuate at the sharply pointed apices. From the lateral aspect the epandrium is moderately high and long, but narrows apically. Cerci small, the epiphallus stout and increasingly stout basally. Dististyli stout, sharply curved, and pointed apically. Apex turned upward. Basistylus moderately stout basally, strongly narrowing on the distal half. Basal apodeme large, except for the greater length from a dorsal aspect and its distally narrowed condition, *Corsomyza* Wiedemann shows very little fundamental difference from *Bombylius* Linné.

Material available for study: I have studied the series of *Corsomyza* Wiedemann in the British Museum (Natural History), from which my illustrations are taken, and Dr. A. J. Hesse has most kindly sent three species for dissection and study.

Immature stages: Unknown.

Ecology and behavior of adults: Hesse (1938) states that the females of *Corsomyza simplex* Wiedemann have a marked resemblance to some tabanids and are more often seen on flowers of composites.

Distribution: Ethiopian: *Corsomyza anceps* Bezzi, 1924; *ancepsoides* Hesse, 1938; *bicolor* Bezzi, 1921; *bipustulata* Bezzi, 1924; *brevicornis* Hesse, 1938; *campicola* Hesse, 1938; *capensis* Hesse, 1938; *clavicornis* Wiedemann, 1819; *depressifrons* Hesse, 1938; *dissimilis* Hesse, 1938; *eremobia* Hesse, 1938; *eremobia braunsii* Hesse, 1938; *fuscipennis* Macquart, 1840; *fusicornis* Hesse, 1938; *gonucera* Hesse, 1938; *hirtipes* Macquart, 1840; *karooana* Hesse, 1938; *longipalpis* Hesse, 1938; *longirostris* (Paramonov), 1954 (as *Pusilla*); *minuscula* Hesse, 1938; *montana* Hesse, 1938; *namana* Hesse, 1938; *nigripes* Wiedemann, 1820 [=*bigotii* Macquart, 1855], *nigripes turneri* Hesse, 1938; *nitida* Macquart, 1840; *ochrostoma* Hesse, 1938; *oneilii* Hesse, 1938; *pallidipes* Hesse, 1938; *pennipes* Wiedemann, 1820; *ruficornis* Bezzi, 1921; *simplex* Wiedemann, 1820; *tricellulata* Hesse, 1938.

As I cannot settle the exact synonymy and relationship of *Pusilla* Paramonov, I add the following comments:

Bowden (1960), in erecting the genus *Zyxmyia* Bowden, speculated that *Pusilla* Paramonov, described from Katanga, is really allied to *Corsomyza* Wiedemann, and more particularly to *Hyperusia* Bezzi. In this view, I believe he is correct. Certainly *Legnotomyia* Bezzi belongs in the Bombyliinae. For *Corsomyza* Wiedemann and its allied and related genera I have set apart the tribe Corsomyzini; *Corsomyza* Wiedemann ranges as far north as Victoria Falls, Northern Rhodesia.

These flies were described from two males taken in the Belgian Congo.

Length: 4 mm.; wing 3 mm.

I have not thus far seen this genus and I quote Dr. Paramonov's description:

Belongs to subfamily Usiinae: hind border of eyes absolutely entire; occiput very slightly prominent, not bilobate above and without a deep central cavity; occipital fringe almost absent; antennae approximate at base; prothorax and metasternum not specially developed; tibiae (femora also) only pilose, without rows of distinct spicules; all bristles on body absent. Cubital fork widely open, venation very similar to *Legnotomyia*, four posterior cells. Proboscis very long, in the type species as long as body. Palpi very thin, cylindrical, as long as mouth cavity, two-segmented, apical segment nearly bare, basal with long hairs. Two basal segments of antennae very short, rather broad and thick, first a little longer than broad, second twice as broad as long, third at least twice as long as the two basal segments together; viewed from the side, of an irregular egg-shape, at least twice as wide as second segment. Mouth cavity reaches to base of antennae, not deep, nearly as broad as sides of face; borders of cavity not high. Face very broad, equal to half of head width, frons in male also very broad, distinctly more than one-third of head width and three times as wide as ocellar triangle.

Head, thorax, scutellum and abdomen covered with very dense, long, erect hairs. Abdomen broad, oval, short, a little longer than and as broad as thorax. Hypopygium of male extremely solid and long. Two submarginal and four posterior cells. All posterior cells and anal cell widely open. Axillary cell (lobe) well developed, alula small, a little rounded, squamae long-haired. Legs without bristles.

Type of genus: *P. longirostris* sp. n.

Ground colour of body black, only basal segment of palpi, veins at base of wings, halteres, tibiae and tarsi yellowish. The long erect hairs on body yellowish-white. Hairs on occiput and face distinctly shorter than on frons. Hairs on frons intermixed with brownish hairs. Hairs on basal two segments of antennae short, third segment bare. Ocellar triangle broader than long. Viewed in profile frons not prominent, face feebly so, also the chin. Sides of frons distinctly divergent. Wings hyaline, veins in basal half yellow, in apical half darkened. Vein r-m situated distinctly beyond, but not far from middle of discal cell. Anal cell at centre only slightly broader than at base or at apex. Ambient vein complete. Alula very sparsely haired. Origin of third longitudinal vein a little nearer to common stem of radial veins than to cross-vein r-m. Legs without bristles, but at apex of tibiae and on tarsi a few black, very short spicules. Pulvilli well developed. Femora with long hairs.

Material available for study: Not seen. This genus was erected and the type deposited after our lengthy stay at the British Museum (Natural History) was concluded.

Immature stages: Unknown.

Ecology and behavior of adults: Not on record.

Distribution: Ethiopian: *Pusilla longirostris* Paramonov, 1954.

Genus *Callynthrophora* Schiner

FIGURES 292, 480, 493

Callynthrophora Schiner, Verh. zool-bot. Ges. Wien, vol. 17, p. 313, 1867. Type of genus: *Callynthrophora capensis* Schiner, 1868, by original designation. One species.

Hesse (1938), pp. 776-777, key to 3 species from South Africa.

A small genus of medium-sized, brownish-black or reddish-brown flies related to *Corsomyza* Wiedemann and also to *Gnumyia* Bezzi. *Callythrophora* Schiner has a very wide head and front and face. There is a peculiar and conspicuous inflation and tumidity on the front which extends into the face. In contrast to *Corsomyza* Wiedemann, but more like *Hyperusia* Bezzi, the antennae are attached only a short distance above the oral recess and as a result the front is even more extensive and prominent. As in *Corsomyza* Wiedemann there is a well-developed, dense, circular, facial brush; *Gnumyia* Bezzi has no distinct facial brush of pile. In *Callynthrophora* Schiner the wide, inflated, tumid front, extending lower downward, the more swollen or thickened first antennal segment, the presence of spinulose pubescence on the second antennal segment all separate it from *Corsomyza* Wiedemann. There are no long, bristly hairs on the last three hind tarsal segments such as found in *Gnumyia* Bezzi and often in *Corsomyza* Wiedemann. Usually with 3 submarginal cells, one species has only 2. The genus *Gnumyia* Bezzi differs further by the presence of a stout, fleshy labellum, very short proboscis, absence of spinules on the second antennal segment, and the presence of feathery pubescence on the hind tibia.

The 3 known species are from South Africa.

Length: 8 mm.; wing a little shorter.

Head, lateral aspect: The head is large and subglobular, with the occiput comparatively prominent but rounded in profile and also sloping strongly inward toward the foramen. Occipital pile short, scanty, bristly, but with a tuft of longer hairs near the vertex, which may also be seen from the front, rising above each eye corner. The front is light reddish brown, and greatly swollen, tumid and broadly rounded. It has numerous very long, erect hairs which continue in the same length downward upon the face but become shorter laterally on each side of the oral recess. The face extends only a short distance below the antenna but the face is as much swollen as is the front. The posterior eye margin is broadly rounded. The proboscis is slender with a slender labellum and the whole structure not longer than the head. Palpus long and slender, at least half as long as the proboscis. The antennae are attached at the middle of the head and are nearly but not quite as long as the head. The first segment is distinctly swollen and bears very long, coarse pile above and below. The second segment is small and beadlike; Hesse (1938) states that it has dense, silvery, spinulelike pubescence upon its inner surface, and it may have a thornlike protuberance above. The third segment is much longer than the first two segments combined; with a knob at the base; the basal half or two-thirds is slender and filiform, the apical part clubbed or spear-shaped, of uniform thickness and nearly twice as thick as the basal part.

Head, anterior aspect: The head is very wide, wider than the thorax, and the front in both sexes much wider than in *Corsomyza* Wiedemann. At the vertex the eyes in the female are separated by 3 times the width in male. The front at the level of the antenna is remarkably wide, each half almost twice as wide as either eye. The antennae are narrowly but distinctly separated. The ocellar tubercle is prominent although low, but the ocelli are large. The oral recess is oval, converging anteriorly below the antenna. The gena on each side of the oral recess is very wide and also wide at the bottom of the head.

Thorax: The thorax is densely covered with long, coarse, erect, yellowish pile which is more abundant anteriorly before the wing or in front of the transverse suture. The anterior suture appears to be curved from the wing in a slightly oblique, forward direction. Scutellum large and thick and convex, with a deep basal crease; its surface bears numerous bristly hairs with similar hair distributed over the margin. The upper half of the mesopleuron is irregularly covered with very dense tufts of long, coarse, brownish-yellow pile. The propleuron and the middle of the sternopleuron bears a little, scanty, scattered tuft of similar pile. Remainder of pleuron bare. Squama short with a dense, thick fringe of fine hairs. There appears to be no anterior collar of pile upon the mesonotum, and bristles are absent on the thorax.

Legs: All the femora are slightly thickened, the tibiae likewise stout. Anterior femur with some long, coarse pile distributed along the ventral surface. Middle femur with a rather dense anterior and posterior fringe of long, coarse pile extending close to the ventral margin. The hind femur has scattered, bristly hair laterally and some longer, similar bristly hair near the apex which is distributed ventrolaterally. The anterior tibia has 3 or 4 dorsal bristles and 5 or 6 minute, posteroventral bristles; middle tibia with four anterior bristles, 6 posterodorsal bristles, and only 2 or 3 minute, posteroventral bristles. The hind tibia has distinct, long bristles laterally and very short medial bristles; it has 12 or 15 dorsomedial bristles some of which are scarcely differentiated from the setate pile, and it has a like number of ventromedial bristles; there are present 12 dorsolateral and 7 ventrolateral bristles. The tarsi end in slender, sharp, gently curved claws and well-developed pulvilli.

Wings: Hyaline, comparatively short and relatively broad apically. There are 3 submarginal cells within the type-species and also in *hastaticornis* Hesse; the species *marginifrons* Bezzi has only 2. The first posterior cell is opened maximally or even widened on wing margin; anal cell wide, acute, closed in the margin or with a short stalk. There is no basal comb. Ambient vein complete and alula narrow.

Abdomen: The abdomen is short oval, or obconical, not quite as wide as the thorax and covered densely along the sides with long, coarse, pale pile which extends along the extreme lateral margin as a brush covering the sternites. The pile upon the dorsum of the abdomen is considerably less extensive and is a little shorter. The abdomen is opaque, dark, brownish or blackish in color and brownish black pollinose.

Hesse (1938) points out that the hypopygium is like that of *Corsomyza;* the basimere shows a posterior lobe, the beak on the dististyle is acute but much less laterally compressed, the band of hair along the upper dorsal part less dense; the aedeagus more acute apically, from the side less broad and without lateral process.

Material available for study: Representatives in the British Museum (Natural History).

Immature stages: Unknown.

Ecology and behavior of adults: Not on record.

Distribution: Ethiopian: *Callynthrophora capensis* Schiner, 1868; *hastaticornis* Hesse, 1938; *marginifrons* Bezzi, 1921, and also *magnifrons*, erratum on plate explanation, p. 180.

Genus *Gnumyia* Bezzi

FIGURES 289, 483, 506

Gnumyia Bezzi, Ann. South African Mus., vol. 18, p. 82, 1921a.
  Type of genus: *Gnumyia brevirostris* Bezzi, 1921, by original designation.

Hesse (1938), pp. 781-782, key to 2 South African species.

Small to medium-size flies that are very remarkable from having a peculiar ensemble of characters. There is a remarkable inflation of the face, sides of face and genae, the more striking since the antennae are attached high up on the head, a great deal nearer the ocelli than to the oragenal cup. The whole head is much widened, wider than the thorax, and most of this unusual head width reflected in the exceptional width of face and front. Male eyes strongly separated. The greater part of the inflation of the face lies below the antennae; moreover, it has the antennae rather widely separated at base, at least as much as width across posterior ocelli and the second segment lacks spinulate pubescence, the third segment is scarcely or not at all clavate at the apex; the proboscis is very short, stout, with fleshy labellum. In its nearest relative, *Callynthrophora* Schiner, the reverse is true; there the most conspicuous and exceptional part of the swelling of the equally wide head lies chiefly on the front above the antennae, which are attached side by side. The antennae have marked spinulose pubescence on the second segment and a dilated apex on the third segment; its longer, more slender proboscis and labellum and its legs lack the dense, feathery pubescence on the hind tibiae, which is found upon *Gnumyia* Bezzi. Known species of *Gnumyia* have 3 marginal cells. The dististyli of the hypopygium are less clawlike than in *Callynthrophora* Schiner.

These flies are part of a very remarkable and characteristically South African section of the subfamily Bombyliinae, including, besides those mentioned above, *Corsomyza* Wiedemann (with the most species), *Megapalpus* Macquart, and *Hyperusia* Bezzi.

Length: 5.5 mm. to 8 mm.; wing a little shorter.

Head, lateral aspect: The head is subglobular in shape due largely to the remarkably swollen front, face, and gena; the occipital part is not extensive. The oragenal cup is not only small but extraordinarily low, and because of this the face is extensive and high; it is at least as high as the length of the antenna. The pile of the face is loose, yet abundant, and is coarse, blackish and comparatively long, the shining, yellowish-brown ground color not obscured. There is no dense brush of pile formed as in other Corsomyzini. Occiput short and poorly developed with rather dense, short, fine, submarginal pile. The eye much narrower than high, appears smaller perhaps as a result of the great development of face and gena; its posterior margin is entire. The thickened proboscis is quite short, extending barely beyond the face, except when thrust downward; labellum short, fleshy, tumid. Hesse (1938) says of the labellum that it is broad and more like a muscid, bears conspicuous, scattered spinules. The palpi in *Gnumyia* Bezzi are shorter than the proboscis. The front is convex and bears pile similar to but shorter and less dense than that of the face. The antennae are attached at the upper two-thirds of the head; they are relatively slender, of uniform thickness, and moderately long, but at most only two-thirds as long as the head. The first segment is a little more than twice as long as wide, the second segment is beadlike with upper apex extended in a styliform process. The first segments are quite widely separated, more so than in other Corsomyzini. The third segment, narrowed at the apex, bears a short, conical microsegment.

Head, anterior aspect: The head is quite wide in consequence of the extensive face and very wide genae, and it is wider than the thorax. The eyes are narrowed anteriorly and quite concave on the medial border. Eyes widely separated in the male, in fact Bezzi, who described the genus from a single individual, thought that he had a female, when in reality he had a male before him. The vertex is elevated and swollen and the ocelli large. Oragenal opening small, narrow, and slitlike and situated at the bottom of the head. The extensive face is separated from the wide gena by a furrow, becoming deeper below. The occiput is broadly and shallowly concave dorsomedially, and there is no medial dorsal fissure.

Thorax: The thorax is rather flattened, with fine, long, erect, dark pile along the sides; bristles absent but margin of scutellum with a few long, bristly hairs. The scutellum is quite convex, large and exceptionally thick. Metapleuron entirely bare, and tuft below the halteres likewise absent; mesopleuron very densely long, coarse, blackish pilose. Squamae with a long, not very dense fringe; no distinct plumula present.

Legs: The femora are slender, the tibiae likewise. Hind femur laterally, medially, and ventrally with rather long, coarse pile, dense below; bristles absent. Middle and posterior tibiae with anterodorsal and posterodorsal rows of fine, long, bristly hairs, but only a posterodorsal row on anterior tibia. Hesse describes some of these elements as "stoutish spicules" on anterior and other tibiae. The tarsi are moderately slender; last segments with more than 3 hairs apically and above. Claws small, sharp, a little curved, the pulvilli well developed with a short, triangular medial lappet.

Wings: The wings are hyaline or with brownish tint, almost identical with *Callynthrophora* Schiner and likewise with 3 submarginal cells; however, the alula, while narrow, is a little wider. The first posterior cell is widely open, in fact wider at the end than at base or middle; the second and third posterior cells widely open. Discal cell relatively short, widened and angulate below; the anterior crossvein lies barely beyond its middle. Upper end of third posterior cell strangely arched and convex. Anal cell closed in the margin or with very short petiole. There is no basal comb; the costa is densely short pilose at base. Ambient vein complete.

Abdomen: The abdomen is quite broad at the base, as wide or a little wider than the thorax and rather strongly narrowed to the end of the fourth segment. The specimen before me seems to have the abdomen a little distorted apically on the left side, but there seems to be a degree of normal, lateral compression of the terminal abdomen. The fifth, sixth, and seventh tergites are nearly parallel-sided. There are 7 tergites with the seventh being three-fourths as long as the sixth. Pile of the abdomen very fine, moderately long, somewhat bristly in character and quite scanty, except along the lateral margins, where it is a little more abundant and, except on the lateral third of the first and second tergites, where it is rather dense. Last sternite of male notched apically.

Hesse found the dististyli rather triangular seen from above, the outer side much produced, the beak distinct and slender, the points not laterally compressed but distinctly compressed dorsoventrally; the basal part on both known species roughened and having fine hairs. Aedeagus like other Corsomyzini and without a ventral process; lateral and basal struts small.

Material available for study: Of this rare genus I have for study only a male of *Gnumyia brevirostris* Bezzi in the British Museum (Natural History) from which my figures are taken. Two species known; the material studied by Malloch may represent a third species. Only males so far known.

Immature stages: Unknown.

Ecology and behavior of adults: Adult habits not recorded.

Distribution: Ethiopian: *Gnumyia brevirostris* Bezzi, 1921; *fuscipennis* Hesse, 1938.

Genus *Hyperusia* Bezzi

FIGURES 62, 295, 492, 500

Hyperusia Bezzi, Ann. South African Mus., vol. 18, p. 84, 1921a. This is the first mention of this genus. Description and type of genus in: Bombyliidae of the Ethiopian Region, p. 107, 1924. Type of genus: *Hyperusia luteifacies* Bezzi, 1924, by original designation.

Hesse (1938), pp. 765–767, key to 6 South African species.

Small, blackish flies, the face sometimes partly and the antenna also in part orange or reddish in some species. It is separated from *Megapalpus* Macquart by the nonprominent buccal rim of the oragenal cavity; this cavity rim does not conspicuously and clearly protrude forward as it does in the latter genus. The antennae are less widely separated, the first segment distinctly shorter, and the proboscis is shorter and noticeably thicker basally. The male eyes are more narrowly separated, the distance between being of the approximate order of *Corsomyza* Wiedemann, to which *Hyperusia* Bezzi is also related.

This genus does have loosely scattered, numerous, long, coarse hairs on all of the front except a narrow medial portion near the antenna and upon the whole sides of the face and the very wide cheeks. Nevertheless, it is loose and scattered, never hiding the surface of the head. Wings hyaline with very pale veins, 2 submarginal cells, the 4 posterior cells all widely open and with the anterior crossvein entering near the outer fourth of the discal cell. These rather small, dark colored flies are covered with rather abundant, long, fine, erect pile but in addition have on the abdomen some sparse, flat-appressed, shining, slightly flattened hairs. The abdomen is broad at the base, slightly narrowed, more so on the last two segments and about 1½ times as long as wide. Vertex separated from the front by a distinct, rather deep, sharp, transverse crease or fovea.

The body of these flies is rather bare and the pale pile sparse, stiff, and almost spikelike upon the face and front and antennae. As Bezzi (1924) points out, there is no facial brush and the whole face and front are relatively bare; the legs are rather stout and short, without bristles.

The femoral hair, while fine, is more abundant than in *Megapalpus* Macquart even in the female. There are 3 long, apical hairs across the apical margin of the last tarsal segment as in *Corsomyza* Wiedemann. The last male sternite is notched at the middle as in both these related genera. The male terminalia are much like *Corsomyza* Wiedemann; unlike that genus the head in front is not conspicuously broadened, the sides of the face, and the gena not conspicuously tumid and prominent. The antennae are inserted much lower down upon the face.

Other characteristics consist of antennae inserted very close to upper oragenal rim; oragenal rim and cup not protruding as it does in *Megapalpus* Macquart, and the proboscis much shorter than in *Megapalpus* Macquart. While the body of the species is blackish the legs and the antenna tend to be in part pale yellow or red, and the contrast in color is heightened by the often stiff, sometimes stubb-tipped and frequently opaque whitish or yellowish pile. A small metapleural tuft is usually present; last tarsal segments with at least 1 long hair.

Hyperusia Bezzi is a small genus of 6 species from South Africa. Hesse (1938) notes that the type-species may be a true *Corsomyza* Wiedemann of the *simplex* Wiedemann series, and notes further that the other species are abundantly distinct.

Length: 3 to 7 mm., of wing 3 to 5.5 mm.

Head, lateral aspect: The head is comparatively short, somewhat rounded behind, more so in front. The occiput is poorly developed posteriorly, although there is a small, deep, central, cuplike recess leaving the intervening area quite wide and extensive when viewed from the rear and covered with dense, short, erect pile growing much shorter toward the eye margin. The face is short in profile but quite wide, bisected by the oragenal cup which reaches to the antenna but is itself short and inconspicuous. Along each outer margin of the oragenal cup between the face and cheeks is a more or less large, sunken area which is apparently the sole remnant of the tentorial pits for there are no fissures above or below. Posterior eye margin almost entire with the merest suggestion of a middle depression.

The face and front are continuous; both bear moderately dense, long erect, coarse, stiff, pale pile. This pile is somewhat longer on the swollen front but is absent medially on the anterior half of the front. The whole of the vertex is occupied by a rather prominent, exceptionally broad, ocellar tubercle or ridge set off anteriorly by a transverse crease or fovea. The ocelli are well developed, though small; the posterior pair is unusually widely separated. The long, loose stiff pile of the face is continued upon the cheeks and the whole of the lateral genofacial depression. The palpus is elongate, slender, and laterally bears a number of conspicuous, long hairs. The second segment is a little widened. Proboscis slender, attenuate at the apex and extending a short distance beyond the antenna. The antennae are attached at the middle of the head; the first segment is twice as long as the beadlike second segment; third segment elongate, widest at the outer third, then a little narrowed to the blunt apex; it is without terminal style.

Head, anterior aspect: The head is much wider than thorax and considerably wider than high, much of which is doubtless occasioned by the remarkably wide face found in this group genera. The oral opening is about one-third the total face width and is 1½ times as wide as long, strongly narrowed on the upper portion extending to the antennae. The antennae are narrowly separated by perhaps the basal width of the first segment.

Thorax: The thorax is about as long as wide excluding the large scutellum. It is more or less shining, usually blackish and loosely covered with long, fine, erect pile, including the humerus and there is similar long tufted pile on all, except the lower portion of the mesopleuron. There are a few long hairs posteriorly on the pteropleuron, the metapleuron appears to be bare. Bristles absent.

Legs: The femora are slightly thickened through the middle. The anterior femur bears minute, appressed setae anteriorly but coarse, long, rather sparse hairs dorsally and posteriorly. The middle femur bears similar long hairs in front and behind and below but short hairs dorsally. Hind femur with short, appressed hairs dorsally and rather fewer, fine, long hairs ventrolaterally. All of the tibiae with fine, scattered, subappressed pile. The dorsal surface of the hind tibia has a longer, more conspicuous fringe; last tarsal segment with 3 long hairs, spicules absent. Claws fine, pulvilli well developed; the arolium forms a minute stub.

Wings: The wings are hyaline with extremely pale, whitish-yellow veins, the surface minutely villose; there are 2 submarginal cells; the 4 posterior cells are all widely open, the anal cell narrowly open, the alula well developed, the ambient vein complete. The second vein arises at the base of the discal cell and the anterior crossvein enters the discal cell near the outer fourth. The wing is rather short and broad, especially basally. The costa is a little expanded triangularly at the base and has a rounded, cup-shaped, somewhat beaklike lobe at the base.

Abdomen: The abdomen is a little longer than the thorax, wide at the base, slightly narrowed and more strongly narrowed at the apex. The tergites bear scattered, quite long, delicate and fine, erect hairs a little more abundant toward the sides of the first three tergites and forming, however, a conspicuous, loose, marginal fringe along the lateral margins of all of the tergites. On the surface of the tergites, more dense and conspicuous on the posterior half of the abdomen, there are fine, flat-appressed, somewhat curled, glittering, smallish hairs.

Hesse (1938) describes the male terminalia:

... much like *Corsomyza* Wiedemann; basal part of basimeres likewise with a rounded lobe-like part; distimeres (beaked apical segments similar) much compressed laterally, claw-like, the acute apex curved downward. Aedeagus similar, also, without a ventral process ending basally above the middle part in a lobe-like process. Basal struts more or less racket shaped, the dorsal margin deeply and angularly incised toward the apex.

Material studied: The type-species *luteifacies* Bezzi, in the British Museum (Natural History).

Immature stages: Unknown.

Ecology and behavior of adults: Not on record.

Distribution: Ethiopian: *Hyperusia luteifacies* Bezzi, 1924; *minor* Bezzi, 1921; *muscoides* Hesse, 1938; *nivea* Hesse, 1936; *soror* Bezzi, 1921; *transvaalensis* Hesse, 1938.

Genus *Megapalpus* Macquart

FIGURES 67, 288, 485, 505, 742

Megapalpus Macquart, Suite à Buffon, vol. 1, p. 394, 1834. Type of genus: *Phthiria capensis* Wiedemann, 1828, by monotypy.
Dasypalpus Macquart, Diptères exotiques, vol. 2, pt. 1, p. 112, 1840. Type of genus: *Phthiria capensis* Wiedemann, 1828, by monotypy, female, as *Dasypalpus capensis*.

Hesse (1938), p. 761, key to 2 South African species.

Small, shining blackish flies, quite bare in appearance with fine, short, silky, depressed, scanty pile on mesonotum and abdomen of the females and the males with scanty erect hairs, and few or no depressed elements. The abdomen is short, apically rounded, and rather convex. There is one weak bristle on the notopleuron and an odd radiating tuft of long, coarse pile on the upper border of the mesopleuron. Face wide but much less wide than in *Callynthrophora* Schiner and *Gnumyia* Bezzi, and not tumidly swollen as in those genera. Moreover, the face is well developed above the moderately high oragenal cup, in contrast to *Hyperusia* Bezzi where the oragenal cup reaches quite to the antenna. The face in *Megapalpus* lacks the curious, striking circular brush of long, dense hair which is so characteristic of *Corsomyza* Wiedemann. Wings small, broad, and hyaline with heavy dark veins and the posterior cells short and wide and widely open; costal comb absent; 2 submarginal cells present; the anterior crossvein enters the discal cell beyond the middle. Hind femur with microsetae below but all tibiae with spicules. Finally the interocular space on vertex is quite wide in both sexes. More closely related to *Hyperusia* Bezzi but sharply differing in the character of the face and oragenal cup and to some extent in the antennae. The third segment of *Megapalpus* Macquart is regularly and gradually dilated to the apex ending in a lipped cup and concealed spine and in *Hyperusia* Bezzi. It is distinctly obliquely truncate and tapered beginning at the outer fourth. In females of *Hyperusia* Bezzi there is considerably more pile above on the body than in females of *Megapalpus* Macquart.

This small genus is confined to South Africa.

Length: 3.5 to 6 mm.

Head, lateral aspect: The head from the lateral view is, apart from the extensive oragenal cup, more or less short oval. The very large eye is not much higher than wide and posteriorly curves strongly inward so that the short occiput is further reduced by this inward retreat of the eye. The occiput is, then shallow and only slightly depressed where the head joins the thorax. The pile is quite short, scanty, fine, and curved outward toward the eye margin throughout. The face is extensive below the antennae but bare, polished and shining, and appears to be divided by a shallow, transverse furrow on the lower two-thirds; beyond this furrow the face forms an added component to the very deep, large, more or less circular, sharp-walled extensive and protrusive oragenal cup. The pile along the wide flat genae is abundant though loose, long, erect, stiff, pale and extends from the bottom of the eye up to the base of the antennae with similar pile continued, though shorter, on the sides of the front.

Posterior margin of the eye entire but not plane. The proboscis is unusually long and unusually slender with minute, long, slender labellum; it is about 3 times as long as the head. The palpus is unusually long and quite slender. Its length from the bottom of the oral opening as long as the antennae. The front is very wide, as wide or a little wider than either eye, somewhat flattened across the middle and bearing very fine, erect, blackish hairs on the dorsal half mixed in with which are some shorter, flattened, curled, subappressed hairs, and laterally on the lower half loosely scattered, long pale hairs like those on the genae. In addition on the lower part of the front at the eye margin there is in females a curious tuft of very dense, matted, subappressed, long, silvery or yellowish scalelike hairs. Middle of front coarsely appressed, scaliform micropubescence in the middle and more densely between the antennae. Remainder of front and the vertex likewise shining and bare. The long black bristles of the front extend onto the vertex. The antennae are attached near the upper third of the head. Antenna slender and about as long as the head or perhaps a little shorter.

The first segment is rather slender, a little widened apically, at least 3 times as long as the apical width and bears several long, bristly hairs and 1 or 2 bristles dorsally along the middle. It also bears several still longer hairs laterally or ventrolaterally, but there is no conspicuous pile or bristles extending downward from below. The second segment is short and beadlike, widened apically. The third segment has a peculiar and characteristic shape; just beyond the base it begins to be narrowed, and on the outer half it is gradually and slightly widened and swollen until it is clublike. However, from the lateral aspect it is not narrowed but only gradually widened. At the apex it bears a short, dorsal, subapical spur rather than a microsegment, which appears to have a small recess within it. The apex of this segment is notched, and the ventral portion is blunt and extends very slightly beyond the dorsal lobe or spur.

Head, anterior aspect: The broad head is much wider than the thorax. The eyes are widely separated in the male, almost as widely separated as in the female. In fact the inner ocular distance in the male across the plane of the anterior ocellus is as great or greater than the width of either eye. The antennae are widely separated at the base by more than twice the thickness of the first antennal segment. The ocellar tubercle is low and small, the ocelli are likewise small. The oral opening is quite large, deep, with the wall thin, knifelike, and protruding a considerable distance down below the eye margin. The interior recess is flattened or very slightly concave and smooth, without conspicuous ridges and without pile. The gena is moderately well developed, about as wide as the thickness of the third antennal segment and shelving inward to the tentorial fissure only along its medial edge.

Thorax: The thorax is quite shallowly convex, except along the anterior margin, which is very abruptly raised above the short pronotum. The whole thorax is shining black with a little faint pollen on the anterior declivity. The mesonotum bears a few, fine, quite loosely scattered, erect black hairs and in addition some equally sparse, short, curled, appressed, brassy hairs. Scutellum similar, the black bristly hairs longer, especially along the margin and also few in numbers. The sides of the scutellum and the prescutellar declivity are gray pollinose. There are no notopleural bristles. The pleuron bears a very conspicuous tuft of dense, long, coarse, slightly flattened pile. About half of this pile is directed anteriorly and dorsally, the other half posteriorly and downward. The upper sternopleuron has a tuft of loosely scattered, fine, short, erect hairs. Pteropleuron, metapleuron, and hypopleuron entirely without pile, except that the hypopleuron has 2 or 3 hairs beneath the spiracle. The propleuron is almost entirely bare.

Legs: The legs are unusually short. The femora are all short and relatively stout with coarse, fine, suberect setae. Hind femur with a few longer stiff hairs below which are scarcely bristly. Anterior tibia without isolated spicules and with only fine setae, a few of which are suberect. Hind tibia with 5 or 6 slender bristly hairs along the anterolateral margin. Tarsi quite short, pulvilli large, long and wide. Claws gently curved from the base.

Wings: The wings are relatively short and moderately wide with the veins thick. There are 2 posterior cells but the second posterior cell is rather short, the anterior branch of the third vein arises rectangularly, and proceeds forward for some distance, then bends strongly, and the remainder of this vein is straight. Marginal cells slightly widened apically, the second vein only gently curved on the outer part. Anterior crossvein enters the wide discal cell at or beyond the outer third. The anterior intercalary vein is long. The first posterior cell is wider on the wing margin than elsewhere. The anal cell is remarkably widely open, the axillary lobe narrow. Alula linear and almost absent. Basal comb weak or absent. The basal spur or hook small. Ambient vein complete.

Abdomen: The abdomen is short and tapered, very slightly convex over the middle, more so laterally. It is shining black with pile quite similar to that of the mesonotum and scutellum. The erect, stiff, or slightly bristly hairs are few in numbers even along the curled side margin and on the second to the fourth tergites, as far as the lateral margin is concerned, they are restricted to the posterior corners, leaving tufts of short, loose, erect, pale colored pile along the remainder of the lateral margin and this pile does not extend inward along the middles of the tergites, except on the first segment. Middles of remaining tergites with a few brassy, slightly flattened, curled, and appressed hairs. I find 7 tergites in the male, the last nearly as long as the preceding. The terminalia are large, tucked into the right beneath the last several segments.

Dr. A. J. Hesse in his work on South African bee flies describes the male hypopygium as follows:

Hypopygium of males like that of *Corsomyza*, the basal parts with a posterior lobe; the beaked apical joints laterally compressed, acutely pointed apically and with a crest of spine-like, bristly hairs along their dorsal (upper) parts; aedeagus much like that of *Corsomyza*-species, without any ventral aedeagal process.

Material available for study: A male of the type-species given to me through the courtesy of Mr. Harold Oldroyd of the British Museum (Natural History) and another through the kindness of Dr. A. J. Hesse. The series of this genus was also inspected at the British Museum (Natural History).

Immature stages: Unknown.

Ecology and behavior of adults: Not on record.

Distribution: Ethiopian: *Megapalpus capensis* (Wiedemann), 1828 (as *Phthiria*); *fulviceps* Bezzi, 1921; *nitidus*, Macquart, 1840.

Genus *Zyxmyia* Bowden

FIGURES 294, 546, 770

Zyxmyia Bowden, Journ. Ent. Soc. South Africa, Pretoria, IV, vol. 23, p. 213, 1960b. Type of genus: *Zyxmyia megachile* Bowden, 1960, by original designation.

Small, black or blackish-brown flies, very similar in general to *Megapalpus* Macquart. The oragenal cup is extensive and prominent, the proboscis long and slender. The face and vertex are quite wide even in males, presently the only captured sex. The antennae are separated by the length of the first two segments. Bowden notes that the short, globular, basal antennal segment is placed close to the upper part of the oragenal cup, and the denser pubescence and absence of stout, alar bristle ally this form to *Hyperusia* Bezzi. Like *Hyperusia* Bezzi and some species of *Corsomyza* Wiedemann it has an excavated third antennal segment; with *Corsomyza* Wiedemann it has a rapidly divergent eye margin; its front, Bowden states, is at the level of the antenna, at least half again as wide as at the vertex. It differs from all other genera in the structure of the legs, especially the incrassate femora, the thorns on the tibiae and the unique, divided, first abdominal tergite. Bowden felt that it stood closest to *Megapalpus* Macquart, and distinguished from it by the position of the antenna, divergent eye margins, more dense pubescence especially upon the pleuron, and by the absence of a bristle at the base of the wing.

This fly is known only from males from Tanganyika at elevation of 4,850 feet.

Length: 6.8 to 8.3 mm.; of wing 6.7 to 8.0 mm.

I have not seen these flies and I quote the author's description:

Allied to *Corsomyza* Wiedemann, *Hyperusia* Bezzi and particularly to *Megapalpus* Macquart. Body with fine erect pubescence on thorax and abdomen and with adpressed hairs, at least in male, denser on abdomen; pleurae with comparatively dense hair on mesopleurae and with short fine hair on pteropleurae and sternopleurae; metapleurae bare, no mesopleural tuft. Head with interocular space very broad in male, frons ridge-like either side of ocellar tubercle, very broad, eye margins divergent, frons notably wrinkled especially near antennae; facial area more or less parallel-sided with sparse, fairly long hair but no facial brush, genae comparatively broad, buccal rim prominently produced; antennae inserted moderately far apart (about ¾ distance between posterior ocelli of genotype) just above buccal rim, first two segments short, subequal, more or less globular, second without or with only very small dorsal process, third segment long relative to first two, broad and strap-like, excavate on inner surface, apex appearing somewhat bifid by reason of a subapical, dorsal pit which contains a minute but stout style; eyes comparatively large, facets equal, at least in *Zyxmyia megachile*; proboscis very long and stout, palps apparently with one segment, long and slender, club-shaped at apices.

Thorax with a small hair tuft on mesopleurae, without any macrochaetae; scutellum clothed as thorax, without any longer, bristly hairs. Abdomen with first tergite split medially, second prominently ridged, other tergites with median transverse depression; abdomen at least as broad at tergite two as thorax, tergites VII and VIII reflexed beneath abdomen, thus only six tergites visible from above; last sternite with small, median, apical notch. Legs with femore markedly incrassate, tibiae and tarsi also stout; without spines or spicules except for some antero-ventrally on hind tibiae; with only very short pubescence; fore and mid tibiae with a postero-dorsal row of very short, stout and heavily sclerotised thorns and similar, but somewhat more slender thorns ventrally on basal tarsal segments of mid and hind legs; last tarsal segments without apical long hairs; claws very stout, widely divergent, strongly bent, pulvilli large, empodium strongly developed, as long as pulvilli. Wings about three times as long as broad, long in relation to body length; with two submarginal cells; discal cell notably broadened in middle, anal cell closed before wing margin and shortly stalked; alula reduced but lobe-like; costa slightly broadened at the base but basal comb absent. Hypopygium of *Corsomyza* type: basal parts rounded, shell-like, with a posterior lobe, beaked apical joints compressed, pointed, without bristly hairs; aedeagus without ventral process, lateral rami asymmetric.

Material available for study: Not seen. Types in the British Museum (Natural History).

Immature stages: Bowden (1960) speculates that it may parasitize species of *Megachile* Latreille.

Ecology and behavior of adults:

The area in which it was captured, northeast of Lolkissale in Northern Tanganyika, contains some of the desert and semi-desert vegetation type of Gillman (1949), the flora of which consists in part of succulents; it would thus appear that *Z. megachile* favors a habitat very similar to the Karoo-type habitat frequented by its southern relatives. It bears a great resemblance to the smaller, brownish species of *Megachile*, such as *gratiosa* Grst., *discolor* Sm., and *semivenusta* Ckll., upon which it may well be parasitic, and to which the trivial name makes reference as well as to the prominently produced buccal rim.

Distribution: Ethiopian: *Zyxmyia megachile* Bowden, 1960.

Genus *Mariobezzia* Becker

FIGURES 291, 488, 496

Mariobezzia Becker, Ann. Mus. Zool. Acad. Imp. Sci., St. Petersbourg, vol. 17, p. 470, 1913a. Type of genus: *Mariobezzia zarudnyi* Becker, 1913, second of two species, by designation of Bezzi, 1924.

Efflatoun (1945), pp. 60-61, key to 2 species.

Small to medium-size flies readily distinguished by the 3 submarginal cells, the very bare aspect with considerable pale coloration, and the unique face. The middle third of the face, which is very extensive dorsoventrally, is extended forward as a prominent, protrusive, nose-shaped extension, separated from the exceptionally wide frontogenal region by a deep crease. On this account Efflatoun (1945) likens the face to that of a conopid. The oral recess in consequence is small and is situated quite low upon the head; from it the short, proboscis is extended; this varies in length from less than the length of the head to about equal to the head length. The male eyes are widely separated and the face and front quite wide in females. The front is shining and waxy white to honey yellow or even light brownish. The face is apt to be blackish in the middle at least. The thorax is mostly black in males; sometimes with pale vittae in females of *catherinae* Efflatoun. The medial keel of the face is separated from the genae, which Efflatoun curiously describes as narrow, by deep fissures in which probably lie the tentorial pits or seam. Antenna elongate; third segment more than twice the length of the basal segments, of nearly uniform width, except that it bears a short, dorsal, preapical style or spine. Second segment at apex in females, sometimes in males, with a short, dorsal spinelike process.

Both *Mariobezzia* Becker and *Corsomyza* Wiedemann have expanded gena but there is no conspicuous facial brush in *Mariobezzia* Becker, its face being almost nude, shining, bare, and polished. The real relationship appears to be with the Corsomyzini; though it is unique it is no more so than other genera within the Bombyliinae, and I can see no reason for creating a separate subfamily for it; I therefore reduce its status to that of a tribe, within that subfamily. Indeed, since both *Mariobezzia* Becker and *Corsomyza* Wiedemann and its allies share in most species and sexes a distinct, thornlike projection anteriorly from the top of the second antennal segment, it would appear that they are closely related. Probably both should be assigned to a single tribe, the name of which would have to be Mariobezzini. Efflatoun (1945) has given a very complete description of both *lichtwardti* Becker and *catherinae* Efflatoun from large series. Among the female variants in his mated series were individuals answering to the description of *zarudnyi* Becker, *pellucida* Paramonov, and *griseohirta* Nurse, all of which he places in synonymy. Although Efflatoun placed *zarudnyi* Becker, the type-species, as a synonym of *lichtwardti* Becker, it appears that the reverse would be true. He listed *lichtwardti* as type-species.

Mariobezzia Becker is a Palaearctic genus of a few, highly variable species, according to Efflatoun, with numerous synonyms. Efflatoun (1945) notes that the type-species has been taken sparsely in Persia, Asia Minor, Egypt, the Sudan, and South India but is common only in Egypt.

Length: 2.3 to 10.8 m., very variable; wings expanded 4.4 to 12.8 mm., with marked differences in color and size in sexes.

Head, lateral aspect: Face rather strongly produced as an exceptionally wide, high, arched, convex ridge, which occupies not quite half the total width between the eyes and which develops a short distance below the antenna and is demarcated by a deep, vertical crease extending all the way to the lower part of the face. This lateral portion, which may be regarded as the expanded gena, ranges from waxy white to honey yellow and it is shining and gently convex. The middle portion may be yellowish brown and shining, with a diffuse, medial, brown stripe. The sides of the face, as well as the middle, have dense, rather long, fine pale pile that is continued onto the anterior half of the front. The oral opening is small, nearly horizontal, 2½ times as long as wide, the opening extends back to the posterior margin of the occiput. The proboscis is short but with the terminal labella large and convex ventrally, plane dorsally, and apically rounded, with lateral and ventral bristly hairs. The palpi are very small, short, cylindrical, composed of one segment. The eye is strongly narrowed below, and posteriorly it is narrowly and gently recessive at a point near the upper third. The occiput is short everywhere and extremely short above the dorsal third, with a broad deep vertical concavity extending from the bottom to the vertex, the latter without medial fissure, and with a crease extending to the corner of each eye.

Head, anterior aspect: The front is rather abruptly narrowed, shining white, or with a diffuse, blackish spot, a little inflated. The ocellar tubercle is rather strongly swollen above the eye, the ocelli reaching nearly to the eye margin and are situated in an obtuse triangle with long erect dorsal pile between the ocelli which is rather abundant. Antennae attached at the upper fifth of the head, rather long and slender, the first two segments are quite short, of equal length and bearing fine terminal bristly hairs. Third segment slender but of nearly uniform width to the middle, then gently tapered to the apex which is sharply pointed with a short, pointed microsegment. The whole dorsal surface is plane. Dorsally at the apex of the second antennal segment, females always and males almost always have a distinct, forward-thrust spine or process.

Thorax: Densely, long, fine, pale pilose, the shining black ground color not obscured. Females of *catherinae* Efflatoun have a whitish to citron yellow thorax with 3 black vittae. Bristles are absent. The scutellum is large, exceptionally thick and convex especially along the margin and there is a deep basal crease and the whole surface and margin have numerous, fine, long, erect hairs. The mesonotal pile is fine and erect, white in some species, darker in others. The mesopleuron and pteropleuron bear dense, long coarse, pale pile; remainder of pleuron without pile; squamae with a fine fringe of delicate hair.

Legs: The legs are rather short, the femur moderately stout without being swollen, the tibiae slender, the hind femur with a ventral fringe of long coarse hair, its tibia with a similar fringe especially laterally. Middle tibia with anterior and posterior dorsal fringe of 10 or more slender pale bristly hairs; these hairs are much shorter, finer, and present only posteriorly on the anterior tibia. Claws very slender, especially at the base, sharp, the pulvilli well developed and with a short, triangular, medial lappet.

Wings: The wings are hyaline, broad, especially so on the basal half. The anal cell is closed and bears a short stalk. The branches from second and third veins which meet may join into a single vertical vein sectioning the first submarginal cell so that there are 3, or the posterior branch of the second vein may be emitted quite obliquely. Axillary lobe quite broad, the alula of medium width. The venation is very similar to *Corsomyza* Wiedemann.

Abdomen: Shining blackish with shining yellow posterior margins to the first three or four tergites of males. The abdomen in females is rather strongly cylindroid, quite convex and gently tapered to the end of the fourth tergite, a little more strongly beyond. The color in females is highly variable but often largely whitish yellow or yellow with black, subbasal bands only on the basal tergites. There are 7 tergites with the seventh two-thirds as long as the sixth; the pile is moderately abundant, unusually fine and long and pale and erect.

Material available for study: The series in the British Museum (Natural History).

Immature stages: Unknown.

Ecology and behavior of adults: Efflatoun (1945) relates that his assistant captured more than 70 males and 100 females of this strange fly in the middle of June in two wadies in Egypt. Because many were taken in copula, Efflatoun was able to establish important synonymy. All specimens were captured on the wing, over a five-day period, their flight resembling omphralids, at a height of about one meter. Always they were found at 1 to 3 P.M., the hottest part of the day.

Distribution: Palaearctic: *Mariobezzia catherinae* Efflatoun, 1945; *ebneri* Becker, 1923; *griseohirta* Nurse, 1922, *griseohirta aegyptiaca* Engel, 1933; *lichtwardti* Becker, 1913; *pellucida* Paramonov, 1929; *zarudnyi* Becker, 1913.

Genus *Acrophthalmyda* Bigot

Figures 18, 251, 437, 450, 797, 798, 799

Scinax Loew, Neue Beiträge II, p. 42, 1855. Type of genus: *Bombylius sphenoptera* Loew, 1855, by original designation. Preoccupied by Wagl., in Reptiles, 1830.

Acrophthalmyda Bigot, Ann. Soc. Ent. France, ser. 3, vol. 6, pp. 573, 583, 1858. Type of genus: *Bombylius sphenoptera* Loew, 1855, as *Cyllenia elegantula* Bigot, 1857, by monotypy.

Ostentator Jaennicke, Abhandl. Senckenberg. Naturg. Ges., vol. 6, p. 348, 1867a. Type of genus: *Bombylius sphenoptera* Loew, by monotypy, as *Ostentator punctipennis* Jaennicke.

A large, peculiar fly of the general aspect of *Bombylius* Linné, but easily distinguished and recognized by the slender wing, margined and spotted with reddish sepia, on which the first posterior cell is open but narrowed and the anterior crossvein lies at the outer third to outer fifth of the discal cell. The alula is absent. Moreover, the completely opaque, reddish-brown thorax and abdomen are relatively bare; the mesonotum is covered with sparse, fine, erect hair, and 5 stout, reddish notopleural bristles, and the abdomen is densely covered with a mat of fine, short, curled, appressed hairs in which a few, long, slender, erect, bristly hairs are intermixed. The base of the second tergite, the sides of the abdomen narrowly, and a narrow, middle stripe down the abdomen have pile of a pale brassy yellow in contrast to the remaining deep brown red pile; sides of the mesonotum likewise contrasting in color. Antennae slender and elongate; oragenal cup prominent, obliquely truncate, its walls thickened, the proboscis elongate, palpus short. Hind femur with stout bristles below; all the tibiae with spiculate bristles.

Edwards (1930) speculated as to the relationship of this fly feeling that it must somehow be related to *Marmosoma* White. It is clearly a member of the Bombyliinae; there are other genera in this subfamily with reduced alula, and far distal anterior crossvein. The spotting of veins and crossveins can be very deceptive. *Marmosoma* White does properly belong in the Toxophorinae; *Tillyardomyia* Tonnoir belongs next to *Eclimus* Loew.

An important characteristic of *Acrophthalmyda* Bigot lies in the very large, unconcealed, terminal genitalia. I believe it is necessary to place these flies in a separate tribe, the Acrophthalmydini.

While I have tentatively accepted the synonymy of *sphenoptera* Loew given by Edwards, I believe that other closely related species may be included within it. Type comparisons will be necessary to resolve this question.

After completing the following description of the strongly reddish type-species with its reddish pile, I have received two other smaller species from Chile, which are gray instead of reddish; in these the anterior crossvein is placed at the outer third of the discal cell; in other respects they are very like the type-species.

These flies are from Chile.

Length: 15 mm. including antennae; wing 14 mm.; wing spread 33 mm.

Head, lateral aspect: The head is elongate due to the strongly produced face and the extensively produced oragenal cup. The occiput is shallow, and not prominent, being only a little convex behind and sloping gradually to the eye margins which are produced outwardly much beyond the occiput. Pile of the pollinose occiput quite loose and scattered and fine, erect, of moderate length; shorter marginally. The face extends forward more than the length of the first antennal segment; it is pollinose and has numerous, but not dense, very fine, rather long, erect hairs. The eye is large, the posterior margin gently rounded, but entire, and it is longer on the upper half. Proboscis quite elongate and slender, 2 to 2½ times as long as the head, but with a short slender palpus, which appears to have a very short, fused facial segment. Front more or less flattened, with a shallow, medial depression running from the ocelli to the anterior one-fourth of the front; its pile is scanty, fine, erect, and moderately long. The antennae are attached very near the top of the head at the upper eighth, they are relatively long and slender, the first segment is more than twice as long as wide, slightly wider than the second segment, which is about two-thirds as long as the first; both of these segments have fine, microscopic, dense, fuzz or microscopic pubescence upon them and each bears some dorsal, lateral, and ventral, long, oblique, bristly hairs, especially the first segment where the hairs are much longer and extend to the base of the segment as well. Third antennal segment longer than the first two combined, slender, slightly narrow at the base, more strongly narrowed on the apical sixth. It bears at the apex a thick, bicolor spine.

Head, anterior aspect: The head is distinctly narrower than the thorax and is not quite twice as wide as high. The head beneath the proboscis between the eyes is somewhat inflated. There are 3 or 4 furrowlike grooves running obliquely from just before the antennae to just behind the antennae. Whole front and face pollinose. The antennae are very narrowly separated, the face is moderately wider than the front of the female. The ocellar tubercle is prominent, the ocellar triangle equilateral, its pile consists of a few erect, long, coarse hairs. The oral opening is large, and relatively deep posteriorly; the lateral walls of the oragenal cup are strongly widened and inflated with fine, oblique grooves. The genae are quite narrow, the tentorial fissure deep and narrow. Sides of the face with a number of loose, scattered, rather long, fine hairs extending forward beyond the oragenal cup but absent immediately beneath the antennae. Viewed from above the head is shallowly bilobate.

Thorax: The thorax is short, slightly longer than wide, distinctly narrowed anteriorly, and very little convex. It is opaque black or sepia widely over the middle including all of the scutellum, except its lower borders, and is light pinkish brown along all of the lateral margin including the humerus and postalar callosity, and the whole of the pleuron is light reddish brown. There are 4 or 5 slender bristles on the notopleuron, several weaker bristles on the postalar callosity and the convex posterior margin of the thick scutellum has numerous, long, fine, bristly hairs. Moreover, the scutellum and mesonotum have abundant, very short, minute, curled, appressed hairs lying on the surface, copper colored on the dark areas, yellowish laterally. Dorsum of the mesonotum with fine, scattered, rather short, erect pile of a reddish color. Pile of the mesopleuron divided into a tuft directed forward along the upper margin and a tuft appressed backward on the posterior margin. In addition there is flattened, scaly pile on the sternopleuron and a dense tuft of scaly pile behind the spiracle. Squama with a wide fringe of short, jetblack, bristly hairs.

Legs: The legs are rather stout and long, the femora distinctly thickened. Anterior tibia with 7 spicules posteriorly, middle tibia with 4 rows, hind tibia likewise with 4 rows of spicules somewhat larger and stouter, each row with 5 or 6 bristles included. Middle femur with 1 bristle below, hind femur with 10 rather stout, ventrolateral bristles. Claws very slender, pulvilli long and spatulate.

Wings: The wings are long and unusually slender. Anterior half of the wing reddish sepia with the costal cell more yellowish brown, the demarcation behind irregular and forming spots at the forking of most veins and likewise just before the apex of the second vein, and anterior branch of the third vein. First posterior cell opened for about one-third its maximal width; medial crossvein long and oblique. Anal cell quite widely open. The anterior crossvein lies at the outer fifth of the discal cell. Axillary lobe narrow, alula absent. The area at the base of the wing and the whole of the basal cell reddish. Costal comb absent. Two submarginal cells only.

Abdomen: The abdomen is rather short, shorter than the thorax, the base is as wide as the thorax to the end of the second segment. Beyond this segment the abdomen is more or less triangular and sharply narrowed. The abdomen is opaque blackish, except the first segment, the lateral margins of all the tergites, the whole of the seventh tergite and a small confluent series of spots in the middles of other tergites, which areas are reddish brown. The pile is comparatively dense, short, curled, scaliform, and glittering. This pile is coppery red on the dark areas, brassy on the posterior border of the first segment, anterior half of the second segment, down the medial row of spots, on all the lateral margins of the tergites and also includes a single row of hairs along the posterior border of the tergites. In addition, there is a considerable number of scattered, fine, erect, bristly hairs over the surface of the tergites and a few along the lateral margins. Anterior half of the first segment with a curious, dense brush of short, erect, pale yellow pile. Last sternite of the female triangular with a tuft of black pile at the apex. The male hypopygium is large and bulbous, attached in such a way as to extend directly backward; not recessed or turned downward.

The male terminalia from dorsal aspect, epandrium removed, show the figure with a very broad base and epiphallus surrounded by a wrapperlike fold on each side, and not meeting medially. The dististylus is quite stout, bean-shaped, with a short, blunt, divergent point at the apex. From the lateral aspect the epiphallus extends as a lobe dorsally above the ejaculatory process.

Material available for study: I have a female of the type-species, and have seen other examples in the collections of the British Museum (Natural History). I also have two other undescribed species lent to me by the California Academy of Sciences. These are predominantly gray pollinose.

Immature stages: Unknown.

Ecology and behavior of adults: Not on record.

Distribution: Neotropical: *Acrophthalmyda sphenoptera* (Loew), 1855 (as *Scinax*) [=*decorata* (Rondani), 1863 (as *Bombylisoma*), =*elegantula* (Bigot), 1857 (as *Cyllenia*), =*paulseni* (Philippi), 1865 (as *Bombylius*), =*punctipennis* (Jaennicke), 1867 (as *Ostentator*)].

Genus *Paratoxophora* Engel

Figures 12, 268, 460, 803, 804

Paratoxophora Engel, Occas. Papers, Rhodesia Mus., no. 5, p. 39, 1936. Type of genus: *Paratoxophora cuthbertsoni* Engel, 1936, by monotypy.

Rather small, peculiar, slender flies, which belong to the Eclimini. The occiput at vertex shows the beginnings of bilobism; the occiput laterally on the upper half of the head is more extensive but is rounded off gradually so that the head, while comparatively short, is subglobular to some extent; the eye extends farther backward below. The males are extensively holoptic with upper anterior facets enlarged and the vertical triangle and ocellar tubercle greatly reduced in size. As in *Eclimus* Loew, the antennae are slender and quite elongate, especially the first segment, which, moreover, is shaggily long pilose, particularly below. Proboscis 3 times as long as the head; face, except below the antennae, with an encircling fringe of long, silvery pile extending down along the narrow gena; sides of oragenal cup thickened above. The femora are somewhat stout, the hind pair with a few bristles below and fine, appressed scales. The tergites have a scanty row of long bristles along the postmargins; males have the hypopygium developed into 2 long, extensive, protruding, dorsal and ventral pieces, separated like the space between pincers; lower process bears the beaked dististyli. Wings hyaline, all cells open; the wing slender, the alula absent, the anterior crossvein a little beyond the middle of the discal cell; 2 submarginal cells. The general picture of this fly may be completed with the fact that it is dull black, with thin pollen or bloom, except on the shining abdomen; the latter has tufts of opaque white, appressed pile on the sides in the corners of the tergites; first segment with band of erect, golden pile.

Only a single South African species is known; it bears no relation to *Toxophora* Meigen.

Length: 6 mm. to 10 mm.; wing 5 mm. to 8 mm.

Head, lateral aspect: Head hemispherical, the occiput moderately prominent, although sloping gradually out to the eye margin, the face projects forward and is bluntly triangular. It extends below the antennae, a length 2 or 3 times that of the second antennal segment. The pile of the face forms a characteristic fringe in 1 or 2 rows on each side of the face next to the eye margin, which consists of long, almost adjacent, fine, shining, white hairs, continued all the way down the gena to the bottom of the head and it is even longer along the gena. Middle of the face without pile. Whole face pollinose. The pile of the occiput is peculiar; on the dorsal half of the head it consists of a loose fringe of some 35 or 40 very long, bristly hairs curved backward. This many are found on each side and it is confined to the upper, outer portion of the upper occiput. Below on the occiput there are rather dense, shining white hairs similar to but shorter than those on the gena. Also, there is a quite wide band of dense, white pollen on the outer part of the lower occiput. Upper occiput shining and metallic. The eye is more or less hemispherical. There is a small cluster of facets dorsally and medially which is much larger than the remaining ones. Posterior margin of the eye entire. Proboscis long and slender, a little more than twice as long as the head, the labellum flat and oval, the palpus small and slender and short. The front in the males is so small it is almost nonexistent. It has a little pollen.

Antennae attached at the upper third of the head, elongate and slender and longer than the head. The first segment is extraordinarily long, at least 6 times as long as wide, the second segment is small, about as long as wide, the third segment is nearly the same width as the second segment, on a little more than the basal half, and then from lateral aspect becomes strongly narrowed. Viewed dorsally this segment is narrowest just before the middle. At the apex it bears a minute spine. The first segment is conspicuous for its long, oblique, shaggy and bristly pile on the whole outer half below, extending forward to the middle of the third segment. Dorsally on the basal half, and also ventrally, it has some short, fine, coarse, or bristly pile.

Head, anterior aspect: The head is a little wider below than above, dominated by the extensive eyes, the enlarged facets show from in front. The head is distinctly wider than the thorax, the eyes meet in the middle from close to the antennae to the small, scarcely raised ocellar tubercle. This tubercle is an equilateral triangle with small ocelli but with a tuft of 4 or 5 long bristly hairs curled forward. The antennae are nearly adjacent at the base, the oral opening is small and triangular and deep behind; its anterior walls are covered with white micropubescence and are curiously inflated and widened anteriorly and narrowing posteriorly. There is a medial fissure on the face beneath the antennae. The genae are narrow and set off by a shallow fissure. From above the postvertex is shallowly bilobate with a shallow fissure.

Thorax: The thorax is elongate, rather compressed laterally, gently arched, the scutellum large and triangular, the mesonotum more or less shining black with bluish reflections and a thin pollen. It is covered with loose scattered, moderately long, erect, shining hairs. Notopleuron with 2 black bristles. Scutellar disc with similar hairs and the margin with a few longer, slender black hairs. Pleuron entirely gray pollinose, the mesopleuron dorsally and posteriorly with much long, fine, tufted white pile. Hypopleuron, pteropleuron, and metapleuron bare. Squamal fringe fine and loose and not very dense. Prothorax no more extensive than in most bombyliids.

Legs: The femora are relatively stout; tibiae and tarsi more or less short but more slender. Anterior tibia with very fine, short spicules, 3 or 4 in each of 3 rows. Middle tibia with 5 or 6 longer stouter spicules in 4 rows. Hind tibia with fewer spicules in 4 rows but they are stouter and longer, most rows with not more than 3 spicules. Hind femur with 3 rather long, stout bristles ventrolaterally, and with dense, minute, short, flat-appressed, iridescent red and green scales, which give way to white scales on this femur as well as the first 4 femora. Base of all femora narrowed slightly. Claws long, sharp; the pulvilli long but quite slender.

Wings: The wings are hyaline, comparatively large, quite slender, especially toward the base. There are 2 submarginal cells. All the cells widely open including the anal cell. The anterior crossvein lies just before the outer fourth of the discal cell. Axillary lobe narrow, alula absent, ambient vein complete.

Abdomen: The abdomen elongate and slender, somewhat compressed laterally, the first two segments are as wide as the thorax and the abdomen is slightly narrowed beyond, more narrowed on the last three segments, and these segments are more strongly compressed. First segment opaque, greenish pollinose with a dense and complete band of rather long, erect, coarse, brassy yellow pile, that changes to long white hairs laterally. Other tergites black, feebly shining with two kinds of pile. Along the posterior margins laterally there are bands or tufts of appressed opaque whitish, somewhat flattened pile, rather matted. Then each tergite has across the middle on the posterior margin less developed on the apical tergites, a fringe of long, slender, black bristles and bristly hairs. They are especially prominent on the fifth to seventh tergites. I find in the male 8 tergites, the last two subequal in length. The hypopygium consists of the ninth tergite and sternite, and it forms a curious, elongate, prominent, laterally compressed beaklike process which has upper and lower valves with a wide gap between. The dorsal component has a fringe of scattered bristly hairs ventrally, the ventral component is divided in the middle. It appears to bear the deep apical segment on each side apically, and has the aedeagus in the middle. Last sternite with considerable fringe of black bristly hairs apically.

The male terminalia from dorsal aspect, epandrium removed, show a very complex structure. The basistylus shows extensive, dorsal membranous areas, which however, are connected and bridged in the middle. The epiphallus and ejaculatory ducts actually arise basally, are turned backward behind the epandrium, and then turn downward as an extremely long, slender, ventral tube. The dististyli are rather curious, long, slender, with odd commalike lobes that turn out apically. It is bilobed, and in the middle visible both dorsally and laterally in what appears as a long tonguelike extension of the basistylus, where one normally finds the epiphallus. This curious condition of an extremely long, slender, coiled, and curved ejaculatory tubelike process has been duplicated and certainly evolved independently in *Amictus* Wiedemann. In *Amictus* the ejaculatory process is also produced from the back part of the hypopygium and turns forward; it runs between the basistyli. The dististyli are larger, with an inward, curved hook at the apex and from the lateral aspect there is a well-developed stout, upwardly curved spout-like epiphallus. The epandrium is quite different from *Paratoxophora* Engel. It has a prominent dorsobasal lobe extended backward, and in front a posteroventral lobe. Cerci arise dorsally at the top of the structure.

Material available for study: A pair of the type-species received through the kindness of Dr. A. J. Hesse of the South African Museum, Capetown.

Immature stages: Unknown.

Ecology and behavior of adults: I have seen no comment upon the adults.

Distribution: Ethiopian: *Paratoxophora cuthbertsoni* Engel, 1936.

Genus *Eclimus* Loew

Figures 15, 258, 454, 572

Eclimus Loew, Stettiner Ent. Zeitung, vol. 5, p. 154, 1844. Type of genus: *Eclimus perspicillaris* Loew, 1844. Designated by Coquillett, p. 536, 1910, the first of 2 species.
Eclimmus Verrall in Scudder, Nomenclator Zoologicus, p. 118, 1882.

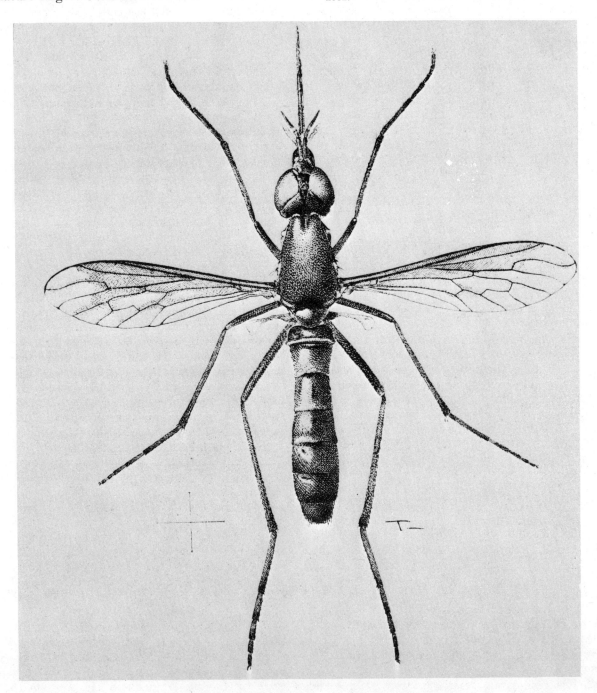

Text-Figure 23.—*Eclimus gracilis* Loew, after Austen.

Engel (1933), p. 55, key to several Palaearctic species.
Maughan (1935), p. 68, key to 3 Utah species.
Hull (1965), pp. 95-97, new western species.

Small, slender flies of blackish coloration characterized by the fact that the wing is very slender and much narrowed at the base. The alula is absent, the axillary cell is very narrow and long, the anal cell very widely opened even more so than the width of the axillary lobe. The wing is dark brown with large but diffused windowlike spots in the 4 posterior cells, the discal, and the apex of the second and third basal cells. The legs are relatively long and slender, the abdomen hemicylindrical, and both thorax and abdomen covered only with microscopic, relatively sparse, appressed setae, each arising from a subtuberculate base, which gives the abdomen and thorax a rugose appearance. Also, the head is quite elongate, the eyes are carried forward, the oragenal area is very extensive, slanted forward with long slender proboscis and long filiform labellum, and a conspicuous 2-segmented palpus which is almost as long as the antennae. The third antennal segment is blunt at the apex, from above the head is semibilobate but there is no cavity. They are almost mosquito-like in appearance except for the spotted wings.

Austen (1937) gives a full complete figure of *Eclimus gracilis* Loew.

These flies are known only from two species found in southern Europe.

Length: 7 mm. excluding the proboscis; 9 mm. including the proboscis. Wing: 5 mm. in length.

Head, lateral aspect: The head is large, especially the eye. The extensive face and oragenal area are slanted forward and almost conical leaving also a very extensive gular area beneath the eye; this wide, shining gular area bears a number of extremely fine, long, dark, downturned hairs, which are therefore erect. The face and the extensive oragenal comb or cup are polished and shining and bare. The gena proper is absent as the eye margin extends quite to the oral cup. The occiput has a clump of rather stiff, blackish, erect bristly hairs on each side situated rather low and opposite the humerus, giving way to more slender pile below. From above the eye is rather strongly and diagonally excavated leaving the dorsal eye profile triangular. The front in the male has a transverse depression nearly midway between the antennae and ocelli. The whole upper part of the front is flat and granulated.

Head, anterior aspect: Ocellar tubercle large but low; the proboscis is 1½ times as long as the maximum head length. The second segment of the palpus is flattened, leaflike, large, conspicuous, twice as wide at the base as at the apex. Second and third antennal segments slightly longer than the slender first segment. First segment rather widely separated at the base. The third segment is blunt apically. The apical part of this segment is scarcely wider than the base.

Thorax: Thorax much more narrow than the head, elongate, high, with parallel sides; at most the mesonotum is very gently convex but a little more convex posteriorly. It is blackish in color, strongly shining, with the pleural sclerites rather reddish-brown centrally. Whereas on the mesonotum and scutellum the microsetae are dense, they are relatively quite sparse on the pleuron, and are confined to the upper metapleuron, upper sternopleuron, and the hypopleuron above the posterior coxae. The hypopleuron above the middle coxae, all the anterior sternopleuron and all the pteropleuron are quite bare, polished and shining. The metapleuron appears to be much reduced in size and is bare with a yellowish spot ventrally; however, there is a small bullose area behind the wing and above the haltere, which is microsetate, and which appears to be metapleuron rather than metanotum. Squama extremely short.

Legs: The legs are quite slender and elongate, especially the hind pair. The slender hind femur is at least 1½ times as long as the anterior femur, however, the anterior coxae is unusually long and slender, fully three-fourths as long as the anterior femur itself. The anterior tibia has numerous oblique, short, setae along the anteroventral margin, and some scattered bristly setae posteriorly. The anterior femur has a few quite long, very slender, erect, ventral, stiffened hairs. Middle femur with a few short, fine hairs ventrally. Hind femur, however, with only very sparse minute appressed setae, placed principally along the lateral margin. Hind tibia even a little longer than the femur, very slender with a few, fine, oblique bristles, not more than 5 in any row, and there are 4 rows. Tarsi extremely slender, the claws fine, scarcely curved at the apex, the pulvilli large.

Wings: The wings are comparatively wide and apically rounded, but remarkably slender and narrow on the basal half. The axillary lobe is very long and slender, the alula absent, the ambient vein complete, the wing fringe long, fine, and dense. The general coloration is rather dark brown but there are large, clear, yet diffusely margined windowlike spots in all of the posterior cells, middle of the discal cell, and apex of the second and third basal cell. Anal cell very widely opened. There are 2 submarginal cells. The anterior branch of the third vein rises abruptly in a rounded arch.

Abdomen: With parallel sides, elongate, hemicylindrical, with 7 tergites visible above, and with the second through the fourth of equal length. The abdomen is largely blackish, the first segment and postmargins and lateral margins of the other segments dark reddish-brown. The abdomen is everywhere shining, rather densely microgranulate with appressed blackish microsetae. All of these tergites have a shallow transverse depression, including the first tergite; they are continuous across the surface.

Material available for study: One female of *Eclimus gracilis* Loew which, because the spotting is somewhat different from the illustration given by Engel, I believe is probably a subspecies.

Immature stages: Unknown.

Ecology and behavior of adults: Not on record.

Distribution: Palaearctic: *Eclimus gracilis* Loew, 1844, *gracilis festae* Bezzi, 1924; *perspicillaris* Loew, 1844.

Genus *Thevenemyia* Bigot

FIGURES 16, 19, 248, 249, 422, 563, 795, 796

Thevenemyia Bigot, Bull. Soc. Ent. France, ser. 5, p. clxxiv, 1875a. Type of genus: *Thevenemyia californica* Bigot, by monotypy.
Thevenemya Bigot, 1875, lapsus.
Thevenetimyia Bigot, Ann. Soc. Ent. France, ser. 7, vol. 2, p. 339, 1892.
Epibates Osten Sacken, Bull. U.S. Geol. Survey, vol. 3, p. 268, 1877. Type of subgenus: *Epibates funestus* Osten Sacken, 1877. Designated by Coquillett, p. 538, 1910, the first of 7 species.
Arthroneura Hull, Ent. News, vol. 77, p. 226, 1966. Type of subgenus: *Eclimus tridentatus* Hull, 1966.

Engel (1933), p. 55, key to several Palaearctic species.
Hall (1969), pp. 12-19, key to 36 world species.

These very distinctive flies have a slender form, with elongate antennae and abdomen, and in some species they are much more slender than in others. The thorax is at most only slightly humped or arched, the abdomen is not decumbent, the head is slightly decumbent in some species, and no scales are present. The proboscis also varies very considerably; it may be 6 times as long as the head or scarcely more than equal the length of the head. The face is shining with some pile or occasionally pollinose with considerable pile; the oragenal cup, however, is large, moderately deep below but filled by a sloping expansion on the upper half and this odd, upper, enclosed part bears a number of long, slender hairs. The oragenal cup may be moderately or may be quite extensively produced; the gena is reduced until it is almost absent. The first segment of the elongate antenna is 2 or 3 times the length of the second; the third is longer than the first segment, gradually and slightly attenuate to the obliquely truncate apex. Pile of first segments varying in length but never long. The tibiae are spiculate, the hind femur has a few short bristles or almost none. The general coloration of these flies is black, shining in some species, dull in others. In some species the mesonotum is nodose spinose in males (subgenus *Epibates* Osten Sacken).

Wings hyaline or with the veins strongly tinged or set out in confluent black margins. All cells are open, the anal cell narrowly, the first posterior cell maximally. The anterior branch of the third vein may be strongly sigmoid or sometimes straight, arising especially abruptly at its base. Male costa sometimes spinose. In *Thevenemyia* Bigot the occiput is very slightly bilobate above. I doubt it is closely related to the Cylleninae; I believe it is more probably an aberrant member of the Bombyliinae. In either case I place it in the tribe Eclimini with *Tillyardomyia* Tonnoir.

The American genus *Thevenemyia* Bigot can readily be distinguished from the European *Eclimus* Loew—of which only 2 species are known—by the strongly narrowed base of the wing in *Eclimus* Loew and also by the curious, much extended, slanting extension of the face and oragenal cup. Also by the type of vestiture which is very minute and short.

Thevenemyia Bigot, while represented by 2 species in the southern Palaearctic region, finds its real home in the western half of the Nearctic area. Several species have been described from Australia (not seen by the author), and one from New Zealand.

Length: 7 mm. to 17 mm.; wing 8 mm. to 13 mm. Proboscis sometimes as much as 9 mm.

Head, lateral aspect: Head is subglobular with both face and oragenal cup extensive but sharply truncate or plane and the oragenal cup receding backward from the lateral aspect. In some individuals the face and oragenal cup are two-thirds as long as the eye. The occiput is prominent, more so in females than in males. The face is polished and bare, except sometimes narrowly pollinose across beneath the antennae and in some species wholly pollinose. It may be totally without pile or have a long fringe of fine hairs projecting forward to the anterior edge. Likewise in some species with pollinose face, they may be covered throughout with rather dense, erect, long, fine pile extending down over the whole of the oragenal cup. In most species the oragenal cup is bare, polished and shining. The occiput bears a rather deep-set, medial, central fringe of long, coarse hairs in several rows, leaving only short, minute setae outwardly, or the entire occiput may be covered with long, fine or coarse pile, becoming shorter toward the margin of the eye. Ventral pile beneath the head and occiput long and fine, either loose or rather dense. The eye is more or less hemicircular, extended outward well beyond the posterior margin which is entire. The eye is large, extending to the bottom of the head, the upper facets slightly enlarged and the eye more or less flattened dorsally.

The proboscis in some species is short and stout with long, flattened, oval, grooved labella which is very little longer than the head in length; here the palpus is half as long as the proboscis, and its second segment is short and lanceolate. In some species the proboscis may be 4 or 5 times as long as the head; in these species the palpus is long and slender, including the distal segment, but no more than one-fifth as long as proboscis; palpal pile always short, and fine and erect. The antennae are attached closely adjacent at the upper sixth of the head. The antennae are elongate and slender with the first segment from 3 to 5 times as long as wide; the second segment from 1 to 1½ times as long as wide; the third segment is as wide at the base as the second segment and gradually tapered to the blunt apex, which may have a minute microsegment. The first segment has either sparse, subappressed setate hairs below and shorter ones above or has rather long, shaggy pile both dorsally and ventrally; second segment with very short, fine setae; the third segment near the apex sometimes has 2 or 3 fine hairs about as long as the segment is wide.

Head, anterior aspect: The head is about 1½ times wider than high, or a little more. The eyes are in contact for a distance approximately equal that of the front, or they may be separated as in *Thevenemyia celer* Cole by not more than half the width of the ocellar tubercle. The antennae are separated at the base by about half the width of the first segment, or in some instances more than the width of the first segment. The ocellar tubercle is small, and from the front is more protuberant than it is laterally, since on each side the eye is raised above the vertex. The ocelli are large, lie adjacent to the eye, and form either an equilateral or short isosceles triangle. The ocellar triangle bears a tuft of fine, black, bristly hairs of varying number, sometimes rather long and sometimes short.

The front is small and triangular in males but flattened across the middle, and in females with a large shallow depression; it is either without pile or with a tuft of short, bristly hairs on each side posteriorly toward the vertex. It is with or without pollen, with or without a transverse groove in the male, and with or without a medial groove. The front in the female has much long, fine, more or less erect pile. The oral opening is either short or long oval, quite deep, except in the middle anteriorly which has an odd, polished, somewhat convex, liplike plate in all species. In a few species this plate is pollinose, and even bears pile anteriorly. The walls of the oragenal cup are knife-sharp and thin and the gena is quite narrow and linear; the tentorial fissure is likewise narrow. Viewed from above the occiput is at most slightly bilobate; it slopes from the swollen interior gradually to the eye margin and sometimes it slopes to the level of the eye or beneath it before reaching the margin, leaving a very distinct vertical rim around the upper part of the eye. Also, ventrally there is, in at least some species, an odd, lobelike bulla placed on each side below the foramen.

Thorax: The thorax is elongate, narrow, compressed laterally, convex when viewed from front or rear, and only very gently convex from the side. The prothorax is distinctly lower than the mesonotum and is much shorter than in *Lepidophora* Westwood and bears a few fine hairs dorsally, a few others laterally. The mesonotum is usually either dull or shining black and may be almost opaque. The pile varies from short, bristly, though fine, mostly subappressed, more nearly erect on the posterior half to others, like *Thevenemyia lotus* Williston, which have two types of pile, short, subappressed, somewhat scaliform golden pile along the whole lateral margins and in front of the scutellum and on the scutellum, and with the dorsum containing mostly erect, numerous, very fine hairs, rather long and becoming still longer on the scutellum. The notopleuron has 3 or 4 strong bristles, either black or pale. The scutellum is thick, convex and large. The mesopleuron has much long, appressed pile in some species, but more scanty, fine hairs and bristly pile in other species which leave this area nearly bare; metapleuron in front of haltere bare, below and behind spiracle with tuft of long pile. Pteropleuron and hypopleuron bare. The sternopleuron has considerable pile and there is a dense, abundant patch of pile situated posteriorly and ventrally to the spiracle. The metanotum has a large, conspicuous patch of long, fine, or bristly pile covering the whole surface; halteres quite large; squamal fringe generally scanty, sometimes in only 1 or 2 rows, sometimes forming a dense, wide border.

Legs: The legs are long and comparatively slender, the hind femur a little more stout or thickened; anterior tibia with 5 minute, slender spicules ventrally, other species with an additional row of 6 posterodorsally. Spicules of middle tibia also weak; on the hind tibia they are short and slender and lie in 4 rows, the ventromedial row contains 7, the dorsomedial row contains 9, the anterodorsal row and the anteroventral row contain only 2 or 3. The hind femur has at most 2 or 3 small, sharp, oblique spicules and sometimes with none; the pile is entirely setate and flat-appressed; the anterior 4 femora sometimes have considerable loose, long, fine, erect pile on most of the ventral side. Also in some species like *Thevenemyia lotus* Williston the pile, instead of being setate, consists of dense, slender, flat-appressed, whitish, scaliform hairs. The claws are short, comparatively stout, strongly curved at the apex with long, spatulate pulvilli. In all species the anterior coxae are unusually long.

Wings: The wings range from hyaline, with the costal cell darkened to more commonly with all the veins margined with brown, sometimes widely, so that the whole wing appears to be more or less tinted. The wing is elongate, the anterior and posterior margins for the most part parallel-sided, and in the males of some species the apex of the wing anteriorly is rather strongly arched backward. Males generally have a double row of tubercles, small or large extending along the costa from the base of the basal cell to or close to the end of the marginal cell. There are 2 submarginal cells and 4 posterior cells with the anal cell open. The anterior branch of the third vein arises abruptly and sometimes even at a right angle, leaving the second submarginal cell very strongly arched at the base and the anterior branch vein sigmoid. The wing of *Thevenemyia* Bigot is a little peculiar in the strong tendency for the anterior crossvein to be shortened pulling the third vein down toward the fourth and both second and third vein at this point may be somewhat thickened. The intercalary vein is often rectangular, and the anterior crossvein lies at the middle of the discal cell. The axillary lobe is relatively narrow, the alula varies from narrow to almost absent as in *Thevenemyia lotus* Williston; ambient vein complete.

Abdomen: The abdomen is decidedly elongate, subcylindrical with the sides nearly parallel and sometimes even compressed laterally. The ground color varies from opaque to shining black. The pile varies from species that have shining, curled, appressed, metallic, somewhat scaliform pile to which may be added moderately numerous, scattered, fine, erect hairs which tend to become longer and more conspicuous laterally, to other species of this genus in which the appressed tomentum is lacking. The pile toward the end of the abdomen may in some species become coarse hair or even fine bristles. Some species have conspicuous postmarginal bands of paler pile on the tergites. Sternites are nearly bare or sometimes rather densely pilose. I find 9 tergites in both males and females. In the females the last 2 tergites are short, recessed with a posterior border of short, stout spines. The dististyli of the male terminalia turn upward at apex of abdomen or extend backward.

The male terminalia from dorsal aspect, epandrium removed, show an oval figure with very large, basally stout, convergent dististyli which are pointed at the apex. In the lateral aspect the epandrium is almost exactly like that of *Sparnopolius* Loew. It has a short, comparatively sharp lobe basally both above and below, and the shape is very similar. The epiphallus is similar but considerably shorter. The ejaculatory process ends quite near the middle. There can be very little doubt these flies belong in the subfamily Bombyliinae, where I have erected a separate tribe for them.

Material available for study: I have before me about 10 species from western North America, and 1 from Mississippi.

Immature stages: Jack Hall, University of California at Riverside, California, recalls having obtained adults of one species from pupa emerging at the holes of anobiid beetle larvae from pieces of deadwood found in that area, and more recently he has obtained a definite record of the flies parasitizing a cerambycid beetle (see life histories).

Ecology and behavior of adults: The ecology of the adults of *Thevenemyia* Bigot is interesting. Various authors have noticed the strong tendency on the part of these flies to hover around the bark of dead trees and to some extent around burnt-over trees.

The rare Mississippi *Thevenemyia* sp. I have never associated with flowers; all others I have collected have been western flower-visiting species; one small species *Thevenemyia culiciformis* Hull was abundant in August in woodland on tall composites like iron weed in Patagonia, Arizona; these flies like *Geron* species bobbed up and down above flower heads before settling down, where because of their small size, long legs, etc., and long slender proboscis they appeared much like giant mosquitoes; the same species also visits low-growing desert flowers.

Distribution: Nearctic: *Thevenemyia californica* Bigot, 1875; *celer* Cole, 1919; *culiciformis* (Hull), 1965 (as *Eclimus*); *funesta* (Osten Sacken), 1877 (as *Epibates*); *halli* (Hull), 1965 (as *Eclimus*); *harrisi* (Osten Sacken), 1877 (as *Epibates*); *harrisi* Johnson, 1914; *laniger* Cresson, 1919; *leechi* Hall, 1954; *lotus* Williston, 1893 [=*auratus* Williston, 1893]; *luctifer* (Osten Sacken), 1877 (as *Epibates*) [=*auripilus* (Bigot), 1892 (as *Amictus*), =*lucifer* Aldrich, 1905, lapsus]; *maculipennis* (Hull), 1965 (as *Eclimus*); *magna* (Osten Sacken), 1877 (as *Epibates*) [=*melanopogon* Bigot, 1892]; *marginata* Osten Sacken, 1877; *melanosa* Williston, 1893; *muricata* (Osten Sacken), 1877 (as *Epibates*); *nigra* (Macquart), 1834 (as *Apatomyza*) [=*aegiale* (Walker), 1849 as *Cyllenia*)]; *ostensackenii* (Burgess), 1878 (as *Epibates*); *sodalis* Williston, 1893; *tridentata* (Hull), 1966 (as *Eclimus*), (subg. *Arthroneura*); *yosemite* Cresson, 1919.

Neotropical: *Thevenemyia auripilus* Osten Sacken, 1887; *fascipennis* Williston, 1901; *halli* (Hull), 1965 (as *Eclimus*); *quadrata* Williston, 1901.

Palaearctic: *Thevenemyia hirta* Loew, 1876; *quedenfeldti* (Engel), 1885 (as *Epibates*); *venosa* Bigot, 1892.

Australian: *Thevenemyia furvicostata* Roberts, 1929; *longipalpis* Hardy, 1922; *nigrapicalis* Roberts, 1929.

Hall (1969) described the following new species from the United States: *accedens, affinis, canuta, melanderi, notata, painterorum, phalantha, speciosa*; and from Australia: *australiensis, mimula, tenta*.

Genus *Tillyardomyia* Tonnoir

FIGURES 17, 250, 440

Tillyardomyia Tonnoir, Rec. Canterbury Mus., vol. 3, p. 101, 1927. Type of genus: *Tillyardomyia gracilis* Tonnoir, 1927, by monotypy.

Relatively slender, medium-sized flies that are readily recognized by the subglobular head and its elongate antenna, together with the patterned, slender wing with 3 submarginal cells, reduced anal area, no alula, and with the odd shape of the second and third posterior cells and discal cell. These flies are related to *Eclimus* Loew in the general form of the head and the quite elongate, shaggy-haired first antennal segment; the third segment is less narrowed and is more spear-shaped, and the occiput is more prominent above. These flies bear straggly, scattered, coarse hairs and stout notopleural bristles.

The only species known is from New Zealand.

Length: 10 mm.

Head, lateral aspect: The head is globular, the appearance heightened by the extensive and tumid occiput which is especially protuberant above, bilobate, and creased in the middle behind. The antennae are set quite low at the middle of the head, hence the face is not very high and extends but little beyond the eye margin. The oral opening is large and extensive, becoming wider below and behind. The proboscis is moderately slender, the apex dilated and leaflike, and the whole perhaps 1½ times as long as the head. The palpus is quite long and slender with narrowed, spatulate apex which appears to be a separate segment. There is also some slight indication that there may be an extremely short basal segment, making 3 segments. First segment of antenna elongate, pilose on all sides, the pile is dense, bushy and black, and longer laterally. The face is bare and shining along the epistoma, shining below the antenna. The occiput has dense, coarse, bristly pile that is rather long, especially above, black, and mixed with some slender black bristles; lower occipital pile fine, long, erect, delicate and white.

Head, anterior aspect: The ocellarium is moderately high with a tuft of long, bristly hairs or slender bristles in the middle. The sides of the front bear numerous, quite long, slender, black bristles leaving the medial part bare, the bristles divergent, and the bristles give way below on each side to a tuft of rather dense, shorter, whitish pile.

Thorax: The thorax is gently arched, mesonotum largely bare with scattered, long, coarse hairs anteriorly, especially across the anterior margin. These hairs are black. There is coarse, opaque, mostly white, mostly appressed, rather straggly pile and hairs on humerus, the lateral margin of the mesonotum, the posterior margin of the scutellum, the upper half of the mesopleuron and a dorsal patch on the sternopleuron. Some of the pile on the upper part of the mesopleuron extends outward, and there is a conspicuous tuft of similar, more or less erect, opaque whitish pile on the sides of the metanotum and on the metapleuron, in both places extensive and abundant and also a small patch in the extreme posterior corner of the hypopleuron. The basal half of scutellum has rather numerous, appressed, flattened, or scalelike black hairs and some similar very scanty hairs on the mesonotum. The notopleuron has 3 conspicuous, quite stout, black bristles; lateral postmargin of the scutellum with 3 pairs of slender, long, black, bristly hairs.

Legs: The legs are rather slender, the femora and tibiae with pale scales; anterior tibiae with 3 or 4 moderately stout, spinous bristles posteriorly; bristles of middle tibiae more conspicuous.

Wings: The wing is relatively slender and the axillary area much reduced and narrowed; there is no alula. There are 3 submarginal cells with the first almost quadrate and the apex of the marginal cell greatly widened, its end nearly truncate. The discal cell is greatly narrowed on the distal half due to the strong forward bend or curve of the third posterior cell tending to occlude the discal cell. The anterior crossvein is near the outer fifth of the discal cell. Anteriorly the wing is rather sharply sepia brownish or blackish, the black color filling basal half of the second basal cell obliquely, all of first basal cell, basal half of second submarginal cell, bordering it anteriorly and distally, and filling all of the first submarginal cell, except the middle and posterior margin, and filling the apex of the wing. There are 4 posterior cells and the anal cell rather widely open.

Abdomen: The abdomen is subcylindrical with nearly parallel sides; it is almost bare but with some rather abundant, black, appressed, slender scales or scalelike hairs on the first two tergites and less abundant, similar hairs on the sides of the remaining tergites. Some of the flattened pile along the sides of the abdomen is white. Probably the entire abdomen is covered in fresh specimens. Apex of abdomen obliquely truncate, the height increased at the apex because of a certain amount of lateral compression. The interior of the apex of the abdomen is densely filled with matted, long, fine pile, so that the terminalia cannot be inspected without dissection.

Material studied: A female of the only known species kindly lent me by Dr. R. H. Painter.

Immature stages: Unknown.

Ecology and behavior of adults: Not on record.

Distribution: Australian: *Tillyardomyia gracilis* Tonnoir, 1927.

Genus *Crocidium* Loew

FIGURES 57, 58, 59, 327, 526, 528, 840, 841, 842

Crocidium Loew, Öfvers. Kongl. Vet. Akad. Forhandl., vol. 17, p. 85, 1861. Type of genus: *Crocidium poecilopterum* Loew, 1860, by monotypy.

Paramonov (1931a), p. 182, treatment of Palaearctic species.
Hesse (1938), pp. 788-792, key to 11 South African species.
Efflatoun (1945), pp. 126-127, key to 2 Egyptian species.

Small, slightly humped, blackish flies, characterized by a conically elongate and more or less tapered and laterally compressed abdomen, which is slightly decumbent. Superficially they resemble the genus *Geron* Meigen. These flies, however, are properly related to the Bombyliinae and, together with *Apatomyza* Wiedemann and *Adelogenys* Hesse, form a rather special group. Overlying the dark ground color there is rather abundant, long, erect, delicate white pile. These flies are a characteristic element of South African fauna. The species fall into two groups, one of which has unspotted hyaline or milky tinted wings, much longer proboscis, longer first antennal segment, the third segment more slender, bare and shining front. The second group has more or less opaque, tinted wings with distinct spots or clouds on crossveins, or bands broken into spots. Front and face dull rather than shining; proboscis shorter and stouter. First antennal segment shorter, that is, much less than 3 times as long as the second. In *Crocidium* Loew the 2 submarginal cells and the anterior branch of the third vein end at the wing apex, in contrast to *Gonarthrus* Bezzi. The thorax is a little arched, but this appearance is merely heightened by the somewhat decumbent abdomen. *Apatomyza* Wiedeman is separated by its very long first and long third antennal segments and both this genus and *Adelogenys* Hesse have, in contrast to *Crocidium* Loew, a more arched thorax and a much longer, more compressed and therevid-like abdomen in appearance. Genae in *Crocidium* Loew distinctly developed, in contrast to *Adelogenys* Hesse where it is absent or virtually so.

Crocidium Loew is a South African Group of flies with 2 additional species in Egypt.

Length: 3 to 6 mm.

Head, lateral aspect: The head is rather subquadrate in profile, the front and the moderately extended face are more or less flattened and also in profile there is little or no occiput visible on the upper half of the eye, although it is comparatively tumid on the lower third of the eye. The genal region is wide and sometimes exceptionally broad; in the very wide species apt to be a little swollen and protrusive. The face is moderately developed, angularly produced since the face is flat and makes a right angle with the low, large, thin walled, sometimes shallow, sometimes deep, obliquely truncate, retreating oragenal cup. The middle of the face is sharply and angularly demarcated from the lateral portion of the face. This middle portion, which reaches to the base of the antenna and is shining, polished and bare, might be regarded as a dorsal extension of the oragenal cup. The lateral face has considerable exceptionally fine, long, erect, delicate, silvery white hairs, which continue to extend over the whole of the wide gena and the ventral occiput. The surface of this area is pollinose. The occiput is without central cavity, but viewed from above each lateral half is slightly convex. The postvertex itself is triangular and a little convex and delineated by lateral creases. Palpus quite fine and slender, apparently of 3 segments, the third quite short; 2 segments are certainly present. Proboscis long and slender and 3 or 4 times as long as the head. Antenna attached near the upper third of the head, elongate and slender, the segments all of about equal width. The first segment varies from at least 3 times as long as the rounded, beadlike second segment to barely more than twice as long.

It bears a few fine hairs. The third segment is nearly twice the combined length of the first 2 segments, slightly narrowed near the apex and with a small, spine-tipped microsegment.

Head, anterior aspect: The head is considerably wider than the thorax. In the male the eyes touch, although for a very short distance, or in other species narrowly separated; they are quite broadly rounded medially so that there is quite a narrow, posterior extension of the front; upper facets of male enlarged. Female eyes may be separated by as much as one-third the width of the head. The front, except for this extension, is quite small and triangular, flattened, with pollen but no pile. The ocellar tubercle is low; the ocelli are small and in the male situated on or almost at the eye margin. The front is separated from the face by a faint diagonal crease; it is not flattened and sunken in either male or female as in *Phthiria* Meigen. From in front the head is broadly oval, and because of the exceptional width of the gena the genofacial space is widely divergent below and viewed as an angle with the front it forms at least a right angle. Oral opening with straight sides, slightly widened toward the proboscis.

Thorax: The thorax is gently convex, distinctly longer than wide and with numerous, long, somewhat flattened, more or less erect, silvery white hairs. There are no differentiated bristles, except that possibly some of the longer hairs on the rounded rim of the rather thick and convex scutellum may be regarded as slender bristles. Mesonotum and pleuron pollinose; mesopleuron with a few, long, scattered hairs; metapleuron bare; hypopleuron with a few hairs.

Legs: The legs are weak. There are no ventral spines on the femora; a rather long apical spur is below upon the middle tibia. The spicules upon the tibiae are longer, and better developed and more conspicuous than in *Phthiria* Meigen.

Wings: The wings are broad at the base, hyaline or milky tinted in one group, clouded or spotted. There are 2 submarginal cells, the second is rather narrow for the anterior branch of the third vein is barely convex. There are 4 posterior cells, all of them quite widely open. The anal cell is closed with a rather long stalk; the axillary lobe is quite large, and the second anal vein quite straight. The anterior crossvein enters the discal cell at the middle or in some species at the outer fourth; the upper intercalary vein is about one-third as long as the medial crossvein. The costal cell is expanded basally; base of costa dilated and with prominent setae; ambient vein complete, the alula of no great width.

Abdomen: The abdomen is elongate but only about 1½ times as long as the thorax, but strongly tapered and even, somewhat compressed laterally. The tergites have numerous, exceptionally long, quite fine, erect hairs, generally silvery white; they also have some appressed, comparatively broad scales or scaliform hairs. Sternites likewise with dense, quite long, erect, white pile growing shorter toward the apex of the abdomen. Eight tergites visible dorsally. Female hypopygium with 3 or 4 pairs of spines. Male hypopygium formed of 2 large, adjacent, mammiform or convex, hoodlike plates turned downward from a dorsal position, and in Egyptian species with on each side a stout recurvate spine.

Hesse (1938) says of the male terminalia:

... the hypopygium is without a long lobe-like process at the base of the basimeres, yet with a distinct process present; dististyles (beaked apical segments) either slender, elongate, cylindrical, sparsely pilose, acutely pointed, or flattened and twisted; aedeagus well developed, without ventral process, joined onto the basal parts by a flange or flap-like ramus; lateral struts either strongly developed, elongate and directed outward or much shorter and directed obliquely downward or towards the base; basal strut racquet-shaped.

The male terminalia from dorsal aspect, epandrium removed, show long, nearly straight attenuate, apically pointed dististyli. What appears to be the aedeagus is long and slender, reaches to the base of the dististyle; it overlies a broader structure arising from the ramus, and deeply divided for nearly two-thirds of its length, and emits 2 long, slender, slightly curved processes of the same thickness as the ejaculatory process. This structure corresponds to the lateral divided claspers in *Phthiria* Meigen. The epandrium is rather similar to that of *Phthiria* Meigen and the basal apodeme similar and large. All of this may indicate a relationship between *Crocidium* and the Phthiriinae. There is a slender bridge from the ramus on each side slightly reminiscent of *Phthiria* Meigen. The base of the aedeagus is large and egg-shaped. From lateral aspect, the large basistyle is bowllike, widening basally, with the dististyle conspicuous and slightly curved.

Material available for study: The series in the British Museum (Natural History) and 3 species kindly sent by Dr. Hesse for dissection and study.

Immature stages: Unknown.

Ecology and behavior of adults: Efflatoun (1945) records capturing a considerable series of *aegyptiacum* Bezzi upon flowers of *Calendula* and other bright colored composites.

Distribution: Palaearctic: *Crocidium aegyptiacum* Bezzi, 1925; *nudum* Efflatoun, 1945.

Ethiopian: *Crocidium chrysonotum* Hesse, 1938; *costilabre* Hesse, 1963; *dasypolium* Hesse, 1963; *depressifrons* Hesse, 1938; *dichopticum* Hesse, 1963; *immaculatum* Bezzi, 1922; *karooanum* Hesse, 1938; *lactipenne* Hesse, 1963; *leucostomum* Hesse, 1963; *melanopalis* Hesse, 1938; *microstictum* Hesse, 1963; *namaquense* Hesse, 1963; *nigrifacies* Bezzi, 1921; *nitidilabre* Hesse, 1938; *pachycerum* Hesse, 1963; *phaenochilum* Hesse, 1938; *phaeopteralis* Hesse, 1938; *pocilopterum* Loew, 1860; *pterostictum* Hesse, 1938; *tinctipenne* Hesse, 1963.

Genus *Mallophthiria* Edwards

FIGURES 341, 522, 532

Mallophthiria Edwards, Dipt. Patagonia and South Chile, pt. 5, p. 169, 1930. Type of genus: *Mallophthiria lanata* Edwards, 1930, by original designation.

Small flies with comparatively short, robust abdomen, with hyaline wings which are virtually identical in venation with *Crocidium* Loew. The face is short and rounded. The males are strongly holoptic and like *Crocidium* Loew there is dense, long, fine pile on head, thorax, and abdomen. They are sharply distinguished by the curious, very much reduced fleshy, stout proboscis with its odd labellum, and by the shorter, different type of antenna. The eye is quite different in shape. Pile of occiput, rather dense, fine, quite long and erect. Ground color black; pile of thorax and abdomen and most of that on face is long, fine, erect, white; some black hair on face and antenna.

Edwards (1930) places this genus within the Phthiriinae probably because of the minute subterminal spine at the apex of the third antennal segment. However, I believe that it is nearer to the genus *Crocidium* Loew and for the present, I leave it next to that genus; only a careful dissection of the genitalia of both sexes and knowledge of the female front can reveal its true relationship.

The face is much shorter than in *Crocidium*; the extremely abundant, long, fine pile on the occiput, face, and gena of *Mallophthiria* is distinctive. While there is some long pile upon these areas in *Crocidium* Loew, it is much more sparse and in that genus the slender, third antennal segment, pointed at the apex, bears a minute microsegment with bristle at tip. *Mallophthiria* Edwards has a somewhat swollen third segment beyond the immediate base and bears at the apex, a minute, subterminal style or spine inserted dorsally; in this respect it approaches *Phthiria* Meigen. All three genera have broad wings, especially on the basal half, with well-developed axillary lobe and alula, and with generalized venation; all of the cells are broadly open except the anal cell which is closed with a short stalk, or rarely narrowly open in *Crocidium* Loew. In all these genera there are 4 posterior cells, and the discal cell widens very considerably toward its outer end, the intercalary vein long, the anterior crossvein lying near, and in *Mallophthiria* Edwards just before the outer third of the discal cell.

This genus is monotypic and at present only known from a single male from Concepción, Chile.

Length: 4 mm.; wing 3.8 mm.

Head, lateral aspect: The face is very short, its length is a little greater than the thickness of the third antennal segment; the face is, however, quite wide below with an extensive space between the eyes and the oral opening. The oral opening extends back to the posterior eye margin and is twice as long as wide; the sides of the face and the entire lateral and ventral margins and cheeks have a dense fringe of exceptionally fine, long, erect, shining, white pile, changing to blackish immediately beneath the antennae. The occiput is rather thick especially in the middle and below; it gradually narrows toward the vertex, and the medial portion is widely and shallowly concave. The vertical triangle is quite small, a little swollen above the eye margin, equilateral, or short and obtuse, and it extends a little backward behind the eye margin. The front is quite small, extremely small in the male. The eyes are holoptic for four-fifths of the total distance and slightly flattened across the top, no more than in *Crocidium* Loew. The proboscis is short and robust; the labellum forms a more slender but obtuse, short, truncate, slightly curved extension extending upward and lying a short distance beyond the face and composed of 2 divided lappets, one on each side.

The palpi are extremely small, cylindrical, concealed within the oral margin, and composed of 2 segments, the second shorter than the first and with a few fine hairs. The antennae are attached at the upper fifth of the head; they are comparatively short, the first segment is a little longer than the second, the second segment beadlike and as long as wide. The third segment is a little compressed laterally, widest on the basal half, slightly narrowed beyond the middle with a short ventral extension; the apex above this extension bears a minute bristly spine. The third segment has an annulate base; the first segment has rather numerous, long, stiff or bristly dark hairs ventrally and laterally; both first and second segments have some slightly shorter coarse hairs dorsally; second segment with a few setae ventrally. The occiput is quite thick, especially below, gradually sloping to the vertex on the upper half, and it bears exceptionally abundant long, fine, white pile and along the middle it has a few short, appressed, shining hairs.

Thorax: Densely long, fine, erect, white pilose especially along the anterior half and on the humeri. The pile is still longer and denser on this mesopleuron and there is almost as long, dense pile on the propleuron; pteropleuron and hypopleuron almost or quite apilose; squama with a rather dense, long, fringe of fine pile, opaque and whitish, tufted laterally. The whole of the mesonotum, except perhaps anteriorly, has a few scattered, more or less appressed, short, shining, curled hairs that are continued on the scutellum where there is also considerable long, erect pile. Bristles are absent; scutellum only moderately thick but convex, moderately large and hemicircular.

Legs: The femur is slightly thickened and bears appressed, slightly flattened, shining, white pile; tibia slender, the hind tibia with a dorsal row of very fine, bristly hairs; the hind femur has an anteroventral fringe of long, fine pile, the anterior femur with a more conspicuous posterior fringe of similar pile, over the whole posterior surface. Tarsi have well-developed pulvilli and a short, stiff, bristlelike empodium.

Wings: Very similar to *Crocidium* Loew, especially *melanopolis* Loew, both of which are here illustrated. The first branch of the fourth vein is slightly arched in *Crocidium* Loew, nearly straight in *Mallophthiria* Edwards. The second submarginal cell is slightly longer, the anterior branch of the third vein plane except at base. The shape of the discal cell and position of the anterior crossvein and the shape, size, and length of all remaining posterior cells approximately the same in both. Alula and axillary lobe are equally wide. The ambient vein is complete in *Mallophthiria* Edwards, the anal cell closed and stalked; wing hyaline.

Abdomen: Long and conical, or strongly tapered beginning at the base. The base is not quite as wide as the thorax. There are 7 tergites in the male, the seventh is half as long as the sixth, which is in turn a little shorter than the fifth. The lateral portion of the tergites is widely covered with quite long, moderately thick, shining white, erect pile, which continues from the base to the apex, and is somewhat more scanty dorsally. Posterior margins of the tergites with some short, shining, curled hairs.

Material studied: The unique type-species, from which my description and illustration are taken.

Immature stages: Unknown.

Ecology and behavior of adults: Not on record.

Distribution: Neotropical (Chile): *Mallophthiria lanata* Edwards, 1930.

Genus *Desmatomyia* Williston

Figures 60, 61, 343, 576, 691, 846, 847, 848

Desmatomyia Williston, Kansas Univ. Quarterly, vol. 3, p. 268, 1895. Type of genus: *Desmatomyia anomala* Williston, 1895, by monotypy.

Small, aberrant, shining black flies with peculiarly plump, thick, slightly decumbent abdomen which at first sight oddly resemble a blackfly gnat, or simuliid. The ends of the first and second and sixth and seventh abdominal segments are narrowly margined behind with shining yellow. This fly is relatively bare; the pile consists of fine, sparse, more or less erect hairs. No bristles or spines on femora or tibiae.

The proboscis is short and stumplike, the palpus also shortened, with a small oval second segment. Oral recess so shallow as to be almost wanting; it has a narrow fissure in the middle. The antenna is remarkable. The first segment is short, the second still shorter, beadlike; the third segment is elongate, of nearly uniform thickness with the basal segments; near the end it has a second beadlike segment, and beyond there is still another segment which is about 3 times as long as the preceding; both these terminal subdivisions are nearly as thick as the principal part of the third segment.

Wings quite generalized; they are hyaline with short alula and all cells open. There are 2 submarginal cells; the second submarginal cell is long and narrow at base and slightly flared at the apex; its veins end about an equal distance above and below the apex of the wing.

This small fly is known only from the States of Colorado and Arizona.

Length: 5 mm.

Head, lateral aspect: The head laterally is short, oval, and not much higher than long. The occiput, however, is moderately prominent, most extensive where the head joins the thorax and slopes gradually outward to the bottom of the eye, to the lateral margin and to the vertex. The pile of the occiput consists of quite scanty, loosely scattered, moderately long, erect, bristly hairs curved a little outward. The front is slightly convex and raised above the eye margin and bears a few bristles scattered and curved downward along the sides of the front and in front of the ocelli. Face very short from the lateral aspect, its height also much reduced. The pile consists of a few scattered, minute setae.

The eye is large, the postmargin entire but broadly rounded almost as much behind as in front. The proboscis is a curious, short, robust, stout structure thrust forward from the bottom of the eye and scarcely extended farther than the first antennal segment. It has a large, curious, inflated, setate, fleshy labellum. The palpus is equally curious. It is quite large and the first segment thick, a little more than 3 times as long as wide and with some 10 or 12 quite long, slender bristles ventrally and laterally, a few others dorsally and apically. The short, plump second segment is a little wider, bluntly rounded, oval, with 6 or 7 bristles shorter than those of the first segment. Both segments have micropubescence. The antennae are attached very near the middle of the head. They are large, robust and long, about $1\tfrac{1}{3}$ times as long as the head. The large, stout, first segment is barely longer than wide, shining and bare, except for some short setae dorsally and a clump of rather long bristles laterally and ventrally. The second segment is equally wide, much shorter, less than half as long as the first segment, and has a few setae above and below. The third segment at the base is a little wider than the second segment. It is not extended above the second segment, but is extended below the second segment. This segment is thick, though a little flattened and curved, largely bare, black and shining, and bears at the apex 2 curious microsegments both quite stout and robust. The first is short, shorter than high, the third is about 3 times as long as wide and blunt at the apex. This segment has a somewhat spoon-shaped or cuplike recess situated ventrally and subapically, and bears a spine.

Head, anterior aspect: The head is as wide as the mesonotum. There is a distinct small depression on the medial eye margin opposite the antennae. The ocellar triangle is small and high, the ocelli occupy nearly an equilateral triangle and there are a few erect, bristly setae within this triangle. Middle of the female front with a conspicuous, broad, concave depression running vertically. The oral opening is extraordinarily shallow, almost nonexistent. There is a vertical, curved depression running down each side of what corresponds to the gena. This genal area is very wide and in fact the whole interocular space in front is very wide. The sides of the oral cavity are delimited by a curious, moderately wide, but concave ridge extending from the base of each palpus about halfway up the head toward the antennae. All the intervening area is quite shallow. Tentorial fissures fused. Pile seems to consist of a few microscopic setae.

Thorax: The thorax is comparatively high and compact, the coxae compacted below. Nevertheless, the mesonotum is not high and humped, it is only moderately elevated and is more abruptly convex behind than in front. It is largely bare and has only a few, fine, scattered, erect hairs on the disc. A few longer ones on the notopleuron, a slightly more dense tuft on the humerus and none along the anterior margin. The scutellum has a few scattered, erect, not very long hairs. Mesopleuron has a tuft of a few scattered hairs on the upper half. Pteropleuron and metapleuron bare. I find no pile at all on even the upper and posterior hypopleuron. There are a few setae on the lower edge of the spiracle. Sternopleuron with pollen only. There is a thin, knifelike narrow plate separating the blackish, shining mesonotum from the blackish pleuron below. This plate is pale yellow. Squama large with a rather short fringe. The haltere is quite large with unusually long club, the upper half curled inward longitudinally.

Legs: The legs are short and quite stout, especially the femora, though they are not swollen toward the base. All the femora are shining with numerous strongly appressed, coarse setae of no great length. The tibiae are similarly covered with coarse, subappressed setae and on the hind tibia there are 1 or 2 slightly longer, erect setae dorsally. Tarsi stout and short. Pulvilli long and wide, rounded apically, claws stout, a little curved at the apex.

Wings: There are 2 submarginal cells, the second submarginal cell is rather long and narrow, widening close to the apex. It arises gradually at the base, and is shaped much as in *Phthiria* Meigen. There are 4 posterior cells which are all long and open maximally. The anterior crossvein enters the discal cell just before the middle. The third vein bends downward at its origin below the second vein. The discal cell is relatively short and wide, strongly angulate below where it is widest. The anterior intercalary vein is one-third as long as the medial crossvein. Axillary lobe broad. Alula 2 or 3 times as long as wide. Ambient vein complete. There is no basal comb.

Abdomen: The abdomen is stout and plump, at the base a little wider than the mesonotum, widest at the end of the third segment and the remainder of the abdomen triangular with the segments rapidly narrowing. There are 7 tergites and a like number of sternites visible. The abdomen is polished, shining, brownish black with the posterior margins of the tergites and the pleuron between tergites and sternites all light yellow. The posterior yellow border of the first and second and the sixth and seventh tergites is much more extensive than that of the others. In some lights there is a faint yellowish white bloom. The pile is very scanty and extraordinarily fine, erect, and not very long. Female terminalia surrounded by strong, curled, flattened hairs turned inward.

The male terminalia from dorsal and lateral aspect, epandrium removed, show dististyli strongly attenuate and strongly divergent. The base of the epiphallus is funnel-shaped. Basistylus bowl-shaped, the epandrium large and long, narrowing apically.

Material available for study: Two females of the type-species. These flies have a curious tendency to become greasy. Dr. Painter gave me one female.

Immature stages: Unknown.

Ecology and behavior of adults: These little flies can be swept up at times from fields of flowers upon open mesa. A single female was collected by Marguerite Hull in middle Arizona in 1959.

Distribution: Nearctic: *Desmatomyia anomala* Williston, 1895; *binotata* Painter, 1968.

Genus *Apatomyza* Wiedemann

FIGURES 97B, 349, 544

Apatomyza Wiedemann, Nova Dipterorum genera, p. 11, 1820.
Type of genus: *Apatomyza punctipennis* Wiedemann, 1820. by monotypy.

Flies of medium size which superficially resemble a large culicid or empidid. The abdomen is tapered, gradually narrowing, densely short appressed pilose and only half as long as a wing. At the base it is as wide as the anterior mesonotum. The thorax is the most prominent part of the fly, distinctly widest across the base of the wings and the mesonotum rather strongly humped. The comparatively small head is only as wide as the anterior part of the mesonotum; it is rather small, decumbent, with a conspicuous cervix so that there is quite a gap between the occiput and the mesonotum. The proboscis is slender, thrust directly forward and not longer than the antenna. The antennae are peculiar for the very long first antennal segments which lie adjacent to one another. The third segments are rather strongly and conspicuously divergent. From the dorsal aspect the eyes are strongly cut away yet rounded on the posteromedial corners, leaving a very extensively exposed occiput, with conspicuous mound-like ocellarium. The mesonotum is covered with short semirecumbent pile. The wings are subhyaline with dense, brownish-yellow villi and an iridescent reflection and narrow, diffused, brownish infuscation at the crossveins, base of the third and fourth posterior cells and of the second submarginal cell. The costal and subcostal cells are light yellowish brown. The ambient vein is complete but weak. The alula is narrow.

This fly is known only from the Cape of Good Hope, and the unique type was collected in October 1817.

Length: 6 mm. exclusive of head. The semidecumbent head including antennae and proboscis measures just over 2 mm., and of about equal length. Wing is 7.5 mm. from apex to extreme base.

Head, lateral aspect: The head is nearly globular but anteroventrally it has an almost rectangular appearance, due to a rather flattened front, a rectangular protrusion of the face beneath the antennae, and the quite plane border from the bottom of the eye to the angle of the face. The occiput from the lateral aspect is quite circularly rounded, not prominent immediately behind the high conspicuous ocellar tubercle, but becoming strongly and almost equally protuberant on the upper half and lower half. The whole posterior profile is quite rounded, and the occiput is also rather extensive at the bottom of the eye, continuing forward as the gena which is almost as wide as the palpus. The upper part of the occiput is blackish and ground color, obscured by yellowish-brown pollen or micropubescence. There are some scattered, pale, brassy, shining, slightly flattened hairs, both above and below, all of which are quite appressed. In addition there are numerous though not dense, long, erect, stiff, brassy, yellow hairs extending almost straight outward. They are shortest on the lower third occiput and much more scanty. The ground color of the occiput toward the middle and also on the ventral fourth or more and upon all of the gena and face, and the lowest half of the front, is rather light brownish yellow. The posterior eye margin is strongly rounded. The facets are large and coarse and tend to appear to be quadrate rather than hexagonal. The upper facets are not larger.

The face is extended forward beneath the antennae, and there is a shallow diagonal fissurelike crease meeting in the midline a short distance below the antennae and then extending a short distance downward and obliquely to the eye margin opposite the proboscis. The face is densely covered with appressed brownish-yellow pubescence and a few short, shining, suberect, yellowish hairs. The front from the lateral profile is visible across the middle but there are two rather prominent bullae, one on each side at the base of and surrounding the side of the antennae which is visible laterally to some extent. Dorsally the tufted pile that lies on the front and eye margin just opposite the anterior ocellar is conspicuous and consists of a broad, wide clump of more or less appressed, somewhat flattened, golden hairs or scalelike hairs, curled forward and over. The ocellar tubercle is quite conspicuous and high, and the ocelli are large. Between the ocelli is a patch of stiff, black, bristly hairs curved forward and of no great length. There are also a few small, reddish-golden ones.

The proboscis is comparatively stout for the size of the fly but it is nowhere swollen. It is slightly wider apically than on the basal third and distinctly pointed above and below near the apex. On the ventral margin there are a few distinct slender, bristly hairs. They are shorter and more numerous on the basal half, larger and more conspicuous apically, and the entire proboscis has several rows of minute, dense, shining, brownish-golden, erect, micropubescence. The palpus is long, and conspicuous and rather stout. The basal segment is longer than the remaining segment. There appear to be two apical segments. One is short and clavate and round, separated from the middle segment by a distinct oblique suture with even a crease and the base of the middle segment narrowed and attached angularly to the long basal segment. The antennae are attached a little above the middle of the head, and are comparatively stout and elongate. The first segment is at least 5 or more times as long as the second, cylindrical, and of about uniform thickness, and at least as thick as the apical part of the proboscis. It is pale, brownish yellow with a few, fine, scanty, appressed hairs dorsally and a few appressed, more stiff or stouter bristly golden hairs, also flat-appressed ventrally and laterally, placed near the apex. The second segment is not quite as long as wide, brownish, with a very few appressed setae. The third segment is slightly widened to the middle and then rather strongly narrowing to the apex. It is more strongly narrowed below. It is brownish yellow, with minute, erect, shining, yellow micropubescence. At the apex there are two distinct and one very minute microsegments, the latter bearing a short spine.

Head, anterior aspect: The head is not quite as wide as the anterior part of the mesonotum. The face below the antennae is distinct, and there is an oblique crease-like fissure circling immediately below the antennae, and extending downward to the eye margin opposite the proboscis base. This narrowing extension laterally below the face in front, probably constitutes the gena, and the crease is equivalent to the tentorial fissure. In any case it leaves a shining, rather bare, quite narrow, cuplike margin along the oragenal cup. From the anterior aspect this oral opening is quite large, wide, extending backward a little beyond the bottom of the eye and moderately deep on each side of the proboscis. The front is quite as wide as the face, the upper part slightly wider. Face or front not quite but almost one-third the total width of the head; this is the male. There is a distinct, horizontal, narrow, linear, brownish band or stripe across the middle of the eye which was probably more distinct in fresh living specimens; eye was probably greenish in life. The antennae are adjacent to one another and continue adjacent throughout the length. The ocellar tubercle is indeed quite large, wide, conspicuous and high. The anterior ocellus is large. The oral opening is large, wide, relatively deep, only slightly narrowed anteriorly. The gena bears minute micropubescence.

Head, dorsal aspect: Viewed from above the semidecumbent head, the elongate antennae are extended horizontally forward, the vertex is slightly wider than the front. There is a distinct, medial concavity or depression over the whole front from the base of the antennae nearly up to the ocellus. The posteromedial corners of the eye are conspicuously cut away so that the occiput is unusually prominent, bulging backward and sloping strongly downward as well, and with fairly numerous, erect bristly hairs. The attachment to the thorax is small, the prothorax is prominent, also with bristly, subappressed back-turned hairs.

Thorax: The thorax is higher than long, rather strongly humped both anteriorly and posteriorly. The prothorax is prominent from all angles and bears a dorsolateral tuft of appressed, backwardly turned, stiff, bristly, golden hairs. It is dark brown laterally on each side, rather light brownish yellow in the middle; this midde portion is covered with extraordinarily minute pollenlike micropubescence. In consequence the head is well removed from the thorax compared to many bombyliids. The mesonotum is black, except laterally and on the humerus, postalar callosity, and the scutellum, all of which are entirely yellowish brown. The yellow color laterally is more extensive in front of the wing and on the notopleuron. The transverse suture is subcircular, tending to curve a little backward near the top of the mesonotum. In the midline there is a rather wide, posteriorly furcate, opaque black stripe. Another stripe not split posteriorly on either side begins about half way between the humerus and the suture, and there is a diffuse, oval, obscurely blackish spot on each side above the base of the wing, well separated from the latter and more or less confluent with the lateral black stripe. The intervening stripe on each side of the midline has a blue or greenish-black appearance. It may be due to the color of the overlying pollen.

All of the mesonotum is opaque or subopaque and there is a fine, light yellow pollen or micropubescence overlying the brownish-yellow areas. The humerus is paler, and has a tuft of stiff, almost bristly pile over most of the surface. The dorsal pile of the mesonotum is rather abundant, quite flat-appressed, brassy yellow, and slightly flattened or subscaliform. On the notopleuron there are numerous comparatively long back-turned, slender, golden bristles, and a few also behind the suture. Postalar callosity with a few, short, bristly hairs. Scutellum semicircular, brownish yellow, bearing dense micropubescence, and some short appressed golden hairs; on the margin there are 5 or 6 pairs of distinctly longer slender bristles; all of its pile and bristles brownish yellow to golden. The mesopleuron has much quite appressed, stiff, bristly, somewhat flattened, golden, subscaliform hairs. There are special tufts turned backward on the posterior margin, backward from the anterior corner and dorsally along the upper margin. Pteropleuron, metapleuron, metanotum, and hypopleuron all bare, except for yellow or brassy micropubescence. There are a few appressed bristly hairs on the upper part of the sternopleuron. The whole pleuron is yellowish brown, except that most of the mesopleuron is dark brown to blackish; the anterior half of the sternopleuron is brown. Halteres conspicuous, the knob elongate oval with a lateral fissure. Squama with a fine fringe of moderately long hairs. I do not find a plumula, but there is a small flangelike extension from the side of the scutellum near its base, to the metanotum beneath the inner corner of the squama.

Legs: The anterior and middle femora are distinctly though slightly swollen, gradually through the middle. All of the legs are light brownish yellow becoming a little paler on the tibia and very densely covered with flat-appressed setae, some of it scalelike, and especially scalelike on the lateral surface of the hind femur where it forms conspicuous, matted, longer, scalelike hairs. There are 3, short, small, blackish bristles on the posterior surface of the middle femur, 1 before the middle and 2 near the apex. On the hind femur, more in evidence when viewed ventrally, there are 4 rows of oblique, sharp, rather short, comparatively small, distinct black bristles. All of them are confined to the outer two-fifths. There are 7 subventrally on the lateral surface, 6 ventrolateral; 4 others are ventral and lie near the apex; 6 others are ventromedial in position. The tarsi are slender, pale brownish yellow, with rather dense, minute, appressed, brassy-yellow bristly hairs; on the anterior pair 5 short black bristles are placed posteroventrally, 5 others placed postdorsally, and anteriorly there is 1 at the apex. Middle tibia similar with 8 bristles in the middle posteriorly, 8 posteriorly along the ventral border, and 7 or 8 posteriorly along the dorsal border; also, there are 6 along the middle of the anterior surface. Hind femur with similar bristles and the same number of rows, and generally with 1 or 2 more bristles in each row. There is also an extra row which is anteroventral. The pulvilli are well developed, the tarsi about as long as the tibia; middle tarsus with a rather long, conspicuous, basally stout, pale-colored bristle ventrally, not present on the other tibia; claws slender.

Wings: Comparatively large, and at least twice as long as the abdomen, with a uniform pale, brownish-yellow tint due chiefly to dense villi of the same color. There are no bare areas. The wing is very slightly broader beyond the end of the anal cell and is broadly rounded apically, not pointed. There are 2 submarginal cells, the second trumpet-shaped, and only a little wider apically, the anterior branch of the third vein arises in a gentle arch with a thickening near its base, and the anterior branch is exactly at the wing apex. The posterior branch ends near the plane of the hind margin. The first and second posterior cells are widely opened, equally wide, and only a little less wide than the apex of the second submarginal cell. Second submarginal cell at apex and the third submarginal cell of about equal width. There are 4 posterior cells; the anal cell is closed with a short stalk, the axillary lobe is a trifle wider than the anal cell, the alula present but narrow and slightly curved. The ambient vein is complete and the posterior fringe is long, fine, and dense. Setae along the costal margin extremely fine and oblique. The posterior crossvein is long, more than half as long as the anterior crossvein. The discal cell ends rectangularly. There is a distinct vein thickening, also, in the middle of both anterior and posterior crossveins, the discal crossvein and again in the lower vein which begins at the middle of the discal cell. The first branch of the radius is thickened and strengthened especially on the outer part. The anterior crossvein enters the discal cell near or just before the outer third of this cell.

Abdomen: The abdomen is distinctly tapered or narrowed posteriorly. There are 8 tergites visible dorsally and laterally, and the ninth forms a short, hoodlike extension over the genitalia; this appears to constitute the epandrium. The abdomen is only half as long as the wing and brownish yellow throughout, except obscurely along the lateral margins, which are blackish. The epandrium is also darker. The pile is flat-appressed, coarse, pale brassy yellow, curved or curled, and bristly. There are a few fine, moderately long, erect hairs along the side margin. The genitalia are terminal and exposed both laterally and apically. The basistylus is large, boatlike, extruded beyond the eighth sternite and distinctly curved upward. The dististyli are a little flattened, curved and diverging outward. From the lateral aspect these structures of the genitalia extend upward as a doubled, clawlike structure.

Material available for study: I have had the good fortune to study the unique type of Wiedemann, the only known specimen, through the courtesy of Dr. Lyneborg of the Copenhagen Museum and of Dr. L. V. Knutsen of the National Museum of Natural History.

Immature stages: Unknown.

Ecology and behavior of adults: Unknown and not on record.

Distribution: Ethiopian: *Apatomyza punctipennis* Wiedemann, 1820.

Genus *Adelogenys* Hesse

FIGURES 223, 301, 564, 568, 1003, 1004, 1005

Adelogenys Hesse, Ann. South African Mus., vol. 34, p. 811, 1938. Type of genus: *Adelogenys culicoides* Hesse, 1938, by original designation.

Hesse (1938), p. 813, key to 3 species.

Small, rather bare, sparsely pilose flies which appear larger because of the length; they are black flies with often much red or yellow on parts of them. The abdomen is attenuate as well as lengthened which gives the flies a somewhat therevid-like appearance; the bare head is strongly subglobular, the wings hyaline and large in proportion to the body of the fly. The thorax is rather strongly humped and somewhat compressed laterally. The ocelli sit upon a relatively large tubercle and the eyes of the male are separated by at least the tubercle width. The antenna is slender and elongate; the first segment is less than 3 times as long as wide, much shorter than in *Apatomyza* Wiedemann and very much more slender than in that genus; the second segment is minute and beadlike. The third segment is lanceolate, widest near the base, elongate and tapering to the apex which bears 2 short microsegments, the last one having a style.

The wings are very long in proportion to size of the insect; all cells open maximally except the anal cell which is closed with a short stalk. The venation is quite generalized and simple; the intercalary vein is reduced, the anterior crossvein lies at the middle of the discal cell; alula narrow.

These flies are related to *Crocidium* Loew, differing in the following particulars. The wings are often spotted. The inner margins of the female eye distinctly diverge anteriorly. The genae are always present, the alula and axillary lobe broader, the thorax less humped and the occiput is more or less concave, in contrast to *Adelogenys* Hesse. From *Apatomyza* Wiedemann it differs in the much shorter first antennal segment which is much thinner, but thick and stout in *Apatomyza* Wiedemann.

These three genera stand off to themselves in this subfamily; it may be noted, however, that the small size of *Adelogenys* Hesse, together with the strong hump on the thorax, may indicate a somewhat distant relationship with *Mythicomyia* Melander with its reduced venation and still more extraordinary hump, and minute size.

This genus is restricted to South Africa.

Length: 2.5 to 7.3 mm.; wing 3 to 8 mm., often longer than the insect.

Head, lateral aspect: The head is almost globular due to the strong, posteriorly developed occiput, which is, however, gradually rounded and sloped outward and not cup-shaped. The face is short but produced forward a little beneath the antennae making a right angle with the large, plane, oragenal cup. The face is without pile except for 1 or 2 minute scattered hairs near the oral margin. There is an extremely microscopic, sparse, scattered pubescence along the outer margin of the oragenal cup, which in profile is almost level with the rounded eye margin against which it abuts; the gena is wanting due to the compression of the eye against the oragenal cup. In profile the occiput is almost as rounded as the anterior half of the head, the pile while still scanty is a little more abundant on the upper half of the occiput where it consists of scattered, metallic, flat-appressed scales. Upper half of the front is similar, with smaller, scattered, appressed scales. The posterior margin of the eye is entire, is very broadly rounded, or rather it is rounded and peaked backward in the middle and less strongly rounded on the upper and lower halves. The proboscis varies from short to long, being in some species at least 1½ times as long as the head and in any case the proboscis is stout and large for the size of the head and the labellum large and conspicuous. The palpus is elongate and slender and conspicuously projecting as in *Crocidium* Loew. It consists of 2 segments, the first one quite elongate and slender, the terminal segment much shorter and broadly ovate or clavate apically with minute hairs. The front is flattened and bare, except on the upper half, which has flat-appressed pile.

The antennae are attached at the middle of the head, relatively slender, almost as long as the head. The first segment is twice as long as wide, 2 to 3 times as long as the short second segment. It is nearly bare, a little widened apically and has a few minute hairs at the apex. Second segment likewise with a more dense micropubescence apically. Third segment is longer than the first two combined, a very little wider than the first segment, and near the outer third it rapidly narrows; the blunt apex bears 2 distinct, narrow microsegments, the first one short, the last one tipped with a bristle. Again, in some species the third segment may be attenuate on more than the apical half, or the segment may be slightly humped, even somewhat spindle-shaped and viewed from above in some species, the basal half is narrowed and the outer half widened toward the middle.

Head, anterior aspect: The head is of approximately the same width as the thorax. The eyes are narrowly separated in the male by slightly less than the width of the ocellar triangle and in females separated by twice the width of the male vertex. The ocellar tubercle is rather large, quite prominent and elevated. The ocelli occupy an equilateral triangle. The oral opening is quite large, shallow throughout, even behind, and it is gradually and slightly widened posteriorly. The genae are virtually obliterated and represented by a scarcely visible line along the eye margin. From a vertical aspect the occiput is completely flat behind and below the ocelli so that there is no trace of groove.

Thorax: The thorax is convex viewed from the side, posteriorly and especially anteriorly and somewhat compressed laterally. The pleuron appears to be comparatively high, due, perhaps, to the somewhat compacted and shortened thorax; this shortening is even more in evidence on the lower half of the thorax with both anterior and middle coxae large and elongate. The pile of the thorax consists of scattered, flat-appressed, metallic scales on the dorsum, some longer, similar hairs on the notopleuron above the wing, and a few shorter scalelike hairs plastered flat and directed backward over the upper half of the mesopleuron. At the upper border of the mesopleuron a few of these are longer and directed upward. Humerus with 2 or 3 long, brassy hairs. Scutellum subtriangular with rounded apex and elevated above the abdomen due to the prominent metanotum over which it hangs; its pile is somewhat similar to that of the mesonotal dorsum. Postalar callosity shelflike and even concave dorsally. Haltere unusually large, the apical club clavate, rounded, elongate, swollen, and nearly half the length of the haltere. Pteropleuron and hypopleuron bare. Metapleuron with a few minute hairs appressed below the haltere.

Legs: The legs are unusually elongate and slender, especially the hind pair. There are no spines on the femur and the femoral pile consists of moderately abundant, flat-appressed, inconspicuous scales or scaliform hairs. Spicules are reduced and inconspicuous, absent on the anterior tibiae, or at least microscopic and appressed. The hind tibia has 7 or 8 widely spaced, small posterodorsal spicules. Hind tarsi quite elongate, longer than the tibia. Pulvilli long in both sexes and the claws well developed but slender and only shallowly curved.

Wings: The wings are unusually large, elongate, hyaline, villose with 2 submarginal cells and 4 posterior cells, all open maximally. The intercalary vein is short, rectangular; the anterior crossvein is at the middle of the discal cell, the anal cell is closed with a short stalk and the stalk may be obliterated. Axillary lobe large, but alula greatly reduced or almost absent. The venation is quite similar to *Crocidium* Loew. The second submarginal cell is shaped more as in *Mallophthiria* Edwards, but in that genus, among other things, the alula is quite well developed. In *Adelogenys* Hesse the ambient vein is weak. In all three of these genera the anterior branch of the third vein is characteristically straight and not sinuous and bent as in many Bombyliinae, and the second submarginal cell is not wide at the base.

Abdomen: The abdomen is elongate but quite strongly tapered from the base to the apex, relatively bare with wide, yellowish, posterior margins on the tergites and the middle of the first three often widely yellowish, with or without central brownish spots. The scanty pile consists mostly of a fringe of long, flat-appressed, brassy hairs along the posterior margin of the tergites plastered back upon succeeding segments and with some shorter hairs in between. The hypopygium of the male is enlarged, clublike and conspicuous, the large dististyli are ventral and fully exposed below. I find 9 tergites in the male, the last one forming a well-developed shield on the genitalia. Female terminalia small with a fringe of fine, long, posteriorly directed hairs from the last tergite.

I quote Dr. Hesse as follows:

Hypopygium of known male very much like those of *Crocidium*; the basal parts similarly shaped; beaked apical joints flattened dorso-ventrally and bifid apically, the outer apical part being produced spine-like, forming a bifid process with beak; aedeagus as in *Crocidium* and without a ventral process;

lateral ramus on each side, joining each basal part to aedeagal part, also broadish and flange-like as in *Crocidium*; lateral struts slightly more rod-like than shown in figure.

Material available for study: Several individuals most kindly furnished for dissection and study by Dr. A. J. Hesse; as I have only seen the type-species, I have supplemented my description with notes from Dr. Hesse.

Immature stages: Unknown.

Ecology and behavior of adults: Hesse states that these flies resemble mosquitos when resting upon the flowers of *Mesembryanthemum* sp.

Distribution: Ethiopian: *Adelogenys braunsii* Hesse, 1938; *culicoides* Hesse, 1938; *namaquensis* Hesse, 1938.

Genus *Semiramis* Becker

FIGURE 325, 535

Semiramis Becker, Ann. Mus. Zool. Acad. Imp. Sci. St. Pétersbourg, vol. 17, p. 486, 1913a. Type of genus: *Semiramis punctipennis* Becker, 1913, by original designation.

Semiramis Becker contains small flies which appear to be closely related to *Crocidium* Loew. I have not seen these flies, but it is apparent from Becker's figure of the wing and profile figure of the head that only slight differences of wing shape and slight differences of cell shape separate it from *Crocidium* Loew. Not only this, but both genera have punctiform spots about the crossveins and vein intersections. Becker, the author of this genus, compared it to *Geron* Meigen, from which it is both obviously and abundantly distinct. Since Becker did not mention *Crocidium* Loew, it can only be assumed that it did not occur to him to compare them.

The wing is a little more pointed apically, and in this respect a little more like *Geron* Meigen; but the cells are remarkably like *Crocidium* Loew; the anterior crossvein is a little closer to the middle of the discal cell, but this, of course, varies some in most genera. The second submarginal cell is longer and narrower in *Semiramis* Becker than in *Crocidium* Loew, and the stalk of the anal cell is longer, the first branch of the medius is perhaps slightly more arched. I have reproduced Becker's figures.

The single known species is from Persia.

Length: 4 mm.; wing 3.5 mm.

I have not seen this fly and I quote Becker's description:

Habitus as in the case of *Geron* Meigen. Thorax and abdomen not broad, with only delicate and short sparse hairs, without bristles. Antennae close together; first and third segments elongated, narrow, the latter with only a short central terminal bristle, without a visible style. Eyes in the female widely separated, in the male without indentation (concavity) on the posterior margin. Lower part of the face short, projecting forward, slightly cone-shaped. Proboscis protuding by a head's length in front of the edge of the mouth, projecting forward horizontally with very long, thread-like, 2-segmented palpus. The posterior margins of the abdominal segments are distinguished by special coloration. The legs are fairly long, without bristles. Venation as in the case of *Phthiria*, with 2 submarginal cells and 4 open posterior cells; the first is widely opened, on the wing edge almost twice as wide as in the middle; anal cell closed and stalked. The junction of the veins are margined with brown.

Type in the museum of St. Petersburg.

Material available for study: None.

Immature stages: Unknown.

Ecology and behavior of adults: Not on record.

Distribution: Palaearctic: *Semiramis punctipennis* Becker, 1913.

Genus *Tamerlania* Paramonov

Tamerlania Paramonov, Trav. Mus. Zool. Kieff, p. 203, 1931a. Type of genus: *Tamerlania grisea* Paramonov, 1931, by original designation.

I have not seen *Tamerlania* Paramonov and I quote the original description, taken by Engel from Paramonov:

Closely related to the genus *Semiramis* Becker, but the venation is somewhat different; the anal cell is opened fairly widely at the margin; the anterior crossvein stands somewhat before or at the middle of the discal cell, and not behind the middle as is the case of *Semiramis* Becker. The branches of the third longitudinal vein are relatively short, only placed a very little closer to the base of the wing than the apex of the second longitudinal vein (very significant in the case of *Semiramis*); the lower branch is straight, the upper one bent in a definite S-form and arises at an almost right angle (in *Semiramis* the two branches are almost straight, and the upper one arises at a very acute angle); the first posterior cell becomes gradually broader over its entire length (without narrowing in the middle). Remaining venation almost the same as with *Semiramis*.

The body is weakly haired and almost without bristles. Upon the thorax, in front of the base of the wings, there is a distinct bristle. The tibiae are short, but distinctly bristled. Abdomen uniformly black. Pulvilli distinct; (in *Semiramis*?). The structure of the antenna differs somewhat perhaps from that of the species of *Semiramis*, but the description of *Semiramis* in Becker is less detailed (Paramonov).

Paramonov describes the type-species in the following words:

The whole body is black, dusted heavily with gray. The face is very slightly protruding in profile; between the mouth and the cheeks there is a groove. Cheeks and chin have sparse but long and whitish hairs. The mouth opening is quadrangular with right angles. The front (frons) is very broad, the vertex of the head takes up considerably more than one-third of the width of the head; toward the bottom the front gradually becomes wider. The ocellar tubercle takes up considerably less than one-third of the width of the vertex; it is covered with extremely short whitish hairs. The front has a distinct transverse groove somewhat above the base of the antennae; below this trough the surface is completely hairless; above it the front is somewhat long, but bears scant, erect, mixed black and yellow hairs; the first segment is almost one and one-half times longer than the second; the third segment is bare, without an easily discernible style or microsegment; it is somewhat blunted at the end, almost one and one-half times longer than the two first segments together; it is somewhat thickened in profile at the basal part, becoming more narrow toward the apex. The indentation on the posterior margin of the eyes is completely missing.

The occiput has sparse, but fairly long whitish hairs; it is fairly flat, without a longitudinal groove or a strongly emphasized central concavity. The proboscis is almost one and a half times longer than the head. The antennae are very long, reaching almost two-thirds the length of the proboscis, somewhat shaped at the apex. Mesonotum with two darker longitudinal lines in the middle, with sparse black and white erect hairs, no scales or bristles, except there is a distinct bristle in front of the base of the wings. The pleura are almost bare, only the mesopleura have sparse but fairly long whitish hairs. The halteres are yellow, the head as long as the stalk. The legs are black, without bristles (on the hind femur there are only long bright colored hairs). The tibiae, however, are short black bristled, the under side of the tarsi likewise. The legs without scales. Wings clouded quite milkily, most blackly bordered along the veins. The costa surrounds the whole edge of the wing. The abdomen on the tergites and sternites is sparsely whitish haired. According to its whole habitus, this fly resembles a species of *Conophorus* Meigen (Paramonov).

Three females (not well preserved examples) Tshangyr, N. W. Buchara, 27 April 1930 (L. Zimin. leg.).

Length: 3-4 mm; wing length: 3-4 mm.

Material available for study: None.

Immature stages: Unknown.

Ecology and behavior of adults: Not on record.

Distribution: Palaearctic: *Tamerlania grisea* Paramonov, 1931.

Subfamily Phthiriinae Becker, 1912

This group has in the past served as a dumping ground for a number of quite disparate elements. Fortunately Hesse (1938) recognizing this, transferred such elements as *Adelogenys* Hess and *Apatomyza* Wiedemann and *Crocidium* Loew back to the Bombyliinae; for them I erect the tribe Crocidini; Hesse also removed the *Geron* Meigen group to a well-marked new subfamily; the genera like *Apolysis* Loew and *Oligodranes* Loew appear to me to show a much greater affinity to the genus *Usia* Latreille, than to *Phthiria* Meigen, though I admit there is some general similarity in the antennae; nevertheless, the *Phthiria*-type antennae has arisen independently among other bee flies, even in the Bombyliinae. Of course, it would be possible to unite the Usiinae and the Phthiriine, and there is some justification for this.

In this subfamily as here constituted the only worldwide representative is *Phthiria* Meigen, which appears to be present on all continents; favoring arid lands, it is abundantly represented in the southwestern United States. In the known species the larvae feed within the egg pods of locusts.

The flies of this subfamily may be characterized by the slightly humped thorax, the 4 posterior cells and 2-branched third vein. While the pile and pubescence is generally sparse and scattered and rather short, there are exceptions in such as the almost woolly *Acreotrichus* Macquart from Australia. Most species are either light in color or have much pale color in the form of vittae or paired spots, and there are interesting sex differences. Thus in some Central American *Phthiria* Meigen the males are almost or quite solid black in contrast to much

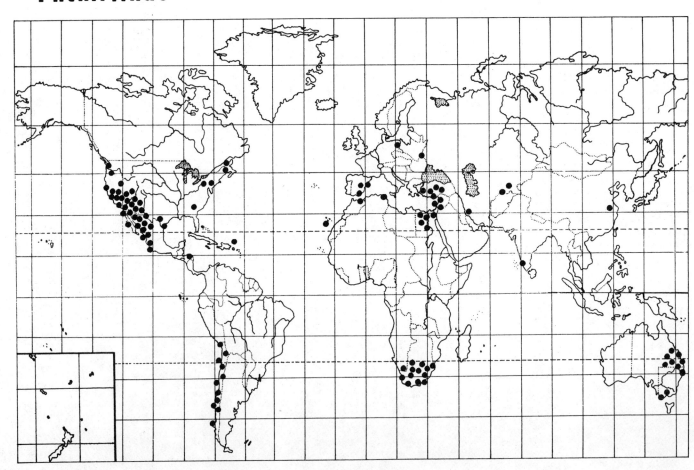

Text-Figure 24.—Pattern of the approximate world distribution of the subfamily Phthiriinae.

lighter females, and I am indebted to Rex Painter for calling my attention to this. Generally, much of the body is bare. The abdomen may be short, plump, compact, barely tapered or in Egyptian species much lengthened and tapered. There is a distinct and characteristic frontal depression before the antenna. Eyes are not infrequently red or green. Male genitalia are large, terminal, conspicuous. All these flies are relatively small.

I recognize four subgenera under *Phthiria*: *Poecilognathus* Jaennicke, distinguished by the conspicuous vein appendices; *Pygocona*, new subgenus, from the Australian area, distinguished by the peculiar conical hypopygium; *Agenosia*, new subgenus, distinguished by the completely bare genofacial area, and the more extended face which is not rounded and turned inward, also, the more extensive frontal area; and *Neacreotrichus* Cockerell, subgenus, new status.

In the Phthiriinae the male terminalia are large, terminal in position, instead of reflexed downward. They are therefore conspicuous and many of the parts and processes can be inspected without dissection. This is very much in contrast to the subfamily Bombyliine, in which the genitalia are only rarely enlarged and terminal, as for example in *Acrophthalmyda* Bigot. In the Phthiriinae there are often conspicuous, elongate, accessory prongs or lobes that we here call claspers. This is suggestive of the situation in Gerontinae, but radically different, because in the Gerontinae we have the entire basistylus undivided and all the processes arising as lobelike extensions. All of the processes and the dististylus arise as lobelike processes rather than as separate movable segments.

KEY TO THE GENERA AND SUBGENERA OF THE PHTHIRIINAE

1. First and second antennal segments of nearly equal length and each approximately as long as wide, and with approximately parallel sides. Epandrium, short, mammiform. Upper lateral face and lateral gena with or without pile . 3
   First antennal segment clearly longer than wide and longer than the second segment; sometimes 2 or more times as long as wide. Third antennal segment angularly widened near the middle, with a dorsal tuft of several bristles or bristly hairs placed at the hump 2
2. Face densely covered with pile of medium length. Bristly hairs on antennal segment 3 or 4 in number and not extraordinarily long. Anterior branch of third vein arising gently. Epipygium formed into a long, V-shaped, gaping, dorsal cone . . . . **Phthiria (Pygocona, new subgenus)**
   Upper lateral face and genal area inflated and densely very long, erect pilose. Anterior branch of third vein arising as a rectangular crossvein with long, backward spur. Dorsal angle of third antennal segment conspicuous, with 7 or 8 conspicuous, long bristles, the longer bristles twisted and bent. Second antennal segment also longer than wide. Whole insect densely covered with very long, dense, erect, fine black hair much as in *Bombylius* Linné. Epandrium short and obtuse . . . . **Acreotrichus Macquart**
3. Angular veins and stumps of veins present in the discal cell; anterior branch of third vein usually rectangular with stump vein . . . . **Phthiria (Poecilognathus) Jaennicke**
   Discal cell without angulated veins and stumps. Anterior branch of third vein always arising gently. Upper genofacial area typically with abundant long hair (if completely bare, *Agenosia*, new subgenus) 4
4. Third antennal segment with conspicuous long, dorsal bristles . . . . **Phthiria (Neacreotrichus) Cockerell**
   Third antennal segment with at most minute bristles, hairs, or none . . . . **Phthiria Meigen, sensu stricto**

Genus *Phthiria* Meigen

FIGURES 99, 100, 215, 308, 566, 570, 849, 850, 851

Phthiria Meigen, Mag. Insektenkunde, vol. 2, p. 268, 1803. Type of genus: *Bombylius pulicaria* Mikan, 1796, by monotypy.
Cyclorhynchus Macquart, Diptères exotiques, vol. 2, p. 114, 1840. Type of genus: *Cyclorhynchus testaceus* Macquart, 1840, by monotypy. Preoccupied Sund., Aves, 1835.
Cyclorrhynchus Bezzi, p. 3, 1924, emendation.
Phtyria Rondani, Arch. Zool. Anat. Fisiol., vol. 3, p. 65 (sep.), 1863, lapsus.
Poecilognathus Jaennicke, vol. 6, p. 350, 1867a. Type of subgenus: *Poecilognathus thlipsomyzoides* Jaennicke, 1867, by monotypy.
Pygocona, new subgenus. Type of subgenus: *Pygocona flavicincta*, new species.
Agenosia, new subgenus. Type of subgenus: *Agenosia vittata*, new species.
Neacreotrichus Cockerell, 1917, subgenus, new status; Proc. U.S. Nat. Mus., vol. 52, p. 377. Type of subgenus: *Acreotrichus atratus* Coquillett, 1904.

Coquillett (1894), pp. 102-103, key to 10 Nearctic species.
Williston (1901), pp. 288-289, key to 9 Mexican species.
Coquillett (1904b), pp. 172-177, key to 12 Nearctic species.
Séguy (1926), p. 234, key to 8 species from France.
Hardy (1933), p. 415, key to 4 Australian species.
Austen (1937), pp. 63-64, key to 8 Palaearctic species.
Engel (1933), pp. 140-142, key to 27 Palaearctic species.
Hesse (1938), pp. 825-831, key to 8 male, 6 female species from South Africa.
Efflatoun (1945), pp. 107-108, key to 4 Egyptian species.
Painter (1962), pp. 28-29, key to 12 Nearctic species.

Compact, rather small flies ranging in size from about 1.5 millimeters to 5 or more times this size. Some Egyptian species have elongate, tapered abdomens 3 times as long as wide. They generally have short, scanty pile but a few have considerable amounts of longer, fine, erect pile. The coloration is commonly pale with or without dark vittae or spots upon face, front, thorax, and abdomen, but some species are entirely black and others have a very fine pale pollen over the thorax. Painter has called my attention to the marked sexual color dimorphism in some species.

In addition to a long, slightly arched proboscis, the genus is especially characterized by the form of the antenna; the third segment is elongate with the apex hollowed out, liplike above, and jutting from the concave pitlike apex is a minute spine or sensory rod. Also in males the eye is distinctly depressed above, a depression that extends to the front of the female and leaves it quite flattened. The face and front are prominently

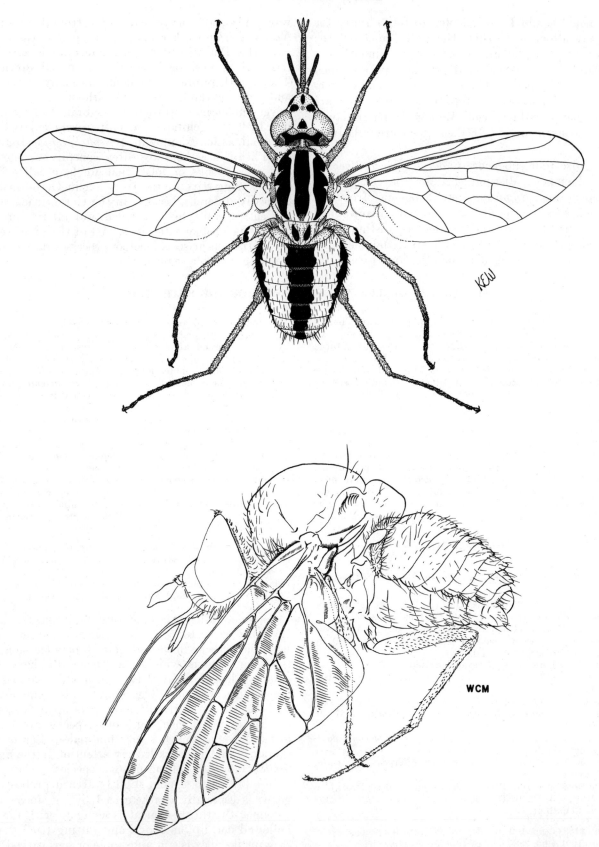

Text-Figure 25.—*Phthiria* species: habitus and lateral aspect.

extended. Wings broad and often quite broad basally. The anterior branch of the third vein and the medial crossvein often have backward spur veins.

This genus is worldwide in extent with rather more species known at this time from the southeastern United States.

Length: About 1.5 mm. to 8 mm., excluding proboscis; wing, 2 to 7 mm.

Head, lateral aspect: The head is sometimes globular and sometimes more or less trapezoidal in shape with the very long front sloping, flattened, or even concave. The occiput varies from short, though moderately and equally extensive above and below, broadly shallowly concave centrally, and this seems to be typical of the genus although there are certainly other species in which the occiput is more prominent and the head subglobular. The occipital pile ranges from scanty, short, stiff, erect, fine, bristly hairs with the eye margin bare to others in which there is rather dense, coarse, bristly pile over almost the entire occiput, the pile below becoming longer and more shaggy. The front in males may be either short, swollen and convex with a complete transverse band of coarse, bristly pile extending all the way around the sides of the antennae and down the sides of the face, or the front may be longer, very little swollen but depressed with a triangular patch of rather dense, appressed, bristly pile reaching from the junctoin of the eye nearly to the antennae. The females are characteristic in having the front wide and very strongly depressed across the middle with transverse foveal crease, and the pile similar in character to the male. The eye is comparatively large, triangular on the upper half and divided in males into an area of larger facets above and smaller facets upon the lower third of the eye. The upper facets are often greatly enlarged. The posterior margin of the eye is entire.

The face is often rather extensive laterally with the short oragenal cup extending to the base of the antennae, sometimes vertically plane and in other species strongly and convexly retreating below the antennae. The oragenal cup does not always reach the base of the antennae. Sides of face either bare and shining, or with considerable long, coarse, bristly or scaliform pile directed forward. The proboscis is always quite long and slender, more attenuate apically, $1\frac{1}{2}$ to 3 times as long as the head and with the labellum quite slender. It is projected straight forward, sometimes a little curled downward. The palpus is extraordinarily slender and long in some species, in others it is lanceolate, shorter and decidedly more broadened before the apex.

The antenna is attached just above the middle of the head, sometimes more nearly at the upper third. It is comparatively elongate, slender, the third segment flattened. All three segments are of nearly uniform width, although the third segment may be a little wider and is sometimes angularly swollen dorsally to a moderate extent in the middle of the segment. This genus is characteristic in the form of the third antennal segment which at the apex bears a dorsal, liplike process, another below the apex, which may be shorter or even longer, leaving the intermediate portion of the apex concave and having a short, conspicuous spine. Third segment either pollinose only or sometimes with several short, stout, oblique bristles in the middle on the dorsal surface. The first segment is a little longer than wide, slightly narrower at the base with strong, long, bristly setae above and a few shorter setae below. Second segment about as long as wide and sometimes shorter, either nearly as long as the first segment, or in other species only half as long, and likewise with a few setae dorsally and at the apex. The third segment varies from 2 to $2\frac{1}{2}$ times as long as the first two segments together.

Head, anterior aspect: The head is ventrally comparatively plane, or sometimes with the suboccipital area a little swollen below. The genofacial area is very wide below becoming in the male slightly convergent toward the antennae, or even with the sides parallel as far as the antennae. The eyes are holoptic for an extensive distance in the male and are depressed where the eyes touch, in addition to a slight overgrowth of the eye on each side. The whole upper side of the eye tends to be flattened. Ocellar tubercle small forming an equilateral triangle, the ocelli resting on the eye margin and the tubercle a little swollen and bearing several fine, long, bristly hairs. Front of female widely separated, slightly convergent toward the vertex and the vertex 3 times the width of the posterior ocelli measured from their outer edges. The antennae are narrowly separated; the first segments touch because they are widened apically. The oragenal cup is rather small, shallow, and short oval, the inner wall is quite wide and slightly sloping inward. Both the lateral face and the gena are quite wide and extensive, with either loose or comparatively dense pile continuing down the side, long and erect. Tentorial fissure absent.

Thorax: The thorax is short and compact, the mesonotum rather high, moderately humped and about equally convex anteriorly and posteriorly. The mesonotal pile varies from scanty and erect or suberect in some species to comparatively dense, rather long, and erect but never hiding the ground color in others. However, some species appear to be either naturally almost bare of pile on mesonotum and abdomen or they become naturally heavily denuded in emergence or when swept up in the net. I have before me a very large series of one very polished black and yellow species in which only an occasional scattered hair can be seen on the mesonotum on any one of the many specimens and in which the abdominal pile is also scanty. The mesonotum may be with or without a fine, quite microscopic, scaliform pollen. The humerus is prominent with a deep 3-cornered recess behind. It seems to always bear long pile, and there is coarse bristly pile on the notopleuron; a few fine hairs on the wide and much flattened posterior callosity, especially on its rear margin. The scutellum is quite large, in many species strongly inflated and convex posteriorly and bears fine, erect pile similar in length and abundance to that found on the mesonotum. If the mesonotum is largely bare, so is the scutellum. The same relationship holds to the mesopleuron and the upper sternopleuron. The pteropleuron, metonotum, the hypopleuron, and the whole of the metapleuron are without pile. Alary portion of the squama large with a scanty fringe. Halteres large with elongate knob and apically widened stalk. Pleuron with or without pollen.

Legs: The legs are moderately long, the femora a little stout, the anterior 4 femora slightly widened toward the base. The hind femur generally slightly widened in the middle. Pile of the femora moderately abundant, obliquely appressed and setate, bristles absent. Tibiae and tarsi a little lengthened and slender. Anterior tibia without bristles or contrasting spicules; the pile consists of appressed setae; other tibiae similarly pilose but in addition they show several rows of oblique, short, stout, spicules. The middle tibia may have as many as 7 or 8 anteroventrally and as many as 15 or 20 dorsally. Hind tibia similar. Claws slender, strongly bent at the apex, the pulvilli long and widely spatulate.

Wings: The wings are hyaline or with spots on the crossveins and wing furcations and sometimes the apices of veins. Characteristically the wing is exceptionally broad at the base. The axillary lobe and the alula are wide. There are 2 submarginal cells and 4 posterior cells, all of them widely open. The anal cell is closed and stalked, the anterior crossvein enters the discal cell always beyond the middle of the discal cell and sometimes at the outer third. The species of *Phthiria* Meigen fall into 2 groups depending on the form and shape of the second submarginal cell and the character of the first section of the anterior branch of the third vein. This first section of the anterior branch of the third vein may be plane and rectangular and simulate a crossvein, with a strong backward spur. In such species there is generally a backward spur at the bend in the lower vein of the discal cell. Painter and Painter (1962) allocate these species to the subgenus *Poecilognathus* Jaennicke. Other species of *Phthiria* Meigen, sensu stricto, have the base of the second submarginal cell arising in a gently rounded fashion, and the lower vein of the discal cell likewise is rounded without a spur. The alula ends at the anal cell.

Abdomen: The abdomen is short and compact and sometimes very short especially in the females, where several tergites are tucked under and recessed below the others. Viewed from above the abdomen is often obtusely conical but in Egyptian species elongate conical, gradually tapered, the sides plane, the whole abdomen 2 or 3 times as long as the basal width. The pile of the abdomen is generally similar to that of the thorax, being very scanty or almost absent on the more bare species, which have a few fine, short, scattered, appressed hairs, and in more pilose species the pile is much more abundant, generally more or less appressed dorsally but forming a conspicuous, erect fringe laterally. In *Phthiria floralis* Coquillett the pile is conspicuous, very fine, long, and erect over the entire abdomen. The female terminalia consist of 2 triangular plates apposed together from each side; from their junction below there are 3 or 4 long, curved, appressed bristles. Male terminalia quite prominent, though not very long, dorsal in position, with the hooks turned downward. The last sternite forms a shelf below with a lateral lobe on each side, the aedeagus lying between.

Since Dr. A. J. Hesse was able to base his conclusions upon a greater number of species, I quote his remarks upon the male hypopygium:

Hypopygium of males with the inner apical part of basal parts in neck region usually prominent or produced into a process, which is broad from side, provided at apex towards the outer side and more dorsally with a crest of shortish spines, with the dorsum of basal parts often provided with long, bristly hairs; beaked apical joints usually slightly curved, hollowed out below, with comparatively few and sparse, or without any, hairs above, but sometimes with a few, usually two, longer, stouter and more spine-like, bristly hairs nearer apex on inner side; aedeagus usually straight or curved upwards apically, without a ventral process, joined on to basal parts on each side by a broadish, flange-like ramus or merely by a strap-like ramus which is under the lateral struts, the aedeagus often produced basally below middle part into a conspicuous process on each side; lateral struts directed outwards and upwards in most of the species.

The male terminalia from dorsal aspect, epandrium removed, show long, slender, apical upturned dististyli which at the apex are apt to bear several very long spikelike spines. In addition, on each side of the epiphallus and in some cases also dorsal medially above the epiphallus, there are apt to be long, accessory, slender, lobelike structures, which in this work we call claspers. The ramus forms a continuous girdle below the base of the aedeagus. In the lateral aspect the basistylus is large, but short oval, the epandrium is triangular with a posterior lobe in the middle, and the basal apodeme quite large and long, with a prominent additional aedeagal apodeme besides the usual lateral apodeme.

Material studied: A large series of several species from the southwestern United States collected by the author and wife, together with several fine pairs exemplifying sexual dimorphism sent to me by Dr. Painter.

Immature stages: The larvae are predators on grasshopper eggs. See discussion in life histories.

Ecology and behavior of adults: These flies are strongly attracted to flowers, especially composites; the author and his wife swept up several species in considerable numbers in the Southwestern States in the fall of 1965. We found one species spread over a hundred-mile area of highway and adjacent fields congregating on sunflower heads in immense numbers. Some of these heads contained as many as 25 individuals, which were greedily at work feeding.

Distribution: Nearctic: *Phthiria aldrichi* (Johnson), 1903 (subg. *Poecilognathus*); *americana* (Coquillett), 1805 (as *Acreotrichus*); *amplicella* (Coquillett), 1904 (subg. *Poecilognathus*); *badia* (Coquillett), 1904 (subg.

Poecilognathus); *bicolor* (Coquillett), 1904 (subg. *Poecilognathus*); *borealis* (Johnson), 1910 (subg. *Poecilognathus*); *cingulata* Loew, 1846; *consors* Osten Sacken, 1887; *coquilletti* (Johnson), 1902 (subg. *Poecilognathus*); *cyanoceps* (Johnson), 1903 (subg. *Poecilognathus*); *diversa* Coquillett, 1894; *egerminans* Loew, 1872; *flaveola* (Coquillett), 1904 (subg. *Poecilognathus*); *floralis* Coquillett, 1894; *humilis* Osten Sacken, 1877; *inornata* (Coquillett), 1904 (subg. *Poecilognathus*); *loewi* Painter, 1965 [=*notata* (Loew), 1863 (subg. *Poecilognathus*)]; *maculipennis* (Cole), 1923 (as *Acreotrichus*); *marginata* (Coquillett), 1904 (subg. *Poecilognathus*); *melanoscuta* Coquillett, 1904; *nubeculosa* (Coquillett), 1904 (subg. *Poecilognathus*); *picturata* Coquillett, 1904; *psi* (Cresson), 1919 (subg. *Poecilognathus*); *punctipennis* (Walker), 1849 (subg. *Poecilognathus*); *scolopax* (Osten Sacken), 1877 (subg. *Poecilognathus*); *similis* Coquillett, 1894; *sulphurea* (Loew), 1863 (subg. *Poecilognathus*); *unimaculata* (Coquillett), 1904 (subg. *Poecilognathus*); *vittiventris* (Coquillett), 1904 (subg. *Poecilognathus*).

Neotropical: *Phthiria albida* Wiedemann, 1821; *alterans* Williston, 1901; *atratus* (Coquillett), 1904 (as *Acreotrichus*; subg. *Neacreotrichus*); *austrandina* Edwards, 1937; *aztec* Painter, 1962; *barbata* Rondani, 1863 [=*barbata* Philippi, 1865, =*vulgaris* Philippi, 1865]; *cana* Philippi, 1865; *chilena* Rondani, 1863; *dolorosa* Williston, 1901 [=*sororia* Williston, 1901]; *exilis* Philippi, 1865; *fasciventris* Twinn, 1928; *fulvida* Coquillett, 1904 (subg. *Poecilognathus*); *lurida* Walker, 1857; *mixteca* Painter, 1962; *olmeca* Painter, 1962; *philippiana* Rondani, 1863; *picta* Philippi, 1865; *pirioni* Edwards, 1937; *pulchella* Williston, 1901; *testacea* Macquart, 1840; *thlipsomyzoides* (Jaennicke), 1867 (as subgenus *Poecilognathus*); *tolteca* Painter, 1962; *tristis* Bigot, 1892.

Palaearctic: *Phthiria albogilva* Séguy, 1941; *atriceps* Loew, 1873; *bicolor* Bezzi, 1925; *cancescens* Loew, 1846 [=*convergens* Loew, 1846, =*pulicaria* Zetterstedt, 1842, =*zimmermanni* Nowicki, 1867, female]; *conspicua* Loew, 1846; *gaedii* Wiedemann in Meigen, 1820 [=*maculata* Wiedemann, 1820, =*punctata* Meigen, 1838]; *incisa* Becker, 1915; *inconspicua* Becker, 1912; *lacteipennis* Strobl, 1909; *minuta* (Fabricius), 1805 (as *Voluccella*) [=*fulva* Meigen, 1804, =*maura* Wiedemann in Miegen, 1820, =*zimmermanni* Nowicki, 1867, male]; *pulchripes* Austen, 1937; *pulicaria* Mikan, 1796 [=*campestris* Fallen, 1815, =*gibbosa* Walker, 1851, not Olivier, 1789, =*nigra* Meigen, 1804, =*pygmaea* (Fabricius), 1805 (as *Voluccella*)], *pulicaria major*, Strobl, 1909; *quadrinotata* Loew, 1873; *rhomphaes* Séguy, 1963; *salmayensis* Efflatoun, 1945; *scutellaris* Wiedemann in Meigen, 1820; *simonyi*, Becker, 1908; *subnitens* Loew, 1846 [=*rustica* Loew, 1846]; *tricolor* Bezzi, 1925; *umbripennis* Loew, 1846 [=*notata* Bigot, 1862]; *unicolor* Bezzi, 1925; *vagans* Loew, 1846, *vagans pallescens* Engel, 1933; *variegata* Austen, 1937; *varipes* Austen, 1937; *virgata* Austen, 1937; *xanthaspis* Bezzi, 1925.

Ethiopian: *Phthiria capensis* Wiedemann, 1828; *cognata* Hesse, 1938; *crocogramma* Hesse, 1938; *fallax* Hesse, 1938; *flavigenualis* Hesse, 1938; *laeta* Bezzi, 1921, *laeta xerophiles* Hesse, 1938; *lanigera* Bezzi, 1921; *nigribarba* Hesse, 1938; *pilirostris* Hesse, 1938; *pubescens* Bezzi, 1922; *pulla* Bezzi, 1922; *simmondsi* Hesse, 1938; *tinctellipennis* Hesse, 1938.

Oriental: *Phthiria gracilis* Walker, 1852.

Australian: *Phthiria albocapitis* Roberts, 1929; *flava* Hardy, 1933; *hilaris* Walker, 1852; *lineifera* Walker, 1857; *nigrina* Hardy, 1933; *pallipes* Bigot, 1892.

Patria Ignota: *Phthiria cingulata* Loew, 1846; *hypoleuca* Wiedemann, 1828.

The following new species may be added to this list:

Nearctic: *Phthiria* (*Agenosia*) *vittata*, new species.

Australian: *Phthiria* (*Pygocona*) *flavicincta*, new species.

Genus *Acreotrichus* Macquart

FIGURES 98, 315, 534, 536

Acreotrichus Macquart, Diptères Exotiques, suppl. 4, p. 121, 1849. Type of genus: *Acreotrichus gibbicornis* Macquart, 1849.

Comparatively small flies which superficially resemble a species of *Sparnopolius* Loew in general shape. The exceptional, long, erect pile of the thorax and still longer, erect, fine pile of the abdomen, spreading outward along the lateral margins and upward dorsally, is of somewhat looser character, and the pile of the face is also very long, erect, and abundant.

They will be quickly recognized by the long, stiff hairs arising from the angulate dorsal margin of the third antennal segment and from the *Phthiria*-like venation in which the anterior branch of the third vein arises rectangularly as a crossvein from which a long spur vein proceeds backward. The face is long pilose in both males and females, in contrast to *Phthiria* Meigen in which only the male face has pile.

These flies are restricted to Australia.

Length: 6 to 8 mm.

Head, lateral aspect: The head is almost triangular at the base because of the remarkable flattened eyes that are depressed along the midline and because of the shortening of the occiput dorsally, a form more or less characteristic of flies in the Phthiriinae. Below, the occiput becomes more extensive from the lateral aspect, but slopes rapidly inward toward the foramen. The pile of the occiput is remarkable for its length. There is a dense band of very long, erect, coarse hair covering almost the whole of the posterior occiput from the vertex to the bottom of the occiput, extending on beneath the eye and up along the sides of the gena from below. In addition and somewhat set apart from the more central band of pile there is a thin, loose fringe of even longer hairs in several irregular rows with attachment close to the eye margin and a short distance behind the eye margin. These hairs are as long as the eye itself and as long as the elongate third antennal segment. The small ocellar triangle is equilateral in shape, the ocelli rest on the eye margin; it has a tuft of very long hairs in a vertical row, some extend up, others extend forward almost to the antenna.

The front constitutes about one-third of the distance from base of antenna to the posterior eye margin. It bears a conspicuous, appressed, matlike tuft of rather long, coarse, pale yellow pile on each side of the medial line divergent outwardly and reaching to the base of the antenna, and directed obliquely away from the front. The face is bulging and rounded on each side, in length nearly as much as the first antennal segment. It bears rather dense clumps of extraordinarily coarse pile, dark reddish brown laterally, changing to slightly shorter, finer more matted, dense, central tufts of pale yellow pile. There is so much pile on the upper and middle part of the face that the ground color is completely obscured. However, laterally the face has gray pollen on the surface. This long extensive pile continues upward and reaches slightly above the antenna and leaves the nonpilose part of the front dark brown pollinose posteriorly and yellowish pollinose anteriorly. This excessively long pile of the face continues equally densely down the sides of the exceptionally wide gena to merge with the pile at the bottom of the head, the only point where the pile is less dense.

The proboscis and labellum are slender and elongate, 1½ times as long as the head. The long, sticklike, almost filiform palpus is half as long as the proboscis and extends beyond the long pile of the face. It bears a few stiff hairs apically. The antennae are attached a little above the middle of the head. They are elongate with the first two segments relatively stout. The first segment is cylindrical, slightly curved from lateral view, and from 4 to 5 times as long as wide. It bears dense, long, erect pile dorsally, somewhat shorter pile laterally, and some long hairs ventrally. The second segment is longer than wide, distinctly widened apically, at least twice as long as its basal width, and bears a remarkably long, rather dense tuft of stiff, almost bristly hair dorsally. There is similar hair ventrally that is not quite as long and not quite as much. The third antennal segment is as long as the first two together and strongly flattened, rather strongly widened dorsally at the middle, and bears at the middle a tuft of 4 long and 3 or 4 very long, bent, curved bristly hairs. They arise laterally just below the high point of the third segment. Apex of the third segment with a short, dorsal, blackish spur and concavity in which there is a yellowish spine, and the ventral lobe that extends conspicuously outward is much thicker and 4 or 5 times as long as the dorsal lobe, and represents the true apex of the third segment.

Head, anterior aspect: Much wider than the thorax. Oval from the anterior aspect because of a shortening and flattening of the head on the lower half. The upper half of the head forms, if it were continued uniformly, half of a circle. Upper eye facets enlarged, but no demarcation between the facets at the bottom of the eye. The eye of the male is conspicuously and tightly holoptic for two-thirds the distance from antenna to occiput. The antennae are quite narrowly separated at the base. The oral cup is obscured by pile, but is rather large, quite short, oval, and deep. Its inner edge is sharp. The genofacial space is strongly widened and expanded, leaving the extensive cheek occupying almost one-third the total width.

Thorax: Subquadrate, only a little wider across the anterior edge of the postalar callosities. The mesonotum is gently convex, a little more so anteriorly. The head is a little drooping. The pile of the mesonotum is opaque black, shining on the scutellum, and densely covered with exceptionally long, nearly or quite erect, coarse, bristly, pale, shining, yellowish-white pile. There is no anterior collar, and pile similar to the mesonotum occurs on the scutellum. There are no bristles. Pleuron with conspicuous pile, extraordinarily so over the whole mesopleuron, radiating out as long, dense tufts. Pteropleuron, anterior hypopleuron, and the metapleuron bare. However, there is a conspicuous tuft of long, erect pile on the posterior hypopleuron above the hind coxa. And there appears to be a knoblike process arising well below the wing attachment between the wing and the base of the haltere bearing radiating pile. This structure is presumably the outer edge of the squama, and the pile is apparently the pile of the plumula extending on medialward. It may also represent a shelflike extension of the squama.

Legs: The legs are rather long and well developed; hind femur stout, but not swollen, a little flattened. The anterior and middle femora bear dense fringes of long pile ventrally, anteroventrally, and posteroventrally. Dorsally on each side they have some appressed, slightly flattened, bristly hairs. Hind femur with a moderate amount of long pile ventrally, especially on the basal half, becoming shorter distally, with considerable appressed, flattened golden pile distally along the lateral margin, and also dorsally and medially. Surface of this femur transversely microrugose. The long tibia have appressed setae on the anterior pair, small, oblique bristles on the others, situated in 4 rows. Posterior tibia with 7 short, stout bristles ventrolaterally, the same number anterodorsally, and about 15 dorsomedially, fewer ventromedially. Claws long, sharp, rounded, but most rectangularly bent apically, and with the pulvilli large and long.

Wings: Nearly hyaline, broad basally, distinctly pointed apically. Almost all the veins are nearly straight with at most slight curvatures, the cubital vein more strongly curved than any. The second vein arises before the end of the second basal cell and is quite straight. Third vein straight to the beginning of the second submarginal cell, which arises by a rectangular, straight, crossvein-like vein and has an equally straight backward spur, which is longer than the basal part of the anterior branch of the third vein. This basal spur presumably represents the remnant of the atrophied third branch of the radius (second branch of second vein). The anterior crossvein enters the discal cell at the outer fourth. The first posterior cell is open maximally, the second posterior cell is widened apically, the third is slightly narrowed, the discal cell is elongate and angular below. Medial crossvein long, nearly straight but slightly wavy, upper anterior intercalary vein straight and longer than the anterior crossvein. Anal cell closed with a long stalk, the lower vein straight, axillary lobe large and alula quite large. The ambient vein ends at the anal cell. The costal comb is not developed, but there is a tuft of hair arising below from the extreme base of the wing.

Abdomen: Short oval and as long as the mesonotum if the scutellum is excluded. The abdomen is blunt, obtuse, and plane apically. There are 7 tergites visible from above, the last three of nearly equal length, and narrowing a little posteriorly. The ground color is opaque black. The pile on the central part of the disc is fine, erect, and quite long, but becoming much more dense gradually toward the sides of the abdomen, which are curled over; it presents a very bushy appearance indeed. This lateral pile is as long or longer than the facial pile. The male terminalia are large. The basal parts are large, rather rounded or subcircular, convex and a little swollen outwardly, covered with moderately long bristly hair, and with the small, attenuate dististyli lying side by side, separated but slightly convergent apically and directed downward. These parts are exposed from the rear aspect and tucked beneath the last tergite. Sternites completely obscured by the long, lateral pile and the curving under of the tergal margins.

Material available for study: Representatives of this genus were found in the collections of the National Museum of Natural History. This well-characterized fly was collected September 9, 1902, near Windsor, New South Wales, by W. W. Froggatt.

Immature stages: Unknown.

Ecology and behavior of adults: Hardy (1922) writes:

... this is the first species of bombyliid to appear in the spring, and it continues on the wing through the summer; it occurs everywhere that wild flowers are abundant, and at times 20 or 30 specimens can be taken with one sweep of the net, and indeed sometimes they are so abundant that they continuously divert one's attention from other insects.

This abundance must be relatively uncommon.

Distribution: Australian: *Acreotrichus fuscicornis* Macquart, 1849; *gibbicornis* Macquart, 1849; *inappendiculatus* Bigot, 1892.

Subfamily Gerontinae Hesse, 1938

This subfamily proposed by Hesse in 1938 appears to be well founded. Although small, it is worldwide or very nearly so. It may be defined by the following 9 characteristics: (1) Flies with never more than 3 posterior cells. (2) In contrast to *Phthiria* Meigen the face below the antenna is extensive. (3) Genal pile, usually present, is stiff, appressed, or curled upward. (4) A discal cell is always present. (5) The first antennal segment is elongate, usually much more than 1½ times as long as the second; it is generally very short in the Phthiriinae. (6) The last sternite of the female is divided, with lappetlike lobes. However, in many species of *Phthiria* Meigen this sternite is deeply concave, with widely separated lobes or prongs. (7) The third antennal segment is elongate, slender, strongly tapered, drawn-out to a fine pointed apex. (8) The wing is broad at base, the anal cell closed and stalked; in this respect they are similar to *Phthiria* Meigen. Proboscis is long and slender in both, the latter has a quite long slender palpus, the *Geron* Meigen species a short slender palpus. *Geron* Meigen has a more narrow, less extended oragenal cup. The venation of the two groups has features in common. It is possible that in the remote past the Gerontinae arose from the *Phthiria* stem by loss of a posterior cell and a different direction to the antennae; both have the front depressed; or possible from a *Gonarthrus*-like fly. (9) The male genitalia are very complex with numerous prongs and spurs. This is evident in *Geron* Meigen, but even more so in the curious South African *Pseudoamictus* Bigot where the terminalia are conspicuous, larger, and more exposed.

In the Gerontinae the male terminalia are much more complex and radically different from similar structures found in either the Bombyliinae or the Phthiriinae. They especially differ in the structure of the basistylus,

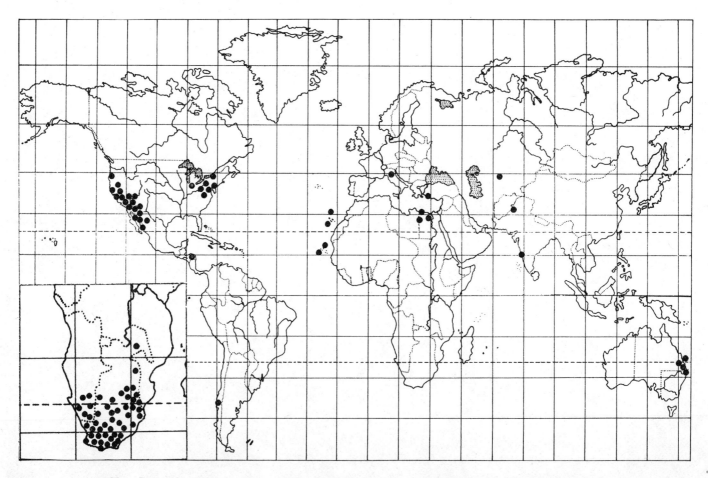

TEXT-FIGURE 26.—Pattern of the approximate world distribution of the species of the subfamily Gerontinae.

which is undivided and not separated into symmetrical halves. Moreover, there are no separate and distinct movable dististyli, this structure being replaced by curious, hornlike lobes or processes from the distal part of the basistylus. Much of the basistylus is membranous. There may be additional hornlike or spinelike prongs on each side of the epiphallus and aedeagal complex.

KEY TO THE GENERA OF THE GERONTINAE

1. Genae very wide, the oral cavity, including the lateral ridge, only about half the width of total space between the eyes on the lower part of head; oral cavity more shallow. Face below antennae pollinose, without pile; genal pile much more extensive on lower third or more of head, long, abundant, bristly, appressed upward. Male eyes rather widely separated or not in contact, or separated by only the width of anterior ocellus, the front therefore much more extensive. First antennal segment as long as others combined, strongly thickened, especially at base, and with dense, long erect, bristly pile extending over its whole length. Whole fly longer, less wide at base, less tapered posteriorly. Wing more elongate, sometimes wholly reddish brown, anal cell closed in the margin, apical cell narrow, the vein closing the middle posterior cell conspicuously sinuous . . . . **Pseudoamictus** Bigot
Genae never extraordinarily wide, usually linear. Oral cavity very deep. Face below antenna with abundant, long, erect pile. Genal pile scanty, largely limited to bottom of eye, curled upward and shorter. Male eye holoptic for a very short distance, depressed at juncture, the male front very small, the ocellar tubercle small but conspicuous. First antennal segment elongate but shorter than remaining two segments, more slender, its pile shorter, and distinctly appressed. Wing broad at base, the anal cell stalked or petiolate, wing almost always hyaline; vein closing middle posterior cell only gently curved. Whole fly shorter, more robust at base distinctly tapered posteriorly . . . 2
2. Male and female face beneath the antenna bare or at most with a few hairs. Pile of first antennal segment long, copious, erect. Anteroventral genal pile abundant. Genae as wide as first antennal segment; third antennal segment pointed but quite long . . . . . **Amictogeron** Hesse
Male and female face with abundant, long erect pile beneath the antenna. Pile of first antennal segment, shorter, scantier, distinctly appressed. Anteroventral genal pile scanty. Genae quite linear. Third antennal segment shorter, pointed, more strongly tapered (*Geron* Meigen, and subgenera) 3
3. Length of last section of fifth branch of radius to the edge of wing about as long as the preceding portion extending to anterior crossvein . . . . ***Geron* (*Empidigeron*)** Painter
Length of last section of fifth branch of radius to the edge of wing never more than two-thirds as long as preceding section to the anterior crossvein, and often much less (*Geron* Meigen, sensu stricto) . . . . . . ***Geron* Meigen**

Genus *Geron* Meigen

FIGURES 104, 105, 309, 523, 543, 754, 755, 859, 860, 861

Geron Meigen, Systematische Beschreibung, vol. 2, p. 223, 1820. Type of genus: *Bombylius gibbosus* Olivier, 1789, as *Geron gibbosus* Meigen, 1820. Designated by Rondani, 1856. Not *Bombylius hybridus* Meigen, 1804, as given by Bezzi, p. 2, 1924, which is a synonym of *B. gibbosus* Olivier.
Empidigeron Painter, subgenus, Trans. American Ent. Soc., vol. 58, p. 143, 1932a. Type of subgenus: *Geron snowi* Painter, 1932a, by original designation.

Bezzi (1924), pp. 112-113, key to 11 Ethiopian species.
Paramonov (1929), p. 185, key to Palaearctic species.
Painter (1932), pp. 144-146, key to 19 Nearctic species.
Engel (1933), pp. 132-133, key to 5 Palaearctic species.
Hesse (1938), pp. 873-883, key to 23 South African species.
Efflatoun (1945), pp. 136-137, key to 7 species and subspecies from Egypt.

Small to medium-sized black flies with only the tibiae and basitarsi and sometimes the femora yellow or brown. They are readily recognized by the presence of only 3 posterior cells, the basally stout, conically tapered abdomen which is a little compressed laterally, the humpbacked mesonotum, and the characteristic antenna. The elongate third antennal segment is strongly tapered from base to apex so that it ends acutely and much attenuate and tipped by a minute, short spine.

The wings are hyaline; the first submarginal cell is short and the anterior branch of the third vein arises in a steep arch; the anal cell is closed. The vestiture of these flies consists of sometimes dense, sometimes sparse, long, fine, erect pile in which there is frequently much, short, glittering, narrowly scaliform and curled or appressed pile on front, mesonotum, and especially the abdomen. Females tend to have more glittering, curled, appressed pile, therefore, the sexes are somewhat dissimilar in appearance. They become very readily denuded.

This genus is distributed through the Palaearctic region on down into Australia; it extends into South Africa, and Chile, and with 22 species in the United States.

Length: 3 to 9 mm. including antenna; wing 2.5 to 8.2 mm.

Geron Meigen, has one subgenus: *Empidigeron* Painter. These flies are characterized by the relative length of the last section of the fifth branch of the radius; this section is relatively long. In *Geron* Meigen in the strict sense, the last section of the fifth branch of the radius is always less than two-thirds the length of the preceding section of that vein as it ends at the anterior crossvein; it is usually much shorter so that the apical cell is relatively short and broad rather than relatively long and narrow as in *Empidigeron* Painter, the flies of which are less numerous and found in the Southwestern United States.

Head, lateral aspect: The head is subglobular, strongly and convexly rounded on the anterior half and with a rather prominent swollen occiput especially in the middle of the head, sloping gradually downward, upward and toward the sides of the eye margin. In the

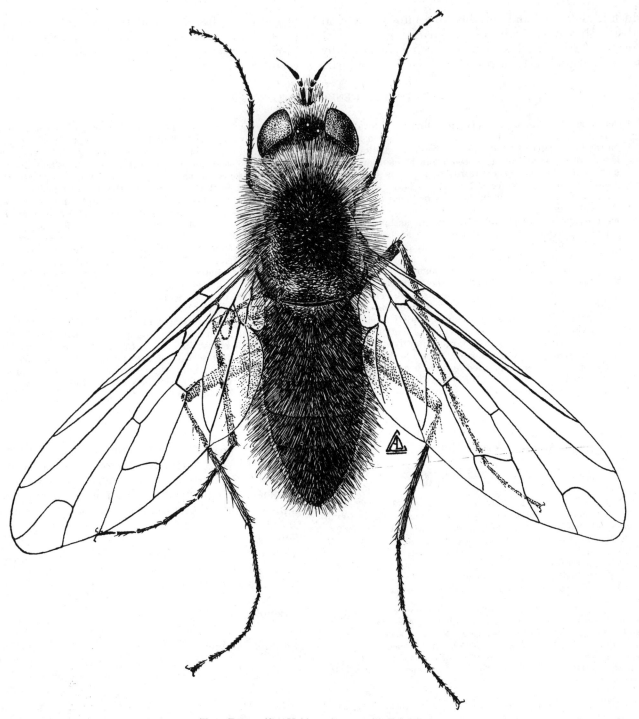

Text-Figure 27.—Habitus, *Geron senilis* Fabricius.

males the upper occiput is more shallow, in the females more extensive. The pile of the occiput is long, dense, and bristly with some shorter, flat-appressed, curled, glittering, scaliform pile intermingled and reaching quite to the eye margin. The front is somewhat flattened, generally plane with the eye, and there is characteristically a low, foveal depression transversely across the lower front shortly in front of the antennae. In males the front and the junction of the eye margins are distinctly sunken with the whole upper occiput flattened and plane from the lateral aspect and the upper half of the eye triangular. The pile of the small male front is pollinose with a few appressed, scaliform hairs in the middle and along the eye margin. The ocellar tubercle is strongly swollen with a tuft of coarse, erect, bristly hairs, and the ocelli rest almost on the eye margin. The pile of the front in the female is pollinose, often dark in the center but silvery on the outside and with either a band of long, coarse, shining bristles down the middle of the front, curved forward, or with flattened scales laterally and smaller, flat-appressed, curled, scaliform pile in the middle or with the long medial bristles wanting.

The face is quite short in profile; both have broad, appressed scales near the eye margin and some longer, slender, generally silvery bristles directed downward along the sides. The walls of the oragenal cup are short, rimlike, following the convex contour of the eye to the bottom; gena absent. The outer margin of this wall is obliquely steep, yet short, is pollinose, and may possibly be considered genal. The posterior eye margin is plane. The proboscis is quite long and very slender, 2 or 3 times as long as the head, the labellum equally slender, the palpus slender but short, bearing several setae at the apex. The antennae are attached near the upper third of the head, they are slender and elongate, the first segment is 2 to 3 or more times as long as the second segment and has quite abundant, long, fine, oblique bristles or bristly setae. The second segment is cylindrical, of uniform width and usually slightly longer than wide. Third segment a little longer than the first two segments together, very strongly tapered and attenuate to the apex, which bears a distinct apical spine. The basal fifth of the first segment is widest, not quite as wide as the second segment.

Head, anterior aspect: The head is quite circular, the eyes widely holoptic in the male for a distance 4 or 5 times as long as the front but in the female the eyes are widely separated and only slightly convergent dorsally. The vertex is about 3 times as wide as the distance between the posterior ocelli. The antennae are narrowly separated at the base, the oragenal cup is more or less rectangular and quite deep with steeply slanting interior walls. Dorsally it begins a long distance from the antennae, and hence the face beneath the antennae is almost as long as wide. There is no tentorial fissure, and the gena appears to be absent except for a minute trace below.

Thorax: The thorax is compact with the mesonotum strongly humped and convex, a little more so anteriorly. The pile of the mesonotum is scanty, fine, and either erect or suberect, very easily denuded, remnants of pile more apt to be preserved along the anterior margin. In females the pile is generally much more abundant, dense, long and bushy; it is better preserved on the anterior third of the mesonotum posteriorly and on the scutellum; mesonotum always pollinose, and rather feebly shining. The humerus is densely pilose, the notopleuron likewise with at most very slender bristles. The pile of postalar callosity is similarly bristly and the scutellum with dense, fine, long, erect pile, except when denuded. In addition to the erect pile both sexes tend to have a varying amount of brilliant, glittering, gold or silver, short, curled, appressed, scaliform pile. The pleuron is pollinose; whole of the mesopleuron and upper sternopleuron densely, long pilose, the latter also with some shining, appressed scales; pteropleuron, metanotum, the anterior metapleuron, and the hypopleuron without pile, except there may be a small patch of scales at the bottom of the pteropleuron, and a patch of scales and some long hairs are on the posterior metapleuron. The alary portion of the squama is moderately long with a radial tuft of long hairs at the lateral corners, and sometimes a scanty fringe over the remainder of the border; halteres large, posteriorly elongate.

Legs: The legs are moderately long and slender, including the femora. The femoral pile consists of matted, flat-appressed scales and a fringe of stiff, shaggy pile ventrally on all three pairs. The anterior tibia has only fine, small, appressed, setate pile; spicules absent, except for perhaps 2 or 3 very minute, short, oblique, slender bristles which are scarcely noticeable. On the middle and hind tibia the spiculate bristles are somewhat longer, still quite slender, oblique, and few in numbers, usually about 3 to 5 dorsally, sometimes as many as 10 on the hind tibia in the anterodorsal row, 6 in the anteroventral row and the same number in the posteroventral row; tarsi rather long and quite slender; claws slender, nearly straight, except at the apex where they are curved; pulvilli long.

Wings: The wings are hyaline, and rather broad at the base, though never as much so as in *Phthiria* Meigen. Most of the veins are weak but distinct. There are only 2 submarginal cells, and it should be noted that there are only 3 posterior cells. The anal cell is closed with a long stalk; vein closing the discal cell either nearly straight or slightly sigmoid. The anterior crossvein enters the discal cell at or just beyond the middle. The second submarginal cell in *Geron* Meigen, sensu stricto, is quite short, but in the subgenus *Empidigeron* Painter it is nearly twice as long as wide.

Abdomen: The abdomen is moderately elongate, at the base nearly or quite as wide as the mesonotum and strongly and characteristically tapered and generally compressed laterally, so that viewed laterally it is rather high. It is only about as long as the thorax with the scutellum included, sometimes slightly longer and occasionally shorter. The pile, except in denuded individuals, is very fine, quite long and erect, varying in quantity and with a considerable amount of curled, flattened, scaliform, glittering pile which is silvery or golden in color. Although it varies, females tend to have more of the glittering pile on both abdomen and thorax and either somewhat less or somewhat shorter, erect, shining pile. The female terminalia consist of the short, conical, or triangular eighth tergite, laterally compressed into 2 shorter, apposed flaps posteriorly, the last sternite making a well-separated, shovellike extension below. The male terminalia consist of an elongate pair of apposed or narrowly separated, spikelike or thornlike protuberances, which represent the apices of the basistyli with the more slender, sharper processes of the dististyli lying between; this is the digitate form of the genitalia of Painter (1932): they also occur in what Painter calls the rosette form. He says that these forms are interchangeable.

Inasmuch as Dr. A. J. Hesse has been able to dissect more species than I have I quote his remarks on the male genitalia:

Hypopygium of male like that of the other *Geroninae*, is more complicated than in any other genus belonging to the *Bombyliidae-Homoeophthalmae*. The hypopygium can only be satisfactorily studied in a liquid medium. As in other *Geroninae* the basal parts form a single structure, not being divided into two symmetrical parts by a dorsal suture or impression and what is designated by me as dorsal is more commonly ventral in position in the intact and set insect, the aedeagus and its attendant structures being more constantly dorsal in position. The basal part is usually more membranous than chitinous on each side towards the apex, the membranous part not having hair or only a few bristly hairs. The dorsum sometimes with hairs towards apex and in some cases with a few or a row of distinct hook-like spines on each side. The distinct and freely movable beaked apical joints of other *Homoeophthalmae* are entirely absent, being represented simply by scarcely movable lobes, by paired or symmetrical extensions, processes or lobes apically on the basal part. These processes in *Geron* assume various shapes in the different species and are thus of taxonomic value in their separation. They are lobe-like, spine-like, hook-like, or dentate and are represented singly, in duplicate or even in triplicate on each side apically. The aedeagal apparatus is complex, composed of a medium straight, curved or even duplicated aedeagus proper ending basally in a spoon-shaped or ladle-like structure, which has a lateral, flattened or flap-like process on each side, corresponding to the lateral struts of other *Bombyliidae* and a dorsally directed bat or racket-shaped basal strut as in other *Bombyliidae*.

The medial aedeagal structure passes through a central guide which is usually produced into a flattened lobe-like process on each side towards base and apically is often pitchfork-like, produced into a long apically directed, straight or curved, spine-like or sabre-like process or prong on each side of the aedeagus either above or below. These prongs, when present, together with the aedeagus, are lodged in or more to and from in a central or median dorsal guide in the form of a partially chitinous or entirely chitinous trough-like or sheath-like structure, usually thinned out towards base or apex or broadened apically or basally, where it is also joined on to the subsidiary apical processes or lobes of the basal part by means of ligaments or membranes. In some species where the central guide is not produced apically into a prong there is nevertheless present on each side a curved or U-twisted spine. Ventral to the aedeagus the central guide abuts, or is joined, on to the basal part apically on each side by a raised strap-like or arch-like ramus, which is sometimes produced apically into a prominent flattened, lobe-like or even spine-like process.

The male terminalia from dorsal aspect, epandrium removed, show a complex dististylus with several acessory lobes all narrow, mostly curved outward and downward. The ramus is small and makes a transverse bridge above the epiphallus. Laterally the basistylus is large, and high, that is, extensive dorsally. The epandrium is subtriangular.

Material studied: A considerable series of many species from Mississippi and the Southwestern United States.

Immature stages: These flies are parasites of tortricid, pyralid, and psychid moths. See discussion under life histories.

Ecology and behavior of adults: These flies come to flowers but are not restricted to composites, though they more often are found there. I have frequently collected them from the heads of composites and have noted a characteristic habit of bobbing up and down several inches above the flower head before settling upon it. They also hover in sunny spots. Painter (1932) states that some species, including the large species *Geron grandis* Painter, are partial to button bush, *Cephalanthus occidentalis*; he also notes that they may be found at the tips of twigs resting much as some asilids do.

On a research collecting expedition to New Mexico, in September 1967, the author collected 99 individuals of a species of this genus in about 20 minutes on a patch of sticky, yellow-flowered composite, *Haplopappus ciliatus* (Nutt.) DC., no more than 50 yards long by 10 yards wide, near a west Texas highway; many others were left; only a small clump of trees was nearby among open fields. Almost without exception these flies selected unoccupied flower heads upon which to alight. One fly was snatched by a small spider as it alighted. Possibly this marked habit of bobbing up and down upon flower heads before alighting is a precautionary instinctive habit. No less than six species or types of spiders were seen on these flower heads besides reduviids; phymatids were absent.

Distribution: Nearctic: *Geron aequalis* Painter, 1932, subg. *Empidigeron*; *albarius* Painter, 1932; *albidipennis* Loew, 1869; *arenicola* Painter, 1932; *argutus* Painter, 1932; *aridus* Painter, 1932, subg. *Empidigeron*; *calvus* Loew, 1863, subg. *Empidigeron* [=*macropterus* Loew, 1869]; *digitarius* Cresson, 1919; *grandis* Painter, 1932; *holosericeus* Walker, 1849 [=*versicolor* Painter, 1932]; *hybus* Coquillett, 1894, subg. *Empidigeron*; *johnsoni* Painter, 1932; *nigripes* Painter, 1932; *niveus* Cresson, 1919; *nudus* Painter, 1932, subg. *Empidigeron*; *parvidus* Painter, 1932; *robustus* Cresson, 1919 (as subsp. of *digitaria*); *snowi* Painter, 1932, subg. *Empidigeron*; *subauratus* Loew, 1863; *trochilus* Coquillett, 1894; *vitripennis* Loew, 1869; *winburni* Painter, 1932.

Neotropical: *Geron albus* Cole, 1923; *albidus* Walker, 1857; *canus* Philippi, 1865; *colei*, nomen nudum; *insularis* (Bigot), 1856 (as *Bombylius*); *litoralis* Painter, 1932; *niveoides* Cole, 1923; *rufipes* Macquart, 1846; *senilis* (Fabricius), 1794 (as *Bombylius*) [=*albidipennis* Loew, 1869, =*vitripennis* Loew, 1869]; *trochilides* Williston, 1901; *insularis* Cole.

Palaearctic: *Geron corcyreus* Frey, 1936; *emiliae* Zaitsev, 1964; *garagniae* Efflatoun, 1945; *gibbosus* Olivier, 1789, *gibbosus erythropus* Bezzi, 1925, *gibbosus halteralis* Wiedemann in Meigen, 1820, *gibbosus subflavofemoratus* Rubio, 1959 [=*capensis* Walker, 1852, =*gibbosus* Meigen, 1820, =*hydridus* (Meigen), 1804, as *Bombylius*), =*olivierii* Macquart, 1840]; *hesperidum* Frey, 1936; *intonsus* Bezzi, 1925 [=*krymensis* Paramonov, 1929]; *longibarbus* Efflatoun, 1945; *longiventris* Efflatoun, 1945; *phallophorus* Bezzi, 1922; *priapeus* Bezzi, 1922.

Ethiopian: *Geron anceps* Hesse, 1938; *australis* Hesse, 1938; *bechuanus* Hesse, 1936; *delicatus* Hesse, 1938; *dichroma* Bigot, 1892; *dissors* Hesse, 1938; *dubiosus* Hesse, 1938; *furcifer* Hesse, 1938; *gariepinus* Hesse, 1938; *lactipennis* Hesse, 1938; *latifrons* Hesse, 1938; *lepidus* Bowden, 1962; *maculifacies* Hesse, 1938; *munroi* Hesse, 1938; *mystacinus* Bezzi, 1924; *nasutus* Bezzi, 1924; *nigerrimus* Hesse, 1938; *nigrifacies* Hesse, 1938; *niveus* Hesse, 1938; *nomadicus* Hesse, 1938, *nomadicus breyeri* Hesse, 1938; *orthoperus* Hesse, 1938; *peringueyi* Hesse, 1938 [=*hybridus* Bezzi, 1921, not Meigen]; *psammobates* Hesse, 1938; *semifuscus* Séguy, 1933; *transvaalensis* Hesse, 1938; *turneri* Hesse, 1938; *parvus* Hesse, 1938.

Oriental: *Geron albescens* Brunetti, 1909; *argentifrons* Brunetti, 1910.

Australian: *Geron australis* Macquart, 1840; *cothurnatus* Bigot, 1892; *dispar* Macquart, 1849; *hilaris* White, 1917; *nigralis* Roberts, 1929; *simplex* Walker, 1859.

Oceania: *Geron umbripennis* Bezzi, 1926.

Patria Ignota: *Geron ?tenuis* Walker, 1857.

Genus *Amictogeron* Hesse

FIGURES 106, 310, 530, 537, 776

Amictogeron Hesse, Ann. South African Mus., vol. 34, p. 918, 1938. Type of genus: *Amictogeron meromelas* Hesse, 1938, by original designation.

Hesse (1938), pp. 921-929, key to males of 15 species, females of 12 species.

Small flies that resemble *Geron* Meigen and with which it is closely related. It may be distinguished by the absence of pile upon the face and the slender, much longer, much less attenuate third antennal segment; also males bear very long, dense bristly pile all around the surface of the long, basally swollen first antennal segment, much in contrast to the males of *Geron* species known to me. The second submarginal cell is long as in *Geron* Meigen.

These flies have a convex and humped mesonotum as in *Geron* Meigen. The head is subglobular, the male eyes extensively holoptic, whereas this genus has long, upturned oragenal pile, none on face; *Geron* Meigen is the reverse, it has facial pile, none on the oragenal strip. Hesse (1938) gives the following additional differences from *Geron* Meigen species known to him:

First antennal segment longer than in *Geron* Meigen, at least 3 and usually much more than 3 times as long as second segment; antennae more widely separated at base than in *Geron* Meigen, but less widely than in *Pseudoamictus* Bigot; third antennal segment more slender than in *Geron* Meigen, more rod-like and more slender, ending in a scarcely visible terminal microsegment which ends in a minute style or spine; palpi very slender, longer than in *Geron* Meigen; upper and lower veins of second submarginal cell generally nearly parallel; alula narrower and less developed than in *Geron* Meigen.

The apical crossvein of discal cell usually more distinctly sigmoid than in *Geron* Meigen; laterally the tergites below do not overlap the sternites as in many species of *Geron* Meigen; spicules of tibiae usually more poorly developed than in *Geron* Meigen; erect pile of the body less dense and shorter even in males than *Geron* Meigen; hairs of first antennal segment and of gena, denser, longer, more bushy, and often very conspicuous in contrast to *Geron* Meigen; no flattened, silvery scalelike pile on the head behind the eyes, on gena, or on frons as is so characteristic of *Geron* Meigen.

Hesse also contrasts minor differences between the terminalia of the genera. He notes a pronounced sexual dimorphism in *Amictogeron*; females having more yellowish markings on pleuron and abdomen, paler or more yellowish femora and tibiae, no long bushy hairs on first antennal segments, no basal pronounced knoblike thickening on first antennal segment, shorter and sparser pubescence on the body, and denser scaling. I have quoted these distinctions from Hesse since the three paratypes of the type-species before me are in very bad condition.

Hesse notes that this genus can only be confused with *Pseudoamictus* Bigot, from which it differs in having a more humped thorax, more globular head, more narrowed oragenal recess, less widely separated antennal segments, narrower genae, narrower interocular space in females, long contact distance of male eyes. He considers *Amictogeron* Hesse a transitional genus between *Geron* Meigen and *Pseudoamictus* Bigot.

The species of *Amictogeron* Hesse are restricted to South Africa.

Length: 3.5 to 7 mm.; wing 4 to 7 mm.

Head, lateral aspect: The head is globular, the occiput extensive and swollen and with scattered, rather scanty, shaggy pile and bristles; in females the occiput is prominent above and below, in males only below, rapidly flattening and receding near the vertex; there is no central cavity, no indentation. The front is rather flattened. A small quadrate depression immediately in front of the antennae bears scattered, stiff, bristly hairs on the upper half which are curled forward. The ocellar tubercle is rather strongly swollen, the ocelli occupying an obtuse triangle with a cluster of 4 or 5 erect bristles. The face, without pile, thinly pollinose, is short in profile, shorter in males, its height is little more than the thickness of the first antennal segment; the antennae rest on the eye margin in lateral aspect, hence the face is not greatly different from the obtusely flattened face of *Geron* Meigen, the surface is pollinose and has 2 or 3 minute hairs.

The lower sides of the oragenal cup and genal area have long, matted, shaggy, flattened hairs directed upward along the sides of the eye; postmargin of the entire eye. The proboscis is long and slender, arising from the bottom of the head and extending forward and is at least twice as long as the head. The palpus is long, but so thin as to be threadlike. The antennae are long and slender, especially the first and third segments. The first segment is 5 or 6 times as long as wide and at least 6 times as long as the short, beadlike second segment. The first segment bears numerous, rather long bristles along the lateral and ventral margins, which are longer and more conspicuous basally; second segment with a few setae laterally at the apex. The third segment is more narrow than the second, straight, nearly cylindrical, and of almost uniform thickness to the apex. Near the apex it becomes strongly narrowed, the sharp apex bears a minute conical spine; inner margin of the third segment with short, suberect setae.

Head, anterior aspect: The head is very short oval, the head nearly as wide as the thorax, the eyes in the male are extensively holoptic for about four-fifths the distance of antennae to ocelli; the line of contact is sunken and the upper facets larger. In the female the eyes are rather strongly separated by somewhat less than in *Geron* Meigen. The ocellar triangle is small, equilateral, and prominent with vertical sides and bears several erect bristly hairs. The vertex is very little wider in the female than the ocellar triangle. The front is small and rhomboidal. The antennae are separated at the base by almost the thickness of one segment. The oragenal cup is ventral in position leaving the face long beneath the antennae, more than half as long as its width. The oragenal cup is quite deep, narrowing anteriorly, truncate beneath the face with the inner walls quite steep and the genal space extensive from the anterior view and bearing a dense tuft of long, shaggy, matted pile directed forward; tentorial fissure or pits absent.

Thorax: The mesonotum is rather strongly humped; it has both erect, long, coarse dark pile intermixed with erect, fine, pale pile and also curled, flat-appressed, glittering hair, nowhere obscuring the ground color. In contrast to *Geron* Meigen, there is always some black pile present. Bristles absent. The scutellum is large, unusually long and exceptionally thick and convex with abundant long pile similar to the mesonotum. The humeri and mesopleuron, sternopleuron and pteropleuron have abundant long fine pile; posthypopleuron with only a few fine, somewhat shorter hairs, metapleuron apilose.

Legs: The legs are quite slender and have scales on the hind femur with an anteroventral fringe of moderately long bristly hair that is rather fine; tibiae and tarsi exceptionally slender and elongate, the hind pair with 8 quite short dorsomedial bristles and 4 or 5 still shorter anteroventral bristles. The anterior tibia has 5 or 6 extremely short, ventral bristles on the outer half, still shorter ones anteriorly; pulvilli well developed.

Wings: The wings are iridescent and hyaline, the subcostal cell yellow, veins pale brown; both first and second veins are nearly straight; the radial sector is short and ends well before the end of the second basal cell; the first and second posterior cells and discal cells gradually diverge. The anterior crossvein is at the middle of the discal cell. The anterior branch of the third vein arises in an oblique, rounded curve almost exactly at the midpoint of the first posterior cell; the remainder of that vein is plane and ends at the apex of the wing. There are 3 posterior cells. the anal cell is closed and shortly stalked, the axillary lobe is large, the alula small, the ambient vein complete; marginal villi moderately long.

Abdomen: The abdomen is short, a little longer than the thorax and somewhat compressed laterally; it is distinctly pinched and compressed laterally. There are 9 segments that may be seen from above, the eighth is shorter and exposed chiefly laterally. The pile of the abdomen is quite scanty but long and bristly and slender and erect.

The pile is intermixed, with which on the dark pollinose background is some scanty, short, curled, shining, somewhat flattened and appressed pile. Female terminalia simple, surrounded by long, bristly hairs on edges of tergites.

Male terminalia very complex and variable between species; of 21 species, Hesse (1938) figures 15; I summarize the comments of Hesse on terminalia in this fashion: like *Geron* Meigen in having the basal part single, not divided into two symmetrical halves by dorsal suture or depression. The greater part of the dorsum is membranous, marked off by a chitinous strip on each side passing into the more rigid chitinous sides; bristly hairs if present placed dorsally on chitinous part; the basal part passes apically into a flattened, lobelike or lappetlike apical process on each side corresponding to the true dististyli (beaked apical segments) of Bombyliinae; these lobes or processes, as in *Geron* Meigen, are not hinged and movable as in the Bombyliini; they are usually flattened, lappet or triangularlike, not boss or fingerlike as in *Geron* Meigen; dorsally each apical process usually has an adpressed spine, at their bases there is ventrally a single spine or recurved process, or a projecting spined process. The medial aedeagal complex is placed in the basal part, attached on each side toward apex by a ramus; the sides of that part joining onto the ramus usually prolonged into a flap or lobe.

The prolongation of this lobe assumes various forms of taxonomic value; on each side the ramus is usually produced apically into a straplike, hooklike, or spinelike process, into a strongly recurved spine or even more complex figure. The ramus on each side meets or is joined onto a medial, central guide, not so well defined as in *Geron* Meigen, which surrounds or into which passes the middle part of aedeagal complex and there may be one or more additional spinelike processes.

Material available for study: The series in the British Museum (Natural History) and also males and females of *meromelas* Hesse sent by Dr. Hesse for study and dissection.

Immature stages: Unknown.

Ecology and behavior of adults: Hesse (1938) found one species settling upon flowers of *Mesembryanthemum* sp.

Distribution: Ethiopian: *Amictogeron anomalus* Hesse, 1938; *barbatus* Bezzi, 1921; *basutoensis* Hesse, 1938; *bezzii* Hesse, 1938; *capicolus* Hesse, 1938; *chellicterus* Hesse, 1938; *consors* Hesse, 1938; *dasycerus* Hesse, 1938; *disparilis* Hesse, 1938; *fuscipes* Hesse,

1938; *karooanus* Hesse, 1938; *lasiocornis* Hesse, 1938; *leptocerus* Bezzi, 1921; *marshalli* Hesse, 1938; *meromelas* Hesse, 1938; *montanus* Hesse, 1938; *namaensis* Hesse, 1938; *nigrifemoris* Hesse, 1938; *peringueyi* Hesse, 1938; *phaeopteris* Hesse, 1938; *waltoni* Hesse, 1938.

Genus *Pseudoamictus* Bigot

FIGURES 103, 311, 548, 555, 862, 863, 864

Pseudoamictus Bigot, Ann. Soc. Ent. France, ser. 7, vol. 2, p. 342, 1892. Type of genus: *Amictus heteropterus* Wiedemann, 1821, not Macquart, by original designation.
Pseudempis Bezzi, Ann. South African Mus., vol. 18, p. 94, 1921. Type of genus: *Amictus heteropterus* Wiedemann, 1821.

Hesse (1938), pp. 960-962, key to 3 South African species.

Curious, blackish flies with brown or gray pollen and with mostly brownish or reddish golden pile that is fine, wiry, partly appressed and upon the whole, rather scanty. The size is larger than the species of *Geron* Meigen and the form somewhat more elongate; the abdomen tends to be cylindroid rather than strongly tapered as in that genus. There is a curious, therevidlike appearance; the head is subglobular and the antennae are elongate. The first antennal segment is as long as the remaining two, strongly widened or swollen, progressively toward the base and the two segments lie adjacent and bear considerable, long, fine, dense, subappressed pile on all sides except the medial aspect; the second segment is beadlike, the third segment progressively tapered to a narrow bristle-tipped apex. The face is extensive ventrally and obliquely and has pollen only but the wide gena bears a conspicuous, vertical, dense, forwardly directed band of stiff, shaggy, or bristly hairs.

The wings are rather long and slender and uniformly tinted with reddish-brown villi. The anterior branch of the third vein arises rather abruptly, and after making a broad, basal arch forward, extends in nearly a straight line to the wing apex. The vein closing the discal cell is strongly sigmoid, the discal cell much longer anteriorly than posteriorly. The anterior crossvein enters the discal cell at or beyond the middle and the anal cell is closed in the margin. As Hesse (1938) notes, there is a tendency for the base of the second submarginal cell to be opposite the apex of the discal cell. It is notable that the eyes in the male are separated, much separated in the type-species. The gena is much broader than *Geron* Meigen or *Amictogeron* Hesse. The male terminalia like the other flies in this subfamily are quite complex; they are much like *Amictogeron* Hesse; it has a single undivided basimere, ending apically on each side in an apical, lappetlike process, directed obliquely, and provided with a dorsal spine.

The few species are entirely South African in distribution.

Length: 5 to 12 mm.; of wing 5 to 12 mm.

Head, lateral aspect: The head is globose, with the face extensive, produced forward a little more than half the length of the eye. It is distinctly sloping beneath the antenna, slightly convex, blunt at the anteriormost ledge above the oragenal cavity. From the lower part of the face, the oragenal cavity slopes obliquely backward. The gena is very wide, and bears conspicuous, long, rather dense, tufts of coarse, bristly, reddish-golden hair or weak bristles, arising near the eye margin well below the middle of the eye and extending as a brush obliquely upward along the sides of the oragenal cavity. The ground color of the entire head is black or brownish black, but completely obscured everywhere by a fine, gray pollen. The occiput is prominent but only toward the middle of the head. It is especially extensive ventrally and is also prominent even below the eye. It slopes gradually inward lengthening all the while, and there is no deep interior cup as found in the division Tomophthalmae. Viewed from above there is no fissure or lobe behind the vertex. The ocelli rest upon a conspicuous, rather high, compact tubercle and are attached to the sides of this tubercle. The posterior ocelli lie behind the eye margin. The occiput is coarsely subpunctate where the sparse pile arises. Outwardly the pile is long, rather fine, stiff but wiry, but the more medial pile is long, matted, coarse, and golden. A fine, wiry, black pile on the upper half of the head extends well beyond the eye margin, although none of it arises at the eye margin. The matted yellow pile scarcely extends beyond the eye margin.

There is not even a depression behind the short column of the ocelli tubercle. The palpus is remarkably filiform and, contrary to the statement of Hesse, it appears to be bent in such a way that there is a distinct long, nearly bare basal segment followed by a longer, equally slender, slightly curved, short, fine, bristly pilose, apical segment. The entire length of the palpus is less than that of the first antennal segment. The proboscis is slender and from extreme base to apex it is only $1\frac{1}{2}$ times as long as the rather long head. The labellum is likewise slender. The antennae are conspicuous and elongate; the first segment is especially long; it is widest and somewhat swollen basally from the point of attachment, and it is as long as the next two segments together, and very hairy on all sides except medially. The medial surface is covered by fine, gray pollen. The hair or pile of the first segment above, below, and laterally is long, rather dense, coarse, bristly, subappressed and reddish brown, extending above to the end of the second segment. The second segment is short and beadlike and of about the same width as the narrowed apex of the first segment. The first segment is only gradually narrowed. The second segment is truncate apically but rounded at the base, and there are a few, short setae. The third segment is a little shorter than the first, of the same width as the second segment, nonannulate from the lateral aspect, with parallel sides on the basal sixth; it then begins to narrow to just beyond the middle where there is a slender, attenuate, stylelike portion quite narrow at the apex, but with a very minute spine.

The pile of the first antennal segment, at least where it rises from the base, is almost as long as the third antennal segment.

Head, anterior aspect: The head is slightly wider than the thorax, oval, with the oragenal and the genofacial space rather rapidly widening below. Below the antenna this space is almost a third of the head's width; at the bottom of the head, it is at least half the head's width. The genae are well-developed below the eye, they are very wide, and the shallow oragenal cup is narrow with a vertical shining clypeal strip in the middle, margined by a narrow, thin, yellowish sidewall of no great depth, on the outside of which is a shallow crease that appears to be part of a tentorial fissure; lateral to this is another pale yellow ridge ascending to the wide pollinose gena. In males the eyes are separated by the width of the ocellar tubercle; in the females the interocular space on the vertex is not broader than twice the width of the ocellar tubercle. Hesse notes that the front of the male is larger, broader, and longer than either *Geron* or *Amictogeron* Hesse. The front in the male is distinctly depressed medially on the lower half above the antenna and extends rhomboidally between the base of the antenna, the segments of which are narrowly separated. Yet, because they extend medially toward each other, they almost touch throughout most of their length. The second and third antennal segments extend obliquely outward and the third segment is conspicuously annulated only when viewed from above.

The upper eye facets are somewhat more coarse and enlarged and distinctly arranged in a circular pattern on each side of the front. The posterior eye margin is quite plane throughout, and there is a very faint, impressed crease on the occiput separating off the bare marginal portion of the occiput which bears no pile.

Thorax: The thorax is slightly compressed laterally and a little convex on the mesonotum but not conspicuously so. The whole thorax is brown or brownish black with a light reddish or sepia-brown pollen. The pile is scanty, straggly, and scattered, rather dense or abundant laterally, but nowhere obscuring the ground color. There are two types of hair; a long, stiff, shining-yellow, curled, erect or partly erect hair, more abundant on the anterior third and slightly more in front of the scutellum, and especially abundant on the lateral margin, forming extremely weak, slender bristles, perhaps no more than bristly hairs on the notopleuron, the postalar callosity, and on the margin and disc of the scutellum. The convex, opaque, brown pollinose, scutellum is almost as long as wide. It forms perhaps more than a hemicircle. Its pile is scanty. The humeral pile is very much like that of the anterior mesonotum. The pleural pile is quite scanty, except dorsally on the mesopleuron where it is loose and more abundant. While the whole pleuron is pollinose and opaque, the pteropleuron, the metapleuron, and the metanotum are bare; the hypopleuron has a few, long scattered hairs beneath the very large haltere which has a curiously elongate knob. The squama has a scanty fringe of fine hair.

Legs: The legs are quite slender and comparatively elongate. The anterior femur has a few stiff, long hairs ventrally near the base. The middle femur has a complete fringe of long, stiff hairs ventrally and similar short hair anteriorly. The hind femur has a ventral lateral fringe of more scanty, long, stiffened hairs. All the femora have scattered, flat-appressed, yellowish, flattened, scalelike hairs. All three pairs of tibia have a few spicules near the apex. Middle tibia with 4 short, stout, spinous setae ventrally at the apex and hind tibia with a circle of at least 12 black spicules on the ventral half. All tarsi long and slender. Ventral spicules are more conspicuous on the hind tarsus. Claws slender, the pulvilli long and spatulate. The arolium acute at the apex.

Wings: Large, much longer than the abdomen and uniformly tinted with brown a little wider at the base, and with a rather distinctive venation. The third vein is slightly arched backward to end well below the apex of the wing. The anterior branch of the third vein arises first rectangularly, then makes a broad curve and is plane throughout much of its length and ends at the wing apex. The anterior crossvein is oblique and enters the discal cell just beyond the middle. There are only 3 posterior cells and the vein closing the discal cell is very strongly sigmoid. Anal cell narrowly open or closed in the margin. The axillary lobe is comparatively wide, but the alula narrow; ambient vein complete. The base of the costa is thickened for a short distance and there is a weak costal comb of oblique setae, with a flattened basal scale without hook.

Abdomen: The abdomen is elongate and comparatively slender with a tendency for the sides to be compressed and it is at least nearly twice as long as the mesonotum, the scutellum excepted. The abdomen is dark, brownish black, opaque, with a yellowish-brown pollen, the eighth tergite is rather light orange brown. The second tergite is a little longer than the next two tergites. The seventh tergite is only apparent laterally. The pile on the side of the first tergite is long, stiff, erect, and golden, but rather scanty. All the remaining tergites except the last one have some scattered, more or less erect, fine, long, yellow hairs together with somewhat more abundant, still scanty, appressed, curled, slightly flattened golden hairs. The rather large terminalia are directed straight backward, or posteriorly, and are fairly complex, with accessory prongs and spurs, which are shining brown or brownish black. Dr. Hesse describes the male terminalia in the following way:

The hypopygium is very much like of *Amictogeron* Hesse. Like that genus it has a single, undivided basal part, ending apically on each side in a lappet-like lobe, which is provided dorsally with a spine as in *Amictogeron* Hesse; with the strongly chitinised strand on each side from base of each apical lappet-like process (which passes) obliquely to the side, sometimes provided with a strong, apically directed spine. The ramus on each side is produced apically into a strap-like process of a bifid process. The central, guide-like part, which abuts on the bases of the rami, with or without an apically directed prong placed on each side and situated dorsal to the aedeagus.

The male terminalia from dorsal aspect, epandrium removed, show a compact, elongate-oval structure in which the dististyli are short, attenuate, and with a rather strong, divergent, curved, hornlike tip. In addition there are no less than 3 pairs of accessory processes, which in this work we term claspers, and these prominent strong, curved, hornlike processes lie dorsal to the epiphallus and aedeagus. From the lateral aspect all of these structures emerge from a single undivided basistylus, and the long flattened leaflike ramus is likewise attached laterally to the basistylus, and is extended obliquely upward on each side of this clasper-complex. The epandrium is large, oval, pointed apically, leaf-shaped, extended almost directly upward, and bears very large cerci.

Material available for study: I am indebted to Dr. A. J. Hesse of the South African Museum, Capetown, for the loan of a male of the type-species.

Immature stages: Unknown.

Ecology and behavior of adults: Not on record.

Distribution: Ethiopian: *Pseudoamictus bezzii* Paramonov, 1930 [=*heteropterus* Bezzi, not Wiedemann]; *heteropterus* Wiedemann, 1821 (as *Amictus*), also (as *Pseudempis*, Bezzi, 1921); *luctuosus* Bezzi, 1921 (as *Geron*).

Subfamily Usiinae Becker, 1912

Small flies which are readily recognized in the type-species by the short thorax and scutellum, the short, very broad, compact abdomen that is much wider than the thorax, the long slender proboscis, the presence of only 3 posterior cells with anal cell always closed and stalked; also the pilosity everywhere is minute, appressed, and setiform. The head is small, much more narrow than the thorax. The cheeks and midlateral face are reduced to a mere line. The greatly swollen occiput is convexly rounded outward to the eye margin, and possessed of a sharply delimited, laterally placed, welt-like, vertical bulla (occipital schwielen of authors); these characters delineate the type-species of the genus. However, there are other species placed within the genus that do not have the occipital bullae. Also, in the past some genera have been placed in this subfamily which have 4 posterior cells, such as *Legnotomyia* Bezzi.

Usia incisa Wiedemann has all the above characteristics except that the pile of front, face, cheeks, occiput, thorax, abdomen, and legs is long, fine, dense, and erect; the abdominal tergites are lined behind with yellow, the face and cheeks are wide laterally and the occiput is poorly developed and lacks the lateral bullae. Generally speaking this species is so similar to species of *Legnotomyia* Bezzi it could easily be mistaken for that genus, except that the venation of the distal part of the radial field is so aberrant in *Legnotomyia* Bezzi, which moreover has 4 posterior cells and very widely open anal cell.

There are so many interesting examples of remarkable convergence in superficial appearance among members of different genera of bee flies that, despite the strong suggestive resemblance of *Legnotomyia* Bezzi to species of the Usiinae, like *Usia incisa* Wiedemann, I reject the idea of any close relationship between them, in spite of the very heavy veins shared by both. Not only does *Legnotomyia* Bezzi have 4 posterior cells, it has a very different type of antenna and a well-developed face below the antenna, and a different type of hypopygial development. I agree with Bowden that *Pusilla* Paramonov finds its real relationship with the *Corsomyza* Wiedemann species. *Psiatholasius* Becker I place in synonymy under *Legnotomyia* Bezzi. I raise *Corsomyza* Wiedemann to tribal status within the Bombyliinae.

Consequently, I restrict the Usiinae to those flies with 3 posterior cells and the other peculiarities outlined above.

Besides *Usia* Latreille and the subgenus *Parageron* Paramonov there are left in this subfamily three other genera: *Oligodranes* Loew and *Apolysis* Loew, and I place *Dagestania* Paramonov near here. I have not seen that genus; Paramonov likens it to *Geron*, *Parageron*, and *Oligodranes*. As for *Geron* Meigen this genus is very different from all these others, and, as Hesse has shown, it amply deserves separate subfamily status. The relationships of these genera, especially *Oligodranes* Loew, is very close to *Usia* Latreille.

I have before me a species of *Usia*, which clearly falls into the subgenus *Parageron* Paramonov.

Melander (1946) has ably discussed the relationship of European type-species of *Oligodranes* Loew and *Apolysis* Loew to their American counterparts; he points out that the European species of both have 2-segmented palpi, whereas the American counterparts have a single segment. He notes that were this character of the number of segments of the palpus regarded as of generic significance, then the American species of *Oligodranes* Loew would fall into *Rhabdoselaphus* Bigot, and the American species of *Apolysis* Loew would require a new name. Hesse (1938) notes that many genera with apparently unisegmented palpus show a plane of separation when treated with potash. Melander further notes that the curious occipital bullae, shared by *Usia* Latreille and *Parageron* Paramonov although not present in all species of either genus, is also present, but not in all species of *Oligodranes* Loew; there are intergrades. Hence, with Melander, I leave all known species of today in *Oligodranes* Loew and *Apolysis* Loew, and *Usia* Latreille, and *Parageron* Paramonov. It seems apparent, as Melander has pointed out, the unisegmented condition of the palpus is a reduction of the palpus from an earlier multisegmented condition, and that *Apolysis* Loew was derived from *Oligodranes* Loew by the loss of the posterior crossvein. And of course it is possible that *Usia* Latreille, a Mediterranean genus, was derived from *Legnotomyia* Bezzi, also a Mediterranean genus, by the loss of a posterior cell; this may well be true.

Melander (1946) found 37 new species of *Oligodranes* Loew in intensive collecting within mountain canyons of the West, the restriction of habit suggesting, in the case of *Oligodranes* Loew (and the very different *Mythicomyia* Melander), a rather youthful taxon.

The Usiinae show a curious, very short, broad figure with several peculiar genera. In *Apolysis* Loew the basal part of the aedeagus is large and bulb-shaped, attenuate, and extended apically. The basistyli are small, the dististyle very large with sharp angles and hooks and with a conspicuous, deep, dorsoapical incision from the lateral aspect. However, these characteristics are entirely lacking in *Oligodranes* Loew. On the whole *Usia* Latreille and *Parageron* Paramonov and *Apolysis* Loew are much more alike than is *Oligodranes* Loew.

Usiinae

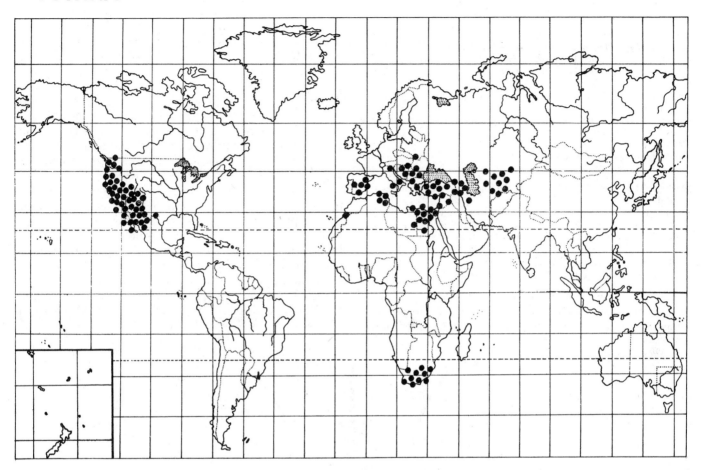

Text-Figure 28.—Pattern of the approximate world distribution of the species of the subfamily Usiinae.

KEY TO THE GENERA AND SUBGENUS OF USIINAE

1. Occiput usually rather strongly swollen, although gradually rounded inwardly and regularly sloping inward from the eye margin. In consequence of the thickness of the occiput the vertical groove or fissure behind the ocelli is prominent. Sides of lower occiput usually with conspicuous vertical ridge of some width. Abdomen wide, more or less subspherical, or subglobose and convex, and at least, short oval. These flies are rather bare, short pilose for the most part, often with a somewhat denuded appearance. Third antennal segment obtuse at apex, with subapical dorsal spine. Flies ranging from 2 mm. rarely, to 10 mm. rarely 2

   Occiput not exceptionally swollen behind and hence the post ocellar vertical fissure not prominent. Abdomen conical and tapered, either long or short. Third antennal segment variable. Flies ranging from minute, three-fourths of a mm. to 4.5 mm. 3

2. Occiput much more poorly developed, less prominent, more flattened; postocellar fissure shallow. Eyes holoptic in male and in contact for a considerable distance and ommatidia on upper three-fourths of male eye, much enlarged and clearly separated from lower facets. Hypopygium generally less swollen . . . . *Usia (Parageron)* **Paramonov**

   Occiput more prominent and usually conspicuously pilose, although the pile is not usually long. Postocellar fissure deep and narrow. Lamellae of epipygium (male hypopygium) grossly swollen, capsular, and conspicuous. Eyes of both sexes widely separated, the ommatidia alike in males . . . . *Usia* **Latreille**

3. The discal cell is complete . . . . *Oligodranes* **Loew**

   Discal cell is confluent with the second posterior cell, the medial crossvein being absent . . . . . . *Apolysis* **Loew**

Genus *Usia* Latreille

Figures 78, 79, 312, 545, 556, 852, 853, 854

Voluccella Fabricius, Entomologica Systematica, vol. 4, p. 412, 1794. Type of genus: *Voluccella florea* Fabricius, 1794. Designated by Latreille, 1802. Error in spelling by Geoffroy, Syrphidae (*Volucella*), 1764. Officially rejected under opinion 441.

Usia Latreille, Hist. nat. Crustacés, pt. 3, p. 430, 1802. Change of name.

Volucella Meigen, Klassifikazion Beschreiburg Insekten, vol. 1, p. 194, 1804, not Geoffroy, 1764.

Parageron Paramonov, Trav. Mus. Zool. Kieff, no. 6, p. 127, 1929c. Type of subgenus: *Parageron orientalis* Paramonov, 1929, by monotypy.

Becker (1906), pp. 196-201, key to 26 males and 27 females of Palaearctic species.
Séguy (1926), pp. 228-229, key to 9 species from France.
Engel (1933), pp. 63-66, key to 30 Palaearctic species.
Austen (1937), p. 55, key to 7 species from Palestine.
Efflatoun (1945), pp. 200-202, key to 13 species from Egypt.
Paramonov (1947c), pp. 207-220, revision, treating many species.
Paramonov (1950), pp. 342-351, key to world species of *Usia*.

Curious flies of very compact, robust form. Most species are rather small. The abdomen is wider, sometimes much wider than the thorax and the flies have either a very bare, denuded appearance as in the type-species, due to the minute appressed setae of the thorax and abdomen, or have long, fine, erect, comparatively dense, pale pile on these areas, as in *Usia incisa* Wiedemann; hence the appearance is variable, but it is always compact. Eyes always holoptic in males, except the species *ignorata* Becker.

These flies have only 3 posterior cells; the anal cell is closed and petiolate and there are only 2 submarginal cells. The veins are strong and stout. Anal cell closed. The proboscis is elongate and slender. The tibiae and femora have fine-appressed setate pile and may have microscopic scales. Male terminalia very prominent and much protruded and conspicuous.

The occiput is never cupulate or hollowed out centrally or bilobed above. Nevertheless, it is peculiar that the occiput in the type-species is so very much swollen and that it bears below on each side a curious, vertical bulla, a thing lacking in some species; this structure is also found in certain species of the genus *Oligodranes* Loew, and sometimes in *Apolysis* Loew.

Usia Latreille is restricted to the Mediterranean and the North African regions. Efflatoun (1945) recorded 13 species from Egypt, 9 of them undescribed.

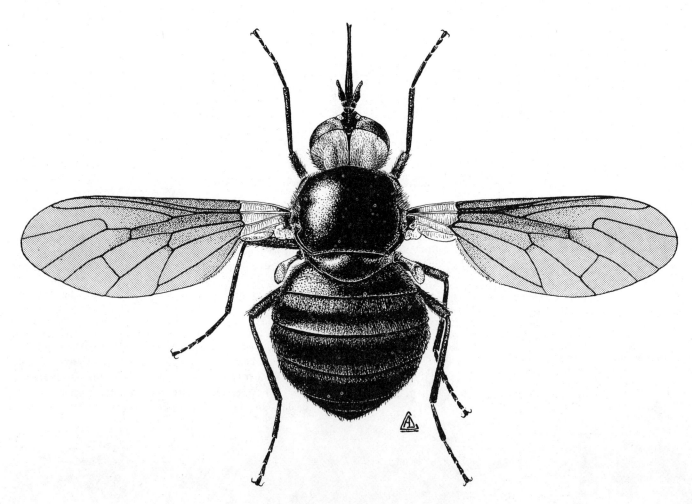

Text-Figure 29.—Habitus, *Usia aenea* Rossi.

Length: Often minute; length of body as little as 1.4 to 3.2 mm., or as much as 8 to 9 mm., excluding proboscis of 3.5 to 4.5 mm. Wing length may reach 7 mm.

Head, lateral aspect: The head is globular, the occiput variable. In males of some species it is greatly reduced in extent immediately behind the vertex, growing more and more prominent ventrally with rounded, convex, sloping sides reaching to the eye margin. The whole occiput is pollinose and covered with rather dense, long, coarse pile outwardly subappressed, the pile becoming shorter toward the vertex and longer in the middle and below. In females, as in the type-species, however, the upper occiput is extensive and swollen dorsally, very rounded and convex and perhaps even a little more extensive than on the lower part of the head. It is also pollinose but with shorter, abundant, relatively short, stiff, erect pile. In both of these types of occiputs the halves are separated below the vertex and the ocellar tubercle marked by a distinct, narrow fascia or sulcus. In the female type-species the posterior ocelli are actually behind the posterior eye corners, and this is due to a recession or curving forward of the eye. The front is either polished and bare with some pollen or micropubescence laterally, and a few short, oblique setae, or may be wholly pollinose, a little inflated, with comparatively dense, long, fine, erect pile. Likewise the face itself varies.

In the type-species the face, as well as the front, is very short in profile, the former restricted to a small triangle on each side just below the base of the antennae, and extended as a mere line down to the middle of the eye. The oragenal cup extends completely to the antennae and because of its dorsal width and breadth occludes the face laterally as well as dorsally. The face margin below the antennae is barely visible in profile as a line along the eye margin. However, in *Usia incisa* Wiedemann both face and front, oragenal cavity and gena are very extensive from the lateral aspect. However, in this species also the oragenal cavity reaches quite to the base of the antennae. The eye is large in *Usia* Latreille with the upper facets twice the size of those below and the posterior eye margin is plane.

The proboscis is very long and quite slender, 2 to 3 times as long as the head, the labellum is even more slender and is about as long as the third antennal segment. Palpus in the type-species minute, cylindrical, quite slender with tiny additional clublike segment at the apex bearing 2 or 3 long hairs. In *Usia incisa* Wiedemann the palpus is large, the second segment strongly clubbed with numerous bristly hairs over the whole surface. The antennae are attached at the upper third of the head, they are relatively short, but slender, the first two segments are nearly equal, both bearing a few setae dorsally in the type-species and in *Usia incisa* Wiedemann the first segment bears numerous, long, coarse, bristly hairs directed forward and outward on the upper half of the segment with shorter pile below. The third segment in this species is a little longer than the first two segments combined, of about the same width though slightly narrowed toward the narrowed apex, and it bears a shallow, spoon-shaped depression at the apex, and also has 4 or 5 long, oblique setae or short bristles dorsolaterally on the outer half. The antennae of the type-species are similar, the third segment is somewhat more robust on the basal half and the spoon-shaped apical depression bears a distinct short, but stout apical spine.

Head, anterior aspect: The head is distinctly wider than high viewed from the front. In the males it is so much broader below that it sometimes appears hemicircular with a rounded dorsal portion, and with the genofacial area steadily widening below. Males strongly holoptic, the front sometimes swollen above the plane of the eye, so that in length the front may be shorter than the eye from the point where it is joined to the posterior eye corners. Ocellar tubercle a little swollen, the large ocelli lie in an equilateral triangle and rest on the eye margin in the male. Front of female in the type-species not quite twice as wide as the width of the ocellar tubercle, which actually lies behind the upper eye corners. And in the type-species the sides of the genofacial area are nearly parallel, being only slightly wider below. The antennae are distinctly separated at the base but curve inward to nearly or quite touch at the end of the first segment. In the type-species the oragenal cup is deep especially opposite the proboscis, but slants forward and upward to the antennae. The inner walls are not vertical but are strongly sloping, and in this species the gena is reduced to a narrow line. In *Usia incisa* Wiedemann the oragenal cup is much more shallow and both face and gena quite extensive. It has a small tentorial pit just opposite the proboscis.

Thorax: The thorax is very short and broad, wider than long, rather evenly convex and while not greatly humped both head and abdomen are somewhat decumbent. The pile in the type-species is comparatively dense, short, bristly or setate and curled either upward or backward. The large, wide scutellum, including its posterior margin, is similarly pilose. Lateral and anterior margins of the mesonotum with dense, microscopic, glittering, appressed, scaliform pile, somewhat pollen-like in character. Humerus bordered above with similar pile and its whole surface bearing erect, bristly hairs like those of the mesonotum. In other species the mesonotum may have pollinose vittae, the scutellum may be strongly inflated across the middle, nonrugose and the whole mesonotum and scutellum with dense, long, matted pile. There is similar long pile on almost the whole of the mesopleuron. In the type-species the whole of the pleuron is pollinose and the upper half of the mesopleuron bears stiff, bristly pile, which is somewhat similar to that on the sides of the mesonotum and only slightly longer. Postalar callosity broad and flattened, the lateral ridge on the posterior half bears numerous, bristly hairs. All the remaining pleuron is without pile in the type-species but a few of the more pilose species may have a few hairs on the metapleuron and upper sternopleuron. The metanotum is extremely short and lacks pile. The squama is large with a fine, scattered fringe of hairs. Halteres are large and short with an unusually large club.

Legs: The legs are stout and relatively short. All of the femora equally but not conspicuously thickened; they are rather more thickened in the type-species which bears rather abundant, appressed, stiff setae with a few slightly longer, stiff, semierect hairs below. Also the femora are shining and rather bare; ground color green black in the type-species. In other species they are covered densely with minute, rounded, microscopic, appressed scales. Tibiae stout but more slender than the femora and clothed with quite abundant, much appressed setae but without outstanding bristles or spiculate setae. The tarsi are relatively short with shorter setae similar to those of the tibiae. Pulvilli large and prominent, widely spatulate, the claws fine and sharp.

Wings: The wings are comparatively wide with stout veins and in the type-species the wing is a little narrowed toward the base with narrow alula and the axillary lobe is no wider than the third basal cell. In other species the alula is much wider and the axillary lobe is also much wider than the third basal cell. Wing always with 2 marginal cells and 3 posterior cells, each of them widely open. The anterior crossvein enters the discal cell either just before or just beyond the middle. The second vein arises upward from the end of the radial sector. The anterior branch of the third vein arises steeply and rectangularly in the type-species with strong outward bend, but in other species it sometimes arises quite gradually. The vein closing the discal cell in the type-species is plane and rectangular, in other species a little curved and oblique. Anal cell closed, the stalk long or short. Ambient vein complete. Villi almost continuous and coarse. Wing usually tinted quite diffusely along the veins; in one species, wings spotted. First vein sometimes rugose.

Abdomen: The abdomen is large, much wider than the thorax and about 1½ times as long. It is very slightly convex across the middle but the tergites are narrowly rolled over laterally. It is widest at the end of the second segment, the second to fourth tergites being practically equal in length, the fourth tergite is only slightly more narrow. The fifth, sixth, and seventh tergites when viewed from above represent an obtuse triangle, and the second, third, and fourth tergites have a nearly complete, transverse, foveal depression near the end of the tergite. Other species lack such depression and have the abdomen somewhat more narrow. In the type-species the pile of the abdomen is rather dense, flat-appressed and setate, very little longer along the sides, and the whole surface is conspicuously rugose, or pitted; the color is uniformly shining greenish black. While there are 8 tergites in both male and female the eighth is extremely short and may be recessed beneath the seventh. Some species have the tergites nonrugose, nonfoliate, but with sublateral depressions in the middle, and each of the 8 tergites bears conspicuous, opaque, linear, yellow postmargins and the pile is comparatively dense, long, fine, and erect and thus in much contrast to the type-species. Female terminalia simple, without spines. The male genitalia are generally large, protruded and conspicuous. The hypopygium often forms a very large, bulblike expansion, directed downward. The dorsal lamella, paired and bipartite, is sometimes segmented at the apex; the ventral, bilobed, and fused pair of lamellae has the upper margin often emarginate or showing several pairs of obtuse teeth; between the ventral lamellae there is a pair of hooks in some species; the recessed and hidden penis (aedeagus) occupies a position between and in front of these hooks.

Material studied: The type-species and several other species.

Immature stages: Xambeu (1898) gave a short description of larvae and pupae but with no illustrations. See discussion under life histories.

Ecology and behavior of adults: Efflatoun Bey (1945) records having captured most of the smaller Egyptian species feeding on flowers of desert Compositae, such as *Astericus, Sonchus, Centaurea, Senecio, Calendula, Launea, Picris*, as well as on *Zilla spinosa*. He states that the adults are excellent hoverers.

Distribution: Palaearctic: *Usia accola* Becker, 1906; *aenea* (Rossi), 1794 (as *Bombylius*) [=*florea* Meigen not Fabricius, 1804]; *aeneoides* Paramonov, 1950; *angustifrons* Becker, 1906 [=*putilla* Becker, 1906]; *anus* Becker, 1906; *atrata* (Fabricius), 1798 (as *Voluccella*) [=*florea* Schiner, 1868, not Fabricius, =*pubera* Loew, nomen nudum in litt. et in coll., =*vicina* Macquart, 1840]; *aurata* (Fabricius), 1794 (as *Voluccella*) [=*taeniolata* Costa, A., 1884], *aurata loewii* Becker, 1906; *bicolor* Macquart, 1855; *calva* Loew, 1869; *carmelitensis* Becker, 1906; *claripennis* Macquart, 1840; *crinipes* Becker, 1906; *deserticola* Efflatoun, 1945; *efflatouni* Venturi, 1945 [=*bicolor* Efflatoun, 1945, not Macquart, 1855]; *elbae* Efflatoun, 1945; *engeli* Paramonov, 1950; *flavipes* Efflatoun, 1945; *florea* (Fabricius), 1794 (as *Voluccella*) [=*cuprea* Macquart, 1834, =*gagathea* Bigot, 1892]; *forcipata* Brullé, 1832; *grata* Loew in Rosenhauer, 1856; *grisea* Efflatoun, 1945; *ignorata* Becker, 1906; *incisa* (Wiedemann), 1830 (subg. *Parageron*) [=*major* Macquart, 1840]; *incognita* Paramonov, 1950; *inornata* Engel, 1932; *lata* Loew, 1846; *manca* Loew, 1846; *minuscula* Efflatoun, 1945; *notata* Loew, 1873; *novakii* Strobl, 1902 [=*pusilla* Macquart, 1840, not Meigen]; *pallescens* Becker, 1906; *parvula* Efflatoun, 1945; *punctipennis* Loew, 1846; *pusilla* Meigen, 1820; *sicula* Egger, 1859 [=*florea* Loew, 1846]; *similis* Paramonov, 1950; *syriaca* Paramonov, 1950; *tewfiki* Efflatoun, 1945; *tomentosa* Engel, 1932; *transcaspica* Paramonov, 1950; *unicolor* Loew, 1873; *vagans* Becker, 1906; *versicolor* (Fabricius), 1794 (as *Voluccella*) [=*hyalipennis* Macquart, 1840]; *vestita* Macquart, 1849; *ornata* Engel.

Oriental: *Usia marginata* Brunetti, 1909; *sedophila* Brunetti, 1909.

Subgenus *Parageron*

Engel (1933), in his key to the species of the genus *Usia* Latreille, incorporated all of those species known to him that fell into Paramonov's taxon *Parageron*. After studying his examples that clearly fall into the subgenus *Parageron*, I agree with Engel that the only valid differences lie in the holoptic eyes of *Parageron* and the relatively smaller hypopygium, the features of which I have summarized from the dissection of this individual that was collected in Spain. Because this specimen was destroyed in return shipment from my technician, I was unable to effect a determination of this species. Nevertheless, it met all of the requirements of *Parageron*.

The male terminalia from dorsal aspect, epandrium removed, show a very broad short structure, somewhat similar to *Apolysis* Loew, except that there is a distinct bridge across from each ramus crossing over the epiphallus. The dististyli are curious, flared, leaflike structures with a thick apical process; this structure with apparently enclosed duct ends in a curious, soft, spongelike structure. The dististyli are strongly flared and divergent laterally. Basistyli blunt and truncate apically. The epandrium large and subquadrate. Laterally the dististyle has a rather deep ventral incision.

Distribution: Palaearctic: *Usia* (*Parageron*) *grisea* Paramonov, 1947; *incisa* Wiedemann, 1830 [=*major* Macquart, 1840]; *lutescens* Bezzi, 1925 [=*orientalis* Paramonov, 1929], *lutescens minor* Efflatoun, 1945; *turkmenica* Paramonov, 1947; *zimini* Paramonov, 1947.

Genus *Apolysis* Loew

FIGURES 81, 329, 332, 549, 560, 855, 856

Apolysis Loew, Öfvers. Kongl. Vet. Akad. Forhandl., vol. 17, p. 86, 1860. Type of genus: *Apolysis humilis* Loew, 1860, by monotypy.

Paramonov (1931), p. 201, treatment of Palaearctic species.
Engel (1933), pp. 128-129, key to 3 species.
Hesse (1938), pp. 849-851, key to 7 species from South Africa.
Melander (1946b), pp. 458-459, key to 9 species from the western United States.

Minute flies, black in general color, or rarely in part yellow. They are strongly humpbacked but less conspicuously than in *Mythicomyia* Coquillett. Grayish pollinose for the most part, they often show spots or vittae upon the mesonotum. The abdomen is usually short and wide, wider than the thorax, often with narrow, yellow postmargins on the tergites. The pile is fine, erect, generally quite scanty and sparse. The proboscis is about twice as long as the subglobular head; the occiput is prominent only below but may show lateral, vertical, weltlike bullae similar to those in some species of *Oligodranes* Loew and *Usia* Latreille. The moderately lengthened antennae have a similar apical pit with spine dorsally placed on the third segment.

These flies are most closely related to *Oligodranes* Loew, from which they are quickly separated by the absence of the discal cell, the crossvein normally closing the discal cell having been lost by specialization. Likewise, the palpi are reduced in American species to one segment by fusion.

Apolysis Loew is primarily a Holarctic genus with several species found in South Africa. In the Old World they are found in the North African and Mediterranean regions as well. All known American species are from California.

Length: 0.75 mm. to 3 mm.

Head, lateral aspect: The head is subglobular but in males the upper part of the occiput is much reduced and is reduced to a thin line behind the ocelli, rapidly widening backward and downward to the point of attachment of head to thorax below the middle of the head. This upper profile of the occiput is plane. The eye is rounded below and recedes forward a little below the middle. The eye is very large occupying most of the lateral aspect of the head. The facets on the lower part of the eye are much smaller than the very large facets above, but there is a more or less gradual transition. The front is extended forward considerably beyond the eye margin. The antennae are attached above the middle of the head; the frontal area is almost as long as the lateral genofacial area. The face itself, in the middle, beneath the antennae, seems to be almost eliminated by the oragenal cavity which extends almost to the antenna. Both face, front, and genae are densely micropubescent. The front has 3 or 4 long, suberect, stiff, bristly hairs on each side near the middle and there is a fringe of still longer, isolated, erect, slender, though stiff, hairs running from near the base of the antennae down the sides of the face and gena, forming several loose rows below and continued over the whole area beneath the eye. The occiput has a sublateral fringe of moderately long, scanty, scattered, bristly hairs and this patch is continued medially almost to the foramen. The posterior surface is scaliform microsetate and the long setae are present over the whole of the occiput, except beneath the ocelli. Posterior margin of the eye entire and plane from the ocelli to the division point of the facets.

The proboscis is elongate, relatively slender, wholly micropubescent with a few fine, short, stiff hairs both above and below, with a rather small labellum. The proboscis extends forward and is about 1½ times as long as the head. The palpus is elongate and slender and not very large. It appears to be composed of 2 segments with the apical segment slightly thicker and bearing apically and dorsally 5 or 6 moderately long setate hairs. The antennae are attached at the middle of the head; they are elongate, and about two-thirds as long as the head. The first segment is beadlike, widest apically, longer than high at the apex and the same length or slightly longer than the second segment. Second antennal segment beadlike, a little wider than the first and both of these segments with 2 or 3 short setae dorsally at the apex. The third segment is flattened, long oval, nearly twice as long as the first two segments, rounded at the apex, the lower margin not quite plane, the whole surface micropubescent and dorsally near the apex. It has a concave incision bearing a short, stout spine.

Head, anterior aspect: The head is as wide or a little wider than the thorax. The face and front are not separated. In the female the front is wide with a slight depression across the middle. It is wholly pollinose and has a few, widely spaced, more or less erect, stiffened hairs. The eyes in the male are in contact for more than half the length from ocellus to antennae. The antennae are narrowly separated. The ocellar tubercle is high and conspicuous, the very large ocelli lie in an equilateral triangle and rest against the eye margin in the male. The oral opening is quite large, elongate oval, extending almost to the base of the antennae and is rather deep. The genae are wide and together with the interior of the oragenal recess are entirely micropubescent, with long, fine, scattered, erect pile on the genae only. The oragenal cup is restricted to a narrow linear strip on the lower half of the oral opening, separated by a distinct tentorial fissure. There is a small hollow in the occiput posteriorly above the junction of the head with the thorax. There are no flanges or tubercles.

Thorax: The mesonotum is high and equally convex and humped anteriorly and posteriorly. The mesonotum is loosely covered with rather long, scanty, fine, erect hairs, and the rather thick, short, convex scutellum has similar scattered discal hairs. Notopleuron without noteworthy bristles or setae. Humerus with 2 or 3 long hairs. The entire thorax is black with usually opaque black vittae and most of the mesonotum and all of the pleuron is densely covered with bluish white or gray or brownish yellow, dense, somewhat scalelike micropubescence. Upper part of mesopleuron with a few scattered hairs. Remainder without any pile, except the micropubescence. Squama rather large and long with a loose fringe of long hairs. Halteres with a large, short oval, closed knob.

Legs: The legs are rather slender, moderately long, the anterior tibia with several rows of short, suberect setae, a little longer and more conspicuous posteroventrally. Anterior femur slightly swollen toward the basal half with a row of short, erect setae above and scattered, long, bristly hairs posteriorly. Middle legs rather similar to the first pair, the femur more slender. Hind legs with the femur much more slender and covered with minute micropubescence that appears scalelike. This femur also has some short, subappressed setae dorsally and medially and has a fringing row laterally and another one ventrally, of rather long, erect, stiff or bristly hairs. Hind tibia quite slender with a dorsal row of bristly hairs similar to that of the femur, perhaps denser and containing 4 long elements in the middle. It also contains several shorter bristly hairs at the base and 1 or 2 at the apex and also a lateral row of 5 or 6 similar hairs that are not quite so long, together with additional, short, medial, and ventral, suberect hairs. It also has micropubescence. Pulvilli large, long, curved and spatulate, the slender claws bent sharply at the apex.

Wings: The wings are relatively large with weak veins. There are 2 submarginal cells, the anterior branch of the third vein arises gradually midway from the wing apex to the anterior crossvein. The whole second submarginal cell is narrow. There are only 3 posterior cells and no discal cell. Anal cell closed with a comparatively long stalk. Axillary lobe broad. Alula large and wide. The ambient vein lies shortly beyond the end of the third vein, the anterior crossvein lies opposite the angulate end of the second basal cell.

Abdomen: The Abdomen is somewhat elongate, about as wide as the thorax or a little more narrow. It is narrowed toward the apex, is generally opaque black, except on the first segment which may be grayish and the pile is fine, long, loose and scattered, and erect, especially along the sides. The abdomen in some species has a great deal of white or yellow color on it. In still other species the posterior margins of the tergites may be linearly whitish. It is much shorter and somewhat appressed along the posterior margins of the tergites. I find 8 segments on the abdomen. The male hypopygium is rather large and forms a troughlike bowl ventrally in which the genitalia lie.

Dr. A. J. Hesse in his work on South African bee flies describes the male hypopygium as follows:

Hypopygium of males with the inner apical angle or part of basal parts considerably produced and prominent, spinulated apically and acting as a sort of guide to the aedeagus; aedeagus also much produced beyond basal parts, lying along middle of scoop-like and conically produced last sternite; beaked apical joints small and provided on the inner side with a membranous, flattened process or flap, sometimes quite broad.

The male terminalia from dorsal and lateral aspect, epandrium removed, show a short, wide, obtuse figure with broad, laterally extensive dististyli which bear several strong, hornlike or tuberclelike processes. The aedeagus and the ejaculatory process extend outward as a long slender process well beyond the strongly divergent dististyli. The position of the dististyli may be an artifact. The basistyle is rather small, though wider basally. The epandrium is quite large, dwarfing the remainder of the structure and completely enclosing the basal apodeme.

Material studied: I have before me paratypes of several species.

Immature stages: Unknown.

Ecology and behavior of adults: Melander (1946) records taking these flies at flowers of *Cryptantha intermediate* and *Eriophyllum confertiflorium*, also of a *Phacelia* sp., and upon *Eriodictyon crassifolium*. All were taken in May in the southern part of California. Hesse (1938) records South African species from in or on the corolla tubes of the red-flowered *Mahernia grandiflora* in the Karoo.

Distribution: Nearctic: *Apolysis aperta* Melander, 1946; *disjuncta* Melander, 1946; *druias* Melander, 1946; *glauca* Melander, 1946; *minutissima* Melander, 1946; *mohavea* Melander, 1946; *petiolata* Melander, 1946; *timberlakei* Melander, 1946.

Palaearctic: *Apolysis andalusiaca* Strobl, 1898; *cinerea* Perris, 1839; *eremophila* Loew, 1873.

Ethiopian: *Apolysis brevirostris* Hesse, 1938; *cingulata* Hesse, 1938; *fumalis* Hesse, 1938; *humilis* Loew, 1860; *lindneri* Hesse, 1962; *maherniaphila* Hesse, 1938; *thornei* Hesse, 1938; *xanthogaster* Hesse, 1938.

Genus *Oligodranes* Loew

FIGURES 77, 231, 338, 550, 552, 553, 554, 737, 857, 858

Oligodranes Loew, Stettiner Ent. Zeitung, vol. 5, p. 160, 1844. Type of genus: *Oligodranes obscuripennis* Loew, 1844, designated by Becker, 1913, p. 484.
Rhabdoselaphus Bigot, Bull. Soc. Ent. France, ser. 6, vol. 6, p. ciii, 1886a. Type of genus: *Rhabdoselaphus mus* Bigot, 1886, by monotypy.
Pseudogeron Cresson, Ent. News, vol. 26, p. 201, 1915a. Type of genus: *Pseudogeron mitis* Cresson, 1915, by original designation.

Paramonov (1929), M. A. Sc. Ukr., vol. XI, p. 190, treatment of Palaearctic spp.
Engel (1933), p. 136, key to 4 Palaearctic species.
Melander (1946), pp. 463-470, key to 57 Nearctic species.

Small flies, sometimes black, polished and shining with yellow borders upon the posterior margins of the tergites; in many other species the general black coloration is heavily modified by dense, brown or silvery-gray pollen, which if it acquires any grease quickly eliminates the patterns upon the thorax and abdomen. These flies are of about the same size range and general appearance as *Phthiria* Meigen and have only 3 posterior cells. They resemble *Phthiria* Meigen in several other respects, such as the broad wing base and alula, the long, slender proboscis, the holoptic males and the obliquely flattened and appressed aspect of the upper eye and front; besides this both *Phthiria* Meigen and *Usia* Latreille have very large, terminal genitalia. Moreover, the antennae are somewhat similar in type but with a differently placed apical spine. A point of similarity between some of the species of *Oligodranes* Loew and the Mediterranean genus *Usia* Latreille is the presence of the curious occipital bullae; while it is possible that these flies were derived from remote *Phthiria*-like ancestors by the loss of one posterior cell, it is more likely that they were derived at one time from a *Legnotomyia*-like fly. Efflatoun (1945) illustrates a wing of *Phthiria salmayensis* Efflatoun in which the entire distal section of the second branch of the medius has dropped out, leaving the wing with only 3 posterior cells in this particular individual.

The genus *Oligodranes* Loew is represented by a very few species in the region of Asia Minor, 2 species from South Africa, and 52 species from the United States.

Length: 1.25 mm. to 6 mm. with most species about 3 mm. in length.

Head, lateral aspect: The head is globular and tends to be elongate. The occiput in the female is extraordinarily swollen and bullose though more prominent on the lower half, convex and sloping sharply outward and forward to the eye margin. In the male it is much more reduced above though still prominent below. The pile is rather dense, fine, and erect, and in males it becomes a rather long fringe along the upper eye borders. The front is flattened in the female on the upper half and plane with the eye but both face and front extend well beyond the eye margin and the lower front is therefore slightly elevated but rather flat. The front of the male is similar to that of the female, but the eyes are much flattened so that the upper half of the eye is distinctly triangular. Posterior eye margin entire and nearly plane; the upper eye facets enlarged. The face has minute, erect setae scantily distributed along the sides, the surface pollinose. In the males this pile is rather longer and more conspicuous, and likewise erect.

The proboscis is quite long and slender, 3 or 4 times as long as the head in most species but not quite twice as long in *Oligodranes sigma* Coquillett. The palpus is rather slender, not very long, although it extends slightly beyond the face. It appears to consist of a short, basal segment and a long, setate, clavate, apical segment which bears some short, fine, bristly hairs below. The antennae are attached near the top of the head at about the upper fifth or sixth. They are comparatively large, the first segment is cylindrical, 4 or 5 times as long as wide, the second segment beadlike, shorter than wide, and approximately one-fourth or one-fifth as long as the first segment. The first segment has some rather conspicuous fine, erect bristles dorsally, as long or longer than the width of the segment and a few others shorter below. The third segment is characteristically broadly oval and somewhat flattened, narrow at the base and minutely attached, then rather rapidly but evenly widening above and below. This segment is more or less long oval, a little plane on the upper surface, strongly convex below and dorsally at the apex it is concave and "hollowed out." At the bottom of the concavity it bears a very short, conical, pubescent microsegment.

Head, anterior aspect: The head is much more narrow than the thorax. In the male the eyes are strongly holoptic, touching and depressed at the junction for a distance equal the length of the front. Actually the front sends a narrow, wedgelike triangle posteriorly in between the eyes for a short distance. The ocellar tubercle is comparatively low, small, equilateral, the ocelli resting on the eye margin. In the female the eyes are rather widely separated. The ocelli do not touch the eye margin and the vertex is half the width of the front at the antennae. The antennae arise narrowly separated by less than the thickness of the first segment. The oragenal recess is curious inasmuch as it begins far back at the bottom of the head, though not as far back as the rounded, posterior eye corners; then it curves steeply and vertically upward, wide and extremely deep, and reaches quite to the base of the antennae. This anterior upper portion is triangular with the inner walls steep and even curled over a little. All of this lateral portion in front of the head is probably to be interpreted as lateral face, for near the bottom of the eye there appears to be a separate, low, marginal border going upward and disappearing, which probably represents the remnant of the gena. I find no trace of the tentorial fissure.

Thorax: The thorax is a little longer than wide, the mesonotum low and only slightly convex. They vary from species in which the mesonotum is almost entirely shining black and the abdomen similar, but with narrow postmargins, to other species in which the thorax and abdomen, even though black, are heavily coated with grayish white pollen. The pile varies from minute, very scanty and short and partly appressed to the group of shining black species which may have abundant, comparatively dense, erect pile over the thorax, the scutellum, and the mesopleuron. Remainder of pleuron and metanotum without pile and usually pollinose. The humerus bears a few scattered hairs or dense, long, pile in the more heavily pilose species, and pile of the scutellum is generally like that of the mesonotum. Sternopleuron sometimes pilose. Squama large with a fine fringe of pile, knob of halteres large, stalk rather short.

Legs: The legs are of no great length, the femora are a little thickened. In pilose species they are rather densely covered with erect, fine pile changing to appressed setae near the apex of the middle and especially the hind pair. In the less pilose species the femora are apt to be pollinose and covered only with scattered, short, stiff, semierect setae, with their tibiae similar in both respects. Tarsi are slender, the pulvilli are broadly spatulate apically, extending nearly to the apex of the relatively short, rather blunt claws, which are strongly curved at the apex.

Wings: The wings are hyaline with the base broad. There are only 3 posterior cells, all widely open. The anterior branch of the third vein arises in a rather low or gentle arch. There are only 2 submarginal cells and the anterior crossvein enters the discal cell slightly before or slightly after the middle. The anal cell is closed with a long stalk; the crossvein closing the discal cell is either plane and rectangular or shows a somewhat sigmoid curve indicating a double origin for this vein. Also, the first longitudinal vein may end rather nearer to the end of the auxiliary vein than to the end of the second vein. Axillary lobe broad basally, the alula comparatively wide. The ambient vein may end at the first posterior cell or at the anal cell.

Abdomen: The abdomen is rather broad at the base as far as the end of the second segment. It may be a little wider than the mesonotum or not quite as wide. In length it is as long as the length of the thorax including the scutellum or sometimes a little shorter. The abdomen may be only as long as wide, nearly circular, viewed from above with all of the tergites, except the first and second, much shortened or it may be elongate conical, somewhat compressed laterally, each tergite narrowed progressively. The species are greatly variable in point of pile present varying from those with only a few subappressed, setate hairs arranged in 1 or 2 transverse rows to other species that are covered with long, erect, rather dense, fine hairs on both tergites and sternites. The female genitalia are surrounded by a fringe of pile, no spines being present. Male terminalia small and short and barely protruded beyond the eighth tergite.

Dr. A. J. Hesse in his work on South African bee flies describes the male hypopygium as follows:

Hypopygium of male based on that of *O. namaensis* n. sp., very small with the outer apical parts or angles of basal parts broad and triangular; beaked apical joints somewhat flattened and with short bristles or setae; aedeagal or middle part with the lateral struts almost vestigial, very small, the basally directed strut more or less lobe-like, with the aedeagus short and apparently in form of two stylet-like processes, the aedeagus produced basally and above middle part into an aedeagal strut on each side; lateral ramus on each side from basal parts joined on to apical aedeagal part.

The male terminalia from dorsal aspect, epandrium removed, show a very short, quite wide, obtuse figure, with the base of the aedeagus and the epiphallus forming curious, short, broad, apically hemicircular plate, all lying far forward. The ejaculatory process seems to be paired or divided. From the lateral aspect the dististyli are small but slender, forming an apical toothlike process. The basistyle is attenuate apically, but very much widened basally, almost completely obscuring the basal apodeme.

Material studied: Several North American species collected by the author, and material in the National Museum of Natural History, the British Museum (Natural History), and the Department of Entomology of the University of California.

Immature stages: Unknown. I have an individual of *Oligodranes cinctura* Coquillett in which the whole posterior part of the abdomen has been eaten away by psocids; to the inner surface of the tergites there are the remains of large, elongate, oval, yellow eggs nearly one-third of a millimeter long. Anteriorly the ovaries show a great mass of very minute eggs.

Ecology and behavior of adults: These flies come to a variety of desert flowers. I have collected several species on flowers in New Mexico. Only one species is known from the eastern United States in temperate rainfall.

Distribution: Nearctic: *Oligodranes acrostichalis* Melander, 1946, *acrostichalis matutinus* Melander, 1946; *albopilosus* Cole, 1923; *analis* Melander, 1946; *anthonomus* Melander, 1946; *ater* (Cresson), 1915 (as *Pseudogeron*); *bicolor* Melander, 1946; *bifarius* Melander, 1946; *bilineatus* Melander, 1946; *bivittatus* (Cresson), 1915 (as *Pseudogeron*); *capax* (Coquillett), 1892 (as *Geron*); *chalybeus* Melander, 1946; *cincturus* (Coquillett), 1894 (as *Geron*); *cinereus* Melander, 1946; *cockerelli* Melander, 1946; *colei* Melander, 1946; *comosus* Melander, 1946; *dissimilis* Melander, 1946; *distinctus*

Melander, 1946; *divisus* Melander, 1946; *dolorosus* Melander, 1946; *eremitis* Melander, 1946; *fasciolus* (Coquillett), 1892 (as *Geron*); *formosus* (Cresson), 1915 (as *Pseudogeron*); *instabilis* Melander, 1946; *knabi* (Cresson), 1915 (as *Pseudogeron*); *lasius* Melander, 1946; *longirostris* Melander, 1946; *loricatus* Melander, 1946; *lugens* Melander, 1946; *maculatus* Melander, 1946; *marginalis* (Cresson), 1915 (as *Pseudogeron*); *mitis* (Cresson), 1915 (as *Pseudogeron*); *montanus* Melander, 1946; *mus* (Bigot), 1886 (as *Rhabdopselaphus*); *neuter* Melander, 1946; *obscurus* (Cresson), 1915 (as *Pseudogeron*); *palpalis* Melander, 1946; *panneus* Melander, 1946; *parkeri* Melander, 1946; *polius* Melander, 1946; *pulcher* Melander, 1946; *pullatus* Melander, 1946; *pygmaeus* Cole, 1923; *quinquenotatus* (Johnson), 1903 (as *Phthiria*); *retrorsus* Melander, 1946; *scapularis* Melander, 1946; *scapulatus* Melander, 1946; *setosus* (Cresson), 1915 (as *Pseudogeron*); *sigma* (Coquillett), 1902 (as *Geron*); *sipho* Melander, 1946; *speculifer* Melander, 1946; *togatus* Melander, 1946; *trifidus* Melander, 1946; *trochilus* (Coquillett), 1894 (as *Geron*).

Palaearctic: *Oligodranes flavus* Paramonov, 1929; *fumipennis* Loew, 1844; *hyalipennis* Séguy, 1941; *modestus* Loew, 1873; *obscuripennis* Loew, 1844; *rubriventris* Paramonov, 1936; *superbus* Engel, 1933.

Ethiopian: *Oligodranes elegans* Hesse, 1938; *flavipleurus* Bowden, 1964; *namaensis* Hesse, 1938.

Genus *Dagestania* Paramonov

Dagestania Paramonov, Trav. Mus. Zool., Kieff, no. 6, p. 133, 1929. Type of genus: *Dagestania pusilla* Paramonov, 1929, by monotypy.

Paramonov (1947), pp. 207-220, key to 2 species.

I have not seen the genus *Dagestania* Paramonov and I quote his description:

It is very close to *Geron* Meigen, *Parageron* Paramonov and *Oligodranes* Loew. Although I have seen only one somewhat damaged female, unfortunately, I have to set up a new genus for this type, since I am not able to classify it under any of the genera known up to now. This genus is distinguished from *Geron* Meigen in the following way: the first antennal segment is not cylindrical and not two and one-half to three times longer than the second, but only slightly longer; and both segments thicken toward the apex. Antennal pile is lacking (the third segment absent in my specimen). The very wide mouth cavity extends to the base of the antenna themselves, while in the case of *Geron* Meigen they definitely do not extend to it. Cheeks almost absent; cheeks very narrow. It is distinct from *Oligodranes* Loew because of the short, single segmented palpus. From *Parageron* Paramonov it is distinct because the anal veins are closed off far from the edge of the wings, and because of the absence of thorax hairs or pile, etc. The usual crossvein is clearly this side of the middle of the discal cell. Pulvilli are present. The structure of the front (frons) is unique.

Paramonov describes the type-species in the following words:

The basic color of the head is dark, only the very narrow cheeks and palpus are almost whitish. The whole head, with the exception of the front (frons), is whitishly dusted. The front (frons) is dusted gold yellow. The structure of the front (frons) is unique; almost at the end of the second third of its length there is a transverse groove, since the front at this point makes a distinct ridge to the base of the antenna. This elevation is three sided: the central triangular plate, which forms a distinct angle with the rest of the surface of the front, is dusted gold yellow, as the latter is; the lateral plates (also triangular), which gradually become the cheeks, are dusted white. The whole head is bare, only the occiput is weakly dusted yellowish. The occiput of the head bulges strongly backward. The posterior border of the eyes are without evagination (indentation). The edges of the front (frons) (stirnrander) viewed from the front, exhibits or takes on breakage or crack in the region where the cross groove is found, after which they diverge more than before.

The basic color of the thorax is dark; the back of the thorax is completely bare, and dusted yellow. On the middle there are two indistinct yellowish longitudinal stripes or vittae, which begin at the front edge and take up two-thirds of the length of the dorsum of the thorax. They are surrounded by a somewhat darker coloration and have a very narrow, dark separating stripe. Between the shoulders and the base of the wings there are a few shining hairs; (The sides of the thorax are bare, partly black and yellowish.) Scutellum dusted yellow, with eight protruding short yellowish bristles, which form a half circle; they are not on the back edge itself, but clearly on the surface of the scutellum. Legs yellow, microscopically pilose, without bristles. All the tarsal segments are black; front femur darkened. Metatarsus very long, almost equal to the length of the second, third, and fourth segments together; the second tarsal segment equal to the third and fourth segments together. Tarsi clearly longer than the tibiae. Haltere bright yellow. Wings transparent. Venation very similar to that of *Parageron* Paramonov but the anal cell is closed off far from the wing edge.

Back part of the abdomen completely yellow beneath, black on top, with narrow, yellow posterior margins on the segments; the pile consists of very short, sparse, dark hairs.

Body length 2.5 mm., wing length 2 mm.

1 female, 16. V. 26. Kumtorkale, Dagestan. Sande. Rjabov leg.

Specimen in Meier Collection

Material available for study: None.
Immature stages: Unknown.
Ecology and behavior of adults: Not on record.
Distribution: Palaearctic: *Dagestania longirostris* Paramonov, 1947; *pusilla* Paramonov, 1929.

Subfamily Heterotropinae Becker, 1912

While I leave these flies all within the subfamily Heterotropinae, it is entirely possible that flies like *Caenotus* Cole and *Apystomyia* Melander, and others, should be placed in a separate family. I leave them in the Bombyliidae because there is so little difference in the venation from that of certain other bombyliids, with the exception that some of these genera have a forked fourth vein. In all the genera included here the eyes in the male are divided into a group of very small facets concentrated very low toward the bottom of the eye. One genus in the subfamily, *Caenotus* Cole, was removed from the Therevidae by Melander (1927); one species has 5 posterior cells and consequently can still be confused with the Therevidae; the proboscis in these flies is very short indeed, fleshy and placed within the deep, triangular oral cup with its sloping sides.

The small, highly polished black flies in *Apystomyia* Melander are as black as shining pitch; their weak, milky wings and petiolate fourth vein are very peculiar. This genus is the same as *Alloxytropus* Bezzi who described it from Egypt (1925c); Efflatoun has redescribed and nicely illustrated the Palaearctic species (1945). The largest genus is *Heterotropus* Loew, abundant in the Middle East, ranging into India, one species of which was described by Melander, collected on April 16 from flowers of desert marigold in Organ Pipe National Monument.

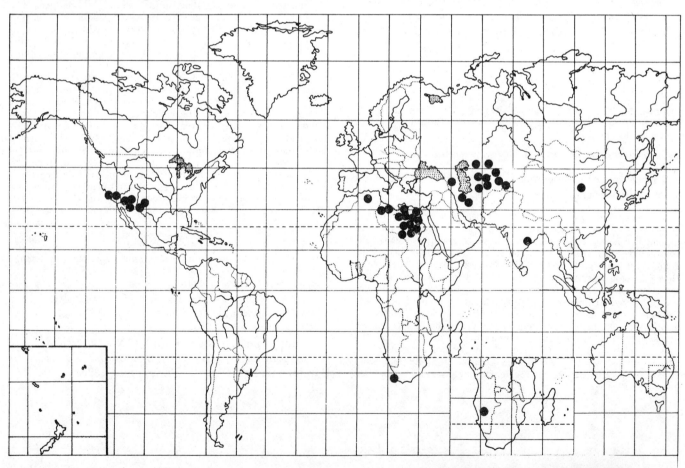

TEXT-FIGURE 30.—Pattern of the approximate world distribution of species of the subfamilies Heterotropinae and Xenoprosopinae.

The subfamily is separated from the remainder of the Bombyliidae by the simple posterior margin of the eye, nonindented, lower facets in male bisected into facets of much smaller size, much as in bibionids; tibiae without seriate spines, the third vein with two branches, and the anal cell petiolate; the occiput is flattened, the anal cell closed and petiolate. Many of these characters are shared by other bombyliids.

The venation is rather unique, the ambient vein lacking in *Prorates* Melander, and ending at the fourth vein in *Apystomyia* Melander. The veins are always dilutely hyaline with little or no pigment or wings milky. The male genitalia are large, exposed, protruded, terminal, nonrecessed as is characteristic in so many lower Brachycera; female often with acanthophorites.

We are acutely in need of some knowledge about their life histories, immature stages, hosts if any, all of which we now lack.

This subfamily shows some relationship with both *Crocidium* Loew, and it might well be reduced to a tribe within the Bombyliinae.

I have been able to dissect males of only one genus, and the genitalia are described under *Heterotropus* Loew.

KEY TO THE GENERA OF HETEROTROPINAE

1. Flies with the characters of the subfamily, especially divided facets in eyes of male, tibiae without spines, anal cell closed and stalked, third vein forked, first posterior cell open, 4 or rarely 5 posterior cells, and proboscis distinctly projecting beyond the oral cavity 2
   With all of the above characters but with the proboscis very short with fleshy labellum and completely confined to the sunken, wide, more or less triangular oral cavity with its sloping walls. Abdomen elongate and not with dense fur-like vestiture 3
2. The intercalary vein arises from the discal cell, between the fourth and fifth veins. Third antennal segment elongate, tapered, slender, style-tipped. Ambient vein present around hind margin of wing. Abdomen either obtuse, or long and tapered. Generally light-colored flies or if dark, with light markings, paired, especially on the abdomen. Anterior face jutting forward as an almost or quite horizontal ledge. Feeds upon flowers . . . . **Heterotropus Loew**
   The intercalary vein arises from the fourth vein. The second posterior cell is petiolate. Ambient vein absent. Third antennal cell conical with a small or microscopic style, weak flies; abdomen a trifle swollen toward the base.
   Prorates Melander
3. Third vein branching near the discal cell; the second posterior cell is sessile; discal cell shorter than the basal cells; thorax opaque pollinose . . . . **Caenotus Cole**
   Third vein branching near the apex of the wing; second posterior cell long petiolate; discal cell longer than the basal cells; whole body, especially thorax, shining pitch black. Wings milky . . . . **Apystomyia Melander**

Genus *Heterotropus* Loew

Figures 96, 527, 547, 774, 782, 785, 837, 838, 839, 976

Heterotropus Loew, Beschreibungen europäischer Dipteren, vol. 3, p. 182, 1873. Type of genus: *Heterotropus albidipennis* Loew, 1873, by monotypy.
Malthacotricha Becker, Ann. Mus. Zool. Acad. Imp. Sci., St. Péterbourg, vol. 12, p. 312, 1907d. Type of genus: *Malthacotricha glauca* Becker, 1907d, original designation. Described as an Empididae.

Paramonov (1929), p. 127, treatment of Palaearctic species.
Séguy (1932), p. 130, key to 7 species.
Engel (1933), pp. 157-159, key to 23 species and subspecies.
Efflatoun (1945), pp. 73-75, key to 8 species from Egypt.

Rather bare, small flies with colorless wing veins and a strange division of the ommatidia into a sharply demarcated region of minute facets below and enlarged facets above; the eyes are very much enlarged in extent with the face correspondingly reduced. The body and head color tend to be largely whitish or pallid yellow, with sharp, blackish markings.

The deep oral recess is quite narrow, largely because the sides of the oral cup are so much widened. The antennae are elongate, slender, of nearly uniform thickness, except that the third segment is attenuate near the apex; the antennal pile is scanty; the first two segments are subequal in length. The entire fly has only fine, scanty pile, the scattered hairs inconspicuous but a little more numerous and appressed upon the abdomen. The anal cell of the wing is closed, all other cells are open; the position of the small crossvein varies, it may be near the middle of the discal cell with the latter drawn up angularly to meet it, or it may be at the outer third of that cell, or a little beyond. The wing tends to be a little narrowed though rounded at the apex with the anterior and posterior branches of the third vein ending about an equal distance above and below the wing apex. The marginal cell is quite narrow; the legs are without bristles; the scutellum has an impressed rim.

Heterotropus Loew is principally a Palaearctic genus largely confined to and characteristic of southern Europe and North Africa, and Egypt. However, there is a species from India, one from South Africa, with exaggerated extension of the face beneath the antennae, and Melander described one species from Arizona.

Length: 4 to 4.5 mm., wing 3.8 to 4.5 mm. The largest species is *elephantinus* Séguy, only 5.5 mm., excluding proboscis.

Head, lateral aspect: The head is large, the eyes extraordinarily developed. In males in lateral aspect the occiput is restricted to the lower half of the eye where it is quite short but bears relatively short, moderately abundant, stiff hairs extending to within a short distance of the eye margin. What I consider to be the thickened wall of the very large, elongate, oragenal cup extends outward in profile on the upper part of the facial area almost immediately beneath the antennae as a shelflike ledge of varying length in different species

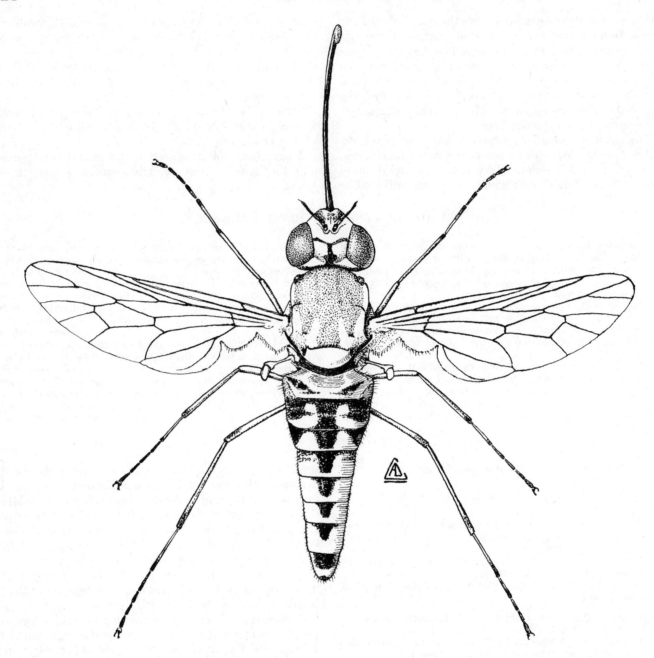

Text-Figure 31.—Habitus, *Heterotropus aegyptiacus* Paramonov.

and farther in males; in *indicus* Nurse it is extended only slightly farther than the end of the first antennal segment. In Egyptian species such as *Heterotropus stigmaticus* Bezzi it extended outward to the end of the second antennal segment or beyond.

The restricted face proper represents a flattened area or even slightly sunken area immediately beneath the antennae, which like the small triangular front is pollinose. The very extensively developed eye is sharply divided obliquely into an area of much small facets on the lower third in front and reduced to the lower fifth posteriorly. Above this section the facets are much larger. The male eye is in contact for only about one-third of the distance from antennae to vertex. The frontal triangle, though relatively small, is long and there is also a small, long, vertical triangle in front of the ocellar tubercle. The posterior margin of the eye is entire, and there are a few, quite short, stout bristles on each side of the upper occiput behind the ocelli and a short distance on each side of the ocellar triangle. The proboscis is comparatively long and rather stout with somewhat fleshy, stout labella and extends either straight forward or is folded up into the long, narrow, quite deep oral recess between the thickened, inflated walls of the large oragenal cup. Palpus small, rather slender and short with a few fine, setate hairs laterally near the apex. The front is very slightly raised above the eye margin and is flat. The antennae are attached at the upper third of the head, the first two segments are quite short, of about equal length and each a little widened and rounded apically with only scattered, minute hairs.

Head, anterior aspect: Head much wider than the thorax; the antennae are adjacent and touching, the ocellar tubercle is low, longitudinally grooved, and forms an equilateral triangle, and the very large ocelli rest on the eye margin. The oral opening is quite large and occupies almost all of the space ventrally between the eyes; its walls are low, very wide, rounded laterally, protrusive to a varying extent, obliquely furrowed anteriorly, infolded and rounded medially. The actual oral opening is long, narrow, and deep. The genae are extraordinarily narrow and sunken along the eye margin and the shallow, tentorial fissure lies on its medial aspect, but the genae expand at the very bottom of the eye as a rather wide triangular area. Viewed from posteriorly the occiput is concave, its lower, outer lateral portion a little inflated.

Thorax: The mesonotum is only moderately high, more convex anteriorly, gently sloping behind, largely greenish or yellowish white in color and opaque with opaque black or brownish vittae on the mesonotum. The pile of the mesonotum is very scanty, fine and erect, and partly subappressed. The entire lateral margin of the notopleuron, however, bears loose, scattered, fine, long hairs. Transverse suture curiously deep and conspicuous. The pleuron is likewise semiopaque white and also bare, not pollinose; mesopleuron with short pile above; remaining pleuron everywhere bare. Scutellum with rounded, impressed margin and moderately abundant, short, erect pile. Squamae with a short fringe. Halteres large, the knob large, elongate and closed.

Legs: The legs are short, the femora a little thickened, especially the hind pair, generally entirely pale in color with the tarsal segments more or less blackish. The anterior tibia has only fine, short pile and 2 or 3 slightly longer hairs below. The anterior femur is almost bare, with a few fine, short, scattered hairs; middle legs similar; hind femur with a dorsal and medial fringe of fine, scattered hairs in length about equal to the apical width of the femur. Laterally and ventrally the hind femur has scattered, rather short, suberect setae; this hind femur is straight in shape. The hind tibia is as long as the femur with a ventral row of 10 or 12 fine, erect, bristly hairs of no great length and a dorsolateral row of 12 or 15 similar more oblique, bristly hairs and additional fine pile; there is also a dorsomedial row of short, suberect setae. The pulvilli are long and rather slender, claws sharply bent at the apex.

Wings: The wings are colorless; the veins are colorless. The costa and subcosta have a slightly yellowish tinge. The costa is expanded basally and has a distinct basal comb; what corresponds to the basal hook is blunt, triangular, and pilose. The second submarginal cell is relatively short and the fork of the third vein is not very wide. The wing is more or less pointed or narrowed apically. There are 4 posterior cells. The anterior crossvein enters the discal cell near the outer fifth and all of the posterior cells are widely open. The intercalary vein is long; discal cell elongate and widened apically and the anal cell is closed with a relatively short stalk. The axillary lobe is quite broad and the rounded alula is rather wide; ambient vein absent.

Abdomen: The abdomen is as long or longer than the thorax, tends to be rather flattened, greenish or yellowish white in color, especially apically, laterally, and on the sternites. There is often a pattern of black, medial, continuous or discontinuous spots on the middles of the tergites. I find 7 segments present in the male with the last one quite short. The hypopygium is large and clublike, obtuse apically. The abdomen has scattered, rather scanty pile, more or less setate, and in the material before me retrorse, except that along the side margin, there is a fringe of fairly abundant, long, coarse, more or less erect pile.

The male terminalia from dorsal aspect, epandrium removed, show a curious type of epiphallus which is sleevelike basally with flared and expanded, narrow prongs distally; these overlie a bridge across the top from the corresponding basistyli. The basal part of the aedeagus is far removed from the center and is small and egg-shaped, and there are 3 quite long, slender, ejaculatory processes that extend beyond the epiphallus in the center of the pygidium. The dististyli are curious and remarkable; they are quite long, club-shaped both basally and apically and concave and much reduced in width in the middle. The entire pygidium is remarkable for being quite oval and the basistylus has a dorsolateral process. From the lateral aspect the epandrium is more or less triangular though somewhat swollen ventrally and distally. The curious dististylus has a long, apical, downturned hook and the general shape of the entire pygidium is different from other genera of Bombyliidae. In the axial system the epiphallus arises in a very strong acute arch from a curious, wide, proximally almost circular, flattened, conspicuous, basal, winglike process.

Material studied: A male of *Heterotropus indicus* Nurse, a gift of the British Museum (Natural History), and a female of *Heterotropus aegyptiacus* Paramonov, sent me through the courtesy of Cairo University, Cairo, Egypt.

Immature stages: Unknown.

Ecology and behavior of adults: Efflatoun (1945), who with his assistants has probably captured more species than anyone, notes that he has only found them during the hottest hours of the day, from 1 to 4, the temperature above 45°C in the shade, a strong south wind. He found them on flowers of *Zygophyllum*, and small desert Compositae, and on flowers of *Statice*. Melander captured his species on flowers of desert marigold, and *Baileya multiradiata*.

Distribution: Nearctic: *Heterotropus senex* Melander, 1950.

Palaearctic: *Heterotropus aegyptiacus* Paramonov, 1929; *albidipennis* Loew, 1873 [=*sulphureus* Paramonov, 1929, female], *albidipennis sudanensis* Becker, 1922; *ammophilus* Paramonov, 1929; *arenivagus* Paramonov, 1929, *arenivagus chivaensis* Paramonov, 1929, *arenivagus flavoscutellatus*, Paramonov, 1929, *arenivagus normalipes*, Paramonov, 1929, *arenivagus repeteki* Paramonov, 1929; *atlanticus* Séguy, 1930, *bisglaucus* Bezzi, 1925; *bourdariei* Séguy, 1938; *elephantinus* Séguy, 1929; *fulvipes* Séguy, 1932; *glaucus* Becker, 1907; *gussakovskiji* Paramonov, 1929; *hohlbecki* Paramonov, 1946; *kazanovskyi* Paramonov, 1925; *longitarsus* Séguy, 1930; *maculatissimus* Villeneuve, 1932; *maculiventris* Bezzi, 1925 [=*tewfiki* Paramonov, 1929]; *magnirostris* Bezzi, 1926; *monogolicus* Paramonov, 1946; *nigrimanus* Séguy, 1931; *sabulosus* Paramonov, 1926, *sabulosus nigritarsis*, Engel, 1933; *stigmaticus* Bezzi, 1926; *trotteri* Bezzi, 1916; *xanthothorax* Efflatoun, 1945; *zarudnyi* Paramonov, 1946; *zimini* Paramonov, 1929, *zimini monticola* Paramonov, 1929.

Ethiopian: *Heterotropus munroi* Bezzi, 1926.

Oriental: *Heterotropus indicus* Nurse, 1922; *pallens* Nurse, 1922.

Genus *Caenotus* Cole

FIGURES 108A, 574, 697, 867, 868

Caenotus Cole, Proc. U.S. Nat. Mus., vol. 62, art. 4, p. 14, 1923. Type of genus: *Caenotus inornatus* Cole, 1923, by original designation.

Melander (1950b), p. 148, key to all known species.

Small flies which were once placed in the Therevidae from which they differ in the second basal cell being pointed apically, the crossvein absent, the absence of bristles, and the short antenna. They are blackish, slender flies with soft, attenuated easily distorted abdomen; the pile of the body is very scanty, fine, and erect though a little more abundant upon the mesonotum. The anal cell is closed but with a short stalk; the first and second branches of the medius arise separately from the discal cell with a very short vein of varying length separating them. The ambient vein, though present, is weak. The first two segments of the antennae are quite short; the third segment is much widened from the base to the middle, then much narrowed and with a short microsegment that is as wide as the apical part of the segment. The eyes are divided into small facets below, large facets above.

Known only from New Mexico, southern Arizona, and California.

Length: 4 to 6 mm.

Head, lateral aspect: The head is approximately hemispherical; in males the occiput does not extend beyond the eye on the upper half and is actually somewhat concave on this portion. In the female the occiput is somewhat similar but extends a little higher toward the vertex and in both sexes it is only moderately developed below the middle and toward the bottom of the head. From the lateral aspect the eye covers almost the entire head and in most specimens the face and front can barely be seen and sometimes in other specimens not at all. The oragenal cup or opening is deeply recessed into the triangular genofacial space, its pile is rather abundant but fine, loose and scattered, long, and extending downward and outward and largely restricted to the lower fourth of the head. Pile is absent below the antennae, except for a continuation of the minute, grayish micropubescence present upon the sides of the front; it continues below the eye to cover all of the gena to the eye margin and all of the oral recess as well. The long, fine pile continues around and upon the occiput on the lower third of the head and on the upper part of the occiput there is, very close to the eye margin, a row of scattered, fine, bristly hairs. The posterior margin of the eye is entire and extends medially inward on the upper half and viewed laterally it is clearly divided, but without a groove or colored line, into an area of minute facets on the lower third, and much larger facets above as in some bibionids.

The proboscis is very short and stout, with a short, stout labellum scarcely extending out from the lower part of the oral opening. Palpus, rather large, with rather numerous, long bristles, and composed of 3 segments. The front is small, flat, somewhat depressed medially in females and with micropubescence only, the middle bare. The antennae are small, the first segment is a little longer than the second which is beadlike, both of these bear several fine bristles dorsally and ventrally; these bristles are nearly equal in length to the vertical width of these segments. The third segment is of about the same length as the first two segments together, and while somewhat flattened laterally, it is angularly widened in the middle to an equal extent above and below, and apically it bears a large, short, flattened microsegment to which is attached a minute, apical microsegment. The antennae are attached a little below the middle of the head and just above the line on which the eye facets are divided into small and large groups.

Head, anterior aspect: The head is a little wider than the thorax; in males the eyes are holoptic and meet extensively, although the medial edge of each is strongly rounded leaving the frontal triangle narrow above and leaving the vertex composed entirely of the large, protuberant ocellar tubercle. In females the front is wide, it begins to be wider and flared outward on the lower third, and its width below the ocelli is more than one-fourth the total head width. As in males, the ocellar tubercle forms a prominent equilateral triangle, but unlike males it is well separated from the posterior eye margin. The ocelli are large. The oral opening is triangular, comparatively large, recessed and only moderately deep. The genae are wide, sloping inward and appear to have only very minute micropubescence. The tentorial fissure is barely evident.

Thorax: The thorax is comparatively high and moderately humped both in front and behind. The prothorax is very short, the vestiture consists of rather abundant, long, quite fine, erect, stiff pile covering the entire mesonotum and the humerus and continued onto the pronotum and as a middle band vertically down the mesopleuron and the sternopleuron, extending as far as the coxa. The sternopleuron and the entire metapleuron and the hypopleuron have only minute, sparse, micropubescence; all pleural pile more delicate than that on the mesonotum. Scutellum with abundant pile similar to that on the mesonotum but longer. Squama with a fine, delicate fringe. Halteres with a quite large, thin-walled, elongate, cup-shaped knob.

Legs: The legs are relatively short and slender, the femora, however, are slightly thickened. Anterior tibia with fine, abundant, subappressed pile only. Anterior femur with a fringe of numerous, long, stiff hairs below in several rows. Middle femur and tibia similar, the latter with a few, short, bristly hairs. Hind femur with abundant, subappressed, shining pile above and laterally and medially, but with the more sparse, short, scattered pile below, among which are 4 or 5 slightly longer, fine, stiff hairs. The hind femur is distinctly arched outward when viewed from above. The hind tibia has 2 or 3 very weak bristly hairs of no great length ventrally and the same number dorsally. Pulvilli long and spatulate. Claws fine and curved slightly at the apex.

Wings: The wings are weak but relatively large and rather wide apically as well as basally. The second submarginal cell is long and narrow, the anterior branch of the third vein forms a narrow fork with the posterior branch. The anterior branch ends a short distance above the apex of the wing. There are 4 posterior cells, all widely open, and the anterior intercalary vein is short. It is worth noting that the anal cell is closed with a long petiole or stalk. The axillary lobe is broad and the rounded alula is relatively narrow. The ambient vein is complete, but like all of the posterior veins behind the radius, it is quite pale. The anterior crossvein enters the discal cell slightly before the middle; this cell is of a peculiar shape, angulate below, short in length, and rather broad across the middle.

Abdomen is peculiar, elongate, tapered, with 8 tergites and sternites and a large, conspicuous, elongate, somewhat clublike hypopygium. The second segment is much longer than the third, and the third, fourth, and fifth segments are equal in length, the remaining segments shorter. Both the tergites and sternites are shining blackish in color and both with moderately abundant, fine, erect pile. In the females the abdomen is even longer and more attenuate, the terminal segments less telescoped and shortened and the female genitalia has on each side a vertical row of 7 or 8 stout, truncate, flattened spines.

The male terminalia from dorsal aspect, epandrium removed, show a very curious elongate structure, the very long tubular epiphallus reaches to the end of the dististyle and is deeply divided near the apex into 2 slender prongs. The dististyle is quite long, slender, curved on the inner and outer margins, convergent apically, the apex sharp. The basal apodeme is curious, very long, widening basally and basally spatulate. From the lateral aspect the dististyle forms a more or less shallow, lunate body curved forward both above and below, and quite thick on each end. The epiphallus is a curious structure, it makes a very strong, downward curve or bend, spoutlike. The basistyle and epandrium are both large. The basal apodeme is also quite prominent from the lateral aspect.

Material studied: Males and females of *Caenotus hospes* Melander, furnished by Dr. Melander from a large series he collected at Organ Pipe National Monument.

Immature stages: Unknown.

Ecology and behavior of adults: Melander (1950) recorded some of these flies coming to windows during the day, a few others attracted to lights; they were captured in April and May.

Distribution: Nearctic: *Caenotus canus* Melander, 1950; *hospes* Melander, 1950; *inornatus* Cole, 1923; *minutus* Cole, 1923.

Genus *Prorates* Melander

FIGURES 97A, 322, 326, 573, 700

Prorates Melander, Ent. News, vol. 17, p. 372, 1906. Type of genus: *Prorates claripennis* Melander, 1906, by monotypy. Described as Empididae.
Alloxytropus Bezzi, Bull. Soc. Roy. Ent. d'Egypte, vol. 8, p. 186, 1925c. Type of genus: *Alloxytropus anomalus* Bezzi, 1925, by original designation.

Melander (1927), p. 376, transfer to the Heterotropinae.
Efflatoun (1945), p. 98, key to 2 Egyptian species (as *Alloxytropus* Bezzi).

Small or very small flies which like *Heterotropus* Loew have colorless veins and generally pallid coloration, the legs especially are wholly pale yellow or whitish. Like *Heterotropus* Loew the eye is much enlarged in extent and sharply divided into a small lower division with minute facets and a more extensive upper division with much larger facets. Eyes in both genera markedly holoptic in males. Melander (1906) originally placed these flies in the family Empididae but in 1927 transferred them to the Bombyliidae.

In *Prorates* Melander the face is even more reduced, to a very small, narrow slitlike recess with small, short proboscis; from below the eyes curve under the head leaving only a small posterior triangle of occiput and the much reduced oral recess. Antennae smaller, and shorter than in *Heterotropus* Loew. In *Prorates* Melander the ambient vein is wanting, the costa ending at the third vein in contrast to *Heterotropus* Loew. Generally speaking the first and second branches of the medius originate from a common stem emitted from the discal cell, and this is true of the American species *claripennis* Melander, which he originally placed in the Empididae, and of the two Egyptian species, except that sometimes in the type-species, *anomalus* Bezzi, these two branches originate separately at the end of the discal cell. The genus is further unique in that the anal cell is closed with a very long stalk, the cubital vein having receded basalward. As Efflatoun (1945) remarks there is a general superficial resemblance to the Scenopinidae; it is just possible that the relationship is more than superficial.

These flies are known only from Egypt, New Mexico, and the Borrego Desert of southern California.

Length: 1.8 to 44 mm.; proboscis .5 to 1.8 mm.

Head, lateral aspect: The head is large, globular, with extraordinary development of the eyes. In the males only a small portion of the occiput is visible in the posterior ventral corner. The front is reduced to a very minute triangle, even smaller than the first antennal segment, and the face is reduced in space until it is not much larger than the front. The eyes extend forward and inward encroaching on space until even the oral recess is greatly narrowed, and the proboscis, which is relatively narrow, extends straight forward beneath the antennae for a short distance. The posterior margin of the eye is entire and the facets are sharply divided into a group of small, black facets on the lower third and much larger light-brown facets on the upper part. The palpus consists of a single, slender, elongate segment. The minute front is flat, lower than the surrounding eye margin, and micropubescent only. The antennae are relatively short; the first two segments are very short and the third segment more elongate and attenuate with a short, stumplike microsegment. The antennae are attached adjacent to one another but the third segment is thrust sharply outward at a right angle at the point of attachment. All of the segments are minutely micropubescent.

Head, anterior aspect: The head is distinctly wider than the thorax. In males the eyes meet for a long distance; the ocellar tubercle is raised, and comparatively large with large ocelli resting at the edge of the eye. The oral opening is narrow, but widens somewhat posteriorly, its depth equal only to that of the proboscis. The genae are quite narrow in front, gradually increasing in width behind.

Thorax: The thorax is relatively high, but only slightly convex over the middle and slopes from the transverse sutures gently back to the scutellum. Prothorax very short, the strongly developed eye tends to rest against both mesonotum, humerus, and prothorax. The mesonotum is opaque and has fine, grayish-white pollen, which under high magnification appears to consist of flat-appressed, scalelike hairs. In addition there are moderately abundant, fine, short, subappressed, slender bristles or bristly hairs. There are 2 distinct, stouter, and longer bristles on the notopleuron and a pair of even longer slender yellow bristles on the margin of the scutellum, otherwise the scutellum and the pleuron are very similar to the mesonotum. The entire pleuron seems to lack pile except for the pollenlike micropubescence and a few isolated hairs on the lower mesopleuron and sternopleuron. Squamae large, thin, with a quite sparse fringe of scanty, stiff hairs of no great length. Halteres large, the knob large, thin-walled, elongate, and cup-shaped or rolled much as a rolled leaf.

Legs: The legs are quite slender, entirely pale, almost translucent in the dried specimen; the femora are very slightly thickened; the pile is everywhere minute, subappressed, rather scanty on the femora, more abundant on the tibiae. The anterior tibiae lack any longer setae that stand out from the abundant, appressed vestiture, but all of the femora have some suberect, fine setae that are a little longer than the remaining pile. Hind tibia similar to the anterior tibia. Pulvilli short, wide, and rounded apically, the sharp, slender claws are strongly bent and curved apically.

Wings: The wings are delicate, the veins colorless throughout, although the costa may be a little darker apically. The second submarginal cell is narrow and the basal fork narrow. There are 4 posterior cells, but the intercalary vein is absent, and the fourth and fifth longitudinal veins have coalesced for some distance away from the discal cell, so that the second posterior cell has a stalklike petiole. The anterior crossvein joins the discal cell near the outer third, and the stalk of the anal cell is even longer than in *Caenotus* Cole and is empidiform in character. The axillary lobe is large, the alula rounded but narrow, the ambient vein is entirely absent.

Abdomen: The abdomen is elongate, relatively narrow and attenuate as in some of the elongate species of *Aphoebantus* Loew. I find 8 segments present. The pile is scanty, fine and erect and relatively short. The hypopygium is large, somewhat bulblike or clublike and rather conical though obtuse.

Material studied: A male of *Prorates claripennis* Melander.

Immature stages: Unknown.

Ecology and behavior of adults: Efflatoun (1945) notes that all these Heterotropini, and likewise the genus *Mariobezzia* Becker, are found only between the hours from 1 to 4, neither he nor his assistants seeing them at any other time. Melander (1950) collected his series in the early morning hours on the windows in his car where they had become trapped during the night; Efflatoun found they were sluggish, slow fliers hovering in the hottest hours of the day only.

Distribution: Nearctic: *Prorates claripennis* Melander, 1906.

Palaearctic: *Prorates anomalus* (Bezzi), 1925 (as *Alloxytropus*); *bezzii* (Paramonov), 1929 (as *Alloxytropus*).

Genus *Apystomyia* Melander

FIGURES 85, 324, 599

Apystomyia Melander, Pan-Pacific Ent., vol. 26, p. 146, 1950b. Type of genus: *Apystomyia elinguis* Melander, 1950, by original designation.

Minute, slender, shiny-black flies with strongly humped mesonotum and with few, scattered, erect, bristly black hairs. The abdomen is elongate and attenuate and approximately twice as long as the thorax. The femora are slightly widened, especially the anterior pair. The wings are of milky-whitish appearance and also slightly iridescent when viewed obliquely. All of the veins are white, except the rather strongly developed costa and second vein.

Length: 3 mm.

Head, lateral aspect: As is so often true with the smaller Bombyliidae the head shrivels up until true proportions in dry material are virtually impossible to interpret. The head appears to be hemispherical, with a smooth, entire, posterior eye margin. The upper eye facets are enormously enlarged throughout most of the eye but without any clear dividing line between the minute facets below. In males the eyes are strongly holoptic for a considerable distance. The ocellar tubercle is triangular, raised above the eye margin by the full depth of the large ocelli, and lying between the ocelli there are 4 comparatively long, stiff, erect, black, bristly hairs. There are 2 other such hairs, even longer, which are placed posterodorsally on the frontal triangle above the antennae and at the confluence of the eyes. The face in the male paratype before me is sunken, probably due to a shriveling effect, however, it appears to be broadly triangular, much wider below. The antennae are large but with the first two segments very short and minute. The third segment is flattened, short oval, lamellate or leaflike, micropubescent, and bears a curious, slender, very short, cylindrical style or microsegment which has minute bristles. The sides of the facial triangle have a row of long, erect, scattered, bristly hairs on each side that are similar to those of the front and vertex. The proboscis is very short and consists of a mere pad or labella-like stub at the base of the head and presumably nonfunctional. On each side of the labellum is a minute, translucent, yellowish white palpus. Lateral margin of the occiput with scattered, erect, bristly hairs.

Head, anterior aspect: From the frontal view the eyes are widely holoptic leaving a comparatively small, approximately equilateral triangle which constitutes the frontal triangle above the antennae. The line of contiguity extends from the large ocelli almost to the antennae, and there are no notches present. The upper eye facets are exceptionally enlarged, but, because the eyes are deeply shrunken inward on both sides in my specimen, I cannot tell if there was a horizontal dividing line present. Melander, however, notes that the lower facets of the eye are abruptly minute. Below the antennae the oral recess or clypeal recess is very broad and large, obliterating the face that bears rather numerous long, erect, black hairs which are at no place dense. There is a blunt stump of a proboscis.

Thorax: Rather strongly humped, less so than in *Mythicomyia* Melander. It is very highly polished black in color, with perhaps a dozen erect, scattered, long, stiff bristles or bristly black hairs lying on each side of the anterior half of the mesonotum. There are 3 or 4 other similar bristly hairs on the posterior half of the mesonotum; among them is a longer element on the postalar callosity. The margin of the flat and rather pubescent scutellum has a conspicuous row of erect, long, black bristles or bristly hairs. The metanotum is large, shiny, and without pile. The entire pleuron appears to lack pile or pubescence of any kind.

Legs: Brownish black. The anterior femur has a dorsal row, a ventral row, and a more conspicuous lateral row of long, black bristly hairs. There are about 10 such hairs placed ventrally, about 7 placed dorsally on the outer half and about 10 placed laterally on the outer two-thirds of this femur. Anterior tibia with a dorsal row containing 6 long, slender, black bristles and near the apex 3 or 4 shorter ones. Middle femur with similar bristles and its tibia with 2 or 3 rows of somewhat shorter, but stouter black bristles; there are 7 or 8 bristles in each row. These tibial bristles become more numerous at the middle of the tibia and beyond. The hind femur has a ventral and a lateral row with slender, whitish, bristly hairs. Its tibia has 1 or 2 rows of weaker, shorter, whitish, bristly hairs. Claws slender and sharp; the pulvilli are well developed. Upon the hind legs the basal half of the first tarsal segment and all of the third and fourth segments are whitish due to the presence of a dense, white micropubescence; and some of the bristles are white. Middle tibia with a curious, black, thick, subapical process or downward projection, composed possibly of fused bristles; the apex of middle basitarsus with a curious apical offset swelling in the direction of the tibial process, and from which the next tarsal segment arises.

Wings: Weak, broad at the base, the axillary lobe especially deep and broadly rounded posterior basally. The alula is present but narrow and in shape gently rounded, both at base and apex, hence with a shallow curvature. The wings have a strong milky whitish cast yet are iridescent from an oblique view. All of the veins are whitish, except the costa and the second vein, both of which are very stout. There are coarse, sharp, oblique, stout setae on the costa. The anterior crossvein is close to the basal third of the very long, slender discal cell. There are 2 submarginal cells but the third vein is almost straight or plane from its origin to the apex of the wing. The anterior branch arises rectangularly, quickly turns toward the apex of the wing, curving over slightly. While there are 4 posterior cells, the discal cell is not quite as long as the cell itself. Second submarginal cell short; its shape hemicircular. The second basal cell is very long, slightly but progressively widened, it is much longer than the first basal cell, and twice as long as the third basal cell. Its length from its base is fully equal the remaining length of the fourth vein to the wing apex.

The costal cell is much widened toward the base, the costa ends at the apex of the wing so there is no ambient vein. Posterior margin of the wing with a distinct fringe and with an especially long fringe in the bend and basal margin of the axillary cell. The anal cell is closed, and the stalk or petiole is almost as long as the last section of the third vein beyond its fork.

Abdomen: Tapered and quite narrow apically. The hypopygium is a little enlarged and clublike. It is dully shining black and has a few, fine, erect, scattered white hairs on all the tergites especially laterally. There is a tuft of scattered, sparse, scanty, long, white pile, laterally on the sides of the first tergite. On the hypopygium there are some quite short, scattered, white hairs.

Because this is a very peculiar fly and because of its relationship within the Bombyliidae, I quote Melander's brief description to supplement my own:

> Eyes of male contiguous from ocelli almost to antennae, not notched, the facets of upper two-thirds coarse, lower facets abruptly minute; front of female very broad above; occiput quite flat; ocelli of male large, on the elevated vertical triangle; mouth opening broad and large, obliterating the face, the cheeks wide, eyes distantly separated below; antennae inserted low on head, basal joints minute, third joint rotund and compressed, microscopically pubescent, the apical style cylindrical, blunt, about one-third as long as the third joint; mouth parts vestigial, fleshy, not projecting. Thorax glistening jet black, with long coarse hairs, scutellar margin setose; pleurae bare. Abdomen slender, tapering, pilose in male, not tomentose, pygidium minute, globular, with two small erect spatulate dorsal palps enclosed by a pair of almost triangular lateral valves tipped with a few setulae, ventral piece small. Legs with coarse hairs, almost setose on femora and tibiae, without tomentum, pulvilli present. Wings very delicate, costa continuing to fourth vein, first vein chitinized, other veins of male thin and translucent, anal lobe very large, alulae moderate, third vein forked near tip of wing, fourth vein forked, the petiole of second posterior cell about as long as posterior crossvein, discal cell narrow, elongate, anterior crossvein at basal third, petiole of anal cell about as long as the arched anal crossvein; calypteres with nearly straight edge, heavily fringed.

Material available for study: A paratype male of the unique type-species furnished through the kindness of Dr. W. W. Wirth and Dr. Karl Krombein of the National Museum of Natural History makes it possible to include this genus within this monograph. Several years ago Dr. Melander, during a personal visit at his home, had offered me a specimen but was unable to locate his series at the time.

Immature stages: Unknown.

Ecology and behavior of adults: Dr. Melander collected his material sweeping vegetation at Wrightwood, California, on the north of San Gabriel Mountains, May 24. He obtained them by sweeping vegetation along a small stream that later entered Sheep Creek Canyon. He also obtained them on Sugarloaf Mt. in the San Bernardino Mts., May to July, and also on Mt. San Jacinto on May 7.

Distribution: Nearctic: *Apystomyia elinguis* Melander, 1950.

Subfamily Toxophorinae Schiner, 1868

This small subfamily of peculiar flies is noted for its humpbacked, robust or sometimes elongate-cylindroid appearance and for the strong bristles upon the mesothorax and prothorax and the legs. These macrochaetae are not only long as a general rule, but in *Toxophora* Meigen they are particularly strongly curved; they are stout basally, sharp apically. All the genera in the subfamily are notable for the unusually prominent, although variable, prothorax, beset with either macrochaetae or spikelike bristles. The wings are variable: small in *Toxophora* Meigen, larger and more elongate in remaining genera. *Lepidophora* Westwood has many scales upon the anterior half of the wing; *Cyrtomyia* Bigot has scales on the basal half. In some genera the wings are mottled or spotted about the crossveins and furcations. The general coloration of the flies is black, relieved in *Toxophora* Meigen by pale pile and dense, matted, scalelike pile placed in bands or spots upon the abdomen. Some species are metallic.

The antennae are elongate, the third segment is sometimes very thin, acutely pointed, as in *Toxophora* Meigen, or shorter in other groups. The first segment is quite elongate, the second usually lengthened but shorter than the first. In *Lepidophora* Westwood the antennae are very long and the first two segments show numerous, conspicuous, largely erect scales, and similar scales adorn the terminal segments of the abdomen.

The femora of *Toxophora* Meigen are swollen and thickened, especially the hind pair, and they are likewise covered with matted scales. There is apt to be a long, troughlike hollow on the ventral sides of the first four femora in this genus in both sexes. The femora in

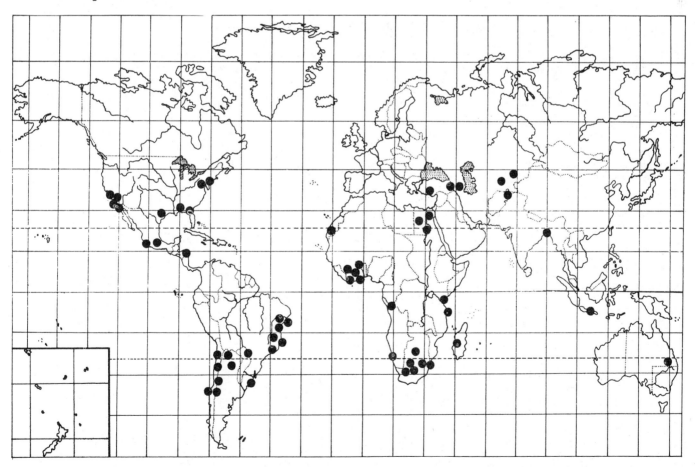

Text-Figure 32.—Pattern of the approximate world distribution of the species of the subfamily Toxophorinae.

this genus frequently, if not generally, lack bristles; whereas, other genera in the subfamily have spinous bristles below in both sexes upon at least the hind femur.

A close study of these flies reveals several peculiar features. In the first place the occiput is not hollowed and cupulate as in Cylleniinae or in the Lomatiinae, and it is not bilobate above behind the vertex as in those groups. Rather, I feel that these flies represent aberrant offshoots of early Bombyliinae stock, and so I place them in the Division Homoeophthalmae; the posterior eye contour is virtually plane.

In *Toxophora* Meigen the anal cell is closed; there are usually 2, but sometimes 3, submarginal cells; in the subgenus *Eniconeura* Macquart the second branch of the medius is partially atrophied. The wings are spotted in *Marmasoma* White, a more elongate fly with tapered abdomen.

Toxophora Meigen parasitizes vespids; it has been reared by Krombein from odynerid-type wasps where it fed on pupae or prepupae and occurs also in the nests of eumenids feeding on either the wasp larvae or the caterpillars they store. Krombein (1967) notes that *Lepidophora* Westwood larvae feed on vespid larvae, and often upon the caterpillars they store and in one instance spiders stored in the nest of a species of *Trypargilum*; see discussion under life histories.

Toxophora Meigen is the only genus in the subfamily that is of worldwide distribution. Has the strong beelike appearance of these flies been a factor in their successful persistence?

The genera within this subfamily are certainly disparate. Hence, I place *Toxophora* Meigen in the tribe Toxophorini and the remaining genera within the tribe Lepidophorini, distinguished by the subglobular head, very swollen occiput and the spikelike occipital bristles.

The subfamily Toxophorinae is characterized by an oval figure in which the dististyli are triangular from dorsal aspect. From the lateral aspect it is very deeply incised dorsally into two subequal sections, in fact, almost cut in half. *Toxophora* Meigen is much more distinct from *Lepidophora* Westwood which is nearly allied to *Cyrtomyia* Bigot. These latter two show a closer relationship, both in the very short, oval form, the large, conical epiphalli, and the dististyli. For this reason I place them in a separate tribe, also pupal characters suggest that they should be separated. After preparation of this work was completed, I revised my views on this subfamily on the basis of pupal characters. I now believe all four genera of the Lepidophorini should be placed as a separate tribe within the Bombyliinae.

KEY TO THE GENERA OF THE TOXOPHORINAE

1. Occiput with dense, rather short, erect pile of no great length. Stout, rather flattened and short, robust flies with strongly arched mesonotum and decumbent head and abdomen. Antennae elongate, especially the first segment; second segment of male antennae with a conspicuous elongate, medial patch of brilliant, silver, microscopic pollenlike microsetae. Third segment slender, elongate, flattened apically, with long attenuate, sharp-pointed style. Only 3 posterior cells. Propleuron and pronotum unusually prominent and bearing, very long, stout, curved macrochaetae, continued along front margin of mesonotum. 2

   Head subglobular, with extensive occiput, not bilobate and not cupulate; at most with faint postvertical crease. Occiput with numerous erect, long, spikelike bristles. Pronotum and anterior mesonotum to a lesser extent with stiff bristly hairs but not with stout, long, curved, scimitarlike macrochaetae. Always with 4 posterior cells; usually with 2 submarginal cells 3

2. Three posterior and 3 submarginal cells.
   Toxophora Meigen
   Three posterior and 2 submarginal cells.
   Toxophora (Toxomyia, new subgenus)
   Three submarginal cells and second branch of medius only partially atrophied . **Toxophora (Eniconeura) Macquart**

3. Antennae unusually elongate, the first two segments both elongate and both covered with dense, lateral fringe of long, broad, erect, conspicuous scales; anterior half of wing, except apex, with numerous smaller scales. Tergites of abdomen laterally with scales, growing longer, wider, tuftlike, and conspicuous on the apex of the abdomen. Strongly humpbacked flies with decumbent head and abdomen. Prothorax very prominent.
   Lepidophora Westwood
   Not such flies. Second antennal segment short and beadlike; prothorax much shorter, its bristles much weaker. Venation and head and antennae very similar to *Thevenimyia* Bigot, but anterior branch of third vein ends above the wing apex and these flies are decumbent. Palpus varies from 2- to 3-segmented 4

4. Wings with scales on basal half. Stout bodied, robust flies, strongly decumbent . . . . **Cyrtomyia Bigot**
   Wings without scales. More slender, less strongly decumbent flies . . . . **Marmasoma White**

Genus *Toxophora* Meigen

FIGURES 91, 92, 316, 317, 558, 577, 865, 866

Toxophora Meigen, Mag. Insektenkunde, vol. 2, p. 270, 1803. No species. Type of genus: *Asilus maculata* Rossi, 1790, designated by Meigen, p. 273, 1804; one species.

Eniconeura Macquart, Diptères exotiques, vol. 2, pt. 1, p. 110, 1840. Type of subgenus: *Eniconeura fuscipennis* Macquart, 1840, by monotypy.

Heniconeura Verrall in Scudder, Nomenclator Zoologicus, p. 159, emendation, 1882.

Toxomyia, new subgenus; type-species: *Toxophora maxima* Coquillett, 1886b.

Coquillett (1891a), p. 199, key to 6 Nearctic species.
Bezzi (1924), pp. 129-130, key to 8 Ethiopian species.
Paramonov (1925), p. 43, key to Palaearctic species.
Engel (1933), p. 82, key to 5 Palaearctic species.
Maughan (1935), p. 71, key to 2 U.S. species.

Hesse (1938), pp. 1032-1036, key to 7 male and 7 female species from South Africa.
Efflatoun (1945), pp. 186-187, key to 3 species from Egypt.
Bowden (1964), p. 53, key to 3 species from Ghana.

Strange, stout, robust flies of small to medium size, which are readily recognized by the characteristic humpbacked form and the long, basally stout, apically pointed, much curved bristles upon the thorax. The legs also have spinelike bristles. They are black insects with bands and spots of matted pale-colored pile on the strongly decumbent abdomen; a few tropical species are metallic green or violaceous. The head is also decumbent and the antennae long with the third segment attenuate and sharp at apex; most males have a brilliant patch of silver pollen on the inner side of the second segment.

The nearest relative is *Marmasoma* White, a Tasmanian and Western Australian genus; it differs from *Toxophora* in the shorter, nonscaly antennae, the less conspicuously drooping head and abdomen, the open anal cell, and the long full row of bristles on each side of the front.

Toxophora species are distributed throughout all world regions but appear to be most abundant in the Southwestern United States and western Mediterranean.

Length: 6 to 12 mm., including antennae; wing, 4 to 7.5 mm.

Head, lateral aspect: The head is subglobular, while the eye is only hemicircular and constitutes the greater part of the head anteriorly. The occiput is more or less swollen posteriorly both in males and females and heightened by dense, spikelike, long pile, or bristles extending backward. The occiput is most extensive centrally, with small, shallow medial depression where the head is joined to the thorax, and from this point it is quite convex and slopes strongly out to the eye margin. Outer part of the occiput covered with dense, appressed scales reaching to the eye margin. Posterior part with dense, erect or obliquely erect fringe of long, somewhat flattened, spikelike bristles. The front in males is reduced to a minute triangle above the antennae bearing a dense, matted tuft of flattened, bristly hairs diverging on each side of the antennae, and in females slightly raised above the eye margin with shallow, central fovea or depression, and with the triangular middle area again with appressed, flattened, bristly pile. The eye is exceptionally large, occupying the greater part of the head, and shows a certain amount of overgrowth posterolaterally, thus reducing the occiput and leaving the greatest eye width at some distance away from the posterior margin. Posterior eye margin laterally not quite plane on the upper half, strongly rounded below; the upper facets in the male are strongly enlarged. From the lateral aspect the face is always quite short, varying a little in extent, but it is relatively long and extensive below the antennae because the large oragenal cup ends some distance away from the antennae. The genae, how-

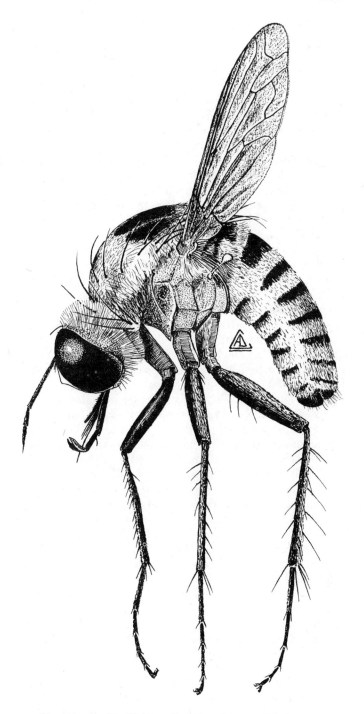

TEXT-FIGURE 33.—Habitus, *Toxophora leucopyga* Wiedemann.

ever, are narrow and rimlike. Face and genae with pollen and minute pubescence only.

The proboscis ranges from 1 to 1½ times the length of the head; it is thrust forward, compressed laterally, with long, narrow, attenuate labellum, which bears laterally many short, erect, spiculate setae. The palpus is quite long and slender and is usually half as long as the proboscis, attenuate, and slightly curved upward; it has a scanty fringe of moderately long, bristly hairs on the outer margin and often shorter bristly hairs on the dorsal and ventral surfaces. The antennae are attached near the top of the head and are extraordinarily long and slender, especially the first segment. The first segment is cylindrical, nearly or quite as long as the head and bears broad, dense scales on all sides. They are, however, sometimes absent below. In addition, there are a few, subappressed, short, black, bristly hairs medially and some setae at the apex. Third segment a little expanded distally with numerous, appressed, short black bristles; in males the medial surface of this segment shows a long, widening patch of silvery pollen extending almost the full length of the segment; and in females there may be a small patch of similar pollen near the apex. The second segment is about one-third as long as the first segment or sometimes a little longer proportionately. The third segment is no wider than the second but is strongly attenuate apically, ending in a long, spurlike style, which is sometimes colored differently and longer in the male than in the female; it is very sharp-pointed, the whole third segment, style excepted, is only a little longer than the second segment.

Head, anterior aspect: The head is almost but not quite as wide as the thorax. It appears to be set in a drooping position partly because of the prominent prothorax and the very high thorax with convex mesonotum. The eyes are strongly holoptic in males, tightly apposed for at least 3 times the length of the front. The ocellar tubercle is comparatively low, the ocelli occupying a somewhat obtuse triangle or sometimes an equilateral triangle and characteristically with a pair of extremely stout, curved, sharp-pointed, spikelike bristles, arising behind the ocelli and directed forward. The eyes in the female are rather widely separated, the front narrowed slightly above, the vertex separated by more than twice the width of the ocellar tubercle. The antennae are adjacent, the head is widest below. The width of the face is about one-fifth the maximum eye width and the oragenal cup peculiar because it appears to have a double wall. The wall consists of an inner portion encircling the oral cavity above and actually ending in the face, since the upper half of the oral recess slants strongly upward and forward until it is level with the facial prominence. This inner wall sinks downward below until it is rather deep within the recess and is quite distinct from a low, vertical rim arising at the bottom of the oral cavity and which extends at least halfway up the oral cavity, leaving the gena narrow but clearly demarcated along the eye margin. Rather near the top of the oragenal recess, between these two walls, there appears to be a shallow tentorial fissure.

Thorax: The thorax is short and compact. The mesonotum is convex although not especially high. The pronotum is unusually prominent and is characteristically large and well developed in this subfamily. The head and especially the abdomen are strongly decumbent in their attachment to the thorax. The mesonotum is almost always strongly denuded in captured individuals but there is generally undisturbed pile along the lateral and anterior margins and narrowly in front of the scutellum and on the brief declivitous portion of the mesonotum. Occasional individuals can be found in which the mesonotal pile is nearly intact. It may be continuous or vittate and consists of dense, lanceolate, flat-appressed scales extending matlike in many directions. In addition, there is much dense, much longer, pointed, delicate, scalelike pile in matted or curled, more or less erect tufts on the anterior and lateral margins, continuing widely over onto almost the whole of the mesopleuron and the whole of the propleuron and pronotum. Along the sides of the mesonotum above the wing, on the postalar callosity and in front of the scutellum and on the margin of the scutellum itself, there is a border of flat-appressed scales intermediate in length between those of the mesonotum and the longer pile on the anterior part of the thorax. In addition, nearly all species of *Toxophora* Meigen have on the mesonotum minute, flat-appressed, black scales which are largely confined to the posterior half of the mesonotum but may form vittate stripes extending forward. These appressed, black scales continue onto the scutellar disc where they are conspicuous and usually larger.

Characteristically, the pronotum and mesonotum have a number of extraordinarily long and basally stout, apically sharp, curved, conspicuous macrochaetae. They form a border on the pronotum of 3 to 7 pairs so long that they curve backward like scimitars over the mesonotum; usually black, sometimes a few are yellow. In addition, there is a row across the anterior part of the mesonotum of similarly long, curved macrochaetae consisting of 2 or 3 pairs. Another is immediately behind the humerus, 3 or 4 between the humerus and the wing base, and a small cluster on the posterior notopleuron, sometimes reduced to as few as 2 or 3; several macrochaetae are above the wing itself; 1 to 3 upon the postalar callosity and several pairs on the margin of the scutellum which are apt to be of varying size and stoutness. There may be a prescutellar pair widely separated, and the outermost pair on the scutellar margin is apt to be much bigger than the others. Upper border of mesopleuron with a horizontal row of several similar macrochaetae. The propleuron is pilose. There is a patch of appressed pile posterodorsally on the sternopleuron, but the pteropleuron, metanotum, hypopleuron, are without pile; except on the latter, a patch of pile is behind the spiracle and below the halteres. Halteral knob with a patch of scales behind in some species. Halteres short but the knob large. The squama is of variable length, sometimes very short, sometimes longer but with more or less plane and thickened margin, always bearing a very conspicuous, marginal fringe of dense, quite long bristles, umbrella-like over the haltere. At the base of the squama medially there is a flattened element that apparently corresponds to the strigula of *Systropus* Wiedemann.

The scutellum of *Toxophora* Meigen is large, wide, and rather flattened becoming thinned and descending backward as a ledge over the whole of the short metanotum. Lateral wall of the metapleuron is thin, extending straight outward behind the haltere. Metasternum short and not sclerotized.

Legs: The legs are unusually stout, especially the femora which are more or less thickened throughout, and the hind femur is the stoutest of the three pairs; it is slightly arched below, more so above. The legs are relatively short, the middle pair a little longer than the front pair and the hind pair a little longer than the middle pair. The femora lack spines or bristles, except rarely the hind femur has a short row ventrally on the basal half. All the femora are densely covered with large, broad, rather short, shinglelike scales, quite flat-appressed, ranging from white, yellow, metallic, or blackish in color. Tibiae with conspicuous, stout, sharp, black bristles. The anterior tibia has a double dorsal row of bristles; the anterior row consists of 6 to 9 bristles, the posterodorsal row has 5 to 8 bristles. There is also a posterior row of 3 to 5 bristles and sometimes additional ventral bristles. Middle tibia with similar and longer bristles, which, however, are more irregular in size and distribution. The hind tibia is stouter than the others and its bristles more erect, large and spike-like. At the base a group of 3 or 4 are placed together transversely, and there are posterodorsal and anterodorsal rows, each consisting of about 4 or 5 bristles. Apex of each tibia with very long, spikelike bristles especially ventrally. Surface of all tibiae covered with loose, shinglelike scales. Tarsi short, slender, especially on the terminal segments, with short setae below and a dense covering of fine, long, flat-appressed setae. Claws long, stout, comparatively blunt and much curved at the apex and the pulvilli long. The last tarsal segments in some species, in males especially, have on each side ventrally a comb of very stout bristles. Lower surface of anterior femur sunken or hollowed and sometimes rugose.

Wings: Wings either hyaline, tinted with brown or yellow, or rarely mottled or spotted. They are relatively short, narrowed at the base. There are 2 or 3 submarginal cells; that part of the vein constituting the second part of the radius is sometimes lost. The anterior branch of the third vein is rectangular; at times it simulates a crossvein; in any case, it is sharply bent outward. The anterior crossvein enters the discal cell at the middle or sometimes at the outer curve; the second branch of the medius may be atrophied completely or partially; hence there are only 3 posterior cells. The anal vein is always closed and petiolate, the cubitus making a strong arch distally. Axillary lobe narrow and the alula of about the same width as the base of the axillary lobe; the ambient vein extends to the end of the first anal cell. There is a thickening of the radial sector. Whole wing villose.

Abdomen: The abdomen is stout, strongly convex, and at the base as wide or even wider than the mesonotum. It may be a little longer and sometimes no longer than the thorax, and it is gradually narrowed and tapered to the end of the first segment. Eight tergites are visible in both sexes and they bear dense, flat-appressed, rather broad, short scales on the basal part of each tergite, except sometimes the last one, and these broad blackish, dark-colored scales often form continuous or paired fascia or a series of spots. The posterior margins and the lateral margins narrowly or widely are covered with matted scales of greater length and in most species yellow or white in color. Sternites generally covered with pale scales throughout and in the middle of each sternite with a curious little tuft of 4 quite fine, somewhat rather long, erect hairs. Female terminalia compressed and without spines. Male terminalia sometimes protrusive and large; in other species completely recessed.

Because of the large number of species that Dr. A. J. Hesse was able to study and summarize, I quote his work on the hypopygium of the males:

Hypopygium of males as in the case of *Systropus*, appears to act in conjunction with the last abdominal segment and the basal part of the hypopygium is constantly ventral in position, the opposing sternite being dorsal or tergal in position. This last sternite is attached to each side of basal part of hypopygium by a stoutish, chitinous process. This process may be roughly club-shaped, thickened at middle and bifid apically or blade-like. It sometimes has a tuft of longish, bristly hairs or it may be smooth. The last sternite ends apically on each side in a terminal plate, probably representing modified last abdominal elements as in other genera. These plates are usually quadrangular or sub-quadrangular, but may also be bilobed, the outer lobe with a comb or crown of short spines. The basal part of the aedeagus is often produced apically into a comb of flattened spines or into a medial process or bilobed process, or it may be without any comb or process. Lodged apically in the basal part is a distinct apical joint (segment) on each sige. This apical joint (segment) may be flanked on the outer side by another process, which has shortish, stoutish, spine-like hairs directed brush-like towards the apical joint (segment). The aedeagal complex is joined on to the basal part beyond the middle on each side by a trace-like or strap-like ramus, which usually forms a flattened process or plate, extending from sides of basal part. This plate really consists of the base of the ramus doubled upon itself, the basal part of which joins on to the process on each side of last sternite. The aedeagal complex consists of a true apical aedeagal part, usually curved towards the basal parts.

The male terminalia from dorsal aspect, epandrium removed, show an oval structure with curious, short, triangular dististyli. The basal ejaculatory apodeme is short. There appears to be an accessory anterior apodeme in addition to the more basal lateral apodeme. From the lateral aspect the epiphallus and ejaculatory process are situated quite high near the top of the epandrium. The dististyli are short, stubby, pointed at the apex but curved backward, and the basistylus has a ventral, apical notch or incision.

Material studied: Numerous Nearctic and Neotropical species.

Immature stages: The larvae are parasitic in wasp nests. See discussion under life histories.

Ecology and behavior of adults: The adults frequent plants of the family Compositae of many genera and species.

Distribution: Nearctic: *Toxophora americana* Guérin-Méneville, with fig. only, 1835; *amphitea* Walker, 1849; *leucopyga* Wiedemann, 1828 [=*fulva* Gray,

1832]; *maxima* Coquillett, 1886; *pellucida* Coquillett, 1886; *vasta* Coquillett, 1891; *virgata* Osten Sacken, 1877.

Neotropical: *Toxophora amicula* Séguy, 1930; *aurea* Macquart, 1848; *aurifera* Rondani in Truqui, 1848 [=*dryitis* Séguy, 1930, =*verona* Curran, 1934]; *cuprea* (Fabricius), 1787 (as *Bombylius*); *lepidocera* d'Andretta & Carrera, 1950 (subg. *Eniconeura*); *leucon* Séguy, 1930 (subg. *Eniconeura*); *pallida* d'Andretta & Carrera, 1950; *travassosi* d'Andretta & Carrera, 1950 (subg. *Eniconeura*); *tristis* Séguy, 1930 (subg. *Eniconeura*); *varipennis* Williston, 1901; *zikani* d'Andretta & Carrera, 1950 (subg. *Eniconeura*).

Palaearctic: *Toxophora aegyptiaca* Efflatoun, 1945; *bezzi* Paramonov, 1933; *deserta* Paramonov, 1933, *deserta tadzhikorum*, 1953; *epargyra* Hermann, 1907; *fuscipennis* (Macquart), 1840 (subg. *Eniconeura*); *lebedevi* Paramonov, 1925; *leyladea* Efflatoun, 1945; *maculata* (Rossi), 1790 (as *Asilus*), *maculata completa* Paramonov, 1933 [=*fasciculata* Villeneuve, 1789, =*maculata* Meigen, 1904]; *psammophila* Paramonov, 1935; *shelkovnikovi* Paramonov, 1953; *turkestanica* Paramonov, 1933, *turkestanica angusta* Paramonov, 1933; *zilpa* Walker, 1849.

Ethiopian: *Toxophora albivittata* Bowden, 1964; *australia* Hesse, 1938; *carcelii* Guerin, 1831; *chopardi* Séguy, 1941; *coeruleiventris* Karsch, 1887; *crocisops* Hesse, 1938; *cyanolepida* Hesse, 1938; *diploptera* Speiser, 1910; *epargyroides* Hesse, 1938; *maculipennis* Karsch, 1886; *obliquisquamosa* Hesse, 1938; *quadricellulata* Hesse, 1963; *punctipennis* Bezzi, 1921; *seyrigi* Séguy, 1934; *trivittata* Bezzi, 1908; *vitripennis* Bezzi.*

Oriental: *Toxophora javana* Wiedemann, 1821.

Australian: *Toxophora compta* Roberts, 1929.

Genus *Lepidophora* Westwood

FIGURES 90, 318, 457, 540, 869, 870, 871

Lepidophora Westwood, London and Edinburgh Philos. Mag. and Journ. Sci., vol. 6, p. 447, 1835. Type of genus: *Toxophora lepidocera* Wiedemann, 1828, by monotypy, as *Ploas aegeriiformis* Gray, 1832.

Painter (1925), p. 120, key to 4 species.
Paramonov (1949), pp. 632-633, key to 7 New World species.
Painter (1962), pp. 50-52, comments on several species.

Quaint and unique flies of medium to large size that are readily recognized by the strongly humped and arched thorax, the decumbent head and decumbent abdomen, also the presence of large, long, broad, black and partly yellow or white scales upon the sides and apex of abdomen, and abundantly on the elongate antennae. The wings are usually smoky, or subhyaline with the veins darker. The second vein and the anterior branch of the third vein make a strong, rounded rectangular bend forward near the apex of the wing. All cells widely open, except the anal cell, which is narrowed; alula well developed; anterior crossvein lies at the middle of the long, narrow discal cell. There are small dark scales on the basal half of the anterior half of the wing and an odd tuft of white or yellow scales at the base of the weak basal comb. The prothorax is strongly developed; it has a spikelike, transverse row of long, stout bristles followed behind in the middle with a large dense patch of erect scales. Head subglobular, the occiput much swollen, especially above, and with numerous spikelike bristles but not bilobate. All femora and tibia with strong, spikelike bristles. Proboscis about 1½ times as long as head; palpus elongate. Oral recess very large, the gena quite narrow but with a row of bristles.

Lepidophora Westwood is a small Nearctic and Neotropical genus ranging into Brazil. Paramonov (1949) described a species from Central Africa, which I have not seen, transferred by François to *Palintonus*, new genus.

Length: Thorax 7 mm., abdomen 6 mm., head 2 mm., antenna 4 to 5 mm., wing 7 mm. to 11 mm.

Head, lateral aspect: The head is globular; occiput unusually tumid but rounded and gradually sloping out to the eye margin. The face projects a short distance near the top of the head; the pile of the face consists of long, wide, erect scales thrust outward near the anterior border of the face in a complete row beneath the antennae. The scales become shorter laterally on the upper corners of the oragenal cup and give way to short, erect bristles below, which arise from the gena. The pile of the occiput consists of numerous, broad, yellow, erect scales, those arising medially are longer, those near the eye margin shorter and more or less appressed to the occiput. Below these scales change to blackish and in other species they may be gray or white, but on the dorsal half of the occiput arising centrally and medially there are numerous, sharp, black, long, erect, spikelike bristles that scarcely rise above the vertex. Pile at the bottom of the occiput and between the eyes consists of long, fine, scanty, scattered hairs. The eye is quite hemispherical and large, the posterior margin entire, almost plane. In males the upper facets are not enlarged. The proboscis is elongate and compressed laterally and from this aspect rather stout. The labellum is similar and one-third the total length of the proboscis. It bears microscopic setae and the whole proboscis is less than 1½ times as long as the head. Palpus large, robust, and rather elongate, extending beyond the oral cavity.

Antennae attached quite at the top of the head, they are remarkably elongate, from 1½ to 2 times as long as the entire head, slender, the first segment extraordinarily long, nearly 3 times as long as the long second segment and the second segment slender, and 5 or 6 times as long as wide; its width is deceptive because of the thick covering of scales. Third segment small, quite slender, knobbed at the base and attenuate at the base and with a medial groove. The first and second segments of the antenna both bear dense, short, suberect, blackish or gray scales dorsally and still longer, equally

---

* Bezzi (1924), p. 20, lists a *Toxophora vitripennis* Bezzi. I have been unable to find this species in print, or listed in *Zoological Record*; it may be a nomen nudum.

numerous, oblique, or subappressed scales ventrally. Those on the second segment tend to be thrust outward in a lateral plane. The third segment bears some short, strong setae laterally on the basal half.

Head, anterior aspect: The head is almost circular, being very little wider than high. The occipital scales and bristles both appear from the front and the head is much narrower than the thorax. Eyes of male narrowly separated by the width of the posterior ocelli at their outer margins. The front where the eyes approach is quite short, this area scarcely longer than wide and pollinose only and likewise flattened. The more extensive area of the front anteriorly before the antennae is somewhat raised and covered with short, broad scales and longer, scattered, slender bristles. Scales often with patches of different color. The antennae are narrowly separated by about half the thickness of the first segment. The ocellar tubercle is quite low, not all protuberant; it is also small and the small ocelli lie in an equilateral triangle from the center of which arise 7 or 8 moderately long, fine, bristly hairs. The oral opening is quite large, oval, and unusually deep anteriorly as well as posteriorly. Its margin is almost knife-thin, except at the extreme rear, where it is a little thickened on each side. The genae are linear and very narrow, the tentorial fissure is completely occluded from above. The postocciput cannot be seen without removal of head.

Thorax: The thorax is rather long and narrow, laterally compressed, strongly compressed anteriorly and posteriorly with both head and abdomen strongly decumbent. The prothorax is extensive and long, convex across the middle, thus creased behind and sloping in front. The pronotum bears long, slender, erect scales, black in the middle, yellow laterally, in some species gray or white, and often intermixed are some coarse hairs of the same color, the hair becoming denser and more in evidence laterally. Anterior margin of the pronotum with a row of long, slender, black bristles, a vertical row laterally of 7 bristles. The elongate, convex humerus bears 10 or 15 slender, black bristles. Mesonotum black with opaque sepia pollen and rather densely covered with short, slender, curved scales, yellow to white to gray in different species. These scales are semiappressed, become longer opposite the wing and on the postalar callosity and on the margin of the scutellum. Laterally it gives way to coarse pile on the corners of the scutellum. The notopleuron bears 5 quite long, rather slender, black bristles. There are other finer, but shorter, bristles. The postalar callosity has 3 long bristles, the scutellar margin has 5 or 6 pairs of black bristles on each side, which are long and some additional slender, black hairs. Vertical outer sides of the postalar callosity without pile, except for dorsal pile that turns over and except, for an anterodorsal, peculiar tuft of long, fine, yellow hairs.

Whole mesopleuron covered with long, coarse, yellow hair slightly appressed backward, some of the hairs becoming flat apically. Pteropleuron with a few hairs anteriorly; hypopleuron with an extensive tuft of hair long and fine on its anterior half, a curious tuft of short, dense hair immediately beneath the haltere and behind the spiracle adjacent to it. Spiracle with a setate rim. Metapleuron bare with a flat, shelflike lobe. Squamal fringe moderately thick but quite fine.

Legs: The femora are stout, the anterior pair rather short, the tibiae and tarsi slender, and the last 4 rather long. Anterior tibiae with stout spicules in 2 rows. There are 6 posterodorsally, 3 posteriorly, only minute setae elsewhere. On the midtibia the spicules are stout and long, in 4 rows, 5 very stout ones anterodorsally, 4 shorter ones posterodorsally, 4 stout ones posteroventrally and 3 or 4 anteroventrally. Hind tibia likewise with 4 rows of stout spicules, 4 or 5 in each row. The femora are not enlarged basally, all 3 pairs bear some stout, spinous bristles. The anterior femur bears a circlet of spines near the apex, midfemur with 4 anteriorly, hind femur with 6 ventrally, and 5 others circling the femur near the apex. Pile of the legs composed of dense, short, wide, flat appressed scales, a few fine, short, erect hairs and sometimes erect, long, slender scales toward the apex. Dorsal scales tend to be dark, ventral scales tend to be pale.

Wings: The wing is long and slender, tinged with brown or gray. Margin of the costa in males with nodulous, spinous tubercles. All the posterior cells widely open, the anal cell narrowly open. Anterior branch of the third vein makes a zigzag bend and at the corner of its first bend often has a backward spur. The second vein makes a rectangular bend toward the costa and often has an outward spur at the corner. Anterior crossvein at or near the middle of the discal cell. The axillary lobe is narrow, the alula large, the anterior half of the wing bears numerous, short, dark scales, except on the outer third. Costal comb moderately well developed. Ambient vein complete.

Abdomen: The abdomen is elongate, subcylindrical with parallel sides; at the base nearly but not quite as wide as the thorax. In length it is not quite as long as the wing. The ground color is shining blackish and is rather densely covered usually by black, short, suberect, more or less appressed, broad scales widely down the middles of the tergites and with large patches of white, gray, or yellow scales sublaterally, which vary from fine and narrow and mixed with fine, erect hairs, to wider scales on the fifth and remaining tergites. Last few tergites likely to have tufts of dense, long, wide, erect black scales. The terminalia are not turned downward but are recessed into the apex of the abdomen. I find 7 tergites in the male. There is no band of short hair beneath the scutellum but there is dense, erect hair laterally outside the scutellum.

The male terminalia from dorsal and lateral aspect, epandrium removed, show a large basally widened aedeagus. The dististyli are short and stout, turned backward medially with an apical notch, and the basistyle apparently has an underlying spoonlike apical process. Epandrium is quite short, high, subtriangular.

Material studied: A considerable series of the type-species and some material of three other species are before me, also pupae obtained by breeding.

Immature stages: These flies live in the nests of wasps. See discussion under life histories.

Ecology and behavior of adults: These flies have a somewhat bouncing flight. I have only seen one attempt at mating. Commonly found in Lafayette County, Mississippi, on flowers of *Rhus* spp., *Rudbeckia hirta* Linné, on sneeze weed, *Helenium tenuifolium* Linné, on *Apocynum canabinum* Linné, and on devil's-walking stick, *Aralia spinosa* Linné. Found once as late as October on *Aster* sp. Robertson (1928) recorded these flies appearing upon *Bidens aristosa* (Michx.) Britt., *Rudbeckia submentosa* Pursh., *Rudbeckia triloba* Linné, and *Coreopsis tripteris* Linné.

Distribution: Nearctic: *Lepidophora lepidocera* (Wiedemann), 1828 (as *Toxophora* [=*aegeriiformis* (Westwood), 1835 (as *Ploas*), =*appendiculata* (Macquart), 1846 (as *Toxophora*)]; *lutea* Painter, 1962.

Neotropical: *Lepidophora acroleuca* Painter, 1930; *culiciformis* Walker, 1850; *cuneata* Painter, 1939; *secutor* Walker, 1857; *vetusta* Walker, 1857.

Genus *Cyrtomyia* Bigot

FIGURES 88, 319, 541, 584, 736, 745, 749, 872, 873, 874

Cyrtophorus Bigot, Ann. Soc. Ent. France, ser. 3, vol. 5, p. 292, 1857. Type of genus: *Cyrtophorus pictipennis* Bigot, 1857, by monotypy. Preoccupied by Le Conte in Coleoptera, 1850.
Cyrtomyia Bigot, Ann. Soc. Ent. France, ser. 7, vol. 2, p. 340, 1892. Change of name.

Large, black flies with beautifully mottled wings, arched thorax, and somewhat decumbent abdomen. They are related to *Lepidophora* Westwood, through the fact that they both have scales upon the wing, a globose head, and a characteristic type of elongate antenna. *Cyrtomyia* Bigot differs by not having scales upon the antenna, and by a more generalized wing. The discal cell in *Lepidophora* Westwood shows a strong, rectangular bend in the fourth vein below, and the second vein and anterior branch of third vein each are very strongly bent near their ends at the apex of the wing. The flies of the genus *Lepidophora* Westwood are much more strongly arched and humped and the abdomen more strongly decumbent. The abdomen of *Cyrtomyia* Bigot is shorter, broader at the base. Both have strong, conspicuous bristles on thorax, scutellum, and upon all tibiae. The posterior femur bears strong, stout bristles; its tibiae with stout, spiculate bristles. Both *Cyrtomyia* Bigot and *Lepidophora* Westwood superficially resemble *Toxophora* Meigen in the decumbent aspect of head and abdomen, but differ conspiciously in the lack of long, curved, stout macrochaetae on the prominent prothorax of *Toxophora* Meigen, in the different venation, in the stout femora which are not swollen as in *Toxophora* Meigen. They also differ in the presence of bristly hairs along the sides of the oragenal cup.

The two known species of *Cyrtomyia* Bigot are from the southwestern part of the Neotropical region.

Length: 13 to 15 mm.

Head, lateral aspect: The head is subglobular, the face but little produced although the oragenal cavity extends well out from the eye margin, it also extends almost but not quite to the antenna, so that the face is quite short in the middle. Front and face continuous and both with rather dense, erect, long, distinctly bristly though slender, blackish pile extending out to the eye margins and continued down upon the whole of the gena leaving a small area beneath the eye which is bare and pollinose only. This bristly pile is not quite as dense upon the front and it is a little shorter on the upper front but a few of these elements are even stouter.

The occiput is quite swollen and prominent but without central cavity. The posterior eye margins are entire, the eye itself convex and inwardly retreating along its posterior border. The occiput is conspicuously covered with dense, long, and fine pile below in the middle which is shorter opposite the ventral eye corners. However, widely over the middle of the lateral occiput, there are long, conspicuous, quite scaliform hairs that arise medially and become shorter, appressed scales near the eye margin. Also there is a tuft of reddish, medial hairs. Dorsally there is a dense clump or cluster of erect, black, spikelike bristles on each side of the dorsolateral occiput. They rise above the ocellar tubercle. Palpus long and conspicuous with at least 2 segments; the outer segment is long and tapered apically. It extends as far as the beginning of the labellum on the proboscis. Dorsally and laterally the palpus bears numerous, moderately long, oblique setae, which are still longer at the apex. The proboscis is large, laterally compressed, with quite large, flattened labellum bearing a dorsal and ventral row of erect, fine setae. The proboscis is about 1½ times as long as the head or twice as long as the eye. The eye itself is almost exactly hemicircular and quite plane posteriorly. Antenna attached near the upper fourth of the head, unusually long and conspicuous. The first two segments have conspicuous, dense, long, bristly pile on all sides. The first segment is 3 or 4 times as long as the second; the second segment is narrowed basally, somewhat dilated quite near the apex and about twice as long as its apical width. Third segment slender, gently narrowed from the base and about 2½ times as long as the second segment; there is no microsegment and no pile is present.

Head, anterior aspect: Genofacial space about one-third the head width with parallel sides or even slightly widened between the lower eye corners. Front flattened, without medial groove, and very slightly raised above the eye margin. On each side in the middle it bears a longitudinal row of 4 or 5 stout, long bristles. The ocellar tubercle is quite low and rather small, but with a few, slender, bristly hairs between the ocelli. The oral opening is wide and very short oval. The head is only about as wide as the posterior thorax.

Thorax: The thorax is distinctly longer than wide, gently arched, and equally arched in front and behind. Mesonotum with rather dense, pale, flat-appressed, scales and scaliform pile, together with considerable, erect, moderately long pile. The pile in the middle is remarkably fine, the pile anteriorly and on the sides becomes coarser and intermixed laterally with long, conspicuous bristles. Some of the bristles on the lower humerus are quite stout but on the notopleuron there are 4 or 5 bristles and on the postalar callosity 3 remarkable, quite long, basally stout, curved, attenuate, reddish bristles. Margin of scutellum with a conspicuous row of long, basally stout, curved, black bristles and the disc is plastered with appressed, white scales. Pleuron shining and mostly bare. Upper mesopleuron and its posterior corner with dense tufts of fine, long, crinkled, white pile, and below it some similar reddish hairs. Metapleuron bare. Hypopleuron with a tuft of quite long, fine, reddish hairs. Squama large with a long fringe of fine hairs. Halteres with cup-shaped excavation apically, the stalk bearing scaliform pile.

Legs: The femora are relatively short but stout with fine setae and more numerous, appressed scales. Especially on the hind femur, ventrally, there are many long, fine hairs. Anterior femur with 2 stout bristles ventrally at the apex, posterior femur with an anteroventral row of some 12 stout, long, oblique, black bristles. All tibiae with long, stout spicules placed in 4 rows; each row contains about 7 elements. Claws slender, pulvilli long, arolium minute.

Wings: The wings are rather broad, narrowed at the base and strongly mottled with sepia brown. There are 2 submarginal cells; the marginal cell is bulbous at the apex. The 4 posterior cells are widely open, the anterior crossvein enters the discal cell at or near the outer third, and the medial crossvein makes a strong but rounded bend basally, otherwise narrowing the long, distal cell much as in *Lepidophora* Westwood. Anal cell closed at the margin, axillary lobe quite long, the alula large, hemicircular with a short fringe. At the apex of the radial sector is a patch of broad, reddish-brown scales and others along the costal cell, and proximal to the arculus.

Abdomen: The abdomen is broad at the base and rather flattened, and as wide or wider than the thorax and gently narrowed toward the blunt, upturned apex of the abdomen which leaves a transverse depression across the middle. Tergites with dense, suberect, bristly pile, and likewise with rather dense scales; these scales are white in a band across the middle, brown or blackish elsewhere.

The male terminalia from dorsal aspect, epandrium removed, show a wide and short-oval structure, with a long, basally expanded, apically clublike, apically incised dististyle which suggests a crab's claw in appearance. The epiphallus is wide and conical, the basal part of the aedeagus also very wide. From the lateral aspect, the basistylus is long and stout, the epiphallus particularly wide, equally wide apically but rounded.

From the lateral aspect, the dististyli are quite short. The epandrium is short, but high, and is triangular.

Material studied: Consists of both the known species; found in the British Museum (Natural History), and material received from a collector.

Immature stages: Unknown. Probably similar to *Lepidophora* Westwood.

Ecology and behavior of adults: Not on record.

Distribution: Neotropical: *Cyrtomyia chilensis* Paramonov, 1931; *pictipennis* Bigot, 1857.

Genus *Marmasoma* White

Figures 89, 313, 542, 578, 878, 879, 880

Marmasoma White, Papers Proc. Roy. Soc. Tasmania, p. 188, 1917. Type of genus: *Marmasoma sumptuosum* White, 1917, by monotypy.

Medium-sized flies with moderately humped thorax. The head is rather more decumbent than the abdomen but both are drooping; the latter is relatively about as elongate as in *Lepidophora* Westwood but a little more slender. Unlike the other genera of the family the antennae are relatively shorter in all segments, especially the second, and the antennae have only slightly widened hairs, not conspicuous scales and scaliform pile. Also, it should be noted that this genus has the anal cell widely open and there are spiny bristles on the femora in both sexes. The female front has a whole row of conspicuous bristles on each side, not just a curved pair as in *Toxophora* Meigen. The wings are spotted. The prothorax is much less prominent than in the other genera. I do not feel that the genus is particularly close to the other genera of the subfamily; certainly the general form and the size and shape of the wings more suggestively resemble *Lepidophora* Westwood.

These flies are known only from Tasmania and Western Australia.

Length: 10 to 12 mm.; of wing 7 to 8 mm.

Head, lateral aspect: The head is subglobular, the occiput very strongly developed posteriorly, especially on the upper part. The occiput bears long, coarse, nonbristly pile, but intermixed with it broadly over the whole upper surface are numerous, black, long, blunt, spikelike bristles and near the eye margin is a fringe of short, appressed, scaliform hairs touching the eye. The front and vertex are plane with the eye, but the small ocellar tubercle is quite raised; the latter bears 2 or 3 strong, long, black bristles curved forward and on each side of the upper half of front and vertex is a regularly spaced row of some 6 or 7 very long, curved, sharply attenuate, stout, black bristles directed forward. The eye is very large, the posterior margin plane, with a slight overgrowth posterolaterally as in *Toxophora* Meigen.

The face is short, but extends downward rather more than twice the length of the second antennal segment. The linear genal area is obscured by an overgrowth of the eye. The proboscis is long and slender, longer than the head, the labellum is likewise slender with a few fine, erect, widely spaced, bristly hairs laterally. Palpus long and slender, the second segment clavate and at least half as long as the first segment. The whole palpus is nearly half as long as the proboscis. The antennae are attached at the middle of the head and are elongate but not as long as in *Toxophora* Meigen. The cylindrical first segment is pollinose, 5 or 6 times as long as wide, and bears long, shaggy, pale pile laterally and below with shorter pile dorsally, and with a few fine, black, bristly hairs dorsally. Second segment short and bead-like, cylindrical, third segment knobbed at the base and at this point much more narrow than the second segment. Third segment knobbed at the base, gradually widened until at the basal fifth it is as wide or slightly wider than the second segment and from this point it is gradually narrowed, the apex is not quite as wide as the base. The apex bears 2 microsegments, the second microsegment bears an apical spine and is slightly longer and more attenuate than the basal microsegment. Third segment of male antenna without the medial, elongate patch of silvery reflecting pollen.

Head, anterior aspect: The head is short oval in females, nearly circular in males, scarcely wider below than above. The eyes are widely holoptic in the male, the front is reduced to a small triangle, the eyes meeting a distance of about 3 times as long as the frontal triangle. Eyes in the female widely separated, the sides of the front are not quite parallel, widening slightly toward the antennae and the vertex a little more than twice as wide as the ocellar triangle. The antennae are very narrowly separated at the base but converge and touch. Face from the frontal aspect nearly as extensive as in *Toxophora* Meigen, pollinose only, and with a deep crease proceeding from near the middle of each antennae obliquely downward toward the eye margin so that the lower part of the face laterally presents a vertical wall, and the crease on the lower part disappears into a narrow, tentorial fissure. Gena occluded, except at the very bottom of the eye where it is linear. Oragenal cup exceptionally deep, especially below, but slanting upward toward the face as in *Toxophora* Meigen, but however without the double wall and enclosed lateral space.

Thorax: The thorax is 3 or 4 times as long as the head, strongly humped and convex both in front and behind. It is even more convex than *Lepidophora* Westwood and the large, rather flattened scutellum is more strongly sloped downward and backward. The head and abdomen are even more decumbent.

The mesonotum is black with the lateral margins brown, quite opaque, the surface densely covered with microscopic scales, the pile consisting of matted, rather long, tangled, slightly flattened hairs, dense anteriorly, more loose behind the humerus and forming a short, transverse, more erect band of pile across the front of the mesonotum and leaving the middle of the mesonotum narrowly black with a few black hairs. The mesonotum bears an extensive patch of scattered, curved, basally stout, apically sharp macrochaetae that cover almost the whole of each lateral half of mesonotum lying in front of the deep, sharply creased, transverse suture. The macrochaetae are smaller than in *Toxophora* Meigen. On the notopleuron is a longitudinal row of 4 much stouter and longer macrochaetae and a similar row of 4 above the wing, 2 on the postalar callosity, a patch of 4 on each side in front of the scutellum, which represent continuation of middorsal rows. The margin of the scutellum has 5 or 6 pairs that are more slender. Discal pile of the large, hemicircular scutellum similar to that of the mesonotum. The postalar callosity is large, pollinose, and with a tuft of pile above and a tuft of pile below.

As in *Lepidophora* Westwood the humerus is unusually elongate, extending far backward from the anterior margin of the mesonotum. The pronotum, however, unlike other members of the subfamily Toxophorinae is very small, not prominent; nevertheless the neck or cervical region seems to be fairly long and the propleuron rather extensive with a pair of very long, strong macrochaetae. Squama quite short, except on the alary portion. The entire border is thick and ridgelike, the alary portion membranous, triangular, with a fine, short fringe of hairs. The thoracic portion is quite long and with a medial strigula. Halteres comparatively small, the knob larger with basal crease above and with cup-like depression apically. The pleuron is largely bare, except for the pollen; the upper half of the mesopleuron has rather dense pile, the lower part a few scattered hairs, and there are a very few hairs on the upper sternopleuron. The lateral metanotum has much long pile and also has 4 or 5 macrochaetae. The metapleuron is bulbously swollen with vertical fringe of pile and 7 or 8 long, slender, black bristles. Likewise, a dense tuft of pile is behind the spiracle, but the hypopleuron and the pteropleuron are bare.

Legs: The legs are comparatively short, the anterior and middle femora distinctly though gradually thickened from apex to base; hind pair stout but not swollen. The anterior femur has 1 bristle posteriorly near the apex and a few fine, long, erect hairs below. It is rather densely covered with large, broad scales above and also below, fewer laterally and behind. Middle femur similar in its fringe of hairs and scales but with 2 stout, subapical spines, one in front and one behind. Hind femur with many scales, almost no fine hairs, but with a ventrolateral row of 8 or 9 oblique and curved, stout, black, spinous bristles. Also there are 12 or more shorter, finer bristles ventrally, 3 dorsolaterally on the outer half, 2 small and 1 very large bristle dorsomedially near the apex. The anterior tibia has many scales and fine, oblique, ventral setae below on the outer half and with 3 rows of prominent, stout, spiculate bristles, somewhat irregularly placed; the anterodorsal row has 8 or 9, the posterodorsal row has 6 or 7 and the posteroventral row has 10; the apical spines are prominent. The middle tibia is similar but the bristles are much stouter and longer. There are at least 6 very large ones anterodorsally and on the basal half there are 3 very large, posteroventral bristles. Hind tibia similar to the middle tibia. There are 5 exaggerated bristles anterodosally ending at the outer third. The tarsi have small scales; basitarsi with a few, very large, spinous bristles on the middle pair. Remaining segments with comblike, stout setae ventrally. Claws long, rather sharp, curved on the outer half, the pulvilli large and long.

Abdomen: The abdomen is elongate but only about 1½ times as long as the thorax; it is much narrower than the thorax beyond the first segment but with nearly parallel sides beyond the third segment so that on the whole the abdomen appears subcylindrical. The tergites are covered with moderately long, flat-appressed, pale scales, which are apt to be largely denuded over most of the tergite, but are apt to persist as a narrow fringe along the posterior margins, and especially below on the lateral margins. On the second to fifth tergites on each side is a row of black, shining, punctiform dots, sometimes a double row. There are some scanty long, fine, erect hairs ventrally on the anterior half of the abdomen. The sternites are apt to be compressed; the whole abdomen is slightly compressed. Male terminalia small, scarcely protruded and not prominent. Last sternite troughlike with triangular apex. There are 8 well-developed tergites present in the males the last ones shorter, but beyond this the ninth tergite comprising the hypopygium is large and elongate, steep laterally with the small genitalic valves appressed.

The male terminalia from dorsal aspect, epandrium removed, show the structure extraordinarily long and slender. The basistyli apically extend beyond the attachment of the long, medially bristly, apically obtuse dististyli, which, however, at the apex have a small process, convergent and turned inward. Epiphallus forming a split sheath, the ejaculatory process spade-shaped. From the lateral aspect, the basistylus is quite bowl-shaped, the dististylus rather small, the apical prong also visible, but the epiphallus forms a very long conspicuous, snakelike, S-shaped process, especially curved downward and then backward at the apex. The epandrium is curious in having a conspicuous, ventral, basal process directed backward rather than downward.

Materal studied: The type-species.
Immature stages: Unknown.
Ecology and behavior of adults: Not recorded.
Distribution: Australian: *Marmasoma sumptusosum* White, 1917.

Genus *Palintonus* François

Palintonus François, F. J., Bull. Ann. Soc. Roy. Ent. Belgique, vol. 100, p. 324, 1964. Type of genus. *Lepidophora austeni* Paramonov, 1949.

Very recently François transferred the Central African fly *Lepidophora austeni* Paramonov to a new genus, which he calls *Palintonus*. I have not seen this genus.

Material available for study. None.
Immature stages: Unknown.
Ecology and behavior of adults: Not on record.
Distribution: Ethiopian: *Palintonus austeni* (Paramonov), 1949 (as *Lepidophora*).

Subfamily Systropinae Brauer, 1880

The flies of this subfamily are unique and very beautiful. They are remarkable examples of mimicry of sphecoid and vespoid wasps and the group extends into all the world regions. Most of these flies are parasitic in the larval stages upon moths of the family Eucleidae, but one species from Central Africa has been reared from cocoons of *Stenomutilla beroe* Pering.

The flies of this subfamily are unique in several respects. They are distinguished by: (1) the long or very long, thin, abdomen, either cylindroid or laterally compressed in females, clavate in males, together with the extraordinarily extensive and lengthened and thickened metasternum in *Systropus* Wiedemann, but ordinary in *Dolichomyia* Wiedemann; (2) the unusually prominent eyes which are often holoptic in both sexes with consequently reduced facial and clypeogenal area; (3) the long, slender, distally flattened, approximate antenna, shorter in some genera, together with the slight but distinctly concave occiput; and (4) the slender wing with only 3 posterior cells and reduced axillary lobe and alula. There are either 2 or 3 submarginal cells.

Other characters of the subfamily that may be noted are the slender wings with axillary lobe either narrowed or completely wanting so that the wing appears to be almost stalked. There is no special pronotal development as in the Toxophorinae but the propleural lobes may be quite swollen. The proboscis varies from as long as the head to much longer. These flies appear to be quite bare but are clothed with fine, minute scales, which may be in regular transverse rows upon the legs, or scattered hairs and pollen. The first posterior cell is

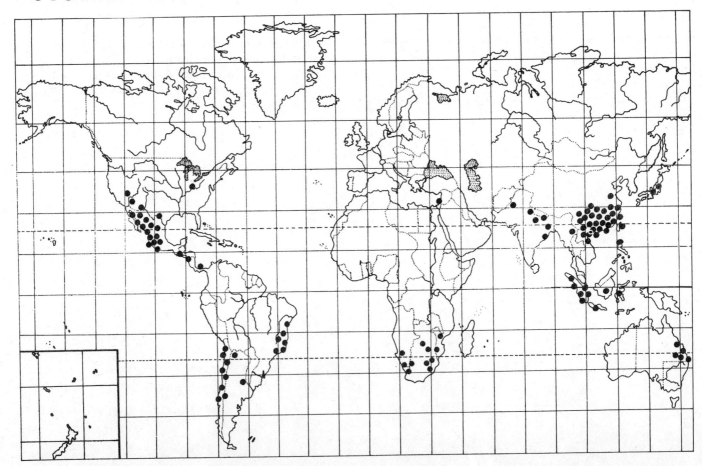

TEXT-FIGURE 34.—Pattern of the approximate world distribution of the species of the subfamily Systropinae.

narrow and is open or narrowly closed, and the anal cell when present is closed and sometimes petiolate. There is a nodelike thickening at the origin of the marginal cell and in some the radial sector has become so shortened that it is very short and vertical. A strigula is characteristically present and prominent, often colored pale. General coloration ranges from smoky blackish species to others black with conspicuous pale yellow or white humeral and propleural spots, others with base of abdomen reddish as in ammophiloid wasps, still others that are a beautiful orange red or brown with paler spots. One South American species has the metasternum pale yellow with 4 black spots. Characteristically there are dorsolateral metasternal spines.

Brues (1939) notes that the East Indian species are more apt to mimic vespoid wasps; whereas, occidental species are more apt to resemble the wasps of the genus *Sphex* (*Ammophila* of authors).

The genus *Systropus* Wiedemann has been subdivided by Enderlein (1926) into a number of minor groups; while it might have been better to leave these subdivisions as "species groups" he has proposed names for 7 subgroups on the relative separation of the eyes in males and females, the number of submarginal cells and the number of segments in the abdominal petiole. While the number of submarginal cells has been appropriately used in many groups of bee flies, and even though it is to a minor extent admittedly unstable in some groups, in many genera it is a very useful character. It does indicate stepwise the strong tendency of the dipterous wing toward simplification and vein reduction.

I leave to the bee fly students of the future the matter of usage of these names proposed by Enderlein as genera and which I leave as subgenera since they are already in the literature; I have indicated their definition as subgenera as included in my key to the subfamily.

There is one well-differentiated additional genus in this subfamily. I refer to *Dolichomyia* Wiedemann with type-species from Colombia. It occurs also in the western United States, in Chile, and Australia, but the species in Australia, of which I have collected 2, are not conspecific and for them I have erected the genus *Zaclava*. All of these flies are much smaller than most species of *Systropus* Wiedemann.

This subfamily is chiefly tropical in distribution. Only 5 species are found north of Mexico and none in Europe proper.

The Systropinae are characterized by a broad hypopygium which is elongate oval without being exceptionally elongate. From the dorsal aspect there is a strong, wide bridge across the middle overlying the epiphallus, from which arises two pairs of peculiar lobelike, apically pointed structures that we here term superior and inferior claspers. The dististyli are quite large, widened apically and blunt. Epandrium very large and subrectangular. In the tribe Dolichomyini, which should possibly be elevated to a subfamily, and this I would be reluctant to do for reasons stated in the section on the general phylogeny, there are none of these accessory processes and lobes and claspers. Also the dististyli are rather curious.

KEY TO THE GENERA AND SUBGENERA OF SYSTROPINAE

1. Metasternum entirely normal. Abdomen laterally compressed throughout its length, the tergites with opaque black spots, saddlelike above. Abdomen not narrowed at base more than at apex. Axillary cell absent. Subcostal cell occluded until the adjacent veins lie almost side by side 2
   Metasternum grossly and remarkably swollen, enlarged, projecting and drawn out posteriorly. The abdomen is long, very slender, and attenuate toward the base, more or less widened posteriorly. Wasplike flies of remarkable appearance. Axillary cell present but alula absent. First antennal segment extraordinarily long, almost threadlike; other segments are also lengthened . . . **Systropus** Wiedemann
2. First antennal segment twice as long as second. Hind femur remarkably swollen distally. No anal vein present. Metasternal plate shelflike, with spine or 1 or 2 teeth.
   Zaclava, new genus
   First antennal segment 4 or more times as long as the second. Hind femur not or scarcely thickened distally. A well-developed anal vein present. Metasternal plate scarcely noticeable . . . . **Dolichomyia** Williston
3. Two submarginal cells.
   Systropus Wiedemann, sensu stricto
   (If eyes of male and female competely holoptic, the subgenus *Cephenius* Enderlein.)
   Three submarginal cells; eyes of female in contact above.
   Systropus (*Coptopelma*) **Enderlein**
   Three submarginal cells; eyes of female separated above.
   Systropus (*Diaerops*) **Enderlein**
   Three submarginal cells; abdominal petiole 2-segmented.
   Systropus (*Dimelopelma*) **Enderlein**
   Three submarginal cells, abdominal petiole with 3 or 4 segments; eyes of female separated above.
   Systropus (*Coptodicrus*) **Enderlein**
   Three submarginal cells; abdominal petiole with 3 or 4 segments; eyes of female in contact above.
   Systropus (*Symballa*) **Enderlein**

Genus *Systropus* Wiedemann

FIGURES 86, 87, 336, 565, 588, 887, 888, 889

Systropus Wiedemann, Nova Dipterorum Genera, p. 18, 1820. Type of genus: *Systropus macilentus* Wiedemann, 1820, by monotypy.
Systrophus Latreille, Familles naturelles du règne animal, p. 496, 1825.

Céphène (Vernacular) Latreille, Familles naturelles du règne animal, p. 496, 1825. "Change of the name *Systrophus*, of Latreille, under the presumption that it was preoccupied by *Systropha*, Illiger 1806, Hymenopt."—Footnote, Bezzi, p. 3, 1924.
Cephenus Berthold, Natürliche Familien des Thierreichs (German translation of Latreille, 1825), p. 506, 1827.
Systrophopus Karsch, Zeitschr. Berliner Ent. Ges. Naturwissensch., vol. 53, p. 656, 1880.

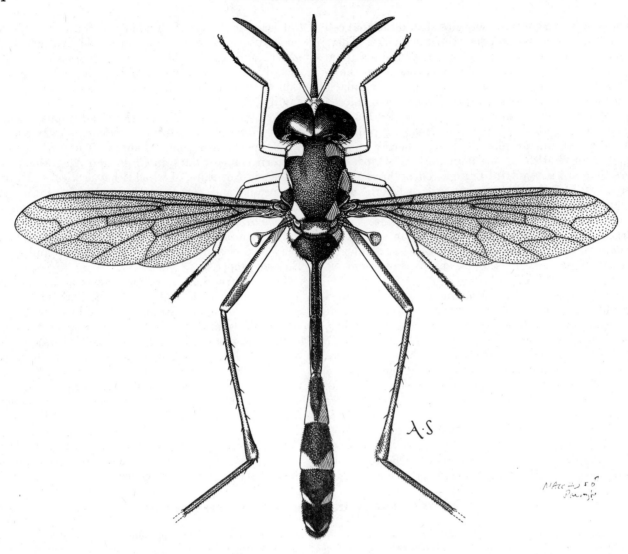

Text-Figure 35.—Habitus, *Systropus* (*Cephenius*) *maccus* Enderlein, from a male paratype.

Cephenus Karsch, Zeitschr. Berliner Ent. Ges. Naturwissensch., vol. 53, p. 656, 1880.

Xystropus Verrall in Scudder, Nomenclator Zoologicus, p. 355, 1882.

Cephenius Enderlein, Wiener Ent. Zeitung, vol. 43, p. 70, 1926b. Type of subgenus: *Systropus studyi* Enderlein, 1926a, by original designation.

Coptodicrus Enderlein, Wiener Ent. Zeitung, vol. 43, p. 91, 1926b. Type of subgenus: *Coptodicrus vespiformis* Enderlein, 1926, by original designation.

Coptopelma Enderlein, Wiener Ent. Zeitung, vol. 43, no. 2, p. 70, 1926b. Type of subgenus: *Systropus* (*Coptopelma*) *schineri* Enderlein, new name for *Systropus macilentus* Wiedemann, as misidentified by Schiner. See Hesse, pp. 1001–1002, 1938.

Diaerops Enderlein, Wiener Ent. Zeitung, vol. 43, no. 2, p. 70, 1926b. Type of subgenus: *Systropus marshalli* Bezzi, 1924, by original designation.

Dimelopelma Enderlein, Wiener Ent. Zeitung, vol 43, p. 90, 1926b. Type of subgenus: *Dimelopelma tessmanni* Enderlein, 1926, by original designation.

Pioperna Enderlein, Wiener Ent. Zeitung, vol. 43, no. 2, p. 70, 1926b. Type of subgenus: *Cephenus femoratus* Karsch, 1880, by original designation.

Symballa Enderlein, Wiener Ent. Zeitung, vol. 43, p. 92, 1926b. Type of subgenus: *Systropus leptogaster* Loew, 1860, by original designation.

Osten Sacken (1887), p. 157, key to 11 species from Mexico.
Williston (1901), pp. 292–297, key to 5 species from Mexico.
Bezzi (1924), p. 117–118, key to 9 Ethiopian species.
Enderlein (1926), pp. 71–75, key to 34 Palaearctic and Oriental species (as *Cephenius*).
Curran (1930), p. 51, key to 18 Neotropical species.
Engel (1933), pp. 87–89, key to 20 Palaearctic and Oriental species (as *Cephenius*).
Hesse (1938), pp. 996–1001, key to 10 male and 7 female South African species.
Painter and Painter (1963), pp. 287–290, key to 24 species and 5 subspecies, Nearctic and Neotropical.

Strange and beautiful flies, which have developed an astonishing pattern of mimicry centered around several groups of elongate, hunting wasps. The American species have centered around caterpillarphile wasps of the sphecid genus formerly called *Ammophila* Kirby, now known as *Sphex* Linnaeus. Certain oriental and Indomalaysian species have been likened in appearance to vespids. Hesse (1938) likens the shape to wasps of the genus *Sphex* Linnaeus, *Sceliphron* Klug, or *Belonogaster* de Saussure.

Flies of the genus *Systropus* Wiedemann are noteworthy for both their extraordinarily slender, laterally compressed, lengthened, petiolate, apically enlarged abdomen, very long hind legs, and very long, slender antennae; together with an extraordinarily bare, denuded appearance due merely to the reduction of pile to the status of minute, flat-appressed, stubby setae. Also, the metasternum is greatly enlarged, heavily sclerotized, and slanted obliquely upward. The coloration of the American forms is more commonly black with yellow spots on propleuron, humerus, notopleuron, postalar callosity, or sometimes a combination of these areas. Nevertheless, a few American species tend to be in considerable part light reddish, or yellowish on the abdomen. One Madagascar species is entirely a beautiful, brownish, orange yellow with 4 conspicuous black spots across the anterior thorax.

The venation in the genus is simple, but noteworthy for its slender wing, narrowed at the base, in which there are only 3 posterior cells; wing hyaline, sometimes smoky. The proboscis is long and quite slender with long, slender labellum, used to probe florets.

The genus is almost worldwide in distribution, more abundant in hot, wet parts of the world. It is peculiar that no species are known from Europe, even in its southern extremes, nor yet North Africa, because the American *Systropus macer* Loew extends northward in the east to Connecticut, at a latitude of about 42 degrees and westward as far as Minnesota. There are 3 species known from North China, and 6 from Sikkim, which at latitude near 30 degrees north is equivalent to Arizona and northern Mexico from where several species are known; the genus appears to be a little more plentiful in tropical and subtropical regions both above and below the equator.

Length: Including antennae 12 to 26 mm. with most species about 15 to 20 mm.

Head, lateral aspect: The head is unique in that it is composed almost entirely of the eye with relatively small frontal and facial areas, and with no occiput visible in lateral profile. The posterior eye margin is nearly plane, perhaps very slightly concave from the top down to the lower third, for the attachment to the cervix of the thorax is distinctly below the middle of the head. Viewed posteriorly the occiput is very slightly and shallowly concave and there is a narrow, vertical groove on the upper part, well below the vertex, and continued by a medial line up to the vertex. The ventrolateral tentorial fissures are mere shallow creases. The head is relatively short, the anterior eye margin is hemicircular in form but the whole head is much less than half a circle. Hence, the head and eye are relatively short. The occiput is pollinose only on the upper three-fourths of the head or more, but sometimes there are a few, fine, suberect hairs ventrally on the lower occiput. The front is small and plane with the eye, and at the base of the antennae extends forward only a very short distance; consequently the upper face is also very limited in extent. The oragenal cup is low and short, but extends nearly to the base of the antennae, leaving a small, medial, facial area.

The proboscis is long and exceptionally slender, always longer than the head, but seldom extending beyond the antennae. It varies from about twice the length of the head to about 3 or 4 times the head length. The labellum is equally slender, a third the length of the proboscis or even more with the lobes divergent, often with curled apices. Palpus visible, minute, elongate, pointed, slender, cylindrical or more or less flattened, with scarcely observable hairs and rather deeply recessed within the oragenal cavity, and composed of a single segment. The antennae are attached at or just above the middle of the head and are remarkably elongate, and quite slender, especially the first two segments. The first segment is longer than the head, sometimes nearly twice as long; the second segment varies from a third to a fifth of the length of the first segment; both segments have short, dense, appressed setae but they are a little longer and more evident on the second segment. The third segment is always strongly flattened, either elongate lanceolate, or elongate oval. The relative proportions of segments may be variable in some species.

Head, anterior aspect: The head is circular, or slightly narrowed from above and much wider than the thorax. The eyes of the male are holoptic for a very considerable distance, leaving a small, low, pollinose, or micropubescent frontal triangle and a still more minute vertical triangle which is composed entirely of the ocellar tubercle. The small ocelli lie on the eye margin and form nearly an equilateral triangle. The ocellar tubercle is a little swollen and raised. The lateral ocelli may be smaller. Oddly, the eyes are often holoptic in the female, though to a more limited extent, leaving the front extending farther upward and forming a very long, slender, wedgelike triangle. In some species the front and face can scarcely be seen from the lateral aspect. Below the antennae the face is about as long as wide, pollinose, with a very few, fine, erect hairs. The oragenal cup is long, quite deep, wedge-shaped, with steep vertical sides on the interior. The outer wall is short, slightly sloping and poorly separated from the long, quite deep sloping upper face and the narrow, ventral gena; middle of gena below with tentorial fissure barely indicated. The antennae are attached adjacent.

Thorax: The thorax is relatively short and compact, the mesonotum very little longer than wide and slightly convex. The wings are attached far to the rear of the thorax, and in fact almost above the haltere. The propleuron and mesopleuron are compacted and compressed below and crowded to the front of the thorax. The metasternum and metapleuron are extremely large, elongate, and obliquely extended upward and heavily sclerotized. The pile of the mesonotum is quite minute, rather abundant, stubby or setate, some of it erect, some of it curled backward. Some species have slightly longer, fine, erect pile over the mesonotum. There is a flangelike lateral plate from the notopleuron resting over the base of the wing. This notopleural flange has a wide, sunken fascia which in some species is densely covered with rather long, erect, bristly pile. The humerus forms a large, triangular, vertical plate, very short in length seen from above, and covered with pile corresponding to that of the pleuron. The scutellum is comparatively small, with curled setae and a little longer marginal pile; it is sometimes bimammillate, or in some species it and the mesonotum as well are coarsely punctate. The metanotum is pollinose only, and there is often a curious, coiled, or basal, scalelike or flatlike extension to the squamae on either side at the base of the scutellum. This curious structure as Painter and Painter (1963) note was called the strigula by Williston (1901), the membranous tubercle by Verrall (1909), the scutellar callosity by Bezzi (1924), and the foliate scutellar callosity by Hesse (1938).

Mesopleuron with erect or appressed pile of no great length. Pteropleuron and metapleuron both pilose. Sternopleuron micropubescent or pollinose only. Propleuron swollen and domelike to a varying extent in species both from the Old and New Worlds. Squamae very short with a short fringe.

Legs: The anterior four legs are nearly equal in size. Both their femora and tibiae are nonspiculate and have dense, appressed micropubescence. Hind legs greatly elongate and much lengthened; they are at least twice as long as the middle pair of legs, sometimes more. The hind femur varies from slightly swollen distally to other species in which it is twice as wide subapically as basally. The pile consists of very minute, appressed setae curiously laid down in transverse rows so that there is a shinglelike effect. In addition there are appressed or erect setae of greater length and stoutness. The hind tibia bears dorsal, lateral, and ventral rows of stout, sharp, spinous bristles of no great length, the number of bristles ranging from 4 to 7; pulvilli large, broad, nearly as long as the claw. The claws are sharp and strongly curved at the apex.

Wings: The wings are elongate, somewhat narrowed at the base, the axillary lobe quite long and evenly curved at base and apex. There are 2 or 3 submarginal cells, usually only 2, and only 3 posterior cells, all open. The anterior crossvein enters the discal cell beyond the middle, the anal vein is closed and stalked and the alula is absent or reduced to a mere linear strip. The base of costa has a round scale. The whole wing is villose, sometimes tinted with smoky brown diffusely about the middle at crossveins or furcations.

Abdomen: The abdomen is elongate and slender, cylindrical or laterally compressed beyond the first segment. The first segment is short and triangular, never as wide as the mesonotum. At the base of the fourth segment the abdomen begins to widen dorsoventrally though often remaining conspicuously flattened and in other species more clavate. The size of the club formed by the seventh to eighth segment varies greatly. The pile is minute, flat-appressed, and rather dense. A little erect, slightly longer pile is on the sides of the first segment. There are 8 visible segments in both sexes. Last female sternite quite variable; "either elongate and scoop-like, or with enclosed elongate process, or with 2 parallel lamellae, or shorter, emarginate apically, with the middle process projecting, or with merely a bifid process at the end of the abdomen" (Hesse, 1938). In the male terminalia the hypopygium is complex and peculiar. The last abdominal segment is so modified as to be closely associated with the true genitalic elements. This is true also of the genus *Toxophora* Meigen, and Hesse (1938) notes that it is also true of *Usia* Latreille. This last abdominal segment is produced on each side into a prong or process, and the genitalic structures are reversed leaving the last sternite dorsal in position, the last tergite ventral.

Hesse (1938) notes from his study of 11 South African species that the hypopygium of this genus is complex and the last abdominal segment modified to such an extent that it is more intimately connected than usual with the hypopygial structures proper, and that it plays a greater part in copulation than similar structures in other bombyliids. As in *Usia* Latreille and *Toxophora* Meigen the last abdominal segment is opposed to the hypopygium and is also produced into a prong or process, and the hypopygial structures are reversed in position so that the usual basal part is ventral, and the usual last sternite, which encloses the aedeagus in most other bombyliids, is dorsal in position and corresponds to the last tergite; apically its apical angle is produced into an elongated process, or prong, which is straight, curved or "hook-like." Hesse finds the apical margin of this tergite is usually emarginate and between the prongs, attached by a membrane, there is a triangular plate. Toward the inner side of each terminal plate he finds an oval, or elongate, sometimes very broad, callus-like structure, its surface roughened, filelike, or faceted; he views these and similar plates in other bombyliid genera as modified terminal abdominal segments. The hypopygium itself, he notes, "consists of a single basal part, more or less feebly divided into two parts by a medial depression, and apically each part ending in an apical segment (joint) which assumes many shapes in different species but usually ends in a spine-like beak directed inward."

The "aedeagus of the aedeagal complex, located in the hollow of the basal part is generally short, more or less concealed by the remainder of the armature, and consists of an aedeagal process which corresponds to the same of other bombyliids, an accessory process and a ramus." The aedeagal process, Hesse finds, is either quite prominent and inflated, or tumid, or more slender and bifid apically; the accessory process either leaf-shaped, twisted, or simply rodlike, or lobelike, and connected on each side to the basally directed aedeagal struts; the aedeagus passes on into the middle part which has the usual lateral strut on each side, and the medial, basally directed basal strut.

The aedeagus is also produced into a basally directed, flattened, strap-like, flattened, or boomerang-shaped aedeagal strut on each side in the hollow of the basal part and usually projecting basally a little beyond the base of the basal part. The whole aedeagal structure is joined on to the basal part on each side by means of the ramus, both together forming a jaw shaped or inverted U-shaped structure, the apical part of which may be broadened or produced into a lobe or spine on each side.

Hesse presents many excellent figures of the complicated genitalia of South African species. Painter and Painter (1964) give excellent illustrations of the male genitalia of five North American species and parts of others and figure females of some.

Dr. A. J. Hesse was able to study a very large assemblage of species in this genus; for this reason I quote his comments on the male genitalia:

Hypopygium of males is complicated and peculiar in that the last abdominal segment is structurally modified to such an extent that it is more intimately connected with the true hypopygial elements than in all the preceding genera. There appears no doubt that it thus plays a greater role in the copulatory act than homologous structures in other Bombyliids in this first division of this family. As in the case of the Palaearctic *Usia* and in the genus *Toxophora* the last abdominal segment, opposed to the hypopygium, is also produced on each side into a prong or process. The hypopygial structures of *Systropus* are also constantly reversed in position in that the usual basal part of the hypopygium is ventral in position and the usual last sternite, enclosing the aedeagal structures in most other Bombyliids, is dorsal in position and corresponding to a last tergite. This latter segment, is usually somewhat sunk in or lower than the tergite before it and is alluded to as the last tergite. Apically on each side its apical angle is produced into an elongated process, prong, or spine which is straight, curved or even hook-like. The apical margin of the last tergite is usually emarginate and on each side between the prongs and attached by a membrane there is a subtriangular or triangular plate. Towards the inner side of each terminal plate above there is an oval or elongate and sometimes very broad, black, indurated, callus-like area, the surface of which is shagreened, file-like or appearing faceted. These and similar plates present in all other Bombyliid-genera probably represent modified terminal abdominal segments.

The last tergite in *Systropus* is attached to the ventrally situated basal part of hypopygium on each side laterally and towards the base and also medially on the inside to the apical part of the hypopygial ramus of the aedeagal complex by a transverse, flattened, strap-like or band-like, chitinous band, extending from the base of one prong to the other and often produced towards the centre into an apically directed process or lobe on each side. The actual attachment to the apical part of the ramus is by means of a tough membrane.

The hypopygium itself consists of a single basal part more or less feebly divided into two parts by a slight medial depression. Apically each part ends in an apical joint which assumes a variety of shapes in the various species and usually ends in a sharp or spine-like beak directed inwards. The aedeagus of the aedeagal complex, lodged in the hollow of the basal part, is usually shortish and more or less hidden by the rest of the armature, consisting of an aedeagal process corresponding to the ventral aedeagal process of some other Bombyliids, an accessory process and the ramus. The aedeagal process is either very prominent, inflated or tumid apically, or more slender and bifid apically. The accessory process is either leaf-shaped and twisted or merely rod-like or lobe-like and is usually connected or joined on each side to the basally directed aedeagal struts. The aedeagus is also produced into a basally directed, flattened, strap-like or boomerang-shaped aedeagal strut on each side in the hollow of the basal part and usually projecting basally a little beyond bases of basal part. The entire aedeagal structure is joined on to the basal part on each side by means of the ramus, both together forming a jaw-shaped or ∩-shaped structure, the apical part of which may be broadened or produced into a lobe or spine on each side.

The male terminalia from dorsal aspect, epandrium removed, show a broad, stout structure in which the epiphallus encloses the ejaculatory process which is much shorter and more slender and emerges below. A strong wide transverse bridge from the ramus emits 2 pairs of prominent lobelike claspers. The superior pair is curved inward at the apex, the inferior pair is curved outward. In the lateral aspect the dististyli are large, wide, blunt, and almost truncate at the apex, and wider apically. The claspers are very prominent and long and the epandrium quite large, subquadrate with a ventral, apical lobe.

Material available for study: The series of species of this genus in the British Museum (Natural History) and of the National Museum of Natural History, together with a representative of *arizonicus* Banks furnished me by Dr. George Byers, and of other material received by purchase from South Asia, South America, or Madagascar, besides several species from the United States and Mexico.

Immature stages: The flies of this genus are mostly parasitic on the larvae of the lepidopterous family Eucleidae (Limacodidae of authors), or in one known instance from the cocoons of the wasp *Stenomutilla beroe* Pering. See section under life histories for further discussion.

Ecology and behavior of adults: Adult flies of this genus are active flies, especially likely to be found around midday in hot sunshine coming to various flowers upon which they feed. Painter and Painter (1964) have found them occasionally as early as 10:30 A.M., as late as 4:00 P.M. Painter and Painter (1964) record Mexican species from the flowers of the following: *Baccharis glutinosa* Pers. (Compositae), the vine *Cissus rhombifolia* Vahl. (Vitaceae), *Melochia pyramidata* L. (Sterculiaceae), *Phyla strigulosa* Mart. and Gal., and the prostrate white verbena *Lippia nudiflora* L. (Verbenaceae), besides more abundant on an undetermined pink-flowered herb, also on *Boerhavia erecta* L. (Nyctaginaceae), and on *Eysenhardtia polystachya* Ortega.

The species *angulatus ammophiloides* Townsend has been found on flowers of *Lippia wrighti*. Nearctic eastern species have been found on *Helianthus* sp. In Mexico, at Milepost 1248, West Pacific Highway, the author and Marguerite Hull, the author's late wife, and son, Sillers Hull, took a large series of *Systropus* of two species on pink flowers of the coral vine *Corculum leptopus* (H & A) Stuntz. Robertson (1928) recorded *macer* Loew from the labiate *Pycnanthemum virginianum*, the verbeniates *Verbena hastata* and *V. urticaefolia*, and the composites *Aster crocoides*, frequently on *Aster turbinellus*, *Bidens aristosa*, *Boltonia asterioides*,

Eupatorium serotinum, Rudbeckia triloba, Solidago nemoralis, and *Vernonia fasciculata.*

Bowden (1964) records 2 species of the genus in Ghana from the flowers of labiate *Hoslundia opposita.*

Painter and Painter (1964) note that there is an even greater resemblance in these flies to wasp models during flight, such resemblance extending to such ichneumonids as *Therion* and *Thyredon.*

Distribution: Nearctic: *Systropus ammophiloides* Townsend, 1901; *angulatus* Karsch, 1880; *arizonicus* Banks, 1909; *macer* Loew, 1863 [=*imbecillus* (Karsch), 1880 (as *Cephenus*), =*infuscatus* (Karsch), 1880 (as *Cephenus*)].

Neotropical: *Systropus acutus* Painter, 1963; *basilaris* Painter, 1963; *bicornis* Painter, 1963; *calopus* Bigot, 1892; *cerdo* Osten Sacken, 1887; *chilensis* Philippi, 1865; *columbianus* Karsch, 1880; *conopoides* Kunckel, 1904; *currani* Carrera and d'Andretta, 1950; *dimidiatus* Curran, 1942; *dolorosus* Williston, 1901; *femoratus* (Karsch), 1880 (as *Cephenus*); *foenoides* Westwood in Guérin, 1842 [=*funereus* Costa, A., 1865]; *fumipennis* Westwood in Guérin, 1842 [=*niger* Walker, 1849]; *geijskesi* Curran, 1942; *lanei* Carrera and d'Andretta, 1950; *lugubris* Osten Sacken, 1887; *mars* Curran, 1942; *nigripes* Painter, 1963; *nitidus* Wiedemann, 1830 [=*brasiliensis* Macquart, 1847]; *oldroydi* Carrera and d'Andretta, 1950; *paloides* Painter, 1963; *pulcher* Williston, 1901; *quadripunctatus* Williston, 1901; *repertus* Carrera and d'Andretta, 1950; *rogersi* Osten Sacken, 1887; *rufiventris* Osten Sacken, 1887; *sallei* A. Costa, 1864; *semialbus* Painter, 1963; *similis* Williston, 1901; *vicinus* Painter, 1963; *willistoni* Curran, 1942; *xanthinus* Painter, 1963.

Palaearctic: *Systropus acuminatus* (Enderlein), 1926 (as *Cephenius*); *annulatus* Engel, 1932; *barbiellinii* (Bezzi), 1905 (as *Cephenius*); *cantonensis* (Enderlein), 1926 (as *Cephenius*); *chinensis* (Bezzi), 1905 (as *Cephenius*); *divulsus* (Séguy), 1963 (as *Cephenius*); *excisus* (Enderlein), 1926 (as *Cephenius*); *exsuccus* (Séguy), 1963 (as *Cephenius*); *flavicornis* (Enderlein), 1926 (as *Cephenius*); *flavipectus* (Enderlein), 1926 (as *Cephenius*); *fodillus* (Séguy), 1963 (as *Cephenius*); *formosanus* (Enderlein), 1926 (as *Cephenius*); *gracilis* (Enderlein), 1926 (as *Cephenius*); *hoppo* Matsumura, 1916; *indogatus* (Séguy), 1963 (as *Cephenius*); *interlitus* (Séguy), 1963 (as *Cephenius*); *laqueatus* (Enderlein), 1926 (as *Cephenius*); *limbatus* (Enderlein), 1926 (as *Cephenius*); *maccus* (Enderlein), 1926 (as *Cephenius*); *melli* (Enderlein), 1926 (as *Cephenius*); *montivagus* (Séguy), 1963 (as *Cephenius*); *mucronatus* (Enderlein), 1926 (as *Cephenius*); *nigricaudus* (Brunetti), 1909 (as *Cephenius*); *nigritarsis* (Enderlein), 1926 (as *Cephenius*); *nitobei* Matsumura, 1916; *polistoides* (Westwood), 1876 (as *Cephenius*); *sauteri* (Enderlein), 1926 (as *Cephenius*); *sikkimensis* Enderlein, 1926 (as *Cephenius*); *studyi* (Enderlein), 1926 (as *Cephenius*); *submixtus* (Séguy), 1963 (as *Cephenius*); *suzukii* Matsumura, 1916.

Ethiopian: *Systropus atratus* Macquart, 1846; *barnardi* Hesse, 1938; *bicoloripennis* Hesse, 1963; *bicuspis* Bezzi, 1924; *buttneri* (Enderlein), 1926 (as *Cephenius*); *clavatus* Karsch, 1880; *crudelis* Westwood, 1876; *cuspidicauda* (Enderlein), 1930 (as *Symballa*); *diremptus* Enderlein, 1926; *fumosus* Hesse, 1938; *gracilis* Hesse, 1963; *hessei* François, 1954; *holaspis* Speiser, 1914; *ichneumoniformis* Hesse, 1958; *leptogaster* (Loew), 1860 (as *Symballa*); *limacodidarum* Enderlein, 1926; *macilentus* Wiedemann, 1828 [=*attenuatus* Macquart, =*capensis* Philippi, 1865, lapsus]; *marshalli* (Bezzi), 1924 (as *Diaerops*); *munroi* Hesse, 1938; *namequensis* Hesse, 1938; *quadripunctatus* Séguy, 1934; *rex* Curran, 1929; *rhodesianus* (Hesse), 1938 (as *Cephenius*); *rufidulus* Bowden, 1962; *rufifemur* (Enderlein), 1926; (as *Cephenius*); *rugosus* Bezzi, 1924; *sanguineus* Bezzi, 1921; *schineri* (Enderlein), 1926 (as *Coptopelma*); *sericeus* Bezzi, 1924; *sheppardi* Hesse, 1963; *silvestrii* Bezzi, 1914; *snowi* Adams, 1905; *subcingulatus* (Enderlein), 1926 (as *Cephenius*), *tenius* (Enderlein), 1926 (as *Cephenius*); *tessmanni* (Enderlein), 1926 (as *Dimelopelma*); *trigonalis* Bezzi, 1924; *vespiformis* (Enderlein), 1926 (as *Coptodicrus*); *zuluensis* Hesse, 1938.

Systropus elegans is proposed as a new name for *Systropus quadripunctatus* Séguy (1934), preoccupied by Williston (1901).

Oriental: *Systropus blumei* Vollenhoven, 1863; *celebensis* Enderlein, 1926; *edwardsi* Brunetti, 1920; *eumenoides* (Westwood), 1842 (as *Cephenius*); *flavicoxa* (Enderlein), 1926 (as *Cephenius*); *furcatus* Enderlein, 1926; *numeratus* de Meijere, 1916; *ophioneus* (Westwood), 1849 (as *Cephenius*); *roepkei* de Meijere, 1914; *sphegoides* Walker, 1860; *tessellatus* Vollenhoven, 1863; *tipuloides* Westwood, 1876; *tricuspis* (Enderlein), 1926 (as *Cephenius*); *udei* (Enderlein), 1926 (as *Cephenius*); *valdezi* Bezzi, 1917; *varipes* Edwards, 1919; *violacescens* (Enderlein), 1926 (as *Cephenius*).

Australian: *Systropus doddi* Roberts, 1929; *flavoornatus* Roberts, 1929.

Genus *Dolichomyia* Wiedemann

FIGURES 93, 94, 335, 579, 589, 881, 882, 883

Dolichomyia Wiedemann, Aussereuropäische zweiflügelige Insekten, vol. 2, p. 642, 1830. Type of genus: *Dolichomyia nigra* Wiedemann, 1830, by monotypy.

Comparatively small flies with an extraordinary increase in the length of the slender abdomen and in the hind legs; in these respects they resemble *Systropus* differing from that genus in the much shorter antennae, the greater anteroposterior length of the head and in the absence of the axillary lobe on the wing. The wing is even more slender and lanceolate than in *Systropus* Wiedemann; the two genera are both bare and with relatively denuded appearance due to only minute pile, and both have a long, slender proboscis. The wings are glassy and hyaline.

This genus, except for the related new Australian genus *Zaclava*, is only known from the southwestern United States and Chile. Thus its present range is much more restricted than *Systropus* Wiedemann.

Length: 5 to 10 mm.

Head, lateral aspect: The head is large, almost entirely consisting of the eye, with very small, short, frontal triangle and smaller vertical triangle. The head is relatively long, nearly as long as high, the occiput in the male is shallow on the upper fifth of the head, dwindling away until none of it can be seen laterally. It is, however, more swollen and extensive on the lower part of the head. And this is true of both sexes. The pile of the occiput consists of pollen or minute pubescence with laterally along the middle a few, scattered, appressed, outwardly directed, short, bristly hairs against a black background and gray pollen. There is a sharp, vertical stripe of yellowish-white pollen running downward from the vertex and in the proper light most of the pollen on the lower part of the head is grayish white. The very small front is gray pollinose without pile. The face is very low and limited by the oragenal cup which extends close to the base of the antennae but which in profile is very short and linear. The pile consists of pollen or minute micropubescence only. The genae are linear.

The proboscis is quite long and slender. The labellum is equally slender and long and it constitutes more than one-third of the total length; it is sharply attenuate, the halves apposed; the whole proboscis is twice as long as the head and extends well beyond the antennae and is thrust straight forward. The palpus is extraordinarily slender but also very elongate and has 3 or 4 fine, oblique, dorsal, bristly hairs. It generally lies close against the proboscis and is not easily seen. The antennae are attached at the middle of the head, elongate and slender but much less so than in *Systropus* Wiedemann. The first segment is cylindrical, 3 or 4 times as long as wide, pale, pollinose, with a few, minute, oblique setae below. The second segment is widest apically and as long as its apical width, and about one-fourth as long as the first segment. The third segment is as long as the first two together, pollinose, the basal half a little thicker than the second segment and at this point narrowed to the blunt apex. The outer half of this segment is a little more than half as wide as the basal part. The whole antennae is perhaps not quite as long as the head.

Head, anterior aspect: The head is circular, the eyes extensively holoptic in the males. The head is much wider than the thorax, the antennae arise adjacent, the vertex is almost restricted to the ocellar tubercle, the latter is a little swollen. The relatively large ocelli lie on the eye margin. There is a narrow, vertical, pollinose wedge extended forward from the ocelli. The oragenal opening is large, elongate, oval, very deep with vertical sides but somewhat more slender on the upper fifth where it is divided by a triangular, sloping, medial ridge. It extends to within a short distance of the antennae. Laterally the walls of the oragenal cup are quite narrow and also short and the upper lateral face and lower genae are linear. The tentorial fissure is deep but consists of a mere line with ventral pit. In a Chilean species the head is curiously long, and globular with the eye cut away on the dorsal half, leaving the occiput much exposed from above.

Thorax: The thorax is short and compact, somewhat similar to *Systropus* Wiedemann but with the metasternum less prominent and not forming fused or semifused, sclerotized, posterior plates, and also the mesonotum is higher and more convex, and is especially steep, vertically abrupt in front above the very short pronotum and curved inward and backward above the pronotum, to a somewhat greater extent than in *Systropus* Wiedemann. The humerus is outwardly swollen and rounded, elongate and extending obliquely backward. The mesonotum is either polished and bare, without a trace of pollen, but with moderately abundant, fine, erect pile or thinly pollinose and vittate with only a very few fine, short, erect hairs. The notopleuron forms a flat, flangelike ledge above the base of the wing somewhat as in *Systropus* Wiedemann but much less prominent.

The scutellum is small, convex with a few fine, short, stiff, erect hairs. The disc is pollinose even on those species which have a bare mesonotum; both halteres unusually large, with large, apical disc or club. The whole pleuron is pollinose and covered with dense, microscopically small, pollenlike scales; upper border of mesopleuron with a few long, stiff hairs appressed upward; remainder of pleuron without pile; squama quite short, without fringe of hairs but with some minute micropubescence. On each side at the base of the scutellum there is a small, circular, cuplike structure at the edge of the squama that corresponds to the strigula found on *Systropus* Wiedemann. There are 2 or 3 setate hairs below and behind the posterior spiracle.

In this genus the dorsal posterior extremity of the metasternum forms a distinct, inflated, rather long, dentate or sometimes bidentate flangelike, lateral extension. It probably corresponds to the small, embedded, oval, or triangular plate at the end of the metasternum situated dorsally in *Systropus* Wiedemann, which is also sometimes toothed.

Legs: Middle pair of legs about 1½ times as long as the anterior pair and only about half as long as the hind pair, which, as in *Systropus* Wiedemann, are greatly lengthened. Beginning near the middle the hind femur is slightly swollen distally. The hind tibia is a little wider apically than at the base. In the new genus *Zaclava* the hind femur is very strongly clavate apically with a row of stout, sharp, ventral spines. The anterior tibia has 2 or 3 fine, short, oblique, bristly setae ventrally and also anterodorsally. The middle tibia has somewhat larger setae anterodorsally, dorsally and posteroventrally. The first 2 rows have about 3 setae, the last one about 7 setae. Hind tibia with prominent bristles, slender but not very long. They are principally concentrated in a dorsal row and a posteroventral row, each containing about 10 elements. The pile of the tibiae consists of a few scattered setate hairs and some scattered, comparatively broad, appressed scales that are more abundant distally on each tibia, and more conspicuous on the hind pair. The anterior femur has a few fine, flat-appressed, scattered microsetae, and this femur is slightly enlarged and thickened in the middle and on the basal half; middle femur more slender with similar pile. The hind femur is narrow at the base, gradually widening until near the apex where it is nearly 3 times as wide. It has some broad, appressed scales, a few small, short, appressed setate bristles ventrally on the outer half and viewed in the proper light it appears to have microscales arranged in transverse rows somewhat as in *Systropus* Wiedemann. The pulvilli are large and elongate with an empodium between. Claws sharp and slender.

Wings: The wings are hyaline, very slender and narrowed on the basal half due to the loss of the axillary lobe and the loss of the alula. There are only 3 posterior cells, each widely open, and the ambient vein is complete. The radial sector is quite short and there is a swelling of the vein where second and third veins meet. The anterior crossvein enters the discal cell close to but not quite at the outer third of that cell. In the type-species the anal vein is quite evident but is set quite far apart from the cubital vein. In the genus *Zaclava* the first anal vein is completely atrophied. There is a knob-like enlargement at the furcation of the second and third veins.

Abdomen: The abdomen in the type-species is 4 or 5 times as long as the thorax and of almost equal width apically as either in the middle or at the base, both when viewed from above and from the side. Thus the abdomen is subcylindrical, but is distinctly a little flattened and compressed laterally, more so at the end of each tergite and especially so upon the second to sixth segments. The posterior corners of each tergite have a more or less oval spot of silvery pollen. Pile of the abdomen minute, appressed and setate, the last tergite with a somewhat conspicuous fringe of bristly hairs apically. Apart from the hypopygium 8 tergites of nearly equal length are visible in the male and female, the eighth being a little shorter. Hypopygial elements above and below are of nearly equal length and apparently the tergite is inverted and reversed as in *Systropus* Wiedemann. The abdomen of the type-species extends far beyond the apex of the hind femur, but in the new genus *Zaclava* it extends only a very short distance beyond the apex of the femur and the abdomen is therefore comparatively shorter.

The male terminalia from dorsal aspect, epandrium removed, show a broadly widened structure toward the base, strongly and gradually narrowing apically with curious dististyli, which in lateral aspect are curved more or less like the letter C, blunt-lobed above and below. The epiphallus forms a broad, dorsal plate, more or less triangular in aspect, occupying the entire outer half of the structure, arising narrowly from the ramus. In lateral aspect the epiphallus is prominent with a curious, attenuate, apical, downturned lobe, and the ejaculatory process has a narrow, prominent incision arising ventrally. The epandrium is triangular, the basistylus plane below, arched into a hemicircle above. None of the intricate accessory claspers are present, such as are so characteristic in *Systropus* Wiedemann. The relationship must be very distant.

Material studied: Several individuals of the type-species and one Chilean species.

Immature stages: Unknown.

Ecology and behavior of adults: In a personal communication Dr. Joseph Bequaert tells me that he collected individuals of the type-species on a species of *Solidago* at an elevation of 6,200 feet, in Mohave County, Arizona, on August 9. R. H. Painter has recorded collecting the same species on *Clematis* on August 23, west of Ft. Collins, Colorado.

Distribution: Nearctic: *Dolichomyia gracilis* Williston, 1894.

Neotropical: *Dolichomyia decta* Schiner, 1868; *nigra* Wiedemann, 1830.

Zaclava, new genus

FIGURES 97, 333, 583

Type of genus: *Systropus clavifemoratus* Hardy, 1922, by present designation.

Differing from the type-species of *Dolichomyia* Wiedemann by the elongate, globular head in which the eye recedes forward leaving the occiput more exposed and with the hind femur strongly swollen on the outer half with numerous, stout, spinous bristles ventrally. The third antennal segment is flat, wide at the base and lanceolate and short. The anterior legs are quite short, proportionately much shorter than in *Dolichomyia* Wiedemann. Proboscis but little longer than the head, stout and robust, the broad, flattened labellum nearly as long as the proboscis. First anal vein completely atrophied.

These flies are known only from Australia.

Length: 5.5 to 10.5 mm.; more commonly 5 or 6 mm.

Material studied: Two new species collected in Australia.

Immature stages: Unknown.

Ecology and behavior of adults: The author collected two Australian species; one at the flowers of a blossoming tree of unknown genus some 10 or 15 miles from Melbourne late in December, and another at Lake George, New South Wales, on *Leptospermum* sp., earlier in that month.

Distribution: Australian: *Zaclava clavifemorata* (Hardy), 1922 (as *Systropus*); *minima* Roberts, 1929; *occidentis* Roberts, 1929.

Subfamily Xenoprosopinae Hesse, 1956

In 1956 Hesse described a very peculiar and aberrant bombyliid from South Africa for which he erected a new subfamily.

A very strange fly immediately distinguished by the large pendulous lobe attached below to the first antennal segment and the greatly reduced mouthparts and lobe-like palpi.

This medium-size fly has, in addition, relatively short legs, no bristles on the body, and no spines on the femora. The short pulvilli are almost spinelike. Blackish in ground color, with fine, white, scaliform pile on the yellowish postmargins of the abdominal tergites. Hesse (1956) notes that it has somewhat the appearance of an asilid or therevid as it rests upon the sand. Wing reduced in size. The wing venation, which is distinctly generalized, shows some relationship to the tribe Cythereini; it is worth noting that the only other bombyliid genera with vestigial mouthparts, *Villoestrus* Paramonov, *Oestranthrax* Bezzi, and *Marleyimyia* Hesse, all belong to the Exoprosopinae to which the Cythereini show a measure of transitional relationship, but which may be only parallel development. The genus *Xenoprosopa* Hesse is even more aberrant. It is known only from a single female, and until males are available I leave it in the new subfamily created for it by Hesse (1956).

Genus *Xenoprosopa* Hesse

Figures 84, 339, 539, 760

Xenoprosopa Hesse, Ann. South African Mus., vol. 35, p. 942, 1956. Type of genus: *Xenoprosopa paradoxa* Hesse, 1956, by original designation.

A very strange fly immediately distinguished by the large pendulous lobe attached below to the first antennal segment and the greatly reduced mouthparts and lobe-like palpi.

This medium-size fly has, in addition, relatively short legs, no bristles on the body, and no spines on the femora. The short pulvilli are almost spinelike. Blackish in ground color, with fine, white scaliform pile on the yellowish postmargins of the abdominal tergites. Hesse (1956) notes that it has somewhat the appearance of an asilid or therevid as it rests upon the sand. Wing reduced in size. The wing venation, which is distinctly generalized, shows some relationship to the tribe Cythereini; it is worth noting that the only other bombyliid genera with vestigial mouthparts, *Villoestrus* Paramonov, *Oestranthrax* Bezzi, and *Marleyimyia* Hesse, all belong to the Exoprosopinae to which the Cythereini show a measure of transitional relationship. This genus *Xenoprosopa* Hesse is even more aberrant. It is known only from a single female, and until males are available I leave it in the new subfamily created for it by Hesse (1956).

These are peculiar and remarkable flies. The most unique feature is the odd antenna coupled with the peculiar oral structure. The proboscis and palpus are much modified, reduced, and transformed and the face depressed. Antennae set close together, with fine hairs on all segments. The first segment has a large, pendulous, prominent, conspicuous, densely haired, inflated, lobelike extension below the segment.

The occiput is rather short, only slightly concave behind, with a distinct, broad, shallow, central, groovelike depression, extending down from behind the ocellar tubercle, and bounded on each side just behind the dorso-posterior angle of the eyes with a slight, but distinct, prominence or lobe.

Hesse (1965) believes that this genus is nearest the Cythereini and that it shows some relationship, though slight, to *Oniromyia* Bezzi.

These flies are readily recognized, Hesse states, by the whitish hair, conspicuous, yellow-ringed abdomen, and yellow legs. They are known from a single female from Pofadder, Bushmanland, taken in October.

Known only from Bushmanland, South Africa.

Length: 10 mm.; wing 7 mm.

As I have not been able to see or obtain material of *Xenoprosopa* Hesse, I quote the author's description of this genus. Hesse states:

Body somewhat elongate. Head broader than thorax, much broader than long; occiput relatively short, only slightly concave behind, but with a distinct, broad, and shallow, central, groove-like depression, extending down from behind ocellar tubercle and bounded on each side just behind dorsoposterior angles of eyes by a slight, but distinct, tumid prominence (or lobe); hind margin of eyes only slightly sinuous, not deeply indented and without a bisecting line; ocellar tubercle broad, only slightly raised, its postero-lateral ocellus on each side well developed, slightly elongate, much larger than small anterior and medial one; frons markedly broad, equally broad throughout, slightly and broadly depressed in front of ocellar tubercle, somewhat transversely prominent a little behind antennae.

The antennae are somewhat close together and with fine hairs on all the joints; joint 1 with a large, prominent and conspicuous, fairly densely haired, inflated or lobe-like extension below; joint 2 transverse, its outer apical angle somewhat sharply pointed; joint 3 stoutish, broadened basally, clubshaped, ending apically in a small or short terminal joint, itself ending in a slender stylet; facial part in front (or below) antennae slightly depressed or excavate, showing and lodging the following mouthparts and structures: a transverse groove demarcating anterior margin of actual face, a central raised boss-like part, separated from eyes and genal part by a deep sulcation or groove (genal groove) and ending medially and anteriorly in a downwardly projecting, somewhat triangular and bluntly pointed process or spine (? anterior or reduced rim of modified buccal cavity) which is bounded on each side and below or behind it by an oval, inflated, hair-covered lobe (? modified palps), these mouthparts in turn being bounded behind by a downwardly projecting lip-like process or extension (? posterior rim of modified buccal cavity). Thorax shortish, subquadrate, but longer than broad, slightly convex anteriorly, appearing slightly humped; scutellum relatively smaller than in other Bombyliid genera, narrowish, rather tumid, leaving on its sides much of postalar calli exposed.

The legs are short, relatively feebly developed, without any spines on femora, only with fine and short hairs, the middle ones however with longish fine ones on outer face; tibia with the front and middle ones scarcely or only a little shorter than femora, the hind ones much shorter, covered with numerous very short spicules, in more than four rows; apical spurs of tibiae short, relatively feebly developed; tarsi shortish, shorter than tibiae, slightly compressed in apical part, especially hind ones, with relatively feebly developed, short spicules, the front tarsi not modified or different from the others; pulvilli present. Vestiture on body relatively poorly developed, composed of fine, shortish, hairs and fine hair-like scales, the former densest on head and the latter on abdomen, without any stiffer bristly hairs or bristles on any part of body; hairs on abdomen, even on sides, fine and very short, absent on venter; pteropleuron, hypopleuron and greater part of metapleuron bare; plumula entirely absent; scaling on legs fine.

The wings are relatively short and narrowish; membrane wrinkled; basal comb wanting; alula wanting; second vein originating some distance before middle cross vein; only two submarginal cells present; discoidal cell relatively short; four posterior cells present, of which the first and third are narrowed apically, the third being very characteristically triangular in shape; anal cell elongate, narrowish; squamae triangular, sparsely hairy along hind border; knobs of halteres rather large. Adbomen elongate, with 7 visible segments in female and in this sex with a posterior genital tuft.

Material available for study: None. This genus is known only from the unique type.

Immature stages: Unknown.

Ecology and behavior of adults: Hesse (1956) states that this unique and remarkable fly has some resemblance to an asilid or therevid when resting upon the sand.

Distribution: Ethiopian: *Xenoprosopa paradoxa* Hesse, 1956.

Subfamily Platypyginae Verrall, 1909

These small anthophilous flies have a characteristic habitus because of their strong humped mesonotum and the globular head. The proboscis while of no great length is usually as long as the head, yet it extends well beyond the confines of the oral cavity and in some species it may be much longer than the head. In the first of two groups the discal cell is complete, as in *Platypygus* Loew, and in the second group the discal cell is open due to the absence of the crossvein; in this second group are found *Cyrtosia* Perris, and also *Onchopelma* Hesse in which the anal cell is closed and stalked, besides the curious Australian genus *Cyrtomorpha* White.

In *Cyrtomorpha* White the size is a trifle larger, the flies have a very compact, subglobular abdomen and the larvae feed within the egg pods of locusts. This is the only genus in the subfamily where we have any knowledge of the larval habits or hosts. The second vein ends close to the first and the proboscis is shorter than the head.

The subfamily *Platypyginae* is characterized and separated from the Mythicomyiinae by the presence of a distinct, well-developed second vein ending independently in the costa beyond the first vein and also by the unbranched third vein. There are present 4 posterior cells, the anal cell usually open, except *Onchopelma* Hesse, and the alula absent or very much reduced.

In *Cyrtisiopsis* Séguy, a genus extending into South Africa, the posterior ventral part of the head is curiously drawn-out and extended backward; probably all those species of *Platypygus* Loew with extraordinarily extensive postocciput should be placed in *Cyrtisiopsis* Séguy. *Ceratolaemus* Hesse though lacking a discal cell also has a remarkable postocciput, raising the question of the comparative value of the discal cell as contrasted with occipital structures. Hesse (1938) calls attention

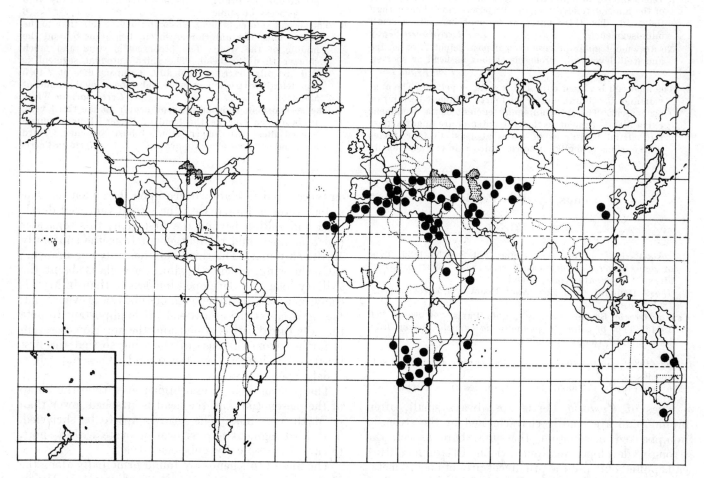

Text-Figure 36.—Pattern of the approximate world distribution of the species of the subfamily Platypyginae.

to the adaptation of this subfamily and flies of the genera *Apolysis* Loew and *Phthiria* Meigen to the matter of floral adaptation. This subfamily has been found on all continents, except South America, and it probably occurs there. These flies seem especially well developed in the Mediterranean and South Africa and probably many more species and unique types are to be found below the equator.

In this subfamily belongs the remarkable genus *Psiloderoides* Hesse. These flies for which I erect the tribe Psiloderini are much like acrocerid spider parasites in general appearance. They are unique also in the marked reduction of the proboscis. They are nearest to the Australian genus *Cyrtomorpha* White in which the proboscis though short is still present, but the discal cell is wanting. Both genera live in and destroy egg pods of locusts.

The subfamily Platypyginae is a disparate group, containing as it does such curious flies as *Psiloderoides* Hesse and *Cyrtomorpha* White.

Those genera with the discal cell absent I include in the tribe Cyrtosiini, including *Ceratolaemus* Hesse, with its quadriarticulate antenna. *Platypygus* Loew and *Cyrtisiopsis* Séguy I would place in the tribe Platypygini.

KEY TO THE GENERA OF PLATYPYGINAE

1. The discal cell is complete, with 3 veins leading from it . . 2
   The discal cell is open, the posterior crossvein absent, consequently the second posterior cell has a long petiole. The anterior limb of the fourth vein has 2 branches, which may sometimes be faint 4
2. Proboscis reduced to a minute, nipplelike structure. Habitus like that of acrocerids; mesonotum humped and abdomen wide, inflated, decumbent. Almost bare flies.
   Psiloderoides Hesse
   Proboscis clearly present 3
3. A remarkable gular process present on the postventral part of the much extended occiput. Proboscis much longer than the head. The third antennal segment has only a small microsegment . . . . **Cyrtisiopsis Séguy**
   No drawn-out gular process even if postocciput is extensive and well developed. Proboscis as long as head or shorter.
   Platypygus Loew
4. The anal cell is closed with a long stalk or petiole. The anal vein and distal branches of the fourth vein may be faint or evanescent. The third antennal segment has a prominent extra distal segment so that the antenna are quadriarticulate. Hind basitarsus of males at least in type-species with curious pronglike process. Third vein extraordinarily straight. Second vein ends close to first vein; male eyes in contact . . . . **Onchopelma Hesse**
   The anal cell is open, at most closed in the margin. Axillary lobe narrow 5
5. Head distinctly elongate due to the remarkable lengthening of the occiput posteriorly which is produced directly backward as a curious spinelike ventral process. Eyes widely separated in both sexes with a minute kink in the margin behind antennal insertion. Antenna also quadriarticulate.
   Ceratolaemus Hesse
   Head rather spherical or slightly oval. If the occiput is prominent posteriorly it is not produced ventrally into any process or spine 6
6. The second vein ends much closer to the first vein than to the third vein. The anterior crossvein lies at or beyond the middle of the wing. The proboscis is stout and much shorter than the head. The third antennal segment is wide basally, narrowed apically. Compact flies of 2 to 5 mm. length with compact, wide, subglobose, semidecumbent abdomen . . . . **Cyrtomorpha White**
   The second vein ends midway or nearer to the third vein. Proboscis as long as the head or longer. Plump, humpbacked flies with longer, stout, robust abdomen. Third antennal segment elongate ovate . . . . . **Cyrtosia Perris**

Genus *Cyrtosia* Perris

FIGURES 321, 593

Cyrtosia Perris, Ann. Soc. Ent. France, vol. 8, p. 55, 1839. Type of genus: *Cyrtosia marginata* Perris, 1839. [Bezzi p. 3, 1924 gives *C. marginata* type-species.]

Cyrthosia Rondani, Arch. Zool. Anat. Fisiol., vol. 3, (sep.) p. 73, 1863, lapsus.

Cephalodromia Becker, Ann. Soc. Ent., Paris, vol. 83, p. 121, 1914a. Type of genus: *Cephalodromia curvata* Becker, 1914, by monotypy (as an empidid).

Engel (1933), pp. 105-106, key to 21 Palaearctic species and subspecies.

Efflatoun (1945), pp. 36-37, key to 9 species from Egypt.

Flies of *Cyrtosia* Perris are always small, often minute, and are readily characterized by the strongly humpbacked mesonotum, the quite bare aspect, the strongly developed occiput, which Efflatoun (1945) calls puffed out, and the prominent pronotum and metanotum. They are usually yellowish flies with conspicuous vittate pattern of blackish or brownish coloration on thorax and fasciate pattern upon the abdomen. The radial sector is forked with the second vein well developed and ending beyond the first vein; this separates *Cyrtosia* Loew and related genera from the subfamily Mythicomyiinae. The third vein has only one branch.

On the wing, which is hyaline, the alula is absent, the axillary lobe well developed but less so than in Mythicomyiinae; the auxiliary vein or subcosta is evanescent, fading away at the apical end. It is important to note that the discal cell is absent and the anterior crossvein is farther from the base of the wing so that the first basal cell is longer than the second. The anal cell is widely open.

The pleuron is high and higher even than the hump of the mesonotum and the head is attached lower than in Mythicomyiinae. The head is apt to be shriveled. The third segment is pyriform or ovate with a short, thick, microscopically pubescent style.

The flies of this genus are found principally along the Mediterranean Coast, in Spain, Sardinia, Asia Minor, Turkestan, etc., and Efflatoun noted 10 species from

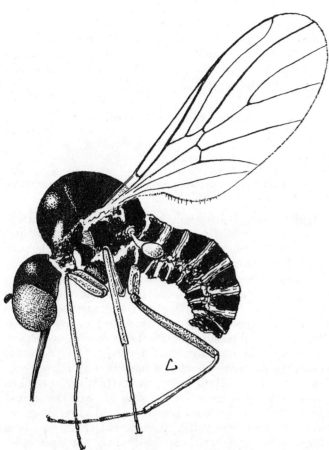

Text-Figure 37.—Habitus, *Cyrtosia nitens* Loew.

Egypt. Hesse has described 2 species from South Africa.

Melander described a species of the related genus *Platypygus* Loew from the southwestern United States. So far the subfamily Mythicomyiinae has a much wider world distribution.

Length: Commonly 1.5 mm., or 2.5 mm., but reaching 4.6 mm. in a few species.

My description taken from the flies in the British Museum (Natural History) has been supplemented by Efflatoun's very excellent description.

Head: Usually ovoid, sometimes as long as, or longer than, broad, and often narrower than the thorax; occiput considerably puffed out and conspicuously produced below, ventrally behind the opening of the mouth, into a triangular or conical projection, which is best seen in profile; frons broader above near the vertex than below where it is always more or less deeply excavated (in most specimens owing to shriveling when dried) and when viewed in profile its sides form almost an angle of 45 degrees with the face which is quite as long (as the frons) and, as in *Empidideicus*, usually more or less produced above, at the upper margin of the mouth; eyes, as in the two preceding genera, with the facets of equal size all over and widely separated in both sexes; genae very narrow, linear; proboscis varying considerably in length, from shorter than the head to as long as the length of the head and the thorax together, in fact it often varies so much in individuals of the same species that it cannot afford a reliable character of distinction; palpi apparently absent, or if present then only very minute and rudimentary (as an extremely minute, microscopic projection bearing a few hairs at the apex); vertex broad, with the ocelli placed on a very feebly raised ocellar tubercle or almost on the same level as the sides of the vertex and the ocelli always form a fairly wide triangle owing to the median ocellus being always placed a little in front of the lateral ocelli.

Antennae inserted about halfway between the vertex and the upper margin of the mouth and consist of a short cylindrical basal segment, a very short somewhat broader second segment and a pyriform to elongate-ovate third segment ending in a short but usually thick style (the thickness due to dense microscopic pubescence); this style, as a rule, is distinctly shorter than in the preceding genus.

Thorax: Very much humpbacked, distinctly more so than in *Leylaiya* and in *Empidideicus*, and with the pronotum and metanotum well developed; the attachment of the head to the thorax is distinctly lower than in the two preceding genera, being much lower than the notopleural suture; humeral and postalar calli stronger than in *Empidideicus*; pleuron well developed and its height usually exceeds the hump of the mesonotum, above the notopleural suture.

Wings: Wings with the alula absent but the anal lobe well developed and with the venation less reduced than in the two preceding genera:

The Sc fades away in the cell, before reaching the costa and the latter is strong and thickened a little beyond the tip of vein R_{4+5}; the radial sector consists of 2 branches, R_{2+3} and R_{4+5}, in addition to the short upper branch R_1, which joins the costa well after the middle; the r-m crossvein is placed further away from the base of the wing than in the other two preceding genera, i.e., toward the middle; the basal portion of vein M_{1+2}, which forms the upper margin of cell 2nd M is usually less strong than the anterior branched fork and 1st M_2 cell is always absent; vein 2nd A is well marked almost to the wing margin, consequently cell 1st A is present and is always more or less widely open; squamae small, with a very sparse, delicate fringe; halteres comparatively short, with a very large knob.

Legs: Moderately long and strong; front and hind coxae long; tarsi, especially the metatarsi, are somewhat long and the apical tarsal segments are usually blackish, partly due to the black pubescence on these parts only; claws and pulvilli small.

Abdomen: Usually not broader than the thorax, cylindrical, feebly bent downward, and often somewhat dorsoventrally flattened; it consists of 7 or 8 segments and the genitalia in both sexes are almost always entirely concealed; the pubescence although sparse and very short is usually distinct.

Material available for study: The series in the collections of the British Museum (Natural History).

Immature stages: Unknown.

Ecology and behavior of adults: Efflatoun (1945) notes that the flies of this genus are usually captured only on or in flower heads; he found them in large numbers in desert Compositae as well as flowers of the common desert plant *Zilla spinosa*. He found hundreds of individuals of *marginata* Perris in February and March on flowers of *Calendula* and *Chrysanthemum*.

Distribution: Palaearctic: *Cyrtosia abragi* Efflatoun, 1945; *aglota* Séguy, 1930; *amnicola* Bowden, 1965; *bicolor* Abreu, 1933; *canariensis* Engel, 1932, *canariensis tenuis* Frey, 1958; *cinerascens* Loew, 1873; *cognata* Engel, 1933, *cognata trisignata* Engel, 1933; *crocea* Séguy, 1949; *flavorufa* Strobl, 1909; *fusca* Séguy, 1938; *gulperii* Efflatoun, 1945; *humeralis* Abreu, 1933; *injii* Efflatoun, 1945; *jeanneli* Séguy, 1949; *luteiventris* Bezzi, 1926, *luteiventris minima* Efflatoun, 1945; *marginata* Perris, 1839; *meridionalis* Rondani, 1864; *nitens* Loew, 1846; *nitidissima* Engel, 1933; *nubila* Bezzi, 1925; *obscura* Fabricius, 1904; *obscuripes* Loew, 1855, *obscuripes maculithorax* Engel, 1933, *obscuripes serena* Becker, 1915; *occidentalis* Rondani, 1863; *opaca* Loew, 1846; *pallipes*, A. Costa, 1885; *perfecta* Becker, 1910; *persicana* Becker, 1903; *pusilla* Loew, 1873; *seia* Séguy, 1963; *separata* Efflatoun, 1945; *tetragramma* Bezzi, 1926, *tetragramma canariensis* Engel, 1933.

Ethiopian: *Cyrtosia curvata* Becker, 1914; *fusca* Séguy, 1938; *ornatifrons* Séguy, 1932; *namaquensis* Hesse, 1967; *stuckenbergi* Hesse, 1967.

Australian: *Cyrtosia parvissima* Roberts, 1929.

Genus *Ceratolaemus* Hesse

FIGURES 334, 585, 592

Ceratolaemus Hesse, Ann. South African Mus., vol. 34, p. 969, 1938. Type of genus: *Platypygus* (*Ceratolaemus*) *xanthogrammus* Hesse, 1938, as a subgenus, by original designation.

Ceratolaemus Hesse is distinguished from its nearest relatives by the lack of the discal cell. It is my belief that it should be put in a separate genus.

These flies are known only from South Africa, and like other members of the Cyrtosiinae they frequent blooms of various flowers in which they may be found feeding.

Length: Of body, about 2¼-2½ mm.; of wing: about 2½-3 mm.

Since I have not been able to see males of this genus I quote the remarks of Dr. A. J. Hesse with respect to genitalia:

> Hypopygium of male with basal parts; the apical process in form of a bidentate structure; aedeagal complex with the aedeagus distinct and slender and joined on to basal parts by a ramus on each side; the basal strut and parts by a ramus on each side; the basal strut and lateral struts cannot be distinctly made out and are minute in contrast with the same structures in *Onchopelma* and *Empidideicus*; the basally directed rod on each side is, however, distinct.

Material available for study: My illustrations are taken from the material of the British Museum (Natural History). I was unable to obtain a dissection of the genitalia.

Immature stages: Unknown.
Ecology and behavior of adults: Unknown.
Distribution: Ethiopian: *Ceratolaemus bilineatus* Hesse, 1967; *longirostris* Hesse, 1967; *montanus* Hesse, 1967; *xanthogrammus* Hesse, 1938.

Genus *Onchopelma* Hesse

FIGURES 352, 538, 580, 582, 590

Onchopelma Hesse, Ann. South African Mus., vol. 34, p. 973, 1938. Type of genus: *Onchopelma pulchellum* Hesse, 1938, by original designation.

Quite small flies; in females there is a superficial appearance to some females of *Heterotropus* Loew and *Phthiria* Meigen which is occasioned by the similar size and yellow color but with radically different wing venation and antenna. The hind basitarsus of males shows a curious, conspicuous, basal hook. The abdomen is plump rather than slender and the mesonotum less strongly humped than in other members of the Cyrtosiinae. The head is subglobular, the male eyes in contact, the occiput short, flattened, normal. The palpus is small, the proboscis shorter, stouter than in *Platypygus* Loew and the labellum longer and better developed. The first basal cell is considerably shorter than the second due to the extreme basal position of the anterior crossvein; the marginal cell is present but there is only a single submarginal cell, and while there are 4 posterior cells there is no discal cell. The auxiliary vein is evanescent distally, and while the first, second, and third veins are strong, the branches of the medius and cubitus are very weak, especially as they approach the wing margin; Hesse (1938) shows the second anal vein as evanescent; my figures from types show it strong, but the anal cell is closed with a quite long stalk. The base of the wing is quite broad in contrast to *Cyrtosia* Perris, due to a very broad axillary lobe; in spite of this the alula is much narrowed or absent. Finally, there is no transverse depression on the front as in *Phthiria* Meigen; front very small in males, broad and wide in females. Curiously the antenna is 4-segmented, the third segment bearing a second segment, as wide as the first, at least half as long, and with small attached stylelike microsegment.

These flies are known only from 3 South African species.

Length: 3 to 3.5 mm.; wing 3 to 3.5 mm.

Head, lateral aspect: Shining yellow flies. The head is distinctly subglobular, the frons, face, and gena bare or nearly so. All pile on these insects very fine, soft, rather sparse. The face is angularly produced forward but only to a very moderate extent. The antennae are attached at or very close to the middle of the head. The occiput is more prominent and tumid and extensive in females; in males it disappears in profile toward the vertex, and without a conspicuously developed central cavity; occipital pile fine, sparse, setiform. The ocellar tubercle, however, large, broad, and conspicuous. The posterior eye margin is slightly but distinctly convex, becoming more so above and below. The palpus is small, not easily discovered, and tucked away hidden in the basal sheath of the proboscis. The proboscis is stout, robust, never longer than the head, and the labellum is prominent and has small spinules, or erect, downturned, stiff bristles. The antennae are rather large, somewhat swollen in *trilineatum* Hesse, more slender in *pulchellum* Hesse; the first and second segments are nearly equal and bear scattered appressed setae; the combined third and fourth segments vary from about twice as long as the first two segments to somewhat less long; the entire length of the antenna is equal the length of the eye; the fourth segment bears a minute spine, or microsegment at the apex.

Head, anterior aspect: The head is nearly the same width as the thorax. The front is broad in females, very small indeed in males; the female front has only a faint indication of a central, vertical depression. The upper eye facets of males are distinctly but not conspicuously larger. The oragenal opening or recess is deep and broad with the upper edge rectangular. The quite large ocelli are situated in a triangle with strong, raised tubercle; in males they rest upon the eye margin. The antennae are attached adjacent.

Thorax: Rounded and convex, less humped than in *Cyrtosia* Perris. The scutellum is transverse. The squamae are small, greatly reduced; halteres short, knobs subglobular or oval.

Legs: The legs are comparatively stout and rather short; there are no spines or bristles beneath the femora. The tibiae are distinctly short and not even the tibia longer than the femur; they have only short, spiny hairs below; there are short apical spurs and the hind basitarus in males has a curious, conspicuous, curved, slender, bristle-tipped spur, or hooklike process; this process has a single, long, spinelike bristle projected upward. The claws and pulvilli are well developed.

Wings: The wings are hyaline. There is a marginal cell but only a single submarginal cell. While there are 4 posterior cells there is no discal cell. The anterior crossvein is far backward toward the base of the wing, placed barely beyond base of second vein. The auxiliary vein is evanescent and so are all the branches of medius, cubitus, anal veins, etc., which normally reach the wing margin. The wing is unusually broad at base but the alula is vestigial or absent.

Abdomen: There are 7 visible segments in the male and 8 in the female. Most of the male hypopygium is exposed; the upper apical angle of the terminal male sternite is angularly and sharply produced. The male hypopygium has distinct bristly hairs basally; according to Hesse (1938) the basimeres are connected to the aedeagal structures by the prolonged basal part; there are apical processes that correspond to the dististyli (beaked apical segments) of the Bombyliinae; they are lobelike and have a stout hook ventrally at the base. As to aedeagal structure, the aedeagus is slender; there is a slender, basally directed rod on each side and a ventrally directed curved rod on each side; lateral struts are well developed; the basal strut has no lateral process on each side, it is very broad, with a flattened laterally extended flange attached dorsally on each side.

Material available for study: The types of both known species in the British Museum (Natural History), from which Arthur Smith made the accompanying illustrations.

Immature stages: Unknown.

Ecology and behavior of adults: Not on record.

Distribution: Ethiopian: *Onchopelma karooanum* Hesse, 1967; *pulchellum* Hesse, 1938; *trilineatum* Hesse, 1938.

Genus *Psiloderoides* Hesse

FIGURE 340

Psiloderoides Hesse, Ann. South African Mus., vol. 50, part 6, p. 121, 1967. Type of genus: *Psiloderoides mansfieldi* Hesse, by original designation.

Very curious flies that greatly resemble small acrocerids. These are small flies of only a few millimeters length. It is rather readily recognized by the minute, nipplelike, or fingerlike lobe representing the much reduced proboscis. The antennae are comparatively short, all segments of about the same length; the third segment has a very thick style, or microsegment with a bristle at tip.

These flies, however, are noteworthy for their strong resemblance to the spider parasitic flies; the abdomen is large, swollen, much decumbent, and subglobular and it is considerably wider than the thorax. The mesonotum is greatly humped and the small head decumbent to a marked degree. Upon the hyaline wing the marginal cell is shortened and there is only 1 submarginal cell. While there are no bristles present the pile of the vestiture is better developed than in related flies. The legs are rather small, a trifle swollen or thickened upon the femora.

These flies have 4 posterior cells and a well-developed discal cell of small, rather wide, and short proportions. They are possibly related to the Australian *Cyrtomorpha* White, which is also subglobose and more or less resembles an acrocerid and, like *Psiloderoides* Hesse, is also a consumer of locust-egg packets in larval stages.

These flies were taken in the Kenhardt District of South Africa in May.

Length: 2.6 to 3.2 mm.; of females 4.2 to 4.4 mm.; of wings 3.2 to 3.88 mm.; of females 4.36 to 4.96 mm.; width of female abdomen, up to 3.6 mm.

Just as the present work was ready to be submitted for publication I was in receipt of Dr. Hesse's paper describing 3 remarkable new genera of bee flies from South Africa. I quote this description:

This new genus, of which the known representative have a striking resemblance to species of the genus *Psilodera* of the spider parasites (dipterous family Acroceridae), is established

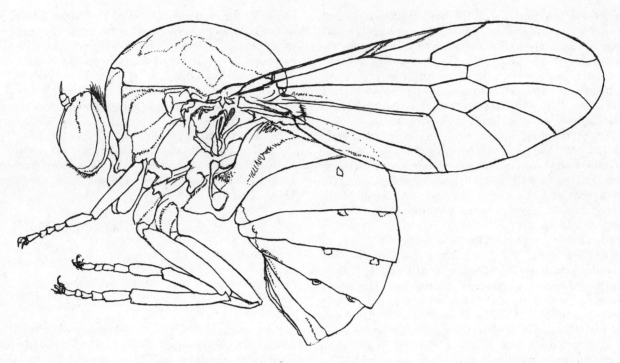

TEXT-FIGURE 38.—Habitus, *Psiloderoides mansfieldi* Hesse. Redrawn after Hesse.

to accommodate a remarkable bombyliid submitted by Mr. R. J. Mansfield and of which the adults were reared in a laboratory of the Department of Agriculture in Pretoria from a batch of bombyliid larvae found both in egg-packets of the brown trek locust (*Locustana pardalina*) and in loose soil in association with such egg-packets.

This bombyliid cannot be referred to any other subfamily of the Bombyliidae but to the Cyrtosiinae [renamed Platypyginae]. With the latter it agrees in certain characters, such as the reduction of the wing-venation, the presence of only one submarginal cell, the peculiar reduced and somewhat triangular marginal cell, the characteristic quadri-articulate antennae, the slight indentation in the inner margin of the eye opposite the antennae, the absence of distinct spines and spicules on the legs, absence of macrochaetae on body, the arched or humped and convex thorax, broad and arched abdomen, and rows of small shiny depressed black spots on abdomen (present in some cyrtosiines [platypygines]).

In certain other characters, such as the very much reduced, rudimentary or vestigal proboscis, the widely separated ocelli of which the middle, slightly anterior, one is remarkably large, the less developed occiput, the sculptured or punctured thorax and abdomen, the excavate venter, and even more convexly humped or arched thorax and abdomen. It, however, differs from other known genera of the Cyrtosiinae [Platypyginae] to such an extent that at least a distinct section or tribe of the latter is indicated to include it.

As the life histories of all the other known South African cyrtosiine [platypygine] bombyliids (if not of the world) are unknown, the discovery of this genus and its host is of great importance.

The genus is characterized as follows:

Body arched and humped, with a striking and marked resemblance to that of the genera *Psilodera*, *Terphis* and *Thyllis* of the spider parasites (fam. Acroceridae), its widest part across between tergites 2 and 3.

Head almost spherical; occiput more flattened, not very prominent, more like that of *Onchopelma*, medially not depressed behind ocellar prominence; eyes large, not tending to be situated far forwards, separated on vertex in both sexes, apparently more widely so in females their inner margin slightly indented opposite antennae, in females somewhat uneven, not uniformly convex, but slightly shallowly depressed groove-like from ocellar corner obliquely down to near middle; ocellar prominence relatively broad, slightly more raised in males, transverse, delimited from frons by a distinct, forwardly-curved, depressed line or suture, not evident in other cyrtosiine genera, the lateral ocellar part higher than middle. The ocelli are widely separated, more so than in the other genera, in a slightly forwardly-curved line, the lateral ones very near or at upper corner of eyes and thus very broadly separated, the ocelli relatively large, especially the middle one which is also more elongated transversely; frons with a slight central depressed line, slightly raised on each side basally in front of each lateral ocellus, broader in females, broader basally than at antennae, and slightly broader than long; face in side view curving down to buccal cavity to the same extent as eyes, longer and narrower than frons, basally separated from antennal insertions by a transverse depression or depressed line, slightly narrowing from base to apex (beginning of buccal cavity); buccal cavity gradually widening to head below, the interocular space on head below being as wide as, or slightly wider than, base of face, the buccal depression not very deep, as long as, or slightly longer than, face; proboscis much reduced, minute or vestigial, represented by a small finger-like lobe or minute nipple; antennae situated close together, quadri-articulate, joint 4 elongate, slender, rod-like, narrower than rest, armed with a terminal style or short bristle, joint 2 cup-shaped; head below and behind broadish, slightly depressed, not sulcate.

Thorax almost globular, very convex above, in side view semicircularly arched or humped above and high above level of vertex, almost or about as high above latter as depth of head itself; anterior sloping part behind head very steep, slightly hollowed and prothoracic part not distinctly separately discernible or prominent as in genus *Cyrtosia*; prothoracic humeral lobes broad, rounded, reminiscent of those of the Acroceridae; sides of thorax above notopleural fold a little anterior to wings not distinctly transversely depressed as in the other genera; postalar calli, owing to convexity of thorax, not so prominently ridge-like; dorsum or discal part of thorax above aerolately punctured, more rugulose posteriorly; scutellum relatively broad, with a slight, but distinct, arcuate depression across basal part; pleurae slightly more convex or bulging than in the other genera, the mesopleuron more triangular and with some setiferous puncturation.

Wings either clear as in males or slightly infuscated as in females; marginal cell reduced, the posterior vein of which joins the costal margin much before apex of wings; one submarginal, a discoidal and four posterior cells present; first basal cell longer than second; anal cell open apically; axillary lobe narrowish, not lobe-like; alula much reduced, narrowish and linear; knobs of halteres tetrahedral in shape.

Abdomen broad, ovate, at broadest part (between tergites 2 and 3), much broader than thorax, arched or humped in appearance; discal part of tergite 1 flattened, slightly depressed, aerolately punctured; rest of tergites above in females also aerolately punctured or sculptured, but only discal basal three-quarters of tergite 2 and to a certain extent narrow discal basal part of 3 (under apical margin of 2) in males, in addition to flattened discal part of 1, aerolately sculptured; rest of tergites in males very finely transversely rugulosely sculptured; dorsum of abdomen in both sexes with two rows of segmental, slightly depressed, shiny, dark or blackish spots on each side from tergite 2 to apex as in some other Cyrtosiinae [Platypyginae], each spot nearer base of the tergite, and in some males often also with an extra central part of tergite 2; venter markedly and characteristically hollowed or excavated, the sides of tergites and plate-like hinder part of metapleurae much inflexed and overhanging venter, and middle of venter with two longitudinal ridge-like elevations.

Legs relatively stoutish and shortish, without any spines on hind femora and without distinct spicules on tibiae; apex of tibiae without distinct spurs, but apparently ending in a minute spine-like point on each side of tarsal insertion.

Vestiture without any macrochaetal elements, the hairs distinctly more developed than in other genera, excepting *Onchopelma*, even the hairs in females, though much shorter than in males, still denser and more evident than in most cyrtosiine genera; hairs on thorax above, scutellum and mesopleuron situated in the areolar crater-like punctures, comparatively dense, longer in males than in females, in latter however apparently equally dense; those on abdomen, sides of tergite 1, greater part of 2 and on rest of tergites in males, though shorter than on thorax, markedly dense, shining silvery whitish, arranged transversely and directed towards centre along the middorsal line of which they form a sort of ridge composed of hairs; hairs on abdomen above in females situated in the crater-like punctures, very much shorter than in males, minute, but also directed towards midline; posterior and lower parts of pleurae bare; hairs on legs relatively longer and denser than in the other genera, slightly longer and finer in males than in females.

Hypopygium of males with the last sternite not spined or very sharply produced at its posterior apical angle as in most other cyrtosiine genera; the basimere of the paramere rather broadish, more saddle-shaped, not shell-like as in the other genera; the telomeres of paramere leaf-shaped, flattened, without any hook or hook-like structure; aedeagal apparatus with the aedeagus proper appearing double at its end, reversed in position, bending towards dorsum instead of towards venter or downwards as in most bombyliids; posterior end of the apodeme of the aedeagal apparatus bulb-like or vesicular, not flattened as in most bombyliid genera; and the paraphyses of the apparatus in form of a ventral hood-like extension of which the hind margin is slightly emarginate medially.

Material available for study: None. This fly was described by Hesse (1967) as this work was being completed. I have reproduced Dr. Hesse's description and figures.

Immature stages: The larvae from which these flies were reared were found in loose soil in association with the egg packets of the brown trek locust of South Africa (*Locustana pardalina*).

Ecology and behavior of adults: Unknown.

Distribution: Ethiopian: *Psiloderoides mansfieldi* Hesse, 1967.

Genus *Cyrtomorpha* White

FIGURES 95, 222, 320, 581, 587, 875, 876, 877

Cyrtomorpha White, Papers Proc. Roy. Soc. Tasmania, p. 185, 1917 (1916). Type of genus: *Cyrtomorpha paganica* White, 1917, by original designation.

Strange, comparatively small, flies of more or less cyrtid-like appearance with rather convex mesonotum, especially in front of the scutellum and with extraordinarily broad, large, inflated abdomen, very much wider than the thorax, which bears minute, appressed, setate pile. The head is small, much more narrow than the thorax and appears to be attached low on the thorax largely because of the elevated mesonotum. The mesonotum is more convex than usually seen in bee flies, both anteriorly and posteriorly but a little more so behind in front of the scutellum. The antennae are minute; the proboscis is short and in some species reduced to a mere stub, not visible from the sides. Oral recess or buccal cavity on the bottom of the head. Male eyes not quite holoptic. The venation is peculiar. The wing is broad; the discal cell is wanting; there are 3 branches to the radius, hence only 2 submarginal cells, and the marginal cell is quite short and narrow. Second vein ends barely beyond the first vein. There are 3 branches to the medius and the anterior 2 branches anastomose a short distance from the radiomedial crossvein. The anal cell is narrowly open; while the wing is broad at the base, the alula is almost absent and is quite linear.

These flies are restricted to Tasmania and Western Australia.

Length: 4 to 5 mm.; wing, 4 mm.

Head, lateral aspect: The head in this aspect is not quite globular because the upper half of the head slopes backward from the middle of the face to the vertex and is less extensively produced than the lower half. The occiput is only moderately extensive but rather strongly convex from the middle outward to the eye margin. Its pile consists of short, stiff, appressed, bristly hairs, those at the bottom of the head more erect. The front is plane with the eye immediately in front of the antennae, gradually rising and swelling toward the vertex, which is rather conspicuously elevated above the eye margin, and there is a small raised ocellar tubercle in the center in females. In males the front is restricted to a small, sunken triangle above the antennae and the vertical triangle is likewise sunken beneath the eye margin and only the equilateral ocellar triangle is visible. The eye is quite large, occupying most of the head in both sexes, the posterior margin entire and in the male shows an overgrowth backward and downward on the upper half, encroaching on the occiput, which in consequence is greatly reduced. Male with the eye almost touching the middle, separated by not more than the thickness of the antennal style, or apparently more widely separated in other species according to Roberts (1929).

Although Roberts describes the proboscis as short, not as long as the head and palpus very small, one-segmented, in the material before me the proboscis is a mere, minute stump, and I find no trace of the palpus. Antennae attached at the middle of the head and quite small. From the lateral aspect the extremely short first segment cannot be seen because of a slight overgrowth of the eye anteriorly on each side. The second segment is also very short and beadlike and both have a few apical setae above and below. The third segment has a somewhat bulblike swelling at the base, not wider than the second segment, beyond which it is rather strongly narrowed, the cylindrical half only about half as wide as the base. It bears a bristle-tipped style of the same thickness as the outer part of the third segment and as long as this segment.

Head, anterior aspect: The head is short-oval being distinctly wider than high, tending to be flattened both dorsally and ventrally and also a little on the sides, so that actually there is a slight subrectangular appearance to the head in the male and a more oval appearance in the female. In the female the eyes are separated by about twice the width of the ocellar tubercle. The frontal pile in the female is scanty, stiff, bristly, suberect, and not very long. The antennae are adjacent at the base. The face is extensive, somewhat convex, the oragenal recess confined to the bottom of the head, broad, nearly circular and since from a central view the eyes diverge posteriorly, the genal area is largely triangular on each side. The oral recess is rather deep with vertical sides behind. Pile of face and gena rather long, fine, more or less curled, curved downward, and of a shining brassy yellow. Tentorial fissure shallow.

Thorax: The thorax is short and compact and strongly humped on the mesonotum. The anterior and posterior part of the mesonotum is almost steep and is rounded. Head and especially abdomen decumbent. Prothorax very small and reduced in extent. The humerus is small, oval, and somewhat flattened with the posterior corner rounded. The mesonotum is yellowish to reddish brown laterally with the central part either black or with 3 wide, black vittae, either shining or partly dulled by pollen that consists of microscopically small rounded, flattened scales. Scutellum thick, convexly rounded posteriorly, comparatively short, its margin and disc and the mesonotum with comparatively dense, coarse, erect, rather bristly pile of no great length. Mesopleuron over nearly its whole surface with pile similar to the mesonotum. Pteropleuron, metapleuron, hypopleuron, and sternopleuron without pile, but with sparse, microscopic scales. Squama short with thickened apilose margin. The outer section of the squama is somewhat larger. Metasternum small, especially reduced behind the scutellum; its most extensive part consists of a narrow, rounded, oblique, lateral fold.

Legs: The legs are relatively short and stout. Middle and hind femora each only slightly longer than the anterior femur and each with a fringe of fine, erect or slightly appressed pile, nowhere longer than the width of the tibia. The femora are loosely covered with microscopic scales. All of the tibiae are similarly pilose but the pile is somewhat shorter than on the femora. The anterior tibia ventrally has a row of oblique, short, numerous, microscopic setae similar to that which extends onto the tarsi. All of the tarsi are quite short including the hind pair and covered with microscopic, appressed setae. Claws and pulvilli small, the former short and sharp, the latter wide, broadly rounded at the apex and as long as the claws.

Wings: The wings are quite broad and relatively short and hyaline, except that the whole surface is covered with coarse, brownish villi. The first vein ends only a little distance beyond the middle anterior wing border; the second vein ends also only a short distance beyond the end of the first vein a distance a little less than that which lies between the ends of the first vein and the auxiliary vein. There is only 1 submarginal cell, the third vein not branched. The radical sector is as long as the first section of the third vein, ending at the anterior crossvein. There is no discal cell and the medius branches or forks a short distance beyond the anterior crossvein equal to approximately the length of the anterior crossvein. There are 4 posterior cells all widely open, and the third vein ends barely before the apex of the wing. Anal cell open, axillary lobe large but alula wanting. The ambient vein is wanting, since the costa ends at the apex of the wing. All of the veins are quite stout, but the branches of the medius are somewhat reduced or even semiatrophied at the wing margin.

Abdomen: The abdomen is very wide, much wider than the mesonotum, especially in the female, and it is short and presents a rather short, rounded appearance from above; in shape and width it reminds one of *Usia* Latreille. There are 7 tergites visible from above with the abdomen tilted forward a little. If viewed directly from above only 5 are visible, the last 2 are a little shorter and strongly turned down due to the convexity of the abdomen. On each tergite there is a prominent, transverse, posteriorly arched, foveate depression on each side near the middle and resting on the the anterior border of the tergites and in addition what appears to be spiracular depression outside of these foveae, yet far from the lateral margin. Viewed from beneath, the brown sclerotized sternites are crowded together, each with a wide, extensive, lateral, yellow, apparently non-sclerotized area, and there is a similar broad, shining, yellow obtuse triangle at the base in the middle of the first tergite, the area much wider than the scutellum. Also, viewed from below, two additional tergites can be discerned in the female, each rounded and cuplike and recessed within each other and both of these recessed within the seventh tergite. Female genitalia sunken and without any spines. Male genitalia small; the 2 main dististyli and basimeres lying side by side and recessed within the ninth tergite.

The male terminalia from dorsal aspect, epandrium removed, show a short, compact rather wide structure with the basistylus quite short and with very peculiar dististyli, wide, obtuse and shaped much as an agaric-type funguslike apex. There is a strong bridge across from the ramus. In the lateral aspect the dististyle is even more peculiar, the basal part large, triangular, produced ventrally downward and pointed, the apical part club-shaped, turned downward and pointed backward and connected by a very narrow neck. The epandrium has unusual depth and is quite short.

Material studied: A male and female of this genus taken in Western Australia.

Immature stages: The larvae of these flies are predators on the eggs of locust. See discussion under life histories.

Ecology and behavior of adults: I have seen no records of the behavior of these flies.

Distribution: Australian: *Cyrtomorpha flaviscutellaris* Roberts, 1929; *paganica* White, 1917.

Genus *Platypygus* Loew

FIGURE 331

Platypygus Loew, Stettiner Ent. Zeitung, vol. 5, p. 127, 1844.
  Type of genus: *Platypygus chrysanthemi* Loew, 1844, by monotypy.
Popsia A. Costa, Atti Roy. Accad. Sci. Napoli, vol. 1, p. 51, 1863.
  Type of genus: *Popsia ridibundus*, 1863, by monotypy.

Paramonov (1929), vol. 11, treatment of Palaearctic species.
Engel (1933), pp. 121-122, key to 8 Palaearctic species.

Quite small flies that are rather strongly humped upon the mesonotum. They are generally blackish with a considerable amount of yellow markings upon head, thorax, or abdomen, or all three. The head is subglobular with the eye large and rounded, but there is much variation in the extent of the occiput, which varies from very short indeed in species like *ridibundus* Costa or *pumilio* Loew, to very extensively produced backward in species like *melleus* Loew and *melinoproctus* Loew. Probably all of those forms with the extraordinarily lengthened occiput should be placed within the genus *Cyrtisiopsis* Séguy, especially because of the variation of the length of the proboscis which is quite long in *Cyrtisiopsis*, but short, not as long as the head in *Platypygus chrysanthemi* Loew, the type-species. Also, the very curiously lengthened occiput of *Ceratolaemus xanthogrammus* Hesse should be noted. Hesse (1938) made this fly a subgenus of *Platypygus* Loew, although the absence of the discal cell and appearance of the wing seem to ally it to *Cyrtosia* Loew. All of the species of *Cyrtosia* Loew appear to have a rounded head and short occiput except *nitidissima* Engel, where it is extensive behind.

Although the discal cell has shown a tendency to be wanting in the subfamily Mythicomyiinae, I am inclined to believe that its presence or absence is more significant than the highly variable postocciput. Because in the fly *Ceratolaemus* Hesse the wing is much more like *Platypygus* Loew in its general form, I leave it as a genus related to *Platypygus*. Note that the third antennal segment has a large, tapered, bristly, setate microsegment ending in a spiny style; palpus and alula are absent. Species of *Platypygus* Loew are apt to have the apex of the abdomen compressed, high vertically, and truncate apically as shown in Engel (1933, Figure 58).

Platypygus Loew is found in Southern Europe, Asia Minor, and South Africa.

Length: 1.8 mm. (in *pumilio* Loew); usually 3.5 to 5.5 mm.

Head: A little narrower than the thorax and almost rounded when seen from above, but when viewed in profile it is much longer than high, owing to the considerable puffed-out occiput (which is even more conspicuously produced below than in the genus *Cyrtosia*), into a large triangular shieldlike projection, one on each side of the posterior mouth opening; frons a little broader above, nearer the vertex than below, where it is somewhat hollowed out above the base of the antennae, and when seen in profile its sides form a weakly convex line with the much shorter, porrect upper part of the face (below the base of antennae), which is then abruptly directed vertically down at an angle of 45 degrees forming the upper side margins of the mouth; eyes widely separated in both sexes and with the facets of equal size all over but always with a minute triangular indention on the anterodorsal margin, a short distance above the base of the antennae; proboscis long (a little longer than the length of the head) or very long (about 3 times the length of the head); palpi minute, very short and often concealed in the opening of the mouth; antennae much resembling that of *Cyrtosia*, the two basal segments very short and equivalent in length and the third segment pyriform, with an apical style that is usually thickened by the presence of microscopic pubescence and always bearing a short apical sensory seta.

Thorax: The thorax is very deeply humpbacked with the pronotum and metanotum strongly developed; the attachment of the head to the thorax is exceedingly low, being immediately above the base of the elongate anterior coxa, so that the summit of the head is on an almost straight line with the notopleural suture.

Legs: The legs are neither strong nor elongate, usually pale except the apical tarsal segments obscured, and even blackish owing to the minute pubescence consisting of microscopic black bristles.

Wings: Wings with a venation much less reduced than in *Cyrtosia* as the 1st M cell is always present and the R sector forms 2 branches; the auxiliary vein does not fade away in the cell before reaching the costa. The costa latter is strong and thickened for a short distance beyond the tip of the third vein; 1st anal cell always open; alula absent; anal lobe well developed; squama almost bare; halteres with a short stalk and a large knob.

Abdomen: The abdomen is distinctly broader than the thorax, but it may be as broad as, or narrower than, the thorax; it consists of 7 or 8 apparent segments and the apical, sternite is large and usually conceals the genitalia; in the male, however, the crossed branches of the forceps and in the female 2 short oval cerci project outward.

Material available for study: I have studied the series of these flies within the British Museum (Natural History), and my illustrations are taken from these flies, supplemented by figures taken from Engel (1933). I have supplemented the description I made in London with quotations from Efflatoun's excellent description of this genus (Efflatoun, 1945).

Immature stages: Unknown.

Ecology and behavior of adults: Efflatoun (1945) notes that *Platypygus* (*Cyrtisiopsis*) *melleus* Loew is common in the eastern desert of Egypt and is an excellent hoverer; he states that he has often seen it hovering over bushes and blooms of *Zygophyllum album*, almost always at the extreme edge of a Wafi; usually at extreme edge and in the shade of a boulder near the grotto housed webs of arachnids.

Distribution: Nearctic: *Platypygus americanus* Melander, 1950.

Palaearctic: Platypygus *algirus* Paramonov, 1929; *bellus* Loew, 1869; *chrysanthemi* Loew, 1844; *depressus* Loew, 1855; *kassanovskiji* Paramonov, 1929; *kurdorum* Paramonov, 1929, *kurdorum persicus* Paramonov, 1934; *lativentris* Loew, 1873; *limatus* Séguy, 1963; *maculiventris* Loew, 1874; *melinoproctus* Loew, 1873; *melleus* Loew, 1856; *pumilio* Loew, 1873; *ridibundus* (A. Costa), 1863 (as *Popsia*) [=*tauricus* Paramonov, 1926]; *turkmenorum* Paramonov, 1929.

Ethiopian: *Platypygus natalensis* Hesse, 1967.

Genus *Cyrtisiopsis* Séguy

FIGURES 330, 600, 997, 998, 999

Cyrtisiopsis Séguy, Ann. Mus. Storia nat. Genova, vol. 55, p. 80, 1930d. Type of genus: *Cyrtisiopsis singularis* Séguy, 1930, by original designation.

Minute, shining, blackish flies with large, opaque, white halteres and sometimes with narrow, yellow postmargins to some of the abdominal segments. The head is quite elongate, and including the eye, longer than high. The occiput is very extensive both above and below but more so ventrally, so there is an extensive grooved gular region which, however, has no setae. The males are widely dichoptic, the proboscis is elongate but rather strongly thickened progressively toward the base. It is at least 1½ times as long as the head length. The genofacial area is rather strongly compressed laterally, because of the extensive anterior development of the eyes. The hyaline wings have black veins, 1 submarginal cell, 4 posterior cells. The anterior intercalary vein at the end of the discal cell is short. The anal cell is widely opened. The auxiliary vein is complete. The ambient vein ends at the apex of the wing.

These flies are known from Italian Somaliland and South Africa.

Length: About 2 mm.; wings about 1½ mm.

Head, lateral aspect: The head is elongate, much longer than high; the occiput is extensive both above and below, more so below, where the long gular region has a medial fissurelike groove. From below, the eyes are obliquely or diagonally excavated. From the dorsal aspect the ocelli are set anteriorly ahead of the posterodorsal corners of the eye, which are broadly rounded. The antennae are set closely adjacent near the anterior part of the head. There is a distinct, comparatively long, though narrow, facial area before the beginning of the plane, nearly vertically, laterally compressed oragenal cup. The gena proper appears to be absent, due to the encroachment of the eye, and consequently there is no tentorial fissure. The proboscis is large, at least 1½ times as long as the length of the head, and it is progressively thickened and widened toward the base. The labellum is quite slender, almost filiform. Palpus filiform.

Head, anterior aspect: The frons has a distinct, extensive, scooplike or shallow depression extending to the base of the large, quite low ocellar tubercle. The ocelli are small, the posterior ocelli set far apart. Antennae adjacent at the base. At the vertex the eyes are separated by a little less than one-third the head width. The oral opening is quite narrow.

Thorax: The thorax is slightly wider than the head, the mesonotum a little longer than wide with a distinct convexed appearance from the side. It is not excessively humpbacked. The surface is dully shining with microscopic, scarcely evident setae, which are more or less appressed. Pleuron shining black, edged with pale brownish yellow, on the propleuron, lateral margins of the mesonotum, posterior margins of the mesopleuron, and with a diagonal whitish spot on the postalar callosity. The shining humeri are brown. The halteres are quite large and opaque yellowish white, and quite conspicuous with a rounded, swollen knob.

Legs: Legs shining black throughout with very microscopic setaelike pile. The claws are very slender; the pulvilli are well developed. All the femora are slightly thickened especially through the middle and the hind femur is only slightly longer than the anterior middle femur.

Wings: Quite hyaline, relatively smooth with iridescent reflection, and brownish-black veins. The wing is elongate oval with rounded apex. There is only 1 submarginal cell but 4 posterior cells, with the upper anterior intercalary vein at the end of the discal cell relatively short. The anterior crossvein enters the discal cell at the middle; the anal cell is widely open and while the axillary lobe is moderately wide, I cannot find an alula on my material.

Abdomen: The abdomen is relatively short, no wider than the thorax, tending to be more or less truncate at the apex of the sixth segment in males. The fifth and sixth tergites in the species before me, *crassirostris* Hesse, paratype, have very thin, narrow, yellow postmargins to these tergites. Other tergites are recessed. The male terminalia are black, rather large, terminal and protrude from this truncated, large, circular terminal ending of the abdomen.

Material available for study: Two male paratypes of *Cyrtisiopsis crassirostris* Hesse furnished me through the generosity of Dr. Hesse.

Immature stages: Unknown.

Ecology and behavior of adults: Found on the heads of composite flowers in South Africa, the long proboscis used in feeding.

Distribution: Ethiopian: *Cyrtisiopsis crassirostris* Hesse, 1967; *singularis* Séguy, 1930.

Subfamily Mythicomyiinae Melander, 1902

Very small to minute flies, some of which are even less than a millimeter in length and in general smaller than the members of the related subfamily Platypyginae. Both subfamilies have a tendency to what Hesse (1967) calls a quadriarticulate antenna. All of them are humpbacked, but some genera more so. This is especially true of *Cyrtosia* Perris, and in the subfamily here discussed, it is particularly true of the species of *Mythicomyia* Coquillett. These flies may be almost entirely shining black as in *Glabellula* Bezzi, or they may have much yellow or pale coloration as in most *Cyrtosia* species, but more generally, *Mythicomyia* species are pollinose and grayish black, and pale color is usually much reduced.

In this subfamily the radial sector never has more than 2 branches, the first of which, the second vein, characteristically always ends in the first vein and does not independently reach the costa; or, the second vein may be vestigial, fused with the first vein, or completely absent.

The third vein is unbranched. In some genera the fourth and sixth veins are apt to be evanescent, faint, or absent. The first and second basal cells are apt to be joined by the loss of the vein dividing them. It should be noted that the discal cell is sometimes wanting as in *Doliopteryx* Hesse, and *Euanthobates* Hesse where it has been lost by fusion with the third posterior cell and lost in the same way in *Empidideicus* Becker, but in *Glabellula* Bezzi lost by fusion with the large second basal cell.

Hesse (1967) has discovered other new genera that clearly belong in this subfamily, such as *Pseudoglabel-*

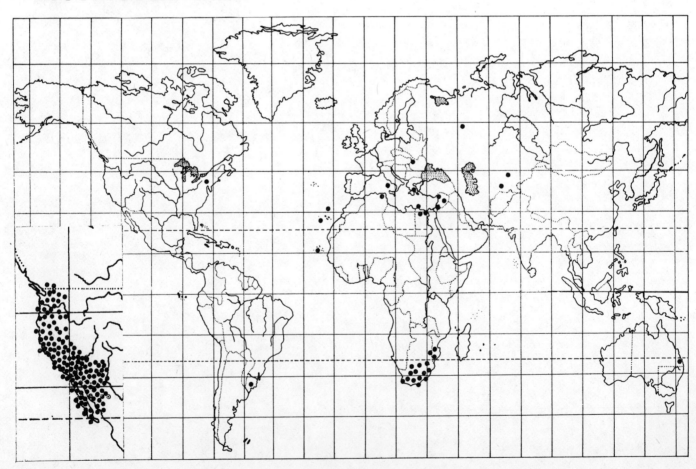

TEXT-FIGURE 39.—Pattern of the approximate world distribution of the species of the subfamily Mythicomyiinae.

lula. Two tribes may be recognized: the Empidideicini with second vein absent, and Mythicomyini in which it is present.

The subfamily Platypyginae is separated from the subfamily Mythicomyiinae by the presence of a distinct and complete second vein ending well beyond the end of the first vein, but it should be noted that the discal cell may be present or absent in both subfamilies. Evolutionally we here rank the medial field as more unstable than the radial field since in so many families of Diptera from simuliids to phorids, etc., this seems to be true. This subfamily, the Mythicomyiinae, appears to have a much wider distribution throughout the world than the Platypyginae. The genus *Mythicomyia* Coquillett has reached an extraordinary development in the southwestern United States. This genus is also known from a single species at Montevideo, Uruguay, collected there by F. Edwards. He notes that the males fly in small swarms in sunlight. The nearest relative of *Mythicomyia* Coquillett appears to be *Aethioptilus* Hesse from South Africa.

Mythicomyia Coquillett has been shifted back and forth from the Rhagionidae (Leptidae) to Empididae to Bombyliidae. Eleven species had been described chiefly from the southwestern United States, another species from Uruguay; as a result of assiduous collecting Melander (1960) decribed 136 new species or varieties from the southwestern United States. The deep canyons emerging from mountains of these western states and particularly from California have resulted in a curious isolation and a species-forming situation, which Melander (1946) likens to Gulick's description of species formation of snails in Hawaiian valleys.

The genus *Mythicomyia* Coquillett is very peculiar. In this genus the basistylus, or as called by others the penis sheath, is absent. This genus is characterized by a large, triangular form in lateral aspect, with extensive, free, basal ejaculatory apodeme and a curious epandrium shaped much like a boomerang. From the dorsal aspect the base of the aedeagus is large and egg-shaped and there are winglike structures on the apical half, which we tentatively call accessory apodemes. There is an especially long, curious aedeagal apodeme lying on each side of the basal ejaculatory apodeme which is shaped like a long hockey stick.

KEY TO THE GENERA OF THE SUBFAMILY MYTHICOMYIINAE

1. The first vein ends near the middle of the wing. The second vein is complete but ends in the first vein, making a small, triangular cell 2
   The scond vein is vestigial or absent 6
2. The discal cell is absent 3
   The discal cell is present 5
3. The second basal cell is much larger than the first due to fusion of the second basal cell with the discal cell; from the combined cell four veins are emitted setting out 4 posterior cells (Holarctic, Ethiopian) . . **Glabellula** Bezzi
   The loss of the discal cell is due to its fusion with the third posterior cell 4
4. No second basal cell present. The anterior branch of the fourth vein atrophied or wanting completely on the basal half, anal vein faint or absent and the fifth vein tending to be atrophied basally (Ethiopian) . **Doliopteryx** Hesse
   Second basal cell distinct and distinctly separated from both discal and first basal cells; consequently only 3 veins proceed from the second basal cell. The first posterior cell is much more narrow and narrowed apically and opens upon the costa above the wing apex; the fourth vein divides into two branches leaving a wide, triangular second posterior and apical cell. There is no projecting pronotal lobe (Ethiopian) . . . . . **Pseudoglabellula** Hesse
5. Vein between first and second basal cell evanescent. Axillary lobe narrow, the anal cell closed and the anal vein faint or evanescent. Whole wing about as wide near the apex as near the base (Ethiopian) . **Aetheoptilus** Hesse
   Vein separating the first and second basal cells strong and distinct. Anal cell open and anal vein distinct. The wing is broad basally, often quite broad, with a prominent, wide axillary lobe (Nearctic, Neotropical).

   Mythicomyia Coquillett
6. Discal cell present. Vein separating first and second basal cells quite faint or absent. Intercalary vein quite short (Ethiopian) . . . . **Anomaloptilus** Hesse
   Discal cell absent, confluent with third posterior cell . . 7
7. Face very short and eye relatively small. Proboscis with basal part as long as the eye but distal part often extremely long, from 2 to 3 times as long as head. Wing with first posterior cell quite narrowed apically. Anal cell narrowly open or closed in the margin. Vein separating first and second basal cells distinct, well formed . . 8
   Face jutting forward as a short, peaklike, triangular elevation. First posterior cell distinctly widened and flared apically. Anal cell widely open. Vein separating first and second basal cells faint or wanting 9
8. Ventral part of head with a "gular" groove, or sulcate; this groove bearing curious, downward projecting, fingerlike, coecalike or straplike processes, arranged much as in a comb on each side of the head (Ethiopian).

   Euanthobates Hesse

   Grooved lower portion of head completely without any trace of protruding, coecalike processes (Ethiopian).

   Euanthobates (**Acoecus**, new subgenus)
9. Style of third antennal segment virtually absent, style minute and visible only under high magnification. Head more elongate due to a more strongly developed occiput. Anterior branch of medius thrust forward, thinned, weak and evanescent distally (Egyptian) . **Leylaiya** Efflatoun
   Style of third antennal segment prominent and well developed. Anterior branch of medius not evanescent, ending below apex of wing 10
10. The anterior ocellus lies in line with the posterior ocellus. Abdomen wider than the thorax and flattened (Palaearctic) . . . . **Empidideicus** (**Cyrtoides**) Engel
    The anterior ocellus lies in front of the others, forming a triangle. Abdomen not wider than thorax (Holarctic, Ethiopian) . . . . **Empidideicus** Becker

Genus *Mythicomyia* Coquillett

Figures 101, 102, 342, 347, 596, 605, 884, 885, 886

Mythicomyia Coquillett, Ent. News, vol. 4, p. 209, 1893. Type of genus: *Mythicomyia rileyi* Coquillett, 1893, monotypy.

Melander (1902), pp. 337-338, treats *Mythicomyia* as an empidid.
Melander (1961), pp. 163-179, key to 136 species and subspecies.

Very minute flies, seldom more than two millimeters in length and often much less. Males are very strongly humpbacked and darker in color, hence sexes are not easily associated. They are distinguished from *Glabellula* Bezzi by the presence of a complete discal cell and by the strongly holoptic eyes of the male. In addition, the abdomen is much more slender, somewhat lengthened and attenuate compared to the short, stout, robust, plump abdomen of *Glabellula* Bezzi. The antennae are longer, more slender and attenuate upon the style of the third segment. The wing is strongly widened basally so that the axillary lobe is triangular with a rounded angle. Anal cell narrowly open, sometimes closed in the margin. In both genera the second vein is very short but well formed, ending in the first vein much like a crossvein resulting in a short, triangular marginal cell.

This genus has had a checkered history. Its author, Coquillett (1893), founded it upon a single specimen, he named *rileyi*, and placed it within the Empididae. Melander (1902) in his review of the Empididae kept it within this family and made it the basis of the subfamily Mythicomyiinae, for flies with abbreviated second vein. Greene (1924) was the first to review the various changed allocation of the genus, noting that Williston, Osten Sacken, and Schiner located the genus within the Rhagionidae (Leptidae of authors). Both

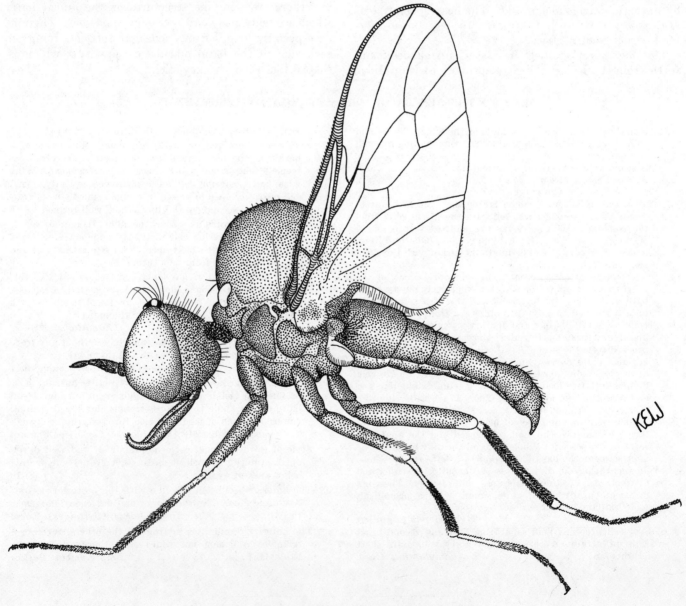

Text-Figure 40.—Habitus, *Mythicomyia monacha* Melander.

Coquillett and Cresson (1915) placed these flies within the Empididae. Aldrich and Kertész in their catalogs located it within the Rhagionidae (Leptidae of authors).

It remained for Greene (1924) to assign it to its present position on the basis of the pupal characters of a fly of the unquestionably closely related *Glabellula* Bezzi, which was reared from a pupa taken from frass of an oak in the District of Columbia. I believe that Greene was correct in his assumptions. Later authors have accepted its assignment to the Bombyliidae on the basis of Greene's study.

The genus *Mythicomyia* Coquillett prior to 1960 was known from 12 species, 2 from Chile and Uruguay respectively, and the other 10 from the southwestern United States and ranging as far east as New Mexico. In 1960 Melander described 127 new species and 9 varieties, of which 92 were from southern California alone (including a few from the bays of Lower California) and 35 other species from southern California and additional western states, usually Arizona or New Mexico. Several species extended northward to Oregon or Washington, a few into Sonora. Two species, *rileyi* Coquillett and *armata* Cresson, are quite widespread in the western United States. The months they fly range from March to September but more commonly in April, May, or June, and again in September. Four species were collected in Arizona only, and 4 in New Mexico alone. Twenty more species were found in both California and Arizona. Nine other species were collected in both California and New Mexico for a total of 13, where I have collected some myself. One species is recorded from Montana, 3 or 4 from Utah, Nevada, Oregon, and Washington.

Length: 0.9 mm. to 3.5 mm., with very few species greater than 2.0 mm. in length.

Head, lateral aspect: The head is subglobular. The occipital component is conspicuous and extensive both above and below, both in males and females but especially in the latter. In lateral aspect the face and oragenal cup are produced beyond the eye. The face in length is about as long as the first two antennal segments and slopes forward and downward. The large oragenal cup is equally produced above and below. The eye is nearly circular, except the portion between the ocelli and the antennae, which is nearly straight, and the portion from the lower front to the face is nearly straight. The facets are minute and all alike. The posterior margin of the eye is entire. The occiput has a few, fine, erect, stiff, short, scattered hairs. The lower portion of the sublateral occiput has considerable loose, long pile. The occiput is at most only slightly concave where it is attached to the thorax. The front is pollinose. The face and oragenal cup are pollinose also and the face and front have a very few fine, scattered hairs.

The proboscis is moderately elongate and about as long as the head. It is slender with long, slender labellum. The palpus is composed of one curious, short, rather flattened, laterally convex, oval segment, which is barely longer than wide and bears a long, slender bristle at the apex; there are 1 or 2 other basal bristles on the outer margin. The palpus appears to be more or less hollow on the median surface. The antenna is somewhat elongate and relatively slender, but considerably shorter than the head. The first segment is very short, scarcely more than half as long as the second; the second segment is about as long as wide; and the third segment is widest just beyond the base where it is a little wider than the second segment. This third segment is somewhat narrowed apically, rather flattened and bears a conspicuous, flattened, rather long microsegment which appears to have a minute, short additional microsegment at its apex. The antenna is everywhere micropubescent. The antenna is attached at the middle of the head.

Head, anterior aspect: The head is about the same width as the thorax or in some species narrower. The face is separated from the front by a transverse groove or furrow in females. The eye in the male is holoptic touching in the middle for a distance of one-third of the total length of front and vertex. The antennae are adjacent and touching. The small front is triangular, pollinose in males and smaller than the triangular ocellar tubercle, which is only moderately raised above the eyes. The ocelli are large, especially the anterior ocellus, and all three rest at the margin of the eye in the male. The female front has parallel sides; its width is a little more than one-fourth the width of the head. The oral opening is triangular, though a little rounded anteriorly and is unusually deep. The medial roof has 3 or 4 rather long, slender, erect hairs on each side. The inner edge or wall of the oragenal cup is quite sharp. I can find no trace of gena or tentorial fissure.

Thorax: The thorax is high with a strongly humped and convex mesonotum, which is relatively short in length. The mesonotum is rather dull, more or less opaque, pollinose, with several rows of very minute, microscopically erect setae. The thorax is generally black or dark brown in color and the humerus, notopleuron and scutellum, and upper mesopleuron are apt to be pale yellow in color. Sometimes only the humerus and postalar callosity are yellow. The hypopleuron may be black or yellow. The upper mesopleuron may have a few long, stiff hairs; the remainder of the pleuron is micropubescent only; scutellum comparatively thin and flattened with several long, erect, stiff hairs on the disc. The squamae are short with a few scattered, stiff hairs; halteres with a short stalk and a very large, thick, slightly elongate club at apex as long as the stalk and divided dorsally by a curved, fissurelike crease.

Legs: The legs are short, often partly yellow, partly black. The femora are a little thickened, the anterior tibiae have several rows of minute, subappressed setae and the tibia is distinctly longer than the femur; posterior femur sometimes a little arched or bent with quite microscopic pubescence and a few scattered, suberect setae. The posterior tibia likewise has some scattered setae. The pulvilli are short, rounded, wide; claws minute and slender, strongly curved at the apex.

Wings: The wings are broad both at apex and base, the base not wider than the middle of the wing. The third vein has 2 branches, the anterior branch is quite short and ends in the first vein. There are 4 posterior cells all very widely open. The anterior crossvein enters the large, relatively long, discal cell at the basal sixth, therefore first and second basal cells are of nearly equal length. The auxiliary vein appears to be reduced and ends in the first vein or simultaneously with it in the costa. The intercalary vein is about half as long as the medial crossvein, the anal cell is open, the axillary lobe large, the alula wanting or reduced to a mere linear band. The ambient vein is absent but the posterior margin of the wing has a fringe of fine hairs and the entire surface of the wing is densely covered with minute, short, microsopic scales or quite flattened scaliform hairs, which are iridescent.

Abdomen: The abdomen is depressed or decumbent, obtuse, and rather plump and is largely black with yellow posterior margins in some species. The pile is setate or bristly consisting of several loose, scattered rows more or less flat-appressed on each tergite, and the last row extends well over and beyond the posterior margin. It is somewhat more dense laterally but equally stiff and bristly. I find 7 segments in the male and female. The male hypopygium is a broadly oval, rather flattened plate with laterally recurved sides containing the slender, elongate genitalia directed forward below.

The male terminalia from dorsal aspect, epandrium removed, show a very peculiar form in which the basistylus or so-called penis sheath is absent, and all of the remaining structures except two apodemes are concentrated forward on the outer apical half. The basal part of the aedeagus is quite large and bulb-shaped or egg-shaped. The epiphallus with enclosed aedeagus and ejaculatory duct, is small, short, and conical. I find no structure equivalent to the dististyli, although there is a very wide lateral structure apically, longer than wide, and extending directly outward and attached to a ramuslike bridge above the epiphallus which is possibly equivalent to the aedeagus. The basal ejaculatory apodeme is very, very long and there appear to be at least accessory apodemes in addition to a prominent, flared, leaflike lateral apodeme. The first of these accessory apodemes lies just across and opposite the base of the aedeagus. It forms an apically widened and rounded, leaflike structure. Arising from the ramuslike bridge across the epiphallus there is an extraordinarily long, backward process, very slender, extending directly by the side of the medial basal ejaculatory apodeme, or basal keel as it is sometimes called. It widens apically and is shaped like a hockey stick. From the lateral aspect the hypopygium is a curious, triangular affair with an odd, very tall or high, dorsally narrowing epandrium, very short in length and ventrally near the middle of the structure, extending obliquely downward and forward as a curious lobe. The cerci are equally peculiar in that they form a very tall, anteriorly convex structure bulging forward and attached anteriorly to all of the epandrium except the downward lobe.

Material available for study: The late Dr. A. L. Melander presented me with several species and I have received several other species through the courtesy of Dr. Wirth and Dr. Karl Krombein of the National Museum of Natural History. Dr. Oldroyd presented me with a paratype of *Mythicomyia pusillima* Edwards, a South American species. I have collected a few species in New Mexico both in 1959 and 1967 while on bee fly research expeditions. In 1959 I swept them from tall masses of goldenrod, *Solidago* sp., flowers in the Ruidoso Mountains. In 1967 they were swept from masses of flowers in White Mt., National Monument, New Mexico.

Immature stages: Unknown; however, Jack Hall, enthusiastic student of bee flies, recently recorded the rearing of this genus, on a multiple basis, from a bee cell.

Ecology and behavior of adults: The flies visit many flowering plants but are partial to some families and genera.

Melander (1960) in his final work on these flies, which he had so assiduously collected for many years, records collecting them upon 49 species of plants distributed among 33 genera of plants found within 19 plant families. They were more commonly found upon various species of *Eriogonum* such as *fasciculatum, inflatum, trichopodium, pusillum, gracile, thomasii, polifolium, nudum, kennedyi parishii,* and *subscaposum,* within the family Polygonaceae. The additional families frequented more sparingly consisted of Euphorbiaceae, Salicaceae, Boraginaceae, Rhamnaceae, Caprifoliaceae, Polemoniaceae, Amaranthaceae, Loasaceae, Apocynaceae, Leguminosae, Zygophyllaceae, Labiatae, Capparidaceae, Rosaceae, Papaveraceae, Liliaceae, Bignoniaceae, and from the genera *Conyza, Baccharis, Chrysothamnus, Guitierrezia, Pectis, Eriophyllum, Isocoma* and *Tetradymia,* all within the family Compositae.

Melander notes that 54 species of *Mythicomyia* were taken from one or more of 11 species of *Eriogonum* and the same species or some of them were also taken occasionally in other plants as well. The full list of plants upon which he collected the flies of this genus is given below.

Including the species already named, Melander collected flies of the genus *Mythicomyia* Coquillett upon the following flowers:

Euphorbia polycarpus and *albomarginata; Croton californicus; Salix exigua; Adenostomum sparsifolium* and *fasciculatum; Conyza coulteri; Cryptantha intermedia; Baccharis pilularis; Chrysothamnus nauseosus; Ceanothus cordulatus* and *integerrimus; Sambucus glaucus; Rhamnus californicum; Lepidospartum squamatum; Hugelia virgata; Baeria aristata; Gutierrezia lucida; Tidestromia* sp.; *Pectis papposa; Apocynum androsaemifolium; Prosopis* sp.; *Petalonix thurberi; Eriophyllum confertiflorum; Larrea divaricata;*

Salvia vaseyi; Wislizenia refracta; Isocoma acradenia; Fallugia paradoxa; Dalea californica, emoryi, spinosa, saundersi; Dendromecon rigida; Nolima parryi; Heliotropium curassavicum; Chilopsis linearis; Tetradymia canescens.

Distribution: Nearctic: *Mythicomyia acuta* Melander, 1961; *agilis* Melander, 1961; *angusta* Melander, 1961; *annulata* Melander, 1961; *anomala* Melander, 1961; *antecessor* Melander, 1961; *apricata* Melander, 1961; *armata* Cresson, 1915; *armipes* Cresson, 1915; *atra* Cresson, 1915; *atrita* Melander, 1961; *aureola* Melander, 1961 [=*flavida* Melander, 1961]; *aurifera* Melander, 1961; *bibosa* Melander, 1961; *bilychnis* Melander, 1961; *bivulneris* Melander, 1961; *bucinator* Melander, 1961; *cala* Melander, 1961 [=*bella* Melander, 1961]; *californica* Greene, 1924; *caligula* Melander, 1961; *callima* Melander, 1961; *calva* Melander, 1961; *carptura* Melander, 1961; *collina* Melander, 1961 [=*actites* Melander, 1961]; *comma* Melander, 1961; *comparata* Melander, 1961; *compta* Melander, 1961; *concinna* Melander, 1961; *concrescens* Melander, 1961; *cressoni* Melander, 1961 [=*rileyi*, authors, not Coquillett]; *cristata* Melander, 1961; *crocina* Melander, 1961 [=*aspilota* Melander, 1961]; *cylla* Melander, 1961; *diadela* Melander, 1961; *dipura* Melander, 1961; *diropeda* Melander, 1961; *enoria* Melander, 1961; *eremica* Melander, 1961; *fasciolata* Melander, 1961; *flavipes* Cresson, 1915; *flaviventris* Melander, 1961; *formosa* Melander, 1961; *frontalia* Melander, 1961; *fulgida* Melander, 1961; *galbea* Melander, 1961; *gausa* Melander, 1961; *gibba* Melander, 1961; *gibbera* Melander, 1961; *gracilis* Melander, 1961; *habra* Melander, 1961; *hamata* Melander, 1961; *hiata* Melander, 1961; *hoplites* Melander, 1961; *hormatha* Melander, 1961; *hybos* Melander, 1961; *illustris* Melander, 1961; *imbellis* Melander, 1961; *indicata* Melander, 1961; *insignis* Melander, 1961; *intermedia* Melander, 1961; *introrsa* Melander, 1961; *irrupta* Melander, 1961; *laticlavia* Melander, 1961; *lenticularis* Melander, 1961; *levigata* Melander, 1961; *liticen* Melander, 1961; *longimana* Melander, 1961; *marginata* Melander, 1961; *minima* Melander, 1961 [=*lucens* Melander, 1961]; *ministra* Melander, 1961; *minor* Melander, 1961; *minuscula* Melander, 1961; *minuta* Greene, 1924; *mira* Melander, 1961; *mirifica* Melander, 1961; *mitrata* Melander, 1961; *modesta* Melander, 1961; *mulsea* Melander, 1961; *murina* Melander, 1961; *mutabilis* Melander, 1961; *napaea* Melander, 1961; *nigricans* Melander, 1961; *nitida* Melander, 1961; *nitidula* Melander, 1961; *ocreata* Melander, 1961; *oporina* Melander, 1961; *optata* Melander, 1961 [=*diloga* Melander, 1961]; *orchestes* Melander, 1961; *ornata* Melander, 1961; *ornatula* Melander, 1961; *pruinosa* Melander, 1961; *pulla* Melander, 1961; *penicula* Melander, 1961; *petena* Melander, 1961; *petes* Melander, 1961; *petiolata* Melander, 1961; *phacodes* Melander, 1961; *phalerata* Melander, 1961; *pharetra* Melander, 1961; *picta* Melander, 1961; *pictipes* Coquillett, 1902; *platycheira* Melander, 1961; *polygena* Melander, 1961; *potrix* Melander, 1961; *pravipes* Melander, 1961; *pruniosa* Melander, 1961; *pulla* Melander, 1961; *pusilla* Melander, 1961; *rhaeba* Melander, 1961; *rileyi* Coquillett, 1893; *robiginosa* Melander, 1961; *salpinx* Melander, 1961; *scapulata* Melander, 1961; *scutellata* Coquillett, 1902; *sorbens* Melander, 1961; *sugens* Melander, 1961; *tagax* Melander, 1961; *tenthes* Melander, 1961; *tibialis* Coquillett, 1895; *trifaria* Melander, 1961; *triformis* Melander, 1961 [=*chitona* Melander, 1961, =*monacha* Melander, 1961]; *tristis* Melander, 1961; *tubicen* Melander, 1961; *tumescens* Melander, 1961; *uncata* Melander, 1961; *vestis* Melander, 1961; *vilis* Melander, 1961; *virgata* Melander, 1961; *virgo* Melander, 1961; *vulnerata* Melander, 1961; *ostentata* Melander, 1961; *parma* Melander, 1961.

Neotropical: *Mythicomyia pusillima* Edwards, 1930.

Genus *Glabellula* Bezzi

FIGURES 85A, 351, 597, 606

Platygaster Zetterstedt, Insecta Lapponica, Dipt., p. 574, 1838. Type of genus: *Platygaster arctica* Zetterstedt, 1838, by monotypy. Preoccupied by Latreille, Hymenoptera, 1809; Schilling, Hemiptera, 1829.

Sphaerogaster Zetterstedt, Diptera scandinaviae, vol. 1, pp. 22, 232, 30, 1842. Change of name. Preoccupied Dejean, Coleoptera, 1821.

Glabella Loew, Beschreibungen europäischer Dipteren, vol. 3, p. 210, 1873. Type of genus: *Glabella femorata* Loew, 1873, by monotypy. Preoccupied Swainson, Mollusc, 1840.

Glabellula Bezzi, Zeitschr. syst. Hymen. Dipt. Teschendorf, vol. 2, p. 191, 1902. Change of name.

Pachyneres Greene, Proc. Ent. Soc. Washington, vol. 26, p. 62, 1924. Type of genus: *Pachyneres crassicornis* Greene, 1924, by monotypy.

Engel (1933), pp. 116-120, treats 5 species.
Melander (1950), pp. 142-143, key to 6 Nearctic species.

Minute, humpbacked, robust little flies, with short, stout, legs and with broad wing showing much reduction of venation. The auxiliary vein is evanescent or reduced, the first vein ends in the costa at the middle of the wing and the second vein, while stout, ends quickly in the first vein and in consequence forms a small, triangular cell; the third vein is unbranched. This genus is related to *Mythicomyia* Coquillett but has the discal cell confluent with the second basal cell, the males are dichoptic and it is much less strongly humped. Moreover, the fifth vein arises from the anal cell without the angulation that is normally found at the base of the discal cell. Engel (1933) notes that these flies look somewhat like small simuliids.

Glabellula Bezzi is a rather small group of flies of widespread occurrence but rather rare in collections; see comments on ecology below. Melander (1950) described 5 additional species, 4 from California, making 6 species from the western United States; he notes that the American species have black legs, whereas those in the Old World have the tibial bases and tarsi brown. The flies are shining black, with often yellow markings over part of the fly. According to Melander (1950) they are also known from Canada, District of Columbia, Lapland, Siberia, Asia Minor, Syria, Turkestan, Palestine and the Philippines. In an earlier paper (1946)

Melander states that they have been taken in India and Australia, but I have not been able to find published references so stating.

Length: 1.0 mm. to 2.0 mm. All of the California species were very small, none over 1.25 mm.

Head, lateral aspect: The head is subglobular but curiously elongate, and the occipital portion is unusually well developed and prominent above and still more so below. The face is not very extensive but is quite high and becomes a little more extended at its lowest point, which is midway between the antennae and the bottom of the genae. Also, the antennae are attached midway between the ocelli and the lowest point of the face and the proboscis arises in the anteroventral corner of the head. The distance from the proboscis to the posteroventral corner of the occiput is as great or a little greater than the distance from the anterior base of the proboscis to the base of the antennae.

The pile of the occiput is restricted to a few, very minute, extraordinarily short, scalelike, submicroscopic setae along the middle and a few sharp, longer, slender setae behind the ocelli and a few fine, short, erect, stiff hairs along the ventral margin of the occiput, some beneath the eye and some placed more posteriorly. The eye is large, filling almost the entire lateral aspect, except the extensive occiput; its posterior margin is entire but broadly rounded. The facets are of uniform size throughout, each with a small central, circular yellow spot. The proboscis is stout and large and may be thrust straight downward or forward. Its length is equal the height of the eye; and the labellum is long but slender.

I can only find the palpus on the specimen where the proboscis is fully extended. What may be the palpus is a very slender linear structure lying immediately against the ventral edge of the gena and the base of the proboscis. I have been able to examine both sides of 2 well-preserved specimens without finding any further indication of the palpus.

Front in female flattened, pollinose, with 1 or 2 minute, erect setae. The antennae are attached at the upper third of the head and are relatively large though short, the first segment is a mere linear ring tucked against the upper edge of a ledge or shelf forming the lower limit of a distinct recessed, frontal depression. The antennae are adjacent. The second segment is 3 times as long as the brief annulus of the first segment, much wider both at base and apex and is broadly and tightly jointed to the large, oval, thick, anteromedially flattened or obliquely truncate third segment. Subdorsally the third segment bears a thick microsegment some 2½ times as long as wide. Both the third segment and the microsegment are densely, coarsely, scaliform microsetate and there is some indication that the microsegment may be composed of 2 parts. Second segment also apically with some microsetae. All of these observations were taken at magnifications 90 × under several strong lights.

Head, anterior aspect: The head is very slightly wider than the thorax, although this may be due to some unnatural compression of the thorax in pinning. The eye is notched medially opposite the base of the antenna, and the upper part of the face has a depression beneath the frontal ledge which in section would be somewhat V-shaped. In males the eyes are dichoptic.

The front above the antennae is a pale yellow, shallow, shallowly excavated, hemicircular depression, bare, except that in the middle there is a large, central brownish spot bearing minute, erect fuzzlike pubescence. Above this excavated portion the vertex proper is shining brownish black with a few very scattered setae on each side. It is very slightly raised above the eye margin. The posterior ocelli are very far apart, the anterior ocellus only a little in front so that the ocellar triangle is very obtuse. The so-called ocellar tubercle is scarcely raised above the adjoining vertex. The posterior occiput is produced quite far behind the ocelli, broadly rounded and sloping inward with almost no trace of depression on either side, and with no tubercles and flanges below.

The oral opening is long and high, twice as wide below as above, with quite sharp medial edge and also the recess is deep. Anteriorly the oral recess has some strong, rather long setae which are somewhat flattened and scalelike. I am unable to determine whether this outer wall from the edge of the oral opening to the eye margin corresponds to gena or oragenal cup, as used elsewhere in this text, for there is not the slightest division of this part of the lower head nor any indication of a tentorial fissure.

Thorax: The mesonotum is rather high and strongly and equally arched both anteriorly and posteriorly, shining brownish black in color; the vestiture consists of rather moderately abundant, short, stiff, suberect setae pointed backward on the anterior half, pointed upward or forward on the posterior half. There are a few longer setae along the transverse suture and 2 distinctly stiffer, slightly longer setae in the posterior corner of the notopleuron. Scutellum thick, short, convex with similar setae to the mesonotum. The pleuron is likewise high and compacted, especially below, the coxae crowded together. It is likewise shining, brownish black with a little yellow on the humerus, the upper and anterior margin of the mesopleuron, laterally upon the very short propleuron, and again on the upper hypopleuron. The only pile upon the pleuron consists of a patch of quite flat-appressed, scaliform microsetae on the lower mesopleuron, the anterior pteropleuron and a small dense patch at the very top on the metapleuron beneath the haltere. Also, there are a few scattered setae on the metapleuron above the hind coxae. Upper mesopleuron with a few fine, short, erect hairs. Squama short with a fringe of short pubescence. Haltere with large, conspicuous elongate oval, closed knob.

Legs: The legs are unusually short, the femora quite stout, all the femora about equally swollen above and below. Both femora and tibiae with numerous, flat-appressed setae and no outstanding spines or bristles. Dorsally on the hind tibia there is a row of very short, fine, erect or suberect short hairs and the apex bears a short, spinous bristle medially. The pulvilli are short and rounded, the claws slender and gradually curved.

Wings: The wings are broad, relatively short, wider basally and almost equally wide apically. The auxiliary vein is vestigial, the first, second, and third veins much stouter and blacker. The first vein ends near or just beyond the middle of the wing, the second vein ends in the first vein about midway from its origin of the costa and the third vein ends a short distance back from the end of the wing and joins the costa near the end of the costa. Ambient vein completely wanting, whole wing villose, the remaining veins colorless. The anterior crossvein lies opposite the end of the first vein. There are 4 posterior cells, the anal vein appears to be vestigial and represented by a fold, in which case the anal cell is open. The discal cell is wanting and there are 3 branches to the medius. Alula absent. Posterior margin of the wing with a fringe of relatively long hairs.

Abdomen: The abdomen is short, broad, rather thick, decumbent, thick and convexly rounded laterally, flattened through the middle, short oval viewed from above and covered with scanty, scattered, appressed, fine setate hairs, which extend beyond the posterior margins of the tergites. I find 8 segments in the female. The sternal area is rather narrow and the last sternite shows 2 long, platelike and pronglike lobes that overlap the apex of the abdomen below and overlap each other apically, leaving a recess within.

Material available for study: Before me are 2 paratypes of *Glabellula crassicornis* Greene (described as *Pachyneres crassicornis*). Also, I have seen Melander's species and the species *nobilis* Kertész, var. *palaestinensis* Engel from material in the British Museum (Natural History).

Immature stages: Largely unknown; Greene (1924) figures the lateral aspect of one pupa of *crassicornis* Greene that was found in the frass from a decaying tree of *Quercus velutina* Lamarck in the District of Columbia; found April 15, the adult emerged April 20.

Ecology and behavior of adults: Melander (1950), who in 50 years collected only 19 individuals, 5 were new species, states that the adults frequent flowers. They have been taken upon flowers of mesquite, on *Rhamnus californica*, and in Manitoba upon strawberry blossoms. Four of the six American species are from California. Engel (1933) found them in flowers of *Polygonum* sp.

Distribution: Nearctic: *Glabellula crassicornis* (Greene), 1924 (as *Pachyneres*); *fasciata* Melander, 1950; *metatarsalis* Melander, 1950; *nanella* Melander, 1950; *pumila* Melander, 1950; *rotundipennis* Melander, 1950; *australis* Malloch.

Palaearctic: *Glabellula arctica* Zetterstedt, 1838; *canariensis* Frey, 1936; *femorata* Loew, 1873; *meridionalis* François, 1955; *nobilis* Kertész, 1912, *nobilis palestinensis* Engel, 1933; *unicolor* Strobl, 1910.

Ethiopian: *Glabellula mellea* Bezzi, 1908; *natalensis* Hesse, 1967.

Genus *Aetheoptilus* Hesse

FIGURE 344

Aetheoptilus Hesse, Ann. South African Mus., Part 6, vol. 50, p. 112, 1967. Type of genus: *Empidideicus* (*Aetheoptilus*) *zuluensis*, Hesse, 1967.

Very minute flies that are clearly related to *Empidideicus* Becker and regarded by Hesse as a subgenus of those flies. This fly is readily separated from *Empidideicus* Becker by the presence of the complete discal cell. Because of the closed anal cell, the shape of the veins at the end of the fused basal cells, I place it in a genus near *Mythicomyia* Coquillett; it appears to me to be the closest of all the South African Mythicomyiinae to the genus *Mythicomyia* Coquillett itself.

Flies that were collected Ndumu Reserve, Ingwavuma District, Zululand.

Length: 1.2 mm.; of wing 1.72 mm.

While Dr. Hesse indicated a subgeneric status for his new bee fly *Aethioptilus* Hesse, 1967, I believe from my study of this subfamily that it deserves equal rank with related forms. His description is quoted below:

The deviation from the normal wing-venation of *Empidideicus*, as described by Becker (1907) and Engel (1933) and as is present in the South African species *Empidideicus turneri* Hesse, has gone a step further in a single female specimen from Zululand in which, in addition to the presence of a discoidal cell as in the subgen. *Anomaloptilus* Hesse there is also present a reduced or vestigial marginal cell as in species of *Glabellula* Bezzi, *Doliopteryx* Hesse and the new genus *Pseudoglabellula*.

To accommodate it a new subgenus *Aetheoptilus* of *Empidideicus* is proposed provisionally, pending the discovery of more material of both sexes.

This new subgenus, as typified by this single female specimen, agrees with *Empidideicus* in most of its generic characters, but differs, apart from the presence of a distinct discoidal cell and a narrowish reduced marginal cell, in having the anal cell angularly acute apically and very shortly stalked, and in its broader frons, face and groove in head below.

From the subgenus *Anomaloptilus*, which also has a discoidal cell, it differs in the presence of a vestigial marginal cell, the apically acute and very shortly stalked anal cell, the more S-curved anterior vein of second posterior cell, more parallelogram-shaped third posterior cell, and in the cephalic characters mentioned above. As far as the wing-venation is concerned this subgenus appears to be even more primitive than *Anomaloptilus*. If the type of wing-venation of *Empidideicus* be considered as a specialization on an ancestral type in which reduction has taken place, a step nearer this ancestral condition is represented by *Anomaloptilus* where a discoidal cell still persists. On this assumption *Aetheoptilus* represents an even more primitive condition in which not only the discoidal cell is still found, but in addition there is also a vestige of the normal marginal cell of the ancestral type.

Material available for study: None. This fly was described by Dr. Hesse (1967) while this work was reaching final stages. *Aethioptilus zuluensis* was described from a single female.

Immature stages: Unknown.

Ecology and behavior of adults: Not on record.

Distribution: Ethiopian: *Aethioptilus zuluensis* Hesse, 1967 (as *Empidideicus*).

Genus *Doliopteryx* Hesse

FIGURE 323

Doliopteryx Hesse, Ann. South African Mus., pt. 3, vol. 35, p. 936, 1956. Type of genus: *Doliopteryx crocea* Hesse, 1956, by original designation.

Minute, humpbacked flies that are unique in possessing the greatest reduction in venation among the Bombyliidae. The antennae are similar to *Empidideicus* Becker, except that the style or microsegment is rather conspicuously pilose or bristly. The wing is broad and has only a single basal cell above which is a small triangular cell just as in *Glabellula* Bezzi, which I consider to be its nearest relative among known genera, and this small closed cell may be regarded as the marginal cell. Of the four remaining longitudinal veins the first one is present only in the apical part of the wing (this is the second vein as styled by Hesse, 1956) and is quite evidently the anterior branch of the medius or fourth longitudinal vein, a view amply supported by a study of the wings of *Leylaiya* Efflatoun and *Empidideicus* Becker. This evanescent vein is clearly part of the nonremigial portion of the wing and is being lost by basal rather than distal atrophy. The next vein present in *Dolipteryx* Hesse is, as I interpret it, the anterior branch of the fifth vein and it, too, is evanescent basally. The last heavy vein is the last branch of the fifth vein and the sixth vein is transparent, the anal cell left widely open. The discal cell and second basal cell are completely wanting. The genus differs further from *Empidideicus* Becker in the presence of a deeper, more conspicuous indentation along the anterior eye margin adjacent to the antenna and having the anterior ocelli placed more to the front, besides hypopygial distinctions described below.

Doliopteryx Hesse is based upon 2 species from South Africa.

Length: 1.5 to 2.0 mm., of wing 1.25 to 1.6 mm.

Head, lateral aspect: Decumbent or lower than mesonotum, and rather similar to *Empidideicus* Becker. The occiput is well developed both above and below but is strongly rounded in lateral profile. It lacks a central groove and it is broad on the sides posteriorly so much that Hesse felt that the eyes appeared shifted forward. The eyes are nearly twice as high as wide and they have a much deeper indentation opposite the antennae than in related genera. The proboscis varies in this genus from short and stout to much longer and more slender, being fully 1 mm. long in *nigrescens* Hesse, the body itself only 1.5 mm. The antennae are attached near or just below the upper third of the head. The first segment is quite short and scarcely discernible; the second segment is twice as long as the first and much broader, especially toward the rounded apex; the third segment is about 4 times as long as the first two segments combined, widest basally and there not quite twice as wide as the basal segments; this third segment is strongly attenuate and has a distinct, rather thick, spine tipped, bristly pilose microsegment. The pubescence ventrally on the third segment is more spinulose. The face is rather long vertically, nearly equal to frontoantennal length; the oragenal part of the face is not quite at right angles to face, contrary to the statement of Hesse, and with a few fine hairs on the inside below the face.

Head, anterior aspect: The front converges anteriorly. The eyes are broadly separated above and equally separated in both sexes. The anterior ocellus lies nearer to the front than in *Empidideicus* Becker; eyes separated by a space as wide as posterior pair of ocelli or slightly more. The front has a linear depression running from ocellus to antennae.

Thorax: High and arched, with the head drooping, the pile moderately abundant but composed of such minute, short setae as to require considerable magnification for study. There are a few slightly longer hairs on humeri, on wing bases and sides of scutellum; pleuron bare. Scutellum quite thick, as long as wide, with fine, moderately long, subappressed black setae.

Legs: All of the femora are distinctly a little thickened; on the hind femur the swelling is entirely dorsal. The tibiae are slender. The legs bear very sparse, fine pubescence, a little more dense and distinct on the tibiae which have no distinct bristles or spicules; apical spurs minute. The tarsi are as long as the tibiae, the hind basitarsus unmodified in males. The claws are small and distinct, curved apically, the pulvilli distinct.

Wings: Greatly reduced, and as Hesse notes, quite unlike Bombyliidae. Nevertheless, if compared with the wings of *Empidideicus* Becker, *Leylaiya* Efflatoun, or *Glabellula* Bezzi, the relationship is clear. The second vein ends in the first vein, as in other Mythicomyiinae, the first branch of the medius ends at the midpoint of the apex of the wing and the basal half of this vein has disappeared. There are 3 branches to the medius and 3 to the radius. The sixth is evanescent but the anal cell is widely open. There is no discal cell and there is only one basal cell, the first being present, the others wanting. The axillary lobe is moderately broad only, the alula and ambient vein wanting.

Abdomen: Short and obtuse. There are 8 tergites with the eighth extremely short dorsally and a little longer laterally. The pile of the abdomen is minute and appressed. The apical angle on each side of last sternite in male is extended inward as a spinelike process. Hypopygium of male minute, made of 2 shell-like basimeres, as in other bee flies, but ending apically in a pointed apical process instead of beaked apical distimeres; each is provided on the lower side with a minute hook.

Since I have not been able to see males of this genus I quote the remarks of Dr. A. J. Hesse with respect to genitalia:

Hypopygium of known male minute, composed of two shell-like basal parts as in the other genera, ending apically in a pointed apical process instead of beaked apical joints, each provided on inner lower side with a minute hook; aedeagal structure, such as the aedeagus and lateral struts, were not discernible after preparation by boiling in dilute NaOH and these have either been destroyed by the NaOH or are composed of easily soluble structures; the basally directed flattened rod on each side is shown in position; the transverse structures shown at base of the apical processes probably represent part of the lateral rami, the aedeagal complex in the natural state probably just ventral to or associated with them; basal strut small, shown from ventral view and from side with its dorsal process blunt.

Material studied: The types of *crocea* Hesse in the British Museum (Natural History).

Immature stages: Unknown.

Ecology and behavior of adults: Hesse notes both species were found in the small white blossoms of a species of *Mesembryanthemum*.

Distribution: Ethiopian. *Doliopteryx crocea* Hesse, 1956; *nigrescens* Hesse, 1956.

Genus *Pseudoglabellula* Hesse

Figure 345

Pseudoglabellula Hesse, Ann. South African Mus., pt. 6, vol. 50, p. 118, 1967. Type of genus: *Pseudoglabellula meridionalis* Hesse, 1967.

Minute flies, the second basal cell is distinct and distinctly separated from both discal and second basal cells; only 3 veins proceed from the second basal cell. They are readily distinguished from *Glabellula* Bezzi and related genera by the much more narrow first posterior cell which is also narrowed apically and opens into the costa above the apex of the wing. The fourth vein divides into 2 branches leaving a wide, triangular second posterior, apical cell. Hesse (1967) notes that there is no projecting pronotal lobe.

The flies are found in the Koup Karoo District of South Africa.

Length: 1.8 mm.; of wing 1.68 mm.

Just as the present work was ready to be submitted for publication, I received Dr. Hesse's paper describing 3 remarkable new genera of bee flies from South Africa. I quote his description:

A female specimen from the Koup Karoo in the collections before me and which was obtained together with other insects by sweeping flowering shrubs, cannot be allocated to any of the known genera of Cyrtosiinae [Platypyginae]. It appears to represent a new and as yet undescribed genus. Certain wing-characters seem to suggest a relationship to the genus *Glabellula* Bezzi, but in most of the other venational characters it shows even closer affinity with the genus *Euanthobates*. If the presence of a distinct, though much reduced, marginal cell be taken as a group character, it is referable to the group of Cyrtosiinae [Platypygine] genera, such as *Cyrtosia, Platypygus, Cyrtisiopsis, Ceratolaemus* and *Glabellula*, in which a marginal cell is present in the wings, even though sometimes much reduced. On the other hand the rest of its wing-characters, as well as certain antennal and cephalic characters, place it in close proximity to *Euanthobates*.

By comparing it with the descriptions and illustrations of species of the Palaearctic *Glabellula*, which also has a similarly reduced marginal cell, its wings, like those of *Euanthobates*, differ in having the first basal cell distinctly very much longer than the second, and this second basal cell is apparently not formed by the fusion of a discoidal and a second basal cell as is suggested in the case of *Glabellula*; only 3 longitudinal veins, not 4, radiate out from this second basal cell; the 4 posterior cells not formed directly by these delimiting longitudinal veins, but the fourth vein bifurcates into two branches, forming the elongate triangular second posterior cell; first posterior cell, unlike that of *Glabellula*, distinctly very much and markedly narrower and narrowed apically, as in *Euanthobates*, to open on anterior or costal margin and not on apical part of it or very near apex of wing as in *Glabellula*; triangular marginal cell comparatively larger than in *Glabellula*.

Other characters which also distinguish it from *Glabellula* are the markedly short face, the great reduction of antennal joint 4, which is minute, scarcely perceptible as in *Euanthobates*, and not elongate, slender or even rod-like, and the distinctly less convexly humped thorax.

From *Euanthobates*, to which it is generically very closely related and with which it shares such wing-characters as the much narrowed first posterior cell, the very short part of fourth vein before base of second posterior cell, and cephalic characters such as the very short face, and much reduced antennal joint 4, it however differs in having a distinct, reduced, triangular marginal cell present, a very much shorter first posterior cell which curves anteriorly, opening on anterior costal margin at a much longer distance before apex of wings, thus reducing the length of the combined marginal and submarginal cells; a much narrower, more parallel-sided anal cell, more like that of *Glabellula*; a very much broader frons, and even shorter face; much larger antennal sockets; comparatively shorter and broader, truncated, vertical, anterior part of head, with the buccal cavity directed more obliquely forward; and the head below broadly hollowed out or excavated, without a well-defined or delimited, central sulcus.

In other characters, such as the absence of a projecting pronotal lobe, it agrees with *Glabellula, Empidideicus* and *Euanthobates*, and differs from *Cyrtosia, Platypygus*, and *Ceratolaemus* where this lobe is present.

Material available for study: None; this fly was described by Dr. Hesse whole this work was undergoing final steps for completion. I have republished Dr. Hesse's description and figure.

Immature stages: Unknown.

Ecology and behavior of adults: Flies taken while sweeping over flowering shrubs.

Distribution: Ethiopian: *Pseudoglabellula meridionalis* Hesse, 1967.

Genus *Empidideicus* Becker

Figures 107, 108, 328, 348, 591, 594

Empidideicus Becker, Zeitschr. syst. Hymen. Dipt. Teschendorf, vol. 7, p. 97, 1907. Type of genus: *Empidideicus carthaginiensis* Becker, 1907, by monotypy.
Cyrtoides Engel, Die Fliegen der palaearktischen Region, vol. 25, p. 100, 1933. Type of subgenus: *Cyrtoides efflatouni* Engel, 1933, by original designation.
Cladella, new subgenus. Type of subgenus: *Empidideicus propleuralis* Melander, 1946.

Engel (1933), p. 101, key to 3 Palaearctic species.
Hesse (1938), pp. 981–982, key to 4 South African species.
Efflatoun (1945), p. 26, key to 3 species from Egypt.
Melander (1946), p. 455, key to 3 species from the southwestern United States.

Very minute, largely black flies with subglobular head, humpbacked mesonotum and stubby abdomen. The proboscis is either very short or in some species even longer than the head. *Empidideicus* Becker is one of several genera, which while having the third vein unforked, likewise have the second vein absent. The vein separating first and second basal cells may be sometimes thinned or evanescent and the discal cell is absent, being confluent with the third posterior cell. There are 4 posterior cells. The anal cell is widely open, the alula wanting as well as the ambient vein. The wing while broad is not so wide as in *Glabellula* Bezzi. *Cyrtoides* Engel is to be distinguished from *Empidideicus* Becker only because the three ocelli are lined up within a straight or nearly straight transverse row. In the related genus *Anomaloptilus* Hesse the discal cell is complete.

Leylaiya Efflatoun is readily distinguished from *Empidideicus* Becker because the anterior branch of the medius in addition to being thrust forward is very weak; its anal cell is closed in the margin. The style of the third antennal segment, well developed; in *Empidideicus* Becker is almost totally wanting.

The species *propleuralis* Melander was placed within *Empidideicus* Becker by Melander but he observed that its relationship was probably closer to *Mythicomyia* Coquillett. I believe this is correct and I transfer it to a new subgenus under *Mythicomyia* Coquillett.

Empidideicus Becker is found along the northern coast of Africa and in or near the Mojave Desert of southern California in the United States. Species characters have to do mostly with coloration and shape and proportion of the segments of the antenna.

Length, in type-species: 0.9 to 1.5 mm., of proboscis 0.25 to 0.45 mm., wing expanse 2.5 to 3.15 mm. of larger species, body 1.6 mm. to 2.1 mm., proboscis 0.4 to 0.6 mm., and wing breadth 3.5 to 4.0 mm.

Head, lateral aspect: The head is globular with the occiput very strongly developed, swollen and convex posteriorly. It is thinly pollinose and has a few fine, scattered, bristly hairs. The front is a little swollen with 3 or 4 slender, bristly hairs of no great length in a vertical row on each side of the front and with a triangular depression in front of the antennae. The vertex is strongly swollen; the ocelli are quite large and between the ocelli there are 2 or 3 slightly stronger, pale bristles of no great length. The face and front are quite pale yellow in contrast to the black occiput. The face is short and rather long vertically, moderately becoming a little more prominent at the dorsal border of the ora-buccal cavity, and there are 2 distinct bristles at the bottom of the face where the buccal cavity begins. Sides of the oragenal cavity very thin, vertical, and while not extensive beyond the eye they are of about the same extent throughout and back to the posterior corner of the eye. The eye is small, the posterior margin is as convex as the anterior margin.

The proboscis is short, very stout, and robust and in an American species extended obliquely forward and downward; in length it is a little more than half as long as the head. The labellum bears conspicuous, ventral, bristly hairs. In European species the proboscis is much shorter and stublike, though held in the same position. The antennae are attached at or below the middle of the head in *humeralis* Melander, and above the middle of the head in European species; also, the proboscis is extraordinarily long and slender in the type-species and in the species *mariouti* Efflatoun and is thrust more or less straight downward. The palpus is apparently absent, but dissected material might show a vestige of it. The antennae arise adjacent, the first segment is extremely short and the second segment likewise short, both of these segments are pale with at most minute setae apically, and the third segment of *humeralis* Melander is cylindrical, about the same width as the second segment or barely more narrow and about 2½ times as long as wide. It is micropubescent and bears at the apex a conical microsegment, sharply pointed, also with long micropubescence. In the European species the third segment is rather larger than the second segment, swollen and pyriform or long oval, with long stylelike microsegment.

Head, anterior aspect: From the anterior aspect the front shows a deep, large, triangular pocket or depression in the female. The ocellar triangle is large and obtuse and the front widens toward the vertex, instead of converging. The eyes are widely separated in both sexes. The antennae arises nearly adjacent, the oragenal recess is placed anteroventrally, is rather small and rectangular, and quite deep. It is barely sufficient to emit the large, fleshy proboscis.

Thorax: The thorax is short and compact, the mesonotum moderately humped and convex, less so than in *Cyrtosia* Perris, but more strongly humped than in *Leylaiya* Efflatoun; it is a little more steep anteriorly than behind. The mesonotum is generally blackish, sometimes vittate with the lateral margins and sometimes prescutellar area pale yellow. The surface is covered with quite microscopic, rounded scales. There are a few erect, scattered, fine bristly hairs on the surface of the mesonotum, including the humerus, the notopleuron, and the postalar callosity. Whole pleuron with pollenlike microscopic scales similar to those of the mesonotum, but otherwise without pile, except on the propleuron and upper mesopleuron, which bears a few long, scattered, fine hairs. The scutellum is somewhat triangular with rounded apex, either shining black or yellow with micropubescence along the margin, the disc feebly pollinose and disc and margin with a few scattered, moderately long, stiff hairs, 10 or 12 on each side. Halteres quite large proportionately on both the stalk and the club; squama very short. The metasternum is short and more or less covered by the broad attachment of the abdomen; the pronotum is also greatly reduced.

Legs: The legs are comparatively short; the anterior femur is distinctly swollen, a little more so toward the base; middle femur slightly swollen; hind femur stout but not as wide as the middle one. All the femora have a few fine, stiff hairs more particularly below and more of them near the base. Dorsally there are a few fine, appressed hairs near the apex. The middle coxae are shorter than the others; whole legs mostly yellowish, the femora sometimes brown and the tarsi often blackish on the terminal segments. The tibiae are slender and bear numerous, subappressed setae, the dorsal row on the anterior tibia has about 12 or more elements in it. Claws slender, sharp, curved at the apex, the pulvilli large, the tarsi short, the last segment appears to be scooped out below.

Wings: The wings are short and broad, the veins rather stout, except the branches of the medius which are weaker. The auxiliary vein is atrophied distally, the end of this vein being free in the middle of the costal cell. The first vein R_1 is stout and strong entering the costa near the middle of the wing. The radial sector is also equally stout and so is its continuation, which is all that remains of the radius. This may be regarded as the third vein. It is unbranched and the second vein (R_2+R_3) may be considered to be completely lost. The discal cell is absent due to the loss of the posterior crossvein. The anterior crossvein is quite basal in position, its position nearer the base of the wing than the end of the first vein; vein separating the first and second basal cells atrophied. There is generally a trace of it basally but it is otherwise extremely weak and hardly detectable. The first branch of the medius is forked. There are 4 posterior cells, the first anal vein is barely indicated, the second one is likewise scarcely distinguishable, so that the first anal cell, while open, can hardly be separated from the axillary lobe. The ambient vein is absent and the alula linear or absent.

Abdomen: The abdomen is broad basally, sometimes broader than the thorax, and either ovoid or conical in shape with 6 or 7 apparent segments. Efflatoun (1945) notes that the genitalia are so much concealed in dried specimens that it is often impossible to distinguish between sexes. Generally the abdomen is in large part yellow with black bands basally of varying extent reaching across the tergites but not extending to the sides and laterally with a black punctiform spot on each side of each tergite.

Since I have not been able to see males of this genus I quote the remarks of Dr. A. J. Hesse with respect to genitalia:

> Hypopygium of males with the basal parts small and shell-like; apical process lobe-like or knob-like, having a prominent hook at its base ventrally; aedeagus shortis and tubular; rods on each side of base of aedeagus very prominent and long; lateral struts prominent and well developed; basal strut peculiar in having a medial dorsal apically directed process and also a dorso-ventrally flattened process on each side.

Material available for study: Consists of the series within the National Museum of Natural History, representatives given me by Dr. Melander, and representatives of the type-species received in exchange from Cairo University.

Immature stages: Unknown.

Ecology and behavior of adults: Efflatoun (1945) comments upon their abundance on the inside of flowers, especially desert Compositae and also in the flowers of *Zilla spinosa*, a common desert plant everywhere in Egypt. He also notes that these flies, like *Cyrtosia* Perris, can be captured in great numbers on the inside of desert flowers. Bezzi (1926) comments on the widespread abundance of all the related genera of this group noting that they have been found in India, Australia, Tasmania, Lapland, Cape of Good Hope, and North America.

Distribution: Nearctic: *Empidideicus flavifrons* Melander, 1950; *humeralis* Melander, 1946; *propleuralis* Melander, 1946; *scutellaris* Melander, 1946.

Palaearctic: *Empidideicus carthaginiensis* Becker, 1907; *efflatouni* Engel, 1933 (as subg. *Cyrtoides*); *hungaricus* Thalhammer, 1911; *insularis* Frey, 1958; *mariouti* Efflatoun, 1945; *nubilus* Bezzi, 1925; *perfectus* Becker, 1910; *serenus* Becker, 1915; *turgestanicus* Paramonov, 1947.

Ethiopian: *Empidideicus beckeri* Bezzi, 1908; *turneri* Hesse, 1938.

Subgenus *Cyrtoides* Engel

Minute, strongly humpbacked flies with short compact abdomen. These flies are very similar to *Empidideicus* Becker and separated from that genus by the alignment of the 3 ocelli into a straight line transversely. *Cyrtoides* Engel is based upon *Empidideicus efflatouni* Engel, which Engel (1933) called a subgenus, and described the species from Egypt and Turkestan. Engel (1933) gave *Cyrtoides* status of subgenus under *Empidideicus* Becker; Efflatoun (1945) made no comment upon it beyond redescribing the species he called abundant in Egypt, but Melander (1946) gave it generic status, which it hardly deserves, and so I have brought it back to subgeneric status inasmuch as, according to Efflatoun (1945), the position or alignment of the ocelli is somewhat variable.

Length: 1.5 to 2.4 mm., wing expanse 2.5 to 4.0 mm.

Distribution: From Egypt and Turkestan.

Genus *Anomaloptilus* Hesse

FIGURES 353, 586, 595

Anomaloptilus Hesse, Ann. South African Mus., vol. 34, p. 983, 1938. Type of genus: *Empidideicus* (*Anomaloptilus*) *cellulifer* Hesse, 1938, by original designation.

Minute flies, which as Hesse (1956) notes are almost indistinguishable from *Empidideicus* Becker except in the constant presence of the distinct discal (discoidal) cell. Since the second vein is wanting, both these genera fall within the subfamily Mythicomyiinae as defined by Melander (1946). Hesse (1938) gave *Anomaloptilus* subgeneric status, but Melander in the above-cited paper raised it to generic status, which I believe it certainly deserves. A separate description is hardly necessary.

These flies are found in several localities of the Cape Province of South Africa.

Length: 1.5 to 1.66 mm., of wing 1.2 to 1.66 mm.

Material studied: The types in the British Museum (Natural History).

Immature stages: Unknown.

Ecology and behavior of adults: Not on record.

Distribution: Palaearctic: *Anomaloptilus atomus* Frey, 1958.

Ethiopian: *Anomaloptilus basutoensis* (Hesse), 1967 (as *Empidideicus*); *brevistilus* (Hesse), 1967 (as *Empidideicus*); *cellulifer* (Hesse), 1938 (as *Empidideicus*); *notatus* (Hesse), 1967 (as *Empidideicus*).

Genus *Euanthobates* Hesse

FIGURES 350, 758

Euanthobates Hesse, South African Animal Life, vol. 11, pp. 482-484, 1965. Type of genus: *Euanthobates pectinigulus* Hesse, by original designation.

Acoecus, new subgenus. Type of subgenus: *Euanthobates mellivorus* Hesse, 1967.

These are very small flies. They are related to *Empidideicus* Becker and less so the *Cyrtosia* Perris yet are quite different. The body is more plump or stout and the thorax is distinctly less humped and is comparatively broader and has no projecting lobelike pronotum. The wings are similar to *Empidideicus* Becker, and much shorter than the body. With other members of the Platypyginae and *Empidideicus* Becker, it has a groove or sulcus beneath the head. They properly belong in the subfamily Mythicomyiinae, and they show the typical venation of this group.

This flower-feeding species, as in the case of most of the known species of both Mythicomyiinae, Phthiriinae and Platypyginae, has become adapted to an anthophilous habitat. The mobile and apparently protrusible, downwardly directed proboscis, and the remarkable, downwardly projecting, chitinous, caecalike or fingerlike processes along the sulcus on the head below are probably adaptations to flower feeding. However, according to Hesse (1967) they are not found among the new species of this genus that he described. These processes, arranged irregularly comblike or brushlike in the gular groove, appear to be unique among known Diptera, if not among insects, and their significance is unknown and can only be guessed at. Their evolutionary origin may be connected with the sulcus or groove below the head that also characterizes other flower-feeding Cyrtosiines, such as the representatives of *Platypygus* Loew, *Cyrtisiopsis* Séguy, *Ceratolaemus* Hesse, and *Empidideicus* Becker, and in which they serve to carry out some unknown and specialized function. These structures do not appear to be connected anatomically with the proboscis, labella, or labrum as are the outgrowths or "pseudotracheal processes" of some other flower-feeding Diptera, such as are found, among others, in representatives of the tabanid genus *Philoliche* Austen.

These flies are from the Kaokoveld area of South West Africa.

Length: 1.6 to 2.04 mm.; of wing 1.16 to 1.4 mm.

Since I have not seen material of this genus I quote Hesse's (1965) description of this genus and comparison with other genera. To the genera cited I have added author's name.

Head perpendicularly truncate in front as in *Empidideicus* Becker, slightly more spherical, but the eyes also situated slightly forward; frons slightly depressed, slightly broader basally than apically; face however markedly short, very much shorter, relative to frontal length, than in either of the other two genera, the antennae thus situated far forwards; antennae differing from those of both *Empidideicus* Becker and *Cyrtosia* Perris in having the third joint more elongate, leaf-shaped or spear-blade-shaped, flattened, ending apically in a minute (not elongate), scarcely perceptible, terminal element, but with joints 1 and 2 however also short as in the other two genera; buccal cavity below the head anteriorly as in *Empidideicus* Becker, not terminal as in *Cyrtosia* Perris; proboscis as in former, with relatively short and stoutish labral part, protrusible, much more so than in *Empidideicus* Becker, and comparatively long; head below grooved or sulcate, the sulcus in this species with remarkable, downwardly-projecting, finger-like, caeca-like or strap-like processes, arranged irregularly comb-like along the sulcus.

Wings as in *Empidideicus* Becker, also much shorter than body, but with the first basal cell distinctly much longer than second; first posterior cell markedly narrow, narrowed apically, and narrowly open on apical margin of wings, its lower or posterior vein thus ending at or before apex of wings, not posterior to it; part of vein between first basal and second posterior cells markedly short, considerably shorter than in *Empidideicus* Becker; anal cell comparatively broad, broader than in latter genus, narrowed apically; alula narrow or wanting as in *Empidideicus* Becker. Vestiture fine and very short, but apparently denser and shorter than in latter genus, especially on the abdomen.

This peculiar species, of which the representatives are unfortunately preserved in alcohol and have thus become distended and discoloured, is characterized as follows:

Body (as far as can be seen in the preserved apecimens) very dark, blackish brown; facial part in front more yellowish; eyes reddish to reddish brown; sides of thorax above and hinder part of border of scutellum reddish brownish; pleurae mainly reddish brown; abdomen above yellowish brownish to mainly dark, dark blackish brownish, with the hind margins of tergites only narrowly, or scarcely, paler and more yellowish, not extensively pale on sides; venter pale or yellowish; femora reddish brownish, their extreme apices more yellowish; tibiae and tarsi more yellowish, but apical parts of tarsi darkened.

Vestiture with the hairs minute, pale.

Head with antennal joint 3 slightly elongate, flattened, spear-blade-shaped or leaf-shaped, broadest nearer base, about three times as long as broad; proboscis with the short, stoutish, labral part about 0.16-0.2 mm. long, the extruded proboscis itself pale, slender, rather soft, much longer than entire body, about 1.14-3.68 mm., its labella small; caeca- or strap-like processes in sulcus below head not all equally long, the longest about 0.16 mm. long, in ventral view appearing slightly broadened straplike.

Wings slightly, but distinctly, tinted yellowish; veins yellowish; first posterior cell markedly narrowed apically; anal cell narrowed apically, but still distinctly open.

Material available for study: The University of Lund lent me specimens of this genus.

Immature stages: Unknown.

Ecology and behavior of adults: "Found on yellow composites near Welwitschia."

Distribution: Ethiopian: *Euanthobates mellivorus* Hesse, 1967; *pectinigulus* Hesse, 1965.

Genus *Leylaiya* Efflatoun

FIGURE 346

Leylaiya Efflatoun, Bull. Soc. Fouad Ier d'Ent., p. 21, 1945. Type of genus: *Leylaiya mimnermia* Efflatoun, 1945, by original designation.

Minute flies, less deeply humped than in either *Empidideicus* Becker or *Cyrtosia* Perris. They are usually entirely orange yellow, except rarely with thorax or third antennal segment obscured by dark brown. They are quickly distinguished from *Empidideicus* Becker by the fact that the style of the third antennal segment is so minute as to be almost nonexistent, and the anterior branch of the medius is so weak that it is nearly effaced and moreover is rather strongly thrust forward. Like *Empidideicus* Becker the auxiliary vein is vestigial distally, the radial sector is unbranched, and the first vein ends in the middle of the wing. Alula, and ambient vein and discal cell are all absent and the anal cell is closed in the margin of the wing.

These curious little flies are known from a single species captured in two desert wadies of Egypt only some 50 kilometers apart.

Length: 1.1 to 1.3 mm., of proboscis 0.1 to 0.5 mm., wing expanse 2.8 to 3.2 mm.

As I have not seen any material of *Leylaiya* Efflatoun, I quote his description:

Head remarkably elongate and somewhat pyriform when seen from above, narrower or much narrower than the thorax and distinctly longer than broad; the head is rounded behind, owing to the very puffed-out and bulging occiput (even more so than in *Cyrtosia* Perris) and then produced forward cone-shaped and gradually narrowing down beak-like apically where it terminates by the short and comparatively stout proboscis; frons a little broader above than below (at the base of the antennae) and narrower in the male than in the female; face short but conspicuously produced vertically upwards below, in a short hump-shaped projection (which is the upper border of the mouth-margin), consequently the very short face itself appears deeply concave, almost semi-circular; from this hump-shaped projection the correct mouth-margin slopes perpendicularly downwards, almost at right angles with the face and frons; genae very narrow, linear; proboscis short, rather stout and about as long as the length of the head (twice as long as the entire mouth-margin).

The palpi are apparently absent; vertex broad and the broad triangle formed by the ocelli (ocellar tubercle) is hardly approximated at the base that they touch, as in *Empidideicus* Becker and *Cyrtosia* Perris and consist of a very short, rounded basal segment, a short cylindrical second segment (about twice as long at the first, as in *Empidideicus* Becker), and a long, broad, pear-shaped third segment whose apical style is so minute (hardly visible even under high magnification) that it may be termed absent.

Material available for study: None. I have quoted the description of Efflatoun.

Immature stages: Unknown.

Ecology and behavior of adults: Not on record.

Distribution: Ethiopian (Egypt): *Leylayia mimnermia* Efflatoun, 1945.

Division TOMOPHTHALMAE
Subfamily Cylleniinae Becker, 1912

This small subfamily contains some remarkable and interesting and unusual bee flies. They are at once characterized by: (1) the deeply cupulate recess within the posterior occiput with its marginal fringe of hair about the rim. (2) The plane, or birectiform, posterior eye margin not indented or bisected either. Generally the antennae and proboscis are elongate, but the former are short within the tribe *Tomomyzini*. (3) The second vein arises basal to the discal cell and arises at an acute angle. The nonindented and nonbisected postmargin of the eye and the place and manner of origin of the second vein separates this subfamily from the Lomatiinae.

After a lengthy and comprehensive study of some sixty characters in almost all of the genera of the two subfamilies, I find these to be the only consistently divisive characters. In fact the two subfamilies are so closely allied they might almost be combined into one with a series of tribes. All included genera have the metanotum bare, except *Peringueyimyia* Bigot. A pilose metanotum is a distinguishing characteristic of the Lomatiinae and it is possible that *Peringueyimyia* Bigot belongs there. This subfamily is almost restricted to the Palaearctic and Ethiopian regions. The unique genus *Neosardus* Roberts is Australian, and so is

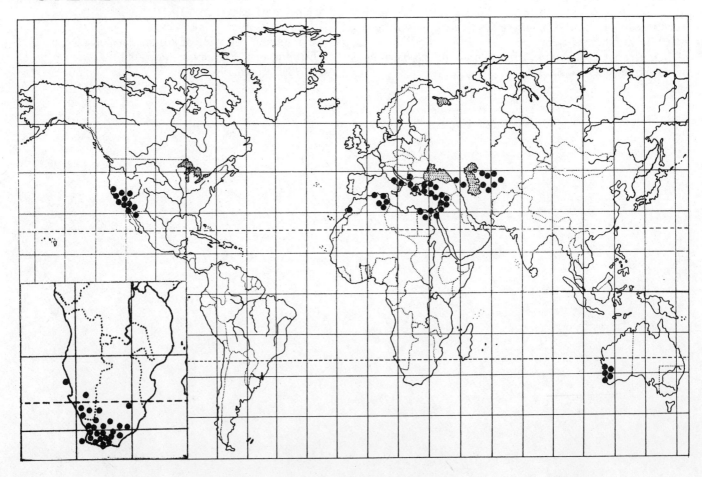

TEXT-FIGURE 41.—Pattern of the approximate world distribution of the species of the subfamily Cylleniinae.

Docidomyia White. Only *Paracosmus* Osten Sacken and its close allies are found in the Nearctic region. Nothing is known concerning the life histories in this subfamily; Bezzi, in Latin, described the pupa of *Cyllenia marginata* without comment as to host. Since many of these flies are commonly found resting upon soil it may be surmised that the hosts are subterranean. The *Henica* Macquart and *Neosardus* Roberts groups both have a well-marked row of acanthophorite spines on each side of the ovipositor much as in soil-ovipositing asilids.

It is my belief that the genus *Eclimus* Loew and allied genera, as *Thevenemyia* Bigot (Ecliminae of Hall, 1969), find their proper relationship within the Bombyliinae where I place them as a special tribal subdivision. I quite concur with Mr. Hall that it deserves special rank. Within the Cylleniinae my studies convince me that I must recognize the following division:

Tribe Cylleniini

This tribe is characterized by thorax and abdomen with strong bristles. The third antennal segment with small bristles especially at apex. Proboscis and antennae varying from long to short. Metapleuron and metanotum bare except for micropubescence. Contains *Cyllenia* Latreille, *Amictus* Wiedemann, and *Sinaia* Becker, and also *Sphenoidoptera* Williston.

Tribe Henicini

(Heniconiinae of Hall, 1969.) Characterized by occipital lobes shallow but narrowly separated; the antennae are long and the proboscis is long; especially characterized by the virtual absence of the metanotum; the abdomen and thorax are so closely jammed that the metanotum is almost obliterated; both metapleuron and metanotum without pile. Included are *Henica* Macquart and *Nomalonia* Rondani.

Tribe Tomomyzini

(Tomomyzinae of Becker, 1913.) Characterized by the projecting face that extends at least beyond the antenna; the face is relatively short but conical and oblique. The ocelli are set far forward from the posterior corner of the eye. The antennae are short and the proboscis is confined to the oral cavity; the condition of the metapleuron and metanotum varies; of the six included genera all have the metanotum bare except *Tomomyza* Wiedemann where it is either pilose or bare; all have the metapleuron pilose; *Tomomyza* Wiedemann has pile on the upper hypopleuron and has a plumula; it should possibly be placed in a separate tribe. Included are: *Paracosmus* Osten Sacken, *Metacosmus* Coquillett, *Amphicosmus* Coquillett, *Pantostomus* Bezzi, *Docidomyia* White, *Tomomyza* Wiedemann, and *Actherosia*, new subgenus. Several South African species formerly placed in *Tomomyza* Wiedemann belong in *Paracosmus* Osten Sacken.

Peringueyimyiini, new tribe

In this tribe I place the single genus *Peringueyimyia* Bigot. It is peculiar. It appears to have all of the characteristics of the Lomatiinae except the indented or bisected post eye margin. The occiput is extraordinarily thick, antennae rather elongate and slender, and the genae are prominent and deeply shelving inward; with the dense long pile beneath the third antennal segment they are much like the Comptosini. The metanotum in front of the haltere is without pile but the metanotum is densely pilose.

Neosardini, new tribe

Here I place the unique genus *Neosardus* Roberts. It has the general characters of the subfamily. The antennae and proboscis are both long and slender. Especially to be noted is the extraordinarily wide face and very wide gena. The occipital cup is present though perhaps smaller, the occipital lobes wide, widely separated and shallow. Both metapleuron and metanotum without pile.

KEY TO THE GENERA AND SUBGENERA OF CYLLENIINAE

1. Genofacial space extraordinarily wide, especially the genae. Front also very broad in females. Males dichoptic. Four submarginal cells present and wing with an *Exoprosopa*-like pattern. Antenna elongate and cylindrical and all three segments conspicuously and densely appressed setate; first antennal segment as long as the third or even longer; last segment tapers sharply from the base and ends acuminate with a minute, bristle-tipped microsegment . . . . **Neosardus Roberts**
   Genofacial space scarcely more than one-third the head width and often much less. Antennae shorter, differently shaped and proportioned. Usually 2 and never more than 3 submarginal cells present 2
2. Thorax, scutellum, hind tibiae, and sometimes the abdominal tergites with strong, long, conspicuous bristles . . 3
   Rather small flies of 4 to 10 millimeters length and often with relatively slender, moderately elongate abdomen. Comparatively bare flies with or without areas of dense, always short, fine or bristly, erect or appressed pile. No strong bristles present anywhere 9
3. Large, robust flies with semidecumbent head and drooping abdomen which bears appressed coarse hair and on the

postmargins of the tergites only appressed rather short, bristly hairs. Femora and tibiae with conspicuous, spine-like bristles. Notopleural and supraalar bristles strong. First antennal segment much swollen with dense downwardly directed pile below and subappressed bristly pile above. Third antennal segment as long as the first, strongly tapered and narrowed. There are 2 submarginal cells; anterior branch of third vein arising as a crossvein with long backward spur; wing hyaline; end of marginal cell bulbous below at apex; first posterior cell closed. Occiput very extensive, lobes tightly apposed.
                                                Peringueyimyia Bezzi

Not such flies. First antennal segment both elongate and heavily swollen. Scutellum with strong bristles 4

4. Postmargins of the tergites with conspicuous, long, erect bristles. Frons depressed and usually with a distinct, medial groove 5
Postmargins of tergites and whole surface with only dense, appressed scaliform pile. Frons elevated, swollen, smooth. First antennal segment short and round, globose. Occipital lobes widely separated. Metanotum almost completed reduced 8

5. Proboscis long and slender, usually much longer than the head; if shortened and scarcely longer than head the wing is hyaline. Antennae comparatively slender and elongate. Flies often brownish yellow in color, vittate thorax, and elongate, cylindroid, distally compressed abdomen. Wings almost always hyaline 7
Proboscis relatively short scarcely as long as head and often shorter. Antennae also rather short; apex of the short-ovate, flattened third antennal segment with a few, slender bristles especially near the apex. Flies rather small, black, the abdomen of moderate length, rather flattened, as wide as mesonotum and with dense, appressed scaliform pile in addition to bristles. Wings with axillary lobe much narrowed, especially toward the base. Wings always with a blackish border or pattern of scattered black or brown spots that may be faint. Intercalary vein plane, usually rectangular to the fourth vein 6

6. Oragenal cup rather extensive both above beneath the antenna and below along the sides, and polished and bare, and convexly rounded outward along its walls; no bristles or pile present upon it. Intercalary vein forms a right angle with fourth vein above. Two submarginal cells. Frons beset with rather long, stiff, erect hairs.
                                                Sphenoidoptera Williston
Oragenal cup very short and with several long, conspicuous, stout bristles and bristly hairs on each side just below the antenna. Frons densely beset with long, erect, bristles. Three submarginal cells and intercalary vein set at an angle. Narrow genae pollinose and not vertically shelving inward along eye margin. . . . . **Cyllenia Latreille**

7. Face quite short, but high, the oral opening reaching only a short distance above the proboscis. Antenna rather like *Amictus* Wiedemann, but a little less slender and the proboscis distinctly shorter and stouter, not much if any longer than the head, and labellum stouter. Wing hyaline, with 2 submarginal cells and all the posterior cells and anal cell widely open. Occiput in profile quite extensive on the upper half of head. . . . . **Sinaia Becker**
Face longer, subconical, distinctly and triangularly peaked forward. Proboscis 2 to 4 times as long as head and together with labellum quite slender. Upper occiput moderately or not at all developed; occiput relatively shallow with the upper lobes separated. First posterior cell open or closed. Submarginal cells usually 2 in number, rarely 3. Wing almost always hyaline; never with conspicuous markings. Males sometimes with curious, very long apical bristles on first tarsal segments. Some males with a double row of short, close-set spines upon lower surface of middle tarsi . . . **Amictus Wiedemann**

8. Marginal cell divided into 3 parts by crossvein. Frons with numerous stiff bristles. . . . . **Henica Macquart**
Marginal cell simple, without crossveins. Frons with a few stiff hairs . . . . **Nomalonia Rondani**

9. Marginal cell at apex simple and not swollen or widened even in forms that have 3 submarginal cells. Wing hyaline. Flies with rather long and slender abdomen. Hind femur with a few fine hairs below 10
Marginal cell distinctly widened at apex and bulbous, protrusive, widened, and rarely even with a radial loop or ending bluntly with the terminal section of the second vein plane 11

10. Three submarginal cells . . . . **Amphicosmus Coquillett**
Two submarginal cells . . . . . . **Metacosmus Coquillett**

11. Wing hyaline with a conspicuous long radial loop; second submarginal cell long and slender and arising as an arch or as crossvein with long backward spur. Eye exceptionally large leaving genofacial and oral area reduced, proboscis very short and muscoidlike. Eyes of male narrowly separately by a long, parallel-sided space, the ocelli set very far forward; slender, compressed, often annulate abdomen and prominent male terminalia much as in *Paracosmus* Osten Sacken. Metanotum bare, hind femur scanty, long erect hairs below . **Docidomyia White**
Not such flies 12

12. Three submarginal cells. Anterior part of wing on basal two-thirds yellowish brown. Anterior branch of third vein arising plane and obliquely. Abdomen club-shaped apically. Whole surface of hind femur densely covered with comparatively long, distinctly appressed, stiff hairs.
Marginal cell wider apically than before the crossvein, but ending plane of genus. Rather large species 12 or 13 mm. in length.
Not such flies 13

13. Wing water clear, the veins yellow. Face scarcely at all produced beyond the antennal insertion. Third antennal segment rather elongate, narrowed basally, dilated apically. Marginal cell dilated or swollen below but ending truncately. Whole insect quite pale yellow or yellowish red except middle of mesonotum and sternum. Abdomen cylindroid but stout, not compressed, basally as wide as mesonotum. Pile very fine, dense, appressed and setate. Ocelli vestigial or reduced. A punctate depression in the middle of frons in front of ocelli.
                                Paracosmus (Actherosia, new subgenus)
Not such flies 14

14. Abdomen slender, compressed laterally, the male terminalia large, protuberant. Abdominal pile very fine, appressed, dense, setate, without any tufts of longer pile in the middle; postmargins of tergites with yellow or white borders widened medially. Frons concave and smooth and without any elevations either toward vertex or on each side of antennal basal base; surface silvery gray pollinose with a few appressed hairs, either fine or bristly. Wing hyaline; 2 submarginal cells; basal arch of second submarginal cell either rounded, or simulating a crossvein with backward spur . . . . **Paracosmus Osten Sacken**
Abdomen comparatively shorter, distinctly wider, rather robust in appearance and always with patches of longer, partly erect, partly curled, and appressed pile lying on each side of the tergites 15

15. Wing with 3 submarginal cells and both second and fourth veins very erratic, and with outward spurs. Wings spotted . . . . **Tomomyza Wiedemann**
Wing with 2 or 3 submarginal cells. Wings hyaline or tinged with brown or black 16

16. Two submarginal cells. First antennal segment twice as long as the second and the face below antenna very short, not extending to the end of the first antennal segment. Ocellarium long and unusually high and well developed. Small, often reddish, rather robust flies with the abdo-

men relatively short like *Tomomyza* Wiedemann, and with the posterior margins of the tergites often strongly raised leaving a corrugated aspect and sometimes with concave or foveate depressions on each side. Anterior branch of third vein usually simulating a crossvein and with a backward spur. Hind femur with appressed setate pile . . . . **Pantostomus** Bezzi
Three submarginal cells. First antennal segment quite short; face below antenna extending much beyond the first antennal segment. Ocellarium raised but rather low and gently raised. Shining black, rather robust flies with abdomen much like the *pictipennis* group of *Tomomyza* Wiedemann. Abdomen without posterior tergal margins strongly raised or lateral foveate depressions. Hind femur with erect or suberect stiff hairs, especially erect and longer below. Venation of second and anterior portion of third vein rather erratic with sharp bends and simulated crossveins . . . . **Tomomyza** Wiedemann

---

Genus *Cyllenia* Latreille

FIGURES 116, 117, 358, 616, 756, 771, 777

Cyllenia Latreille, Histoire nat. Crustacés Insectes, part 3, p. 429, 1802. Type of genus: *Cyllenia maculata* Latreille, 1802, monotypy.

Engel (1933), p. 11, key to 5 Palaearctic species.

Small flies, distinguished by the subglobular head, a little peaked in front beneath the antenna, the conspicuously swollen occiput and the presence of 3 submarginal cells in the wing; also, there are strong bristles beneath the hind femur and on the postmargins of the abdominal tergites. The abdomen is somewhat compressed, convex, subcylindrical, and slightly arched or decumbent.

There are other characteristic features. The anal cell is widely open and the third posterior cell arises abruptly and very convexly from the fourth posterior cell; the wing is slender basally, the alula wanting or much reduced; the wing is usually patterned with brown spots more or less sharply demarcated. The occi-

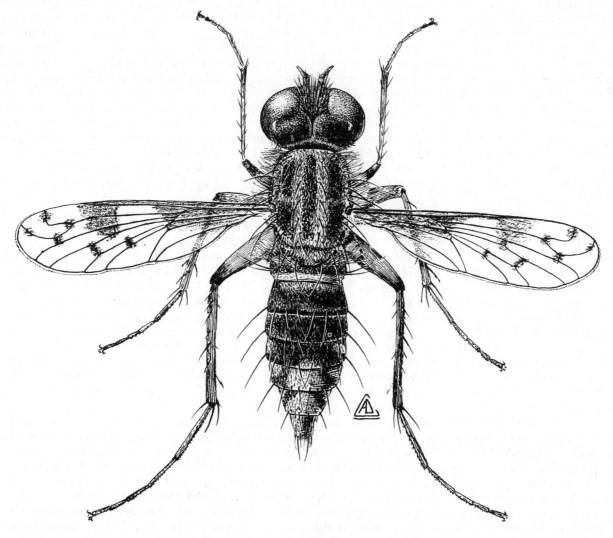

TEXT-FIGURE 42.—Habitus, *Cyllenia maculata* Latreille.

put is strongly bilobate above with a small but deep central cavity; viewed laterally it is conspicuous and wide on the upper two-thirds; posterior vertical margin of eye very slightly concave but without middle indentation or bisecting line. This is the genus upon which the subfamily Cylleniinae is founded. The proboscis is stout and of medium length, and the antennae remind one of the Lomatiinae, the laterally compressed and thinned and offset third segment has apical bristles.

Cyllenia Latreille comprises a small group of Palaearctic flies distributed through southern Europe, Turkestan, and Asia Minor.

Length: 3.5 to 8 mm., usually about 5 mm.

Head, lateral aspect: The head is subglobular in shape or even somewhat triangular, partly due to the angular peak of the face extending out nearly to the end of the first antennal segment and partly due to the prominent, extensive occiput, which is especially extensive from the dorsal aspect; the occiput is also bilobate and has a well-developed central cuplike recess or cavity. However, the posterior margin of the eye is very slightly concave and there is no medial indentation or bisecting line. The face is only moderately prominent and the peak is almost exactly opposite the midlevel of the head; it bears a conspicuous fringe of slender bristles down the sides of the oragenal cavity, but the face immediately beneath the antennae is bare. The oragenal recess is wide, narrowed above, and the gena is as wide as the basal antennal segment. The palpus is elongate and quite slender and clearly composed of 2 segments; it is nonhairy.

The proboscis is relatively short and stout with large fleshy labellum bearing a number of fine, stiff, erect hairs on its ventral surface. It projects a little beyond the oral cavity and slightly beyond the antenna, but it is not as long as the head. The antennae are attached a little above the middle of the head; they are comparatively short and the first segment is widened distally, bears a few short, fine bristles dorsally and is broadly attached to the second segment; it has rather more bristles ventrally, but they are not numerous nor long. Viewed dorsally the second antennal segment is obliquely subtruncate apically both medially and again laterally; it is to the lateral apex that the third segment is attached; viewed dorsally it has a dense patch of minute setae, some longer ones below. The third segment viewed laterally is stout, except at the immediate apex, which is turned downward, ends truncately, bears an apical spine and 4 or 5 distinct bristles; dorsally there is also a cluster of appressed bristles in the middle of the segment; viewed dorsally it is greatly thinned, becoming attenuate near the base. The third segment is not quite as long as the combined length of the first two segments.

Head, anterior aspect: The eyes in the male are separated at the vertex by a distance not quite twice the width of the ocellar tubercle; the front widens rapidly below. The ocellar tubercle is conspicuous, high, with vertical sides and lobiform. Between the small ocelli there is a tuft of long, coarse hairs. The upper eye has rounded corners viewed above and is separated from the ocellar tubercle. Viewed above the two sides of the occiput meet behind the ocellarium and leave a deep pocket. The pile of the occiput is dense, except near the eye margin, but short and fine, and directed backward. The eyes are bare. There is almost no development of the head beneath the eyes although the cheeks are moderately wide seen from below. The tentorial groove is deep and ends opposite the upper point of the oral recess. The antennae are attached close to the upper end of the oral recess. Both face and front are pollinose. The face has a preoral fringe of about 9 or 10 moderately long, stout, silvery bristles on each side.

The middle of the front has a large, triangular patch of long, rather dense, coarse hairs; the dorsal pile is black, the lower pile above the antenna is silvery.

Thorax: The thorax is relatively narrow and slightly convex with 5 irregular, indistinct rows of flattened, white scales; also there is a short, abbreviated row of 5 or 6 fine, acrostical, bristly hairs. There are some similar scattered hairs laterally, 2 long, but quite slender, bristles on the notopleuron, 2 above the wing, 2 on the postalar callosity, 3 pairs on the scutellar margin that are rather long, and some shorter hairs, 2 or 3 on each side of the scutellum. The basal half of the disc of the thick, posteriorly convex scutellum is rather densely covered with pale scales, and a line of scales divides a shallow, medial, posterior depression. The humerus has scales and some coarse hairs. The pleuron is relatively bare; there is a horizontal row of rather long, conspicuous, outwardly directed, whitish scales on the mesopleuron culminating in a posterior patch on this sclerite. Above the scales is a row of scattered, fine hairs and a small patch of scattered scales is on the upper part of the sternopleuron, otherwise the pleuron is bare.

Legs: The femora are moderately stout, slightly enlarged through the middle, rather densely covered with whitish scales; tibiae with more scattered similar scales. The hind femur has a conspicuous row of ventromedial, stout, long bristles that give way basally and ventrally to equally long but more bristly hairs. The hind tibiae has 3 rather long, anterodorsal bristles and 2 or 3 similar, posterodorsal bristles and has 1 anteroventral bristle and 2 or 3 setae; apex with 5 bristles, 3 or 4 of which are long and conspicuous. The anterior femur is narrowed distally, a little swollen basally, without bristles but with a ventral row throughout of dense, fine, suberect, very abundant setae; its tibia has 4 or 5 fine, slender, rather short dorsal bristles.

Wings: The wings are hyaline, in most species with rather light brownish spots about the crossveins, and sometimes with a band extending to the costa. There are characteristically 3 submarginal cells, the vein R 4 simulating a crossvein. In the middle the discal cell is strongly occluded from below by the strongly arched base of the third posterior cell; the intercalary vein simulates a crossvein. The anal cell is widely open, the axillary lobe slender, the alula quite narrow or absent.

The anterior crossvein enters the discal cell at the outer fourth; the marginal cell is strongly widened apically.

Abdomen: The abdomen is compressed laterally; only the first segment is as wide as the thorax; the apex of the abdomen is a little widened, and it is convex and slightly recumbent. The posterior margins of the segments on each side have conspicuous, quite long, erect, black bristles, 3 or 4 on each side. Lower down on the sides of the abdomen there is much long hair varying from coarse to fine, white in color and the bristles black. The whole abdomen is covered rather densely with scales, pale in color, or of two colors, mostly white below, mostly light brown above. The lower scales are large, the upper scales narrow and more like flattened hairs. Terminalia with a pair of convex plates closely adjacent, lower corners of the last sternite curved upward and backward as a flap or lobe. Beak of terminalia extended downward.

Material available for study: I am deeply indebted to Dr. R. H. Painter for lending material in this genus.

Immature stages: Unknown.

Ecology and behavior of adults: Not on record.

Distribution: Neotropical: *Cyllenia unicolor* Jaennicke, 1867.

Palaearctic: *Cyllenia globiceps* Loew, 1870; *laevis* Bigot, 1888; *maculata* Latreille, 1802 [=*cyllenia* (Lamarck), 1816 (as *Ploas*)]; *marginata* Loew, 1846; *obsoleta* Loew, 1846; *rustica* (Rossi), 1790 (as *Asilus*) [=?*maculata* Latreille, 1802]; *turgestanica* Paramonov, 1929.

Genus *Sphenoidoptera* Williston

FIGURES 235, 364, 601, 610, 1024, 1025, 1026

Sphenoidoptera Williston, Biologia Centrali-Americana, Dipt., vol. 1, supplement, p. 295, 1901. Type of genus: *Sphenoidoptera varipennis* Williston, by monotypy.

Small, robust flies with a slightly decumbent head and abdomen, and very large but rather short and wide oral opening which is oblique, leaving the face produced for a short distance beneath the antennae. The proboscis is short, and stout with a large, oval, flattened labella; it is distinctly longer than the antenna and extends beyond it. The head is long with the occiput especially prominent dorsally and becoming quite short along the lower eye margins. Thorax and abdomen and legs with scales. Anterior and posterior tibiae with spines.

The whole sides and middle of the face, both of which are narrow, are bare and the upper occiput is strikingly prominent and produced. The slender wing with narrow axillary lobe, open anal cell and absent alula is suggestive of *Cyllenia marginata* Loew, but in that genus, although the dark pattern in *marginata* Loew is suggestively similar, the shape of the third posterior cell is quite different and there are 3 submarginal cells, moreover, its face bears a vertical row of distinct bristles.

This genus is known from Mexico.

Length: 7 to 8 mm.

Head, lateral aspect: The face is short, polished, bare, and produced obliquely forward and downward from the antennae for a distince not quite as long as the first antennal segment. At the epistomal margin it is obliquely truncate leaving a large wide opening, at most 1½ times as long as wide, with thin, narrow walls that extend a short distance from the eye margin on the lower portion of the head but not as great as the length of the face. The cheeks form a short triangle posteriorly. Eye prominent without lateral indentation and from the dorsal aspect strongly extended beyond the occiput but only over the lateral and ventral portion of the head. Occiput especially thick and tumid dorsally. The high ocellarium at the vertex lies in a deep fissure, and also between the posterior eye corners, the occiput recedes strongly below. The pile is short, fine and scanty with a few longer, bristly hairs dorsally. The proboscis is slightly widened distally, extended obliquely downward, the terminal lamella large, long, oval, strongly compressed with lateral bristly pile. The palpus is long, quite slender, extending nearly half the length of the proboscis. The antennae are attached at the middle of the head, or moderately elongate and slender; the first segment is nearly 3 times as long as the second, the second segment is short and beadlike; third segment approximately as long as the first and bearing a short, small microsegment beyond which is a second conical microsegment with short, stylate apex.

Head, anterior aspect: The front is wide, only slightly narrowed at the vertex with shallow medial groove, some pollen, and laterally it has numerous, rather long, distinctly bristly, blackish hairs; vertex with 3 or 4 pairs of similar long slender bristles or bristly hairs all lying on the ocellar tubercle behind the ocelli. Slopes of the ocellarium vertical.

Thorax: With scattered, rather short, black, bristly hair, or slender bristles, and similar pile upon the humerus, the lateral margins have rather long, comparatively slender bristles; 2 notopleurals, 2 to 3 supraalars, 2 on postalar callosity. There are 2 pairs on scutellar margin with 2 shorter pairs laterally. The thorax is quite convex and thick with a deep crease before scutellum and without pollen or pile. The propleuron has a vertical fringe of scanty, long, stiff pile; the upper half of the mesopleuron bears numerous, rather long, fine bristles or bristly hairs, together with below these bristles a fringe of short, narrow, rather dense, flattened, whitish pile. The sternopleuron, pteropleuron, hypopleuron and metapleuron and the metanotum are all without pile. Ventral metasternum narrow, chitinized, with long, stiff pile. The postmetacoxal area is membranous; squamae without a fringe of pile.

Legs: The hind femur is distinctly thickened chiefly through the middle and beyond and a little compressed laterally. The anterior femur is slightly thickened toward the base. The femora and tibiae, the anterior and posterior coxae all have scales. Hind femur with a rather stout bristle ventrolaterally at the apical fifth; its tibia has 9 dorsomedial, 5 long, stout, and 2 short dorsolateral, 8 mostly shorter ventrolateral and 10 quite short, fine, ventromedial bristles. The anterior femur has a subapical anteroventral bristle; its tibia has 10 posterodorsal, 10 posterior and 9 minute, posterior ventral bristles. Terminal tibial spines strong. Claws short, sharp, curved from the middle and the pulvilli well developed, and with only an exceptionally minute medial lappet.

Wings: Slender, hyaline, with a sharply demarcated blackish pattern of spots and the foreborder blackish to the end of the first vein. There are 2 submarginal cells. The four posterior cells are open, the first posterior cell of uniform width and scarcely wider than the discal cell. The anterior crossvein is situated at the outer third of the discal cell. The anal cell is widely open, the axillary lobe narrow, and while the ambient vein is complete the alula is completely absent. There is a rounded blackish spot at apex of second vein, at base and apex of anterior branch of third vein, on each side of intercalary vein, a band across middle of discal cell widening toward the costa and connected with basal dark border which ends with it; basal to it a clear hyaline area extends to and ends sharply at the second vein.

Abdomen: Broad and robust and longer than wide; it is moderately convex with nearly parallel sides to the end of the fourth segment and with numerous, broad, white scales especially over the middle and posterior margin of the first segment, the basal portions of the remaining segments and with almost as many black scales intermixed. Other pile is very scanty, but the posterior margins of all of the tergites especially laterally have a fringe of quite long, conspicuous, slender, black bristles, which are continued onto the posterior margin of the sternites. Seven tergites present, the seventh is as long or longer than the sixth and strongly compressed laterally, so that its walls are pinched together.

Material available for study: A female of the type-species of the Biologia Centrali-Americana material in the British Museum (Natural History). Also, a male given me by my friend, Dr. Rex Painter.

Immature stages: Unknown.

Ecology and behavior of adults: Not on record.

Distribution: Neotropical: *Sphenoidoptera varipennis* Williston, 1901.

Genus *Amictus* Wiedemann

FIGURES 109, 357, 611, 641, 741, 752, 779, 890, 891, 892

Amictus Wiedemann, Zoologisches Magazin, vol. 1, pt. 1, p. 58, 1817. Type of genus: *Bombylius oblongus* Fabricius, 1805, by monotypy.

Thlipsomyza Wiedemann, Nova Dipterorum Genera, p. 12, 1820a. Type of genus: *Bombylius compressus* Fabricius, 1805, by monotypy.

Thlypsogaster Rondani, Arch. Zool. Anat. Fisiol., vol. 3, p. 72, 1863. Type of genus: *Thlypsomyza castanea* Macquart, 1840. Rondani had 2 species. Bezzi, p. 4, 1924, gives *T. castanea* as type-species.

Truquia Rondani, Arch. Zool. Anat. Fisiol., vol. 3, (sep.), p. 72, 1863. Type of genus: *Truquia insularis* Rondani, 1863, by original designation.

Thlipsogaster Coquillett, Trans. American Ent. Soc. Philadelphia,, vol. 21, p. 108, 1894a. Error of identification and lapsus.

Engel (1933), pp. 16-18, key to 21 species.
Austen (1937), p. 78, key to 5 species from Palestine.
Efflatoun (1945), p. 158, key to 5 species from Egypt.

Flies of small or medium size, usually pale pollinose, and readily distinguished by the strong macrochaetae upon thorax, scutellum, and abdomen. The abdomen is narrow, somewhat tapered and either cylindroid or laterally a little compressed. The mesonotum is often vittate. Face nearly bare or with a few, stiff, scattered hairs. The face is rather extensive and extended obliquely forward, the oragenal or buccal cavity sloping obliquely backward, the orafacial angle rectangular. The proboscis is quite long and slender and the antennae are distinctive; the first segment is lengthened, the second beadlike, the third segment moderately elongate with the outer half strongly attenuate, bearing short, stout oblique bristles or setae, its apex with small short, distinct microsegment and bristle. The first posterior cell is open or closed. *Sinaia* Becker is a monotypic genus related to *Amictus* Wiedemann, found in the same region and distinguished by the extraordinarily prominent upper occiput, particularly in the dorsal aspect.

This genus is restricted to the southern Palaearctic region, ranging along North Africa from Morocco to Egypt, the southern border of Europe, numerous in Turkestan and Asia Minor; of 5 species found by Austen in Palestine, 4 were undescribed.

Length: 6 to 11 mm., wings shorter.

Head, lateral aspect: Head hemicircular but with the face rather strongly produced triangularly forward. The occiput is not very extensive but is a little longer dorsally, and, in fact, the occiput is rather flattened and shelves rapidly inward to the margin of the interior cup or recess which, while distinct, is only moderately deep. The fringe on the border of the occipital cup is long, erect, and dense on the upper half, the hairs apparently somewhat scaliform or flattened and with the pile decreasing in length laterally to the eye margin and becoming flat-appressed at the upper eye margin. The front is flat, slightly sunken beneath the eye margin, pollinose, only the lower half bearing some stiff bristly hairs directed forward. Middle of front on the lower half with a deep, medial groove. The face is conically peaked forward, shining and apt to be light yellow or brown with some pollen laterally and with a few bristly hairs of varying lengths never as long as the first antennal segment. In females the entire front may have scattered bristly hairs. The posterior margin of the eye on the upper two-thirds is not quite plane in some species, more nearly so in others. Upper eye facets of the male enlarged only narrowly near the eye margins.

The proboscis is long and slender, 2 to 3 times the length of the head. The labellum likewise slender. Palpus very slender, not extending beyond the oral recess. Antenna attached a little above the middle of the head, moderately long and slender. The first segment is a little wider apically, about 4 times as long as wide, and reaching as far as the apex of the face in most species. This segment bears only a few short appressed setae dorsally. The second segment is short and beadlike. The third segment is as long as the first two segments together, may have a narrow knob at the base, is never wider than the second segment, and is very gradually attenuated on the outer half. It may become quite narrow on the apical third and often bears some short, distinct, bristles medially. Shorter bristles may extend dorsally nearly to the base. Apically there are 2 very small beadlike microsegments with a bristle at the tip of the last one.

Head, anterior aspect: From the anterior view the head is almost circular; the eyes are distinctly separated in the males by a distance a little greater than the ocellar tubercle, and, perhaps, nearly equal to twice the width of the second antennal segment. The front at the antenna is 2 or 3 times as wide as the vertex, the whole front opaque and pollinose, the ocellar tubercle unusually high and prominent, elongate, extending back behind the ocelli, and with the eye on each side rising above the base of the tubercle. The ocelli are placed in a short isosceles triangle. The antennae are separated by a distance equal the basal width of the antenna. Viewed anteriorly the large, oblique, subventral oral recess occupies the lower third of the head. It forms a wide, broadly oval, elongate recess with deep vertical walls, even anteriorly, and it is recessed backward at the base below. The proboscis is usually thrust directly forward or a little downward. Gena quite narrow but widening a little at the bottom of the eye and the tentorial fissure practically closed. From above the lobes of the occiput are separated by about half the width of the ocellar tubercle. The interior recess has an opaque membranous pillowlike extension arising from the cervix interiorly.

Thorax: The thorax is considerably longer than wide, much narrower than the head, rather opaque brownish or blackish, but with very dense, opaque, flat-appressed, somewhat flattened tomentum or pile that forms regular patterns directed in definite directions, mostly posteriorly but near the scutellum radiating outward, and anteriorly broken up into a more loose arrangement, semiappressed and nondirectional. Many species show distinct vittate strips upon the thorax. Pile of the scutellum quite similar to that of the mesonotum, the ground color, however, apt to be paler, especially apically. The humerus and postalar callosity are apt to be light in color, the humerus with humerus erect, long bristly hairs. The entire lateral margin back to the wing often has curled shining bristles. The notopleuron with 2 or 3 quite long, stout bristles, 3 others above the wing, 3 or more upon the postalar callosity, 1 or 2 pairs in front of the scutellum, 3 pairs of especially long, curled, stout, pale bristles on the scutellar margin, weaker in some species, and with a dorsal central row of 6 or 7 bristles on the posterior part of the mesonotum. The mesonotum is rather strongly convex, the pronotum high, abrupt and vertical, upper mesopleuron with a band of loose upwardly directed bristles, and along its middle longitudinally, a band of matted, long, pale, scaliform pile. Whole pleuron generally pale and pale pollinose. The entire hypopleuron, metapleuron, metanotum, and pteropleuron, and all of the sternopleuron, except the upper border, without pile or bristles. The squama is almost bare. The plumula likewise, except for dense micropubescence. Halteres large, the stalk widened apically. The posterior spiracle is unusually large, oval, and widely open with a short pubescent fringe anteriorly, and there is a curious, flattened pit, posterodorsally next to the haltere.

Legs: Femora distinctly swollen on all three pairs, more stout than almost any bombyline except *Toxophora* Meigen. Anterior femur ventrally with a cluster of spines anteroventrally on the outer half as many as 8, and as many as 10 spinous bristles posteriorly. Middle femur with 10 spines posteroventrally. The hind femur with 4 or 5 short and 1 or 2 long spines anteroventrally, but with 4 or 5 quite long, quite stout spines or spinous bristles ventromedially and sometimes some shorter ones. Anterior tibia with numerous, short, oblique, spinous setae and short bristles anterodorsally, but with almost none elsewhere. Middle tibia somewhat similar, hind tibia, however, very long, stout, of uniform thickness with 4 groups of long, often white, stout bristles in pairs along the dorsal surface, and with an additional cluster at the apex. The tarsi are remarkably long and have a fringe of very stout, blunt spines along the ventral margin forming a comblike structure. These are on the anterior margin of the middle and hind tibia, and on the posterior margin the spines are somewhat longer and distinctly sharper. Austen (1937) called attention to the fact that several species of *Amictus* have one or more remarkably elongate bristles on the underside of the first segment of the front tarsus just before the tip. The claws in *Amictus* are unusually long, sharp, barely curved at the tip, and the long slender pulvilli extend almost to the apex of the claw.

Wings: Wing elongate and slender with wrinkled, iridescent surface. It may be totally hyaline or, as in species like *minor*, all of the crossveins may be diffusely margined with blackish brown. The furcation of the third vein and anterior intercalary vein is blackish brown as well. The first posterior cell may be closed with a long stalk as in *validus* Loew, closed to the margin as in *tigrinus* Austen, or this cell may be open in maximal width as in *minor* Austen. In *variegatus* Meigen the first posterior cell is open, and there are 3 submarginal cells. The anal cell is always widely open, the ambient vein quite stout and complete, the alula absent, the costa thickened at the base, but with minute setae and with a furry lappet in place of a hook.

Abdomen: Decumbent, distinctly compressed laterally so that the abdomen is nearly cylindrical, quite so on the second segment, and slightly more narrowed on the remaining segments. First segment as wide as the thorax with a conspicuous anterior brush of dense, pale, long, rigid, spikelike pile, which, however, is confined to the outer third of the segment, leaving none beneath the scutellum. The remainder of the first segment and all the remaining tergites are densely covered with a curious, flat-appressed mat of opaque or feebly shining, flattened pile or tomentum of different colors all so pale so that characteristic patterns are formed in different species. In addition, each tergite has on the posterior margin a single row of quite long, erect, conspicuous pale bristles, usually 5 on each side. The male terminalia are quite large and long, placed vertically and truncately at the end of the abdomen, the last segments increased in depth thereby, the dististyli exposed from above, symmetrical, directed backward, and the whole hypopygium laterally with a dense fringe of long, bristly hair.

The male terminalia from dorsal aspect, epandrium removed, show a pygidium that is wide basally, narrowing apically with a long slender ejaculatory process. In the lateral view, this process arises at the extreme base of the genitalia and curves downward and forward the full length of the genitalia and beyond. It is curiously suggestive of the condition found in *Paratoxophora* Engel, which also has this curious backwardly turned snakelike or tubelike ejaculatory process of extraordinary length, and the process also goes backward and turns downward and forward just as in *Amictus* Wiedemann. The two genera from their external morphology do not suggest a relationship, however, and there are significant differences in the genitalia as well. However, the basistylus of *Amictus* does not turn up proximally, and the epiphallus and dististyle are quite different, and the epandrium is also quite different. Numerous differences appear both from the dorsal and lateral aspect.

Material available for study: The species in this genus from both the National Museum of Natural History and the British Museum (Natural History). I also have material of *Amictus aegyptiacus* Paramonov received through the courtesy of the Department of Entomology of Cairo University in exchange for American bee flies.

Immature stages: Unknown.

Ecology and behavior of adults: Engel (1932) states that these flies are usually taken in April or May and again in autumn and are found in southern countries for the most part. They are taken from the flowers of steppe-loving plants.

Distribution: Neotropical: *Amictus cinerascens* Bigot, 1892.

Palaearctic: *Amictus aegyptiacus* Paramonov, 1931; *castaneus* (Macquart), 1840 (as *Thlipsomyza*); *compressus* (Fabricius), 1805 (as *Bombylius*); *firjuzanus* Paramonov, 1931; *funebris* Paramonov, 1931; *gebeli* Efflatoun, 1946; *heteropterus* (Macquart), 1840 (as *Thlipsomyza*); *insignis* Loew, 1870; *insularis* (Rondani), 1863 (as *Truquia*); *latifrons* Loew, 1873; *minor* Austen, 1937; *nobilis* Loew, 1870; *obliquenotatus* Austen, 1937; *oblongus* (Fabricius), 1805 (as *Bombylius*); *pictus* Loew, 1869; *pulchellus* Macquart, 1849 [=*striligatus* Loew, 1869]; *scutellaris* Loew, 1869; *setosus* Loew, 1869; *shafiki* Efflatoun, 1945; *similis* Paramonov, 1931; *tener* Becker, 1906; *tigrinus* Austen, 1937; *validus* Loew, 1869 [=*compressus* Wiedemann, 1828, without descr.]; *variegatus* (Meig. in Waltl's Reise), 1835 (as *Thlipsomyza*); *virgatus* Austen, 1937; *zinamominus* Becker, 1906.

Genus *Sinaia* Becker

FIGURES 212, 365, 604, 617, 638, 759, 973, 974, 975

Sinaia Becker, Ann. Mus. Nat. Hungarici, vol. 14, p. 65, 1916.
Type of genus: *Sinaia kneuckeri* Becker, 1916, by original designation.

Flies of medium size and of about the same size as *Amictus* Wiedemann to which it is closely related. It is readily separable by the much shorter face, which is gently rounded, the less attenuate antenna, shorter proboscis, and the characteristic shape of the male hypopygium, which I illustrate from Efflatoun. They are grayish black flies with a sheen of silvery pubescence or pollen not hiding the ground color. The proboscis is about as long as the head, and the bristles are even longer and stronger than in *Amictus* Wiedemann. Efflatoun (1945) notes that the statements of Engel with regard to *Sinaia* Becker are misleading because of the very bad preservation of the unique female type. He has been able to correct this because he had obtained a series of both males and females in perfect condition.

Sinaia Becker is known only from Sinai and from the Northern Sinai region; collected between April 13 and 24.

Length: 8 to 9.2 mm., breadth 10 to 12 mm.

Head: Head a little broader than the thorax; face feebly produced forward "beak-like"; proboscis as long as the length of the head; antennae as in *Amictus* Wiedemann, and the third segment bearing a minute cylindrical joint ending by a microscopic sensory rod.

Wings: Wings of the same shape and with the same venation as in *Amictus* Wiedemann and in the type of the genus, *kneuckeri* Becker, which is the only species known so far, cell R_5 is widely open and the r-m crossvein is placed well after the middle of 1st M_2 cell; 1st A cell open; alula obsolete.

Thorax: Thorax with weak but distinct rows of d.c. bristles; pre-alar, supra-alar and post-alar bristles well developed; scutellum with marginal bristles. Posterior femora with ventral rows of bristles and the tibiae with much fewer and shorter bristles except on the posterior tibiae where the short bristles are fairly numerous.

Abdomen: Abdomen with distinct marginal bristles particularly on the sides and toward the apex.

Material available for study: None. I have quoted Efflatoun's excellent and penetrating description, based upon a series of fresh individuals of both sexes; previously this genus has been only known from a single, much damaged female.

Immature stages: Unknown.

Ecology and behavior of adults: Not recorded.

Distribution: Sinai Desert (Ethiopian): *Sinaia aharonii* Engel, 1937; *kneuckeri* Becker, 1916.

Genus *Henica* Macquart

FIGURES 226, 406, 620, 973, 978, 1030

Enica Macquart, Hist. nat. Insectes. Dipt., Suite à Buffon, vol. 1, p. 399, 1834. Type of genus: *Anthrax longirostris* Wiedemann, 1819, p. 134, also 1828, p. 281, by monotypy.

Henica, emendation. See Kertész p. 69, 1909.

Lagochilus Loew, Öfvers, Kongl. Vet. Akad. Förhandl., vol. 17, p. 87, 1860. Type of genus: *Anthrax longirostris* Wiedemann, 1819, as *Cyllenia afra* Wiedemann, 1828, p. 358, by monotypy, not *Cyllenia afra* Macquart, 1840; see *Nomalonia*.

Alonipola Rondani, Arch. Zool. Anat. Fisiol., vol. 3, p. 71 (sep.), 1863. Type of genus: *Anthrax longirostris* Wiedemann, 1819, as *Cyllenia pluricellata* Macquart, 1855, by original designation.

We prefer *Henica* on basis of usage.

An odd fly, robust, of medium size, and brownish color with yellowish-white coloration on the head. The pile of the front, the sides of the thorax and scutellum and all tibiae and femora is decidedly bristly and its color black. The mesonotum and the whole abdomen, however, is covered with matted, appressed pile, slightly scaliform, of mixed colors, pale brown and brownish white or yellow.

The wings are quite distinctive; the costal cell is yellowish and the remainder of the wing brown, though paler on the posterior half; it has 8 distinct and conspicuous pale whitish fenestrate spots centered about the crossveins and vein furcations, and each of considerable size. There are 3 submarginal cells and the marginal cell is divided into 3 parts by 2 supernumerary crossveins. The second vein arises as a crossvein with backward spur. The wing is narrow basally, the alula narrow.

The occiput is bilobed and cupulate but much more shallow than *Peringueyimyia* Bezzi and the lobes are separated. Its nearest allied genus is *Nomalonia* Rondani, which has a similar occiput and similarly has what Hesse describes as a muscoid appearance; *Nomalonia* Rondani differs principally in lacking the extra crossveins in the marginal cell; also the front and especially the sides of the face are bare in contrast to *Henica* Macquart which has not only a very bristly front but also has a conspicuous extension of pile down the side of each antenna.

Henica Macquart is a characteristic element of South African Diptera fauna, of which only one species is known.

Length: 4.5 to 11 mm., wing, 4.5 to 9.5 mm.

The exceptional confusion that has centered around *Henica* Macquart is because Wiedemann described a species as *Anthrax longirostris* Wiedemann, another as *Cyllenia longirostris* Wiedemann, and a third species as *Cyllenia afra* Wiedemann. The first two species were distinct, but placed in the wrong genera. It was later shown that his third species, *Cyllenia afra* Wiedemann, is really a strict synonym of his *Anthrax longirostris* for which Macquart erected the genus *Henica*, but not before additional genera had been proposed for it. Macquart, as a result of misidentification, also redescribed *Anthrax longirostris* Wiedemann as *Cyllenia pluricellata* (1855). Rondani added to the confusion by creating a new genus *Alonipola* (1863) for it. Loew, realizing that Wiedemann's third species, *Cyllenia afra*, did not belong to this genus created the genus *Lagochilus* for it. *Lagochilus* Loew falls into synonymy under *Henica* Macquart.

Head, lateral aspect: The head is approximately hemicircular from lateral aspect, only a little produced in the middle at the attachment of the antennae. The occiput is only moderately produced but is slightly more extensive on the upper half, leaving the eye margins almost but not quite plane. The eye is circular in front but the entire eye forms less than half a circle. The pile of the occiput consists of a broad, wide band of scanty, fine, erect hairs on the wide posterior rim of the occiput. Pile is absent on the lateral margin of the occiput adjacent to the eye where only pollen is present, but on the inner margin of the occiput this scanty fringe is directed backward. The front is less convex than in *Nomalonia* Rondani, except along the eye margin it is covered with loosely scattered, rather long, stiff, black, slender bristles curved downward toward the antennae. Eye margins of the front with pollen.

The proboscis slender, about as long as in *Nomalonia* Rondani, generally held obliquely upward. The labellum may show a rugosity due to minute spinules in some lights. Palpus quite small and slender of no great length, with a few fine hairs near the apex, and composed of a single segment. The antennae are attached quite in the middle of the head; they are small and slender. The first segment is not quite as long as its apical width, a little narrowed at the base with a tuft of 6 or 8 setae ventrally, not extending to the end of the second segment; second segment cylindrical, as long as wide, with 3 or 4 black, dorsal setae at the apex. Third segment more slender, of about equal width to beyond the middle, beyond which it is reduced to about half the basal width. At the apex it ends in a small, oblique, medial, cup-shaped depression, bearing minute, enclosed spine.

Head, anterior aspect: The head as wide or wider than the thorax, the oragenal cup is less protrusive than in *Nomalonia* Rondani, but likewise reaches to the base of the antennae and is separated from the sides of the face by a deep, wide furrow, which on the lower fifth of the head merges into a long, quite deep, narrow, tentorial fissure. The front is large and triangular, the eyes of the male separated by a space equal to the space between the posterior ocelli. All 3 of the ocelli are quite large and set in a short isosceles triangle. The ocellar tubercle is convex and raised and bears a tuft of numerous slender black bristles, those behind the ocelli erect, those in front curved forward. Front of the female much wider. The antennae are narrowly separated by a space equal to the width of the second antennal segment. The oragenal cup is quite deep below and although quite elongate is more shallow above with the inner wall sloping, both sides of the upper furrow covered with moderately dense, long, coarse pile curved downward. Pile is absent on the oral side of the tentorial fissure but abundantly present on the side opposite next to the eye. The genae, while extending downward below the eye, are not as extensive as in *Nomalonia* Rondani. From a vertical aspect, the two halves of the occiput do not meet above but leave a rather wide, deep sulcus or fissure behind the vertex.

Thorax: The thorax is moderately convex, more so than in *Nomalonia* Rondani, and the mesonotum is densely covered with appressed, coarse, narrowly scaliform pile, partly arranged in vittae of different colors. Very little of the black ground color shining through. Mesonotum anterolaterally and in front of the scutellum with some scattered, moderately long, fine, erect, bristly hairs. Sides of mesonotum, the scutellum, prescutellar area, and the pleuron pale, brownish-clay color, on the mesonotum and scutellum partly translucent. Humerus with pile and bristly hairs similar to that of the mesonotum. Notopleuron with a tuft of 7 long, sharp, curved, basally stout, black bristles. Four similar more slender bristles in a row above the wing; 3 or 4 longer ones on the postalar callosity and the scutellum with 5 pairs on the margin, besides a few other shorter, fine hairs. Greater basal part of the scutellum with appressed, pale, scaliform pile like that of the mesonotum.

Upper border of the mesopleuron with a conspicuous border of upwardly directed, rather numerous, black, bristly hairs. Metapleuron pollinose only. Metanotum likewise. Upper posterior hypopleuron behind the spiracle with a patch of appressed scales. Lower half of the pteropleuron with a large, dense patch of appressed scales directed forward. Entire ventral half of the mesopleuron densely covered with long, pale, flattened scales. Some similar scales on the sternopleuron. Squamae somewhat vesicular and with a fine fringe of marginal hairs. Haltere with the stalk dilated and flattened apically and bearing scaliform pile.

Legs: The legs are unusually stout, especially the femora. The anterior 4 femora are somewhat swollen toward the base. Hind femur stout. Anterior femur with a posteroventral row of 10 or 12 short bristles and 7 or 8 irregularly placed along the anterior surface. All coxae with 4 to 10 long, black, downwardly directed bristles. The whole surface otherwise covered with appressed, pale-colored scales. Middle femur similar to the anterior pair. Hind femur with a row of 8 or more long, slender, downwardly directed black bristles on both anteroventral and ventromedial surfaces, and with a cluster of bristles dorsoapically in several rows. Anterior tibia with 9 prominent, black bristles posterodorsally and 5 or 6 shorter ones anterodorsally. Middle tibiae with 8 long, anterodorsal bristles and 12 or more fine, short posterior bristles, besides 5 long, stout, posteroventral bristles, and 8 anteroventral bristles. Hind femur with still longer bristles in 4 rows of 6 or 7 bristles each. All tibiae with 2 or 3 very long, stout, apical ventral bristles. Last 3 segments of tarsi in male with curious pads ventrally. Pulvilli minute. Claws short and rather stout.

Wings: The wings narrowed toward the base, largely brown villose except for large fenestrate, subquadrate spots with whitish villi on each side of all crossveins, 8 spots in each wing. Costal cell paler. There are 3 marginal cells and 3 submarginal cells. The second vein arises rectangularly simulating a crossvein or sometimes a double crossvein with backward spur. The rectangular anterior crossvein lies close to the outer fifth of the discal cell. The intercalary vein is long and straight. Anal cell very widely open. Axillary lobe narrow and the alulae quite narrow, ambient vein complete.

Abdomen: The abdomen is broad, wider than the thorax and only gently convex. Whole abdomen rather densely covered with narrow, appressed, mostly pale, scaliform pile with a little darker pile of a fine, bristly character forming borders along the posterior margin and the pale scales often of different colors. Ground color black.

Dr. A. J. Hesse in his work on South African bee flies describes the male hypopygium as follows:

Hypopygium of male with the basal parts elongate, not separated into two parts by a dorsal suture, the line of suture being indicated only by a slight longitudinal depression towards apical part, with only very short and scattered fine hairs above and very fine pruinescence, but no long, bristly hairs; beaked apical joints, with the apical part more or less flattened and bifid, with a transverse, raised, keel-like ridge above, behind which the dorsum is covered with dense bristly hairs; aedeagus with the slender apical part curved upwards, then broad and tubular for some distance and then appearing very broad and really constituting part of the ventral inner part of basal part, the actual tube of the aedeagus being visible along its center; middle part of aedeagal complex prominent, inflated, helmet-like and, together with basal part of aedeagus, connected on each side by means of ramus to basal part and also to last sternite, with the ramus on each side continued on inner side of basally directed projection of basal part as a narrow, strap-like process; lateral struts shortish.

Material available for study: Through the courtesy of Dr. A. J. Hesse I have been able to study a pair of the type-species. The British Museum (Natural History) also gave me a specimen of that species.

Immature stages: Unknown.

Ecology and behavior of adults: I have no record of habits and behavior or flower preferences of the adults.

Distribution: Ethiopian: *Henica longirostris* (Wiedemann), 1819, 1828 (as *Anthrax*) [=*afra* (Wiedemann), 1828 (as *Cyllenia*), =*pluricellata* (Macquart), 1855 (as

Cyllenia), also=*pluricellata* (Macquart in Rondani), 1863 (as *Alonipola*), also =*afer* (Wiedemann in Loew), 1860 (as *Lagochilus*)].

Genus *Nomalonia* Rondani

FIGURES 224, 225, 363, 609, 612, 696, 728, 994, 995, 996

Nomalonia Rondani, Arch. Zool. Anat. Fisiol., vol. 3, p. 71, 1863. Type of genus: *Cyllenia afra* Macquart, 1840, not Wiedemann, by original designation.

[*Nomalonia* Rondani, like *Henica* Macquart, has suffered from a confusion due to two species with great superficial similarity—which in early collections were rarely seen—and were poorly understood. Macquart (1840) described a species as *Cyllenia afra* Wiedemann, which proved to be not that species, but an undescribed species which Rondani later made the type of a new genus, *Nomalonia afra* Macquart.]

Hesse (1956), pp. 34-36, key to 6 Ethiopian species.

Brownish flies with appressed pile on abdomen and mesonotum. Closely related to *Henica* Macquart in general form, coloration, and prominence of the black bristles on the sides of thorax, on the scutellum, and the legs. It is distinguished primarily by the absence of the additional crossveins in the marginal cell and the wing is less conspicuously infuscated and fenestrate. The sides of the face opposite each antenna and the face itself below the antenna are quite bare, rather polished and shining in contrast to the dense subappressed pile that is present in *Henica* Macquart.

Nomalonia is a South African genus with at least 6 species.

Length: 6.5 to 12 mm.; of wing 6 to 10 mm.

Head, lateral aspect: The head is more or less triangular in profile, more especially in the male. In both male and female the front is convex, swollen, and somewhat protuberant, but in the male the margin of the oragenal cup appears to be more plane. The occiput is only moderately developed, far less extensive than in *Peringueyimyia* Bigot. It is almost equally long above and below and the posterior margin of the eye is almost but not quite plane. From near the top to shortly above the lower corner there is an almost imperceptible curve or excavation. Pile of the occiput very short and scanty, almost nonexistent close to the eye but becoming longer and forming a distinct fringe backward, of no great length along the inner margin of the occipital cup. The halves of the occiput do not meet above behind the vertex and leave a rather wide gap between the occiput almost equal to the space between the posterior ocelli. The front in the male is pale waxy yellow as is the entire head, except the inner rim of the occiput. It is rather strongly swollen and a little convex from the lateral aspect and bears scattered, bristly black hairs down the middle, and in addition upon a triangular area above the antennae there is abundant, long, subappressed, white scales. These scales extend more scantily to the ocelli. The ocellar tubercle is swollen and conspicuous. The ocelli are large, especially the posterior ones. In the female there are rather more black bristles than yellowish scales on the front.

The proboscis is elongate and quite slender, about 1½ times as long as the head but the part that extends beyond the buccal cavity is no longer than the head; it tends to be held obliquely upward and outward. The palpus is long and quite slender and composed of one segment only; it is much more slender than in *Peringueyimyia* Bigot. It is deeply recessed, composed of a single segment and colored so like the background of the cavity that it is not easily perceived. The antennae are relatively small but slender, except the first segment, which is a little swollen, very short, distinctly beadlike and pale yellow. The second segment is no longer, but is cylindrical, scarcely more than half as wide as the first segment, and the third segment is a little more narrow at the base and attenuate on the outer two-thirds, with a short, stublike, apical microsegment. Whole apex of the third antennal segment colored black. The first antennal segment has only a few minute setae below, the second segment has a few setae above at the apex, and the third segment has 3 or 4 setae dorsally. The third antennal segment is twice as long as the first two segments combined.

Head, anterior aspect: Viewed from in front the head is narrowly oval with the genae swollen and prominently extended down beneath the eyes. The head is as wide as the thorax. The eyes in the male are separated by a space equal the width of the first antennal segment. There is a distinct sulcus separating the face from the front and continuing laterally in the direction of the eye and ending in a deep tentorial pit, and then continued obliquely downward separating the genae from the sides of the oragenal cup. In the female the front is quite wide both above as well as below. The ocelli occupy a triangle not quite equilateral and are separated from the eye on each side by a distance equal to the width of this triangle. The antennae separated by a space equal to the length of the second antennal segment. The oral opening is large, wide, elongate, and quite deep. It extends to within a short distance of the antennae forming a somewhat nose-shaped or peaked extension beneath the antennae, so that the midfacial area below the antennae is quite short. Sides of the oragenal cup and face with pollen only and shining waxy yellow. It is therefore in great contrast with *Peringueyimyia* Bigot. There tends to be a brownish spot on each side near the bottom of the eye.

Thorax: The mesonotum is rather short and moderately convex, especially on the anterior half. The pile is composed of comparatively dense, appressed, brownish-yellow scales, usually more or less denuded over the middle of the mesonotum, and more abundant on the anterior lateral margins and in front of the scutellum and on the base and sides of the latter. Middle of the mesonotum broadly polished black bearing pollinose vittae when viewed from the front. There is a scanty, anterior fringe of pile around the borders of the extremely short prothorax. The humerus is prominent, and somewhat swollen; it bears pale scales and a few black bristles. Notopleuron with a double row of strong, black bristles, 4 above and 3 much stouter and longer below. Above the wing there are 3 slender black bristles, and there are 5 long, stout, curved black bristles in a single row on the postalar callosity. Postmargin of the scutellum with 4 curved black bristles, not quite as long as those of the postalar callosity. The humerus, lateral margins of the mesonotum, prescutellar area and the whole of the scutellum are light yellowish brown. This yellow area in front of the scutellum is bisected with black. Metapleuron and sides of the metanotum without pile; hypopleuron and pteropleuron likewise bare. The mesopleuron on the upper border has a wide longitudinal row of stout, black, upturned bristles. The lower part of mesopleuron has a few scattered, pale, brownish yellow, lanceolate scales similar to those of the mesonotum. There is a small patch of similar scales in the middle of the sternopleuron.

Legs: All coxae have a vertical row of stout, black bristles, the legs are stout but comparatively short, the anterior femur is a little swollen toward the base and bears a few black bristles below in the middle and posteriorly. Middle femur with 3 or 4 black bristles ventrally, and with appressed, linear, black scales above and shorter, broader, pale scales below. Hind femur stout and rather thick throughout with 6 stout, black, anteroventral bristles; there are 3 others at the apex laterally and 2 above, and with the scales abundant and almost entirely pale. The anterior tibia has irregular, dorsal rows of stout, spiculate, black bristles, 8 or 10 in each row, and 7 or 8 posteriorly. The apex of this tibia has 4 long, stout, sharp, black bristles. The 2 ventral bristles are more prominent. Middle tibiae with similar apex, the hind tibia with a circlet of 8 or 9 stout bristles, and with 6 long, anterodorsal bristles, 4 or 5 shorter anteroventral bristles and 9 long, stout, posterior bristles. Tarsi rather short, the pulvilli greatly reduced, the claws almost straight, only slightly curved toward the apex.

Wings: The wings are rather conspicuously brown villose and subhyaline with whitish, opaque, conspicuous spots on the crossveins, apex of the second basal cell and base of third vein. There are 3 submarginal cells and the base of the anterior branch of the third vein arises steeply and plane and obliquely, simulating a crossvein. The second vein arises abruptly and rectangularly and often has a distinct spur vein directed backward. The significance of this spur is uncertain. The first posterior cell is widely open, the rectangular, anterior crossvein is near the outer fifth of the discal cell, the lower marginal vein of the discal cell is straight except at the base where it is curved. Anal cell very widly opened, ambient vein complete, alula rather narrow. Base of costa a little swollen, the scales and fine setae and a triangular lappet present instead of a costal hook. The first vein ends no great distance from the end of the second vein. Marginal cell narrow at the apex.

Abdomen: The first and second segments of the abdomen in the male are as wide as the mesonotum and the first four segments in the female are as wide or even wider than the thorax. The abdomen is very slightly convex, moderately narrowed and compressed laterally in the male. The color tends to be more or less shining black, on the basal parts of the tergites becoming diffusely reddish brown along the posterior margin and sometimes over most of the posterior tergites. The pile consists of flat-appressed, black setae, rather densely mixed with appressed, reddish brown, short, wide scales that become brownish white along the posterior margins especially toward the sides of the tergites. On the posterior tergites toward the sides of the setae give way to longer, slender, flat-appressed, bristly hairs. Male terminalia elongate, large, and prominently protruded. Hesse (1956) points out that the basal portion of the hypopygium, like that of *Henica* Macquart, is fused dorsally and ending basally on each side in a lobelike prolongation. The dististyli are excavated or hollowed on the sides below and the inner excavation is often produced apically into a toothlike lobe.

Because of the large number of species that Dr. A. J. Hesse was able to study and summarize, I quote his work on the hypopygium of the males:

Hypopygium of the male is also very similar to that of *Henica* in some respects; basal parts also fused dorsally, elongate, ending basally on each side in a lobe-like prolongation; beaked apical joints excavated or hollowed on the sides towards the lower aspect, the inner excavation often produced apically into a tooth-like or prong-like lobe which gives the joints a bifid appearance from above, dorsally these joints are covered with shortish, sometimes spine-like, bristles; apical joints usually also end in a sharp point, but they may be blunt and, when viewed from above, their shapes are variable; aedeagus tubular as in *Henica* and also constitutes part of ventral aspect of basal parts, being fused ventrally with the basal part; middle part of aedeagal complex helmet-shaped and intimately connected with the last sternite by membranous or chitinous connection; lateral struts usually broadish, ladle-shaped, ovoid or tongue-shaped; basal strut racket-shaped or chopper-shaped.

Material available for study: I have a pair of the type-species most kindy sent by Dr. Hesse for study and dissection.

Immature stagse: Unknown.

Ecology and behavior of adults: I have seen no published record of the behavior of the adults, or its preference for flowers.

Distribution: Ethiopian: *Nomalonia afra* Macquart, 1840 [=*Cyllenia afra* Wiedemann, in Marquart, misid.]; *clavicornis*, Hesse, 1956; *henicoides* Hesse, 1956; *imitata* Hesse, 1956; *sporanthera* Hesse, 1956; *syrticola* Hesse, 1956.

Genus *Tomomyza* Wiedemann

FIGURES 113, 115, 362, 603, 613, 1021, 1022, 1023

Tomomyza Wiedemann, Nova Dipterorum Genera, p. 9, 1820a.
  Type of genus: *Tomomyza anthracoides* Wiedemann, 1820, by monotypy.

Hesse (1956), pp. 74-77, key to 11 Ethiopian species.

Small flies of relatively bare aspect due to dense, minute, appressed pile. They are only moderately elongate; the abdomen is cylindrical and long ovoid, rather than elongate, in the same extent as in *Paracosmus* Osten Sacken. *Tomomyza* Wiedemann is more like *Pantostomus* Bezzi, from which it differs in the generally mottled wings and the erratic venation in the submarginal part of the radial field. The hind femur tends to be somewhat thickened distally and lacks bristles or spines; the face and front are long conical or triangular and the proboscis does not extend beyond the oral recess; lobes of occiput tightly apposed.

Tomomyza is a small genus of South African flies.

Length: 4 to 11 mm., usually about 5 to 8 mm.; wing, 3.5 to 8 mm.

Head, lateral aspect: The head is subglobular though with strongly peaked and forward production of face and front at the middle of the head. The occipital component is unusually prominent, especially above, but also at the bottom of the head; it is least extensive near the lower sixth. The posterior occiput laterally is rather steeply sloping, a little convex, so that the interior cup or recess is quite large, circular and deep. The pile of the occiput is dense, minute, and appressed, extending all the way to the eye margin and the inner margin of the occiput has a fine, erect, sparse fringe. The front and face are quite prominent as a result of their peaked forward extension. The oragenal cup, which reaches quite to the antennae is even more extensive, extending beyond the first antennal segment. The front is plane across the eye on the upper half, well developed below and bears considerable, fine, moderately long, subappressed pile running from the whole of the vertex widely down the middle of the front to the base of the antennae. The whole of the front and all the vertex is densely micropubescent, except the low ocellar tubercle, which is nodular behind and, except for a medial stripe on the upper half of the front and vertex, and except a small spot along the eyes in the middle of the front, the sides of the frontofacial area opposite the antennae are flat and covered only with dense, silvery micropubescence. Similar pubescence continued on down the sides of the face to a point just below the middle of the eyes. The oragenal cup is largely bare, and shining, separated from the face by a very shallow, oblique depression. The eye is large, occupying the greater part of the head; the posterior margin is nearly plane. The upper anterior facets are somewhat enlarged.

The proboscis is short and rather stout, laterally compressed, the labellum is elongate but narrow, curved upward and attenuate to the apex. The palpus is quite long and slender, curved, with short, apical, clublike segment. The long, curved basal portion has a few long, scattered, ventral hairs. The antennae are attached at the middle of the head; the first segment is stout but not swollen, widened apically, longest medially, and this segment is less than twice as long as its basal width. The second segment is short and beadlike, rather longer medially, so that the flattened pyriform third segment which is shaped much as in *Pantostomus* Bezzi is held strongly outward. The first two segments bear only dense, minute, appressed pubescence. The third segment is pollinose and bears a minute, apical spine.

Head, anterior aspect: The head is circular and distinctly wider than the thorax. The eyes in the male are separated in both sexes, narrowly in the male, more widely in the female, the extent variable, but generally about 3 times the width of the ocellar tubercle, but in some instances twice the width of the ocellar tubercle. The antennae are widely separated by a strong, triangular production of the front similar to *Pantostomus* Bezzi, yet the first two segments lie adjacent at the apex. The ocellar tubercle is short isosceles and situated a short distance forward from the posterior eye corners. The oragenal cup is deep and long and narrow, extending quite to the base of the antennae from the bottom of the eye. Its inner wall is vertical, the middle portion of the outer wall is obliquely sloping but steep. Genae reduced to a mere line of pubescence, and similarly reduced at the bottom of the head. Tentorial fissure absent. From vertical aspect the occiput is bilobed and has a dorsal fissure. It has a conspicuous sunken depression where the occipital lobes meet the posterior vertex; they meet at a point still some distance behind the eye margin. Sides of occipital lobes touching.

Thorax: the thorax is slightly convex, the mesonotum generally opaque blackish with paler borders laterally. The humerus and scutellum are more or less pale. Pile of mesonotum dense, fine, curled, and flat-appressed giving way to slightly longer, coarse hairs directed outward along the lower margin of the notopleuron and tufts of erect, slightly flattened hairs on the humerus. The pile of the scutellum is similar to the mesonotum; the curled hairs of the margin are a little longer. Metapleuron pilose, hypopleuron with micropubescence only, pteropleuron without pile. The pile of the mesopleuron on the upper part consists of only appressed micropubescence.

Legs: The legs are relatively short, the anterior 4 femora are slender, the posterior pair distinctly swollen a little toward the apex. The pile of the femora and tibiae is minute, rather dense, and quite flat-appressed, pale and glittering at least in part on the tibiae. Hind tibia and femur sometimes banded, due to darker pile apically. Pulvilli slender and linear, as long as the claw. Claws very fine, sharp, and slightly curved at the apex.

Wings: The wings have erratic venation with numerous dark, brownish spots on the crossvein, the whole of the first basal cell, except near the apex, the whole of the costal cell and basal half of the marginal cell. There are 3 submarginal cells. The first submarginal cell is indented on each side by a rectangular apex to the marginal cell and by a rectangular bend in the last section of the anterior branch of the third vein. This section of the third vein also arises as a crossvein placed rectangularly and connects with the second vein by a rectangularly bent vein, which often emits a spur basalward. Other bends in veins above described often emit spurs. Hesse (1956) figures one species with a venation very little different from that of *Paracosmus* Osten Sacken. The numerous crossveins and right angle bends are absent. First posterior cell open but narrowed, lower vein of the discal cell with one or more spurs at bends and the anterior crossvein lies well beyond the middle of the discal cell. Anal cell open, ambient vein complete, alula absent.

Abdomen: The abdomen is moderately elongate, the tergites rather strongly convex from side to side. Seven tergites are present; the pile is short, dense, fine, not flattened and quite flat-appressed. On the sides of the first segment the pile is slightly longer and directed outward and appressed. Female terminalia concealed within compressed seventh tergite and border of short pubescence. Hesse (1956) points out that the last sternite in the male is dorsal in position, the basal part composed of a single, undivided part generally covered with dense hairs, ending on each side basally with a lobelike process. The dististyli form a more or less curved clawlike structure. Aedeagus as in *Pantostomus* Bezzi but apex sometimes much recurved.

Dr. A. J. Hesse in his work on South African bee flies describes the male hypopygium as follows:

Hypopygium of male also situated on the ventral aspect of abdomen as in *Pantostomus*, the last sternite of which is dorsal in position; basal part of hypopygium also composed of a single and undivided part, usually covered with fairly dense hair, ending basally on each side in a sort of process or lobe and in two known species at least also with an outwardly projecting process at base of this lobe, with the outer apical angle of basal part usually slightly produced and rounded; beaked apical joints usually curved claw-shaped; aedeagus shaped much as in *Pantostomus*, but sometimes with the apex much recurved.

Material available for study: Four species available through the generosity of Dr. A. J. Hesse of the South African Museum, Capetown.

Immature stages: Unknown.

Ecology and behavior of adults: Hesse (1956) records species of *Tomomyza* Wiedemann captured on flowers of *Carpobrotus edulis* and *Carpobrotus acinaciformis* and also resting on the sand between such plants. Another species of this genus was found resting or flying among branches and dried twigs of *Mesembryanthemum*. Still a third species was found resting on bare stones or ground or sand between bushes. A fourth species was caught on the sandy banks of a dry river course and also of the flowers of *Mesembryanthemum*.

Distribution: Ethiopian: *Tomomyza anomala* Hesse, 1956; *anthracoides* Wiedemann, 1820; *barbatula* Bezzi, 1922; *guillarmodi* Hesse, 1956; *karooana* Hesse, 1956; *nitidula* Hesse, 1956; *pallipes* Bezzi, 1922; *pantostomoides* Hesse, 1956; *philoxera* Hesse, 1956; *pictipennis* Bezzi, 1921; *stenolopha* Hesse, 1956.

Genus *Pantostomus* Bezzi

FIGURES 111, 356, 602, 614

Pantostomus Bezzi, Ann. South African Mus., vol. 18, p. 474, 1921b. Type of genus: *Pantostomus gibbiventris* Bezzi, 1921b, by monotypy. This is the first mention of this genus. Also Broteria (Ser. Zool.), vol. 20, fasc. 2, pp. 69, 79, 1922a.

Hesse (1956), pp 53-55, key to 9 Ethiopian species.

Small flies with slightly elongated abdomen and with a very bare aspect due to the presence of appressed, scaliform pile. Closely related to *Paracosmus* Coquillett, and with similar venation, it differs in the presence of a very long, high, ocellar mound or tubercle. The ocelli are set almost halfway between front and posterior eye corners. Also, in *Pantostomus* Bezzi the lateral frontofacial area forms a strong, rugose, pillowlike support on each side of the antenna, and there is a deep foveal groove or sulcus setting off the upper rim of the genofacial cup or buccal cavity.

In *Pantostomus* Bezzi species that I have seen, the abdomen is strongly bullose on the dorsal tergites and corrugate when seen from the sides, much in contrast to *Paracosmus* Coquillett.

This genus is also related to *Tomomyza* Wiedemann; that genus differs in having the wings generally mottled with brown spots and the venation, especially in the anterior apex or submarginal area, much more erratic. In *Pantostomus* Bezzi the wings are generally glassy hyaline but sometimes tinged with pale brown or yellow.

Pantostomus Bezzi is a South African genus which is a characteristic element of that fauna; Hesse in his remarkable work on this bee fly fauna has added 8 new species to the one partially described by Bezzi.

Length: 5 to 10 mm.; wing, 3.5 to 7.5 mm.

Head, lateral aspect: Head more or less triangular from the lateral view; the occiput is rather strongly developed on the upper half of the head, a little less so below. The central interior of the occiput is very widely and deeply cuplike and deeply recessed. The occipital pile is distinctly sparse and consists mostly of scattered, slender, scaliform hairs quite appressed and directed outwardly toward the eye margin. Along the inner rim of the occiput there are only a few, quite short, fine erect hairs. The front is plane across the eyes for only a very short distance at a point about one-third of the distance from the antennae to the posterior eye corners. The posterior parts of the front and the vertex are a little swollen, and the unusually long ocellar triangle long isosceles and set anteriorly a considerable distance from the eye corners. The lower third of the front, like the face, is prominently extended forward and on each side, adjacent to the antennae, there is a somewhat pillowlike, convex, transversely striate, moundlike swelling. The pile of the front consists of sparse, curled, flat-appressed, shining scales and on the ocellar tubercle there is some short, stiff, pale hairs curled forward in a medial row. The oragenal cup extends obliquely upward and forward to the base of the antennae, at which point it extends forward nearly the length of the first antennal segment and is laterally separated from the face by a rather deep sulcus or crease. The eye is quite large, occupying by far the greater part of the head. The anterodorsal facets are very little enlarged, the whole anterior contour is hemicircular. The posterior margin is shallowly concave about the middle.

The proboscis is short, and relatively stout; the labellum is elongate and strongly tapered from the base to the rather sharp apex. The outer margin straight, the inner margin curved. Its length is a little less than that of the head. The palpus is quite long and slender. It is often curved and the apex is rather abruptly and slightly enlarged and swollen forming a very short terminal segment. The lateral margin has some fine, long, erect hairs, the remaining portion with minute, short hairs. The antennae are situated at the middle of the head; the first segment is short, ranging from barely longer than the second to slightly more than twice as long. It is more or less cylindrical, narrow at the base, a little widened at the apex, and bearing minute, appressed hairs. The second segment is quite short, slightly widened apically, and similarly pilose. The third segment is basally nearly twice as wide in some species; in others scarcely wider. It gradually narrows toward the apex leaving the whole segment long or short pyriform, thus rounded at the base though broad and narrowed apically. Dorsally at the apex it bears a minute, upturned spine.

Head, anterior aspect: The head is considerably wider than the thorax, the eyes separated in the male by a distance fully equal to the width of the antennae at the base. In the female the eyes are separated by only about twice the distance between outer margins of the posterior pair of the ocelli. The antennae are rather widely separated at the base, although this is not immediately apparent since they often touch one another. There is a strong, triangular extension of the front produced between the bases of the antennae. The oragenal cup is quite long, extending from the bottom of the eye to the base of the antennae. It is also deep, both inner and outer walls steep and vertical. The gena is reduced to a linear strip, except at the bottom of the eye, and the tentorial fissure is absent. From a dorsal aspect the occiput is bilobed, the halves touching.

Thorax: The thorax is gently convex, the mesonotum covered with dense, very short, curled appressed and flattened pile which becomes longer and still appressed along the lateral margins. The metapleuron has a large tuft of long, coarse, flattened, appressed shining pile. Posterior hypopleuron with a patch of flattened pile behind the haltere, and the anterior hypopleuron with a similar patch of appressed pile. The pteropleuron is shining and bare, except for a small, posteroventral patch of micropubescence; upper sternopleuron with similar micropubescence; upper border of the mesopleuron with scattered, sparse, appressed, flattened hairs.

Legs: The legs are rather short, the anterior 4 femora rather slender, the hind femur stout and a little swollen through the middle and beyond. Femora and tibiae with sparse, flat-appressed, glittering, scaliform pile. Pulvilli large, nearly as long as the claw, broad apically, the claws long and sharp, curved at the apex.

Wings: The wings are hyaline; 2 submarginal cells are present; the anterior branch of the third vein arises rectangularly at the crossvein and has a backward spur. Marginal cell rather strongly swollen and bulbous at the apex. First posterior cell slightly narrowed on the margin. Anal cell widely open; alula absent, ambient vein complete.

Abdomen: The abdomen is comparatively robust, moderately elongate, somewhat drooping; the first segment is short, shorter than in *Paracosmus* Osten Sacken. The second, third, and fourth segments have a wide but rather strongly raised, bullose ridge down the middle with a strong depression on each side and an outlying bulla. Pile rather dense but short, flat-appressed, glittering, and scaliform. I find 5 long tergites and 2 very short ones. The male terminalia are strongly recessed within the abdomen in contrast to *Paracosmus* Osten Sacken. Hesse (1956) points out that the last sternite is dorsal in position; the basal part is undivided but covered with fine, dense hairs. The dististyli are claw-shaped from lateral aspect and much flattened or laterally compressed. Apical part of aedeagus curved upward a little; basal strut in some species with a distinctly visible lateral process on each side basally. He notes that the hypopygium of male is very uniform and similar in all the species.

Material available for study: A pair of the type-species kindly furnished by Dr. A. J. Hesse, and a paratype of *capensis* Hesse presented to me by the British Museum (Natural History).

Immature stages: Unknown.

Ecology and behavior of adults: Hesse (1956) comments on the habits of these flies. One species frequents sandy patches between shrubs. Still another species was caught resting on sand, and a third species was found on the dried twigs of *Mesembryanthemum;* when resting, their wings lie beside the abdomen and their reddish color harmonizes with the twigs and foliage of this plant.

Distribution: Ethiopian: *Pantostomus bullulatus* Hesse, 1956; *capensis* Hesse, 1956; *fruticicolus* Hesse, 1956; *gibbiventris* Bezzi, 1922; *mallochi* Hesse, 1956; *melanotidus* Hesse, 1956; *pilosulus* Hesse, 1956; *psammophilus* Hesse, 1956; *tinctellipennis* Hesse, 1956.

Genus *Amphicosmus* Coquillett

FIGURES 110, 114, 361, 608, 618

Amphicosmus Coquillett, West American Scientist, vol. 7, p. 219, 1891b. Type of genus: *Amphicosmus elegans* Coquillett, 1891b, by monotypy.

Glycophorba, new subgenus. Type of subgenus: *Amphicosmus cincturus* Williston, 1901.

Hall (1957), p. 142, key to 3 species.

Small slender, shining flies of bare or denuded aspect due to the minute pile. The glassy, hyaline wing has 3 submarginal cells and the end of the marginal cell is simple, not bulbous or swollen as in *Paracosmus* Osten Sacken; consequently *Amphicosmus* Coquillett is more closely related to *Metacosmus* Coquillett and like it has weak legs and the female front is convex, never concave.

There appears to be some sexual dimorphism; I took what I consider to be a pair at the same spot in Arizona; the female has the lower front, lower face, and lower occiput, besides tibiae, bright canary yellow; the male has these parts black or steel gray, the tibiae brown.

Amphicosmus Coquillett is a Nearctic genus known from arid western lands.

Length: 7 to 9 mm.

Head, lateral aspect: The head is globular, the occiput is moderately prominent, a little more so above. Viewed from the rear the occipital margins are wide and rather strongly convex and sloping to the eye margin and covered with dense, short, micropubescence and some pollen, the ground color shining through. The inner margin of the deep, interior cup has a very fine fringe of short, scanty hairs directed backward. The face is conically or triangularly produced forward much as in *Metacosmus* Coquillett. The pile of the face consists of a curious, oval downward extension of the dense micropubescence of the front; this reaches to the anterior margin of the oragenal cup well below the polished, bare upper part of the triangular face and extends obliquely forward beneath the antennae. Margin of the oragenal cup with a few minute hairs. The front is plane with the eye on the upper half, but visible in profile on the lower third at least where it extends forward a space equal to the length of the first two antennal segments. Its pile is moderately abundant, fine and suberect. The vertex and the very low ocellar triangle have equally fine, scanty hairs. The eye is quite large, although proportionately smaller than in *Metacosmus* Coquillett; the posterior margin is almost but not quite plane, and the oragenal cup extends in lateral aspect distinctly beyond the eye margin. The proboscis is short, but is nearly as long as the eye and is extended straight forward with comparatively slender though apically attenuate, flattened, divergent labellae that are curved on both inner and outer margins. The palpus is minute and slender, elongate, lying laterally along the sides of the proboscis, with 1 or 2 apical hairs.

The antennae are small, the first segment and the second subequal, shorter than high, pollinose, with 3 or 4 minute short, pale hairs apically both above and below. Third segment long conical in the male, a little longer than the first two segments combined and with minute apical spines. In the female it is a little more widened through the middle and more abruptly narrowed toward the apex. The antennae are attached a little above the middle of the head.

Head, anterior aspect: The head is circular, the genae and oral opening very slightly produced below in the female, but scarcely at all in the male. The ocellar tubercle is also more prominent in the female. In the male the eyes are divided or separated by a space equal to the width of the third antennal segment. The ocelli occupy a long isosceles triangle and are set a considerable distance forward in front of the posterior eye corners, but are much closer to the posterior eye corners than in *Metacosmus* Coquillett. The female front is 3 or 4 times as wide as the front in the male, and the antennae are quite narrowly separated. Oragenal recess elongate, rather narrow, quite deep with steep vertical sides medially, and the vertical wall also steep laterally. There is a small, tentorial pit, rather shallow, at the eye margin about halfway up along the middle of the oral cup. Genae completely occluded by the development of the eye margin, except near the bottom of the eye where it is possible that the lower vertical wall of the oragenal recess, where it protrudes below the eye, is probably part of the genae. From the dorsal aspect the occiput is bilobed with a very narrow, linear fissure visible behind the vertex, the sides of the occiput appear to be almost or quite touching.

Thorax: The thorax is gently convex, the mesonotum polished black with rather sparse, fine, erect pile, and some pollen on the notopleuron. Humeri rather lobate with strong recess immediately behind and with a flared, liplike outer rim, which is pollinose, that separates mesonotum and mesopleuron. The upper border of the mesopleuron is also pollinose. Pleuron polished black, except sometimes the upper border of the mesopleuron, anterior metapleuron, the humerus, a notopleural spot, and the upper border of the postalar callosity, all of which are pale, polished yellow. Scutellum polished black with much erect, rather long fine hairs. Metapleuron with a dense tuft of long, bushy, often pale, coarse, flattened hairs; its surface is pollinose only. Metanotum shining and bare; hypopleuron with coarse micropubescence. Pteropleuron without pile, and the sternopleuron dorsally with a patch of appressed setae. Haltere with a very large club at the apex.

Legs: The legs are rather short, especially the femora which are moderately stout but not swollen. Both femora and tibiae without bristles, covered rather densely with short, appressed, shining, sharp, somewhat flattened or scaliform hairs that become longer on the hind femur. Pulvilli large, as long as the claw. Claws strongly curved at the apex.

Wings: The wings are hyaline, glassy, with 3 submarginal cells. First posterior cell widely open, the anterior crossvein lies a little beyond the middle of the discal cell. The anal cell is widely open, the ambient vein complete, the alula quite narrow.

Abdomen: The abdomen is elongate, slender, and laterally compressed. It is largely shining in color, generally light reddish brown, black on the first segment, except for a pale yellow, posterolateral spot, and black at the base of the second segment and on the terminal segments. Female terminalia enclosed by a fringe of large, flattened hairs arising from the inner surface of the last tergite. The male hypopygium as in *Paracosmus* Osten Sacken is large, strongly protruded, and the basimeres and dististyli are turned upward and flared outward, sharp, curved, and hooklike.

Material available for study: Representatives of the type-species at various museums and also a pair collected by the author in the western part of Arizona.

Immature stages: Unknown.

Ecology and behavior of adults: I have found this species resting upon hot, desert-crusted soil among weeds and herbs in lower Arizona in August; this habit is very similar to that of *Paracosmus* Osten Sacken.

Distribution: Nearctic: *Amphicosmus arizonensis* Johnson and Johnson, 1960; *cincturus* Williston, 1901; *elegans* Coquillett, 1891; *vanduzeei* Cole, 1923.

Glycophorba, new subgenus

Type of subgenus: *Amphicosmus cincturus* Williston, 1901.

Flies with club-shaped abdomen, a conical face that extends well beyond the end of the second antennal segment, and which have the wing containing 3 submarginal cells as in *Amphicosmus* Coquillett. They are related to *Amphicosmus* Coquillett and differ in the much larger size, the noncompressed, club-shaped abdomen, and in the venation. The anterior branch of the third vein arises in a long, rounded arch, and the marginal cell is much more widely swollen. Also, the lower surface of the hind femur is densely covered with rather long, dense, oblique stiffened hairs. The dense patches of microspinules present on the sides of the tergites in *Amphicosmus* Coquillett are much more poorly developed.

Flies from central Mexico.

Length: 14 mm.; of wing 10.5 mm.

Genus *Metacosmus* Coquillett

FIGURES 118, 355, 615

Metacosmus Coquillett, West American Sci., vol. 7, p. 220, 1891b.
 Type of genus: *Metacosmus exilis* Coquillett, 1891, by monotypy.

Melander (1950b), p. 156, key to 3 species. All Nearctic.

Small, slender, shining black flies of bare aspect, with the abdomen rather elongate, especially in males, and often somewhat compressed. The postmargins of the tergites are linearly yellow or whitish. The head in *Metacosmus* Coquillett is larger than the entire thorax and wider. The ocelli are set far forward, the anterior ocellus almost midway between antenna and occiput. There are 2 submarginal cells and 4 posterior cells. The anal cell is as widely open on the wing margin as the first posterior cell; axillary lobe present and whole wing glassy hyaline. The genus is more closely related to *Amphicosmus* Coquillett than to *Paracosmus* Coquillett; in the latter genus the female front is concave and the hind legs more stout; in *Metacosmus* Coquillett the legs are weak and the front convex; also the marginal cell is swollen, or bulbous at the apex in *Paracosmus* Coquillett and the anterior branch of the third vein arises as a simulated crossvein, with backward spur. This vein arises simply and obliquely without a spur in *Metacosmus* Coquillett. All three genera have the males dichoptic.

Metacosmus Coquillett is a Nearctic genus. One species is known from Pennsylvania, and I have taken it in New Jersey; the other two species are desert frequenters and known from Arizona, New Mexico, and California.

Length: 4 to 5 mm.

Head, lateral aspect: The head is nearly globular and quite elongate; the eye is enormously developed and constitutes the major part of the head. Its anterodorsal facets are much enlarged. The entire head is larger than the thorax, much wider and equally long. The occiput is rather strongly developed on the dorsal half of the head, viewed from the rear it is wide but it only gradually and gently slopes down to the eye margin. The interior cup is quite deep and abruptly recessed within. Pile fine, sparse, and scanty. It is more or less appressed outwardly, except along the inner edge of the occipital recess, where this scanty fringe is directed backward.

The front is quite narrow in the male with parallel sides from the posterior eye corners behind the vertex to within a short distance of the antennae, where it widens slightly. Surface of front and vertex mostly pollinose, plane with the eye margins, except for a very small area above the antennae, which is slightly raised. The ocelli are barely visible in profile and form a long, acute triangle. The upper front and vertex bear 4 or 5 short, scattered hairs, and in the female the lower half of the front has some scanty, suberect pile. Female front about twice as wide as that of the male. Posterior eye margin not quite plane. The proboscis is quite short, relatively stout, somewhat divergent apically with short, rimlike, and conspicuous labellum. Palpus small, short, more or less concealed in the oral recess. Antennae attached at the middle of the head, small and short, the first segment is so short as to form a low, circular, rimlike cup from which the second segment arises. The second segment is very short, about twice as wide as long. Both the first and second segments bear 1 or 2 short, flattened shining hairs at their apices. Third segment about twice as long as the first two segments; the basal half is wide and stout, nearly as wide as the second segment, and this expanded portion is about as long as wide. The outer half of the third segment is sharply reduced at the middle of the segment and very much narrower; it ends in a minute spine.

Head, anterior aspect: Head very much wider than the thorax; the antennae are adjacent or at most very narrowly separated. The face is prominent, conically or triangularly peaked and produced forward, often pale yellow in color and with only minute micropubescence along the sides. The genofacial recess is elongate, comparatively shallow in front, deeper behind and below, and confined to the lower fifth of the head. It begins some distance anteriorly to the posteroventral corners of the eye, hence there is a considerable anterior extension of the occiput, which is, however, nearly or quite plane with the eye. The genae are quite nonexistent because of the excessive medial expansion of the eye.

Thorax: The thorax is polished black, very little convex, the mesonotum with scattered, fine, subappressed hairs. Margin and disc of scutellum with a few, fine, long, erect hairs. Humerus chiefly pollinose with only a few minute hairs behind. Metapleuron rather widely long pilose. Metanotum and hypopleuron pollinose. Pteropleuron and sternopleuron without pile but with a little pollen ventrally and posteriorly. Mesopleuron with a dorsal border of fine, erect, slightly flattened hairs. Knob of halteres large and elongate.

Legs: The legs are short and weak, the anterior femur is slightly enlarged basally. Middle and posterior femora with a row of quite fine, moderately long hairs anteroventrally. All of the tibiae lack spicules or bristles but are covered with relatively sparse, appressed, sharp, somewhat flattened hairs. Pulvilli large, broad and rounded, the claws fine, sharp, strongly bent at the apex.

Wings: The wings are hyaline, narrowed at the base and have 2 submarginal cells. The marginal cell is rather wide and still wider near the apex as it is gently curved but swollen backward. First posterior cell widely open, the anterior crossvein lies in the middle of the discal cell. Anal cell extremely widely open, almost fully as wide on the margin as the first posterior cell. Ambient vein complete, the alula absent.

Abdomen: The abdomen is elongate, slender, somewhat laterally compressed in the middle and posteriorly; the ground color is shining black with narrow, pale postmargins to the tergites and the sternites. The pile is sparse and scanty, fine and appressed. There are longer, upright tufts of pile on the first tergite. Male hypopygium bulbous with dististyli concealed.

Material available for study: All three species through the courtesy of Dr. R. H. Painter and Mr. Jack Hall; I have a small series of an undescribed species I collected.

Immature stages: Unknown.

Ecology and behavior of adults: I have collected these flies on two occasions. I took an individual of *Metacosmus exilis* Coquillett in June hovering about a forest path in New Jersey. I have also taken several of these flies of a different species flying in very hot sunshine about clumps of prickly desert plants in the California desert not far from Desert Center; time, late August. The range of ecological habitats is remarkable.

Distribution: Nearctic: *Metacosmus exilis* Coquillett, 1891; *mancipennis* Coquillett, 1910; *nitidus* Cole, 1923.

Genus *Paracosmus* Osten Sacken

FIGURES 112, 119, 360, 619, 896, 897

Allocotus Loew, Berliner Ent. Zeitschr., vol. 16, p. 82, 1872.
    Type of genus: *Allocotus edwardsii* Loew, 1872, by monotypy.
    Preoccupied Fisch., Pisces; Mayr Hemiptera, 1864.
Paracosmus Osten Sacken, Bull. U.S. Geol. Survey, vol. 3, p. 262, 1877. Change of name.
Actherosia, new subgenus. Type of subgenus: *Paracosmus rubicundus* Melander, 1950.

Melander (1950b), p. 154, key to 4 species.

Small, slender flies easily recognized by the bare aspect and the elongate abdomen, compressed laterally. Like *Metacosmus* Coquillett and *Amphicosmus* Coquillett, the head is subglobular and wider than the thorax; these three genera, together with *Pantostomus* Bezzi and *Tomomyza* Wiedemann, all have not only a wide head but a relatively elongate subcylindrical abdomen; in the three American genera there appears to be a tendency for the abdomen to be more or less compressed. Moreover, in all these genera, and in the strange and very different genus *Peringueyimyia* Bezzi, the occipital lobes are very tightly apposed. Oddly in all five the palpus is extremely long and slender, coiled snakelike, with short apical clublike segment.

In *Paracosmus* Osten Sacken the flies have a very bare aspect due to very short, dense, appressed, setate pile and even some bare and polished areas on face and pleuron. Like *Metacosmus* Coquillett the flies are generally black, largely shining, with pale-yellow postmargins on the abdominal tergites; it differs from the latter genus in the larger head and venation; the anterior branch of the third vein arises as a rectangular crossvein which usually emits a spur directed toward the base of the wing.

In the male of *Paracosmus* Osten Sacken the hypopygium is unusually prominent and protruded and directed obliquely upward.

Paracosmus Osten Sacken is a Nearctic genus that is characteristic of the southwestern United States with some species ranging into Mexico. I have taken the type-species commonly in northeastern Mexico, in Harlingen, Texas, near Beeville, Texas, and in southern California near Pasadena.

Length: 6 to 10 mm.

Head, lateral aspect: The head is more or less triangular, with a moderately produced occiput and a rather peaked and strongly produced front and face. The occiput is dorsally about as long as the third antennal segment and in the middle and below a little longer than the second antennal segment. The part of the occiput is hollowed out into a large, deep cavity. From the outer ridge it slopes sharply forward to the eye margin. The whole outer surface is densely micropubescent but the pubescence runs in different directions. That part on the upper third or more runs backward and is silvery shining from the lateral aspect. That part in the middle runs forward and laterally, this part allows the brassy black ground color to shine through. Lower occiput thinly pollinose. In addition, along the rim of the central cavity, there is a row of fine, short, erect hairs. The front in both sexes is flat with a very shallow but quite wide transverse concavity or depression deepest across the middle of the front. Surface pollinose with a separated band of fine, erect pile on each side of the front in the male; in the female it is less clearly separated and the color is apt to be different.

The face is peculiar in that the unusually long oragenal cup extends obliquely upward to the base of the antennae and outward almost to the end of the first antennal segment and is demarcated from the lateral face by a crease. Lateral face micropubescent and anteriorly with a row of hairs. Oragenal cup bare, polished, and shining. The eye is hemispherical with the posterior margin nearly plane on the upper half and making a very low, slight bend in the middle and this lower portion gently curved; the postmargin is entire. The proboscis is slender and about as long as the head. Palpus long and slender with a few rather long hairs laterally and at the apex, and with a very small clavate end segment. The proboscis is capable of being rather strongly withdrawn into the head, in some insects it is not as long as the eye. The antenna attached just above the middle of the head, short and relatively small. The first segment is less than twice as long as wide, the second, viewed medially, is as long as wide and slightly wider apically. It has a tuft of stiff setae below on the apical half. The third segment is oval, nearly as long as the first two segments together; it is narrowed briefly at the base, widest on the basal fourth and then narrowed to the truncate apex, and this apex dorsally bears a blunt tubercle or spine. The medial surface of the third segment is covered, except at the extreme base, by a depressed area of micropubescence.

Head, anterior aspect: The head is much wider than the thorax. The eyes in the male are well separated by almost the width of the ocellar tubercle. The ocellar tubercle itself is long and narrow, but the posterior ocelli are set considerably forward near the middle, and the tubercle bears a number of fine, erect hairs of no great length. The female front is not quite twice the width of the male and the ocellar tubercle much lower. The antennae are separated at the base by more than half the thickness of the first segment. The oral opening is quite long, extending from the bottom of the head to the antennae. Interiorly it is rather deep and concave with the lateral wall of the oragenal cup folded inward in the middle with a marked ridge where it ends. The walls of the oragenal cup are moderately sharp throughout lapping over on the lower fourth.

Gena extremely narrow but detectable as a narrow strip along the eye and separated by the tentorial fissure.

Thorax: The mesonotum is low and very gently convex, except in the aberrant species *Paracosmus rubicundus* Melander. The mesonotum has vittae of pollen or coarse micropubescence that is rather conspicuous. Mesonotum also covered with rather abundant, in part dense, appressed, curled, short, wiry pile. Laterally along the side margins the pile becomes longer, still strongly appressed, shining and a little flattened and also curled. Scutellum similar to the mesonotum. Humerus densely micropubescent and rather densely covered with erect pile. Curiously the pleuron is very largely bare and shining, with an oblique or horizontal stripe of fine, pollenlike micropubescence running across the upper sternopleuron and the lower pteropleuron and the whole of the metapleuron. The posterior hypopleuron above the hind coxa is also pubescent. Upper mesopleuron with a narrow band of scanty, curled hairs. Metapleuron with rather dense, long, erect, silvery, flattened hairs and an area of shorter pile behind the hind coxa; another beneath and behind the haltere. Squama short with a fine, short fringe. Haltere relatively small, the knob flattened and concave below.

Legs: The legs are relatively short, the femora are a little thickened but only very slightly enlarged toward the base and the rather longer hind femur is distinctly narrowed and spindly toward the base. Whole tibiae and femora rather densely covered with short, appressed, setate pile on all surfaces, none of it upstanding; the short tarsi are similar. Pulvilli large, as long as the claw. The claws are slightly bent at the apex.

Wings: The wings are hyaline and unusually glassy; the costa and subcostal cells are also hyaline. The marginal cell is strongly swollen near the apex, bent backward, broadly rounded, the second vein meeting the costa at a right angle, or more unusually obliquely backward. The anterior branch of the third vein arises rectangularly, makes a rectangular bend outward at which point it usually has a backward spur and then tends near the apex of the wing to make a sharp bend forward nearly as abrupt as the second vein. The anterior crossvein enters the discal cell just before the outer fourth. First posterior cell long and very slightly narrowed. Anterior intercalary vein long. The anal cell is widely open, axillary lobe rather narrow, the alula quite narrow and linear. Ambient vein complete.

Abdomen: The abdomen is elongate, at least 1½ times as long as the thorax and rather strongly compressed laterally and black, except in *rubicundus* Melander, and with sharp, pale yellow, bare, posterior margins to all of the tergites, more extensive in the middles and more extensive and wider laterally on the first three. I find 7 tergites, the last one short conical in the female, and in the male there is an eighth tergite laterally. The male, likewise, has a deep, transverse, groovelike depression near the base of the second segment. In the female it is much closer to the base and less pronounced. The pile of the abdomen is moderately dense, flat-appressed and setate. The sternites have longer and suberect setae. The male terminalia are large, clublike, the dististyli lie on the upper part of the large hypopygium and are extended upward.

The male terminalia from dorsal aspect, epandrium removed, show a rather simplified body, with long sausagelike dististyli which originate a considerable distance back from the basistylus. The epiphallus is split and divided into 2 ejaculatory processes. In the lateral aspect the figure is quite unique, the very large basistylus is almost bowl-shaped, strongly arched beneath, much higher basally. The dististyli that arise from the end of the basistylus are curved upward. The epiphallus and included ejaculatory process are long and rather straight attenuate. Epandrium comparatively long, bluntly lobelike apically.

Material available for study: A considerable series of the type-species and some material of all the known species. I am indebted to Dr. Frank Cole for material of *rubicundus* Melander.

Immature stages: Unknown.

Ecology and behavior of adults: I have collected these flies sometimes rather abundantly resting upon the hot surface of caked sandy places in hot semidesert country; also on sandy banks beside western streams. They rest on the very hot sand in strong sunlight. I have taken them from eastern Mexico, southern Texas, and Los Angeles County, California.

Distribution: Nearctic: *Paracosmus edwardsii* (Loew), 1872 (as *Allocotus*); *insolens* Coquillett, 1891; *morrisoni* Osten Sacken, 1887; *rubicundus* Melander, 1950; *similis* Hall, 1957.

Actherosia, new subgenus

Small, pale flies with comparatively robust unconstricted abdomen; everywhere pale yellow or reddish, including face and frons and only the disc of the mesonotum and the sternopleuron black. Wings water-clear hyaline.

These pretty little flies are related to *Paracosmus* Osten Sacken but differ in the more extensive occiput, the longer, basally tapered third antennal segment, shorter first antennal segment and the vestigial or rudimentary or absent anterior ocellus. Finally in *Paracosmus* and related genera the abdomen is characteristically compressed laterally and the abdomen is narrow, less stout, rather longer.

These flies are regarded as a subgenus of *Paracosmus* Osten Sacken, with type-species *Paracosmus rubicundus* Melander. I want to thank Dr. Frank Cole for providing me with material.

Flies are found in southern California.

Length: 9 mm., including antenna; wing 6.5 mm.

Genus *Peringueyimyia* Bigot

Figures 120, 359, 607, 627, 642, 893, 894, 895

Peringueyimyia Bigot, Bull. Soc. Ent. France, ser. 6, vol. 6, p. cx, 1886b. Type of genus: *Peringueyimyia capensis* Bigot, 1886b, by monotypy.

A peculiar genus of medium-sized black flies with more or less drooping head and abdomen. The tergites are linearly margined behind with brownish yellow; their pile is matted and appressed, white or yellowish white. The mesonotal pile is erect, fine, bristly, and chiefly black. All femora and tibiae with spiny black bristles. The proboscis extends beyond the antennae, and the third segment of the latter has a ventral brush of long, yellowish-white pile.

The head is subglobular with the occiput extremely prominent, strongly and convexly sloping backward and inward toward the deep central cavity; the lobes behind the vertex are tightly apposed in contrast to *Nomalonia* Rondani and *Henica* Macquart. The female vertex is much less widely separated and the wing venation is quite different. Glassy hyaline, the wing has 3 tiny brown spots and the anterior branch of the third vein arises as a simulated crossvein with long backward spur vein; the marginal cell is bulbous at the apex.

The single known species is a unique element in the South African bee fly fauna.

Length: 8.5 to 13 mm.; wing: 8.5 to 12 mm.

Head, lateral aspect: The head from the dorsal aspect is subglobular but this is chiefly due to the enormous extent of the occiput, which is about four-fifths as long as the eye itself, and which is strongly sloping and rounded backward both from above and from the side. The halves of the occiput meet closely behind the vertex and form a comparatively shallow but distinct, elongate groove visible from the interior when the head is removed. The occiput is deeply cuplike and there is a linear crease where the two lobes of the occiput are fused above. Pile of occiput scanty, short, rather fine, appressed anteriorly and forming around the rim of the cup an erect, short, sparse fringe. Inner, lower tentorial pits large, deep and obliquely elongate. From the lateral aspect the occiput is greatly reduced in extent below and only about one-third as long as its dorsal length. The front in the female is flattened and has a medial groove; on the upper half it is only as wide as the base of the swollen antennae, and the pile is bristly, comparatively sparse, curled forward; these hairs are not quite as long as the first antennal segment. The front of the male is small, and triangular and bears minute, gray micropubescence only. From the lateral aspect the face is scarcely visible, nonprotrusive and bears near the eye margin a fringe of coarse long hairs merging with similar, equally long, dense pile on the lower half of the first antennal segment. It also continues, though shorter, downward to the beginning of the genae.

The eye is quite large and high, a little shorter below than above, in spite of the fact that the occiput extends farther forward on the dorsal half. The posterior margin of the eye, while not indented, shows a distinct break or bend at the middle, where the occiput expands forward. Eye facets without pile and only slightly enlarged above in the male. The proboscis is extended obliquely upward or forward and is only a little longer than the head and actually shorter when it is extended upward. It is rather slender, with long, slender labellum. Palpus large, elongate, flattened apically, with sparse, bristly hairs laterally, the inner surface particularly flattened and the whole deeply recessed in the oragenal cavity, and consisting of a single segment. The antennae are comparatively small; however, the first segment is swollen, not only below but medially and it is nearly or quite as long as the slender, attenuate third segment. The antennae are separated narrowly at the base; the first segment bears sparse, short, fine, bristly black pile dorsally and on the entire upper half, changing to long, bushy, pale-colored, fringelike pile on the lower half. The second segment is small, short, beadlike, with stubby, black setae above and a few longer, pale hairs below. The third segment is 3 or 4 times as wide on the basal fourth as at the apex; it is never as wide as the second segment; it bears medially a deep, elongate pit on the basal third and the apex is capped with a minute, short, pale, conical spine.

Head, anterior aspect: The head is somewhat more narrow than the thorax, quite circular from the anterior aspect. Eyes in male virtually holoptic for a short distance, the vertical triangle small, longer than wide; the rather small, convex ocelli are not quite equilateral. The oragenal cup is large and extensive, deep above and below. The edge of the cup is rather sharp and linear and separated from the short face by a deep groove completely hidden from the lateral view. This deep furrow leaves the inner margin of the eye strongly sloping and the fringe of pile that ranges down the eye arises from this slope. The genae are reduced to a linear trace by the expansion of the oragenal cup. The lateral furrow extends ventrally into a deep, very narrow, tentorial fissure. Deep inner recess of the occiput almost bare with pollen or minute pubescence only.

Thorax: The mesonotum is a little longer than wide, slightly convex and sloping, especially anteriorly; it is faintly pollinose but subshining and bears comparatively sparse, fine, subappressed pale hairs and equal or greater amounts of fine, erect, not very long, black bristly hairs. The pile becomes more dense, matted, and coarse along the sides and the anterior margin of the mesonotum, where it forms a rather dense fringe cupping around the occiput of the head. Notopleuron with 4 rather long, basally stout, finely tapered, curved, black bristles. There are 6 others that are conspicuous, though more slender, situated above the base of the wing and 3 or 4 still longer bristles curved backward from the postalar callosity. The shining brown scutellum has some short, pale, coarse, curled, appressed hairs, especially toward the base; there is a border of fine, erect, black, bristly hairs on the outer half and a few at the base and on the lower margin it bears a loose fringe of long, pale, yellowish bristles. The humerus is rather inconspicuous, slightly elongate and convex, and thickly covered with bushy pile. Metapleuron with a conspicuous tuft of pile extending upward and entirely situated above the haltere. Lower metapleuron, hypopleuron, and pteropleuron bare. Upper half and posterior margin of the mesopleuron with much coarse pile, and a small tuft on the upper half of the sternopleuron. Squama semicircular but short with a wide band or fringe of rather dense, fine pile.

Legs: The legs are comparatively long and unusually stout, especially on the femora, yet not swollen. Anterior femur a little enlarged at the base with 4 or 5 short bristles posteroventrally and a similar number anteroventrally and a few, short, appressed, black setae behind intermixed with dense, wide, appressed grayish-white scales. Middle femur with similar scales in front and behind, and similar setae. However, the anterior surface on the outer two-thirds bears along the middle anteroventrally 4 long, basally stout, black bristles, 2 equally conspicuous bristles near but not at the apex, and posteriorly there are 8 outer bristles likewise near the apex. Hind femur with loose, brownish scales and with 10 moderately long, basally stout, apically sharp, posteroventral bristles and on each side near the apex 2 others arranged vertically and quite stout. The anterior tibia bears an anterodorsal and a posterodorsal row of about 12 short, well-developed spines or spicules and posteriorly a row of 6 or 8 spines, 2 or 3 of which are unusually long and stout. Middle tibia with 7 rather long, rather oblique spinous bristles dorsally, 3 or 4 shorter ones anteriorly, 12 or 14 short spines posteriorly and several longer posteroventral spinous bristles. Hind tibia with 4 rows of more sparse spinous bristles, 7 or 8 anteroventrally, 6 anterodorsally and a like number posterodorsally, and likewise ventrally. Pulvilli well developed, tibiae and tarsi with scales, nearly as long as the claw. Claws curved from the base.

Wings: The wings are glassy hyaline; the costal and subcostal cells are pale yellow. The costal, subcostal, and the first longitudinal vein yellow, the remaining veins chiefly dark brown. The wing is large, elongate, not greatly narrowed at the base and there are small, diffuse, blackish spots on the crossveins, especially the radiomedial crossvein and also at or below the extremely short radial sector. There are 2 submarginal cells, but there is a long but quite characteristic spur vein directed backward arising at the end of the basal section of the anterior branch of the third vein. This basal section of the third vein is quite rectangular, simulating a crossvein. The end of the marginal cell is broadly rounded; the second vein is recurrent at the apex. The third vein arises from the second vein very near to the base of the latter, hence the radial sector is extremely short. The anterior crossvein lies rectangularly at the outer fifth of the discal cell. The vein on the lower side of the discal cell is much more sigmoid than in either *Nomalonia* Rondani or *Henica* Macquart. The first posterior cell is closed with a short stalk and the other half of this cell is triangular. The medial crossvein at the end of the discal cell is not quite as long as the anterior crossvein. Anal cell widely open, ambient vein complete, the alula short and narrow with plane margin and nearly as wide apically as basally. The costa is somewhat expanded at the base with weak comb or brush of setae and no hook.

Abdomen: The abdomen is comparatively short oval but narrowing a little apically. Near the end of the third tergite it becomes as wide as the mesonotum and it is rather densely covered with matted, curled, shining, brownish yellow, somewhat flattened hairs, with, in addition, a postmarginal fringe of rather long, shining, slender, brownish yellow bristles. These bristles lie more or less appressed along the abdomen, and laterally along the posterior parts of the segment, there are some long, fine, black, bristly hairs. Ground color of abdomen shining black with the posterior margins yellowish brown. I find 8 tergites in the male, the last two considerably more narrowed, and I find 7 in the female, the last one longer than the preceding ones. Female terminalia with a circlet of 5 or 6 blunt, shining, reddish spines on each side apically. Male terminalia only shortly protrusive beyond the last narrow sternite and tergite. The dististyli are apposed, slightly divergent, arising below and turning upward. The hypopygium is thus reversed in position. The last tergite, which Hesse (1938) states is the last sternite, is notched in the middle. Dististyli elongate and flattened with sharp apex and flared subapically.

Material available for study: I am deeply indebted to Dr. A. J. Hesse for furnishing material of this genus.

Immature stages: Unknown. However, the spiny armature on the female genitalia suggests that the eggs are laid in the soil.

Ecology and behavior of adults: I have seen no recorded comment on the behavior of adults or the flowers they visit.

Distribution: Ethiopian: *Peringueyimyia capensis* Bigot, 1886.

Genus *Neosardus* Roberts

FIGURES 121, 398, 621, 639, 900, 901, 902

Neosardus Roberts, Proc. Linnean Soc. New South Wales, vol. 54, p. 560, 1929. Type of genus: *Neosardus principius* Roberts, 1929, by original designation.

Roberts (1929), p. 560, key to 4 species.

Strange flies of small to medium size, robust in form and at once peculiar in the extraordinary width of the front, vertex, and inflated face. The antennae are strikingly elongate, cylindrical, and relatively slender. The wings have a brown pattern. All species seem to have emerald or malachite green scales on the thorax. The wing has 4 submarginal cells. Finally, it is to be noted that although the occiput is shallow and short laterally and dorsally, it is definitely bilobate, and it has a deep central cavity. There are no closely related genera. It must be regarded as a member of the Cylleninae, which is relatively generalized with respect to the occiput and highly specialized in the matter of the front and face and antenna.

This genus is only known from Australia but is found on both sides of the continent.

Length: 6.5 to 12.5 mm.; wing, 5.5 to 10 mm.

Head, lateral aspect: The head is more or less subglobular, very strongly inflated, in consequence of which the eye appears to be somewhat reduced although the eye is not small. The occiput is not at all extensive from the lateral aspect, except below, where in some species it is even more pendulous than in *Nomalonia* Rondani, and thus there is considerable variation in this respect. The occiput from a posterior aspect shows a deep, cuplike depression such as is characteristic of the Cylleninae; nevertheless its area is central and small so that the outer part of the occiput laterally between the recess and the eye margin is very broad, moderately convex, gently sloping from the middle to the eye margin, and the halves, while bilobed above, have a usually wide or deep sulcus behind the ocelli. In some respects these features and the long, slender antennae suggest a remote derivation from the Bombyliinae. The pile of the occiput is fine, erect, and rather scanty and becoming stiff and almost bristly on the upper fourth near the vertex. The eye is comparatively long compared to the type, hemicircular in front, strongly rounded below with the posterior margin approximately plane only over the middle of the head. The front is extremely wide even in the male of the type-species where it forms a large triangle, widely divergent below, everywhere pollinose, and covered with quite long, loose, scattered, rather fine hairs, except along the eye margin. In *Neosardus paramonovi*, new species, it is even wider, inflated shining yellow with loose, scattered, erect, bristly hairs and the whole translucent. In the type-species and in *Neosardus paramonovi*, new species, each side of the face is considerably wider than the eye, its width extraordinary, and its sides parallel.

The oragenal cup is rather strongly developed, but laterally its wall on the upper half is very broad, gently sloping and convex, and demarcated from the face and genae by a rather distinct crease or fissurelike sulcus, which never descends into a deep, slitlike, tentorial fissure as in *Henica* Marquart. This creaselike fissure extends to the base of the antennae so that the upper medial wall of the oragenal cup is long and convex and this part of the face bears a tuft of bristly hairs beneath the antennae. Outer wall of the oragenal cup without pile, except beneath the antennae, but the extremely broad space between the furrow and the eyes and the equally extensive genae below bear numerous, long, fine, erect, stiff, scattered hairs, which are pale in color. The proboscis is quite elongate and slender and twice as long as the head. It originates at the extreme rear of the head almost behind the eye. The labellum is long, microspinulose. Palpus slender, elongate, and clearly composed of 2 segments, the apical segment short, and the whole palpus bearing scattered, fine hairs of no great length. The antennae are attached at the upper third of the head and are elongate and slender; the first segment is especially elongate, cylindrical, twice as long as the second segment, which is nearly the same width. Both of these segments bear appressed setate hairs dorsally and ventrally. Third segment with disclike rim at the base, not quite as wide as the second segment, rapidly widening to the middle, then more narrow to the apex, which may or may not bear a short microsegment but in any case has a minute apical spine. The third segment is nearly or quite as long as the second segment.

Head, anterior aspect: The head is much wider than the thorax. The eyes in the male are widely separated by a distance slightly greater than the ocellar tubercle. The ocellar tubercle is prominent, rather swollen, obtuse, with all 3 ocelli large, the posterior pair almost touching the eye margin, and the tubercle bears a tuft of long, fine hairs, sometimes bristly, which stand erect. The oral opening is relatively small when compared to the width of the face and it is relatively shallow with its inner wall sloping as much as its outer wall. The genae are excessively wide in consequence of the extraordinary width of the face.

Thorax: The thorax is only slightly convex; the mesonotum is black in ground color, thinly pollinose, and rather densely covered in many species with emerald green or coppery scales. The scales are rather limited in the type-species. The humerus, lateral margin, and the apical half of the scutellum and the whole pleuron in some species are light-brownish orange. In the type-species the whole thorax is blackish, with rather dense, gray pollen. Notopleuron with 2 or 3 long, slender bristles, postalar callosity with 1 or 2, and the scutellar margin with 2 or 3 pairs of bristles. Metanotum and metapleuron bare, upper posterior hypopleuron below the haltere with a patch of pile. Pteropleuron pollinose only. Mesopleuron rather densely and sternopleuron rather scantily pilose.

Legs: The legs are slender and rather short. Anterior femur with fine pile only. The type-species has fine pile only on the remaining femora but in other species with 3 or 4 slender bristles ventrally on these last 4 femora. Anterior tibia with a few minute spicules, 10 or 12 anterodorsally, 3 or 4 posterodorsally, or in the small, delicate type-species these spicules at the most are represented by short, stiff hairs. Bristles of hind tibia present in 4 rows of 4 to 6 elements, all slender and of no great length. Pulvilli slender, but not as long as the claws, which are fine, sharp, and gently curved.

Wings: The wings are largely mottled with brown with some clear areas along the posterior border, sometimes at apex of wing. Auxiliary cell often clear and a large hyaline spot distally in the second basal cell. First posterior cell widely open, discal cell short or long, and the anterior crossvein lying at the middle of the discal cell. Ambient vein complete; alula hemicircular.

Abdomen: The abdomen is elongate in the type-species, and only the base of the abdomen is as wide as the thorax. In some other species the abdomen widens beyond the base, is oval and drooping, and wider than the mesonotum. Color black, shining, and somewhat metallic, or in other species shining orange brown or with black spots or vittae medially. The pile is rather scanty, erect, and short, or may be appressed, fine and bristlelike and short. First abdominal segment with a rather dense brush of rather long, erect pile laterally. Hypopygium partly recessed, the long basal lobes adjacent, the dististyli concealed by curled hairs from the last sternite.

Material available for study: Two species, the type-species, given me by the late Dr. A. J. Nicholson of the CSIRO, and a larger, undescribed species collected by the author in Western Australia, here illustrated and named in honor of Dr. Paramonov, who truly liked the bee flies, and to whom we are all indebted for his astute observations.

Immature stages: Unknown.

Ecology and behavior of adults: Very little is known about the ecology of this genus. In January 1954, I collected a female of a large species resting on hot, sandy ground amid weeds, grass, and herbs in Western Australia near Bull Pond out from Perth. It was a sunny day with a very strong wind blowing. Females have a well-developed fringe of spines around the ovipositor, clearly suggesting that the eggs are deposited in soil.

Distribution: Australian: *Neosardus circumdatus* Roberts, 1929; *lepidus* Roberts, 1929; *nigratus* Roberts, 1929; *paramonovi* Hull, new species; *principius* Roberts, 1929.

Subfamily Lomatiinae Schiner, 1868

In the Lomatiinae we find several characteristic tribes. The Lomatiini are best developed in the Old World, especially southern Europe and South Africa. The Comptosiini are greatly concentrated in Australia and in the Chilean subregion of the neotropics, areas to which they are restricted and where there are many species. I believe the troublesome group of genera centered around *Xeramoeba* Hesse and *Petrorossia* Bezzi are annectant, though distant, with the subfamily Anthracinae; we find that not only is the manner of origin of the second vein highly variable but so is its point of origin. The squamal fringe in the tribe Xeramoebini and throughout the whole subfamily is composed of only fine, soft hairs; the relationship between the genera in this tribe and certain genera in several other tribes is so close as to almost amount to hairsplitting. From a comparative study of more than sixty characters within virtually all of the known genera of the subfamily, I conclude it is necessary for a preservation of values to reduce the subfamily Aphoebantinae to tribal rank and to separate other distinct groups as tribes. I have not seen the genus *Edmundiella* Becker. Tribal characters are set forth below:

Tribe Lomatiini

This tribe is characterized by indented eyes, with or without a bisecting line. The metanotum is pilose without exception, but bare within 15 other genera in other

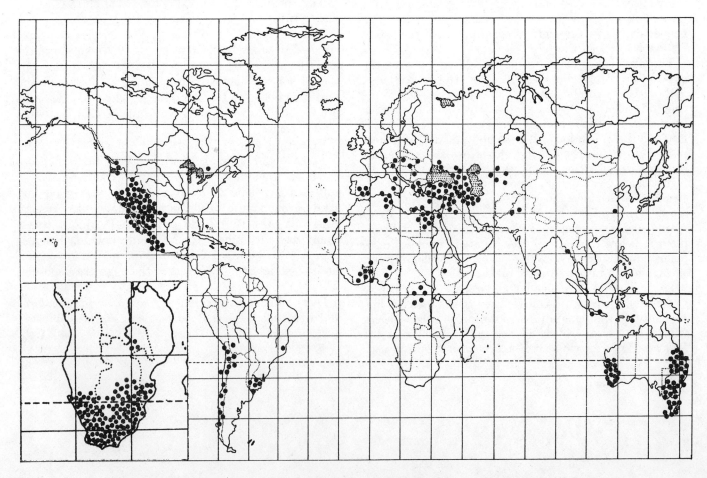

TEXT-FIGURE 43.—Pattern of the approximate world distribution of the species of the subfamily Lomatiinae.

tribes, and in this tribe the metapleuron in front of the haltere may or may not have a tuft of pile. The second vein is basal in origin and arises gently, that is, acutely. The gena is shelving inward. The marginal cell is at most slightly widened and often ends obtusely, and the second vein ends at right angles but may end even acutely; hence there is no radial loop. One or two Australian species do have a slightly protrusive marginal cell, but yet are far from the radial loop found in the Comptosiini; see figure. The tribe contains *Lomatia* Wiedemann and its subgenera. It also contains *Anisotamia* Macquart, *Ogcodocera* Macquart, and *Bryodemina* Hull. I also place *Edmundiella* Becker here.

Myonematini, new tribe

This tribe is erected for the unique genus *Myonema* Roberts, which I believe more properly belongs within the Lomatiinae. The eyes are indented, the genae are shelving inward, which is an important character of Lomatiinae, and finally not only are the ocelli set very far forward indeed, but there is an extraordinarily deep, curious, fossa or trench between the eyes posterior to the ocelli.

Antoniini, new tribe

Here the eyes are indented and bisected, but there is no postocellar fovea or trench. The ocelli are placed posteriorly; the genae shelve inward but are very narrow. The front is strongly protuberant; the eyes have a unique, strong, greenish to reddish reflection. There are 3 submarginal cells, the first posterior cell is closed and stalked, and more important the anterior crossvein is at the outer tenth of the discal cell. It contains the one genus *Antonia* Loew, a partial mimic of such flies as the syrphid *Xanthogramma* Schiner. It is perhaps more correct to call this a parallel development.

Tribe Aphoebantini

A group in which the eyes are both indented and bisected, but its principal distinguishing characteristics are the short, vertical or retreating, nonconical, nonhoodlike face together with the gentle, acute origin, and basal position of origin of the second vein; the third vein arises from the second vein, dipping downward, swollen, and often with a knotlike swelling at point of origin. The metanotum is bare. Contained are the 4 genera: *Aphoebantus* Loew, *Pteraulax* Bezzi, *Epacmus* Osten Sacken, and *Pteraulacodes* Hesse. Male terminalia terminal, large, and of considerable size.

Tribe Prorostomatini

In general appearance a group of flies not greatly dissimilar to *Aphoebantus* Loew. Both have the eyes indented and bisected. However, they are very distinct in three particulars. The second vein arises abruptly with rounded curve, and its point of origin lies between the base of the third vein and the anterior crossvein, rarely close to the crossvein; the metanotum is pilose in all the included genera, except the European type-species of *Plesiocera* Macquart. Included are *Prorostoma* Hesse, *Conomyza* Hesse, *Stomylomyia* Bigot, *Exepacmus* Coquillett, *Eucessia* Coquillett, *Epacmoides* Hesse, *Coryprosopa* Hesse, and *Plesiocera* Macquart.

Xeramoebini, new tribe

The second vein arises abruptly though with rounded base; its point of origin is rather close to the anterior crossvein. The face is short, not conical or hoodlike. The proboscis is short and contained within the oral cavity and the labellum muscoidlike. The metasternum is bare. A plumula is present (in *Xeramoeba* Hesse and *Chionamoeba* Sack), or absent. Contained are *Petrorossia* Bezzi, *Desmatoneura* Williston, *Xeramoeba* Hesse, *Chionamoeba* Sack, and *Chiasmella* Bezzi.

Tribe Comptosiini

These are large flies confined to Australia and southern South America. Without exception all have the metanotum and the upper hypopleuron pilose. The South American *Lyophlaeba* Rondani, and all of its subgenera, including *Macrocondyla* Rondani, are sharply distinguished from the Australian *Comptosia* Macquart, and its subgenera by the dense tuft of pile upon the metapleuron. The separate *Ylasoia* Speiser and *Doddosia* Edwards do not follow this rule. All these genera in this tribe have a protrusive, expanded, fingerlike or otherwise complex marginal cell. Submarginal cells range from 2 to 4 with those genera having 4 such cells certainly representing the more archaic forms of these relict genera. In Australia I have collected annectant species of *Lomatia* Meigen that merge into *Comptosia* Macquart, but I have collected only females.

In this subfamily the hypopygial figure is characteristically elongate and slender in contrast to the conspicuously widened base found in the Exoprosopinae. Also, the epandrium is rather characteristic, being elongately triangular, therefore rather narrowed in terms of maximal width of thickness. Quite often it is lobed or sharply angulate dorsobasally and almost always with a prominent, sometimes quite conspicuous, narrow, basiventral, downturned lobe or extension. Again the dististylus is usually quite large, rather elongate, upturned, with pointed but not incised or hooked apex. An outstanding exception, to this general character in the subfamily is the genus *Epacmus* Osten Sacken, which fits none of the above descriptions. Possibly *Epacmus* belongs in the Cylleniinae.

KEY TO THE GENERA AND SUBGENERA OF THE LOMATIINAE

1. The second vein arises abruptly and at right angles or almost at right angles to the third vein, stem vein, but with the curve broadly rounded, usually at some point between the base of the discal cell and the anterior crossvein, rarely opposite the base. Squama always with soft, fine-hair fringe only; scales absent. Antennal style never with a well-developed whorl or brush of hairs at apex, 2 or 3 submarginal cells present. Proboscis quite short and contained within the oral cavity 2
   The second vein arises gently at or before the base of discal cell and always at an acute angle. Squamae likewise without scales on margin. Two to 4 submarginal cells present 16
2. Wing with 2 submarginal cells 3
   Wing with 3 submarginal cells 14
3. Middle of posterior eye margin without a distinct concave indentation or concavity, and without any medial fissure or groove on the middle of eye. Upper half of eye margin anteriorly receding leaving the occiput prominent. Face obliquely conical, hoodlike. Plumula present, but weak . 4
   Middle of posterior eye margin with a distinct indentation or concavity of varying depth, and with or without an accompanying fissure or groove on the middle of eye. Upper half of posterior eye margin sometimes recessive to a varying degree, or not at all receding . . 5
4. Face produced and bluntly rounded in front, not extending to the apex of the antenna or beyond it. Surface of face pollinose with dense, short, appressed, stiff pile. Anterior branch of third vein arises gently in an arch.
   ***Prorostoma* Hesse**
   Face strongly produced and peaked, rounded above, plane on the oragenal cup and face extended well beyond the apex of the antenna. Surface of face polished, with scattered setae. Anterior branch of third vein arises gently. Second vein arises rectangularly as a simulated crossvein with basal spur ? . . . . ***Coryprosopa* Hesse**
5. Front strongly swollen; face also bulging forward, at least on the upper part. Face vertical or receding, not extended beneath the antenna; not peaked, hoodlike, or conical. Xeramoebini 7
   Front not swollen 6
6. Face very short, both vertically and horizontally, even at base of antenna 8
   Face distinctly produced below the antenna, prominent, rounded in front or peaked and conical 10
7. Face vertical, the oragenal cup separated from the face by a distinct sulcus. Front of male with dense mat of subappressed silver or golden micropubescence. Abdomen slender and tapered. Anterior tibia with minute bristles.
   ***Desmatoneura* Williston**
   Face shorter and retreating below, the oragenal area not separated. Front with longer pubescence but not matlike or extraordinarily dense. Abdomen shorter, wider at base. Anterior tibia with slender spicules. Plumula present. Labellum muscoidlike . . . . ***Chionamoeba* Sack**
8. Third antennal segment onion-shaped, the short style thickened and the microsegment with a crown or circlet of a few long hairs at the apex, and also an apical spine; plumula present; in other respects differing in only specific characters from *Petrorossia* Bezzi. Proboscis-labellum of muscoid type (Xeramoebini).
   ***Xeramoeba* Hesse**
   Third antennal segment more conical, longer, the style more slender, the apical spine at end of microsegment without a circlet of long hairs. Plumula absent 9
9. Anterior crossvein at least midway between base and middle of discal cell. Abdomen broad at base and elongate oval, without scales. Alula and squamae short but distinct. Pile of front and face dense and soft and rather long. Eyes well separated. Metanotum bare. Apex of third segment often with a pencil of hairs and a style. Beaked dististylus produced into a short twisted or curved, or scroll-like process . . . . ***Petrorossia* Bezzi**
   Anterior crossvein quite at the middle of discal cell. Abdomen elongate, slender, cylindrical, covered with scales. Eyes approaching closely in both sexes. Alula and squama greatly reduced. Metanotum pilose.
   ***Petrorossia (Pipunculopsis)* Bezzi**
10. Face strongly produced but bluntly rounded. Antennae set quite far apart, separated by at least their own length. Proboscis confined to the oral recess; second vein arises a considerable distance proximal to the anterior crossvein. Squama with only flattened hairs. (See Exoprosopinae.)
    ***Synthesia* Bezzi**
    Antennae not widely separated. Face either conical or with liplike extension 11
11. Proboscis distinctly extended beyond the face. Second vein arises at the anterior crossvein, sometimes slightly before or after. Squama with a fringe of hair.
    ***Chiasmella* Bezzi**
    Proboscis contained within the oragenal recess 12
12. Alula greatly reduced, the axillary lobe narrow. Oragenal cup rather conical and peaked, or triangular, but not extending beyond the antenna. Indentation of posterior eye margin usually deep and conspicuous 13
    Alula well developed; axillary lobe wide. Oragenal cup prominent but shorter, more rounded and convex dorsally and bearing appressed scaliform pile. Indentation of eye margin weak and shallow. Anterior branch of third vein simulates a crossvein. Pile of the tapered, cylindrical, compressed, slender, elongate abdomen is largely appressed and scaliform in character . ***Epacmoides* Hesse**
13. Occiput rather long behind the ocelli, the fissure deep and narrow, the lobes almost or quite touching. Much erect, long, dense pile present anteriorly on mesonotum and anteriorly on base and sides of the abdomen.
    ***Plesiocera* Macquart**
    Occiput short, not extensive behind the ocelli, the fissure wide and correspondingly shallow . . ***Conomyza* Hesse**
14. Proboscis twice as long as head or longer. Alula absent or nearly so and axillary lobe narrow. Anterior margin of wing blackish, the black margin irregular behind. Face conical and with scales, the sides not converging below. Sides of abdominal tergites with scales, the abdomen relatively broad and oval. The second vein arises abruptly midway between base and middle of discal cell. Last section of second vein gently sigmoid at base, more strongly sigmoid apically. Squama with only flattened hairs. (See Exoprosopinae.) . . . . . ***Isotamia* Bezzi**
    Proboscis not extending beyond the apex of face. Alula and axillary lobe well developed 15
15. Middle of posterior eye margin with a deep, conspicuous indentation together with accompanying horizontal, anterior, fissure. Anterior tibia without spicules. Face conical and extending beyond the antenna.
    ***Stomylomyia* Bigot**

Middle of posterior eye margin with a shallow indentation. Face conical, extending at least to end of antenna. Second vein arising midway between base and anterior crossvein of discal cell. Indentation of posterior eye margin quite shallow . . . . **Exepacmus Coquillett**

16. Second vein with a strong radial loop giving the marginal cell a protrusive, fingerlike ventroapical lobelike form of varying prominence. Hence the second vein is strongly recurrent. Two to 4 submarginal cells present. Wings elongate and slender, the base and axillary cell either wide or whole base of wing sometimes greatly narrowed; wings usually banded or wholly brownish or veins suffused and apex sometimes chalky white in some groups, especially in males. Oragenal recess extending almost or quite to the base of antenna. Flies which are rarely less than medium size; these flies usually large and often very large. Abdomen usually elongate with only a slight taper; it is only rarely broad, wide, and flattened and such species are relatively flat with lateral, flattened tufts of pile on the tergites. Metapleuron in front of halteres bare, or pilose. Male genitalia always small and recessed within and overlapped by the hoodlike seventh and eighth tergites. Comptosiini 32

    Marginal cell at most broadened apically, the end obtuse, gently rounded or acutely angulate, the second vein meeting costa usually approximately at a right angle; apex of marginal cell commonly gently swollen below like the end of a finger. In this division the flies are either small, with obtusely tapered and narrowed or even laterally compressed abdomen with large conspicuous terminal male genitalia, and wings vitreous hyaline, rarely tinged, or flies are larger, with wider abdomen with nearly parallel sides and wings clear, basally infuscated, or suffusely clouded, banded, or spotted. Male genitalia smaller and visible only from below. Metapleuron in front of haltere either pilose or bare. A few giant species have a very wide abdomen, either flattened, with lateral tufts or pile or a convex abdomen with bushy pile laterally and posteriorly. Oragenal recess not extending to the base of the antenna 17

17. Small, slender flies for the most part; abdomen elongate and gently tapered or occasionally short oval; abdomen sometimes compressed through the middle. Male genitalia usually large, terminal, and conspicuous leaving the apex of the abdomen obtuse. Wings hyaline or rarely brownish, yellowish, or smoky at base 18

    Flies of small or large size; usually medium sized but occasionally very large and such giant species usually have conspicuous tufts of long pile in front of the haltere. Whole thorax including scutellum, always black or brownish black. Abdomen usually as wide as the thorax or wider and oval and often with bright, shining yellow postmarginal bands on the tergites; tergites thinly covered with fine, more or less erect pile or with post marginal bands of either straight or curled golden, flat-lying hairs; lateral margin of abdomen with fringe of uniformly colored pile either yellow or black or of alternating colors, sometimes formed into conspicuous flattened tufts. Male genitalia of the more or less recessive and concealed type, larger than in Comptosiini but much smaller than the Aphoebantini and nonterminal in type. Lomatiini . . 25

18. Three submarginal cells 19
    Two submarginal cells 21

19. Anterior crossvein enters the discal cell almost at the end of this cell. Wing very long and slender, about 4 times as long as wide. Anal cell widely open. First posterior cell closed and with a long stalk. Radial sector vein scarcely evident, the junction of veins almost pointlike. Front of these flies swollen and face peaked forward into a long cone of varying extent. Much opaque or feebly shining canary-yellow color on head, thorax, and abdomen (*Xanthogramma* Schiner mimics); abdomen moderately elongate, with more or less parallel sides and lateral tergal margins scalloped or crenulate. Male terminalia rather large. Male eyes separated. Antoniini.

    Antonia Loew

    Not such flies. The anterior crossvein enters the discal cell at, near, or before the middle. Male eyes holoptic . . 20

20. First posterior cell closed with a distinct stalk or petiole. Male genitalia large, terminal, obtuse. Face in lateral aspect only slightly convex or tumid, the hair upon it equally dense or distributed all over its surface and not confined to an apical brush. The first antennal segment is longer and not cuplike; third antennal segment distinctly more quickly broadened basally, therefore the base more bulblike, or onion-shaped. The vestiture of the front, face, and antenna is longer and denser and takes the form of erect, bristly hairs. Bristly hairs on the hind margins of the tergites long, erect, prominent especially on last segments of both male and female. Hind femur comparatively stout, with prominent bristles below, especially in males. Squama large, broad, earlike . . **Pteraulax Bezzi**

    First posterior cell open. Squama poorly developed. Face in lateral aspect more subconically prominent, the apical part of the oragenal cavity deep and the apical part of the face appearing to project over it due to the presence of a conspicuous brush or dense tuft of bristly hairs across apex of face; the remainder of the face without long hairs. The first antennal segment is distinctly shorter and cup-shaped; third segment more gradually broadened basally, more conical, hence the base not bulblike. The vestiture of the front consists of short hairs, still shorter and more sparse on the antenna. Pile of face sparse and short; most of face and front with dense cover of flattened scales. Bristly hairs across hind margins of tergites poorly developed . . . . . **Pteraulacodes Hesse**

21. Proboscis long and slender, distinctly extending beyond the apex of the oral cavity and thrust upward except when feeding. Third antennal segment bulblike at base. Suddenly constricted into a slender, columnlike style with small bristle-tipped microsegment twisted or hooked upward or sideways. Face protruding below; genae narrow, oral opening elongate. Labellum slender, corneous.

    Epacmus Osten Sacken

    Proboscis almost always contained within the oragenal cavity at or most barely extended beyond; labellum enlarged, fleshy, muscoidlike. Third antennal segment pyriform, conical or whole apical part thickened, sausagelike. Face vertical, more usually retreating. If hoodlike and projecting see *Plesiocera* Macquart 22

22. Third antennal segment evenly and regularly conical, little longer than wide and gradually narrowing. The apex is capped with minute, bristle-tipped microsegment. Proboscis barely extends beyond oral recess, the labellum fleshy. Abdomen short oval. Face bare above. Coloration of type-species light red. Circular margin of upper face bears a fringe in one row of strong red bristles. Middle of first tergite strongly depressed.

    Eucessia Coquillett

    Third antennal segment distinctly constricted subbasally and distinctly longer than wide. Third segment of varying length and width especially with respect to the narrowed part 23

23. Third antennal segment bearing a two-segmented style, the second segment tipped with a bristle.

    Aphoebantus (Cononedys) Hermann

    Third antennal segment with a one-segmented style . . 24

24. Vestiture of abdomen composed of abundant scales of rather slender form, or flattened, appressed hairs; hairs sparse. Metapleuron in front of haltere bare. Vestiture consists mostly of rather densely appressed scales or hair and is alike in the two sexes . . . . **Aphoebantus Loew**

    Vestiture of male abdomen, less so in females composed of abundant, long, very fine, erect, usually pale or whitish hair. Few or no scales present on body. Vestiture of males and females, however, often different. Metapleuron in front of haltere with appressed tufts of fine twisted pile . . . . . . **Aphoebantus (Triodina, new subgenus)**

25. Metapleuron in front of haltere bare; at most covered with pollen 26
Metapleuron in front of haltere with a distinct, conspicuous tuft of soft pile or bristly hair 29

26. Wings very long and slender, at least twice as long as the abdomen. Only 2 known genera and 2 known species included here 27
Wing at most 1½ times as long as the abdomen. Many species in Europe and South Africa included 28

27. Wings and whole fly quite large. Wings uniformly brown. Third antennal segment small but much constricted at the base and onion-shaped, the stylar portion long. Whole insect gray pollinose. A single species from Canary Island.
Lomatia (Canariellum) Strand
Small, gray flies not or scarcely more than 6 mm., in length.
Edmundiella Becker

28. First posterior cell closed and stalked. Unusually large, flies with straw-yellow pile and reddish-yellow antenna. Venation typically like *Lomatia* Wiedemann; the base of the anterior branch of the third vein arises rectangularly at least in some individuals and there is a stump vein. Genus monotypic . . . . **Anisotamia Macquart**
First posterior cell open though sometimes characteristically narrowed especially in South African species. Wings clear with anterior border smoky to the middle or with a suffused middle band; abdomen shining black with sulphur-yellow postmarginal bands on the tergites, which are sometimes interrupted. South African species with wings sometimes spotted, the abdomen black with postmarginal fascia of appressed golden hair; wings usually banded. Flies of medium to large size. Anterior crossvein enters the discal cell well beyond the middle . **Lomatia Meigen**

29. Giant flies with strongly convex abdomen and bushy pile laterally and posteriorly, or short pile and long flat tufts laterally. Third antennal segment long conical and slender. Proboscis short, the labellum large. First posterior cell greatly narrowed or closed and stalked 30
Quite small flies, with flattened abdomen, only slightly convex. Pile fine, erect, sparse dorsally on abdomen, with conspicuous lateral border of pile, flattened into a mat, and again at apex of abdomen. Genitalia minute. First posterior cell open maximally. Third antennal segment short conical . . . . **Ogcodocera Macquart**

30. First posterior cell closed and stalked. Proboscis slender throughout but labellum leaflike. Lateral pile all black and discal pile black except pinkish posteriorly at apex of abdomen. Very large, robust flies.
Bryodemina Hull, new genus
First posterior cell greatly narrowed. Large flies with strongly flattened, broadened abdomen, the pile appressed, black, except on postmargins and part of tufted, flattened, lateral border where it is yellow. Proboscis expanded distally and the labellum quite fleshy and muscoidlike.
Bryodemina (Brachydemia, new subgenus)

31. Small flies of about 8 mm. length with quite slender, compressed or subcylindrical abdomen and very large subglobular head. Femora and tibiae with only soft, minute, short, fine, more or less appressed pile. The anterior branch of the third vein arises by a subrectangular, simulative, false crossvein, or as in the type-species arises as an abruptly rounded arch. Wing narrowed at base, vitreous hyaline, the alula absent, the anal cell and first posterior cell widely open; second vein strongly recurrent at apex. Males dichoptic, very narrowly, but with parallel sides; ocelli set extremely far forward as in *Myonema* Roberts but with no deep trench or fossa behind the ocelli as in that genus. (See Cylleniinae.)
Docidomyia White
Not such flies 32

32. Metapleuron in front of haltere bare 33
Metapleuron in front of haltere with a distinct tuft of long, fine hair or bristly pile 42

33. A tuft of bristly hairs on the lower posterior corner of the pteropleuron. Metanotum bristly pilose laterally, the plumula wanting. Third posterior cell very much narrowed on the wing margin 45
Without such a tuft of bristly hairs on the pteropleuron. Third posterior cell not conspicuously narrowed . . . 34

34. Alula quite large and anal vein sinuous. Large flies with very wide, short-oval, shallowly convex, dull, short-pilose abdomen, the tergites with prominent, flattened tufts of rather long pile. Third antennal segment long and attenuate. Wings maculate . . . . . **Oncodosia Edwards**
Alula either reduced or absent. Anal vein straight . . 35

35. Alula completely absent. Male eyes narrowly but distinctly separated. Wing markings form a special type; chalklike bands never present in either sex; second section of third branch of the medius as long or usually longer than the first; fourth branch of radius more strongly recurved and the upturned part distinctly and characteristically sinuous. Axillary lobe quite narrow. Three submarginal cells.
Comptosia (Aleucosia) Edwards
Alula present. Male eyes usually in contact. Axillary lobe not reduced 36

36. Two submarginal cells 39
More than 2 submarginal cells 37

37. Four submarginal cells. Wings with a clear apical non-chalky band and a clear band across the middle. Axillary lobe wide, alula clearly present, excluding *plena* Edwards.
Comptosia (Epidosia, new subgenus)
Three submarginal cells 38

38. Wings speckled and clouded; medium-sized, usually blackish flies, in which the veins evidence a tendency to stubs, becoming very advanced in *norrisi* Paramonov, the selected type of *Chelina*. Alula present. Proboscis slender, distinctly longer than the oral cavity; labellum small, slender, corneous. Third antennal segment sagittate, the true style short . . **Comptosia (Chelina, new subgenus)**
Large flies, the color generally dark brown or even light red; scutellum generally red. Wings long and reddish brown or rufus, sometimes wholly reddish; often with chalky white apical or preapical band, accompanied by an elongate, red, black-vittate abdomen and micronodulate costa, or very large species with short-oval, blackish abdomen modified by dense, curled appressed pile and a pilose vitta and fasciate bands. The costa of wing sometimes micronodulate. Alula present, axillary lobe wide. Antennae always sagittate. Male eyes touching or separated . . . . **Comptosia Macquart**

39. Antennae distinctly sagittate or spearlike. Large flies with more or less strongly developed bands of brown upon the wings. Axillary lobe well developed. Alula present.
Comptosia (Alyosia) Rondani
Antennae bulbiform, the basal bulb of third segment small, the abruptly formed style long and slender 40

40. Anterior branch of third vein arising very close to the end of the discal cell. Small species with dark brown wings in which there are a few clear spots. Alula present. Axillary lobe only moderately narrowed. Proboscis extends beyond oral cavity; labellum short, slender. Species: *fenestrata*, new species.
Comptosia (Opsonia, new subgenus)
Anterior branch of third vein arising far beyond the end of discal cell 41

41. Large flies with broad abdomen wider than the thorax, convex, more or less decumbent; wings large with two distinct brown crossbands and remainder of veins more or less suffused. Alula quite wide. No lateral mats of flattened pile upon the abdomen and wing not narrowed apically, the base greatly widened and anal vein only slightly sinuous. Species: *duofasciata*, new species.
Comptosia (Paradosia, new subgenus)

Small to medium-sized flies with elongate-oval abdomen, blackish, rather uniform in color and wings barely smoky subhyaline except on the anterior border, which is more blackish. Proboscis contained within the oral cavity; labellum large, muscoidlike, sometimes longer than the the basal stem. Species: *casimira*, new species, figured.
Comptosia (Anthocolon, new subgenus)

42. Metapleuron with a strongly developed tuft of long, fine pile in front of the haltere. Two submarginal cells. Radial loop short, alula present; posterior part of wing entirely hyaline, except for a diffuse light-brown middle band; 2 of these bands, the basal and the apical are incomplete. Abdomen elongate, with parallel sides; antennae sagittate, male eyes touching. Proboscis stout and shorter than the head, barely extending beyond the oral cavity.
Doddosia Edwards
More than 2 submarginal cells. Proboscis long and slender; always longer than the head. (*Lyophlaeba* Rondani, including subgenera *Macrocondyla* Rondani and *Compsella* Hull, new subgenus) 43

43. Four submarginal cells present and the third antennal segment, long, conspicuous, flattened and swordlike. Pulvilli vestigial, stublike . . . . . . **Lyophlaeba Rondani**
Three submarginal cells. Pulvilli well developed. Third antennal segment spikelike, long slender and attenuate. 44

44. First posterior cell undivided.
Lyophlaeba (Macrocondyla) Rondani
First posterior cell divided by a crossvein.
Lyophlaeba (Compsella, new subgenus)

45. Long winged, black flies, the mesonotum yellowish vittate. Wings with quite blackish bands, the hyaline areas with chalk-white villi. Wing of normal shape at apex. Radial loop long. Two submarginal cells present. Third posterior cell long but narrowed apically . . . **Ylasoia Speiser**
Reddish-brown flies with dense, very short, bristly, reddish pile; appearance rather bare. Apex of wing distinctly hook-shaped. Wings hyaline with complex pattern of reddish sepia. Three submarginal cells and radial loop short; third posterior cell short, rounded, and club-shaped above, narrowed below . . . **Ulosometa Hull, new genus**

Genus *Lomatia* Meigen

FIGURES 130A, 139, 144, 145, 146, 216, 354, 701, 733, 903, 904, 905

Stygia Meigen, Systematische Beschreibung, vol. 2, p. 137, 1820. Type of genus: *Bibio sabaea* Fabricius, 1781, designated by Becker, 1913a, p. 460, but given as *Lomatia sabaea* Fabricius. Preoccupied by Latreille, Lepidoptera, 1804.
Lomatia Meigen, Systematische Beschreibung, vol. 3, Vorwort, 1822. Change of name. Type of genus: *Bibio sabaea* Fabricius, 1781.
Canaria Becker, Ann. Mus. Zool. Acad. Sci. Imp. St. Pétersbourg, vol. 17, p. 462, 1913a. Type of subgenus: *Anthrax brunnipennis* Macquart, 1838, by original designation. Preoccupied by Partington (1835), also others.
Canariellum Strand, Arch. Naturgesch., vol. 92, Abt. A., pt. 8, p. 48, 1928. Change of name.

Bezzi (1924), pp. 142-143, key to 17 Ethiopian species.
Paramonov (1924, 1925, 1926), keys and descriptions to many Palaearctic species.
Séguy (1926), p. 213, key to 3 species from France.
Engel (1934), pp. 363-368, keys to 17 species.
Austen (1937), pp. 95-96, key to 9 species from Palestine.
Hesse (1956), pp. 149-181, keys to 90 Ethiopian species.

Rather large flies of characteristic form and appearance. They are shining black, or brownish-black flies with the abdomen broad, slightly oval, sometimes a little wider than the thorax even in males, the surface only a little convex so that they appear to be somewhat flattened. Palaearctic species have conspicuous, narrow or wide, postmarginal, pale yellow or orange fascia on the tergites, first tergite sometimes excepted, and these fasciae may be separated in the middle so that there is a series of lateral bandlike spots instead of continuous fascia. The numerous South African species lack these yellow bands, the abdomen being wholly black or nearly black. The wing of Palaearctic species shows a characteristic, diffusely edged, brown band of infuscation covering the costal, first basal, sometimes the second basal cell, base of marginal cell and in various species extending to, beyond, or just before the origin of the second submarginal cell or downward as a crossband over the whole end of the discal cell; South African species besides having similar patterns often have the wings spotted.

The wing shows only 2 submarginal cells with the marginal cell wide apically, recurrent at apex, sometimes a little swollen but never conspicuously swollen. The first posterior cell and the anal cell are always open.

The head is hemiglobular, the short, robust proboscis barely extends beyond the confines of the oragenal recess. Front and face are short, the latter receding. The first antennal segment is large, flattened above, swollen toward the apex so that the beadlike second segment and the bulbose, apically attenuate third segments are strongly divergent. Occiput strongly indented near midpoint of eye margin. Upper occiput extensive, the lobes touching. The lower occiput is generally about half as long as the upper occiput. Pile of thorax and abdomen abundant, long, coarse, erect or semierect, forming shaggy tufts alongside of abdomen which, however, are not flattened as is so characteristic of certain other genera.

I agree with Hesse (1956) that *Lomatia* Meigen represents a highly plastic genus. The genus as represented in South Africa is certainly dissimilar to the representatives in Europe. Probably *Lomatia* Meigen is a recent genus or one that is now reaching a peak of speciation; what little we know of life histories (European) indicates predation on grasshopper eggs. Certainly there are enough species of locusts in South Africa and enough varied ecological situations to provide for an explosion of species.

Lomatia Meigen is a genus characteristic of and confined to the southern half of the Palaearctic, especially the Mediterranean, North Africa, Asia Minor, and extending below the Sahara into South Africa where Hesse (1956) lists 87 species in his remarkable memoir.

Apparently, there are a few Australian flies near *Lomatia* Meigen, and I place them here provisionally. I have collected some in Australia myself, but I have

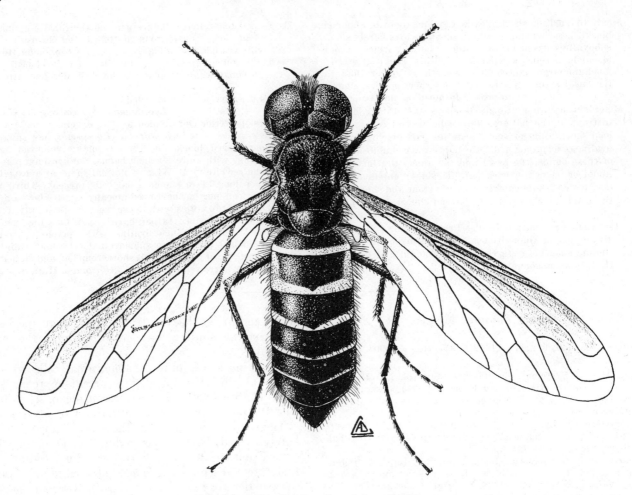

Text-Figure 44.—Habitus, *Lomatia lateralis* Meigen.

no males. It is probable that these are distinct from their European relatives, but a decision must depend upon the comparisoin of male genitalia.

The African species of *Lomatia* Meigen, as Bezzi (1924) notes, are quite dissimilar from the Palaearctic species in a number of ways. Generally speaking, the discal cell in the African species ends acutely; this is almost the rule, but there are a few exceptions such as *uniplaga* Hesse and *pterosticha* Hesse; moreover, where the European species are polished black with divided fasciae of pale yellow or at least medially narrowed bands upon the tergites, the African species are dull black without yellow bands; the tergites posteriorly have bands of appressed, curled, glittering, often golden hair; in some, the wings are mottled with obscure spots. It should be noted further that while the posterior margin of the eye is typically deeply indented in the middle, I have one South African species with completely hyaline wings and almost no indentation upon the posterior eye margin.

Included in *Lomatia* Meigen, I place *Canariellum* Strand as a subgenus.

Canariellum Strand, with type of subgenus *Anthrax brunnipennis* Macquart (1838) was described from the Canary Islands by Becker (1913a) under the name *Canaria* Becker. I am unable to pass upon the validity by which Strand changed the name. Frey illustrated this fly by a photograph in one of his papers. It may be recognized by the unusually long, very dark wings, and the antenna; third segment small, base small and bulbous, style long and quite slender; first segment strongly rounded.

Length: The flies in the genus *Lomatia* Meigen vary greatly in size. Body length: 5.5 to 16.5 mm. Wing length: 5.5 to 19.5 mm. Many of the South African species are rather small, 6 to 7 mm. in length; some species vary much in size; other species are characteristically large flies, such as the European *superba* Loew and the South African *longitudinalis* Loew, besides still other species.

Head, lateral aspect: The head is subglobular, conspicuously convex on the anterior half and equally so above and below the antennae. The occiput is very prominent above both in males and females; in females it is sometimes more than twice as extensive behind the vertex as it is at the lower eye corners, and the posterior eye margin is deeply and triangularly or sometimes roundly indented in the middle, but never with bisecting line. Pile of the occiput sparse, short, rather bristly, somewhat appressed or directed forward toward the eye margin; in some species it is much more dense, and always with a dense fringe of short, erect pile along the inner margin of the large, deep, central, occipital cup. The front is convex, a little swollen. The front has dense, shaggy, coarse or bristly pile over the entire front and vertex, except upon a small medial triangular space immediately in front of the antennae, which is pollinose only, or sometimes only shining and marked by a characteristic V-shaped depression or pit. The ocellar tubercle is prominent with pile similar to that of the front, the ocelli set in an isosceles triangle.

Below the antennae the face is short, extending no further than the front above the antennae, the oragenal cup is also short and extends to the base of the antennae. The lateral face and gena are pollinose and bear dense, rather long, shaggy or bristly pile, extending down to the base of the proboscis. The proboscis is short and stout with short, fleshy, muscoidlike, rugose labellum, which occasionally is long and pointed as in *Lomatia infuscata* Bezzi. The proboscis is thrust forward or somewhat upward, and in some species it is no longer than the oral recess, in others it may extend a little beyond, and in a few species it is long. The palpus is elongate, slender, clavate and consists of only one segment; it is sometimes almost bare, with only minute pubescence; in others it bears long, shaggy pile laterally near the apex.

The antennae are attached at the middle of the head; the whole antennae are relatively small. The first segment is large, swollen medioapically, generally longer than the second segment. The first segment has short, setate pile in the middle dorsally. It also has conspicuous, long, brushlike tufts of rather bristly pile, which are long laterally and below and equal in length to the thickness of the segment but with shorter, medial pile where the segments converge. The small, beadlike second segment is firmly attached at the outer lateral margin of the first segment, hence this segment and the third segment are strongly divergent. The third segment is either pyriform at the base and usually drawn out into strongly attenuate, stylelike lobe bearing a bristle at the apex. The swollen, bulblike base is as wide or sometimes less wide than the second segment, and the swollen part is approximately as long as wide before narrowing abruptly into the apical process. The apical process, when viewed from the inner side, is strongly asymmetrical and arises from one edge of the third segment; again, in a few species, as in *Lomatia acutangula* Loew, the whole third segment is very gently and gradually tapered to the apex, which usually bears a small spine or minute microsegment.

Head, anterior aspect: The head is more narrow than the thorax. The eyes in the male always separated, though sometimes very narrowly by as little as the width of one ocellus as in the type-species, or sometimes by as much as the width of the ocellar tubercle. The eyes of the female are separated by less than or a little more than twice the space between the outer margins of the posterior ocelli and occasionally 3, or more than 3, times this distance. The oragenal cavity is obliquely directed upward and is comparatively deep with the inner walls vertical, or even overhanging at the bottom of the head. Near the middle the oragenal recess is pinched-in somewhat toward the middle and is triangular above, leaving a bare, longitudinal ridge extending to the base of the antennae. The width of the facial space is about equal to the width of the eye, but widens a little above toward the antennae. The facial part of the oral recess and the deeply sunken border of the narrow genal portion all bear dense, rather long pile. Tentorial fissure quite deep. The posterior part of the buccal excavation may descend well below the eye and back beyond the posterior eye corners. From the posterior aspect the strongly lobed occiput has a very narrow fissure above, the two lobes usually touching.

Thorax: The thorax is short and compact, widest posteriorly across the postalar callosity. The mesonotum is not longer than wide; it is comparatively low; slightly more raised anteriorly. It tends to be somewhat flattened in the middle and behind. The color of the entire thorax is black, polished and shining on the mesonotum and scutellum but sometimes becoming reddish brown on the lower part of the pleuron. Pile of the mesonotum dense, fine, erect and moderately long, nowhere obscuring the ground color. It is equally long on the humeri, rather stiff between them, with pile extended continuously over the notopleuron. There are several weak, notopleural bristles that do not rise above the pile. Considerable dimorphism may be seen in the females of some species with respect to the pile. In males the thoracic pile is entirely black, except that of the squama, but in females the mesonotal, upper mesopleural, and scutellar pile is golden, the remaining pleural pile and squamal pile white.

Postalar callosity with 2 or 3 weak bristles. Scutellar disc with a few long, fine, erect, scattered hairs, and the margin with similar long hairs scarcely or not at all stouter. Scutellum rather large, thinning toward the posterior margin. The pleuron is pollinose. Almost the whole of the mesopleuron bears bushy, long, coarse pile along the upper border that is directed upward. On the lower half the pile is finer, directed outward, or in females the pile is coarser and the pile is directed downward. Upper half of sternopleuron with fine, erect pile, or with subappressed tufts of bushy pile. Pteropleuron pollinose only. Hypopleuron bare, except for 1 or 2 hairs. Metanotum, however, with the characteristic, dense band of stiff pile. There is no hair in front of the posterior spiracle but behind the spiracle there is a conspicuous, large tuft of appressed, coarse pile. Squama peculiar in the presence of a very wide, broad, extensive band of long, plumelike pile. Propleuron with long, collarlike pile.

Legs: The legs are comparatively stout without being swollen, the anterior femur may be slightly enlarged toward the base. All of the femora bear appressed, small scales, somewhat larger and longer dorsally on the hind femur. Anterior femur with a few fine, bristly or stiffened hairs ventrally of no great length. Middle femur similar with a few slender bristles anteroventrally near the apex. Hind femur with 3 or 4 rather stouter bristles anteroventrally on the outer half, sometimes as many as 7 or 8. Anterior tibia slender with 12 tiny, short, oblique, spiculate hairs dorsally, a few others posteriorly. Middle tibia with longer, oblique, slender bristles in 4 rows. There are 10 anterodorsally, the same number posterodorsally, and about the same number on each side ventrally. Hind tibia with 10 or 12 dorsal bristles, 7 or 8 anteroventral bristles, 6 or 8 much more slender posteroventral bristles but only minute, erect setae dorsomedially. All of the tibiae are progressively longer, the hind pair being the longest. All tarsi short, claws slender, rather strongly but gradually curved on the apical half, the pulvilli long and slender.

Wings: The wings are elongate, narrowing a little toward the base, and characteristically tinged with light reddish sepia brown to a varying extent but generally extending over the whole anterior border to a little beyond the anterior crossvein. This somewhat diffuse brownish color includes the whole of the first basal cell, anterior border of the discal cell, and sometimes extends diffusely along the posterior border of the second basal cell. In some species the whole wing is somewhat lighter on the basal part and the ground color tends to extend somewhat as a crossband over the whole outer half of the discal cell and a little below it. In several South African species the wings bear distinct but somewhat diffuse spots subapically over the base of the second submarginal cell, end of the discal cell and beyond it, base of the third posterior cell, end of second basal cell and base of discal cell. In *Lomatia neavei* Bezzi the entire wing is blackish except the costal cell and the posterior basal angle, a few individuals showing a hyaline streak in the center of the discoidal cell and in the posterior cell. There are only 2 submarginal cells and there is a characteristic form to the end of the marginal cell and the submarginal cell. Marginal cell slightly widened apically but forming a rounded never accentuated bulge. The second vein returns to the costa either rectangularly or slightly recurrent. The anterior branch of the third vein rises abruptly, makes a rounded bend and then turns up strongly toward the costa so that the first marginal cell, which is of nearly equal width throughout, makes a rather strong upward bend to the foremargin of the wing.

There are 4 posterior cells, all open, the first posterior cell always open though sometimes a little narrowed. Anal cell widely open, the anal vein straight for most of its distance, but curved apically. The anterior crossvein is oblique and enters the discal cell at the outer fifth of the type-species and sometimes either a little before or a little beyond the outer fifth. The axillary lobe is long and slender, equally narrowed basally and distally. The alula is quite narrow and linear with a plane margin. Ambient vein complete.

Abdomen: The abdomen is broadly oval, as wide basally as on the thorax and in both males and females wide apically. The whole abdomen is only very gently and slightly convex. In African species the abdomen is entirely black. In the type-species and in Palaearctic species the abdomen is characteristically marked with narrow or wide posterior yellow margins on the tergites, sometimes broken into pairs of fasciate or triangular spots by these bands being interrupted in the middle, as in *Lomatia suberba* Loew. The pile of the abdomen is rather long and fine and erect on the first 4 tergites, very dense laterally on the first tergite but becoming rather curled and appressed on the last 4 tergites. Also, the lateral pile of the tergites is more dense, stiff and bushy but never with long, flattened tufts as in some related genera. On the abdomen of males the pile on the first segment is rather whitish and straw colored, but golden yellow in the females. The marginal pile of the tergites in the males is black but in the females yellowish white. I find 9 tergites in the males, the last one small and overhanging the genitalia. In females I find only 7 tergites, the seventh tergite and sternite each bearing a dense fringe of curled pile surrounding the apex of the female abdomen.

Dr. A. J. Hesse was able to study a very large assemblage of species in this genus; for this reason I quote his comments on the male genitalia:

Hypopygium of males usually reversed in position, the true dorsal side being directed to the sides or more frequently to the ventral side, and its opposing sternite (last unmodified sternite) dorsal in position; neck region of the two symmetrical basal parts more or less narrowed and covered with hairs which are sometimes dense, longish, conspicuous and situated in distinct punctures; outer apical angle of each basal part sometimes angularly, or even sharply, produced; beaked apical joints remarkably uniform in shape, ending in a slightly outwardly and downwardly directed pointed beak and usually covered with fine shortish hairs; aedeagus shaped; lateral struts from middle part very uniform in shape, but sometimes tending to be broadish or narrowed apically or even narrow and rod-like; basal strut variable in shape, very frequently with its dorsal edge incised or emarginate or sharply pointed and in many forms with a distinct flattened, ledge-like, triangular extension on each side basally.

From my own study, I find the male terminalia from dorsal aspect, epandrium removed, show a long slender body with peculiar dististyli, which have a large, wide, curved tubercle. In lateral aspect it has a noselike tubercle at the apex in the type-species. The epiphallus is quite long, and relatively slender. From the lateral aspect the basistylus is long, a little thicker though widened basally and the epiphallus and ejaculatory process are long and spoutlike. Epandrium quite large and rhomboidal in shape. Cerci large. Basal apodeme small.

Material available for study: I have representatives of European species presented to me by the Museum of Comparative Zoology and by Professor J. Timon-

David of Marseille. Also, material of South African species received through the courtesy of Mr. B. R. Stuckenberg of the Natal Museum.

Immature stages: One species has been associated in the larval state with the egg pods of locusts. See discussion under life histories.

Ecology and behavior of adults: Hesse (1956) records one species taken on the flower clusters of *Berzelia*.

Distribution: Neotropical: *Lomatia caloptera* Macquart, 1834 (probably belongs in another genus).

Palaearctic: *Lomatia abbreviata* Villeneuve, 1912; *alecto* Loew, 1846; *armeniaca* Paramonov, 1924; *bella* Loew, 1873; *bella tibialis* Loew, 1873; *belzebul* (Fabricius), 1794 (as *Anthrax*), *belzebul lowei* Paramonov, 1931; *brunnipennis* (Macquart in Webb), 1838 (as *Anthrax*); *comata* Austen, 1937; *conspicabilis* Austen, 1937; *erinnis* Loew, 1869, *erinnis erinnoides* Paramonov, 1928, *erinnis obscuripennis* Loew, 1869; *fasciculata* Loew, 1869; *fuscipennis* Portschinsky, 1887; *graeca* Paramonov, 1928, *graeca erinnoides* Paramonov, 1931; *grajugena* Loew, 1869; *gratiosa* Loew, 1869; *hamata* Austen, 1937; *hamifera* Becker, 1915; *hecate* Meigen, 1839; *infernalis* Schiner, 1868 [=*infernalis* Loew, 1869, =*tisphone* Lowe, 1869]; *innominata* Austen, 1937; *lachesis* Egger, 1859, *lachesis graeca* Paramonov, 1928 [=*limbata* (Megerle in litt. in Meigen), 1820 (as *Stygia*), =*tysiphone* (Loew in litt. in Schiner), 1862]; *lateralis* (Meigen), 1820 (as *Stygia*), *lateralis atropos* Egger, 1859 [=*belzebul* (Panzer), 1797 (as *Anthrax*)]; *lepida* Austen, 1937; *montana* Paramonov, 1924; *multifasciata* Austen, 1937; *pallida* Tsacas, 1962; *persica* Paramonov, 1924; *polyzona* Loew, 1869, *polyzona meridiana* Paramonov, 1931 [=*hecate* Loew, 1846 (not Meigen), =*polygona* Bezzi, 1903, lapsus]; *rogenhoferi* Nowicki, 1867, *rogenhoferi caspica* Paramonov, 1926; *rufa* Wiedemann, 1818; *sabaea* (Fabricius), 1781 (as *Bibio*); *shelkovnikovi* Paramonov, 1924; *superba* Loew, 1868; *taurica* Paramonov, 1925, *taurica completa* Paramonov, 1925; *variegata* Paramonov, 1925.

Ethiopian: *Lomatia acutangula* Loew, 1860, *acutangula transvaalensis* Hesse, 1956; *albata* Hesse, 1936; *albicincta* Hesse, 1956; *albicoma* Hesse, 1956; *albizonata* Hesse, 1956; *albulata* Hesse, 1956; *apicalis* Hesse, 1956; *arenaria* Hesse, 1956; *asaphoedesma* Hesse, 1956; *atrella* Hesse, 1956; *basutoensis* Hesse, 1956; *bembesiana* Hesse, 1956; *berzeliaphila* Hesse, 1956; *bevisii* Hesse, 1956; *brunnitincta* Hesse, 1956; *canescens* Hesse, 1956; *chraecoptera* Hesse, 1956; *cinereola* Hesse, 1956; *citraria* Hesse, 1956; *cognata* (Walker), 1849 (as *Anthrax*); *compsocoma* Hesse, 1956; *conicera* Hesse, 1956; *conocephala* (Macquart), 1840 (as *Anthrax*); *consors* Hesse, 1956; *conostoma* Hesse, 1956; *crossodesma* Hesse, 1956; *desmophora* Hesse, 1956; *dimidiata* Hesse, 1956; *eremia* Hesse, 1956; *fasciolaris* Walker, 1857; *flavifrons* Hesse, 1956; *fucatipennis* Hesse, 1956; *fulva* Hesse, 1956; *fulvipleura* Hesse, 1956; *gigantea* Bezzi, 1912; *glauciella* Hesse, 1956; *grahami* Hesse, 1956; *grisealis* Hesse, 1956; *halteralis* Hesse, 1958; *hemichroa* Hesse, 1956; *heterocoma* Hesse, 1956; *hylesia* Hesse, 1956; *infuscata* Bezzi, 1921; *inornata* Loew, 1854; *jansei* Hesse, 1956; *kalaharica* Hesse, 1956; *koakoana* Hesse, 1956; *latifrons* Hesse, 1956; *latiuscula* Loew, 1860; *lawrencei* Hesse, 1956; *leucochlaena* Hesse, 1956; *leucophasia* Hesse, 1956; *leucopsis* Hesse, 1956; *liturata* Loew, 1860; *loewi* Bezzi, 1912 [=*inornata* Loew, 1860 (not Loew, 1854)]; *longitudinalis* Loew, 1860; *marleyi* Hesse, 1956; *matabeleensis* Hesse, 1956; *melampogon* Loew, 1860; *melanoloma* Hesse, 1956; *melanthia* Hesse, 1956; *mesoleuca* Hesse, 1956; *mitis* Lowe, 1860; *mollivestis* Hesse, 1956; *monticola* Hesse, 1956; *mozambica* Hesse, 1956; *natalicola* Hesse, 1956; *namaqua* Hesse, 1956; *neavei* Bezzi, 1924; *nigrescens* Richardo, 1901, *nigrescens aterrima* Hesse, 1956, *nigrescens bulawayoensis* Hesse, 1956; *nivosa* Hesse, 1956; *oreoica* Hesse, 1956; *orephila* Hesse, 1956; *ovamboensis* Hesse, 1956; *pedunculata* Hesse, 1956; *phaenostigma* Hesse, 1956; *pictipennis* (Wiedemann) 1828 (as *Anthrax*) [=*aurata* (Macquart), 1846 (as *Anthrax*), =*centralis* (Macquart), 1840 (as *Anisotamia*)]; *plocamoleuca* Hesse, 1956; *pseudofasciata* Hesse, 1956; *pterosticta* Hesse, 1956; *pulchriceps* Loew, 1860, *pulchriceps linganaui* Hesse, 1956, *pulchriceps ogilviei* Hesse, 1956, *pulchriceps tinctella* Hesse, 1956; *punctifrons* Bezzi, 1924; *purpuripennis* Hesse, 1958; *salticola* Hesse, 1956; *semiclara* Hesse, 1956; *septoptera* Hesse, 1956; *sericosoma* Hesse, 1956; *simplex* (Wiedemann), 1819 (as *Anthrax*); *sinuosa* Hesse, 1956; *spiloptera* Bezzi, 1924; *stenometopa* Hesse, 1956; *stenoptera* Hesse, 1956; *subcaliga* Hesse, 1956 *tenera* Loew, 1860; *thysanomela* Hesse, 1956; *uniplaga* Hesse, 1956; *vicinalis* Hesse, 1956.

Oriental: *Lomatia brunnipennis* (Wulp), 1868 (as *Comptosia*).

Autralian: *Lomatia anthracina* (Thomson), 1869 (as *Comptosia*); *australensis* Schiner, 1868; *sorbicula* (Walker), 1857 (as *Anthrax*), *sobricula*, erratum.

Genus *Anisotamia* Macquart

Anisotamia Macquart, Diptères exotiques, vol. 2, pt. 1, p. 81, 1840. Type of genus: *Anisotamia ruficornis* Macquart, 1840. Designated by Coquillett, p. 507, 1910a, the first of 2 species. The second species has been transferred to *Lomatia* Meigen.

Since I have not seen *Anisotamia* Macquart, I give a translation of Engel who describes this genus as follows:

In its plastic characteristics this genus scarcely differs at all from *Lomatia* Meigen, from which it is distinguished only by the closed first posterior cell, the prevalence of yellow body coloration, and its size.

The eyes of the male are separated by the ocellar tubercle and diverge immediately below it; in the female the vertex of the head is somewhat broader. The frons is thickly pilose almost up to the ocellar tubercle, and with hairs of the same length as the face. The thorax has some yellow parts in basic color too. The pile on the metapleuron is wanting or absent, as in the case of *Lomatia* Meigen, but the sides of the metanotum are similarly pilose. On the yellow legs, the pulvilli are fairly narrow. The abdomen is the same width as the thorax, and is about one and a half times as long, primarily yellow in its basic color, with black anterior parts on the tergites and sternites. The epipygium is small, and like the probably spined ovipositor, concealed in the last segment.

Metamorphosis unknown.

The only species is spread over all of north Africa as far as Asia. The genus *Oncodocera* Macquart, cited in the Kertész catalogue as a questionable synonym for *Anisotamia* Macquart, is a distinct genus with a completely different body and antenna.

Only a single species is known. Engel describes the unique type-species in the following terms:

Male and female in large part predominantly straw yellow. The head is straw yellow pilose on a reddish yellow basal color of face and front; the back of the head is black, but also with short yellow hairs. The antennae are reddish yellow, of the same form as in *Lomatia* Meigen; a few short little bristles on its basal segments are black. The eyes of the male are separated from each other by the ocellar tubercle, and closer to each other for only a short distance before it; in the female, on either side next to the ocellar tubercle, there is an intermediate space, which is not quite as long as the distance of the front one from the side ocellus, and the area around the ocellar tubercle is red yellow in base color. The facets are slightly enlarged in the male in a band or area next to the front. The thorax is black in basic color, somewhat shiny on the mesonotum, dull on the pleuron; the base of the yellow scutellum is also more or less black. In the female the sides of the mesonotum at the level of the alar callosity are also red yellow in basic color. The pile of the prosternum, pleura and coxa is yellowish white; that of the notopleural suture, of mesonotum and scutellum is straw yellow. The legs are yellow with short, black little bristles and darkened tarsi. The pulvilli are short and narrow. The wings weakly gray yellow, with yellow-brown bordered veins, of which only the tips of the second and third radius and the forked branches of the fourth and fifth and the anal vein are unbordered.

The venation corresponds to that of *Lomatia* Meigen, only the first posterior cell is closed off and stalked. The discal cell is shorter and wider than in *Lomatia* Meigen; the diagonally situated anterior crossvein, however, is in the same position as in the latter. The wings of the female are colored somewhat paler. Squama with white fringe; haltere ochre yellow. On the abdomen the sides of the first tergite are protruding and bear longer straw yellow hairs than the others. The anterior margins of all tergites are black on half their segment length, which color extends with an arc in the middle, into the reddish yellow back half. In other words, the back edges of the tergite are very broad, those of the sternite also broad and colored yellow. On the latter, only the front edge remains dull black for a third of the length of the sternite. The pile of the abdomen is also straw yellow.

Material available for study: There is a single female individual of *A. ruficornis* Macquart in the collections of the British Museum (Natural History), which was recorded from Palestine by Austen (1937).

Since *Anisotamia* Macquart is so much like *Lomatia* Meigen in every respect, except for the closed first posterior cell, I have not redescribed it; instead I quote Engel's description.

Immature stages: Unknown.

Ecology and behavior of adults: Austen recorded the above-mentioned female as apparently investigating and possibly ovipositing in the nest of fossorial Hymenoptera. The species is well known in Egypt.

Distribution: Palaearctic: *Anisotamia ruficornis* Macquart, 1840.

Genus *Ogcodocera* Macquart

FIGURES 122, 713, 724, 911, 912

Ogcodocera Macquart, Diptères exotiques, vol. 2, pt. 1, p. 83, 1840. Type of genus: *Mulio leucoprocta* Wiedemann, 1828, by monotypy, as *Ogcodocera dimidiata* Macquart, 1840.
Oncodocera Osten Sacken, Bull. U.S. Geol. Surv., vol. 3, p. 247, 1877, emendation.

Small flies of black color and broad, rather flattened abdomen. They can be quickly recognized by this broad, rather flattened, short oval abdomen, which is quite wide apically in males, with the last tergite an equilateral triangle in females, but more especially by the peculiar distribution of pile. The surface of the tergites has fine, scattered, long, erect hairs, but the sides of the tergites have a dense flat brush of matted hair extending in a flat platelike formation out directly away from the sides; this lateral hair band is very long and at least half as long as each lateral half of the tergites. Lateral pile in the type-species is entirely brownish black, but the posterior margin of the last tergite has an equally long fringe of opaque snow-white pile forming a shelf above the small genitalia.

Wings with basal half obliquely brownish, the outer part hyaline. There are 2 submarginal cells, the anterior branch of the third vein is gently sinuous and the marginal cell is scarcely widened apically. Four posterior cells are present, all of them open widely; the anal cell is closed in the margin or near it. The anterior crossvein enters the discal cell at or a little beyond the outer third of that cell.

The head is hemiglobular, but the occiput is longer in females and at least twice as long as in males. The upper occipital lobes are tightly apposed, and the indentation near the midpoint of the posterior eye margin is comparatively deep. The antennae are small and set at or slightly below the middle of the head so that the front, while very short in extension, is of great length from vertex to antenna and strongly convex. The first antennal segment is relatively large, widened apically, the second segment small, the third elongate conical and strongly tapered to a fine point and bearing a minute apical spine.

Ogcodocera Macquart is certainly related to *Lomatia* Meigen, an Old World genus. It differs in the smaller size, the closed anal cell, widely open in *Lomatia* Meigen, and in the exceptionally wide flattened bands of long pile along the sides of the abdominal tergites. *Ogcodocera* Macquart is a genus of few species restricted to the Nearctic and Neotropical regions.

Length: 5.5 to 8 mm.; wing, 6 to 9 mm.

Head, lateral aspect: The head is comparatively short with the upper anterior half very convex and the lower anterior half below the antennae more nearly plane due to the receding face and the receding eye. The occiput is very extensive on the dorsal half, even more so in the middle, at which point the posterior eye margin is angularly indented and has a small notch. The upper eye facets are considerably enlarged. The pile of the occiput is dense but quite short and appressed forward and reaches the eye margin. There is a fringe of short erect hairs along the inner margin of the circular occipital cup. The front is convex, densely covered with long pile, erect along the eye margins, somewhat appressed in the middle of the front. This pile is scaliform, silvery, and matted in the males of some species, black and erect in other species; this pile continues to extend down between the widely separated antennae. The face is very short, the oragenal recess lying below the eye margin. The eye is extensive, somewhat reniform.

The proboscis is short, flattened laterally with a short, flattened, lanceolate labellum turned upward. The proboscis is slightly longer than the oragenal recess. The palpus is quite slender and composed of one segment and about one-third as long as the proboscis. It bears a very few fine, long, appressed hairs. The antennae are small, the first segment short, broad, almost completely concealed by the matted pile of the front overlapping its own extensve pile. The first segment bears a tuft of long, silvery, flattened hairs above, another below and medially with rather less pile laterally. The second segment is only about half as long as wide, the third segment is a little longer than the first 2 segments combined. It is long conical being gradually tapered from the base and has a minute spine at the apex. At the base it is not quite as wide as the second segment.

Head, anterior aspect: The head is circular, and not quite as wide as the thorax. At first sight it appears to be much more narrow than the thorax due to the deceptive, long, bushy pile on the mesonotum and pleuron of the thorax. Eyes in the male holoptic, in contact for about one-fourth of the distance from ocellus to antennae. The ocellar tubercle is small but high and the ocelli are set in a short isosceles triangle. A tuft of fine, stiff hairs arise between the ocelli. The front is quite wide forming an almost equilateral triangle. In females the eyes are separated a space equal to the first two antennal segments. The small, low ocellar tubercle lies at the posterior corner of the eye and is not set slightly forward as in the male. The anterior part of the vertex is swollen, but the front is rather sunken actually below the level of the eye; its pile is similar to that of the male but less extensive. Most of the antennal pile is brownish or black. The oragenal cup is elongate, rather deep, the inner wall vertical or even partly cupped over; near the middle it is somewhat compressed inward, and it is separated from the eye margins by a deep foveal recess. This shelving outer wall bears a fringe of coarse, erect pile of no great length. From the posterior aspect the occiput is bilobate with the dorsomedial fissure and accompanying depression extending only half the length of the occiput. The occipital lobes are tightly apposed.

Thorax: The thorax is quite short but distinctly widened posteriorly; the mesonotum rather flat and low. The whole thorax is of very little depth. The pile of the mesonotum is dense, erect, fine and quite bushy and rather long. In the material that I have seen, these flies are usually more or less denuded on the posterior half of the mesonotum, scutellum, and generally rather widely over the middles of the tergites. However, it appears that normally this fine, long, erect pile is comparatively abundant over the whole of the mesonotum and the tergum of the abdomen and much of the disc of the scutellum, but certainly this pile is much more dense and bushy on the anterior third of the mesonotum, the humerus and the notopleuron, and perhaps even longer and denser on the upper mesopleuron. There are fewer hairs on the lower mesopleuron and sternopleuron; the pteropleuron is without pile, but there is long, dense pile on the metanotum, and a large, oval, conspicuous tuft of long pile on the metapleuron in front of the halteres, a smaller, shorter tuft behind. The scutellum is rather large, wide, slightly triangular but obtuse and rather flattened. The squama is small and quite short but bears a peculiar, very wide, dense brush of pile widened out dorsoventrally. In this respect it is quite like *Lomatia* Meigen. In *Ogcodocera* Macquart there is a sexual dimorphism in the color of the pile. In females the mesopleural, squamal, scutellar and basiabdominal pile, including the lateral tufts, is snow white. In males these areas have reddish or brownish-black pile.

Legs: The legs are weak; the femora are slender and loosely covered with flat-appressed, rather long, dark scales. The small anterior femur is shorter than the middle pair and has only a few fine hairs below, some 8 or 9 altogether. The middle femur is considerably shorter than the hind femur, and contains 15 or 20 long, fine, ventral hairs. Hind femur with a similar number of hairs below and 2 or 3 quite slender, apical, bristly hairs ventrally, a few others dorsally. Middle tibia similar, the stiff dorsal hairs a little longer; the 7 or 8 oblique, short, bristly hairs ventral in position. Hind femur with 8 or 9 very slender, short bristles anteroventrally and with 10 to 12 bristles anterodorsally, and nearly as many dorsomedially. There are 6 ventromedially. Tarsi short and slender, claws long and straight but sharply curved at the apex; the slender pulvilli are long.

Wings: The wings are rather broad basally with the basal half obliquely dark reddish sepia but scarcely extending into the axillary lobe and not all of the anal cell. The alula and base of the axillary lobe, however, are dark. The sepia color fills out the costal cell and extends farther back basally in the marginal cell, still farther back in the first submarginal cell and colors the base of the discal cell and extends downward posteriorly along from the base of the fourth posterior cell. The remainder of the wing is hyaline and villose. There are 2 submarginal cells with the anterior branch of the third vein arising obliquely and rather sigmoid. Marginal cell is not at all widened apically. All the posterior cells are widely open and the anal cell is closed in the margin, or more usually has a short stalk. Ambient vein complete, the alula moderately wide.

Abdomen: The abdomen is quite broadly oval, usually wider than the thorax and rather strongly flattened so that it is only a little convex. I find 7 tergites in the male and 7 in the female. Unabraded, fresh material has a considerable amount of long, fine, erect pile on the middle of the tergites with some shorter, flat appressed, fine, nonscaliform pile arising at the base of the tergites and posteriorly directed and appressed on the tergites. The abdomen, however, is conspicuous and noteworthy for the extraordinarily long fringes arranged along the sides of the tergites as a flat bank of outwardly directed, overlapping tufts of pile. In length these tufts of pile are almost as long as half the width of the abdomen; hence they add materially to the apparent width of the abdomen. Female terminalia are surrounded and concealed on each side by a dense, matted tuft of pile leaving the barest longitudinal slit visible. Male terminalia small, the basimeres and dististyli turned ventralward side by side beneath a curiously notched and hoodlike, apical tergite.

The male terminalia from dorsal aspect, epandrium removed, show an unusually slender body with a very long, epiphallus progressively widened toward the base. The dististyli are unusually large and viewed from above are stout, obtuse, long, wide, and conspicuous. Viewed laterally, however, they are even more conspicuous, and strongly curled upward. The basistylus is long, but slender, the basal apodeme, small and short, the epiphallus is long and spoutlike but curved upward on the outer part. The epandrium is small, triangular, with a very long basiventral lobe or process.

Material available for study: I have collected 2 species in this small genus: one Nearctic species, not common in Mississippi; another species in southern Mexico. I have a third species from south Brazil.

Immature stages: Unknown.

Ecology of adults: These flies can be collected on flowers such as fleabane and other composites in early summer.

Distribution: Nearctic: *Ogcodocera analis* (Williston), 1901 (as *Oncodocera*); *leucoprocta* (Wiedemann), 1828 (as *Mulio*) [=*dimidiata* Macquart, 1840, =*leucotelus* (Walker), 1852 (as *Anthrax*), =*terminalis* (Wiedemann), 1830 (as *Anthrax*)].

Bryodemina Hull, new genus

FIGURES 207, 375, 695, 898, 899

Bryodema Hull, Ent. News, vol. 77, p. 227, 1966, preoccupied Fieber, 1853. Type of genus: *Anthrax valida* Wiedemann, 1830.
Brachydemia, new subgenus. Type of subgenus: *latisoma*, new species.

Very large, giant flies of exceptionally stout, robust form. The are related to *Ogcodocera* Macquart, which are very much smaller, with flattened abdomen accentuated by the remarkably long, flattened tufts of pile along the sides of the tergites, which are scarcely noticeable in *Bryodemina* Hull. In *Bryodemina* the first posterior cell is closed and stalked, the anterior crossvein, which is quite oblique, enters the discal cell at the outer fifth of the discal cell, not near the middle of the cell. The marginal cell is widened apically and ends obtusely. The third antennal segment is long, and gradually attenuated, narrow, and spikelike.

Flies are abundant in central Mexico. They are nearly related to another broad, wide, somewhat similar fly from Brazil, which I place in the new subgenus *Brachydemia*. Here the anterior crossvein is in the same relative position, very oblique, and the first posterior cell is greatly narrowed. The marginal cell is similar and the antennae are similar. Nevertheless, these very large flies have a thinned abdomen with very conspicuous radiating tufts of dense, flattened tufts of pile along the lateral margins of the tergites. Figured in Williston (1908).

Distribution: Central Mexico and southern Brazil.

Length: 22 mm.; rarely as little as 13 mm.; wing 17 to 20 mm.

I have before me 3 species of *Ogcodocera* Macquart, the type-species *leucoprocta* Wiedemann, *analis* Williston and a third species from South Brazil; all are almost equal in size, small, flattened, with lateral tufted pile and basally swollen third antennal segment. The anterior crossvein lies quite close to the middle of the discal cell, the first posterior cell is open maximally.

Elsewhere, I proposed the name *Bryodemina* for the giant bee flies with very robust, thickened bodies that formerly went under the name *Ogcodocera valida* Wiedemann. These differ from *Ogodocera* Macquart, in the strict sense, of which I have 3 species before me, in the very thick body and giant size, in the closed and stalked, or narrowly open first posterior cell, and the position of the anterior crossvein, which is very strongly oblique and placed at the outer sixth of the discal cell. All of the 3 species of true *Ogcodocera* Macquart before me, including the type-species and one ranging to south Brazil, are of nearly uniform size, flattened body, basally swollen third antennal segment, the first posterior cell open maximally; the anterior crossvein is very near the middle of the discal cell and almost rectangular.

Head, lateral aspects: The head is subglobular and the occipital component is quite extensive with the dorsal lobes tightly apposed and with shallow crease. The posterior eye margin is only slightly indented and angularly indented with a very short bisecting notch. The lower half of the occiput is almost as strongly produced as the upper half; from the dorsal aspect it slopes strongly backward. The pile of the occiput is dense and fine, very short, and setate; the fringe around the occipital cup is quite short but dense and in the type-species pale in color in contrast to the black pile of the occiput. The face extends downward below the antenna a distance nearly equal the length of the first two antennal segments. In profile the extreme upper face and antennal base are scarcely visible and extend only a little way beyond the eye. The prominent frontal area is somewhat convex and bullose and is distinctly seen from the lateral aspect. On the lower part of the head, as well as the upper part above the antenna, the head is strongly rounded and retreating. Since the labellum of the proboscis extends above the apex of the oragenal recess and reaches the antennae, in consequence, this area rests against the antenna. This area is without pile. On each side, however, the pile is dense, coarse, directed outward, and extends a little beyond the second antennal segment. It becomes shorter below along the sides of the deep tentorial fissure; the lowermost sides of the gena and bottom of the head are without pile. The frontal area is opaque, and like all of the remainder of the head blackish, and the pile is dense, fine, erect, and nearly as long as the first two antennal segments.

The proboscis extends as far as the base of the antenna. The labellum is large, oval, flattened, and compressed on the outer half. It makes up not quite half the length of the proboscis. Palpus small, quite slender, deeply recessed, and lying against the base of the proboscis; it appears to have a minute, apical segment. The antennae are attached at the middle of the head; the first segment is quite large; swollen, twice as long and twice as wide as the small second segment; it bears some scattered, bristly pile laterally and also some medially toward the apex. The second segment is short, beadlike, with a row of stiff setae dorsally. Third segment at the base not quite as wide as the second segment and about twice as long as the first two segments together. It is gradually and distinctly but slightly tapered to the rather thick apex which bears a small, short spine; both upper and lower surfaces are nearly plane.

Head, anterior aspect: The head is nearly circular but much more narrow than the thorax. It is slightly flattened laterally and again at the vertex. The eyes in the male are strongly holoptic; they join at a point a little less than halfway from the posterior eye margin to the antennal base. The ocellar triangle is a short isosceles triangle and in the males makes up about one-half of the rather small vertical triangle. The ocellar triangle is quite low and scarcely visible in profile. The frons forms a widely expanding equilateral triangle, its length a little more than half the distance from the posterior eye margin to the antenna. In the female the occiput is even more extensive and may be approximately four-fifths as long as the eye itself; the junction point of the dorsal lobes in both sexes is set back a considerable distance from the corners of the eye and from it radiates slender grooves or striations. The oral opening is wider below, narrowing slightly above, and its walls are thin and knifelike on both sides.

Thorax: The thorax is quadrate and scarcely if any longer than the width across the postalar callosity; it is rather flat especially on the posterior half with the pile extraordinarily dense, black, fine, and erect, with a dark, reddish-brown cast, over the middle of the mesonotum when viewed laterally. The scutellum is quite large and hemicircular, slightly convex, but strongly convex on the margin, and shining brown or blackish. The pile is dense, fine, erect, and blackish. Largely bare or denuded in the middle the pile is distinctly more abundant on the sides of the scutellum; this seems to be in part a continuation of the pile arrangement of the abdomen, which is much more dense on the sides of the abdomen leaving a narrow, longitudinal stretch down the middle which has only scanty pile. There are 4 or 5 quite slender bristles on the notopleural area which barely rise above the basic pile. There are several still more slender bristles on the postalar callosity. Pile on upper mesopleuron quite dense, fine and long, thinning out and becoming much more scanty below; it is still more scanty on the sternopleuron; the hypopleuron appears to lack pile. There are some scattered hairs on the pteropleuron, and much long fine pile is on the metanotum and also on the metapleuron above and in front of the spiracle. The haltere is small and slender; the squama has a very dense, long, fine fringe.

Legs: Rather small, and short on all 3 pairs. Femora stout without being swollen. Above, they bear coarse, appressed pile scarcely if at all flattened, and with short, fine, ventral fringe posteriorly on the first four femora. Each of these first four femora are slightly widened toward the base, and slightly bent at the apex; middle femur with 8 or 9 short, oblique, black bristles; hind femur with the appressed pile, black but shining and slightly flattened. The anteroventral margin bears 8 to 10 slender, oblique bristles and a few fine hairs. Anterior and middle tibiae quite slender, with short, fine, appressed setae and a few very fine, short, stiff, oblique bristlelike hairs dorsally. The hind tibia is at least twice as thick as the middle tibiae and very little longer; it is slightly compressed, and bears both dorsally. The hind tibia is at least twice as thick as the middle tibiae and very little longer; it is slightly compressed, and bears both dorsally and ventrally a dense, oblique fringe of moderately long, bristly hairs which lie in one plane and slightly longer dorsally and scarcely longer than the width of the tibiae. The fringe disappears toward the apex of the tibiae. Hind basitarsus narrow and tapered toward the apex and distinctly longer than the four remaining segments; pile short, fine oblique and setate. Pulvilli well developed; claws sharp; they are bent in a rectangular curve.

Wings: The wings are large, and narrow apically, quite wide basally. The axillary lobe is especially prominent. The alula has the postmargin plane or slightly concave and is about 3 times as long as wide; the ambient vein is continued strongly to the base of the alula. The basicosta is strongly widened and flattened in both sexes and bears dense, appressed, slightly flattened pile. I do not find any hook or spur. The veins are stout and the anterior crossvein is strongly oblique entering the discal cell at the outer sixth. The first posterior cell is closed and stalked; the two veins on the outer half of this cell are plane and the intercalary vein is long. The second vein makes a rectangular bend entering the costa at a right angle leaving the marginal cell widened and obtuse at the apex. The anterior branch of the vein arises a gentle arch; it is bent before the middle; all the remainder is quite plane and it ends well before the apex of the wing. Anal cell closed with a short stalk; wing hyaline, with dark brown villi and with foreborder tinged with brown back as far as the discal cell and a little beyond the anterior crossvein.

Abdomen: Very large, thick and convex and wider than the thorax, appearing much wider because of the lateral mats of pile. There is much extreme, dense, long, fine, mostly shaggy and appressed pile over the tergites; this pile is a little more erect basally more matted and appressed posteriorly; it forms a wide, flaring, lateral mat. Narrowly, longitudinally the pile on the medial part of the tergites is thinned. The whole abdomen is faintly shiny black obscured by the dense pile. In type-species the lateral basal pile is black with the apical pile forming 2 widely separated expanding mats of brownish, cream-colored pile; in other species there is a transverse band of pile. Male terminalia small, posterior, recessed. There are 8 tergites visible above in males.

The male terminalia from dorsal aspect, epandrium removed, show a figure in general similar to that of *Ogcodocera* Macquart, to which it is related. The dististyli are wider apically, the lateral apodeme is longer and larger. The basal part of the aedeagus is much larger and bulblike. The epiphallus does not extend to the end of the basistylus. The epandrium is similar, but shorter and the cerci are larger.

Material available for study: A large series of both sexes of the type-species collected by the author and wife in central Mexico in the late summer of 1962. Some also collected by my son, Clovis Sillers Hull. Still others collected when in Mexico in 1951 a few miles from Cuernavaca. Two females of *Brachydemina* from Chapada, Brazil, presented to me through the courtesy of Dr. Curran.

Immature stages: Unknown.

Ecology and behavior of adults: The author found a series hovering in early afternoon beneath an apple tree on the outskirts of Cuernavaca; there was much high grass. Most of those collected in 1962 were in a lush meadow-like area about forty miles east of Guadalajara; others were in wet, dark woodland areas where there were patches of sunlight; a few were found on comparatively open hilly slopes. All were hovering and none were seen alight.

Distribution: Neotropical: *Bryodemina fasciata* (Williston), 1901 (as *Oncodocera*); *hedickei* (Paramonov), 1930 (as *Oncodocera*); *latisoma*, new species (subg. *Brachydemia*), figured in Williston (1908); *valida* (Wiedemann), 1830 (as *Anthrax*) [=*eximia* (Macquart), 1850 (as *Anisotamia*)].

Genus *Edmundiella* Becker

Edmundiella Becker, Wiener Ent. Zeitung, vol. 34, p. 347, 1915.
Type of genus: *Edmundiella niveifrons* Becker, 1915, by monotypy.

I have not seen *Edmundiella* Becker. I give a translation of Becker, who describes this genus in these terms:

This diptera species has the characteristics of the group: the antennae which are close together, the non-protruding face, the small, round head with the indentation on the posterior margin of the eyes, and the bisection of eyes as well as the wing venation. It differs in the posterior body, which is strongly and heavily covered with tufts of hair on the sides, whereby it looks much broader than it really is.

This is strongly suggestive of *Ogcodocera* Macquart, it is not impossible that it is the same.

A medium sized, entirely dull black species. Thorax short and distinctly pilose. Head, viewed from the front, circular, viewed from above somewhat broader than long, but not wider than the thorax. Eyes in the male touching or holoptic for a short distance, situated high, bare, indented or excavated on the posterior margin; of the head; the head (occiput) protruding conspicuously (wulstartig). The upper border of the mouth extends to the antenna, so that a facial surface is not present; the distance of the edges of the eyes from each other is scarcely more than one eye width, narrowing toward the bottom. Antennae approaching each other at the base, but otherwise diverging, almost at a right angle; the second segment thickly and heavily pilose, the third lengthened or elongate in the form of a cone, with a small, fine, blunt apex and a delicate short terminal bristle, which is located not quite in the center. Front (frons) densely pilose to below the antennae. Proboscis short, in the example at hand stretched perpendicularly downward, with narrow, insignificant palpi; the base of the proboscis extends to the antenna. Abdomen no wider than the thorax; about twice as long as it is wide, the lateral margins are distinguished by bristle-like tufts of hair, standing far out, so that the abdomen with its fairly dense pile takes up a great width and thus approaches a circular shape. Legs simple, the under side of the femora without bristles. Pulvilli are present; wings long, narrow; the second and third branch of radius forms a slender arch as it opens into the anterior border; it is arched and right angled. The forked vein of the radius takes a course completely parallel to the other. The first posterior cell is opened widely, with parallel longitudinal veins, the small crossvein lies somewhat diagonally and distinctly beyond the middle of the discal cell.

Male. Thorax completely black, the pile somewhat long, gray brownish to black, sides of the pleuron black haired. Haltere brown with a bright stem. Head: back of the head dull black, bare; the whole forehead is covered densely with silver white to gray glittering hairs which are directed downward. At the edges of the eyes there are small black hairs. The upper mouth opening and the second antennal segment are also covered with dense white hairs. Antennae black, in the form as stated above; proboscis and palpus also black. Abdomen completely black, dull, every segment with thick side strands, which consist of black brown hairs accompanied by a few whitish ones, particularly on the anterior segments; on the upper side there are only few hairs, the last segment is black pilose. Sternites bare. Legs completely black. Pile and bristles only delicate. Wings with the aforementioned venation; the basal half is brown, the limitation of the color runs from the end of the first longitudinal vein diagonally over the base of the discal cell and still borders the fifth longitudinal vein.

Length of the body and the wings 6 mm.
A specimen from Greece.

Material available for study: None.
Immature stages: Not on record.
Ecology and behavior of adults: None recorded.
Distribution: Palaearctic: *Edmundiella niveifrons* Becker, 1915.

Genus *Myonema* Roberts

Figures 135, 140, 219, 388, 626, 629, 757

Myonema Roberts, Proc. Linnean Soc. New South Wales, vol. 54, p. 563, 1929. Type of genus: *Myonema humile*, 1929, by original designation.

Flies of medium size, characterized by a wing in which there are several features worthy of note. The second vein is almost plane throughout, showing a mere trace of curve; the anterior branch of the third vein shows very little curve at base or apex but does turn up slightly on the outer half; the posterior branch of the third vein is slightly curved backward, the first branch of the medius slightly curved forward so while the first posterior cell is narrowed it is still rather widely open. The hyaline, generalized wing has an anterior border of dark brown extending downward irregularly, in what Roberts (1929) called three "waves." The anterior crossvein lies before the middle of the broad discal cell; the second vein arises acutely at opposite the base of the discal cell, and the anal cell is widely open. The head in profile is subtriangular, with the eye reniform and extending to the full height of the head, but it does so behind the ocelli and opposite the bilobate occiput. The ocelli are set very far forward, leaving a deep fossa extending for a considerable distance backward to the posterior eye corner. The slightly concave occiput has a deep, central, cuplike recess, but there is no anterodorsal recession of the eye exposing the upper occiput, and there is no indentation and bisecting line at the middle posteriorly.

Finally, it should be noted that the entire head and antennae are much more like that of the Lomatiinae, a fact which Roberts (1929) astutely observed; the short first antennal segment is swollen and has long bristly pile below; the second is of about the same length, a little convex above and below and much less wide; the third segment is conical, a little constricted before the narrow apex, which bears a minute spine. The oragenal cup is moderately prominent and produced above a little beyond the second antennal segment.

This fly is known only from the type-species from Brisbane, Australia.

Length: 10 mm.; wing 9 mm.

Head, lateral aspect: The head is approximately subtriangular in profile because of the oblique anterior extension of the face and oragenal cavity jutting forth beneath the antenna. From its farthest forward point the face rather strongly recedes backward and downward. The eyes reach the full height of the head but only behind the ocelli which are placed anteriorly and upon a raised flattened tubercle; therefore, the eyes extend to a point opposite the bilobate occiput. The face is comparatively small from upper border of oragenal recess to the antennae, but its extent is best gauged from the accompanying figure; it narrows a little below, with the narrowing of the recess.

The antennae are relatively small. The first segment is obtuse and swollen, strongly convex above on which side it bears a few setae and the upper margin is a little longer than the lower, nearly plane margin; below, it bears numerous, dense, slender bristles that are almost as long as the first two segments combined. The second segment is not quite as long as wide, but much less wide than the first segment into which it is rather snugly fitted; it is also tightly attached to the third segment, and it bears a few short setae dorsally. The third segment is as long as the first two segments combined and shaped more or less like a slightly convex cone which strongly narrows beginning at the middle and ends in narrow, slender apex slightly constricted subapically; at the apex is a trace of a very minute stublike microsegment, which bears a small slender, apical spine. Roberts called it spindle-shaped when viewed laterally, but with its inner surface concave when viewed from above. The palpus is slender and curled and about half as long as the proboscis; it has a short apical segment, the both bear short, fine, erect hairs.

The proboscis is stout and rather short, barely longer, if as long, as the oral recess; the labellum is large; the whole structure may be invisible laterally if tucked into the recess. Its length not longer than the head length. The front and vertex bears a considerable amount of scattered, stiff, erect, fine, bristly hairs perhaps as long as the first two antennal segments; this pile continues opposite the antenna as a downturned fringe of similar hair and continues down along the sides of the upper face and the genal area below to the bottom of the eye.

The occiput is slightly concave but the concavity extends from near the vertex to near the lower point of the eye and is therefore spread out over most of the vertical height of the head; there is no dorsal recession of the eye exposing the occiput as in many other genera of this group. There is no sharp middle posterior indentation or bisecting line on the eye. The occiput bears only a few fine, scattered hairs, the central fringe short.

Head, anterior aspect: The head is nearly as broad as the thorax. Viewed from above, the occiput is raised and bilobate; it has a large, central cavity. In males the eyes are dichoptic, separated by about half the width of the large, rounded, ocellar tubercle, which is situated forward and well below the vertex of the head; ocelli large. Antennae rather close together; first segment distinctly cupuliform.

Thorax: The thorax is longer than broad, almost flat, and clothed with erect and also appressed pile; notopleural bristles are present but weak. The scutellum is wide and semicircular. The mesopleuron has dense, fine, long, bushy pile. The haltere is small and slender and the squama short with a long fringe.

Legs: The legs are slender; the anterior femur has a long, fine fringe of pile; the anterior tibia has 12 fine, black bristles and there are 10 to 12 similar, more distinct bristles on the midtibia; Roberts (1929) calls them spicules. The hind femur is moderately elongate, distinctly dilated distally and bears scattered, long, fine hairs below. On the distal two fifths, there is a dense brush all around for long, black, subappressed pile, which is longer dorsally and laterally. Pulvilli are present.

Wings: Comparatively narrow and distinctly narrowed basally with both axillary lobe and alula narrow; Roberts (1929) calls the alula absent. There are 2 submarginal cells, but a feature of the wing is the almost plane second vein, and the anterior branch of the third vein is scarcely curved, barely turned up at apex; its posterior branch turns down strongly so that the second submarginal cell is short and rather triangular. The distal cell is broad, the anterior crossvein entering before the middle, the first posterior cell narrows apically from both sides yet is still widely open. The anal cell is widely open. The second vein arises acutely opposite the base of the discal cell. The hyaline wing has a conspicuous anterior border of dark brown extending irregularly downward at three places: basally to fill the whole of first and second basal cells and base of the third basal cell; narrowly over base of discal and fourth posterior cell and whole basal half of first posterior cell, except for a small posterior subbasal clear spot; over all of marginal cell, except apex, and all of first submarginal cell, except apex and a posterior middle clear spot.

Abdomen: The abdomen is obtusely conical; it bears long pile on the sides on the basal half, and with longer, finer, sparser pile posterolaterally and with appressed pale tomentum dorsally. The extreme lateral margins of segments 4 to 7 have thick, long, downturned tufts of short, black, brushlike pile. The eighth segment bears the genitalia; the abdomen broadens to the fifth segment, the remaining ones are decumbent.

Material studied: The series of male types in the Queensland Museum through the courtesy of the Director, Mr. Mack. My illustrations were taken from this series. I also have two females given me by Prof. F. A. Perkins of the University of Queensland.

Immature stages: Unknown.

Ecology and behavior of adults: Not on record. This fly obviously appears early in the spring months, in August and September, mostly August.

Distribution: Australian: *Myonema humile* Roberts, (1929).

Genus *Docidomyia* White

FIGURES 142, 372, 623, 633

Docidomyia White, Papers Proc. Roy. Soc. Tasmania, p. 204, 1917 (1916). Type of genus: *Docidomyia puellaris* White, 1917, by original designation.

Paramonov (1951), pp. 745-746, key to 5 species from Australia.

Small or medium size flies, black and shining in general color. The head is large and more or less globular and very largely given over to the development of the eye. Posterior eye margin quite plane, the occiput not long, but the occipital lobes contiguous, and the occipital recess deep but small. In the males the eyes are narrowly separated, the vertex quite long, as long as the front, and the small ocellar tubercle situated far forward. The first two segments of the antennae are large, rather bulbous and short, each with tufts of bushy pile. The third segment with small, conical bulb at the base and a slender styliform process a little longer than the base. Below the antennae the face is moderately wide but not protrusive. The oragenal recess is very shallow and triangular above and narrow and extending nearly to the antennae, but both the sides of the face and the oragenal recess slope backward into a deep, steep-walled recess behind. Proboscis stout with short, fleshy labellum. This genus will be rather readily recognized by the slender, attenuate abdomen of reduced size, the hyaline wings narrowed at the base, with the strong apical bend and loop in the second vein.

These flies are restricted to Tasmania and Western Australia.

Length: 6 mm. to 9 mm.

Head, lateral aspect: The head is subglobular, the occiput moderately developed above and below and bears fine, minute micropubescence on both the upper and lower half, and a very scanty fringe of hairs along the circular inner margin of the occipital recess. The eye is very large, occupying most of the head, the posterior margin is entire and plane, the upper facets a little enlarged. The front is convex, distinctly swollen, rather smooth, slightly flattened above the antennae; it has a large, circular, pale colored, hemicircular arch above each antennae. Pile of the front fine and erect and only moderately abundant. The face is very short from the lateral aspect, no longer anywhere than is the lower front, immediately in front of the antennae, and it slopes rapidly and backward.

The proboscis is extremely short, stumplike with large, fleshy, spinuolse, muscoidlike lobes. The palpus is small, elongate, somewhat clavate on the second segment which is about as long as the first segment. The antennae are attached a little below the middle of the head, they are relatively small and short, with the first segment large, obtuse, about as long as wide with dense, short, bushy pile dorsally and medially and with a tuft of longer, dense, bristly pile dorsolaterally and below. The second segment is short and beadlike, or barrel-shaped, almost as wide as the first segment and with a very dense tuft of long bristles dorsolaterally and an equally conspicuous tuft below. Both tufts are directed obliquely outward and downward. Mediolateral surface of this segment only with short setae. Third segment at the base as wide as the second segment, the basal portion somewhat flattened, held obliquely, strongly divergent outwardly and the swollen, basal part about as long as wide. It is abruptly narrowed into a long, stylelike process with a minute, apical spine. This process of the third segment arises dorsally and asymmetrically and is about 1½ times as long as the bulbar basal portion. Whole third segment densely microsetate.

Head, anterior aspect: The head is more narrow than the thorax and is circular. The eyes are separated in the male and the vertex is extraordinarily long with narrow parallel sides. The width between the eyes is no greater than the width of the third antennal segment. This narrow vertex occupies fully half the distance from the antennae to the posterior eye corners and the short isosceles ocellar tubercle is set very far forward. Front in the female wider than in the male. The face is strongly convergent below, the distance between the eye margins at the bottom of the head a little more than 3 times the width of the very narrow male vertex. The sides of the face, which are lateral to the narrow V-shaped or triangular, upward, wedgelike extension of the oragenal recess, slants sharply beneath the eye margin at the middle of the lower head, so that the eye margins on either side of the short proboscis are deep and quite vertical. The gena appears to be absent, for the sides on the lower part of the oragenal cavity press directly against the eye margin. The slanting part of the face bears considerable long, dense, pale pile up to the point where it drops down below the eye margin. From the posterior view the occiput is strongly bilobed, the dorsal fissure deep, the two lobes touching.

Thorax: The thorax is polished and shining black with rather dense, quite fine, comparatively long, erect pile over the whole of the mesonotum, the notopleuron, the humerus and the disc and margin of the scutellum. There is similar pile on the upper half of the mesopleuron, and such pile is extensive, long, and erect over the whole of the metapleuron leaving the hypopleuron, the pteropleuron, and the sternopleuron bare, except for appressed pollenlike micropubescence. The squama is quite short with a short fringe.

Legs: The legs are comparatively slender, the anterior four femora are a little thicker than the hind pair and all 3 pairs of femora have a ventral fringe of long, erect, fine, stiff hairs. The middle femora also have very abundant similar pile all over the posterior surface. Anterior tibia with numerous, appressed, setate hairs and posteriorly with a row of more or less erect, longer, slender, bristly hairs and a few dorsally. Middle tibia similar. Hind tibia with similar appressed setae and with only a few, short, fine, erect, additional, bristly hairs located principally anteriorly. Pulvilli large, and widely spatulate distally and somewhat notched or bilobed apically. Claws long, sharp, curved at the apex, swollen basally.

Wings: The wings are hyaline, vitreous, narrow at the base with alula absent and axillary lobe narrowed. The first vein ends close to the end of the auxiliary vein. This vein and the costa beyond the costal cell are somewhat thickened and stout. The second vein arises obliquely before the end of the second basal cell and near the apex makes a sharp, downward, or backward fingerlike loop. The anterior branch of the third vein arises rectangularly as a simulated crossvein with a long backward spur, and the remainder of this vein makes a strong, rounded bend, backward and then forward ending before the apex of the wing. Anterior crossvein enters the discal cell at the outer third. The first posterior cell is open but slightly narrowed, the anal cell is widely open and the ambient vein is complete.

Abdomen: The abdomen is rather small and reduced in size, the first segment is not quite as wide as the thorax, the second much more narrowed, the remaining segments tend to be a little more narrowed with the sides of the abdomen laterally compressed. The pile of the abdomen is scanty, rather fine, and erect with some additional, rather minute, appressed setae over the middle and dorsal portion of the tergites. Male terminalia conspicuous, comparatively elongate, bulbous, and directed obliquely upward.

Material available for study: I have a male of the type-species from Tasmania presented to me by the Commonwealth Scientific and Industrial Research Organization at Canberra, and also other species from Western Australia.

Immature stages: Unknown.

Ecology and behavior of adults: I have seen no record of the behavior of adults or of the flowers visited.

Distribution: Australian: *Docidomyia froggatti* Paramonov, 1951; *norrisi* Paramonov, 1951; *puellaris* White, 1917; *swani* Paramonov, 1951; *victoriana* Paramonov, 1951.

Genus *Antonia* Loew

Figures 136, 137, 138, 384, 624, 637, 977, 979, 981

Antonia Loew, Neue Beiträge IV, p. 30, 1856. Type of genus: *Antonia suavissima* Loew, 1856, by monotypy.
Dimorphophora Walker, Entomologist, vol. 5, p. 272, 1871. Type of genus: *Dimorphophora syrphoides* Walker, 1871, by monotypy.
Dimorphaphus Walker, Entomologist, vol. 5, p. 225, 1871; lapsus.
Dimorpaphus (Walker).—Loew in Scudder, U.S. Nat. Mus. Bull. 19, part 1, p. 108, and part 2, p. 99, 1882.
Antoniaustralia Becker, Ann. Mus. Zool. St. Pètersburg, vol. 17, p. 458, 1913a. Type of genus: *Antoniaustralia hermanni* Becker, 1913, by monotypy.

Bezzi (1924), p. 137, key to 4 species from Ethiopia.
Paramonov (1929), p. 67, treatment of Palaearctic species.
Engel (1934), p. 358, key to 5 Palaearctic species.
Hesse (1956), pp. 134-143, descriptions, discussion, key to 2 species.

Small to medium-sized flies of rather slender, elongate abdomen, which is more or less tapered and subcylindrical. They are comparatively bare and sparsely pilose and most if not all species show bright reddish and sulphur-yellow markings.

The head is subglobular with deep indentation on the middle of posterior eye margin together with bisecting line among eye facets; reflection of eyes green and red; occiput bilobate above. Antenna short, third segment long and conical with minute spical spine. One of the most remarkable features in this genus is the sharply conical, anterodorsal extension of the oragenal cup leaving the upper surface beneath the antenna horizontal and the oragenal cup strongly oblique. Equally interesting is the long, slender hyaline wing with the very long, first basal cell, the anterior crossvein located almost at the end of the discal cell; also the second vein and the third vein arise before the end of the second basal cell, with almost no prefurca or radial sector present. There are 3 submarginal cells. The last 3 or 4 tergites are laterally extended and flaplike, the preceding ones rolled tightly around the sternites.

Antonia Loew is a widely distributed genus of the Old World, with 4 or 5 species known from Ethiopia, the same number from the Palaearctic, and 2 species from Australia.

Length: 8 to 16 mm.; of wing 8 to 11.5; maximum wing span about 26 mm.

Head, lateral aspect: The head is hemiglobular, the occiput is dense and comparatively long, shorter, flat-appressed adjacent to the eye margin, but becoming more and more erect down to the central fringe around the very large, central cavity. The eye is slightly reniform, although the indentation on the middle of the posterior border, while conspicuous, is not deep. The upper and lower angles of this indentation are equal and there is an unusually long, linear incision horizontally among the eye facets at the bottom of the indentation. Upper eye facets are only slightly enlarged and the whole eye has a brilliant green and red reflection, even in dried material. Front strongly swollen on the lower part to produce a prominent, rounded, villose tubercle bearing posteriorly a shallow, medial fovea or furrow. Pile on the upper part of the front and at the vertex scanty, fine, and erect and not longer than the first antennal segment. Middle of the tubercular swelling with shorter, appressed pile, and lateral flanks of the tubercle with longer pile directed downward. Tubercle quite bare above the antenna. From the lateral aspect the face is short but produced at least as much as the upper part of the frontal tubercle. The oragenal cup, however, is extraordinarily produced, much more than in *Plesiocera* Macquart, and is sharper, conical, and extends nearly as far as the antennal style. Its sides are bare, except for a narrow, oblique band of pile; it is micropollinose below. Oragenal cup sharply separated by a crease from the limited, lateral triangle of the face. The oragenal cup has an indentation laterally at the middle opposite the beginning of the deep, knife-thin, tentorial fissure.

The proboscis is slender, barely extending beyond the epistoma or more usually confined within it; the labellum is long, slender, and pointed. Palpus rather short, slender, composed of 2 segments. It seems to bear only micropubescence. The antennae are attached a little above the middle of the head and occupy a deep cavity between the prominent frontal tubercle and the projecting face. The first segment is large, strongly widened apically, to the greatest extent medially. It is about 1½ times as long as the beadlike second segment, which is distinctly recessed into the cuplike apex of the first segment. Third segment as long or slightly longer than the first two together, not so wide as the second segment, the base distinctly bulblike, abruptly narrowed into a long, thin style. In the material I have seen, the style ends in a comparatively short, slender, spine-tipped microsegment. The first segment bears pile only ventrally and along the sides. The second segment has a tuft of short pile laterally.

Head, anterior aspect: The head is circular though the eyes are a little flattened laterally, and about as wide as the thorax. The eyes are separated in the male by barely more than the width of the strongly swollen, elongate, ocellar tubercle, and anterior to the tubercle the eyes are still more narrowed. Sides of ocellar tubercle vertical. Ocelli are placed in an isosceles triangle distinctly anterior to the posterior eye margin. Posterior eye corners a little rounded backward from the ocellar tubercle. The antennae are widely separated by at least the width of the basal segment. The oragenal cup is quite long, narrow, and very deep anteriorly, scarcely deeper, although strongly widened on a little more than the posterior third. At the point where the oragenal cup is widened the walls are thin and vertical, but in front of this they are strongly inflated inward and flattened and grooved. The walls are also vertical on the anterior part of the oragenal recess. Tentorial fissure long, deep, and very thin. Subocular part of the gena small and triangular. From above, the occipital lobes are distinctly but narrowly separated.

Thorax: The thorax is short, scarcely wider posteriorly, dull black, with, in some species, light gray pollinose vittae, and a grayish triangle of pollen in front of the scutellum. Characteristically the swollen humerus, the somewhat swollen notopleuron and postalar callosity are bright, lemon yellow; mesonotum in some species, including the type-species, with continuous, yellow, lateral margins. The scutellum is wholly yellow or has a quite black basal band, sometimes the corners black. Pile of the mesonotum dense, quite long, brassy yellow, as long as the anterior collar but becoming more sparse posteriorly; also, there is some short, scattered brassy yellow, curled, appressed pile on the disc. Postalar callosity with long, slender, stiff, or slightly bristly golden hairs. Scutellum with similar hairs, which are quite long; no macrochaetae present. Squama short with a fine, short, somewhat banked fringe. Pleuron of no great height, pollinose, lemon yellow with black spots. Mesopleuron chiefly black with dense, small, bristly pile on most of its surface, thinning out anteroventrally. The posterior border of the sternopleuron has long pile. Pteropleuron and hypopleuron without pile but metanotum and the metapleuron in front of the haltere with a large clump of long, erect pile. Curiously there is no pile behind the spiracle. While the propleuron anteriorly has a large, conspicuous clump of long, yellow pile, the prosternum is without pile.

Legs: The legs are slender, only the middle femur and hind femur comparatively stout. Anterior femur, however, slightly widened basally. The first four femora bear a fringe of dense pile posteriorly of no great length and no bristles. Hind femur ventrolaterally with a fine fringe of not very long, stiff hairs but over on its ventromedial surface the hairs are more numerous, oblique, and while slender are almost bristlelike. There are no true bristles present, except 1 or 2 dorsal bristles near the apex on each side, these are yellow in some species. Almost the whole of the lateral surface of the hind femur is covered with flat-appressed, coarse pile. Tibiae long and slender, the anterior tibia with flat-appressed setae; there are 5 of these very small setae posterodorsally that stand out obliquely, and 6 situated anterodorsally. Middle tibia similar, the spiculate setae scarcely longer. Hind tibia with flat-appressed, black pile, and with quite small, oblique, yellow bristles of which there are 4 anteroventrally, 5 anterodosally, and no other rows. However, ventromedially there is a continuation of the curious, black, bristlelike pile found on the inner surface of the hind femur. Anterior tarsi extraordinarily slender, without spicules below. Middle tarsi not quite so slender with 2 or 3 weak spicules on the basitarsi. Hind basitarsi rather strongly spiculate. Claws long and slender, the pulvilli long and spatulate, the arolium minute.

Wings: The wings are hyaline, exceptionally long and slender, narrowed toward the base, the humeral crossvein beyond the midpoint between the base of the wing and radial sector. Radial sector extremely short, almost nonexistent. The second vein arises from a knotlike swelling at the base of the third vein. Auxiliary and first vein very long, the auxiliary vein extending nearly to the outer fifth of the wing. There are 3 submarginal cells, the first posterior cell is rather strongly narrowed apically, the anterior crossvein enters the discal cell at the outer tenth of that cell. Anal cell very widely open. Axillary lobe long and slender but equally wide basally and apically. Alula absent, the ambient vein complete.

Abdomen: The abdomen is about 1½ times as long as the thorax, rather strongly flattened, the sides either nearly parallel and the apex of the male blunt, or somewhat tapered, and ending quite narrowly as shown by Bezzi (1924) in the figure of *Antonia suavissima* Loew. This appears to be the abdomen of a female, although the head is that of a male and it is labeled a male. The abdomen is rather largely light yellow or red and nearly opaque but in all species with a certain amount of black in the way of opaque black spots, which are either small and triangular and confined to the medial, basal part of the segment, or larger, quadrate, rectangular, extended laterally as a fascia or connecting with lateral black spots, but always with the posterior margins of the tergites linearly yellowish. There are 7 prominent tergites observed from above with an eighth tergite recessed beneath the seventh and the ninth forming a notched, flaplike lid over the genitalia. Last tergite sometimes with a posterior prong or lobe partly encircling the genitalia. Side margins of the tergites extended downward laterally as rounded, thin, overlapping plates. Pile of the abdomen consists of rather abundant, minute, flat-appressed, golden pile with, especially along the posterior half, scanty yet conspicuous, long, erect, partly black, partly golden, bristly hairs, which become even longer but no more abundant laterally. Lateral corners of the first tergite with an unusually loose clump of quite long, erect, stiff, yellow pile. Male terminalia small, the two basimeres lying side by side and exposed apically and ventrally.

Since I have not been able to see males of this genus I quote the remarks of Dr. A. J. Hesse with respect to genitalia:

Hypopygium of males (as based upon the known males of the two South African species *xanthogramma* and *cercoplecta*) with the basal part more or less divided into two distinct parts, each somewhat shell-like in appearance, ending apically in an outer apical process, the apex of which is bent slightly inwards, each basal part also provided with longish and fairly dense bristly hairs in apical half and on the process; beaked apical joints subtriangular in outline, very much flattened or laterally compressed, the outer face being somewhat concave, ending apically in a fairly sharp point, the lower edge of which is recurved and spine-like, with about 3-4 (or more) longish and stoutish bristles along the upper margin of each, and sometimes also with a few shorter hairs; aedeagal complex with the aedeagus itself short, sometimes provided below with a parrot-beak-like process; lateral ramus, from each basal part on each side, produced together ventrally into a V-shaped ventral process; dorsal part of aedeagal complex produced on each side into a basally directed process; lateral struts shoe-horn-shaped; basal strut remarkable in being four-vaned, being produced on each side into a shelf-like or flange-like plate.

Material studied: The type-species in the British Museum (Natural History) and an Australian species, *Antonia rieki* Paramonov. Also a third species presented to me by John Bowden.

Immature stages: Unknown. But Hesse (1956) gives strong evidence that they concern *Bembex* species or masarids, as adults were captured in close association with these wasps.

Ecology of adults: Bezzi (1924) notes the fact that Kneucker captured the type-species in the Sinai Desert on flowers of *Zygophyllum coccineum* Linné. Hesse captured 2 species settling on the damp sand of river beds. There is a harmonizing of color as they rest on some flowers.

Distribution: Palaearctic: *Antonia arenacea* Paramonov, 1934; *armeniaca* Paramonov, 1929; *bouillonae* Séguy, 1932; *fedtschenkoi* Loew, 1873; *persicana* Paramonov, 1936; *suavissima* Loew, 1856 [=*syrphoides* (Walker), 1871 (as *Dimorphophora*)].

Ethiopian: *Antonia aurata* Bowden, 1959; *bella* Curran, 1927 *cercoplecta* Hesse, 1956; *cirrhata* Bezzi, 1924; *nigrifrons* Bezzi, 1924; *occidentalis* Bowden, 1959; *xanthogramma* Bezzi, 1924.

Australian: *Antonia decorata* Paramonov, 1950; *hermanni* (Becker), 1913 (as *Antoniautralia*); *rieki* Paramonov, 1953; *roddi* Paramonov, 1950.

Genus *Aphoebantus* Loew

Figures 126, 205, 213, 227, 369, 405, 706, 719, 906, 907

Aphoebantus Loew, Berliner Ent. Zeitscher., vol. 16, p. 77, note, 1872. Type of genus: *Aphoebantus cervinus* Loew, 1872, by monotypy.
Triodites Osten Sacken, Bull. U.S. Geol. Surv., vol. 3, p. 245, 1877. Type of subgenus: *Triodites mus* Osten Sacken, 1877, by monotypy; preoc. Gistel, 1848.
Triodytus Riley, American Ent., vol. 3, p. 279, 1880b; lapsus.
Aphoebantes Loew, in Scudder, U.S. Nat. Mus. Bull., part 1, p. 27, 1882; lapsus.
Aphochantus Hyslop, Proc. Ent. Soc. Washington, vol. 14, p. 101, 1912; lapsus.
Cononedys Hermann, Zeitschr. fur system. Hymenop. und Dipt., vol. 7, p. 197, 1907. Type of subgenus: *Anthrax stenura* Loew, by original designation. *Conogaster* Hermann =*Cononedys* Hermann, see Bezzi 1924.
Triodina, new subgenus. Type of subgenus: *Triodites mus* Osten Sacken, by present designation.
Meganedys, new subgenus, type of subgenus, *petiolata*, new species.

Coquillett (1891c), pp. 254-256, key to 24 species from southwestern U.S.
Coquillett (1894a), pp. 105-107, key to 28 species of southwestern U.S.
Bezzi (1925), pp. 203-205, key to 9 Palaearctic species.
Engel (1934), pp. 397-398, key to 10 Palaearctic species.
Maughan (1935), p. 69, key to 4 species from western U.S.
Melander (1950a), pp. 15-21, key to 64 species and subspecies from U.S.

The genus *Aphoebantus* Loew was based upon an American species, *cervinus* Loew, which is before me, together with many other Nearctic species. These flies are relatively small and characterized by the rather slender, somewhat elongate, and always tapered abdomen. The general coloration is almost always black with thin grayish white pollen although legs may be in part yellowish and the appressed pile of slenderly flattened scalelike hairs or tomentum and other more erect hairs is usually white or yellowish brown. The antennae are short, the third segment with bulbus or sometimes subtriangular base that is produced into a slender bristle-tipped style. Generally the males have a more silvery head. The head is large, subglobular, easily detached, and males are holoptic; the occiput is only moderately developed. Scutellum sometimes bilobed. They are distinguished from the related *Epacmus* Osten Sacken by the quite short mouthparts and the less slender antennal style; in *Epacmus* Osten Sacken the stylar part of the third segment is abruptly formed and very slender, often with a twisted or semihooklike apex, and pulvilli are absent. In *Aphoebantus* Loew pulvilli are generally well developed. Characteristically the male hypopygium is large and prominent, protrusive posteriorly; they furnish excellent specific characters. Wing hyaline, or rarely smoky toward base, with only 2 submarginal cells; apex of marginal cell slightly and characteristically dilated ventrally; base of second submarginal cell sometimes rectangular, with basal spur-vein. This genus appears to have its greatest development in the southwestern part of the United States. It is also known from the Mediterranean and Ceylon.

Length: 3 to 10 mm., most species in between. Wing 4 to 10 mm.

Included in *Aphoebantus* Loew is the distinct new subgenus *Triodina*, in which the males and often the females have rather dense, erect, very fine, long, upstanding pile upon the abdomen and no flattened hairs or tomentum. Pulvilli present or absent.

Included also is the new subgenus *Meganedys*, with first posterior cell closed and with a long stalk; proboscis slender and elongate; pulvilli short but present, rounded, abdomen rather wide; antenna as in *Aphoebantus* Loew. The type will be *petiolata*, new species, southwestern United States, the name derived from the stalk of the first posterior cell.

Melander (1950) has ably discussed the phylogeny of the genus with its related forms including *Eucessia* Coquillett. Even though in many respects there is an overlapping of characters, these genera, or subgenera, are very useful. Possibly a more extended study, with dissected genitalia may show the need for further subdivision. Also the presence of so many species in the California and Arizona area suggests that here we have a generic complex of recent origin undergoing perhaps considerable phylogenetic differentiation at this time.

Also I place *Cononedys* Hermann (1907) as a subgenus of *Aphoebantus*. It is distinguished by the presence of 2 distinct apical microsegments at the end of the third antennal segment.

American species of *Aphoebantus* Loew may be described in the following way.

Head, lateral aspect: The head is subglobular with very prominent occiput in both sexes, but especially in the female, the occiput is bilobate above with narrow, paper-thin fissure. Posterior eye margin distinctly indented with minute anterior notch, with the extent of the indentation variable. Occiput either thinly or densely pollinose, the pile largely consists of numerous, flattened scales scattered over the lateral border and a scanty fringe of short, erect hairs along the inner rim. The front is convex, nearly plane, with the eye near the vertex, very slightly swollen in some species. In other species the front is more strongly swollen on the lower half in front of the antennae. The front is generally pollinose and the pile erect and abundant or sparse and bristly with sometimes patches of flattened, scalelike hairs laterally. In other species the whole front is covered with comparatively numerous bristles, at least as long as the first segment of the antennae and all of them curved forward. In *Aphoebantus vittatus* Coquillett the scales on the sides of the front are rather thick, broad, matted, and silvery white. The face in profile beneath the antennae is long vertically and extends a considerable distance beneath the antennae.

The proboscis is extremely short and stout with robust, stout, swollen, fleshy labellum. In no species before me is it longer than the oragenal recess. Palpus slender, long and cylindrical, with comparatively long bristly hairs laterally and at the tip. The antennae are small, attached to the middle of the head. The first segment is very short, at least twice as wide as long, it bears a few long setae ventrally, shorter setae dorsally and on the sides. The second segment is even shorter, only about half as long as the first. The third segment is bulblike at the base, swollen until it is distinctly wider than the second segment, then suddenly and abruptly narrowed medially so that on the outer edge there is a long, conical process, rather slender and about twice as long as the basal part and bears a spine at the apex. If the fly is rotated the asymmetry of this antennal process is not apparent, and it appears to be abruptly narrowed from both sides of the segment. The third segment bears a small spine apically.

Head, anterior aspect: The head is circular, the head is wide or slightly wider than the thorax. The eyes in the male are holoptic for a distance which may be as much as half the length of the front or as little as one-fourth the length of the front. Eyes a little sunken and depressed where they meet. The female front is wide, converging above, and the vertex is not quite 3 times as wide as the width of the ocellar tubercle. The large ocelli rest on the sides of a swollen tubercle with vertical, lateral, and posterior margin placed in a short isosceles triangle and with a tuft of fine, bristly hairs arising on the tubercle. The antennae are quite widely separated by a distance equal to the length of the first two antennal segments and sometimes even more widely separated. The oragenal recess is elongate oval extending back below nearly to the posterior eye margins so that much of the buccal cavity is ventral. In profile the buccal cavity is gently oblique. The cavity is deep with the inner walls vertical, the genae are narrow and linear and with a shallow tentorial fissure near the middle.

Thorax: The thorax is short and compact. The mesonotum is low, very little convex, a little more so in some species. The scutellum is quite large, more or less flattened on the disc, sometimes very thick and convex apically, usually shining, and more shining in the mesonotum and varying from faintly pollinose to polished black. Pile of mesonotum varies from rather dense, flat-appressed, somewhat elongate, scaliform hairs, which are very easily rubbed off leaving a denuded surface. This appressed vestiture may be intermixed with scattered, fine, long, erect, or semierect, bristly pile, and there is generally a collar of erect pile across the anterior margin of the mesonotum. The humerus varies from faintly to densely pollinose and from sparsely to quite densely pilose. Notopleuron with several bristles of varying stoutness but rarely as strong as the macrochaetae sometimes found on the apex of the scutellum. Upper half of mesopleuron with long, coarse, or slightly flattened pile curled upward, the lower half with appressed pile or scales. Sternopleuron with appressed scales or with erect, coarse pile as in the type-species. Pteropleuron, metanotum, anterior metapleuron pollinose only. There is a dense tuft of pile on the posterior metapleuron behind the spiracle. Propleuron laterally with a tuft of long pile and the prosternum with pile and scales. Squama moderately large with a short, fine, sparse fringe.

Legs: The femora are relatively short, the hind femur is stout and sometimes very slightly thickened through the middle. The middle and anterior femur may be slightly thickened but is more slender toward the apex. All the femora often with comparatively dense, broad, distinct, flat-appressed scales. Anterior femur with or without a posteroventral fringe of 3 to 8 slender bristles located principally on the outer half. On the middle femur the bristles are stouter, more distinct, apt to be entirely restricted to the outer half, and in addition to those on the posterior row there are 1 to 3 bristles weak or stout on the anteroventral surface. Hind femur with 3 to 8 stout bristles posteroventrally, sometimes restricted to the outer half, again extending closer to the base. Anterior tibia quite slender, as long as the femur, with smaller, flattened scales or with only appressed setae. In the type-species there are present 3 or 4 stout, anterodorsal bristles and 4 or 5 stout posterior bristles, but in some species there is only a row of 6 to 10 small, suberect setae dorsally but with a few stout bristles posteriorly. Middle tibia with quite stout and long dorsal and posteroventral bristles and with shorter, posterior, and anteroventral bristles in the type-species. In other species the tibial bristles are short and much less conspicuous. Again on the hind tibia in the type-species there are 3 prominent rows of bristles, an anteroventral, an anterodorsal, and a posterodorsal, each containing 5 elements. Similarly the number and strength of these bristles vary in other species. Tarsi comparatively short, the pulvilli nearly as long as the claw and rather slender. Claws slender, sharp, curved from the base. The legs in *Aphoebantus* Loew vary from dark reddish brown to black on the femora with the tibia and tarsi usually very dark brown and sometimes brown or even light yellow.

Wings: The wings are vitreous hyaline. There are 2 posterior cells. Sometimes, as in the type-species, a strong spur extends backward from the end of the plane, crossveinlike basal section of the anterior branch of the third vein. Absent, or only faintly indicated in the type-species, there is usually a weak but distinct bend in the second vein a little beyond the furcation of the third vein so that the apex of the marginal cell is elongate, rounded, and bent downward like the shape of the apical segment of a finger. The anterior crossvein enters the discal cell at or slightly before or slightly after the middle of the discal cell. There are 4 posterior cells. The first posterior cell is distinctly narrowed at the apex and slightly widened in the middle. Anal cell widely open, axillary lobe and alula wide, base of costa distinctly expanded with a costal comb of setae and with a prominent, flattened, apical spur. The second vein arises at a sharp angle near the base of the discal cell or across from it. Ambient vein complete.

Abdomen: The abdomen is usually elongate and distinctly tapered, its length a little greater than that of the thorax including the scutellum, but in some species actually less in length. Viewed from above the sides of the abdomen tend to be plane, slightly concave, or slightly convex. The color is usually black and shining covered with a variable amount of flat-appressed, scaliform pile of sometimes contrasting colors, often together with a postmarginal fringe of fine, semierect, bristly hairs, which sometimes instead of being erect tend to be curled downward and backward and not readily distinguished from the remaining pile. They may be lacking or restricted to a few hairs near the posterior corners. Characteristically there is a band of dense, coarse, erect, spikelike pile on the outer third of each side of the first segment. Sternal pile rather similar to the tergal pile. Abdomen sometimes more or less flattened, or distinctly cylindrical or compressed laterally. In females there are 7 tergites with the seventh generally compressed laterally, and there is a dense fringe of shining, bristly or slightly flattened pile curled inward around the female terminalia. Male terminalia distinctly large and protrusive.

The male terminalia from dorsal aspect, epandrium removed, show a figure that is broad basally. The basistylus narrows conspicuously on the outer half. The epiphallus is very short. The basal apodeme is extraordinarly large and long, and at least as large as the whole epandrium. The dististyli are small with a curved, apical hook, which viewed from above is curved outward, viewed laterally is curved upward and backward. The epiphallus is strongly separated dorsally from the ejaculatory process. Epandrium is more or less triangular.

Material studied: Besides a female of the type-species given me many years ago by R. H. Painter, I have a large series of many species collected by me and by Marguerite Hull in the southwestern United States.

Immature stages: The larvae of this genus consume the egg pods of locust, and at least one species is known to be a parasite of *Tiphia* wasps in India. See discussion under life histories.

Ecology and behavior of adults: The adults are usually found resting on the ground, on the sand, and sometimes on rocky surfaces, but may be uncommonly found on flowers. Melander (1950) has recorded them on creosote bushes, *Larrea divaricata*, *Wislizenia refracta*, *Heliotropium* species, *Gilia*, *Phycelia*, and *Cryptantha* species, on *Adenostoma*, wild buckwheat or *Eriogonum fasciculatum*, *Lotus* and *Hugelia* species.

Distribution: Nearctic: *Aphoebantus abnormis* Coquillett, 1891; *altercinctus* Melander, 1950; *arenicola* Melander, 1950; *balteatus* Melander, 1950; *bisulcus* Osten Sacken, 1887; *borealis* Cole, 1926; *brevistylus* Coquillett, 1891; *capax* Coquillett, 1891; *carbonarius* Osten Sacken, 1887; *catenarius* Melander, 1950; *catulus* Coquillett, 1894; *cervinus* Loew, 1872; *concinnus* (Coquillett), 1892 (as *Epacmus*); *contiguus* Melander, 1905, *contiguus separatus* Melander, 1950; *conurus* Osten Sacken, 1887; *cyclops* Osten Sacken, 1887; *denudatus* Melander, 1950; *desertus* Coquillett, 1891; *eremicola* Melander, 1950; *fucatus* Coquillett, 1894; *fumidus* (Coquillett), 1891; *fumosus* Coquillett, 1892 (as *Epacmus*); *gluteatus* Melander, 1950; *halteratus* Melander, 1950; *hians* Melander, 1950; *hirsutus* Coquillett, 1886; *inermis* Coquillett, 1891 (erratum for *abnormis*); *interrputus* Coquillett, 1891; *inversus* Melander, 1950; *leviculus* Coquillett, 1894; *maculatus* Melander, 1950; *marcidus* Coquillett, 1891 [=*squamosus* Coquillett, 1891, males only]; *marginatus* Cole, 1923; *micropyga* Melander, 1950; *mixtus* Coquillett, 1891; *mormon* Melander, 1950; *mus* (Osten Sacken), 1877 (as *Triodites*; subg. *Triodina*), *mus barbatus* Melander, 1950; *obtectus* Melander, 1950; *parkeri* Melander, 1950; *pavidus* Coquillett, 1886; *pellucidus* (Coquillett), 1892 (as *Epacmus*); *poedes* Osten Sacken, 1887; *rattus* Osten Sacken, 1887; *scalaris* Melander, 1950; *schlingeri* Hall, 1957; *scriptus* Coquillett, 1891; *sperryorum* Melander, 1950; *squamosus* Coquillett, 1891, females only; *tardus* Coquillett, 1891; *timberlakei* Melander, 1950; *transitus* (Coquillett), 1886 (as *Leptochilus*); *ursula* Melander, 1950; *varius* Coquillett, 1891; *vasatus* Melander, 1950; *vittatus* Coquillett, 1886; *vulpecula* Coquillett, 1894.

Neotropical: *Aphoebantus argentifrons* Cole, 1923; *dentei* d'Andretta & Carrera, 1952.

Palaearctic: *Aphoebantus albicinctus* Paramonov, 1925; *armeniacus* (Paramonov), 1925 (as subg. *Cononedys*); *armenicus* (Paramonov), 1929, nomen nudum (as subg. *Cononedys*) [=*armeniacus* Paramonov, 1929]; *bituberculatus* (Becker), 1915 (as subg. *Cononedys*); *claripennis* Becker, 1903; *clauseni* Aldrich, 1928; *costalis* Paramonov, 1929, *costalis subcostalis* Paramonov, 1929; *dichromatopus* Bezzi, 1925; *efflatouni* Bezzi, 1925; *erythraspis* (Hermann), 1907 (as subg. *Cononedys*); *escheri* Bezzi, 1908; *latifrons* Paramonov, 1929; *persicus* Becker, 1912; *pusillus* Paramonov, 1929; *scutellatus* (Meigen), 1838 (as *Anthrax*); *seratus* Aldrich, 1928; *sternurus* (Loew), 1870 (as *Anthrax*) [type of *Cononedys*]; *transcaspicus* Paramonov, 1929; *turkmenicus* Paramonov, 1929; *wadensis* Bezzi, 1925.

Ethiopian: *Aphoebantus bilobatus* Bezzi, 1924.

Oriental: *Aphoebantus ceylonicus* (Brunetti), 1909 (as *Argyromoeba*).

Genus *Epacmus* Osten Sacken

FIGURES 141, 143, 628, 631, 913, 914, 915, 916, 917

Leptochilus Loew, Berliner Ent. Zeitschr., vol. 16, p. 78, nota. 1872. Type of genus: *Leptochilus modestus* Loew, 1872, by monotypy.

Epacmus Osten Sacken, Biologia Centrali-Americana, Dipt., vol. 1, p. 142, 1887. Change of name.

Coquillett (1892a), p. 9, key to 5 species from the western United States.

Melander (1950a), pp. 5-6, key to 12 species from the southwestern United States.

Small or medium-sized flies with tapered, cylindrical, or compressed abdomen, which are very similar in general appearance to *Aphoebantus* Loew and are distinguished by the produced face, the relatively long proboscis that to a variable extent extends beyond the confines of the oragenal recess and has a longer, slender, nonfleshy-type labellum.

Several South African genera within this tribal complex also have a peaked or produced face, but without exception they are separated by the manner of origin of the second vein, which arises abruptly, far distal to the base of the discal cell.

The antennal style in *Epacmus* is as a rule unusually slender and long, strongly differentiated from the basal bulblike portion, and the thornlike spine or bristle at the apex is frequently curved up after a characteristic fashion. The pulvilli are variable but are very often much reduced, almost vestigial.

Epacmus is a characteristic element of the bombyliid fauna of the western United States; the 12 species range from Washington to California, a few as far east as New Mexico or extreme western Texas.

Length: 9 to 9 mm.; wing 5 to 8 mm.

Head, lateral aspect: The head is subglobular, the occiput only moderately extended above and behind the vertex and equally prominent below. It bears dense, broad, flattened scales that extend to the eye margin over a pollinose surface. The marginal fringe along the inner edge of the occipital recess is not only quite fine but very scanty. The posterior eye margin is distinctly though rather shallowly indented and has a small, triangular notch in the eye facets. The eyes are more prominent on the upper half than below. The front is convex, not at all extensive and bears scattered, scaliform hairs and is generally densely pollinose, often silvery, or may be golden in color. The face is triangularly produced forward but somewhat obtusely, and is generally covered rather densely with shining, pale pollen or micropubescence, some of it scalelike. Sides of the oragenal cup rather prominent, polished and bare. Proboscis elongate and slender, longer than the head and distinctly much longer than the oragenal cup. The palpus is very slender, threadlike, elongate with 2 or 3 fine hairs on the apical segment, which is only about one-fourth as long as the basal segment. The antennae are small, the first segment is much shorter than wide, considerably wider apically, rather cup-shaped, not greatly produced medially, a little more so laterally and the second segment, which is about the same length, is widely and solidly attached. The second segment is more than twice as wide as long with the sides rounded and convex. The first segment has a tuft of setae dorsolaterally, a few minute setae ventromedially and below, the second segment has similar pile and setae similarly placed. The third segment is formed as a low, compact bulb sharply contracted into an asymmetrical, long, slender, pronglike lobe which arises on the lateral edge of the third segment. It bears apically an oblique, long, conical microsegment.

Head, anterior aspect: The head is not quite circular; it is somewhat subquadrate. The head is about the same width as the thorax. The eyes are holoptic in the male for a short distance, which is usually not much greater than the length of the ocellar tubercle, hence the front is large as well as convex. The eyes are a little sunken where they meet and the rather small, conspicuous ocellar tubercle is nearly equilateral. The ocelli rest on the eye margin and the eyes viewed from above are curved forward on their upper posterior margins. The female has the eyes more widely separated. The antennae are widely and clearly separated by a distance nearly equal to the width of the first antennal segment. The oragenal cup is comparatively short, slightly widening anteriorly, the upper half of its inner margin slopes strongly and the lower portion has vertical walls. The ventral and dorsal ends of this recess are both distinctly triangular, the one extended forward, the other extended backward. The genae are confined to a narrow, pollinose line along the eye margin. Tentorial fissure absent. The occiput viewed posteriorly is bilobate with the adjacent lobes touching; the fissure between is deep but without any additional pocket anteriorly.

Thorax: The thorax is short and compact; the mesonotum is as wide as long, rather low and flattened, sometimes moderately shining and in some species partly or wholly opaque pollinose. The mesonotum is widely and rather densely covered with loose, comparatively wide, long, lanceolate scales. They are apt to be rubbed off over the middle part of the mesonotum, but they adhere to the whole notopleuron, humerus, and postalar callosity. Scales are generally present also on the scutellum, unless rubbed off. Pollen of mesonotum often vittate. There are sometimes a few, very slender, erect, long hairs on the notopleuron and always 2 or 3 long, slender bristles. Margin of scutellum with several irregularly placed hairs or slender bristles and some other long hairs on the disc. A few species have the postscutellar margin with 4 pairs of comparatively stout, long bristles, and the margin indented in the middle and bearing scales. Postalar callosity sometimes with 3 or 4 long, conspicuous bristles. Pleuron pollinose, the upper half of the mesopleuron with a rather dense band of upwardly turned bristles and scales and in the posterior corner with a conspicuous, backwardly turned tuft containing mostly scales but with some hairs. Lower part of mesopleuron with a few loose scales; upper half of sternopleuron with a large, dense patch of appressed scales. Pteropleuron, metanotum, and hypopleuron pollinose only. However, the metapleuron in front of the halteres, and posteriorly behind the spiracle have either a small patch of scales or a dense, wide patch of long, coarse pile. Squama large with a long, fine fringe on the medial portion and the outer part with a wide, plumelike border of long, extensive hairs which varies greatly in different species.

Legs: All of the femora are rather stout, especially the last four, the anterior pair varying in thickness. All pairs covered rather densely with long, appressed, more or less truncate, shinglelike scales. The anterior femur has a few semierect, bristly hairs of no great length posterodorsally and sometimes with 5 or 6 rather stout bristles anteroventrally. Middle femur similar with a row of 4 or 5 posteroventral and a like number of somewhat stronger anteroventral bristles. Hind femur with 3 or 4 strong, ventral bristles on the basal half, and on the outer half of the hind femur there are 3 to 5 stout, anteroventral and 1 to 2 posteroventral bristles. Sometimes there are 2 or 3 dorsoapical bristles. All of the tibiae are quite long and slender. They have small, appressed scales, the anterior pair has a few quite appressed, small setae and 3 or 4 posteroventral, short bristles; in some species the bristles are more conspicuous. Middle tibia with 3 or 4 stout bristles anteriorly, 4 or 5 posteriorly. Hind tibia with 4 rows of stout, rather long, oblique bristles. There are 3 or 4 anteroventrally, 5 or 6 anterodorsally, 5 or 6 dorsomedially and a like number ventromedially. Tarsi long and slender with short, ventral bristles and longer ones apically. The claws are only slightly and gently curved from base to apex. They are long, relatively blunt at apex, and the pulvilli are greatly reduced.

Wings: The wings are hyaline, elongate, and rather slender. There are 2 submarginal cells, the anterior branch of the third vein arises usually plane and simulating a crossvein, with or without a backward spur. The marginal cell is slightly but gradually widened apically, has at most a slight preapical bend, but is usually blunt at the apex. The second vein meets the costa at a right angle. First posterior cell open but to a varying extent; in those species where it is more widely open the first bend of the anterior branch of the third vein tends to be rounded. The anterior crossvein enters the discal cell at or slightly beyond the middle. The second and third veins fork acutely opposite or before the base of the discal cell. Anal cell widely open, ambient vein complete, the alula small but rather wide.

Abdomen: The abdomen is longer than the thorax, in males subcylindroid, but rather compressed laterally, especially through the middle so that viewed from above the sides are often concave. In females the abdomen tends to be elongate, conical, and distinctly tapered, the third and fourth tergites only slightly more narrow than the first two, the fifth to seventh more strongly narrowed. I find 7 tergites in the females and 8 in the males with the eighth tergite quite short. The pile consists of rather dense, large, coarse, appressed or partly erect scales which are more prominent on the sides of the tergites and more apt to be rubbed off dorsally and medially. In both males and females there is a conspicuous band of long, dense scales or scaliform hairs; appressed along the the posterior margin of the first segment and changing to a wide tuft of dense, long, erect, spikelike pile laterally. Females of some species have dense, black scales on the tergites with a narrow white or yellow fringe of scales along the posterior border overlapping the next tergite. In addition both sexes have a considerable amount of fine, erect, long pile laterally on the tergites. Female terminalia surrounded on all sides by a dense fringe of shining, bristly pile curled inward. Male terminalia somewhat enlarged and clublike, enclosing the two prongs of the aedeagus.

The male terminalia from dorsal aspect, epandrium removed, show a figure similar to *Aphoebantus* Loew. It is much broader, wider, short and compact. The dististyli are similar and turn outward. The epiphallus is shortened, the epandrium is quadrate. The basal apodeme much smaller and shorter from the lateral aspect, more or less concealed by the epandrium. Just as in *Aphoebantus* Loew, the ejaculatory process is well separated from the overlying epiphallus.

Material studied: A male of *Epacmus litus* Coquillett lent me by Jack Hall and males and females of several species recently collected by the author in the Southwest.

Immature stages: Unknown. Probably similar to *Aphoebantus* Loew.

Ecology of adults: Similar to *Aphoebantus* Loew, generally found resting on sandy or rocky ground.

Distribution: Nearctic: *Epacmus cirratus* Melander, 1950; *clunalis* Melander, 1950; *connectens* Melander, 1950; *labiosus* Melander, 1950; *litus* Coquillett, 1886; *modestus* (Loew), 1872 (as *Leptochilus*); *morsicans* Melander, 1950; *nebritus* Coquillett, 1894; *nitidus* Cole, 1921; *pallidus* Cresson, 1919; *ponderosus* Melander, 1950; *pulvereus* Melander, 1950; *tomentosus* Melander, 1950.

Genus *Pteraulax* Bezzi

FIGURES 123, 382, 625, 630

Pteraulax Bezzi, Ann. South African Mus., vol. 18, p. 117, 1921a. Type of genus: *Pteraulax flexicornis* Bezzi, 1921, by original designation.

Hesse (1956), pp. 336-338, key to 7 species.

Small flies strongly resembling *Aphoebantus* Loew; however, there are 3 submarginal cells and the first posterior cell is closed; and the hyaline or nearly hyaline wing is strongly wrinkled, the anal cell open, and pulvilli present. The eyes of the male are holoptic. These dark colored flies are characterized further by pale pile, and a narrow, elongate, cylindrical abdomen; however, the abdomen is no longer than the thorax. The abdomen has conspicuous, flat-appressed, broadly scalelike pile; the posterior margins of the tergites bear lateral fringes of long, stout, bristly hairs; legs with conspicuous bristles.

Pteraulax Bezzi is a small genus of at least 7 South African species.

Length: 7 to 11 mm.

Head, lateral aspect: The head is more or less hemispherical, the occiput is only moderately developed posteriorly; as in related genera there is a deep central recess bordered with only extremely short hairs. It is bilobate above the narrow, medial fissure. Laterally the occiput bears flat-appressed, narrowly scaliform hairs. There is a deep, conspicuous indentation on the posterior eye margin accompanied by a bisecting line or shallow crease. The face is rounded and very little produced forward. Its surface, together with the front, is pollinose and broken by tiny, punctate bare spots. The face bears only loose, scattered, erect, moderately long sometimes bristly pile, but among the similar coarse pile of the front there are some 40 to 50 conspicuous, rather wide, pale, brownish yellow scales which are suberect. The ocular margin of both front and face are narrowly bare, except for pollen. The oragenal cup is scarcely in evidence at all in the profile aspect, and it does not extend onto the face proper. The lower face is bluntly and broadly rounded and pollinose, extending over into the oral cavity. However, just above the proboscis there is a narrow, cuplike rim of very limited proportions set almost against the eye margin and separated from it by the usual crease. Palpus exceptionally fine, long and threadlike and composed of 2 segments; the apex is a little swollen and both segments bear a few long, scattered hairs. The stout proboscis is subcylindrical, shining, nearly bare, a little arched upward and slightly longer than the oral cavity. The short, stout labellum is wider, more or less muscoidlike.

The antenna is attached at the middle of the head, or barely above it; the antenna is short and stout, the first segment 2 or 3 times as long as the second and nearly twice as wide at its apex as the second segment. This segment bears the characteristic medial extension or protuberance found in related genera. The third segment is wide, broad basally, pyriform and produced on the outer half into a rather stout styliform piece, which carries a minute, short microsegment, which is truncate apically.

Head, anterior aspect: The head is wider than the thorax with the face and front continuous and relatively narrow, comprising little more than one-fourth the head width. The oral opening is approximately 2½ times as long as wide, a little narrowed in the middle, since the walls of the very low oragenal cup bend inward. The eye margins very slightly converge so that the sides of the face are not quite parallel. There is a small area of conspicuously enlarged facets on the eye where the eyes meet. The antennae are separated by the apical width of the first segment. The ocelli are distinct and set upon a low, anteriorly segregated tubercle.

Thorax: The thorax is slightly longer than wide with a rather short, scanty collar of coarse pile along the anterior margin but including the humerus. The notopleural and nearly all of the mesonotal pile is subappressed, or flat-appressed and more or less coarse and wiry with a few much finer, erect hairs intermixed. The scutellum is covered with similar pile, flat-appressed also, but distinctly more flattened and somewhat scale-like, but not forming the conspicuous scales present on the abdomen; the margin of the scutellum with a few longer, stiff hairs or even bristles. The notopleuron has 1 or 2 unusually stout, long bristles; postalar callosity with 2 more slender bristles; the outer squamae are extensive, thin and flaplike. The pleuron is bare, except for a tuft of bristly hairs along the dorsal border, a more dense tuft of less bristly hairs in the upper anterior corner which grades into finer hairs below. Oddly, the mesopleuron bears a few scattered scales, the sternopleuron has a conspicuous, dense patch of flat-appressed scales. The pteropleuron, the metapleuron in front of haltere, and metasternum, and all of the hypopleuron above the middle coxae usually bare; there is a dense tuft of scales below the spiracle, another smaller tuft in front of the posterior coxa.

Legs: All of the femora are somewhat stout and slightly swollen and are 3 or more times as thick as their corresponding, uniformly slender tibiae. In addition the hind femur bears a double row of conspicuous bristles, 5 on the ventrolateral margin, 3 on the ventromedial margin confined to the basal half; otherwise the pile of the legs is distinctly composed of rather broad, plastered, flat-appressed, pale scales. The anterior tibia has a posterior row of 6 long spicules and an anterior row of some 10 short spicules; middle tibia has similar spicules; hind tibia has 4 rows of rather longer, bristle-like spicules, each row containing only 3 or 4 elements. The claws are long, quite slender, the pulvilli well developed; arolium minute and stublike; dorsal apex of last tarsal segment with a stout, black, spinal seta; ventral surface of last 4 tarsal segments with dense, erect, short, black, spinous setae.

Wings: The wings are rather small, a little wider apically, the anterior branch of third vein sigmoid; 3 submarginal cells present, the first posterior cell is closed and stalked; anal cell is open, alula well developed, the ambient vein is complete. The wings are hyaline but the surface bears fluted ripples especially anteriorly. The subcostal vein ends close to the end of the second vein and the anterior crossvein enters the discal cell near the middle; costal setae small.

Abdomen: Although the abdomen is elongate, tapered, cylindrical and blunt at the male apex, it is rather small and is shorter than the thorax if the large scutellum is included. The first segment is nearly twice as wide as the apex of the abdomen, and bears a band of rather dense, erect, long, quite coarse pile, which is, however, confined to the outer third and is long laterally as well. The posterior margin of this tergite has an appressed fringe of flattened pile. Remaining tergites rather densely covered, especially laterally, with flat-appressed, narrow scales directed more or less medially. Also laterally the posterior margins bear 4 to 6 conspicuous, long, erect slender bristles or bristly hairs; sternites with a few similar bristles, and the posterior margin of the last tergite and sternite bears a conspicuous fringe of 7 or 8 long, dark, bristly hairs, or even stout, long bristles.

Male terminalia enlarged, directed backward and plane with the axis. I quote Dr. Hesse's description of the terminalia based on males of many species.

Dr. A. J. Hesse in his work on South African bee flies describes the male hypopygium as follows:

Hypopygium with the sternite opposite it well developed, broad, its apical angles more or less rectangular, sometimes rounded, and its hind margin not incised, sometimes broadly emarginate; basal parts of hypopygium with a small triangular tergite dorsally between their bases, dorsal part in neck region almost without or with only very fine and short hairs, the dorsal apical part sometimes with a lobe-like process, the outer lower edge or margin of each basal part sometimes produced into a lobe-like or even triangular process; beaked apical joints laterally compressed, either elongate, almost blade-like, or bird-head-shaped from side, covered with shortish hairs, the apex usually curved downwards; a distinct, conspicuous, ventral aedeagal process or apparatus present and formed by the lateral ramus on each side from each basal part coalescing apically to form a projecting rod-like or beak-like process on which there is a transverse slit-like groove or incision through which or in which the apex of aedeagus opens; basal part of ramus on each side where it joins each basal part produced into a process or broad leaf-like extension; lateral struts comparatively broad, shoe-horn-shaped or scapula-like.

Material studied: Representatives of *flexicornis* Bezzi in British Museum (Natural History) and a pair of *setaria* Hesse kindly sent by Dr. Hesse for study and dissection.

Immature stages: Unknown.

Ecology and behavior of adults: Not on record.

Distribution: Ethiopian: *Pteraulax ausana* Hesse, 1956; *braunsi* Bezzi, 1922; *cinctalis* Hesse, 1956; *eremophilia* Hesse, 1956; *eurymetopa* Hesse, 1956; *flexicornis* Bezzi, 1921; *latifacies* Hesse, 1956; *setaria* Hesse, 1956.

Genus *Pteraulacodes* Hesse

Pteraulacodes Hesse, Ann. South African Mus., p. 348, 1956a.
Type of genus: *Pteraulacodes karooensis* Hesse, 1956, by original designation.

This is a genus of small, *Plesiocera*-like flies that have a relatively elongate, tapered abdomen; consequently they also resemble the Holarctic genus *Aphoebantus* Loew. Hesse (1956) notes that it resembles both *Plesiocera* Macquart and *Stomylomyia* Bigot, especially the latter genus, and may be distinguished from both by the absence of a well-marked facial cone, the bisected hind margins of the eyes, the longist and slender terminal element and style of the third antennal segment, the slender and bristlelike spines on the hind femur, and bristlelike spicules on the outer part of the tibiae, etc. Moreover, it is separated further from *Plesiocera* Macquart by the presence of 3 submarginal cells.

Pteraulacodes Hesse is one of a small group of genera characteristic of South Africa; there are related forms in the southwestern United States.

Length: 5 to 6 mm., of wing 4 to 5 mm.

As I have not been able to see or obtain material of *Pteraulacodes* Hesse, I quote Hesse's description of the genus:

Body similarly shaped; abdomen in males also cylindrical and in females conically pointed, also predominantly black, but with the sides of abdomen, hind margins of sternites, the femora and the tibiae yellowish or yellowish reddish to a variable extent. Vestiture with the erect hairs on body less developed, relatively sparser and shorter; hair on frons distinctly shorter and sparser; that on face for the greater part wanting, present only as a conspicuous brush-like tuft anteriorly and overhanging apex of buccal cavity; that on disc of thorax very much shorter, finer and sparser; vestiture on mesopleuron also sparser and shorter, composed of scales and bristly hairs, greater part of pleurae even barer than in *Pteraulax* Bezzi; metanotal tuft also wanting; prealar, postalar and scutellar bristles however similarly developed, but apparently fewer in number; hair on abdomen, excepting only dense brush on sides of tergite 1, also sparse; transversely arranged bristly hairs across hind margins of tergites short, not developed to the same extent as in *Pteraulax* Bezzi, conspicuous and longish only on last tergite in males; scaling slightly more developed and denser than in *Pteraulax* Bezzi, very dense on frons and face, mostly lanceolate in shape, finer on body above; that on pleuron, mesopleuron, sternopleuron, hind margin of metapleural part and on coxae dense and cretaceous whitish; that on venter also very dense and that on legs as in *Pteraulax* Bezzi.

The wings have the membrane also wrinkled; basal comb wanting; three submarginal cells present; first posterior cell however open apically, not acute and closed as in *Pteraulax* Bezzi; squamae much smaller, more normal. Head large, subglobular, broader than thorax; eyes large, indented in hind margin and also with a short bisecting line extending from indentation, also in contact for some distance in front of ocellar tubercle in males, separated on vertex in females; occiput like that of *Pteraulax* Bezzi, the medial sulcus tending to be narrower; frons also convex in males and anteriorly in females; face from side however distinctly subconically prominent or snout-like, not gradually sloping into buccal cavity as in *Pteraulax* Bezzi, but overhanging the buccal cavity, the brush-like tuft of hair emphasizing this character; buccal cavity deep; genae very narrow and linear, the genal furrows not so distinct as in *Pteraulax* Bezzi; antennae with joint 1 relatively much shorter, cup-like and with much fewer and sparser hairs; joint 3 conical, gradually broadened basally, ending apically in a slender joint-like terminal joint bearing a relatively long style; proboscis confined to buccal cavity or only projecting slightly beyond buccal cavity; palps slender, longish, biarticulate, the apical joint clavate and very much shorter than basal joint.

The legs without any dense fine or longish hairs on front and middle femora below; middle and hind femora also with slender bristle-like spines below as in *Pteraulax*, those near base on hind femora also long and slender and more or less disposed irregularly in pairs, in males especially; tibiae also with the spicules on outer part long and bristly-like, and with the 2 or 3 lower spurs on middle and hind ones also markedly long and slender; claws with their apices bent downwards and the pulvilli well developed. Genital brush in females also terminal. Apical angles of last sternite in males slightly produced, more lobe-like and not rectangular as in *Pteraulax* Bezzi. Hypopygium of male very much like that of *Pteraulax* Bezzi; dorsal parts of basal parts also finely and relatively poorly covered with hair; beaked apical joints compressed, curved and elongate; a ventral aedeagal process in the form of a rod-like process also present; apex of aedeagus also ending in a transverse slit-like aperture in aedeagal process; lateral struts broadish, shoehorn-shaped; basal strut with triangular lateral extension on each side basally and with its apical margin along dorsal aspect also with a lateral ledge-like extension.

Since I have not been able to see males of this genus I quote the remarks of Dr. A. J. Hesse with respect to genitalia:

Hypopygium of male very much like that of *Pteraulax*; dorsal parts of basal parts also finely and relatively poorly covered with hairs; beaked apical joints compressed, curved and elongate; a ventral aedeagal process in the form of a rod-like process also present; apex of aedeagus also ending in a transverse slit-like aperture in aedeagal process; lateral struts broadish, shoe-horn-shaped; basal strut with a triangular lateral extension on each side basally and with its apical margin along dorsal aspect also with a lateral ledge-like extension.

Material available for study: None. I have not been able to personally examine these flies as they are known only from the unique types.

Immature stages: Unknown.

Ecology and behavior of adults: Not on record.

Distribution: Ethiopian: *Pteraulacodes karooensis* Hesse, 1956.

Genus *Prorostoma* Hesse

FIGURES 130, 389, 622, 640

Prorostoma Hesse, Ann. South African Mus., Pt. II, vol. 35, p. 121, 1956a. Type of genus: *Plesiocera integra* Bezzi, 1922a, by original designation.

Small flies which in shape and length of abdomen and vestiture strongly resemble *Epacmoides* Hesse. The wing is essentially the same also; it is hyaline, the second vein arises abruptly in nearly the same position, the anal cell is open, the marginal cell has a more conspicuous kink subapically, and the end portion of this cell is a little more conspicuously swollen downward. The antennae are similar but whereas *Epacmoides* Hesse has a rather sharply triangular face, *Prorostoma* Hesse has the produced face rounded.

In comparison with *Epacmoides* Hesse, *Prorostoma* Hesse has the hind margin of the eye only slightly indented and emarginate; the scutellar margin is normally rounded and not slightly bidentate or bimammiform as in *Epacmoides* Hesse. Hesse contrasts this genus with *Epacmoides* Hesse in his key couplets; he states further that *Epacmoides* Hesse is almost without tiny bristles beneath the femora, but that they are numerous in *Prorostoma* Hesse; in *Prorostoma* Hesse I find 4 beneath the anterior femur, 5 beneath the middle femur and 5 below the hind femur (none in the female before me); Hesse states that there is double row; all of these bristles are minute, however.

Prorostoma Hesse appears to be a weakly characterized monotypic genus from South Africa.

Length: 4 to 9.5 mm.; wing, 4 to 8.5 mm.

Head, lateral aspect: The head is subglobular, the occiput moderately prominent above, and bears dense, short, erect pile extending almost to the eye margin. Only the outer row of hairs is curved forward; it is more extensively appressed at the indentation of the eye. Whole surface of the occiput pollinose or appressed micropubescent. The fringe along the inner occipital recess is rather dense. Eyes with a shallow indentation near the middle. Front densely pale pollinose, with abundant, fine, bristly pile obliquely and slightly curved forward and covering the whole front. Front very slightly convex. The face is prominent, but is bluntly rounded and does not extend as far as the apex of the antennae. In both of these respects it differs from *Coryprosopa* Hesse. Moreover, the surface of the face is pollinose with dense, short, appressed, stiff pile.

The proboscis is short, and rather stout, not longer than the oragenal recess, in which it is usually tucked away, and the labellum quite long, fully as long as the proboscis and with microspinulose surface. The palpus appears to consist of a short, basal segment that bears 1 or 2 long, slender bristly hairs and of a more flattened, moderately lengthened, apically blunt, micropubescent segment carrying 4 long, lateral, slender bristles and 2 or 3 at the apex. Its total length is about one-third the length of the proboscis. Antennae are small; the first segment is wide, short, distinctly wider than long, black with a thin, pale, apical rim of somewhat more extensive and extended cuplike appearance on the medial side. Second segment shorter than wide, distinctly convex and rounded on each side, nearly as wide as the first segment but broadly attached laterally, hence the entire antennae are strongly divergent. The third segment is pyriform, bulblike on the basal half, the thick, apical process somewhat asymmetrical, its outer margin plane, except near the base, the inner margin distinctly concave. The third segment bears a short, cylindrical microsegment no longer than wide with minute, concealed spine. The second segment laterally bears several long, stiff bristly hairs and a few fine, short hairs medially. Also, the first segment has a few short setae medially and dorsally and a tuft of somewhat longer hairs ventrally.

Head, anterior aspect: The head is circular, wider than the thorax, the eyes widely separated in the male by slightly more than the thickness of the first antennal segment and the ocellar tubercle in the male is quite prominent with vertical sides; the ocelli set in a nearly equilateral triangle. In the female the ocellar tubercle is a little less prominent and the vertex is almost as wide as the front, the sides of the front and vertex being only slightly convergent. Moreover, the upper part of the front in the female, immediately in front of the ocellar tubercle, is rather strongly and medially depressed or sunken, and distinctly concave. This wide fovea is characteristic. The antennae are separated at the base, though rather narrowly. The first two antennal segments touch apically and medially due to the medial extension of the first segment. The oragenal cup is distinctly ventral in position, at most very slightly oblique. It is elongate oval, rather deep, slightly pinched in the middle, and this is the only point at which the inner walls are vertical. Both below and above the inner walls are rather strongly sloping. The gena is quite linear, the only deep depression is a short one between the eye and the oragenal cup at the point where it is compressed inward. Here there is a moderately deep tentorial fissure. The sides of the face are not only densely pollinose but bear numerous, stiff, bristly hairs curled downward and forming a fringe that extends along the edge of the whole upper half of the oragenal cup. From the posterior aspect the lobes of the occiput are in contact, except that there is behind the vertex between the posterior eye corners a deep, continuous, triangular, vertical shaft or pocket, which appears to extend through to the occipital cavity.

Thorax: The thorax is black, the mesonotum low and gently convex, covered loosely with abundant, coarse, suberect, shining yellow pile, in places rather matted and more appressed on the posterior half of the mesonotum. Notopleural and humeral pile more tangled and erect and longer. Anterior mesonotal collar of pile long. Notopleuron with 2 stout, pale bristles. Postalar callosity with 2 long, slender bristles. Scutellum more shining than the mesonotum, especially on the apex, the disc faintly pollinose and covered with pile similar to that of the mesonotum. There are a few fine, long, but not conspicuous hairs across the mesonotum and in front of the scutellum. Scutellar margin with 2 or 3 pairs of slender bristles. Propleuron with erect, vertical fringe of long hairs. Whole pleuron silvery gray pollinose. Mesopleuron almost entirely covered with coarse, flattened, subappressed hairs which are tangled above and with longer, bristly hairs on the posterior half. A very large triangular area of the sternopleuron bears matted, appressed, scaliform pile with, at the extreme ventral part a fringe of bristles in one row. Pteropleuron and hypopleuron bare. Lateral metanotum and anterior metapleuron each with a rather conspicuous tuft of coarse, partly erect, partly appressed or curled hairs. There is no patch of pile behind the spiracle. Squama short with a short, thin fringe of pile.

Legs: The legs are rather slender and comparatively short. Anterior femur gradually thickened a little toward the base; middle femur likewise somewhat stouter in the middle and toward the base. Hind femur stout, anterior femur with 4 or 5 stout spicules ventrally, middle femur with 10 anteroventrally and 7, a little longer, posteroventrally, also with some spines on the lower, inner aspect of the hind femur. Anterior tibia much shorter than the middle pair; slender, with appressed pile and with 6 or 7 minute, fine, suberect setae dorsally in each of 2 rows and 7 or 8 setae posteroventrally. Middle tibia with equally fine, slender, oblique, bristly setae, 3 or 4 anteroventrally, 5 or 6 on each side dorsally and 7 or 8 posteroventrally. Apical spines long and stout. Hind tibia not longer than the middle pair, only slightly thicker with 4 or 5 setae in each dorsal row, 4 ventrolaterally and 6 weaker ones ventromedially. Apex with one spinous bristle much longer than the others. Tarsi short, slender, sharp, strongly curved at the apex, pulvilli long.

Wings: The wings are hyaline, long and slender and villose. The auxiliary cell yellow. Second vein with a strong bend anteriorly before the apex. End of marginal cell broadly rounded backward. The anterior branch of the third vein arises abruptly but is rounded at the bend and curves forward near the middle. First posterior cell long and slender, a little narrowed at the apex, the anal cell equally widely open. The anterior crossvein enters the discal cell before the middle; the second vein arises abruptly but not quite rectangularly and has a rounded bend without spur. It arises a little before the midpoint between the beginning of the discal cell and the anterior crossvein. Ambient vein complete. Alula well developed. Costa with a lappetlike spur.

Abdomen: In the male the abdomen is elongate and subcylindrical, slightly longer than the thorax, dully shining and thinly pollinose, the pile is rather dense, brownish yellow, partly flat-appressed with a few, scattered, long, fine hairs arising posteriorly along the outer part of the tergites, especially the first four. Sides of the first tergite with dense, broad band of coarse, erect, pale pile. The fringe of long, fine, erect hairs along the outer margins of the anterior tergites, while consisting of only scattered hairs, is conspicuous. I find 9 tergites in the male, the ninth quite short, the tenth larger, slanted downward and forming a curled hood enclosing the remainder of the genitalia. The female abdomen is slenderly oval. At the end of the third segment a little wider than the thorax and densely covered with flat appressed, long, brownish yellow, scalelike pile. I find 7 well-developed tergites in the female and a very short eighth tucked under the seventh. In males the basimeres are large and lie close together ventrally.

Dr. A. J. Hesse sums up the hypopygium of these flies in the following words:

Hypopygium of males with the apical angles of the opposing last sternite also slightly angularly prominent; basal parts of hypopygium also divided into two parts by an impressed suture, each covered dorsally with shortish hair, the base of each produced into a lobe-like process and the apical neck-like part narrow and well marked off from basal two-thirds, the outer apical angle not prominently produced as in *Plesiocera;* beaked apical joints elongate, curved outwards near apex, without any distinct or visible hairs above; aedeagus shortish apically, passing basally and on dorsal aspect into a broadish flattened process on each side, with the lateral ramus on each side from each basal part fused together bandwise, across aedeagal complex, to form a ventral aedeagal process, prolonged on each side into a ventrally directed process the apical part of which is flattened and broadened.

Material available for study: A pair of the typespecies sent to me through the courtesy of Dr. A. J. Hesse.

Immature stages: Unknown.

Ecology and behavior of adults: I have seen no comment upon the behavior of adults.

Distribution: Ethiopian: *Prorostoma integrum* (Bezzi), 1922 (as *Plesiocera*).

Genus *Coryprosopa* Hesse

FIGURES 132, 386, 632, 647

Coryprosopa Hesse, Ann. South African Mus., vol. 35, p. 118, 1956. Type of genus: *Coryprosopa lineata* Hesse, 1956, by original designation.

Small flies superficially appearing like *Aphoebantus* Loew, with elongate tapered abdomen, a little compressed and subcylindrical in the male and with blunt apex, but rather quickly recognized by the long, scattered, coarse, bristly hairs arising from the posterior half or more of the tergites and the posterior borders on the female, and similar shining mesonotal hairs, again more conspicuous in the male. Whole surface everywhere pollinose, the general coloration primarily pale, with black stripes on the abdomen and the mesonotum centrally blackish, and both of these areas with opaque, matted, pale, scaliform pile.

In addition, in this genus, the face is extremely prominent and conically extended beyond the pyriform, bulbous, third antennal segment. Moreover, the chiefly hyaline wings mostly tinted in the costal and auxiliary cells; it is noteworthy for the rectangular origin of the second vein with basal stump and for the anterior bend in the second vein on the outer fourth of this vein. Finally, the lobes of the occiput are separated.

Coryprosopa Hesse is a monotypic South African genus related to *Plesiocera* Macquart, and *Conomyza* Hesse from which it is distinguished by the above characters.

Length: 5.5 to 9 mm.; wing 4.5 to 6.5 mm.

Head, lateral aspect: The head is subglobular, with the occiput only moderately extensive above in either sex, and from the lateral aspect very little indeed shows from the lower half. Middle of the eye margin shallowly indented. The front is rather flattened, the vertex sinking below the eye margins. Whole front and face pale yellow and densely pale pollinose, the pile in the male consists of 2 or 3 rows of slender, pale, bristly hairs directed forward. In the female the pile is darker, more nearly erect and more irregular. The pile of the occiput is fine and short and appressed outward. Inner margin of the occiput with a scanty fringe around the circular recess. The face in *Coryprosopa* Hesse is very characteristically extended forward into a long, beaklike, triangular cone, nearly equilateral and reaching beyond the apex of the antennae. The proboscis is short, comparatively stout with flattened, pointed labellum; while Hesse (1956) notes the labellum is nearly half as long as the rest of the proboscis, it appears to me to be a little less than one-third as long. And in the material before me the proboscis is tucked in the oral recess lying flat against the head. Labella surface spinulose. The palpus is quite long and slender, cylindrical, curved, with a few fine, long, erect, bristly hairs at the base. It appears to have a short, second apical segment.

The antennae are small and short, the first segment is short and quite wide, scarcely half as long as its width viewed from above. The second segment is less wide, only half as long as the first segment and this segment is also only half as long as wide. The third segment is pyriform but somewhat asymmetrical, the basal bulblike portion is attached somewhat broadly and medially on the second segment, bulging out medially to a moderate extent and only gradually tapering to the blunt apex which bears a short, spine-tipped microsegment. This segment is covered with pale scalelike micropubescence. The second segment has a fringe of bristly setae, except medially. The first segment has a tuft of shining, scaliform pile laterally, and it has also a shelving, more or less cuplike ledge developed medioapically, and touching the ledge of the opposite antennae.

Head, anterior aspect: The head is circular and distinctly wider than the thorax. Eyes of the male strongly separated by a distance equal the breadth of the first antennal segment. Eyes more widely separated in the female. The vertex is scarcely narrowed over the front, and it is at least 3 times as wide as the space between the outer edges of the ocelli. The antennae are widely separated, although the first two segments touch at the apex, usually, because of the lobelike outgrowth of this segment. Because of the position of the second and third segments the outer part of the antennae is strongly divergent. The ocellar tubercle is equilateral, rather convex and prominent, especially in the male. The oragenal recess is set low upon the head and is slightly oblique, hence it is readily inspected only from the bottom. The face is very strongly conically projected forward. The gena is a mere linear strip along the eye margin and there is a paper-thin, tentorial fissure near the bottom of the eye. The face is pale pollinose with scattered, fine, pale bristles in the male and rather stiffer, stouter, reddish bristles in the female. From the posterior aspect the occiput is bilobate, deeply cupulate and dorsally the lobes fail to meet by the distance between the ocelli.

Thorax: The thorax is longer than wide, the mesonotum densely pollinose and also densely covered with matted, flat-appressed, coarse pile. Anterior mesonotal fringe of erect pile short and scanty. Humeri rather inflated, pale yellow, and pollinose and with an extensive tuft of long, coarse, somewhat flattened hairs. Notopleuron with scattered, slender bristles and in the posterior corner with 2 very stout, long, pale bristles. Postalar callosity with a few slender bristles, prescutellar margin with a row of bristles and the scutellar margin with 5 or 6 pairs of bristles. The surface of the scutellum has flat-appressed, coarse, long pile. Middle of mesonotum blackish in ground color but yellow in front of the scutellum and yellow throughout the entire lateral margins. Scutellum yellow, except for a medial, blackish stripe. Pleuron yellow and pollinose, mesopleuron with only scanty, scattered, long hairs and a few others on the sternopleuron. Pteropleuron and hypopleuron without pile. There is a tuft of hairs on the metapleuron in front of the haltere and the lateral metanotum bears a large, bushy tuft of coarse, erect pile. Halteres small, squama short with a fine, short fringe in one row only. All the thoracic pile is shining where erect but rather opaque where flat-appressed.

Legs: The legs are yellow. The femora rather stout, especially the hind pair. Anterior femora much shorter than the middle pair and the last two pairs nearly equal in length. All the femora covered rather loosely with large, flat-appressed, yellow scales. Anterior femur with 7 or 8 minute, erect, black setae ventrally on the swollen basal half or more. Middle femur with very fine, short, erect bristles, 6 anteroventrally, 3 posterior ones. Hind femur with fine, erect bristles that are a little more stout and also short. There are 6 anteroventral, 2 ventromedial ones apically, a dorsal bristle on each side at the apex and 1 or 2 other dorsal bristles near the apex. Anterior tibia short and quite slender, much shorter than the others; this tibia has an anterodorsal row of about 10 minute setae and 3 others equally minute posteriorly. Middle tibia with fine, short, oblique, slightly longer, scarcely more stout. This hind tibia has 4 rows of bristles, 6 anteroventral, 8 anterodorsal, 10 or more dorsomedial, and 3 or 4 ventromedial. Apical spur longer, sharp and stouter than the other bristles of the tibiae. Tarsi short, claws strongly curved at the apex and fine and sharp at the apex. The pulvilli are slender, but extend to the apex of the claws.

Wings: The wings are hyaline, vitreous with iridescent tints and the whole wing is elongate and slender. The costal and auxiliary cell are entirely brownish yellow and this yellow continues opposite the base of the discal cell as far as the third vein and includes the third vein. There are 4 diffuse, brownish spots, one just before the origin of the third vein, one at the lower end of the second basal cell, one over the basal end of the first submarginal cell, and one over the anterior crossvein. These seem to be best developed in the female, males may lack them or may have them much reduced. There are 2 submarginal cells, the second vein is rather strongly bent subapically, the marginal cell is rounded apically below. The second vein arises rectangularly with basal spur and its origin lies about halfway between the base of the discal cell and the anterior crossvein. All 4 posterior cells are open, the first one a little narrowed, and the anal cell is as widely open as the first posterior cell. Axillary lobe narrow, ambient vein complete, alula quite short, costa with a stout basal swelling and weak comb of bristles. There appears to be a bullose thickening in the first basal cell below where the radial sector begins. Anterior crossvein enters the discal cell in the middle.

Abdomen: The abdomen is elongate, about 1½ times as long as the thorax, subcylindrical, a little compressed in the middle on each side, brownish orange or brownish yellow, opaque yellow pollinose. There is a black vittae on each side originating in a rather large, subquadrate, blackish spot on each side of the second segment in both males and females, and slightly narrowing and fading away on the seventh tergite. I find 7 well-developed tergites in the female with 2 minute, additional ones beyond. Males with 7 tergites, each of nearly equal length and with 2 other shorter ones slanting downward; the male terminalia are distinctly recessed within these and a little more exposed below. Pile of the male abodmen dense, opaque yellow, flat-appressed, rather coarse and flattened but also with a rather conspicuous, loose, irregular band or double band of long, slender, shining yellow bristles or bristly hairs curled backward from along the posterior margins of the tergites. Female abdomen with more scanty, appressed pile that seems to be finer in character and also with scattered, long, fine hairs along the posterior margins of the tergites.

Dr. A. J. Hesse in his work on South African bee flies describes the male hypopygium as follows:

Hypopygium of male not so exposed as in *Plesiocera*, withdrawn into apical part of abdomen; last opposing abdominal sternite relatively shorter than in latter genus, its apical angles also slightly pointed and subprominent; basal parts of hypopygium itself divided into two parts by a dorsal dividing suture, each part with a basal tongue-like lobe and apical angle of each basal part not produced to the same extent as in *Plesiocera*; beaked apical joints entirely different, broadened in basal half, produced apically into a outwardly curved beak, the outer apical angle angular, the dorsum covered with hairs, especially towards base; aedeagus with the apical part shortish, produced basally on dorsal aspect into a prominent process on each side, the lateral ramus on each side from each basal part coalescing and forming an elongate ventral process, the ventral part of which is hollowed out and the apex slightly recurved; lateral struts very well developed, more strongly developed than in *Plesicera*; basal strut also longer and broader.

Material available for study: A pair of type-species sent by Dr. A. J. Hesse for dissection and study.

Immature stages: Unknown.

Ecology and behavior of adults: I have seen no comment on the behavior of adults.

Distribution: Ethiopian: *Coryprosopa lineata* Hesse, 1956.

Genus *Epacmoides* Hesse

FIGURES 129, 380, 646, 650

Epacmoides Hesse, Ann. South African Mus., Part II, vol. 35, p. 125, 1956. Type of genus: *Plesiocera biumbonatum* Bezzi, 1922, by original designation.

Hesse (1956), pp. 127-128, key to 4 Ethiopian species.

Small flies with short, appressed pile, except on the sternites, mesopleuron, and anterior margin of the mesonotum. The abdomen is as long as the thorax, the scutellum included, but in males is rapidly narrowed to a cylindroid form beyond the second segment; females elongate oval with the last segment triangular. The wings are hyaline, and these flies are readily distinguished from *Aphoebantus* Loew, which they resemble in general aspect, by the formation of the second vein; this vein arises abruptly from the third vein at a point a little beyond the middle of the space between the base of the discal cell and the anterior crossvein. The face is triangularly produced and in this respect resembles *Epacmus* Osten Sacken from which it is again readily separated by the abrupt origin of the second vein. Moreover, it should be noted that in addition to rather strong bristles on the margin of the scutellum there is a transverse row of equally well-developed bristles lying in front of the scutellum. Notopleuron also with similar bristles and anterior mesonotal margin with a collar of pile.

The metanotum and upper and posterior metapleuron have much coarse tufted or partly appressed pile. Proboscis limited to the oragenal recess.

The anterior branch of the third vein is strongly sigmoid, the end of the marginal cell ovally rounded and swollen downward.

This small genus contains only 4 South African species.

Length: 3.6 to 8 mm.; of wing 4 to 8 mm.

Head, lateral aspect: The head is subglobular, occiput only moderately developed and with rather dense, appressed, scaliform pile reaching to and making a matted fringe along the upper half of the eye margin; it begins at the indentation of the eye, which is deeper and more conspicuous than in *Prorostoma* Hesse. Inner margin of the occiput with a very scanty fringe of scattered hairs. The front is flat on the upper half below the ocelli but is distinctly swollen in the male in *Epacmoides* Hesse, much more so than in *Prorostoma* Hesse. Also the entire front, except immediately anterior to the ocelli, is shining and bare of pollen, whereas in *Prorostoma* Hesse the front and vertex are uniformly and densely silvery pollinose. The pile of the male front is rather dense, bristly, and projected forward. The face is triangularly produced almost as far as the apex of the antennae, it is more pointed and acute than in *Prorostoma* Hesse and bears scattered, appressed scales, the surface polished. The proboscis is slender with a short, flattened, oval, somewhat pointed labellum. It is shorter or in any case not longer than the oral recess. Palpus rather long and slender with a few long hairs near the apex and laterally in the middle, and apparently with a short basal segment. It is about one-third as long as the proboscis. The antennae are similar to *Prorostoma* Hesse, the first segment flared and widening both medially and apically with apical fringe of stiff bristly hairs longer laterally. Second segment quite short, less than half as long as wide, the third segment smaller, bulblike at the base with a long slender, cylindrical process, not quite twice as long as the basal part.

Head, anterior aspect: The head is circular and a little wider than the thorax. The eyes are separated in the male by a distance equal the apical width of the first antennal segment. The vertex is sunken beneath the eye margin, the ocellar tubercle prominent with vertical sides and very large, wide, anterior ocellus. The ocelli lie in an equilateral triangle and there is a row of coarse, bristly hairs extending forward over the middle of the ocellar tubercle. Pile of anterior front distinctly scaliform. In the female the eyes are much more widely separated and the sides of the front converge only slightly. The ocellar tubercle is equally prominent, likewise with vertical sides and from the anterior ocellus to the middle of the front there is a marked, concave depression or concavity. While the antennae are rather widely separated at the base, nevertheless the first segments touch medially and apically. The oragenal cavity is quite deep, elongate, with vertical sides and slightly pinched in along the middle. The genae are quite linear along the middle of the oragenal cavity and are slightly wider below. Viewed from behind the occiput is bilobate and very different from *Prorostoma* Hesse. Where the lobes touch in *Prorostoma* Hesse with a deep triangular recess, the lobes in this genus touch but lack any deep anterior pocket.

Thorax: The thorax is short and black, the mesonotum is low, a little convex, rather higher than in *Prorostoma* Hesse. It is loosely covered with coarse, flat appressed, scaliform pile. The anterior collar is rather high, and narrow, the coarse bristly hairs rather long and the flat appressed pile begins immediately behind the collar. Humerus with a dense, tangled tuft of erect and partly appressed pile. Notopleuron densely pilose and with a large, loose clump of 8 or 9 slender, red bristles, 2 of them more stout. Postalar callosity with at least 3 long, slender, red bristles, and there is a band of similar bristles in front of the scutellum which are curved backward. Scutellum black, with matted, appressed pile on the disc and the lateral corners, and in the medial fovea on the posterior margin of the scutellum. Scutellum is rather thick apically, polished black, rather strongly bimammillate. The margin has 4 or 5 pairs of long, red bristles. Pleuron pollinose, whole mesopleuron and almost the entire sternopleuron with dense mat of flat-appressed, scaliform pile. The upper and posterior mesopleuron have a few long, erect, bristly hairs. Propleuron with a narrow, long fringe. Metanotum and the anterior metapleuron in front of the spiracle and an equally conspicuous patch behind the spiracle made up of flat-appressed, scaliform pile. However, that on the metanotum is more erect. Halteres short but the knob large and short. Squama short with a weak fringe.

Legs: Anterior and middle femora distinctly but not strongly thickened over the middle and more so at the base. Hind femur stout but not swollen. All femora covered with dense, yellowish white, large, flat-appressed scales. Middle femur with 2 quite stout, anteroventral bristles and 2 or 3 minute setae posteriorly. Hind femur with 5 bristles ventrolaterally on the apical half, 3 of them stout; 1 stout bristle and 2 or 3 smaller ones dorsolaterally near the apex. Anterior tibia with fine, bristly spicules, 3 extremely minute spicules dorsally, 5 which are a little longer posteriorly, and 5 still larger ones posteriorly ventrally. Middle tibia with only minute bristles dorsally, but with 5 stouter, longer bristles ventrally. Hind tibia with 4 rows of bristles, the anteroventral row contains 4 moderately long, stout bristles, the 2 dorsal rows contain 5 or 6 quite small bristles and the ventromedial row contains 6 slender bristles. Tarsi slender, short, but rather strongly spiculate below. Claws slender, only slightly curved with the pulvilli rather long.

Wings: The wings are hyaline, elongate, but comparatively broad, wholly villose. Auxiliary cell yellowish. Second vein with a strong, preapical bend forward so that the marginal cell is rounded outward posteriorly. The anterior branch of the third vein rises abruptly, not quite rectangularly and makes a rounded, strong bend outwardly and then is bent forward a little beyond the middle so that the second submarginal cell has nearly or quite parallel sides on both its basal half and its outer half. First posterior cell open widely, the anal cell opened half as wide. The anterior crossvein enters the discal cell a little before the middle. The second vein arises abruptly, has a rounded bend, its point of origin is halfway from the beginning of the discal cell to the anterior crossvein. Ambient vein complete, alula rather wide. Costa at the base thickened, and with a large lappetlike spur.

Abdomen: The abdomen in the male is short, scarcely as long as the thorax, gradually narrowed beyond the base of the second segment, subcylindrical, slightly compressed laterally and apically and densely covered with flat-appressed, pale scales and scaliform pile and each tergite with a subapical fringe of fine, bristly, reddish hairs, of no great length and appressed backward. Color of abdomen black with the posterior margins obscurely and rather widely reddish brown. Middle of the first segment with on each side of the midline a large, more or less semicircular tuft of flat-appressed, yellow scales, the outer third of this segment with dense band of long, coarse, erect, spikelike pile. Extreme lower margins of the second, third and fourth tergites with considerable, long, erect, coarse pile. Female abdomen elongate conical with slightly rounded sides and gently convex on the disc. It is densely covered with matted, appressed, scaliform pile partly yellowish or reddish brown, especially on each side of the middle and on the posterior part leaving a distinct, medial, pale yellow stripe and a less distinct row of disjointed spots laterally. I find 7 well-developed tergites, the last one is compressed laterally and has a heavy fringe of bristly hairs surrounding the female terminalia. I find 7 well-developed tergites in the male and a short eighth turned downward and almost concealed beneath the seventh. It encloses the small genitalia, the two pieces of which lie side by side recessed and turned to one side.

Since Dr. A. J. Hesse was able to base his conclusions upon a greater number of species, I quote his remarks upon the male hypopygium:

Hypopygium of males with the lateral apical angles of opposing last sternite also somewhat angular; basal parts of hypopygium itself differing from that of *Prorostoma* in not having the apical part so well marked off; beaked apical joints more like those of *Coryprosopa*, not narrowish throughout, but with an outer angular apical part; aedeagus without a complex, ventral, aedeagal process, the process, if present, in form of a forwardly projecting ledge or a pair of stylets; basal strut sometimes with a distinct, shelf-like, lateral ledge present on each side.

Material studied: A pair of the type-species sent by Dr. A. J. Hesse to me.

Immature stages: Unknown.

Ecology and behavior of adults: I have seen no comment on the behavior of adults.

Distribution: Ethiopian: *Epacmoides albifrons* Hesse, 1956, *albifrons pallidulum* Hesse, 1956; *biumbonatum* (Bezzi), 1922 (as *Plesiocera*); *cryptochaunum* Hesse, 1956; *xerophilum* Hesse, 1956.

Genus *Stomylomyia* Bigot

Figures 131, 385

Stomylomyia Bigot, Bull. Soc. Ent. France, ser. 6, vol. 7, p. xxxi, 1887a. Type of genus: *Tomomyza europaea* Loew, 1869, as *Stomylomyia leonina* Bigot, 1887, by monotypy.

Small, comparatively slender flies, varying from yellowish brown to blackish in ground color with some brownish or yellowish or rarely silver gray pollen, and also with scattered, slightly flattened pile or even narrow scales in some species. They are rather readily recognized by the sharp and conically produced face and the clear, hyaline wings, which may be tinged with brown or yellow near base and costa. Also, note the origin of the second vein which is abrupt and rounded, and lying halfway between anterior crossvein and base of radial sector.

All of the species of this genus have 3 submarginal cells, separating them at once from *Plesiocera* Macquart which also has a much less sharply produced face. Engel (1934) made *Stomylomyia* Bigot a subgenus of *Plesiocera* Macquart placing 5 species within it. I have had for study the type-species of *Stomylomyia* Bigot, received through the courtesy of the British Museum, and I have been able to study the type of *Exepacmus nasalis* Melander. Dr. A. J. Hesse kindly sent several South African species of *Plesiocera* Macquart for study, and I have the type-species as well. *Stomylomyia* Bigot is distinguished from *Plesiocera* Macquart by the oblique, basal origin of the second vein; its face is long and conical extending well beyond the antennae. It is separated from *Exepacmus* Coquillett by the deep, conspicuous indentation of the posterior eye margin. All of these small genera have a general similarity in appearance, and *Desmatoneura* Williston belongs in the same group, differing, as does *Plesiocera* Macquart, in having only 2 submarginal cells and distinctions; however, it is worth noting that the peculiar front found in *Desmatoneura* Williston is matched by the front of *Plesiocera flavifrons* Becker (Engel 1934, fig. 159); the latter species differs sharply in facial ledge and indentation of the eye.

As now formed, *Stomylomyia* Bigot is restricted to Asia Minor and southern Europe and North Africa.

Length: 3.5 to 11 mm.; wing 3.0 to 9 mm.

Head, lateral aspect: The head is subglobular, the occiput is short above and below. There is a very shallow indentation at the middle of the posterior eye margin without accompanying notch, hence this genus is difficult to separate from the Cylleniinae. The pile of the occiput consists of erect, micropubescence and a fine, scanty fringe along the sides of the interior occipital cup and the pile becomes a little longer and more conspicuous dorsally behind the vertex on each occipital lobe. The front is short but distinctly protuberant above the antennae. The pile is loose, scanty, bristly, of no great length and scattered over the whole of the front. The ground color is shining black with some pollenlike micropubescence down the middle of the front, linearly along each eye margin and just above the antennae. The face in profile is strongly and conically peaked forward. In this respect it is somewhat like *Coryprosopa* Hesse, from which it differs in the presence of 3 submarginal cells. The face is bare and shining, the conical part, which represents an extension of the oragenal cup, is sharply delimited from the rest of the face by a narrow, almost rectangular crease. Face with scanty micropubescence laterally, and a few short, bristly hairs below on the edge of the peak of the oragenal recess. The eye is only moderately large, distinctly long on the upper half, the upper facets are a little enlarged.

The proboscis is slender, not longer than the oragenal recess, and the labellum slender, rather flattened, and nearly one-third the total length of the proboscis. The palpus is quite long and slender and almost filiform. I can only find one segment; it bears a few, fine hairs laterally. The antennae are attached near the upper third of the head, the first two segments are subequal in length and both are short. Viewed from below the second segment is a little longer than the first. Ventrally the occiput descends rather strongly beneath the eye, so that the genal region is extensive at the bottom of the head, though very short and linear above.

Head, anterior aspect: The head is circular and much wider than the thorax. The front in the female is widely separated by a distance somewhat greater than the space across the first antennal segment measured from the outside. The ocellar tubercle is slightly obtuse, large, low, a little flattened. The antennae are narrowly separated by the width of the first segment in the female. The oragenal recess is unusually long, quite deep throughout, and strongly narrowed anteriorly. The crease separating the sides of the oragenal recess from the gena is distinct and rather deep. The gena is narrow, except below the eyes where it is extensive. The ventral occiput extends even deeper below the eyes. From the posterior view the occipital lobes are broadly, gently rounded, widely separated, deep, and divided by a rather wide fissure.

Thorax: The thorax is rather convex though not conspicuously so. Mesonotum and scutellum are blackish with brownish yellow pollen and scanty, more or less appressed setate pile dorsally and a considerable amount of narrow, appressed scales which are more numerous along the anterior margin of the mesonotum, above the wing, on the notopleuron and humerus. The upper mesopleuron bears in addition some appressed scales, and bears a patch of long, scattered, bristly hairs dorsally. There are a few, appressed scales on the lower mesopleuron and upper sternopleuron and there is a collar of scanty, erect pile along the anterior margin of the mesonotum. There are 2 moderately stout, long, curved, red bristles on the notopleuron and 4 pairs of long, more slender red bristles on the margin of the scutellum. Pteropleuron, hypopleuron, and the metapleuron, with the exception of the metanotum, bare and pollinose only. The metanotum bears an extensive, conspicuous tuft or clump of long, erect scaliform pile and in this respect is like *Coryprosopa* Hesse. There is also a radiating tuft of appressed scales behind the spiracle and below the haltere. The squama is short.

Legs: The femora are loosely covered with scattered, flat appressed scales, the hind femur has 2 minute, slender bristles in the middle laterally, 2 others much longer and much thicker on the outer half. The anterior tibia bears some weak spicules, middle and hind femur with more distinct bristles, slender and oblique. On the hind femur there are 3 rows, each with 4 or 5 bristles.

Wings: The wings are hyaline and microvillose. The second vein arises abruptly but with rounded bend about halfway between the upper end of the second basal cell and the anterior crossvein. The anterior crossvein enters the discal cell at the middle. There are 3 submarginal cells. The anterior crossvein arises almost rectangularly as a plane, simulated crossvein and at the end of the crossvein it sends a backward spur to join the second vein, this spur vein represents the third branch of the radius. The apical part of the second vein and the anterior branch of the third vein both bend upward, ending before the apex of the wing. First posterior cell widely open, the anal cell also widely open. axillary lobe and alula both moderately narrow. Ambient vein complete.

Abdomen: The abdomen is elongate, about as long as the head and thorax together, subcylindrical, blackish, feebly shining, with thin pollen and with loose, scattered, moderately abundant, appressed scales, which are somewhat wider laterally, more narrow dorsally and there are wide, conspicuous scales on the sternites. Female terminalia with a dense fringe of hairs at the apex.

Material available for study: Material for the type-species in the collections of the British Museum (Natural History). Also the type of *Exepacmus nasalis* Melander. And for comparison with the related genus *Plesiocera* Macquart, the type-species given me by the British Museum (Natural History) and several South African species furnished through the generosity of Dr. A. J. Hesse.

Immature stages: Unknown.

Ecology and behavior of adults: Not on record.

Distribution: Palaearctic: *Stomylomyia europaea* (Loew), 1869 (as *Tomomyza*), *europaea aegyptiaca* Bezzi, 1925, *europaea nigrirostris* Bezzi, 1925 [=*leonina* (Bigot), 1887 (as *Stomylomyia*)].

Genus *Plesiocera* Macquart

FIGURES 125, 127, 381, 643, 648

Plesiocera Macquart, Diptères exotiques, vol. 2, pt. 1, p. 82, 1840.
    Type of genus: *Plesiocera algira* Macquart, 1840, by monotypy.
Calledax, new subgenus. Type of subgenus: *Plesiocera psammophila* Hesse, 1956.

Engel (1934), p. 391, key to 7 Palaearctic species.
Hesse (1956), pp. 99-102, key to 7 species.

Medium-sized flies in the type-species, with tapered, nonconstricted abdomen, the ground color black with grayish-ochre pollen overlaid, thinner on the mesonotum and abdomen, more in evidence elsewhere. The pile is rather yellowish, nearly opaque and erect on the anterior third of the mesonotum and with much appressed brassy yellow pile upon the remainder of the mesonotum, scutellum, and abdomen; basal corners of abdomen with dense, conspicuous tuft of erect, golden pile. The face is projecting forward and separated by a distinct crease, sharper laterally, and crossing a short distance below the antenna. Marginal cell at end with a distinct, thumblike bend downward; wings largely hyaline, faintly brownish on foreborder, except at apex. The anal cell is widely open. The second vein arises opposite the base of the discal cell and at a gently acute angle, or sometimes a little more abruptly. The legs in the type-species are pale yellow; anterior tibia with only ultramicroscopic black microspinules; hind tibiae with distinct, regular, slender, black spines throughout. Claws and pulvilli small but well developed. In the type-species the posterior border of the eye is deeply indented and clearly bisected, and affinity appears to be with *Epacmus* Osten Sacken. I have only females.

Plesiocera Macquart contains one subgenus, *Calledax*, new subgenus, which is described herein.

Head, lateral aspect: The head is globular with the occiput very extensive dorsally but strongly rounded and sloping inward. The central fringe is long and dense, the cavity large and very deep. The entire outer surface of the occiput is pollinose and densely covered with short, nearly erect, golden pile which tends to curl forward. There is a tuft of longer, golden pile extending outward opposite the indentation of the eye. The front is convex due to the produced and greatly rounded head. It is scarcely visible above the eye profile, bears golden pollen and a mixture of fine, black and golden, erect hairs. The eye is reniform, the indentation deep and broad, the lower angle of the indentation more acute than the one above, and there is rather long, distinct, horizontal notch in the eye at the point of indentation.

The face is a little extended forward beneath the antennae but rounded and retreating to the point where the very extensive oragenal cup begins to jut out. This dorsal and lateral anterior wall of the oragenal cup is sharply demarcated from the face by a conspicuous, circular crease. Whole face and oragenal cup on the upper half is golden pollinose, and bears scattered, fine, more or less erect, black and yellow hairs, especially along the sides. This pile is not quite as long as the first antennal segment. The proboscis is slender, the labellum likewise, not longer than the oragenal recess into which it may be completely hidden when at rest. Palpus slender and small with a few fine, long, golden hairs. Antennae attached at the middle of the head. They are small, the first segment is the largest, 2 or 3 times as long as the second and this segment is strongly widened medially toward the apex. It bears bushy, long, bristly pile laterally and below with shorter setae medially and dorsally. The second segment is wider than long and bears long, fine setae on all sides. Third segment basally as wide as the second segment, rounded at the base and progressively narrowed to the apex. The stylar portion is thick and bears a distinct, cylindrical, spine-tipped microsegment. This third segment is about as long as the first two together, or slightly longer.

Head, anterior aspect: The head is circular and considerably wider than the thorax. Eyes in the male slightly separated, or by the width of the ocellar tubercle in Palaearctic species, or in Ethiopian species also separated by the breadth of the ocellar tubercle. In the female of the type-species the eyes are separated at the vertex by nearly 3 times the width of the small ocellar tubercle, which posteriorly and posterolaterally is very high, steep, and vertical. The width of the female eye is also equal to the length of the third antennal segment. In African species it varies, but it is usually little more than twice the width of the first segment. This first segment is not very wide at the base but expands considerably on the medial side toward the apex; remaining two segments are divergent, the second segment not more than half as long as the first, more narrow, shorter than wide and cylinder-like rather than bead-shaped. First segment with a few short setae above but with a conspicuous covering of fine, black bristles over the whole lateral surface which extends slightly beyond the second segment. There are shorter bristles below. Second segment with fine setae on all sides. Third segment more narrow than the second, plane at the base and nearly plane dorsally. Below, this segment rapidly narrows and tapers uniformly to the thick apex, which bears a distinct, short, spine-tipped microsegment.

The oragenal cup is rather wide, but about 3 times as long as its inner width. Anteriorly a short distance beneath the antennae the oragenal cup is very strongly produced. Both the upper margin of the produced portion and laterally the upper half are rounded and convex or bulging outward a little. It is sharply separated from the face by a deep crease and the height gradually diminishes until it is quite low just before the base of the proboscis. Since in the middle there is a slight diagonal, foveate depression, it would seem that this strong, jutting forward of the oral walls has an upper facial component and a posterogenal component. The tentorial fissure is a mere line but quite long, leaving a noticeable border along the lower eye which is pollinose, and contrasting because the oragenal cup is more thinly pollinose. Sides of both the pollinose face above and the sides of the whole anterior margin of the oragenal cup covered with long, scattered, bristly hairs. The head viewed from above shows the occipital lobes to be adjacent but with a deep furrow between and with a large, deep triangular pocket behind the vertex.

Thorax: The thorax is slightly longer than wide, the scutellum excepted, and also it is very slightly wider behind than in front. The mesonotum is low and slightly convex, black and rather densely olive-brown pollinose. The pile consists of dense, shining, brassy, appressed, slightly flattened hairs, which become more strongly appressed and more abundant and conspicuous on the posterior half of the mesonotum and likewise shorter but which anteriorly change to longer, more erect, dense pile merging into the anterior collar of long, dense, erect pile. The humerus is swollen and bears long pile like that of the anterior collar. Notopleuron with a clump of weak bristles, one of them longer and stouter than the others. Upper postalar callosity with bristly pile and 7 or 8 slender bristles, 2 or 3 of them much longer. There is also a considerable tuft of matted pile of no great length on the anterior, outer wall of the callosity. Scutellum large, bluntly subtriangular, distinctly thin, bearing pollen and pile similar to that of the mesonotum. The scutellar margin has about 5 pairs of long, slender bristles. Also, the mesonotum posteriorly in front of the scutellum bears a row of slender bristles. Squama short with a scanty fringe, which, however, seems to be banked in several rows. Pleuron pollinose with very dense, conspicuous, long, bristly pile over the whole of the mesopleuron, a large upward tuft turned upward and the remainder more or less erect but slightly curved backward or downward. Pteropleuron and hypopleuron without pile. The metanotum and the metapleuron in front of the haltere also without pile, but behind the spiracle there is a conspicuous tuft of somewhat appressed pile.

Legs: The legs are slender but with the femora distinctly stout. Anterior and middle femora slightly thickened toward the base, the hind femur throughout. All of the legs are pale yellow in the type-species, the last tarsal segment on the last four legs brownish. The femora are densely covered with flat-appressed, long, yellow, scaliform pile. The middle femur bears 2 black setae toward the apex, the hind femur bears 7 on the outer half on the anteroventral aspect; the last 4 are distinct bristles. Anterior tibia without spicules, except of the most microscopic type. Middle tibia, however, with 4 rows short, fine, oblique, sharp, slender bristles, 7 anterodorsally, 5 anteroventrally, 10 posterodorsally, and 7 posteroventrally. The quite long, slender, hind tibia also with 4 rows, 6 anteroventrally, 13 anterodorsally, 10 dorsomedially and these are the smallest, and 12 ventromedially. Claws small, slender, curved on the outer half, the pulvilli long and broad.

Wings: The wings are tinted with brown on the anterior half, the whole wing villose, the dark brown villi giving the posterior half of the wing a slight brownish tint also. The auxiliary vein ends near the base of the second submarginal cell and not very far from the end of the first vein. The second vein arises at a strong but not abrupt angle opposite the base of the discal cell. It has a distinct bend forward subapically so that the end of the marginal cell is widened and rounded posteriorly, the anterior branch of the third vein arises as an oblique, simulated crossvein with a backward spur. Outwardly it makes a very strong, deep bend turning forward to end before the apex of the wing. First posterior cell widely open, anal cell also widely open. The anterior crossvein enters the discal cell a little before the middle. The humeral crossvein is placed far out from the base of the wing. This basal section before the humeral crossvein is nearly a fifth of the length of the wing, hence the wing is strongly narrowed and lengthened at the base. The axillary lobe is narrow, especially basally, the anal vein curved, the alula short, the ambient vein complete and the costa with a strong lappetlike spur at the base, and the costa is also widened at the base and bears only minute setae.

Abdomen: The abdomen in the female type-species is elongate and a little tapered from the base to apex, the sides of the abdomen nearly or quite plane; the abdomen is a little longer than the thorax and of the same width at the base; it tends to be rather strongly flattened and covered very densely indeed with flat-appressed, matted, golden pile in the type-species, with a few, additional, fine, suberect, black or yellow bristly hairs in the posterior corners laterally. The sides are a little turned over and are yellow with long, golden pile, whereas the disc from above is entirely black, except for a yellow hind margin on the sixth and seventh tergites and a concealed eighth. The first segment has extraordinarily long, dense, erect, bristly pile extending almost to the midline but leaving the anterior margin widely bare as far as the base of the squama. Female terminalia small and concealed by bristles.

Dr. A. J. Hesse sums up the hypopygium of these flies in the following words:

Hypopygium of male usually subject to slight torsion, situated on side opposite last sternite, usually visible at end of abdomen; basal part divided dorsally into two separate parts by a suture or impressed suture, usually covered with fine hairs above, each basal part with a more or less well-marked-off apical part and the outer apical part or angle usually produce or angularly prominent; beaked apical joints usually elongate and curved, usually ending apically in a sharp point or upturned spine, sometimes with a row of spinules or spinule-like hairs along outer aspect nearer apical part; aedeagus with the apical part usually slender, often directed slightly upwards, the aedeagal part produced basally on dorsal aspect into a distinct, lobe-like or tongue-like process on each side which is joined on to lateral ramus on each side, the aedeagus also with a distinct ventral process, formed as a medial, apically directed structure by the fusion and coalescence of each lateral ramus, the apical part of this process assuming various and complex shapes; lateral and basal struts, the latter without a lateral shelf-like flange, but a short process on each side basally may be present.

Material available for study: I have a female of the type-species given to me by the British Museum (Natural History). I also have three other species sent to me through the generosity and kindness of Dr. A. J. Hesse of the South African Museum, Cape Town.

Immature stages: Unknown.

Ecology and behavior of adults: I have seen no record of the behavior of adults.

Distribution: Palaearctic: *Plesiocera algira* Macquart, 1840 [=*fornicata* (Loew), 1857 (as *Anthrax*), =*inaequalis* (Becker), 1906 (as *Anthrax*)]; *araxana* Paramonov, 1925; *flavifrons* Becker, 1915; *pusilla* Bezzi, 1925; *tenella* (Loew), 1869 (as *Tomomyza*); *turkestanica* Paramonov, 1929.

Ethiopian: *Plesiocera curvistoma* Hesse, 1956; *flavilabris* Hesse, 1956; *pernotata* Hesse, 1956; *philerema* Hesse, 1956; *psammophila* Hesse, 1956; *rhodesiensis* Hesse, 1956; *rufiventris* Hesse, 1956.

Calledax, new subgenus

Type of subgenus: *Plesiocera psammophila* Hesse.

Small, slender, the abdomen somewhat constricted, blackish flies, more or less shining. The second vein origin is variable in position, but always clearly and abruptly at right angles and usually quite beyond the base of the discal cell. The posterior eye margin is receded forward on the upper half of the head much as in *Tomomyza* Wiedemann, or *Amphicosmus* Coquillett, but is not indented or bisected. It includes the South African species placed in *Plesiocera* Macquart. In *philerema* Hesse the second vein arises at right angles opposite base of discal cell. In *curvistoma* Hesse it arises halfway between.

Flies of *Plesiocera* Macquart are Palaeartic and Ethiopian in distribution.

Plesiocera Macquart is related to *Stomylomyia* Bigot and to *Exepacmus* Coquillett. Engel included *Stomylomyia* Bigot as a subgenus of *Plesiocera*.

Length: 3.5 to 10 mm.; wing 3.5 to 9.5 mm. The type-species appears to be the largest species.

Genus *Conomyza* Hesse

Conomyza Hesse, Ann. South African Mus., part 2, vol. 35, p. 114, 1956. Type of genus: *Conomyza semirufella* Hesse, 1956, original designation.

Small flies that are related to *Plesiocera* Macquart. The occiput is shorter behind the ocellar tubercle and the occipital indentation or gap is greater, more distinct, and wider. The scales on the upper part of the body are distinctly finer and more hairlike. The front tibiae bear distinct, minute spicules and spurs; the tarsi, especially the female tarsi, are distinctly shorter than the tibia, and they bear distinct spinules below on the front tarsi. In this genus the front claws are almost as large as the middle claws, not nearly or quite vestigial as in *Plesiocera* Macquart. A weak genus, possibly only a subgenus.

This genus is known from the type-species and one well-marked variety from Namaqualand, the variety from Koup Karoo.

Length: 4 to 7 mm.; of wing 3.5 to 5 mm.

As I have not been able to see or obtain material of *Conomyza* Hesse, I quote the author's description of this genus. Hesse states:

This new genus is erected to contain a species which on account of certain distinct characters cannot be retained in the genus *Plesiocera* Macquart, as defined in this memoir. In quite a number of characters it is, however, almost indistinguishable from the latter genus. Its chief characters, when compared with the latter, are as follows:

Body similarly shaped. Vesiture also composed of hairs and addressed scaling, the former also comparatively or moderately sparse; the finer and longer ones disposed in the same way and on the same parts of body, those on abdomen also fine, shortish and inconspicuous; scaling on body, however, distinctly finer, more hair-like or pile-like, the individual scales, even along sides of thorax above, on pleurae, abdomen and on legs finer and more hair-like, giving the insects a distinctly more pily appearance than in species of *Plesiocera* Macquart; scaling on sides of thorax, even if denser than on disc, not contrastingly cretacious whitish or ochreous yellowish and that on pleurae also not contrastingly white.

The head is as in *Plesiocera* Macquart, broader across eyes than across thorax; occiput, however, distinctly shorter or narrower behind ocellar tubercle, the occipital lobes thus less prominent, the medial sulcation distinctly broader, more gap-like, the two lobes more broadly separated, the sulcus nearly as broad as width of ocellar tubercle; eyes with the hind margin as in *Plesiocera* Macquart, not indented or subangularly emarginate, only broadly sinuous, separated above on vertex in both sexes, narrower in males and about as broad as ocellar tubercle, appearing slightly narrower than in *Plesiocera* Macquart, there being no small space in females a little less than 2 times width of tubercle; facial part also produced, distinctly cone-like, the conical part, however, less sharply pointed, slightly blunter, more rounded; antennae as in *Plesiocera* Macquart, joint 3 gradually narrowed apically, more rapidly along lower part, no distinct slender apical part visible as in the case of some species of *Plesiocera* Macquart, ending apically in an almost indistinguishable terminal style, bearing a stylet; proboscis also short, confined to buccal cavity, its labella blunter apically. Wings as in *Plesiocera* Macquart; alula also much reduced and axillary lobe narrowish; middle cross vein at about or near middle of discoidal cell; second vein originating somewhat obliquely quite a distance away from base of third vein also kinked.

The legs have some spines on outer lower and apical aspect of hind femora and also with a row of distinct spines on inner lower aspect from base to beyond middle in males; tibiae always with distinct, though small, spicules and spurs on front ones; front tarsi scarcely as long as front tibiae, usually shorter, with distinct spicules like those on middle tarsi, though smaller; front claws not markedly reduced or vestigial as in *Plesiocera Macquart;* pulvilli well developed. Hypopygium of males rather large and conspicuous, with the opposing last sternite rather more elongate than in *Plesiocera* Macquart, its apical angles also angular; basal part divided dorsally into two by a suture, the apical part of each hairy and the base of each drawn out into a lobe-like process; beaked apical joints narrowish, curved inwards, more or less hollowed above, the edges more or less ridge-like, the inner one bearing an upwardly directed dentate process or spine near apex, the hinder part of each joint provided with hairs above (side and dorsal views in middle); aedeagus very elongate, slender, cylindrical and curved as shown in the side view, with a long ventral process (Ae. Pr.) arising near its base from middle part of aedeagal complex, the aedeagal part also produced basally and dorsally on each side into a lobe-like process (seen in outline in left-hand figures and also in right-hand one); ramus on each side from each basal part as shown in figures; lateral and basal struts well developed, shaped as shown in figures.

Since I have not been able to see males of this genus I quote the remarks of Dr. A. J. Hesse with respect to genitalia:

Hypopygium of males rather large and conspicuous, with the opposing last sternite rather more elongate than in *Plesiocera*, its apical angles also angular; basal part divided dorsally into two by a suture, the apical part of each hairy and the base of each drawn out into a lobe-like process; beaked apical joints narrowish, curved inwards, more or less hollowed above, the edges more or less ridge-like, the inner one bearing an upwardly directed dentate process or spine near apex, the hinder part of each joint provided with hairs above; aedeagus very elongate, slender, cylindrical and curved, with a long ventral process arising near its base from middle part of aedeagal complex, the aedeagal part also produced basally and dorsally on each side into a lobe-like process; ramus on each side from each basal part; lateral and basal struts well developed.

Material available for study: None seen. Known only from the unique types in the Cape Town Museum.

Immature stages: Unknown.

Ecology and behavior of adults: No information on record.

Distribution: Ethiopian: *Conomyza semirufella* Hesse, 1956, *semirufella karooana* Hesse, 1956.

Genus *Exepacmus* Coquillett

FIGURE 293

Exepacmus Coquillett, Trans. American Ent. Soc. Philadelphia, vol. 21, p. 101, 1894a. Type of genus: *Exepacmus johnsoni* Coquillett, 1894, by monotypy.

I have compared the type of *Exepacmus nasalis* Melander with the type-species of *Stomylomyia* Bigot and I find the only significant difference lies in the very deep excavation of the posterior occiput in the one species as against the very shallow indentation in the other species. For this reason I leave *Exepacmus* as a distinct genus, and I have included it in the key to the genera of the Lomatiinae, but I consider a detailed description unnecessary.

Distribution: Nearctic: *Exepacmus johnsoni* Coquillett, 1894 [=*nasalis* Melander, 1950].

Genus *Eucessia* Coquillett

FIGURES 124, 374, 635, 653

Eucessia Coquillett, Canadian Ent., vol. 18, p. 82, 1886a. Type of genus: *Eucessia rubens* Coquillett, 1886, by monotypy.

Small flies with hyaline wings and with rather short, slightly tapered, cylindrical abdomen as in some *Aphoebantus* Loew. The large globular head is a little wider than the thorax. The only known species has a blackish mesonotum with the abdomen predominantly light brownish orange and the legs of the same color. The genus differs from *Aphoebantus* Loew principally in the short, bulblike third antennal segment which is very little longer than wide. The dense, silver pollen on the slightly inflated front of the male is rather similar to *Desmatoneura* Williston, but in *Eucessia* Coquillett the face is a little produced forward, slightly more so in the female, and it has a very weakly indicated furrow across the face beneath the antennae. Also from this genus it differs in the position of the anterior crossvein, which is at the middle of the discal cell in *Eucessia* Coquillett, and moreover the second vein originates obliquely before the end of the second basal cell as measured at the anterior angle. These little flies also have the scales on the mesonotum rather wide, dense, and conspicuous, especially laterally and again in patches along the sides of the tergites and in a dense, posterior border on the first antennal tergite.

This little genus is restricted to a single species hitherto known only from southern California. I have a single female of what appears to be a different species from Texas, taken 20 miles west of Toyah, August 4, 1954.

Length: 4.5 mm. to 6 mm.

Head, lateral aspect: The head is subglobular, the occiput relatively short above and below and the posterior eye margin has a deep but broad and rather shallow indentation a little below the middle of the head. There is only a minute anterior incision on the eye margin at this indentation. The pile of the occiput is dense, scaliform, and appressed and extends abundantly to the eye margin. The inner fringe of the occipital recess is short, fine, and sparse. The front is short, very slightly protuberant dorsally, densely covered with silvery white, pollenlike micropubescence in the male and also the female. Both sexes have an isolated, longitudinal band of stiff, bristly, oblique pile beginning near the posterior part of the front, well separated from the corresponding band on each side and also separated from the eye margin. The face is slightly projecting, but not all conspicuously. It is a little more projecting in the female from Texas, and it is fringed and bordered along the oragenal recess with coarse, pale, bristly pile. There is only the faintest indication of a transverse, circular furrow beneath the antennae.

The eye is quite large, occupying most of the head, the upper facets only a little enlarged. The proboscis is short, robust, the labellum comprising more than half the length of the proboscis with fleshy lobes at the apex that bear a few stout bristles ventrally. The palpus is long and slender and composed of 2 segments, the first segment curved and arched outward, the second segment shorter, also slender and only about one-third as long as the first segment. The antennae are small, they are attached a little above the middle of the head, the first segment, though quite short, is twice as long as the second segment. The third segment viewed laterally is wide and only a little longer than this basal, vertical width and the styliform extension is short and blunt. It bears a distinct microsegment with bristle at the tip. In the Texas female, viewed laterally, the third segment is equally narrowed above and below so that the whole segment is entirely conical. In both of these individuals the segment is more narrow when viewed from above and it is 2 to 3 times as long as wide from this view. The pile of the antennae is minute and scanty.

Head, anterior aspect: The head is circular, distinctly wider than the thorax, the eyes are separated in the male by an extent equal to the basal width of the third antennal segment viewed from above. In the Texas female it is rather more than twice as wide. The ocellar tubercle is somewhat swollen with vertical sides and the ocelli occupy an equilateral triangle, and there is a clump of stiff, reddish bristles in the middle of the tubercle, curved forward and not longer than the tubercle itself. The sides of the eye rise above the vertex. The antennae are separated by a distance equal to about half the width of the first segment. The antennae are only slightly divergent. The oragenal recess is comparatively small, elongate oval, rather shallow anteriorly, deeper below and almost equally wide at both ends. The genae are present but quite linear, and there is no tentorial fissure. Viewed from behind the occipital recess is deep but rather small. Viewed from above the occipital lobes are narrowly separated.

Thorax: The mesonotum is shining blackish, the scutellum likewise and without pollen, but in the Texas female the mesonotum has a thin, light pollen over the black part, the ground color changing to light brownish orange laterally as indeed over the entire remaining part of the female, except for the lower pleuron, which has some blackish brown diffusely present, and the scutellum of the female is almost entirely orange brown. The pile of the mesonotum, scutellum, and pleuron consists of matted, dense, broad, conspicuous scales from among which arise a very few, fine, scattered, curved, bristly hairs of no great length. However, the notopleuron has a tuft of 3 rather long, stout, red bristles and 4 or 5 other more slender ones. Postalar callosity with 2 or 3 long, red bristles, margin of scutellum with 3 or 4 pairs. The pollinose humerus in the male of the type-species bears scaliform pile and also stiff bristly pile. In the female scales are lacking on the humerus, except perhaps the posterior border but some bristly pile is present. Moreover, along the anterior margin of the pronotum of both sexes there is a collarlike fringe of pile. Disc of scutellum with scales and an odd patch on each side at the immediate base. Upper mesopleuron with appressed scales and rather dense, erect, bristly pile. Lower mesopleuron and upper sternopleuron with a patch of appressed scales. Pteropleuron, metanotum, metapleuron, and hypopleuron bare, pollinose only, except that there is a small tuft of erect, scalelike pile posterior to the spiracle. The squama is quite short on the thoracic part, larger on the alar portion, the short fringe scanty.

Legs: The legs are small but the femora distinctly a little thickened, especially the second and third pairs. They bear scattered, appressed scales. The anterior tibia has a few microsetae dorsally and posteriorly, which stand out from the rather dense cover of smaller, flat-appressed setae. Middle tibia with 5 or 6 short, oblique, pale, spinous bristles posteriorly and a few dorsally. On the hind tibia the spinous bristles are a little longer, pale in color, 5 or 6 in each row including those at the apex and there are 3 rows present. The pulvilli are vestigial, claws sharp and a little curved.

Wings: The wings are hyaline, glassy, there are 2 submarginal cells and all the posterior cells are open. The anterior crossvein lies at the middle of the discal cell, the second vein arises in a straight line with the radial sector, the third vein arises obliquely, both of these veins arise before the end of the second basal cell as measured at the anterior basal angle. The anterior branch of the third vein arises steeply but makes a rounded bend outward. The marginal cell is wide on the apical half and has a mere suggestion of a bend below, opposite the furcation of the third vein. Anal cell widely open, axillary lobe wide and alula wide and the ambient vein complete. Base of the costa with strong, sharp, scalelike spurs.

Abdomen: The abdomen is only a little longer than the thorax, cylindrical, sometimes slightly compressed laterally but in the Texas female rather flattened, and less narrowed apically. In the type-species there is a band of broad, flattened, appressed scales across the middle of each tergite but not extending quite to the lateral margin. In addition there are a few scattered, subappressed, stiff, pale, reddish hairs becoming more numerous on the last tergite. First tergite with a fringe of rather long, erect, coarse pile basally and a narrow band of appressed scales basally. The male terminalia are large and conspicuous, inverted and form a somewhat narrowed, conical, truncate protuberance. Female terminalia surrounded by a dense fringe of inwardly turned, bristly hairs.

Material available for study: The type-species male, and the female of a second species from Texas.

Immature stages: Unknown.

Ecology and behavior of adults: Taken by Timberlake on flowers of *Euphorbia albomarginata* at Riverside, California, in May (Melander, 1950).

Distribution: Nearctic: *Eucessia rubens* Coquillett, 1886.

Genus *Xeramoeba* Hesse

FIGURES 128, 655

Xeramoeba Hesse, Ann. South African Mus., part II, vol. 35, p. 356, 1956. Type of genus: *Xeramoeba apricaria* Hesse, 1956, by original designation.

Small flies of dark color and very thin, grayish pollen. The pile is short, scanty, largely appressed and such few bristles as are present are thin, short, and scanty. They are very much like *Aphoebantus* Loew in general appearance with subglobular head, wider than the mesonotum, and similarly the eyes are indented and bisected and the antennae are very similar; in both the face is rounded and retreating in lateral aspect. However, the oragenal cavity is much narrowed when viewed from below and the proboscis very short, stumplike, the labellum is large and muscoidlike. Like *Chionamoeba* Sack, the metanotum is quite short and there is no pile on metapleuron, metanotum, or hypopleuron and in *Xeramoeba* Hesse, none on the pteropleuron. In both the plumula is present but there is no strigula. Unlike *Chionamoeba* Sack the front above antenna is not swollen. As in many of the Lomatiinae the second vein arises abruptly not too far distant from the anterior crossvein. Bowden (1964) believes that these two genera belong in the Anthracini; while granting that to some extent they may be annectant, they may just as well be convergent and it seems to be better to create for them a separate tribe within the Lomatiinae, the more so, since there are nine other genera within the Lomatiinae with a second vein which shows abrupt origin. Bowden (1964) takes note of the microspicules upon the front tibiae. While they certainly serve to differentiate this genus from *Prothaplocnemis* Bezzi, inasmuch as most bombylids showing a varying development of spines upon the legs, this appears to be one of the most uncertain characters upon which to base phyletic relationships.

Hesse (1956) compares these flies to *Villa* Lioy or *Thyridanthrax* Osten Sacken as far as superficial aspect is concerned. He notes that *Xeramoeba*, in comparison with *Chionamoeba* Sack, is less elongate, the pile and bristly hair slightly longer, those on front and face denser, longer (and stiffer), the eyes of the male more widely separated. I can see little difference in the upper occipital lobes of the two genera. Hesse finds the basal wing comb better developed, the costal cell shorter, the tibial spicules better developed and conspicuous than in *Chionamoeba* Sack; he finds the hypopygium differs from that genus and also from *Anthrax* Scopoli. He finds that the tubelike aedeagal process suggests *Chionamoeba* and the downwardly directed apical hook of the aedeagus resembles that present in some *Anthrax* species.

These small flies are known only from South Africa.

Length: 4.3 to 7.5 mm.; wing 4.5 to 8.5 mm.

Head, lateral aspect: The head is large and nearly globular; the occiput is comparatively long dorsally and considerably reduced in length below. The surface is thinly pollinose, the pile is short and rather scanty, except on the rather isolated, central fringe where the pile is dense. The indentation of the eye is deep, without a bare triangle but with the lines separating upper and lower divisions of the eye quite long and slender. Eye greatly extended forward so that the whole head is very broadly rounded in front both above and below the antennae. Both front and face extend only a very short distance outward; the cheeks are not visible below the eye margin, and the middle base of the head behind the proboscis barely shows. The front is rounded and bears short, oblique, stiff, rather abundant pile; that of the face similar in length and character. The antennae are attached at the middle of the head; they are quite small, the first and second segments equal in length, the first segment only slightly wider, and both segments more than twice as wide as long. Third segment strongly bulblike at the base, abruptly swollen below to a short, thick style, which bears a very short microsegment. The proboscis is extremely short and exceptionally stout and robust. The large, flat lobes of the labellum are muscoid in character and flared outward. Palpus small and minute.

Head, anterior aspect: The head is slightly wider than the thorax. Eyes in the female widely separated by 3 times the width of the knoblike, prominent, ocellar tubercle the sides of which are vertical, and bear 2 or 3 short hairs directed forward. Eyes in male also widely separated but less than in females. The antennae are separated by a little more than half the width of the first segment. The face is extensive and slightly retreating below, nearly vertical above, the antennal sclerite distinct. There is only a faint diagonal or crescentic line extending from the antenna obliquely down to the eye margin. The eyes are green in reflection, the oragenal recess is small and strongly oblique, narrow, deep in front and behind with an odd, deeper, medial fossae. The oragenal recess is so narrow that posteriorly it touches the sides of the proboscis, hence the face and recess are strongly convergent. The tentorial fissure is linear and slitlike, the space between it and the eye quite linear and this bare, linear stripe continued on upward to opposite the antennae. There is a rather strong overgrowth of the eye ventrally along the margin of the oragenal recess and again on the upper sides of front and vertex and upper occiput. Viewed from above the lobes of the occiput are narrowly separated; the fissure is deep and rounded on each side with a deep, triangular pocket behind the vertex.

Thorax: The thorax is black and compact and triangular below the mesonotum. The mesonotum is low, a little more strongly convex anteriorly. It is faintly pollinose, the pile consisting of scattered, curled, appressed, rather short and slightly flattened hairs including all of the anterior and lateral pile, except the anterior collar and except the humeral pile. Notopleuron with 1 or 2 fine, long bristles. Postalar callosity with 2 or 3, and the scutellar margin with about 4 pairs of more slender, long, bristly hairs and others in front of the scutellum. Pile and pollen of the scutellum similar to that of the mesonotum. Plumula present, the tuft of hairs moderately long but not numerous. Pleuron pollinose, upper mesopleuron with a conspicuous tuft of scattered, sparse, upturned, slender, bristly hairs. Posterior half of mesopleuron with some flat-appressed, scaliform pile below and a loose clump of bristly hairs curled backward above. There is a large, dorsal triangle on the sternopleuron densely covered with flat-appressed scales. Metanotum, pteropleuron, hypopleuron, and the anterior metapleuron pollinose only. There is a small tuft of scales behind the spiracle. Halteres small and short.

Legs: The legs are slender and rather short. Anterior femur are at most very slightly thickened toward the base; all the femora have flat-appressed, pale scales. The anterior femur has 2 or 3 small, fine setae anteriorly; the middle femur likewise. The posterior fringes of pile are lacking on the first four femora. The hind femur bears 1 or 2 comparatively strong, attenuate, sharp bristles ventrolaterally near the apex, 3 smaller ones dorsally above. All the tibiae are slender; anterior tibia has 7 tiny, fine setae dorsally, 2 ventrally and 4 or 5 posteriorly. The middle tibia bears scarcely stronger setae; there are 4 anteroventrally, 6 anterodorsally, 6 posterodorsally and 3 or 4 posteroventrally. On the hind tibia the setae are a little longer, a little stouter but still slender; there are 5 ventrolateral, 6 or 7 dorsolateral, and 4 very minute, slender ones ventromedially. Claws quite slender, a little curved at the apex, the pulvilli wide and rounded at the apex and only about half as long as the claws.

Wings: The wings are hyaline, the surface wrinkled and iridescent. There are only 2 submarginal cells. The second vein shows only a very slight, scarcely noticeable kink or bend opposite the base of the second submarginal cell. The anterior branch of the third vein arises plane but not quite at a right angle; generally without a stump or trace of stump. The second vein likewise arises plane, not at a right angle, its origin is near the anterior crossvein but basal to it and separated from it by the length of the anterior or less. The anterior crossvein enters the discal near the basal fourth, the first posterior cell is quite widely open, almost maximally; the anal cell is open; the ambient vein is complete, and the alula is well developed. The costa is expanded at the base but bears only weak setae. The wing is large and long, at rest extending far beyond the apex of the abdomen. Costal hook large and scalelike. Prehumeral part of costa as long as the last section of the third vein.

Abdomen: the abdomen is black, as long as the thorax, slightly wider than the mesonotum to the end of the second tergite. Second tergite more than twice as long as the third; 7 tergites are visible dorsally. The abdomen as a whole is quite broad and broadly oval apically with the apex of the abdomen rather triangular; the last tergite in the female conical. The whole abdomen is gently convex, thinly pollinose; the pile consists of flat-appressed, rather loose and scattered, slender, slightly flattened hairs. Sides of the tergites with some longer, erect, bristly hairs. Sides of the first tergite with a dense brush or band of dense, erect, coarse pile becoming much shorter medially.

Since I have not been able to see males of this genus I quote the remarks of Dr. A. J. Hesse with respect to genitalia:

> Hypopygium of male with a rather conspicuous basal process to the shell-like basal parts which also have some hairs dorsoapically; beaked apical joints compressed and twisted; aedeagal apparatus in form of a tube-like or funnel-like process lodging the aedeagus proper and ending apically below in a downwardly directed recurved hook or process. Hypopygium of this genus is different from that of the preceding genus *Chionamoeba* and the following genus *Anthrax*. The tube-like aedeagal process, however, is reminiscent of that of the former and the downwardly directed apical hook of this structure also resembles that present in some species of Anthrax.

Material available for study: Representatives of the genus presented to me through the courtesy of John Bowden.

Immature stages: Unknown.

Ecology and behavior of adults: Bowden (1964) notes that *Xeramoeba gracilis* Bowden was taken settling upon bare ground in company with species of *Petrorossia* Bezzi.

Distribution: Ethiopian: *Xeramoeba apricaria* Hesse, 1956; *gracilis* Bowden, 1964.

Genus *Chionamoeba* Sack

FIGURES 228, 387, 645, 651

Chionamoeba Sack, Abhandl. Senckenberg. Naturf. Ges., vol. 30, pt. 4, p. 543, 1909. Type of genus: *Bibio nivea* Rossi, 1790, by original designation.

Engel (1934), p. 415, key to 5 Palaearctic species.

Small flies with broad and only slightly convex abdomen; in South African species, however, the abdomen is more elongate, but on the basal half fully as wide as the mesonotum and distinctly and gently tapered. The face is very much rounded throughout and retreating below, scarcely protruded beyond the eye margin. The second vein arises either at the anterior crossvein, or a short distance before. Pile everywhere dense, fine, silky, matted, and appressed, except that on the mesopleuron there are 2 large bands or clumps of coarse, long, bristly, pale hairs, one directed upward, the other backward, and both subappressed; also the postmargins of the abdominal tergites have a row of short slender bristles and there are 3 more distinct bristles on the notopleuron with others on postalar callosity and on the margin of the scutellum.

These flies are found in southern Europe and 2 species are described from South Africa which are somewhat different.

Bezzi (1924) and Hesse (1956) note that in certain respects this genus appears to be more or less annectant with the Anthracini; this is perhaps more nearly true of the present genus and of *Xeramoeba* Hesse, both of which have a more or less muscoidlike labellum; Bezzi considered *Chiasmella* Bezzi to show such a relationship. Bowden (1964) notes that the aedeagus of his South African species *Chionamoeba choreutes* Bowden has the aedeagal sheath more extensive, in this respect being more like Anthracini. The front above the antenna is bossed much as in *Desmatoneura* Williston.

However, my studies lead me to believe that in spite of several anthracine characteristics the greater relationship is with the Lomatiinae and for them I erect a separate tribe, the Xeramoebini, separated from other groups by the presence of the plumula and the reduced size of the hypopygium; see further discussion under the subfamily Lomatiinae.

Hesse (1956) notes that *Chionamoeba* Sack differs from *Anthrax* Scopoli in the following ways: it has a more elongate abdomen, either no bristles, or fewer, on the body, less dense, erect pile, a not concavely, saucer-shaped second antennal segment, a terminal microsegment attached to segment three which bears no apical circlet of hair (such a circlet may indeed be a parallel development), a second vein not originating at right angles nearer opposite the anterior crossvein, and the base without an appendix, fewer spines on the femora, shorter and less conspicuous spicules on the tibia. From *Petrorossia* Becker it differs in the broader wings, longer costal cell, presence of a plumula, etc.

Length: 4.5 to 9.5 mm.

Head, lateral aspect: The head is nearly globular with the eyes considerably lengthened, quite circular in front, the entire front, especially the lower part, strongly produced forward and swollen conspicuously, gradually and regularly across from each eye margin. The face is somewhat less produced, but retreats slightly immediately below the antennae, then it is slightly raised or produced slightly above the oragenal margin. Pile of front and vertex abundant, but not dense and of no great length. On the front the pile is shorter and curled downward and is subappressed but erect near the vertex. The occiput is only moderately extensive dorsally. The indentation of the posterior eye margin near the middle is deep but much less so than in *Petrorossia* Bezzi. There is a nonpollinose triangle near the bottom of the indentation and the line of separation of the eye facets quite short. Upper eye facets moderately enlarged. Green reflections from the eye are only faint.

The proboscis is very short, not extending to the apex of the oragenal recess, the labellum large, somewhat flattened, nearly as long as the basal part of the proboscis. Palpus very small, slender, with a few long, fine hairs. Antennae attached a little below the middle of the head in profile, quite small, the first segment very short, especially dorsally, scarcely visible and apparently overlapped by the extra development of the lower front. It is somewhat longer below since the swollen front gives way between the antennae. The first segment bears some long, numerous, bristly hairs ventrally and laterally. The second segment is rather large, apically widened and rounded, short pyriform in shape with the small end embedded in the first segment. The whole segment is as long as the conical, bulblike part of the third segment. The second segment bears conspicuous, long, scattered, bristly hairs dorsolaterally, and ventrally and laterally, none medially. The third segment is bulblike at the base with thick, rather short, micropubescent style.

Head, anterior aspect: The head is wider than the thorax, nearly circular with all 4 sides slightly flattened in their middles. The front is very extensive; in the male the eyes are separated by about twice the width of the high, knoblike, ocellar tubercle, which has vertical sides, and forms an equilateral triangle and has a few fine, erect hairs. The front is very strongly widened to the middle of the head and the face and oragenal recess almost equally strong and narrowed below and at the bottom of the head scarcely wider than the vertex. There is a faint, impressed, fossalike division extending circularly outward on each side below the antennae. Antennal sclerites prominent and differently colored. The oragenal recess is narrow, quite deep posteriorly, rather deep anteriorly but a little widened and a fringe of fine, facial hairs lies around its edge. The tentorial fissure is linear leaving only a narrow stripe next to the eyes. Gena very much reduced, almost absent, except for the flattened, triangular base on each side behind the oral recess. Antennae quite widely separated. Viewed from above the occipital lobes almost touch. The triangular pocket behind the vertex is deep.

Thorax: The thorax is black and compact, the lower part triangular, the mesonotum low, and flattened posteriorly. The extreme lateral margin, humerus, and postalar callosity are reddish brown, the whole surface of these parts pollinose. The pile is rather dense, fine, long, and erect over the whole surface, but it is coarser and more bristly on the anterior collar and perhaps even longer. There appears to be a deep fossalike recess medial to the thin, high, vertically ridgelike postalar callosity. Opposite it the vertically raised mesonotum bears shaggy tufts of pile. The notopleuron has 1 or 2 bristles, the postalar ridge has a tuft of pile and 1 or 2 bristles. The alary squamae are quite short, the fringe long but consisting of only 1 or 2 ranks of hairs. Thoracic squama very conspicuous with no hairs. Plumula with only a few hairs. Scutellum black and covered like the mesonotum. Pleuron pollinose, the upper mesopleuron with dense, long, bristly, upturned pile on the upper half and with more scattered, long, slender, bristly hairs below and posteriorly. Anterior pteropleuron with 2 tufts of hairs. Sternopleuron on the upper half with scattered, stiff, subappressed hairs. Metanotum and the whole of the metapleuron and hypopleuron pollinose only.

Legs: The legs are slender including the femora. All the femora have rather numerous, flat appressed, conspicuous scales. Middle femur with conspicuous fringe of rather long hairs posteroventrally. None of the femora bear bristles. The anterior tibia bears minute, fine, suberect, slender spicules; there are 3 or 4 dorsally, 4 or 5 posteroventrally. On the middle tibia the spicules are slightly stouter and longer; there are 5 anteroventral, 6 or 7 posterodorsal, and 6 to 8 equally developed posteroventral ones. Bristly spicules of hind tibia equally slender, scarcely longer; there are 2 anteroventral, 6 to 8 anterodorsal, and 4 or 5 ventromedial ones. Front tarsi quite short, hind tarsi quite long. Claws slender, curved a little at the apex, the pulvilli long and spatulate.

Wings: The wings are quite large, long, and rather broad basally and at rest extending far beyond the abdomen. The surface of the wing is wrinkled and iridescent. There are 2 submarginal cells. The marginal cell is scarcely widened apically and there is no kink or bend in the second vein. The anterior branch of the third vein arises rather gently and is rounded at the first bend. The second vein arises plane, abruptly but not rectangularly, near and basal to the anterior crossvein, removed from it by a distance nearly equal or equal to the length of the anterior crossvein, the first bend of this vein is slightly rounded. The anterior crossvein enters the discal cell near the basal fourth; the first posterior cell is widely open, the anal cell likewise. The ambient vein complete, the alula moderately well developed. Costa expanded at the base but with only a few minute setae. Costal hook scalelike.

Abdomen: The abdomen is elongate, a little longer than the thorax with the scutellum included, at the base not wider than the thorax and distinctly tapered or narrowed posteriorly, rather strongly convex dorsally, more flattened below. The abdomen is black, shining, with a very faint pollen and with dense pile of two types. Tergites beyond the middle of the second densely covered with matted, flat-appressed scales, with a conspicuous, lateral fringe of long, erect, pale, bristly pile on the sides of the abdomen, becoming even longer and denser, and coarser on the sides of the first tergite. Lateral pile of first tergite not extending in the middle of this segment. Apex of last tergite with a fringe of long hairs. Second tergite not twice as long as the third. Third and fourth tergites equal in length. Seven tergites and a very short eighth are visible in the male. The male terminalia often twisted to the left side, the basimeres broad basally, attenuate apically, but lying adjacent so that each is somewhat triangular.

Since I have not been able to see males of this genus I quote the remarks of Dr. A. J. Hesse with respect to genitalia:

> Hypopygium of male of South African species without any process to basal parts on outer side, but with the inner apical part produced; beaked apical joints elongate, cylindrical and curved; aedeagal structure in form of a funnel lodging the aedeagus; lateral struts small. Terminal lamellae at posterior end in last sternite with their dorso-apical angles not produced.

Material available for study: The representatives of this genus in the collections of the British Museum (Natural History). Also, material sent through the courtesy of John Bowden in exchange.

Immature stages: Unknown.

Ecology and behavior of adults: Bowden (1964) notes that while males were seen hovering over a sand heap, the females were found investigating small holes in the timber of buildings.

Distribution: Palaearctic: *Chionamoeba erythrostoma* (Rondani), 1873 (as *Anthrax*); *frontalis* (Wiedemann), 1828 (as *Mulio*) [=*Cytherea frontalis* Wiedemann, 1828, in Engel]; *lepida* (Hermann), 1907 (as *Argyromoeba*); *nivea* (Rossi), 1790 (as *Bibio*), *nivea lioyi* Griffin, 1896; *sabulonis* (Becker), 1906 (as *Argyromoeba*), *sabulonis rubicunda* Bezzi, 1925; *semirufa* Sack, 1909.

Ethiopian: *Chionamoeba choreutes* Bowden, 1964; *meridionalis* Hesse, 1956.

Genus *Petrorossia* Bezzi

FIGURES 134, 234, 383, 634, 652, 918, 919, 920

Petrorossia Bezzi, Zeitschr. syst. Hymen. Dipt. Teschendorf, vol. 8, p. 35, 1908c. Type of genus: *Bibio hesperus* Rossi, 1790, by original designation.

Pipunculopsis Bezzi, Bull. Soc. Roy. Ent. d'Egypte, vol. 8, p. 211, 1925c. Type of subgenus: *Pipunculopsis bivittata* Bezzi, 1925, by monotypy.

Bezzi (1924), pp. 151-152, key to 6 Ethiopian species.
Bezzi (1925), pp. 209-211, key to 8 species.
Engel (1934), p. 409, key to 5 Palaearctic species.
Hesse (1956), pp. 312-315, key to 9 Ethiopian species.
Bowden (1964), pp. 60-61, key to 9 Ethiopian species.

Small, dark colored flies with faint grayish-white pollen; the abdomen in some species may be reddish laterally, or terminally. In one species the first two antennal segments are white, the third black. The pile is scanty, rather short, flat-appressed, more erect laterally and as in *Chionamoeba* Sack it is apt to be slenderly scaliform on the abdomen. The face is rounded and retreating, but varies a little in prominence; the proboscis and labellum, short and contained within the oral recess. The wings are hyaline or tinged with brown anterobasally. The second vein arises abruptly between the base of the discal cell and the anterior crossvein. Anal cell widely open. End of marginal cell with thumblike enlargement, but this may be wanting in one or two species. The legs are pale; the black front tibial spinules present but minute. The plumula is absent. The occipital lobes are in contact but the crease is deep, the sides rounded. In general appearance these flies are apt to suggest a small species of *Villa* Lioy. There is more than a passing resemblance among not only *Petrorossia* Bezzi, but its relatives *Xeramoeba* Hesse and *Chionamoeba* Sack to a few uncommon species of *Anthrax* Scopoli, which are more elongate and slender than usual. While some species of *Petrorossia* Bezzi have an apical microsegment attached to the end of the third antennal segment, it is surmounted by a thick spine and there may be a circlet of a few fine hairs, as in *fulvipes* Loew. It is related to *Prothaplocnemis* Bezzi, which also has a terminal tuft of hairs on the antenna. Engel makes *Pipunculopsis* Bezzi a subgenus.

This genus is represented by several species in the South Palaearctic region and by more numerous species in South Africa. It also occurs in Australia.

Length: 4.8 to 10.5 mm., of wing 4.6 to 12 mm.

Head, lateral aspect: The head is nearly globular, the occiput quite short immediately behind the vertex but lengthening almost immediately and quite long, and deep at the midpoint of the eye laterally. The incision of the eye at this point is extremely broad and deep; both dorsally and behind the vertex the occiput slopes steeply inward toward the comparatively small, cuplike recess. The whole surface of the occiput is covered with dense, pollenlike, micropubescence, and dense, erect pile becoming longer at the central internal fringe. Pile and pollen reach to the eye margin, except at the bottom of the indentation there is a bare triangle and the horizontal line separating upper and lower facets at this point is unusually long. The eye is reniform, the upper facets only a little enlarged; the entire eye has a greenish reflection. The front at the point of attachment of the antennae is very short, nevertheless, because the antennae are attached at or slightly below the middle of the head, its length from antennae to vertex is great, and since the length of the head is considerable, the front is strongly rounded in profile.

The face extends only a short distance beyond the eye margin; it is rounded and strongly recessive, the oragenal cup reaching only two-thirds of the distance from its posterior margin to the base of the antennae. Proboscis is very short, not extending beyond the oragenal recess. The entire proboscis, including the labellum, is quite stout and robust; the latter is somewhat flattened and leaflike and nearly or quite as long as the basal part of the proboscis. Palpus short, robust, almost barrel-shaped with some fine hairs at the apex and laterally. Antennae placed at or just below the middle of the head, small, the first segment only a little larger than the second and slightly longer. The first segment bears short, stiff hairs medially, a conspicuous tuft of longer hairs dorsally and ventrally, which are confined to the middle of the segment. Second segment a little more than half as long as high but longer in the middle on the medial aspect than either above or below; it bears a few long setate hairs dorsally and ventrally and laterally but none medially. Third segment with a strong, bulblike base nearly as wide as the second segment and abruptly narrowed into a thick, stout style about twice as long as the swollen basal part. The style bears a short, stout microsegment with a thick, apical spine and 2 or 3 short apical hairs, which are minute and not all conspicuous.

Head, anterior aspect: Head a little wider than the thorax, nearly circular but slightly flattened on all 4 sides. There are no grooves or fossae separating the face or curving from side to side beneath the antennae, although the antennal sclerite is distinct. The eyes in the male are separated by a distance fully equal to the high, knoblike, ocellar tubercle, the sides of which even curve under a little. The posterior eye corners are quite broadly rounded and curve outward and away from the tubercle, which is short isosceles; the ocelli lie on the edge of the tubercle and bear a few moderately long, stiff hairs centrally. The eye margin is raised above the level of the upper front and vertex, the whole surface finely pollinose and the lower two-thirds of the front bears dense, slightly flattened, curved, subappressed, shining hairs over the whole surface, except for a linear, medial, nonpilose stripe and a linear, nonpilose, marginal stripe continued on down to the gena below. The antennae are separated by nearly the basal width of one segment and the antennae are strongly divergent. Face extensive dorsoventrally, pollinose and uniformly covered with similar pile of the same color, length and appression as the pile of the front. The oragenal recess is comparatively small and narrow, slightly pinched in the middle, deep anteriorly and posteriorly, its total width a little less than one-fifth the total head width. The tentorial fissure is linear and deep and begins at the point where the oragenal recess narrows anteriorly. Walls of oragenal cup quite thin and vertical. Genae decidedly linear, the eyes extended below the genae in profile. Viewed from above the occipital lobes are distinctly separated with a narrow fissure between and with a deep, triangular pocket behind the posterior vertical wall of the ocellar tubercle.

Thorax: The thorax is compact, almost triangular below the mesonotum. The mesonotum is low, very little convex, widened only behind the wings where the postalar callosity is rather prominent and is raised with a triangular troughlike depression dorsally. The thorax is black, thinly pollinose, the pile of the mesonotum consists of glittering, flat-appressed, narrow, scaliform pile with rather dense, fine, long, erect pile, longer on the anterior third and merging directly into the anterior collar. This pile is equally abundant on the swollen humerus, the posterior corner of which is vertical; equally abundant over the whole notopleuron and the sides of the mesonotum, but, however, shorter in the middle of the mesonotum between the wings. There is one long bristle on the notopleuron, usually 2 or 3 on the postalar callosity, sometimes a row of fine hairs in front of the scutellum and the flattened, subtriangular scutellum covered with the same pollen and pile as the mesonotum and its margin at most with a few long, slender, bristly hairs in addition to appressed pile. Squama short, plumula absent. Pleuron pollinose, the mesopleuron is almost wholly covered by dense, quite long, bristly pile, these elements on the upper border are stouter, all of those which lie on the upper half are curled upward and the remainder are strongly curled backward and downward. Upper half of sternopleuron and sometimes the whole of it covered with appressed, shorter pile directed backward. Pteropleuron, metanotum, anterior metapleuron, and the hypopleuron pollinose only. There is a tuft of pile behind the spiracle.

Legs: The legs are slender, the anterior femur, however, distinctly a little widened basally beginning at the outer third. All of the femora and the hind tibia are covered with flat appressed, narrow scales. These scales are quite dense in some species, in others sparse. The anterior and middle femora have a row of short hairs posteroventrally. Hind femur with 1 or 2 fine, weak bristles ventrolaterally near the apex and sometimes 2 or 3 shorter ones dorsolaterally on the outer half. All of the tibiae quite slender; anterior tibia with 1 or 2 minute, fine setae posterodorsally and 4 or 5 others posteroventrally. Middle tibia also with a few slender, short, oblique bristles. Hind tibia with 8 fine, oblique, slender, black bristles dorsolaterally and 7 ventrolaterally, besides 3 or 4 ventromedial bristles. Claws quite slender, sharp, strongly curved near the apex, the pulvilli long and spatulate.

Wings: The wings are hyaline, large, and at rest extend far beyond the abdomen. The wing surface is wrinkled and iridescent. The second vein arises between the anterior crossvein and the base of the discal cell and almost halfway. It arises abruptly but not rectangularly. There are only 2 submarginal cells, the second vein with a strong, preapical bend so that the marginal cell is bent backward and is thumblike in outline. The anterior branch of the third vein arises obliquely, but not quite rectangularly, and there is a strong backward bend in this vein corresponding to the bend of the second vein. The anterior crossvein enters the discal cell at or before the basal third. The first posterior cell is a little narrowed apically and the anal vein is open. Ambient vein complete, the alula short but distinct. Costal comb weak, but costa strongly expanded. The costal hook or scale is well developed.

Abdomen: The abdomen is a little wider than the thorax at the base, obtuse apically, often rather high and thick or subcylindrical, the sides tapered or narrowed, the whole abdomen a little longer than the thorax with scutellum included. Tergites covered with dense, flat-appressed, slightly flattened pile and sometimes with additional sparse, very fine, bristly pile, likewise more or less appressed. Sides of the tergites sometimes with longer, pale bristly hairs which are more or less erect. Ground color often reddish or orange brown laterally, the tergites widely blackish in the middle, the apical tergites may be wholly red. Hypopygium large with the basimeres triangular and turned to one side or dorsal in position.

Since Dr. A. J. Hesse was able to base his conclusions upon a greater number of species, I quote his remarks upon the male hypopygium:

Hypopygium of males with the basal parts convex, shell-like, covered dorsally and along outer apical margin with fairly stoutish, conspicuous and longish, bristly hairs, the base usually drawn out into a sort of scoop-like process; beaked apical joints either more or less twisted, curved or scroll-like, the dorsal or outer apical part excavated or hollowed, produced apically into a spine-like process or into upper and lower processes or even three processes, the upper one usually directed outwards over lower one, the dorsum of these joints usually with hairs or a tuft; aedeagus with a ventral process below, formed by the union of a ramus on each side from sides of basal parts, the apical part of this process either pick-like, blade-like, or ending in a bidentate or bified process which is directed downwards; lateral struts shoe-horn-shaped, sometimes broadish and long; basal strut ham-shaped, sub-racket-shaped, or chopper-shaped, its apical margin sometimes with a ledge-like extension.

The male terminalia from dorsal aspect, epandrium removed, show a short, broad, robust figure, narrowed apically. The dististyli are blunt or indented apically. In the lateral view there are curious, basally expanded structures arising at or near the ramus or the upper wall of the basistyle, which curve forward as strongly convergent, lobelike or clasperlike structures, overlying the apex of the epiphallus. They appear to have no relationship to the epiphallus or ejaculatory process.

Material available for study: Material in the collections of the British Museum (Natural History), and other examples sent through the kindness of John Bowden.

Immature stages: Unknown, although Bowden (1964) suggests that the probable hosts are small halictids, nomadids, or even megachilids, or crabronids.

Ecology and behavior of adults: Bowden (1964) comments on the habits of certain species. Females of *Petrorossia letho* Wiedemann are taken hovering over hot, bare soil or settling on it near the nest holes of small halictic and nomadid bees. *Petrorossia consobrina* Bowden was found hovering around the entrance holes of bee nests presumably megachilids and crabronids in the timbers of buildings. *Petrorossia phthinoxantha* Bowden was taken hovering at cracks and crevices in a bank frequented by various species of small bees; one female was taken at the flowers of *Tagetes* sp.; another species was found hovering over a path in rather dense forest; another about the bank of small stream; still others about the base of an uprooted tree.

Distribution: Palaearctic: *Petrorossia albifacies* (Macquart), 1840 (as *Bombylius*) [=*rufipes* Macquart, 1840]; *appendiculata* (Macquart), 1849 (as *Lomatia*); *bivittata* Bezzi, 1925 (subg. *Pipunculopsis*); *hesperus* (Rossi), 1790 (as *Bibio*), *hesperus tropicalis* Bezzi, 1921 [=*albipectus* (Walker), 1849 (as *Anthrax*)]; *latifrons* Bezzi, 1924; *letho* (Wiedemann), 1928 (as *Anthrax*), *letho liliputiana* Bezzi, 1924; [=*longitarsis* (Becker), 1902 (as *Anthrax*)]; *sceliphronina* Séguy, 1935; *stenogaster* (A. Costa), 1884 (as *Anthrax*).

Ethiopian: *Petrorossia angustibasalis* Hesse, 1956, *angustibasalis buziana* Hesse, 1956; *chapini* Curran, 1927; *consobrina* Bowden, 1964; *deducta* Bowden, 1964; *flavicans* Bowden, 1964; *fulvipes* (Loew), 1860 (as *Anthrax*); *fumipennis* Hesse, 1956; *fuscicosta* Bezzi, 1924; *gratiosa* Bezzi, 1912; *imbutata* Hesse, 1956; *karooana* Hesse, 1956; *masieneensis* Hesse, 1956; *media* Séguy, 1931; *obscurior* Bowden, 1964; *phthinoxantha* Bowden, 1964; *plerophaia* Hesse, 1956; *vinula* Bezzi, 1921.

Genus *Desmatoneura* Williston

FIGURES 133, 379, 636, 649, 908, 909, 910

Desmatoneura Williston, Kansas Univ. Quart., vol. 3, p. 267, 1895. Type of genus: *Desmatoneura argentifrons* Williston, 1895, by monotypy.

Small, rather slender flies with the abdomen tapered and longer than the thorax, tending to be a little compressed laterally and both thorax and abdomen covered with scattered, appressed, scaliform pile and longer, more erect pile along the sides of the thorax, pleuron, and sides of the abdomen. Closely related to *Aphoe-*

bantus Loew and *Euccesia* Coquillett, the venation is distinctive inasmuch as the second vein arises abruptly but not rectangularly at a distance in front of the anterior crossvein 1½ to 2 times the length of that crossvein. Apically the marginal cell is swollen as rather more commonly seen in the South African genera of the Aphoebantini, but also present in some American *Aphoebantus* Loew. This apical swelling is due to an apical bend in the second vein. Another quite distinctive characteristic of *Desmatoneura* Williston is the striking, swollen and inflated front. The antennae are set at or slightly below the middle of the head in profile, hence the front is quite extended. The face is vertical, and there is a distinct transverse or curved fascia running midway between the antennae and the anterior edge of the oragenal recess. Finally, it may be added that the third antennal segment has the swollen basal part distinct from the short, apical styliform part and the eyes are dichoptic in the male. Females always 2 or 3 times as large as the males.

Desmatoneura Williston as presently known is restricted to the arid region of the southwestern United States.

Length: 4 to 9 mm.

Head, lateral aspect: The head is subglobular, the occiput is short, and in the male of about equal length above and below, although at the middle of the posterior margin of the eye the occiput is a little more extensive because of the prominent, rectangular indentation of the eye margin accompanied by a short, horizontal, creaselike excision of the margin. In the female the upper occiput is a little longer than in the male, the lower occiput distinctly much shorter than in the male and the middle indentation along the eye margin deep, broad and angular but not rectangular. The pile of the occiput is quite short and minute and appressed near the eye margin but changes to a rather dense, moderately long fringe along the inner margin of the occipital recess. The front is distinctly swollen and inflated in both sexes in a characteristic fashion. And in the males it is completely covered with dense, silvery white, flat-appressed micropubescence directed forward and downward, extending from the ocelli to the antennae and out to the eye margin, among the micropubescence are a few, fine, scattered, erect hairs. In the females, however, the front is largely bare down the middle, shining, blackish in the type-species with some coppery reflection which appears to proceed from minute, scalelike elements and beginning in the middle of the front there are loose, appressed, scattered, scalelike hairs. In the females above the antennae and along the eye margins the ground color becomes yellowish and on the eye margins at least there is similar, minute, silvery micropubescence similar to that of the male. Also, the erect, scattered hairs are somewhat more numerous and extensive over the whole surface of the front, except linearly upon the eye margins. The face and front extend anteriorly forward from the eye a distance nearly the length of the third antennal segment.

The circular, transverse fascia beneath the antennae is characteristic of this genus. The pile of the face consists principally of dense, fine, micropubescence with some additional, short, scattered, erect pile, more abundant in the male, and more abundant diagonally below each antennae. The fine fringe of facial pile extends as a fringe below the edge of the oral margin. The eye is large and extensive, quite hemicircular on the anterior half, the upper facets distinctly enlarged in the male. The proboscis is short, not longer than the oragenal recess, and not reaching to its apex either. It is rather thick and stout with a short, fleshy, muscoidlike development which bears a number of rather stiff hairs. The palpus is composed of one short, slightly swollen segment and a second, much clavate and apically swollen and rounded second segment, about 3 or 4 times as long as the first segment, and very narrowly attenuate and attached at its base. The first segment has minute micropubescence, the second bears similar micropubescence and laterally a few, short, stiff, bristly hairs. The antennae are attached at just below the middle of the head and are thrust strongly out laterally and are therefore much divergent apically. The first segment viewed from below is much longer than the second segment, viewed from above they are of about equal length and they are both short. Second segment with 3 or 4 long, stiff hairs dorsally. The third segment is bulbous at the base, rather abruptly narrowed on the lower half, less conspicuously on the dorsal part; the whole bulbar base not longer than the second segment of the antennae. The styliform part of the third segment is thick and of about equal width and at the apex bears a short, blunt, conical microsegment. The first segment has a fringe of short pile laterally and ventrally.

Head, anterior aspect: The head is slightly wider than the thorax, almost circular from the front. In the male the eyes are separated by a distance equal the width between the antennae and in the female by twice this much distance. The antennae are separated widely by a distance rather greater than the length of the first two antennal segments. The ocellar tubercle is prominent, high, with vertical sides, small, the ocelli in an equilateral triangle. It is separated from the eye margin in the male and in the female separated on each side by almost the distance between the posterior ocelli. Anterior ocellus twice as large as the front ocelli. The oragenal recess is rather short, oval, comparatively shallow with sloping sides. From the front it is obliquely sloping and retreating backward. The genae are reduced to a mere line on the oragenal recess. The head viewed from the posterior aspect shows a very large, deep, occipital recess. Viewed from above the occipital lobes are separated by a distinct fissure behind the ocellar tubercle.

Thorax: The mesonotum is black, feebly shining, loosely covered with appressed, short, flattened scales or hairs. There is a tuft of long, stiff, bristly hairs on the notopleuron and the scutellum has 2 pairs of long, slender, convergent bristles in the male, and in the female more numerous bristles. Disc of scutellum with similar scales to that of the mesonotum. The pleuron more or less yellowish or brownish, especially below; the mesopleuron has a conspicuous clump or band of stiff, pale bristles directed upward. Propleuron with a vertical fan of long, slender bristles directed outward. Whole surface of pleuron pollinose. The middle and lower mesopleuron and upper sternopleuron with appressed pile. Metanotum, metapleuron, and hypopleuron bare. There is, however, a tuft of flattened hairs immediately behind the spiracle and below the haltere.

Legs: The legs are slender, the middle and especially the hind pair much longer than the anterior legs. All of the femora plastered with closely appressed, small scales, rather more dense below. Anterior femur with a row of microsetae anteriorly, middle femur with 1 or 2 bristles anteriorly, hind femur with 4 or 5 sharp, stout, short, spinous bristles laterally on the posterior half and 2 or 3 others ventrolaterally and distally. Anterior tibia with 2 or 3 rows of minute, short setae; middle and hind tibia each with 3 rows of sharp, oblique, spinous bristles of no great length, and hind tibia with 4 rows of similar bristles. On the hind tibia these rows contain 8 to 12 bristles. Pulvilli long and slender, the claws sharp, slender and curved.

Wings: The wings are hyaline, the auxiliary cell and to a lesser extent the costal cell tinted. There are 2 submarginal cells. The anterior branch of the third vein arises obliquely but abruptly but never rectangularly. It is rather broadly rounded at the first bend and is strongly bent forward at the outer half. The second vein has an abrupt bend distally so that the apex of the marginal cell is widened. All the posterior cells are open, the anterior crossvein lies at or close to the basal fourth of the discal cell. The second vein arises very abruptly, in contrast to *Aphoebantus* Loew, but it is rounded and not rectangular. Anal cell widely open, alulae large, the ambient vein complete, the base of the wing with a hooklike spur. The second vein arises comparatively close to the anterior crossvein, removed from it by about 1½ to 2 times the length of the anterior crossvein.

Abdomen: The abdomen is longer than the thorax, gently tapered posteriorly and more or less cylindrical. It is generally reddish or yellowish brown in color but tends to be obscurely blackish on the basal tergites, except laterally. The pile consists of rather loose, scattered, flat-appressed, rather long scales and flattened hairs. The first tergite, however, has a fringe of dense, erect, rather long, bristly hairs on the outer half. Male terminalia conical, triangular, moderately large and protrusive with inverted recess.

The male terminalia from dorsal aspect, epandrium removed, show a broad, short structure in which there is a wide, prominent, basally widened epiphallus with truncate, snoutlike apex. The basal part of the aedeagus is extraordinarily large and bulblike. Dististyli flared outward like long, slender, undulate, apically pointed, leaflike structures. From the lateral aspect, these are also slender, undulate and pointed and the epiphallus is bifid. The large epandrium is elongate, yet triangular with rather strong, sharp, ventral lobe at the base. Basistylus also triangular and apically pointed.

Material available for study: I have before me a pair of the type-species given my by Dr. R. H. Painter. I have a series of individuals of the same species and one additional species collected by my wife, Marguerite Hull, and me the day after a heavy desert rain at Sells, Arizona. I also have a female I collected at Las Cruces, New Mexico.

Immature stages: Unknown.

Ecology and behavior of adults: These flies are found resting upon desert sand or soil and are often attracted to the edges of temporary ponds. I have never seen them upon flowers.

Distribution: Nearctic: *Desmatoneura argentifrons* Willison, 1895.

Genus *Chiasmella* Bezzi

FIGURES 378, 644, 654

Chiasmella Bezzi, The Bombyliidae of the Ethiopian Region, p. 156, 1924. Type of genus: *Chiasmella brevipennis* Bezzi, 1924, by original designation.

Medium-sized flies, rather similar in appearance to *Aphoebantus* Loew. The head is large and more or less globular, the posterior eye margin has a strong deep indentation. They are dark colored flies with fine grayish-white pollen and similar colored pile. They have hyaline wings and gently tapered abdomen.

The only known species is from Arabia.

Length: 11 mm.

Head, lateral aspect: The head is subglobular; the face is rather strongly produced, gently increasing to the oral margin; its height is nearly equal the length of the short antennae. The sides of the face have pollen above on the upper half, extending obliquely down to the eye margin; oral opening long but narrow, 4 times as long as wide. The eye has a prominent, deep indentation posteriorly, the occiput is tumid but chiefly swollen submedially, and it abruptly slopes to the eye margin where it is nearly plane with the eye. The middle of the occiput has a very large circular depression or bowllike opening; there is a deep, medial fissure leading to the vertex; the pile is short, coarse, and flat-appressed but becomes erect, submedially along the rim of the opening. The palpus is short and concealed within the oral opening; the first segment is quite slender in comparison to the second, a little shorter, the second segment wide with more or less parallel sides, subtriangular in section and the apex rounded and bearing a number of rather short, bristly hairs apically, dorsally, and laterally. The short proboscis is extended forward, it bears a few stiff hairs ventrally near the apex. The antennae are attached at the middle of the head and are short; the first two segments are quite short and nearly equal in length; the third segment is long, conical with nearly parallel sides on the attenuate distal third and bears 2 quite short, ringlike or annulate microsegments.

Head, anterior aspect: The front is slightly narrowed toward the vertex with scattered, flattened, appressed hairs; the ocellar triangle is quite small, situated just a little before the posterior corner of the eye; in the male the front is a little more narrow; it is about half the width of the female.

Thorax: With fine, scattered, suberect pile over the dorsum, which is rather short but this pile becomes a little longer laterally; along the anterior margin, it is more dense and appressed. The humerus is densely long pilose with a dense, long, pronotal collar of pile, coarse but nonflattened. The propleuron, laterally, and the whole upper half of the mesopleuron have dense, long, coarse pile. The sternopleuron dorsally and posteriorly, the posthypopleuron, and all the coxae have numerous, rather slender, whitish scales. The dorsal pteropleuron and the metapleuron have a rather dense patch of shorter pile; on the metapleuron it is curly. Squamae with a fringe of long pile. The scutellum is large, slightly thickened, convex along the margin, the disc and especially the margin with dense slightly flattened and flat-appressed pile; its margin has 2 or 3 pairs of slender, bristly hairs. The notopleuron has 1 long slender bristle, the postalar callosity has 3 pairs.

Legs: Slender; the femora and tibiae are covered with broad white scales. The hind femur has 5 or 6 short bristles placed laterally, and at the outer half, there are 6 ventrolateral bristles; hind tibia with 12 quite short, posterodorsal, 7 or 8 stouter anterodorsal, 5 or 6 anteroventral, and 6 or 7 posteroventral bristles. Middle femur with 5 anteroventral bristles; its tibia has 5 or 6 posterodorsal, 7 posteroventral, 4 or 5 anterodorsal, and a like number of anteroventral bristles. All the femora have scales. The anterior femur has a few short bristles; they are placed in the anteroventral and posteroventral, and the anteroventral position and are distributed along the middle; its tibia has 4 or 5 short, posteroventral bristles; pulvilli well developed; the arolium forms a minute short lappet.

Wings: The wing is hyaline. There are 2 submarginal cells and the first posterior cell almost maximally open, the anal cell narrowly open. The axillary lobe is narrow, the alula present but small, so that at the base the wing is about as wide as the wing apex. The anterior crossvein lies just beyond the basal third of the discal cell, and the second vein arises abruptly but rounded just before the anterior crossvein and removed from it by not more than the length of this crossvein. The discal cell is comparatively wide, with on the outer half the vein above and the vein below about equally convex and slightly narrowing the intercalary vein. The anterior branch of the third vein arises abruptly but plane and there is a strong backward bend in this vein, the outer section plane; there is a similar subapical bend in the second vein. The ambient vein is complete.

Abdomen: With eight segments in the male, with dense, flat-appressed, flattened whitish pile on all of the segments which on the terminal part of the abdomen becomes almost scalelike. There is also a conspicuous lateral fringe of abundant, long, fine, white pile becoming a little shorter near the midline; the sternites are hidden by the curled tergite. The middles of the tergites on the second, third, and fourth segments have a blackish basal fascia that becomes a little wider in the middle.

Material studied: The type-species, a pair in the British Museum (Natural History).

Immature stages: Unknown.

Ecology and behavior of adults: Not on record.

Distribution: Palaearctic (Arabia): *Chiasmella brevipennis* Bezzi, 1924.

Genus *Comptosia* Macquart

FIGURES 150, 156, 158, 209, 366, 424, 693, 735, 921, 922, 923, 924, 925, 926

Comptosia Macquart, Diptères exotiques, vol. 2, pt. 1, p. 80, 1840. Type of genus: *Comptosia fascipennis* Macquart, 1840, by monotypy.
Neuria Newman, Entomologist, vol. 1, no. 14, p. 220, 1841. Type of genus: *Neuria lateralis* Newman, 1841. Designated by G. H. H. Hardy, p. 54, 1922b. Edwards, p. 82, 1934, footnote, states: "I cannot trace an earlier selection of a genotype for *Neuria*; consequently Bezzi's 1924 reference to *praeargentata* as the genotype is invalid."
Alyosia Rondani, Arch. Zool. Anat. Fisiol., vol. 3 (sep.), p. 54, 1863. Type of subgenus: *Comptosia maculipennis* Macquart, 1846. Designated by G. H. H. Hardy, p. 54, 1922b.
Aleucosia Edwards, Encycl. Ent., Paris, ser. B. II., Dipt., vol. 7, p. 95, 1934. Type of subgenus: *Neuria corculum* Newman, 1841, by original designation.
Epidosina, new subgenus. Type of subgenus: *Comptosia vittata* Edwards, 1934.

White, in Hardy (1924), pp. 77-78, key to 18 species from Australia.
Edwards (1934), pp. 81-112, key to groups and description of 45 species.
Hardy (1941), pp. 223-225, key to 34 Australian species.

Large and sometimes very large flies ranging up to 56 mm. in wing spread. They are characterized by a relatively long, slender wing of comparatively uniform breadth in which the marginal cell is drawn backward and outward into a long, rounded, fingerlike lobe, sometimes called the radial loop, which cuts deeply into the first submarginal cell. Submarginal cells vary from 2 or 3, to occasionally 4 as in *Comptosia plena* Walker. Most species, however, have only 2 submarginal cells but the type-species has 3. A few species, moreover, have the wing narrowed toward the base with reduction of the axillary lobe. While a few species have the wings hyaline, the great majority of species have wings that are variously patterned with brown accentuated along the veins and crossveins and especially upon the foreborder of the wings and often leaving the apical fifth or the whole posterior border of the wings clear. Thus the species fall into many pattern groups. In the type-species the wing is a uniform, beautiful, rufous brown in color extending nearly to the end of the marginal cell, appearing again at the apex of the wing and interrupted in the male by a transverse band of opaque white. Several species have the apex of the wing white or with preapical band. This band, more chalky in males, may be absent in females.

This complex of genera centering around the genus *Comptosia* Macquart has received a great deal of study by Edwards (1934), Hardy (1922, 1933), and Paramonov (1931), each of whom has made valuable and important contributions to our understanding of these genera. Because of the erratic, almost distorted, highly peculiar venation one is inclined to regard these flies as rather archaic; the loops and curves of the veins are probably mere specializations, but the extra crossveins suggest a condition that might have obtained a long time ago.

Rondani (1863) first attempted to subdivide the genus.

He restricted *Comptosia* Macquart to the species with 3 submarginal cells, and the antenna with a long style. He characterized *Macrocondyla* Rondani as having 3 submarginal cells and an indistinct antennal style. *Alyosia* Rondani was created for those species with 2 submarginal cells, and *Lyophlaeba* Rondani for the species with 4 submarginal cells. Because of misunderstanding of genera by several authors considerable synonymy resulted, which has been cleared by Edwards in his fine illustrated paper of 1934. Moreover, Edwards sought for and appears to have found other supplementary characters of considerable value. I have followed him to a large extent in this paper, but I have adopted *Lyophlaeba* Rondani as a genus for the South American forms because it has priority over *Macrocondyla* Rondani, and I have retained the latter only in a subgeneric sense.

Also, the abdomen is elongate oval, yet always considerably shorter than the wing. The pile of these flies is fine, short, and subappressed in most species but is somewhat long, erect, dense, and woolly in a few species; the majority have a strongly bare aspect.

The legs are slender and bear fine bristles on the hind femur. It should be noted that the proboscis varies from relatively short and stout with fleshy labellum, to others more slender in both respects but never greatly longer than the oral recess. The antennae are not greatly different from other Lomatiinae; the first segment is swollen and medially flared, generally a little longer and larger, and remaining segments much smaller. The face is short, the oral cavity strongly retreating.

The numerous species of *Comptosia* Macquart are restricted to Australia; the closely allied flies in the Chilean subregion fall in the related genus *Lyophlaeba* Rondani and the subgenus *Macrocondyla* Rondani; they differ from *Comptosia* Macquart by the presence of a large tuft of soft pile on the anterior metapleuron.

Length: 10 to 22 mm.; wings 10 to 56 mm.

The *Comptosia* Macquart complex of related genera and subgenera may be expressed in the following way:

Comptosia Macquart: No pile in front of haltere and none on pteropleuron.

- *Comptosia* Macquart in the strict sense: 3 submarginal cells; alula narrow; anal vein straight; radial loop deep; male costa tuberculate; wing near apex in male, sometimes female, with opaque white band. Type-species: *Comptosia fascipennis* Macquart, by monotypy.

- *Alyosia* Rondani, subgenus: 2 submarginal cells; radial loop strongly developed and fingerlike. While either the distal crossvein, which breaks up the radial field into 4 submarginal cells, or the branch connecting the radial loop to the third vein (sectorial crossvein) may be regarded as a true supernumerary crossvein, I interpret the other veins in the radial field as being probable coalescent branches of the radius; I regard the whole *Comptosia* complex as constituting an apparently archaic relict group of bee flies which tend to reflect an early generalized venation. Therefore I leave *Alyosia* Rondani intact as a convenient name already in use. The presence of additional submarginal cells (Schiner, notwithstanding) is very useful in places within the family. The majority of genera have moved in the direction of a still more simplified venation, some even having the radial field reduced to a single submarginal cell. Type-species: *Comptosia maculipennis* Macquart, by designation of Becker.

Edwards (1934) presents no less than 4 species groups within this subdivision: *fasciata* Fabricius, *preargentata* MacLeay, *apicalis* Macquart, and the *ocellata* Newman group; 17 species included in all.

- *Epidosina*, new subgenus. This name I propose for those species of the *Comptosia* complex with 4 submarginal cells. Radial loop strong and fingerlike. Anal vein straight. Axillary lobe wide and alula narrow. Type-species: *Comptosia vittata* Edwards.

- *Oncodosia* Edwards. A genus distinguished by the large, very broad abdomen without prehalteral pile. Sides of abdomen with conspicuous flattened tufts of hair. Two submarginal cells present, except in occasional freakish individuals. Alula wide and large and anal vein sinuous. Wing characteristically very broad basally. Radial loop not extensive; commonly collected by Hull and his sons in Western Australia in 1954. Australian only. Type-species: *Anthrax patula* Walker, by original designation.

- *Aleucosia* Edwards, subgenus: Proposed by Edwards for those 15 species having the following ensemble of characters: three submarginal cells in the type-species, alula completely absent; wing markings of a different type, chalky white markings never present; second section of third branch of medius longer than or at least as long as the first, this rarely true in *Comptosia* Macquart; the fourth branch of the radius more strongly recurved, the upturned part distinctly sinuous, rarely true in *Comptosia* Macquart. Type-species: *Neuria corculum* Newman, by original designation, contains 15 species.

Doddosia Edwards. A related genus with this ensemble of characters: Metapleuron with a conspicuous tuft of hair; pteropleuron bare in front of spiracle; distinct bristles present on middle and hind tibiae; abdomen with parallel sides; wings very broad, costal cell a little widened at base; venation essentially like *Comptosia* Macquart, the radial loop well marked; third posterior cell not narrowed at tip; anal vein straight; alula small. Type of genus: *Doddosia picta* Edwards. Found abundantly by Hull in northern Queensland.

However, I do not agree with certain earlier writers that the number of submarginal cells is of no significance whatsoever; we certainly find it invaluable in differentiating the species of *Ligyra* Newman (*Hyperalonia* of authors) from the more numerous forms of *Exoprosopa* Macquart; nor have I heard any one suggest abandoning it. I have retained the number of submarginal cells as part criterion for certain groups I recognize as useful subgenera, but I have superimposed this concept upon the pleural characters discovered by Edwards.

Head, lateral aspect: The head is globular with very strongly developed occiput, which, however, is steeply and convexly sloping to the eye margin down to the comparatively small but deep central cavity of the occiput. In some species the occiput is more extensive dorsally than it is below, in still others it is approximately equally extended dorsally and ventrally. However, in any case there is a characteristic, deep indentation at the middle of the posterior eye margin, with or without accompanying notch, but without bisecting line. The occiput is pollinose and the pile consists of minute hairs, more or less appressed or curled. There is an erect fringe of pile around the inner margin bordering the central occipital cup which varies from sparse to dense in different species. The front is convex, forming in males almost an equilateral triangle and extending about half the distance from the antennae to the posterior eye corners. This pile is erect or suberect, fine, bristly, and rather dense and in some species with subappressed, scaliform, glittering hairs. Typically there is a bare, narrow, foveal groove in the middle of the front which may be linear or as wide as an ocellus. It may extend as in the type-species, with crimped margin all the way between the antennae, or it may end in front of the antennae. In any case, however, there is a bare, triangular, or rhomboidal space in front of the antennae, bare of pile and containing pollen only. The front of the female is similar with pile a little more sparse.

The eye is quite large, slightly reniform, equally developed above and below, the upper and lower halves therefore symmetrical. The face in the male of the type-species is not or scarcely at all visible in profile as it begins to recede from the upper margin of the antennae, and the oragenal cup likewise sinks below the eye margin, except at the bottom of the head near the base of the proboscis. This is likewise true of the female, except that the receding face is a little more extensive and visible beneath the antennae. The sides of the face bear dense coarse pile becoming shorter but extending down along the sides of the gena nearly to the base of the proboscis. The proboscis is stout, sometimes confined to the oragenal recess into which it is tucked. In other species it extends beyond the oragenal recess by one-third its length but when thrust upward, as it often is, it comes to rest between the antennae. In the type-species the proboscis is relatively short, and even with the fleshy lobes of the labellum apposed it would barely fit into the oragenal recess, and possibly a little beyond. In this species the labellum is short, broad, fleshy, with ruguse or transverse channels much as in a muscoid. The two lobes are often divergent backward as large, more or less triangular flaps. The palpus is clearly composed of a short, basal segment, sometimes not longer than wide, again 2 or 3 times as long as wide, again 2 or 3 times as long as wide and with a long, slender distal segment, which bears numerous, fine, long, subappressed hairs. The apex is pointed, the whole segment cylindrical.

The antennae are attached near the upper third of the head. The first segment is strongly swollen, widened apically especially medially and also somewhat produced apically on the medial aspect, so that the apex is more or less obliquely truncate. Moreover, the first segment is rather strongly swollen and rounded below in the type-species and bears a tuft of rather long, oblique, bristly pile of the same length as the pile of the upper face. This segment is at least twice as long as its basal width and its apical width at least 1½ times as long as its basal width. Thus, it appears to be somewhat flattened dorsoventrally and somewhat flared medially toward the apex. It is rather densely beset with stout, black, setate bristles on the whole medial surface, the dorsal surface, except at the base, and longer, finer bristles laterally proceeding chiefly from the basal half. The second segment is small, quite beadlike with rounded, convex sides medially and laterally; it is attached near the outer edge of the first segment so that the second and third segments diverge strongly. It has a few short setae dorsally. The third segment is more narrow than the second, widest at the base, rather strongly tapered to a point just before the middle from which the remainder of the segment is drawn out into a style of uniform thickeness.

Head, anterior aspect: The head is circular and distinctly more narrow than the thorax. The eyes in the male vary from being almost holoptic and separated by only half the width of the ocellus, each upper eye contour being strongly rounded as in the case of the type-species, to others in which the eyes are separated by a little more than the thickness of the ocellus. In a species before me the eyes are truly touching. In females the front is comparatively narrow and the eyes are separated at the vertex by a space sometimes as little as half the lower width of the front, and sometimes it is a little wider. The ocellar tubercle is never conspicuous, it has a clump of 8 to 20 long, fine, slender hairs curved forward and lying in a medial row and the ocelli occupy a short, isosceles triangle.

In the type-species the second antennal segment is as long as wide, in some species only half as long. The third segment in some species has the bulbous, basal part of the third segment much reduced in length, thus more sharply attenuate, leaving a proportionate, longer, slender, stylelike process on this segment. The antennae are separated at the base by a distance nearly equal the basal width of the first segment. From the anterior aspect the face is rather wide along the sides of the upper part but rapidly sinking down below the eye level near the upper third of the oragenal cavity. It continues to drop down into a long, narrow, deep, tentorial fissure which lies between the lower part of the oragenal cup and the genal part of the head.

The facial component may be considered to stop at the bottom of the facial slope. The whole sides of the face bear dense, long, somewhat oblique pile and the genal sides below along the eye margin bear a narrow fringe of pile. Walls of the oragenal cup folded inward near the middle. The head from the posterior view shows a strongly bilobed occiput with deep medial crease extending almost to the vertex in the type-species, in other species only half the dorsal length of the occiput.

Thorax: The thorax is short and compact, rather strongly widened posteriorly, the posterior width of the mesonotum is as great as its length, and the humeri extend outward behind as a transverse, vertically flattened lobe. The mesonotum is densely pollinose and in some species feebly shining. It is comparatively low and only gently convex, equally convex in front and behind. The pile consists of dense, short, oblique, very fine, bristly hairs arising from minute tubercles in some species; in others the fine hairs are longer, erect, and without tubercles. In still other species it is composed of coarse, flat-appressed, pale hairs intermixed with extraordinarily fine, erect, black hairs. Pile of scutellum similar to that of the mesonotum. The humeri have dense pile also similar to the mesonotum. Notopleuron with 3 or 4 and postalar callosoity with 1 or 2 stout, rather long bristles, pale in color. Scutellar margin usually only with long, bristly hairs. The scutellum is large, distinctly subtriangular with the apex rounded, the pollen thin, and the whole scutellum flattened toward the apex.

The pleuron is pollinose; the upper half of the mesopleuron bears dense pile, which is fine, bristly and curled and extending upward. The upper sternopleuron has a little fine pile but the pteropleuron and hypopleuron and metapleuron in front of the spiracle are all bare. The lateral metanotum bears a conspicuous band of long pile, generally pale, with some black hairs intermixed, and there are a few hairs on the posterior metapleuron above the hind coxa. The propleuron has tufts of long pile. Prosternum with long pile; the alary squama is short but bears the usual *Lomatia*-like, wide, plumose band of pile. The halteres are relatively small.

Legs: The legs are well proportioned to the body without being large. The first 4 femora are slightly enlarged over the middle and base, the hind pair stout without being large. The femora vary from black to yellow or red, with the tibia always pale. The anterior femur of the type-species is covered with minute, flat-appressed, setate hairs and 4 or 5 spines anteroventrally. On the middle pair the pile is similar, but somewhat larger, and the bristles of the same number, and on the hind femur in addition to the small, flat-appressed hairs, there are 7 short, setate bristles anterodorsally on the outer half, and on the whole anteroventral length of the hind femur there are 12 somewhat longer, oblique, stout, black bristles, and in a few species even more. In others there are a few slender bristles restricted to the outer half. In some species of *Comptosia* Macquart the anterior 4 femora have scattered, fine, long, erect, bristly hairs on the posterior surface and they may lack bristles on the anterior femur and the pile of the posterior femur may be rather matted and scalelike.

Anterior tibia anteroventrally with a dense brush of appressed, golden setate pile in the type-species and a similar brush of darker color in blackish species. The spicules or bristles of the anterior tibia are small, fine and in some species scarcely differentiated. In the type-species there are 3 anteriorly, 5 or 6 posterodorsally. Other species have 8 to 10 anterodorsally, 6 to 8 posterodorsally, and a like number posteroventrally. The middle tibia has 4 rows of bristles of variable length and stoutness. The anterodorsal row has 7 to 10; the posterodorsal row has 8 to 12; the posteroventral row has 3 to 8; and the anteroventral row has 6 to 12. Hind tibia likewise with 4 rows of bristles of about the same numbers. Tarsi short, strongly spiculate below, except that there are none upon the anterior tarsi, and only 1 row upon the hind tarsi. Claws and pulvilli usually small, the latter extending to the end of the claw with rounded apex. In the type-species the claws and pulvilli are quite large, the former rather slender but strongly curved at the apex, the wide spatulate pulvilli are rather truncate at apex, and with the arolium rather large, but smaller in other species.

Wings: The wings are elongate, slender in most species but in some species as in the type-species, they are extraordinarily long and slender. Wing patterns fall into several groups; in the type-species it is a uniform, beautiful, reddish brown throughout in the females, similarly colored but with a conspicuous, opaque, chalky white, preapical band in males. This preapical band is characteristic of a number of species in this genus. Some equally large species have the veins of the wing diffusely margined with dark brown, the interior of the cells lighter. Still a third group of species have the anterior half of the wing dark brown without preapical spots. The posterior border of the brown color is irregular, usually filling all of the first and part of the second and third basal cells and cutting across cells to the apex. This group of species tends to have the wings strongly narrowed toward the base. The axillary lobe is extremely narrow, especially basally, and the alula absent or nearly so (subgenus *Alyosia* of Edwards).

In all members of the genus the radial loop is long and fingerlike, the marginal cell greatly extended downward, the end of the second vein strongly recurrent. All 4 posterior cells are open, the anal cell is widely open, the anal vein nearly straight, never bowed backward, and the anterior crossvein enters the discal cell at the outer seventh or outer eighth of the discal cell. Costal margin denticulate in the males of some species including the type-species, and there is a prominent, precostal, triangular nodule or spur. The alula in *Comptosia* Macquart ranges from narrow with a plane margin to species where it is absent; the ambient vein is complete.

Abdomen: The abdomen varies from species where it is short oval, basally as wide or wider than the width of the thorax behind the wings, to other species where it is more elongate, 1½ to 2 times as long as the thorax and with the lateral margin nearly plane, the whole abdomen tapered. In males I find 8 tergites with the ninth forming a medially notched shield placed dorsally over the small and quite recessed terminalia. In the type-species the eighth and ninth male tergites are likely to be completely recessed beneath the seventh, the last sternite likewise recessed; ninth male tergite bilobed; in some males there appear to be 10 tergites. In females I find 8 tergites with the female terminalia recessed within a posterior pocket and surrounded by strong, bristly hairs. Postmargin of last sternite always plane.

The abdomen of *Comptosia* Macquart is usually blackish; in the type-species, however, it is brownish red laterally with the broad, medial, black, continuous vittae becoming narrower posteriorly and extending outward along the posterior margin laterally but never reaching the corners. The pollen of the abdomen is generally faint but may be thick and dense, often leaving the abdomen shining. The abdominal pile consists of scattered, fine, short, largely appressed, bristly hairs, or with similar, even finer hairs erect and extending upward over disc and lateral margin, and with a variable amount of fine, short, curled, pale, flat-appressed hairs generally concentrated along the posterior margin and more sparse on the anterior part of the tergites. Anteriorly the lateral border of the first tergite shows a brush of erect, unusually fine hairs, not coarse and spikelike as in other genera. It may extend all the way across the segment and in a few species forms a narrow, complete band of bushy, tangled pile on the posterior border of this tergite. Oddly enough, some species have only a few hairs basolaterally on the first tergite.

The male terminalia from dorsal aspect, epandrium removed, show an elongate figure with more or less parallel sides much as in *Lomatia* Meigen. Here again the dististyli are either leaflike viewed from above or twisted outward; in any case they are stout and blunt. The epiphallus and aedeagal complex are long and spoutlike. Basistylus elongate and dorsally ventrally widened apically.

Material available for study: In my own collection there are about 20 or more species, many of them collected by the author and his sons while in Australia in 1953 and 1954, in both the eastern and western sections of the country. I have also studied the material in the collection of the Commonwealth Scientific and Industrial Research Organization at Canberra, through the courtesy of Dr. A. J. Nicholson and his associates, especially Dr. S. J. Paramonov. I have also had the advantage of studying the material in the British Museum (Natural History) and in the United States National Museum (now National Museum of Natural History).

Immature stages: Unknown.

Ecology and behavior of adults: These flies commonly feed on various species of the genus *Leptospermum*, a flowering genus of the family Myrtaceae, ranging from low, herblike forms to shrubs and small trees; few plants attract Diptera in such quantities or such great variety. At Canberra I collected a series of the beautiful type-species on these flowers.

Distribution: Oriental: *Comptosia indecora* (Wulp), 1885 (as *Neuria*).

Australian: *Comptosia albofasciata* Thomson, 1869; *angusta* Edwards, 1934 (subg. *Aleucosia*); *apicalis* Macquart, 1848; *atherix* (Newman), 1841 (as *Neuria*) (subg. *Aleucosia*) [=*geometrica* Macquart, 1847]; *aurifrons* Macquart, 1849; *bancrofti* Edwards, 1934; *basalis* (Walker), 1849 (as *Anthrax*); *bicolor* Macquart, 1849; *biguttata* Edwards, 1934; *brunnea* Edwards, 1934; *calophthalma* Thomson, 1869 (subg. *Aleucosia*); *cincta* Edwards, 1934 (subg. *Aleucosia*); *corculum* (Newman), 1841 (as *Neuria*) (subg. *Aleucosia*); *costalis* Edwards, 1934 (subg. *Aleucosia*); *cuneata* Edwards, 1934 (subg. *Aleucosia*); *decedens* (Walker), 1849 (as *Anthrax*); *dorsalis* (Walker), 1849 (as *Anthrax*) (subg. *Aleucosia*); *edwardsi* Hardy, 1941; *extensa* (Walker), 1835 (as *Anthrax*); *fasciata* (Fabricius), 1805 (as *Anthrax*) [=*nigrescens* (Newman), 1841 (as *Neuria*)]; *fascipennis* Macquart, 1840; *fulvipes* Bigot, 1892 (subg. *Aleucosia*); *gemina* Hardy, 1941, [=n.n. for *cognata* Walker, 1852 (as *Anthrax*) not *cognata* Walker, 1849]; *hemiteles* (Schiner), 1868 (as *Neuria*) (subg. *Aleucosia*); *inclusa* (Walker), 1849 (as *Anthrax*); *lateralis* (Newman), 1841 (as *Neuria*) [=*ducens* (Walker), 1852 (as *Anthrax*); =*grandis* (Schiner), 1868 (as *Neuria*), =*rufoscutellata* Jaennicke, 1867]; *maculipennis* Macquart, 1846 (subg. *Alyosia*); *maculosa* (Newman), 1841 (as *Neuria*) (subg. *Aleucosia*), [=*fulvipes*,

Bigot, 1892]; *moretonii* Macquart, 1855; *murina* (Newman), 1841 (as *Neuria*); *obscura* (Walker), 1852 (as *Anthrax*); *ocellata* (Newman), 1841 (as *Neuria*); *partita* (Newman), 1841 (as *Neuria*) (subg. *Aleucosia*); *plena* (Walker), 1849 (as *Anthrax*) (subg. *Aleucosia*); *praeargentata* (MacLeay in King), 1828 (as *Anthrax*); *quadripennis* (Walker), 1849 (as *Anthrax*); *rubrifera* (Bigot), 1881 (as *Ligyra*); *sobria* (Walker), 1849 (as *Anthrax*); *serpentiger* (Walker), 1849 (as *Anthrax*) (subg. *Aleucosia* Edwards); *stria* (Walker), 1849 (as *Anthrax*); *sylvana* (Fabricius), 1775 (as *Bibio*); *tendens* (Walker), 1849 (as *Anthrax*); *tricellata* Macquart, 1847 (subg. *Aleucosia*); *tripunctata* Edwards, 1934 (subg. *Aleucosia*); *vittata* Edwards, 1934; *walkeri* Edwards, 1934; *wilkinsi* Edwards, 1934; *insignis* Walker; *subsenex* Walker.

Genus *Lyophlaeba* Rondani

FIGURES 147, 151, 154, 157, 161, 162, 368, 656, 673, 988–993

Lyophlaeba Rondani, Arch. Zool. Anat. Fisiol., vol. 3, p. 55 (sep.), 1863. Type of genus: *Lyophlaeba lugubris* Rondani, 1863, by original designation; 1 species.

Macrocondyla Rondani, Arch. Zool. Anat. Fisiol., vol. 3, p. 55 (sep.), *Macrocondyla* is regarded in this work as a subgenus of *Lyophlaeba* Rondani. 1863. Type of subgenus: *Macrocondyla pictinervis* Rondani, 1863, by original designation; 1 species.

Tritoneura Schiner, Verh. zool.-bot. Ges. Wein, vol. 17, p. 312, 1867. Type of genus: *Comptosia lugubris* Philippi, 1865, by monotypy.

Liophleba Bezzi, The Bombyliidae of the Ethiopian Region, p. 4, 1924, emendation.

Compsella, new subgenus. Type of subgenus: *Macrocondyla transandina* Edwards.

Paramonov (1931), pp. 1-218, treatment of *Lyophlaeba* Rondani, including *Macrocondyla* Rondani.

Edwards (1934), pp. 103-110, treatment of *Macrocondyla* Rondani and *Lyophlaeba* Rondani.

Paramonov (1948), pp. 146-148, key to 19 species (including *Macrocondyla*), from Bolivia, Chile, and Patgonia.

Large flies with comparatively long, tapered, flattened abdomen bearing generally a medial vitta or series of medial spots and pile generally of two types, with very fine, erect, black hairs and soft, matted, appressed tufts of pale pile outlining the tergites; these tufts are sometimes lacking.

Adopting the principle of the first reviser of the group, we have followed Paramonov (1931), and so we subordinate *Macrocondyla* Rondani as a subgenus of *Lyophlaeba* Rondani. We accept *Lyophlaeba* Rondani for all of the South American species related to *Comptosia* Macquart because of this principle; this prevents the matter of priority being a question for future consideration. Because the number of submarginal cells in these and related flies is stable within the species, I cannot agree that it is without value; coupled with the characteristics of the antenna and the pleuron, it forms a useful character for the division of *Lyophlaeba* Rondani into subgenera. Edwards (1934) noted that all the flies in this genus have a large, dense tuft of pile on the anterior metapleuron, much similar hairs on the pteropleuron, third posterior cell not marginally narrowed, radial loop prominent, anal vein straight, axillary area moderately large, alula small with plane margin. The eyes of the male are either narrowly separated or almost touching. Edwards (1934) proposed treating *Lyophlaeba* Rondani as a subgenus. Paramonov considered that *Lyophlaeba* Rondani should take precedence over *Macrocondyla* Rondani, and we have decided to follow Paramonov for the reasons given above. The type-species of subgenus *Macrocondyla* is *pictinervis* Rondani. I divide the genus as follows:

The genus *Lyophlaeba* Rondani, with type-species *lugubris* Rondani, has 4 submarginal cells. The first antennal segment is long, without medial hair tuft at apex. Third antennal segment quite long, not styliform apically, in males shaped like a sword with a curved blade, in females more plane; apex bifid, the short spine articulated and dorsal and subapical. The proboscis is long and pulvilli are quite rudimentary.

The subgenus *Macrocondyla* Rondani has 3 submarginal cells. The first antennal segment is widened medially at apex and has a conspicuous tuft of long pile; third antennal segment strongly attenuate with minute spine at apex. Pulvilli conspicuous, more than half as long as claw. Wings clouded or banded.

Compsella, new subgenus, with 4 submarginal cells and first posterior cell divided by a crossvein. Type of subgenus: *Macrocondyla transandina* Edwards. These would be the least simplified in terms of venation and presumably the oldest of the group. Radial loop long, wing mottled and banded.

This genus with its subgenera is entirely South American and almost restricted to the Chilean subregion.

Length: 13 to 22 mm.; of wing, 12 to 22 mm.

Head, lateral aspect: The head is subglobular, the occipital component very prominent on the upper half and from the eye margin rounded and sloping backward leaving the occipital cup small but deep, and the marginal fringe of the recess dense but short. The upper half of the occiput varies from species in which it is shining, and uncommonly without pollen, to most species where the pollen is dense. The pile on this part of the occiput is rather scanty, consisting of short, stiff hairs curled and appressed forward and not reaching the eye margin. This condition extends to the bottom of the eye, or lower. The front in males is triangular and pollinose and in both sexes covered rather densely with shaggy pile appressed downward. From the lateral aspect both face and front are quite short, not extending beyond the eye margin. The face bears dense, long pile laterally in considerable amount and the whole surface is pollinose. The eye is kidney-shaped with deep but broad indentation a little below the middle, the indentation is equal in angle, above and below its greatest indentation. The proboscis is elongate and slender with very small labellum, often thrust obliquely downward, but in the resting position it extends obliquely up between the antennae. It is shortest in those species of the subgenus *Macrocondyla* Rondani, sensu stricto. Generally the proboscis extends well beyond the head and antennae. Palpus slender and elongate, composed of a short, somewhat fused basal segment and a much longer apical segment. The palpus bears a number of long, fine hairs.

The antennae are attached at the upper third of the head, the first segment is large, much swollen, barrellike but not greatly asymmetrical. It is 4 or 5 times as long as the much smaller, much more narrowed, beadlike second segment. Third segment nearly as long as the first two segments together but sometimes a little shorter. It may be flattened, lanceolate, and leaflike, or cylindrical and tapering on the outer third where it is much more narrow. Apically it bears a small, stubby, minute spine.

Head, anterior aspect: The head is circular, not quite as wide as the thorax at its greatest width. The eyes in the male are narrowly separated, sometimes by as little as the width of one ocellus, sometimes slightly more. The ocellar tubercle is set almost at the posterior eye corner, is rather low, and the lateral ocelli touch the eye margin in the males. The front forms an almost equilateral triangle, and the vertex is extended forward in front of the ocelli with nearly parallel sides for an additional distance rather greater than the length of the ocellar triangle; this triangle is isosceles in shape. In the females the eyes are separated at the vertex by a distance equal 3 times the space between the ocelli. In both sexes the front is densely pollinose with a narrow, bare, longitudinal, foveate depression extending from the vertex and widening immediately in front of the antennae and extended as a tonguelike, triangular extension between the antennae. The antennae are separated quite widely in some species, enough for the proboscis to fit between them. The oragenal opening is quite wide below reaching almost to the eye margins but strongly and progressively narrowed anteriorly, ending immediately beneath the antennae. The walls of the oragenal recess are thick and inflated, except near the bottom of the eye where they become much thinner, and the recess is much deeper. However, this recess is quite deep throughout the lower four-fifths, becoming shallow only a short distance from the antennae.

Moreover, the tentorial fissure is wide and extremely deep, so that at no great distance from the antennae this fissure begins and leaves the head adjacent to the eye margin itself, shelving deeply inward. It ends at the beginning of the gena which extends deeply beneath the eye. Anterior genal margin polished and bare. The head viewed from above shows that the very extensive occipital lobes are tightly apposed behind the vertex.

Thorax: The thorax is longer than wide, widest posteriorly; the mesonotum is low and quite gently convex, its pile consists of dense, fine, rather short, slightly appressed hairs both pale and black and in some species vittae of flat-appressed, straggly, dully white, tangled hairs. The more erect, bristly hairs, though fine, tend to be longer on the posterior third of the mesonotum and in front of the scutellum. There is a well-developed anterior collar of stiff pile on the mesonotum which extends over and includes the prominent humeri. The latter have a posterolateral blunt spur. Scutellum large with circular, posterior margin somewhat flattened apically, the pile similar to the mesonotum, with weak, long bristles on the margin and sometimes laterally. Notopleuron with 4 or 5, and postalar callosity with 5 or 6 bristles, which are rather stout and conspicuous. Pleuron pollinose with a conspicuous tuft of fine, tangled, and radiating pile and some bristles on the upper half of the mesopleuron; these may extend more loosely over the whole pleuron. The sternopleuron is more than usually pilose, the whole upper half has a bushy mass of dense, long pile appressed downward. In a few species the quantity is reduced. Propleural and prosternal pile extensive and long. The metanotum is both pilose and bristly. Both the anterior metapleuron in front of the haltere and the posterior metapleuron behind the spiracle have conspicuous tufts of rather dense, long, fine pile. Hypopleuron bare. Halteres small. Alary squama short but with dense, plumelike fringe. Pteropleuron without pile.

Legs: The legs are well developed. Anterior and middle femora slightly but distinctly swollen from the apex to the base. Hind femur stout without being swollen. All the femora have flat-appressed pile, sometimes fine, black and setate, on others pale, shining and scaliform, and in some the pile is a mixture of these two types. The appressed scaliform pile is more common. Anterior femur with a few very fine, erect hairs below and generally with 4 to 7 slender, anteroventral bristles. Middle femur with 4 or 5 much stouter, sharp, anterior bristles and the hind femur characteristically with 6 to 8 prominent, rather long, slender, black, anteroventral bristles, and 4 or 5 anterodorsal setae on the outer half. The anterior tibia is shorter than the middle tibia, the middle tibia nearly if not quite as long as the hind tibia. Anterior tibia with distinct, bristly spicules; there are 5 or 6 anterodorsal ones, 8 to 10 anteroventral, sometimes 6 to 8 additional bristles between, 10 to 12 smaller, posterodorsal ones, and 8 to 10 posteroventral ones. Lower surface of anterior tibia with a slender brush of dense, appressed setae. Middle tibia with 4 rows of bristles; 8 to 10 anterodorsally, about the same number of posterodorsal bristles which are rather smaller and about 10 posteroventral bristles, and there are 5 or 6 anteroventral bristles. Hind tibia with well-developed bristles, all of them short, the dorsal bristles a little stouter. There are 10 to 12 anteroventral ones, the same number of anterodorsal ones, and 4 to 7 posterodorsal bristles, and 7 to 10 posteroventral bristles.

Anterior tarsi quite short without spicules below, remaining tarsi, however, strongly spiculate ventrally. Claws slender, sharp, and strongly curved apically, and with a long pulvillus which is quite slender. There is a short, triangular arolium.

Wings: The wings are long and slender, longer than the abdomen. There are 3 submarginal cells, except in the genus *Lyophlaeba* Rondani, which has 4. The subgenus *Macrocondyla* has a woolly, densely pilose mesonotum and abdomen and the first posterior cell almost closed. Wing with a strong brown pattern somewhat diffuse, darker on the basal third of the wing in front of the second basal cell. The brown pattern runs down each side of the cubital vein, as a spot across the anal vein and with a second concentration of brown color subapically across the wing, leaving the veins widely margined with brown; the centers of the cells more or less diffusely hyaline; in some species, as in *Lyophlaeba lugubris* Philippi, almost the whole wing is brown with very faint, subhyaline spots in a few cells. The radial loop is prominent, although its extent varies. The first posterior cell is more strongly narrowed than in *Comptosia* Macquart. The anterior crossvein enters the discal cell from the outer fourth to the outer sixth. Alula short and linear. Costa widened at the base with a basal, costal spur. Margin of the costa never denticulate.

Abdomen: The abdomen is elongate, distinctly longer than the thorax, rather strongly flattened and long conical or slightly oval in shape. I find 9 tergites in the male, the last one forming a short hood over the small recessed genitalia. In the females I note only 7 tergites, any others being completely recessed. The pile of the abdomen varies from dense, matted, fine, short, flat-appressed, pale pile among which there is considerable amount of longer, quite fine, bristly, black pile over the whole tergite, and curled obliquely backward. The pale pile in some species is restricted to narrow, posterior borders, and in some species there is similar, matted, appressed, slightly longer pile forming a continuous vitta or stripe down the middle of the abdomen and again narrowly along the lateral margins. The erect, bristly pile tends to be a little stouter along the lateral borders, but there are no banked, flattened tufts of lateral pile. Lateral third of the first tergite usually with a rather dense brush of erect, coarse pile.

Material studied: I have 5 species, including several sent to me by Alfredo Faz from Chile in 1923 and also the type-species.

Immature stages: Unknown.

Ecology and behavior of adults: I have seen no record of the flowers these flies frequent.

Distribution: Neotropical: *Lyophlaeba argentinae* Paramonov, 1940 (as *Lyophlaeba*); *bifasciata* (Macquart), 1849 (as *Comptosia*); *bigoti* Edwards, 1934; *blanchardi* Paramonov, 1931 (as *Lyophlaeba*) [=*bifasciata* Blanchard, 1852, not Macquart]; *boliviana* Paramonov, 1931 (as *Lyophlaeba*); *canescens* Philippi, 1865; *chilensis* Paramonov, 1931 (as *Lyophlaeba*); *consobrina* (Philippi), 1865 (as *Comptosia*); *edwardsi* Paramonov, 1948 (as *Lyophlaeba*) [=*montana* Edwards, 1934, not Philippi]; *haywardi* Edwards, 1934; *infumata* (Philippi), 1865 (as *Comptosia*); *koslowskyi* (Edwards), 1934 (as *Ligyra*); *landbecki* (Philippi), 1865 (as *Comptosia*); *lugubris* Rondani, 1863 [=*lugubris* (Philippi), 1865 (as *Comptosia*)]; *manca* Edwards, 1934; *minuta* Paramonov, 1931 (as *Lyophlaeba*); *montana* (Philippi), 1865 (as *Comptosia*); *pallipennis* Paramonov, 1940 (as *Lyophlaeba*); *parbifasciata* Paramonov (as *Lyophlaeba*), 1940; *philippii* Paramonov, 1931 (as *Lyophlaeba*) [=*bifasciata* Philippi, 1865, not Macquart]; *pictinervis* (Rondani), 1863 (type of subgenus: *Macrocondyla*), [=*vulgaris* (Philippi), 1865 (as *Comptosia*)]; *setosa* Paramonov, 1940 (as *Lyophlaeba*); *transandina* Edwards, 1934.

Genus *Ylasoia* Speiser

FIGURES 148, 367, 659, 670, 985, 986, 987

Ylasoia Speiser, Zool. Jahrb., Jena, vol. 43, p. 213, 1920. Type of genus: *Anthrax pegasus* Wiedemann, 1828, by original designation.

Long-winged, black-winged flies of medium size in which there are 2 irregular hyaline crossbands on the wing, one near the base, one near the apex, in addition the middles of some posterior and apical cells are faintly paler; the color of the wing is much blacker than commonly seen. There are only 2 submarginal cells and the radial loop is long. There is a faint, narrow, gray, medial vittae down the abdomen, a similar one on the mesonotum besides on each side a sublateral, wider, orange vitta.

The occiput is very long and bilobate above, the third antennal segment long and attenuate, the proboscis very short, quite stout with large muscoidlike labellum.

The pteropleuron bears a tuft of hair posteroventrally and there is a small tuft on the metapleuron behind the spiracle but none in front. Lateral metanotum bristly.

The most distinctive character of this genus apart from the patch of pile on the pteropleuron lies in the third posterior cell which marginally is greatly narrowed.

Three species have been assigned to this south Neotropical genus; two of them may perhaps be subspecies. Males have the hyaline areas more reduced and the second hyaline band more of a chalky white. The affinities of the genus seems to be nearer the Australian genus *Oncodosia* Edwards, than to the South American genera like *Lyophlaeba* Rondani.

Length: 10 to 12 mm., of wing 13 to 15 mm.

Head, lateral aspect: The head is hemiglobular with the occiput on the upper part of the head and unusually reduced below. The occiput is thinly covered with brown pollen, the pile consists of abundant but not dense, fine, quite short, erect, black setate hairs, curved a little forward and extending to the eye margin. The eye is kidney-shaped, very deeply indented, the lower angle much more acute than the upper angle. There is a small, impressed rim behind the eye on the upper half. Central fringe of the occipital recess short and scanty, the cavity unusually large. The front is flattened, except on each side in front of the antennae, where it is slightly raised. Pile abundant but not dense, moderately long, fine, erect, and black. The face in lateral aspect is not at all produced and is scarcely visible.

The proboscis is short, stout, and robust but with an exceptionally large, fleshlike labellum, the inner surfaces of which have transverse grooves. The labellum is at least as long as the basal part of the proboscis; it does not reach quite to the antennae. Palpus stout with a short basal segment more or less fused to the slightly longer clavate, apical segment. The latter bears a few ventral hairs. Antennae attached a little above the middle of the head, they are elongate, the first segment stout but much more slender than in related genera; they are slightly widened at the apex medially and bear apically at the outer edge a small, short, beadlike segment. First segment with numerous appressed, fine bristles and bristly setae above and much longer, appressed, coarse, black bristly pile below. Second segment with fine setae on all sides. Third segment at the base no wider than the second segment, slightly widened just beyond the base, then gradually tapered to near the apex where the thick styliform portion has parallel sides and bears a small, short spine. The third segment is slightly longer than the first two segments combined.

Head, anterior aspect: The head is circular, as wide as the thorax, the eyes in the females are separated at the vertex by about twice the width of the ocellar triangle. Ocellar triangle short isosceles; the tubercle is a little raised and bears fine, moderately long, black hairs similar to that of the front. Whole front sepia brown pollinose. Antennae rather widely separated at the base with a triangular, tonguelike extension of the front between the antennae. The oragenal recess is wide but rather shallow. It extends to and between the antennae, and is deeper posteriorly. The lower sides of the face shelve inward almost vertically beginning on the lower third of the eye. The tentorial fissure is quite deep and the lower wall of the oragenal recess quite thin, and not extending upward to the level of the eye. Sides of the face throughout with a few scattered, long, fine, erect hairs. Gena small and narrow and neither occiput nor gena at all extensive below the eye. The occiput viewed from above shows the lobes to be quite tightly apposed with a punctate depression in front of the fissure and behind the ocellar tubercle; the latter extends a short distance behind the posterior eye corners, which are, however, rounded forward.

Thorax: The thorax is dark brown with clay colored pollen on most of the pleuron and remaining pleuron with darker pollen. The thorax is wider posteriorly, the mesonotum low, slightly convex posteriorly and anteriorly, flattened before the scutellum and without any intermediate fissure. The mesonotum is black with pale, strikingly colored vittae of quite dense pollen, besides a median, linear, gray vittae there are 2 widely separated, rather wide, posteriorly divergent, bright, brownish yellow vittae running from the anterior margin and disappearing in an area of brown pollen just before the scutellum. In between and on each side of these yellow vittae there are 3 black stripes bearing very dark pollen, and the lateral margin is light brown. The pile of the mesonotum consists of scanty, fine, erect or suberect hairs of no great length, but the anterior mesonotal collar consists of spikelike black, erect bristles extending over the humerus, especially anteriorly. The humerus is large and swollen, with the posterior surface high and vertical and bearing a posterolateral tubercle. Notopleuron with 2 stout, long bristles. Postalar callosity with 2 long bristles and a few bristly hairs. Scutellum large, the posterior margin evenly rounded, the whole opaque black, the sparse discal pile like that of the mesonotum but longer, the margin with numerous, longer, slightly stiffer bristles and a little short, curled, yellow pile. Whole mesopleuron pilose with scattered, fine, erect, shining black hairs becoming slightly bristly above. Upper sternopleuron with a few similar, appressed hairs. Metapleuron in front of the haltere without pile but metapleuron behind the spiracle with a tuft of 10 or 12 hairs and the posteroventral pteropleuron with a tuft of about 25 long, erect, slender, bristly hairs. Hypopleuron without pile. Squama with a wide band of pile but very loose and bristly in character. Halteres long, the elongate club flat.

Legs: The legs are stout, the anterior coxa quite long, three-fourths as long as the anterior femur. All the femora have appressed, shining, blackish setae and a scanty fringe of fine hair below. Anterior femur shorter than the middle one and the middle one shorter than the third femur. The femur bears conspicuous, sharp, stout, rather strongly appressed, anteroventral bristles. They vary somewhat in numbers but there are usually 6 or 7 present. Other femora with at most 1 or 2 bristles, sometimes none. Anterior tibia stout with appressed setae and with rather well developed, short, oblique bristles. There are 10 or 12 anterodorsally, 10 slightly longer ones posterodorsally, 8 posteroventral, and 6 or 8 ventral or anteroventral. Middle tibia similar, the bristles slightly longer and slightly stouter, and with 5 rows, the anterodorsal row with 14, the others 6 to 8. Hind tibia similar, the bristles slightly longer, and with 4 rows, the anterodorsal row with 10, the dorsomedial row with the same number, the ventromedial row with 5, and the anteroventral row with 5 bristles. Tarsi long but stout, strongly spiculate below. Claws sharp, strongly arched, pulvilli long, rather broad and the arolium long.

Wings: The wings are rather broad but long, much longer than the abdomen, somewhat narrowed at the base, in part hyaline but more extensively blackish. The black color is due to very dark, sepia brown coloration, and dark brown villi, and appears even blacker because of the dense, nearly white villi of the sharply contrasted pale bands and spots. There are only 2 submarginal cells. The radial loop and prolongation of the marginal cell is extensive. The first posterior cell is narrowed apically, the third posterior cell is equally narrowed apically, and the anal cell is as widely open as both these posterior cells. The anterior crossvein enters the discal cell at the outer sixth, the axillary lobe is narrow, the alula wanting, the ambient vein complete. There is a continuous, pale, vertical stripe across the wing on the outer third, some small, diffuse, pale spots in the centers of most cells, a large, elongate, pale spot just before the apex and lying in the outer half of the second basal cell. There is a larger spot occupying the whole of the middle of the anal cell, continued across the anal vein in wider extent to occupy the middle of the axillary lobe.

Abdomen: The abdomen is elongate oval and in the middle wider than the thorax. It is opaque black with dense, very dark brown pollen, becoming greenish when viewed from the rear and with a narrow, medial vitta down the abdomen composed of confluent, wedgelike spots of pale brownish or greenish-yellow pollen. Pile of the abdomen scanty, fine, black, suberect, bristly in character. There is similar pile a little more abundant, longer, and erect at the lateral margin of the first tergite and tufts of appressed, similar pile, some of it shorter, some of it longer, on the lateral margins of the remaining tergites but never comprising massive tufts on the tergites themselves. I find 7 well-developed tergites visible above in the females, but arising from the remaining recessed tergites and sternites, there is below the seventh tergite, a great mass of tangled, dense, pale pile forming an enclosure about the female terminalia; the amount that is exposed varies greatly, sometimes very extensive and again tucked within the terminal recess of the abdomen.

Material available for study: A series of individuals of the type-species collected in southern Brazil.

Immature stages: Unknown.

Ecology and behavior of adults: I have seen no record of the behavior of these flies.

Distribution: Neotropical: *Ylasoia pegasus* (Wiedemann), 1928 (as *Anthrax*).

Genus *Doddosia* Edwards

Figures 149, 155, 431b, 698, 722, 931, 932

Doddosia Edwards, Encycl. Ent., Paris, ser. B. II, Dipt., vol. 7, p. 103, 1934. Type of genus: *Doddosia picta* Edwards, 1934, by original designation.

Australian bee flies of medium size which show a somewhat greater resemblance to *Lyophlaeba* Rondani than to *Comptosia* Macquart and which possess the large tuft of long, soft, white pile on the anterior metapleuron in front of the spiracle, a characteristic of the South American group of related genera. The sides of the abdomen are parallel, the costa is slightly widened before the middle. The attenuate third antennal segment is styliform at the apex, the proboscis is more than 1½ times as long as the oral recess. Wings are banded but the brown color is very diffuse; radial loop of only medium length; only 2 submarginal cells present; anal vein straight, except at apex, the alula short and plane.

Known only from Northern Queensland. The type material came from Townsville; the material I have collected came from near Dimbula, Queensland.

Length: 12 to 14 mm.; wing length 12 to 14 mm.

Head, lateral aspect: The head is hemiglobular; the occiput is comparatively extensive above, strongly sloping inward and twice as long dorsally as below. The pile of the occiput is fine, flat-appressed, sericeus, slightly flattened and reaching the eye margin. The interior collar is quite short and rather scanty. The front is short and, together with the face, does not extend beyond the eye in the lateral aspect. The frontal pile is abundant and coarse and obliquely turned downward and rather more dense in females. The eye is reniform with a quite deep, broad indentation below the middle; the angle above is equal to the angle below. The palpus is composed of a very short, fused basal segment and a much longer, curved, clavate, apical segment, which has a scanty fringe of long, fine hairs below. The proboscis is short and quite stout, the labellum large, flat, and elongate oval with interior rows of small tubercles; proboscis a little longer than the head and at rest extends upward obliquely between the antennae. The antennae are attached at the middle of the head. The first segment is large and swollen, widened inwardly at the apex and bears dense, long hairs below, and shorter setae above at the apex, besides longer hairs dorsolaterally. The second segment is much smaller, beadlike, strongly divergent; the third segment is not as wide as the second but gradually attenuate from the base to the styliform apex. This segment is as long as the first two segments together and bears a minute, apical spine.

Head, anterior aspect: The head is circular, as wide as the thorax. The eyes in the male are holoptic or virtually so for a distance a little greater than the vertical triangle. They are separated at most by half the width of an ocellus. The male front is an equilateral triangle, pollinose, with scanty, long, yellow pile laterally and a bare streak down the middle. Front extended forward as a triangular tongue between the antennae. In the females the eyes are separated by about 3 times the distance between the posterior ocelli. The front in females has similar pile to the male. Ocellar tubercle small, low, long-isosceles. The oragenal recess is deep throughout and is slightly narrowed anteriorly, and it ends a considerable distance from the antennae but the inner foveate recess is widest anteriorly and narrows below. This is because the walls of the oragenal recess are thick and wide but have an odd, curious, oval hollow on each side on the lower half. Tentorial fissure moderately deep, beginning on the lower half of the head, and the whole sides of the face densely obscured by long, extensive pile. The face medially is more extensive and vertically it is nearly but not quite as long as the first antennal segment. The antennae rest upon a placodelike, basal sclerite. The gena is moderately extensive below the eye, polished, shining, without pile, whereas the pile on the sides of the face is strongly directed downward. The pile on the sloping part of the face, adjacent to the eye margin, and along the middle of the oragenal recess is directed straight outward. Viewed from above the occipital lobes are tightly adjacent. The occipital recess is rather large.

Thorax: The thorax is compact, the mesonotum low and gently convex, faintly pollinose, densely covered with glittering, golden, curled, flat-appressed, slightly flattened pile. The thorax also has a considerable amount of fine, erect, brownish-golden pile which anteriorly becomes quite dense, especially in the male. There is similar, long, brownish-golden pile distributed very densely along the anterior mesonotal margin, on the whole of the humerus, which is much obscured, the whole notopleuron, and shorter but densely on the postalar callosity. The notopleuron has 2 bristles. Scutellum large, triangular, with pile corresponding to the mesonotum. Pleuron everywhere densely pollinose and conspicuously pilose. Whole of the mesopleuron and almost the whole of the sternopleuron with tangled mats of long pile. Only the pteropleuron and hypopleuron lack pile. The lateral metanotum, the anterior metapleuron in front of the haltere especially, and the metapleuron behind the spiracle are all rather densely, long pilose. Squama with a wide, plumose fringe.

Legs: The legs are well developed but with the first 4 femora scarcely thickened basally. All the femora bear flat-appressed yellow scales and a few fine, suberect hairs posteroventrally; the anterior femur has 1 or 2 setae anteriorly, middle femur likewise; the hind femur has 2 or 3 weak anteroventral bristles and a few minute, anterodorsal setae on the outer half. The anterior tibia has 8 fine, oblique, slender bristles anterodorsally, 6 others posterodorsally and 8 equally minute bristles posteroventrally. The middle tibia has 6 slender bristles anterodorsally, 6 others slightly more stout posterodorsally, and with 6 posteroventrally. The hind tibia is long and slender; it bears 12 short, moderately stout anterodorsal bristles, 9 fine, anteroventral bristles, 8 fine, dorsomedial bristles, and 18 fine, oblique, ventromedial bristles. The anterior basitarsi are as long or longer than the remaining segments and with oblique, spiculate setae below. The claws are slender, but wide at the base, sharp at the apex, rectangularly curved at the apex and with long, slender, spatulate pulvilli, and short, triangular arolium.

Wings: The wings are much longer than the abdomen, large, with the anterior border brown extending to the end of the second vein. The brown color extends obliquely across the wing to just beyond the third vein at the origin of its anterior branch and again extends all the way across the wing on the outer half of the discal cell to the end of the third branch of the medius. Also, the whole of the first and most of the second basal cell is brown. The second vein forms a well-developed radial loop extending to the end of the marginal cell outward and downward. There are only 2 submarginal cells. The first posterior cell and anal cell are widely open. The anterior crossvein enters the discal cell at the outer fifth. Axillary lobe well developed, ambient vein complete, the alula 3 times as long as wide, its posterior margin plane.

Abdomen: The abdomen is elongate, longer than the thorax, and with parallel sides in both sexes. It is feebly shining brownish black with 7 well-developed large tergites in the male and a short eighth tergite recessed under the seventh but a little longer laterally, in addition there appears to be an equally short, hood-like, ninth tergite overhanging the terminalia. Females with 7 well-developed tergites. Any additional ones are completely concealed and the female terminalia densely enclosed and laced about by long, matted, flattened, golden hairs. The pile of the abdomen consists of rather numerous, quite long, erect, golden hairs on the first three tergites, never dense, becoming shorter and fewer posteriorly and with a few very fine, blackish hairs appearing on the last three tergites. In addition, however, there is a conspicuous, posterior border of flat-appressed, curled, golden, flattened pile on the first seven tergites which is somewhat more extensive near the lateral margin, especially more extensive on the fourth tergite and beyond. The male terminalia are exposed ventrally, the basistyle large and adjacent.

The male terminalia from dorsal aspect, epandrium removed, show a figure rather similar to *Comptosia* Macquart. The dististyli are quite large, very blunt, and truncate apically. The epandrium is reduced in size and is narrow. In all of these the epandrium shows a strong basiventral lobe.

Material available for study: A series of males and females of the type-species that were collected by the author near Dimbula, North Queensland, September 1953.

Immature stages: Unknown.

Ecology and behavior of adults: The flies that I have collected were alighting on damp sand in a small, otherwise dry, stream bed.

Distribution: Australian: *Doddosia picta* Edwards, 1934.

Ulosometa, new genus

Figures 431c, 929, 930

Type of genus: *Ulosometa falcata*, new species.

Large, relatively bare flies with peculiar wings. They are reddish brown with black mesonotum, its borders reddish. The abdomen is rather flattened, wider than the thorax, and narrowed only beyond the fifth tergite; 8 tergites are visible above in males. The pile is fine, short, yellowish, scanty, and curled like wool, and appressed.

These flies are related to *Ylasoia* Speiser and possibly to *Oncodosia* Edwards, but differ in the wing and wing venation. The apex of the wing is hook-shaped as in certain nemestrinids and has a somewhat similar pattern. They are sharply maculate with a reddish-brown foreborder and also with an irregular broken series of vertical bands: one subbasal, one beyond the middle, and a third along the apical border to the wing apex. Venation peculiar and erratic There are 3 submarginal cells; the anterior crossvein lies almost at the end of the long, widened, posteroapically hollowed discal cell; first posterior cell greatly narrowed on the wing margin; third posterior cell also much narrowed on the margin from both directions, leaving it short and club-shaped above and obtruded forward into the end of the discal cell. Anal cell open. Uncolored areas clear except for brownish-yellow villi. In males the eyes are narrowly separated by the thickness of the second antennal segment; first segment large, short, swollen; third segment elongate, slender, tapered, narrow. Male dististyli elongate and protruding to the left.

These flies are known from a single species from Tucuman, Argentina.

Length: 17 mm.; wing 16 mm.

Head, lateral aspect: The head is comparatively short from the lateral aspect. The occiput while prominent is greatly rounded and sloping backward and inward to the rim of the moderately large circular cup. The eye is reniform, indented, but not deeply indented, and without a bisecting line. The upper eye facets in the male are scarcely enlarged. The eye is more than twice as high as long. The face does not protrude forward beyond the eye margin, but the entire face and the upper genae are densely covered with a very coarse, erect, reddish, bristly pile, with similar pile intermixed upon the lower surface of the first antennal segment. The occipital pile is similarly colored but scanty, short, and flat-appressed.

The proboscis is short and stout with a long, robust labellum and extends a little beyond the apex of the oral cavity, to rest between the large, stout, first antennal segment. Palpus composed of 2 segments but of no great length, its pile minute. The front is flattened and depressed down the middle and bears a lateral patch of dense, gray, bristly pile diverging outward along the eye margin. The antennae are attached just above the middle of the head. The first segment is 3 times as long as the second segment and more than twice as wide as the second segment. It bears dense, reddish, bristly pile laterally, medially, and still longer dense, red pile ventrally. The third segment is less wide than the second, tapers rapidly from the base to the apex which has a minute, spine-tipped microsegment. The third segment is very little longer than the first segment.

Head, anterior aspect: The head is much narrower than the thorax and even narrower than the humeral space. The males are dichoptic and eyes separated by nearly the width of the ocellar tubercle. The ocellar triangle is elongate, distinctly raised above the eye margin, and the posterior ocelli are set forward a short distance from the posterior eye corner, and this tubercle bears a dense patch of reddish, bristly pile curving forward. The antennae are separated by rather more than the thickness of the second antennal segment and has a distinct groove between them. The oragenal opening is elongate and comparatively deep with vertical and hollowed sides and the genae are shelved deeply inward along the conspicuous tentorial fissure. Behind the proboscis the ventral occiput is excavated and striated. Upper genal pile short, erect, bristly becoming longer as it merges with the pile of the face. From above the occipital lobes are adjacent with shallow crease between.

Thorax: The thorax is quadrate, slightly wider across the prominent, swollen, postalar callosity. The humeral callosity is prominent, sharply margined laterally, vertically platelike behind. The pile of the mesonotum is scanty, erect, bristly, and short. There are 2 or 3 distinct, short, black bristles on the notopleuron. Center of mesonotum broadly opaque black with reddish border on all sides and a large, dull, red rectangle in front of the scutellum. The scutellum has a broad, deep, transverse fossa at the base which is broadly rounded below. It is barely notched medially at the apex, the whole surface reddish with median stripe of pale, brownish-yellow or whitish pollen, and a few scattered, appressed, short, bristly hairs, especially on the margin. The pleuron is almost entirely brownish red, dully black in the lower anterior corner of the metapleuron and black above the coxae. There is a small tuft of long hairs in the lower corner of the pteropleuron; the metapleuron is bare, but the metanotum has dense, red pile. There is a tuft of red pile below the haltere which is situated some distance behind the spiracle. The plumula is absent, but there is a curious tuft of rather dense, long, fine pile extended outward from an oval patch on the anterolateral portion of the postalar callosity. Squama with a broad, wide border of long, fine pile.

Legs: Legs are rather small, the tarsi short, especially on the middle and hind pairs, where they are distinctly shorter than the tibia. The femora are shining and without scaliform pile but with scattered, appressed, short, bristly hairs. The tibiae likewise have appressed short, bristly, reddish yellow hairs and also several rows of short, strongly appressed, fine, black bristles, 15 to 20 in each row. The black ventral spicules of the tarsi are sharp, conspicuous, and appressed. Pulvilli prominent; arolium stublike and distinct.

Wings: The wings are elongate, widest just beyond the middle, narrowing toward base and apex; they are sepia brown in color, etched by yellowish veins, and the edges of the mottled pattern are rather sharp. The auxiliary vein and the first vein end close together. The costal border beyond is strongly thickened; marginal cell rounded and slightly protuberant at the apex. There are 3 submarginal cells. The anterior crossvein enters the discal cell almost at the end of that cell;

first posterior cell much narrowed at the wing margin; third posterior cell almost equally narrowed; anal cell widely open. The wing is densely covered with yellowish villi on the hyaline areas. The reddish sepia color covers the anterior border of the wing backward to the anterior border of second basal and discal cells; the wing is more yellowish in the middle, and more brownish obliquely across the base of the second and third basal cells, and the axillary lobe. There is also an irregular, vertical stripe at the end of each of these 3 cells, which extends over narrowly into the base of the discal and fourth posterior cells. There is a more prominent, oblique, crossband covering the outer third of the wing and edging both sides of the third posterior cell, the anterior border of this cell, the whole apex of the discal, also the basal border of the first posterior cell and its anterior border; it continues to the anterior border of the wing, and then borders the anterior branch of the third vein, and the apex of the second vein. Base of the costa with a lobe instead of a hook.

Abdomen: Elongate, oval, a little convex, a little wider than the mesonotum; rather bare but possibly partly denuded; the pile is fine, flat-appressed, and reddish. The whole abdomen is reddish brown with obscure blackish bands basally on the third to fifth tergites becoming narrower and reduced on each succeeding tergite. There are 6 tergites, which are of nearly equal width. These are only slightly reduced progressively in length. The seventh and eighth tergites, while conspicuous, are more narrowed and form a shovellike hood for the terminalia. The male terminalia are terminal, large, conspicuous, toothlike and extended to the left.

The male terminalia from dorsal aspect, epandrium removed, show a body which is long, and somewhat wider than *Bryodemina* Hull. Likewise, it has nearly parallel sides and therefore with very little taper. The base of the aedeagus is large, the lateral apodeme is much larger and triangular. The epiphallus is constricted in the middle and like *Bryodemina* Hull not reaching the end of the basistylus. The basistylus itself is quite long but stout. Epandrium triangular, the basal ventral lobe much thicker.

Material available for study: Two males, the unique type and paratype, from Tucuman, Argentina, May 7, 1918. Species description incorporated in the generic description; type in the National Museum of Natural History.

Immature stages: Unknown.

Ecology and behavior of adults: Not on record.

Distribution: Neotropical: *Ulosometa falcata*, new species, as described above.

Genus *Oncodosia* Edwards

FIGURES 152, 153, 427, 699, 723, 933, 934, 935

Oncodosia Edwards (Encycl. Ent., Paris, ser. B. II, Dipt., vol. 7, p. 101, 1934. Type of genus: *Anthrax patula* Walker, 1849, by original designation.

Large flies rather readily recognized by the pattern of the wing and its pointed apex and relatively broad base. Moreover, the alula is large and extensive, and the anal vein is distinctly bowed backward in the middle. There are normally only 2 submarginal cells, the anal loop is short, but in typical species the discoidal cell is acutely pointed apically. As to the wing pattern there is a brown foreborder ending with the first vein and extending obliquely backward to fill the base of the first posterior cell, whole of first basal cell but only narrowly along the edge of the discal cell; in addition there are 4 isolated spots along the posterior part of the wing.

Characteristically the abdomen quite broad, broader than the thorax, and has dense, rather flattened tufts of long hair along the lateral margins of the tergites in contrast to the short, appressed pile over the remaining dorsal surface. The antenna is short and attenuate but with a very fine stylelike, unsegmented extension at the apex.

These flies are only known from Western Australia. Length: 15 to 18 mm.; of wing 18 to 19 mm.

Head, lateral aspect: The head is hemiglobular with the occiput quite extensive and prominent above but strongly sloping inward. The pile of the occiput consists of very minute, short, erect or subappressed setae in scanty quantities. In addition, the central fringe on the border of the occipital recess is very short and appears to consist of only one or two rows of hairs. There is a curious, impressed collar quite near the ocular margin extending from near the vertex to the lateral indentation of the eye. The front in the male is a little swollen but from the lateral aspect scarcely extends beyond the eye. This is likewise true of the face. Pile of the front is dense, fine, bristly, and erect. Within the middle, between the widely separated antennae, there is a wide, curious patch of flat-appressed, golden pile. The eye is reniform and in spite of the shorter occiput on the lower half of the eye, the eye is shorter on this lower half because of the receding face.

Proboscis slender, a little widened at the base, the labellum flat and oval and somewhat pointed apically. The proboscis is longer than the head and in repose extends up and between the antennae. Palpus small, slender, a little clavate apically and with fine hairs below. The antennae are attached slightly above the middle of the head; the first segment is large, robust, barrellike, and although truncate apically and slightly oblique, it is not swollen inward at the apex. It bears very dense, bristly pile, especially bristly medially and above but a little longer and quite coarse ventrally. Second segment attached on the outer edge of the first segment, small and beadlike with numerous, short, black setae dorsally and laterally only. Third segment at the base almost as wide as the second segment. It is a little compressed laterally but from the lateral view it is short-conical, and bears a short, apical spine.

Head, anterior aspect: The head is circular, much more narrow than the thorax and relatively small. The eyes are separated in the male by a distance equal the width of the posterior ocelli from their outer edges. The ocellar tubercle is a little swollen and short isosceles. The posterior ocelli very narrowly separated from the eye. The antennae are unusually widely separated and far apart; they are actually closer to the eye margin than they are to each other, the distance between being almost equal the maximum width of the greatly swollen first segment. The oragenal opening is shallow, except below where it is moderately deep; it is also narrowed above and fails to reach the antennae by a distance equal the width of the first antennal segment. This middle part of the face is bare, but on either side there is a large, extensive patch or clump of numerous, long, fine, erect, rather stiff hairs, black ones above and yellow ones below. At the lowest point of the face where it shelves deeply inward to the deep tentorial fissure, there is a dense patch of long, erect pile. The genae are narrow, a little more extensive at the bottom of the eye but much less so than in *Comptosia* Macquart. The eye is pilose and conspicuously pilose on the lower half. There are only a few scattered hairs on the upper third of the eye and these upper facets are scarcely enlarged in the male. The head viewed from above shows the occipital lobe to be very tightly appressed and the depression shallow. The ocellar tubercle extends behind the eye.

Thorax: The thorax is large, the mesonotum quite low, rather strongly flattened on the posterior half and only slightly convex anteriorly. It is much wider posteriorly than anteriorly. It is pollinose with narrow, medial, linear victae, feebly shining behind and very densely covered with short, suberect, or slightly appressed, coarse, pale pile, which becomes more matted and appressed laterally, but changes along the anterior margin to a dense collar of long, erect, bristly pile, which extends over the broad humerus. Whole lateral margin from humerus to wing also has many reddish bristles mixed with pile. The notopleuron has 5 more stout, slightly longer reddish bristles. Postalar callosity with an extraordinarily long, bushy tuft of dense, fine pile on the sides of the vertical surface behind the wing, with shorter pile in front above and with 8 or 10 red and some black bristles on the dorsolateral border of the callosity.

The scutellum is quite large tending to be very obtusely triangular and quite rounded apically. At the base, it and the preceding mesonotum are both sharply declivitous with deep, trenchlike furrow. Postmargin of the mesonotum with long, fine, curled pile. Scutellar disc with dense, fine, erect, mostly black, bristly pile. The margin of the scutellum has short reddish and black bristles which are distinctly slender. Squama with a wide band of plumoselike pile. Pleuron quite thinly pollinose, the pteropleuron and hypopleuron bare, the whole of the mesopleuron with dense, long, fine pile and a cluster of slender red bristles dorsally curled upward. Upper sternopleuron with scattered, tangled pile, propleuron and prosternum with much long, bristly pile. The lateral metanotum is very short dorsally; it bears dense, long pile laterally. The anterior metapleuron is completely without pile in front of the haltere, but there is a conspicuous patch of long, fine pile behind the spiracle. Halteres small and slender.

Legs: The legs are stout, the tibiae are almost as thick as the femora. The first 4 femora are only slightly enlarged toward the base; all 3 pairs of femora are covered densely with fine, flat-appressed, shining, slender, reddish brown, flattened pile. Anterior femur with 3 or 4 fine, anteroventral bristles, and posteroventrally with some 10 still finer and shorter bristles and some additional fine pile. Middle femur with 10 longer, stouter, anteroventral bristles, but posteriorly and ventrally with only very fine, suberect hairs. Hind femur with 7 or 8 fine, slender oblique, anteroventral bristles, smaller than those on the middle femur and a few small setae ventrally. The anterior tibia is shorter than the middle tibia, the middle tibia shorter than the hind tibia. The hind tibia is as long as its femur and is slightly curved. Anterior tibia with several rows of minute, bristly spicules; there are 12 anterodorsally, 8 posteriorly, 6 or 8 ventrally, and the same number of much smaller ones posterodorsally. Middle tibia with similar bristles in length and number. Hind tibia likewise with the bristles short and slender, the numbers are also about the same as the middle tibia. There are 12 on the ventromedial row; those dorsally are scarcely longer than the numerous, fine, oblique setate hairs. There are at least 12 that stand out as weak bristles. Tarsi rather short, terminal segment slender. Claws slender, strongly bent on the apical half, the pulvilli long and slenderly spatulate. Arolium distinct but short.

Wings: The wings are quite large and long, much longer than the abdomen, very broad at the base with a large, extensive alula. The costa at the base is greatly thickened and expanded and has a conspicuous comb of setae in the male. The costal cell is wide both before and after the humeral crossvein. The wing is rather pointed, the marginal cell widened apically and bulging outward at the apex, although the radial loop is not deep. There are only 2 submarginal cells, the first posterior cell is narrowed at the margin; the anal cell is likewise open and also of about the same width as the first posterior cell. The anal vein is distinctly bowed back in the middle, therefore slightly sinuous. The anterior crossvein is very much oblique, entering the discal cell near the outer fifth. Anterior border of wing reddish brown, the red ending on the foreborder opposite the base of the second submarginal cell, then extending obliquely backward, it extends a little beyond the first basal cell, narrowly borders the discal cell in front and fills out the basal half of the second basal cell, extreme base of the third basal cell, and the whole of the alula. There are prominent brown spots below at the end of the second basal cell, along the discal cell at the furcation of the fifth vein, below at the end of the discal cell, and a round spot at the base of the second submarginal cell. Remainder of wing hyaline, thinly villose, ambient vein complete.

Abdomen: The abdomen is extremely broad, only a little convex, tending to be flattened especially on the posterior half, short oval. It is much wider than the thorax, shining brownish yellow with a series of small, blackish triangles basally in the middle of the second to the sixth tergite. The first tergite entirely black, except for linear, yellow hind border. This segment bears dense, long, fine, yellow pile extending even under the scutellum, subappressed across the middle, remarkably dense, longer, erect but fine laterally where it looks like cottony tufts. Wide middle portion of the tergites covered with fine, appressed, short, black and partly yellow pile, the black pile denser along the anterior margin of the second tergite and on all of this central area beyond the base of the second tergite the pile is quite sparse. Laterally, however, far out on the edge of the tergites there are very dense, lateral, flattened tufts of conspicuous pile; the pile of the posterior corners pale yellow tending to be more appressed backward and tending to narrowly follow the posterior margins inward on the outer third and the anterior corners with equally conspicuous tufts of black pile which tends to be less strongly turned backward. I find 7 well-developed, visible tergites in the male from above. There is an eighth completely recessed beneath the seventh and in addition a smaller, apically notched, recessed, ninth tergite forming a hood over the small genitalia. The apices of the two pieces can barely be seen beneath.

The male terminalia from dorsal aspect, epandrium removed, show short, almost triangular dististyli; the base of the aedeagus seems undifferentiated from the epiphallus. The lateral aspect shows a long, rather stout basistylus. The epandrium and the entire structure is very similar to *Doddosia* Edwards. It differs principally in the dististyli and the epiphallus. Like the other members of the Comptosini, the epandrium has a prominent basiventral lobe.

Material available for study: I have a male of *Oncodosia plana* Walker, which I collected in Western Australia, January 6, 1954. According to the figures given by Edwards the wing of *Oncodosia plana* Walker is almost identical with the wing of the type-species, *Oncodosia patula* Walker.

Immature stages: Unknown.

Ecology and behavior of adults: This species can be collected on low-growing flowers of *Leptospermum*.

Distribution: Australian: *Oncodosia ampla* (Walker), 1852 (as *Anthrax*); *patula* (Walker), 1849 (as *Anthrax*); *plana* (Walker), 1849 (as *Anthrax*).

Subfamily Exoprosopinae Becker, 1912

The flies of this very large subfamily fall within some 35 genera and many subgenera centered around the enormous and worldwide genus *Exoprosopa* Macquart with its 443 described species.

The outstanding characteristics of the subfamily lie within the condition of the posterior eye margin, which is strongly indented but is variable with respect to the presence or absence of a sharp notch or bisecting line. All of these eye characters need to be correlated with genitalic characters as well as the extraordinarily numerous divergences of wing venation. This involves a special morphological study of the genus *Exoprosopa* Macquart in particular, but also of related genera. Dissection of many genera of this subfamily has shown a distinctly unique characteristic of the epiphallus: from the dorsal aspect it is curiously a broad, apically rounded, tonguelike process proceeding from a swollen base.

I recognize 3 tribes: the Exoprosopini, with the majority of species; the Villini, with a very large number of species and like the former group, also worldwide; and the Villoestrini, a small group and a probable offshoot from the *Villa*-like ancestry, which are peculiar for the remarkable and complete loss of effective mouthparts. This is total in *Villoestrus* Paramonov, and virtually so in the other genera that fall here. These are found on 4 continents.

Tribe Exoprosopini

These flies of worldwide distribution are characterized by great diversity and have appropriately been divided into many subgenera. Most species are large and some are veritable giants among the bee flies,

Exoprosopinae

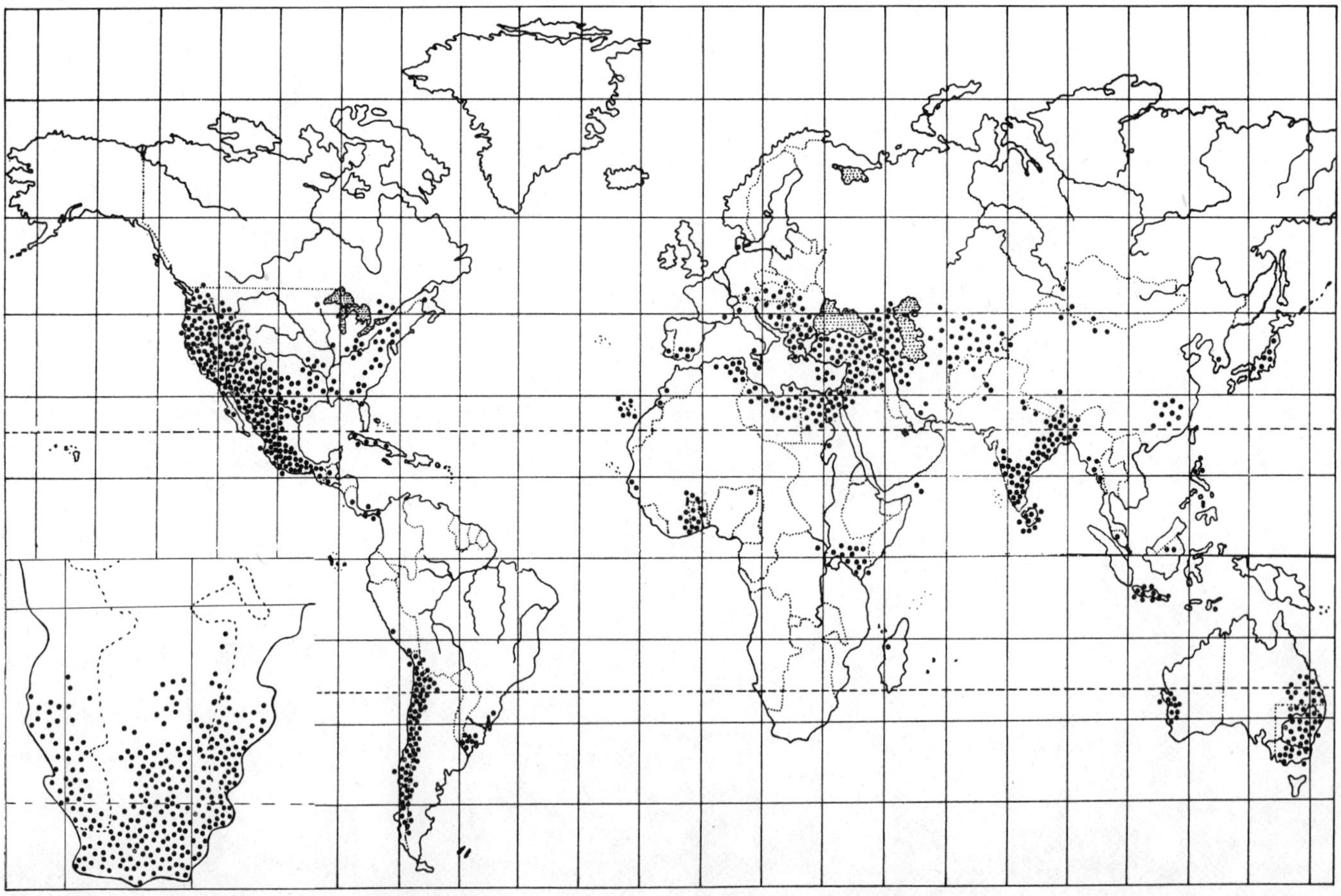

TEXT-FIGURE 45.—Pattern of the approximate world distribution of the species of the subfamily Exoprosopinae.

reaching a wing expanse as in *goliath* Bezzi up to 64 mm., in others with a wing spread of as little as 12 to 14 mm. In this tribe the more distinctive genera are *Isotamia* Bezzi (described under three names), *Litorrhynchus* Macquart, *Ligyra* Newman, and *Colossoptera* Hull. *Diatropomma* Bowden also is unusual. The genus *Exoprosopa* Macquart is one of the most remarkable in the entire family. It is worldwide in distribution, and several hundred species have been described and undoubtedly many more remain unknown. Also, much work remains to be done in the analysis of this one genus, and I believe when all of the terminalia can be completely dissected and delineated and compared, that it will be necessary to recognize additional genera and subgenera. There appears to be immense variety in the patterns of the wing and in all accompanying features.

Villini, new tribe

Bee flies ranging in size from 4 to 5 mm. in length to 16 or 17 mm. Antennae widely separated and the third antennal segment bulbose at base, the style short to long, the thickness variable and the style nonarticulate except rarely. Metanotal and metapleural pile tufts present, often dense. Squamal and alary scales narrow. Oral opening in some forms greatly reduced. Proboscis generally shorter than oral recess and with fleshy labellum but in some genera distinctly longer than head, more slender throughout and at apex. Anterior tibia varying from smooth, without bristles, and reduced in size, to normal in size, and with well-developed slender bristles or smaller spicules. Wings with 2 to 3 submarginal cells. Costal comb usually prominent. Pulvilli generally absent, occasionally present.

It is not easy to find a dependable means of separating the Villini group of genera from the Exoprosopini group of flies. Points of overlap occur in the number of submarginal cells; there are a few species of *Thyridanthrax*, like *ternarius* Bezzi and *laetus* Loew, which have 3 such cells yet are so clearly like *Thyridanthrax* in other respects that there is little doubt of their position. Many of the Villini are very distinctive: such are *Cyananthrax* Painter, *Diplocampta* Roeder, *Lepidanthrax* Osten Sacken. Also *Poecilanthrax* Osten Sacken, *Dipalta* Osten Sacken, and *Hemipenthes* Loew, and *Paravilla* Painter are each rather distinctive.

American species of *Thyridanthrax* Loew show the distinctively triangular face of the type-species, and in Europe this genus often shows a very blunt face and these I assign to *Tauropsis*, new subgenus, with *irrorellus* Klug as type. Also in European species the third antennal segment varies from the typical form with the stylate part abruptly formed and much narrowed and attenuate but far from linear, to those like *misellus* Loew and *semifuscus* Engel where it is long, scarcely tapered, and at the tip end thick and stout. Still others have the segment form typical but with a short, yet distinct microsegment as in *Exoprosopa* species, such as *stigmulus* Klug and *afer* Fabricius, which, however, are readily distinguished from *Exoprosopa* species by the untoothed claw and other differences. The variability of the third antennal segment is also well shown in such species as *Exoprosopa subfasciata* Engel, where it is quite long, of uniform thickness, not at all tapered, and lacks an apical style; this species should be assigned to *Corycetta*, new subgenus.

Four characters distinguish these two tribes, and the allocation of an unknown fly to one or the other depends upon the preponderance of characters. For example, in those rare species of *Thyridanthrax* which have 3 submarginal cells (the *ternarius* Bezzi, *laetus* Loew group) claws lack the basal tooth, ocelli are remote, or antennal style is undivided.

Host selections where known are quite diverse. Larvae of: *Rhynchanthrax* Painter parasitize species of *Tiphia*, *Tiphia* parasitizing white grubs; *Paravilla* Painter, bembecid wasps; *Chrysanthrax* Osten Sacken, scoliid wasps in turn preying on white grubs; *Thyridanthrax* Osten Sacken parasitic on tsetse fly puparia, and other muscoid flies, such as *Calliphora*, and living also more commonly in the egg cases of locusts; *Hemipenthes* Loew behaving as hyperparasites attacking species of *Ophion*, *Banchus*, and tachinids such as *Masicera*, which in turn attack nocturnal Lepidoptera; *Villa* Lioy, primarily parasites of nocturnal Lepidoptera, but also tenebrionid beetles in North America and northeastern Europe; *Dipalta* Osten Sacken parasitic on myrmeleonids; *Poecilanthrax* Osten Sacken parasitic on a variety of nocturnal Lepidoptera such as cutworms, sod worms, etc.

There has been much discussion by various authors of the taxonomic limits of *Villa* Lioy, sensu stricto. Much of this is due to overlapping in different taxa of characters that must necessarily be used in defining the genera and subgenera of what might be called the *Villa* complex in the wide sense (and for which I here erect the tribe *Villini*). In this tribe there are some 10 to 15 genera, of which 6 are of particular interest: *Villa* Lioy, *Hemipenthes* Loew, *Thyridanthrax* Osten Sacken, *Chrysanthrax* Osten Sacken, *Paravilla* Painter, and *Rhynchanthrax* Painter. Painter (1965) places all of these as subgenera of *Villa* Lioy. Engel treated *Thyridanthrax* Osten Sacken and *Hemipenthes* Loew as entirely separate genera, treating 48 Palaearctic species of the former genus and 10 species of the latter. Bezzi (1924) in his Bombyliidae of Ethiopia took the same view and so has Hesse (1956) in his monumental study of South African bee flies. Still more recently Bowden (1964) segregated these forms away from *Villa* Lioy; in spite of what appears to be overlapping, I believe Bowden's is the correct view.

While admitting the relatively close relationship of these several taxa, my own studies lead me to believe that systematic and bionomic studies are better served by leaving them as separate genera until such time as a complete study of the genitalia of all or nearly all species can be made. This provision reduces the use of trinomials. Certainly there is no doubt in my mind that the very interesting taxon *Rhynchanthrax* Painter is distinctive, separated by the characters discussed under that genus. And I feel that the same can be said of *Paravilla* Painter with its conical face. Probably facial contours, length of antennal style, and length of proboscis on the whole provide more valid relationships than wing patterns, which, to some extent, may tend to repeat themselves in certain genera. To assist in clarifying the above genera I have illustrated the faces, wings, and antennae of typical species of each.

Villoestrini, new tribe

In this group I place four remarkable genera, all of which show profound reduction of the mouthparts. These flies seem to be nearest related to the genus *Villa* Lioy and a probable offshoot from this ancestral line. The male terminalia suggest *Villa*, and Hesse (1956) finds a resemblance to the *leucochila* species group of *Villa*. These flies are widespread, ranging from South Africa, where three of the four genera are found, to Palestine and to the southwestern United States and Brazil. Besides the reduced mouthparts they are notable for the small antenna, the blunt, short, rounded, and retreating face, absence of pulvilli, large costal comb, etc. The hosts for one species proves to be cossid larvae.

The hypopygium in the Exoprosopinae is rather readily characterized by the extraordinarily long, quite wide, tonguelike epiphallus. Nearly all of the genera have the hypopygium wider on the basal half, distinctly narrowed apically, usually the whole apical half dominated by the conspicuous epiphallus. This will be especially noted in genera like *Oestranthrax* Bezzi, *Astrophanes* Osten Sacken, *Dipalta* Osten Sacken, and many others. Also the dististylus is characteristically subapical in position, almost always small, with a distinct dorsoapical incision and upturned hook. This is less conspicuous in *Astrophanes* Osten Sacken and a few other genera. Such genera as *Chrysanthrax* Osten Sacken, *Hemipenthes* Loew, and *Lepidanthrax* Osten Sacken have a large, wide, rather deep, rounded, anteroventral incision on the epandrium.

KEY TO THE GENERA AND SUBGENERA OF THE EXOPROSOPINAE

1. Mouthparts greatly reduced and vestigial and sometimes entirely absent (Tribe Villoestrini) 2
   Mouthparts normal and functional 4
2. Mouthparts completely absent, the oral opening reduced to a minute punctate slit, and even the palpus wanting; pile of face setae. Occiput reduced near vertex.
                                         ***Villoestrus* Paramonov**
   Minute vestiges of the proboscis remain and sometimes traces of the palpus as well 3
3. Head reduced in size, narrower than the mesonotum at the level of the wings. Third posterior cell relatively short, its base only a little way before the middle vein separating it from the fourth posterior cell, and usually with a short stump of vein projecting into the discal cell from this base. Second vein strongly recurved apically and with a trace of a stump vein basally. Third antennal segment with swollen base, small, short, onion-shaped, the apical piece long and narrow. Occiput quite prominent and tumid near vertex . . . . ***Oestranthrax* Bezzi**
   Head as wide or wider than the mesonotum at level of wing. Third posterior cell relatively long, its base near the apex of the second basal cell and usually without a stump vein into the discal cell. Second vein not strongly recurved apically and without a stump. Third antennal segment more conically stout, the base not strongly constricted, the apical piece not long and narrow.
                                          ***Marleyimyia* Hesse**
4. Second vein arising abruptly at a right angle from main stem vein and arising a long distance basal or proximal to the anterior crossvein and arising obtusely rounded anterobasally. Squama fringed with hairs, no scales present. Wings entirely or almost entirely hyaline. (Tribe Xeramoebini) . . . . LOMATIINAE
   The second vein arises quite rectangularly from the stem vein, but usually arises exactly at, or barely before, or barely beyond the anterior crossvein, with which it often appears confluent; distance out of alignment seldom more than the vein's thickness; sometimes as much as the length of vertical part of second vein. Origin of second vein either rounded anterobasally or with a backward stump vein. In about three-fourths of all species the wings bear characteristic, distinctive patterns of spots and bands, with at least the basal part of wing brown . 5
5. Third antennal segment with a small, short, suture-separated microsegment always bearing a tuft, usually a conspicuous tuft, or circlet of apical hairs. Face always strongly retreating and rounded below. Oragenal cup short, its walls thin. Gena with a wide, deep tentorial pit or pocket on each side of the reduced upper part of oragenal cup. Proboscis always short and of no great length or size. Metapleuron without pile. Squamal fringe composed of hairs. Claws without basal teeth.
                                            ANTHRACINAE
   Attenuate part of third antennal segment with or without suture separated, spine-tipped style, but never with a pencil or circlet of hairs at the apex of style or microsegment. Many genera and species fall into each division. The metapleuron is pilose and the fringe of the squama is composed of scales, narrow or wide. Claws with or without a basal tooth. Two, 3, or sometimes 4 submarginal cells present. Oragenal cup comparatively long and narrow, the proboscis enclosed with its recess except when projecting beyond the face; proboscis often much longer than the oragenal cup. No deep tentorial pocket present above the genae. (Exoprosopinae) 6
6. Third antennal segment with a distinctly separated spine-tipped or bristle-tipped style or microsegment, or variable length and thickness, which is separated by a suture, distinct from the base of the third segment. Claws with a basal tooth, long, or short. Three or sometimes 4 submarginal cells present. Ocellar tubercle far from vertex. If a separated style is present together with only 2 submarginal cells, the claws lack teeth; seen in couplet 5B. (Tribe Exoprosopini) 27

Third antennal segment without any sort of a separated style whatever; it bears only a short spine of variable thickness present at stylate apex. Two submarginal cells or 3 with great rarity. Claws without a basal tooth, a small obtuse tubercle rarely present. Ocelli placed near the vertex. Proboscis long or short. (Tribe Villini) . . 7

7. Anal cell closed, in or before the margin of the wing. Eye of male separated by less than the width of the ocellar tubercle. Swollen base of third antennal segment very short and wide and saddle-shaped, apical part slender. Face and occiput with broad scales; apex of abdomen broadly silvery scaled . . . . **Astrophanes Osten Sacken**
Anal cell open. Eyes of male separated by much more than width of the ocellar tuberacle 8

8. Alula absent; axillary lobe also greatly narrowed and reduced. Anal cell widely open. Large flies with very large, long, entirely dark blue wings that are disproportionate to the body. Discal cell narrow, acute apically, with spur vein below. These flies are dark, shining blue, black pilose, the anterior mesonotal pile orange. Face strongly conical, abdomen elongate oval, quite bristly on the sides and scales absent . . . . **Cyananthrax Painter**
Flies with the alula present or if narrowed or reduced they are bee flies entirely different in appearance 9

9. Three submarginal cells present 10
Two submarginal cells present 15

10. Wings and other features of face, claws, and antennae as in *Thyridanthrax* Osten Sacken, but regularly with 3 submarginal cells; group *ternarius* Bezzi, *lactus* Loew, etc.
Thyridanthrax Osten Sacken
Wings venation and antenna quite different from *Thyridanthrax* 11

11. Second vein very strongly sigmoid on its outer section, both basally and apically 12
Second vein narrowly curved and marginal cell slightly or moderately bulbous at the apex 13

12. Wing narrowed at base somewhat as in *Mancia* Coquillett, but alula and axillary lobe well developed, but both curves of the last section of third vein broadly rounded. Third antennal segment very strongly onion-shaped at base and the style rising abruptly . **Diplocampta Schiner**
Wing not conspicuously narrowed at base. Both curves of last section of the second vein rectangular, the base of third submarginal cell with a stump vein above a rectangular supernumerary crossvein. Face bluntly conical, with flattened hairs . . . . **Dipalta Osten Sacken**

13. The third antennal segment is conical at base and gently tapered into the antennal style. Face with scales . . 14
Base of third antennal segment small, short, onion-shaped, the style quite long and thin, of uniform thickness. Face with dense, coarse hairs among which are a few flattened hairs. Abdomen with coarse, dense hairs along the sides and a few flattened hairs and with flattened hairs along the posterior margins of the posterior tergites. Basal half of wing generally dark sepia, the apex with one or more spots . . . . **Stonyx Osten Sacken**

14. Face bluntly rounded, with short, erect hairs and scaliform hairs. Upper angle of third posterior cell with a long stump vein into the discal cell. Abdomen with scaliform pile. Anterior tibia quite without spicules. Wings hyaline.
Atrichochira Hesse
Face conical and with glittering scales. Upper angle of third posterior cell without stump vein. Abdomen with patches of silvery scales along the sides of the tergites. Anterior tibia with weak spicules or setae. Wing black with hyaline base. Discal cell relatively long.
Pseudopenthes Roberts

15. Face very strongly conical and pointed apically and covered with sparse, short, scattered setae. Anterior branch of third vein strongly bent backward and approximately at a right angle. Base of third antennal segment gently tapered and conical. If wing very large and abdomen unusually elongate, the subgenus *Agitonia* Hull.
Neodiplocampta Curran
Face rounded or conical. Curve of anterior branch of third vein rather shallow, at least not rectangularly rounded. Third antennal segment of various shapes 16

16. Abdomen, especially at the base and sides, and to a lesser extent on the dorsum with matted, broad, conspicuous scales which are often silvery. Proboscis projects well beyond the oral opening. Third antennal segment small, short, onion-shaped, the style long and slender. Contact of discal cell and fourth posterior cell as long or longer than base of fourth posterior cell.
Lepidanthrax Osten Sacken
Abdominal scales narrow, if present at all, and rarely silvery. Proboscis and antenna variable in form and length. Contact of discal cell and fourth posterior cell usually shorter than base of fourth posterior cell . . 17

17. Third segment of antenna short, small, onion-shaped, with long abruptly formed styliform portion which is linear, of uniform thickness, even threadlike. Face and proboscis, each long or short 18
Third antennal segment long or short conical, attenuate beyond the wide base; also rarely, long and thick and scarcely tapered. Face and proboscis likewise variable, long or short 21

18. Alula much reduced, almost wanting; axillary lobe very narrow. Face conical but blunt at apex. The proboscis extends a short distance beyond the oral cavity. Quite small, dark brown or black flies with hyaline wings and brownish yellow, and white or silvery scales on face, frons, thorax, and abdomen . . . . . **Mancia Coquillett**
Not such flies 19

19. Face produced, bluntly rounded or short conical and the proboscis distinctly extended well beyond the oral opening. Abdomen, especially at the base and sides and to a lesser extent upon the dorsum, is covered with matted, broad, conspicuous scales which are often silvery. Style of antenna long and slender. Contact of discal and fourth posterior cell as long or longer than base of fourth posterior cell . . . . **Lepidanthrax Osten Sacken**
Abdominal scales narrow, if present at all, rarely silvery. Face and proboscis of varying form. Contact of discal and fourth posterior cells usually shorter than base of fourth posterior cell 20

20. The proboscis extends beyond the face and oral opening by at least half the length of the oral opening. Styliform part of third antennal segment long and slender. Hypopleuron bare or nearly so. Wing generally with the basal half dark . . . . **Rhynchanthrax Painter**
Proboscis not projecting beyond oral opening. Face retreating and generally bluntly rounded. Hypopleural pile usually present. Front tibia generally spiculate. Base of wings blackish on the basal half or more with irregular margin to the black pattern posteriorly. Discal cell at the apex typically expanded, above and below, bulblike and obtuse. Color and wing pattern generally suggestive of *Anthrax* Scopoli species. Pulvilli sometimes present; metapleural fringe scanty and basal comb small.
Hemipenthes Loew

21. Wings hyaline, or very nearly so. Face rounded and retreating and the quite short proboscis does not project beyond the oral opening. Third antennal segment conical but very short, the style long and slender. The pulvilli are absent, the metapleural fringe dense, the basal comb large and often conspicuously silvery. Discal cell acute at apex. Anterior tibia with setae or even spines.
Villa Lioy
Wings partly or wholly colored, or if nearly hyaline the face is either prominent, or the third antennal segment is long conical. Face always more or less extended, bluntly rounded or conical or even sharply conical 22

22. Front tibia spiculate in 3 rows. Face strongly projecting and conical, usually acutely conical. Third antennal segment long conical . . . . ***Paravilla*** Painter
    Front tibia relatively smooth. Claws small, rarely large. Face rounded anteriorly. Face bluntly triangular, or if subconical the wings are large and long . . 23
23. Front with rather dense, erect, coarse, bristly hairs; face subconical, with similar hairs but more sparse. Both face and front usually without any appressed, flattened hairs. Wings usually with extensive, mottled pattern covering whole wing, with some hyaline spots; a few species with posterior margin and immediate apex hyaline. Wings unusually large and elongate; first posterior cell characteristically narrowed toward the margin. Lower vein of discal cell bowed, with an angle, and usually with a long, characteristic, ventroapical spur vein or an anterobasal spur vein. End of discal cell acute. Anterior tibia smooth; anterior claws minute. Pulvilli absent.
    Poecilanthrax Osten Sacken
    At least the apical half or two-thirds of wing clear. Wings relatively much smaller. Pattern of wing on basal half more or less solid and uniform, or with one to several subhyaline flecks, or dilutely hyaline with only the veins and costa of basal half blackish or brownish, or yellow. Discal cell below without angle and lower spur vein . 24
24. Face bluntly but distinctly triangular (sensu stricto), or shorter, subvertical, bluntly rounded (subgenus *Tauropsis*, type *irrorellus* klug). Wings dark at base, but with windowlike hyaline spots about the crossveins and furcations of veins (*fenestratus* Fallen group). Wings sometimes entirely hyaline (as in *lloydi* Austen group), or with the base only narrowly brownish (*afer* Fabricius group). Third posterior cell varying from very short to very long. Third antennal segment usually short and distinctly conical. Pulvilli absent; front tibia smooth.
    Thyridanthrax Osten Sacken
    Face more prominent, triangular but never acutely conical, the apex more or less rounded. Style of antenna much more slender near the base of the third segment not at all conical and not onion-shaped either 25
25. Face prominently protrusive, triangularly projecting and bluntly rounded at the apex. Anterior tibia smooth; pulvilli long and well developed. The second vein arises well before the anterior crossvein. Discal cell at apex wide and truncate. Veins merely tinged with brown at the base.
    Synthesia Bezzi
    Second vein arising at anterior crossvein or beyond. Intercalary vein at apex of discal cell much shorter and discal cell not rectangularly truncate. Pulvilli absent . . . 26
26. Face triangularly projecting but rounded and blunt. Wings dark at base or more or less hyaline, but always darker about the veins, if any color is present. Third antennal segment usually long conical and stylate part quite slender. Contact of discal cell and fourth posterior cell rather generally punctiform or nearly so. Discal cell rather pointed apically. Second posterior cell large. Proboscis confined to oval cavity but, together with labellum, quite slender. Anterior tibiae smooth.
    Chrysanthrax Osten Sacken
    Face conically projecting, but not acutely pointed. Wings very broad, bluish purple. On the basal half, clear hyaline beyond. Anterior tibiae with microsetae. Proboscis short, stout, the labellum quite short and muscoidlike. Third antennal segment short conical. Second posterior cell much reduced. Discal cell as in *Hemipenthas* Loew.
    Deusopora Hull
27. Wings with 2 submarginal cells. Claws without basal tooth (*laetus* Loew, *afer* Fabricius group of *Thyridanthrax*).
    Thyridanthrax Osten Sacken
    Wing with 3 or more submarginal cells 28
28. Wing with 3 submarginal cells 29
    Wing with 4 submarginal cells . . . . . ***Ligyra*** Newman
29. Wing partly hyaline and much narrowed toward the base, with both alula and axillary lobe greatly narrowed; thorax short; pleuron relatively bare. Genal grooves indistinct. Proboscis much longer than in *Exoprosopa* Macquart and extending beyond the prominent face by at least the length of the head 30
    Wing of normal shape, or if rarely narrowed basally the alula is distinct or the proboscis is much shorter . . . 31
30. The second vein arising distal to the anterior crossvein. Eyes distinctly pubescent under low power. Third antennal segment stout and elongate, with a curious, short-conical cap, and sclerotic apex . ***Diatropomma*** Bowden
    Aberrant flies with the origin of the second vein basal to the anterior crossvein by at least double the length of the crossvein. Wings much narrowed toward the base, alula and axillary lobe both quite narrow and wing blackish on anterior half with irregular margin behind. Proboscis extending beyond apex of the face at least a distance equal the length of the head, and distinctly longer than in *Exoprosopa* Macquart, but the palpi are shorter. The buccal cavity is relatively longer, the pleuron more bare, the thorax shorter, the front claws very short, but front claws less reduced. Postvertex with a very deep transverse furrow. Third antennal segment more like *Anthrax* with almost onion-shaped third segment and with the thin, separated style almost twice as long as third segment. Pulvilli absent . . . . ***Isotamia*** Bezzi
31. Wing entirely iridescent hyaline; at most with very slight, indistinct basal infuscation. Dense, brilliantly shining scales present on several parts of body; hairs pectinate to the base (Bowden, 1964); eyes of male almost touching behind the ocelli. Dististyli deeply U-shaped and curved up at apex. Small species. Bristles on body and legs reduced; integument shining . . ***Micomitra*** Bowden
    Wings always infuscated even if only costal cell and parts of basal cells; usually with extensive patterns of brown, yellow, red or blackish, and otherwise differing . . . 32
32. Large flies with extraordinarily large, long wings, at most feebly shining, more so disproportionate to the body, narrowed basally, much like either *Poecilanthrax* Osten Sacken or *Cyananthrax* Painter, and completely, uniformly, dark sepia in color. Abdomen rather narrowed and doubly as elongate in proportion to *Exoprosopa* Macquart. Second posterior cell more than twice as wide basally as distally; third posterior cell in reverse correspondingly widened on the margin. Tooth of claw long.
    Colossoptera Hull
    Flies with wing and abdomen normally shaped; venation like *Exoprosopa*, sensu latus 33
33. Face rounded and retreating. Anterior tibia beset with spicules. Proboscis longer than oral cavity and projecting. Wing venation and pattern characteristic and peculiar . . . . ***Litorrhynchus*** Macquart
    Face more or less projecting and generally quite conical. Anterior tibia generally smooth 34
34. Marginal cell divided into two parts by an anterior, supernumerary crossvein. Wing with numerous, isolated, eyelike spots becoming confluent basally. Small flies.
    Heteralonia Rondani
    Marginal cell undivided, the wing with only 4 apical cells including the marginal cell. Wings without eyelike spots. 35
35. First posterior cell divided into two parts by a supernumerary crossvein 36
    First posterior cell simple, open, or closed 37
36. Second and third posterior cells of unique shape, each long, reclinate, and encroaching upon each other.
    Exoprosopa (Exoptata) Coquillett
    Second and third posterior cells of normal shape, very little wider at base than on wing margin, not elongate and encroaching upon each other.
    Exoprosopa (Zygodipla) Bezzi

37. Discal cell on lower margin, before the tip, with a very prominent angle, emitting an appendix. Terminal vein of discal cell more or less sinuous and running at an angle or perpendicularly to longitudinal axis of wing . . . 38

    Discal cell without apical projecting angle below, and if sometimes dilated, it is rounded, not angular, and always without appendix 41

38. Five posterior cells present 39

    Only 4 posterior cells present 40

39. Among posterior cells, the third alone closed and provided with a long stalk . . . . *Exoprosopa (Mesoclis)* **Bezzi**

    Among posterior cells, the fourth alone closed and provided with a short stalk . . . *Exoprosopa (Metapenta)* **Bezzi**

40. Front tibia short and thick, beset with well-developed spicules. First posterior cell always open.
    Exoprosopa (Acrodisca) **Bezzi**

    Front tibia long and smooth and thin. First posterior cell often closed and stalked.
    Exoprosopa (Cladodisca) **Bezzi**

41. Vein between discal and second posterior cells long, deeply and regularly S-shaped, and in the same line with longitudinal axis of wing, or nearly so. Wings never reduced in size. Discal cell narrow and long, not expanded above or below near its apex 42

    Vein between discal and second posterior cells short, straight or only slightly sinuous; if deeply sinuous, then the vein is short, not S-shaped, or forming an angle with long axis of the wing. Wing sometimes reduced in size. Discal cell shorter, and broader, and sometimes expanded above or below the apex 43

42. First posterior cell closed and ending bluntly, and provided with a long stalk. Hind tibia with very short spicules. Wings with basal half yellow and margined with brown.
    Exoprosopa (Trinaria) **Mulsant**

    First posterior cell usually open, or if closed, terminating in a narrow apex and with a short stalk. Hind tibia with long, stout spicules. Basal half of wing not yellow, not margined with brown . . *Exoprosopa (Defilippia)* **Lioy**

43. Legs feathered. Discal cell contracted in the middle. Upper branch of third vein not widely divergent from main stem. Marginal cell very narrow at the end.
    Exoprosopa (Pterobates) **Bezzi**

    Legs not feathered and venation different. Marginal crossvein straight or sinuous and recurrent. Second vein originating opposite or before or beyond middle crossvein. First posterior cell typically open, but sometimes closed. Wings never fenestrate, or if so the spots not hyaline. Abdomen with or without silvery bands (including *Argyrospila* of Bezzi, 1924) . . . *Exoprosopa* **Macquart**

Genus *Villa* Lioy

FIGURES 181, 182, 390, 667, 681, 953, 954

Villa Lioy, Atti Inst. Veneto, ser. 3, vol. 9, p. 732, 1864. Type of genus: *Anthrax abaddon* Fabricius, 1794, as *Anthrax concinnus* Meigen, 1820. Designated by Coquillett, p. 619, 1910, the second of 7 (as 8) species.

Hyalanthrax Osten Sacken, Biologia Centrali-Americana, Dipt., vol. 1, p. 134, 1887. Type of genus: *Anthrax faustina* Osten Sacken, 1887. Designated by Coquillett, p. 553, 1910, the first of 5 species.

Aspiloptera Künckel, Bull. Sci. France et Belgique, vol. 39, p. 145, nota, 1905. Type of genus: *Anthrax flavus* Meigen, 1820. See Bezzi, p. 6, 1924.

Coquillett (1887), pp. 160-163, key to 52 species (as *Anthrax*).
Coquillett (1887), pp. 170-177, key to 78 species (as *Anthrax*).
Coquillett (1894a), pp. 97-98, key to 16 species (as *Anthrax*).
Becker (1913), pp. 17-67, treatment of Palaearctic species.
White in Hardy (1924), pp. 74-76, key to 13 species.
Bezzi (1924), pp. 182-183, key to 15 species.
Séguy (1926), pp. 192-193, key to 35 species from France.
Painter (1926b), pp. 206-208, key to 9 *Villa* (s. str.) species and 7 related species.
Engel (1936), pp. 571-576, key to 35 males and 32 females of species.
Austen (1937), p. 133, key to 7 species from Palestine.
Hesse (1956b), pp. 472-480, key to 24 males and 24 females species and subspecies.

Small to rather large flies widely variable in many respects but characterized in the following terms: the head is globular, the face varying from entirely rounded and retreating below the antennae to others in which it is very slightly projecting and not quite vertical. The proboscis is always short, not, or at most very slightly, extending beyond the rather ventrally located orageneal recess. The face itself is extensive in dorsoventral length. The widely separated antennae have the third antennal segment with bulblike base but the abruptness with which the style is formed varies considerably. The long, stylate apical part of the third segment is nonarticulate. With rare exceptions most genera in the tribe Villini can in this way be separated from the tribe Exoprosopini. The anterior tibia varies from completely smooth to species that have one or more rows of distinct, fine, sharp, appressed, small spinules.

The type-species of *Villa* Lioy has clear, hyaline wings, except for a varying amount of pale yellow or brownish infuscation at the base and narrowly along the anterior margin; all species, except as noted below, have clear wings with the brown color extending only into costal and marginal cell. In a few species—*fulviana* Say, *occulta* Wiedemann, and a few others—the infuscation, which may be dark, extends diffusely into the cells immediately behind the subcostal; Austen has described a Palaestine species that has *Thyridanthrax*-like pale pattern on the basal third of the wing. Even in those species in which the costal cell is water clear, the subcostal cell is yellow as far as the end of the subcostal vein. Hesse (1956) notes that 2 South African species—*leucophila* Bezzi, etc.—have distinct *Anthrax*-like wing infuscations, but have nevertheless been referred to *Villa*. Hesse further notes that Painter (1930, 1933) in his study of some American representatives of these two genera has relegated *Hemipenthes* as a subgenus of *Villa*. Representatives of *Thyridanthrax* and *Oestranthrax* on the other hand, however, show certain distinct characters that distinguish them respectively and generically from species of *Villa*.

Males of some species have a brilliant mat of silvery patagial scales.

The genus *Villa* Lioy in the strict sense is found on all the continents.

Length: 5 to 17 mm.

Head, lateral aspect: The head is nearly globular, the occiput well developed and long both above and below, but slightly longer dorsally and strongly rounded and sloping backward on all sides toward the large, central, occipital cup. The indentation at the midpoint of the eye margin is shallow but distinct, angulate, and equally extensive above and below the linear division of the upper eye facets. The pile of the occiput is appressed, scaliform, abundant, reaching almost to the eye margin and the interior fringe about the margin of the occipital cup is short and dense. The eye is quite large, sometimes with green reflections, hemicircular in front. The oral cavity and gena are not or only very slightly extended below the eye margin. The extensive frontal area is quite low, convex, and rounded, following the eye margin. Its pile generally is comparatively dense, fine, and erect, varying somewhat in length but usually shorter than the first antennal segment. In addition to the erect pile there is often a considerable amount of flat-appressed scales or scaliform pile, usually of contrasting color.

The face in lateral aspect is short in most species but it varies from nearly vertical on the one hand, to either slightly and bluntly produced, to still other species that are quite rounded, convex, and receding. The pile of the face predominantly consists of suberect, dense, long scales mixed with a certain amount of stiff more or less appressed hairs. The palpus is long and slender, with one segment a little flattened, pointed at the apex, and bears a few fine hairs. The proboscis is always short, usually contained within the oral recess, sometimes slightly protruding beyond. In some species the proboscis seems to be distinctly shorter than the oragenal recess. The antennae are placed at the middle of the head and are relatively small, the first segment is a little widened apically, sometimes nearly twice as long as wide and at least twice as long as the short second segment. The first segment has a tuft of long, stiff, bristly hairs dorsolaterally, a still longer tuft medially, but only a few fine, short setae dorsally. This segment is bare in the middle below and some of the medial hairs may be scalelike. The third segment is bulblike at the base, generally a little wider than the second segment, and apically it bears a stylelike extension, which is not only long but may be formed either abruptly, the quite narrow style rising abruptly from the basal part of the segment, or the swollen basal part may be somewhat conical and only gradually attenuate; generally plane on the dorsal margin but curved and concave below; apex with minute spine.

Head, anterior aspect: The head is as wide as the thorax, sometimes a little wider. The eyes of the male are always separated by approximately the width of the small, raised, ocellar tubercle, which almost touches the eye laterally. This ocellar tubercle varies distinctly in height in species; in some the sides may be quite vertical. In females the eyes are separated by 3 or more times the width of the ocellar tubercle. The ocellar tubercle is small, the ocelli lying either in an equilateral or a short isosceles triangle and they are placed very close to the posterior eye corners, which are a little rounded. The antennae are widely separated by a distance nearly equal to twice the first segment. The oragenal recess is narrow but quite deep in front as well as behind; it varies in length and is generally wedge-shaped with the sides vertical and plane, but it may be slightly constricted in the middle. The gena consists of a narrow rim at the bottom of the eye and the oragenal recess is sunk below and within the eye margin on the lower half. Viewed from above the occipital lobes are tightly apposed without anterior pit.

Thorax: The general coloration of the thorax is black. The thorax is short, rather flattened especially posteriorly but low throughout and only slightly convex anteriorly. The pile consists of comparatively dense, coarse or fine, erect pile so dense in some species as to obscure the ground color, always dense anteriorly in the region of the collar and along the lateral margins, notopleuron, and above the wing, but often thinning out considerably over the middle of the dorsum, which generally bears a number of flat-appressed, or sometimes loosely suberect, small scales varying from gold to black in color. The notopleuron and postalar callosity at most have very poorly developed, slender, bristly hairs. The scutellum varies in length but is distinctly triangular, its color, pile, and scales corresponding to that of the mesonotum. The plumula is well developed; squama with a long fringe of fine hairs in a narrow plane. The pleuron is densely covered with masses of long, fine, tufted pile, except upon the hypopleuron.

Legs: The femora and tibiae are slender. The femora and especially the hind tibia bear numerous, flat-appressed scales. Anterior and middle femora with a fringe of fine hairs ventrally and posteroventrally, the middle femur often with a distinct row of bristles; hind femur always with a few spines or bristles and often with many. The anterior tibia either with or without bristles or spines; middle and hind tibiae with 4 rows. All the claws are small especially the anterior ones and there are no basal teeth present on claws; pulvilli absent, in contrast to *Anthrax* Scopoli.

Wings: The wings are large, broad at base, pointed apically and are hyaline, glasslike, with wrinkled membrane and slightly iridescent reflections. Quite rarely, as in *Villa fulviana* Say, a faint brown tint from villi is perceptible over the whole of the wing and the costal cell, subcostal cell, and with base of wing in front of alula dark brown; also with diffuse, pale brown color filling out most of the first basal cell and diffusely margining the first vein posteriorly and the anterior crossvein. In *Villa decipula* Austen there is a *Thyridanthrax*-like pattern on the wing, but the face is rounded, the front tibia spinose. Most species of *Villa* Lioy have the costal cell hyaline with only the subcostal cell yellow, but in many species both the costal cell and the subcostal cell are pale yellow, the remainder of the wing hyaline. There are only 2 submarginal cells, but the anterior branch of the third vein, in those species where it rises plane and nearly rectangularly, sometimes has a short basal spur. In other species this vein is broadly rounded at the first bend. The second vein rectangular exactly at or slightly before the anterior crossvein. The marginal cell is a little widened before the apex. The anterior branch of the third vein has a sharp upward bend near the middle, the vein ending well before the apex of the wing.

The anterior crossvein enters the discal cell at the middle or sometimes a little before the middle. The discal cell is acute and pointed apically in contrast to *Hemipenthes* Loew where it is usually swollen apically and obtuse, and this cell is strongly constricted before the middle. The axillary lobe is large and wide, the alula large but about twice as long as wide, its margin bearing scales. The costa at the base is very greatly expanded with strong, bristly comb, its width about twice that of *Hemipenthes* Loew. Also, the costal hook is spinelike, and the patagium in some species is quite large and conspicuous forming a flat, radiating fan of considerable proportions. In other species it is small and inconspicuous. In those species where it is enlarged it will be yellowish or whitish in the females but brilliantly silvered in the males and this silver pile sometimes extends onto the base of the costa.

Abdomen: The abdomen is relatively short, broadly oval, wide even at the end of the sixth tergite and the whole abdomen wider than the thorax, sometimes of equal width. The abdomen is only gently convex, the pile varying in different groups in the genus. The erect, long pile may be quite dense as in forms like *Villa pretiosa* Loew or *Villa fulviana* Say with, however, one or sometimes two types of flat-appressed pile lying on the tergites. Most species show transverse bands of appressed scaliform hairs or scales on the anterior margins of the second and remaining tergites. These bands are of varying width and color and sometimes narrowly interrupted in the middle. Some species have in addition a rather dense coating of flat-appressed black scales or scaliform hairs over the remainder of the tergites. Lateral margins of the segments either only with appressed, scaliform hair or in the species of longer pile with a very dense lateral border of long, plushlike pile. In some species the pile of the abdomen is not distributed in bands. Sternites largely bare in some species, in others with a dense, matted coat of appressed, scaliform pile. Males of some species have often a dense, shelflike, projecting mat of brilliant silver scales extending out from the posterior margin of the last tergite. Seventh tergite in males of some species sometimes forms a recessed collar around the genitalia. In the more woolly species there are 7 tergites visible from above in both sexes. Female terminalia have a row of small spines on each side.

Because of the large number of species that Dr. A. J. Hesse was able to study and summarize, I quote his work on the hypopygium of the males:

Hypopygium of males markedly uniform in series of species: dorso-apical parts of clasper-like basal parts always covered with fairly dense hairs; beaked apical joints very uniform in shape and not of much use in the separation of the species, more or less laterally compressed in apical parts and directed slightly outwards, ending apically in an upper tooth-like process or lobe and a shorter lower one; aedeagal apparatus consisting of a slightly curved and pointed aedeagus and a prominent and conspicuous, apically directed, ventral, aedeagal process which is usually in form of a broad, flattened, scoop-like process of which the apical part may either end in two separated or contiguous downwardly directed spines, processes or hooks, or it may be without spines and merely scoop-like or with a central, keel-like ridge dorsally and apically; middle bulb-like part of aedeagus usually small; lateral struts small; basal strut usually scarcely projecting posteriorly, bat-shaped, ham-shaped or boomerang-shaped. Ovipositor of females with a series of spines on each side, increasing in size from above to below and each with its apex bent sub-hook-like.

The male terminalia from dorsal aspect, epandrium removed, show a figure that is broad basally, narrowed apically, and while the epiphallus is large and quite wide basally, it is strongly attenuate regularly narrowing toward a point apically. It has a curious, backward-turned hooklike process on each side of the apex. From the lateral aspect the epandrium is triangular, not incised apically, although there is a lobelike extension at the top. The dististyle does not have a deep incision, although there is a very slight incision dorsally.

Material available for study: Numerous American species in my own collection, other species in the collections of the National Museum of Natural History and of the British Museum (Natural History). Prof. J. Timon-David of Marseille kindly sent me several European species for study. I also have representative material from South America and Australia.

Immature stages: Flies of this genus are parasites of nocturnal Lepidoptera. Some species attack coleopterous larvae of the family Tenebrionidae. See remarks under life histories.

Ecology and behavior of adults: The adults may be captured on a variety of flowers, especially composites. Painter (1926) records *Villa lateralis* Say from the flowers of *Tetraneuris linearis*, *Lepadenia marginatum* (Pursh) Niewel, and from *Rhus* species. He states that they are found in Texas from March to November. I have taken species of *Villa* commonly on *Rhus* species and on sneeze weed, *Helenium tenuifolium* Linné. I have frequently found them hovering for long periods of time in forested areas, and I have also seen species apparently ovipositing in soil.

Distribution: Nearctic: *Villa adusta* (Loew), 1869 (as *Anthrax*); *aenea* (Coquillet), 1887 (as *Anthrax*); *agrippina* (Osten Sacken), 1887 (as *Anthrax*); *albicincta* Cole, 1923, s.l.; *albovittata* (Macquart), 1850 (as *Anthrax*, s.l.); *alternata* (Say), 1823 (as *Anthrax*) [=*albipectus* (Macquart), 1848 (as *Anthrax*), =*bastardi* (Macquart) 1840 (as *Anthrax*), =*consanguineus* (Macquart), 1840 (as *Anthrax*)], *alternata nigropectus* Cresson, 1916; *anna* (Coquillett), 1887 (as *Anthrax*, s.l.); *caprea* (Coquillett), 1887 (as *Anthrax*, s.l.); *cautor* (Coquillett), 1887 (as *Anthrax*, s.l.); *chromo-*

lepida Cole, 1923; *compressa* Painter, 1926; *connexa* (Macquart), 1855 (as *Anthrax*); *consessor* (Coquillett), 1887 (as *Anthrax*); *costata* (Say), 1824 (as *Anthrax*, s.l.); *faustina* (Osten Sacken), 1887 (as *Anthrax*); *fissa* (Bigot), 1892 (as *Anthrax*, s.l.); *flavocostalis* Painter, 1926; *fulviana* (Say), 1824 (as *Anthrax*), *fulviana nigricauda* Loew, 1869; *fumicosta* Painter, 1962; *fumida* Coquillett, 1887; *gemella* (Coquillett), 1892 (as *Anthrax*, s.l.); *gracilis* (Macquart), 1840 (as *Anthrax*); *handfordi* Curran, 1935; *harveyi* (Hine), 1904 (as *Anthrax*); *hircina* (Coquillett, 1892 (as *Anthrax*); *hypomelaena* (Macquart), 1840 (as *Anthrax*); *inculta* (Coquillett), 1892 (as *Anthrax*, s.l.); *lateralis* (Say), 1823 (as *Anthrax*) [=*fulvipes* (Coquillett), 1887 (as *Anthrax*)], *lateralis arenicola* (Johnson), 1908 (as *Anthrax*), *lateralis atra* Painter, 1926, *lateralis johnsoni* Painter, 1926, *lateralis nigra* Cresson, 1916, *lateralis semifulvipes* Painter, 1962; *levicula* (Coquillett), 1894 (as *Anthrax*, s.l.); *livia* (Osten Sacken), 1887 (as *Anthrax*); *meridionalis* Cole, 1923, s.l.; *miscella* (Coquillett), 1887 (as *Anthrax*, s.l.); *molitor* (Loew), 1869 (as *Anthrax*); *moneta* (Osten Sacken), 1887 (as *Anthrax*); *mucorea* (Loew), 1869 (as *Anthrax*); *muscaria* (Coquillett), 1892 (as *Anthrax*); *nebulo* (Coquillett), 1887 (as *Anthrax*); *pretiosa* (Coquillett), 1887 (as *Anthrax*); *psammina* Cole, 1960, s.l. [=*arenicola* Cole, 1923, s.l., not Johnson]; *sabina* (Osten Sacken), 1887 (as *Anthrax*); *salebrosa* Painter, 1926; *scrobiculata* (Loew), 1869 (as *Anthrax*); *shawii* (Johnson), 1908 (as *Anthrax*); *sini* Cole, 1923, s.l.; *sodom* (Williston), 1893 (as *Anthrax*, s.l.); *squamigera* (Coquillett), 1892 (as *Anthrax*); *stenozona* (Loew), 1869 (as *Anthrax*); *supina* (Coquillett), 1887 (as *Anthrax*); *telluris* (Coquillett), 1892 (as *Anthrax*, s.l.); *terrena* (Coquillett), 1892 (as *Anthrax*, s.l.); *vacans* (Coquillett), 1887 (as *Anthrax*); *vanduzeei* Cole, 1923, s.l.; *vestita* (Walker), 1849 (as *Anthrax*).

Neotropical: *Villa abbreviata* (Wiedemann), 1830 (as *Anthrax*); *albicincta* Cole, 1923; *albicollaris* Cole, 1934; *albifacies* (Rondani), 1863 (as *Anthrax*); *albissima* Oldroyd, 1938; *amasia* (Wiedemann), 1828 [=*vicinus* (Macquart), 1840 (as *Anthrax*)]; *ambigua* (Lynch Arribálzaga), 1878 (as *Anthrax*); *angulata* Brèthes, 1909; *antica* (Walker), 1852 (as *Anthrax*); *argentiflua* (Philippi), 1865 (as *Anthrax*); *argentosa* Painter, 1933; *ariditata* Cole, 1923; *balteata* (Philippi), 1865 (as *Anthrax*); *barbiventris* (Rondani), 1868 (as *Anthrax*); *bellula* (Philippi), 1865 (as *Anthrax*); *binotata* (Macquart), 1846 (as *Anthrax*); *bipartita* (Jaennicke), 1867 (as *Anthrax*); *bistella* (Walker), 1850 (as *Anthrax*); *brachialis* (Thomson), 1869 (as *Anthrax*); *calogaster* (Philippi), 1865; *caloptera* (Philippi), 1865 (as *Anthrax*); *calvescens* (Philippi), 1865 (as *Anthrax*); *castanea* (Jaennicke), 1867 (as *Anthrax*); *ceria* (Williston), 1901 (as *Anthrax*); *chilensis* (Philippi), 1865 (as *Anthrax*); *chimaera* (Osten Sacken), 1887 (as *Anthrax*); *cinerea* Cole, 1923; *conclusa* (Walker), 1857 (as *Anthrax*); *connexa* (Bigot), 1856 (as *Anthrax*); *conopas* (Philippi), 1865 (as *Anthrax*); *constituta* (Walker), 1852 (as *Anthrax*); *convexa* (Walker), 1857 (as *Anthrax*); *corrigiolata* (Rondani), 1863 (as *Anthrax*); *costalis* (Wiedemann), 1828 (as *Anthrax*); *crepuscularis* (Lynch-Arribálzaga), 1878 (as *Anthrax*); *cuniculus* (Osten Sacken), 1886 (as *Anthrax*); *curvirostris* (Thomson), 1869 (as *Anthrax*); *decemmacula* (Walker), 1857 (as *Anthrax*); *delicatula* (Walker), 1849 (as *Anthrax*); *detecta* (Walker), 1852 (as *Anthrax*) [=*bipenicillatus* (Bigot), 1892 (as *Anthrax*)]; *diana* (Williston), 1901 (as *Anthrax*); *diminutiva* (Macquart), 1846 (as *Anthrax*); *ditaenia* (Wiedemann), 1828 (as *Anthrax*); *divisa* (Walker), 1852 (as *Anthrax*); *dorsalis* (Walker), 1857 (as *Anthrax*); *duodecimpunctata* (Philippi), 1865 (as *Anthrax*); *durvillei* (Macquart), 1840 (as *Anthrax*); *edwardsi* Oldroyd, 1938; *ephebus* (Osten Sacken), 1886 (as *Anthrax*); *epilais* (Wiedemann), 1828 (as *Anthrax*); *eurrhinata* (Bigot), 1892 (as *Anthrax*); *excisa* (Walker), 1852 (as *Anthrax*); *extremitis* (Coquillett), 1902 (as *Anthrax*); *fenestrellus* (Wiedemann), 1828 (as *Anthrax*); *fenestralis* (Wiedemann), 1830 (as *Anthrax*); *festiva* (Philippi), 1865 (as *Anthrax*); *flavicincta* Cole, 1923; *flavipilosa* Cole, 1923; *fulvago* (Philippi), 1865 (as *Anthrax*); *fulvipeda* (Rondani), 1863 (as *Anthrax*); *funebris* (Macquart), 1840 (as *Anthrax*); *fusca* (Fabricius), 1805 (as *Cytherea*); *galathea* (Osten Sacken), 1886 (as *Anthrax*); *gayi* (Macquart), 1840 (as *Anthrax*); *gorgon* (Fabricius), 1805 (as *Anthrax*); [=*maimon* (Fabricius), 1805 (as *Anthrax*)]; *gradata* (Macquart), 1847 (as *Anthrax*); *hilarii* (Macquart), 1840 (as *Anthrax*); *hirsuta* (Williston), 1901 (as *Anthrax*); *hyalacra* (Wiedemann), 1828 (as *Anthrax*); *hyalinipennis* (Blanchard in Gay), 1852 (as *Anthrax*); *ignea* (Macquart), 1846 (as *Anthrax*); *imitans* (Schiner), 1868 (as *Argyramoeba*); *inexacta* (Walker), 1857 (as *Anthrax*); *ingloria* (Philippi), 1865 (as *Anthrax*); *inordinata* (Rondani), 1863 (as *Anthrax*); *ioptera* (Wiedemann), 1828 (as *Anthrax*); *latifimbria* (Walker), 1852 (as *Anthrax*); *lelia* (Williston), 1901 (as *Stonyx*); *lemniscata* (Philippi), 1865 (as *Anthrax*); *lepidota* (Osten Sacken), 1887 (as *Anthrax*); *leucomallus* (Philippi), 1865 (as *Anthrax*); *leucothoa* (Wiedemann), 1830 (as *Anthrax*); *lineata* (Walker), 1857 (as *Anthrax*); *maria* (Williston), 1901 (as *Anthrax*); *melaleuca* (Wiedemann), 1828 (as *Anthrax*); *melanogaster* (Bigot), 1892 (as *Anthrax*); *mendozana* Brèthes, 1909; *micromelaena* (Bigot), 1892 (as *Anthrax*); *midas* (Fabricius), 1805 (as *Anthrax*); *minerva* (Wiedemann), 1828 (as *Anthrax*); *miniata* Oldroyd, 1938; *moerens* (Philippi), 1865 (as *Anthrax*); *moneta* (Osten Sacken), 1887 (as *Hyalanthrax*); *murina* (Philippi), 1865 (as *Anthrax*); *mutua* (Walker), 1849 (as *Anthrax*); *nero* (Fabricius); 1805 (as *Anthrax*); *nigricosta* (Schiner), 1868 (as *Anthrax*); *nigrita* (Fabricius), 1775 (as *Bibio*); *nigrofimbriata* (Williston), 1901 (as *Anthrax*); *nitida* Cole, 1923; *nivea* Cole, 1923; *nudiuscula* (Thomson), 1869 (as *Anthrax*) [=*lateralis* (Thomson), 1869 (as *Anthrax*)]; *obliqua* (Macquart), 1834 (as *Anthrax*); *obscuripes* (Bigot), 1892 (as *An-

thrax); *orbitalis* (Williston), 1901 (as *Anthrax*); *pallipes* (Bigot), 1892 (as *Anthrax*); *peninsularis* Cole, 1923; *perimele* Wiedemann, 1828; *philippii* (Rondani), 1863 (as *Anthrax*); *pleuralis* (Williston), 1901 (as *Anthrax*); *pluricella* (Williston), 1901 (as *Anthrax*); *polyphemus* (Wiedemann), 1821 (as *Anthrax*); *porteri* Oldroyd, 1938; *praeterita* Oldroyd, 1938; *primitiva* (Walker), 1849 (as *Anthrax*); *procedens* (Walker), 1852 (as *Anthrax*); *propinqua* (Schiner), 1868 (as *Argyramoeba*); *pusio* (Macquart), 1840 (as *Anthrax*); *pusio* (Philippi), 1865 (as *Anthrax*); *quadricincta* (Rondani), 1863 (as *Anthrax*); *quadricincta* (Philippi), 1865 (as *Anthrax*); *quadripunctata* Cole, 1923; *quinquepunctata* (Thomson), 1869 (as *Lepidanthrax*); *rava* Painter, 1933; *recta* (Walker), 1852 (as *Anthrax*); *reperta* (Walker), 1852 (as *Anthrax*); *restituta* (Walker), 1852 (as *Anthrax*); *ruficollis* (Bigot), 1892 (as *Anthrax*); *sackeniana* (Williston), 1901 (as *Isopenthes*); *scylla* (Osten Sacken), 1887 (as *Anthrax*); *sejungenda* (Rondani), 1863 (as *Anthrax*); *semitincta* (Schiner), 1868 (as *Anthrax*) [=*semicinctus* (Bigot), 1892 (as *Glossista*), lapsus]; *solita* (Walker), 1857 (as *Anthrax*); *sonorensis* Cole, 1923; *spiloptera* (Wiedemann), 1828 (as *Anthrax*); *stheno* (Wiedemann), 1828 (as *Anthrax*); *subaequalis* (Lynch Arribálzaga), 1878 (as *Anthrax*); *tenuirostris* (Macquart), 1849 (as *Anthrax*); *tincta* (Thomson), 1869 (as *Anthrax*); *translata* (Walker), 1852 (as *Anthrax*); *trifigurata* (Walker), 1860 (as *Anthrax*); *trimacula* (Walker), 1849 (as *Exoprosopa*); *trimaculata* (Macquart), 1848 (as *Anthrax*); *una* Oldroyd, 1938; *vastitatis* Cole, 1923; *verdensis* Oldroyd, 1938; *vicina* (Macquart), 1846 (as *Anthrax*); *antipoda*, n.n., for *vicina* (Blanchard in Gay), 1852, not *vicina* (Macquart), 1846; *viduata* (Loew), 1860 (as *Anthrax*); *villicus* (Philippi), 1865 (as *Anthrax*); *vitripennis* (Philippi), 1865 (as *Anthrax*); *vulpecula* (Philippi), 1865 (as *Anthrax*).

Palaearctic: *Villa abbadon* (Fabricius), 1794 (as *Anthrax*) [=*abbadon* (Meigen), 1820 (as *Anthrax*), =*concinnus* (Meigen), 1820 (as *Anthrax*)]; *aegyptiaca* (Macquart), 1840 (as *Anthrax*); *albida* Becker, 1916; *albiventris* Frey, 1936; *albula* (Loew), 1869 (as *Anthrax*) [=*quinta* Becker, 1916]; *arabica* (Macquart), 1840 (as *Anthrax*); *atricauda* Austen, 1937; *baluchiana* Brunetti, 1920; *bicingulata* (Macquart), 1840 (as *Anthrax*); *bimacula* (Walker), 1849 (as *Anthrax*); *bivirgata* Austen, 1937; *bizonata* Becker, 1916; *bombiformis* Becker, 1916; *brunnea* Becker, 1916; *ceballosi* Rubio, 1959; *cingulata* (Meigen), 1804 (as *Anthrax*) [=*hottentotta* Jaennicke, 1867, =*cingulatus* (Zetterstedt), 1842, not Meigen (as *Anthrax*)]; *cingulum* (Wiedemann in Meigen), 1820 (as *Anthrax*); *circumdata* (Meigen), 1820 (as *Anthrax*); [=*cingulatus* (Walker not Meigen), 1851 (as *Anthrax*), =*flavus* (Curtis not Meigen), 1824 (as *Anthrax*), =*hottentottus* (Schellenberg), 1803 (as *Anthrax*), =*stoechades* (Jaennicke), 1867 (as *Anthrax*)], *circumdata algeciras* Strobl, 1909, *circumdata fulvimaculata* Abreu, 1926; *claretta*, n.n., for *claripennis* Becker, 1923; *claripennis* (Kowarz), 1867 (as *Anthrax*); *clarissima* (Loew), 1857 (as *Anthrax*) [=*cyprignus* (Rondani), 1863 (as *Anthrax*)]; *combinata* (Walker), 1857 (as *Anthrax*), *decipula* Austen, 1937; *distincta* (Meigen), 1838 (as *Anthrax*); *doriae* Bezzi, 1922; *euzona* (Loew), 1869 (as *Anthrax*); *fallax* Austen, 1937; *fasciata* (Meigen), 1804 (as *Anthrax*); *fasciculata* Becker, 1916; *fasciventris* (Macquart), 1849 (as *Anthrax*) [=*fusciventris* Loew, 1860, lapsus]; *haesitans* Becker, 1916; *halteralis* (Kowarz), 1883 (as *Anthrax*) [=*circumdatus* (Meigen), 1820]; *hottentotta* (Linné), 1758 (as *Musca*) [=*flavus* (Meigen), 1820 (as *Anthrax*), =*perfecta* Becker, 1906, =*suprema* Becker, 1916]; *hottentotta modesta* (Meigen), 1820 (as *Anthrax*); *humilis* (Ruthe), 1831 (as *Anthrax*) [=*mucidus* (Zeller), 1840 (as *Anthrax*)]; *insignis* Austen, 1937; *ixion* (Fabricius), 1794 (as *Anthrax*); *laevis* Becker, 1915 [=*inconstans* Becker, 1916]; *latifascia* (Walker), 1857 (as *Anthrax*); *leucostoma* (Meigen), 1820 (as *Anthrax*); *limbata* (Coquillett), 1898 (as *Anthrax*); *longipennis* (Macquart), 1840 (as *Anthrax*); *longitarsis* (Becker), 1902 (as *Anthrax*); *luculenta* Séguy, 1941; *marginalis* (Wiedemann in Meigen), 1820 (as *Anthrax*); *melanura* (Loew), 1869 (as *Anthrax*); *micrargyra* (Walker), 1871 (as *Anthrax*); *multibalteata* Austen, 1936; *muscaria* (Pallas in Meigen), 1818 (as *Anthrax*); *nigriceps* (Macquart in Webb and Berthelot), 1838 (as *Anthrax*), *nigriceps abdominalis* Abreu, 1926; *nigrifrons* (Macquart in Webb and Berthelot), 1838 (as *Anthrax*); *niphobleta* (Loew), 1869 (as *Anthrax*) [=*nana* Becker, 1916]; *nomas* (Eversmann), 1854 (as *Anthrax*); *obscura* (Weber), 1801 (as *Anthrax*); *occulta* (Wiedemann in Meigen), 1820 (as *Anthrax*); *ovata* (Loew), 1869 (as *Anthrax*); *palumbii* Rondani, 1877; *paniscus* (Rossi), 1790 (as *Bibio*) [=*bimaculatus* (Macquart), 1834 (as *Anthrax*), =*cingulatus* (Meigen), 1820 (as *Anthrax*), =*hottentottus* (Walker not Linné), 1849 (as *Anthrax*)]; *perfecta* (Becker), 1906 (as *Anthrax*); *persica* (Macquart), 1840 (as *Anthrax*); *plagiata* (Walker), 1849 (as *Anthrax*); *punctata* (Macquart), 1834 (as *Anthrax*) [=*punctulatus* (Macquart), 1835 (as *Anthrax*); *punctum* (Loew), 1854 (as *Anthrax*); *pygarga* (Loew), 1868 (as *Anthrax*); *quinquefasciata* (Wiedemann in Meigen), 1820 (as *Anthrax*) [=*blanda* Loew, 1869, =*cana* Meigen, 1804, =*inquieta* Becker, 1916, =*mus* Becker, 1916]; *ruficeps* (Macquart), 1840 (as *Anthrax*); *scrutata* (Wiedemann in Meigen), 1820 (as *Anthrax*); *sardous* (Macquart), 1849 (as *Anthrax*); *satanas* Becker, 1916; *senecio* (Loew), 1869 (as *Anthrax*); *squamifer* (Jaennicke), 1867 (as *Anthrax*); *stenozona* (Loew), 1869 (as *Anthrax*) [=*disjuncta* Becker, 1916]; *suffusa* (Walker), 1849 (as *Anthrax*); *svenhedini* Paramonov, 1934; *syphax* (Fabricius), 1798 (as *Anthrax*); *tankerina* (Bigot), 1892 (as *Anthrax*); *tashkentica* Paramonov, 1925; *tenuis* (Walker), 1871 (as *Anthrax*); *tomentosa* Becker, 1916; *varipennis* (Macquart), 1849 (as *Anthrax*); *ventruosus* (Loew), 1869 (as *Anthrax*); *venusta* (Meigen), 1820 (as *Anthrax*) [=*dolosus* (Jaennicke), 1867 (as *An-*

thrax), =*margaritifer* (Dufour), 1833 (as *Anthrax*), =*turbidus* (Loew), 1869 (as *Anthrax*)]; *vitripennis* (Loew), 1860 (as *Anthrax*).

Ethiopian: *Villa albescens* (Loew), 1860 (as *Anthrax*); *anthophoroides* Hesse, 1956; *apiformis* Hesse, 1956; *apparens* (Walker), 1852 (as *Anthrax*); *argentina* Bezzi, 1921; *aspiculata* Hesse, 1956; *atrisquama* Bezzi, 1924; *aurocincta* (Bigot), 1892 (as *Anthrax*); *biflexa* Loew, 1852; *biguttata* Macquart, 1834; *bravae* Bezzi, 1922; *candidata* Bezzi, 1924; *chionalis* Hesse, 1956; *chrysothrix* Bezzi, 1924; *commiles* (Walker), 1857 (as *Anthrax*); *dissimilis* Hesse, 1956; *dizona* (Loew), 1860 (as *Anthrax*); *eucnemis* François, 1961; *fenestralis* (Macquart), 1840 (as *Anthrax*); *filicornis* Hesse, 1956; *flammigera* (Walker), 1849 (as *Anthrax*); *flavalis* Hesse, 1956; *flavescens* (Loew), 1860 (as *Anthax*); *flavipes* (Loew), 1860 (as *Anthrax*); *fulvipleura* Hesse, 1956; *gariepina* Hesse, 1956; *harroyi* François, 1961; *hybrida* Hesse, 1956; *kaokoensis* Hesse, 1956; *karasana* Hesse, 1956; *karooensis* Hesse, 1956; *lasia* (Wiedemann), 1824 (as *Anthrax*); *leptopus* (Thomson), 1869 (as *Anthrax*); *leucochila* Bezzi, 1921 [=*leucostomus* (Weidemann), 1821 (as *Anthrax*)]; *leucoproctus* (Loew), 1860 (as *Anthrax*); *lloydi* Austen, 1914; *loewii* Hesse, 1956; *madagascariensis* (Macquart), 1847 (as *Anthrax*); *nilotica* (Jaennicke), 1867 (as *Anthrax*); *niphobletoides* Hesse, 1956; *nivearia* Hesse, 1956; *pachystyla* Hesse, 1956; *paniscoides* Bezzi, 1912; *phaeotaenia* Bezzi, 1922; *punctulata* (Macquart), 1840 (as *Anthrax*); *pusilla* (Wiedemann), 1821 (as *Anthrax*); *sexfasciata* (Wiedemann), 1821 as *Anthrax*); *sokotrae* (Ricardo in Forbes), 1903 (as *Anthrax*); *submacula* (Walker), 1849 (as *Anthrax*); *terminus* (Walker), 1849 (as *Anthrax*); *turneri* Hesse, 1956; *unifasciata* (Macquart), 1840 (as *Anthrax*); *validicornis* Bezzi, 1924.

Oriental: *Villa abeilla*, n.n., for *Anthrax antecedens* (Walker), 1859, not (Walker), 1852 (as *Anthrax*, U.S.); *absalon* (Wiedemann), 1824 (as *Anthrax*); *albofulva* (Walker), 1852 (as *Anthrax*); *angustata* (Doleschall), 1858 (as *Anthrax*); *antecedens* (Walker), 1860 (as *Anthrax*); *aperta* (Walker), 1852 (as *Anthrax*); *argentilatus* (Walker), 1857 (as *Anthrax*); *aterrima* (Doleschall), 1858 (as *Anthrax*) [=*proferens* Walker, 1860]; *aureohirta* Brunetti, 1909; *aygula* (Fabricius), 1805 (as *Anthrax*); *carbonaria* Walker, 1852 (as *Anthrax*); *clara* (Walker), 1852 (as *Anthrax*); *confirmata* (Walker), 1861 (as *Anthrax*); *congrua* (Walker), 1852 (as *Anthrax*); *demonstrans* (Walker), 1860 (as *Anthrax*); *devecta* (Walker), 1861 (as *Anthrax*); *dia* (Wiedemann), 1824 (as *Anthrax*); *emarginata* (Macquart), 1840 (as *Anthrax*); *emissa* (Walker), 1864 (as *Anthrax*); *emittens* (Walker), 1861 (as *Anthrax*); *instituta* (Walker), 1852 (as *Anthrax*); *leucopyga* (Macquart), 1840 (as *Anthrax*); *limpida* (Walker), 1852 (as *Anthrax*); *manifesta* (Walker), 1852 (as *Anthrax*); *melasoma* (Wulp), 1882 (as *Anthrax*); *nigricostalis* (Guérin-Ménéville), 1838 (as *Anthrax*); *praedicans* (Walker), 1860 (as *Anthrax*); *pretendens* (Walker), 1860 (as *Anthrax*); *satellitia* (Walker), 1857 (as *Anthrax*); *semifuscata* Brunetti, 1920; *semihyalina* (Meijere), 1907 (as *Anthrax*); *terminalis* (Wulp), 1868 (as *Anthrax*) [=*leucostigma* Wulp, 1898]; *trimaculata* (Wulp), 1868 (as *Anthrax*); *troglodyta* (Fabricius), 1775 (as *Bibio*) [=*hyalinus* Wiedemann, 1821, =*lucens* Walker, 1852].

Australian: *Villa albata* Roberts, 1928; *albirufa* (Walker), 1857 (as *Anthrax*); *albobasalis* Roberts, 1928; *alternans* (Macquart), 1849 (as *Anthrax*); *alterna* (Walker), 1849 (as *Anthrax*); *angularis* (Thomson), 1869 (as *Anthrax*); *apicifera* (Walker), 1865 (as *Anthrax*); *aprica* Roberts, 1928; *argentipennis* White, 1917; *australis* (Walker), 1852 (as *Anthrax*); *brunea* Roberts, 1928; *commista* (Macquart), 1849 (as *Anthrax*); *consimilis* Thomson, 1869 (as *Anthrax*); *flaveola* (Macquart), 1849 (as *Anthrax*); *fumea* Roberts, 1928; *incisa* (Macquart), 1847, (as *Anthrax*); *incompta* (Walker), 1849 (as *Anthrax*); *maculata* (Macquart), 1846 (as *Anthrax*); *marginata* (Walker), 1852 (as *Anthrax*); *minor* Macquart, 1849 (as *Anthrax*); *nigricosta* (Macquart), 1849 (as *Anthrax*); *obscura* Macquart, 1846; *pellucida* (Walker), 1852 (as *Anthrax*); *proconcisa* Hardy, 1942 [=*concisa* (Macquart), 1849 (as *Anthrax*)]; *quinqueguttata* Roberts, 1928; *rava* Roberts, 1928; *resurgens* (Walker), 1849 (as *Anthrax*); *semimacula* (Walker), 1849 (as *Anthrax*); *serpentiger* (Walker), 1849 (as *Anthrax*); *sobricula* (Walker), 1857 (as *Anthrax*); *subobscura* Hardy, 1942 [=*obscurus* Macquart, 1846]; *subsenex* (Walker), 1857 (as *Anthrax*); *trivincula* Roberts, 1928; *varipennis* Roberts, 1928; *vitre* (Walker), 1852 (as *Anthrax*); *lucidus* Walker; *leucotelus* Walker.

Oceania: *Villa basalis* (Macquart), 1855 (as *Anthrax*); *biappendiculata* Macquart, 1855.

Ignota Patria: *Villa brunipennis* Macquart, 1840; *confluens* (Macquart), 1840 (as *Anthrax*); *diana* (Walker), 1849 (as *Anthrax*); *fervidus* (Walker), 1849 (as *Anthrax*); *fumipennis* (Wiedemann), 1828 (as *Anthrax*); *fuscicostata* (Macquart), 1846 (as *Anthrax*); *gnata* (Walker), 1852 (as *Anthrax*); *ignifera* (Walker), 1852 (as *Anthrax*); *illata* (Walker), 1852 (as *Anthrax*); *insularis* (Walker), 1849 (as *Anthrax*); *noctiluna* (Walker), 1849 (as *Anthrax*); *notabilis* (Macquart), 1840 (as *Anthrax*); *purpurata* (Wiedemann), 1828 (as *Anthrax*); *reducta* (Walker), 1852 (as *Anthrax*); *semilimpida* (Wiedemann), 1828 (as *Anthrax*); *sexnotata* (Macquart), 1855 (as *Anthrax*); *subannulus* (Walker), 1849 (as *Anthrax*); *succedens* (Walker), 1852 (as *Anthrax*); *trivittata* (Macquart), 1834 (as *Anthrax*).

Genus *Astrophanes* Osten Sacken

FIGURES 172, 395, 660, 666, 927, 928

Astrophanes Osten Sacken, Biologia Centrali-Americana, Dipt., vol. 1, p. 106, 1886. Type of genus: *Astrophanes adonis* Osten Sacken, 1886, by monotypy.

Queer little flies of black color, the males with the last half of the abdomen covered with dense, silvery or creamy-white pile laid down in a mat in which the shining hairs are longitudinally placed. The head is large, subglobular, distinctly wider than the thorax. The front, except a smaller, upper, triangular part, is densely covered with shining white hair; this hair is appressed forward and only silvery in appearance when viewed from above. Wings hyaline.

In addition to the general appearance described above there are several other noteworthy characters. These small flies have the male eyes separated by less than the width of one ocellus. The occiput is unusually short and poorly developed but there is a shallow indentation and a distinct bisecting line. The adjacent collar of pile on the anterior mesonotum is high, conspicuous in contrast to the remaining pile. The face is short, nearly vertical, not protruding, but rounded from side to side. The antennae are quite short; the third segment is so short that Osten Sacken described it as a mere disc and its short stylar portion ends in a minute bristle. The anal cell is closed and bears a short stalk.

The type-species ranges from northern Mexico into Arizona and Utah on the west, into Texas and Kansas on the east; I have collected it in several localities. Brèthes (1909) described the species *andina* from Mendoza which he placed here.

Length: 4 to 5 mm.

Head, lateral aspect: Large and subglobular. The occiput, while relatively short, shortens still more toward the vertex, especially in the male. And the pile of the occiput in both sexes is scanty, fine, and erect near the vertex but flat-appressed, matted, and scalelike at and above the indentation. Because the very deep occipital cup or recess is so large the short, dense, silvery, marginal fringe is relatively close to the eye margin, especially in the male; in the female the occiput is wider and more extensive. The front is quite extensive and convex, almost or completely level with the eye, scarcely protruding even at the antenna. It is very slightly convex from side to side. The male front has dense, matted, subappressed, silvery scales and a small, upper triangle with fine, erect, black hairs. Female front with broad, silvery scales in a narrow band just in front of the antennae and in a small, lateral tuft near the middle. The face is prominent, extensive vertically, and extended forward at the oral recess, a distance fully equal the length of the antennal style. It is somewhat shorter beneath the antenna and is rather more protrusive in the female. The pile of the face consists of a few, scattered, silvery scales, a tuft of similar scales along the eye margin representing an extension from the front and reaching halfway down the face. Also the face has scattered, fine, erect blackish and whitish hairs. The eye is extremely large. The posterior margin has a moderately deep, acute indentation of little width or breadth, with a small sunken triangle from which proceeds a long, bisecting line directed forward almost half the length of the eye. Upper facets strongly enlarged. The eye is almost as long as high.

The proboscis is small, short, slender, the labellum small, short oval, not extending to the apex of the oral recess. Palpus quite slender, clavate, and pointed apically and about half as long as the proboscis. The antennae are attached at the middle of the head, and are medium sized, but exceptionally short on all three segments. The very short first segment has a lateral fringe of coarse, bristly pile extending from the middle above almost to the ventral surface. The third segment is much wider medially than the second segment and has the form of a flat, short disc with rounded edges. It bears subdorsally a short, thick style in length perhaps as much as the width of the basal disk.

Head, anterior aspect: Head considerably wider than the thorax. In the male the eyes are linearly separated by a little less than the width of the anterior ocellus. The upper posterior corners of the eye are broadly rounded and each eye is more or less recessed forward so that, viewed from above, the posterior eye margin is not plane. The front in the male is wide at the antenna, the whole front forming a very short isosceles triangle. In the female the eyes are separated by rather more than the width of the second antennal segment and less than the third. Ocellar tubercle moderately high, the ocelli set in an equilateral triangle at the posterior eye corners. The antennae are quite widely separated by nearly twice the amount of the distance from antenna to eye. The oral recess is small, long, and narrow with a deeper medial groove or pocket, extending to the apex. The eye margins almost reach to the wall of the oral recess but recede ventrally so that the gena is in evidence only at the bottom of the eye. There is a quite linear fissure along the eye margin, midway between the antenna and the base of the proboscis. Viewed from above the lobes of the occiput are shallow and narrowly separated, more strongly divergent behind the vertex.

Thorax: Shining black with more or less scattered, erect, slightly flattened hairs over the mesonotum and with tufts of longer, distinctly scaliform pile on humerus and notopleuron. The anterior collar is composed of a single row of long, stiff, white hairs. Scutellum subtriangular, bluntly pointed apically, slightly rounded near the base on each side. The mesopleuron has only a little pile near the top, and it is stiff and forms scaliform bristles. Pteropleuron with a few bristly hairs, hypopleuron bare, but metapleuron with an extensive tuft of bristly hair. Scales of squamae broad, conspicuous and white.

Legs: Rather small, black in color. Anterior femur with a few minute setae ventrally, middle femur with a conspicuous fringe of slender, bristly hairs posteroventrally. However, the hind femur has only 4 to 6 very slender, oblique bristles ventrally. Anterior tibiae smooth above, the tarsi small. Middle tibia with 2 or 3 bristles anterodorsally, but with 7 to 9 considerably stouter, though slender, oblique, sharp bristles posterodorsally. Middle basal tarsi obliquely spiculate. Hind tibia with 3 or 4 small ventral bristles but with 12 to 14 more conspicuous, longer, stout yet slender bristles posterodorsally. Claws small, slightly curved, pulvilli absent. All the femora and tibiae and the tarsi as well covered with flat-appressed, iridescent scales.

Wings: Relatively large, quite broad, completely hyaline, all the posterior cells widely open. There are 2 submarginal cells; the second vein has a gentle bend near the apex so that the marginal cell is gently rounded below apically. The anterior crossvein enters the middle of the discal cell. The second vein arises opposite it rectangularly, usually with a rounded bend, but sometimes with a rectangular bend. Anal cell closed in the margin or with a short stalk. Alula large, ambient vein complete. Basal comb wide with stiff black setae.

Abdomen: At the base as wide as the thorax, at the end of the third tergite even wider. Males with erect, coarse, brushy pile laterally on the first tergite and with a band of whitish, subappressed pile across the base of the second tergite, which is somewhat scaliform and is apt to be more conspicuous in the females and matted and appressed. In males the remainder of the second tergite and all of the third tergite are covered with small purplish to greenish-black scales that are fully appressed, but with the fourth and remaining tergites covered by a dense, appressed mat of silvery or sometimes cream-colored scales, extending over the posterior margin and each band overlapping and the last band forming a hood over the small, recessed, asymmetrical genitalia. Female with shorter, yellowish scales along the base of all the tergites but forming no conspicuous silvery mat. Remainder of the female tergites with the small, greenish or coppery scales.

The male terminalia from dorsal aspect, epandrium removed, show a basally wide, apically narrowing structure. The basal part of the aedeagus is very large and bulb-shaped, narrowing apically where it reaches under the remarkably prominent, wide, tongue-shaped epiphallus. Dististyli small, attenuate apically with a hooklike divergent process. From the lateral aspect, the basistyle is long and stout but the epandrium is relatively short and is subquadrate, but with prominent, conical, downturned basal lobes.

Material available for study: The type-species collected by author and wife in New Mexico, in 1926.

Immature stages: Unknown.

Ecology and behavior of adults: These flies are generally found resting on the sandy soil of semiarid regions. I have collected them on several occasions in the southwestern United States.

Distribution: Neotropical: *Astrophanes adonis* Osten Sacken, 1886; *andinus* Brèthes, 1909.

Genus *Chrysanthrax* Osten Sacken

FIGURES 170, 179, 663, 936, 937

Chrysanthrax Osten Sacken, Biologia Centrali-Americana, Dipt., vol. 1, p. 121, 1886. Type of genus: *Anthrax cypris* Meigen, 1820, as *Anthrax fulvohirta* Wiedemann. Designated by Coquillett, p. 523, 1910, the fourth of 4 species.

Small to medium-sized, *Villa*-like flies in which the oblique basal band or triangle of the wing is sometimes dark brown and sharply delimited, as in the type-species and a few other species but is more often pale, much reduced in color, more yellowish, the margin diffuse or reduced further to a few, small spots about crossveins and basal furcations.

Also, in this genus, the pile is uniformly short and entirely or almost entirely flat-appressed; margins of segments of the abdomen without tufts or extensive pile; general color of pile yellowish or golden, sometimes straw-colored or whitish, doubtless providing the name selected by the author of the genus.

The proboscis is slender, the labellum thin vertically, its shape lanceolate ovate and usually protruding a short distance beyond the oragenal recess. Anterior tibia quite smooth, reduced in size; pulvilli absent. The extent of contact of the fifth vein with the discal cell varies from short to commonly punctiform.

Chrysanthrax as at present known is characteristic of the southwestern United States; 17 of the 22 species included in the genus by Painter and Painter (1965) are from California; 2 or 3 species are found in the eastern and southeastern United States.

Length: 6 to 15 mm.

Head, lateral aspect: Head hemiglobular, the occiput strongly developed, shorter below, and the pile consists of appressed scaliform hairs, and a central fringe which becomes denser dorsally. The eye is large, the posterior indentation is quite shallow, although extensive above and below, and the bisecting line at the midpoint of the indentation extends inward over the eye for a considerable distance. The eye often has a strong greenish reflection. The front in profile is slightly protuberant or raised over the eye margin and covered with scattered, fine, erect, bristly pile. And in addition either with suberect, scaliform hairs or with numerous, slender, bright-colored scales. Similar hair and scales extend downward rather densely over the middle and sides of the face and form a loose fringe over the margin on the anterior part of the oral cup. The face is characteristically extended forward as a sharp, or sometimes lightly blunted triangle or cone, but not so sharp and extensive as in *Dipalta* Osten Sacken or *Stonyx* Osten Sacken. The proboscis is slender and may be confined to the oral cup or the labellum may extend beyond the cup. Palpus short and cylindrical with a scanty ventral fringe of hairs. The antennae are attached slightly above the middle of the head. They are small and slender. A rather tiny first segment is almost twice as long as wide, wider apically with numerous, short, fine setae on all sides, except ventrally. The second segment is small and beadlike, nearly as long as wide, with minute setae. The third segment is elongate conical, tapering into a slender style which occupies the outer half of the segment and bears a minute spine at the apex.

Head, anterior aspect: A little narrower than the posterior width of the mesonotum, nearly circular from the front. The eyes are separated by slightly more than the length of the first two antennal segments and by about 3 times the width of the ocellar tubercle. The ocellar tubercle is quite small, almost equilateral, sometimes nearly twice as long as wide. Front covered with pollen in addition to the pile. The antennae are quite widely separated and are not more distant from the eye margin than the length of the first antennal segment. They may be even closer. The oragenal cup is long, rather narrow, deep posteriorly with strongly thickened sides along the middle, and the recess is distinctly narrowed anteriorly. The oral cup extends downward below the eye margin to a moderate extent, most of which may be considered gena. Anteriorly the gena is linear, shelving inward for an extensive distance as a deep, conspicuous fissure. Laterally there is a bare circling band reaching from the eye margins to each antenna, on which pile is absent. Viewed from above the occipital lobes are extensive and nearly horizontal, although they slope downward to some extent. The ocellar tubercle is set a considerable distance forward from the slightly rounded posterior eye corners.

Thorax: Slightly wider posteriorly, the mesonotum opaque, pollinose, with blackish ground color, though generally lighter along the lateral margin, postalar callosity, and scutellum. The pile, except as denuded, is comparatively dense, fine, appressed, and scaliform. The anterior collar is dense and rather extensive in depth; it is comparatively short, but extends backward grading into the mesonotal pile. Notopleural bristles weak, postalar callosity with fine bristles or as many as 4 rather strong, long bristles. The rather flattened scutellum bears a fringe of numerous, rather fine bristles. There are dense tufts of long coarse pile on the propleuron and the whole of the mesopleuron and metapleuron. There is some scattered, stiff pile on the pteropleuron and a tuft of hair beneath the haltere. Plumula likewise pilose. Hypopleuron bare, except for 3 or 4 hairs above the posterior coxa. There is a fringe of short stubby hair on the spiracle. Squamal scales short.

Legs: Slender and weak, especially the anterior pair. Anterior tibia smooth. The tarsus has only fine hair beneath. Middle femur with 2 or 3 rather stout bristles at the middle or beyond. Hind femur with 5 to 7 rather strong bristles ventrally. Middle and hind tibia with a few sharp, rather slender, oblique bristles in 4 rows. The hind tibia has 6 or 7 bristles anterodorsally, the same number posterodorsally, 5 or 6 anteroventrally, and the same number smaller and finer posteroventrally. The more minute setate bristles have been omitted from this count. Hind tarsi strongly attenuate, the terminal segments are minute and slender, but setate below. Claws small, fine, very slightly curved, pulvilli absent. Anterior claws quite minute.

Wings: The wings vary from obliquely and irregularly but sharply reddish brown on the basal half or more as in the type-species to a pattern somewhat more common where the color is much more dilute and pale, but where there are smaller diffuse, darker spots on the crossveins. There are only 2 submarginal cells. The anterior crossvein enters the discal cell at or often beyond the middle of that cell and the second vein arises at or before the anterior crossvein. The contact of the discal cell with the fourth posterior cell is characteristically punctiform or nearly so. Anal cell open but only to a moderate extent. Alula moderately wide. Ambient vein complete. Costa rather strongly expanded at the base, but with minute setae and the costal hook sharp and thornlike.

Abdomen: Short oval. Generally wider than the thorax at least at the base. Color black, but usually with the sides of the tergites broadly and sometimes very broadly reddish yellow or orange and the sternites similarly pale, except in the middle. The pile is dense, flat-appressed, and distinctly scaliform or even forming lanceolate scales; generally some shade of yellow, orange, or red, there may be a few black scales intermixed and commonly there are some scattered black hairs along the posterior margins of the tergites, quite fine and suberect, or even entirely erect in the middles of the tergites, but becoming a little more numerous and conspicuous near the lateral margins of the tergites. The dense pile on the sides of the first tergite is conspicuous and blunt-tipped and often extends backward over the sides of the second tergite. There are 7 tergites visible from above. The last sternite may form a clasperlike lappet on each side. The acanthophorites have a fringe of 5 red spines on each side. Male terminalia large but rather strongly recessed with a tendency to be enclosed by sternal lappets.

The male terminalia from dorsal aspect, epandrium removed, show a broad, wide base, the whole figure narrowing apically. Dististyli are small, the epiphallus prominent and spade-shaped apically. From the lateral aspect, the small dististylus is incised dorsally and apically, with the apex sharp, hook-shaped and turned upward. It arises dorsoapically from the extensive basistylus. The epiphallus forms a curious, narrow, slender structure with a small dilated cap at the apex above and a triangular process below. The epandrium is odd; it forms a large, long, curved structure somewhat like an inverted U with the basal arm strongly extended downward.

Material for study: Many species from the southwestern United States, collected by the author and his wife on various expeditions—1926, 1949, 1954, 1959, and 1962.

Immature stages: Larvae of these species where the habits are known attack white grubs.

Ecology and behavior of adults: The adults are abundant and readily captured on low-growing flowers, especially composites. They may be found in temperate regions of high rainfall as well as in semiarid country.

Distribution: Nearctic: *Chrysanthrax adumbratus* (Coquillett), 1887 (as *Anthrax*); *altus* (Tucker), 1907 (as *Anthrax*); *arenosus* (Coquillett), 1892 (as *Anthrax*); *arizonensis* (Coquillett), 1887 (as *Anthrax*); *cinefactus* (Coquillett), 1892 (as *Anthrax*); *crocinus* (Coquillett), 1892 (as *Anthrax*); *cypris* (Meigen), 1820 (as *Anthrax*) [=*conifacies* (Macquart), 1850 (as *Anthrax*), =*fulvohirta* (Wiedemann), 1821 (as *Anthrax*)]; *dispar* (Coquillett), 1887 (as *Anthrax*); *edititius* (Say), 1829 (as *Anthrax*) [=*impiger* (Coquillett), 1887 (as *Anthrax*), =*sabulosa* (Coquillett), 1892 (as *Anthrax*)]; *eudorus* (Coquillett), 1887 (as *Anthrax*); *hircinus* (Coquillett), 1892 (as *Anthrax*); *junctura* (Coquillett), 1887 (as *Anthrax*); *lepidotoides* Johnson, 1919 [=*lepidota*, authors, not Osten Sacken]; *mirus* (Coquillett), 1887 (as *Anthrax*); *nebulosus* (Coquillett), 1894 (as *Anthrax*); *pallidulus* (Coquillett), 1894 (as *Anthrax*); *scitulus* (Coquillett), 1887 (as *Anthrax*); *tantillus* (Coquillett), 1892 (as *Anthrax*); *turbatus* (Coquillett), 1887 (as *Anthrax*) [=*comparata* (Tucker), 1907 (as *Anthrax*)]; *vanus* (Coquillett), 1887 (as *Anthrax*); *variatus* (Coquillett), 1892 (as *Anthrax*); *vulpinus* (Coquillett), 1892 (as *Anthrax*).

Neotropical: *Chrysanthrax astarte* (Wiedemann), 1830 (as *Anthrax*).

Genus *Cyananthrax* Painter

FIGURES 175, 373, 675, 938, 939, 940, 941

Cyananthrax Painter, Journ. Kansas Ent. Soc., vol. 32, p. 73, 1959. Type of genus: *Anthrax cyanoptera* Wiedemann, 1830, by original designation.

Rather large, peculiar flies of moderately lengthened abdomen, of brown or blue-black coloration, which have very large, broad, uniformly dark wings and the alula absent or quite linear. The face is quite conical, the proboscis confined to the oral recess, and the front tibiae are smooth; pulvilli vestigial. These characters are sufficient to identify this genus which is probably distantly related to *Poecilanthrax* Osten Sacken; like the latter genus *Cyananthrax* Painter has a distally directed spur vein from the lower outer corner of the discal cell. Unlike *Poecilanthrax* Osten Sacken the discal cell is separated from the fourth posterior cell by the presence of the posterior crossvein.

Only one species is known. I collected it around Tepotzlan in southern Mexico.

Length: 9 to 11 mm.; of wing 11 to 15 mm.

Head, lateral aspect: Head hemiglobular with strongly developed occiput, except near the bottom of the head where it is much shorter. Occipital pile bristly, subappressed in a forward direction. The central fringe about the exceptionally large postoccipital recess is dense and short. The front is slightly raised, covered with sparse, bristly hairs of no great length, especially dorsally and toward the sides. These hairs are slightly curled forward and the lower middle part of the front is covered with some appressed, short bristly hair. The face is rather acutely conical; it extends forward in a long triangle almost as far as the apex of the antennal style. The eye is slightly reniform. The posterior indentation is rather shallow but quite long vertically. The eye bulges upward and away from the occiput on the dorsal third. Upper facets are small, the anterior facets only slightly more enlarged. The face bears scattered, strongly appressed, short bristly hairs and at the peak of the face above the narrowed oral cup, there are 3 groups of short, stout, matted bristles. The proboscis is comparatively slender and is confined to the oral recess; its lower surface is grooved. The filiform palpus is only a fourth as long as the proboscis but with 6 or 7 moderately long, quite slender, bristly hairs along its ventral lateral margin.

The antennae are attached at the upper third of the head. The first segment is a little more than twice as long as wide. The second segment is short and annulate, quite plane apically. At most it is as long as wide and the third segment is not attached tightly. The third segment is elongate conical with nearly plane base beginning to narrow just before the middle into an apically very slender style with very minute spine at tip. The whole third segment is very little longer than the first two segments combined. The first segment bears some short, shaggy bristly pile.

Head, anterior aspect: The head is a little more narrow than the posterior part of the thorax, nearly circular from the front. Not quite as wide as high because of the prominent ventral production of the thin-walled oral cup, which also extends as far backward as the posteroventral plane of the eye. The face and front are relatively narrow. The front at the vertex is about 3 times as wide as the ocellar triangle and at the antenna the face is scarcely twice as wide as the vertex. The ocellar tubercle is small but high and prominent, higher than the ocelli, an equilateral triangle in shape and set well forward from the rounded posterior corners of the eye. The antennae are widely separated, divergent, and set distant from the eye margin by no more than the width of the second antennal segment. The oral cup is large and quite elongate, very deep, with thin walls posteriorly which are vertical and are distinctly thickened on the anterior half of the recess and likewise narrowed anteriorly. The gena is quite linear and there is a tiny tentorial fissurelike tip. Viewed from above the occipital lobes have a horizontal fissure rather than a vertical one.

Thorax: The mesonotum is distinctly wider posteriorly, shining black, with slender, short, flat-appressed, scanty pile over much of the disc, and also some scattered, erect, exceptionally fine, slender hairs. There are tufts of rather dense, erect bristly hair between the humerus and the wing. The notopleuron has 2 long, rather stout, curved, attenuate, sharp-pointed bristles. There are 3 bristles which are more slender above the wing and three, strikingly long, slender, black bristles along the postalar callosity, which are curved backward. Anterior mesonotal collar composed of a band of quite stiff, blunt-tipped, reddish-golden pile, continued across the front of the anterolaterally swollen humerus, and with a similar tuft directed upward from the dorsal half of the mesopleuron. Scutellum polished bluish black with long, very fine, erect hairs on the disc and longer slender bristles on the margin. The posteroventral mesopleuron, the upper sternopleuron, and the whole pteropleuron have scattered, fine, long, black, bristly hairs. Metapleuron with a vertical fringe of bristly pile and likewise some in front on the dorsal part. Hypopleuron entirely bare and no pile beneath the haltere. The plumula consists of coarse, long, white pile.

Legs: All of the femora are comparatively stout and the last four tibiae and tarsi likewise stout. However, the anterior tibiae and tarsi are very slender, completely smooth above; the tarsi reduced in size and bears fine, erect, nonglandular hairs. The anterior tibiae posteroventrally has an extremely dense, almost comblike band of short, stout setae. Middle femur with 1 or 2 bristles in front. Hind femur with at least 4 rather stout, long bristles on the ventral aspect of the distal half and 2 subapical bristles dorsally. Middle tibia with 5 to 7 weak, slender bristles in each row. Hind tibia with longer, slightly more stout bristles that are oblique. There are 4 anteroventrally, 6 anterodorsally, about the same number posterodorsally, and about 5 posteroventrally. All the femora are covered with flat-appressed, shining, black scales and some appressed setae. Claws very slightly curved, but curved backward, slender even at the base, and at most with minute stublike pulvilli.

Wings: Extraordinarily large, broad, blunt at apex, strongly narrowed at the base. The alula is wanting or at most quite linear and the axillary lobe greatly reduced throughout and attenuate basally. The entire wing is uniformly rather dark, sepia brown with a violet reflection above and below. There are 2 submarginal cells. There is sometimes a spur at the base of the second submarginal cell. The discal cell is very long and slender, and posteriorly the second submarginal cell is extended backward toward the wing base leaving the apex of the discal cell narrowed and pointed and the upper intercalary vein sigmoid. From the base of the second posterior cell there is a rectangular bend with a spur vein along the lower edge of the discal cell. Moreover, the discal cell lacks contact with the fourth posterior cell, and the posterior crossvein is generally present. The anal cell is widely open, the ambient vein complete. The costal comb is weak, but the basal spur is quite sharp and thornlike.

Abdomen: Abdomen elongate, and at the base barely or sometimes not quite as wide as the thorax. The abdomen is slightly narrowed posteriorly. The sides are nearly plane in both sexes to the end of the fifth segment, and the abdomen is at least 2½ times as long as wide in males, a trifle shorter in females. There are 7 tergites visible. In the female the seventh tergite is large, triangular. The abdomen is shining, bluish black in color, the blue reflection strong. The pile consists entirely of sparse, scattered, exceptionally fine, erect, black hairs of moderate length, and along the sides of the tergites, in addition to the longer hairs. They are particularly dense along the sides and apex of the last tergite. Female terminalia with a row of nine, rather long, blunt-tipped, black-tipped, slender, golden-red spines on each acanthophorite. Male terminalia small, asymmetrical, more or less recessed. Male individuals smaller than females.

The male terminalia from dorsal aspect, epandrium removed, show a triangular structure. The epiphallus is more narrow than usual but still the basal part of the aedeagus is extraordinarily large and prominent, bulblike-shaped at the apex, and the dististyle is small, arising subapically from the basistyle and with an apical hooklike process and incision as seems to be typical of the subfamily. From the dorsal aspect the dististyle is elongate and slender. From the lateral aspect, the basistyle is quite long; the dististyle shows the characteristic incision, the epandrium is triangular with rather sharp basiventral lobe.

Material available for study: A series of the typespecies collected by the author and his daughter in Mexico in 1951 and by the author and his late wife in Mexico in 1962. I have seen a second species.

Immature stages: Unknown.

Ecology and behavior of adults: I have captured individuals of both sexes hovering about and alighting upon shrubbery in Tepotzlan, Mexico. The locale was close to the borders of a mountain stream.

Distribution: Neotropical: *Cyananthrax cyanopterus* (Wiedemann), 1830 (as *Anthrax*).

Genus *Deusopora* Hull

Figures 173, 277, 664

Deusopora Hull, Journ. Georgia Ent. Soc., vol. 6, no. 1, pp. 1-3, 1971. Type of genus: *Deuspopora sapphirina* Hull, by original designation.

Medium-sized flies, recognized by the large, very broad base of the wings, the triangular scutellum, and the peculiar color pattern. The basal two-thirds of the wings in both sexes is sepia brown with a strong violaceous reflection, the outer third water-clear and sharply demarcated. The venation is more like *Rhynchanthrax* Painter, the projection of the face and the antenna more like *Thyridanthrax* Osten Sacken. The humeral cell is slightly widened but the setate brush narrow and instead of a curved spine, there is a broadly widened hook as in *Rhynchanthrax* Painter. The alula is narrow, the anal cell exceptionally widely opened. Mesopleuron with an anterior tuft and propleuron with a collar of orange pile. Abdomen elongate, rather narrow, shining black with emerald green scales and appressed, short, black, bristly or setate pile. All claws small but of nearly equal size.

This peculiar fly from southern Brazil is so aberrant and so unlike other members of the tribe *Villini* that I find it necessary to place it in a separate genus.

Length: 10 mm., wing 10 mm., base of wing 4 mm.

Head, lateral aspect: Head hemiglobular. The occiput is moderately prominent above, but shorter at the bottom of the head. The pile is flattened, brassy yellow, curled, and appressed flat. The central fringe is dense and short. The eye is large, rather long horizontally, with angulate, shallow, but extensive depression posteriorly, which extends from near the top of the eye to near the bottom. The eye is extensively bisected at the bottom of the indentation. The front is barely visible in profile, slightly more so on the lower half, with a shallow depression midway between ocellus and antenna, the pile erect and bristly. There are also some scattered, metallic-green scales on the front. The face is triangularly produced though not extensively produced. Medioapically the face, over a narrow, small space, is bluntly obtuse. Sides of face with a few pale scales and some appressed black setae and the anterior fifth of the oral cup bears a dense fringe of stout, black bristly setae overhanging the edge. A bare diagonal line goes from the eye margin to each antenna. Proboscis and labium distinctly shorter than the oral recess. The proboscis is rather stout, the labium thick. The palpus is long and cylindrical with a few fine, long hairs laterally and ventrally. It is more than half as long as the proboscis, labium excluded. The antennae are attached a little above the middle of the head, small, divergent, the first segment expanded medioapically and bearing moderately long setae on all sides including below. The second segment is distinctly shorter than wide, broadly attached to the third segment. The third segment is short conical and reaches almost maximum attenuation at the middle of the segment. Whole segment scarcely longer than the first two segments. There is a minute spine at the tip.

Head, anterior aspect: Wider than the thorax, circular, the vertex narrowly separated, almost equally narrow in both sexes. In the male the ocelli rest on the eye margin, and in the females it is separated from the eye margin by not more than the thickness of a posterior ocellus. The lower eye facets are pilose. The front at antenna is about 4 or 5 times as wide as at vertex. The antennae are widely separated and distant from the eye margin by the length of the first segment. The oral recess is long and slender, a little compressed anteriorly at the bottom of the eye, deep posteriorly, and extending quite back to the posterior plane of the eye. The genal suture along the anterior half of the oral cup is quite linear, the fissure itself threadlike. The oral cup is slightly produced below the eye. From above, the lobes of the occiput, though prominent and more or less horizontal, are distinctly separated with a considerable gap between them. The ocellar tubercle is placed at the point where the corners of the eye are rounded backward and outward.

Thorax: A little longer than wide with parallel sides, except for the postalar callosity. The humerus is not greatly protuberant, and the posterior edge and surface are rounded. Mesonotum black with thin, grayish-brown pollen and with numerous, metallic, malachite-green scales, which are violet, placed over each wing. There is also some minute, fine, short, suberect hair, a few longer hairs in front of the scutellum. Anterior collar narrow, rather long, the coarse pile sharp apically, and bright yellow, and covering the anterior part of the humerus. Scutellum clothed with green scales like the mesonotum, with brown pollen and a few fine, long hairs on the margin. Propleuron with tufts of long yellow pile. Upper anterior mesopleuron with tuft of long, yellow pile. Whole posterior half of mesopleuron and upper sternopleuron with silvery white pile. The pteropleuron has some scattered dark hairs, and similar dark reddish-brown hairs are abundant on the upper half of the metapleuron. The entire hypopleuron is bare and there is a tuft of pile beneath the haltere.

Legs: Moderately well developed, the front pair is scarcely reduced, the anterior femur and tibia are smooth. Middle femur without spines or bristles anteriorly and no ventral hair. Hind femur likewise without spines or fringe of hair below. All these femora are covered with dark, flat-appressed, shining, slender scales, and with a few very minute, appressed, fine, setae dorsally and laterally along the hind femur. Hind tibia with 5 or 6 fine, oblique, short, stiff, bristly hairs ventrolaterally. Dorsally, on both sides, the hind tibia has 20 or more short, somewhat more stout, oblique, spiculate setae. Anterior claws a little reduced, but no more than the posterior claws which are very small, sharp, scarcely curved, with no pulvilli.

Wings: Extraordinarily broad, a little less than twice as long as the maximum breadth. The wings are dark, sepia brown with strong violet to greenish reflection, this dark color is vertically delimited very close to the outer end of the discal cell. There are 2 submarginal cells. The second vein arises obliquely, distant from the anterior crossvein, by more than half the length of that crossvein. First posterior cell open maximally. Second posterior cell quite small, narrowed basally. Contact of the discal cell with the fourth posterior cell as long as the anterior crossvein. This crossvein enters the middle of the discal cell. Anal cell very widely open by a length two-thirds the length of the anterior crossvein. Second anal vein straight. Axillary lobe broad. Alula 2 or 3 times as long as wide.

Abdomen: Elongate oval, no wider than the thorax, with almost parallel sides for the first 4 or 5 tergites.

The length of the abdomen is 1½ to 2 times the length of the mesonotum, the scutellum excepted. The abdomen is shining black, rather densely covered with brilliant green, flat-appressed scales. The lateral sixth of the first tergite and anterior corners of the second have erect, coarse, moderately dense reddish-brown pile; the curled-over margins of the tergite and the last three tergites posteriorly have a few fine, black, appressed hairs. Eight tergites are visible from above. The eighth tergite is large, triangular, sloping or compressed downward or inward at the end of which the terminalia bear 6 pairs of black spines and laterally a tuft of fine yellow hair. These appear to be rather commonly exposed in the female by a beaklike gape of the last sternite. Male terminalia asymmetrical, small, and recessed.

Material available for study: The type series.
Immature stages: Unknown.
Ecology and behavior of adults: Unknown.
Distribution: Neotropical: *Deusopora sapphirina* Hull.

Genus *Dipalta* Osten Sacken

Figures 186, 370, 662, 665, 944, 945

Dipalta Osten Sacken, Bull. U. S. Geol. Survey, vol. 3, p. 236, 1877. Type of genus: *Dipalta serpentina* Osten Sacken, 1877, by monotypy.

Flies of medium size with conical face and 3 submarginal cells which are rather readily recognized by the mottled wings and the erratic venation. As Osten Sacken pointed out the second vein is greatly contorted, bent into the form of a recumbent S-like figure situated near the point of contact with the anterior branch of the third vein; however, the exact shape of this vein is very variable and generally there are one or more spur veins directed either backward or outward; however, spur veins may be absent.

The genus is related to the *Villa* Lioy complex rather than to *Exoprosopa* Macquart from which it is easily separated by the absence of the terminal stylelike segment attached to the third segment. It is most closely related to *Diplocampta* Schiner. Like that genus it has 3 submarginal cells, but the wing base in *Diplocampta* is very narrow, the bend in the third vein, anterior branch, very deeply kinked and sigmoid.

Dipalta Osten Sacken is known from one species ranging over northern Mexico and the western United States and one species from Virginia and Ohio. It is quite possible that there is only a single highly variable species among which a number of subspecies may be distinguished. I have collected it from nine localities in seven states including three from Mexico.

Length: 9 to 10 mm.; wing 10 to 13 mm.

Head, lateral aspect: The head is hemiglobular and rather triangular in shape because of the strongly peaked or conical and triangular face. The occiput is extensive, especially on the upper part of the head. At the bottom of head, it is comparatively short. The pile is scanty, scaliform, and completely flat-appressed; the central fringe around the recess is dense, but short. The front is very slightly raised above the eye margin, flat across the middle, or even slightly depressed. Its pile is erect, not longer than the first antennal segment, abundant without being dense, and extending from vertex to the base of the antenna. In addition there are numerous, slender, flat-appressed scales on the front; the vertex, however, is pollinose only in addition to the hair. Face produced forward as an almost equilateral triangle. Its pile consists of scanty, scattered, short, subappressed hairs, a fringe of somewhat stiffer, more bristly hairs in the middle along the edge of the oral cup, and also numerous appressed scales like those of the front. There is also a narrow, scarlike, bare band or creaselike band, extending from near the middle of the oral recess laterally around to the base of each antenna. The eye is large, upper facets small, anterior facets enlarged somewhat in the middle of the eye. There is a deep, but very broad or extensive indentation, and a bisecting line at the middle of the eye posteriorly which extends a short distance forward. The eyes have a strong, greenish reflection.

The proboscis is short, rather slender, a little flattened, grooved laterally, and not extending beyond the oral recess. The labellum is rather long, slender, and flattened. The palpus is long and filiform with a few, fine, erect hairs laterally. The antennae are attached at the upper third of the head and are comparatively small. The first segment is twice as long as its basal width, but it is a little widened apically, and there are some short setae dorsally, slightly longer setae laterally. Second segment narrow at the base, rather rounded and beadlike, and not as long as wide. The third segment forms a short, bulblike base, triangular outwardly; it is slightly wider than the second segment and has a short, abruptly formed, slightly attenuate style, carrying a minute spine at the tip.

Head, anterior aspect: The head is not quite as wide as the posterior mesonotum. It is nearly circular from the front. The eyes in the male are separated by about twice the width of the ocellar tubercle. Ocellar tubercles are strongly and abruptly raised and set distinctly forward from the posterior eye corners, and the ocelli occupy an equilateral triangle. The posterior eye corners are only slightly rounded. The antennae are comparatively widely separated, but are distant from the eye margin by the length of the first antennal segment. The oral opening is long, large, quite deep, with vertical sides in the middle, overhanging edges posteriorly. The margin is paper-thin posteriorly, a little thickened on the anterior half. Gena linear. Tentorial pits absent. Viewed from above, the lobes of the occiput have a vertical fissure.

Thorax: The mesonotum is distinctly widened posteriorly but the humerus is prominent, vertically truncate posteriorly, and extends outward considerably, and the dense, stiff, erect hair of the anterior collar extends over the humerus in full. Prosternum with long tufts of equally stiff and bristly hair. The mesonotum is more or less opaque black, generally with dark brown pollen and loosely covered with flat-appressed, short, pale, slightly scaliform hair, which extends also over the whole scutellum, but both areas are apt to be largely denuded. The notopleuron has several long, slender bristles and others extend up to the base of the humerus. Postalar callosity with long slender bristles. The margin of the scutellum may have 4 or 5 pairs of still more slender bristles. Upper half of mesopleuron with numerous, long, stiff or bristly hairs, pale in color, extending outwardly and upwardly. Remainder of mesopleuron with more slender or shorter pile appressed backward or downward. The pteropleuron has an anterior tuft of stiff pile. The whole of the metapleuron is covered with long, stiff hairs, and there is an anterior and posterior tuft of appressed, stiff hair beneath the haltere. Hypopleuron bare, front and behind. Squama with a very short fringe of scales.

Legs: Comparatively strong without being stout or swollen. Anterior and middle femur, slightly narrowed apically. The middle femur has 2 or 3 rather strong bristles anteroventrally at the middle. Hind femur with 5 or 6 more slender, short bristles along the posteroventral margin; however, the last one distally is a little more stout. Anterior tibia slender and characteristically quite smooth without bristles or spicules. Anterior tarsi reduced in length, and the whole of the anterior pair of legs somewhat reduced in size. Anterior tarsus bears vertical, glandular-tipped hairs vertically. Middle tibia with some quite fine, short, slender bristles, 7 or 8 anterodorsally, the same number posterodorsally and posteroventrally, but only 4 or 5 anteroventrally. Hind tibia long, with 15 or more bristles in the anterodorsal row. There are about 10 in the posterodorsal row, 7 or 8 in the posteroventral row, but only 5 or 6 anteroventrally; all these bristles are only slightly stronger than those of the tibia. The hind femur may bear 7 or 8 oblique spicules of bristles dorsally at the middle and on the outer half. Claws sharp, curved from the base, without pulvilli.

Wings: The wings are large and elongate, characteristically mottled with comparatively sharp, dark brown spots and irregular brown group of coalesced spots crossing the wing at the middle where the second vein arises. A still more irregular wide group of coalesced spots are located subapically on the outer third of the wing. The venation is quite erratic especially with regard to the second vein, and to a less extent with the third vein. The second vein arises abruptly near or very close to the anterior crossvein, but makes a rounded bend outwardly. There is a deep loop at the end of the marginal cell, sometimes with a spur outward from this loop, and there is a second broad loop forward, located in front of the base of the second submarginal cell, of very variable shape and sometimes with a backward spur. There are 3 submarginal cells formed, as I interpret it, by the posterior branch of the second vein, going backward to join the anterior branch of the third vein. This seems to me a logical interpretation of this wing. Base of the second submarginal cell sometimes with a backward spur. Contact of the discal cell with the fourth posterior cell comparatively short. Anal cell open, usually widely. Axillary lobe long and narrow. Alula present with circular posterior margin. Ambient vein complete. Costal comb large, the setae minute, the basal spur sharp and thornlike.

Abdomen: The abdomen is as broad as the thorax, slightly elongate in most forms. The abdomen is usually a dully shining black color, in some forms yellowish along the sides. The whole surface is rather abundantly covered with short, flat-appressed, yellow, scaliform pile with a little black hair in the middle. In addition, there is generally a loosely formed, very fine, obliquely erect band of fine, black, bristly hairs of no great length along the posterior margin of the tergites. Normally 7 tergites are visible. Occasionally, the eighth and ninth can be seen, very small and short. Male terminalia asymmetrical, more or less recessed and of no great size.

The male terminalia from dorsal aspect, epandrium removed, show a figure very broad on the basal half and the epiphallus very large, wide apically, with rounded, tonguelike structure. From the lateral aspect the small, rather elongate, slender dististyle is incised apically with upturned hook. The epandrium is rectangular with a rather prominent, downturned lobe basally.

Material available for study: A considerable series collected by Marguerite Hull and me in Western States and in Mexico. Several new species may be involved. What bee flies may parasitize the giant *Palpares* in South Africa?

Immature stages: Flies of this genus parasitize antlions (Myrmeleonidae).

Ecology and behavior of adults: I have found these flies comparatively abundant on low-growing flowers in desert areas. I have no record of the plants, but they were found frequently on a greenish-yellow flower in the Ruidoso Mountains of New Mexico.

Distribution: Nearctic: *Dipalta banksi* Johnson, 1921; *serpentina* Osten Sacken, 1877.

Genus *Diplocampta* Schiner

Figures 195, 391, 671, 674

Diplocampta Schiner, Verh. zool.-bot. Ges. Wien, vol. 17, p. 312, 1867. Type of genus: *Diplocampta singularis* Schiner, 1868, by original designation.

Curran (1931), pp. 7-8, key to 3 West Indian and Neotropical species.

Small flies of *Villa*-like aspect, with 3 submarginal cells, but at once distinguished by the characteristic S-shaped bends in the anterior branch of the third vein and apical part of second vein. Moreover, the wing is strongly narrowed toward the base, with widely open anal cell, which together with the scales on face and abdomen, the small size, etc., relate them to *Mancia* Coquillett which shares these peculiarities, except that its venation is normal. The face of *Diplocampta* Schiner is moderately produced but blunt and for a portion of its height nearly vertically plane. Anterior tibia without spicules. These flies do not appear to be closely related to *Neodiplocampta* Curran, which, instead of an onion-shaped third antennal segment, has a long conical one, has 2 submarginal cells and only gently sigmoid radial veins, and is besides larger, with the wing normally wide at base and scales rather more poorly developed.

This small genus is known only from southern South America.

Length: 4 to 6 mm.

Head, lateral aspect: The head is subglobular, the face prominent but blunt anteriorly, hence there is no conical peak. The face is covered loosely with beautiful broad, oval, rather pale scales, together with loosely scattered, not very long, erect, or subappressed, black, bristly hairs. The front has the same kind of covering of scales and hairs, the bristly pile somewhat more erect, especially on the upper half of the front where it is also somewhat longer. The occiput is comparatively short, with large, central cavity that has a dense fringe of short, erect pile along its margin. Lateral portion of occiput conspicuously but loosely covered with short, pale scales. Occiput bilobate above with deep medial fissure. Posterior eye margin very shallowly indented but obtusely angular in the middle with short, medial notch. The eye is almost as long as high. Palpus long and slender, curled upward with long, scattered, very fine hairs ventrally. Proboscis short, not extending beyond the oral opening. The antenna is attached at the upper third of the head and all 3 segments quite short. The first segment is a little longer than the second; it bears numerous, short, erect, black setae laterally and also medially. The second segment is 2 or 3 times wider than long, third segment wider than the second at the rimlike base which is no longer than the second segment and then abruptly expands into a long, stylelike process of approximately 2 or more times the length of the basal, bulbous part. Apex with a minute spine and possibly 1 or 2 very minute microsegments.

Head, anterior aspect: The head is distinctly wider than the thorax, the vertex of the female is narrow, about one-third as wide at the front as the antenna. The antennae are widely separated, each is separated from the eye margin by a little less than the width of the first segment. Ocellar tubercle low, of an equilateral triangular shape and with vertical sides.

Thorax: The thorax is short, the mesonotum dully shining black, very slightly convex and rather loosely covered with narrow scales which may be brownish in the middle and white anteriorly. Anterior margin with a narrow collar of erect, not very long, coarse pile. The protuberant humerus bears long, coarse hairs and similar scaliform pile which is continued down over the notopleuron. The notopleuron has 3 delicate, slender, black bristles. The postalar callosity has 2 or 3 longer, somewhat stouter, sometimes reddish bristles. Margin of the scutellum with 3 or 4 pairs of longer, slender bristles which are black or sometimes yellow, the inner pair crossed. Squama with a fringe of slender scales. Mesopleuron with scanty, loosely scattered, long, bristly hairs curled upward along the dorsal border, and a few appressed scaliform hairs posteroventrally. Pteropleuron likewise with a few long bristles or bristly hairs curled backward. Metapleuron with a vertical posterior band of long, fine, erect pile. Hypopleuron bare.

Legs: The legs are slender, short, and delicate. The femora and tibiae bear rather dense, matted, pale, glistening scales. Middle femur with a fringe of loose pile posteriorly. Anterior tibia without spicules of any kind, but middle and hind tibia with well-developed spicules though unusually slender and of no great length; these are situated in 4 rows containing from 7 to 16 elements. Claws exceptionally slender, pulvilli absent.

Wings: The wings are characteristically narrowed toward the base and a little broadened and rounded apically at least by contrast with the base. The shape is much as found in *Mancia* Coquillett. The apex of the wing is sometimes tinged with brown, otherwise hyaline. The venation is characteristic, the second vein arises at or barely before the anterior crossvein; it is rectangular with basal spur. The second vein beyond the fork on its anterior branch is remarkably sigmoid and contorted, quite S-shaped. The anterior branch of the third vein forms a short, plane, rectangular crossvein. Anal cell widely open, alula short, base of the costa with a small, flangelike expansion and comb, the basal lobe spinelike. The proximal vein at the base of the third posterior cell has an inward spur into the discal cell.

Abdomen: The abdomen is about 1½ times as long as the thorax and not any wider at the base. From the base it is gently and regularly narrowed and rounded at the apex. Surface covered with rather dense, flat-appressed, pale scales often of two or more colors. Posterior margin of the first tergite with a conspicuous row of radiating scales and on the anterior half with a few fine, scattered, erect hairs changing in the anterolateral corners to dense, coarse, short pile. Posterior margin of the tergites also with a scanty fringe of more or less appressed, very fine, slender, black, bristly hairs. Sternites similar to the tergites.

Material studied: Examples in the collection of the British Museum (Natural History).

Immature stages: Unknown.

Ecology and behavior of adults: Not on record.

Distribution: Neotropical: *Diplocampta secunda* Paramonov, 1931; *singularis* Schiner, 1868; *subsinuata* Oldroyd, 1938.

Genus *Hemipenthes* Loew

FIGURES 177, 178, 183, 185, 187, 413, 672, 678, 942, 943

Hemipenthes Loew, Berliner Ent. Zeitschr., vol. 13, p. 28, nota 1869. Type of genus: *Musca morio* Linné, 1758. Designated by Coquillett, p. 550, 1910; two species as one.
Isopenthes Osten Sacken, Biologia Centrali-Americana, Dipt., vol. 1, pp. 80, 96, 1886. Type of genus: *Anthrax sinuosa* Wiedemann, 1821, as *Isopenthes jaennickeana* Osten Sacken, 1886, subspecies. Designated by Coquillett, p. 556, 1910, the second of two species.

Engel (1936), pp. 563-564, key to 7 Palaearctic species.
Painter (1962), pp. 89-90, key to 12 species from the United States.

Small to rather large flies with the basal half, and sometimes more than half or even the whole wing, dark sepia brown or blackish; the blackish color is sharply defined, diagonal or oblique, either with irregular, steplike edge or smooth, and often extending no farther backward than the end of the second submarginal cell. The face is either rounded and retreating, or at most blunt and vertical, apt to be rather heavily covered, together with the front, with erect stiff pile. The third antennal segment is onion-shaped at base, rather abruptly narrowed with long, fine style as in *Rhynchanthrax* Painter, but never so strongly reduced to a bulblike base. Anterior tibia smooth, without spinules or at most fine, sharp, small ones scarcely noticeable. The flies of this genus, as Bezzi has pointed out are rather readily distinguished from *Villa* Lioy by the wide, obtuse ending of the discal cell compared with the acute apex in *Villa* Lioy. Bezzi calls attention to the small costal comb in comparison with *Villa* Lioy; he calls the metapleural fringe sparse, and this may be true of some European species, but I do not find it so in my material.

Hemipenthes Loew is abundant in the Palaearctic and Nearctic regions, less so elsewhere and apparently absent from South Africa.

Length: 5 to 14 mm.

Head, lateral aspect: The head is hemiglobular, the occiput moderately developed, a little shorter below and the pile of the occiput is fine and appressed forward toward the eye and in one species before me, from Mongolia, some of the short setate hair is nearly erect. In some species it is nearly scaliform. The eye is reniform, the facets of nearly equal size, the posterior indentation is shallow, but quite broad, extending some distance above and below the middle of the eye. The eye is distinctly bisected by a line extending a considerable distance inward over the eye from the midpoint of the excavation. The front is comparatively wide, covered with dense, conspicuous, erect, black, bristly pile and a few small, short appressed scales in some species. These scales are virtually lacking in some species.

The face varies from quite bluntly protuberant, nearly or quite vertical beneath the antenna, to others in which it is even rounded and retreating. The face is covered with similar, dense, bristly pile and scales like that of the front, and this bristly pile forms a distinct fringe, extending over the margin of the oral cup on the anterior half of the cup. The proboscis is short characteristically, varying from species where it extends just to the apex of the oral recess to others in which it is scarcely more than half as long as the oral recess. This marked reduction in the length of the proboscis is perhaps a trend in the same direction as such species as *Anthrax* Scopoli and *Oestranthrax* Bezzi. The proboscis is quite stout and the labellum large and fleshy as in a muscoid. Palpus slender, cylindrical or filiform, often curled or coiled, and with a fringe of fine hairs ventrally and laterally. Sometimes the labellum though large is elongate oval and flattened. The antennae are attached at the middle of the head and are comparatively small and short. The first segment is about twice as long as wide, a little widened apically or extended medioapically and bears especially on the outer half numerous, moderately long, stiff, bristly hairs, never quite as long as the segment itself. They may extend almost to the base. The second segment is short and beadlike, rounded in the middle, and narrowly attached to the third segment. The third segment is as wide as, or very slightly wider than, the third segment, bulbiform, onion-shaped, or very short conical, to which is attached a rather slender style. The style is approximately as long as the first two antennal segments or very slightly longer, and it bears a minute spine at the apex.

Head, anterior aspect: The head is approximately the same width as the mesonotum between the wings. It is nearly circular from the anterior view, or may be very slightly flattened on all four sides. The eyes are narrowly separated in the male by a distance approximately equal to the length of the first anterior segment. They are a little more widely separated in the female. The ocellar tubercle is nearly equilateral, has vertical sides, is almost hidden by the overgrowth of the eye on each side. The antennae are less widely separated than in some related genera, the distance between the antennae being little more than one-third the width of the front at this point. The oral cup is rather wide, about 2½ times as long as its width. It is deep anteriorly and especially posteriorly. In some species the sides are quite wide anteriorly, in others narrow and sloping. The gena is rather narrow but the tentorial fissure is deep midway between the antenna and the posterior end of the recess. This fissure is conspicuously deep. Viewed from above the lobes of the occiput are swollen posterodorsally and consequently the fissure is more or less horizontal with a pitlike depression behind the ocellar column.

Thorax: Wider posteriorly, blackish in color with loosely scattered, fine, appressed hair or scaliform pile. The humerus is rather strongly swollen out laterally and the wide anterior collar of stiff pile extends over and includes the whole of the humerus. In both American species and species from Central Asia there is often a dense, lateral band of appressed pale pile running down the sides of the mesonotum and extending over at least part of the postalar callosity. In some species there is additional pile on the disc of the mesonotum, consisting of exceptionally fine, erect hairs. Notopleuron with a few bristles and sometimes they may be quite stout. Similarly there may be several long stout bristles on the postalar callosity and some bristles, either fine or stout, on the margin of the scutellum. The disc of the scutellum, unless denuded, is covered very much like the mesonotum. The pleuron is rather densely pilose, with dense, coarse tufts on the propleuron, most of the sternopleuron, whole of the mesopleuron and metapleuron, and most of the pteropleuron. The hypopleuron is bare above the middle coxa. There is a little pile above the posterior coxa and a distinct short, appressed tuft of pile beneath the haltere. Squamal scales short.

Legs: Comparatively stout, but not swollen. Anterior femur with a fringe of stiff hairs ventrally on both sides in some species. The anterior tarsi are nearly smooth in some species but may have a few oblique slender setae or bristly hairs dorsally. Middle femur with stiff hairs beneath in some species, hind femur similar but may have 3 or 4 long slender bristles anteroventrally of varying proportions, rather short in some species. Basal half of hind femur with some long, fine, stiff hairs ventrally. All the femora and tibiae have some scales which may be conspicuous and long in some species, especially where they are pale in color. Middle tibia with a few, fine, oblique bristles. Hind tibia with slightly stronger bristles in 4 rows. Including small elements there are 15 to 25 bristles and spiculate setae in the anterodorsal row, and 10 to 15 in the posterodorsal row. There are 5 to 7 in each ventral row. In at least one species the hind tibia is a little flattened and strongly feathered above and below with conspicuous blackish scales. The tarsi are slender and attenuate posteriorly. Claws small, sharp, nearly straight, with small pulvilli. Anterior claws very minute.

Wings: Comparatively uniform in width. The wings are wide apically and basally with a characteristic black or reddish-brown pattern, obliquely, but irregularly and sharply distributed on the basal half of the wing. Sometimes this dark pattern leaves as much as half of the anal cell and axillary lobe clear and sometimes with this area almost entirely blackish and with the dark color sometimes ending at or even before the middle of the wing and in others more or less filling completely the marginal cell and beyond. There are only 2 submarginal cells. The second vein arises abruptly at just before or just beyond the anterior crossvein. Marginal cell generally bent backward like a thumb at the apex. Usually there are some faint spots about the crossveins within the dark basal portion of the wing. The anterior crossvein enters the middle of the discal cell. The upper anterior intercalary vein may be oblique or rectangular, and apically the discal cell may be relatively narrow or may be widened above and below. Contact of discal cell with the fourth posterior cell usually not longer than the length of the anterior crossvein. Anal cell widely open. Alula narrow. Ambient vein complete. Costal comb weak and narrow. The costal hook sharp and thornlike.

Abdomen: Short oval, sometimes no wider than the thorax, again distinctly wider though not greatly. The color is black. There are 7 tergites, which are clothed with moderately dense hair either appressed or subappressed, fine and curly, or matted and scaliform, black, whitish, yellow or brownish orange in color, and often mixed. Sides of the tergites with a rather inconspicuous fringe of coarse hair and sometimes larger tufts. In some species the dense lateral fringe of the first tergite extends entirely down the sides of the second tergite, and there may be alternating tufts of black and pale pile. Females with a few slender bristly spines along the acanthophorites. Male terminalia small, conical and asymmetrical, partly exserted beneath the last tergite. The remaining tergites beyond the seventh are completely tucked in under the seventh.

The male terminalia from dorsal aspect, epandrium removed, show a basally wide, apically narrowed figure with broad, long, spatulate or tonguelike epiphallus with concave sides, very blunt at apex and the dististyli are nearly apical, as they arise from the basistylus. Basal part of the aedeagus large, very short oval, and rounded both basally and apically. From the lateral aspect the epiphallus has a ventral extension in the middle, the dististylus is comparatively short, basistylus long and the epandrium has a wide, archlike or concave incision ventrally from the anterior to the posterior corners.

Material available for study: In addition to the series at several museums, I have material sent me by the late Dr. Timon-David and also considerable material collected by myself and my late wife in the western United States.

Immature stages: The larvae of *Hemipenthes* Loew behave as hyperparasites of parasitic Hymenoptera such as the genera *Banchus* and *Ophion* and also hyperparasites of such tachinids as *Masicera* Macquart, which in turn have attacked nocturnal *Lepidoptera*.

Ecology and behavior of adults: Species of this genus are abundant on various types of flowers, mostly composites.

Distribution: Nearctic: *Hemipenthes bigradata* (Loew), 1869 (as *Anthrax*); *castanipes* Bigot, 1892; *catulina* (Coquillett), 1894 (as *Anthrax*); *celeris* (Wiedemann), 1828 (as *Anthrax*) [=*floridensis* Macquart, 1849]; *chimaera* (Osten Sacken), 1887 (as *Anthrax*); *comanche* Painter, 1962; *curta* (Loew), 1869 (as *Anthrax*); *edwardsii* (Coquillett), 1894 (as *Anthrax*); *eumenes* (Osten Sacken), 1887 (as *Anthrax*); *floridana* (Macquart), 1850 (as *Anthrax*); *incisiva* Painter, 1962 [=*incisa* (Walker), 1852 (as *Anthrax*)]; *inops* (Coquillett), 1887 (as *Anthrax*);

lepidota (Osten Sacken), 1887 (as *Anthrax*); *mobile* (Coquillett), 1894 (as *Anthrax*); *morio* (Linnaeus), 1758 (as *Musca*) [=*morioides* (Say), 1823 (as *Anthrax*)]; *pima* Painter, 1962; *pullata* (Coquillett), 1894 (as *Anthrax*); *sagata* (Loew), 1869 (as *Anthrax*) [=*orbitalis* (Williston), 1901 (as *Anthrax*)]; *scylla* (Osten Sacken), 1887 (as *Anthrax*); [=*succincta* (Coquillett), 1894 (as *Argyromoeba*)]; *seminigra* Loew, 1869; *sinuosa* (Wiedemann), 1821 (as *Anthrax*) [=*assimilis* (Macquart), 1846 (as *Anthrax*), =*concisa* (Macquart), 1840 (as *Anthrax*), =*gideon*, authors, not Fabricius, =*nycthemerus* (Macquart), 1840 (as *Anthrax*)], *sinuosa blanchardiana* (Jaennicke), 1867 (as *Exoprosopa*), *sinuosa jaennickeana* (Osten Sacken), 1886 (as *Isopenthes*); *webberi* Johnson, 1919; *wilcoxi* Painter, 1933; *yaqui* Painter, 1962.

Neotropical: *Hemipenthes minas* Macquart, 1848.

Palaearctic: *Hemipenthes alata* Becker, 1907; *argentifrons* Becker, 1907; *chorassani* Becker, 1913; *circumdata* Becker, 1912, *circumdata quinquefasciata*, Becker, 1912; *gaudanica* Paramonov, 1927; *hamifera* (Loew), 1854 (as *Anthrax*); *intermedia* Paramonov, 1927; *latipennis* Paramonov, 1927; *maura* (Linné), 1758 (as *Musca*) [=*bifasciatus* (Meigen), 1804 (as *Anthrax*), =*daemon* (Panzer), 1797 (as *Anthrax*), =*denigratus* (Linné), 1767 (as *Anthrax*), =*hirsutus* (Villers), 1789 (as *Anthrax*)], *maura flavotomentosa* Paramonov, 1926; *melaena* Bowden, 1965; *mischanensis* Paramonov, 1926; *morio* (Linné), 1758 (as *Musca*) [=*morioides* Say, 1823, =*semiater* (Meigen), 1820 (as *Anthrax*), =*seminiger* (Loew), 1869]; *nitidofasciata* (Portschinsky), 1891 (as *Anthrax*); *praecisa* (Loew), 1869 (as *Anthrax*); *rufipes* Becker, 1912; *subarcuata* (Loew), 1870 (as *Anthrax*); *transcaspica* Paramonov, 1926; *uncina* Loew, 1869; *velutina* (Meigen), 1820 (as *Anthrax* [=*bicinctus* (Wiedemann in Meigen), 1820 (as *Anthrax*), =*holosericeus* (Meigen, not Fabricius), 1804 (as *Anthrax*), =*nycthemerus* (Wiedemann in Meigen), 1820 (as *Anthrax*)].

Genus *Lepidanthrax* Osten Sacken

FIGURES 211, 567, 702, 946, 947

Lepidanthrax Osten Sacken, Biologia Centrali-Americana, Dipt. vol. 1, p. 107, 1886. Type of genus: *Anthrax disjunctus* Wiedemann, 1830. Designated by Coquillett, p. 559, 1910, the first of 3 species.

Podolepida, new subgenus. Type of subgenus: *Exoprosopa lutzi* Curran, 1930.

Curran (1930), pp. 1-2, key to 12 North American species.

A genus of comparatively small flies readily distinguished by the length of the slender proboscis which extends a considerable distance beyond the end of the oral recess and is nearly or sometimes quite twice the length of the head. Another important character concerns the antenna; the third segment is minute, abruptly onion-shaped, the style quite long, almost threadlike and without suture and ends in a minute, short bristle at the apex; the antennae are unusually widely separated.

The front tibia bears distinct, slender spinulate bristles dorsally. There are no pulvilli. The wings are completely hyaline in some species with only the subcostal cell yellowish. In other species the costal cell is tinted and a small spot about the anterior crossvein, and in still others there is a dark brown costal cell and a distinct pattern of small, more or less isolated blackish spots. First posterior cell in some species with supernumerary crossvein near the end of the discal cell. Contact of discal cell with posterior cell unusually long. The tergites of the broad, short, rather flattened abdomen have numerous appressed scales, and the lateral margins have extending fringes of scales which are sometimes quite wide. These mats of scales are sometimes brilliantly silver or pale golden.

This group of flies is found widely in the southwestern United States and Mexico and ranging east to Texas and New Mexico and south to Panama. Two species from India have been assigned by Brunetti to this genus and two from Australia by Roberts; I have not seen these.

Length: Most species 5 to 9 mm., proboscis excluded; 1 species reaches 11 mm.

Head, lateral aspect: Head semiglobular, the occiput unusually prominent, especially in the middle on account of the long, quite deep, angular indentation of the eye. Eye sometimes with a short bisecting line, or sometimes only with the facets curving forward at the bottom of the indentation. The pile of the occiput is dense and composed of conspicuous, short, broad, flat-appressed scales; the posterior fringe around the central recess is rather long and dense. The front is flattened or in some species very slightly raised in front of the ocelli and flattened or even a little depressed on the lower half. Pile of the front and vertex, consisting of comparatively short, erect, fine hairs which are scanty and sparse, but stiff and bristly in the larger species. Characteristically the front is also covered densely with a conspicuous mat of broad, pale-colored, flat-appressed scales. The face is conically or triangularly extended forward conspicuously, extending well beyond the end of the third antennal segment if the style be excepted. Pile of face, like that of the front, with scattered, fine, bristly hairs and numerous broad, conspicuous, pale appressed scales.

The proboscis is long and slender, extending far beyond the oral recess. Its total length is about 1½ times the head length; the labellum is small and slender, and the surface microspinulose. Palpus long, slender, and filiform with fine, erect, stiffened, ventral hairs. The antennae are attached at the middle of the head. They are small, the first segment slightly longer than the second, both these segments a little widened apically. The third segment a little rounded at the base, the whole segment is small and more or less globular or bulblike, with an abruptly formed, long, slender, filiform style, rather like that of *Stonyx* Osten Sacken. The style bears a minute spine at the apex.

Head, anterior aspect: The head is not quite as wide as the posterior width of the mesonotum. It is nearly circular, slightly flattened above, below, and laterally, the eyes in the male are separated by a little more than twice the width of the ocellar tubercle, a little more widely in females, and the entire face and front comparatively narrow. Upper eye facets small. Anterior eye facets opposite the antenna enlarged. The eyes often have a greenish tinge. They are large, high, and reniform. The antennae are quite widely separated from the eye margin by not more than the length of the first segment. The oral recess is quite long, slender, deep, with vertical sides, even anteriorly, the margin thin, the cavity a little wider posteriorly and the oral cup extending a little downward beneath the eye. The tentorial fissure is rather anterior to the middle of the recess, and the gena appears to be linear. There is a small bare triangle at the bottom of the posterior eye corner. The ventral eye corners are broadly rounded in some species, angular in others, but the vertical eye corners are broadly rounded in all species.

Thorax: Short, almost quadrate, a little wider posteriorly as in *Stonyx* Osten Sacken. The mesonotum is feebly shining with faint pollen, densely covered with appressed, slender scales, and the scutellum similarly covered. There may be a few long, slender, bristly hairs in front of the scutellum and in front of the wing. Notopleuron and postalar callosity with comparatively stout, moderately long bristles. Margin of the subtriangular scutellum with rather long, slender bristles. The anterior collar is rather extensive, comparatively long and stiff anteriorly, grading off into shorter pile posteriorly. Only the upper half of the mesopleuron has bristly pile. The lower half in most species has numerous, quite broad, conspicuous, pale scales, occasionally more slender in some species. Both pteropleuron and sternopleuron with patches of scales and some stiff hair. The pteropleuron may have hair only. Metapleuron with a conspicuous, vertical fringe of stiff, bristly hair. Posterior hypopleuron below the haltere with a patch of scales, and there is a tuft of hair immediately beneath the haltere, another one posterodorsally behind the spiracle. Haltere small with a small knob. Squamal fringe composed of slender scales.

Legs: Neither stout nor slender, the anterior femur a little thickened toward the base, and all the femora and tibiae densely covered with broad, short scales. Anterior femur sometimes with a row of long, stiff hairs ventrally, bristles absent. Middle femur with 2 to 4 stout bristles anteroventrally along the middle. Hind femur with 4 or 5 similar, slightly larger bristles along the anteroventral margin. Anterior tibia with 5 or 6 minute, fine bristly spicules on the anterodorsal segment and as many as 10 on the posterodorsal margin. Middle tibia with 5 to 7 stronger and longer bristles on the anterodorsal margin, 8 or 10 on the postero- dorsal margin and 6 or 7 on the posteroventral margin and almost as many weaker bristles on the anteroventral margin, making 4 rows. Hind tibia similar, the bristles equally large but fewer in number. All of the tarsi are quite weak, poorly developed, short and slender and attenuate. Claws scarcely curved, pulvilli absent.

Wings: Characteristically slender, varying from hyaline with only the subcostal cell tinted yellow to species with 1 or 2 small brownish spots to species with numerous, sharply defined, brown spots. There are 2 submarginal cells. One group of species has an extra crossvein across the first posterior cell near the end of the discal cell. These are much larger than the usual *Lepidanthrax* and may be placed in the subgenus *Podolepida*, new subgenus, with *lutzi* Curran as type. The discal cell is in contact with the fourth posterior cell for a very considerable distance. The first posterior cell is a little narrowed apically. The anal cell is very widely open apically. Alula well developed. Ambient vein complete. The costal comb varies from small to wide and extensive. The costal hook is sharp and thornlike.

Abdomen: Wider than the thorax in most species, comparatively flattened. There are 7 tergites visible. In most species the abdomen is feebly shining black, but in some species the sides of the tergites, and as many as the last five tergites, are pale brownish yellow. Characteristically the pile of the abdomen consists of dense mats of scales, often silvery, which in a few species are confined to the lateral margins of the tergites. The last two tergites have the pile elsewhere consisting of flat-appressed, scaliform hair. There is a difference between the sexes in some species. The males may have mats of silver scales on the terminal half of the abdomen, which are appressed in an outward direction laterally. Females of such species may have the silver scales replaced by oblique mats of slender, whitish scales. However, in addition to the scales and the appressed scaliform hair, most species with a few exceptions have a single row of long, slender, blackish, bristly hairs along the posterior margin of the tergites. Sternites likewise covered with scales and with long, slender, erect, bristly hairs. Male terminalia comparatively small and recessed.

The male terminalia from dorsal and lateral aspect, epandrium removed, show a figure basally broad, narrowed apically but relatively short. The epiphallus is a quite wide, tonguelike structure, barely rounded or curved on the blunt apex. The sharp, basally stout dististyle arises subapically, and has the usual dorsoapical incision and hook. Epandrium also with a deep, broad, wide, ventral incision, and the dorsal margin plane.

Material studied: Many species which the author has collected in the southwestern United States, and other material given him by Mr. Sperry.

Immature stages: Unknown.

Ecology and behavior of adults: Commonly found about isolated clumps of desert flowers, also resting on the foliage or on the ground. The flies are active in the morning in brilliant, hot sunshine, and I have seen the males in pursuit of females around such clumps between Riverside and Desert Center, California.

Distribution: Nearctic: *Lepidanthrax agrestis* (Coquillett), 1887 (as *Anthrax*); *angulus* Osten Sacken, 1886; *campestris* (Coquillett), 1887; *disjunctus* (Wiedemann), 1830 (as *Anthrax*); *inauratus* (Coquillett), 1887 (as *Anthrax*); *lautus* (Coquillett), 1887 (as *Anthrax*); *lutzi* (Curran), 1930, subg. *Podolepida*; *morna* Curran, 1930; *proboscideus* (Loew), 1869 (as *Anthrax*); *rufolimbatus* (Bigot), 1892 (as *Epacmus*),

Neotropical: *Lepidanthrax hyalipennis* Cole, 1923; *indecisus* Curran, 1930; *panamensis* Curran, 1930.

Oriental: *Lepidanthrax compactus* Brunnetti, 1920; *transversus* Brunetti, 1920.

Australian: *Lepidanthrax albifrons* Roberts, 1928; *linguatus* Roberts, 1928.

Genus *Mancia* Coquillett

FIGURES 168, 396, 661, 738

Mancia Coquillett, Canadian Ent., vol. 18, p. 159, 1886b. Type of genus: *Mancia nana* Coquillett, 1886, by monotypy.

Odd, quite small, blackish, *Villa*-like flies noteworthy for the character of the axillary cell which is very slender, much reduced in size because of the strongly narrowed condition of the wing toward the base. The anal cell is very widely open, almost or quite as wide at wing margin as the second or third posterior cell. Second vein arising, with backward spur, a little before anterior crossvein. The wing is hyaline, except the extreme base and the subcostal cell and a minute brown cloud at base of anterior crossvein and middle of first basal cell. The face is bluntly, obtusely conical with a stripe of silvery whitish scales running down between the widely separated antenna from the lower part of the front to the anterior edge of the oral recess; proboscis slender, a little longer than the oral recess. The abdomen bears scattered, light reddish brown, narrow, appressed scales and a white band at base of abdomen.

A single species is known from California.

Length: 3.5 to 4.5 mm.

Head, lateral aspect: Nearly globular, the occiput moderately long, but rather strongly rounded inward, the occipital cup large and deep with short, dense fringe which becomes longer dorsally. The pile of the occiput consists of flat-appressed, rather broad, pale, brownish-yellow scales. The front is about twice as wide as the long, swollen ocellar tubercle in the female, and there is a faint groove running forward on each side of the tubercle so that the upper middle part of the front is very slightly depressed. Remainder of front almost flat.

The upper eye corners are strongly rounded and the ocelli are placed a little forward from the posterior margin of the eye. The pile of the occiput consists of scattered, erect and somewhat curled, bristly, black hairs; and anteriorly, especially in front of each antenna, there is a patch of subappressed, brownish-orange scales with a larger patch of silver scales placed transversely in front of the antenna and continuing on down the middle of the face almost to the oral margin. The face from the lateral aspect is rather prominent, projecting triangularly forward, not quite as far as the tip of the antennal style. In addition to the scales in the middle of the face, scarcely seen laterally, there are scattered, appressed, black and yellow, bristly hairs along the sides of the anterior part of the oral cup. The eye is large, the posterior indentation is rather low, but extensive above and below the midpoint. The facets of the eye diverge at the midpoint of the excavation, but there is no bisecting line. Proboscis rather short, but slender, with slender labellum, the whole proboscis is very little longer than the head and the palpus is quite slender and threadlike. The antennae are placed just above the middle of the head. They are small and short, the first segment quite short and disclike with a few short setae dorsally. The second segment is about twice as long as the first, shorter than wide. The third segment is not quite as wide as the second segment, forming a very small, short, knoblike or bulblike base with a very abruptly formed, long, quite slender style of uniform width with a bristle at the tip.

Head, anterior aspect: Head almost circular, a little wider than the thorax, the face at the antenna comprises a little less than one-third, slightly more than one-fourth of the head width. From the anterior view the ocellar tubercle is high, prominent, abruptly formed, and the vertex sinks a little below the eye margin. The antennae are widely separated and arise very close to the eye margin. The face below the antenna is very little wider than the front. The oral cup is unusually long and slender with thick walls, pinched in laterally at the middle and comparatively deep throughout, and without medial grooves or ridges anteriorly. Gena narrow, tentorial pit or fissure shallow. Head, viewed from above, with the occipital lobes adjacent and the fissure rather deep.

Thorax: Short and broad with large wide, convex scutellum. The mesonotum is opaque, blackish on the anterior half with thin gray pollen. Posterior half of the scutellum shining. In material I have seen there is a furrowlike medial crease on the anterior half; it may be abnormal. Pile of the mesonotum on the anterior half consists of abundant, erect, shining, white, flattened pile of medium length, but with the very sparse anterior collar composed of longer pile. In addition there is a considerable amount of broad, flat-appressed, yellowish and whitish scales, especially laterally and on the humerus and notopleuron, and the scales on the posterior part of the mesonotum, and on the base side and margin of the scutellum are a little more slender, appressed, and brownish yellow. Mesopleuron with a tuft of radiating bristly pile. Metanotum, metapleuron with a conspicuous tuft of erect, bristly pile. Hypopleuron and pteropleuron bare. Scales of squama narrow.

Legs: There is a peculiar tuft of very large, long, broad, white scales, extending downward on each side of the front coxa. The legs are small and weak, the hind femur a little thickened and flattened, more stout than the other legs, but in no way swollen. The anterior tibia is smooth, except for 1 or 2 minute, bristly hairs and 3 or 4 even smaller similar hairs ventrally. Middle and hind tibia with 4 or 5 slender, sharp, oblique bristles dorsally on each side.

Wings: Wings hyaline with dark veins and the extreme base and the auxiliary cell yellowish. There is a tiny trace of brown yellow below the origin of the third vein and at the anterior crossvein. The wing is characteristic in shape, being strongly narrowed basally. The axillary lobe is small, short and narrow, and the anal cell open by almost the full length of the axillary lobe, being extraordinarily wide. There are 2 posterior cells. There is scarcely any trace of bend on the second vein. It arises rectangularly a little before the anterior crossvein and has a small backward spur. The anterior crossvein enters the discal cell at the middle. There are only 2 submarginal cells. The first posterior cell, and the others as well, are widest at the wing margin. Ambient vein complete. Alula rudimentary or vestigial. There is a small basal comb.

Abdomen: Short oval, the seventh tergite in the female triangular. Abdomen at the base as wide as the thorax. First tergite with a small patch of erect pile laterally and the posterior margin bearing a fringe of broad, brownish-yellow scales. Second tergite with flat-appressed, radiating, broad, brownish-white scales; it has a few brown scales on the posterior margin. Third to sixth tergites with numerous brownish-orange, broad, appressed scales laterally which are apt to be denuded centrally. Terminal, concealed tergites yellowish orange with fringe of long, shining whitish hairs. I have seen only the female. The next to the last sternite has a curious, conspicuous postmarginal band or fringe of long, broad, silvery or yellowish-white scales.

Material available for study: A representative of the type-species.

Immature stages: Unknown.

Ecology and behavior of adults: Not on record.

Distribution: Nearctic: *Mancia nana* Coquillett, 1866.

Genus *Neodiplocampta* Curran

Figures 167, 189, 392, 657, 658, 669, 743

Neodiplocampta Curran, Families and Genera of North American Diptera, pp. 193, 200, 1934. Type of genus: *Diplocampta roederi* Curran, 1931, by original designation.

Agitonia Hull, Ent. News, vol. 77, pp. 225-227, 1966. Type of subgenus: *Agitonia sepia* Hull, 1966.

Flies rather less than medium size in this family and with the face short to long conical or even bluntly protuberant and slightly convex below the antenna. They can usually be readily distinguished by the brown or yellowish-brown anterior border upon the wing with brown mottlings upon crossveins and vein furcations coupled with the strong sigmoid bend of the second vein. At least one species has the wing unusually broad, almost entirely sepia brown, the bend in the second vein unusually strong. From *Dipalta* Osten Sacken, to which it is related, it is separated by the presence of only 2 submarginal cells. From *Chrysanthrax* Osten Sacken, to which it also seems related, it is separated by the distinct double bend in the second vein.

There is one subgenus, *Agitonia* Hull (1966), distinguished by the very broad or wide, large wing, larger size, second vein more strongly contorted, more elongate abdomen, etc. The wing is almost entirely deep sepia in color. Found in southern Brazil.

Neodiplocampta is a small genus ranging from southern South America to Mexico, the West Indies, and the southwestern United States.

Length: 5 to 11 mm.; wing 5 to 11 mm.

Head, lateral aspect: Head hemiglobular, the occiput only moderately long in lateral profile and a little shorter ventrally. The pile consists of long, slender, flat-appressed scales of a yellowish color. The central fringe around the occipital cup is dense and short. The eye is reniform with the posterior indentation shallow, but broad. The eye facets are rather uniform in size, and there is a very short bisecting line extending forward from the bottom of the indentation.

In some species the scales of the occiput are more sparse, and intermixed with them are some fine, black, more or less erect setae. The front is long and slightly raised in front of the antennae, covered with numerous, fine, nearly erect, bristlelike hairs and also with numerous, broad, more or less appressed, spear-shaped, yellowish scales, which vary considerably in number and size in different species. The face is strongly peaked triangularly and conically forward with black bristly hairs similar to those of the front, but more or less appressed, and with, in addition, scales similar to those of the front but of varying density. Sometimes the face is completely covered with a mat of these scales and the bristly hairs may form a fringe overlapping the edge of the oral cavity. In addition there is a bare, pollinose band, rather narrow, free of pile and scales, encircling the upper face and reaching to the base of each antenna, but not extending across the middle of the face.

The proboscis is comparatively slender, the labellum long, slender and pointed, and usually the tip of the labellum barely extends beyond the peak of the oral recess. However, in one species the entire labellum extends forward beyond the peak of the face. The palpus is rather long and filiform, often curled, with a row of stiff, erect setate hairs ventrally and laterally. The antennae are attached a little above the middle of the head. They are comparatively small, the first segment is a little widened apically, twice as long as its apical width, and bears numerous bristly setae or short, bristly hairs. Second segment wider than long, broadly attached to the third segment, but with rounded lateral margins. Third segment with a bulbiform, conical base stretching out into an attenuated style, the whole segment only slightly longer than the first two combined, and the thickness of the style varies in different species, and represents a prolongation of the dorsal part of the third segment.

Head, anterior aspect: Head as wide as the posterior part of the thorax. The eyes are rather narrowly separated in the males by a distance at the vertex not more than twice the width of the second antennal segment. Of this narrow space, the prominent convex ocellar tubercle, with nearly vertical sides, occupies at least four-fifths. The ocellar tubercle is set near, but not at, the posterior eye corners and it is not quite equilateral in form. From the lateral aspect it is almost hidden by the upward overgrowth of the eyes; posterior eye corners slightly rounded. Face and lower front comparatively wide, the face strongly convergent below. The head is nearly circular from the front view. The antennae are widely separated, but distant from the eye margin by a little less than the length of the comparatively long first antennal segment. The oral recess is long and slender and quite deep, the posterior wall is a little thickened, although its edge is sharp, and the drop down below the eye is a short distance only. The wall of the oral cup is thickened in the middle, and again thinned and narrowed anteriorly. The gena is linear and there is a very shallow, linear, tentorial crease. Viewed from above the occipital lobes have a vertical fissure rather than a horizontal one and the lobes are scarcely touching.

Thorax: The mesonotum is slightly convex, black, more or less shining with brown pollen which may be quite faint or quite dense. The mesonotum is a little wider posteriorly and either sparsely or densely covered with fine-appressed pale hair in some species, and in other species, distinct scaliform hair. The anterior collar of pile is stiff, blunt-tipped, but comparatively short and loose, and grading back into the pile of the mesonotum. Notopleuron and sides of mesonotum with very weak, slender bristles, sometimes yellow, sometimes black and those of the notopleuron occasionally more stout. Postalar callosity and the margin of the scutellum with a few slender bristles and bristly hairs. Disc of scutellum colored and clothed like the mesonotum. The whole of the mesopleuron covered with stiff bristly hair. The pteropleuron, the metapleuron, the sides of the metanotum, and the plumula, all with abundant, long, stiff pile. There is a tuft of hair beneath the haltere. The whole of the hypopleuron is bare. Scales of the squamal fringe short.

Legs: The legs are only moderately strong, the front pair reduced, and the front tarsi nearly smooth, having only fine, minute appressed setae dorsally. The anterior femur may have a few fine, short bristles dorsally. Middle femur has 1 to 3 rather stout bristles in the middle anteriorly, and the hind femur has 3 to 6 bristles of varying size, some of them stout, on the anteroventral margin. Middle and hind tibia with fine, slender, oblique bristles, stouter on the hind legs, placed in 4 rows. There are 6 or 8 bristles anterodorsally on the middle tibia, 6 to 10 on the posterodorsal row, 6 to 10 on the posteroventral row, but only 5 or 6 anteroventrally. Hind tibia with the numbers similar, but the bristles stouter, and in some species with somewhat greater numbers on the dorsal rows, claws small, sharp, almost straight and pulvilli absent.

Wings: Of moderate width, and almost as wide apically as basally. There are 2 submarginal cells and the distinctive feature of the wing lies in the very strong, double bend or sigmoid kink in the second vein. The first posterior cell is a little narrowed. The anal cell is widely open in most species, narrowly open in some. The alula is present but not very wide. The ambient vein is complete. Costal comb present but narrow. Costal hook sharp and thornlike. The second vein arises opposite or shortly in front of the anterior crossvein or shortly beyond. The anterior crossvein enters the discal cell before or at the middle.

In some species the wing is broadly oval or even extremely wide and large and the abdomen distinctly more elongate than the type-species from the Nearctic and the second vein much more strongly contorted. Such species have been relegated to the subgenus *Agitonia* Hull.

Abdomen: Short oval, varying from almost entirely brownish yellow or red with the first tergite and a triangle on the second blackish to other species in which there is greater proportion of black down the middle of the abdomen or species in which the abdomen is wholly black. The pile consists of a dense covering of slender, appressed scales, or scaliform pile, intermixed with subappressed, black setae. These setae form a scanty fringe along the lateral margins of the tergites. The stiff pale pile on the first tergite may extend completely across the segment or be absent in the middle, or may extend on to the basal corners of the second tergite, or be absent from this area and replaced by a few, stiff, black hairs. There are 7 tergites visible. The male terminalia are small, asymmetrical, short conical. The females have a fringe of spines on each acanthophorite.

Material available for study: Several species collected by the author and by Dr. W. G. Downs in Mexico and in the southwestern United States.

Immature stages: Unknown.

Ecology and behavior of adults: These flies frequent low-growing flowers in arid regions. They can also be found resting upon the ground.

Distribution: Neotropical: *Neodiplocampta paradoxa* Jaennicke, 1867; *roederi* Curran, 1931; *sepia* Hull, 1966 (as subg. *Agitonia*). I have 6 new species before me that are included in a new manuscript.

Genus *Paravilla* Painter

FIGURES 176, 184, 407, 707, 717, 951, 952

Paravilla Painter, Journ. Kansas Ent. Soc., vol. 6, p. 10, 1933a. Type of genus: *Paravilla edititoides* Painter, 1933, by original designation.

This group was erected by Painter (1933) for all those species of the old *Villa* Lioy complex that have the front tibia spinose and the face conically projecting. In *Paravilla* the face is quite conical, distinctly more conical than in either *Hemipenthes* Loew or *Chrysanthrax* Osten Sacken. The front tibiae are distinctly spinose; the pulvilli are absent. The basal part of the wing is diagonally dark and never with lighter fenestrate spots about the crossveins and furcations. Finally, it should be noted that the fourth posterior cell has only punctiform or nearly punctiform contact with the discal cell and the latter is obtuse at the apex.

As far as is known, *Paravilla* Painter is a New World group of rather numerous species.

Length: 5 to 15 mm., mostly flies about 10 mm. in length.

Head, lateral aspect: The head is hemiglobular, however, with the face sharply and conically produced. The occiput is extensive, both above and below, but strongly retreating inward from the eye margin at a strong, rounded, oblique slant so that the curvature of the eye is continued back over the occiput. The pile is dense and consists of appressed scaliform pile or slender scales or both; central fringe around the occipital recess is short. The eye is large and only slightly concave posteriorly since the posterior indentation is shallow though extensive. There is a short bisecting line at the bottom of the indentation along the posterior eye margin. The front is comparatively flat, often a little sunken midway between the vertex and antenna and slightly raised in front of the antennae. The pile is very coarse and bristly and erect, generally black, and characteristically there are numerous scales that are mostly slender, ranging from erect to appressed or subappressed. Both pile and scales are continued onto the sharply peaked conical face. The bristly setae, however, on the face become rather strongly appressed and also form a scanty fringe extending beyond the margin of the oral cup and generally somewhat concentrated medioapically at the tip of the face. The face is as extensive as in *Dipalta* Osten Sacken or *Stonyx* Osten Sacken or perhaps even more prominent. It may extend almost to the apex of the antennal style, and the total length of the face from eye margin to apex is nearly equal to the total length of the antenna. The antenna is set forward from the eye margin a distance equal to the length of the second segment.

The proboscis is slender with long, slender, pointed labellum, and extends beyond the length of the oral cavity by not more than the length of the labellum; more commonly merely the tip of the labellum protrudes beyond the length of the apex, or is even shorter than the oral cavity. The palpus is slender, long, cylindrical, often curled with a fringe of stiff hairs laterally and ventrally. The antennae are attached distinctly above the middle of the head, not far from the upper third of the head. They are short; the first segment is equal to twice as long as wide, and it is somewhat swollen and widened apically. The second segment is beadlike, scarcely as long as wide, rather broadly attached to the base of the third segment. The third segment is elongate conical in which the more narrow unsegmented stylelike portion varies in length from one-third to one-half of the total length of the segment. The first segment bears numerous short bristles on all sides, sometimes absent, sometimes present below. In length these bristles are usually a little longer than the second antennal segment. Second segment with short setae dorsally and laterally.

Head, anterior aspect: The head is not quite as wide as the posterior width of the thorax. It is nearly circular. In males the eyes are separated by the length of the first two antennal segments or about twice the width of the antennal tubercle. The antennal tubercle is placed near the slightly posterior eye corners, and laterally the eye rises above the tubercle. At the antennae the face is slightly more than twice as wide as the width of the vertex of the head. The antennae are widely separated and are distant from the eye margin by not more than and usually less than the width of the first segment. The oragenal cup is long, slender, especially anteriorly, and quite deep, becoming shallow only at the apex of the face. On the anterior half the walls are thickened. Anteriorly the gena is completely linear with a very long linear tentorial crease that extends backward nearly to the bottom of the eye, at which point the gena is a little more extensive, extending downward from the eye. Viewed from above, the occipital lobes are tightly apposed, the fissure more vertical than horizontal and with a wider fissurelike pit behind the ocelli.

Thorax: The thorax is slightly wider posteriorly, the postalar callosities wide and shelflike, ground color black, but usually with brown, gray, or yellowish pollen present and not infrequently greenish pollen. The anterior collar varies from loose and sparse with mostly pointed hairs to other species where it is more dense with blunt hairs. Pile of the mesonotum is matted, appressed, scaliform with more bushy pile lying appressed over the wing and anteriorly to the humerus. In addition, in many species there are very fine, erect and suberect, black or yellow bristly hairs. Notopleuron, the postalar callosity, and the margin of the scutellum with weak bristles of increasing length. The scutellum is clothed like the mesonotum, comparatively flat with nearly hemicircular rim. The propleuron, the mesopleuron, and almost the whole of the metapleuron have conspicuous tufts of long, coarse, bristly pile, usually pale, but sometimes black in color. Pteropleuron with considerable pile, a conspicuous patch of shorter pile above the posterior costa. There is a tuft of hair beneath the haltere; the anterior hypopleuron above the middle coxa is bare, and the squamal scales are rather long and conspicuous.

Legs: Relatively strong including the front pair which are smaller but not greatly reduced. First four femora slightly widened toward the base. Hind femur stout and of uniform thickness. The anterior femur has 3 to 5 bristles near the middle anteroventrally. They may be quite stout. Hind femur with 3 to 5 bristles, sometimes stout and sometimes rather small, along the anteroventral margin. Surface of femora and tibiae with rather broad, conspicuous scales. Characteristically the anterior tibiae bear spicules or short bristles, usually 5 to 7, along the anterodorsal and posterodorsal margins. Middle and hind tibiae with more conspicuous, sharp, oblique bristles in 4 rows together with some small setae. There are usually 6 to 8 bristles in these rows excluding the shorter setae. Claws large, even anteriorly, slightly and gradually curved from the base, pulvilli absent.

Wings: Large, comparatively elongate and generally slender, a little pointed apically, sometimes broader at the base. Approximately the outer half of the wing diagonally is hyaline or subhyaline, with or without an occasional small spot at the more distal furcations of veins. Usually the dark pattern and spots are gently suffuse instead of sharply marked; however, this varies considerably and in species like *castanea* Jaennicke, there is an exceptional pattern of 3, irregular, sharply-defined spots across the wing, one at the base, one at the middle, one apical with supplementary spots. Characteristically, the contact of the discal cell with the fourth posterior cell is nearly punctiform or at least short. The anal cell is open. First posterior cell rarely narrowed strongly, usually only slightly narrowed. The anterior crossvein enters the discal cell at the middle or before the middle of the cell, and the second vein arises at shortly before or shortly after the anterior crossvein. The ambient vein is complete and the alula is relatively narrow.

Abdomen: As wide, or sometimes distinctly wider than the thorax, especially in females. Ground color usually black with terminal tergites sometimes yellowish. The pile varies from dense mats of flat-appressed hair, in some species slightly scaliform hair, and other species in which appressed pile is more sparse, although sometimes forming distinct bands along the posterior margins of the tergites. Most species have considerable amounts of fine, erect pile, a little longer toward the base of the abdomen, yellowish, or sometimes intermixed with black. Males of some species have the last two tergites with appressed silvery scales. In species like *castanea* Jaennicke there may be more bushy pile than appressed pile on the dorsum of the abdomen. Female terminalia with 8 pairs of slender, slightly spatulate spines on the sides of the acanthophorites. Male terminalia asymmetrical, more or less recessed, obtusely conical.

The male terminalia from dorsal aspect, epandrium removed, show a structure wide basally, narrowed apically; the epiphallus is extremely broad, long and tonguelike. The basal part of the aedeagus is large and subquadrate; lateral apodemes leaflike, and paired. From the lateral aspect, the dististylus is distinctly subapical, has a dorsoapical incision and hook. The epiphallus is conspicuous, and thick and long. The epandrium triangular with extensive downturned basal lobe.

Material available for study: Nearly 40 species, mostly collected by the author and his wife.

Immature stages: The host preferences for only a few species are known. See discussion under life histories.

Ecology and behavior of adults: These flies are readily attracted to a number of species of flowers, mostly to the composites.

Distribution: Nearctic: *Paravilla apicola* Cole, 1952; *castanea* (Jaennicke), 1867 (as *Anthrax*); *cinerea* Cole, 1923; *consul* (Osten Sacken), 1886 (as *Anthrax*); *coquilletti* Painter, 1965 [=*obscura* (Coquillett), 1894 (as *Anthrax*)]; *cunicula* (Osten Sacken), 1886 (as *Anthrax*); *diagonalis* (Loew), 1869 (as *Anthrax*); *edititoides* Painter, 1933; *emulata* Painter, 1962; *epheba* (Osten Sacken), 1886 (as *Anthrax*); *extremitis* (Coquillett), 1902 (as *Anthrax*); *flavipilosa* Cole, 1923; *fulvicoma* (Coquillett), 1887 (as *Anthrax*); *fumida* (Coquillett), 1887 (as *Anthrax*); *lacunaris* (Coquillett), 1892, (as *Anthrax*); *mercedis* (Coquillett), 1887 (as *Anthrax*); *nigronasica* Painter, 1933; *palliata* (Loew), 1869 (as *Anthrax*); *perplexa* (Coquillett), 1887 (as *Anthrax*); *separata* (Walker), 1852 (as *Anthrax*) [=*nemakagonensis* (Graenicher), 1910 (as *Anthrax*)]; *spaldingi* Painter, 1933; *syrtis* (Coquillett), 1887 (as *Anthrax*); *tricellula* Cole, 1952; *vastus* (Coquillett), 1892 (as *Anthrax*); *vigilans* (Coquillett), 1887 (as *Anthrax*); *xanthina* Painter, 1933.

Genus *Poecilanthrax* Osten Sacken

FIGURES 180, 190, 198A, 417, 710, 720, 948, 949, 950

Poecilanthrax Osten Sacken, Biologia Centrali-Americana, Dipt., vol. 1, p. 119, 1886. Type of genus: *Anthrax alcyon* Say, 1825. Designated by Coquillett, p. 593, 1910, the first of 7 species.

Cole (1917), pp. 67-78, key to 22 species (as *Anthrax*).
Maughan (1935), pp. 53-56, key to 13 species from Utah and western states.
Johnson and Johnson (1957), pp. 2-5, key to 27 species and subspecies.
Painter and Hall (1960), pp. 19-25, key to 36 species and subspecies.

A genus of large flies; only an occasional individual is reduced in size. They may be recognized by an ensemble of characters: the most outstanding are the large, elongate, pictured wings, the presence of 2 or rarely 3 submarginal cells, the emission of vein M2 from well in front of the lower outer angle of the discal cell, and the lower outer angle usually bearing a prominent spur vein. To the characterization may be added the fact that the abdomen is moderately lengthened, the face is sharply to bluntly conical, and the proboscis projects a little way beyond the oral recess or is entirely confined to it. The front tibia is smooth and its tarsi reduced, the pulvilli absent. Vestiture consists of sparse, rather stiff hair, scales absent. While the wing patterns are characteristic and conspicuous, they tend to be distinctly diffuse along the edges of the patterns; a few species have merely the foreborder of the wing darkened with yellow or brown to the end of the marginal cell.

Poecilanthrax Osten Sacken is primarily a Nearctic genus with most species western and not a few species extending down into northern Mexico.

Length: 8 to 14 mm.; wing, 8 to 17 mm.

Head, lateral aspect: Head hemiglobular, a little shorter than in some related forms, and the occiput, though well developed, is shorter and slopes inward much more rapidly at a strong slant. Hence the occipital fissure is vertical rather than horizontal. The surface of the occiput is strongly pollinose, usually pale gray, and the pile short, rather fine, suberect and directed forward to the eye margin. The pile is rather loose and scanty. The eye is slenderly reniform with an exceptionally broad excavation that extends from near the bottom of the eye to close to the top of the eye. At the bottom of the indentation the eye facets curve forward toward the front of the head for a short distance, but there is scarcely any trace of a definite bisecting line. The front is slightly raised and a little more so in front of the antenna. It is dully shining without evident pollen, except across the vertex, and the pile is abundant, sometimes fine, sometimes coarse, erect, and never as long as the first antennal segment. The ground color varies from pale yellow to dark reddish brown and changing to black at the vertex. Face produced forward as a short, triangular cone in most species, but with a varying degree of obtuseness; thus, for example, in *californicus* Cole it is less sharply peaked than in *demogorgon* Walker or as in *flaviceps* Loew. The pile of the face which is pale in ground color consists of scattered bristly hairs, black or pale or both, a little more curled and appressed and less erect than the pile of the front, and less numerous beneath the antenna. There is in fact a narrow, encircling band beneath the antenna that is bare of pile. Some hairs extend over the margin of the oral recess and there is often a cluster of dense setae, black or pale, concentrated medially at the apex of the peak of the face.

Representative definitive distribution patterns and ranges of species of *Poecilanthrax* Osten Sacken. Adapted and redrawn from Painter and Hall.

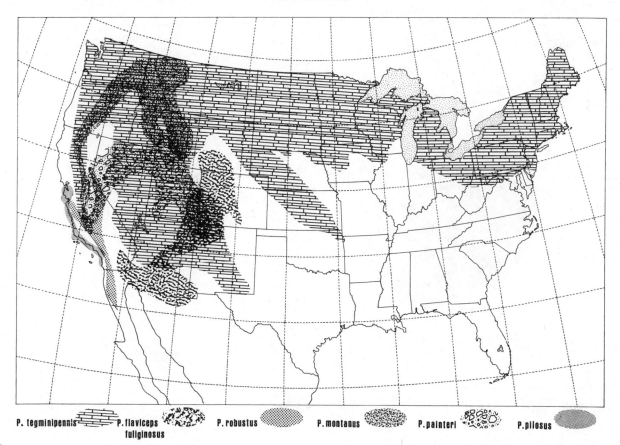

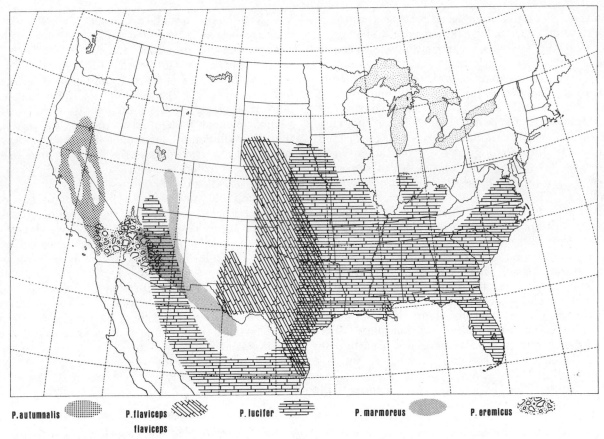
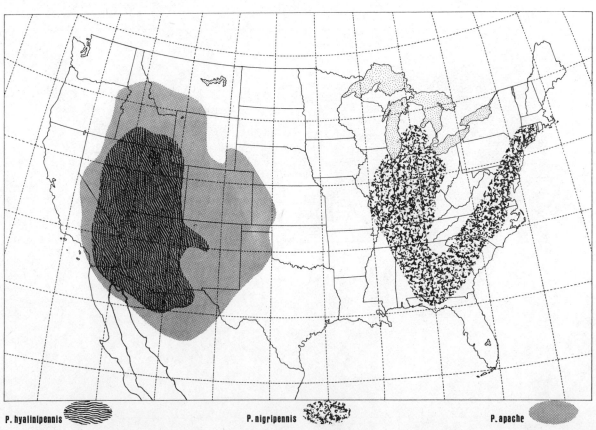

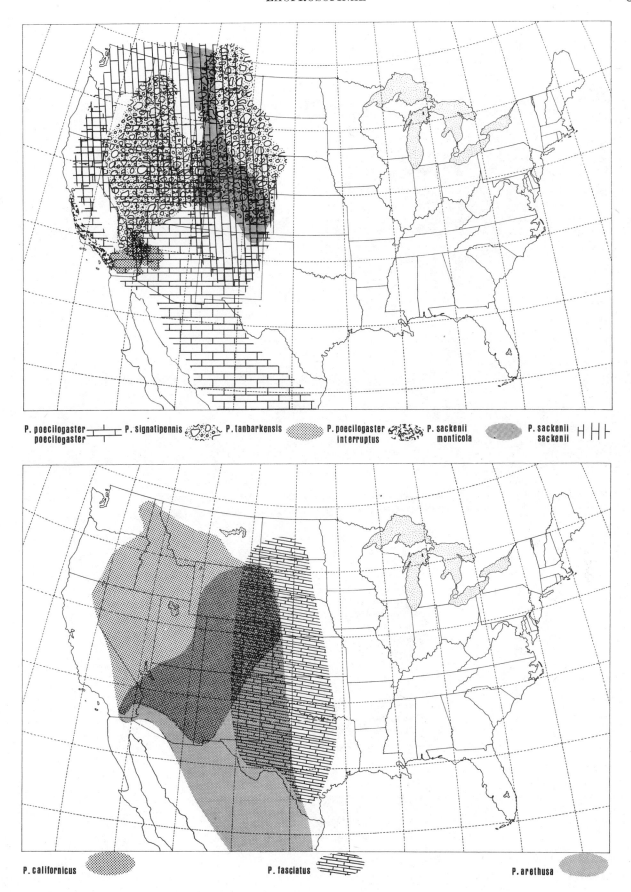

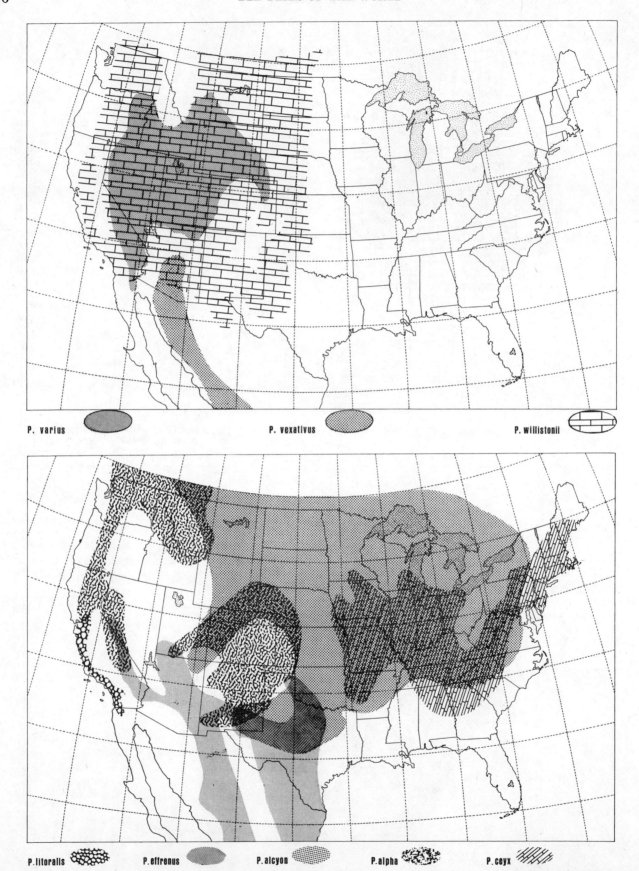

The proboscis and labellum are slender. The tip of the latter extends just beyond the apex of the face. Palpus slender, almost filiform, not quite half as long as the proboscis and bearing only extremely fine hairs scattered along the lower and lateral margin. The antennae are attached slightly above the middle of the head and are relatively small and slender. The first antennal segment is very distinctly widened medially at the apex. It is about 2½ times as long as its basal width, and 3½ to 4 times as long as the small beadlike second segment. The first segment bears considerable coarse, bristly hair, black or yellow, over the whole lateral, dorsal and medial surfaces; but these tufts are accentuated on each side and the segments are bare ventrally. The second segment, however, has a complete ring of appressed setae. Third segment not wider than the second, elongate conical with the base bulblike and extending rather gradually into the attenuate style, so that beyond the swollen base the middle third is considerably thicker than the apical third. Apex with a minute spine. The third segment, style included, is very slightly longer than the first two segments; in a few species it is of equal length, and the style is an attenuate continuation of the dorsal part of the segment.

Head, anterior aspect: The head is distinctly wider than the anterior thorax, but about the same width as the posterior part. It is nearly circular, slightly flattened on all four sides with the eye and at the vertex distinctly raised above the vertex, almost obscuring the ocelli when viewed from the side, and the small ocellar triangle has vertical sides, is extended backward as a low ridge. The ocelli themselves occupy a short isosceles triangle and are characteristically placed far forward from the posterior eye corners. Eyes in the male separated by a little more than 3 times the width of the ocellar tubercle and in the female by 4 or more times the width. The antennae are set rather far apart. They appear a little closer together because the first segment is convergent and they are separated from the eye margin by nearly the length of the first antennal segment. The oral recess is narrow and deep, especially posteriorly, with vertical walls, and on the anterior half the lower edge is rounded and inflated, the margin sharp beneath the eye. The genae are narrow, separated medially by a deep though narrow crease, and they become a little more extensive at the bottom of the eye. From above, the occipital lobes vary in development and also in apposition. The lobes and the tight fissure are almost horizontal or are only a little slanted, whereas in a few species there is a distinct gap between the lobes that do not touch.

Thorax: Shining black with reddish or sometimes bluish-gray pollen. The pile is rather short, flat-appressed, and scaliform, merging anteriorly into a broad collar of stiff, rather pointed pile. The pile is longer in some species, extending fully over upon the humerus, sometimes extending back to the notopleuron, and merged with the dense tufts of pile on the upper mesopleuron. The pile above the wing is coarse, sometimes rather long and matted, in other species consisting only of a few, short, curled, bristly hairs as in *demogorgon* Walker. The notopleuron has some short, stiff bristles. There are 4 to 6 bristles on the postalar callosity, and the margin of the scutellum has short, slender, weak bristles, longer in a few species. Disc of scutellum coated with appressed hair similar to the mesonotum, but apt to be denuded posteriorly. Mesopleuron, pteropleuron, and metapleuron, and the upper sternopleuron with much dense pile, finer and shorter on the sternopleuron, longer, coarser, almost bristly on the upper mesopleuron. There is a patch of erect hairs on the hypopleuron above the posterior coxa, and a tuft of hair beneath the haltere. Anterior hypopleuron bare. Plumula pilose, scales of the squama quite short, poorly developed, a few extending onto the alula.

Legs: Moderately developed, the anterior pair reduced in size, and its tibia smooth. Middle femur with a row of slender bristles along the middle anteriorly, and usually with a few bristles also posteriorly, 3 to 6 in each row and variable in size. They are rather prominent in *californicus* Cole. Hind femur with 6 to 9 bristles anteroventrally and with 4 or 5 dorsally near the apex in some species. Hind tibia with conspicuous, long, sharp, oblique bristles in 4 rows. There are 7 to 9 anteroventrally, about 13 in the anterodorsal row excluding the shorter setae, and about the same number dorsomedially, but fewer ventromedially. Claws slender, sharp, long, with scarcely a trace of basal tooth and no pulvilli. Anterior claws exceptionally minute.

Wings: The wings are characteristic, large and quite long, extending far beyond the abdomen, not broadened at the base, sometimes slightly pointed at the immediate apex. They are notable for the fact that they are never hyaline but colored by several types of patterns. In some species the apical half of the wing is irregularly pale, tinged by dense villi. In most species the wings are mottled. They may be dark sepia brown as in *demogorgon* Walker and often the cells, especially apically, and the posterior cells become grayish hyaline in the centers, leaving most of the veins diffusely margined with reddish yellowish or blackish brown. In *lucifer* Fabricius the entire wing is nearly uniformly brown. The patterns, while confusing, are distinctive for each species. The second vein arises at or a little before or even beyond the anterior crossvein and the anterior crossvein enters the discal cell usually at, sometimes slightly before the middle of, the discal cell. The contact of the discal cell with the fourth posterior cell is usually about as great as the length of the anterior crossvein, not more, sometimes less, and in one species punctiform. Generally there are 2 submarginal cells, occasionally 3, formed by an erratic crossvein, connecting the posterior branches of the third vein. Generally there are 4 posterior cells, but sometimes 5, formed by an erratic crossvein extending all the way through the third posterior cell. The first posterior cell is narrowed at the wing margin and may be almost closed. Anal vein widely open.

The alula varies from 3 to 4 times as long as wide to quite narrow in a few species, 8 or more times as long as wide. The ambient vein is complete with short fringe of hair. The costal comb is only moderately expanded, bearing short, fine-appressed setae, and the costal hook small, slender, and sharp.

Abdomen: The abdomen with 7 tergites visible from above. It is narrowly oval, distinctly longer and wider than the thorax, the color varying from black with only small, obscure reddish spots or triangles at the sides of the second and third tergites; even these may be lacking. In other species where a slender triangle of red extends inward from the lateral margin a varying degree toward the middle, leaving the segment blackish in front and behind, except on the margin. In one species the abdomen is entirely fox red, except for a blackish spot of decreasing size on the first 4 tergites. Sides of the female genitalia with a circlet of small slender spines. Male terminalia large with rather long coarse hairs dorsally and apically. The dististyle has a broad base tapering to an apical hook curved sharply inward. I cannot do better than quote Painter and Hall's excellent description of the aedeagus:

The aedeagus is slightly curved from base to apex reaching to beaked apical segments; base broad, gonopore in the middle of aedeagus, elevated projecting inward and upward, the aedeagal struts dorsad of middle part and projecting laterad and above lateral struts; middle part rectangular and not prominent, the lateral struts flange-like, short and broad, the basal strut fan-shaped in profile, prominent, outer edge minutely serrated.

Material available for study: Besides the collection in the National Museum of Natural History, I have extensive material acquired by collecting in six Western States and also in Mississippi.

Immature stages: Hosts are known for 7 species and consist of some 15 species of cutworms, sod worms and armyworms, all from the family *Noctuidae* (*Phalaenidea*). One species has been recorded as a hyperparasite of white grubs. See discussion under life histories.

Ecology and behavior of adults: These flies are abundant on flowers where they feed upon nectar and pollen. Painter and Hall point out that the association is with the group of flowers that happen to be blooming at the time the flies emerge. They note a partiality in the West to various species of rabbit bush, *Chrysothamnus*, and to a lesser extent on *Gutierrezia*, *Grindelia*, and *Senecio*. They note an association in the Eastern Rocky Mountains on resinweed, *Silphium*, and *Asters*, and on the plains species associated with gayfeather, *Liatris*, puncture vine, *Tribulus*, smartweed, *Persicaria*, and goldenrod, *Solidago*. He notes a frequent association with sagebrush, *Artemisia*. In Mississippi species are especially attracted to *Helenium tenuifolium*, and to a lesser extent on ironweed, goldenrod, *Rudbeckia*, *Coreopsis*, fleabane, and sumac.

Distribution: Nearctic: *Poecilanthrax alcyon* (Say), 1824 (as *Anthrax*) [=*halycon* Wiedemann, 1828]; *alpha* (Osten Sacken), 1877 (as *Anthrax*); *apache* Painter and Hall, 1960; *arethusa* (Osten Sacken), 1886 (as *Anthrax*); *autumnalis* (Cole), 1917 (as *Anthrax*); *californicus* (Cole), 1917 (as *Anthrax*); *colei* Johnson and Johnson, 1957; *demogorgon* (Walker), 1849 (as *Anthrax*), [=*ceyx* (Loew), 1869 (as *Anthrax*)]; *effrenus* (Coquillett), 1887 (as *Anthrax*); *eremicus* Painter and Hall, 1960; *fasciatus* Johnson and Johnson, 1957; *flaviceps* (Loew), 1869 (as *Anthrax*) [=*butleri* Johnson and Johnson, 1957], *flaviceps fuliginosus* (Loew), 1869 (as *Anthrax*); *hyalinipennis* Painter and Hall, 1960; *ingens* Johnson and Johnson, 1957; *johnsonorum* Painter and Hall, 1960; *litoralis* Painter and Hall, 1960; *lucifer* (Fabricius), 1775 (as *Bibio*) [=*fumiflamma* (Walker), 1852 (as *Anthrax*)]; *marmoreus* Johnson and Johnson, 1957; *moffitti* Painter and Hall, 1960; *montanus* Painter and Hall, 1960; *nigripennis* (Cole), 1917 (as *Anthrax*); *painteri* Maughan, 1935; *pilosus* (Cole), 1917 (as *Anthrax*); *poecilogaster* (Osten Sacken), 1886 (as *Anthrax*), *poecilogaster interruptus* Painter and Hall, 1960; *robustus* Johnson and Johnson, 1957; *sackenii* (Coquillett), 1887 (as *Anthrax*), *sackenii monticola* Johnson and Johnson, 1957; *signatipennis* (Cole), 1917 (as *Anthrax*) [=*marginatus* Johnson and Johnson, 1959, =*yellowstonei* (Cole), 1917 (as *Anthrax*)]; *tanbarkenis* Painter and Hall, 1960; *tegminipennis* (Say), 1824 (as *Anthrax*) [=*fuscipennis* (Macquart), 1834 (as *Anthrax*)]; *varius* Painter and Hall, 1960; *vexativus* Painter and Hall, 1960; *willistonii* (Coquillett), 1887 (as *Anthrax*); *zionensis* Johnson and Johnson, 1957.

Genus *Prothaplocnemis* Bezzi

FIGURE 214

Prothaplocnemis Bezzi, Bull. Soc. Roy. Ent. d'Egypte, vol. 8, p. 220, 1925c. Type of genus: *Argyramoeba anthracina* Becker, 1902, by original designation.

Bezzi based this genus upon the species *Argyramoeba anthracina* Becker, and maintained that this species contained transitional forms to *Aphoebantus* Loew and *Petrossia* Becker. Bezzi placed it, however, as a subgenus to *Anthrax* Scopoli (as *Argyramoeba* Schiner), to which he believed it shows even closer connections, for this species has a plumula. The scutellum is almost triangular with a weakly developed apex, and the abdomen in the female is scarcely longer than mesonotum and scutellum together. The epipygium stands free and is relatively large, but in its parts, it approaches more the characteristic structure for the species of the genus *Anthrax* (*Argyramoeba*). The appendages of the dorsal lamella are still simply hooklike, but the basal lamellae are stretched out and not broad or provided with side lobes. As in the case of *Aphoebantus* Loew, the hairing and scaling of the body is sparse, so that the shining black basic color is very conspicuous. In my figure of the wing the delineation of the brownish black color over most of the wing and the basal hyaline wing areas were omitted.

Material available for study: I have not seen this fly.
Immature stages: Unknown.
Ecology and behavior of adults: Not on record.
Distribution: Palaearctic: *Prothaplocnemis anthracina* (Becker), 1902 (as *Argyramoeba*); *persica* Paramonov.

Genus *Pseudopenthes* Roberts

Figures 196, 393, 668, 676

Pseudopenthes Roberts, Proc. Linnean Soc. New South Wales, vol. 53, p. 132, 1928a. Type of genus: *Pseudopenthes fenestrata* Roberts, 1928, by original designation.

A small to medium-sized, shining black fly, the wings almost wholly dark brownish black but with the greater part of the second basal cell and the basal half of the third basal cell quite hyaline. The front and face has iridescent whitish scales and on the abdomen the anterior corners of the second and third tergites and the posterior corners of the remaining tergites have patches of white scales giving way abruptly to violaceous black scales.

This genus has the third antennal segment much as in *Villa* Lioy, sensu lato, but has 3 submarginal cells as in *Stonyx* Osten Sacken; I do not consider it closely if at all related to these. *Pseudopenthes* Roberts also has a long, sharp, pale colored, basally stout tooth at the base of the claw as in *Exoprosopa* Macquart and *Litorrhynchus* Macquart. The face is bluntly conical with apex rounded, and the ocellar tubercle is set a considerable distance forward from the rear of the vertex. Proboscis quite short, rather stout, labellum fleshy, both confined to the oral recess. The relationship is perhaps closest to *Exoprosopa* Macquart.

A single species is known from South Australia, New South Wales, and Queensland. I collected a single male about midday north of Queensland on a hot, sun-baked open space about a large anthill in bottle-tree country.

Length: 10.5 to 12 mm.; wing 11 to 11.5 mm.

Head, lateral aspect: Head subglobular, the occiput comparatively short, the occipital cup deep, but relatively small. In the male the front is slightly swollen and raised above the eye margin, rather flattened in the middle, and the pile consists of rather loose comparatively short, erect, stiff, black hairs, more numerous on the lower half. Pile of the occiput, minute, subappressed and setate, but with scattered, completely appressed, small silvery scales in addition to the setae and more numerous at the indentation. The face is triangularly produced forward, slightly rounded at the apex of the oral cup. It bears scanty, scattered, stiff, sharp, subappressed black setae, and also appressed silver scales. The lower front, likewise, has appressed silver scales. Proboscis short, stout, with stout, fleshy labellum, which extends slightly beyond the oral recess. The long, slender palpus is cylindrical, of equal width throughout, and has some very short erect pile. The antennae are attached a little above the middle of the head. They are small and short. The first segment is twice as long as wide with distinctly abundant, fine, oblique or subappressed setae, the hairs about as long as the second antennal segment. Second antennal segment beadlike, slightly less than half as long as the first segment. Third segment triangular at the base, rather strongly attenuated, especially on the lower part, into a thick style with minute bristle, no microsegment, the length of the style about equal to the length of the first segment. The eye is slightly reniform. The posterior indentation is shallow but quite extensive above and below and with a short bisecting line. Facets of the male eye quite small above and below, but wider in the middle anteriorly, and again along the posterior margin on the lower half.

Head, anterior aspect: Nearly circular, wider than the thorax, the antennae widely separated by nearly twice the distance of the first segment to the eye margin. Oral recess wedge-shaped, narrowed anteriorly, long and slender, exceptionally deep, the walls thick, particularly on the anterior half where they are even slightly widened or inflated, though with an anterior or ventral ridge. The tentorial crease, or fissure is long and distinct, the gena quite narrow. The head from above has the occipital lobes slightly separated, rounded and divergent with a 5-grooved, radiating sunken depression behind the vertex.

Thorax: Thorax short and broad, narrowed anteriorly with large, long scutellum, the mesonotum only gently convex, shining black, with dense, small, wide, short, flat-appressed, purplish, black scales, and with a long, lateral, matted border of subappressed, coarse, white pile including the humerus and running to the base of the scutellum. Anterior collar of pile, rather dense, coarse, white, of several rows, and set rather far forward. There are no bristles on the mesonotum. Mesopleuron with a large, dense tuft of erect, coarse, black, bristly pile directed outward. Pteropleuron and hypopleuron bare. Metapleuron with a tuft of bristly hair. Squama with narrow white scales.

Legs: All the femora are a little stout, especially middle and hind pairs, but not swollen. The first four have only a few fine hairs ventrally, hind femur with 3 short, oblique, small, spinous bristles on the outer half. Anterior tibia smooth. The tarsi small. Middle tibia with 4 fine, oblique bristles ventrally, and with 6 or 7 small setae above. Hind tibia with 6 or 7 small, oblique bristles ventrally, but with about 20 short, small anterodorsal, spinous setae, and 12 or 14 posterodorsally. Claws nearly straight, slightly curved at the apex.

Wings: Large and unusually broad, strongly tinged with sepia brown, the middles of the cells faintly paler, but with most of the second, half of the third, and half of the axillary lobe completely hyaline. There are 3 submarginal cells and the posterior branch of the second vein makes a strong rectangular, but rounded, bend outward. The apex of the marginal cell is greatly swollen backward so that the first submarginal cell is foot-shaped. The second vein arises abruptly with rounded bend opposite the anterior crossvein, and the anterior crossvein enters the discal cell distinctly before the middle. Anal cell widely open, the axillary lobe large and the alula large. Costal comb wide. Ambient vein complete.

Abdomen: Elongate, 1½ times as long as the thorax, the sides nearly parallel and only slightly narrowed posteriorly. The abdomen is a little flattened in the middle on the first four tergites, but more strongly convex on the side, and 7 tergites are visible in the male. The abdomen is shining black, thickly plastered with minute, purplish, black scales, apt to be denuded in the middle of the segments. The base of the first tergite laterally has a band of erect, stiff, white pile, which is confined to the outer third. The base of the second segment has on each side, laterally a narrow band of white, appressed scales. Remaining tergites have a triangular spot of similar scales, which on the third tergite is basal and is the largest spot, and on the fourth tergite it is in the middle of the lateral margin, and on the remaining tergites it lies in the posterior corner. In the male the terminalia are rather small, asymmetrical, more or less recessed and turned toward the right side. I have not seen the female.

Material available for study: A representative of the type-species.

Immature stages: Unknown.

Ecology and behavior of adults: I captured a single male resting on the side of an anthill in the hot sunshine about a hundred miles north of Brisbane, Queensland.

Distribution: Australian: *Pseudopenthes fenestrata* Roberts, 1928.

Genus *Stonyx* Osten Sacken

FIGURES 200, 429, 708, 718, 960, 961

Stonyx Osten Sacken, Biologia Centrali-Americana, Dipt., vol. 1, pp. 80, 94, 1886. Type of genus: *Stonyx clelia* Osten Sacken, 1886. Designated by Coquillett, p. 609, 1910, the second of 2 species.

Flies of medium size, with 3 submarginal cells. This genus is readily distinguished from *Exoprosopa* Macquart by the absence of a divided antennal style. The third antennal segment is small, onion-shaped, and the stylar portion remarkably long and slender and needlelike with minute apical bristle; in the character of the third antennal segment they remind one of *Lepidanthrax* Osten Sacken; both of these genera have the slender proboscis extending well beyond the oral recess. The anterior tibia has a distinct row of spinules or bristles and the pulvilli are absent.

All the known species have the wings sharply variegated with brown and the discal cell is more or less constricted in the middle. The anal cell is open, and may be widely open. The pile of abdomen and mesonotum is in large part fine, short, and strongly appressed. Face bluntly projecting.

Three species are known; they range from Panama to Mexico and into Arizona. I have collected 2 species in Mexico.

Length: 7 to 12 mm.

Head, lateral aspect: The head is hemispherical, the occiput rather strongly developed, especially in the middle with the posterior surface plane and flattened, rather than rounded, as far back as the margin of the central recess. The stiff, bristly pile is quite dense, but also quite short, and closely appressed to the surface. The inner marginal fringe is dense, but short. The front is very slightly raised at the eye margin, flattened across the middle, bearing comparatively dense, stiff, erect pile, which is blackish in the type-species and almost as long as the first two antennal segments. In the male the anterior and ventral eye facets are larger than those at the vertex. The eye has a very long, comparatively shallow indentation and is bisected by a line at the middle of the eye. At the deepest point of the indentation there is a small bare triangle. The face is extended forward in a low obtuse triangle, a distance somewhat greater than the length of the first three antennal segments, style excepted. In the type-species the pile of the face is entirely similar in length, color, and relative stiffness to the pile of the front. The pile is a little shorter and somewhat appressed in other species. It is rather thick, extending on down the sides of the face, and reappearing more sparsely on the gena.

Proboscis and labellum slender, the latter pointed, all the surface microsetate under high power. It is at least 1½ times as long as the head and generally thrust straight forward. Palpus long, slender, filiform with conspicuous, loose, erect, scattered stiff hairs. The antennae are attached a little above the middle of the head and are comparatively small. The first segment is a little longer than wide, the second wider than long, and not quite as long as the first segment. The third segment is small, scarcely as wide as the second segment, its base plane, but not tightly attached to the preceding segment. The third segment is distinctly shorter than long, shaped much as a flat onion or bulb in the type-species but forming a short triangular cone in other species. Attached to it near the middle is a very long, slender, abruptly formed, needlelike style, with a short spine at the apex. This long style is 1½ times the length of all the three segments, and is very much like that of *Lepidanthrax* Osten Sacken.

Head, anterior aspect: The head is wider than the anterior part of the mesonotum, but a little more narrow than the posterior part. From the anterior aspect the vertex and sides of the eyes are slightly flattened. In males the eyes are separated by a little more than the length of the first three antennal segments, style excepted, or by more than 3 times the width of the small ocellar tubercle. This tubercle is distinctly raised, isosceles in shape, set forward a little from the posterior eye corners, and the whole surface of the front and vertex is pollinose. All species bear some flattened, slender scales on the front and mixed in with the pile of the face. The type-species has few such scales, but they are dense in other species. The antennae are unusually widely separated, distant from the eye margin by not more than the length of the first segment. The oral recess is large, long, and quite deep, with vertical sides for at least the posterior three-fourths of its length. The lateral margin is thin, set close to the eye, with the gena extending downward, not quite the thickness of the proboscis. There is a pale, bare, circular, pollinose line curving down the sides of the upper part of the face and in some species it can be seen curving medially above and between the antennae, and apparently forming the demarcation of face and antennae. The tentorial pit forms a short, deep fissure, lineally separated from the eye margin at the midpoint of the oral recess. Viewed from above the occipital lobes are deeply sulcate posterovertically and not horizontally as in those species where the dorsal occiput is more extensive.

Thorax: Short, distinctly narrowed anteriorly, becoming broadest across the postalar callosities. The mesonotum is comparatively flat in the middle, but a little raised and convex near the wing and only slightly convex viewed laterally. The scutellum is quite wide, but comparatively short, very slightly triangular, each half a little rounded on the margin, both margin and base rather strongly convex. The mesonotum is opaque, blackish, with a conspicuous anterior fringe, narrow, high, dense and very coarse, the bristly hairs with rather blunt apices. Mesonotum of males sometimes distinct from the female as in the type-species. The pile of the mesonotum in the male may consist entirely of rather short, stiff, pale, erect hairs, longer in front of the scutellum, the anterior collar in the male of one color and a different color in the female. This is also true of the pleural pile. Disc of mesonotum in the female rather densely coated with completely appressed, short, coarse, pale bristly hair, the scutellum likewise. Margin of scutellum with slender bristles and notopleuron with a clump of 5 or 6 short, stout bristles. Pile of mesopleuron quite dense, long and bristly and erect. The anterior pteropleuron bears bristly hair, and there is a dense, vertical fringe of hair on the metapleuron; hypopleuron bare, and no pile behind the posterior spiracle. However, there is a small, curious tuft of hair immediately below the haltere which is appressed downward. Metanotum very short. Squamal fringe quite short, dense, and scalelike.

Legs: The middle and posterior femora are moderately stout, anterior femur rather slender, but both anterior and middle femora a little widened toward the base. The anterior femur sometimes has a few slender, bristly hairs below, quite close to the base. Anterior trochanter inserted in a socketlike joint on the coxa. Middle femur with 2 or 3 distinct, sometimes stout bristles in the middle on the anterior surface. Hind femur always with 3 or 4 distinct bristles on the anteroventral margin of the distal half. They may be rather stout. All femora with scales. Anterior tibia with 8 to 10 slender, oblique bristles anterodorsally, posterodorsally, and ventrally. Middle tibia with about the same number of somewhat longer dorsal bristles, but with sometimes fewer ventral bristles. Posterior tibia with about the same number of bristles as the middle tibia, a little longer, also oblique, sometimes more stout. Claws sharp, gently curved from the base with short pulvilli.

Wings: Wings large and broad basally, blunt apically with 3 submarginal cells and a strong rectangular posterior bend subapically in both the second vein and the anterior branch of the third vein. First posterior cell slightly narrowed apically, the anterior crossvein rectangular, entering the distal cell distinctly before the middle. Contact of discal and first posterior cells never greater than the length of the anterior crossvein. Anal cell widely open. Costa very strongly widened at the base. Setae of costal comb long, bristly and appressed outward. Costal hook long, sharp, and thornlike. Alula comparatively wide, the ambient vein complete, with a long fringe of villi. Whole wing villose, and in all the species known to me, the basal half of the wing, including all or most of the axillary lobe is obliquely dark sepia in color with a varying amount of additional sharp markings of sepia brown on the outer half of the wing, the pattern differing in different species. The second vein arises either before or after the anterior crossvein.

Abdomen: Abdomen short and broad; gently convex. Surface either shining black or opaque brown, the spikelike erect fringe on the first segment may extend completely across this segment, or be much reduced beneath the scutellum, leaving dense, long tufts only laterally. Surface of abdomen with comparatively dense, slightly flattened pile, but in some species becoming distinctly scalelike on the sixth and seventh tergites. Male terminalia large, asymmetrical, extended downward as a prominent cone. Lateral margins of the tergites with moderately prominent tufts of long, coarse, bushy pile, which sometimes shows alternately contrasting colors.

The male terminalia from dorsal aspect, epandrium removed, show a structure with the epiphallus unusually long, narrowing a little near the middle and with the middle part of the aedeagus large, wide and short oval, and much rounded apically. From the lateral aspect it is much like *Rhynchanthrax* Painter except that the ejaculatory process is well separated from the epiphallus. The epandrium is quite similar. The dististyle is subapical with dorsoapical incision and hook.

Material available for study: I have representatives of the type-species given me by Dr. C. H. Curran, also additional species collected by my late wife and me in our last expedition to Mexico.

Immature stages: Unknown.

Ecology and behavior of adults: Very similar to other species of the *Villa* Lioy and *Exoprosopa* Macquart genera. Found on composites or resting upon the ground.

Distribution: Nearctic: *Stonyx clelia* Osten Sacken, 1886 [=*keenii* (Coquillett), 1887 (as *Anthrax*)].

Neotropical: *Stonyx clotho* (Wiedemann), 1830 (as *Anthrax*); *lacerus* (Wiedemann), 1830 (as *Anthrax*); *melia* Williston, 1901.

Genus *Synthesia* Bezzi

FIGURES 229, 715, 726

Synthesia Bezzi, Ann. South African Mus., vol. 18, p. 130, 1921a.
  Type of genus: *Synthesia fucoides* Bezzi, 1921, by original designation.

A comparatively small, blackish, *Villa*-like fly of aberrant character, which is distinguished by the yellowish-brown legs, face, lower front, and apex of abdomen, and coarse appressed pile of the same color, and by several characters that set it apart from most other Exoprosopinae. The pulvilli are slender but long and well developed, the anterior tibia has spinules only ventrally. While the posterior eye margin is quite shallowly indented, and there is no bisecting line, many American species of *Villa* Lioy approach this condition and differ little. What is more distinctive, perhaps, is the relatively long and prominent face which bulges outward at the eye margin and is prominent and rounded, not conical; the proboscis is shorter than the oral cavity, stout with stout, fleshy labellum; the antennae do not differ greatly from certain species of *Villa* Lioy before me, but they are set quite far apart.

The second vein arises at a right angle well in front of the anterior crossvein, but removed from the crossvein by less than the length of the crossvein; this position while unusual is matched by other species and genera in the subfamily and is not unique. The basal half of the wings are deeply tinged with yellow.

One species is known from South Africa.

Length: 5 to 10 mm.; if wing approximately 5 to 9.5 mm.

Head, lateral aspect: The head is approximately hemispherical. The occiput is short with the central cup or recess large and deep. Pile of the occiput moderately dense, flat-appressed, and consisting of narrow scales densely clustered along the eye margin; the central fringe is short but rather dense on the upper half of the eye and almost wanting below. The front is flattened, slightly raised above the eye margin on the lower part, and densely covered with appressed and subappressed, long, slender, pale, scaliform hairs. There is also some nearly erect, bristly pile mixed in. The face is bluntly protuberant, rounded in front, extending forward at least the length of the third antennal segment. Its pile is moderately abundant, shining, golden reddish, and flattened. Some of it appears to be merely stiff bristles. All of it is short and subappressed. The eye is longer just above the middle. The posterior margin bears a very shallow, long (or wide), indentation without any bisecting line.

Proboscis short, thin, and laterally compressed but with a stout, somewhat triangular, short, wide labellum. The whole structure is confined to the oral recess. Palpus quite long and slender. A little flattened and clavate apically with a few, short, bristly hairs toward the apex and some long, white, fine, erect hairs on the basal half. Antenna attached a little above the middle of the head, rather small and short. The first segment is considerably widened apically, particularly toward the medial aspect where it forms almost a rounded lobe. Apically it is as wide as long. The first segment bears a tuft of short, stiff, bristly hairs at the apex both medially and laterally in between. Second segment quite short, wider than long, but with a rounded margin. Third segment bulblike at the base, no wider than the second segment and with an asymmetrical, dorsally placed, bristle-tipped, micropubescent style. The style is 3 times as long as the base of the segment and is a little thickened on the basal third.

Head, anterior aspect: The head is slightly flattened dorsally and likewise on each side; it is fully as wide as the thorax. The eyes are separated in the male by a little less than 3 times the ocellar tubercle and the vertex is slightly narrowed in front of the ocellar tubercle. The ocellar tubercle is rather high, the ocelli placed marginally and in an isosceles triangle. The antennae quite widely separated, distant from the eye margin by no more than the length of the first segment. The width of the face at the level of the antenna is a little greater than one-third the total head width. From the anterior aspect the oral recess occupies the lower fourth of the head and is relatively ventral in position. It is elongate with the anterior half comparatively shallow with sloping sides. Posteriorly the walls are vertical. The genae are narrow, separated from the wall of the oral cup by a fissurelike crease with a deep tentorial pit anteriorly. Viewed from above the occipital lobes are adjacent but rounded inward (forward) from behind and the fissure is deep.

Thorax: Comparatively short. The mesonotum is distinctly convex anteriorly, feebly shining blackish in color with dense, flat-appressed, matted, slightly flattened, pale pile and with a few suberect hairs in front of the scutellum, elsewhere on the dorsum and with rather dense tufts of somewhat longer, shining reddish-yellow pile on the whole of the humerus extending back to the base of the wing. Bristles are absent, but the tuft of pile on the notopleuron is dense, more distinctly flattened. The scutellum is quite large, largely pale reddish brown in color with the base blackish and with pile similar to the mesonotum. The marginal hair slightly bristly on the sides, very short in the middle. The anterior collar is distinct but narrow, coarse, moderately long and dense. Propleuron, upper half of sternopleuron, whole of mesopleuron with much coarse, rather long, shining, bristly pile thrust upward on the mesopleuron and outward collarlike on the propleuron. Pteropleuron and hypopleuron with thin pollen only. Metapleuron, however, with a scanty tuft of long, coarse, hair over the whole posterior half. Squamal scales scanty and narrow. There is a fanlike fringe of conspicuous pile beneath the postalar callosity and the lateral metanotum is pilose.

Legs: The legs are weak, especially the front pair. Anterior femur with 1 or 2 tiny setae. Middle femur with 2 distinct, slender bristles anteroventrally on the outer half and hind femur with a row of 5 to 7 equally slender, oblique bristles anteroventrally. Anterior tibiae smooth, their tarsi small and short, middle tibia with 4 or 5 small anteroventral bristles, slender and sharp, and the same number in each dorsal row and posteroventrally. Hind tibia similar, the dorsal rows with 7 or 8 tiny, slender bristles; middle and hind tibiae with strong spiculate ventral bristles. Claws slender only slightly curved with pulvilli long and slender.

Wings: Comparatively large, broad basally, hyaline on the outer half and posterior margin and similar in both sexes. The whole of the costal cell, auxiliary cell, base of marginal cell, whole of second and third basal cells light reddish brown. Also the basal part of the discal, fourth posterior, and anal cells and alula are tinged with reddish brown diffusely merging into the hyaline part of the wing. The marginal cell subapically is rather strongly widened backward, the second vein has a spoonlike loop. The posterior cells and the anal cell are all widely open. The anterior crossvein enters the discal cell at the middle. The second vein arises rectangularly with rounded bend distinctly basal to the anterior crossvein at least as much as the length of that crossvein. Ambient vein complete, alula relatively narrow, basal comb weak, the basal spur sharp but lappet-like.

Abdomen: Obconical with somewhat rounded sides in both sexes, the base as wide as the thorax, and the length approximately the same as the thorax. The color is chiefly light yellowish brown in the male, which has a wide, basal, black band over the middle of tergite, not reaching the apex of the sides, a smaller band on the third tergite and a medial spot on the fourth, likewise black. The female, however, is much more extensively blackish, only the sides of the tergites, the posterior margin of the fifth and the sixth and the whole of the seventh being reddish. There are 7 tergites observable in both sexes, the last one being comparatively long and more or less triangular. The pile of the abdomen is pale whitish yellow or cream-colored, forming a broad, wide brush of erect dense pile on the lateral third of the first tergite, continuing down the lateral margin of the abdomen, less dense, somewhat shorter and subappressed backward. Dorsum of the tergites with comparatively abundant similar pile, mostly flat-appressed, more abundant along the base of the margins and more likely to be denuded posteriorly. Female terminalia with about 4 or 5 quite slender, long, reddish, acanthophoric bristles or spines on each side. Male terminalia small rather strongly recessed, quite ventral in position and tending to be obscured by overlapping pile.

Hesse states:

Of the hypopygium that the apical angles of the shell-like basal parts are slightly produced, bases of latter also produced and lobe-like; beaked apical points somewhat flattened from side to side, slightly twisted; aedeagal apparatus with the aedeagus itself very short and the ventral aedeagal process prominently projecting, its apical part flattened and broadened spoon-like, ending apically below in a recurved lip-like or ledge-like process which is usually not much evident or is depressed against the surface in dried specimens.

Material available for study: A pair of the type-species sent for study and dissection by the courtesy of Dr. A. J. Hesse of the South African Museum, Capetown.

Immature stages: Unknown.
Ecology and behavior of adults: Not on record.
Distribution: Ethiopian: *Synthesia fucoides* Bezzi, 1921.

Genus *Thyridanthrax* Osten Sacken

FIGURES 169, 171, 188, 217, 218, 220, 221, 705, 725, 958, 959

Thyridanthrax Osten Sacken, Biologia Centrali-Americana, Dipt., vol. 1, p. 123, 1886. Type of genus: *Anthrax selene* Osten Sacken, 1886. Designated by Coquillett, p. 615, 1910, the first of 4 species.
Exhyalanthrax Becker, Ann. Mus. Nat. Hungarici, vol. 14, p. 44, 1916. Type of subgenus: *Anthrax vagans* Loew, 1862. See Bezzi, p. 7, 1924 for designation.
Tumulus, new subgenus. Type of subgenus: *Anthrax misellus* (Loew), 1869.
Oriellus, new subgenus. Type of subgenus: *Anthrax stigmulus* (Klug), 1832.

Bezzi (1924), pp. 193-195, key to 20 Ethiopian species.
Engel (1936), pp. 524-527, key to 33 species.
Austen (1937), pp. 147-148, key to 9 species from Palestine.
Hesse (1956b), p. 528, key to 55 male and 51 female species and subspecies.
Bowden (1964), pp. 101-102, key to 11 species from Ghana.

Medium-sized flies falling into several disparate groups, especially as far as the Palaearctic and Ethiopian species are concerned. The typical members of the genus are exemplified by the type-species, *selene* Osten Sacken; this and a number of other species from both the above regions, together with Nearctic species—I have at least 6 before me—have the wings dark at the base with characteristic hyaline spots centered about the crossveins and vein furcations. Some other characteristics of the restricted group *fenestratus* Fallen are: the end of the discal cell is pointed, not transverse; those species with windowlike spots strongly developed have a very short contact of the fourth posterior and discal cells; the abdomen has stiff, shaggy, appressed, bristly pile both dorsally and especially laterally and terminally. Within the genus the face varies from bluntly convex (slightly protuberant below) to obtusely conical, the peak quite short. The antennae are bulboconical with the stylar portion slender and generally much shorter than in *Chrysanthrax* Osten Sacken.

In 1924, Bezzi widened the scope of the genus as far as Palaearctic and Ethiopian species are concerned to include several quite different groups. He recognized four groups. He included the clear-winged species that Becker (1916) had relegated to his *Exhyalanthrax* and who regarded them as a subgenus of *Villa* Lioy. Still earlier Becker had placed *Thyridanthrax* Osten Sacken, as conceived by Osten Sacken, as a subgenus of *Hemipenthes* Loew; Paramonov (1927) adopted the same view. Bezzi in his broadening of the concept in 1924 included the group *afer* Fabricius, species with conical face and extensive, nonfenestrate wing pattern; this group and the clear-winged *Exhyalanthrax* Becker group include some parasites of *Glossina* Wiedemann flies among the South African member species.

Hesse (1956) notes that most authors have followed Bezzi in his broadened concept of *Thyridanthrax* Osten Sacken, but Hesse makes an excellent case for the restriction of *Thyridanthrax* Osten Sacken to the shaggy, fenestrate group of species, some of which are found in all three world regions cited above.

However, I feel that the old name *Exhyalanthrax* Becker may well be revived for the inclusion of those Old World species which have almost entirely hyaline wings. I, therefore, utilize it in a subgeneric capacity in this work. I also find it necessary to segregate species like *semifuscus* Engel and *misellus* Loew with their curious apically stout third antennal segments to separate status. For these I propose the subgenus *Tumulus*, with type *misellus* (Loew), 1869. I also cannot avoid the conclusion that the species *stigmulus* Klug and *vagans* Loew with their *Exoprosopa*-like articulated style should also be separated. For them I propose the subgenus *Oriellus*, with type *stigmulus* (Klug), 1832. I would agree with Hesse that much further subdivision of this large group of flies must be left for a time when some worker can attempt to elucidate large segments of the genus on a basic study of the genitalia of the Palaearctic and Nearctic species. Such a study might be attempted in the relatively near future, but it will be a long time before life histories can be added for more than a few species, since this type of study proceeds slowly and is often based upon fortuitous discoveries.

It is possible that the more conical-faced European species now in *Thyridanthrax* Osten Sacken, if critically studied, would fall into *Chrysanthrax* Osten Sacken. Painter (1933) separates these 2 primarily on the shape of the face.

It is perhaps not inappropriate to take note of the general similarity of some species of *Thyridanthrax* Osten Sacken to *Exoprosopa* Macquart. The *fenestratus* Fallen group has several species, such as *ternarius* Bezzi and *laetus* Loew, with 3 submarginal cells commonly and typically present; they are separable from *Exoprosopa* Macquart only by the lack of a third antennal segment, articulated style, and the absence of the characteristic tooth at the base of the claw. Some species of *Exoprosopa* Macquart, like *divisa* Coquillett, have a strongly fenestrate wing pattern. Although Engel (1936) gave the type-species as *Anthrax fenestratus* Fallen, Coquillett (1910) had made a previous designation.

Species have been placed in this genus from Chile, and if *afer* Fabricius is included, it is recorded by Kertész from India and Oceania. Painter (1964) lists 9 Nearctic species. The majority are from southern Europe, Asia Minor, North and South Africa.

Length: 4 to 15 mm., most species 7 to 13 mm.

Head, lateral aspect: Head hemiglobular with very extensive occiput. The occipital pile consists of flat-appressed, loosely distributed scales and a quite short, dense fringe around the central cup or recess. The eye is reniform with a shallow, extensive indentation. From the lateral aspect the lower corners of the eye are rounded so that the eye bulges somewhat backward and the upper eye corners are rather strongly rounded. The facets at the lower corner of the eye are quite small, and there is only a slight enlargement of the facets on each side of the front. There is a short bisecting triangle at the bottom of the indentation and no bisecting line. The front is covered with abundant, coarse, erect pile, and some flat-appressed scales which become more dense on the face. While there are some fine hairs upon the face, they are scanty, and even the fringe which extends over the edge of the oral recess is scanty. The proboscis and labellum are slender, reaching slightly beyond the apex of the oral recess. The palpus is cylindrical and not quite half as long as the proboscis. It has only very fine hairs fringing it ventrally and laterally. The antenna is attached at the upper third of the head. The first segment is swollen apically and sometimes medially. The first antennal segment bears considerable, rather long bristly pile, not longer than this segment. Second segment small and bead-shaped. The third is elongate conical, the short pile becoming quite slender and bearing a minute bristle.

Head, anterior aspect: As wide as the thorax, nearly circular from the front, but slightly flattened above and on the sides. The genae extend down below the eye rather conspicuously, though not more than the thickness of the proboscis, yet more so than in related genera. The antennae are widely separated, distant from the eye margin by rather less than the first antennal segment. The oral recess is chiefly ventral, but slightly oblique, narrow, deep posteriorly with vertical sides, and strongly narrowed anteriorly toward the apex of the face where the sides are somewhat widened or obliquely sloping inward. From above, the lobes of the occiput are tightly apposed, rather horizontal in position, but with a neat, deep, narrow, cleftlike fissure behind the ocelli. The ocellar tubercle is a little elongate, quite prominent and protrusive, yet not rising above the uppermost level of the eye. In males the eyes are rather narrowly separated by not more than and sometimes less than the width of the ocellar tubercle.

Thorax: The thorax is broad anteriorly, at least as wide anteriorly as the mesonotum is behind the wing, but the mesonotum is a little wider across the postalar callosities. The ground color is black with faint, dark brown pollen, and the scutellum may be almost entirely reddish or yellow in some species, or the basal half may be black. Pile of the mesonotum consists of fine, appressed, bristly or scaliform hair with considerable amounts of erect or suberect, stiff, pale pile, becoming more dense, anteriorly, and merging into an anterior collar which may be extensive, and which may form a wide, broad band along the anterior margin. In other species there is no progressive gradation of this anterior pile into the remainder of the mesonotum. The propleuron and mesopleuron are extensively pilose with tufts of very coarse pile. There are few hairs on the pteropleuron. The metapleuron is densely and widely pilose. There is a tuft of hair beneath the haltere, and a conspicuous tuft appressed downward above the posterior coxa. The anterior hypopleuron is bare, but the plumula bears much long pile and the scales of the squama and alula are long and conspicuous and extend as shorter scales onto the axillary lobe.

Legs: Anterior pair of legs much reduced, the anterior tibia smooth. The tarsus is small and the claws minute. The middle femur bears anteroventrally a fringe of long, slender bristles, 5 to 7 in number. Hind femur with 5 to 12 slender, oblique bristles, a little more stout in some species, and a few short bristles of spicules dorsally at the apex. All the femora, especially the hind femora, and the last four tibiae with conspicuous, narrow, appressed scales. Middle and posterior tibiae with numerous, fine, oblique bristles in 4 rows, and some shorter setae intermixed. Most of these rows have 10 such bristles, but the anterodorsal row on the hind tibia of *nugator* Coquillett has more than 20 such bristles. Claws small, quite slender, sharp, and scarcely curved. Pulvilli absent.

Wings: The wings tend to be broad at the base and pointed apically. Characteristically they have rather more than the basal half diagonally colored with brown or brownish black of varying intensity. Always with irregular margin and sometimes diffuse. This basic black basal color in a few species is reduced to a series of diffuse spots. But in those species with a more intense pattern of brown, very characteristically there are windowlike spots, one extending on the anterior crossvein and above it, one extending on all four sides of the contact point of the discal and fourth posterior cell. Another on all sides at the upper end of the second basal call and generally at the base of basal cells. The anterior crossvein enters the discal cell at the middle or often distinctly beyond the middle, though never as far as the outer third. In some of the European species, placed by European authors in this genus, the anterior crossvein may be relatively basal, even located at the basal third of the discal cell; and at least 4 or 5 European species lack the windowlike spots and have only a pale fleck within the brown area located where the thyridium is normally placed. It is notable that these European species that lack the fenestrate wing likewise have a long contact of the discal cell with the first posterior, and the typical species of the genus show a tendency for the contact of discal and posterior cells to be either punctiform or at least short.

The type-species of this genus is *selene* Osten Sacken and not *fenestratus* Fallén as given by Engel. The shape of the discal cell varies widely, sometimes long, sometimes quite short. The first posterior cell is a little narrowed, the anal cell widely open, the alula about 3 times as long as wide, and there are only 2 submarginal cells. The ambient vein is complete. Another interesting character lies in the extraordinary width of the costal comb. The costal hook is long and thornlike.

Abdomen: The abdomen is short oval, mostly black in ground color, but some species at the sides of the basal tergite and the last two tergites are more or less brownish yellow, or with reddish brown that extends inward toward the midline. The pile is dense and consists of slender matted scales or scaliform pile, sometimes with bands of varying color along the posterior margins, and often in many species with a considerable amount of fine, erect, or suberect pile which may form a bristly terminal fringe. Males may have a little silvery pile on the last tergite. Females with the terminalia recessed and bearing several pairs of spines on the lateral margin of the acanthophorite. Male terminalia comparatively large but recessed and asymmetrical, enclosed by the last sternite which pinches together to form a protecting cover.

Dr. A. J. Hesse was able to study a very large assemblage of species in this genus; for this reason I quote his comments on the male genitalia:

> Hypopygium of males usually with a dorsal ridge on each shell-like or clasper-like basal part, this ridge usually hairy, usually projecting freely in the form of a short process posteriorly, the freely projecting part sometimes fused with its neighbour, forming a U-shaped crest on the two basal parts; aedeagus proper usually short, sinuous; aedeagal apparatus always with a conspicuous, projecting, scoop-like, aedeagal process which differs in shape in the various species and is usually provided apically and ventrally on each side with a hooklet or spine.

The male terminalia from dorsal aspect, epandrium removed, show a remarkably broad, wide, long, extensive epiphallus, quite tonguelike. The basal part of the aedeagus short oval and egg-shaped. In the lateral aspect, the dististyle is apical with dorsoapical incision and hook. The ejaculatory duct is separated below from the epiphallus. The epandrium has a strong basal, downward lobe, leaving a wide ventroapical incision, much as in *Rhynchanthrax* Painter.

Material available for study. All the material in the National Museum of Natural History, the Museum of Comparative Zoology at Harvard, the British Museum (Natural History), and considerable personally collected material.

Immature stages: The larvae of these flies consume pods of locust eggs; they also parasitize *Glossina* and *Calliphora*.

Ecology and behavior of adults: The adults are abundant on many types of low-growing flowers, especially composites.

Distribution: Nearctic: *Thyridanthrax atratus* (Coquillett), 1887 (as *Anthrax*); *bifenestratus* (Bigot), 1892 (as *Anthrax*); *fenestratoides* Coquillett, 1892 (as *Anthrax*) [=*macula* (Cole), 1919 (as *Anthrax*)]; *melasoma* (Wulp), 1882 (as *Anthrax*); *nugator* (Coquillett), 1887 (as *Anthrax*); *pallidus* (Coquillett), 1887 (as *Anthrax*); *pertusus* (Loew), 1869 (as *Anthrax*); *selene* (Osten Sacken), 1886 (as *Anthrax*) [=*otiosa* (Coquillett), 1887 (as *Anthrax*)]; *utahensis* Maughan, 1935.

Neotropical: *Thyridanthrax hypoxanthus* Macquart, 1840 [=*blanchardii* (Philippi), 1865 (as *Anthrax*)]; *semilugens* Philippi, 1865 [=*unicinctus* (Bigot), 1892 (as *Anthrax*)]; *semitristis* Philippi, 1865.

Palaearctic: *Thyridanthrax afer* (Fabricius), 1794 (as *Anthrax*), [=*fimbriatus* (Meigen), 1804 (as *Anthrax*), =*sirius* (Hoffmannsegg in coll. in Meigen), 1820 (as *Anthrax*)]; *agnitionalis* Austen, 1937; *albicingula* Austen, 1936; *albinus* Becker, 1913; *albosegmentatus* Engel, 1936; *amoenus* Austen, 1937; *angusteoculatus* (Becker), 1902 (as *Anthrax*); *anus* (Wiedemann), 1828 (as *Anthrax*); *argentifer* (Becker), 1916 (as *Exhyalanthrax*); *autumnalis* (Becker), 1916 (as *Exhyalanthrax*); *chinophorus* Bezzi, 1925; *circe* (Klug), 1832 (as *Anthrax*); *dagniensis* Paramonov, 1934; *elegans* (Wiedemann in Meigen), 1820 (as *Anthrax*), [=*variegatus* Jaennicke, 1867]; *fenestratus* (Fallen), 1814 (as *Anthrax*), [=*fenestralis* Wahlgren, 1907, =*nigritus* (Fabricius), 1781 (as *Bibio*), =*ornatus* (Curtis), 1824 (as *Anthrax*), =*variegatus* (Pallas in Wiedemann), 1818 (as *Anthrax*)]; *fenestratus montanus* Paramonov, 1926; *fulvifacies* Austen, 1937; *graecus* Paramonov, 1926; *griseolus* (Klug), 1832 (as *Anthrax*); *heteropterus* Paramonov, 1926; *hispanus* (Loew), 1869 (as *Anthrax*); *inauratus* (Klug), 1832 (as *Anthrax*); *incanus* (Klug), 1832 (as *Anthrax*) [=*testaceus* (Macquart), 1840 (as *Anthrax*); *indigenus* Becker, 1908(as *Anthrax*); *innocens* Austen, 1936; *intermedius* (Paramonov), 1927 (as *Hemipenthes*); *irrorellus* (Klug), 1832 (as *Anthrax*) [=*inconspicuus* (Loew), 1856 (as *Anthrax*)]; *kushkaensis* (Paramonov), 1927 (as *Hemipenthes*); *latipennis* Paramonov, 1927; *latona* (Wiedemann), 1828 (as *Anthrax*); *lepidulus* Austen, 1937; *leucotaeniatus* Engle, 1936; *lineus* (Loew), 1860 (as *Anthrax*); *lotus* (Loew), 1869 (as *Anthrax*) [=*confusus* Becker, 19—]; *macrops* (Portschinsky), 1887 (as *Anthrax*); *melanchlaenus* (Loew), 1869 (as *Anthrax*); *mervensis* Paramonov, 1926; *minutus* (Macquart), 1849 (as *Anthrax*); *misellus* (Loew), 1869 (as *Anthrax*); *montanorum* Austen, 1936; *mutilus* (Loew), 1869 (as *Anthrax*); *nebulosus* (Dufour), 1852 (as *Anthrax*) [=*nubilus* Loew, 1871, lapsus, =*occipitalis* Loew, 1869]; *nitidifrons* Austen, 1936; *noscibilis* Austen, 1936; *obliteratus* (Loew), 1862 (as *Anthrax*); *pauper* Becker, 1916 (as *Exhyalanthrax*); *perpusillus* Austen, 1937; *perspicillaris* (Loew), 1869(as *Anthrax*) [=*fenestratus* (Meigen), 1820 (as *Anthrax*), =*gallus* (Loew), 1869 (as *Anthrax*)]; *polyphemus* (Wiedemann in Meigen), 1820 (as *Anthrax*), *polyphemus pumilio* Austen, 1937; *punctum* (Loew), 1869 (as *Anthrax*); *rohdendorfi* (Paramonov), 1925 (as *Villa*); *rotundifacies* Paramonov, 1926; *rubriventris* Paramonov, 1926; *samarkandicus* Paramonov, 1926; *semifuscus* Engel, 1936; *stigmulus* (Klug), 1832 (as *Anthrax*); *tabaninus* Bezzi, 1925; *timurensis* Paramonov, 1926; *trancaspicus* (Paramonov), 1924 (as *Exhyalanthrax*); *turcomanus* Paramonov, 1926; *unctus* Loew, 1869; *unicolor* (Becker), 1902 (as *Anthrax*); *vagans* (Loew), 1869 (as *Anthrax*) [=*marginalis* Loew, 1869], *vagans beckeri* (Paramonov), 1926 (as *Hemipenthes*), *vagans contrarius* (Becker), 1916 (as *Exhyalanthrax*), *vagans parvus* (Paramonov), 1926 (as *Hemipenthes*), *vagans unistriatus*, Engel, 1936; *venustulus* Austen, 1936; *vetulus* (Wiedemann), 1828 (as *Anthrax*).

Ethiopian: *Thyridanthrax aberrans* Hesse, 1956; *abruptoides* Hesse, 1956; *abruptus* (Loew), 1860 (as *Anthrax*); *alliopterus* Hesse, 1956; *anisospilus* Hesse, 1956; *arenicolus* Hesse, 1956; *argentifrons* Austen, 1929; *argyrolophus* Hesse, 1956; *atriventris* Hesse, 1956; *bechuanus* Hesse, 1956; *beckerianus* Bezzi, 1924; *beneficus* Austen, 1929; *bolbocerus* Hesse, 1956; *brevifacies* Hesse, 1956; *burtii* Hesse, 1956; *caffrariae* Hesse, 1956; *calochromatus* Bezzi, 1921; *cidarellus* Hesse, 1956; *griseifrons* Hesse, 1956; *idolus* Hesse, 1956; *incipiens* Bezzi, 1924; *lugens* (Loew), 1860 (as *Anthrax*); *luteolus* Bezzi, 1924; *lutulentus* Bezzi, 1921; *macquarti* Bezzi, 1912; *melanopleurus* Bezzi, 1912; *monticolus* Hesse, 1956; *nitidifrons* Hesse, 1956; *niveifrons* Hesse, 1956; *occiduus* Hesse, 1956; *phileremus* Hesse, 1956; *pseudoflammiger* Bezzi, 1924; *salutaris* Austen, 1929; *semilautus* Hesse, 1956; *simmondsi* Hesse, 1956; *stylicornis* Hesse, 1956; *subperspicillaris* Bezzi, 1924; *ternarius* Bezzi, 1921, *ternarius speciosus* Hesse, 1956; *thyridus* Hesse, 1956; *transiens* Bezzi, 1921; *triangularis* Bezzi, 1924; *uroganus* Hesse, 1956; *vicinalis* Hesse, 1956; *zinnii* Hesse, 1956.

Oriental: *Thyridanthrax unicinctus* Guérin, 1838.

Genus *Rhynchantrax* Painter

FIGURES 232, 401, 716, 727, 964, 965

Rhynchanthrax Painter, Journ. Kansas Ent. Soc. vol. 6, p. 6, 1933a. Type of genus: *Anthrax parvicornis* Loew, 1869, by original designation.

Painter (1933), p. 7, key to 5 species.

Small to medium-sized flies with approximately the basal half of the wing dark brown but sharply and irregularly and diagonally delimited and enclosing paler windowlike spots. From other genera of the *Villa* Lioy complex it is separated by the sharply narrowed bulblike base of the third antennal segment with its very long, slender style. In the long, slender style and in the long, slender proboscis and labellum, which is about twice as long as the head, it is quite similar to

Lepidanthrax Osten Sacken; from it separated by the broad wing, different type of wing pattern, and the absence of the patches of matted scales on the sides of the abdomen. In *Rhynchanthrax* Painter the bristles of the anterior tibia vary even within the species; I have individuals of *parvicornis* Loew from Kansas with strong, stout bristles, others from Mississippi with weak, fine bristles. Pulvilli are absent; this character and the 2 submarginal cells separate these flies from *Stonyx* Osten Sacken.

The range of this genus extends from Colorado, Kansas, and Illinois on the north to Mississippi on the east, Arizona on the west, and Texas and northern Mexico on the south.

Length: 5 to 14 mm.

Head, lateral aspect: Head is hemiglobular, but the occiput is only moderately produced and developed backward and the occipital pile is appressed, short and bristly with rather narrow scales, the central fringe short. The indentation of the eye is unusually shallow, although the occiput is extensive above and below, and the eye facets at this point curve forward leaving a small triangular indentation, but no bisecting line. The front has much apparently short, fine, erect, bristly hair, some faint pollen and numerous flat-appressed scales especially on the anterior half. Occasionally they are reduced in quantity, becoming more numerous, though loose and scattered in appearance on the face. Characteristically the face has a considerable amount of fine, erect, or suberect bristly pile, in length scarcely longer than the second antennal segment. The face may be described as bluntly extended forward, maximum length from eye being slightly more than the length of the three antennal segments if the style be excepted. The face is not peaked or conical but blunt anteriorly.

The proboscis in this genus is very slender, the labellum also, and it is characteristically elongate, extending far beyond the apex of the oral recess. It is nearly twice as long as the oral recess. Palpus very slender and cylindrical with fringe of stiff, bristly hairs ventrally and laterally. The antenna is attached near or just below the upper third of the face and all the segments of the antenna are small, the first segment scarcely longer than wide with a few short bristly hairs on all sides, except below. Second segment beadlike, a little wider apically, broadly attached to the base of the third antennal segment. The third segment is a quite small onion-shaped or bulb-shaped segment and short with an abruptly formed, very slender style of considerable length. The whole segment is very much like that of *Lepidanthrax* Osten Sacken which, incidentally, also has a long proboscis. The style of the antenna is a little longer than the total length of the three segments.

Head, anterior aspect: The head is large, as wide as the thorax. The eyes are rather widely separated, even in males, by 3 times the width of the ocellar tubercle and by at least the length of the three antennal segments. In the female it is at least 4 times as wide as the ocellar tubercle. The antennae are unusually widely spaced apart, the space from the eye margin is rather less than the length of the first segment. The oral recess is wider than in related genera, deep posteriorly, and narrowed only quite near the anterior peak or face, and this narrowed area is widened by obliquely sloping inward. The palpus is unusually long and filiform and the proboscis has a double bulbiform base. Viewed from above the ocellar tubercle is rather low, but placed distinctly forward from the anterior corners of the eye, which are broadly rounded. Lobes of the occiput are vertical and not touching.

Thorax: the thorax is subquadrate, widened only across the prominent postalar callosities. It is black with dark brown pollen, the apical half of the scutellum more or less obscurely reddish brown. The pile of the mesonotum is coarse, short and appressed or subappressed. There is a row of very fine, erect hairs in front of the scutellum. The anterior marginal band of pile is coarse, wide in depth, and the pile comparatively short. There are 3 or 4 bristles on the notopleuron, some bristly pile anterior to these, 4 or 5 longer bristles on the postalar callosity and 7 or 8 pairs on the margin of the scutellum which are long and conspicuous. Propleuron and mesopleuron with extensive, long, coarse, dense, appressed pile. There are a few hairs on the pteropleuron. The metapleuron has dense pile, but the hypopleuron is bare anteriorly and posteriorly. There is a tuft of hair beneath the haltere and the plumula is densely pilose. Squamal scales are conspicuous and other long, extending onto the alula and the axillary lobe.

Legs: The legs are stout, including the anterior pair which are well developed. The anterior pair has only the tarsus reduced and with a fringe of fine hair rather than setae. The front tarsal claws are rather large. The middle femur has several stout bristles placed in the middle anteriorly. The hind femur has 8 to 10 bristles anteroventrally; in some species it is slender, in other species stout. Characteristically the anterior tibia is spiculose in several rows, but these bristles are fine, evident but not conspicuous. Hind tibia with 4 rows of bristles, moderately long and oblique. There are 7 or 8 anteroventral, 10 to 14 anterodorsally, including the smaller setae and rather small finer bristles in the other two rows. Claws stout, not very long, and pulvilli quite rudimentary or absent.

Wings: Broad at the base, pointed apically, the basal half of the wing is diagonally covered with dark sepia brown, the margin sharp, but irregular and with fenestrate spots at and above the anterior crossvein at the end of the second basal cell and below it and likewise narrowly at the base of the basal cells. The depigmented fleck on the thyridial area is large. The first posterior cell is scarcely narrowed, the anal cell is widely open. There are only 2 submarginal cells. The alula is 3 times as long as wide. The ambient vein is complete and fringed, and the costal comb while not as conspicuous as in *Thyridanthrax* is wide and has many stout setae, and the costal hook is sharp and strongly thornlike.

Abdomen: Short and obtuse, wider than the thorax, black in some species, reddish orange in others with medial black bands on the second and third and fourth tergites of decreasing width, but the first tergite is always entirely black. The pile consists of dense, appressed scales for the most part, which are black and yellow or intermixed, and which laterally in *texana* Painter are orange red. Viewed laterally a few fine, black hairs are suberect, and these become more in evidence laterally and on the last two tergites. Sides of first two tergites with copious long, dense, pale, coarse pile. The acanthophorites have a row of spines on the side. The male terminalia are large and asymmetrical and rather strongly recessed.

The male terminalia from dorsal aspect, epandrium removed, show a very wide, long, apically rounded epiphallus with nearly parallel sides. The basal part of the aedeagus is large and short oval, the lateral apodemes are paired. From the lateral aspect the dististyle is thick, only moderately long, with dorsoapical incision and hook. The epiphallus is long and thick without distinctly separated ejaculatory process. The epandrium is much rounded, especially basally or posteriorly. It has a deep, wide, anterior incision on the lower half.

Material available for study: A series of the type-species and one more species; also, the series in the National Museum of Natural History.

Immature stages: Parasites of white grubs. See life histories.

Ecology and behavior of adults: These flies are not uncommon on a variety of flowers and are especially attracted to *Rudbeckia* species and other composites.

Distribution: Nearctic: *Rhynchanthrax parvicornis* (Loew), 1869 (as *Anthrax*); *quivera* Painter, 1933; *rex* (Osten Sacken), 1886 (as *Anthrax*) [=*plagosa* (Coquillett), 1887 (as *Anthrax*)]; *texanus* Painter, 1933.

Genus *Paranthracina* Paramonov

Paranthrax Paramonov, Trav. Mus. Zool., Kieff, no. 11, p. 57, 1931. Type of genus: *Paranthrax africanus* Paramonov, 1931, by monotypy. Preoccupied Bigot, Bombyliidae, 1876.
Paranthracina Paramonov, Trav. Mus. Zool. Kieff, no. 12, p. 52, 1933. Change of name.

I have not been able to see this genus.

Distribution: Ethiopian: *Paranthracina africanus* Paramonov.

Genus *Villoestrus* Paramonov

FIGURES 410, 682, 685

Villoestrus Paramonov, Trav. Mus. Zool. Kieff, no. 11, p. 93, 1931. Type of genus: *Villoestrus uvarovi* Paramonov, 1931, by original designation.

Large, robust, broad, very compact, obtuse, greasy flies in which the oragenal cavity is reduced to a mere slit or narrow cleft. The proboscis and palpus are completely absent. Related to *Oestranthrax* Bezzi and several other Exoprosopinae with reduced mouthparts, they represent a much more extreme specialization. The head is large, subglobular and together with face and front, it is very strongly convex. There are some other peculiarities. The auxiliary vein extends far toward the wing apex. There appears to be a vena spuria immediately behind the first vein. There are 2 submarginal cells and the 4 posterior cells are widely open. The discal cell is shaped like a shoe or slipper and the anterior crossvein enters at the middle; the second vein arises almost rectangularly just in front of the anterior crossvein. The anal cell is closed with a short stalk. Wings sometimes strikingly dimorphic in color and pattern; males may have a strong brownish blotch across the middle or lack it; female of *dimorphus* Hesse has a very dark, conspicuous anteromedial blackish spot on the wing; female of type-species unknown. The abdomen is especially thick and high. The general color varies from pale ochraceous brown to reddish.

These rare flies are known from Palestine and from Strandfontein dunes in the Cape Flats near Capetown, South Africa.

Length: 16 to 16.5 mm., in both species; wing 12 to 13.5 mm.

Head, lateral aspect: The face is rather prominent but strongly convex and receding; it ends below in a deep, narrow, oblique crease or fissure which does not quite extend to the middle of the face. Just before its medial end, which is deep punctate, it extends shallowly and longitudinally a short distance back toward the occiput. The cheeks are wide but do not extend below the eye, except for a faintly linear margin; they are barely visible at the bottom of the eye. The oral opening is confined to a short, narrow cleft or slit; proboscis and palpus absent. The eye is much longer dorsally, therefore only gently convex above but strongly convex below; the posterior margin in the middle has a low indentation and a long continued transverse crease upon the eye. The occiput is thick and prominent but gently sloping medially and with a very large, abrupt, deep, central concavity; the pile of the occiput is extremely scanty, minute, fine, and laterally flat-appressed; upper occiput without a dorsal fissure. The antennae are attached at the middle of the head and short; the first segment is as long as the second and third with the style omitted. This first segment is greatly swollen along its medial aspect and is therefore much wider than the second, the second segment is quite short and disclike; the third segment is curiously and rather strongly compressed, flattened on the outer surface; it is convex medially, pyriform, the style narrow, slightly longer than the segment itself, very slightly dilated at its apex and with a minute conical spine.

Head, anterior aspect: The face bears dense, stout, stiff, subappressed pile with black and yellow hair intermixed; the latter is somewhat flattened or scalelike and with distinct rather broad scales on the upper half between the antennae and on the front. Antennae separated by a distance nearly equal to the combined length of all three segments, style omitted. The front also has considerable short, suberect, coarse, black pile. The ocellar triangle is small, low, nearly equilateral, the ocelli small.

Thorax: Rather densely short, matted, fine pilose; especially matted with pale pile along the lateral margins, much of the mesonotum is bare; this is almost certainly due to being denuded. Scutellum large, wide, rather convex and somewhat gently swollen over the middle and abrupt at the base, with densely tufted, matted, fine, pale pile laterally and along the margin and with shorter appressed pile of the same color on the disc. There is scarcely any trace of anteromedial depression marking off the large swollen postalar callosity. Pleuron with a dense tuft of pile of rather fine pale color anteriorly below or on the anterior margin of the humerus and across the anterior margin forming a collar; also over almost the whole of the mesopleuron and with an especially dense tuft on the metapleuron. Sternopleuron with abundant but less dense pile, a scanty tuft on the pteropleuron and the posthypopleuron; lateral metanotum pilose.

Legs: Small, short and weak with small, fine scales on the femur and tibia; there is no long pile. It has minute fine setae on the hind femur, more numerous fine, short bristly hairs on the anterior femur; the hind tibia has a lateral, ventrolateral, dorsolateral and dorsomedial row, medial row and ventromedial row of fine, short setae or minute bristles; anterior femur with 2 rows posteriorly and 1 dorsal or posterodorsal row.

Wings: Slightly reduced in size, and relatively short, but strong and broad. Mostly hyaline, they may differ in color within the sexes and in some females have a strong central black pattern. The basal costal hook is stronger than in *Oestranthrax* Bezzi and broader. The second vein has no basal stump, and is not as sharply curved upward apically and, moreover, the second submarginal cell is longer; also, the second and third posterior cells are decidedly longer; the discal cell is shaped like a shoe or slipper and is somewhat constricted in the middle; the anterior crossvein enters at the middle and the second vein arises just basal to it and arises abruptly. The axillary lobe is broader, the alula wide, the ambient vein complete.

Abdomen: Short, thick, obtuse, with a dense tuft of more or less erect, fine, brownish-white pile laterally on the first segment and equally dense, flat-appressed, matted pile laterally on the second and the third tergites; on the third tergite most of this pile becomes small, slender scales; the third and the very short fourth, fifth, sixth, and seventh tergites and their sternites have rather numerous, slender, brownish-yellow scales. There are visible from above 7 tergites in the male with a short eighth tergite concealed beneath the seventh. Terminalia quite short and inconspicuous.

Material available for study: The type-species in the British Museum (Natural History).

Immature stages: Unknown.

Ecology and behavior of adults: Unknown. Hesse (1956) remarks that the South African species was overlooked by dipterists and captured by a lepidopterist looking for butterflies! He comments on the "short seasonal occurrence" this genus must have.

Distribution: Palaearctic: *Villoestrus uvarovi* Paramonov, 1931.

Ethiopian: *Villoestrus dimorphus* Hesse, 1956.

Genus *Oestranthrax* Bezzi

FIGURES 166, 230, 233, 394, 683, 686, 740, 775, 783, 955, 956, 957

Oestranthrax Bezzi, Ann. South African Mus., vol. 18, p. 130, 1921. Type of genus: *Anthrax obesus* Loew, 1863, by original designation; first mentioned here. Described in: Voyage Alluaud et Jeannel Afrique orientale, pt. 6, p. 326, 1923.

Engel (1936), pp. 556-561, descriptions of 4 species, 2 subspecies, Palaearctic.

Hesse (1956), pp. 511-512, key to 4 South African species.

Strange flies of large size, and rather obese and bloated form. This fly is unique because of the much reduced mouthparts; the mouth opening itself is small and the proboscis is rudimentary; cheeks broad and separated from the face region by a deep fissure or furrow. Palpus short and distinct. No macrochaetae on thorax. Wings shorter than the body of fly and the second basal cell is short and dilated. Wings hyaline. Bezzi (1924) relates this fly to the *Villa* Lioy complex of genera, and notes a similar trend toward reduction of mouthparts in such flies as oestrids, the tabanid *Adersia* Austen, and the muscoids of the group Trixinae; he correlates the reduction with high degree of specialization in its parasitic relationship.

Oestranthrax Bezzi is related to *Villoestrus* Paramonov, which has complete reduction of mouthparts and loss of palpus, and *Marleyimyia* Hesse, intermediate between the other two genera. Hesse (1956) comments upon the peculiar oily exudate that comes from the surface of all these genera and suggests that the host of these flies is probably excessively rich in fats and oils. Could they be hyperparasites upon late stages of meloids? Related genera have been reared from logs containing Cossidae.

Oestranthrax Bezzi contains 8 species from Asia Minor and South Africa; there are also recently discovered species in Utah, U.S.A.

Length: 9.5 to 16 mm.; wing 8 to 14 mm.

Head, lateral aspect: The head is subglobular, the face moderately projecting, extensive, and of considerable depth but broadly rounded and retreating. Only the lower third of the front extends beyond the eye. The eye itself rather short and consequently much higher than long, the face is rather densely covered with conspicuous, more or less appressed or suberect, broad, pale scales extending nearly to the eye margin. The occiput is rather prominent and tumid, bilobate above with slender fissure and large, central, cuplike cavity. Pile of occiput scanty, fine, short, erect and located on the inner border around the cavity, except that the remaining outer portion tends to have flat-appressed, matted, slender scales becoming dense at the eye margin. Posterior eye margin broadly and shallowly indented and with a bisecting crease in the middle. Pile of the front moderately abundant and consisting of slender, erect scales which are apt to be black, changing over to appressed yellow scales in front along the eye margin and also with a number of scattered, quite fine, erect, delicate, black hairs no longer than the scales and present throughout. The proboscis is very greatly reduced and vestigial and is snugly recessed within the small, shallow, oral cavity lying between the posterior eye corners. The palpus is present and short and appears to consist of 2 segments with the basal segment minute. It also is more or less recessed and lies on each side of the proboscis. The genae are moderately wide, although the oragenal space is narrowed across the proboscis.

The genofacial fissure is conspicuously deep and rather flared, overlapped by pile from the face. There is no oragenal cup in the usual accepted sense. Antennae attached slightly above the middle of the head. All antennal segments short, the first is about twice as long as the second, narrow at the base, and expanded somewhat ventrally at the apex. Second segment a little compressed and attached obliquely to the first segment. First segment with a tuft of long, coarse setae dorsolaterally and another one ventrally, third segment consists of a minute basal, bulblike portion abruptly followed by a slender, stylelike process, which is about as long as the first two segments together.

Head, anterior aspect: The head, no wider than the thorax or a little more narrow. The vertex of the female is narrow, approximately one-fourth of the width across the antenna. Ocellar tubercle quite small but with vertical sides and sharply demarcated. The ocelli are large, the anterior ocellus quite vertical, the others less so. They occupy a short isosceles triangle and the tubercle bears 2 or 3 setae of no great length. The antennae are widely separated at the base by a distance nearly equal to twice the length of the first segment.

Thorax: The thorax is broad and short, shallowly convex, the mesonotum feebly shining, rather densely covered with quite slender, appressed scales apt to be denuded across the middle but more in evidence as a band in front of the scutellum, or scattered along the anterior portion. Anterior margin of mesonotum with a moderately dense, not very long collar or band of erect, coarse, bristly, pale pile. Bristles are quite absent, the notopleuron has a dense, bushy tuft of extensive, erect, more or less radiating pile in front of the wing, and there is a narrow fringe or erect pile above the wing. Scutellum more or less triangular, wide at the base, the posterolateral angles broader than the anterobasal angle. Surface covered with scales and a few fine hairs along the margin. Squama with a fringe of dense, long hairs barely flattened. Mesopleuron with a very strongly developed, quite broad, dense, radiating tuft of pile below the haltere, another tuft lower down on the lateral metasternum. Hypopleuron bare.

Legs: The legs are rather short, all the femora a little thickened, the anterior tibiae and tarsi quite short, the former arcuate. Pile of the femora and tibiae consists entirely of appressed scales, the middle femur lacks any posterior fringe. Hind femur with 2 or 3 small setae below and a few still smaller setae laterally. All of the tibiae have conspicuous, quite fine, sharp spicules. They are comparatively well developed though of no great length or thickness. The anterior tibia has 4 rows of spicules about equally developed each with some 8 or 10 in number; it is slightly curved, with a fine brush of hair below. Spicules of hind tibia very little if any larger except ventrally. There are 5 rows that have 8 to 17 elements. Claws long and slender without basal teeth, curved from the base. Pulvilli absent.

Wings: The wings are hyaline, comparatively small, the costal cell and to some extent the first basal cell tinged with yellowish brown. Auxiliary vein ends obliquely but comparatively close to the end of the first vein. There are only 2 submarginal cells, but the anterior branch of the third vein has rather characteristic pattern being perhaps a little bit more abrupt in its origin and broadly rounded as it turns to the apex. Second vein arises a little before the anterior crossvein but rectangularly and without spur. The anterior crossvein enters the discal cell near or just beyond the middle. The upper vein of the third posterior cell has a shallow, angulate bend with a spur directed proximally in the discal cell. The discal cell is convexly widened both anteriorly and posteriorly; second basal cell quite short and wide. The alula is moderately large bearing a fringe of narrow scales; ambient vein complete. The base of the costa is expanded with prominent comb and a sharp basal spur.

Abdomen: The abdomen is very short oval, distinctly broader on the first three tergites than the thorax and covered densely with appressed scales, tending to be pale basally on the tergites and black posteriorly; first tergite, however, with dense, erect hairs, except along its posterior margin where the hairs are replaced by appressed scales.

The dististyli of the male are compressed laterally, rather twisted, and bifid as in *Villa* Lioy. The aedeagus is relatively short and according to Hesse (1956) has a well-developed scooplike process of which the central, dorsal, apical part is raised keel-like. The lateral struts project straight outward, the basal strut does not project beyond the base.

Since Dr. A. J. Hesse was able to base his conclusions upon a greater number of species, I quote his remarks upon the male hypopygium:

Hypopygium of male very much like that of *Villa*, especially of the *leucochila*-section; beaked apical joints compressed apically, more or less twisted, bifid apically as in *Villa*; aedeagal apparatus with a relatively short aedeagus and a well-developed scoop-like aedeagal process of which the central, dorsal, apical part is raised keel-like; lateral struts projecting straight out laterally; basal strut not projecting beyond bases of basal parts.

The male terminalia from dorsal aspect, epandrium removed, show a basally wide, apically narrowed, elongate structure in which the epiphallus is very large, wide, apically widened and dilated, with apex rounded, and quite tonguelike in appearance. The basal part of the aedeagus is rather large, egg-shaped or bulblike. From the dorsal aspect the ramus proceeds backward as 2 long, curved, convergent, lobelike extensions extending as far backward as the end of the basal ejaculatory apodeme. From the lateral aspect, the epandrium is large and triangular, the basistylus large, stout or robust both apically and even more so above the base. The dististyli are short, stubby, dorsoapically incised and hooked and subapical in position. The epiphallus from lateral aspect is large, long, and tonguelike.

Material available for study: I am deeply indebted to Dr. A. J. Hesse of the South African Museum, Cape Town, for sending a pair of the type-species for dissection and study. I have also previously seen the species in the British Museum (Natural History), and I have seen the species *farinosus* Johnson and Maughan in the National Museum of Natural History.

Immature stages: Unknown.

Ecology and behavior of adults: Johnson and Maughan (1953) write of the fly *farinosus* Johnson and Maughan; this species inhabits the eastern edge of the Pahvant Valley of Utah, a broken terrain with intermittent areas of sand and loam and dominant vegetation sagebrush and greasewood, and they stated further that the sandy areas produced many kinds of bee flies, the loamy areas nearly barren yet it was in these more or less barren areas the flies of this genus were found. They were nearly always resting near the ground on dead twigs or stems, their pale color blending well with the weather-whitened wood; most of his 13 specimens were seen by walking into the sun and watching for the glint of sunlight on the heads and the epaulets of white scales on the wings; capture was difficult and uncertain, their darting flight being extremely rapid. They often returned to the same general locality, had to be captured in flight and were very easily abraded; a year later the same locality produced none.

Distribution: Nearctic: *Oestranthrax farinosus* Johnson and Maughan, 1953.

Palaearctic: *Oestranthrax arabicus* Paramonov, 1931; *brunnescens* (Loew), 1857 (as *Anthrax*); *goliath* Oldroyd, 1951; *karavajevi* Paramonov, 1931; *rubriventris* Paramonov, 1934 (1936).

Ethiopian: *Oestranthrax disparilis* Hesse, 1956; *obesus* (Loew), 1863 (as *Anthrax*), *obesus olfieri* Paramonov, 1931, *obesus pallifrons* Bezzi, 1926; *speiserianus* Bezzi, 1923; *zimini* Paramonov, 1934 (1936).

Oestrimyza, new genus

Figures 415, 703

Type of genus: *Oestrimyza fenestrata*, new species.

Large flies which are related to *Oestranthrax* Bezzi. The proboscis is reduced to a minute bilateral spur that protrudes downward from a small, sunken, oval recess.

The wing is large and entirely dark yellow brown, with 9 small, yellow, barlike and windowlike spots. The wing is so large that the fly suggests *Poecilanthrax* Osten Sacken in a superficial way and like it the wings are much longer than the abdomen. There are only 2 submarginal cells; the first posterior cell is scarcely narrowed; the anal cell is open although narrowly.

These flies are found in Brazil.

Length: 14 mm.; wing 15 mm.

Head, lateral aspect: Hemiglobular, the occiput prominent, although strongly sloping inward, and the occipital pile scanty, appressed, flat, and rather wiry. There is a small outward swelling of the occiput, at the deepest point of the long, comparatively deep indentation. Eye facets not bisected. The upper occiput is twice as long as the lower occiput. Central fringe rather long and loose and longer above. The front is a little swollen, distinctly visible above the eyeline and more prominent below. Its pile is quite bristly, short, erect, or curled forward. The lower part of the front protrudes almost the length of the two short first antennal segments. The face is strongly swollen forward, convex and rounded, and abruptly rounded below at the lower fifth of the head where there are deep, ventrally widening fissures, rather triangular in form, which constitute the tentorial fissure and pit. The pile of the face is dense and quite bristly, a little longer below, and with the more or less narrow, bare encircling band or line, curving up to the base of each antenna. Proboscis is reduced to a minute, bilateral spur protruding downward from a small, sunken, oval recess; at the posterior end of the spur is a pair of small, seedlike, adjacent, bulbous structures, which may be part of the proboscis or part of the palpus. The antennae are attached slightly above the middle of the head.

Head, anterior aspect: Eyes of the male are rather widely separated by 3 times the width of the low yet abrupt ocellar tubercle. Ocellar triangle short isosceles, and placed far forward from the posterior eye corners, which are unusually broadly rounded, and almost obliquely truncate. The antennae are widely separated by nearly the length of the long third segment and distant from the eye margin by the length of the first two segments. There is an impressed line down the middle of the face. Viewed from above the occipital lobes have a deep, vertical gap or fissure between them, and therefore they are not in contact.

The antennae though conspicuous are not exceptionally large. The first segment is short, a little widened medially, and a little shorter than its apical width. It bears a hemicirclet of short, stiff, bristly setae, but none below. Second segment quite short, small, beadlike, the edges rounded, attached narrowly to the third segment. The third segment has a rounded, short, bulblike base, very gradually tapering above and below into the long, quite thick, bristle-tipped style. Viewed from above the style, which is dorsoventrally flattened, is more than half as wide as the base. Viewed laterally it is less than half as wide, and the apex is bluntly truncate and contains the stubby, asymmetrically placed spine. Head strongly subquadrate from the anterior view.

Thorax: A little longer than wide, the scutellum excepted. A little widened across the prominent postalar callosities, which are not set off by more than the most shallow crease. Color black, surface greasy. The humerus and postalar callosities are brown, the scutellum likewise. Pile of mesonotum fine, short, partly erect, and partly curled downward. The anterior collar is quite short and narrow, but includes the humerus. There are 3 short bristles on the notopleuron, and a patch of short bristly pile between the notopleuron and the humerus and a similar patch on the upper mesopleuron, which is somewhat convexly swollen outward, leaving a flat, narrow, wedge-shaped depression between it and the slightly convex, pilose pteropleuron. Metapleuron with pile. Hypopleuron bare. There appears to be a minute hair tuft under the haltere. Squamal scales short.

Legs: Rather poorly developed and comparatively slender, the anterior pair only a little reduced. All of the legs are rather spiky with short, oblique, numerous, black, spiculate bristles. There are fine bristly setae on the anterior femur. Anterior tibia with 15 or more distinct dorsal bristles, nearly as many anteriorly and posteriorly. The short bristles in the middle femur with 8 or 10 bristles anteroventrally. Hind femur also with 8 or 10 bristles, mostly on the outer half, some of them finer, and all of them less stout than the short bristles of the tibia. Middle and hind tibiae with numerus, conspicuous, stout, short bristles. On the hind tibia there are 3 or 4 rows dorsally and extending somewhat downward laterally, and with 2 ventral rows. Only those of the dorsomedial row are shorter and finer. The more lateral of the dorsal rows contains 20 or more bristles. All tarsi are strongly setate below. Claws fine, slender, sharp, only slightly curved, with no basal tooth, and pulvilli rudimentary or wanting.

Wings: The wings are very large and broad, pointed only at the immediate apex. Like *Poecilanthrax* Osten Sacken they are much longer than the abdomen. The entire wing is dark yellowish brown, rather flat or scarcely wrinkled. The wing has 9, small, yellow, windowlike spots, a bar on each side at the lower end of the second basal cell, again on each side at the upper end of the second basal cell, on each side of the anterior intercalary vein which are extended proximally on the distal portion of the medial crossvein and again on each side of the basal, plane part of the medial crossvein; also spots on both sides of the anterior crossvein and the base of the second vein, on both sides of the base of the anterior branch of the third vein and again on the outer bend of this vein. There is a trace of this yellow color at the base of the second basal cell. There are only 2 submarginal cells, the first posterior cell is scarcely narrowed. The anterior crossvein enters the discal cell at the middle, and the second vein arises a little before the anterior crossvein. Anal cell open, but narrowly. The ambient vein is complete. The alula is about 3 times as long as wide. The basal comb is of moderate width, with numerous short, oblique setae, and the costal hook is stout basally, sharp at apex.

Abdomen: The abdomen is rather elongate, nearly twice as long as the mesonotum, the scutellum excepted. It is distinctly wider than the thorax and has parallel sides to the end of the sixth tergite. Third to sixth tergite of nearly equal length. The abdomen is greasy, blackish, with the lateral margin of the tergites extending over, with rounded edges, but not curled over. The tergites laterally are dark reddish brown. The pile of the abdomen is minute, flat-appressed and setae, and very little longer along the lateral margins of the tergites. First tergite with slightly longer pile, but confined to the outer edges only. Seven tergites are visible from above, the seventh is nearly or quite as long as the sixth, but rounded laterally. The male terminalia extend downward as an asymmetrically placed cone enclosed by the last sternite.

Material available for study: The unique male type found among flies studied from miscellaneous collections in the National Museum of Natural History.

Immature stages: Unknown.

Ecology and behavior of adults: Not on record.

Distribution: Neotropical: *Oestrimyza fenestrata*, new species.

Genus *Marleyimyia* Hesse

FIGURE 174

Marleyimyia Hesse, Ann. South African Mus., part 3, vol. 35, p. 521, 1956b. Type of genus: *Marleyimyia natalensis* Hesse, 1956, by original designation; one species.

Large, obese flies, related to *Oestranthrax* Bezzi, larger in size, with shorter, stouter legs and much longer hairs on the femora. Basal tarsal segment more thickened and claws without basal tooth. Separation of male eyes much less than in *Oestranthrax* Bezzi; the genae are very much narrower, according to Hesse (1956). The buccal cavity is broadly visible and not slitlike. The proboscis is more distinct, about equal in length to the third antennal segment. The head is much larger and broader, the indentation of hind eye margin scarcely noticeable. Third antennal segment larger at base and the stylate portion thicker. Hesse says that the buccal cavity is a little less reduced than in *Oestranthrax* Bezzi but much better developed than in *Villoestrus* Paramonov.

A unique character of the only known species is the presence of dense, golden yellow hairs on the posterior sternites, which Hesse (1956) notes is somewhat similar to, perhaps a mimic of, the scopa of a gastrilegous bee such as *Megachile*.

All of these several genera, including *Oestrimyza* Hull, are closely related and form a distinct subgroup of the Exoprosopinae.

Marleyimyia Hesse is known from a single male from Natal. Type in South African Museum.

Length: 17 mm.; of wing 15.5 mm.

As I have not been able to see or obtain material of *Marleyimyia* Hesse, I quote Hesse's description of the genus:

From both *Oestranthrax* Bezzi and *Villoestrus* Paramonov this new genus differs in the following respects:

Body larger, mesopleuron distinctly much more convexly bulging and prominent. Head relative to body much larger, more like that of *Villoestrus* Paramonov being quite broad as across disc of thorax at level of wings; eyes as in *Villoestrus* Paramonov more than twice as high as long at level of indentation; facets much coarser, those in upper part in male distinctly coarser than in lower part and coarser than in *Villoestrus* Paramonov; interocular space on vertex in male much narrower than in *Oestranthrax* Bezzi, but like that of *Villoestrus* Paramonov: face slightly less convex and also less so than in latter genus; buccal cavity slightly less reduced than in most species of *Oestranthrax* Bezzi, but distinctly much more developed than in *Villoestrus* Paramonov and in form of a broad and deepish depression bounded on each side by somewhat prominent rims or lips; genae very much narrower than in *Oestranthrax* Bezzi and also narrower than in *Villoestrus* Paramonov; rudimentary proboscis like that of former, its labellar part however pointed, not ending in two vestigal lobes; palps also very similar; antennae more like those of latter genus and joint 3 much more broadened at base than in *Oestranthrax* Bezzi, its slender part also relatively shorter than in latter and not so long and filiform. Vestiture with the hairs and scales on first antennal joints and face distinctly longer and very much and markedly denser than in both the other two genera; hair on thorax above discally and anterolaterally, on pleurae, coxae, in metapleural tuft, on scutellum, first tergite, basal half of tergite 2 and base of venter distinctly very much longer, denser, more shaggy in appearance than in former two genera; hairs on sides of abdomen also distinctly longer and denser; those on venter markedly dense, much denser and longer than in any other Exoprosopine and having a marked resemblance to the scopa of a gastrilegous bee; scales on face narrowish as in *Villoestrus* Paramonov, but very much longer; that in form of whitish transverse bands across bases of tergites in narrower bands than in *Oestranthrax* Bezzi, more like those of *Villoestrus* Paramonov, but with a more conspicuous band across base of tergite 3, but none across base of 2 as in former genus; rest of scaling on abdomen above dark or blackish, without any yellowish or ochreous ones.

Legs: Shorter and stouter than in *Oestranthrax* Bezzi, more like those of *Villoestrus* Paramonov, but with much longer hairs on femora; femora, tibiae, and tarsi more or less equal in length as in latter genus; femora without any spines or spinelets below; front tibiae with more and denser spicules than in *Oestranthrax* Bezzi; basil joint of tarsi more thickened than in latter genus; claws without a basal tooth, slightly stouter and more bent down apically than in former two genera. Hypopygium of male very similar to that of *Oestranthrax* Bezzi and those of the *leuochila*-section of *Villa*, but aedeagus much shorter, the scoop-like aedeagal process more rounded knob-like apically and the basal strut more slender.

Wings: More like those of *Villoestrus* Paramonov, with similar venational characters, but not infuscated along costal part or with a dark cross band; second vein slightly less recurved apically than in *Oestranthrax* Bezzi; second submarginal cell longer and narrower basally; second and third posterior cells much shorter, the latter extending basally to much nearer apex of second basal cell; fourth posterior cell tending to be very broad apically, broader than in the other two genera; anal cell closed apically and with a very short stalk in this unique male at least; basal hook of wings longer, more developed than in either genus, more prong-like; squamae with the apical demarcated part considerably shorter than in *Villoestrus* Paramonov.

Since I have not been able to see males of this genus I quote the remarks of Dr. A. J. Hesse with respect to genitalia:

Hypopygium of male very similar to that of *Oestranthrax* and those of the *leucochila*-section of *Villa*, but aedeagus much shorter, the scoop-like aedeagal process more rounded knob-like apically and the basal strut more slender.

Material available for study: None.

Immature stages: Unknown largely. However, the collector of the single male type stated that the fly emerged from an old log containing cossid larvae (Hesse, 1956).

Ecology and behavior of adults: Not recorded.

Distribution: Ethiopian: *Marleyimyia natalensis* Hesse, 1956.

Genus *Exoprosopa* Macquart

Figures 197, 198, 208, 210, 376, 680, 747, 971, 972

Exoprosopa Macquart, Dipteres exotiques, vol. 2, pt. 1, p. 35, 1840. Type of genus: *Bibio capucina* Fabricius, 1781, as *Anthrax pandora* Macquart, 1826. Designated by Coquillett, p. 544, 1910, the ninth of 41 species.

Mima Megerle in literature, in Meigen, 1820, p. 175.

Trinaria Mulsant, Mem. Acad. Sci. Lyon, ser. 2, vol. 2, p. 20, 1852. Type of subgenus: *Anthrax rutila* Pallas in Wiedemann, 1818, as *Anthrax interrupta* Mulsant, 1852. Designated by Coquillett, p. 617, 1910, the first of 2 species.

Argyrospyla Rondani, Dipterologiae Italicae Prodromus, vol. 1, p. 162, 1856. Type of subgenus: *Anthrax jacchus* Fabricius, 1805 (*jaccus*), by original designation; one species.

Defilippia Lioy, Atti. Instit. Veneto, ser. 3, vol. 9, p. 733, 1864. Type of subgenus: *Anthrax minos* Meigen, 1804. Designated by Griffini, p. 42, 1896.

Argyrospila Verrall in Scudder, Nomenclator Zoologicus, U.S. Nat. Mus. Bull. 19, part 1, p. 32, 1882; emendation.

Exoptata Coquillett, Canadian Ent., vol. 19, p. 13, 1887a; as a genus, now regarded as a subgenus. Type of subgenus: *Exoprosopa divisa* Coquillett, 1887, by monotypy.

Mesoclis Bezzi, The Bombyliidae of the Ethiopian Region, p. 229, 1924. Type of subgenus: *Anthrax pygmalion* Fabricius, 1805, by original designation.

Metapenta Bezzi, The Bombyliidae of the Ethiopian Region, p. 230, 1924. Type of subgenus: *Exoprosopa pentala* Macquart, 1840, by original designation.

Acrodisca Bezzi, The Bombyliidae of the Ethiopian Region, p. 234, 1924. Type of subgenus: *Exoprosopa angulata* Loew, 1860, by original designation.

Cladodisca Bezzi, The Bombyliidae of the Ethiopian Region, p. 243, 1924. Type of subgenus: *Exoprosopa suffusa* Klug, 1832, by original designation.

Pterobates Bezzi, The Bombyliidae of the Ethiopian Region, p. 273, 1924. Type of subgenus: *Anthrax apicalis* Wiedemann, 1821, by original designation.

Zygodipla Bezzi, Bull. Soc. Roy. Ent. d'Egypte, vol. 9, p. 268, 1926c. Type of subgenus: *Anthrax algira* Fabricius, 1794, by original designation.

Corycetta, new subgenus. Type of subgenus: *Exoprosopa subfasciata* Engel, 1936.

Osten Sacken (1886), p. 87, key to 3 species.
Coquillett (1892b), p. 123, key to 9 species from western United States.
Williston (1901), pp. 269-270, key to 11 species from Mexico.
White in Hardy (1924), p. 74, key to 4 Australian species.

Bezzi (1924), pp. 231-236, key to 18 species; p. 246, key to 3 species; pp. 250-252, key to 21 species; p. 274, key to 3 species; pp. 277-286, key to 105 species.
Séguy (1926), pp. 184-185, key to 15 species from France.
Paramonov (1928), pp. 181-303, 6 pl., key to Palaearctic species.
Curran (1930), pp. 2-4, key to 37 species from Mexico and the United States.
Maughan (1935), pp. 34-38, key to 12 species from Utah.
Engel (1936), pp. 454-464, key to 67 Palaearctic species.
Austen (1937), pp. 162-164, key to 14 species from Palestine.
Hesse (1956b), pp. 658-726, key to 171 males and 187 females of species and subspecies.
Johnson and Johnson (1958), pp. 84-85, key to 28 species from western United States.
Bowden (1964), pp. 118-119, key to 15 species from Ghana.
Painter and Painter (1969), pp. 5-33, key to species of north neotropical sector.

This is a genus of large flies; occasionally depauperate individuals as little as 6 mm. in body length and 14 mm. in wing expanse, but most species are much larger and some of these species comprise the largest known bee flies. The head is large, subglobular, and loosely attached to the thorax; the occiput is extensive but slopes strongly backward to the margin of the large, deep, central recess. The face is either bluntly protuberant, or very frequently acutely conical. The antennae are relatively small, very widely separated. The third segment is characteristically attenuate and has an articulate, separated, apical, bristle-tipped style; this character with rare exceptions separates *Exoprosopa* Macquart and allied genera from the *Villa* Lioy group of related genera. The antennal style is without an apical whorl of hairs such as found in the Anthracinae. The proboscis in *Exoprosopa* Macquart is slender, varying from no longer than the oral recess to others where it is almost twice as long.

The flies of this genus can usually be selected out of any assortment of bee flies by the combination of the globular head, conical or subconical face, small, slender, relatively short, widely separated antennae, broadly oval, moderately convex abdomen, together with the large wings, which within the genus show a variety of patterns. The three principal types are species in which: (1) the whole wing is diffusely brown, paler posteriorly; (2) the anterior foreborder is rather sharply brown, the vein closing the second basal cell infuscated; and (3) the wing shows 2 major, generally sharply defined, irregular crossbands, the first one occupying the basal half of the wing and itself sometimes broken into 2 bands, and the second major band proceeding chiefly backward from the apex of the costal cell. This basic pattern of markings assumes many variations, often filling out the whole of the auxiliary cell, often having clear sinuses about the crossveins; the color of this maculation varies from very dark sepia to brownish yellow or combinations of these colors. Presumably all species with completely hyaline wings such as *chionea* Bezzi, *ancilla* Bezzi, *chrystallina* Bezzi, and others should be assigned to *Micomitra* Bowden.

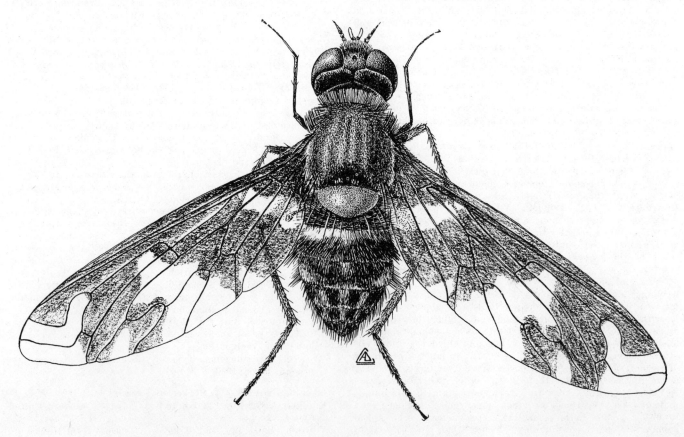

Text-Figure 46.—Habitus, *Exoprosopa rhea* Osten Sacken.

In *Exoprosopa* Macquart the squama and generally the alula as well have lanceolate scales on the margin instead of hairs and the metapleuron is hairy; these two characters, together with the absence of hairs on the antennal style separate *Exoprosopa* Macquart from the subfamily Anthracinae.

Face, front, mesonotum, and abdomen with numerous scales, which are narrow or wide, large or small. The front legs are generally reduced in size and sometimes greatly reduced; their tibiae and attenuate tarsi are often smooth and nonspiculate or short, thick, and spiculate.

Also tibiae are sometimes pennate. It should be noted that the claw of the pretarsus has a basal tooth, either blunt or long and sharp, and the main claw itself is scarcely curved; whole pretarsus of front tarsus greatly reduced.

As for the classification of this very large, worldwide, important group of flies I have adopted the basic subdivision of the genus into the subgenera recognized by Bezzi (1924), together with the subgenera added by later writers such as Bowden (1964). The subdivision of this genus was begun by Rondani (1856, 1864), Mulsant (1852), Lioy (1964) and Coquillett (1887) and continued by Bezzi (1924). While I quite agree with Bezzi in the usefulness of these subdivisions, nevertheless, Paramonov (1928) has demonstrated that because of overlapping many of these subgroups are only species groups. These subgroupings necessarily have to be based on the numerous variations in wing venation, rather than mere color pattern, as was attempted by Becker. Useful as an arbitrary division in the color patterns may be in the formation of keys to species, the alterations in the venation itself form a slightly more substantial basis for an understanding of phylogenetic origins.

South Africa is especially rich in this remarkable genus. Hesse (1956) treats 135 species and in his key has 209 couplets; he finds that 8 of the so-called subgenera of Bezzi are present, but in attempting to bring together near-related species he has been obliged to take note of 33 additional species groups: three of these groups contained 9, one contained 8, five contained 7, two contained 6, four contained 5, three contained 3, seven contained 2, and eight groups contained only 1 species. As Hesse notes, these groups are artificial and are set up for purposes of identification only within a very large genus, where it is extraordinarily difficult to expedite and clarify identification; because of a scarcity of males he was able to present figures of terminalia for only 53 species; of wings he showed 17 species; it is possible that a complete atlas of terminalia and wings and abdominal patterns, etc., might make it possible to completely delimit groups save for occasional transitional forms.

Hesse (1956) also notes that *Exoprosopa* Macquart can only be confused in South Africa with the species of *Litorrhynchus* Macquart, which it closely resembles and is differentiated by the constantly more produced face, relatively much shorter proboscis, not or rarely longer than the head, not or only slightly projecting beyond the oragenal cavity, front tibia nonspiculate and slender in most species, a more slender spinelike basal tooth on claw, rarely with a *Litorrhynchus*-like clear spot in middle of wings and a different arrangement of pale white scales on abdomen. From *Thyridanthrax* Osten Sacken it is at once distinguished by the 3 submarginal cells, different wing pattern, basal tooth on claw, long stylar element on the third antennal segment. From the perhaps superficially similar American *Poecilanthrax* Osten Sacken, they are distinguished by the 3 submarginal cels, biarticulate antennal style, presence of tooth on claw, etc. The American *Stonyx* Osten Sacken, has among other things a very different antenna, and lacks tooth on claws.

There is a certain amount of sexual dimorphism in certain species of *Exoprosopa* Macquart. Hesse (1956) shows wing variation in the sexes of two species on his Plate III, the brown color is less extensive in the marginal, submarginal, and first posterior cells of the male. Also in some American species the white scales of the abdomen become bright and silvery on the last tergite of the male. In addition to the presently utilized subdivision of the genus, a great many additional species groups can be formed by an intensive study of the species, particularly from a regional basis.

Exoprosopa Macquart species are found in all of the world regions, and the genus seems to have reached its highest development in the Mediterranean and South African regions. The great variations in forms and patterns suggest that we are dealing with a rather strongly plastic, expanding group of flies, forming a dominant element in the bee fly fauna in temperate and arid regions.

Length: Body 6 to 22 mm.; wing 6 to 27 mm.; wing expanse 14 to 64 mm.

For a better understanding of past groupings into subgenera I quote Bezzi's arrangement (1924):

1 (2). Marginal cell divided into two by a supernumerary cross-vein; wings with eye-like spots . . . .
. . . . i. *Heteralonia*, Rondani.
2 (1). Marginal cell simple; wings devoid of eye-like spots.
3 (4). First posterior cell divided into two by a supernumerary cross-vein (thus five posterior cells present); discoidal cell devoid of a projecting lower apical angle, but with a sinuous, horizontal terminal vein; front tibiae thick and beset with spicules . . . .
. . . . vi. *Exoptata* Coquillett.
(now *Zygodipla* Bezzi—see Bezzi, 1925.)
4 (3). First posterior cell simple.
5 (12). Discoidal cell on lower margin, before tip, with a very prominent angle, emitting an appendix; terminal vein of discoidal cell more or less sinuous, and running at an angle or perpendicularly to longitudinal axis of wing.
6 (9). Five posterior cells present.
7 (8). Among posterior cells, third alone closed and provided with a long stalk . . . ii. *Mesoclis*, subgen. nov.
8 (7). Among posterior cells, fourth alone closed and provided with a short stalk . iii. *Metapenta*, subgen. nov.
9 (6). Only four posterior cells present as a rule.

10 (11). Front tibiae short and thick, beset with well-developed spicules; first posterior cell always open . . . . . .
. . . . iv. *Acrodisca*, subgen. nov.
11 (10). Front tibiae long and thin, smooth; first posterior cell often closed and stalked . v. *Cladodisca*, subgen. nov.
12 (5). Discoidal cell without apical projecting angle below, and if sometimes dilated, dilation rounded, not angular, and at any rate devoid of appendix.
13 (16). Vein between discoidal and second posterior cells long, deeply and regularly S-shaped, and in same line with longitudinal axis of wing, or nearly so; wings never dimidiate; discoidal cell narrow and long, not expanded above or below before distal extremity.
14 (15). First posterior cell closed, terminating bluntly, and provided with a long stalk; hind tibiae with very short spicules; wings with the basal half yellow and margined with brown . . . . vii. *Trinaria* Mulsant.
15 (14). First posterior cell usually open, or, if closed, terminating in a narrow apex and with a short stalk; hind tibiae with long and stout spicules; basal half of wings not yellow nor margined with brown . . . . .
. . . . viii. *Defilippia* Lioy.
16 (13). Vein between discoidal and second posterior cells short, straight or only faintly sinuous (if deeply sinuous, then short, or not regularly S-shaped, or forming an angle with longitudinal axis of wing); wings sometimes expanded above or below before distal extremity.
17 (18). Legs feathered; discoidal cell contracted in middle; upper branch of cubital fork not widely divergent from main stem; marginal cell very narrow at end . .
. . . . ix. *Pterobates*, subgen. nov.
18 (17). Legs not feathered.
19 (20). Marginal cross-vein straight and not recurrent; second longitudinal vein orginating beyond middle cross-vein; first posterior cell typically closed and provided with a short stalk; wings with quite hyaline "windowpanes"; abdomen with silvery bands, at least in female . . . . x. *Argyrospila* Rondani.
20 (19). Marginal cross-vein sinuous and recurrent; second vein originating opposite or before middle cross-vein; first posterior cell typically open, but sometimes closed; wings never fenestrate, or if so "windowpanes" not hyaline; abdomen devoid of silvery bands . . . . xi. *Exoprosopa* Macquart.

Head, lateral aspect: The head is more or less hemiglobular, its apparent length is increased by the forward extension of the face which ranges from blunt species, like *rostrifera* Jaennicke (compare *Litorrhynchus* Macquart), to those in which the face is conically and acutely produced, like *jacchus* Fabricius. There are large numbers of both the blunt-nosed and conical-nosed species. The occiput dorsally is rather extensive but slopes strongly downward toward the deep, central, cuplike recess. The degree of indentation on the eye varies strongly from species, like *ingens* Cresson, where the excision is so gradual and so slight that the occiput is very little longer opposite the middle of the eye than it is dorsally, to others, like *jacchus* Fabricius, where the indentation is more distinct and angulate with upper and lower angles approximately equal. The pile of the occiput varies from a uniform covering of flat-appressed, scaliform pile directed outward toward the eye margin, generally touching and overlapping the eye margin, to others in which there is a distinct, pollinose, pile-free rim about the margin of the eye. This occipital pile gives way medially and centrally to the always short and often scanty interior fringe about the occipital recess. Eyes only slightly reniform. There is a bare triangle at the bottom of the indentation and the line dividing the upper and lower eye facets varies from short to long. Upper eye facets slightly enlarged in the male; some of the lower eye facets tend to be short pilose.

The oragenal opening is long, narrowed in the middle, a little widened anteriorly but much more so posteriorly, with thin, vertical margin posteriorly and the inner wall strongly inflated at the beginning of the tentorial fissure below. The tentorial fissure is linear and deep, leaving only a narrow, sloping margin adjacent to the eye. Cheeks below the eye comparatively prominent. The oragenal recess is quite deep but near the anterior apex sloping up to the facial margin. The head viewed from above shows that the occipital lobes are quite tightly apposed but with a fissure anteriorly and the vertex always shallowly sloped.

The face in *rostrifera* Jaennicke is quite bluntly rounded, extending out about twice the length of the first antennal segment; but in *jacchus* Fabricus the face is more than 3 times as long as the first antennal segment and this segment is about 2 to 3 times as long as its basal width and bears comparatively short, stiff, numerous setae on all sides, except ventrally. The antennal sclerite is distinct but narrow. Second antennal segment disclike, short, from a third to a fourth as long as the first segment and with setae on all sides, except medioventrally. The third segment basally is less wide than the second, somewhat bulbose at the base but with the dorsal surface plane, hence most of the constriction is ventral, to other species that are quite long and conical ending in a more or less attenuate stylar portion, and in all species with an attached segmented style, which varies from one-fourth to one-half as long as the length of the third segment.

Head, anterior aspect: The head is as wide as the thorax, is subquadrate, being rather distinctly flattened on all four sides. The eyes in the male are widely separated by almost as much space as there is between the antennae, seldom as little as two-thirds of the space between the antennae. Females with the eyes not much more separated than in the males. The antennae are very widely separated, the middle space more than twice the distance from the outer edge of the antennae to the eye. The ocellar tubercle is small, quite low, inconspicuous, short isosceles or equilateral, in some species extended backward beyond the ocelli, in others ending abruptly. It is placed well forward beyond the posterior eye corners, which are strongly rounded and the tubercle bears a few short setae in most species. Front sometimes with a sunken depression in the middle on the upper third. Vertex with or without scales.

Thorax: The thorax is comparatively low on the ventral half beneath the wing. It is also rather long. The mesonotum is very little convex, rather flattened posteriorly, pollinose, generally blackish, but some species have postalar callosity and scutellum pale; the pile varies from dense, fine, flat-appressed hairs, scarcely widened, to others in which the hair is distinctly setate, in still others where they have broad, short scales. In the scalose species there are generally some fine, erect hairs of no great length.

Humerus and the margin as far back as the notopleuron with longer, coarse pile. The anterior collar is very distinct, composed of blunt, dense, long spikelike pile, often of contrasting color. Notopleuron with 3 bristles, often quite stout; postalar callosity with 3 or 4 stout and several more slender bristles; scutellum with 6 to 8 pairs of slender or stout, long bristles. Scutellar disc covered like the mesonotum. The pleuron is thinly pollinose, the mesopleuron densely covered with long, coarse pile, the pile on the upper half sometimes distinctly bristly, and turned upward. Sternopleuron with appressed scales posteriorly and sometimes over nearly the whole surface; they may be very dense. Pteropleuron with bristles and pile and sometimes with scales extensively over the anterior and dorsal part and sometimes with a posteroventral, isolated tuft of setae. The metanotum behind and beneath the scutellum is extremely short, due to the close attachment of the abdomen, laterally there is a tuft of short pile and the swollen metapleuron posteriorly bears a very conspicuous brush of long, coarse pile, sometimes separated into a vertical band posteriorly and an oblique band dorsally. Plumula long, extensive and conspicuous, composed of coarse hairs generally contrasting with the remainder of the pile when the remaining pile is dark. Hypopleuron entirely without pile, except sometimes a few hairs in front of the spiracle.

A conspicuous tuft of flattened scales or hairs is diagonally placed below and behind the spiracle in species like *iota* Osten Sacken, but is completely lacking in other species like *ingens* Cresson. Squama divided by a longitudinal crease with a prominent rim and a fringe of scales that are narrow or sometimes lanceolate and extended onto the alula and base of the axillary lobe. Halteres relatively quite small.

Legs: The femora are stout; the anterior femur slightly thickened basally. Middle and hind tibiae quite stout. The femora are covered with appressed scales varying from jet black to yellowish with besides scattered, suberect, fine setae as a rule, which may become a short, posterior fringe. Middle femur with some stout, spiculate bristles of varying length and thickness and numbers. Hind femur in a species like *ingens* Cresson with as many as 9 or 10 short, quite stout, ventrolateral bristles, in others with 5 or 6, and sometimes with a few quite slender bristles.

The scales on the hind femur and tibia may be so dense and wide as to be quite shinglelike. The anterior tibia bears a dorsal and ventral row of fine, suberect spicules in species like *rostrifera* Jaennicke, in others like *jacchus* Fabricius the anterior tibia is quite slender and bears only very dense, finely appressed microsetae, and, moveover, it is strongly attenuate apically in contrast to a group of species like *ingens* Cresson or *rostrifera* Jaennicke where the anterior tibia is actually widened apically. In general all of the tarsi are small, weak, slender, gradually attenuate and short. The anterior tarsi bear a fringe of fine, erect hairs instead of stiff bristles. The anterior claws are minute, the remaining claws large, long, scarcely curved at the apex, sharp apically with a basal tooth and the pulvilli absent.

Wings: Wings large, usually broad at the base, but with exceptions as *ceuthodonta* Hesse, etc., always much longer than the abdomen when at rest. The costal cell ends nearly opposite the base of the first marginal cell. There are 3 submarginal cells, those species having 4 submarginal cells are placed in the very closely related genus *Ligyra* Newman. The pattern of the submarginal group of cells is variable. The kink or bend where these veins are upturned toward the foremargin is either gentle or deep; the veins closing the second submarginal cell and the marginal cell are particularly erratic. From a check of the wings of 90 species, I find that 47 show the second vein arising before the anterior crossvein, and 37 show it precisely arising at the anterior crossvein, while 6 species show it arising beyond this crossvein; in none of the species where it lies before or after the crossvein is it separated from it by more than the length of the crossvein itself, usually less. This second vein usually arises rectangularly, but it is oblique in a few species as *goliath* Bezzi and *argentifrons* Macquart and very gradual in origin in *lacerata* Engel.

In some species the marginal cell is divided into two parts by an extra crossvein and the wing has eyelike spots. This is the basis for the genus *Heteralonia* Rondani. Again the first posterior cell may be divided into two by a supernumerary crossvein, thus leaving 5 posterior cells as in *Exoprosopa divisa* Coquillett. This is the basis for the subgenus *Exoptata* Coquillett, further distinguished by the perfectly smooth, slender, anterior tibia.

Another group of species shows that the discal cell on its lower margin before the apex has a strong, distinct, angular bend, frequently emitting an appendix vein, and the terminal vein of the discal cell sinuous and placed at an angle or perpendicularly or longitudinally to the axis of the wing. Within this group the species having the fourth posterior cell closed and stalked represent the subgenus *Metapenta* Bezzi. Those with the third posterior cell closed and stalked constitute the subgenus *Mesoclis* Bezzi. Each of these have 5 posterior cells. Those with only 4 posterior cells have the first posterior cell always open and the front tibia short and thick and spiculate; these belong to the subgenus *Acrodisca* Bezzi. Those species that have the front tibia long, thin, and smooth and the first posterior cell often closed and stalked belong to the subgenus *Cladodisca* Bezzi. There are several other distinct venational groups; for example, those species that have the vein between the discal and second posterior cells long and deeply and regularly S-shaped, the discoidal cell itself narrow and long constitute another group of subgenera.

I am obliged to agree with Bezzi (1924), who states: "These characters have at least the same value and constancy as the crossvein uniting the 2 branches of the cubital fork in *Hyperalonia* Rondani (*Ligyra* Newman), and may therefore be employed for the purpose of generic or subgeneric division."

In *Exoprosopa* Macquart the anal cell is open, the anal vein is sometimes straight and sometimes strongly sinuous, the alula is moderately extensive and characteristically bears scales similar to those found upon the squama. A few scales are continued basally along the edge of the axillary lobe. The distal end of the alula is always rounded, the ambient vein is complete and in both sexes the costa at base is strongly expanded and flattened, but the comb of spicules which this area bears are minute or coarse. There is a well developed basal hook. The pattern types and coloration of the wing have been dealt with earlier, but it should be noted that the posterior third of the wing is very strongly rippled or fluted transversely.

Abdomen: The abdomen is broadly oval, a little wider than the thorax itself, rather gently convex and in a few species more or less flattened, usually black; it may be sometimes light brownish orange with black medial spots. The pile characteristically consists of dense, slender, linear scales that lie flat-appressed, frequently forming pale bands along the anterior part of one or more and sometimes all of the tergites beyond the first, leaving the posterior margin with black scales. In *jacchus* Fabricius patches on each side of 6 tergites bear dense, flat-appressed, silvery scales making this species look very much like a *Lepidanthrax* Osten Sacken. Sides of the first tergite with the usual bombylid-like brush of dense, erect, coarse pile. Sides of the remaining tergites either with very short, appressed fringes of pile or sometimes with long tufts of erect and suberect, bristly pile, as in *calyptera* Say. There are 7 tergites visible above in the male and female, a very short eighth recessed beneath the seventh. Female terminalia have about 7 pairs of strong spines upon the acanthophorites. The male terminalia are large, elongate, recessed beneath the last tergite but generally visible from below.

Dr. A. J. Hesse was able to study a very large assemblage of species in this genus; for this reason I quote his comments on the male genitalia:

Hypopygium of males very uniform and almost similar in a large number of species, often however strikingly different, suggesting generic differences; the shell-like basal parts invariably hairy on dorsal aspect, sometimes with a few characteristic curved or hook-like, dark, flattened, spine-like setae medially and dorsally between the partly fused basal parts, the apical angle of basal parts sometimes prominently projecting and bases sometimes also produced; beaked apical points always more or less curved, their apical parts directed upwards, their apices usually sharp and in many forms with a slight indentation, giving them a bifid appearance, their outer lateral angles usually prominent, angular, or even bluntly spine-like, and with some hairs, the dorsal surface usually hollowed out or depressed to a variable extent; aedeagus rarely long; a ventral aedeagal process invariably present and well developed, assuming various shapes which are of specific value in the separation of the species; lateral struts fairly uniform, but sometimes broad or much reduced; basal strut assuming various shapes, much reduced and small in a few species.

The male terminalia from dorsal aspect, epandrium removed, show a figure broad basally, much narrowed on the outer third or less. The epiphallus is rather prominent, narrowing apically, the sides not quite parallel, and the sides nearly plane. In a genus as large as this, the figures must vary greatly. From the lateral aspect the dististyle is quite subapical and incised just before the upturned, hooklike apex. The end of the aedeagus shows a curious cap located vertically over the rather narrow preceding neck; it is notched below. Epandrium large, plane above with a blunt, short, downturned lobe basally.

Hesse finds the hypopygium and terminalia within the male very uniform and almost similar in a large number of species, however, sometimes "strikingly" different, thus suggesting generic differences; the shell-like basal portion he finds invariably with hair dorsally, sometimes with a few curved or hooklike, flattened, spiny setae medially and dorsally between the partly fused basal parts, and with the apical angles of the basal parts sometimes prominently projecting and the base also sometimes produced; he found the dististyli (beaked apical segments) always more or less curved, with their apices directed upward, the apices sharp, often with a slight indentation, which produces a bifid appearance, their outer lateral angles usually prominent, angular or even spinelike. The aedeagus was rarely long, a ventral aedeagal process always present and well developed and assuming various shapes of specific value; the lateral struts were uniform but sometimes broad, or much reduced; basal struts much reduced and small in a few species.

Material studied: Many Nearctic and Neotropical species and some Palaearctic, Ethiopian, and Australian material.

Immature stages: Such stages are known for several species of this genus and such data as we have indicate hyperparasitism on bees, wasps, and beetle larvae that live in the soil. See discussion under life histories.

Ecology and behavior of adults: The adults can be captured either resting upon the ground in sandy places or upon many species of flowers, especially composites, and while arid regions are greatly favored by this genus, about 16 percent of the bee fly fauna of Lafayette County, Mississippi, a temperate region of considerable rainfall, are comprised within the genus *Exoprosopa* Macquart. Certainly many known species have been found in the Mediterranean and in Asia Minor, and the South African fauna is extraordinarily rich in this genus. The wing patterns of over a hundred species have been illustrated by various authors. The flies of this genus have an extraordinarily powerful flight, and the marked strengthening of the costa at its base doubtless has something to do with this and perhaps the fluted wrinkling of the wing as well. Even the most wary species, however, may be captured by quietly stalking them.

Distribution: Nearctic: *Exoprosopa agassizii* Loew, 1869; *albifrons* Curran, 1930; *anomala* Painter, 1934; *arenicola* Johnson and Johnson, 1959; *bifurca* Loew, 1869; *butleri* Johnson and Johnson, 1959; *californiae* (Walker), 1852 (as *Anthrax*); *caliptera* (Say), 1823 (as *Anthrax*); *celeris* Cole, 1916; *clarki* Curran, 1930; *decora* Loew, 1869; *divisa* (Coquillett), 1887 (as *Exoptata*); *dodrans* Osten Sacken, 1877; *dodrina* Curran, 1930; *dorcadion* Osten Sacken, 1877; *doris* Osten Sacken, 1877 [=*pallens* Bigot, 1892]; *eremita* Osten Sacken, 1877 [=*grata* Coquillett, 1892]; *fasciata* Macquart, 1840 [=*americana* (Wulp), 1867 (as *Mulio*), =*longirostris* Macquart, 1850, =*rubiginosa* Macquart, 1840]; *fascipennis* Say, 1824 [=*coniceps* Macquart, 1850, =*melanura* Bigot, 1892, =*philadelphica* Macquart, 1840], *fascipennis albicollaris* Painter, 1962, *fascipennis noctula* (Wiedemann), 1830 (as *Anthrax*); *fumosa* Cresson, 1919; *hulli* Painter, 1930; *hyalipennis* Cole, 1923; *ingens* Cresson, 1919; *iota* Osten Sacken, 1886; *jonesi* Cresson, 1919; *junta* Curran, 1930; *meigenii* (Wiedemann), 1828 (as *Anthrax*) [=*emarginata* Macquart, 1840]; *painterorum* Johnson and Johnson, 1960 [=*cingulata* Johnson and Johnson, 1959]; *pardus* Osten Sacken, 1886; *pueblensis* Jaennicke, 1867; *rhea* Osten Sacken, 1886; *rostrifera* Jaennicke, 1867; *sharonae* Johnson and Johnson, 1959; *sima* Osten Sacken, 1877; *socia* Osten Sacken, 1886; *sordida* Loew, 1869; *texana* Curran, 1930; *tiburonensis* Cole, 1923; *titubans* Osten Sacken, 1877; *utahensis* Johnson and Johnson, 1959; *xanthina* Painter, 1934.

Neotropical: *Exoprosopa anthracoides* Jaennicke, 1867 [=*trabalis* Loew, 1869]; *argentifasciata* Macquart, 1864; *atripes* Cole, 1923; *bipartita* Bigot, 1892; *bizona* Walker, 1850; *blanchardiana* Jaennicke, 1867; *brevirostris* Williston, 1901; *brevistylata* Williston, 1901; *castilla* Painter, 1930; *cubana* Loew, 1869; *extensa* Wulp, 1888; *filia* Osten Sacken, 1886; *flavinervis* Macquart, 1846; *fumosa* Cresson, 1919; *harpyia* (Wiedemann), 1828 (as *Anthrax*); *interrupta* Wiedemann, 1828; *limbipennis* Macquart, 1846; *maldonadensis* Macquart, 1849; *melanura* Bigot, 1892; *mus* Curran, 1930; *nubifera* Loew, 1869; *orcus* Walker, 1849; *panamensis* Curran, 1930; *parva* Loew, 1869; *pavida* Williston, 1901; *phlegethon* Walker, 1849; *procne* Osten Sacken, 1886; *prometheus* Macquart, 1855; *punctata* Macquart, 1849; *rufiventris* (Blanchard in Gay), 1852 (as *Paranthrax*); *sackeni* Williston, 1901; *sanctipauli* Macquart, 1840; *sola* Painter, 1939; *spadix* Painter, 1933; *stymphalis* Wiedeman, 1828; *subfascia* Walker, 1849; *uruguayi* Macquart, 1840; *varicolor* Macquart, 1846.

Palaearctic: *Exoprosopa aberrans* (Paramonov), 1928 (as *Zygodipla*); *adelphia* (Becker), 1906 (as *Zygodipla*); *aeacus* (Meigen), 1804 (as *Trinaria*) [=*bombyciformis* Dufour, 1833, =*livida* Pallas in Wiedemann, 1818, =*lutea* Macquart, 1840]; *aegina* (Wiedemann), 1828 (as *Trinaria*) [=*bovei* Macquart, 1840, =*bovis* Becker, 1906]; *aegyptiaca* (Paramonov), 1928 (as *Defilippia*); *albicincta* Macquart, 1840; *algira* (Fabricius), 1794 (as *Anthrax*, later as *Zygodipla*) [=*archimedea* Bigot, 1860, =*pygmalion* Macquart, 1834, not Fabricius, =*sicula* Macquart, 1834, =*singularis* Macquart, 1840]; *ammophila* (Paramonov), 1931 (as *Trinaria*); *antica* Walker, 1871; *anus* (Wiedemann), 1828 (as *Anthrax*); *apicalis* Klug in Loew, 1860; *araxana* Paramonov, 1828; *arenacea* (Becker), 1906 (as *Zygodipla*); *bagdadensis* (Macquart), 1840 (as *Zygodipla*) [=*clausa* Becker, 1913]; *beckeri* Austen, 1913; *bezzi* Paramonov, 1928; *bucharensis* Paramonov, 1929; *busiris* Jaennicke, 1867; *capucina* (Fabricius), 1781 (as *Argyrospyla*) [=*caloptera* Pallas in Wiedemann, 1818, =*dorcadion* Osten Sacken, 1877, =*nigrata* (Wiedemann), 1828 (as *Anthrax*), =*nigrita* (Fabricius), 1775 (as *Bibio*), =*pandora* Macquart, 1826 (not Fabricius)]; *chalybaea* (Roeder), 1887 (as *Pterobates*), *chalybaea chrysogaster* Bezzi, 1926; *chan* Paramonov, 1928; *circeoides* (Paramonov), 1928 (as *Defilippia*), *circeoides nigrofasciata* Paramonov, 1928; *cleomene* (Egger), 1859 (as *Argyrospyla*); *conspicienda* Austen, 1937; *completa* (Loew), 1873 (as *Zygodipla*); *decrepita* (Wiedemann), 1828 (as *Defilippia*) [=*albifacies* Paramonov, 1928, =*flava* Paramonov, 1928]; *dedecor* (Loew), 1870 (as *Defilippia*), *dedecor minoides* Paramonov, 1928; *dedecoroides* Paramonov, 1928; *deserta* Paramonov, 1928; *dispar* (Loew), 1869 (as *Cladodisca*) [=*rivularis* Griffin, 1892], *dispar interrupta* Paramonov, 1926; *disrupta* Walker, 1871; *dives* Walker, 1849; *dulcis* Austen, 1936; *efflatoun beyi* Paramonov, 1928; *efflatouni* (Bezzi), 1925 (as *Defilippia*); *exoprosopoides* Paramonov, 1928; *fallaciosa* (Loew), 1873 (as *Zygodipla*) [=*evanescens* Becker, 1913]; *ferruginea* Klug, 1832; *fusconotata* Becker, 1913; *grandis* Wiedemann in Meigen, 1820 [=*turcomana* Portschinsky, 1877, =*fasciata* Dufour, 1850]; *griseipennis* Macquart, 1849; *guttipennis* Bezzi, 1924 [=*punctipennis* Ricardo in Forbes, 1903]; *herzi* Portschinsky, 1892; *hyalipennis* Paramonov, 1928; *insularis* (Ricardo in Forbes), 1903 (as *Cladodisca*); *italica* (Weidemann in Meigen), 1820 (as *Argyrospyla*), *italica megaera* Wiedemann in Meigen, 1820; *jacchus* (Fabricius), 1805 (as *Argyrospyla*) [=*italica* Rossi, 1792, =*jocchus* Fabricius, 1805, lapsus, =*pandora* Meigen, 1820, not Fabricius, =*picta* Wiedemann in Meigen, 1820], *jacchus baccha* Loew, 1869, *jacchus maenas* Loew, 1869, *jacchus quadripunctata* Paramonov, 1928; *kirgizorum* Paramonov, 1928 [=*uzbekorum* Paramonov, 1928]; *lacerata* Engel, 1936; *latiuscula* (Loew), 1873 (as *Zygodipla*) [=*farinosa* Becker, 1913]; *lucidifrons* Becker, 1913; *lugubris* Macquart, 1840; *megerlei* Meigen, 1820 [=*campicola* Eversman, 1854, =*mayeti* Bigot, 1888, =*turanica* Paramonov, 1928, =*vespertilio* Wiedemann in Meigen, 1820], *megerlei consanguinea* Macquart, 1840, *megerlei deserticola* Paramonov, 1928, *megerlei vesperugo* A. Costa, 1893; *melaena* Loew, 1874, *melaena abbreviata* Paramonov, 1928; *melanoptera* Pallas in Wiedemnan, 1818; *meridionalis* Paramonov, 1928; *minoana* Paramonov, 1928; *minois* (Loew), 1869 (as *Defilippia*); *minos*

(Meigen), 1804 (as *Anthrax*, later as *Defilippia*) [=*albiventris* Macquart, 1840, =*germari* Wiedemann in Meigen, 1820, =*semialba* Wiedemann, 1818, =*semiflavida* Becker, 1906, =*senilis* Klug, 1832, =*siderata* Pallas in Wiedemann, 1818], *minos pharaonis* Paramonov, 1928; *mucorea* (Klug), 1832 (as *Zygodipla*) [=*bagdadensis* Becker, 1902, not Macquart, syn. of *tephroleuca* Loew, =*tephroleuca* Loew, 1856, =*olivierii* Macquart, 1840, =*sauvipennis* Macquart, 1849]; *munda* Loew, 1869, *munda rivularis* Griffin, 1896; *nigrifera* Walker, 1871; *noctilio* Klug, 1832 [=*marginalis* Walker, 1871]; *nonna* Becker, 1913; *normalis* (Loew), 1869 (as *Zygodipla*); *occlusa* (Loew), 1873 (as *Zygodipla*), *occlusa nubeculosa* Loew, 1870; *occlusoides* (Paramonov), 1928 (as *Zygodipla*), *occlusoides lactea* Paramonov, 1928; *onusta* Walker, 1852 (as *Anthrax*) [=*pectoralis* Loew, 1937, =*truquii* Rondani 1863]; *pallasii* Wiedemann, 1818 [=*dionysii* Bigot, 1860, =*rhymnica* Eversmann, 1854]; *pallidisetigera* Austen, 1937; *pandora* Fabricius, 1805; *pauper* Walker, 1871 (as *Anthrax*); *phaeoptera* Meigen, 1820 [=*titanus* Megerle in litt. in Meigen, 1820]; *pictilipennis* Austen, 1936; *pleskei* Paramonov, 1928; *portshinskiji* Paramonov, 1928; *pulcherrima* Paramonov, 1928; *punctinervis* (Becker), 1913 (as *Cladodisca*); *pygmalion* (Fabricius), 1805 (as *Anthrax*, later as *Mesocalis*) [=*delineata* Becker, 1906, =*varinervis* Macquart, 1840, =*zona* Bigot, 1860]; *rivularis* (Meigen), 1820 (as *Cladodisca*) [=*argyrocephala* Macquart, 1840, =*rivularis* Griffin, 1896, =*sabaea* Meigen, not Fabricius, 1804], *rivularis munda* Loew, 1869; *rivulosa* (Becker), 1902 (as *Zygodipla*); *rutila* (Pallas in Wiedemann), 1818 (as *Trinaria*) [=*daubei* Guerin, 1835, =*hilaris* Eversman, 1854, =*interrupta* Mulsant, 1852, =*krugeri* Bezzi, 1926, =*miegi* Dufour, 1850]; *seniculus* Wiedemann, 1828 [=*lugens* Paramonov, 1928]; *serpentata* (Loew); 1854 (as *Argyrospyla*); *shelkovnikovi* (Paramonov), 1928 (as *Pterobates*); *similis* Coquillett, 1898, *spiloneura* Bezzi, 1926; *squamea* Mulsant, 1852; *subfasciata* Engel, 1936; *suffusa* (Klug), 1832 (as *Cladodisca*) [=*conturbata* Loew, 1869]; *sytshuana* Paramonov, 1928; *tamerlan* (Portschinsky), 1887 (as *Trinaria*), *tamerlan bezzii* Paramonov, 1928; *telamon* Loew, 1869 [=*loewi* Paramonov, 1928]; *transcaucasica* Paramonov, 1928; *turkestanica* (Paramonov), 1925 (as *Argyrospylia*), *turkestanica mongolica* Paramonov, 1928; *uzbekorum* Paramonov, 1928, female of *uzbekorum* = syn. of *kirgizorum* Paramonov, 1928; *vassiljevi* Paramonov, 1928; *vlasovi* Paramonov, 1928; *volitans* Wiedemann, 1828; *zimini* Paramonov, 1929; *zanoni* (Bezzi), 1922 (as *Argyrospyla*); *zarudnyji* Paramonov, 1928.

Ethiopian: *Exoprosopa acrodiscoides* Bezzi, 1921; *acrospila* Bezzi, 1923; *albata* Bezzi, 1924, *albata novaeformis* Bezzi, 1924; *albofimbriata* Bezzi, 1924; *albonigra* Bezzi, 1924; *altaica* Paramonov, 1925; *angulata* Loew, 1860; *angusta* Bezzi, 1924; *aphelesticta* Hesse, 1956; *apicalis* Wiedeman, 1821; *apiformis* Hesse, 1956; *arcuata* Macquart, 1847; *argentifrons* Macquart, 1855, *argentifrons scalaris* Bezzi, 1924; *argillocosmia* Hesse, 1956; *argyrophora* Bezzi, 1912; *aridicola* Hesse, 1956; *atrata* Hesse, 1956; *atrella* Hesse, 1956; *atrinasis* Speiser, 1910; *atrisquama* Hesse, 1956; *aululans* Bezzi, 1924; *balioptera* Loew, 1860; *barnardi* Hesse, 1956; *basalis* Ricardo, 1901; *basifascia* Walker, 1849 (as *Anthrax*); *batrachoides* Bezzi, 1912; *bolbocera* Hesse, 1956; *brachipleuralis* Hesse, 1956; *brevinasis* Bezzi, 1924; *cadericina* Bezzi, 1924; *caffra* Wiedemann, 1821; *caffrariana* Hesse, 1956; *campestris* Hesse, 1956; *capensis* Wiedemann, 1921; *capnoptera* Bezzi, 1912; *cervina* Bezzi, 1921; *ceuthodonta* Hesse, 1936; *cingulalis* Hesse, 1956; *claripennis* Hesse, 1956; *clathrata* Bezzi, 1923; *clausina* Bezzi, 1924; *compar* Bezzi, 1924; *connivens* Bezzi, 1924; *conochila* Bezzi, 1924; *contorta* Bezzi, 1924; *corvina* Loew, 1860; *corvinoides* Hesse, 1956; *costalis* Macquart, 1846; *curvicornis* Bezzi, 1924; *damarensis* Hesse, 1956; *decastroi* Hesse, 1950; *decipiens* Bezzi, 1924; *decolor* Bezzi, 1924; *didesma* Hesse, 1956; *dilatata* Bezzi, 1921; *diluta* Bezzi, 1924; *dimidiata* Macquart, 1846; *discriminata* Bezzi, 1912; *dolichoptera* Bezzi, 1924; *dubia* Ricardo, 1901; *dux* Wiedemann, 1828; *elipsis* Bezzi, 1924; *elongata* Ricardo, 1901; *eluta* Loew, 1860; *engyoptera* Hesse, 1956; *erebus* Walker, 1849 (as *Anthrax*); *eremochara* Hesse, 1956; *exigua* Macquart, 1855 [=*tenuis* Macquart, 1855]; *fastidiosa* Bezzi, 1924; *ferreirae* Hesse, 1950; *fimbriatella* Bezzi, 1921, *fimbriatella furvalis* Hesse, 1956; *flavicans* Bezzi, 1924; *formosula* Bezzi, 1921; *fulviops* Szilady, 1942; *furvipennis* Hesse, 1956; *fuscula* Bezzi, 1924; *gentilis* Bezzi, 1924; *goliath* Bezzi, 1924; *gonioneura* Hesse, 1956 [=*spoliata* Bezzi, 1922]; *griqua* Hesse, 1956; *guillarmodi* Hesse, 1956; *hamata* Macquart, 1840; *hamula* Hesse, 1956; *haustellata* Bezzi, 1924; *hemiphaea* Hesse, 1956; *heterocera* Bezzi, 1912; *heros* Wiedemann, 1819, *heros litoralis* Hesse, 1956, *heros melanthia* Hesse, 1956, *heros protuberans* Bezzi, 1924; *hyalodisca* Bezzi, 1923; *hyaloptera* Hesse, 1956; *hypargira* Bezzi, 1921; *hypargyroides* Hesse, 1956; *hypomelaena* Bezzi, 1912; *ignava* Loew, 1860; *inaequalipes* Loew, 1852 [=*peociloptera* Bezzi, 1924, =*hirtipes* Loew, 1860]; *indecisa* Walker, 1849 (as *Anthrax*); *infumata* Bezzi, 1924; *infumata* Bezzi, 1921; *inornata* Loew, 1860; *jacchoides* Bezzi, 1912; *jubatipes* Hesse, 1956; *kaokoensis* (Hesse), 1956 (as *Heteralonia*); *karooana* Hesse, 1956; *laeta* Loew, 1860; *latifrons* Bezzi, 1924; *lepidogastra* Bezzi, 1912; *leucothyrida* Hesse, 1956; *linearis* Bezzi, 1924; *loewiana* Bezzi, 1924; *loxospila* Hesse, 1956; *luctifera* Bezzi, 1912; *luteicincta* Hesse, 1956; *luteicosta* Bezzi, 1921, *luteicosta metapleuralis* Hesse, 1956; *luteocera* Hesse, 1956; *macroptera* Loew, 1860 [=*longipennis* Loew, 1863, sine descr.]; *maculifer* Bezzi, 1921; *maculosa* Wiedemann, 1819; *madagascariensis* Macquart, 1849; *magnipennis* Bezzi, 1924; *major* Ricardo, 1901; *majuscula* Hesse, 1956; *mara* Walker, 1849 (as *Anthrax*); *marleyi* Hesse, 1956; *masienensis* Hesse, 1950; *melanaspis* Bezzi, 1924; *melanostola* Hesse, 1956; *melanozona* Hesse, 1956; *merope* Wiedemann, 1824; *mesopleuralis* Bezzi, 1924; *metopargyra* Hesse, 1956; *mimetica* Hesse, 1956; *mira* Hesse,

1936; *monticola* Hesse, 1956; *morosa* Loew, 1860; *mozambica* Hesse, 1956; *mydasiformis* Bezzi, 1924; *nebulosa* Hesse, 1956; *nigerrima* Bezzi, 1924; *nigrovenosa* Bezzi, 1924; *nemesis* Fabricius, 1805 [=*nox* Walker, 1849]; *nephoneura* Hesse, 1950; *neurospila* Bezzi, 1921; *nigrifimbriata* Hesse, 1956; *nigrina* Bezzi, 1924; *nigripennis* Loew, 1852; *nigrispina* Bezzi, 1924; *nigritella* Bezzi, 1924; *notabilis* Macquart, 1840 [=*tewfiki* Paramonov, 1931]; *nova* Ricardo, 1901; *nuragasana* Hesse, 1956; *obscurinotata* Hesse, 1956; *obscuripennis* Hesse, 1956; *obtusa* Bezzi, 1924; *offuscata* Bezzi, 1921; *ogilviei* Hesse, 1956; *othello* Szilady, 1942; *ovamboana* Hesse, 1956; *pallida* Bezzi, 1924; *pallidifacies* Hesse, 1956; *pallidipes* Hesse, 1956; *palustris* Bezzi, 1924; *parvula* Bezzi, 1921; *paucispina* Hesse, 1956; *pentala* Macquart, 1840; *penthoptera* Bezzi, 1956; *personata* Bezzi, 1921; *perpulchra* Bezzi, 1924; *plerosticta* Hesse, 1956; *pleroxantha* Hesse, 1936; *polyspila* Bezzi, 1924; *polysticta* Hesse, 1956; *porectella* Hesse, 1956; *poricella* Hesse, 1956; *praefica* Loew, 1860; *pterosticha* Hesse, 1936; *punctifrons* Bezzi, 1924; *punctulata* Macquart, 1840; *pusilla* Macquart, 1840 [=*jacchoides* Bezzi, 1924]; *rasa* Loew, 1860; *rectifascia* Bezzi, 1924; *recurrens* Loew, 1860; *referta* Bezzi, 1924; *reticulata* Loew, 1860; *retracta* Bezzi, 1924; *rhodesiensis* Hesse, 1956; *robertii* Macquart, 1840; *rubella* Bezzi, 1924; *rubescens* Bezzi, 1924; *rubicunda* Hesse, 1956; *rutiloides* Bezzi, 1924; *sabulina* Becker, 1913; *saskae* Szilady, 1942; *scaligera* Bezzi, 1912; *schmidti* Karsch, 1887; *senegalensis* Macquart, 1840; *sigmoidea* Bezzi, 1912; *simillima* Hesse, 1956; *sisyphus* Fabricius, 1805; *spectrum* Speiser, 1910; *spoliata* Bezzi, 1924; *stannusi* Bezzi, 1912; *stenometaena* Bezzi, 1924; *stevensoni* Hesse, 1956; *strenua* Loew, 1860; *suffusipennis* Bezzi, 1923; *tabanoides* Bezzi, 1924; *thoracica* Bezzi, 1924; *tollini* Loew, 1863; *tricolor* Macquart, 1840, *tricolor orientalis* Bezzi, 1924; *trigradata* Hesse, 1956; *triloculina* Hesse, 1956; *tripartita* Hesse, 1956; *triplex* Bezzi, 1924; *tuckeri* Bezzi, 1921; *umbrosa* Loew, 1860; *unifasciata* Ricardo, 1901; *venosa* Wiedemann, 1819; *villaeformis* Bezzi, 1912; *villosa* Bezzi, 1924; *violaceo* Bezzi, 1924; *vumbuensis* Hesse, 1956; *zambesiana* Hesse, 1956; *zonata* Hesse, 1936; *parva* Ricardo.

Oriental: *Exoprosopa abjecta* Nurse, 1922; *abrogata* Nurse, 1922; *affinissima* Senior-White 1924; *albida* Walker, 1852; *alexon* Walker, 1849; *annandalei* Brunetti, 1909; *aurantiaca* Guerin, 1835; *auriplena* Walker, 1852; *bengalensis* Macquart, 1840; *binotata* Macquart, 1855; *brahma* Schiner, 1868; [=*bramah* Wulp, 1896, lapsus]; *collaris* Wiedemann, 1828 [=*ruficollis* Saunders, 1841]; *dissoluta* Nurse, 1922; *flammea* Brunetti, 1909; *flavipennis* Brunetti, 1909; *flaviventris* Dolschall, 1857; *fuscipennis* Macquart, 1848; *gujaratica* Nurse, 1922; *insulata* Walker, 1852; *interstitialis* Nurse, 1922; *iridipennis* Nurse, 1922; *javana* Macquart, 1840; *lateralis* Brunetti, 1909; latipennis Brunetti, 1909; *maculiventris* Brunetti, 1920; *niveiventris* Brunetti, 1909; *obliqua* Macquart, 1840; *pennata* Nurse, 1922; *pennipes* Wiedemann, 1821; *puerula* Brunetti, 1920; *punjabensis* Nurse, 1922; *purpuraria* Walker, 1852; *retrorsa* Brunetti, 1909; *semilucida* Walker, 1852; *sipho* Paramonov, 1928; *siva* Nurse, 1922; *stylata* Brunetti, 1920; *tristis* Wulp, 1868; *umbrifer* Walker, 1849; *vitrea* Bigot, 1892; *vitripennis* Brunetti, 1912.

Australian: *Exoprosopa adelaidica* Macquart, 1855; *bicellata* Macquart, 1847; *dimidiata* Roberts, 1928; *latelimbata* Bigot, 1892; *obliquebifasciata* Macquart, 1849; *stellifer* Walker, 1849.

Oceania: *Exoprosopa alternans* Macquart, 1855; *sinuatifascia* Macquart, 1855.

Country Unknown: *Exoprosopa argentifera* Walker, 1849; *marginicollis* Gray in Griffiths, 1833; *pulchra* Walker, 1852; *thomae* Fabricius, 1805; *umbra* Walker, 1849; *undans* Walker, 1849.

Genus *Atrichochira* Hesse

FIGURES 206, 431A, 711, 721

Atrichochira Hesse, Ann. South African Mus., vol. 35, p. 911, 1956. Type of genus: *Exoprosopa pediformis* Bezzi, 1921, by original designation.

A rather curious, yellowish to reddish-brown fly of medium size with black mesonotum and related to *Exoprosopa* Macquart. The stylate part of the third antennal segment is thickish, and the division into the apical segment of the style is not easily found; this microsegment is apically pointed, and about 5 times as long as wide; there is a spinous bristle at the end of the style. The face is more rounded and tumid than *Exoprosopa* Macquart, and the mesonotum and scutellum have the bristles very weak or absent; unlike *Exoprosopa* Macquart the head is entirely light red, the proboscis quite short. Hesse (1956) notes further that the legs have the spines very weak and reduced and the basal tooth of the claw minute, short and spinelike. The anterior tarsi are appressed setate instead of hairy. There is a sharp and characteristic bend in the vein on the lower side of the discal cell, this vein arises rectangularly from the last branch of the fifth vein, makes a sharp bend outward, and emits a backward spur at the angle. Hesse notes that the dististyli of the male genitalia differ from most *Exoprosopa* Macquart by being hook-shaped, the beak curved upward and outward, not bifid or notched.

Two species have been placed here. One, the type-species, occurs in the southern part of the Cape Province, Zululand, and Natal, and a second species, *inermis* Bezzi, is from Nyasaland.

Length: 10 to 13 mm.; wing a trifle shorter.

Head, lateral aspect: Subglobular, the occiput is comparatively prominent although strongly rounded backward and inward. The occipital cup is deep but relatively small; occipital pile scanty, flat-appressed, and a little flattened. The front is long and convex and moderately protuberant especially on the lower half, its pile scanty, rather short, stiff, and erect. The face is quite prominent, bluntly or obtusely conical, its length above the oral recess is at least as long as the small antenna; facial pile also scattered and scanty, bristly and suberect. The eye is large, the upper facets in the male scarcely enlarged, and the indentation in the middle of the posterior eye margin, although quite broad, is quite shallow also. There is a bisecting line in the middle of the eye posteriorly. The proboscis is quite short but also slender, the slender labellum comprises more than the outer third and the slender, long palpus reaches almost to the labellum and has a fine, apical, and lateral fringe of short hairs. The proboscis is enclosed in the narrow oral recess and does not reach to the apex of the face. The antennae are attached a little above the middle of the head in profile. They are small and the first segment is nearly twice as long as the second; the second segment is about half as long as wide; the third segment forms a triangular cone at the base, broadly attached to the second segment and not wider than this segment. Apically the third segment is drawn out into a narrow style with bristle at tip and no additional microsegments; the style is not quite as long as the remaining antenna. The first antennal segment bears only a few short setae.

Head, anterior aspect: Head not quite as wide as the thorax, almost circular and at most slightly flattened laterally and at the vertex. The vertex is level with the eye margin; the minute ocellar triangle is very low, equilateral or obtuse, and is set a considerable distance forward from the posterior eye corners. The eyes in the male are separated at the posterior corners by about twice the width of the ocellar tubercle. The front becomes much wider at the antenna and is a little wider than either eye at this point. Antennae widely separated and removed from the eye margin by no more than the length of the first two antennal segments. The oral recess occupies only the lower fourth of the head and is almost entirely ventral in position. It forms a long, triangular opening, quite narrow anteriorly and not very wide posteriorly. The gena is demarcated laterally by a longitudinal tentorial fissurelike crease. It occupies about half the space from the eye to the inner wall of the oral cup. The head from above shows that the occipital lobes are very tightly apposed and the fissure quite shallow. From it arises some scaliform pile.

Thorax: Thorax a little longer than wide, gently convex, feebly shining, largely blackish over the mesonotum with the entire lateral margin including the humerus, the entire scutellum, and 2 medial vittae in front of the scutellum yellowish or reddish brown. Pile over the middle of the mesonotum scanty, short, flat-appressed, more or less flattened. Lateral pile longer and much more dense and subappressed or even a little matted. None of this lateral pile is of any great length. The anterior collar of pile forms an erect band of moderately long, stiff hair becoming shorter posteriorly. There are no bristles anywhere on mesonotum or scutellum. The mesopleuron and upper sternopleuron, and upper half of pteropleuron, each have considerable comparatively long, stiff pile. Also the metapleuron in front of the haltere has much long, dense pile over almost its entire surface. There is a tuft of long, flattened hair beneath the base of the haltere, another tuft above the posterior coxa. Metanotum extremely short, apparently bare, greatly overhung by the scutellum. Halteres quite small. Squama with a fringe of long scales.

Legs: Entirely yellow and quite weak. All the femora are short and slender and covered with fine, rather closely appressed, sharp, black setae. In addition the hind femur bears 3 or 4 oblique, fine, short bristles on the posteroventral margin. Anterior tibia smooth, except for 3 or 4 dorsal, small, flat-appressed setae. The front tarsi are quite short, anterior claws without pulvilli. The middle tibia has 4 rows of very fine, sharp, small, black bristles, 4 or 5 in each ventral row and 6 or 7 in each dorsal row. Hind tibia similar, the dorsal bristles smaller and appressed. The anterodorsal margin with 20 or more closely adjacent, small, oblique, short, black bristles. Hind tarsi short, the claws scarcely bent even at the apex, yellow at the base, black at the apex and without pulvilli. Hesse (1956) notes that the basal tooth of the claw is very much reduced, short and spinelike.

Wings: Uniformly tinged with light yellowish brown. There are 3 submarginal cells. The anterior branch of the third vein arises quite rectangularly, cuts halfway across the second submarginal cell before dividing it into two. The united branches of the second and third vein make a very strong rectangular bend, and the first submarginal cell has the shape of a "foot." The anterior crossvein arises at the middle of the discal cell and the second vein arises gently or sometimes almost rectangularly just in front of this crossvein. The discal cell is of curious shape. It is deeply occluded through the middle by a long, rectangular branch from the lower division of the medius; this rectangular vein has a backward spur. Anal cell widely open. Axillary lobe large. The wing is wide at the base, but the alula is rather slender and elongate oval. Ambient vein complete. Basal comb of the costa exceptionally wide. The basal spur is prominent.

Abdomen: Short, oval, and obtuse, as wide as the thorax and 7 tergites are visible in the male. The first tergite laterally bears an appressed, scanty, fringe of long, pale, bristly pile; middle of the dorsum of the abdomen covered especially along the posterior borders of the tergite with flat-appressed, blackish scales and some pale scales laterally. Moreover, pale scales and hair predominate on the last two tergites and form a conspicuous fringe on the posterior margin of the sternites. Male terminalia prominent, extruded, directed obliquely and diagonally downward and not symmetrical in position.

Hesse states "that the hypopygium differs from that of most forms of *Exoprosopa* in having the beaked apical joints hook-shaped, the slender beak curved upwards and outwards and its apex not bifid, is indented or notched."

Material available for study: A pair of the type-species kindly furnished through the courtesy of Dr. A. J. Hesse.

Immature stages: Unknown.

Ecology and behavior of adults: Not on record.

Distribution: Ethiopian: *Atrichochira inermis* (Bezzi), 1912 (as *Exoprosopa*); *pediformis* (Bezzi), 1924 (as *Exoprosopa*).

Genus *Diatropomma* Bowden

FIGURES 194A, 199, 404

Diatropomma Bowden, Journ. Ent. Soc. South Africa, Pretoria, vol. 25, no. 1, p. 116, 1962. Type of genus: *Diatropomma carcassoni* Bowden, 1962, by original designation.

Diatropomma Bowden consists of comparatively small *Exoprosopa*-like flies in which the base of the largely blackish wing, with midposterior clear spot and tinged apex, is rather strongly narrowed. This reduction of the axillary lobe and alula is shared by *Isotamia* Bezzi and these flies differ from that genus and also from *Exoprosopa* Macquart by the distinct pubescence of the eyes, visible at low magnification. Bowden observes that the form of the genitalia, the short, stiff pubescence, form of discal cell, and the sharp upward bend of the anterior branch of the third vein all ally this fly to the *Heteralonia* Rondani or the *Acrodisca* Bezzi groups within the Exoprosopinae.

Only 2 species are known, the type-species from Kenya, and *claudia* François from Urundi.

Length: 7.8 to 10.1 mm., wing 9.9 to 11.3 mm., proboscis 2.3 to 2.8 mm.

One can find nothing to add to Bowden's analytical comparison, which I quote:

This new genus appears to have most in common with *Isotamia* Bezzi (=*Francoisia* Hesse, *Ogilviella* (Paramonov). With *Isotamia* Bezzi it shares several features, such as reduced axillary lobe and alula, anal cell of about equal width, short thorax, relatively bare pleura, short, stiff and blunted hairs on pleura and at base of abdomen, and also the depressed (although not grooved) face, long proboscis and indistinct genal grooves. It differs conspicuously from *Isotamia* Bezzi and *Exoprosopa* Macquart in the distinct pubescence of the eyes (although not previously reported, many species of Bombyliidae possess microscopic pubescence on the eyes, which is usually only visible in good light at high magnification, whereas in *Diatropomma* Bowden the pubescence is easily visible at low magnification), the peculiar structure of the third antennal segment, the greatly reduced fringes to the squama and alula and the remarkable development of cilia all round the wing. The genitalia of the male are unremarkable, except that the aedeagal sheath is considerably broadened and rounded apically, and are generally reminiscent of species of *Heteralonia* Macquart or *Acrodisca* Bezzi rather than other subgenera and species groups within *Exoprosopa* Macquart. This relation is further suggested by the sharp upward bend of R_{2+3}, the form of the discal cell, head structure and to a certain extent the short stiff pubescence. The general appearance of the type species also suggest a *Heteralonia* Rondani or a species of *Defillipia* Lioy such as *consanguinea* Macq. or *nigerrima* Bezzi.

As I have not seen material of *Diatropomma* Bowden, I quote his description in full:

Body with vestiture generally reduced and sparse; thoracic hair in collar, along notopleura and on pleura, as well as that in lateral tuft of first abdominal segment, stiff, short and coarse. Head with occiput about level with vertex, occipital pit deep but occipital cleft more or less closed; ocellar tubercle set well forward of vertex; frons of approximately equal width in both sexes; eyes with small indentation and weak bisecting line, with equal facets and with very short, sparse but quite distinct pubescence; face conically produced, somewhat depressed medially and indented at apex, genal furrows almost obliterated; antennae with third segment stout, with a small conical apical sub-segment which bears a minute apparently sclerotized cap; under high magnification third antennal segment densely hairy; proboscis long, projecting well forward of face; palpi slender only slightly shorter in length than antennae.

Thorax: Short relative to scutellum, barely twice length of latter, and broad, quite distinctly broader than long; humeral calli prominent, postalar calli not or barely marked off from disc of thorax and represented by projecting posterior angles of notum; meso- and metapleural sclerites relatively prominent, particularly the latter, plumula sparse; scutellum broadly rounded; thoracic bristles well developed, especially postalar bristles, scutellar weaker.

Abdomen: Long and narrow; bristles greatly reduced, absent except for extreme sides and small ones at apex. Legs with fore coxae long, little shorter than fore femora (excluding trochanters); fore femora, tibiae and tarsi subequal, the fore legs thus relatively unmodified, fore femora without bristles, fore tibiae with well developed and numerous spicules; tarsi with strong ventral bristle rows, fore tarsi also with some longish hairs in both sexes; claws of fore tarsi less markedly reduced than in *Exoprosopa* Macquart, other claws well developed and all with a blunt basal tooth; pulvilli absent.

Wings: Long relative to body, markedly pedunculate, axillary lobe reduced and linear, about as wide as anal cell; alula reduced, lobelike; squamae much reduced, auriform; basal comb small; R_{2+3} arising in a smooth obtuse curve at about or slightly beyond r-m, the latter slightly before middle of discal cell which is relatively long, apical vein of discal cell short, regularly S-shaped, more or less parallel to wing margin; first posterior cell widely open, second and third posterior cells of equal width at wing margin, anal cell of approximately equal width throughout, widely open at wing margin; cilia of costal margin extending right round wing, becoming longer at base of axillary lobe and on alula, which is without scales; squamae with sparse narrow scales and also with some bristly hairs at angle with alula; wings conspicuously infuscate with an extensive pattern like that of *Littorrhynchus* Macquart, which has, in type species, a prolongation into second submarginal cell and a subsidiary, less intense, apical infuscation separated from main infuscation by a narrow hyaline band; hyaline or subhyaline fenestrae present over main bifurcations and cross veins. Hypopygium with basimeres long, subquadrate; telomeres long and narrow, with a reflexed apical point; aedeagus short and slender, ventral aedeagal processes (aedeagal sheath) large, broadly rounded; paraphyses broad, relatively pointed at apices, apodeme simple.

Material available for study: None. I have had to rely upon Bowden's description and excellent illustrations.

Immature stages: Unknown.

Ecology and behavior of adults: Not on record. Bowden states that it seems to be at home in the semiarid country of the Rift Valley.

Distribution: Ethiopian: *Diatropomma carcassoni* Bowden, 1962; *claudia* (François), 1960 (as *Exoprosopa*).

Genus *Heteralonia* Rondani

FIGURE 194

Heteralonia Rondani, Arch. Zool. Anat. Fisiol., vol. 3, p. 57, sep., 1863. Type of genus: *Exoprosopa oculata* Macquart, 1840, by original designation, as *Exoprosopa occulta* Rondani, 1863, lapsus.

A curious little fly, about 6 mm. long, that is quickly distinguished by the peculiar venation of its small speckled wing. It is black and feebly shining, the wing sepia brown obliquely on the anterior half with still darker spots on all crossveins, vein furcations and additional spots near the posterior margin. There are 3 or 4 submarginal cells, and the first submarginal cell is closed and stalked and has a peculiar shape.

An erect, plane, rectangular crossvein can be interpreted as by Bezzi (1924) as dividing the marginal cell into two parts, or as by Hesse (1956) as constituting an extra oblique crossvein dividing the anterior, apical, submarginal cell into two parts. Both the two known species show this peculiarity constantly. Hesse suggests that since a somewhat similar situation exists in *Ligyra* Newman, there is adequate basis for leaving *Heteralonia* Rondani a separate genus; with this view I am inclined to agree. It is possible to interpret all of these so-called crossveins as normal branches of the radial system.

Two species are known, one from Senegal and Ghana, another from South West Africa.

Length: 6 to 8 mm.; wing 6 to 9 mm.

Head, lateral aspect: The head is hemiglobular but peaked in front on account of the triangular, conical face. The occiput is comparatively short, both at the vertex of the head and also at the bottom of the head. Dorsally the occiput is convex near the eye margin, but rapidly curved vertically downward so that it has not much thickness at this point and the medial fissure is vertical. At the bottom of the head, however, the occiput slopes very gradually backward, and it is also produced distinctly below the eye, and bears chiefly pollen on the bottom, but on the middle and dorsal part of the occiput there is a matlike cover of rather dense, flat-appressed, outwardly directed yellowish scales. The eye rises abruptly and conspicuously above the occiput on the dorsal half of the eye. The eye is reniform, the posterior indentation deep and extensive above and below with a rather pronounced, bisecting line between the eye facets at the bottom of this indentation. The front is slightly raised, very convex in profile, pollinose over a black ground color, with scattered, strong, erect or curled, bristly hairs, those at the vertex longer than the first antennal segment.

The face is conically or triangularly produced forward as a not quite equilateral triangle. The pile of the face is similar to that of the front, bristly or long setate, rather strongly appressed, also black in color and with the long, bristly setae forming a fringe, overlapping the edge of the oral recess; the elements are somewhat stouter in the middle of the face. There are also a few, appressed, golden scales. The proboscis is rather short, extending barely beyond the oral recess. It is slender with long, slender, pointed labellum. Palpus small, cylindrical and short with a few, stiff hairs below and laterally. The antennae are attached at the upper third of the head. The first segment is widened apically with very short setae dorsally and laterally and a few ventrally. This segment is about twice as long as the basal width. The second segment is beadlike, wider than long. The third segment is elongate and conical. At the apex of the cone it bears an articulated, slender style, bristle-tipped, the style no longer than the remainder of the segment.

Head, anterior aspect: The head is a little wider than the thorax, circular, the eyes separated in the male by a distance scarcely more than the length of the first two antennal segments. Female vertex slightly wider. A rather unique feature appears to be the extreme forward position of the small, low, isosceles ocellar tubercle which lies distant from the posterior eye margin at least one-third the length of the front. Posterior eye corners rounded and the eye facets of uniform size. The antennae are separated widely and are distant from the eye margin by less than the length of the first segment. Face and oral cup narrow, the latter is triangular. Posteriorly behind the proboscis, the oral cup is thin-walled and moderately deep, but this only begins midway along the sides. Anteriorly the wall is inflated and thickened with rounded instead of sharp edge and this thickened area reaches to half of the anterior division of the oral recess. Gena linear. The oral cup is moderately produced downward from the eye margin.

Thorax: Almost quadrate, but a little narrowed across the anterior margin. Humerus swollen with numerous, long, black, bristly hairs. Surface of mesonotum and scutellum shining with faint pollen and scattered, fine, sparse, erect hairs, and scattered, flat-appressed setae above the wing and more widely on the scutellum. Notopleuron with some short, stout, black bristles. Postalar callosity with 3 or 4 long, slender, black bristles and the scutellar margin with 3 or 4 pairs of more slender, black bristles. Mesopleuron with a bushy tuft of very stiff, black, blunt, bristles or bristly hairs. Anterior collar composed of similar elements, but rather short. The anterior pteropleuron has a tuft of bristly hairs. The sternopleuron has some appressed, bristly hair and long, bristly hairs, very dense on the metapleuron. Hypopleuron bare. No tufts of pile beneath the haltere.

Legs: The legs are comparatively short and stout, but not swollen. The anterior tarsi has strongly appressed, short, stout setae. Middle femur with some long, slender bristles. There are 2 in the middle on the front side, 3 shorter ones posteriorly. The middle tarsus has some rather conspicuous, though slender, bristles, which are few in number. There are 3 anterodorsal and 2 or 3 posterodorsal bristles. The hind femur bears 4 moderately long, slender bristles anteroventrally and 2 dorsally at the apex. Its tibia also has the bristles scanty. There are 4 or 5 anterodorsal bristles, the same number anteroventrally, and 5 or 6 posterodorsally. Claws long, slender, sharp, slightly curved toward the apex and with a distinct, basal tooth.

Wings: Comparatively large, considering the small size of the fly, slender, with a characteristic dark sepia, ocellated pattern in spots especially on the crossveins and becoming confluent anterobasally. There are 3 submarginal cells and 2 marginal cells. The shape of the second submarginal cell is peculiar. It is somewhat scapulate, extending as far basally from the base of the first submarginal cell as it does outward into the first submarginal cell. First posterior cell strongly narrowed at the apex. Anal cell open almost maximally. Wings slender at the base, the axillary lobe comparatively narrow, the alula narrow, the ambient vein complete. The costal comb narrow, and the costal hook thornlike.

Abdomen: Longer than the thorax and distinctly though not greatly wider. It is shining black, narrowly oval, and densely covered with flat-appressed, slender, scaliform black pile and considerable amounts of brownish-yellow wider scales. There are a few, scattered, fine, sparse, black hairs along the lateral margin, and subapically on the last tergite. Posterior margins of the sternites with similarly slender, somewhat more numerous hairs. Male terminalia moderately large, visible from the sides or below, conical and thrust toward the left of the insect. More in evidence from below, the lateral margin of most of the tergites and posterior edge of the last tergite have a conspicuous single row or fringe of short, broad, black scales.

Material available for study: The type-species presented to me through the courtesy of Mr. John Bowden.

Immature stages: Unknown.

Ecology and behavior of adults: I have seen no record of adult behavior.

Distribution: Ethiopian: *Heteralonia kaokoensis* Hesse, 1956; *oculata* (Macquart), 1840 (as *Exoprosopa*) [=*cosmoptera* Bezzi, 19—].

Genus *Isotamia* Bezzi

FIGURES 193, 402, 684, 688

Isotamia Bezzi, Trans. Ent. Soc. London, p. 627, 1912 (1911). Type of genus: *Isotamia daveyi*, by original designation.
Francoisia Hesse, Bull. Inst. Sci. nat. Belgique, vol. 28, no. 42, p. 1, 1952. Type of genus: *Francoisia sulcifacies* Hesse, 1952, by original designation.
Ogilviella Paramonov, Proc. Roy. Ent. Soc. London, (B), vol. 23, p. 26, 1954b. Type of genus: *Ogilviella tridentata* Paramonov, 1954b, by original designation.

These are peculiar, comparatively small flies, which are rather bare; they have short scales on head, thorax, and more abundantly upon the abdomen. The slender, basally narrowed wing, has 3 submarginal cells, and a sharply delimited irregular anterior, blackish pattern superficially very much like some species of *Anthrax* Scopoli. The third vein arises, however, well before the anterior crossvein, and while obtuse at its origin it is not rectangular. The third antennal segment is conical with long, slender apex and in addition a long terminal style. The proboscis is long and slender; posterior eye margin is indented together with short bisecting line; occiput only moderately tumid or conspicuous but with a deep central cavity. Anterior and middle tibiae with weak spicules, hind tibia with somewhat more prominent ones; hind femur with 3 spines below near apex; pulvilli absent.

Though placed by Bezzi in the Lomatiinae, I agree with Hesse concerning its true relationship within the Exoprosopinae. It differs from *Exoprosopa* Macquart by the reduced axillary lobe and alula, less pile on pleuron, less reduced front claws, dorsal groovelike depression of the face, short palpus, long proboscis, and stiff, stout, bristly pile on pleuron and base of abdomen.

Known from a single species found in central Africa. Length: 7 to 8 mm.; wing 9 mm.

Head, lateral aspect: The head is hemispherical. The face is rather conspicuous and angularly produced; however, since it is demarcated laterally and obliquely by a crease running just inside the antenna to the eye margin, and this crease or sulcus is continued transversely across just below the antenna, possibly the extended portion should properly be regarded as part of the oragenal cup. The oragenal cup is deep but narrow, divergent below with the outer margin sharply marked but with the walls thick and expanded on each side. Gena virtually absent, confined to an indistinct groove near the eye at the lower part of the eye. The pile of the face consists of flat-appressed, minute, broad and quite short scales. There are 2 shallow, central depressions on the front, 1 over the middle and 1 above the antenna. Also, there is a curious, V-shaped depression in the middle at the oral margin. The sides of the front and the middle of the face and sides of the oragenal cup each have a few, scattered, stout, short, oblique setae. Occiput only moderately developed but a little longer above though slanting backward and it is also bilobate with a narrow fissure. On the lower half the eye extends farther backward and also the occiput extends farther downward behind the proboscis. The occiput has a large, central cavity bordered with abundant, short, erect pile. The remainder and anterior portion is covered with small, appressed, scattered scales. The palpus is moderately long and slender with long, slender, setate hairs laterally, especially toward the base.

The proboscis is quite long and slender and about twice as long as the head. The antennae are attached near the upper fourth of the head or slightly beneath it; they are minute, the first segment is nearly twice as long as the second and bears above and below a few fine setae; second segment with some shorter setae dorsally and 1 or 2 setae below. The third segment is strongly and triangularly bulbous at the base, abruptly drawn out into a slender, stylelike portion only as long as the basal part; it bears an equally slender style which is a little longer than the length of the 3 segments combined and likewise has a microscopic spine at the tip.

Head, anterior aspect: The head is as wide as thorax, the female vertex is very little more than half the width of the front at the antenna. The antennae are quite widely separated, and separated from the eye margins by less than the width of 1 segment. The ocellar tubercle is small and so shallow as to be almost absent; the shallow, convex ridge, however, is creased laterally and extends backward to the lobes of the occiput.

Thorax: The thorax is short and broad, the scutellum quite wide but rather short and without strongly developed posterolateral angles. The mesonotum is shining black, the scutellum likewise, both covered with moderately abundant, flat-appressed scales, some of them minute, most of them moderately large. The anterior margin of the mesonotum has a narrow band of coarse, pale yellow, erect, bristly pile of moderate length changing to black on the swollen, protuberant humerus on its outer portion. Notopleuron with some short bristles, postalar callosity with 3 or 4 longer, slightly stouter, tuberculate bristles. Scutellar margin with 2 or 3 slender, not very long, black bristles. There is a conspicuous tuft or band of long, moderately stout, black bristles upon the upper margin of the mesopleuron and the more anterior elements are better developed, the greater, lower part of the mesopleuron bare. The upper pteropleuron has a few setae; upper metapleuron above the haltere and in front of it with a tuft of numerous, long, slender, black bristles; hypopleuron bare; the squama has only a very few fine, short hairs on it, and no scales.

Legs: The legs rather slender but comparatively short, they bear scattered, flat-appressed, dark scales, and also scanty, minute, appressed setae. The hind femur has 2 long, stout spicules below and 1 smaller and 1 or 2 others dorsally on each side near the apex; anterior tibia has 2 or 3 minute spicules dorsally. The hind tibia, however, has 3 or 4 conspicuous spicules especially below and also 5 or 6 shorter ones above; there are 4 rows altogether. The tarsi are quite short, the claws small, pulvilli absent.

Wings: The wings are rather long and slender and characteristically greatly narrowed toward the base. The alula is absent or reduced to a narrow, thickened fold. There are 3 submarginal cells, the anal cell is quite widely open, the basal vein anteriorly on the third posterior cell is undulate, the anterior crossvein enters the discal cell beyond the middle and the second vein arises a considerable distance proximal to the anterior crossvein by 2 or more times the length of the anterior crossvein. The anterior half of the wing is blackish sepia, irregularly but sharply demarcated; base of costa without conspicuous setae but there is a sharp, spinelike lobe.

Abdomen: The abdomen is moderately elongate, not quite twice as long as the thorax but broader than the thorax and with nearly parallel sides to the end of the sixth segment. Abdomen shining black, covered with conspicuous, pale, broad, flat-appressed scales, which are apt to be largely denuded and with some black scales intermixed especially along the posterior margin of the sixth and seventh tergites where they extend out free and are broad and conspicuous. The sides of the first tergite have a dense band of moderately long, erect, coarse, black pile which, however, does not appear to extend under the scutellum in the middle of the segment. Sternites with scales only.

Material available for study: Type of *Ogilviella tridentata* Paramonov in the British Museum (Natural History).

Immature stages: Unknown.

Ecology and behavior of adults: Not on record.

Distribution: Ethiopian: *Isotamia daveyi* Bezzi, 1912 [=*sulcifacies* (Hesse), 1952 (as *Francoisia*), =*tridentata* (Paramonov), 1954 (as *Ogilviella*)].

Genus *Litorrhynchus* Macquart

FIGURES 191, 408, 687, 689

Litorhynchus Macquart, Diptères exotiques, vol. 2, pt. 1, p. 78, 1840. Type of genus: *Bibio lar* Fabricius, 1781. Designated by Oldroyd, 1940, as type by remainder and correction of error in Coquillett, 1910.

Litorrhynchus Bezzi, The Bombyliidae of the Ethiopian Region, p. 3, 1924, emendation.

Coquillett (1910, p. 562) designated *Litorrhynchus hamatus* Macquart, the second of 4 species, as type-species. Coquillett did not realize that 3 of the species Macquart included—*Anthrax seniculus* Wiedemann, *A. collaris* Wiedemann, and his own new species *L. hamatus*—were as shown by Loew (1860) and by Ricardo (1901) actually equivalent to Exoprosopa, sensu stricto.

Bezzi (1924), pp. 212-214, key to 25 Ethiopian species.
Hesse (1956), pp. 625-631, key to 16 South African species.

While dipterists at the National Museum of Natural History suggest that I change the name *Litorrhynchus* Macquart to read *Litomyza*, new name, because of strict adherence to the rules, I strongly believe the purpose of taxonomy is ill served by changing a name in use for 130 years. I decline to do so, hence I leave the fixation as Oldroyd made it 100 years after the naming of the genus.

Large, beautiful flies related to *Exoprosopa* Macquart and generally with a handsome and very characteristic pattern of brown or yellow or both colors on the wing. Bezzi, who in 1912 redescribed the genus and in 1924 commented upon it further, emphasized the rounded face, with longer proboscis extending beyond the oral cavity. Bezzi particularly called attention to the long, sharp, basal tooth on the claws, shared by *Exoprosopa* Macquart, and also the spicules present on the anterior tibia. Bezzi admitted that some species of *Exoprosopa* Macquart do have front tibial spicules, and he noted that a few species like *ignava* Loew have a similar wing pattern; see also *decipiens* Bezzi, *curvicornis* Bezzi, and *pterosticha* Hesse. The very different American genus *Stonyx* Osten Sacken likewise has a basal tooth on the claw, but a different wing pattern and other differences. While admittedly there is some overlapping of characters, it is at once apparent when comparing a number of species that *Litorrhynchus* Macquart constitutes a rather unique group within the tribe Exoprosopini and that this genus should be restricted to those *Exoprosopa*-like flies that combine blunt, rounded face with tooth on the claws, spicules on the anterior tibia, and characteristic wing pattern. The very rounded face alone will usually separate them from the species of *Exoprosopa* Macquart, which have a conically pointed face.

Macquart in erecting this genus noted that the Indian species *Bibio lar* Fabricius from the Orient was to be included within the genus and contained its characters; but then he also assigned to it a South African species and another Oriental species both of which subsequently proved to be true members of the genus *Exoprosopa* Macquart in the strict sense. Oldroyd (1940) chose the only sensible course and designated the Indian species *lar* Fabricius as the type-species; this species differs slightly from the African species. I have illustrated *lar* Fabricius through the courtesy of Mr. Oldroyd and also have been able to study a number of African species through the courtesy of Dr. A. J. Hesse.

Litorrhynchus Macquart is a characteristic element of the bee fly fauna of that part of Africa lying below the equator. Bezzi (1924) keys 25 species and treats 18; Hesse (1956) treats 16 species.

Length: 6.5 to 18 mm., more commonly 10 to 12 mm.; wing 9 to 22 mm.

Head, lateral aspect: The head is subglobular with strongly developed and tumid occiput and the face, while it is moderately prominent, is broadly rounded. The face extends a little farther outward than the length of the first two antennal segments; it is characteristically bluntly rounded as compared with the conical face of *Exoprosopa* Macquart; the oragenal cup is narrow, but deep, except in front, its walls are much widened or thickened and rounded with the surface pollinose. The pile of both face and front is abundant, yet not dense. It is fine, erect, stiff, and rather short with a considerable number of long, flat-appressed scales intermixed. The posterior eye margin very shallow and broadly concave, but at the middle it has a short, bisecting line. The occiput is about as long as the face at its greatest length. The pile is short and erect giving away along the eye margins to a rather broad border of quite dense, matted, or flat-appressed, usually pale scales. The occiput behind the ocelli is bilobate; the fissure is quite narrow since the lobes are touching. The palpus has 1 segment, is quite long and quite slender with numerous, rather long, slender, bristly hairs on the ventral surface, shorter ones above. The proboscis extends distinctly beyond the oral cavity. The nonoral part of the genae is exceptionally narrow, the crease is deep but the sides close.

The antennae are attached at the upper third of the head, the first segment is longer and may be about twice as long as the second and 3 or 4 times as large. It bears numerous, moderately long, coarse, bristly setae on all sides, except the upper half of the medial surface, which is more or less bare. Third segment either conical or clublike, its base broader and more bulbiform; this segment is often long; terminally it ends in a slender, apically dilated stylelike segment, itself bearing an apical spinelet.

Head, anterior aspect: The head is as wide as the thorax, the female vertex is only half as wide as the front at the base of antennae and the ocellar tubercle quite low and undeveloped. The small ocelli are set in a short isosceles triangle. The front is relatively smooth but may be slightly swollen in the middle. The oral opening is narrow, rapidly widening at the base of the proboscis.

Thorax: The thorax is short and broad and only slightly convex. The scutellum is quite large, the mesonotum tends to have rather dense, flat-appressed, often shining, scaliform pile in which, however, are intermixed some subappressed, quite fine, posteriorly directed hairs. Anterior margin of the the mesonotum with a conspicuous fringe of long, slender, erect, dense, pale bristles. The humerus is sharply swollen and often bears both black and yellow bristles. Notopleuron with 2 or 3 rather long, conspicuous, stout bristles and anteriorly a row of shorter bristles. Postalar callosity with 2 or 3 quite stout, conspicuous bristles and several more slender ones. Margin of the scutellum with 6 to 8 pairs of rather stout, moderately long bristles, which tend to be curved across one another in the middle. The disc bears abundant, fine, suberect pile and some yellow scales or scaliform pile, flat-appressed especially on the base. Squama with a fringe of broad, often brownish scales. Almost the whole of the mesopleuron is covered with dense, coarse, bristly or wiry pile often of 2 colors, the upper, coarser, more bristly pile is longer, directed upward and the pteropleuron especially posterodorsally has a large, subappressed clump of coarse, bristly pile. Metapleuron with dense, erect, long, coarse pile on a vertical band down its middle. Also, there is a tuft of pile at the base of the halteres below. Hypopleuron bare. Propleuron with tufts of long pile on each side.

Legs: The legs are relatively short and stout, especially the femora. Anterior femur swollen from base to near the apex. The pile consists of dense, large, broad, flat-appressed scales together with minute, fine, appressed, setate hairs. The pile of the tibiae is similar but the setae are finer and more appressed, the scales perhaps smaller. The hind femur has 2 lateral spicules but none below. The anterior tibia has a posterodorsal row of 8 or 9 quite small spicules and a few much smaller spicules anteriorly, also a few much more slender posteroventral spicules. The spicules on the hind tibia are appressed and moderately well developed with about 10 to 12 in each row; claws rather sharp but slender with a basal tooth; pulvilli absent.

Wings: The wings are large and tending to be a little longer than the fly. They are apt to have a very characteristic pattern of sepia brown, or reddish brown and yellowish brown with some fenestrate spots and the apex often hyaline as well as a posterior triangular spot in the middle. The discal cell is broad with the second posterior cell wide toward its base. Base of costa with well developed comb of bristles and with a sharp, long, hooked or beaklike process.

Abdomen: The abdomen is comparatively long oval but quite broad across the first two tergites and shallowly oval. The ground color is apt to be light brown, reddish or yellowish and the surface is covered densely with scales that are mostly dark but sometimes these are intermixed with scales or patches of scales that are opaque white, and there are also considerable, short, fine, strongly appressed, bristly hairs. The female terminalia bear a row of curved spines on each side. The male terminalia, according to Hesse, have a crest of dense hairs or spiny hairs dorsally in apical part of each basal part; distimeres (beaked apical segments) slightly curved outward and up, the apex of each ending in a slightly longer, upper, outer, curved hooklike point and in a lower, inner, shorter, blunt process; each distimere has a prominent, raised, blunt, projecting process on the outer side just before middle of joint; the aedeagus has a prominent, ventral, scooplike process which usually forms a broadened apical, central keel; the aedeagus has a prominent, ventral, scooplike process which usually forms a broadened apical, central keel; the aedeagus itself is short; lateral struts shaped much as a shoe horn; posteriorly projecting strut racquet-shaped.

Dr. A. J. Hesse in his work on South African bee flies describes the male hypopygium as follows:

Hypopygium of males with a crest of dense hairs or sometimes almost spine-like hairs along dorsal part in apical half of each basal part; beaked apical joints slightly curved outwards and upwards, the apex of each ending in a slightly longer, upper, outer, curved, hook-like point and in a lower, inner, shorter, blunter process; each beaked apical joint also with a prominently raised, blunt, projecting process on outer side just before middle of joint; aedeagal apparatus with a prominent, ventral, scoop-like aedeagal process, which is usually broadened apically and centrally keel-like on dorsal aspect (side view); aedeagus itself relatively short; lateral struts of middle aedeagal part strongly developed, broadly shoe-horn-shaped; posteriorly projecting strut chopper-shaped or racket-shaped.

Material studied: The series in the British Museum (Natural History) and an example of the type-species lent by this museum. Dr. A. J. Hesse kindly sent several species from South Africa for dissection and study.

Immature stages: One species has been bred in South Africa from the mud nests of a wasp. See comments under life histories.

Ecology and behavior of adults: Very few notes are on record. Hesse (1958) notes that they appear more abundantly in grass-steppe, savannah, and broken-bush country, and are not common in the Karoo and arid types of country; he says the wing pattern, striking when in open grass veldt, renders them almost invisible in the shadows of broken thornbush.

Distribution: Ethiopian: *Litorrhynchus argyrolepis* Bezzi, 1912; *atricapillus* Hesse, 1956; *basalis* (Ricardo), 1901 (as *Exoprosopa*); *bechuanus* Hesse, 1936; *corticeus* Bezzi, 1924, *corticeus corticalis* Bezzi, 1924; *damerensis* Hesse, 1956; *dentifer* Bezzi, 1912; *dilatatus* Bezzi, 1912; *ectophaeus* Hesse, 1956; *erythraeus* Bezzi, 1906, *erythraeus allothyris* Bezzi, 1923; *infuscatus* Bezzi, 1924; *kaokoensis* Hesse, 1956; *macropterus* (Loew), 1860 (as *Exoprosopa*) [=*longipennis* Loew, 1863 (as *Exoprosopa*) (no descr.), =*tollini* Bezzi, 1924, not Loew]; *maurus* (Thunberg), 1827 (as *Tanyglossa*) [=*collaris* Macquart, 1840, not Wiedemann, *erythraea* Speiser, subg. *Exosoma*, =*rostrata* Loew, 1860]; *metapleuralis* Bezzi, 1924; *nyasae* Ricardo, 1901 [=*argyrolepis* Bezzi, 1921, in part]; *obumbratus* Bezzi, 1924 [=*macropterus* Ricardo, not Loew]; *perplexus* Bezzi, 1912; *phloeochromus* Bezzi, 1922; *productus* Bezzi, 1924; *pseudocollaris* Bezzi, 1924 [=*maurus* Bezzi, 1921, in part]; *repletus* Bezzi, 1912; *ricardoi* Bezzi, 1912; *rostratus* (Loew), 1860 (as *Exoprosopa*); *siccifolius* Bezzi, 1924; *suberosus* Bezzi, 1924; *suspensus* Bezzi, 1924; *tollini* (Loew), 1863 (as *Exoprosopa*); *vernayi* Hesse, 1956 [=*nyasae* Hesse, 1936, not Ricardo].

Oriental: *Litorrhynchus lar* (Fabricius), 1781 (as *Bibio*).

Genus *Micomitra* Bowden

FIGURES 203, 421, 694

Micomitra Bowden, Mem. Ent. Soc. Southern Africa, Pretoria, no. 8, p. 147, 1964. Type of genus: *Exoprosopa parvicellula* Bezzi, 1921, by original designation.

The genus *Micomitra* Bowden was erected to contain a group of species of South Africa *Exoprosopa* Macquart, sensu lato, which displayed a combination of characters not found among other species. According to Bowden they show a shining integument, generally brilliantly shining, iridescent scales on several parts of the body, greatly reduced chaetotaxy, much more reduced than in other *Exoprosopa* species, and with thoracic bristles especially weak or wanting. Also Bowden points out that the males have U-shaped telomeres and greatly expanded ventral aedeagal process.

This genus contains the species that were included by Bezzi (1924) in his *stupida* Rossi group, and by Hesse (1956) in his *parvicellula* Bezzi group. They have the wings completely hyaline, shining and iridescent, even without infuscation basally or in the costal cell. First posterior cell usually broadly open according to Hesse, with the integument of the frons shining. Like the *Acrodisca* and *Heteralonia* groups of *Exoprosopa* the ocelli are set rather far forward. Note that the antennal style of the third antennal segment is minute and spinelike. Also the species are of small size, rarely exceeding 10 mm. in length. Bezzi noted the slight development of the basal tooth on the claw, the short wings without praediscoidal spot.

Micomitra is distributed through the Mediterranean and Ethiopian regions, more abundantly in South Africa. Bowden placed 12 species here: *parvicellula* Bezzi (type-species), *stupida* Rossi, *iris* Loew, *famula* Bezzi, *chrystallina* Bezzi, *pharao* Paramonov, *bella* Austen, *iridipennis* Hesse, *opalina* Hesse, *perlucida* Hesse, *aerata* Hesse, and *anthracina* Bowden. Bezzi included *serva* Bezzi, *erronea* Bezzi, *brachycera* Bezzi, *ancilla* Bezzi, *chionea* Bezzi, and *latissima* Bezzi.

Length: 5.5 to 11 mm.; wing 4.8 to 9 mm. Most species less than 10 mm.

Head, lateral aspect: Semiglobular, the face bluntly triangular, forming almost a rectangle with the oragenal cavity. The face is covered with broad, iridescent, pale scales which are flatly plastered against the surface. Continued narrowly down the sides of the gena and with a few short, very fine, quite erect, white hairs anteriorly on the gena. Face and gena and the basal segments of the antenna often pale yellow. There may be a few minute setae medially on the peak of the face. The occiput is quite prominent, rounded, and sloping to the eye margin, plastered over the flat or tightly appressed silvery scales and the large central cavity bearing a short fringe around its border. Occipital lobes tightly apposed. Posterior eye margin broadly but deeply indented below the middle and with a bisecting line in the facets. The upper facets are twice as large as the ventral ones. The palpus is slender, but short, cylindrical, with a few fine hairs tending to be curved and less than half as long as the proboscis. The proboscis is rather short, often directed straight forward, but not extending farther than the oragenal recess. It is strongly compressed laterally with microspinules ventrally which are arranged in rows. The labellum is also compressed, a little more than twice as long as wide, a little pointed apically, and a little widened when viewed from above.

The antennae are attached just above the middle of the head. They are small with the first two segments nearly equal, the third is longer than the first two segments combined; its width at the base equal to that of the short second segment and beyond the base it is progressively attenuate to the apex. The apex consists of an extremely small, short, slender, articulated microsegment.

Head, anterior aspect: The head is distinctly wider than the thorax, very globular when viewed from above. The oragenal recess is quite narrow and deep, both above and below, widened posteriorly where the proboscis emerges. The walls of the oragenal recess are widened and strongly sloping outward and also backward to the eye margin. They are only slightly inflated medially just below the middle. There is a long, sharp, tentorial fissure separating the quite narrow gena throughout its length. The antennae are very widely separated and removed from the eye margin by less than the thickness of the first segment. The eyes in the male are very narrowly separated behind the ocelli by the thickness of one ocellus. The ocellar tubercle is prominent, though small and is set rather far forward from the posterior eye corner. The male front rapidly widens to form a rather wide isosceles triangle completely covered with wide, iridescent, silvery, obliquely appressed scales that completely obscure the black ground color. This dense mat of scales ends abruptly between and opposite the first antennal segment. The first antennal segment bears some very short, fine white pile in the species *opalina* Hesse.

Thorax: The thorax is distinctly longer than wide, gently convex, feebly shining-black, with a few scattered broad, coppery scales in *opalina* Hesse, completely flat-appressed. The anterior collar is composed of dense, long, erect, shining white pile, and there is a lateral border of dense, matted, opaque, white pile on the sides of the metanotum and the anterior postalar callosity. This callosity is set apart by a deep medial crease. The scutellum is large but only gently convex, subtriangular in shape with a few slender, white bristly hairs along the margin and a deep basal crease. The mesopleuron has a dense mat of long, opaque, white pile dorsally extending backward over the pteropleuron. Only the ventral margin of the pteropleuron has a few scattered, white hairs. Posteriorly the sternopleuron has some green scales in *opalina* Hesse. It is set apart along the middle by an angular ridge, leaving the upper part depressed and concave. The hypopleuron is bare, and there is a small tuft of pile on the metapleuron above the spiracle. Above the posterior coxa is a dense mat of appressed silvery scales in *opalina* Hesse. Squama with a dense border of opaque, white pile.

Legs: The legs are comparatively short. The hind pair are, however, longer. The middle and posterior femora are stout without being conspicuously thickened or swollen. They bear broad, appressed, white scales and the hind femur especially has a few, fine, scattered, appressed, black setae. The hind tibiae are much longer than the hind femora and have several rows of oblique, numerous black setae. Claws slender; the basal tooth is quite short; pulvilli absent. The arolium is stublike. Anterior tibiae without spicules. The tarsi are quite short.

Wings: The wings are completely hyaline, strongly iridescent. The anterior veins are pale yellow, the posterior veins are light brown. The second vein arises abruptly but not rectangularly just beyond the anterior crossvein. The anterior crossvein is rectangular entering the discal cell a little before the middle. There are 3 submarginal cells. The first and third are equally wide at the base; the marginal cell is a little expanded and rounded posteriorly apically. Apically the discal cell is expanded anteriorly, slightly narrowing the first posterior cell, and the discal cell is apically pointed and drawn-out distally. The intercalary vein is parallel with the wing margin; the second and third posterior cells are equally long on the wing margin and the third posterior cell is angulate and plane basally and anteriorly but without spur veins. Anal cell widely open, axillary lobe wide; the alula twice as long as wide and the ambient vein complete. The costal comb is broad and conspicuous with microsetate hairs; there is a small spinelike spur at the extreme base.

Abdomen: Abdomen is elongate, as long as thorax and scutellum combined, rather convex, with 7 tergites visible from above. In *opalina* Hesse it is shiny black with numerous, small, metallic, black scales and large very wide, flat-appressed, white scales, which tend to form a conspicuous anterior border on the third tergite, extending also onto the base of the second and more widely and conspicuously on the last two tergites. Sternites with conspicuous anterior or posterior borders of white scales. Male terminalia exposed ventrally and twisted to the right. Hesse in describing the male terminalia of this group of species, which were included by Bowden in his genus *Micromitra* Bowden, notes that the male of *opalina* Hesse is very similar to that of *iridipennis* Hesse and *parvicellula* Bezzi, differing from both in having the apical angle of basal parts distinctly more angular and not rounded; basal strut neither broad nor very narrow; posterior processes of aedeagal complex distinctly much shorter; lateral apical angles of ventral aedeagal process more spinelike than in *parvicellula* Bezzi.

Material available for study: Through the generosity of Dr. A. J. Hesse I have received a male of *opalina* Hesse for dissection and study.

Immature stages: Unknown.

Ecology and behavior of adults: Not on record.

Distribution: Palaearctic: *Micomitra aerata* Hesse, 1956; *ancilla* Bezzi, 1924; *anthracina* Bowden, 1964; *bella* Austen, 1937; *brachycera* Bezzi, 1924; *chionea* Bezzi, 1924; *chrystallina* Bezzi, 1924; *erronea* Bezzi, 1924; *famula* Bezzi, 1923; *iridipennis* Hesse, 1956; *iris* Loew, 1869; *latissima* Bezzi, 1924; *opalina* Hesse, 1956; *parvicellula* (Bezzi), 1921 (as *Exoprosopa*); *perlucida* Hesse, 1956; *pharao* Paramonov, 19—; *serva* Bezzi, 1924; *stupida* Rossi, 1790 [=*chalcoides* Pallas in Wiedemann, 1818, =*rhadamanthus* Meigen, 1804, =*vitreicosta* Walker, 1849], *stupida pharao* Paramonov, 1928, female.

Colossoptera, new genus

Figures 201, 418, 714, 969, 970

Colossoptera, new genus. Type of genus: *Exoprosopa latipennis* Brunetti, 1909, by present designation.

An *Exoprosopa*-like fly with extraordinarily large, extensive, broad wing, of uniform sepia brown color; readily mistaken at first sight for *Cyananthrax* Painter, they have articulated antennal style and 3 submarginal cells, as in *Exoprosopa* Macquart. The wings dwarf the elongate, parallel-sided abdomen. The third antennal segment is tiny, but the style is long and slender. The first posterior cell is narrowly yet distinctly open. The third posterior cell on the wing margin is 3 times, or more, as wide as the second posterior cell on the margin; there are no spur veins. Face conical.

Found in northern India.

Length: 16 mm.; wing 19 mm.; abdomen 10 mm.; its width 4 mm.

Head, lateral aspect: Head hemiglobular; the occiput is moderately prominent, wider above than below, and strongly sloping inward. The pile is flat-appressed and consists of narrow scales; the central fringe is short. The eye is reniform, the indentation broadly rounded, extensive above and below, and the eye is bisected at the middle posteriorly. The front is a little swollen, arising above the eye margin, and not any more extensive at the antenna than it is in the middle. The pile of the front is bristly, erect, and black, and the front is opaque black with dark brown pollen. There are a few pale, scattered scales on the lower front and few on the face. The face is triangularly produced forward, but not acutely peaked. It bears considerable erect bristly pile, concentrated medially at the apex and forming a fringe down the sides of the oral recess. The proboscis and labellum are slender, only a little longer than the oral recess, the palpus slender, nearly half as long as the proboscis with a few long, stiff hairs below. The antennae are attached just above the middle of the head and are rather small. First segment is cylindrical, $2\frac{1}{2}$ times as long as wide. Second segment quite beadlike, as wide as long with rounded sides, narrowly attached to the third segment. The first segment bears some long bristly setae or short bristles on all sides, except below. The second segment has minute setae. The third segment is small, forming an elongate, attenuate cone, nearly as long as the first segment, nearly plane on all sides and with a segmented style a trifle longer than the segment.

Head, anterior aspect: Circular, wider than the thorax. The ocellar tubercle is minute, abrupt, slightly longer than wide, and with the space between the eyes at least $4\frac{1}{2}$ times as great as the width of the ocellar tubercle. The antennae are widely separated by nearly the length of the first three segments and separated from the eye margin by little more than half the length of the first segment. The oral recess is long and slender with thickened sloping walls on the anterior half; it is shallow posteriorly and not enclosing the proboscis or palpus. Genal space very narrow. The tentorial crease is long and very narrow. Gena a little more extensive beneath the eye perhaps. The occipital lobes are tightly apposed.

Thorax: Subquadrate and small, opaque black with brown pollen. The pile of the mesonotum consists of very fine, erect, blackish hairs becoming longer in front of the scutellum, with a narrow, anterior collar of deep yellow pile continued over the whole of the humerus and with tufts of pile continued laterally along the sides of the mesonotum, yellow and matted and long, with a conspicuous tuft over the wing and on the postalar callosity. Scutellum clothed like the mesonotum. Pleuron black with brown pollen, the pile black, except dorsally. The upper mesopleural pile and the dense pile of the metapleuron is long and yellow. The pteropleuron also bears some blackish pile. The sternopleural and propleural pile is all brownish black. Hair tuft beneath the halteres apparently absent. Squamal scales slender. There are 3 quite stout, long, black bristles on the notopleuron, 4 or 5 on the postalar callosity; the margin of the scutellum has only fine long hairs, the plumula with yellow hairs, the humerus with a rather sharp lateral spur.

Legs: Slender and long. Anterior tibia nearly smooth. Middle femur with several stout bristles anteriorly along the middle. Hind femur with a double row of long, slender, distinct bristles ventrally, 7 bristles ventrolaterally. There are 3 or 4 other long, fine hairs, also 6 ventromedial. Also there are several additional long hairs distally, and 5 bristles anterodorsally on the outer half. All the femora and tibiae with numerous, flat-appressed, shining, brownish-black scales. The long hind tibiae have slender bristles placed in 4 rows; there are 9 ventrolateral, 18 dorsolateral, and 10 or more dorsomedial, besides 10 ventromedial. All tarsi spiculate below. Anterior claws slightly reduced. Other claws long, sharp, slender, slightly bent at the apex, and with a strong, long, basal tooth.

Wings: Very large and broad, giving this fly the appearance of the genus *Cyananthrax* Painter, for which I first mistook it. The wing is uniformly dark sepia brown, unrelieved by spots or windows and without a thyridial fleck. The venation is a little peculiar. There are 3 submarginal cells. The first submarginal cell is strongly drawn-out toward the base of the wing, and the anterior branch of the second vein is strongly sigmoid, and the posterior branch is also sigmoid. The first posterior cell is narrowed apically. The second posterior cell is much narrowed, leaving the third posterior cell very wide on the margin. Medial crossvein with a double bend, one at the base and one at the apex. The very long, upper anterior intercalary vein is also bent at each end. Contact of the discal cell with the fourth posterior cell short, and shorter than the anterior crossvein; this crossvein enters the discal cell before the middle and the second vein arises opposite the anterior crossvein. Alula narrow. Ambient vein complete. Costal comb narrow, the hook long and sharp.

Abdomen: Abdomen elongate; it is nearly 3 times as long as the mesonotum if the scutellum is excepted. Color black, more or less shining, with faint pollen, the sides of the second and third tergites a little yellowish. The abdomen bears numerous, flat-appressed, black scales with some yellow scales laterally, and it also bears conspicuous, coarse, matted, straw-yellow, long, backwardly appressed pile on the sides of the abdomen, except the last tergite which has only black. There are 7 tergites visible. The last one has a fringe of bristly black hair laterally and posteriorly. Male terminalia large, the dististyli extruded posteriorly.

The male terminalia from dorsal aspect, epandrium removed, show a short, oval structure with quite wide, tonguelike or shovellike epiphallus. The base of the aedeagus is pear-shaped, rather wide basally. From the lateral aspect the basistylus is stout, especially basally; it is more than twice as long as wide; it is bluntly rounded apically, with the dististyli rising quite subapically. The epiphallus is prominent; the epandrium is sharply triangular, more or less rectangular dorsobasally, but hooked back at that angle, and with a very prominent extensive, attenuate lobe turned downward basally.

Material available for study: The unique type.
Immature stages: Unknown.
Ecology and behavior of adults: Unknown.
Distribution: Oriental: *Colossoptera latipennis* (Brunetti), 1909 (as *Exoprosopa*).

Genus *Ligyra* Newman

FIGURES 192, 202, 425, 704, 731, 962, 963

Ligyra Newman, Entomologist, vol. 1, no. 14, p. 220, 1841. Type of genus: *Anthrax bombyliformis* MacLeay, 1827, by monotypy.
Hyperalonia Rondani, Arch. Zool. Anat. Fisiol., vol. 3, p. 58, 1863. Type of genus: *Stomoxys morio* Fabricius, 1775, as *Anthrax erythrocephala* Fabricius, 1805. Designated by Coquillett, p. 554, 1910, the 5th of 13 species. In this work regarded as a subgenus.
Velocia Coquillett, Canadian Ent., vol. 18, p. 158, 1886b. Type of genus: *Anthrax cerberus* Fabricius, 1794, by monotypy. Preoccupied by Robineau-Desvoidy, 1863.
Ligira, of Authors.

Bezzi (1924), pp. 361-363, key to 19 Ethiopian species (as *Hyperalonia*).
Roberts (1928a), p. 96, key to 9 Australian species (as *Hyperalonia*).
Paramonov (1929a), pp. 188-189, comments on the *venus* Karsch group (as *Hyperalonia*).
Engel (1936), pp. 455-456, key to 6 Palaearctic species (as *Hyperalonia*).
Paramonov (1953a), pp. 220-222, comments on the *venus* group (as *Ligyra*).
Hesse (1956), pp. 914-917, key to 10 Ethiopian species (as *Ligyra*).
Bowden (1964), p. 151, key to 5 West African species (as *Ligyra*).
Paramonov (1967), pp. 124-127, key to 22 known species from Australia (as *Ligyra*).
Painter (1968), pp. 107-121, review of *Hyperalonia* Rondani, segregates as a separate South American genus.

Large robust flies which cannot be effectively distinguished from *Exoprosopa* Macquart except in the constant presence of 4 or more submarginal cells and the short, blunt tooth at the base of the claw. The wing is usually long; sometimes extraordinarily long and slender as in the *venus* Karsch group of species from South Africa. In other species it is broad as in *kaupii* Jaennicke.

The wings show a striking pattern of sepia brown, sometimes a very dark color, and in some species the base of the wing, the costa, and subcostal cells are yellowish. There are many types of wing patterns such as that of *chilensis* Rondani, *proserpina* Wiedemann, *bombyliformis* MacLeay, *gazophylax* Loew, *tantalus* Fabricius, *venus* Karsch, and *atricosta* Bezzi; each is different. Species groups can be more effectively grouped around the type of venation, which varies in several particulars; but more especially the discal cell; this cell in one group is long or short, strongly constricted below in the middle, in another group it is lengthened and pointed at the apex, in a third group, wide and obtuse at apex with the closing vein rather long. The anterior crossvein usually enters the discal cell at the middle, but in *venus* Karsch and a few others it enters before the middle. The shape of the anal vein offers useful specific characters. Unfortunately, even when taken with other characters such as the relative width and length of the wing and the abdominal and leg vestiture, there is so much overlapping of species that species groups still remain ill defined.

In Chile they extend as far south as Araucania, but are common at the latitude of Conceptión. These species were relegated back to *Hyperalonia* Rondani by Painter (1968).

I regard the basal crossvein dividing the submarginal area as being the anterior branch of the third vein, meeting the posterior branch of the second vein in partial confluence and then departing for the edge of the wing.

In species of *Exoprosopa* Macquart the posteriorly directed branch closing the upper half of the third submarginal (basal) cell has often been regarded as a crossvein (sectorial vein); I believe a more accurate interpretation is to regard it as the posterior branch of the second vein meeting the anterior branch of the third vein; in support of this I point out that it is strongly oblique in many species. The additional transverse veins found in *Ligyra* Newman, always one, sometimes two, may be regarded as remnants of extra veins similar to those found in the Nemestrinidae. In any case since *Exoprosopa* Macquart shows more simplified venation, *Ligyra* Newman must be regarded as more generalized and ancient in type; the former genus more successful, more recent, with far wider distribution.

Length: Body 10 to 20 mm.; width of body 5 to 8 mm.; length of wing 10 to 24 mm.; width of wing 6 to 7 mm.; wing spread 24 to 56 mm.; in *L. venus* Karsch the wing is 6 mm. wide and 25 mm. long.

The genus *Ligyra* Newman differs so little from *Exoprosopa* Macquart that a detailed and lengthy description would merely duplicate the description of the latter genus in nearly every particular. I have therefore chosen to comment on the genus in only general terms and include remarks on distribution and species groups.

Head, lateral aspect: Similar to *Exoprosopa* Macquart. The proboscis is contained within the oral recess or projects only a short distance beyond it except in *albiventris* Macquart where it is twice as long. There appear to be numerous specific differences in the ornamentation of the proboscis and labellum in the species before me. The face is bluntly conical in all species with which I am acquainted except *L. chilensis* Rondani and *albiventris* Macquart where it is bluntly rounded but protuberant. Antenna similar to *Exoprosopa* Macquart.

Head, anterior aspect: Oragenal recess narrow, inflated in the middle, the inflated part usually rounded, sometimes ridged.

Thorax: Similar to *Exoprosopa* Macquart. Disc of mesonotum rarely with scales. Bristles of notopleuron strong, variable in number; stout and long on the postalar callosity.

Legs: Like *Exoprosopa* Macquart. Anterior tibia usually smooth, more rarely spiculate. Hind legs at least with scales except in *chilensis* Rondani. Tooth at base of claw very short, scarcely discernible in some species; always blunt.

Wings: Similar to *Exoprosopa* Macquart in size, shape and frequency of strong patterns of brown or black. Constantly different from *Exoprosopa* Macquart in the presence of 4 or sometimes 5 submarginal cells. Costa strongly expanded at base.

Abdomen: Similar to *Exoprosopa* Macquart. Scales present on the abdomen in most species. Female terminalia with spines on each side near apex. Hesse (1956) notes that the genitalia do not differ much from those of *Exoprosopa* Macquart in the structure and shape of the dististyli and the aedeagal process.

Bezzi (1924) and Hesse (1956) recognized the following groups of South African species:

1. The *Ligyra nigripennis* Loew group

Wings relatively broad. Venation rather similar to *Litorrhynchus* Macquart or *Exoprosopa* Macquart, subgenus *Defilippia* Lioy. Discal cell of characteristic shape, the constriction lying below the anterior crossvein and the upper part of the distal extremity, elongate, rounded, pointed at apex; wings uniformly infuscated, in color yellow brown, brown or even purplish, cell middles paler. Third antennal segment relatively short and more conical, the style longer than half the length of the segment. Abdomen and legs with yellow or golden scalose pile.

Species included: *vittata* Ricardo, *niveifrons* Bezzi, *evanida* Bezzi, *helena* Loew, *paris* Bezzi, *coleoptrata* Bezzi, *nigripennis* Loew, and according to Bezzi the Oriental species *tantalus* Fabricius, *leucanoe* Jaennicke, *oenomaus* Rondani, and *chrysolampis* Jaennicke.

2. The *Ligyra sisyphus* Fabricious group

Wings narrower than in the preceding group and the general coloration of the legs and body more reddish. Wings varying from hyaline to others with a much reduced brown or yellow pattern narrowly restricted to the foreborder or occasionally with broad crossbands. Venation of these species quite different, the discal cell regular, not strongly constricted in the middle, the apex wide and nearly or quite plain, not pointed and elongated; vein between second and third posterior cells nearly straight.

Species included: *Ligyra sisyphus* Fabricius, *atricosta* Bezzi, *alula* Bezzi, *virgo* Bezzi, *thyridophora* Bezzi, *doryca* Boisduval, *transiens* Bezzi.

3. The *Ligyra venus* Karsch group

Bezzi speaks of these as the most beautiful of all African Diptera; those I have seen are certainly remarkable. Wings very long and narrow with characteristic and extensive pattern of dark brown bands extending backward from the dark foreborder. Discal cell wide, gradually and particularly widened apically, its end vein oblique and a little shortened; this cell is only slightly constricted. The front tibia is spiculate and the hind legs are feathered with rather long suberect scales. Abdomen with conspicuous patches of golden scales, generally two basal, one at apex, arranged as a triangle. Head sometimes red, sometimes with collar, and notopleural pile tufts bright orange or red.

Species included: *Ligyra venus* Karsch, *cupido* Bezzi, *mars* Bezzi, *mars-vulcanus* Bezzi, *enderleini* Paramonov. This group is more tropical in its distribution and is absent from South Africa.

While I am familiar with 9 of the 14 New World species, I am not able to fit them well into the Old World species groups. The three species *hela* Erichson, *albiventris* Macquart, and *gazophylax* Loew all have such similar wing venation, wing spot pattern, and abdominal scale coloration and distribution pattern that they would seem to have a very close relationship; yet *albiventris* Macquart has a very blunt, rounded face and proboscis twice as long as the oral cavity, in great contrast to the other two species; this seems to indicate a more important and stable relationship for the wing characteristics, pattern included, than for type of face and length of proboscis.

Since Williston (1901, p. 273), and Painter (1930) point out that the Chilean group of species, *morio* Fabricius, *chilensis* Rondani, *erythrocephala* Fabricius, differs from typical *Ligyra* in lacking a segmented style, I here resurrect the name *Hyperalonia* Rondani as a subgenus to include only these three species; they are further notable for the absence of true scales and for the greatly swollen apical half of the discal cell. Painter (1930) says the metapleuron has only 1 or 2 hairs; this is not true of my material, which does not differ in this respect. This new use of the name *Hyperalonia* is apart from the erroneous use by many early authors who were unaware of the priority of *Ligyra*. I find 5 submarginal cells constantly present in *Ligyra kaupei* Jaennicke and in my material of *Ligyra tenebrosa* Paramonov.

Abdominal patterns may be characteristic; of species before me 8 have a band of pale scales at the base of the second and rarely of the third tergite; 6 species have either apical tergites wholly pale scalose or the last 2 or 3 tergites each with an isolated spot of pale scales on each side; *Ligyra venus* Karsch and its allies are noted for patches of golden scales at base and apex of abdomen.

The male terminalia from dorsal aspect, epandrium removed, show a figure moderately wide basally and narrowed apically. The epiphallus widens from the base to the apex where it is quite wide, barely curved at the blunt apex. Base of aedeagus bulblike and almost circular. From the lateral aspect the apically incised, hooded dististyle is small, lobelike and subapical to only a moderate extent. Epandrium distinctly triangular, but with the anterior basal margin broadly but only shallowly incised.

Prior to 1934 *Exoprosopa*-like flies with 4 submarginal cells were placed under the name *Hyperalonia* Rondani (1864). Edwards (1934) called attention to the older name *Ligyra* Newman (1841) and also to the fact that Newman in naming his genus *Ligyra* selected *Anthrax bombyliformis* MacLeay as the type-species. Hardy (1941, p. 233; 1942, p. 203) has indicated a belief that *bombyliformis* MacLeay is the same as *Bibio sylvanus* Fabricius (1787), but as Hardy can hardly have seen the Fabricius type, I feel that the statement wants confirmation. This synonymy is here adopted with reservation. *Ligyra* Newman is a genus of much fewer species than *Exoprosopa* Macquart; nevertheless, they are widespread. Two species are found in southwestern United States: *gazophylax* Loew from southern Arizona and California, which I have taken as far north as San Lucas; and *pilatei* Macquart, which occurs in Peru and extends into southern Arizona. From the Palaearctic region, species are known only from China, Japan, Formosa, India, Arabia, Egypt, Erythrea, Ussuri, and one species from Greece and the Island of Rhodes. About an equal number of species are known from the tropical part of the New World and the tropical part of Ethiopia, and a larger number than from any other area.

Material available for study: Thirteen species in my own collection together with the material in the British Museum (Natural History) and the National Museum of Natural History.

Immature stages: Hyperparasites; as far as known parasites of scoliids, such as *Tiphia*, which in turn are parasites of subterranean scarabaeid larvae. See life history section for further details.

Ecology and behavior of adults: Bowden (1964) furnished some interesting observations on several species he found hovering about small clearings in scrub or hawking along the sides of hedges, hovering about buildings; he found *L. paris* Bezzi hovering about the ends of projecting branches of trees and that the species showed a strong territorial attitude, driving intruders away from its chosen spot, or returning if chased away. He found *sisyphus* Fabricius to be a ground-frequenting species. The species I have taken in California, Mexico, and Australia have all been found in stream beds resting warily upon rocks or sand.

Painter (1930) collected 2 species, *albiventris* Macquart and *latreillii* Wiedemann, flying about sandy ocean beaches in Honduras.

Distribution: Nearctic: *Ligyra gazophylax* (Loew), 1869 (as *Exoprosopa*); *pilatei* (Macquart), 1846 (as *Exoprosopa*).

Neotropical: *Ligyra albiventris* (Macquart), 1848 (as *Exoprosopa*) [=*hela* (Erichson), 1848 (as *Anthrax*)]; *cerberus* (Fabricius), 1794 (as *Anthrax*); *chilensis* Rondani, 1863; *dido* Osten Sacken, 1886; *flavosparsa* Bigot, 1892; *gargantua* Knab, 1915; *guerini* (Macquart), 1846 (as *Exoprosopa*); *kaupii* (Jaennicke), 1867 (as *Exoprosopa*); *koslowskyi* Edwards, 1930; *latreillii* (Wiedemann), 1830 (as *Anthrax*); *morio* (Fabricius), 1775 (as *Stomoxys*) [=*coeruleiventris* (Macquart), 1846 (as *Exoprosopa*), =*erythrocephala* (Fabricius), 1805 (as *Anthrax*)]; *proserpina* (Wiedeman), 1828 (as *Anthrax*) [=*cerberus* (Macquart), 1840 (as *Exoprosopa*), =*klugii* (Wiedemann), 1830 (as *Anthrax*), =*rufescens* (Walker), 1849, (as *Anthrax*)]; *servillei* (Macquart), 1840 (as *Exoprosopa*, no descr.); *surinamensis* Rondani, 1863.

Palaearctic: *Ligyra ferrea* (Walker), 1849 (as *Anthrax*) [=*versicolor* Loew in litt.]; *flavofasciata* (Macquart), 1855 (as *Exoprosopa*); *formosana* Paramonov, 1931; *gebleri* (Loew), 1854 (as *Exoprosopa*); *helena* (Loew), 1854 (as *Exoprosopa*) [=*gloriosa* (Walker), 1871 (as *Exoprosopa*)]; *monacha* (Klug), 1832 (as *Anthrax*); *orientalis* Paramonov, 1931; *shirakii* Paramonov, 1931; *similis* Coquillett, 1898; *tantalus* (Fabricius), 1794 (as *Anthrax*) [=*coeruleopennis* (Doleschall), 1858 (as *Anthrax*)]; *ussuriensis* Paramonov, 1936.

Ethiopian: *Ligyra alula* (Bezzi), 1906 (as *Exoprosopa*); *atricosta* Bezzi, 1924; *coleoptrata* Bezzi, 1921; *cupido* Bezzi, 1924; *mars* Bezzi, 1924, *mars-vulcanus* Bezzi, 1924; *melanoptera* Bowden, 1964; *minerva* Paramonov, 1953; *nigripennis* (Loew), 1852 (as *Exoprosopa*); *niveifrons* Bezzi, 1914; *paris* Bezzi, 1924; *sisyphus* (Fabricius), 1805 (as *Anthrax*); *thyridophora* Bezzi, 1912; *transiens* Bezzi, 1924; *venus* Karsch, 1887 [=*enderleini* Paramonov, 1953]; *virgo* Bezzi, 1924; *vittata* Ricardo, 1901.

Oriental: *Ligyra albicincta* (Macquart), 1931 (as *Anthrax*); *argyura* de Meijere, 1924; *celebesi* Paramonov, 1931; *chrysolampis* (Jaennicke), 1867 (as *Exoprosopa*); *curvata* de Meijere, 1911; *doryca* (Boisduval), 1835 (as *Anthrax*) [=*audouinii* (Macquart), 1840 (as *Exoprosopa*), =*leuconoe* (Jaennicke), 1867 (as *Exoprosopa*), =*pelops* (Walker), 1859 (as *Anthrax*), =*ventrimacula* (Doleschall), 1857 (as *Anthrax*)]; *enderleini* Paramonov, 1929; *evanida* Bezzi, 1920; *flavotomentosa* de Meijere, 1929; *macassarensis* Paramonov, 1931; *oenomaus* Rondani, 1875, *oenomaus flora* Frey, 1934; *paludosa* de Meijere, 1911; *semifuscata* Brunetti, 1912; *sphinx* (Fabricius), 1787 (as *Bibio*) [=*imbuta* (Walker), 1849 (as *Anthrax*)]; *suffusipennis* Brunetti, 1909; *sumatrensis* de Meijere, 1911; *tantalus* (Fabricius), 1794 (as *Anthrax*) [=*coeruleiventris* (Doleschall), 1858 (as *Anthrax*)].

Australian: *Ligyra bombyliformis* (Macleay), 1827 (as *Anthrax*); [=*albiventris* (Thomson), 1869 (as *Exoprosopa*), =*punctipennis* (Macquart), 1849 (as *Exoprosopa*)]; *burnsi* Paramonov, 1967; *calabyana* Paramonov, 1967; *campbelli* Paramonov, 1967; *chinnicki* Paramonov, 1967; *cingulata* (Wulp), 1885 (as *Exoprosopa*); *commoni* Paramonov, 1967; *contrasta* Paramonov, 1967; *dentata* (Roberts), 1928 (as *Hyperalonia*); *doryca* (Boisduval), 1835 (as *Anthrax*); *eyreana* Paramonov, 1967; *fasciata* Paramonov, 1967; *hemifusca* (Roberts), 1928 (as *Hyperalonia*); *inquinita* (Roberts), 1928 (as *Hyperalonia*); *macraspis* (Thomson), 1868 (as *Exoprosopa*) [=*argenticincta* (Bigot), 1892 (as *Hyperalonia*), =*sinuatifascia* (Hardy not Macquart), 1924 (as *Hyperalonia*)]; *orest* Paramonov, 1967; *pilad* Paramonov, 1967; *robertsi* Paramonov, 1967; *satyrus* (Fabricius), 1775 (as *Bibio*) [=*funesta* (Walker), 1849 (as *Exoprosopa*), =*insignis* (Macquart), 1855 (as *Exoprosopa*)]; *septentrionis* (Roberts), 1928 (as *Hyperalonia*); *sinuatifascia* (Macquart), 1855 (as *Exoprosopa*); *tenebrosa* Paramonov, 1967.

Subfamily Anthracinae Latreille, 1804
(As Anthracii)

All of these flies have a number of interesting features in common. Many species are quite large. Many, and in fact the majority, are a dull somber black relieved sometimes by spots of silver pile on the abdominal tergites or on the apex of abdomen or with a wide band of silver pile. Wings with 2 or 3 submarginal cells. The second vein arises abruptly, at a right angle, at a point located at or very near the anterior crossvein; a backward spur vein may be present. First segment of antenna always quite short; second segment likewise short. The third segment is usually strongly bulblike or onion-shaped at base with an attenuate style; however, the third segment may be conical. This style bears a rather short apical microsegment of the same thickness as the stylate part of the third segment. At the apex this microsegment bears a characteristic circlet or brush of hairs except in *Walkeromyia* Paramonov. Male genitalia large and placed symmetrically.

As pointed out by Bowden (1964), Bezzi was mistaken in thinking this group lacked spines on the ovipositor. The ovipositor does have a circlet of spines which are recessed because of the invagination of the acanthophorites. The face is always rounded, the proboscis quite short; pulvilli well developed; besides these,

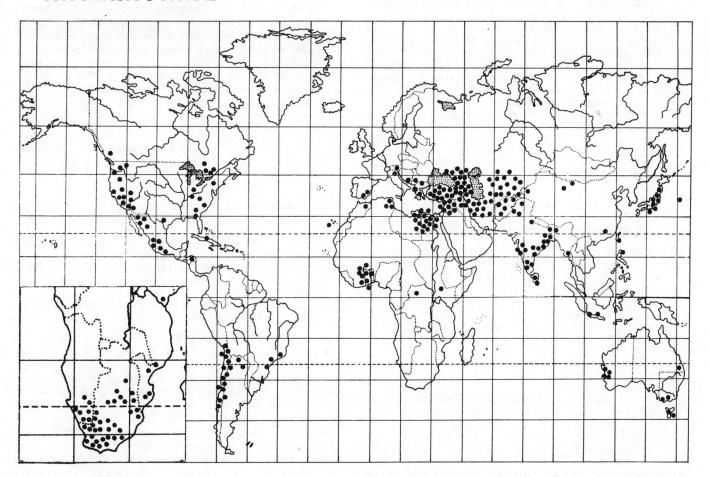

TEXT-FIGURE 47.—Pattern of the approximate world distribution of the species of the subfamily Anthracinae.

several other characters separate this group from the Exoprosopinae. The metapleuron is constantly bare and without pile on front of the haltere and the squama has a fringe of hairs instead of scales. The male genitalia are large and placed symmetrically. It appears to be connected to the Lomatiinae through the tribe Xeramoebini.

The question arises as to what genera may be properly included in this subfamily and what should be left in the Lomatiinae. It is the opinion of the author that *Chionanoeba* Sack, *Petrorossia* Becker, and *Xeramoeba* Hesse should remain in Lomatiinae; perhaps also *Prothaplocnemis* Bezzi; in these genera the second vein does indeed arise rather abruptly and at a varying distance from the fork of the third vein; the face is short, the proboscis short, but so it is in *Aphoebantus* Loew. These features may be parallel developments rather than annectent relationships. And so I restrict the Anthracinae to *Anthrax* Scopoli (together with certain subgenera), *Dicranoclista* Bezzi, and the curious and aberrant *Walkeromyia* Paramonov.

The distribution is interesting. *Anthrax* Scopoli is of worldwide distribution; *Walkeromyia* Paramonov is found in Trinidad and also the Chilean subregion and is practically a phylogeront or relict type; *Dicranoclista* Bezzi is found in central Africa and the southwestern United States.

Based on the North American species I have dissected, the genus *Anthrax* Scopoli shows a structure in which the epandrium is extraordinarily long and rather slender with a downturned basal lobe itself long and prominent and suggestive of a condition often seen in both the Exoprosopinae and the Lomatiinae. There may be a short, blunt lobe in the Bombyliinae but never a long slender lobe; the basal stylus also is peculiar, it is quite wide at the base, long, slender, and narrowed apically. The epiphallus extends well beyond the dististyli, and this structure is narrowed in the middle.

KEY TO THE GENERA OF THE ANTHRACINAE*

1. Wings with 2 submarginal cells, the second having a basal spur. Third antennal segment long, conical, gradually tapered from the base, the separated microsegment has reduced apical hairs and with apical spine, or the hairs even sometimes absent or denuded. Face below antenna more than usually extensive due to the shortened, reduced oral recess and shortened proboscis. Face very blunt and bullose. Very large flies; in the male the long hind legs bear an extraordinary fringe of quite long scales extending from the outer third of the hind femur to the second tarsal segment . . . . **Walkeromyia** Paramonov
   Apical fringe or circlet of hairs at the end of the third antennal segment, microsegment well developed. Male hind femur without curious fringe of scales 2
2. Third antennal segment very loosely attached to the second segment. Wings with 2 submarginal cells. Anterior branch of third vein with or without a spur vein.
   Anthrax Scopoli
   Third antennal segment very tightly apposed and joined to the second segment. Wings with 3 submarginal cells; rarely with 2 3
3. Three submarginal cells formed below by a supernumerary crossvein as in *Ligyra* Newman, the base of anterior branch of third vein with a long basal stump vein, and origin of second vein likewise with a stump vein. First posterior cell closed and bearing a long stalk. No scales or bristles on abdomen . . . . **Dicranoclista** Bezzi
   Three submarginal cells formed above by extension of the stump vein at base of anterior branch of third vein. No supernumerary crossvein below. First posterior cell open. (*Argyramoeba* of authors). The subgenus *Spogostylum* in the Macquart sense.

Anthrax (Spogostylum) Macquart

* See also *Coniomastix* Enderlein (Pamirs, Central Asia).

Genus *Anthrax* Scopoli

FIGURES 159, 160A, 163, 337, 409, 712, 734, 966, 967, 968

Anthrax Scopoli, Entomologia Carniolica, p. 358, 1763. Type of genus; *Musca anthrax* Schrank, 1781 (misident. as *Musca morio* Linnaeus, 1758). See Aldrich, 1926, p. 12.
Chalcamoeba Sack, Abhand. Senckenberg. Naturf. Ges., vol. 30, p. 522, 1909. Type of genus: *Anthrax virgo* Egger, 1859, by original designation.
Chrysamoeba Sack, Abhand. Senckenberg. Naturf. Ges., vol. 30, p. 522, 1909. Type of genus: *Chrysamoeba vulpina* Sack, 1909, by original designation.
Leucamoeba Sack, Abhand. Senckenberg. Naturf. Ges., vol. 30, p. 520, 1909. Type of genus: *Bibio aethiops* Fabricius, 1781, by original designation.
Molybdamoeba Sack, Abhand. Senckenberg. Naturf. Ges., vol. 30, p. 519, 1909. Type of genus: *Anthrax tripunctata* Wiedemann in Meigen, 1820, by original designation.
Psamatamoeba Sack, Abhand. Senckenberg. Naturf. Ges., vol. 30, p. 536, 1909. Type of genus: *Anthrax isis* Meigen, 1820, by original designation.
Satyramoeba Sack, Abhand. Senckenberg. Naturf. Ges., vol. 30, p. 517, 1909. Type of genus: *Anthrax etrusca* Fabricius, 1794, by original designation.
Spogostylum Macquart, Diptères exotiques, vol. 2, pt. 1, p. 53, 1840. Type of subgenus: *Spogostylum mystaceum* Macquart, 1840, by monotypy.
Spongostylum Williston [not Macquart], Manual of North American Dipt., edition II, p. 65, 1896 (emendation).
Argyromoeba Schiner, Wien. Ent. Monatschr., vol. 4, p. 51, 1860. Type of genus: *Musca anthrax* Schrank, designated by Sack in 1909.
Argyramoeba E .and L. Coucke, Ann. Soc. Ent. Belgique, vol. 38, p. 286, 1894 (emendation).

Coquillett (1894), p. 95, key to 18 species (as *Argyramoeba*).
White in Hardy (1924), pp. 76-77, key to 3 species (as *Argyramoeba*).
Bezzi (1924), pp. 160-161, key to 12 Ethiopian species.
Bezzi (1924), pp. 168-169, key to 11 Ethiopian species (as *Spongostylum*).
Séguy (1926), pp. 205-206, key to 9 species from France.
Engel (1936), pp. 422-424, key to species.
Engel (1936), pp. 439-440, key to species.
Austen (1937), pp. 109-110, key to 3 species from Palestine (as *Anthrax*).
Austen (1937), pp. 114-115, key to 8 species from Palestine (as *Spongostylum*).

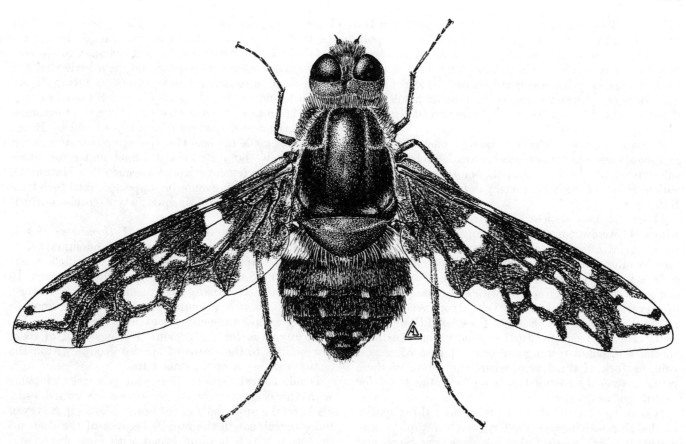

Text-Figure 48.—Habitus, *Anthrax tigrinus* DeGeer.

Hesse (1956), pp. 366-384, key to 42 male and 44 female species (as *Anthrax*).
Hesse (1956), pp. 384-387, key to species.
Paramonov (1957), pp. 126-148, key to 19 species.
Bowden (1964), p. 81, key to 6 species from Ghana (as *Anthrax*).
Bowden (1964), pp. 87-88, key to 5 species from Ghana (as *Argyramoeba*).
Osten Sacken (1886), pp. 100, key to 8 species (as *Argyramoeba*).
Marston (1970), key to all New World species except *albofasciatus* group.

Dull black flies, which are sometimes feebly shining. These are usually medium sized flies. Many species are rather large; a few are small. Quite commonly the species show a striking wing pattern of sharply delimited spots of varying number. Still others have only the anterior margin more or less black, and in a few species the wing is almost entirely hyaline or tinged with pale yellow. All species are characterized by a divided style, which bears upon the apex of the terminal section a characteristic circlet or brush of hairs arranged about an often blunt or truncate tip. All species likewise bear the following group characteristics: metapleuron in front of halteres bare, face short and rounded and proboscis always short, squamae with a fringe of hairs but no scales, etc.

Probably no genus in the family Bombyliidae, with the possible exception of *Exoprosopa* Macquart, has been involved in as much confusion. Many diverse views and opinions have been given with regard to its limitations and what should be included within it. Nevertheless there is a small base of sensible conclusions upon which to build.

This genus is worldwide in distribution, with an immense number of species for the Palaearctic and Ethiopian regions.

Length: 4 to 20 mm.

Anthrax Scopoli as the oldest name cannot be abandoned. Its type is *anthrax* Schrank, misidentified as *Musca morio* Linnaeus. It is curious that Sack (1909) overlooked *Anthrax* Scopoli in his excellent paper. *Anthrax anthrax* Schrank has 2 submarginal cells; the anterior branch of the third vein arises rectangularly, bends rectangularly, and has a well-developed backward spur which is never complete; the second vein also arises rectangularly, bends rectangularly, and has a backward spur. All species that have the above combination of characters unquestionably belong to *Anthrax* Scopoli. Beyond this point there are certain aspects of variability in the characters found within the group.

Argyromoeba Schiner has as its type, *Musca anthrax* Schrank, clearly designated by Sack in 1909, a year before Coquillett selected *Anthrax tripunctata* Wiedemann. It is then an exact synonym of *Anthrax* Scopoli and further use is invalid. Also the wing of *Anthrax tripunctata* Wiedemann (Engel, 1936, fig. 113) is identical in venation with *Anthrax anthrax* Schrank; since Sack designated *tripunctata* as the type of his *Molybdamoeba*, this likewise becomes a synonym of *Anthrax* Scopoli, sensu stricto.

Beyond the above characterization one may quite reasonably ask what characters are available for further subdivision of *Anthrax* Scopoli, sensu lato, and into which group of such characters does the type-species fall.

There are two choices: First, the venation of the wings. If we are correct in assuming that the trend of wing specialization is in the main toward simplification, then we must consider the reduction from 3 submarginal cells to 2 such cells to be in the direction of simplification; and second, those species with constant spur (base of posterior branch of second vein atrophied) may be reckoned intermediate. Each of these conditions might well form the basis for further subdivision. In those species without a rectangular bend at base of second vein, or fork of third vein, where the origin of these veins are evenly rounded out, we have the basis for useful species groups.

It may be noted that there are at least three species in the Eurasian area with 3 regularly complete submarginal cells: *ocyale* Fabricius, *flavescens* Sack, and the latter is the type of *Anthracamoeba* Sack, a name that might well be used for the species with 3 submarginal cells provided it can be demonstrated to be different from *Spogostylum* Macquart, a South American species which Bezzi (1924) notes has the submarginal cell divided in quite a different way from that of the three above-named species. The type-species of *Spogostylum* Macquart (*Spongostylum* of authors) is *mystaceum* Macquart and not *tripunctatum* Wiedemann as given by Engel (1936). It is not possible to completely change the definition of *Spogostylum* Macquart as Engel has done. Hence I reserve a decision as to the disposition of *Spogostylum* Macquart until sufficient material is available to determine (1) the manner of subdivision of the submarginal cell, (2) the regularity of the presence of 3 submarginal cells, and (3) the type of structure of the third antennal segment. Provisionally, either *Spogostylum* Macquart or *Anthracamoeba* Sack is available for species with 3 submarginal cells; but not *Argyromoeba* Schiner because of this synonymy given above, as it is an exact synonym of *Anthrax* Scopoli. For this reason I have had to reject the use Hesse (1956) has made of this name. I prefer to leave *Anthrax* an unsplit group with many species groups, pending a worldwide study of *Anthrax* Scopoli, in the wide sense.

Apart from the venation of *Anthrax*, sensu lato, a rather creditable attempt has been made to split up this large assemblage on the basis of the form of the second and third antennal segments and their manner of attachment. Unfortunately it is applied with some difficulty. These two types consist, first, of those species in which the third antennal segment, at its rounded base, projects a little ways outward beyond the apex of the second segment which correspondingly is a little rounded on its outer margins, the two segments not fitting closely together, and second, the group in which these two segments fit closely into one another; this second arrangement is the cup-shaped attachment described by Bezzi, and it was made the basis for the separation of a group of species by both Bezzi and Engel under the name *Spongostylum* (emendation of *Spogostylum* Macquart). This interpretation cannot be supported until such time as adequate material of Macquart's type-species is available.

There have been efforts to split *Anthrax* species still further upon the basis of wing pattern, abdominal coloration, and shape of the stylate part of the third antennal segment (*Satyramoeba* Sack); it does not seem to me profitable to divide each possible section further, inasmuch as such a course would be tantamount to creating a separate genus for each species group.

In any case, for the present, I relegate all of these distinctions to the status of species groups within the genus *Anthrax* Scopoli, senus latus.

Head, lateral aspect: The head is nearly globular with the occiput strongly developed backward especially on the upper half of the head. The occiput is even more prominent in the middle because of the deep indentation which is quite broad with both upper and lower angles nearly equal. There is a distinct bisecting line on the posterior eye facets. The pile of the occiput is scanty and consists of fine appressed hairs or scales or both; the circular fringe about the interior cup is short and dense. Some species have special tufts of scales along the eye margin. The front is convex, very little swollen and uniformly covered in most species with dense, long, bristly hair. However, in some *Anthrax* species the bristly pile is concentrated in patches of more dense character and there may be numerous appressed scales mixed in with the pile. The face is short and convex, pollinose, and vertically it may extend down the head as much as the length of the third antennal segment. There are a few species in which the face is slightly though bluntly protuberant beneath the antenna. There is a characteristic, conspicuous, transverse band of dense, black, bristly pile above the oral recess and the immediate area beneath the antennae is apt to be bare of pile, and these bare areas may meet in the middle and extend outward almost to the eye margin. The eye is reniform because of the deep indentation posteriorly.

The proboscis is always quite short, the labellum equal to half the length of the proboscis, long oval, but comparatively thin. The proboscis is confined to the oral recess. Palpus rather small and slender and short with a few long, bristly hairs at the apex. The antennae are attached at the middle of the head. The first two segments are quite short; the first segment is twice as long as the second and with a circular fringe of bristly hairs on all sides in several rows. The second and third segments are either closely attached together or may be disclike with the adjacent ends of each segment rounded so that the area of attachment is reduced, and generally speaking in this latter situation, the rounded base of the third segment is apt to be extended outward, making this segment wider than the second segment. The third segment varies greatly in shape. Generally it is strongly constricted near the base, leaving an onion-shaped appearance, with abruptly narrowed style, divided into a terminal microsegment of varying length. There are other species, however (subgenus *Satyramoeba* Sack), in which the third segment is drawn out into a long tapered cone. All species of *Anthrax* Scopoli have at least a small microsegment at the apex of the third segment which bears a brush or circlet of hairs.

Head, anterior aspect: The head is more narrow than the thorax in some species such as *pluto* Wiedemann, but in the *gideon* Fabricius group it is fully as wide as the thorax. The eyes are separated in both sexes by a distance 3 times as great as the ocellar tubercle. The ocellar tubercle is quite small and obtuse. The antennae are widely separated and are rather closer to the eye margin than they are to each other in some species, in others equally spaced across the face. The oral opening is small, quite ventral in position and divided into a quadrate portion bordered by a narrow flangelike production of the gena. The gena itself is either moderately wide or linear. The sides of this part of the oral recess quite deep and vertical and this lateral ridge or wall abruptly descends interiorly so that there is a wider transverse or semicircular excavation beneath the lower margin of the face into which the labellum fits. This anterior recess has three deep, longitudinal grooves or troughs. Tentorial fissure on each side deep. Viewed from above the occipital lobes are tightly apposed and have a deep, longitudinal fissure behind the vertex.

Thorax: The thorax is slightly narrowed anteriorly, widest on the posterior half, opaque or feebly shining black or brownish black and the mesonotal pile consists of sparse, fine, short, erect, scattered hairs over the disc which in some species may be comparatively dense and the pile is longer and more abundant laterally from the humerus to the postalar callosities. This pile may be soft, matted and appressed, and even somewhat curled, but more commonly it consists of stiff, erect, bristly hairs. Bristles on the mesonotum vary from a few, weak, short bristles on the notopleuron to some species which have comparatively long, moderately stout bristles. The anterior collar of pile is distinct and may consist of only 1 or 2 rows of stiff, long hairs or may be widened into a distinct band of pile. Scutellum clothed like the mesonotum. The marginal bristles are long or short, but slender in character. Mesopleuron with much coarse, stiff pile, especially on the upper half. Pteropleuron covered densely with stiff pile. The hypopleuron and the entire metapleuron in front and behind the haltere are bare. Metanotum extremely short. Squama with a fringe of hairs.

Legs: Legs well developed. The anterior pair is only slightly smaller than the middle pair. Anterior tibia with a double row of 12 or 14 slender bristles and about the same number of short, slightly stouter bristles, posteroventrally. Anterior femur with a few slender bristles basally. Middle tibia and legs similar with 4 rows of bristles on the tibia, somewhat stouter in character and the anterodorsal with fewer elements. Hind femur stout but not swollen with an anteroventral row of stiff, sharp, pointed bristles, which may contain as few as 4 or 5 on the outer half, or 10 or 12 reaching nearly to the base. The hind tibia is stout and may have only the same number and character of bristles as on the middle tibia. In *Anthrax tigrina* De Geer the anterior and middle tibiae are almost smooth, even the posterior bristles being minute, but the hind tibia in this species has a very dense band of appressed, stout, bristly setae, both dorsally and ventrally. Moreover, this species has a mat of dense, appressed scales on all of the femora, minutely on the anterior foretibia, but large and conspicuous on the hind tibia. Claws sharp, strongly bent at the middle with large, oval pulvilli.

Wings: Wings large and elongate, sometimes quite broad at the base with a very wide extensive alula and wide axillary lobe or with the wings strongly narrowed toward the base with narrow lobe and alula. In the matter of patterns there are 4 main types of wings in the species: (1) The apical part of the wing is hyaline and the base, to a varying extent, is marked off by uniform, unbroken black, as in the *gideon* Fabricius group. In this group sometimes the entire wing is blackish with only the middles of the cells dilutely paler (*aterrimus* Bigot). (2) The wing is hyaline, except for a few small spots around the basal crossveins and the furcation of the third vein. (3) A comparatively uncommon group in which the wing is entirely hyaline or covered with a uniform tinge of pale brownish yellow. (4) The spots of the wing are very numerous, extending to the apex and posterior border, and with larger areas blocked out in black. Such is the *irrorata* Say group.

The venation within the genus *Anthrax* Scopoli, sensu lato, presents several distinct types. There are those species with 3 submarginal cells formed as in *mystaceum* Macquart, which is the generic type of the genus *Spogostylum* Macquart, and those that have 3 submarginal cells differently formed as in *ocyale* Wiedemann. Another group has only 2 submarginal cells and no trace of spur veins arising either from the second or third veins as in *Anthrax virgo* Egger and *Anthrax aethiops* Fabricius. There is a third group with a constantly present, well-developed spur vein extending backward at the origin of the second vein, which in this case is always rectangular and not rounded as in *stictica* Klug, and which, furthermore, has a strong spur vein at the base of the anterior branch of the third vein. The anal cell is narrowly open or closed in the margin of the wing. The ambient vein is complete, and there is a small basal comb on the costa, which is quite narrow in some species, strongly developed in others. Finally, the aberrant condition of *Anthrax tigrina* De Geer should be noticed in which the outer bend of the anterior branch of the third vein is sharply rectangular with a short spur and also the curious fifth posterior cell is cut off near the base of the discal cell. Possibly *tigrina* should be placed in a separate subgenus.

Abdomen: Abdomen always short and robust, wider than the thorax, generally rather opaque black, sometimes feebly shining. There are 7 tergites visible from above in both sexes. The pile consists of fine, stiff, erect, bristly hairs of varying length, often arranged in distinct bands, concentrated on the posterior part of the tergites and with the lateral margins having more extensive and more dense, bushy, tufted mats of pile, sometimes of contrasting colors, and sometimes with white or black scales intermixed. These scales may extend over much of the posterior border of the tergites but are apt to be concentrated laterally and there may even be a few scales on the posterior margin of the mesonotum and scutellum. In one group of species the abdomen is covered with a mat of appressed, soft, curled or slightly flattened pile; in another group all the setae are appressed instead of erect; also a rather characteristic group has the whole apex of the abdomen, the last 2 or 3 tergites, covered completely by a dense mat of broad silver scales in the males. Female terminalia largely recessed and hidden by a dense mass of matted curly pile. Male genitalia large in many species. Usually the dististyli are turned straight downward symmetrically side by side. They are quite long, which together with ventral elements make a conspicuous cone-shaped hypopygium.

Inasmuch as Dr. A. J. Hesse has been able to dissect more species than I have, I quote his remarks on the male genitalia:

Hypopygium of males with the two symmetrical clasper-like basal parts always produced basally into a flattened process, their apical part in most cases also produced into a blade-like or prong-like process; beaked apical joints very variable in form and shape, usually much flattened and twisted, or leaf-like, their upper and lower margins extended to a variable extent and the pointed apex or beak usually recurved outwardly and backwardly to a variable extent, sometimes bifid apically or with more than one projection or tooth, their surfaces or upper basal part rarely with conspicuous hairs; aedeagal complex usually with the apical part of the guidepart, which lodges the aedeagus proper, produced ventralwards into a recurved hook or plate, flanked basally on each side by a subsidiary process or tooth, sometimes also with another smaller process on each side, sometimes with a medial flattened key-like or wattle-like extension or even a disc-like expansion below apical part; lateral struts and medial basal strut usually large and well developed, broad, the former usually tongue-shaped, ladle-shaped or shoehorn-shaped, the latter with its dorsal and basal edge often extended on each side into a flange or wing-like extension, its posterior margin sometimes also with a flattened lateral extension. Last sternite opposite hypopygium sometimes slightly notched or indented apically, rarely with its posterior angles produced hook-like. Terminal lamellae in last sternite variable in shape, their dorsoapical part produced into a hook-like or spine-like process in some species.

The male terminalia from dorsal aspect, epandrium removed, show a peculiar figure extremely broad on the basal half, much narrowed on the apical half. The dististyli are small, curious structures, which emit dorsally a hornlike process, curved forward and bluntly pointed. The epiphallus is constricted in the middle and extends beyond the dististylus as a flattened, tonguelike process. From the lateral aspect three things are apparent: the basistylus is club-shaped, strongly expanded at the base, much narrowed apically. Also, there is a long backward triangular lobe from the base, the epiphallus and the ejaculatory processes are both very long, slender, and quite separated. Basal apodeme large and extended backward. The epandrium is curious. It is quite long, apt to be very bristly, almost equally blunt basally as well as apically, with prominent cerci apically, and with a downwardly directed lobe at the base reminiscent of Comptosiini, to which of course it bears no relation.

Material available for study: The series of species in the National Museum of Natural History, British Museum (Natural History), Museum of Comparative Zoology, and extensive material in my own collection from the United States, South America, Europe, and Australia.

Immature stages: Immature stages are known for a number of species that parasitize wasps that build mud nests. Also, one species parasitizes a tiger beetle, *Cicindela* species. See under life histories section.

Ecology and behavior of adults: The adults in this genus spend very little time feeding and a great deal of activity given to searching for the nests of their hosts. These flies display much curiosity concerning people, often hovering about the neck or arms of people and alighting on them.

Distribution: Nearctic: *Anthrax albofasciatus* Macquart, 1840 [=*analis* Macquart, 1834], *albofasciatus cascadensis* Marston, 1963, *albofasciatus piceus* Marston, 1963; *albosparsus* (Bigot), 1892 (as *Argyromoeba*); *analis* Say, 1823; *antecedens* Walker, 1852; *argentatus* (Cole), 1919 (as *Spogostylum*); *argyropygus* Wiedemann, 1828 [=*contigua* (Loew), 1869 (as *Argyromoeba*)]; *aterrimus* (Bigot), 1892 (as *Argyromoeba*) [=*slossonae* (Johnson), 1913 (as *Spogostylum*)]; *aureosquamosus* Marston, 1963, *aureosquamosus chaparralus* Marston, 1963; *cedens* Walker, 1852; *cintalapa* Cole, 1957; *cybele* (Coquillett), 1894 (as *Argyromoeba*); *daphne* (Osten Sacken), 1886 (as *Argyromoeba*); *delila* (Loew), 1869 (as *Argyromoeba*); *fur* (Osten Sacken), 1877 (as *Argyromoeba*); *georgicus* Macquart, 1834; *grossbecki* (Johnson), 1913 (as *Spogostylum*); *irroratus* Say, 1823; *latelimbatus* (Bigot), 1892 (as *Hemipenthes*); *limatulus* Say, 1829 [=*obsoleta* (Loew), 1869 (as *Argyromoeba*)], *limatulus artemesia* Marston, 1963, *limatulus colombiensis* Marston, 1963, *limatulus larrea* Marston, 1963, *limatulus vallicola* Marston, 1963; *melanopogon* (Bigot), 1892 (as *Argyromoeba*); *nidicola* Cole, 1952; *occidentalis* (Johnson), 1913 (as *Spogostylum*); *pauper* (Loew), 1869 (as *Argyromoeba*); *plesius* Curran, 1927; *pluricellus* Williston, 1901 [=*capucina* Fabricius of Rau, 1940]; *pluto* Wiedemann,

1828; *seriepunctatus* (Osten Sacken), 1886 (as *Argyromoeba*); *stellans* (Loew), 1869 (as *Argyromoeba*); *tigrinus* (De Geer), 1776 (as *Nemotelus*) [=*scriptus* (Say), 1823 (as *Anthrax*), =*simson* (Fabricius), 1805 (as *Anthrax*)]; *varicolor* (Bigot), 1892 (as *Argyromoeba*); *vierecki* Cresson, 1919 (as *Spogostylum*).

Neotropical: *Anthrax acroleucus* Wiedemann, 1828 [=*gideon* Macquart, not Fabricius, 1840]; *angustipennis* Macquart, 1840; *calopterus* (Schiner), 1868 (as *Argyromoeba*); *cephus* Fabricius, 1805; *crinitus* (Bigot), 1892 (as *Argyromoeba*); *dimidiatus* Wiedemann, 1819; *euplanes* (Loew), 1869 (as *Argyromoeba*); *gideon* Fabricius, 1805; *guianicus* Curran, 1934; *imitans* Schiner, 1868; *inappendiculatus* Bigot, 1892, subg. *Spogostylum*; *leucocephala* Wulp, 1882; *luctuosus* Macquart, 1840; *macquarti* d'Andretta and Carrera, 1952, n.n. [=*leucopyga* Macquart, 1855, not 1840]; *mexicanus* Cole, 1957; *minimaculatus* Oldroyd, 1938; *mystaceus* Macquart, 1840, subg. *Spogostylum*; *oedipus* Fabricius, 1805 [=*aequa* Walker, 1852, =*irrorata* Macquart, 1840, =*punctum* Walker, 1849]; *plurinotus* Bigot, 1892; *poecilophorus* Schiner, 1868; *propinqus* Schiner, 1868; *selene* Osten Sacken, 1886 [=*obtiosus* Coquillett, 1887]; *seriepunctatus* Osten Sacken, 1886, also Nearctic; *squalidus* Philippi, 1865; *subandinus* Philippi, 1865.

Palaearctic: *Anthrax acroleucus* (Bigot), 1892 (as *Argyromoeba*); *actuosus* Paramonov, 1936; *aegypticola* Paramonov, 1936; *aethiops* (Fabricius), 1781 (as *Bibio* and as *Leucamoeba*) [=*punctata* Meigen, 1820], *aethiops bezzii* Paramonov, 1957; *alagoezicus* Paramonov, 1936; *alashanicus* (Paramonov), 1957 (as *Spogostylum*); *algirus* (Villeneuve), 1910 (as *Argyromoeba*); *anthrax* (Schrank), 1781 (as *Musca*) [=*morio* (Linné), 1758 (as *Musca*), =*sinuata* Meigen, 1804]; *antiopa* (Bezzi), 1924 (as *Spogostylum*); *appendiculatus* Macquart, 1855; *arenivagus* Austen, 1937 (as *Spogostylum*); *aernophilus* Paramonov, 1936; *armeniacus* Paramonov, 1936; *bezzianus* Paramonov, 1936; *binotatus* Wiedemann, 1820 [=*subnotata* Wiedemann, in Meigen, in part, 1820]; *bisniphas* Bezzi, 1925; *boninensis* (Matsumura), 1916 (as *Argyromoeba*); *brevis* Becker, 1913; *brunnicosus* (Becker), 1913 (as *Argyromoeba*); *bucharensis* (Paramonov), 1936; *Paramonovella*, n.n., for *bucharensis* Paramonov, 1957, not 1936; *bucharensis* Paramonov, 1957 (as *Spogostylum*); *cairensis* Paramonov, 1936; *candidus* (Sack), 1909 (as *Psamatamoeba*); *candidapex* Austen, 1937; *caucasicus* (Zaitsev), 1961 (as *Spogostylum*); *chionanthrax* (Bezzi), 1926 (as *Argyromoeba*); *chionostigma* Tsacas, 1962; *chivaensis* Paramonov, 1936; *cinereus* (Zaitsev), 1961 (as *Spogostylum*); *comatus* Austen, 1937 (as *Spogostylum*); *comptus* Paramonov, 1936; *crosi* (Villeneuve), 1910 (as *Argyromoeba*); *desertorum* Paramonov, 1936; *distigma* Wiedemann, 1828 [=*tripunctata* Wulp, 1868]; *doctus* (Paramonov), 1957 (as *Spogostylum*); *dubius* (Zaitsev), 1961 (as *Spogostylum*); *efflatouni* Paramonov, 1936; *sergeiae*, n.n., for *efflatouni* Paramonov, 1957, not 1936; *efflatouni* (Paramonov), 1957 (as *Spogostylum*); *erythrogaster* Paramonov, 1936;

etruscus (Fabricius), 1794 (as *Satyramoeba*) [=*formosa* Dufour, 1852, =*hetrusca* Fabricius, of authors, =*rubiginipennis* Macquart, 1840, =*satyrus* (Rossi), 1790, not Fabricius (as *Bibio*)]; *flavus* (Sack), 1929 (as *Leucamoeba*); *flavescens* (Sack), 1909 (as *Spogostylum*); *flavipennis* (Sack), 1909 (as *Spogostylum*); *flavipes* Roeder, 1896 (as *Spogostylum*); *grisescens* Paramonov, 1936; *hassani* Paramonov, 1936; *helenae* (Paramonov), 1957 (as *Spogostylum*); *heteropyga* (Sack), 1909 (as *Argyromoeba*); *hippolyta* (Wiedemann), 1828 (as *Spogostylum*) [=*dedecor* (Hermann), 1907 (as *Argyromoeba*), =*villosa* Klug, 1832]; *incitus* Paramonov, 1936; *indigenus* (Becker), 1913 (as *Argyromoeba*); *irrorellus* (Klug), 1832 (as *Spogostylum*); *isis* (Meigen), 1820 (as *Psamatamoeba*) [=*binotata*, Wiedemann in Meigen, 1820, =*ixion* Megerle in litt. in Meigen, =*subnotata* Walker, 1871]; *jazykovi* Paramonov, 1936; *jezoensis* Matsumura, 1916; *karavaievi* Paramonov, 1924, subg. *Spongostylum*; *kiritshenkoi* Paramonov, 1936; *koshunensis* Matsumura, 1916; *kozlovi* (Paramonov), 1957 (as *Spongostylum*); *laetus* Paramonov, 1936; *laevipennis* Paramonov, 1936; *lucidus* (Becker), 1902 (as *Argyromoeba*); *macrocerus* Paramonov, 1936; *maculosus* (Sack), 1909 (as *Argyramoeba*); *massinissa* Wiedemann, 1828; *melanistus* (Bezzi), 1925 (as *Argyramoeba*); *mendax* Austin, 1937; *monacha* (Sack), 1909 (as *Chalcamoeba*); *mongolicus* Paramonov, 1936; *montanus* (Paramonov), 1957 (as *Spogostylum*); *monticola* (Paramonov), 1957 (as *Spogostylum*); *nanellus* Paramonov, 1936; *nanus* Paramonov, 1936; *nigrus* Austen, 1937 (as *Spogostylum*); *niphas* (Hermann), 1907 (as *Psamatamoeba*), *niphas bilineatus* Engel, 1934; *nitidus* Austen, 1937 (as *Spogostylum*); *obuchovae* (Paramonov), 1957 (as *Spongostylum*); *ocyalus* (Wiedemann), 1828 (as *Anthrax*), (subg. *Spogostylum*); *ogasawarensis* Matsumura, 1916; *oophagus* Paramonov, 1936, *oophagus parvus* Paramonov, 1936; *obscurus* (Sack), 1909 (as *Anthracamoeba*); *pallipes* (Loew), 1869 (as *Spogostylum*); *pamirensis* (Paramonov), 1957 (as *Spogostylum*); *pargrisescens* Paramonov, 1936; *Paramonovella*, n.n., for *bucharensis* Paramonov, 1957; *parobscurus* (Paramonov), 1957 (as *Spongostylum*); *perfectus* Becker, 1923; *perpusillus* Austen, 1937 (as *Spogostylum*); *persicus* Paramonov, 1936; *pharaonis* Paramonov, 1936; *pilosulus* Strobl, 1902, *pilosulus tadzhikorum* Paramonov, 1936; *punctipennis* (Zaitsev), 1961 (as *Spogostylum*); *putealis* Matsumura, 1905; *ramsesi*, Paramonov, 1936; *repetekianus* Paramonov, 1936; *rufulus* (Zaitsev), 1961 (as *Spogostylum*); *sacki* Paramonov, 1936; *sergeiae*, n.n., for *efflatouni* Paramonov, 1957; *semirufus* Sack, 1909 [=*semifura* Sack, lapsus, 1909]; *shelkovnikovi* Paramonov, 1936; *sordidus* Sack, 1909, subg. *Spongostylum* [=*antiopa* Bezzi, 1925]; *stackelbergi* (Paramonov), 1957 (as *Spongostylum*); *stepensis* Paramonov, 1936; *stricticus* Klug, 1832 [=*polystigma* Sack, 1909]; *submacrocerus* Paramonov, 1936; *tenebrosus* (Paramonov), 1957 (as *Spongostylum*); *transcaspicus* Paramonov, 1936; *trifasciatus* Meigen, 1804 [=*aethiops*

Laboulbene (not Fabricius), 1873, =*capitalata* Mulsant, 1952], *trifasciatus leucogaster* Wiedemann in Meigen, 1820; *trinotatus* Dufour, 1852, [=*trimaculata* Becker, 1908, Sack, 1909, Engel, 1936], *trinotatus areolatus* Abreu, 1926; *tripunctatus* (Wiedemann in Meigen), 1820 (as *Modybdamoeba*) [=*difficulis* Wiedemann in Meigen, 1820]; *turanicus* Paramonov, 1936; *turcmenicus* Paramonov, 1926; *turkmenicola* (Paramonov), 1957 (as *Spongostylum*); *turkmenorum* Paramonov, 1936; *ultrareolatus* Abreu, 1926; *uralensis* Paramonov, 1936; *varius* Fabricius, 1794; *velox* Loew, 1862; *virgo* (Egger), 1859 (as *Chalcamoeba*), *virgo pedemontanus* Griffin, 1896, *virgo pilosulus* Strobl, 1902; *volitans* Wiedemann, 1828; *vulpinus* (Sack), 1909 (as *Chrysamoeba*); *yamashiroensis* Matsumara, 1916; *zonabriphagus* Portschinsky, 1895; *fulvipes* Loew.

Ethiopian: *Anthrax aetheocoma* (Hesse), 1956 (as *Argyramoeba*); *aygulus* Fabricius, 1805 [=*biflexus* Loew, 1852]; *badius* Hesse, 1956; *bifarius* Hesse, 1956; *busonicus* François, 1960; *caffer* Hesse, 1956; *camptocladius* Bezzi, 1912; *candidulus* Hesse, 1956; *chalciurus* Hesse, 1956; *consobrinus* Hesse, 1956, *consobrinus suffusipunctis* Hesse, 1956; *cunctator* Hesse, 1956; *cuthbertsoni* Hesse, 1956; *dagomba* (Bowden), 1964 (as *Argyramoeba*); *decipiens* (Bezzi), 1912 (as *Molybdamoeba*); *diffusus* Wiedeman, 1824 [=*maculipeennis* Macquart, 1840], *diffusus decusys* Bezzi, 1924, *diffusus fuscopurpuratus* Hesse, 1956, *diffusus hybridus* Hesse, 1956, *diffusus majusculus* Bezzi, 1924, *diffusus pallidulus* Hesse, 1956, *diffusus suffusipennis* Hesse, 1956; *dimidiatipennis* Hesse, 1956; *doliops* Hesse, 1956, *doliops fulviventris* Hesse, 1956, *doliops gamka* Hesse, 1956; *dolomatus* Bowden, 1964; *dubius* Macquart, 1840; *elutus* Bezzi, 1921; *eremobius* Hesse, 1956; *eurypterus* Hesse, 1956; *fracidus* (Bowden), 1964 (as *Argyramoeba*); *fratercula* (Bowden), 1964 (as *Argyramoeba*); *fulvastra* (Bowden), 1964 (as *Argyramoeba*) [=*ventrale* (Bezzi), 1924 (as *Spongostylum*)] *fulvipes* Loew, 1860; *furvus* Hesse, 1956; *fuscipennis* Ricardo in Forbes, 1903 [=*dentata* (Becker), 1906 (as *Argyramoeba*), =*muscaria* Klug, 1832]; *hessii* Wiedemann, 1828; *homogeneus* Bezzi, 1912; *immaculatus* (Bowden), 1964 (as *Argyramoeba*); *incisuralis* Macquart, 1840, *incisuralis aridicola* (Hesse), 1956 (as *Argyromoeba*), *incisuralis fumosus* (Hesse), 1956 (as *Argyramoeba*), *incisuralis glaucescens* (Hesse), 1956 (as *Argyramoeba*), *intermedius* Hesse, 1936; *kaokoensis* Hesse, 1956; *lemimelas* Speiser, 1910; *leucopogon* (Bezzi), 1912 (as *Molybdamoeba*); *leucurus* Hesse, 1956, *massauensis* Jaennicke, 1867; *mimetes* Hesse, 1956; *mixtus* Loew, 1860; *munroi* Hesse, 1956, *munroi willowmorensis* Hesse, 1956; *muticus* Bezzi, 1921 (as *Spogostylum*); *namaensis* Hesse, 1956; *nanus* Hesse, 1956; *nigerrimus* Bezzi, 1924, *nigerrimus ocellatus* Bezzi, 1924; *nubeculosus* Hesse, 1956; *phaeopteralis* Hesse, 1956; *pithecius* Fabricius, 1805 [=*confusemaculata* Macquart, 1855, =*conspurcata* Wiedemann, 1828, =*spectabilis* Loew, 1860]; *plumipes* Hesse, 1956; *princeps* Bezzi, 1924; *punctipennis* Wiedemann, 1821; *puncturellus* Hesse, 1950; *punicisetosus* (Hesse), 1955 (as *Argyromoeba*); *pycnopeltis* (Hesse), 1956 (as *Argyromoeba*); *quinquemaculatus* Bezzi, 1924 (as *Spogostylum*); *rhodesiensis* Hesse, 1956; *robustalis* (Hesse), 1956 (as *Argyromoeba*); *saturatus* Bezzi, 1924 (as *Spogostylum*); *simmillimus* Hesse, 1956; *spathistylus* Hesse, 1956; *sticticalis* Hesse, 1956; *subanthrax* Bezzi, 1924; *tetraspilus* Hesse, 1956; *triatomus* Hesse, 1956 [=*trimaculatus* Bezzi, not Wulp, 1921]; *triguttellus* Hesse, 1956; *trisinuatus* Hesse, 1956; *trixus* Bowden, 1964; *xerozous* Hesse, 1956.

Oriental: *Anthrax appendiculatus* (Bigot), 1892 (as *Argyromoeba*); *argentiapicalis* (Brunetti), 1912 (as *Argyromoeba*); *austeni* Brunetti, 1920; *bipunctatus* Fabricius, 1805; *carbo* Rondani, 1875; *ceylonicus* Brunetti, 1909, *claripennis* Brunetti, 1909; *clausus* Brunetti, 1909; *distigma* Wiedemann, 1828 [=*argyropyga* Doleschall, 1857, =*tripunctata* Wulp, 1868]; *duvaucelii* Macquart, 1840; *fallax* de Meijere, 1907; *fletcheri* Brunetti, 1920; *fulvulus* Wiedemann, 1821 [=*degener* Walker, 1856, =unnamed variety Wulp, 1880]; *gentilis* Brunetti, 1909; *gestroi* Brunetti, 1912; *guttatipennis* Brunetti, 1920; *himalayensis* Brunetti, 1909; *indicatus* Nurse, 1922; *intermedius* Brunetti, 1920; *limitarsis* Brunetti, 1909; *leucopygus* Macquart, 1840; *melania* Wulp, 1885; *nigrofemoratus* Brunetti, 1909; *niveicauda* Brunetti, 1920; *niveisquamis* Brunetti, 1909; *obscurifrons* Brunetti, 1909; *semiscitus* Walker, 1857.

Australian: *Anthrax arcus* de Meijere, 1913; *ater* Roberts, 1928; *confluensis* Roberts, 1928; *lepidiota* Roberts, 1928; *proconcisus* Hardy [=*concisus* Macquart, 1849, not Macquart, 1840]; *prosimplex* Hardy [=*simplex* Macquart, 1847]; *velox* White, 1917.

Ignota Patria: *Anthrax degenera* Walker, 1852a.

Genus *Dicranoclista* Bezzi

FIGURES 160, 403, 679, 753, 1018, 1019, 1020

Dicranoclista Bezzi, The Bombyliidae of the Ethiopian Region, p. 178, 1924. Type of genus: *Dicranoclista simpsoni* Bezzi, 1924, original designation.

Coquillettia Williston, Manual North American Dipt., Ed. II, p. 65, 1896b. Type of genus: *Spogostylum vandykei* Coquillett, 1894, by original designation. Preoccupied by Uhler, 1890.

Shining black flies, with dense, erect, rather long, plushlike, reddish yellow to yellowish-brown pile upon thorax and abdomen and easily recognized by the peculiar venation of the wing. The vein that is the anterior branch of the third vein coalesces and is confluent for quite a distance with the posterior branch of the second vein, but this branch of the third vein rejoins the last branch of the third vein so that there is an extra submarginal cell corresponding to the fourth cell in *Ligyra* Newman. Moreover, the attachment of the second branch of the second vein is atrophied and directed backward as a spur from the point of coalescence so that there is in fact only 3 submarginal cells. First posterior cell is also closed and stalked. The second vein arises quite rectangularly and has a strong backward spur. The wing is quite hyaline on the outer half, the base being wholly brown to yellow or subhyaline with the veins narrowly brown and a vertical spot of brown of varying size on the anterior crossvein and extending up to the wing margin.

Antennae typical of the Anthracinae; the third segment is very short and broad, the style slender, with 2 segments and a pencil of hairs at the apex. Pile of front fine and erect, of face coarse and erect, short beneath antennae, long and mustache-like above the oral margin; appressed scales lie along the middle of the front. Squama and alula with fine pile, the former dense and bushy. Anterior tibia long and slender with numerous, fine, long spines.

Three species are known, one from Gambia, one from Utah, and a third species, *vandykei* Coquillett, from California, Utah, and Texas. Johnson and Johnson (1960) note that the latter species peculiarly resembles *Chrysanthrax fulvianus* Say in general appearance.

Length: 10 to 12 mm.; wing: 9 mm.

The antennae are attached a little below the middle of the head. The first segment is widened apically, is only a little longer than the second, and bears a few, fine hairs of no great length laterally and dorsolaterally and some shorter ones medially. The second segment is shorter than wide, fully and broadly attached to the base of the third segment. The third segment is quite short and onion-shaped with a short, stylar, asymmetrical portion directed laterally, slightly knobbed apically, and bearing a short microsegment with apical brush or circle of hairs.

Head, lateral aspect: Head subglobular, the occiput is very prominent and extensive dorsally, even more so in the middle of the eye at the point of indentation, but ventrally the occiput is only about half as long as it is dorsally. Pile of the occiput consists of narrow, flat-appressed, scattered scales, more dense at the point of indentation, but also with a tuft of erect hair at this point. Interior cup of the occiput large and deep with a dense fringe of short pile on the border. The front is convex, quite extensive, scarcely raised above the eye margin, but with numerous scattered, appressed, long, slender scales and with abundant, coarse erect pile extending from the ocellar tubercle nearly to the antenna. The scales overlap the base of the antenna. From the lateral aspect, very little of the face shows. It retreats downward slightly below the eye margin, the total depth of the face is almost equal the length of the antenna. There is a conspicuous, circular, transverse band of dense, coarse, almost bristly, reddish pile, reaching from the eye margin across the lower face. The eye is reniform with a deep, wide excavation, the upper and lower angles are approximately equal, and there is a bisecting line at the midpoint of the eye. Upper facets of the eye in the male very slightly enlarged. The proboscis is short and stout and confined to the oral recess and the labellar portion comprises at least four-fifths of its length. The palpus is quite short, small, and clavate with a few stiff hairs near the apex.

Head, anterior aspect: The head is not quite as wide as the thorax. The eyes are rather widely separated in the male by 3 times the width of the ocellar tubercle. Their width also a little more than the width of the antennae. The ocellar tubercle is quite small but prominent and the ocelli occupy an equilateral triangle. Antennae widely separated, the space between equals the space between antenna and eye. Below the face the eyes are slightly convergent with deep, sloping sides almost reaching to the bottom of the oral recess. Oral recess narrowed a little before the middle, the posterior part quite deep with vertical walls. Gena very narrow. Tentorial pits deep. From above, the lobes of the occiput are tightly apposed, with a comparatively deep fissure.

Thorax: The thorax is considerably longer than wide. The mesonotum and the scutellum are shining black with very dense, reddish-yellow, plushlike pile which in nowise obscures the ground color, and this pile becomes even more dense broadly along the anterior margin and on the humerus. There are 3 or 4 distinct bristles on the notopleuron. The pile on the propleuron, mesopleuron, and pteropleuron is quite dense. The hypopleuron and metapleuron are bare, but there is a tuft of long pile posteriorly behind the spiracle. The metanotum appears to be exceptionally short and appears to be bare. The squama has a fringe of dense, matlike hair. The plumula has a wide band of long pile.

Legs: The legs are stout, all of the femora being strengthened without being swollen. The first four femora are, however, a little wider basally. First four femora also with a fringe of hairs distally. Hind femur with a ventrolateral row of 12 to 15 oblique, sharp, slender bristles and at least 10 bristles ventromedially besides scattered appressed setae dorsally on at least the outer half. All of the femora are densely covered with very broad, loosely appressed, large, brownish-yellow scales. Anterior tibia slender, elongate with some 16 slender prominent though short spiculate bristles anterodorsally and 20 bristles posterodorsally. There are 12 or 14 similar bristles posteroventrally. Middle tibia similar with an additional true ventral row containing few bristles. Hind tibia with 6 or 7 bristles anteroventrally and 10 to 12 bristles in each dorsal row. Claws large, long, sharp, strongly bent at the apex, the pulvilli quite long and spatulate.

Wings: Hyaline in the type-species with the basal third completely brownish yellow filling the anal cell, axillary lobe and alula, filling the second basal cell and extending beyond as a triangle into the discal and fourth posterior cell, and extending upward to the first vein and moreover filling almost all of the costal cell and with a broad spot covering the anterior crossvein and reaching up to the first vein. In the American species *vandykei* Coquillett the wing is in every respect similar to the type-species, except that the brownish-yellow color is found only at the base of the wing, borders the posterior margin of the second and anterior border of the third basal cells, fills out the whole of the costal and subcostal cells, and has only a small spot on the anterior crossvein and base of second vein. In *fasciata* Johnson and Johnson the dark color of the wing is almost restricted to the prehumeral area of the wing and to the costal cell. This genus is very unique in the presence of an extra submarginal cell formed much as in *Ligyra* Newman, except that there is an addition and incomplete spur constantly present extending back toward the second vein. As I interpret this wing the anterior branch of the third vein and posterior branch of the second vein coalesced at some time in the past to present this formation. The first posterior cell is closed and stalked. There is a well-developed basal comb and a stout basal hook. Ambient vein complete. Alula large.

Abdomen: Abdomen oval, narrowed somewhat posteriorly. Bezzi (1924) says of the type-species that the abdomen is a little pointed. It is a little wider basally than the thorax, rather flattened, shining black with dense, erect pile of considerable length, somewhat more loose and sparse medially, but certainly very dense laterally and upon the lateral margins. The ground color is in no way obscured and there are no bristles. In the American species the posterior margins of the second, fifth, and sixth tergites have a complete band of conspicuous, leaf-shaped, brownish-yellow scales; curiously these scales on the third and fourth tergites are restricted to a short row on the lateral margin. Male genitalia symmetrical, pale in color. The dististyle is directed downward.

Material available for study: The type-species in the British Museum (Natural History) and also material of *vandykei* Coquillett.

Immature stages: Unknown.

Ecology and behavior of adults: I have seen no record of adult behavior, except that Johnson and Johnson (1960) record taking a female of *fasciata* Johnson and Johnson on a twig before sunrise by the Provo River, Utah.

Distribution: Nearctic: *Dicranoclista fasciata* Johnson and Johnson, 1960; *vandykei* (Coquillett, 1894 (as *Spogostylum*).

Ethiopian: *Dicranoclista simpsoni* Bezzi, 1924.

Genus *Walkeromyia* Paramonov

Figures 164, 165, 411, 426, 677, 709, 730, 746, 748, 750, 751

Walkeromyia Paramonov, Konowia, vol. 13, p. 23, 1934c. Type of genus: *Anthrax luridus* Walker, 1857, by original designation.

Large, robust, brownish, or brownish-yellow, or reddish-black, shining flies, which have a strongly swollen front and face, much reduced fleshy proboscis and labellum, and correspondingly shortened, ventral oral recess. Related most closely to Anthracinae with which it agrees in the type of mouth, wing venation and especially in the bare metapleuron in front of the haltere, and in the simple pile of the squama. It is aberrant in that the long, conical, third antennal segment has a minute, short style without any hair tuft at the apex. A further peculiarity lies in the fact that the hind legs of the male have dorsally and ventrally a long dense band of slender black scales extending from basal third of the femur to include the first 3 segments of the tarsus. Wing generally uniformly brownish yellow; 2 submarginal cells; anterior branch of third vein arising plane, not quite at right angles but with long backward spur.

Two species, from southern part of South America and from Trinidad.

Length: 14 to 19 mm.; wing 16 to 21 mm., the wing longer than the body.

Head, lateral aspect: Head almost globular, with the occiput exceptionally produced and strongly projected backward. The length of the occiput at the vertex is even greater than the width of the male vertex, and even greater in extent if measured from the deep indentation of the posterior eye margin. The occiput is slightly more extensive in males than in females, and, viewed from above, shows 3 or 4 faint transverse grooves. Instead of being produced directly backward, the occiput slopes posteriorly almost as much as the eye slopes anteriorly. The central indentation on the posterior eye margin is very broad and deep, the angle above equal to the angle below; posterior part of eye with bisecting line. Pile of occiput scanty, short, appressed and setate, and forming longer tufts of stiff hair in the hollow of the indentation, these bristly hairs directed forward. The fringe about the deep occipital cup is scanty below, dense above.

The front is very large and extensive, a little swollen, and at the level of the antennae, almost twice as wide as the width at the vertex. The pile of the front consists of scattered, fine, erect hairs of no great length, but with 6 patches of similar denser hair, 3 distributed on the upper part of the front and 3 distributed across the front on its lower third. The face is quite extensive, swollen like the front, rounded and convex, produced forward, its length equal the conical part of the third antennal section, pile of face comparatively dense, coarse, black and bristly, its length slightly greater than the pile of the front. There is a distinct bare space beneath each antenna. Oral recess small and deep, confined to the lower eighth of the head. It is deep and the posterior part in which the proboscis is imbedded has vertical walls. This swollen ridgelike genal sidewall, almost as wide as the oral recess, curves abruptly inward anteriorly so as to form a deep, transverse recess and extension of the posterior part and has on each side a deep tentorial pocket. The proboscis is shorter than in *Anthrax* Scopoli and fully two-thirds of the proboscis is composed of the swollen clublike labellum. Palpus clavate with minute setae and as long as the basal part of the proboscis. Antennae attached at the middle of the head, widely separated, the first and second segments are both quite short, the first segment is twice as long as the second, the third segment is a little wider than the second, closely attached to it, short conical with a narrow stylar portion not more than one-fifth the total length of the segment. Attached to the apex of the third segment is a very small, thick, short, pale, bristle-tipped microsegment. There is no circlet of hairs at the apex of the segments.

Head, anterior aspect: The head is slightly flattened above, below and on each side, and it is slightly wider below. The vertex in males and females is of almost equal width and is about three-fourths as wide as the genal space between the eyes below. The antennae are widely separated, each attached at the outer third of the face. Ocellar tubercle low, the ocelli placed in a short isosceles triangle, oral opening deep and short rectangular, widening laterally in front; the genae are pollinose and without pile. Viewed from above, the occiput has a deep, long fissure, the sides touching, the cleft deeply recessed.

Thorax: The thorax is a little longer than wide, widest posteriorly, and wider than the head. It is feebly shining, very dark brown in color. The pile is fine, erect, and rather short in the middle of the mesonotum, but there is a considerable amount of curled, appressed pile laterally and posteriorly and the anterior collar pile is considerably longer and erect. Bristles are absent. Scutellar pile appressed and consisting of narrow scales. Upper mesopleuron with a dense tuft of pile, the anterior half appressed forward, the posterior half appressed backward. Pteropleuron with quite abundant pile. The hypopleuron and the metapleuron in front of the haltere are bare. Metanotum beneath the scutellum very short and apparently bare. Squama with a short fringe of hair.

Legs: Weak on the anterior forelegs. The hind legs, however, are quite long and strongly developed although slender. In the male there is a curious, dense, wide, rather dense band of brownish black, long, slender scales, which begin a little before the middle of the hind femur and form a dorsal and ventral feathered band on the hind femur, both dorsal and ventral surfaces of the hind tibia, and both surfaces of first two segments of the hind tarsus. The band of scales on the hind femur is broadest below, but on the tibia and tarsus it is much more extensive dorsally than below. Hind basitarsus more than twice as long as the remaining segments. Claws and pulvilli well developed.

Wings: The wings are quite large and long, extending far beyond the apex of the abdomen, uniformly tinged with brownish-yellow color. There are 2 submarginal cells and a backward spur vein arises near the base of the anterior branch of the third vein. The second arises rectangularly, and the anterior crossvein is placed at the middle of the discal cell. There is a short spur vein backward from the medial crossvein below the discal cell. Anal vein narrowed apically and also narrowly open. Alula comparatively narrow. The ambient vein is complete. There is a small basal comb on the costa; also a distinct, flattened, apically rounded lobe.

Abdomen: Abdomen robust, shorter than the thorax in males, of about equal length in females. It is strongly convex across the middle. I find 6 tergites visible in the male, 7 in the female. The abdomen is rather densely covered with matted, slender, appressed, shining black scales, and a few wider scales laterally and along the posterior edges of the tergites and also laterally some stiff, bristly hair which tends to become more conspicuous and often matted into tufts along the posterior border of the last tergite and the last sternite. Male terminalia deeply recessed and enclosed by bands of stiff long black hair. Female terminalia similarly concealed.

Material available for study: A pair of *luridus* Walker lent by the National Museum of Natural History and additional material purchased from Fritz Plaumann.

Immature stages: The host and larval forms have not been recorded, as far as I am aware, but in the National Museum of Natural History there is a female from Trinidad that was reared in a nest of *Xylocopa submordax* by F. D. Bennett, who deserves much credit for having preserved this record.

Ecology and behavior of adults: Not on record. Probably very similar to *Anthrax tigrinus* Fabricius, which also parasitizes bees of the genus *Xylocopa* Latreille. I believe these anthracine flies should fall within a separate tribe, the Walkeromyini. This requires that *Anthrax tigrinus* Fabricius be placed within a genus separate from other North American *Anthrax* Scopoli. I propose *Stymphalina*, new name.

Distribution: Neotropical: *Walkeromyia lurida* Walker, 1857; *plumipes* Philippi, 1873.

Genus *Coniomastix* Enderlein

FIGURE 412

Coniomastix Enderlein, Deutsche Ent. Zeitschr., 1934, p. 140, 1934. Type of genus: *Coniomastix montana* Enderlein, 1933, by original designation.

I have not seen *Coniomastix* Enderlein and I quote his description.

Genus of the subfamily Anthracinae. Eyes somewhat indented or excavated (posteriorly). Antenna flagellum (microsegment) extraordinarily short and in the form of a minute cone; it bears on the point scattered little hairs. In front of the ocelli, there is a sharply impressed frontal furrow or indentation reaching as far as the ocellar tubercle. The second and third branches of the radius at the base with vein stumps. Base of radius two and radius three (appearing as a crossvein) and also a spur situated proximal before the radiomedial crossvein. Fourth longitudinal vein strongly bent forward towards the front; the second and third branches bent more weakly. The fifth band of radius and first of medius end separately, likewise the second cubitus and the two middle branches of the medius. The crossvein between the second radius and the third and fourth is complete.

Distribution: Palaearctic: *Coniomastix montana* Enderlein, 1933.

Supplemental Text Figures

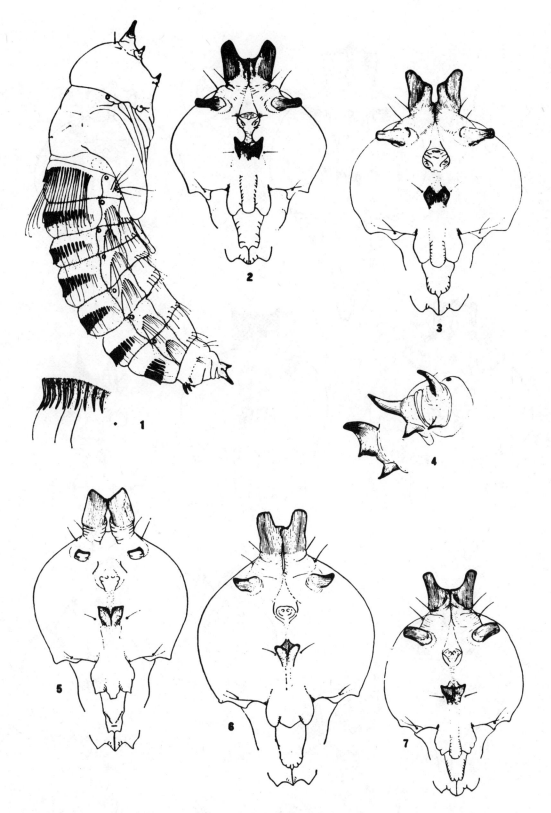

TEXT-FIGURES 1–7.—Bee fly pupae: 1, *Villa alternata*, lateral view, with enlargement of dorsal armature of seventh abdominal segment. 2,3,5–7. Heads of pupae, frontal view: 2, *Villa* species; 3, *V. handfordi;* 5, *V. fulviana;* 6, *V. alternata;* 7, *V. molitor.* 4. Apical segment of *V. alternata*, posterolateral view of variations of tubercles. After A. R. Brooks.

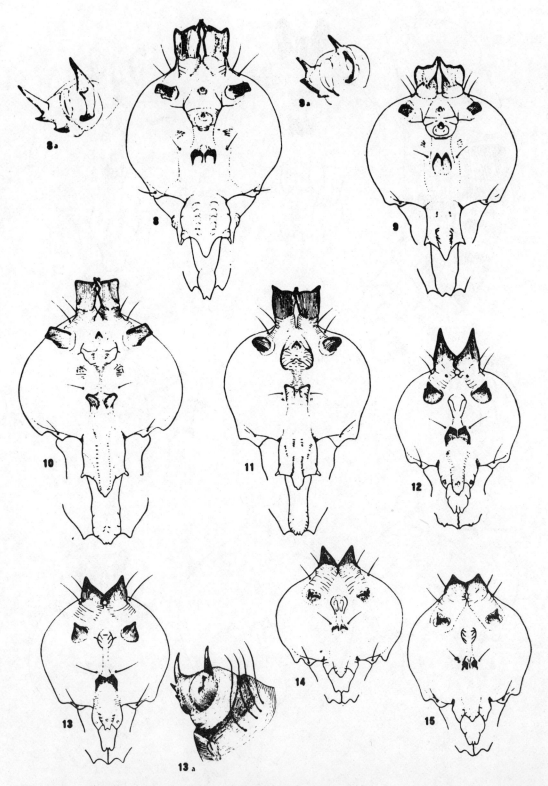

TEXT-FIGURES 8–15.—Heads of pupae, frontal view: 8, *Poecilanthrax sackenii;* 9, *P. terminipennis;* 10, *P. willistonii;* 11, *P. alcyon;* 12, *Hemipenthes* species; 13, *H. morioides;* 14, *H. sinuosa;* 15, *H. catulina.* Apical segment of pupa, posterolateral view: 8a, *P. sackenii;* 9a, *P. tegminipennis;* 13a, *H. morioides.* After A. R. Brooks.

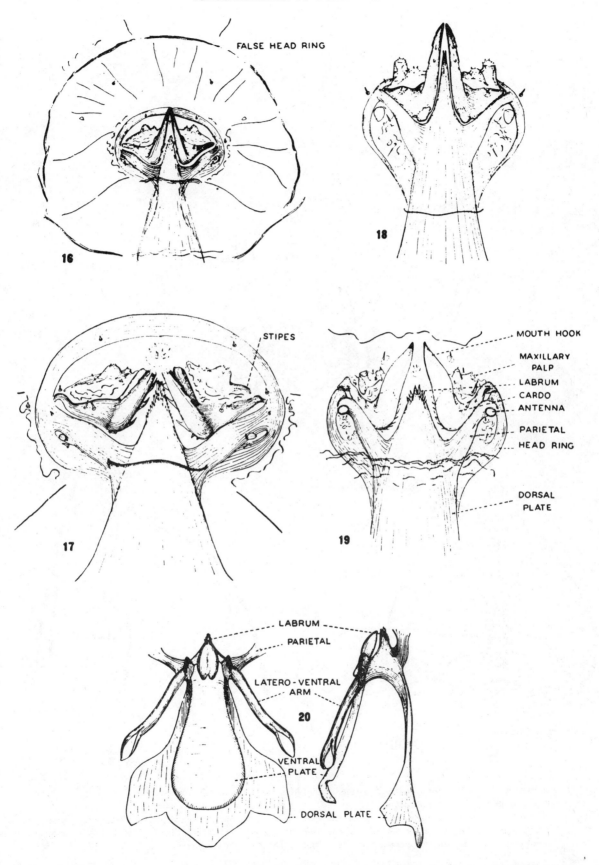

TEXT-FIGURES 16–20.—Bee fly larval head structure: 16–19, anterior head capsule, dorsal view: 16, *Poecilanthrax willistonii;* 17, *P. tegminipennis;* 18, *Hemipenthes morioides;* 19, *Villa alternata.* Posterior head capsule, ventral and ateral views: 20, *Poecilanthrax willistonii.* After A. R. Brooks.

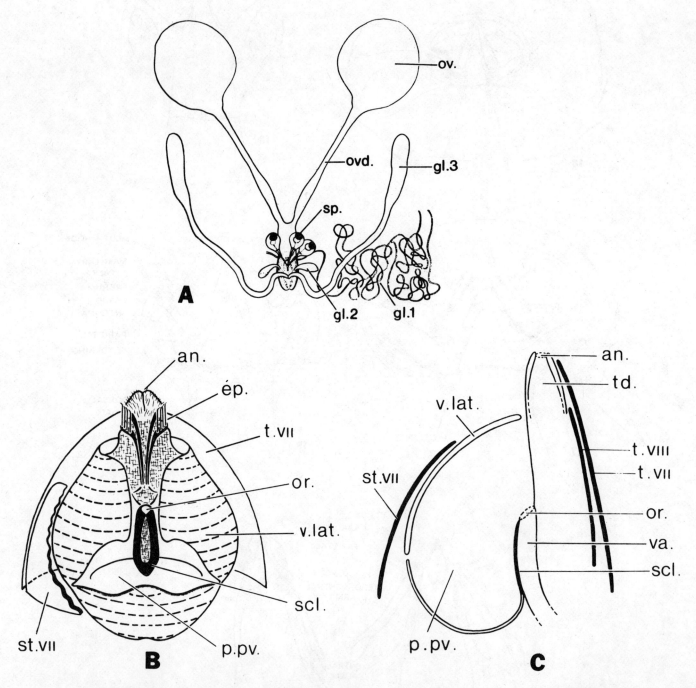

Text-Figure 21.—*Villa quinquefasciata* Wiedemann apud Meigen. A. Female genital structures: gl. 1, 2, 3 accessory glands; ov. ovary; ovd. oviduct; sp. spermatheca. B. Anal extremity of a female. Ventral view of seventh sternite, major parts of the lateral valves slightly separated from one another: an. anus; p.pv. perivaginal pouch; st. sternite; ep. anal spines; or. genital orifice; scl. sclerites enclosing the vagina; t. tergite; td. digestive tube; va. vagina; v. lat. lateral valves. C. Longitudinal cross section. After Biliotti, Demolin, and du Merle.

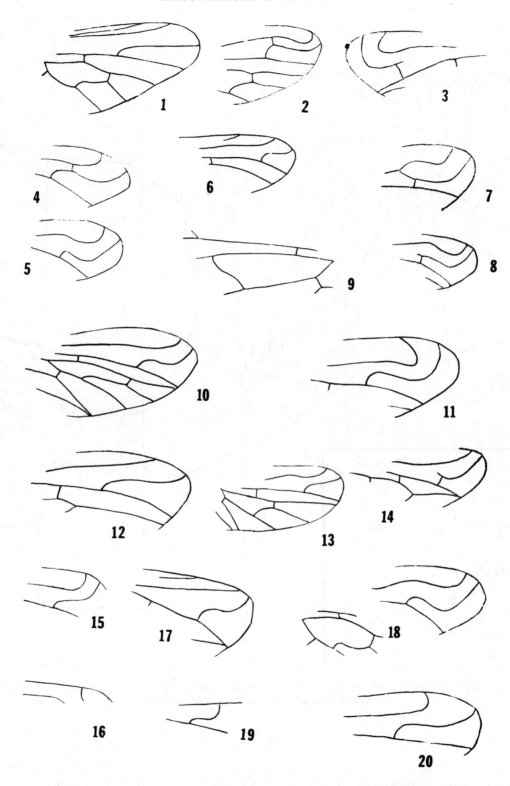

TEXT-FIGURE 22.—Wings of fossil bee flies: 1, *Acreotrichus atratus*, from type; 2, *Aldrichia ehrmanni*, from type; 3, *Alepidophora pealei*; 4, *Amphicosmus elegans*, from type; 5, *Aphoebantus squamosus*, from type; 6, *Corsomyza crassirostris*; 7, *Exepacmus johnsoni*, from type; 8, *Epacmus nebritus*, from type; 9, *Geronites stigmalis*, discal cell; 10, *Lithocosmus coquilletti*; 11, *Megacosmus mirandus*; 12, *Melanderella glossalis*; 13, *Metacosmus exilis*, from type; 14, *Pachysystropus rohweri*; 15, *Paracosmus insolens*, from type; 16, *Paracosmus morrisoni*; 17, *Pachysystropus condemnatus*; 18, *Triodites mus*, discal cell and apex of wing; 19, *Toxophora virgata*, end of discal cell; 20, *Verrallites cladurus*. After T. D. A. Cockerell.

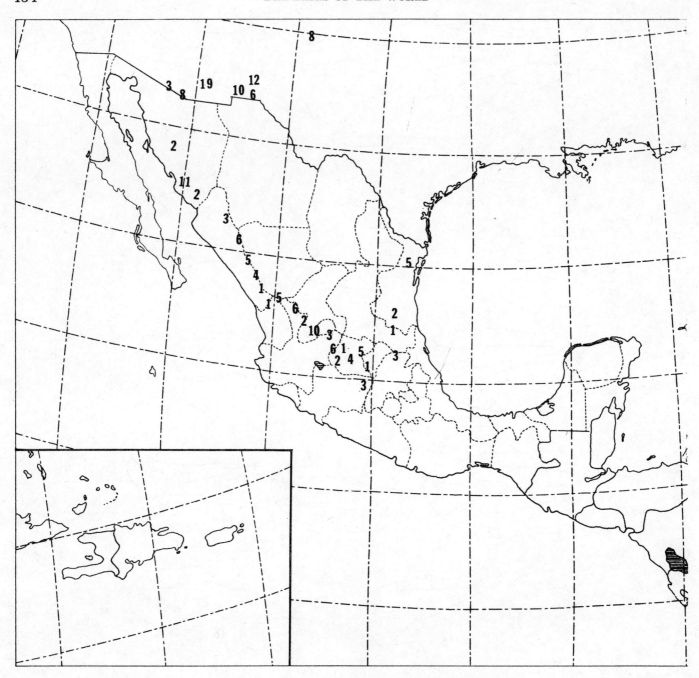

TEXT-FIGURE 23.—Number of bee fly species obtained by three collectors in August (1962) at each of 32 stops in upper Mexico and the southwestern border of the United States.

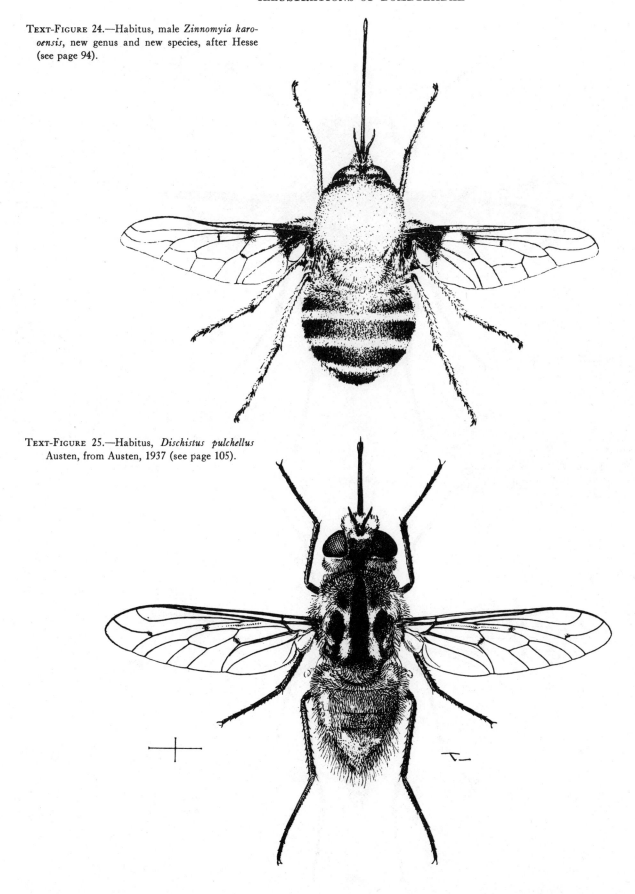

Text-Figure 24.—Habitus, male *Zinnomyia karooensis*, new genus and new species, after Hesse (see page 94).

Text-Figure 25.—Habitus, *Dischistus pulchellus* Austen, from Austen, 1937 (see page 105).

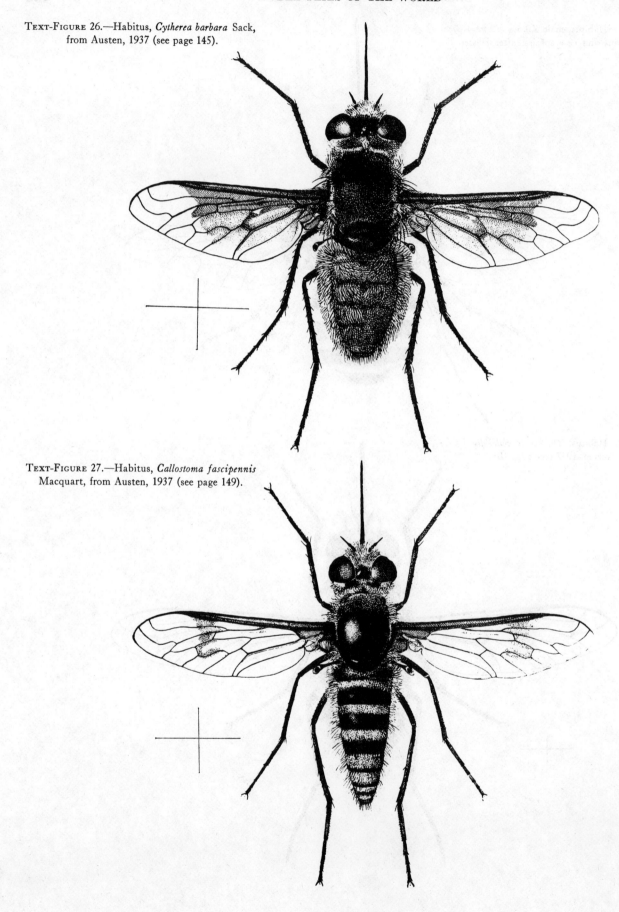

Text-Figure 26.—Habitus, *Cytherea barbara* Sack, from Austen, 1937 (see page 145).

Text-Figure 27.—Habitus, *Callostoma fascipennis* Macquart, from Austen, 1937 (see page 149).

TEXT-FIGURE 28.—Habitus, *Legnotomyia trichorhoea* Loew, from Austen, 1937 (see page 162).

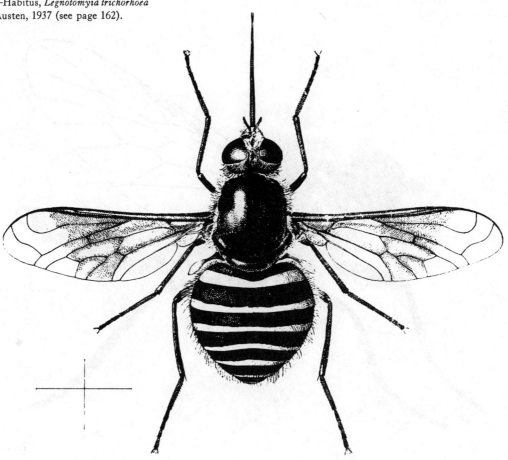

TEXT-FIGURE 29.—Habitus, female paratype of *Heterotropus nigrithorax*, new species, after Frans J. J. Francois (see page 223).

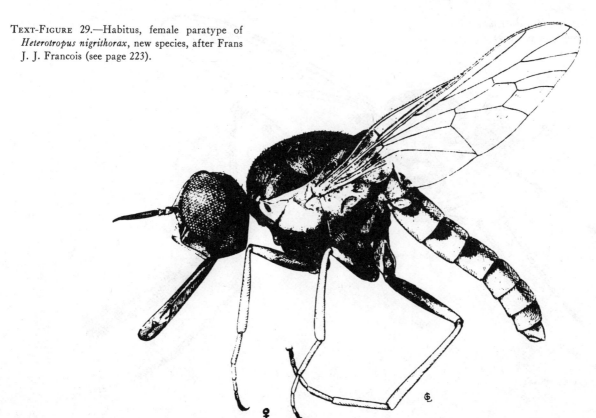

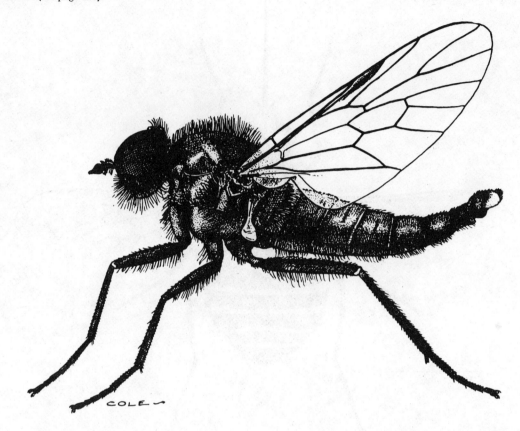

Text-Figure 30.—Habitus, *Caenotus minutus* Cole, after Cole (see page 226).

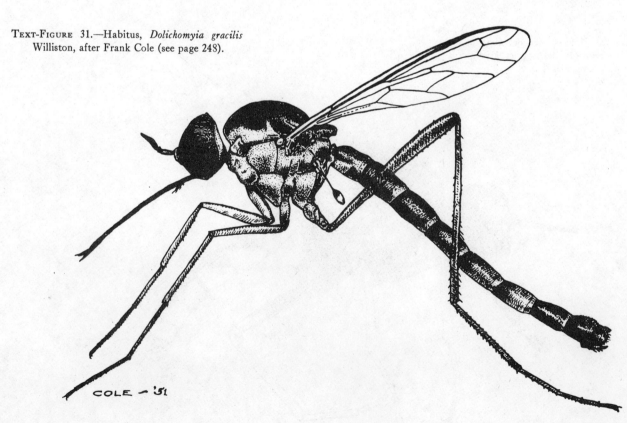

Text-Figure 31.—Habitus, *Dolichomyia gracilis* Williston, after Frank Cole (see page 248).

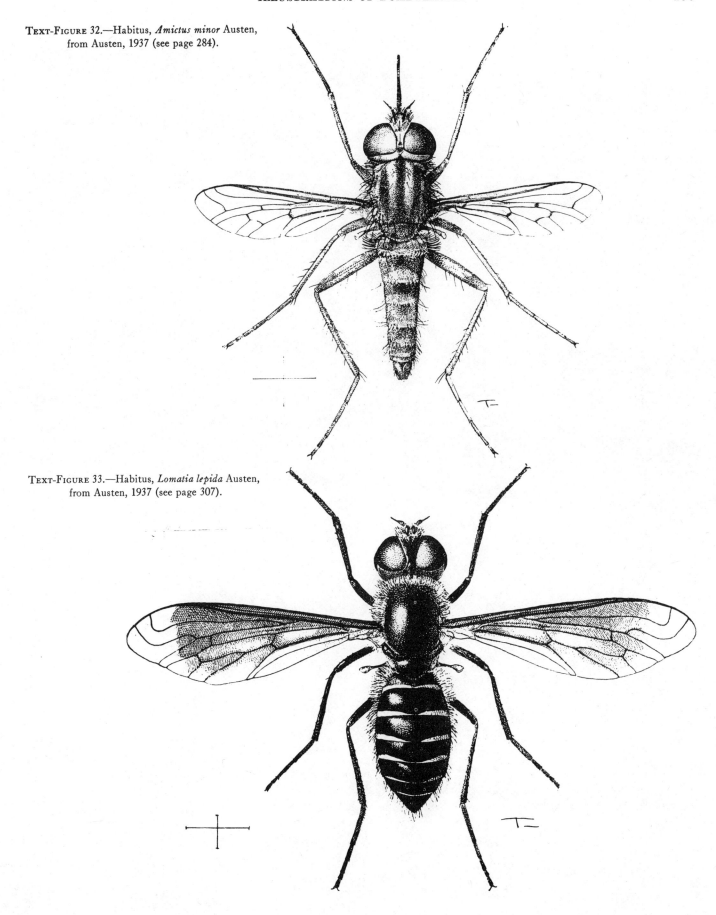

Text-Figure 32.—Habitus, *Amictus minor* Austen, from Austen, 1937 (see page 284).

Text-Figure 33.—Habitus, *Lomatia lepida* Austen, from Austen, 1937 (see page 307).

Illustrations of Bombyliidae

	Figures	Pages
Antennae	1– 235	463–484
Wings	236– 431c	485–500
Heads	432– 769	501–540
Miscellaneous	770– 785	541
Genitalia	786–1029	542–572

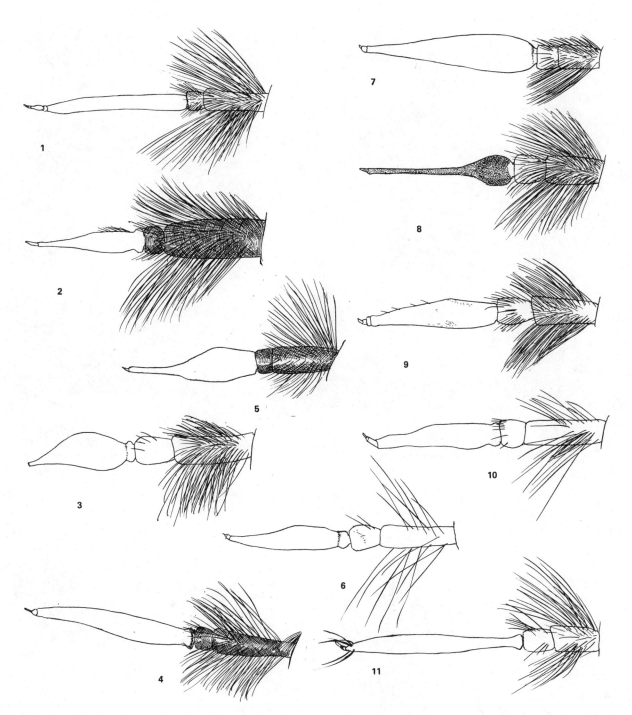

Figures 1–11.—1, *Bombylius major* Linné. 2, *Acanthogeron senex* Wiedemann. 3, *Zinnomyia karooensis* Hesse, paratype. 4, *Bombylius (Triplasius)* sp., Algeria. 5, *Bombylodes eximius* Becker. 6, *Systoechus oreas* Osten Sacken. 7, *Bombylius punctatus* Fabricius. 8, *Anastoechus trisignatus* Portschinsky, female. 9, *Parabombylius coquilletti* Williston. 10, *Systoechus ctenopterus* Mikan. 11, *Sisyromyia decorata* Walker.

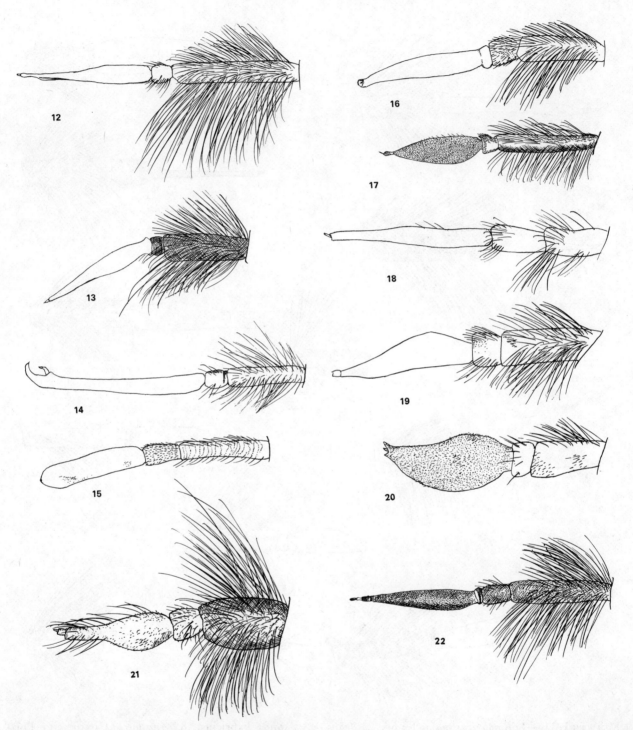

FIGURES 12–22.—12, *Paratoxophora cuthbertsoni* Engel. 13, *Lissomerus niveicomatus* Austen, holotype. 14, *Brychosoma pulchra*, new species. 15, *Eclimus gracilis* Loew. 16, *Thevenemyia magnus* Osten Sacken. 17, *Tillyardomyia gracilis* Tonnoir. 18, *Acrophthalmyda sphenoptera* Loew, male. 19, *Thevenemyia (Arthroneura) tridentata* Hull. 20, *Nectaropota setigera* Philippi. 21, *Platamomyia depressus* Loew. 22, *Eusurbus nigrocinctus* Roberts.

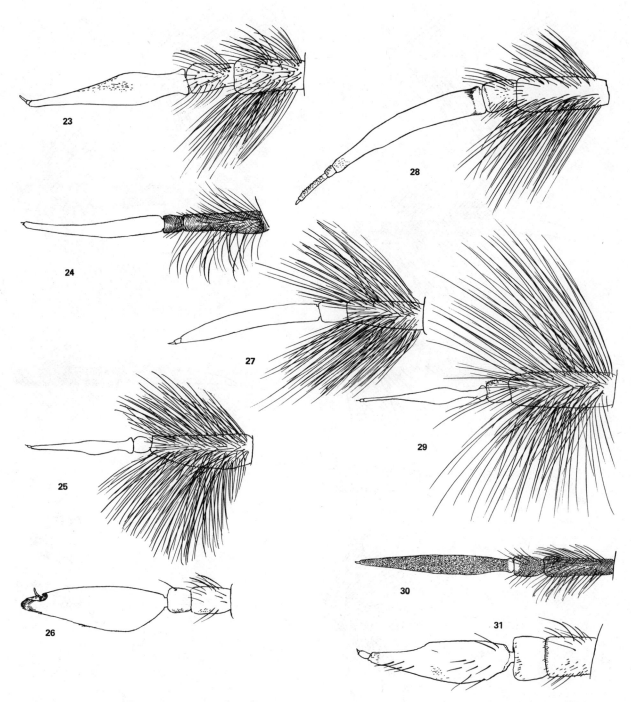

Figures 23–31.—23, *Laurella auripila* Hull. 24, *Gonarthrus cylindricus* Bezzi. 25, *Dischistus plumipalpis* Hesse. 26, *Neodischistus collaris* Painter. 27, *Bombylisoma minimus* Schrank. 28, *Pilosia flavopilosa*, new species. 29, *Dischistus mystax* Wiedemann. 30, *Sericusia lanata* Edwards. 31, *Legnotomyia fascipennis* Bezzi.

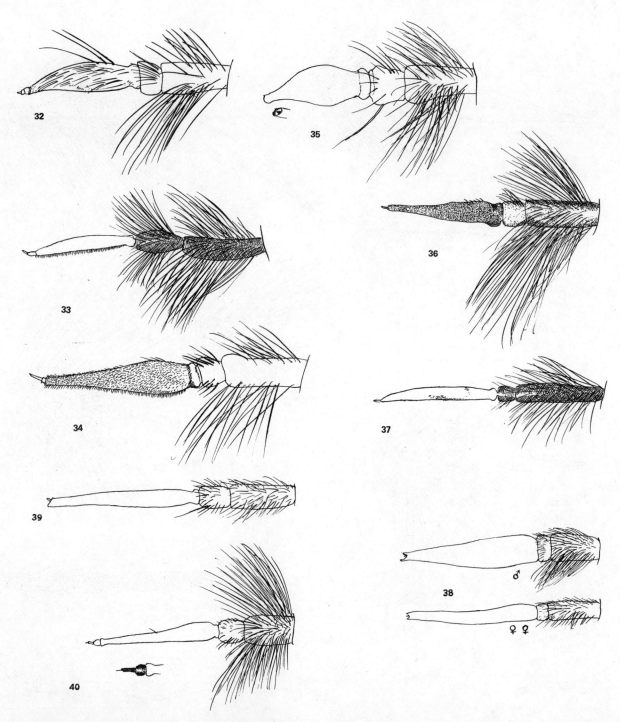

FIGURES 32–40.—32, *Sosiomyia carnata* Bezzi, male. 33, *Conophorina bicellaris* Becker, male. 34, *Adelidea anomala* Wiedemann. 35, *Geminaria canalis* Coquillett. 36, *Hallidia plumipilosa* Hull. 37, *Doliogethes aridricolus* Hesse, male, paratype. 38, *Bromoglycis robustus* Hull. 39, *Cryomyia argyropila*, new species. 40, *Sericosoma squamiventris* Edwards.

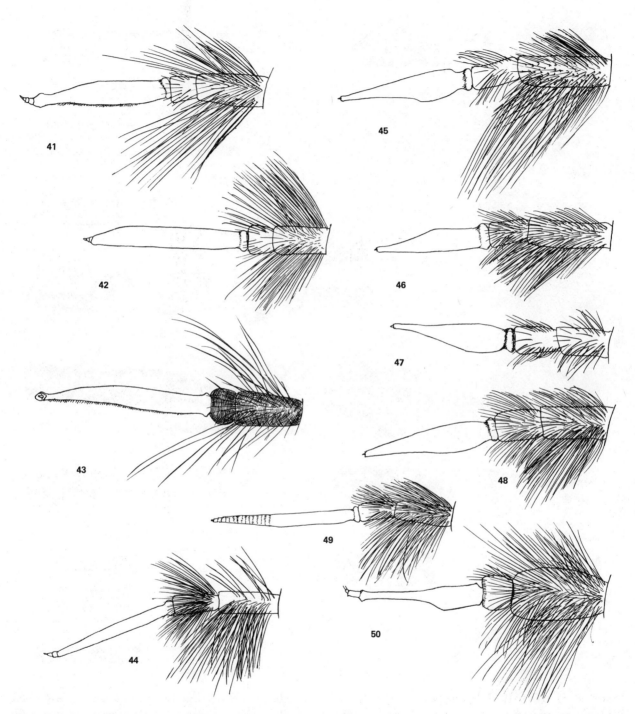

FIGURES 41–50.—41, *Chasmoneura argyropygus* Wiedemann. 42, *Chasmoneura pectoralis* Loew. 43, *Lepidochlanus fimbriatus* Hesse, male, paratype. 44, *Cacoplox griseata* Hull. 45, *Lordotus gibbus* Loew. 46, *Lordotus zonus* Coquillett. 47, *Lordotus bipartitus* Painter, paratype. 48, *Lordotus diversus* Coquillett. 49, *Euprepina nuda* Hull. 50, *Sericusia* sp.

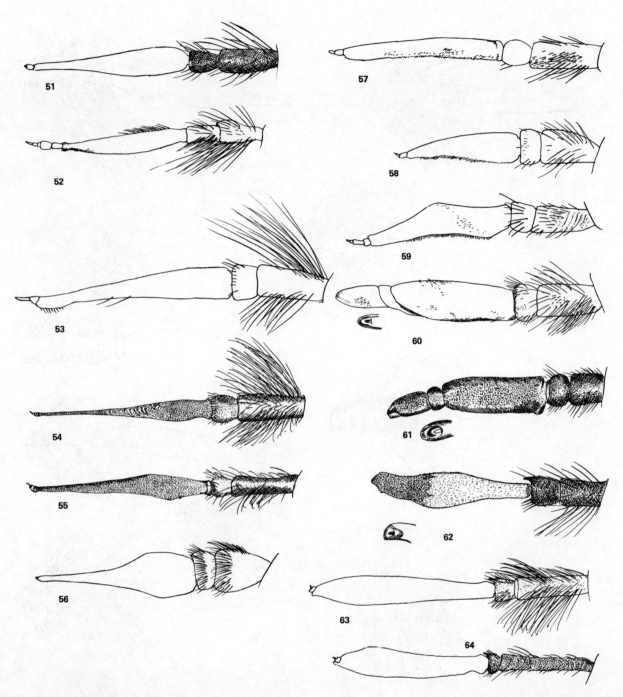

Figures 51–64.—51, *Eurycarenus laticeps* Loew. 52, *Eurycarenus dichopticus* Bezzi. 53, *Triploechus heteroneurus* Macquart. 54, *Heterostylum croceum* Painter, paratype. 55, *Heterostylum robustum* Osten Sacken. 56, *Gyrocraspedum pleskei* Becker. 57, *Crocidium* sp. 58, *Crocidium namaquense* Hesse. 59, *Crocidium phaeopteralis* Hesse. 60, *Desmatomyria binotata* Painter, paratype. 61, *Desmatomyia anomala* Williston. 62, *Hyperusia soror* Bezzi. 63, *Corsomyza brevicornis* Hesse. 64, *Corsomyza simplex* Wiedemann.

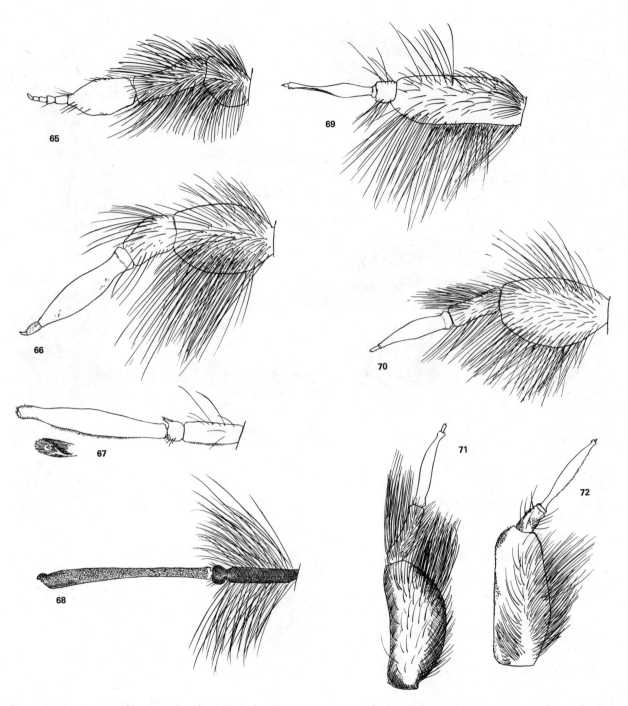

FIGURES 65–72.—65, *Aldrichia erhmanii* Coquillett. 66, *Conophorus cristatus* Painter. 67, *Megapalpus capensis* Wiedemann. 68, *Corsomyza nigripes* Wiedemann, male. 69, *Conophorus virescens* Fabricius. 70, *Conophorus fenestratus* Osten Sacken. 71, *Conophorus fenestratus* Osten Sacken, male, dorsal aspect. 72, *Conophorus virescens* Fabricius, male, dorsal aspect.

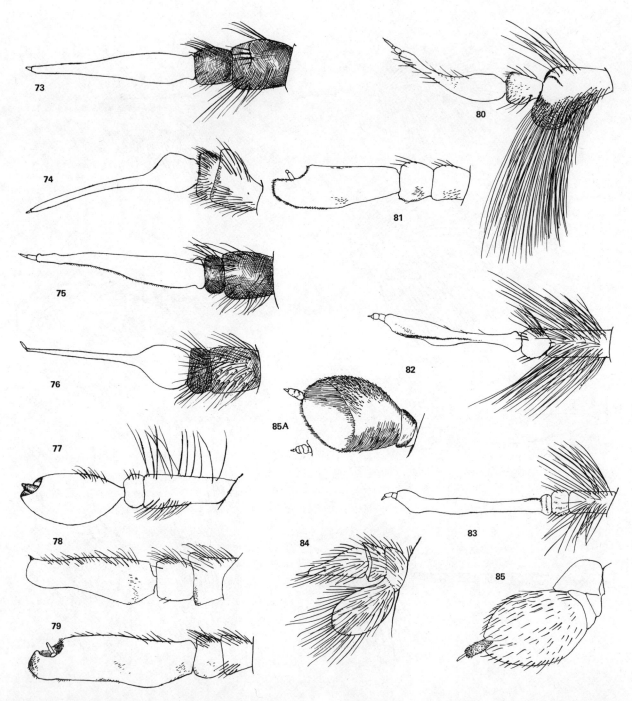

FIGURES 73–85A.—73, *Cytherea obscura* Fabricius, medial aspect. 74, *Callostoma fascipenne* Macquart, dorsal aspect. 75, *Cytherea (Chalcochiton) pallasii* Loew. 76, *Callostoma fascipenne* Macquart, lateral aspect. 77, *Oligodranes cincturus* Coquillett. 78, *Usia florea* Fabricius, male. 79, *Usia aenea* Rossi. 80, *Oniromyia pachyceratus* Bigot. 81, *Apolysis druias* Melander. 82, *Pantarbes capito* Osten Sacken. 83, *Pantarbes willistoni* Osten Sacken. 84, *Xenoprosopa paradoxa* Hesse, after Hesse. 85, *Apystomyia elinguis* Melander, paratype. 85A, *Glabellula* sp.

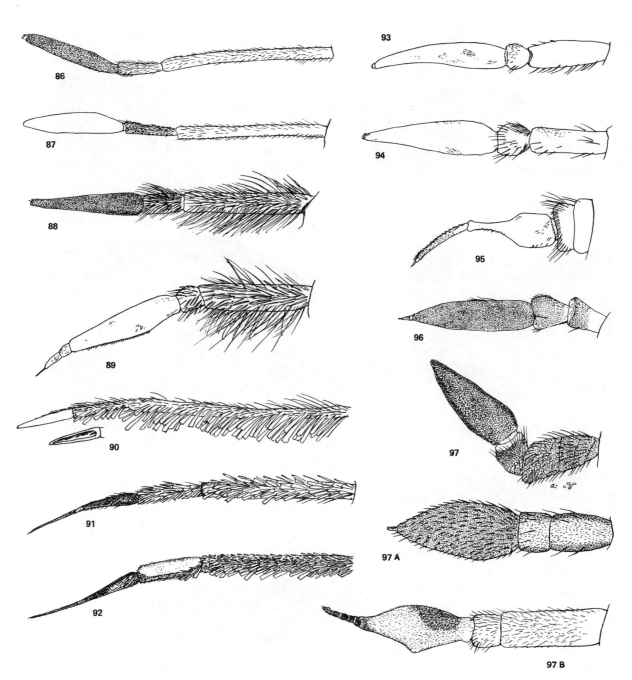

Figures 86–97b.—86, *Systropus quadripunctata* Seguy. 87, *Systropus angulatus* Karsch. 88, *Cyrtomyia chilensis* Paramonov. 89, *Marmasoma sumptuosa* White. 90, *Lepidophora lepidocera* Wiedemann. 91, *Toxophora amphitea* Walker, male, lateral aspect. 92, *Toxophora amphitea* Walker, male, dorsal aspect. 93, *Dolichomyia gracilis* Williston. 94, *Dolichomyia* sp., Chile. 95, *Cyrtomorpha paganica* White. 96, *Heterotropus aegyptiacus* Paramonov. 97, *Zaclava* sp. 97a, *Prorates claripennis* Melander. 97b, *Apatomyza punctipennis* Wiedemann, holotype.

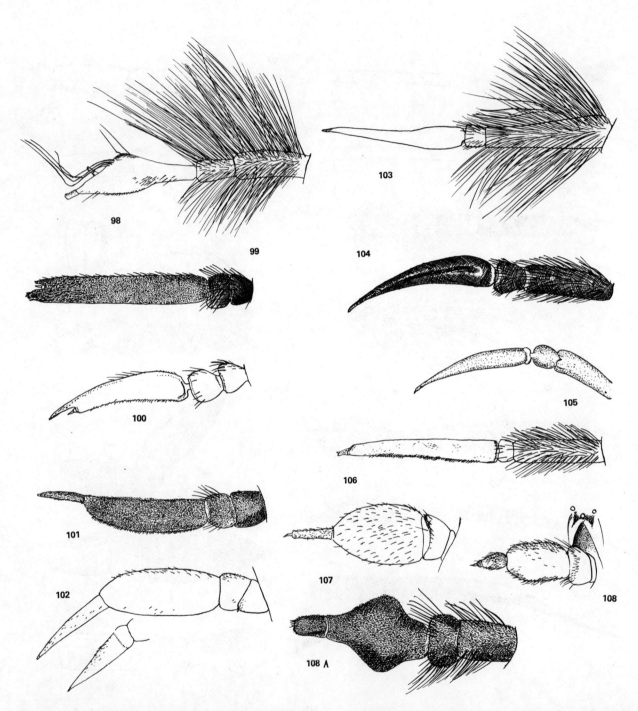

FIGURES 98–108A.—98, *Acreotrichus gibbicornis* Macquart. 99, *Phthiria* sp. 100, *Phthiria gaedei* Wiedemann, after Engel. 101, *Mythicomyia monacha* Melander, paratype. 102, *Mythicomyia armada* Cresson, paratype. 103, *Pseudoamictus heteropterus* Wiedemann. 104, *Geron* sp., New Mexico. 105, *Geron gibbosus* Olivier. 106, *Amictogeron meromelas* Hesse, paratype. 107, *Empidideicus efflatouni* Engel. 108, *Empidideicus humilis* Melander. 108A, *Caenotus hospes* Melander.

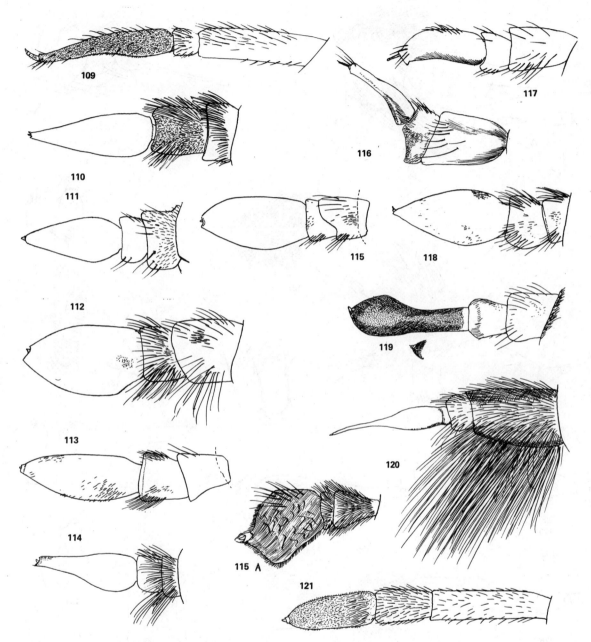

FIGURES 109–121.—109, *Amictus validus* Loew. 110, *Amphicosmus cincturus* Williston. 111, *Pantostomus tinctellipennis* Hesse, paratype. 112, *Paracosmus morrisoni* Osten Sacken. 113, *Tomomyza anthracoides* Wiedemann. 114, *Amphicosmus elegans* Coquillett. 115, *Tomomyza philoxera* Hesse, paratype. 115A, *Proracthes longirostris* Bezzi. 116, *Cyllenia marginata* Loew, dorsal aspect. 117, *Cyllenia marginata* Loew, lateral aspect. 118, *Metacosmus nitidus* Cole. 119, *Paracosmus (Actherosia) rubicundus* Melander. 120, *Peringueyimyia capensis* Wiedemann. 121, *Neosardus principius* Roberts.

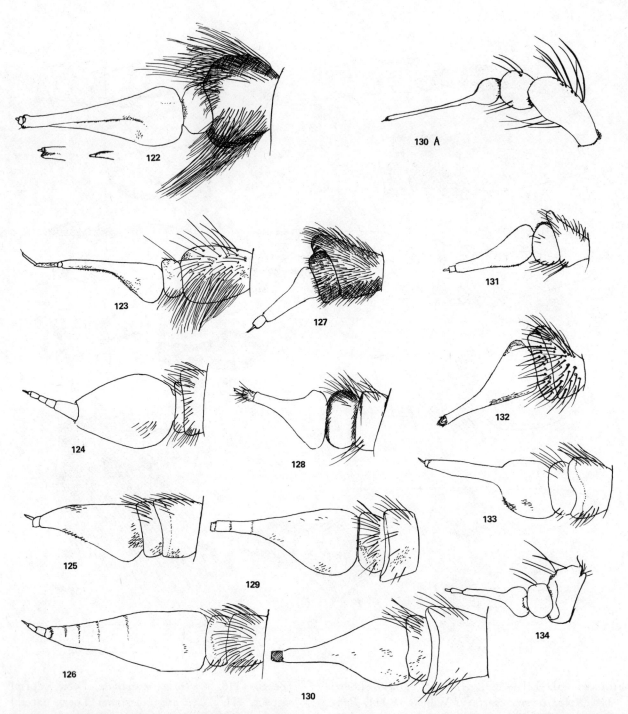

Figures 122–134.—122, *Ogcodocera leucoprocta* Wiedemann. 123, *Pteraulax setaria* Hesse. 124, *Eucessia* sp. 125, *Plesiocera psammophila* Hesse, paratype. 126, *Aphoebantus* sp., Texas. 127, *Plesiocera algira* Macquart. 128, *Xeramoeba gracilis* Bowden, paratype. 129, *Epacmoides biumbonatum* Bezzi. 130, *Prorostoma integrum* Bezzi. 130A, *Lomatia (Canariellum) brunnipennis* Macquart, after Engel. 131, *Stomylomyia europae* Loew. 132, *Coryprosopa lineata* Hesse, paratype. 133, *Desmatoneura argentifrons* Williston. 134, *Petrorossia hesperus* Rossi, after Engel.

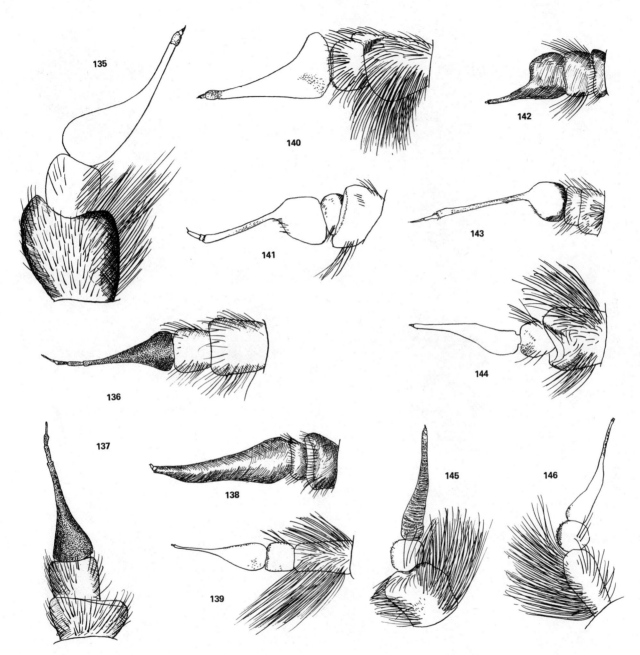

FIGURES 135–146.—135, *Myonema humilis* Roberts, dorsal aspect. 136, *Antonia suavissima* Loew. 137, *Antonia suavissima* Loew, dorsal aspect. 138, *Antonia* sp., Australia. 139, *Lomatia* sp., So. Africa. 140, *Myonema humilis* Roberts. 141, *Epacmus* sp. 142, *Docidomyia puellaris* White. 143, *Epacmus litus* Coquillett. 144, *Lomatia belzebul* Fabricius. 145, *Lomatia belzebul* Fabricius, dorsal aspect. 146, *Lomatia* sp., So. Africa, dorsal aspect.

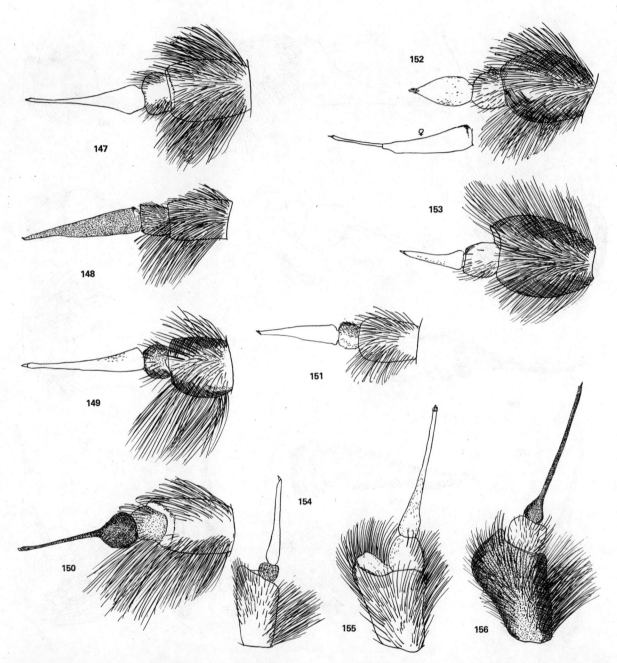

Figures 147–156.—147, *Lyophlaeba* sp. 148, *Ylasoia pegasus* Wiedemann. 149, *Doddosia picta* Edwards. 150, *Comptosia (Paradosia) tendens* Walker. 151, *Lyophlaeba (Macrocondyla) montana* Philippi. 152, *Oncodosia planus* Walker, male, female. 153, *Oncodosia planus* Walker, dorsal aspect. 154, *Lyophlaeba (Macrocondyla) montana* Philippi, dorsal aspect. 155, *Doddosia picta* Edwards, dorsal aspect. 156, *Comptosia fascipennis* Macquart, dorsal aspect.

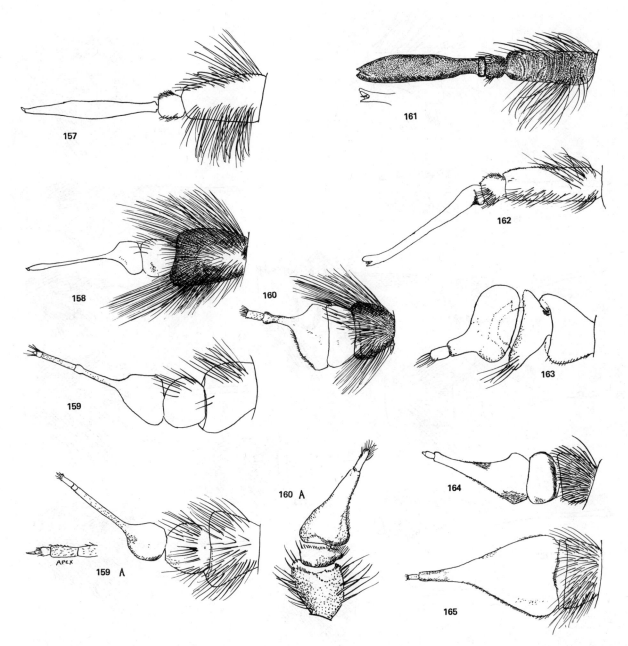

FIGURES 157–165.—157, *Lyophlaeba lugubris* Rondani. 158, *Comptosia ocellata* Newman, male. 159, *Anthrax argyropygus* Wiedemann. 159A, *Anthrax analis* Say. 160, *Dicranoclista vandykei* Coquillett. 160A, *Anthrax (Satyramoeba) etrusca* Fabricius, after Engel. 161, *Lyophlaeba* sp. 162, *Lyophlaeba* sp., dorsal aspect. 163, *Anthrax aethiops* Fabricius, after Engel. 164, *Walkeromyia luridus* Walker. 165, *Walkeromyia* sp., male, Nova Teutonia.

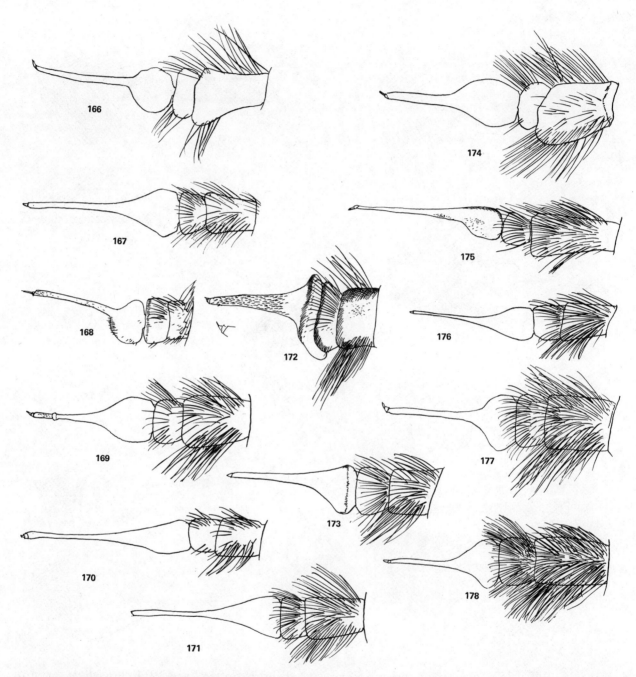

Figures 166–178.—166, *Oestranthrax obesus* Loew. 167, *Neodiplocampta paradoxus* Jaennicke. 168, *Mancia nana* Coquillett. 169, *Thyridanthrax afer* Fabricius. 170, *Chrysanthrax cypris* Meigen. 171, *Thyridanthrax perspicillaris* Loew. 172, *Astrophanes adonis* Osten Sacken. 173, *Deusopora sapphirina* Hull. 174, *Marleyimyia natalensis* Hesse, after Hesse. 175, *Cyananthrax cyanoptera* Wiedemann. 176, *Paravilla castanea* Jaennicke. 177, *Hemipenthes morio* Linné. 178, *Hemipenthes maurus* Linné.

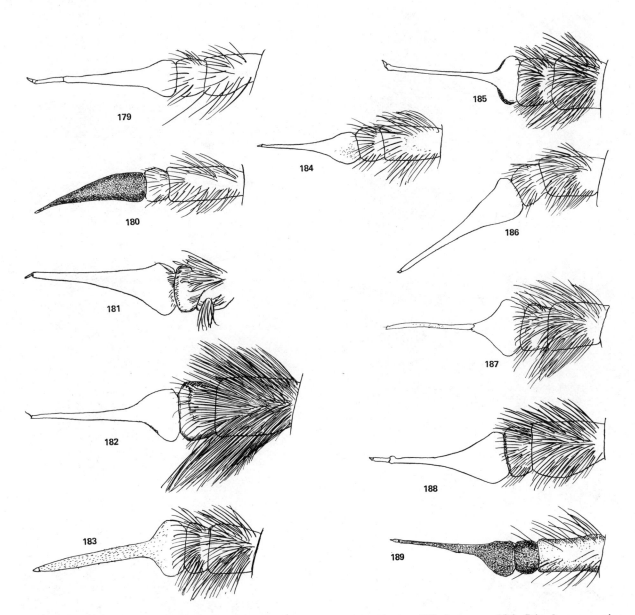

FIGURES 179–189.—179, *Chrysanthrax edititia* Say. 180, *Poecilanthrax effrenus* Coquillett. 181, *Villa cingulata* Wiedemann. 182, *Villa alternata* Say. 183, *Hemipenthes lepidotus* Osten Sacken. 184, *Paravilla xanthina* Painter. 185, *Hemipenthes* sp., near *sinuosa* Wiedemann. 186, *Dipalta serpentina* Osten Sacken. 187, *Hemipenthes velutinus* Meigen. 188, *Thyridanthrax pertusa* Loew. 189, *Neodiplocampta* (*Agitonia*) *sepia* Hull.

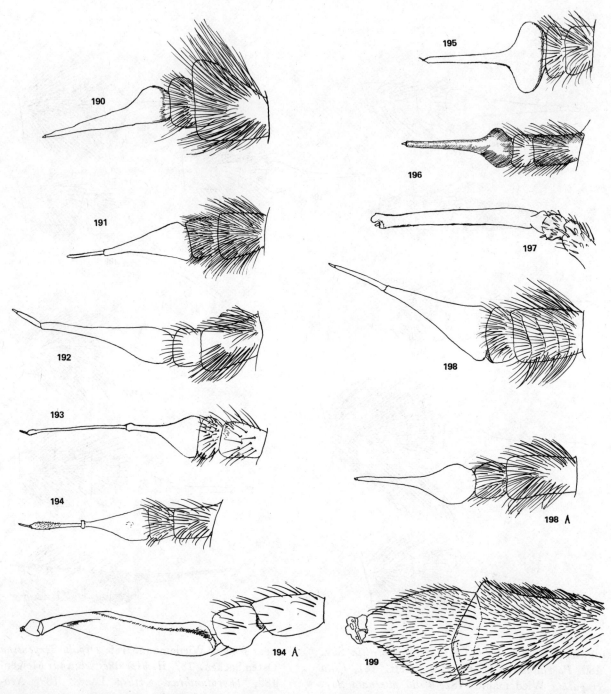

FIGURES 190–199.—190, *Poecilanthrax californicus* Cole, dorsal aspect. 191, *Litorrhynchus lar* Fabricius. 192, *Hyperalonia morio* Fabricius. 193, *Isotamia daveyi* Bezzi, paratype of *Ogilviella tridenta* Paramonov. 194, *Heteralonia kaokoensis* Hesse. 194A, *Diatropomma carcassoni* Bowden, after Bowden. 195, *Diplocampta secunda* Paramonov. 196, *Pseudopenthes fenestrata* Roberts. 197, *Exoprosopa subfasciata* Engel, after Engel. 198, *Exoprosopa jacchus* Fabricius. 198A, *Poecilanthrax californicus* Cole, lateral aspect. 199, *Diatropomma carcassoni* Bowden, after Bowden.

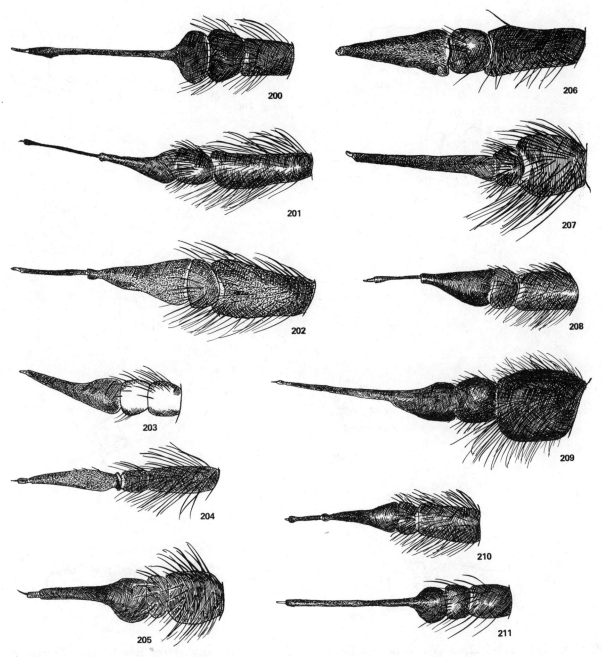

FIGURES 200–211.—200, *Stonyx* sp. 201, *Colossoptera latipennis* Brunnetti. 202, *Ligyra tenebrosa* Paramonov. 203, *Micomitra opalina* Hesse. 204, *Sparnopolius lherminierii* Macquart. 205, *Aphoebantus* sp. 206, *Atrichochira pediformis* Bezzi. 207, *Bryodemina valida* Wiedemann. 108, *Exoprosopa minos* Meigen. 209, *Comptosia fascipennis* Macquart. 210, *Exoprosopa apicalis* Klug. 211, *Lepidanthrax* sp.

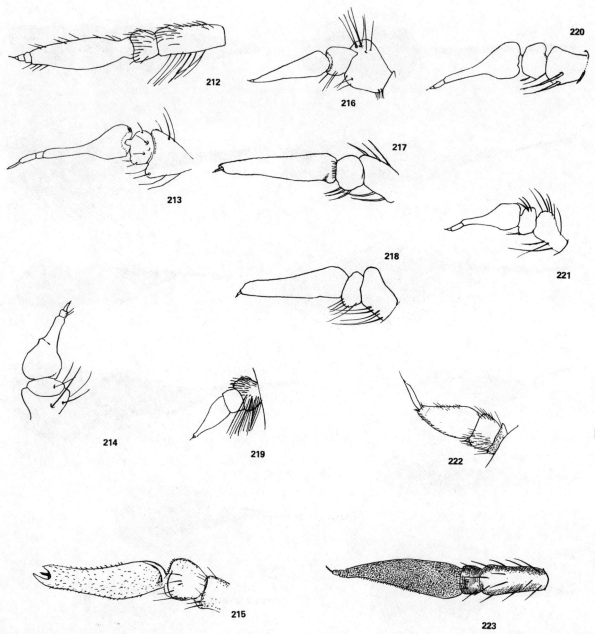

Figures 212–223.—212, *Sinaia kneuckeri* Becker, after Efflatoun. 213, *Aphoebantus (Cononedys) erythraspis* Hermann, after Engel. 214, *Prothaplocnemis anthracina* Becker, after Engel. 215, *Phthiria pulicaria* Mikan, after Engel. 216, *Lomatia bella* Loew, after Engel. 217, *Thyridanthrax misellus* Loew, after Engel. 218, *Thyridanthrax semifuscus* Engel, after Engel. 219, *Myonema* sp. 220, *Thyridanthrax vagans* Loew, after Engel. 221, *Thyridanthrax stigmulis* Klug, after Engel. 222, *Cyrtomorpha* sp. 223, *Adelogenys culicoides* Hesse.

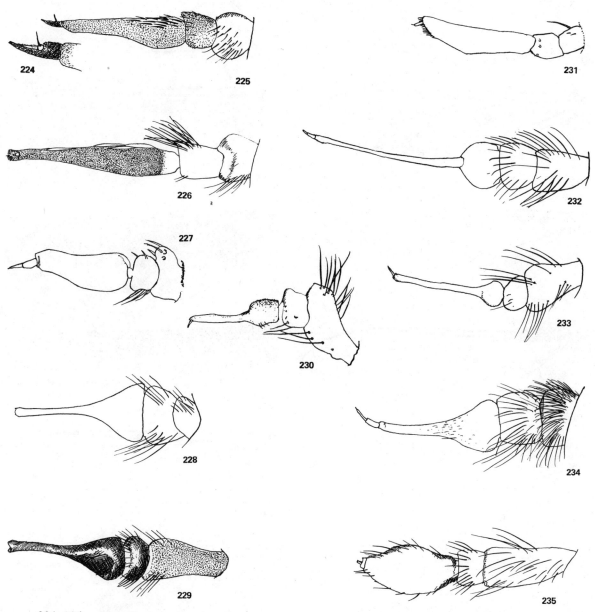

FIGURES 224–235.—224, *Nomalonia afra* Macquart, apex of antenna. 225, *Nomalonia afra* Macquart. 226, *Henica longirostris* Wiedemann. 227, *Aphoebantus claripennis* Becker, after Engel. 228, *Chionamoeba choreutes* Bowden, paratype. 229, *Synthesia fucoides* Bezzi. 230, *Oestranthrax karavajevi* Paramonov, after Engel. 231, *Oligodranes flavus* Paramonov, after Engel. 232, *Rhynchanthrax parvicornis* Loew. 233, *Oestranthrax brunnescens* Loew, after Engel. 234, *Petrorossia phthinoxantha* Hesse. 235, *Sphenoidoptera varipennis* Williston.

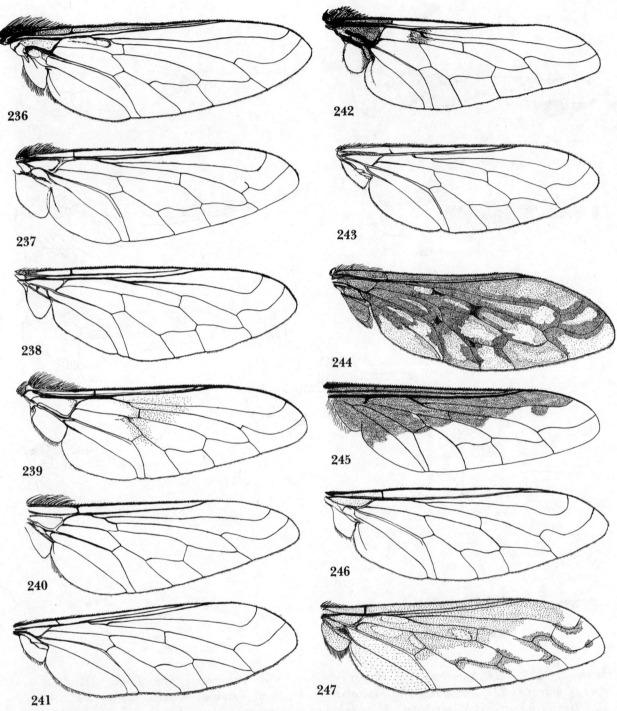

Figures 236–247.—236, *Systoechus vulgaris* Osten Sacken. 237, *Acanthogeron senex* Wiedemann. 238, *Lissomerus niveicomatus* Austen, holotype. 239, *Anastoechus trisignatus* Portschinsky, female. 240, *Anastoechus nitidulus* Fabricius. 241, *Sisyrophanus pyrrhocerus* Bezzi, type. 242, *Zinnomyia karooensis* Hesse, paratype. 243, *Sisyrophanus abdominalis* Bezzi, type. 244, *Bombylius* (*Triplasius*) sp., Atlas Mts. 245, *Bombylius major* Linné. 246, *Isocnemus nemestrinus* Bezzi, type. 247, *Bombylius* (*Triplasius*) *bivittatus* Loew.

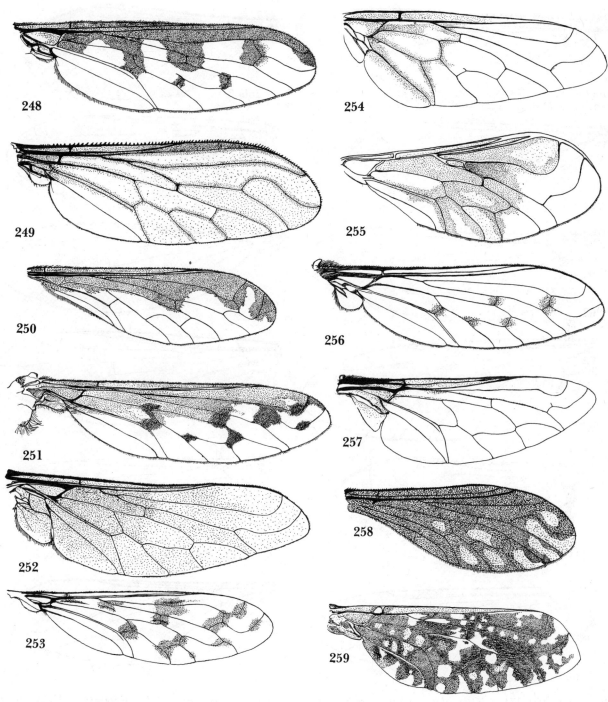

FIGURES 248–259.—248, *Thevenemyia (Arthroneura) tridentatus* Hull. 249, *Thevenemyia* sp. 250, *Tillyardomyia gracilis* Tonnoir. 251, *Acropthalmyda sphenoptera* Loew. 252, *Sisyromyia auratus* Walker. 253, *Nectaropota setigera* Philippi. 254, *Legnotomyia cinerea* Austen, type. 255, *Legnotomyia trichorhoea* Loew. 256, *Adelidea anomala* Wiedemann. 257, *Eusurbus nigrocinctus* Roberts. 258, *Eclimus gracilis* Loew. 259, *Proracthes longirostris* Bezzi.

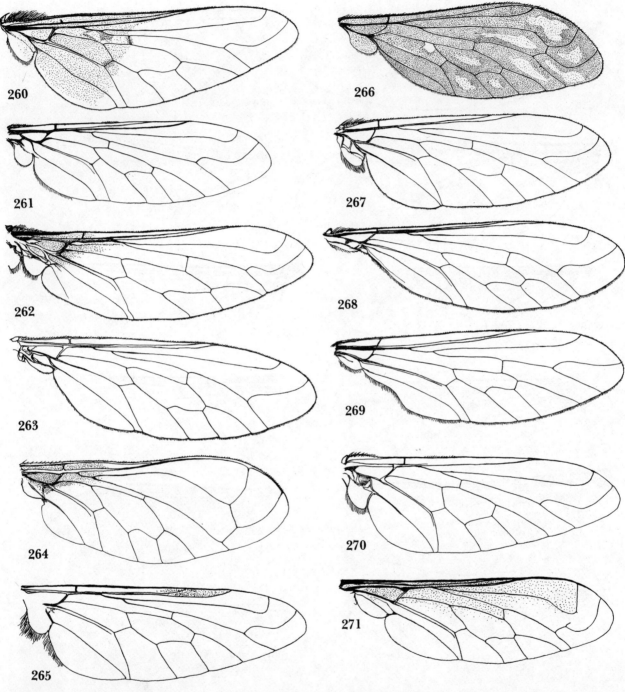

Figures 260–271.—260, *Doliogethes aridicolus* Hesse, paratype. 261, *Neodischistus collaris* Painter. 262, *Bombylodes eximius* Becker. 263, *Gonarthrus cylindricus* Bezzi. 264, *Othniomyia tylopelta* Hesse, type. 265, *Cryomyia argyropila*, new species. 266, *Adelidea pterosticta* Hesse, paratype. 267, *Sericusia lanata* Edwards. 268, *Paratoxophora cuthbertsoni* Engel. 269, *Conophorina bicellaris* Becker. 270, *Lepidochlanus fimbriatus* Hesse, paratype. 271, *Bromoglycis robustus* Hull.

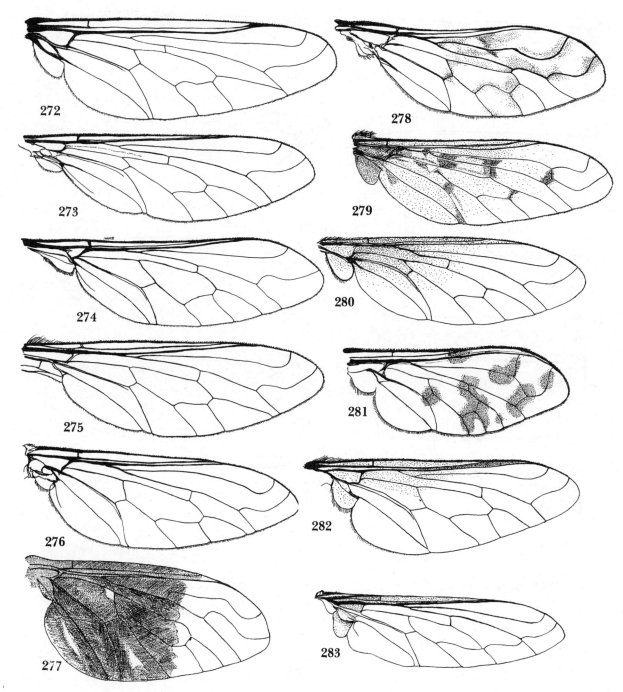

Figures 272–283.—272, *Parabombylius coquilletti* Williston. 273, *Bombylisoma minimus* Schrank. 274, *Laurella auripila* Hull. 275, *Dischistus mystax* Wiedemann. 276, *Cacoplox griseata* Hull. 277, *Deusopora sapphirina* Hull. 278, *Hallidia plumipilosa* Hull. 279, *Sosiomyia carnata* Bezzi. 280, *Chasmoneura argyropygus* Wiedemann. 281, *Geminaria canalis* Coquillett. 282, *Pilosia flavopilosa*, new species. 283, *Heterostylum robustum* Osten Sacken.

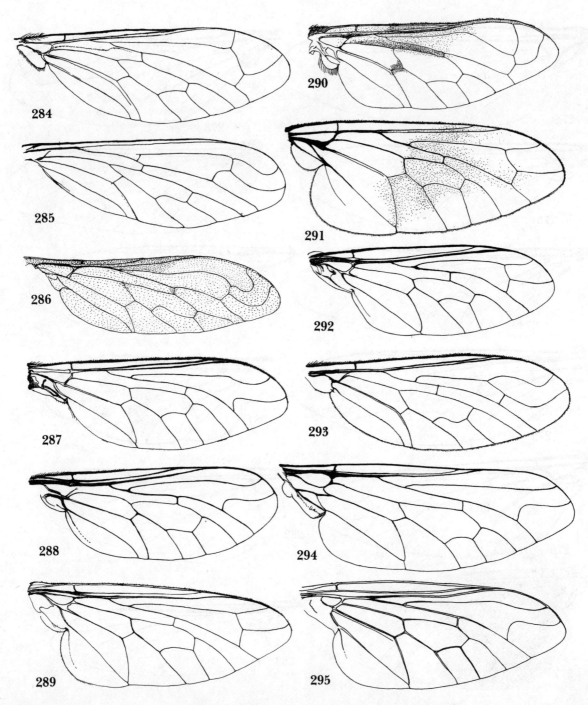

FIGURES 284–295.—284, *Conophorus fuliginosus* Wiedemann. 285, *Aldrichia ehrmanii* Coquillett. 286, *Platamomyia depressus* Loew. 287, *Corsomyza nigripes* Wiedemann. 288, *Megapalpus capensis* Wiedemann. 289, *Gnumyia brevirostris* Bezzi. 290, *Lordotus gibbus* Loew. 291, *Mariobezzia griseohirta* Nurse. 292, *Callynthrophora capensis* Schiner. 293, *Exepacmus johnsoni* Coquillett. 294, *Zyxmyia megachile* Bowden, after Bowden. 295, *Hyperusia luteifacies* Bezzi, type.

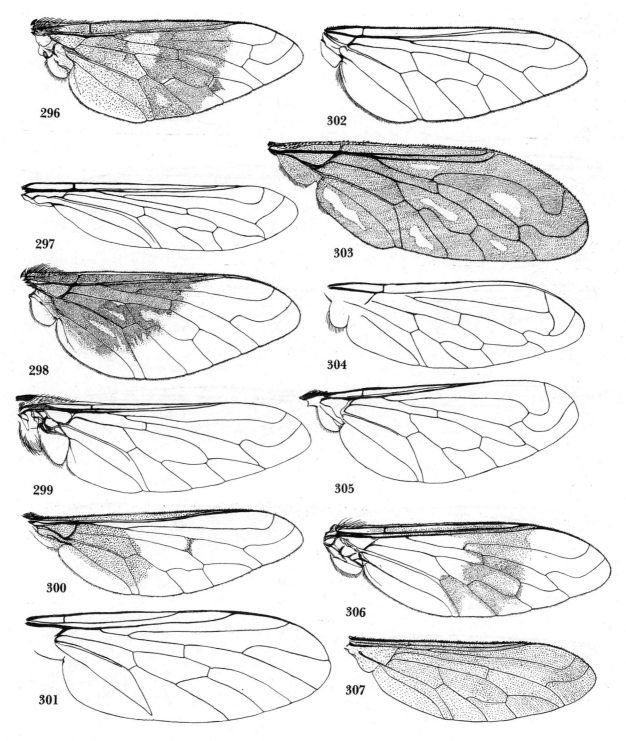

FIGURES 296–307.—296, *Cytherea obscura* Fabricius. 297, *Oniromyia pachyceratus* Bigot. 298, *Cytherea (Chalcochiton) pallasii* Wiedemann. 299, *Eurycarenus laticeps* Loew. 300, *Efflatounia* sp. 301, *Adelogenys culicoides* Hesse. 302, *Sericosoma squamiventris* Edwards, type. 303, *Gyrocraspedum pleskei* Becker. 304, *Triploechus heteroneurus* Macquart. 305, *Pantarbes pusio* Osten Sacken. 306, *Callostoma fascipenne* Macquart. 307, *Karakumia nigra* Paramonov, after Paramonov.

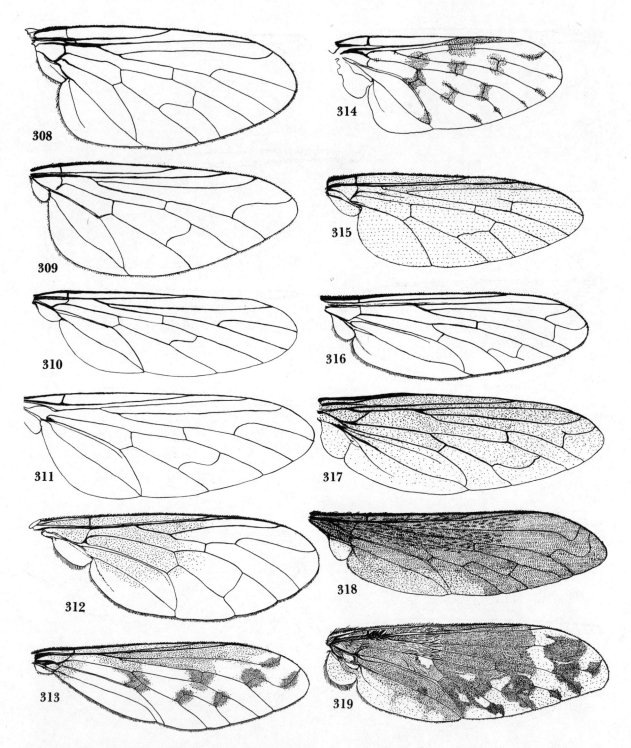

FIGURES 308–319.—308, *Phthiria* sp. 309, *Geron* sp. 310, *Amictogeron* sp. 311, *Pseudoamictus heteropterus* Wiedemann. 312, *Usia atrata* Fabricius. 313, *Marmasoma sumptuosa* White. 314, *Phthiria (Poecilognathus) thlipsomyzoides* Jaennicke. 315, *Acreotrichus gibbicornis* Macquart. 316, *Toxophora (Eniconeura) amphitea* Walker. 317, *Toxophora* sp. 318, *Lepidophora lepidocera* Wiedemann. 319, *Cyrtomyia chilensis* Paramonov.

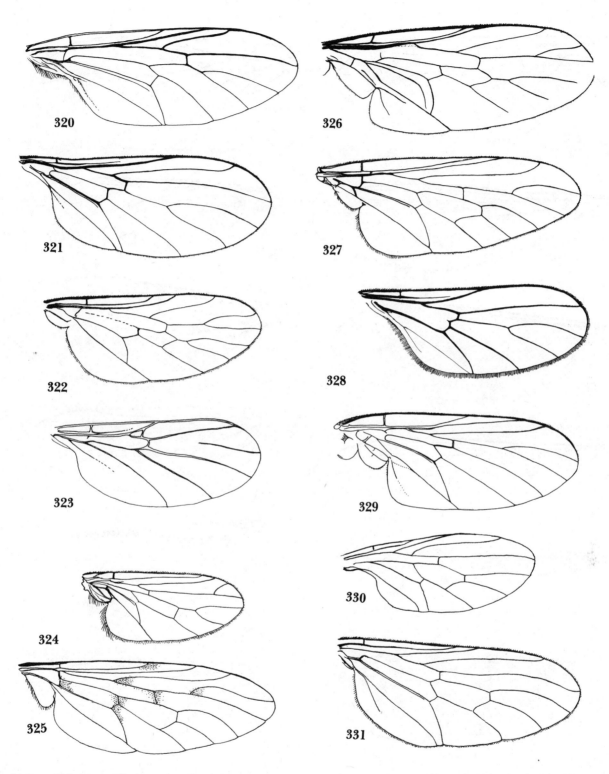

FIGURES 320–331.—320, *Cyrtomorpha paganica* White. 321, *Cyrtosia* sp. 322, *Prorates claripennis* Melander. 323, *Doliopteryx crocea* Hesse, paratype. 324, *Apystomyia elingius* Melander. 325, *Semiramis punctipennis* Becker, after Engel. 326, *Prorates anomalus* Bezzi, after Efflatoun. 327, *Crocidium melanopalis* Hesse, paratype. 328, *Empidideicus turneri* Hesse, paratype. 329, *Apolysis maherniaphila* Hesse, paratype. 330, *Cyrtisiopsis crassirostris* Hesse, paratype. 331, *Platypygus melleus* Bezzi, after Efflatoun.

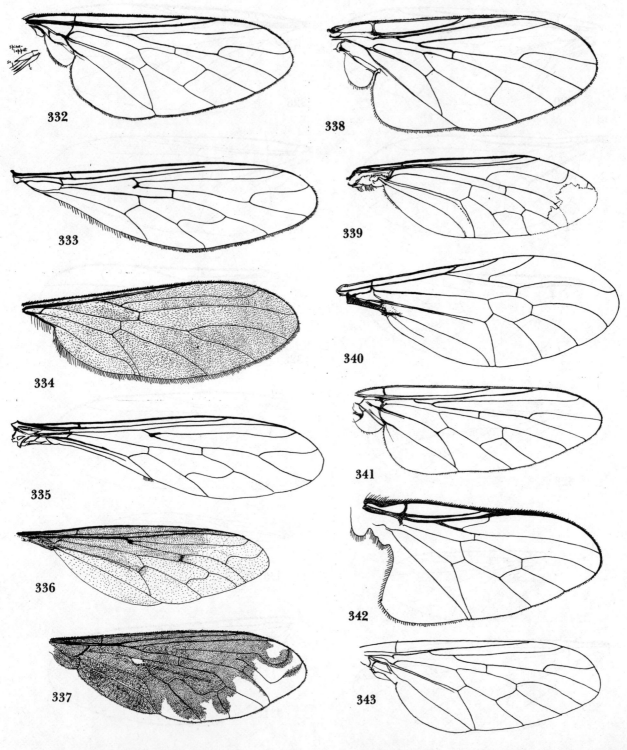

Figures 332–343.—332, *Oligodranes cincturus* Coquillett. 333, *Zaclava* sp. 334, *Ceratolaemus xanthogrammus* Hesse, paratype. 335, *Dolichomyia gracilis* Williston. 336, *Systropus angulatus* Karsch. 337, *Anthrax anthrax* Schrank. 338, *Oligodranes elegans* Hesse, type. 339, *Xenoprosopa paradoxa* Hesse, after Hesse. 340, *Psiloderoides mansfieldi* Hesse, after Hesse. 341, *Mallophthiria lanata* Edwards, type. 342, *Mythicomyia monacha* Melander, paratype. 343, *Desmatomyia anomala* Williston.

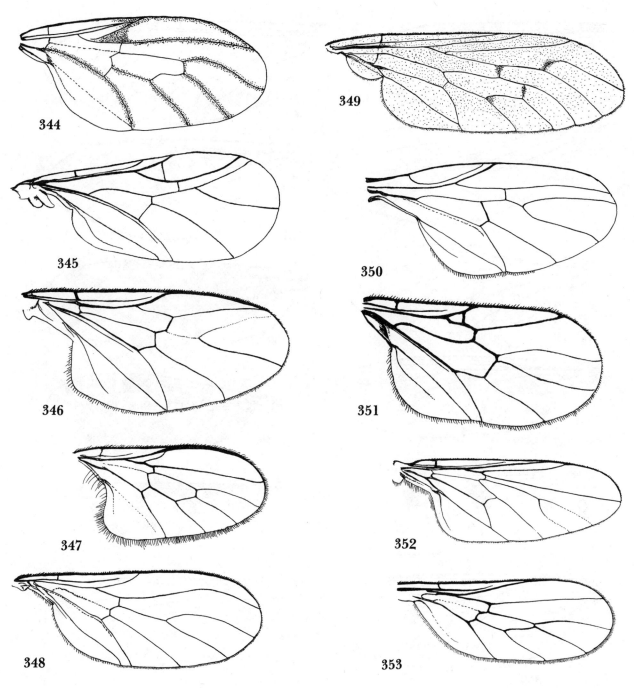

FIGURES 344–353.—344, *Aetheoptilus zuluensis* Hesse. 345, *Pseudoglabellula meridionalis* Hesse, after Hesse. 346, *Leylaiya mimnervia* Efflatoun, after Efflatoun. 347, *Mythicomyia pusillima* Edwards, type. 348, *Empidideicus carthaginiensis* Becker, after Efflatoun. 349, *Apatomyza punctipennis* Wiedemann, from the type. 350, *Euanthobates mellivorus* Hesse, after Hesse. 351, *Glabellula nobilis* Kertesz. 352, *Onchopelma trilineata* Hesse, type. 353, *Anomaloptilus celluliferus* Hesse, paratype.

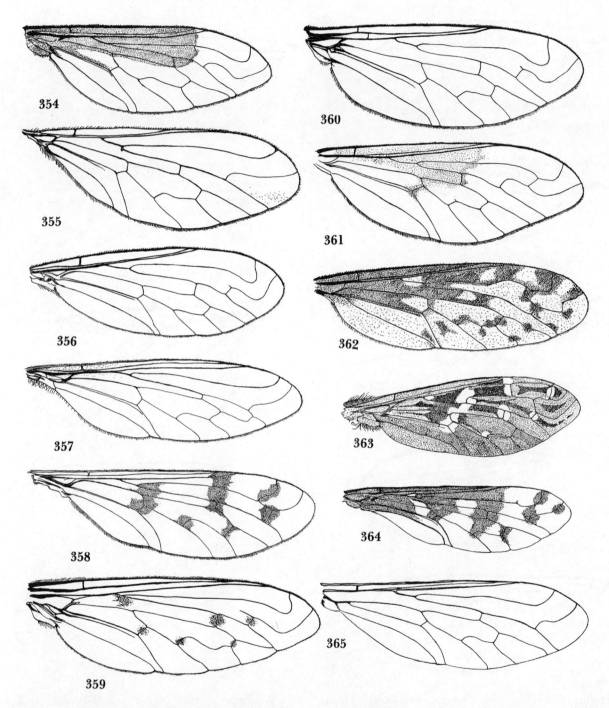

Figures 354–365.—354, *Lomatia sabaea* Fabricius. 355, *Metacosmus nitidus* Cole. 356, *Pantostomus tinctellipennis* Hesse. 357, *Amictus validus* Loew. 358, *Cyllenia marginata* Loew. 359, *Peringueyimyia capensis* Bigot. 360, *Paracosmus morrisoni* Osten Sacken. 361, *Amphicosmus cincturus* Williston. 362, *Tomomyza pictipennis* Bezzi. 363, *Nomalonia afra* Macquart. 364, *Sphenoidoptera varipennis* Williston, type. 365, *Semiramis punctipennis* Becker, after Engel.

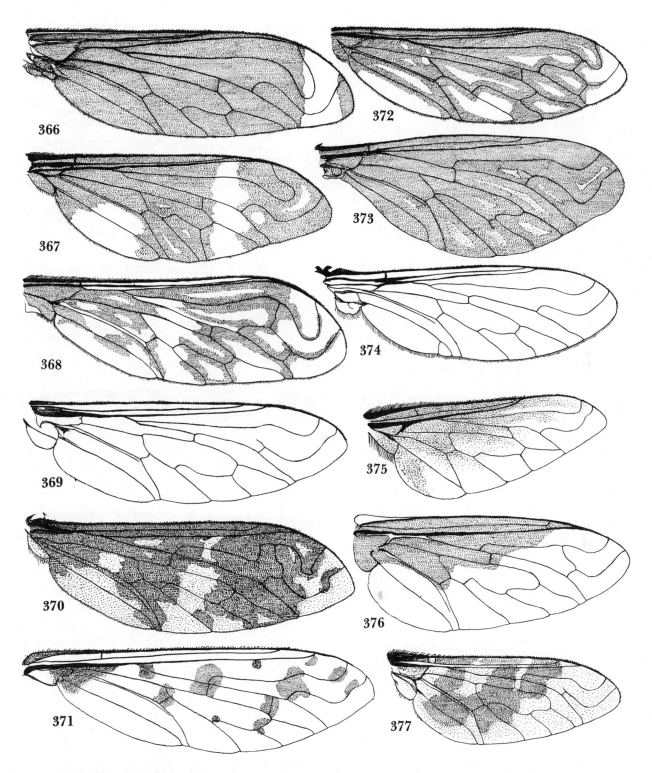

FIGURES 366–377.—366, *Comptosia antipodesia*, new species. 367, *Ylasoia pegasus* Wiedemann. 368, *Lyophlaeba lugubris* Wiedemann. 369, *Aphoebantus* sp. 370, *Dipalta serpentina* Osten Sacken. 371, *Lepidanthrax* sp. 372, *Lyophlaeba* (*Macrocondyla*) sp. 373, *Cyananthrax cyanopterus* Wiedemann. 374, *Eucessia rubens* Coquillett. 375, *Bryodemina valida* Wiedemann, new name. 376, *Exoprosopa minos* Meigen. 377, *Thyridanthrax fenestratus* Fallen.

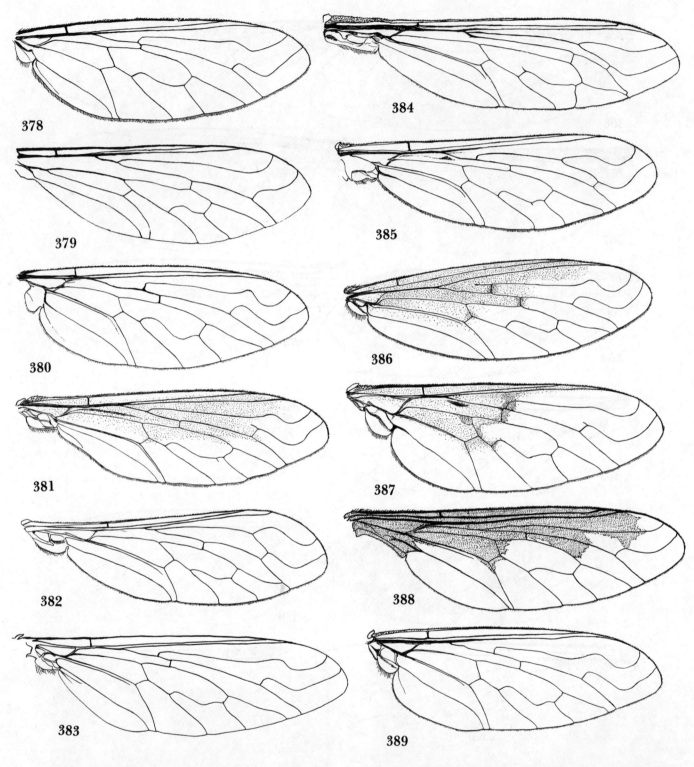

Figures 378–389.—378, *Chiasmella brevipennis* Bezzi, type. 379, *Desmatoneura argentifrons* Williston. 380, *Epacmoides albifrons* Hesse. 381, *Plesiocera algira* Macquart. 382, *Pteraulax setaria* Hesse, paratype. 383, *Petrorossia letho* Wiedemann. 384, *Antonia suavissima* Loew. 385, *Stomylomyia europae* Loew. 386, *Coryprosopa lineata* Hesse. 387, *Chionamoeba sabulonis* Becker. 388, *Myonema humilis* Roberts. 389, *Prorostoma integrum* Bezzi.

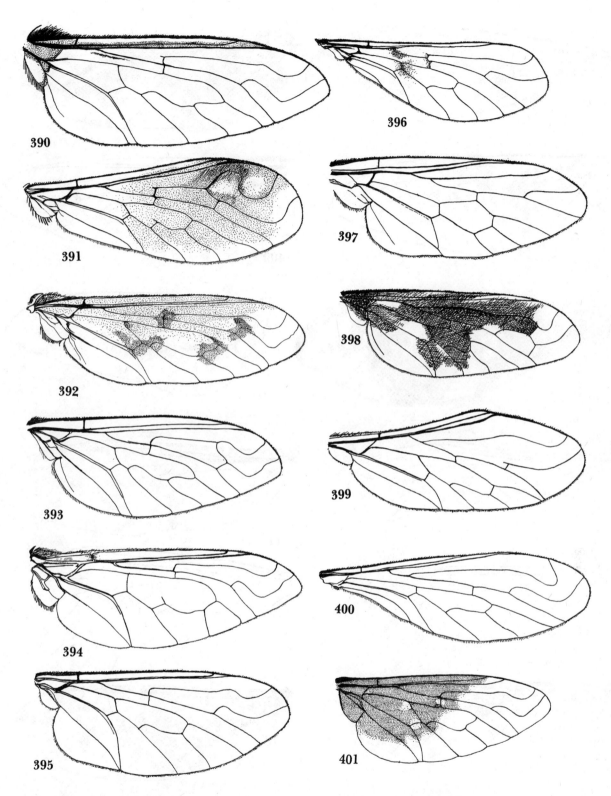

Figures 390–401.—390, *Villa alternatus* Say. 391, *Diplocampta secunda* Paramonov. 392, *Neodiplocampta paradoxa* Jaennicke. 393, *Pseudopenthes fenestrata* Roberts (color omitted). 394, *Oestranthrax obesus* Loew. 395, *Astrophanes adonis* Osten Sacken. 396, *Mancia nana* Coquillett. 397, *Heterotropus indicus* Nurse. 398, *Neosardus principius* Roberts. 399, *Anthrax isis* Meigen. 400, *Docidomyia puellaris* White. 401, *Rhynchanthrax parvicornis* Loew.

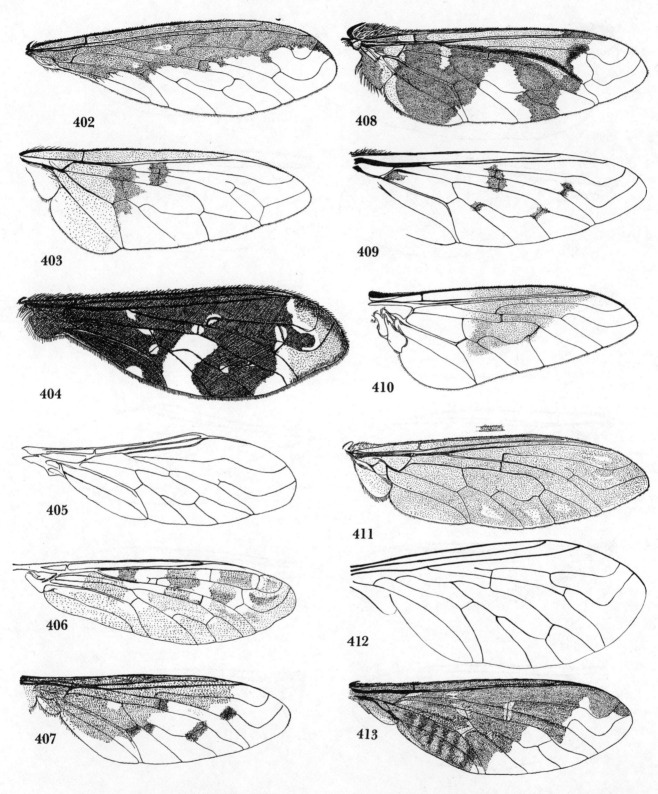

Figures 402–413.—402, *Isotamia daveyi* Bezzi, type. 403, *Dicranoclista simpsoni* Bezzi, type. 404, *Diatropomma carcassoni* Bowden, after Bowden. 405, *Aphoebantus* sp. 406, *Henica longirostris* Wiedemann. 407, *Paravilla* sp. 408, *Litorrhynchus lar* Fabricius. 409, *Anthrax punctipennis* Wiedemann. 410, *Villoestrus uvarovi* Paramonov, from type. 411, *Walkeromyia luridus* Walker. 412, *Coniomastix montana* Enderlein, after Enderlein. 413, *Hemipenthes sinuosa* Wiedemann.

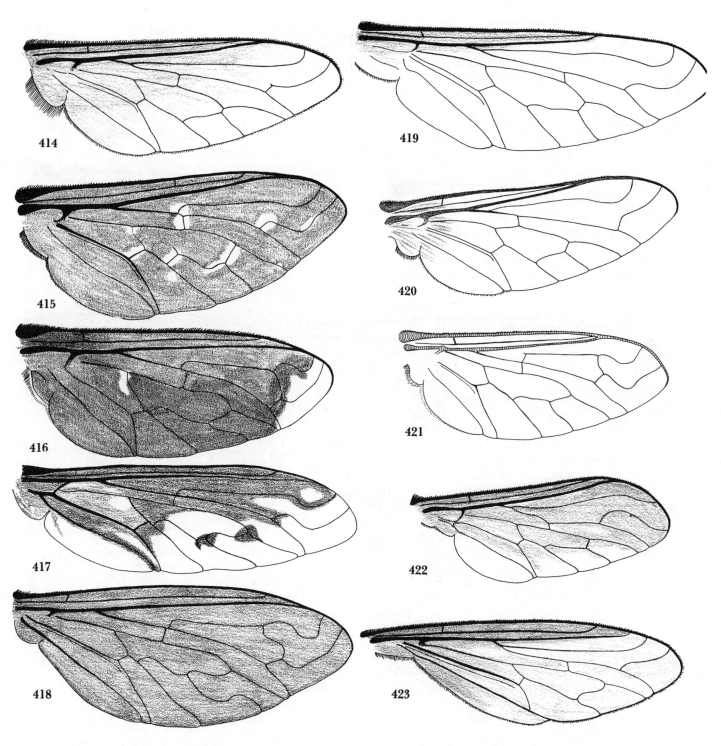

Figures 414–423.—414, *Chasmoneura argyropyga* Wiedemann. 415, *Oestrimyza fenestrata*, new species. 416, *Exoprosopa apicalis* Klug. 417, *Poecilanthrax californicus* Cole. 418, *Colossoptera latipennis* Brunnetti. 419, *Euprepina nuda* Hull. 420, *Sparnopolius lherminierii* Macquart. 421, *Micromitra opalina* Hesse. 422, *Thevenemyia* sp. 423, *Comptosia* (*Anthocolon*) *Casimira*, new species.

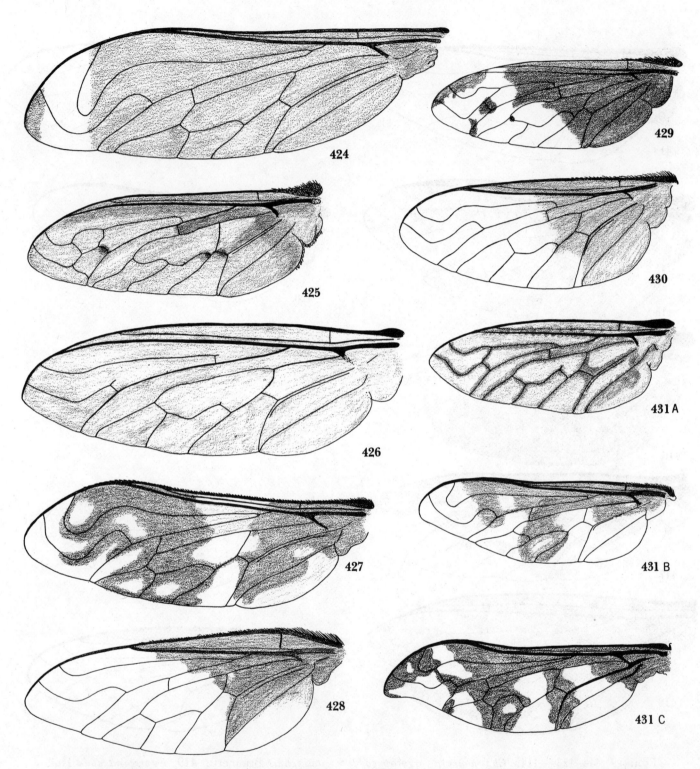

Figures 424–431c.—424, *Comptosia fascipennis* Macquart. 425, *Ligyra tenebrosa* Paramonov. 426, *Walkeromyia* sp. 427, *Oncodosia planus* Walker. 428, *Ogcodocera leucoprocta* Wiedemann. 429, *Stonyx clotho* Wiedemann. 430, *Synthesia fucoides* Bezzi. 431a, *Atrichochira pediformis* Bezzi. 431b, *Doddosia picta* Edwards. 431c, *Ulosometa falcata*, new species.

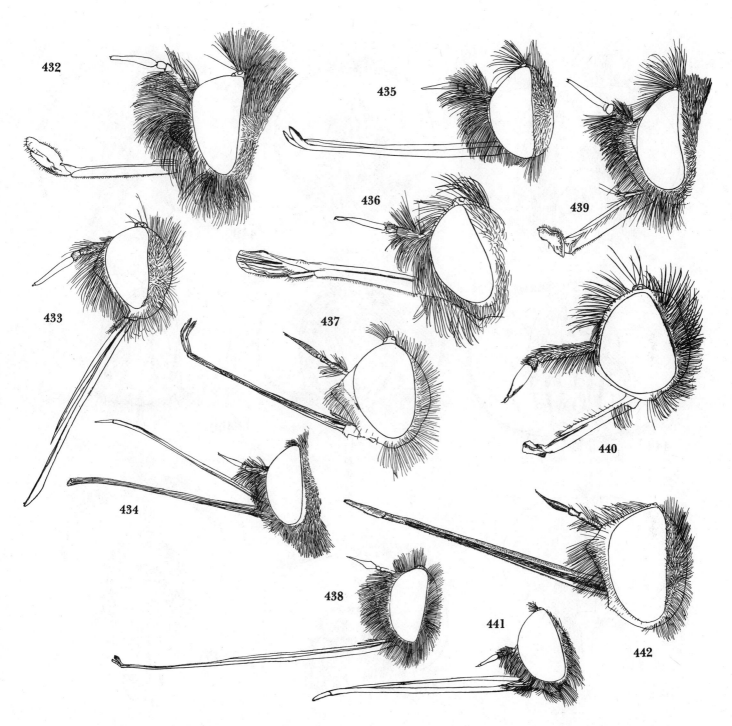

Figures 432–442.—432, *Bombylius bifrons* Walker. 433, *Bombylius (Triplasius) bivittatus* Loew. 434, *Sisyrophanus abdominalis* Bezzi, male. 435, *Lissomerus niveicomatus* Austen, type. 436, *Sisyromyia brevirostris* Macquart. 437, *Acrophthalmyda sphenoptera* Loew. 438, *Bombylodes eximius* Becker. 439, *Eusurbus crassilabris* Macquart. 440, *Tillyardomyia gracilis* Tonnoir. 441, *Isocnemus nemestrinus* Bezzi, from type. 442, *Systoechus vulgaris* Loew.

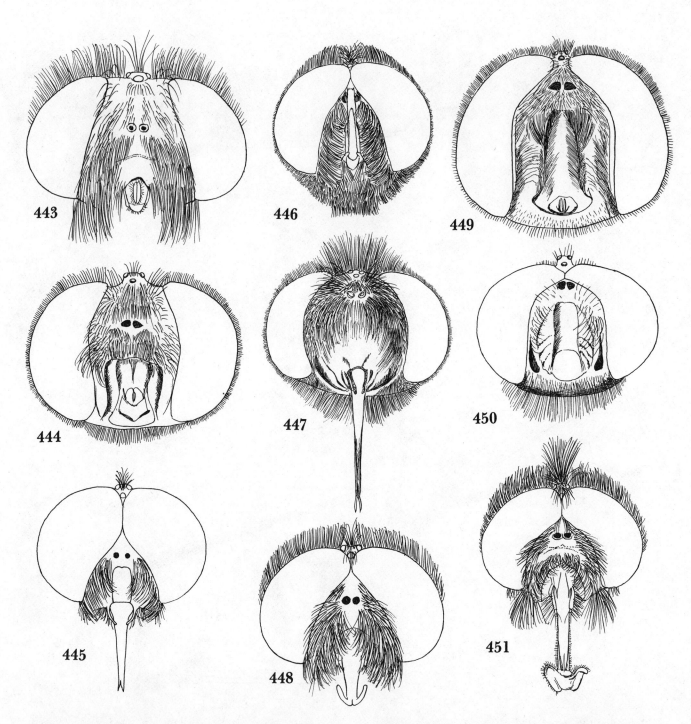

FIGURES 443–451.—443, *Sisyromyia brevirostris* Macquart. 444, *Systoechus vulgaris* Loew. 445, *Isocnemus nemestrinus* Bezzi, type, male. 446, *Sisyrophanus abdominalis* Bezzi, type, male. 447, *Anastoechus trisignatus* Portschinsky. 448, *Lissomerus niveicomatus* Austen, type. 449, *Zinnomyia karooensis* Hesse. 450, *Acrophthalmyda sphenoptera* Loew. 451, *Eusurbus crassilabris* Macquart.

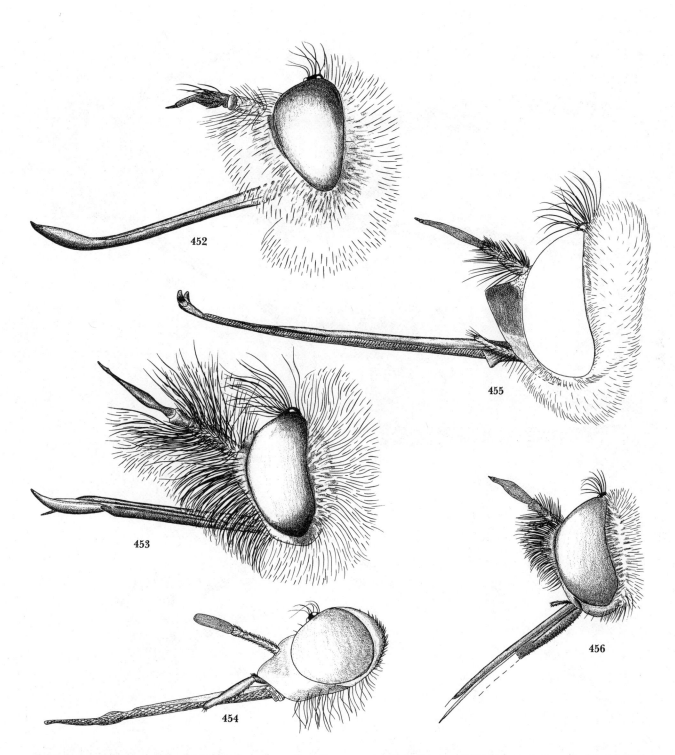

FIGURES 452–456.—452, *Sosiomyia carnata* Bezzi. 453, *Dischistus mystax* Wiedemann. 454, *Eclimus gracilis* Loew. 455, *Euprepina nuda* Hull. 456, *Parabombylius coquilletti* Williston.

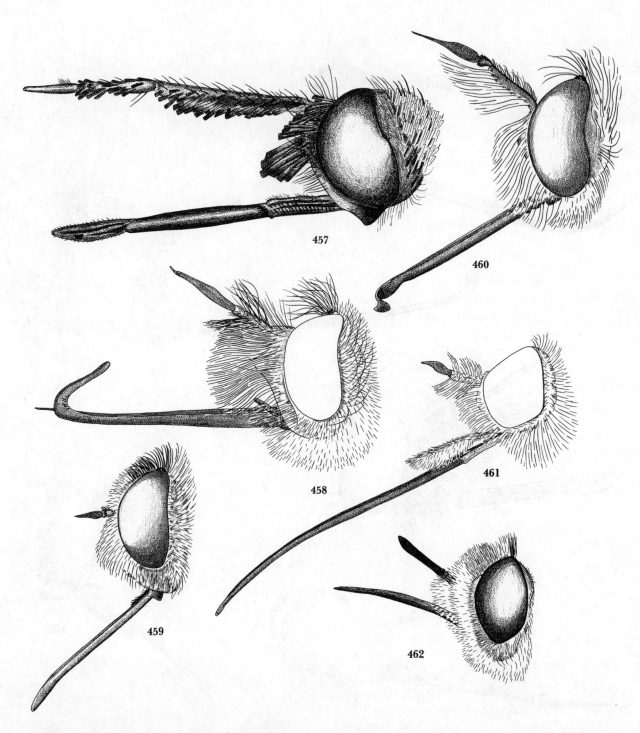

FIGURES 457–462.—457, *Lepidophora lepidocera* Wiedemann. 458, *Bombylius major* Linné. 459, *Cytherea obscura* Fabricius. 460, *Paratoxophora cuthbertsoni* Engel. 461, *Geminaria canalis* Coquillett. 462, *Corsomyza*.

Figures 463–470.—463, *Zinnomyia karooensis* Hesse, paratype male. 464, *Lepidochlanus fimbriatus* Hesse, paratype, male. 464A, *Nectaropta setigera* Philippi. 465, *Anastoechus trisignatus* Portschinsky. 466, *Sisyrophanus pyrrhocerus* Bezzi. 467, *Acanthogeron senex* Wiedemann. 468, *Doliogethes aridicolus* Hesse, paratype. 469, *Legnotomyia trichorhoea* Loew. 470, *Othniomyia tylopelta* Hesse, from type.

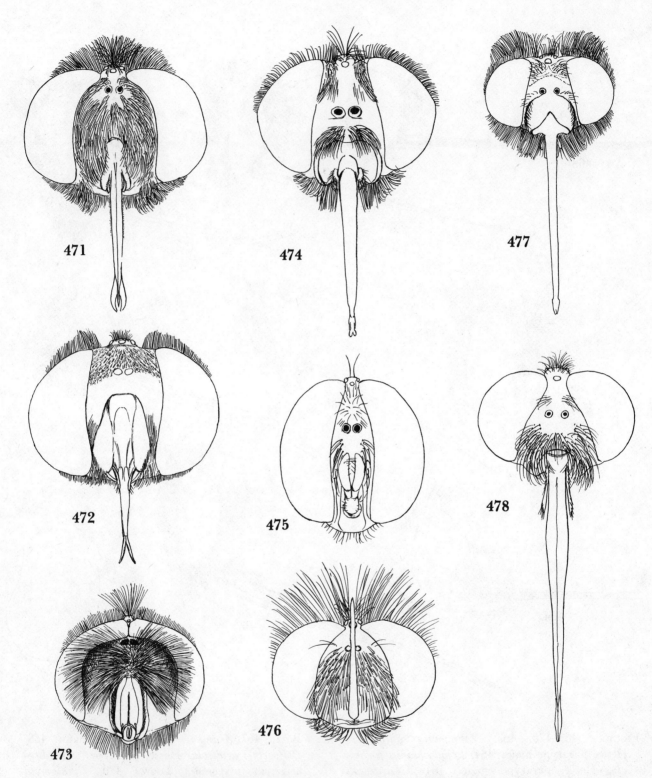

FIGURES 471–478.—471, *Doliogethes aridicolus* Hesse. 472, *Neodischistus collaris* Painter. 473, *Bombylisoma minimus* Schrank. 474, *Legnotomyia trichorhoea* Loew. 475, *Othniomyia tylopelta* Hesse. 476, *Lepidochlanus fimbriatus* Hesse, paratype male. 477, *Sisyrophanus pyrrhocerus* Bezzi. 478, *Adelidea pterosticta* Hesse, paratype.

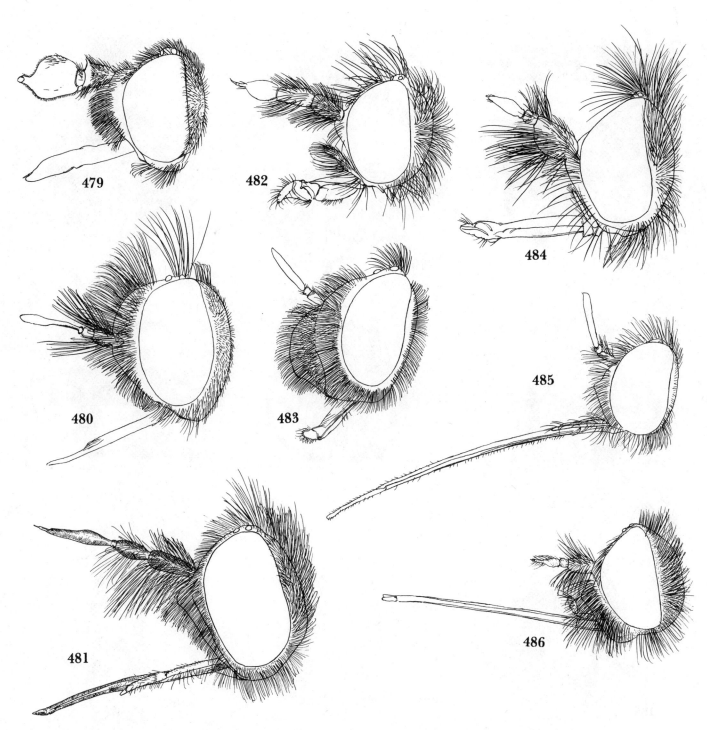

FIGURES 479–486.—479, *Prorachthes longirostris* Bezzi. 480, *Callynthrophora capensis* Schiner. 481, *Conophorina bicellaris* Becker. 482, *Aldrichia ehrmanii* Coquillett. 483, *Gnumyia brevirostris* Bezzi. 484, *Conophorus fuliginosus*. 485, *Megapalpus capensis* Wiedemann. 486, *Platamomyia depressus* Loew.

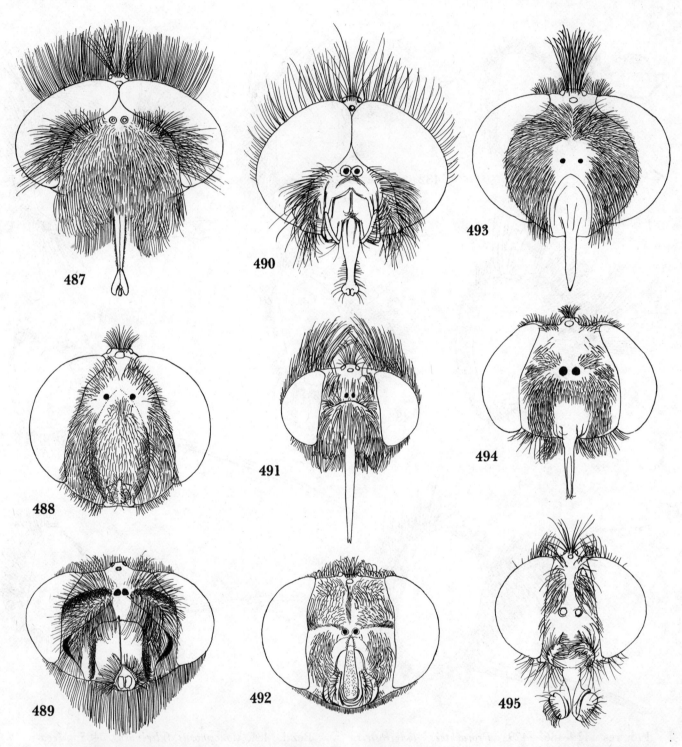

Figures 487–495.—487, *Bombylius bifrons* Walker, from type. 488, *Mariobezzia griseohirta* Nurse. 489, *Cacoplox griseata* Hull. 490, *Conophorus fuliginosus* Wiedemann. 491, *Sericusia lanata* Edwards, type. 492, *Hyperusia luteifacies* Bezzi. 493, *Callynthrophora capensis* Schiner. 494, *Prorachthes longirostris* Bezzi. 495, *Aldrichia ehrmanii* Coquillett.

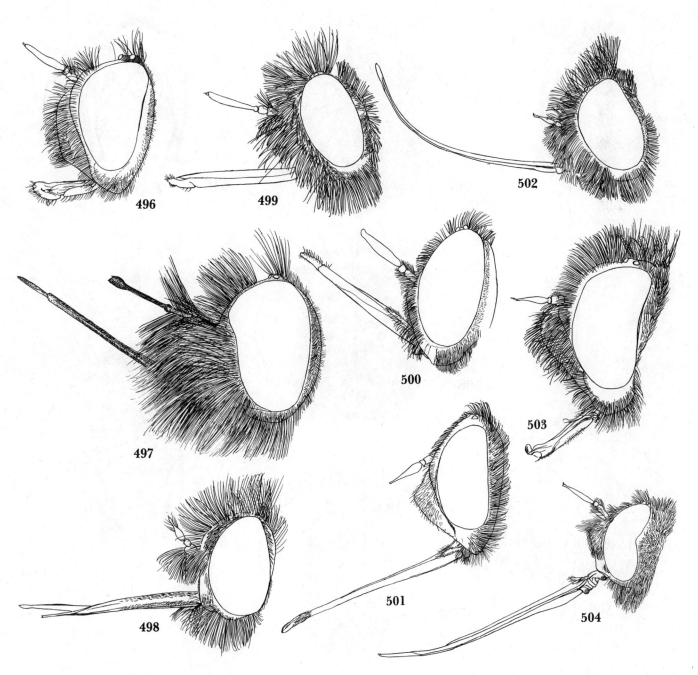

FIGURES 496–504.—496, *Mariobezzia griseohirta* Nurse, from type. 497, *Corsomyza nigripes* Wiedemann. 498, *Oniromyia pachyceratus* Bigot. 499, *Pantarbes pusio* Osten Sacken. 500, *Hyperusia luteifacies* Bezzi. 501, *Gyrocraspedum pleskei* Becker. 502, *Sericosoma squamiventris* Edwards, type. 503, *Cytherea (Chalcochiton) pallasii* Loew. 504, *Heterostylum robustum* Osten Sacken.

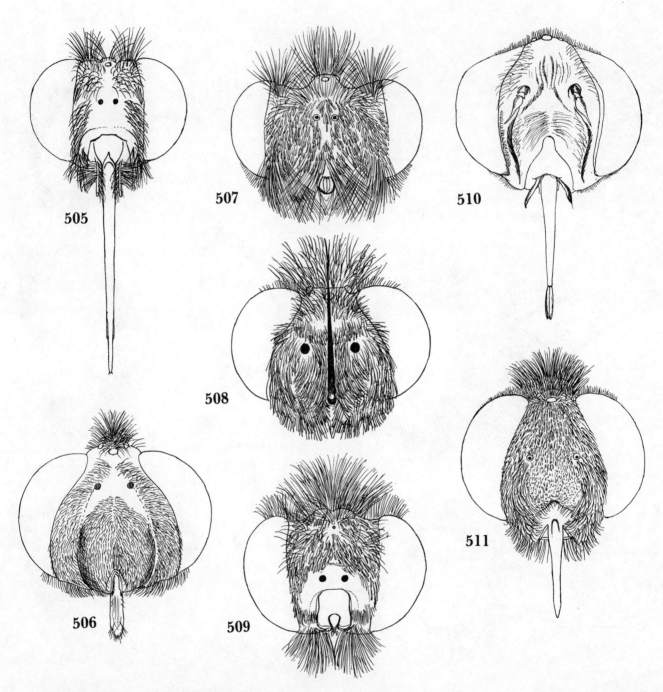

FIGURES 505–511.—505, *Megapalpus capensis* Wiedemann. 506, *Gnumyia brevirostris* Bezzi, from type. 507, *Pantarbes pusio* Osten Sacken, type. 508, *Sericosoma squamiventris* Edwards, type. 509, *Oniromyia pachyceratus* Bigot. 510, *Gyrocraspedum pleskei* Becker. 511, *Callostoma fascipenne palaestinae* Paramonov.

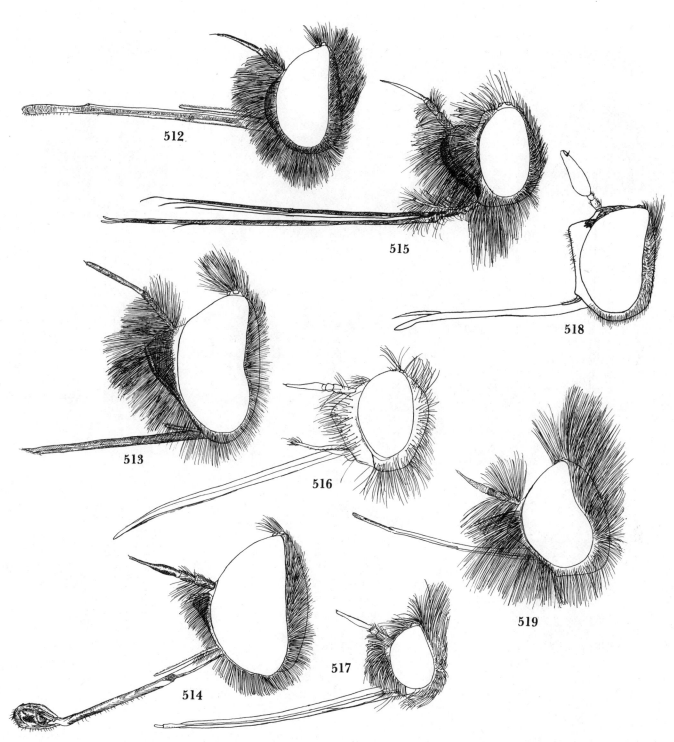

Figures 512–519.—512, *Pilosia flavopilosa*, new species. 513, *Bombylisoma minimus* Schrank. 514, *Laurella auripila* Hull, type. 515, *Cacoplox grieseata* Hull, type. 516, *Adelidea pterosticta* Hesse, paratype. 517, *Sericusia lanata* Edwards, type. 518, *Neodischistus collaris* Painter, type. 519, *Hallidia plumipilosa* Hull, from type.

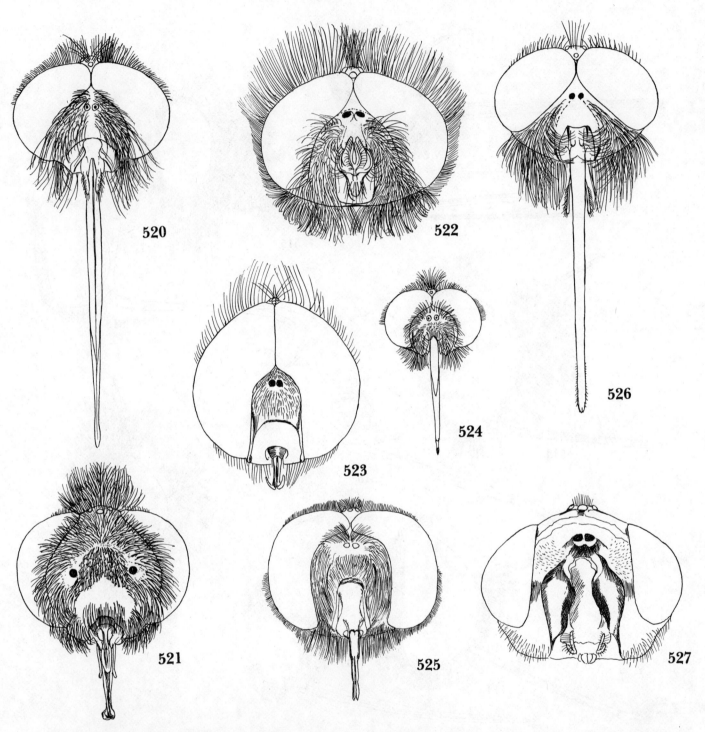

FIGURES 520–527.—520, *Bombylius (Triplasius) bivittatus* Loew. 521, *Cytherea (Chalcochiton) pallasii* Loew. 522, *Mallophthiria lanata* Edwards, holotype. 523, *Geron* sp. 524, *Triploechus heteroneurus* Macquart. 525, *Eurycarenus laticeps* Loew. 526, *Crocidium melanopalis* Hesse. 527, *Heterotropus aegyptiacus* Paramonov.

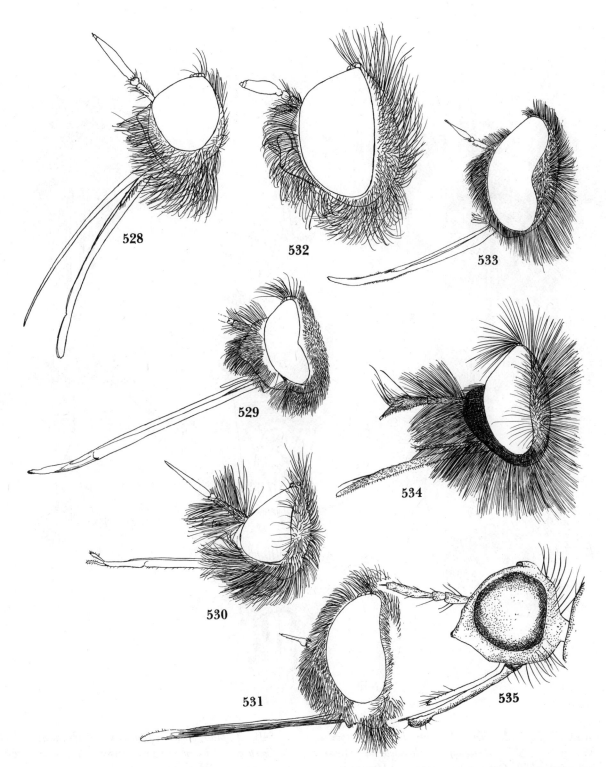

FIGURES 528–535.—528, *Crocidium melanopalis* Hesse. 529, *Efflatounia* sp. 530, *Amictogeron lasiocornis* Hesse, holotype. 531, *Callostoma fascipenne (palaestinae)* Paramonov. 532, *Mallophthiria lanata* Edwards, holotype. 533, *Eurycarenus laticeps* Loew. 534, *Acreotrichus gibbicornis* Macquart. 535, *Semiramis punctipennis* Becker, after Becker.

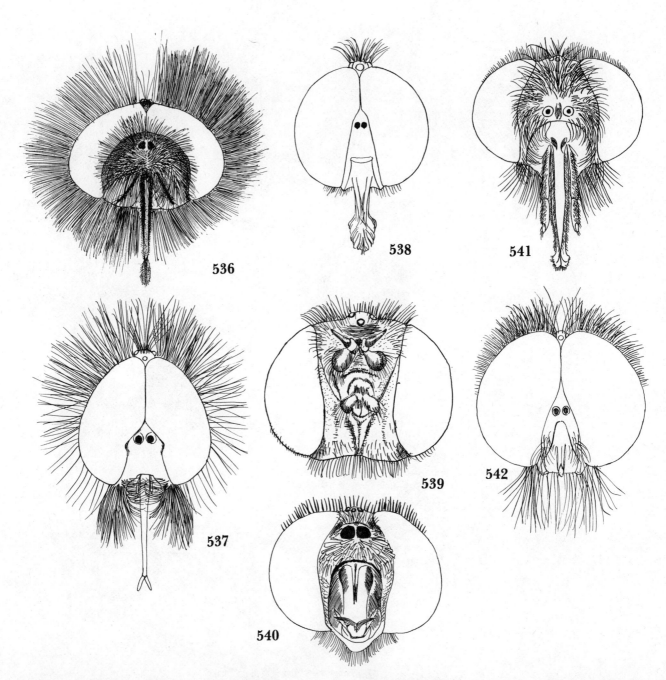

FIGURES 536–542.—536, *Acreotrichus gibbicornis* Macquart. 537, *Amictogeron lasiocornis* Hesse, holotype. 538, *Onchopelma pulchella* Hesse, type. 539, *Xenoprosopa paradoxa* Hesse, after Hesse. 540, *Lepidophora lepidocera* Wiedemann. 541, *Cyrtomyia chilensis* Paramonov. 542, *Marmasoma sumptuosa* White.

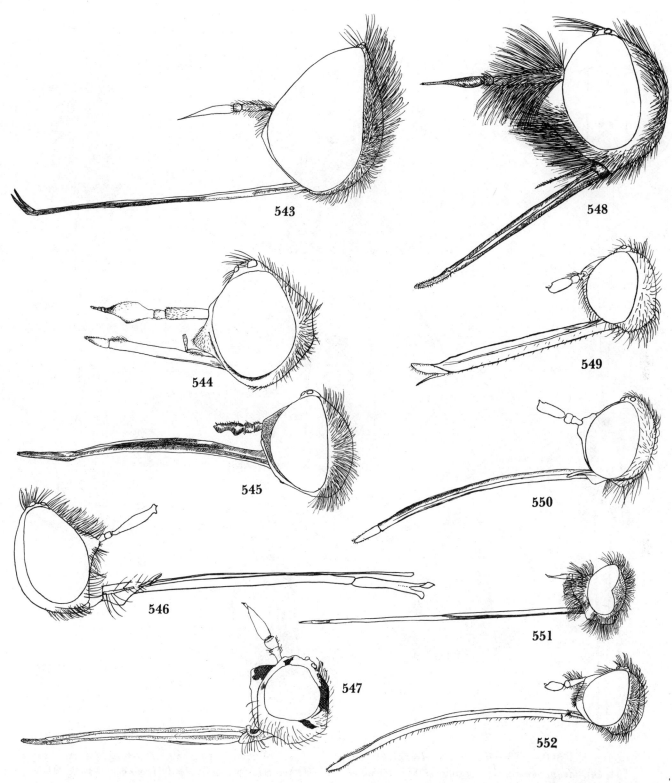

FIGURES 543–552.—543, *Geron* sp. 544, *Apatomyza punctipennis* Wiedemann, from type. 545, *Usia atrata* Fabricius. 546, *Zyxmyia megachile* Bowden, after Bowden. 547, *Heterotropus aegyptiacus* Paramonov. 548, *Pseudoamictus heteropterus* Wiedemann. 549, *Apolysis maherniaphila* Hesse. 550, *Oligodranes elegans* Hesse. 551, *Triploechus heteroneurus* Macquart. 552, *Oligodranes cincturus* Coquillett.

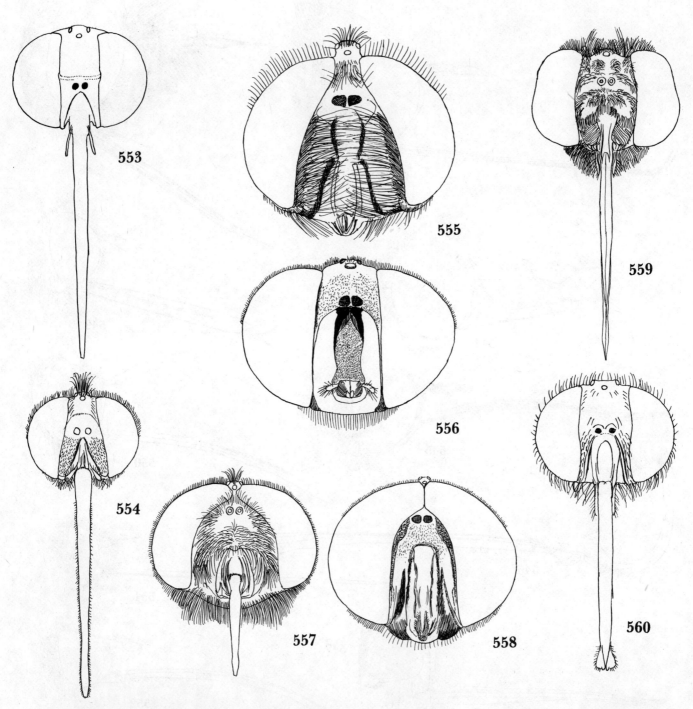

FIGURES 553–560.—553, *Oligodranes elegans* Hesse. 554, *Oligodranes cincturus* Coquillett. 555, *Pseudoamictus heteropterus* Wiedemann. 556, *Usia aenea* Rossi. 557, *Efflatounia* sp. 558, *Toxophora leucopyga* Wiedemann. 559, *Heterostylum robustum* Osten Sacken. 560, *Apolysis maherniaphila* Hesse, type.

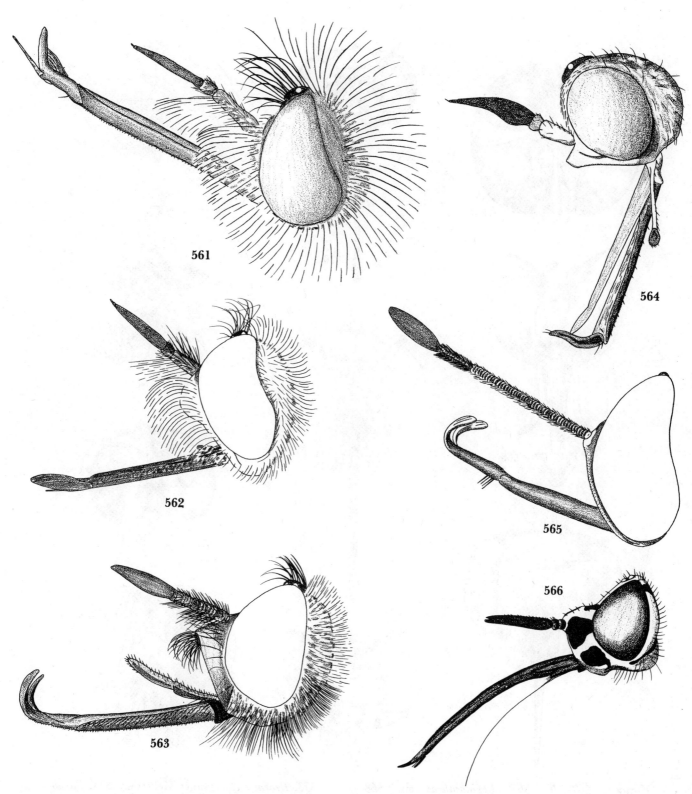

FIGURES 561–566.—561, *Gonarthrus cylindricus* Bezzi. 562, *Chasmoneura argyropygus* Wiedemann. 563, *Thevenemyia magnus* Osten Sacken. 564, *Adelogenys culicoides* Hesse. 565, *Systropus angulatus* Karsch. 566, *Phthiria* sp.

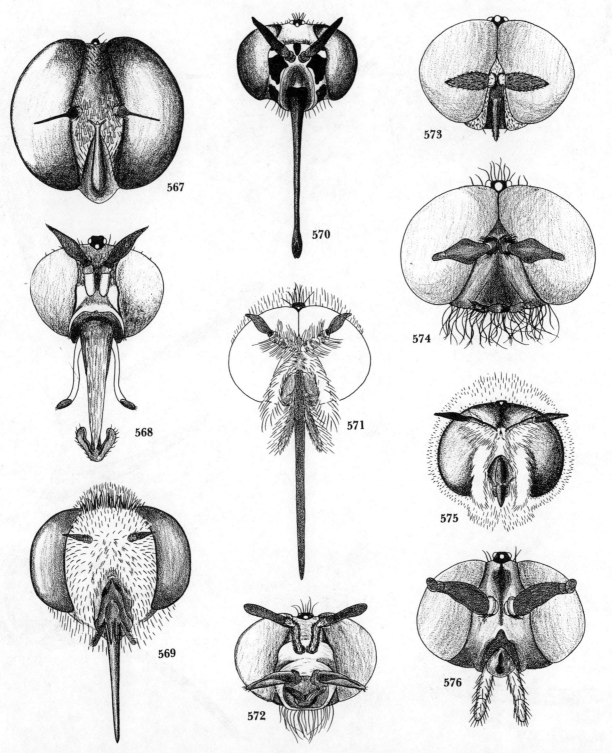

FIGURES 567–576.—567, *Lepidanthrax* sp. 568, *Adelogenys culicoides* Hesse. 569, *Cytherea obscura* Fabricius. 570, *Phthiria* sp. 571, *Geminaria canalis* Coquillett. 572, *Eclimus gracilis* Loew. 573, *Prorates claripennis* Melander. 574, *Caenotus hospes* Melander. 575, *Lordotus gibbus* Loew. 576, *Desmatomyia anomala* Williston.

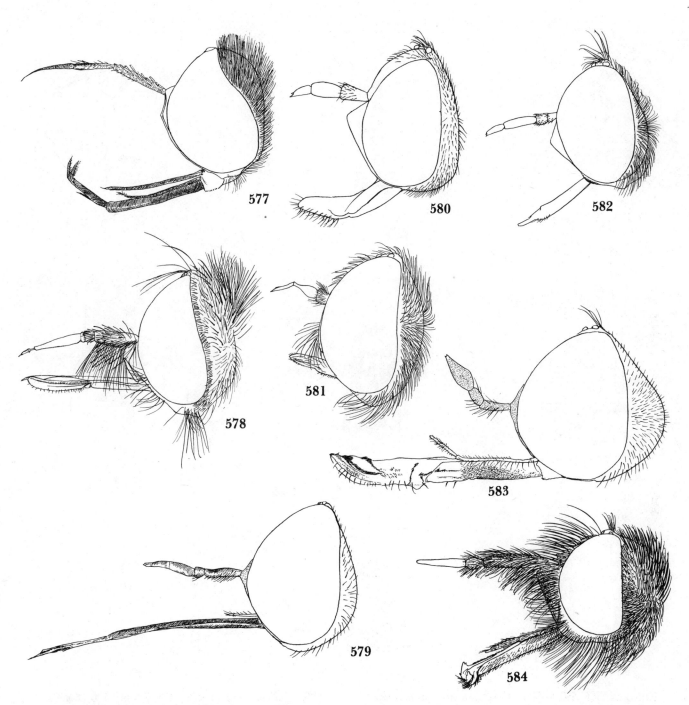

Figures 577–584.—577, *Toxophora leucopyga* Wiedemann. 578, *Marmasoma sumptuosa* White. 579, *Dolichomyia gracilis* Williston. 580, *Onchopelma trilineata* Hesse, type. 581, *Cyrtomorpha paganica* White. 582, *Onchopelma pulchella* Hesse, type. 583, *Zaclava* sp. 584, *Cyrtomyia chilensis* Paramonov.

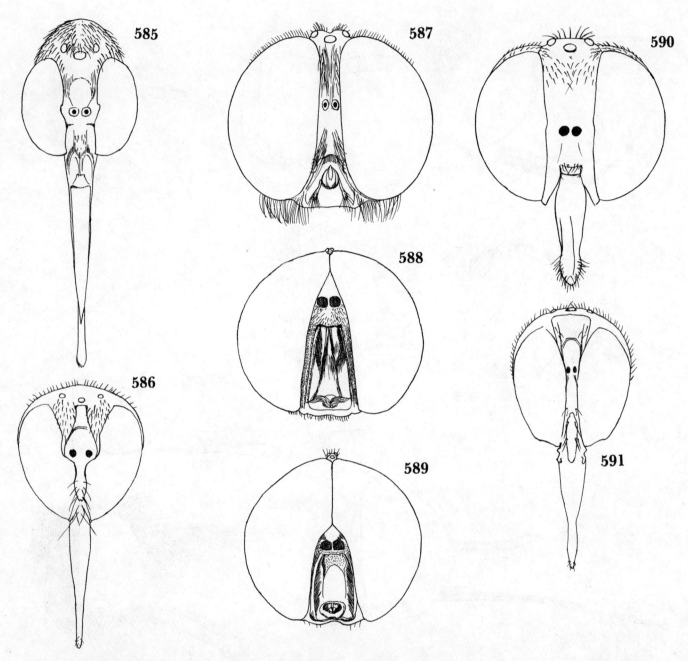

FIGURES 585–591.—585, *Ceratolaemus xanthogrammus* Hesse, female paratype. 586, *Anomaloptilus celluliferus* Hesse, allotype. 587, *Cyrtomorpha paganica* White. 588, *Systropus angulatus* Karsch. 589, *Dolichomyia gracilis* Williston. 590, *Onchopelma trilineata* Hesse, type. 591, *Empidideicus turneri* Hesse, allotype, female.

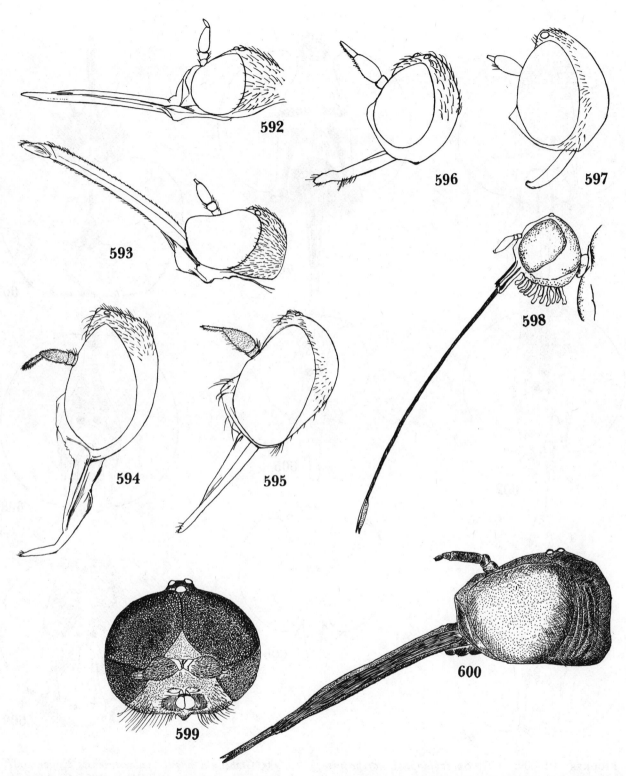

Figures 592–600.—592, *Ceratolaemus xanthogrammus* Hesse, female, paratype. 593, *Cyrtosia* sp. 594, *Empidideicus turneri* Hesse, female, allotype. 595, *Anomaloptilus celluliferus* Hesse, allotype. 596, *Mythicomyia pusillima* Edwards, type. 597, *Glabellula nobilis palaestinensis* Engel. 598, *Euanthobates pectinigulus* Hesse. 599, *Apystomyia elinguis* Melander. 600, *Cyrtisiopsis crassirostris* Hesse.

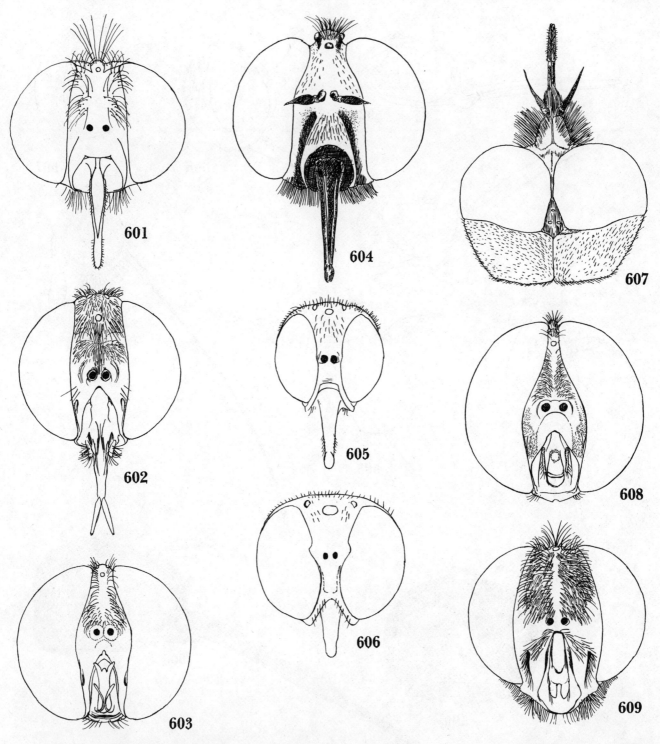

FIGURES 601–609.—601, *Sphenoidoptera varipennis* Williston, type. 602, *Pantostomus tinctellipennis* Hesse, paratype. 603, *Tomomyza pictipennis* Bezzi. 604, *Sinaia kneuckeri* Becker, after Efflatoun. 605, *Mythicomyia pusillima* Edwards, type. 606, *Glabellula nobilis palaestinensis* Engel. 607, *Peringueyimyia capensis* Bigot. 608, *Amphicosmus cincturus* Williston. 609, *Nomalonia afra* Macquart.

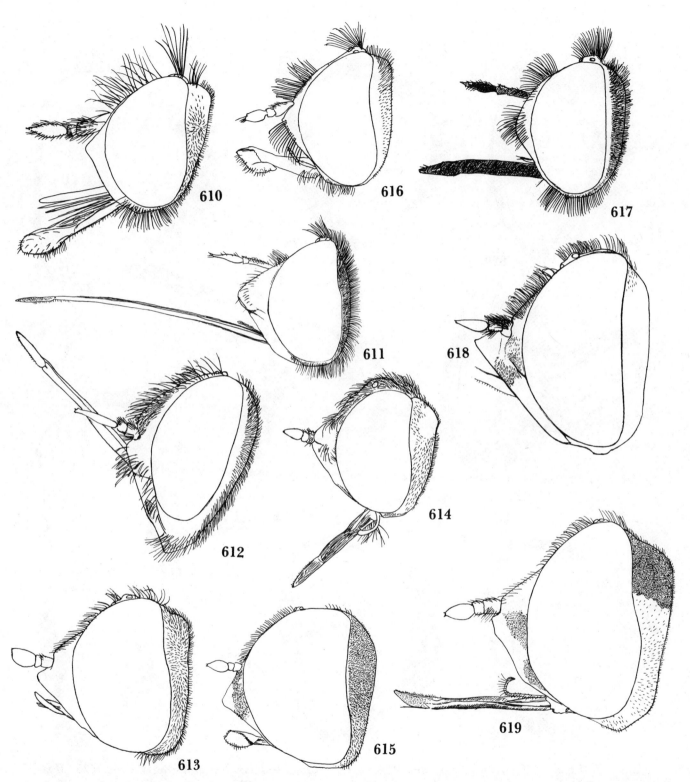

Figures 610–619.—610, *Sphenoidoptera varipennis* Williston, type. 611, *Amictus validus* Loew. 612, *Nomalonia afra* Macquart. 613, *Tomomyza pictipennis* Bezzi. 614, *Pantostomus tinctellipennis* Hesse, paratype. 615, *Metacosmus* sp. 616, *Cyllenia marginata* Loew. 617, *Sinaia kneuckeri* Becker, after Efflatoun. 618, *Amphicosmus cincturus* Williston. 619, *Paracosmus morrisoni* Osten Sacken.

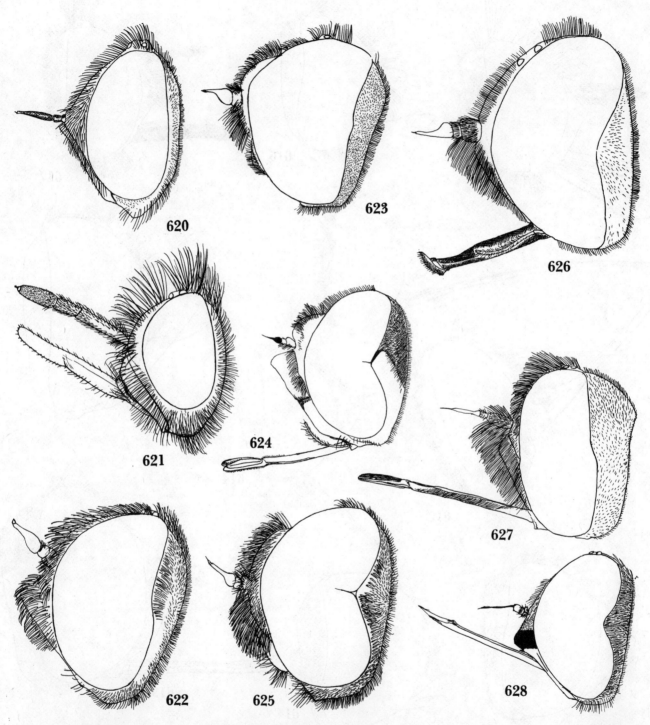

FIGURES 620–628.—620, *Henica longirostris* Wiedemann. 621, *Neosardus principius* Roberts. 622, *Prorostoma integrum* Bezzi. 623, *Docidomyia puellaris* White. 624, *Antonia suavissima* Loew. 625, *Pteraulax flexicornis* Bezzi, male. 626, *Myonema* sp. 627, *Peringueyimyia capensis* Bigot. 628, *Epacmus litus* Coquillett.

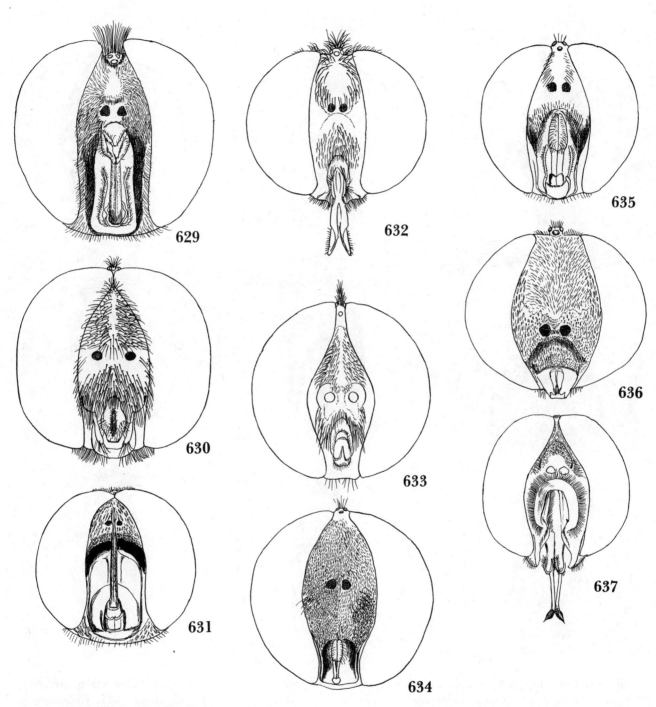

FIGURES 629–637.—629, *Myonema humilis* Roberts. 630, *Pteraulax flexicornis* Bezzi. 631, *Epacmus litus* Coquillett. 632, *Coryprosopa lineata* Hesse, paratype. 633, *Docidomyia puellaris* White. 634, *Petrorossia letho* Wiedemann. 635, *Eucessia rubens* Coquillett. 636, *Desmatoneura argentifrons* Williston. 637, *Antonia suavissima* Loew.

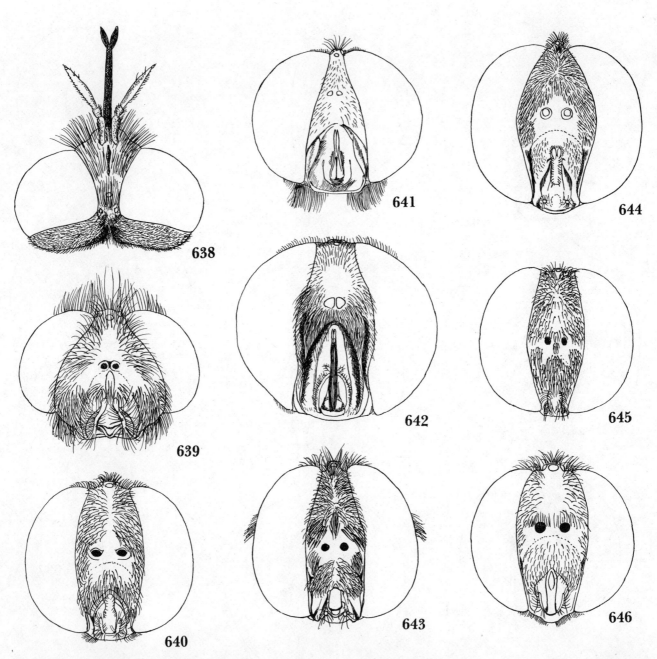

FIGURES 638–646.—638, *Sinaia kneuckeri* Becker, after Efflatoun. 639, *Neosardus principius* Roberts. 640, *Prorostoma integrum* Bezzi. 641, *Amictus validus* Loew. 642, *Peringueyimyia capensis* Bigot. 643, *Plesiocera algira* Macquart. 644, *Chiasmella brevipennis* Bezzi. 645, *Chionamoeba sabulonis* Becker. 646, *Epacmoides albifrons* Hesse, paratype.

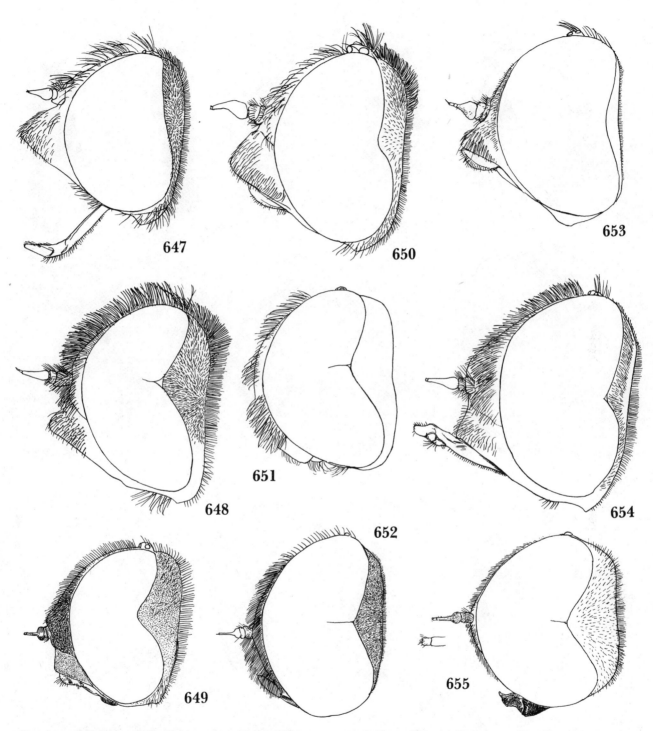

FIGURES 647–655.—647, *Coryprosopa lineata* Hesse, paratype. 648, *Plesiocera algira* Macquart. 649, *Desmatoneura argentifrons* Williston. 650, *Epacmoides albifrons* Hesse, paratype. 651, *Chionamoeba sabulonis* Becker. 652, *Petrorossia letho* Wiedemann. 653, *Eucessia rubens* Coquillett. 654, *Chiasmella brevipennis* Bezzi. 655, *Xeramoeba gracilis* Bowden.

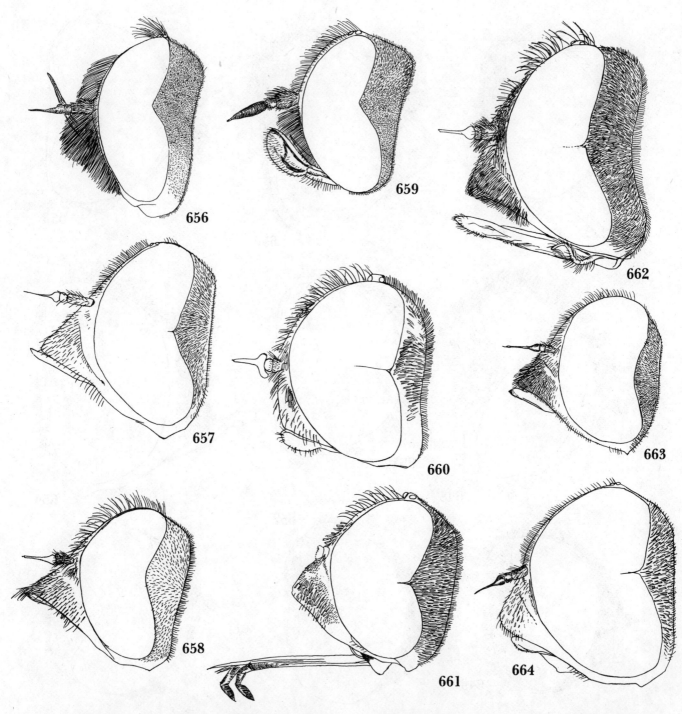

FIGURES 656–664.—656, *Lyophlaeba* (*Macrocondyla*) *montana* Philippi. 657, *Neodiplocampta* (*Agitonia*) *sepia* Hull. 658, *Neodiplocampta paradoxa* Jaennicke. 659, *Ylasoia pegasus* Wiedemann. 660, *Astrophanes adonis* Osten Sacken. 661, *Mancia nana* Coquillett. 662, *Dipalta serpentina* Osten Sacken. 663, *Chrysanthrax cypris* Meigen. 664, *Deusopora sapphirina* Hull.

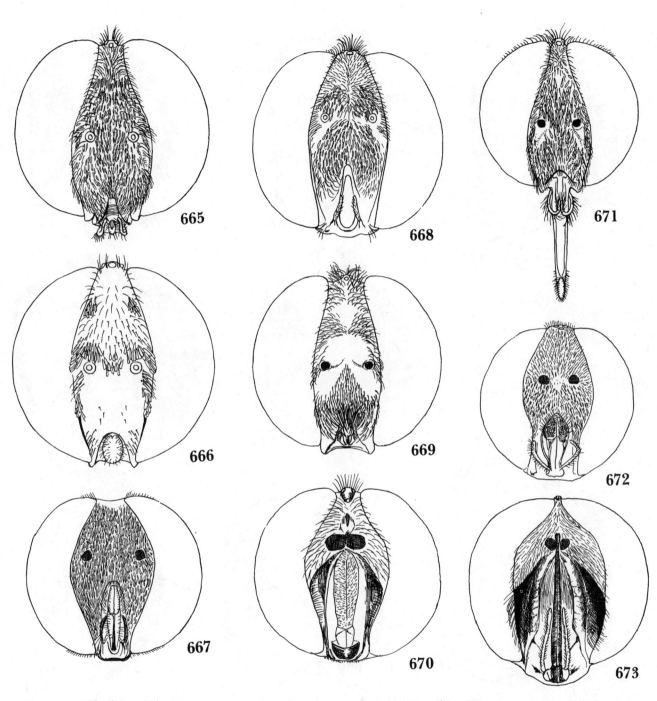

FIGURES 665–673.—665, *Dipalta serpentina* Osten Sacken. 666, *Astrophanes adonis* Osten Sacken. 667, *Villa pretiosa* Loew. 668, *Pseudopenthes fenestrata* Roberts. 669, *Neodiplocampta paradoxa* Jaennicke. 670, *Ylasoia pegasus* Wiedemann. 671, *Diplocampta secunda* Paramonov. 672, *Hemipenthes sinuosa* Wiedemann. 673, *Lyophlaeba (Macrocondyla) montana* Philippi.

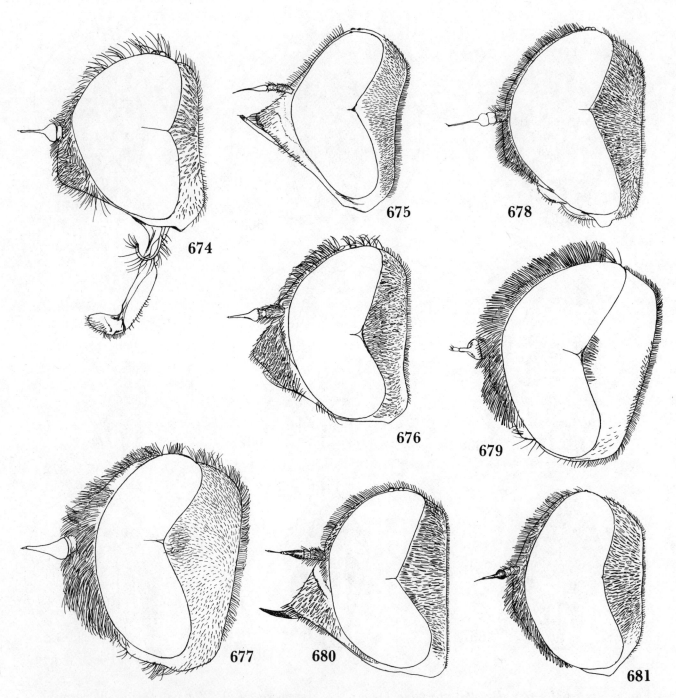

Figures 674–681.—674, *Diplocampta secunda* Paramonov. 675, *Cyananthrax cyanoptera* Wiedemann. 676, *Pseudopenthes fenestrata* Roberts. 677, *Walkeromyia luridus* Walker. 678, *Hemipenthes sinuosa* Wiedemann. 679, *Dicranoclista simpsoni* Bezzi, type. 680, *Exoprosopa jacchus* Fabricius. 681, *Villa pretiosa* Loew.

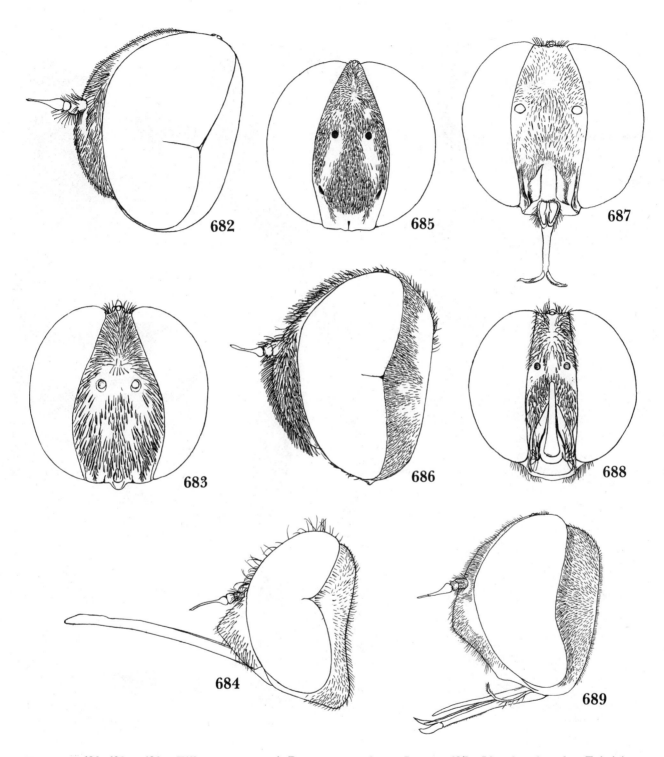

Figures 682–689.—682, *Villoestrus uvarovi* Paramonov, from type. 683, *Oestranthrax obesus* Loew. 684, *Isotamia daveyi* Bezzi, type. 685, *Villoestrus uvarovi* Paramonov, from type. 686, *Oestranthrax obesus* Loew. 687, *Litorrhynchus lar* Fabricius. 688, *Isotamia daveyi* Bezzi, type. 689, *Litorrhynchus lar* Fabricius.

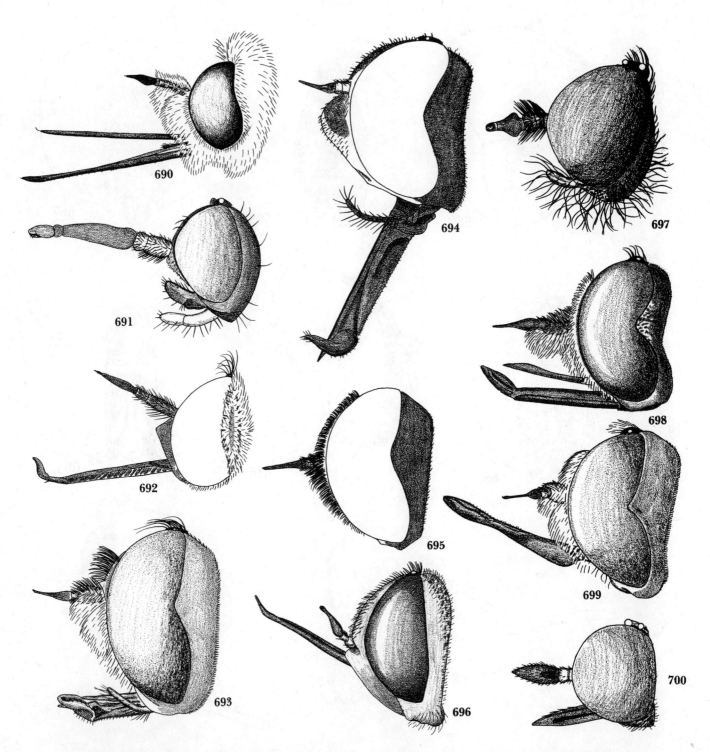

FIGURES 690–700.—690, *Lordotus gibbus* Loew. 691, *Desmatomyia anomala* Williston. 692, *Sparnopolius lherminierii* Macquart. 693, *Comptosia* sp. 694, *Micomitra opalina* Hesse. 695, *Bryodemina valida* Wiedemann. 696, *Nomalonia afra* Macquart. 697, *Caenotus hospes* Melander. 698, *Doddosia picta* Edwards. 699, *Oncodosia planus* Walker. 700, *Prorates claripennis* Melander.

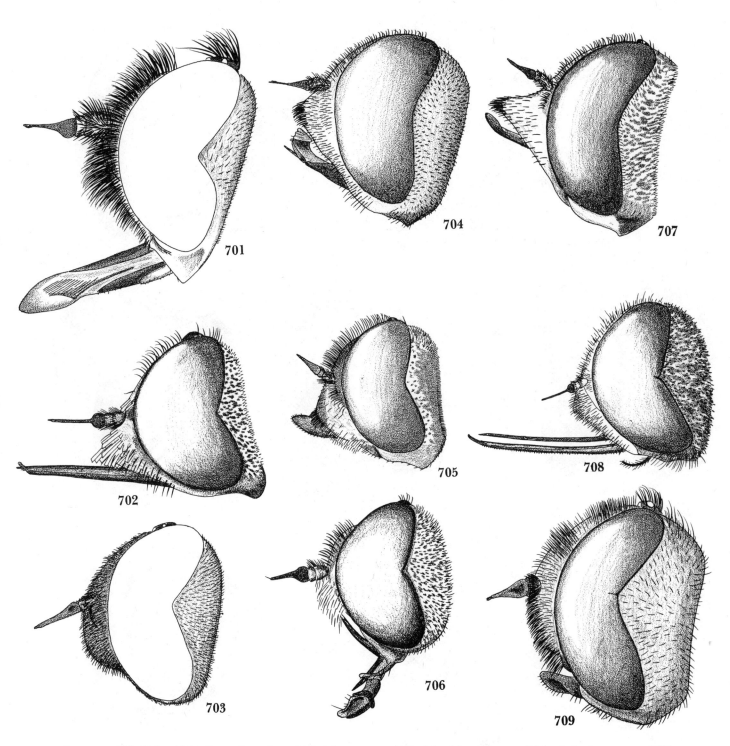

FIGURES 701–709.—701, *Lomatia sabaea* Fabricius. 702, *Lepidanthrax* sp. 703, *Oestrimyza fenestrata*, new species. 704, *Ligyra tenebrosa* Paramonov. 705, *Thyridanthrax* sp. 706, *Aphoebantus* sp. 707, *Paravilla* sp. 708, *Stonyx clotho* Wiedemann. 709, *Walkeromyia luridus* Walker.

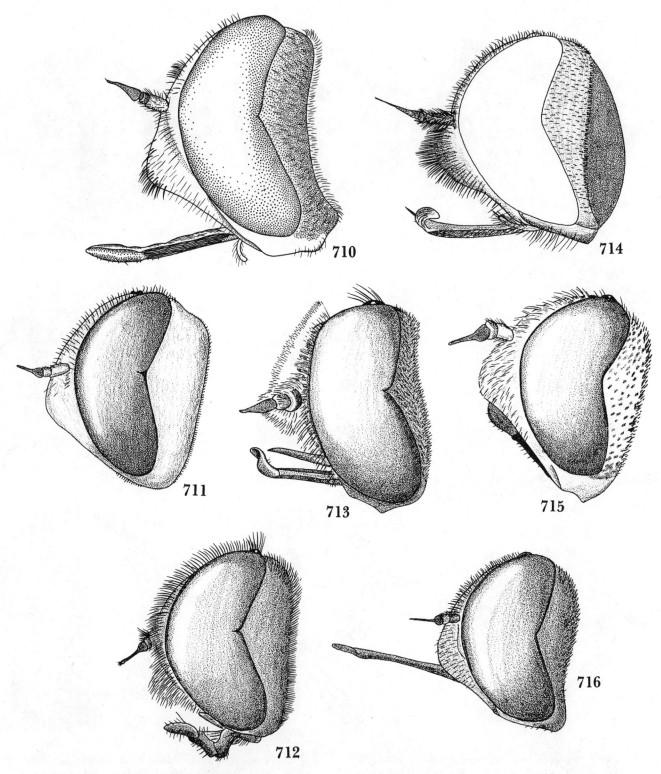

Figures 710–716.—710, *Poecilanthrax californicus* Cole. 711, *Atrichochira pediformis* Bezzi. 712, *Anthrax tigrinus* De Geer. 713, *Ogcodocera leucoprocta* Wiedemann. 714, *Colossoptera latipennis* Brunetti. 715, *Synthesia fucoides* Bezzi. 716, *Rhynchanthrax parvicornis* Loew.

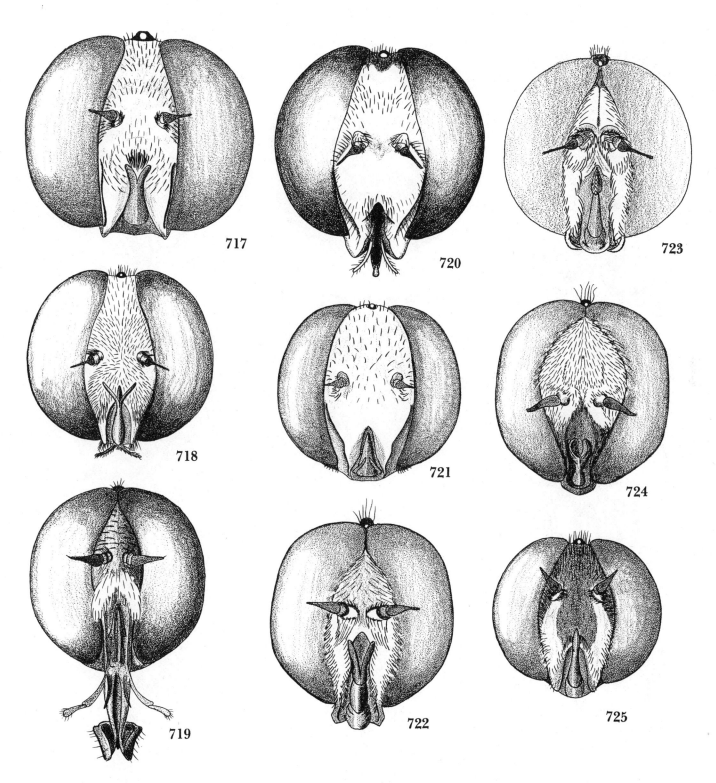

FIGURES 717–725.—717, *Parvilla* sp. 718, *Stonyx clotho* Wiedemann. 719, *Aphoebantus* sp. 720, *Poecilanthrax californicus* Cole. 721, *Atrichochira pediformis* Bezzi. 722, *Doddosia picta* Edwards. 723, *Oncodosia planus* Walker. 724, *Ogcodocera leucoprocta* Wiedemann. 725, *Thyridanthrax fenestratus* Fallen.

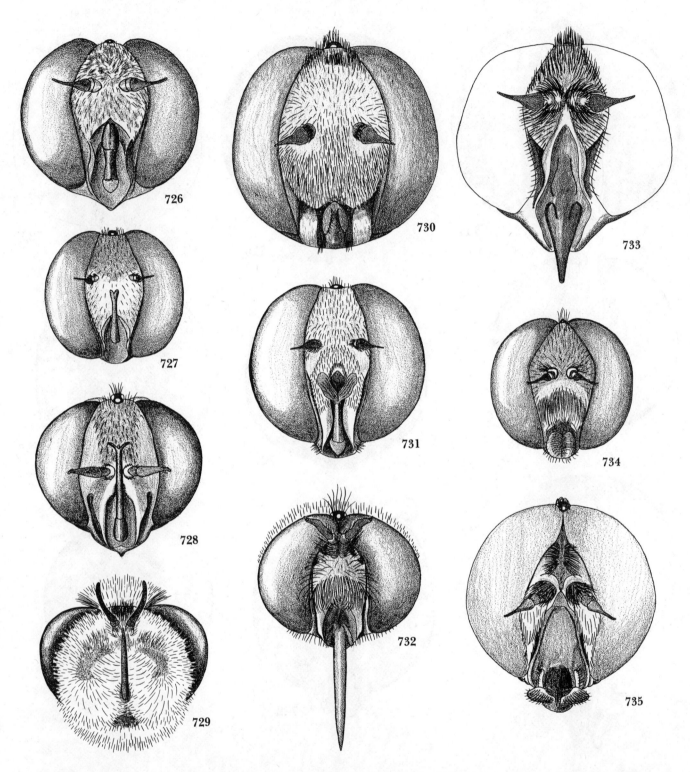

FIGURES 726–735.—726, *Synthesia fucoides* Bezzi. 727, *Rhynchanthrax parvicornis* Loew. 728, *Nomalonia afra* Macquart. 729, *Corsomyza nigripes* Wiedemann. 730, *Walkeromyia luridus* Walker. 731, *Ligyra tenebrosa* Paramonov. 732, *Parabombylius coquilletti* Williston. 733, *Lomatia belzebul* Fabricius. 734, *Anthrax tigrinus* De Geer. 735, *Comptosia fascipennis* Macquart.

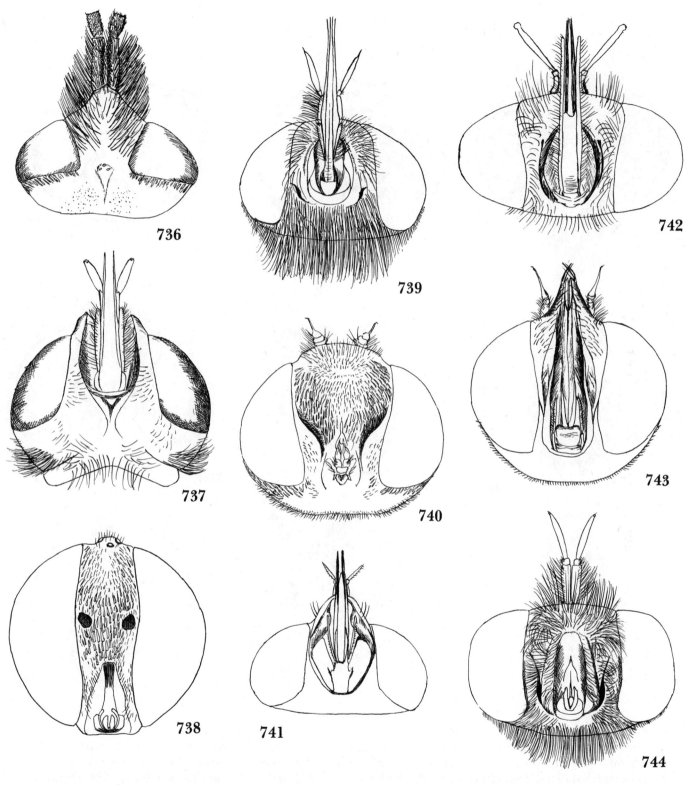

FIGURES 736–744.—736, *Cyrtomyia chilensis* Paramonov, dorsal aspect. 737, *Oligodranes cincturus* Coquillett, ventral aspect. 738, *Mancia nana* Coquillett, frontal aspect. 739, *Adelidea anomala* Wiedemann, ventral aspect. 740, *Oestranthrax obesus* Loew, ventral aspect. 741, *Amictus validus* Loew, ventral aspect. 742, *Megapalpus capensis* Wiedemann, ventral aspect. 743, *Neodiplocampta paradoxa* Jaennicke, ventral aspect. 744, *Doliogethes* sp., ventral aspect.

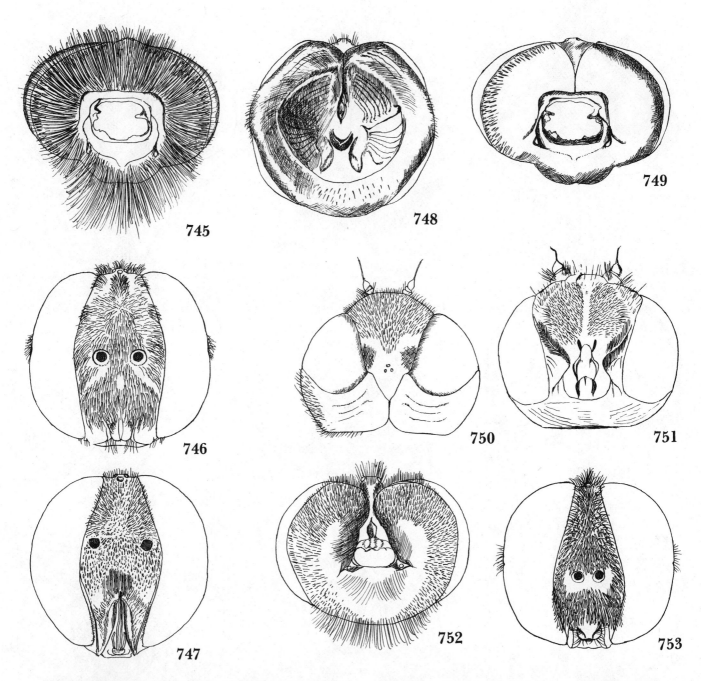

Figures 745–753.—745, *Cyrtomyia chilensis* Paramonov, occiput. 746, *Walkeromyia luridus* Walker, frontal aspect. 747, *Exoprosopa (Argyrospila) jacchus* Fabricius, frontal aspect. 748, *Walkeromyia luridus* Walker, occiput. 749, *Cyrtomyia chilensis* Paramonov, occiput depiled. 750, *Walkeromyia luridus* Walker, dorsal aspect. 751, *Walkeromyia luridus* Walker, ventral aspect. 752, *Amictus validus* Loew, occiput. 753, *Dicranoclista simpsoni* Bezzi, type, frontal aspect.

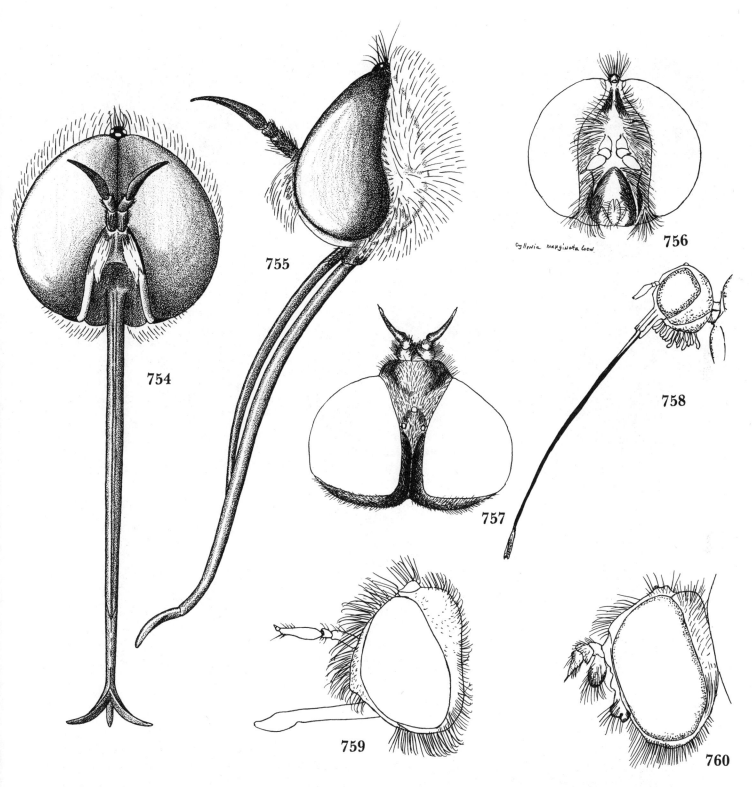

Figures 754–760.—754, *Geron* sp., New Mexico. 755, *Geron* sp. New Mexico. 756, *Cyllenia marginata* Loew. 757, *Myonema humilis* Roberts. 758, *Euanthobates pectinigulus* Hesse, after Hesse. 759, *Sinaia kneuckeri* Becker, after Engel. 760, *Xenoprosopa paradoxa* Hesse, after Hesse.

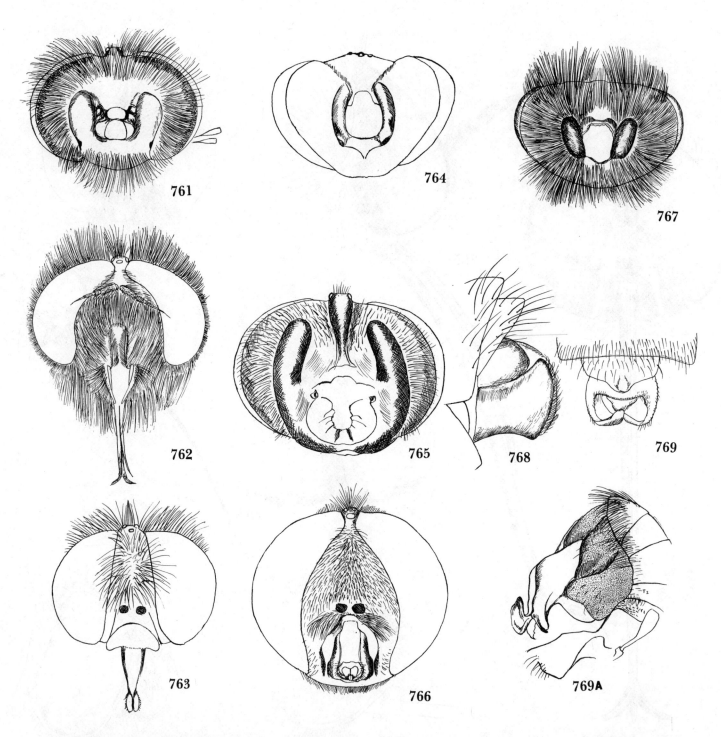

FIGURES 761–769A.—761, *Lepidochlanus fimbriatus* Hesse. 762, *Acanthogeron senex* Meigen. 763, *Tillardomyia gracilis* Tonnoir. 764, *Megapalpus capensis* Wiedemann. 765, *Oligodranes cincturus* Coquillett. 766, *Henica longirostris* Wiedemann. 767, *Sericusia lanata* Edwards, from type. 768, *Acrophthalmyda sphenoptera* Loew, male genitalia. 769, *Antonia rieki* Paramonov, female genitalia. 769A, *Usia florea* Fabricius, male genitalia.

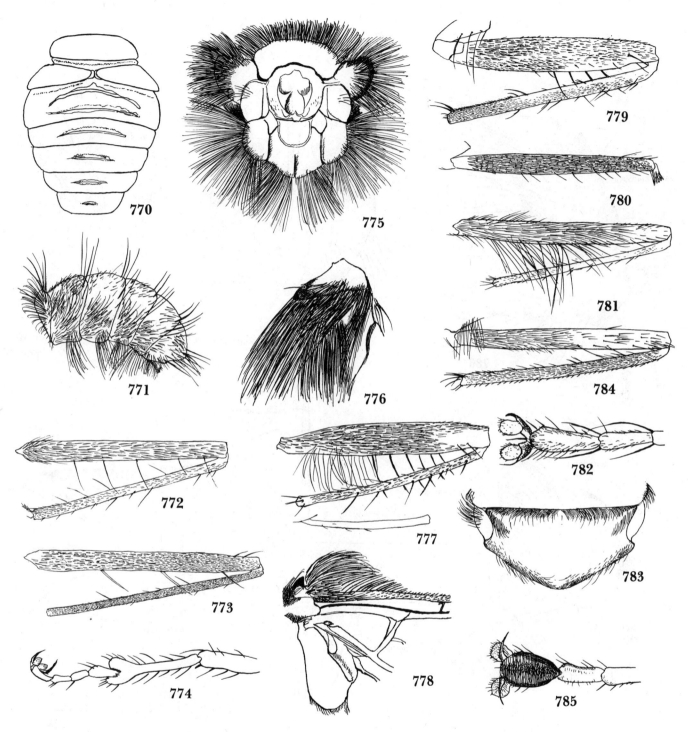

FIGURES 770–785.—770, *Zyxmyia megachile* Bowden, dorsal aspect abdomen, after Bowden. 771, *Cyllenia marginata* Loew, lateral aspect abdomen. 772, *Doliogethes aridicolus* Hesse, paratype. 773, *Lepidochlanus fimbriatus* Hesse, paratype. 774, *Heterotropus arenivagus* Paramonov, middle metatarsus, after Engel. 775, a Homeophthalmae type of occiput. 776, *Acanthogeron senex* Wiedemann, squama. 777, *Cyllenia marginata* Loew, hind femur. 778, *Anastoechus trisignatus* Portschinsky, costal hook and comb. 779, *Amictus validus* Loew, hind femur. 780, *Anastoechus trisignatus* Portschinsky, hind femur. 781, *Lissomerus niveicomatus* Austen, type, hind femur. 782, *Heterotropus arenivagus* Paramonov, hind tarsus, after Engel. 783, *Oestranthrax obesus* Loew, scutellum. 784, *Acanthogeron senex* Wiedemann, hind femur. 785, *Heterotropus arenivagus* Paramonov, anterior tarsus, after Engel.

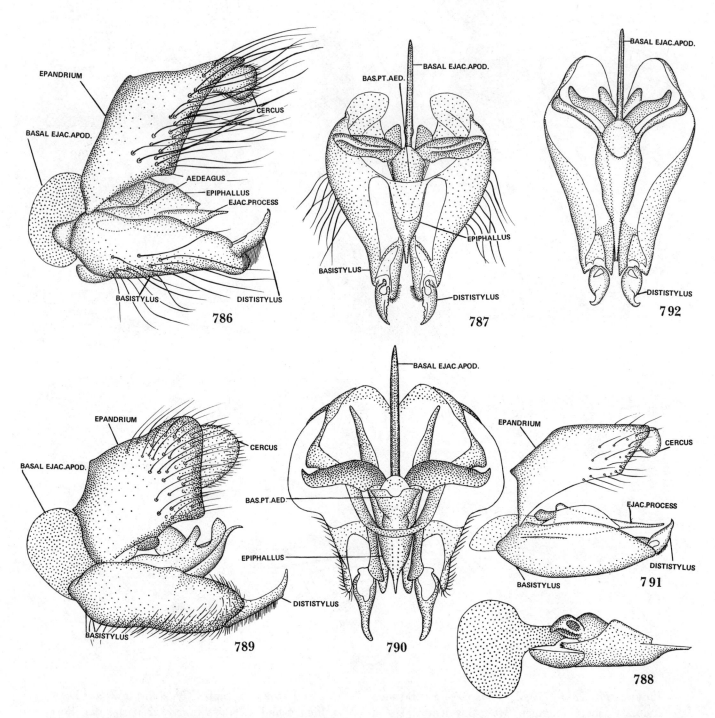

FIGURES 786–792.—786, *Bombylius major* Linné, lateral aspect. 787, *Bombylius major* Linné, dorsal aspect. 788, *Bombylius major* Linné, axial system. 789, *Systoechus vulgaris* Osten Sacken, lateral aspect. 790, *Systoechus vulgaris* Osten Sacken, dorsal aspect. 791, *Anastoechus barbatus* Osten Sacken, lateral aspect. 792, *Anastoechus barbatus* Osten Sacken, dorsal aspect.

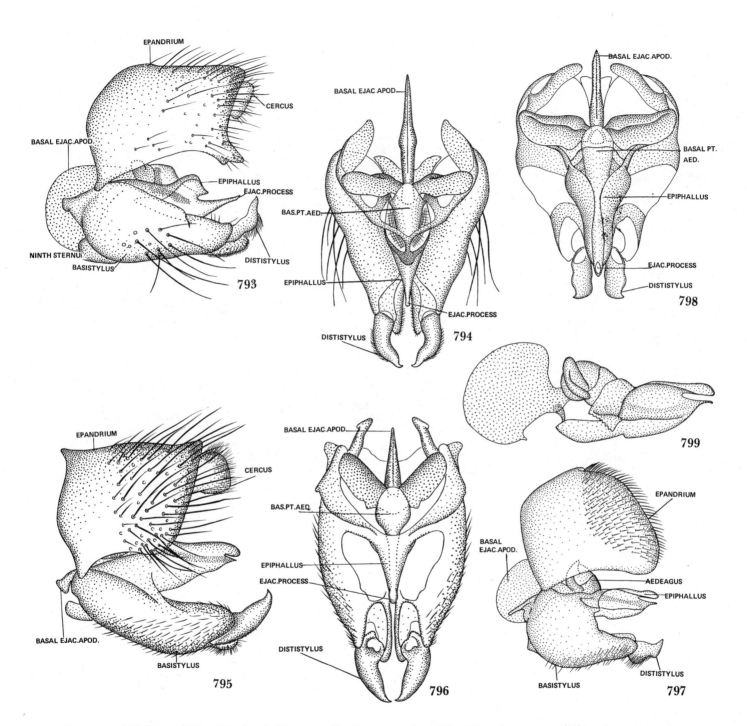

FIGURES 793–799.—793, *Parabombylius coquilletti* Williston, lateral aspect. 794, *Parabombylius coquilletti* Williston, dorsal aspect. 795, *Thevenemyia celer* Cole, lateral aspect. 796, *Thevenemyia celer* Cole, dorsal aspect. 797, *Acropthalmyda sphenoptera* Loew, lateral aspect. 798, *Acropthalmyda sphenoptera* Loew, dorsal aspect. 799, *Acropthalmyda sphenoptera* Loew, axial system.

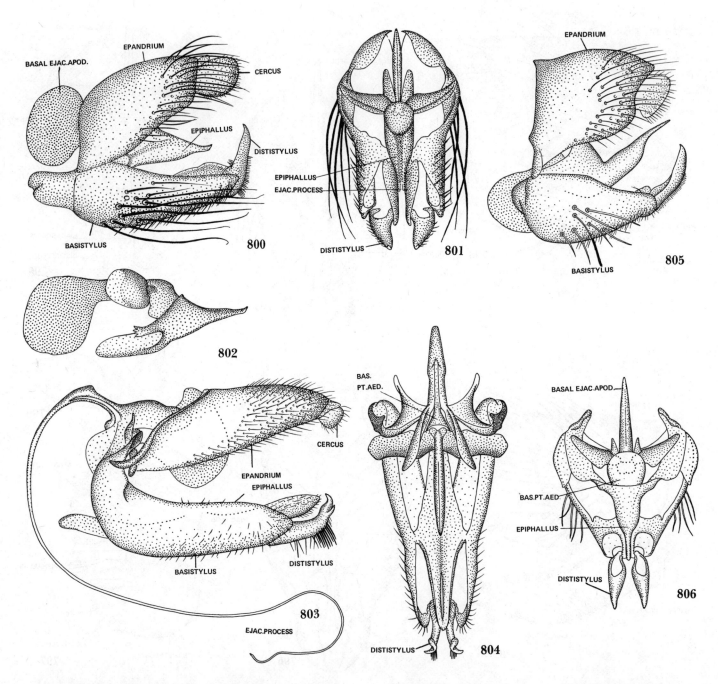

FIGURES 800–806.—800, *Legnotomyia trichorhoea* Loew, lateral aspect. 801, *Legnotomyia trichorhoea* Loew, dorsal aspect. 802, *Legnotomyia trichorhoea* Loew, axial system. 803, *Paratoxophora cuthbertsoni* Engel, lateral aspect. 804, *Paratoxophora cuthbertsoni* Engel, dorsal aspect. 805, *Sparnopolius lherminierii* Macquart, lateral aspect. 806, *Sparnopolius lherminierii* Macquart, dorsal aspect.

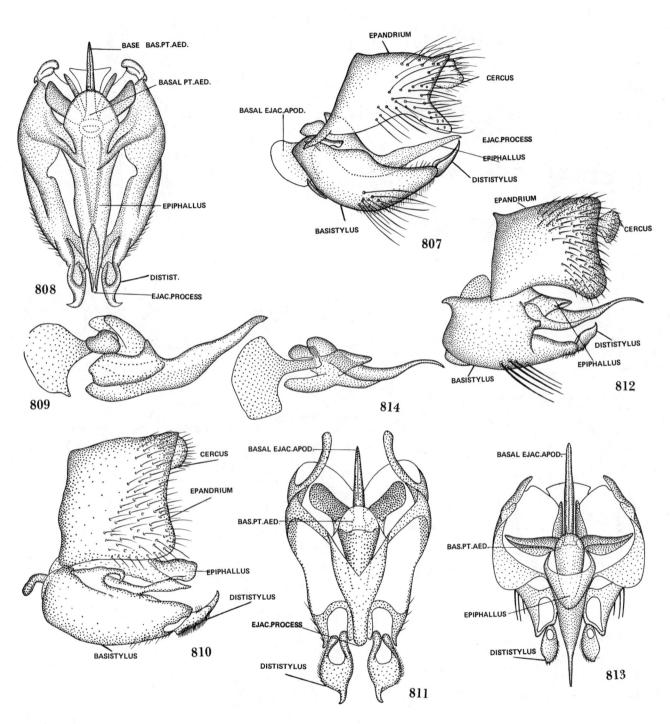

Figures 807–814.—807, *Lordotus gibbus* Loew, lateral aspect. 808, *Lordotus gibbus* Loew, dorsal aspect. 809, *Lordotus gibbus* Loew, axial system. 810, *Sosiomyia carnata* Bezzi, lateral aspect. 811, *Sosiomyia carnata* Bezzi, dorsal aspect. 812, *Hallidia plumipilosa* Hull, lateral aspect. 813, *Hallidia plumipilosa* Hull, dorsal aspect. 814, *Hallidia plumipilosa* Hull, axial system.

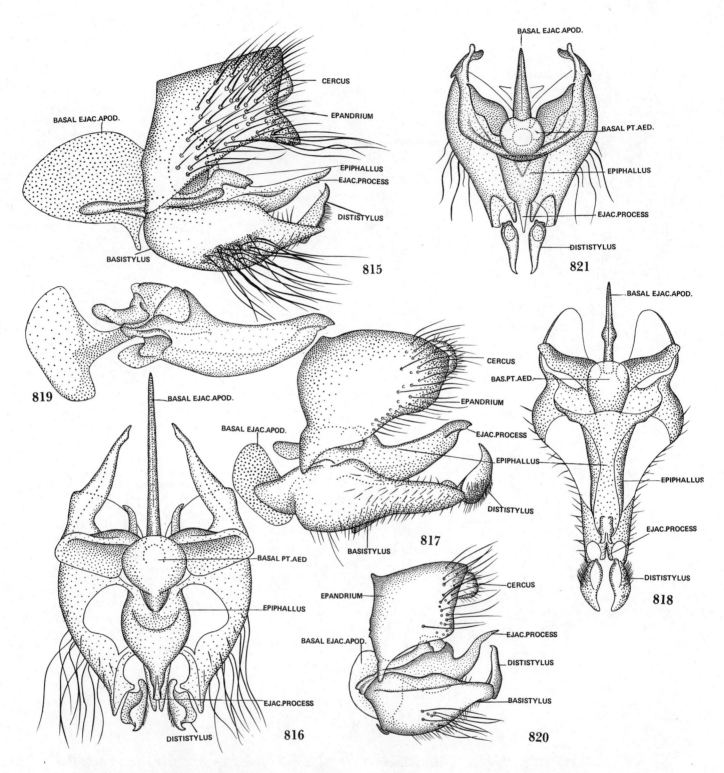

Figures 815–821.—815, *Euprepina nuda* Hull, lateral aspect. 816, *Euprepina nuda* Hull, dorsal aspect. 817, *Corsomyza simplex* Wiedemann, lateral aspect. 818, *Corsomyza simplex* Wiedemann, dorsal aspect. 819, *Corsomyza simplex* Wiedemann, axial system. 820, *Conophorus nigripennis* Loew, lateral aspect. 821, *Conophorus nigripennis* Loew, dorsal aspect.

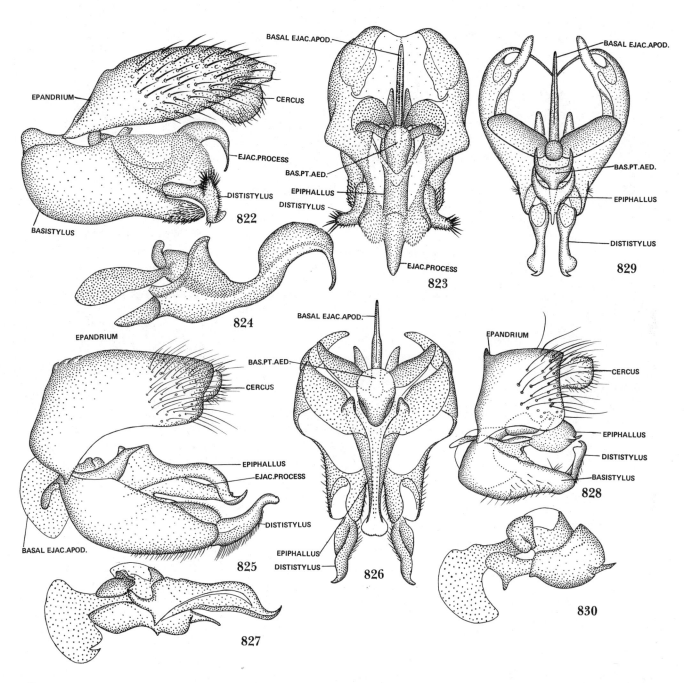

FIGURES 822–830.—822, *Oniromyia pachyceratus* Bigot, lateral aspect. 823, *Oniromyia pachyceratus* Bigot, dorsal aspect. 824, *Oniromyia pachyceratus* Bigot, axial system. 825, *Cytherea obscura* Fabricius, lateral aspect. 826, *Cytherea obscura* Fabricius, dorsal aspect. 827, *Cytherea obscura* Fabricius, axial aspect. 828, *Pantarbes willistoni* Osten Sacken, lateral aspect. 829, *Pantarbes willistoni* Osten Sacken, dorsal aspect. 830, *Pantarbes willistoni* Osten Sacken, axial aspect.

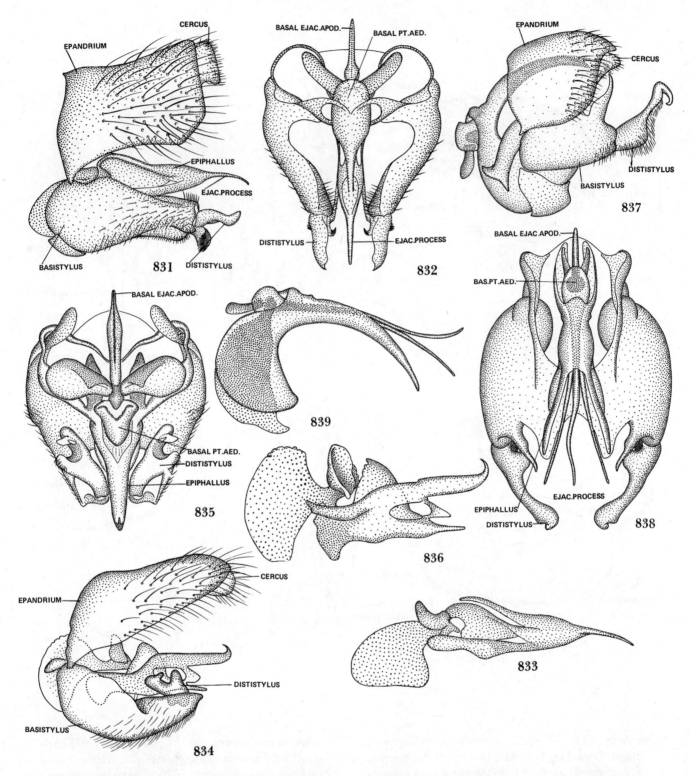

FIGURES 831–839.—831, *Eurycarenus dichopticus* Bezzi, lateral aspect. 832, *Eurycarenus dichopticus* Bezzi, dorsal aspect. 833, *Eurycarenus dichopticus* Bezzi, axial system. 834, *Heterostylum robustum* Osten Sacken, lateral aspect. 835, *Heterostylum robustum* Osten Sacken, dorsal aspect. 836, *Heterostylum robustum* Osten Sacken, axial system. 837, *Heterotropus indicus* Nurse, lateral aspect. 838, *Heterotropus indicus* Nurse, dorsal aspect. 839, *Heterotropus indicus* Nurse, axial system.

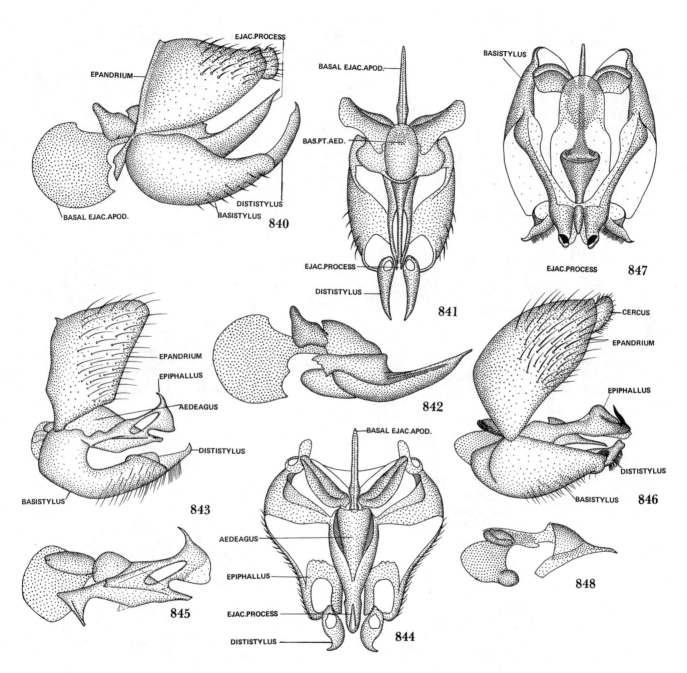

FIGURES 840–848.—840, *Crocidium melanapolis* Hesse, lateral aspect. 841, *Crocidium melanapolis* Hesse, dorsal aspect. 842, *Crocidium melanapolis* Hesse, axial system. 843, *Triploechus heteroneurus* Macquart, lateral aspect. 844, *Triploechus heteroneurus* Macquart, dorsal aspect. 845, *Triploechus heteroneurus* Macquart, axial system. 846, *Desmatomyia anomala* Williston, lateral aspect. 847, *Desmatomyia anomala* Williston, dorsal aspect. 848, *Desmatomyia anomala* Williston, axial aspect.

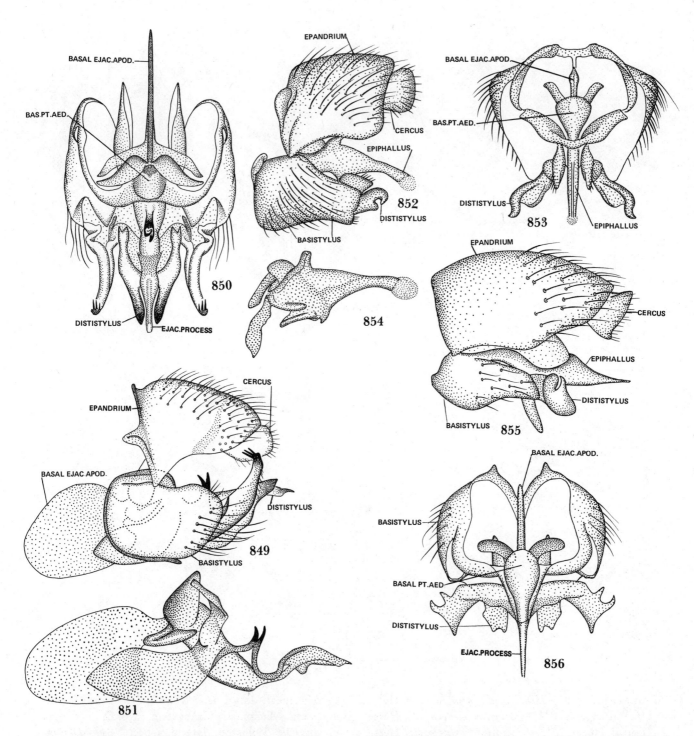

Figures 849–856.—849, *Phthiria* sp., lateral aspect. 850, *Phthiria* sp., dorsal aspect. 851, *Phthiria* sp., axial system. 852, *Usia* (*Parageron*) sp., lateral aspect. 853, *Usia* (*Parageron*) sp., dorsal aspect. 854, *Usia* (*Parageron*) sp., axial system. 855, *Apolysis druias* Melander, lateral aspect. 856, *Apolysis druias* Melander, dorsal aspect.

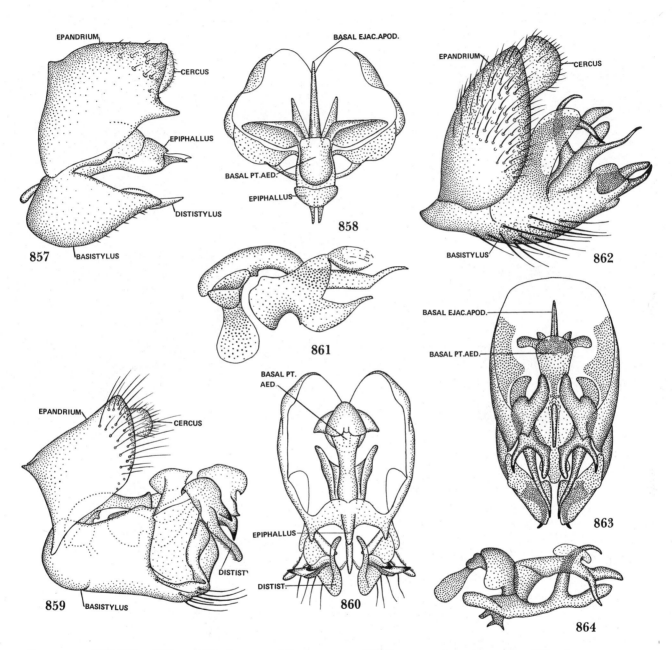

FIGURES 857–864.—857, *Oligodranes bifarius* Melander, lateral aspect. 858, *Oligodranes bifarius* Melander, dorsal aspect. 859, *Geron* sp., lateral aspect. 860, *Geron* sp., dorsal aspect. 861, *Geron* sp., axial system. 862, *Pseudoamictus heteropterus* Wiedemann, lateral aspect, 863, *Pseudoamictus heteropterus* Wiedemann, dorsal aspect. 864, *Pseudoamictus heteropterus* Wiedemann, axial system.

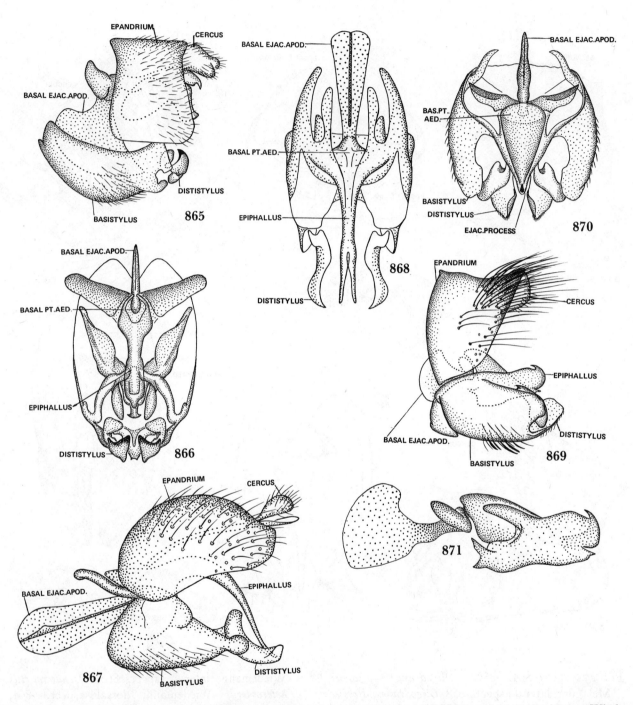

FIGURES 865–871.—865, *Toxophora amphitea* Walker, lateral aspect. 866, *Toxophora amphitea* Walker, dorsal aspect. 867, *Caenotus hospes* Melander, lateral aspect. 868, *Caenotus hospes* Melander, dorsal aspect. 869, *Lepidophora lepidocera* Wiedemann, lateral aspect. 870, *Lepidophora lepidocera* Wiedemann, dorsal aspect. 871, *Lepidophora lepidocera* Wiedemann, axial system.

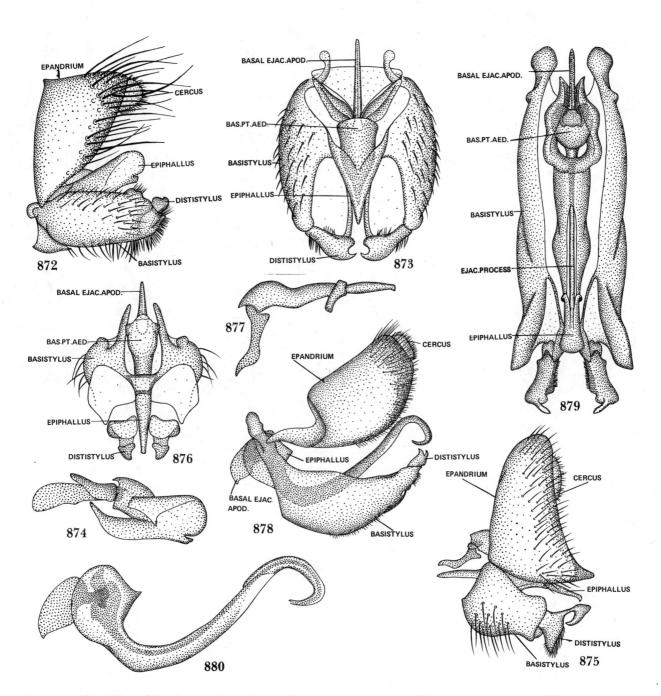

Figures 872–880.—872, *Cyrtomyia chilensis* Paramonov, lateral aspect. 873, *Cyrtomyia chilensis* Paramonov, dorsal aspect. 874, *Cyrtomyia chilensis* Paramonov, axial aspect. 875, *Cyrtomorpha paganica* White, lateral aspect. 876, *Cyrtomorpha paganica* White, dorsal aspect. 877, *Cyrtomorpha paganica* White, axial system. 878, *Marmasoma sumptuosa* White, lateral aspect. 879, *Marmasoma sumptuosa* White, dorsal aspect. 880, *Marmasoma sumptuosa* White, axial system.

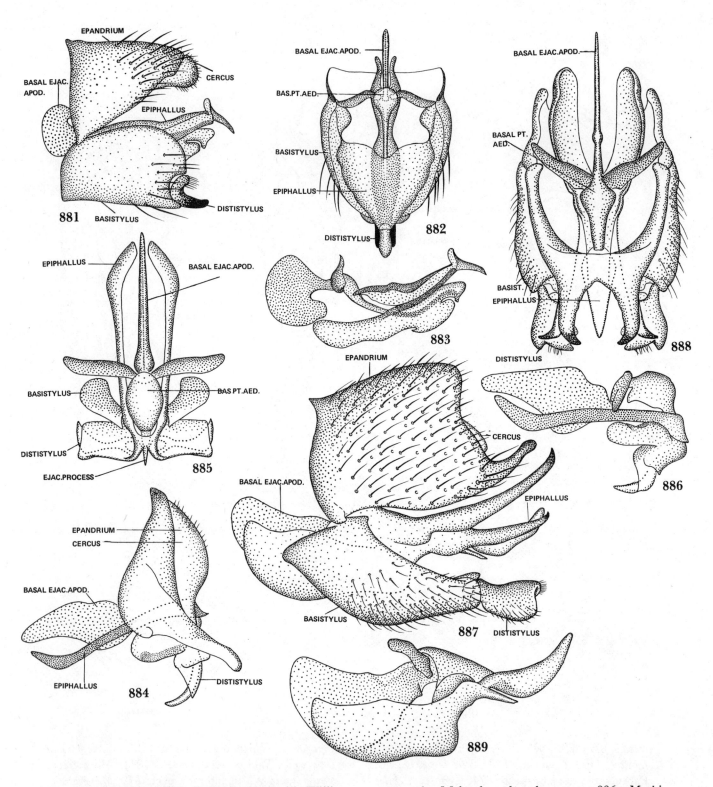

Figures 881-889.—881, *Dolichomyia gracilis* Williston, lateral aspect. 882, *Dolichomyia gracilis* Williston, dorsal aspect. 883, *Dolichomyia gracilis* Williston, axial system. 884, *Mythicomyia monacha* Melander, lateral aspect. 885, *Mythicomyia monacha* Melander, dorsal aspect. 886, *Mythicomyia monacha* Melander, axial system. 887, *Systropus angulatus* Karsch, lateral system. 888, *Systropus angulatus* Karsch, dorsal system. 889, *Systropus angulatus* Karsch, axial system.

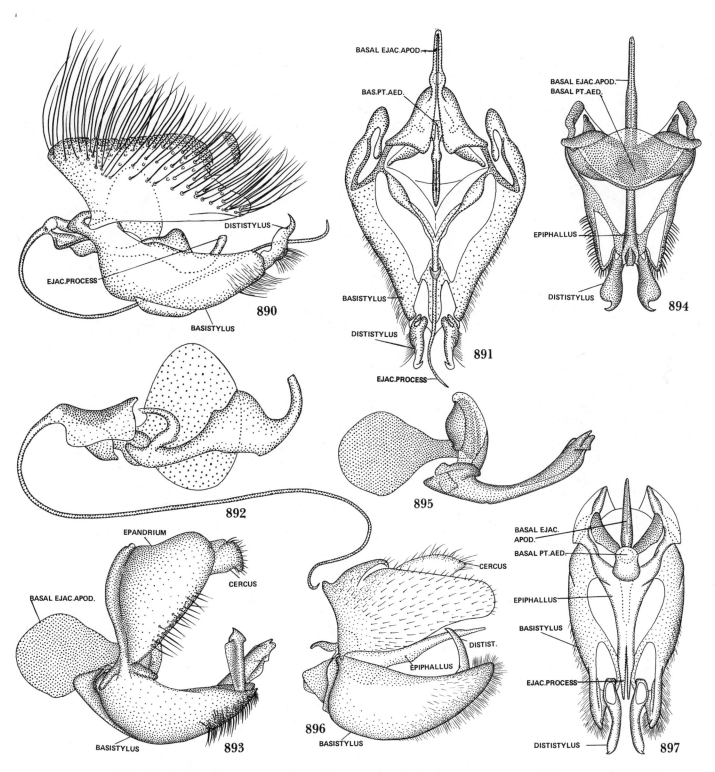

FIGURES 890–897.—890, *Amictus validus* Loew, lateral aspect. 891, *Amictus validus* Loew, dorsal aspect. 892, *Amictus validus* Loew, axial system. 893, *Peringueyimyia capensis* Bigot, lateral aspect. 894, *Peringueyimyia capensis* Bigot, dorsal aspect. 895, *Peringueyimyia capensis* Bigot, axial system. 896, *Paracosmus morrisoni* Osten Sacken, lateral aspect. 897, *Paracosmus morrisoni* Osten Sacken, dorsal aspect.

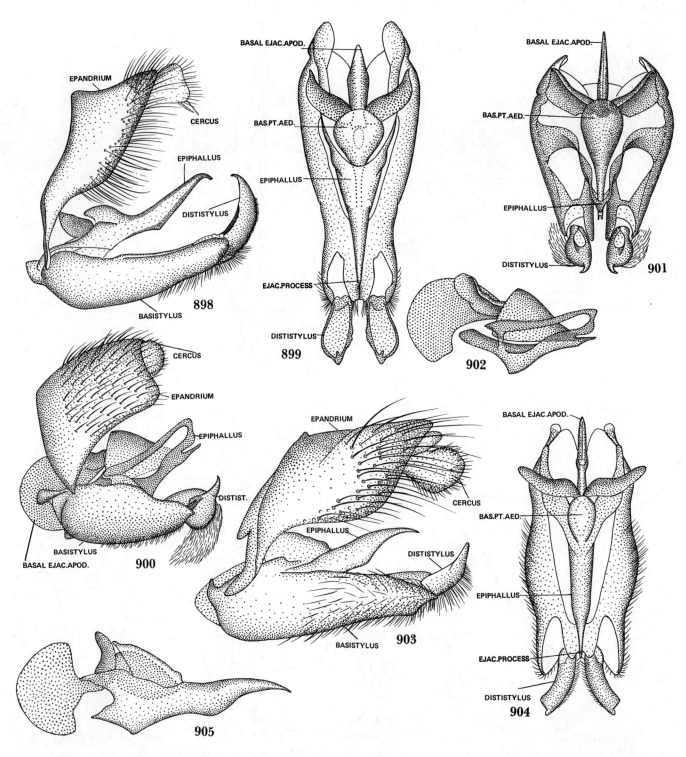

Figures 898–905.—898, *Bryodemina valida* Wiedemann, lateral aspect. 899, *Bryodemina valida* Wiedemann, dorsal aspect. 900, *Neosardus principius* Roberts, lateral aspect. 901, *Neosardus principius* Roberts, dorsal aspect. 902, *Neosardus principius* Roberts, axial system. 903, *Lomatia sabaea* Fabricius, lateral aspect. 904, *Lomatia sabaea* Fabricius, dorsal aspect. 905, *Lomatia sabaea* Fabricius, axial system.

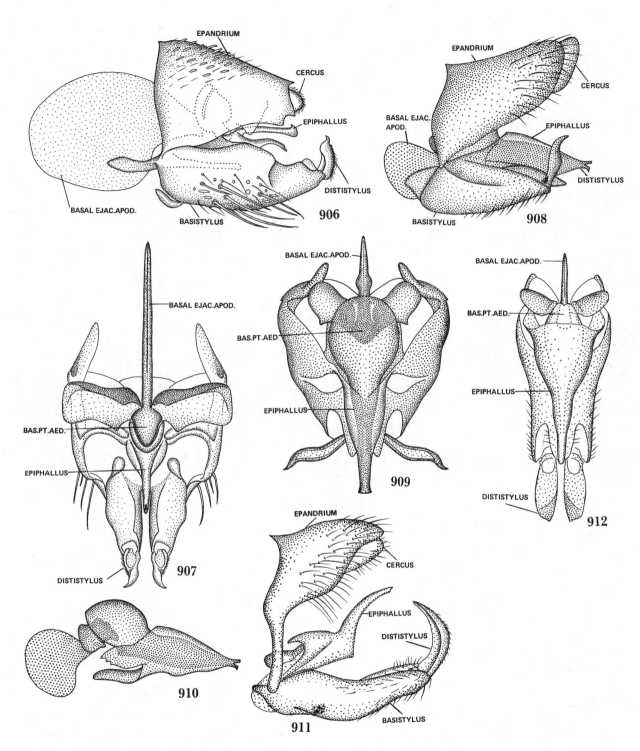

Figures 906–912.—906, *Aphoebantus* sp., lateral aspect. 907, *Aphoebantus* sp., dorsal aspect. 908, *Desmatoneura argentifrons* Williston, lateral aspect. 909, *Desmatoneura argentifrons* Williston, dorsal aspect. 910, *Desmatoneura argentifrons* Williston, axial system. 911, *Ogcodocera leucoprocta* Wiedemann, lateral aspect. 912, *Ogcodocera leucoprocta* Wiedemann, dorsal aspect.

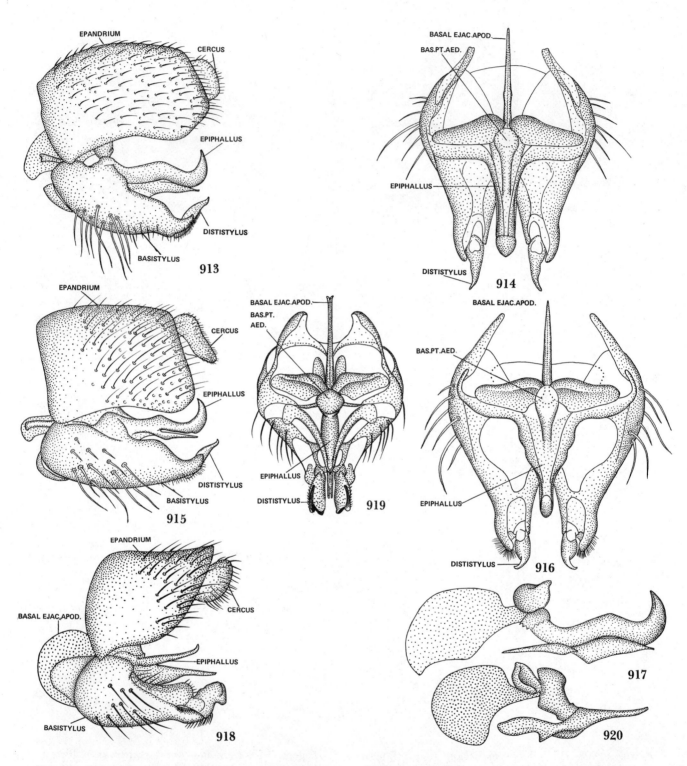

Figures 913–920.—913, *Epacmus* sp., lateral aspect. 914, *Epacmus* sp., dorsal aspect. 915, *Epacmus* sp., lateral aspect. 916, *Epacmus* sp., dorsal aspect. 917, *Epacmus* sp., axial system. 918, *Petrorossia letho* Wiedemann, lateral aspect. 919, *Petrorossia letho* Wiedemann, dorsal aspect. 920, *Petrorossia letho* Wiedemann, axial system.

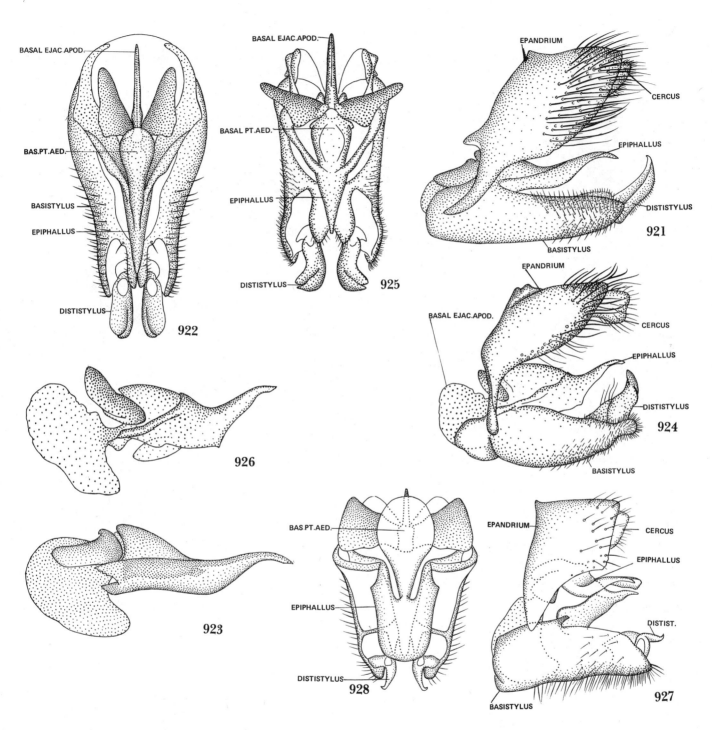

Figures 921–928.—921, *Comptosia fascipennis* Macquart, lateral aspect. 922, *Comptosia* sp. 923, *Comptosia fascipennis* Macquart. 924, *Comptosia* sp. 925, *Comptosia* sp. 926, *Comptosia fascipennis* Macquart. 927, *Astrophanes adonis* Osten Sacken, lateral aspect. 928, *Astrophanes adonis* Osten Sacken, dorsal aspect.

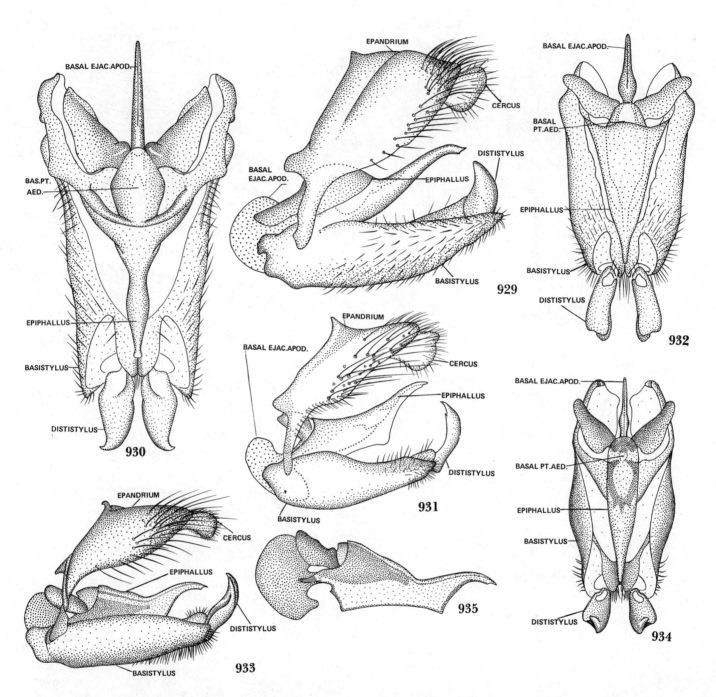

Figures 929–935.—929, *Ulosometa falcata*, new species, lateral aspect. 930, *Ulosometa falcata*, new species, dorsal aspect. 931, *Doddosia picta* Edwards, lateral aspect. 932, *Doddosia picta* Edwards, dorsal aspect. 933, *Oncodosia ampla* Walker, lateral aspect. 934, *Oncodosia ampla* Walker, dorsal aspect. 935, *Oncodosia ampla* Walker, axial system.

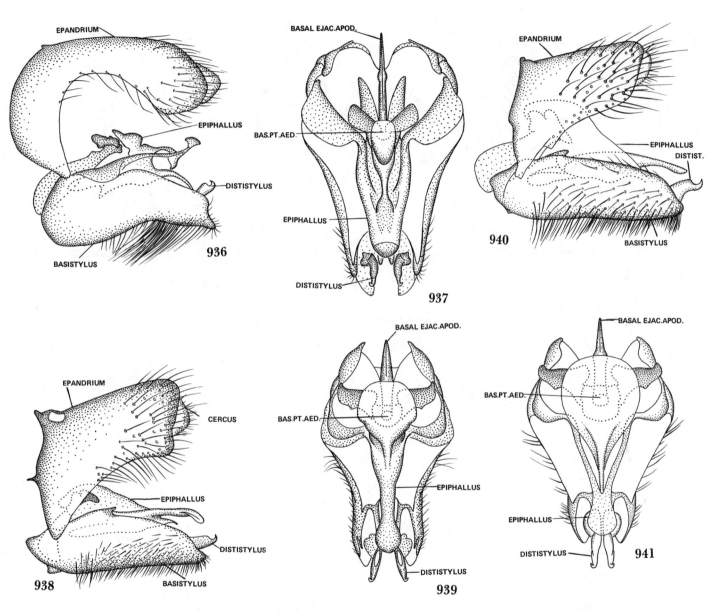

Figures 936–941.—936, *Chrysanthrax cypris* Meigen, lateral aspect. 937, *Chrysanthrax cypris* Meigen, dorsal aspect. 938, *Cyananthrax* sp., lateral aspect. 939, *Cyananthrax* sp., dorsal aspect. 940, *Cyananthrax cyanopterus* Wiedemann, lateral aspect. 941, *Cyananthrax cyanopterus* Wiedemann, dorsal aspect.

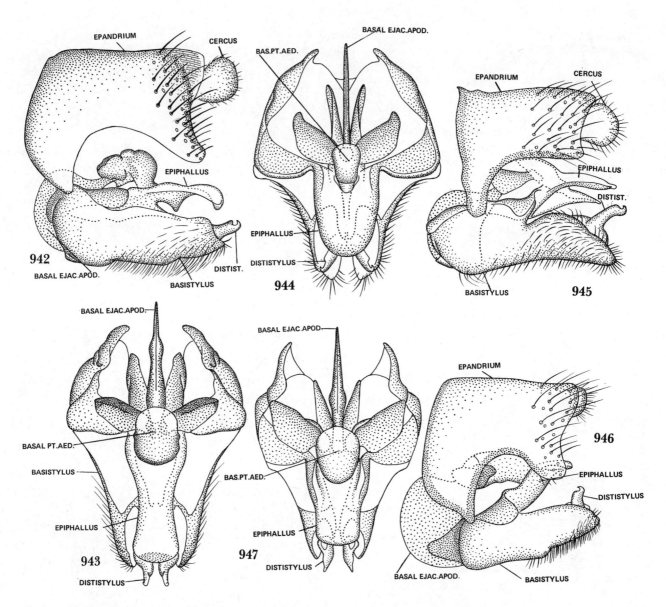

FIGURES 942–947.—942, *Hemipenthes eumenes* Osten Sacken, lateral aspect. 943, *Hemipenthes eumenes* Osten Sacken, dorsal aspect. 944, *Dipalta serpentina* Osten Sacken, lateral aspect. 945, *Dipalta serpentina* Osten Sacken, dorsal aspect. 946, *Lepidanthrax* sp., lateral aspect. 947, *Lepidanthrax* sp., dorsal aspect.

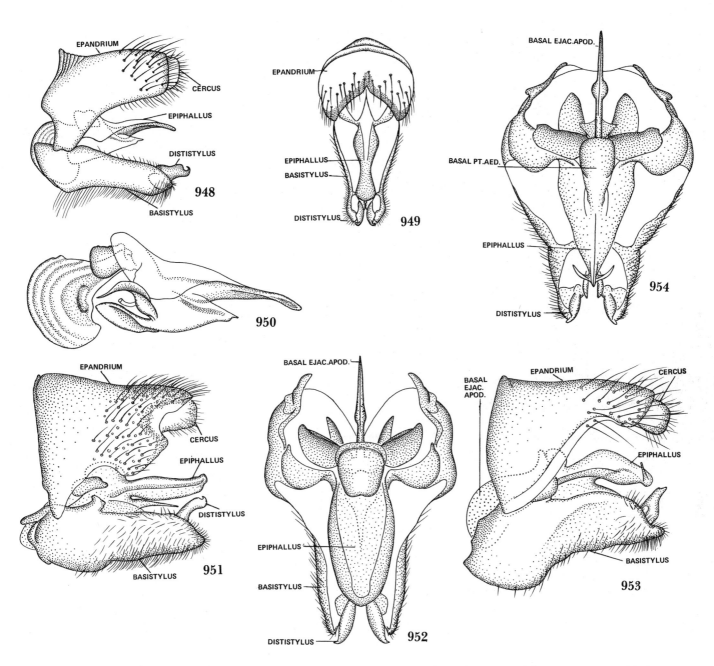

FIGURES 948–954.—948, *Poecilanthrax californicus* Cole, lateral aspect. 949, *Poecilanthrax californicus* Cole, dorsal aspect. 950, *Poecilanthrax californicus* Cole, axial system. 951, *Paravilla edititoides* Painter, lateral aspect. 952, *Paravilla edititoides* Painter, dorsal aspect. 953, *Villa aggripina* Osten Sacken, lateral aspect. 954, *Villa aggripina* Osten Sacken, dorsal aspect.

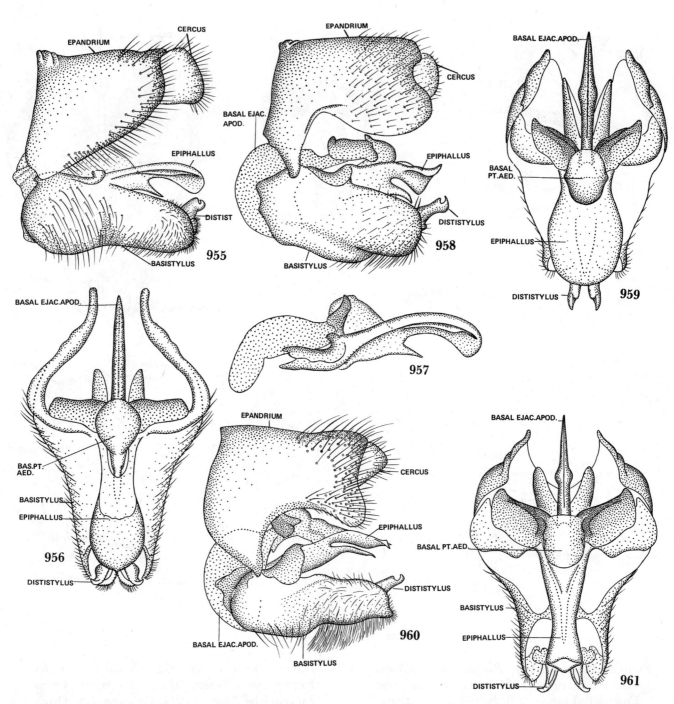

FIGURES 955–961.—955, *Oestranthrax obesus* Loew, lateral aspect. 956, *Oestranthrax obesus* Loew, dorsal aspect. 957, *Oestranthrax obesus* Loew, axial. 958, *Thyridanthrax fenestratus* Fallen, lateral aspect. 959, *Thyridanthrax fenestratus* Fallen, dorsal aspect. 960, *Stonyx clotho* Wiedemann, lateral aspect. 961, *Stonyx clotho* Wiedemann, dorsal aspect.

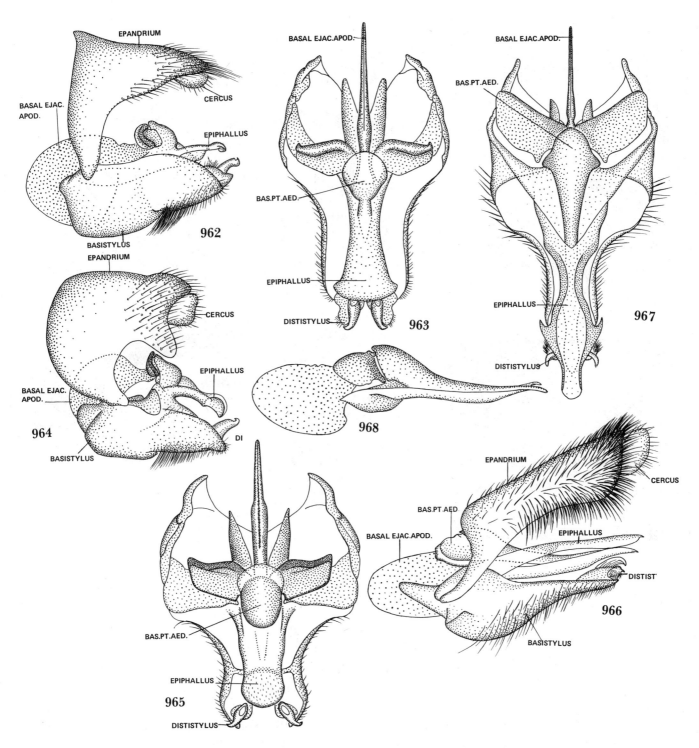

FIGURES 962–968.—962, *Ligyra tenebrosa* Paramonov, lateral aspect. 963, *Ligyra tenebrosa* Paramonov, dorsal aspect. 964, *Rhynchanthrax parvicornis* Loew, lateral aspect. 965, *Rhynchanthrax parvicornis* Loew, dorsal aspect. 966, *Anthrax aterrima* Bigot, lateral aspect. 967, *Anthrax aterrima* Bigot, dorsal aspect. 968, *Anthrax aterrima* Bigot, axial system.

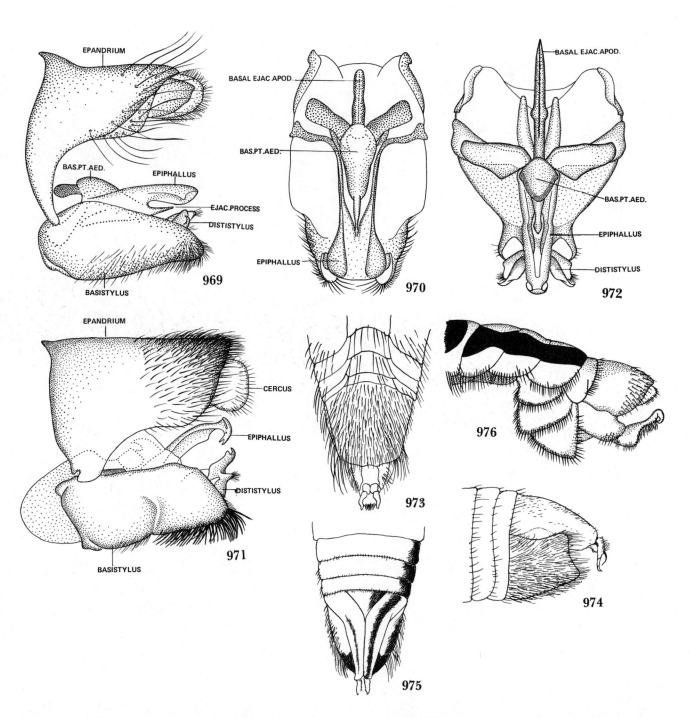

Figures 969–976.—969, *Colossoptera latipennis* Brunetti, lateral aspect. 970, *Colossoptera latipennis* Brunetti, dorsal aspect. 971, *Exoprosopa ingens* Cresson, lateral aspect. 972, *Exoprosopa ingens* Cresson, dorsal aspect. 973, *Sinaia kneuckeri* Becker, ventral aspect of pygidium, after Efflatoun. 974, *Sinaia kneuckeri*, Becker, lateral aspect of pygidium, after Efflatoun. 975, *Sinaia kneuckeri* Becker, dorsal aspect of pygidium, after Efflatoun. 976, *Heterotropus indicus* Nurse, lateral aspect of pygidium.

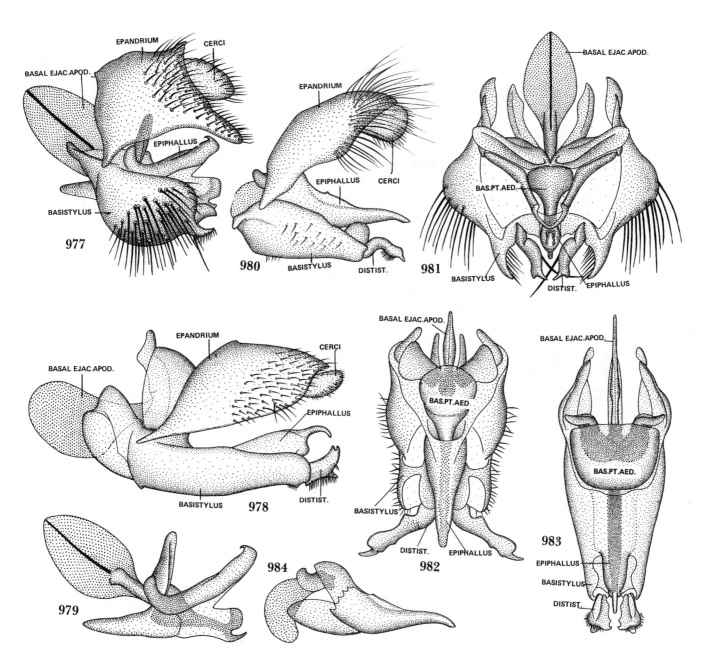

Figures 977–984.—977, *Antonia xanthogramma* Bezzi, lateral aspect. 978, *Henica longirostris* Wiedemann, lateral aspect. 979, *Antonia xanthogramma* Bezzi, dorsal aspect. 980, *Dischistus mystax* Wiedemann, lateral aspect. 981, *Antonia xanthogramma* Bezzi, axial system. 982, *Dischistus mystax* Wiedemann, dorsal aspect. 983, *Henica longirostris* Wiedemann, dorsal aspect. 984, *Dischistus mystax* Wiedemann, axial system.

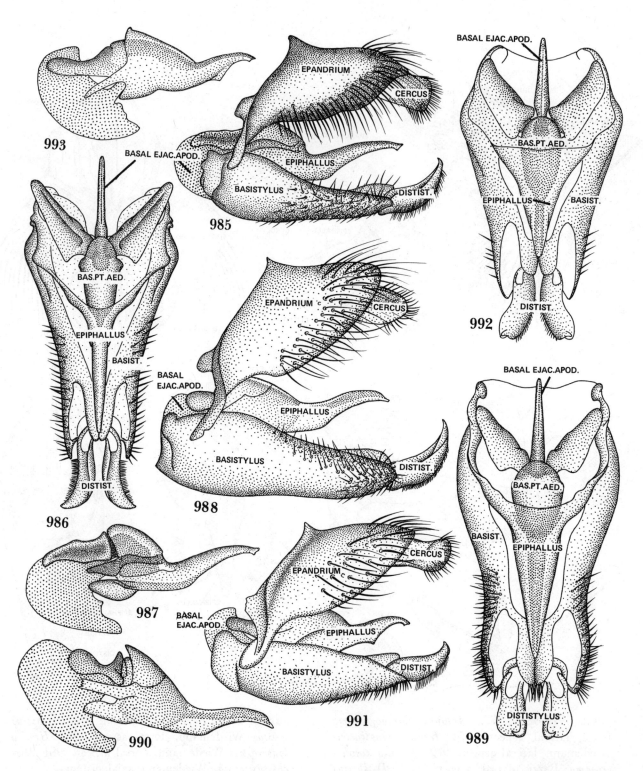

FIGURES 985–993.—985, *Ylasoia pegasus* Wiedemann, lateral aspect. 986, *Ylasoia pegasus* Wiedemann, dorsal aspect. 987, *Ylasoia pegasus* Wiedemann, axial system. 988, *Lyophlaeba bifasciata* Macquart, lateral aspect. 989, *Lyophlaeba bifasciata* Macquart, dorsal aspect. 990, *Lyophlaeba bifasciata* Macquart, axial system. 991, *Lyophylaeba lugubris* Rondani, lateral aspect. 992, *Lyophlaeba lugubris* Rondani, dorsal aspect. 993, *Lyophlaeba lugubris* Rondani, axial system.

ILLUSTRATIONS OF BOMBYLIIDAE 569

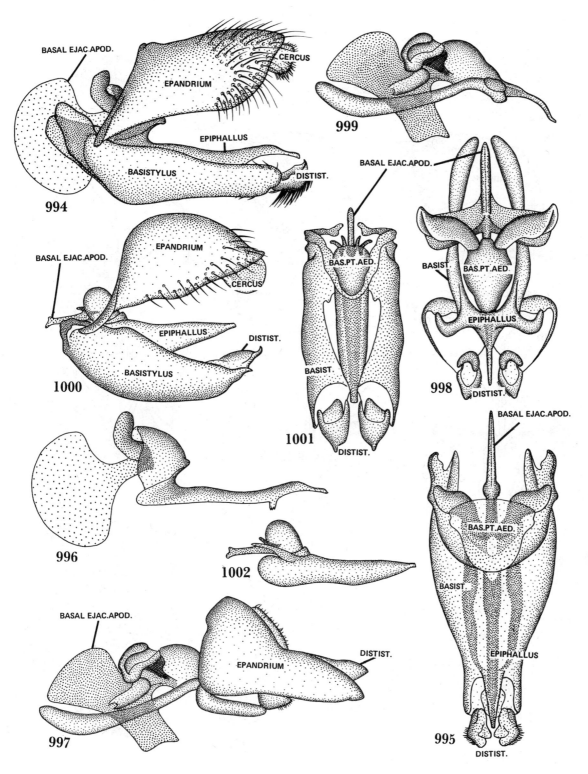

FIGURES 994–1002.—994, *Nomalonia clavicornis* Hesse, lateral aspect. 995, *Nomalonia clavicornis* Hesse, dorsal aspect. 996, *Nomalonia clavicornis* Hesse, axial system. 997, *Cyrtisiopsis crassirostris* Hesse, lateral aspect. 998, *Cyrtisiopsis crassirostris* Hesse, dorsal aspect. 999, *Cyrtisiopsis crassirostris* Hesse, axial system. 1000, *Sericosoma squamiventris* Edwards, lateral aspect. 1001, *Sericosoma squamiventris* Edwards, dorsal aspect. 1002, *Sericosoma squamiventris* Edwards, axial system.

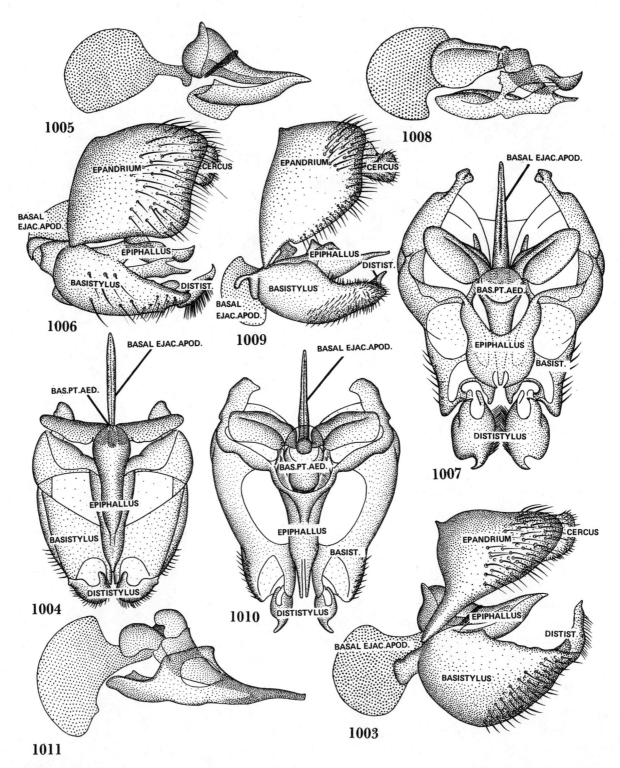

FIGURES 1003–1011.—1003, *Adelogenys culicoides* Hesse, lateral aspect. 1004, *Adelogenys culicoides* Hesse, dorsal aspect. 1005, *Adelogenys culicoides* Hesse, axial system. 1006, *Adelidea maculata* Hesse, lateral aspect. 1007, *Adelidea maculata* Hesse, dorsal aspect. 1008, *Adelidea maculata* Hesse, axial system. 1009, *Platamomyia depressa* Loew, lateral aspect. 1010, *Platamomyia depressa* Loew, dorsal aspect. 1011, *Platamomyia depressa* Loew, axial system.

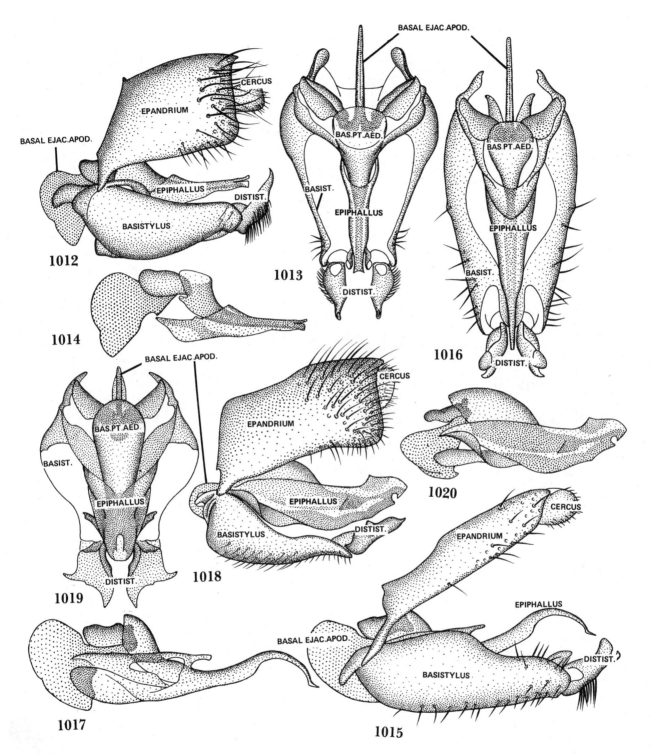

Figures 1012–1020.—1012, *Prorachthes conspersipennis* Hesse, lateral aspect. 1013, *Prorachthes conspersipennis* Hesse, dorsal aspect. 1014, *Prorachthes conspersipennis* Hesse, axial system. 1015, *Gonarthrus* sp., lateral aspect. 1016, *Gonarthrus* sp., dorsal aspect. 1017, *Gonarthrus* sp., axial system. 1018, *Dicranoclista fasciata* Johnson and Johnson, lateral aspect. 1019, *Dicranoclista fasciata* Johnson and Johnson, dorsal aspect. 1020, *Dicranoclista fasciata* Johnson and Johnson, axial system.

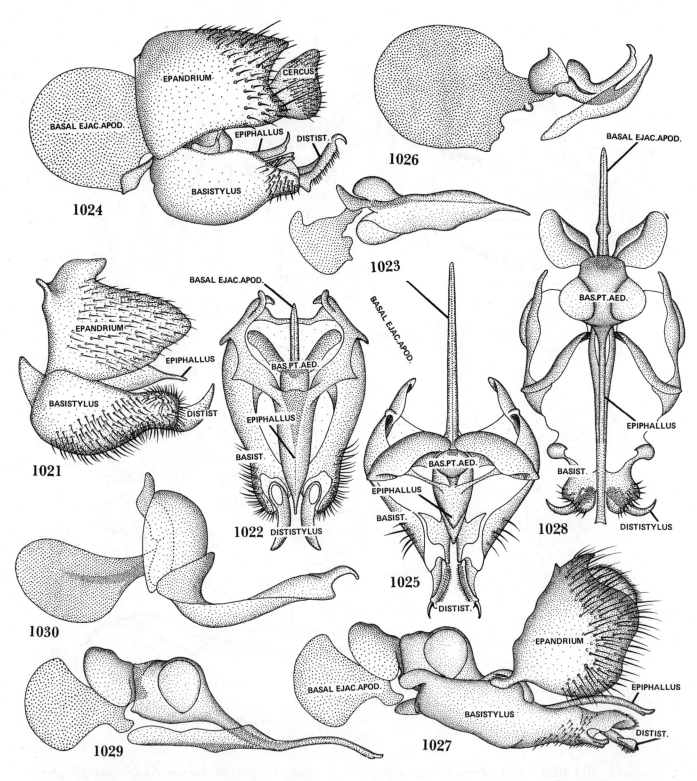

Figures 1021–1030.—1021, *Tomomyza pictipennis* Bezzi, lateral aspect. 1022, *Tomomyza pictipennis* Bezzi, dorsal aspect. 1023, *Tomomyza pictipennis* Bezzi, axial system. 1024, *Sphenoidoptera varipennis* Williston, lateral aspect. 1025, *Sphenoidoptera varipennis* Williston, dorsal aspect. 1026, *Sphenoidoptera varipennis* Williston, axial system. 1027, *Nectaropota setigera* Philippi, lateral aspect. 1028, *Nectaropota setigera* Philippi, dorsal aspect. 1029, *Nectaropta setigera* Philippi, axial system. 1030, *Henica longirostris* Wiedemann, axial system.

Bibliography

ABBASSIAN-LINTZEN, ROSEMARIE
　1968. Bombyliidae (Diptera) of Iran: IV. Species of the Subfamily Cythereinae. Journ. Natur. Hist., vol. 2, no. 2, pp. 231–238.

ABREU, E. SANTOS
　1926. Monografía de los Bombylidos de las islas Canarias. Mem. Roy. Acad. Barcelona, ser. 3, vol. 20, pp. 1–67, 1 pl., 3 figs.

ADACHI, TSUNAMITSU, AOKI, AKIRA, and SHIRAKI, T., and many others
　1954. Iconographia Insectorum Japanicorum. Edition 2. Pp. 1–1736; index pp. 1–203; 15 color pls.; figs. 1–4962, featuring 4962 species. (Bombyliidae pp. 1593–1597; figs. 4577–4588.)

ADAMS, CHARLES CHRISTOPHER
　1905. An ecological study of prairie and forest invertebrates. Bull. Illinois State Lab. Nat. Hist., vol. 11, art. 2, pp. 33–280, 63 pls., 18 text-diagrams. (Bombyliidae pp. 120, 185–186, 198, text-fig. 16.)

ADAMS, CHARLES FREDERICK
　1905. Diptera Africana. I. Kansas Univ. Sci. Bull., vol. 3, no. 6, pp. 149–208. (Bombyliidae pp. 155–156.)

ADELUNG, EDUARD VON
　1895. Note: Summary of Portschinsky's papers of 1894 and 1895. [In German.] Zool. Centralbl., vol. 2, p. 285.

AGASSIZ, JEAN LOUIS RODOLPHE, and others
　1842–1847. Nomenclator zoologicus. In 12 parts. Also 1848, pp. 1–1135.
　　　1846. Nomina systematica generum dipterorum, by Agassiz and Loew, pp. 9–10, 1–42.

AHRENS, AUGUST
　1812–1814. Fauna insectorum Europae. Halle, Kümmel. Books 1 and 2 by Ahrens; other books by Germar. Each book with 25 pages and 25 plates.
　　　1812. Book 1, pp. 1–25, pls. 1–25. (Bombyliidae p. 25, pl. 25.)

ALDRICH, JOHN MERTON
　1905. A catalogue of North American Diptera. Smithsonian Misc. Coll., vol. 46, no. 1444, 680 pp. (Bombyliidae pp. 218, 221–246.)
　1907. Additions to my catalogue of North American Diptera. Journ. New York Ent. Soc., vol. 15, pp. 2–9. (Bombyliidae p. 4.)
　1916. Dipterological notes II. A study of the *lateralis* group of the bombyliid genus *Villa* (*Anthrax* of authors, in part). Ent. News, vol. 27, pp. 439–444.
　1926. On the status of the generic name *Anthrax* Scopoli. Insecutor Inscitiae Menstruus, vol. 14, nos. 1–3, pp. 12–15.
　1928. Three new species of two-winged flies of the family Bombyliidae from India. Proc. U.S. Nat. Mus., vol. 74, art, 2, no. 2747, 3 pp.

ALLEN, F. J., and UNDERHILL, H. M. J.
　1875. Notes on the Diptera. April: Humble-bee flies. Science Gossip. London. (Hardwicke, editor.), 1875, pp. 79–81, figs. 46–51.
　1876. Notes on the Diptera. VI. Conopidae. (This includes note on Bombylii and Sitaris.) Science Gossip, 1876, p. 171.

Allen, H. W.
- 1921. Notes on a bombylid parasite and a polyhedral disease of the southern grass worm, *Laphygma frugiperda*. Journ. Econ. Ent., vol. 14, pp. 510–511.

Allred, D. M., Johnson, D. E., and Beck, D. E.
- 1965. A list of some beeflies of the Nevada test site. 6R. Basin Naturalist, vol. 25, pp. 5–11, 2 figs.

Andres, Adolf
- 1913. Note additionnelle à la communication de M. Boehm sur *Anthrax griseola* Klug (Dipt.). Bull. Soc. Roy. d'Egypte, vol. 3 (1912–1913), p. 31.

d'Andretta, Maria A. V., and Carrera, Messias
- 1950. Sobre as espécies Brasileiras de Toxophorinae (Diptera, Bombyliidae). Dusenia, vol. 1, pt. 6, pp. 351–374, figs. 1–45.
- 1952. Resultados de uma expedição scientífica ao Território do Acre. Diptera. Papéis Avulsos Dept. Zool., São Paulo, vol. 10, no. 17, pp. 293–306, 8 figs. (Bombyliidoe pp. 296–297.)

Andreu Rubio, Jose M.
- 1959. Bombilidos marroquies del Instituto Español de Entomologia (Diptera). Eos Rev. Española Ent., vol. 35, no. 1, pp. 7–19.
- 1961. Los dipteros bombilidos espanoles y su distribucion geografica. Murcia. 65 pp.

Androic, M.
- 1956. Contribution a l'etude de *Cnethocampa pityocampa* Schiff. Rev. Path. Veg. Ent. Agric., vol. 35, pp. 251–262.

Angus, James
- 1868. Habits of the carpenter bees. Amer. Natur., vol. 1 (1867), p. 157.
- 1869. Bee parasite. Amer. Natur., vol. 2 (1868), pp. 48–49.

Aoki, Akira [see Adachi, Aoki, Shiraki, and others]

Austen, Ernest Edward
- 1909. Ruwenzori Expedition reports. Part 10. Diptera. Trans. Zool. Soc. London, vol. 19, pp. 85–100, pl. 3. (Bombyliidae p. 86.)
- 1913. On Diptera collected in the western Sahara by Dr. Ernst Hartert, with descriptions of new species. Part 1. Bombyliidae. Nov. Zool. Tring, vol. 20, pp. 460–465.
- 1914. A dipterous parasite of *Glossina morsitans*. Bull. Ent. Research, London, vol. 5, pp. 91–93, fig. 1.
- 1929. The Tsetse fly parasites belonging to the genus *Thyridanthrax* (Diptera. Family Bombyliidae), with descriptions of new species. Bull. Ent. Research, London, vol. 20, pp. 151–164, text-figs. 1–10.
- 1936. New Palaearctic Bombyliidae (Diptera). Ann. Mag. Nat. Hist., London, ser. 10, vol. 18, pp. 181–204, pl. 5, figs. 1–7.
- 1937. The Bombyliidae of Palestine. British Museum (Natural History), pp. i–ix, 1–188, 3 pls., 1 color frontispiece, 72 text-figs., 1 map. (Immature stages pp. 2–3, 14, 38, 40, 46, 52, 53, 54, 55, 62, 75–76, 77, 91, 94, 95, 113, 130, 133, 146–147, 162.)

Austen, Ernest Edward, and Hegh, E.
- 1922. Tsetse-flies, their characteristics, distribution and bionomics, with some account of possible methods for their control. Imper. Instit. Ent., London, 188 pp., figs. (Bombyliidae pp. 120, 130, 131, figs, 9, 10.)

Baer, W.
- 1920–1921. Die Tachinen als Schmarotzer der schädlichen insekten. Ihre Lebensweise, wirtschaftliche Bedeutung und systematische Kennzeichnung. Zeitscher. Angew. Ent.
    - 1920. Vol. 6, pt. 1, pp. 185–246. (Bombyliidae pp. 187, 235–237, figs. 32-33.)
    - 1921. Vol. 7, pt. 7, pp. 97–163, 349–423, 63 figs.

BAKER, GEORGE W. (editor)
- 1948. Index to the literature on flies (other than mosquitoes) and the diseases they convey, abstracted in the *Review of Applied Entomology, Section B., medical and veterinary*, volumes 1–34, 1913 to 1946 inclusive. In 2 volumes. Vol. 1, pp. 1–296; vol. 2, pp. 297–597. (Bombyliidae: Vol. 1 contains references to bombyliid parasites ot tsetse flies (*Glossina* species).

BALDUF, W. V.
- 1962. Life of the carpenter bee, *Xylocopa virginica* (Linn.) (Xylocopidae, Hymenoptera). Ann. Ent. Soc. America, vol. 55, pp. 263–271. (Bombyliidae p. 270.)

BANKS, NATHAN
- 1909a. A new species of *Systropus* (Bombyliidae), Ent. News, vol. 20, p. 18.
- 1909b. Directions for collecting and preserving insects. U.S. Nat. Mus. Bull. 67, 135 pp., 188 figs. (Bombyliidae p. 30, fig. 58.)

BARRAND, PHILLIP J. [see Gibbs and Barrand]

BATRA, SUZANNE W. T.
- 1965. Organisms Associated with *Lasioglossum zephyrum* (Hymenoptera: Halictidae). Journ. Kansas Ent. Soc., vol. 38, pp. 367–389.

BAUDOT, E. [see Séguy and Baudot]

BECHER, E.
- 1882. Zur Kenntniss der Mundtheile der Dipteren. Denkschr. Akad. Wiss. Wien, vol. 45, pp. 123–162, pls. 1–4. (Bombyliidae p. 145, pl. 3, fig. 3 a-d, fig. 4 a-g, fig. 5 a-f.)

BECKER, THEODOR
- 1887. Beiträge zur Kenntniss der Dipteren-Fauna von St. Moritz. Berliner Ent. Zeitschr., vol. 31, pp. 93–141. (Bombyliidae pp. 106–107; 6 spp.)
- 1889. Altes und neues aus der Schweiz. Wiener Ent. Zeitung, vol. 8, pp. 73–84, pl. 1. (Bombyliidae pp. 73–75.)
- 1891. Neues aus der Schweiz. Ein dipterologischer Beitrag. Wiener Ent. Zeitung, vol. 10, pp. 289–296. (Bombyliidae pp. 294–296.)
- 1900. Beiträge zur Dipteren-Fauna Sibiriens. Nordwest-Sibirische Dipteren gesammelt vom Prof. John Sahlberg aus Helsingfors im Jahre 1876 und vom Dr. E. Bergroth aus Temmerfors im Jahre 1877. Acta Soc. Sci. Fennicae, vol. 26, no. 9, pp. 1–66, 2 pls. (Bombyliidae pp. 14–16.)
- 1902–1903. Aegyptische Diptern. Mitt. Zool. Mus. Berlin.
    - 1902. Vol. 2, pt. 2, pp. 1–66, 1 pl. (Bombyliidae pp. 11–32.)
- 1903. Die paläarktischen Formen der Gattung *Mulio* Latreille (Dipt.) Zeitschr. syst. Hymen. Dipt. Teschendorf, vol. 3, pt. 1, pp. 17–32, 89–96, 193–196, 197–198.
- 1906a. *Usia* Latr. Berliner Ent. Zeitschr., vol. 50 (1905), pp. 193–228, pl. 6, figs. 1–7 (color).
- 1906b. Notiz zu *Usia taeniolata* Ach. Costa. Ann. Mus. Zool. Univ. Napoli, n. ser., vol. 2, no. 8, 1 unnumbered page.
- 1906–1907. Die Ergebnisse meiner Dipterologischen Frühjahrsreise nach Algier und Tunis. 1906. Zeitschr, syst. Hymen. Dipt. Teschendorf.
    - 1906. Vol. 6, pp. 1–16, 97–114, 145–158, 273–287, 353–368. (Bombyliidae pp. 13–16, 97–114, 145–158, 273–275.)
    - 1907. Vol. 7, pp. 33–61, 97–128, 225–256, 369–407. (Bombyliidae pp. 97–98.)
- 1907a. *Legnotus* Lw. (Dipt.). Zeitschr. syst. Hymen. Dipt, Teschendorf, vol. 7, pp. 452–454, 3 figs.
- 1907b. Dipterologische Ergebnisse. Berichtigung. Zeitschr. syst. Hymen. Dipt. Teschendorf, vol. 7, pp. 454–455. (Bombyliidae p. 455.)
- 1907c. *In* Becker, Stein, Treplin, Villeneuve: Dipteren aus Südarabien und von der Insel Sokotra. Denkschr. Math.-Naturw. Klasse Kaiser Akad. Wiss., vol. 71, pp. 131–160, 5 figs. (Bombyliidae pp. 131–135.)
- 1907d. Zur Kenntnis der Dipteren von Central-Asien. 1. Cyclorrhapha schizophora holometopa und Orthorrhapha brachycera. Ann. Mus. Zool. Acad. Imp. Sci., St. Pétersbourg, vol. 12, pp. 253–317, pls. 1–2. (Bombyliidae pp. 312–314.) Reprint pp. 1–65, pls. 1–2. (Bombyliidae pp. 60–62, pl. 2, figs. 13, 18.)

BECKER, THEODOR—Continued
- 1908. Dipteren der Kanarischen Inseln und der Insel Madeira. Mitt. Zool. Mus. Berlin, 4, pp. 3–206, 4 pls. (Bombyliidae pp. 12–22.)
- 1909. Collections recueillies par M. Maurice de Rothschild dans l'Afrique orientale angalaise. Insectes: diptères nouveaux. Bull. Mus. Hist. Nat. Paris, 1909, pp. 113–121. (Bombyliidae p. 115; 2 spp.)
- 1910a. Dipterologische Sammelreise nach Korsika. Part 1. Orthorrhapha brachycera. Deutsche Ent. Zeitschr., 1910, pp. 635–665. (Bombyliidae pp. 636–638.)
- 1910b. Voyage de M. Maurice de Rothschild en Ethiopie et dans l'Afrique orientale. Ann. Soc. Ent. France, vol. 79, pp. 22–30. (Bombyliidae pp. 23–24; 2 n. spp.)
- 1911. Die Loew'schen Typen in der Rosenhauerschen Dipteren-Sammlung. Wiener Ent. Zeitung, vol. 30, pp. 71–76. (Bombyliidae pp. 71–72.)
- 1913a. Genera Bombyliidarum. Ann. Mus. Zool. Acad. Imp. Sci. St. Pétersbourg, vol. 17 (1912), pp. 421–502, 36 figs., 1 chart.
- 1913b. *In* Becker and Stein: Persische Dipteren von den Expeditionen des Herrn N. Zarudny, 1898 und 1901. Ann. Mus. Zool. Acad. Imp. Sci. St. Pétersbourg, vol. 17 (1912), pp. 503–654, pls. 12–14 with figs. 1–39 (figs. 20–26 and 30 are omitted), text-figs. 1–27. (Bombyliidae pp. 506–507, 553–587, pl. 13, figs. 17–19, text figs. 20–27.)
- 1913c. *In* Becker and Stein: Dipteren aus Marokko. Ann. Mus. Zool. Acad. Imp. Sci. St. Pétersbourg, vol. 18, pp. 62–95, 2 figs. (Bombyliidae pp. 83–84; 21 spp.)
- 1914a. Diptères nouveaux recoltés par Ch. Alluaud et R. Jeannel en Afrique orientale. Ann. Soc. Ent., Paris, vol. 83, pp. 121–130. (Bombyliidae [as Empididae] p. 121.)
- 1914b. Ergebnisse einer von Prof. Franz Werner im Sommer 1910 . . . ausgeführten zoologischen Forschungsreise nach Algerien. IV. Dipteren. Sitz. Akad. Wiss. Wien. Math.-nat. Klasse, vol. 123, sect. 1, pp. 605–608. (Bombyliidae p. 607.)
- 1915a. Dipteren aus Tunis in der Sammlung der ungarischen National-Museums. Ann. Mus. Nat. Hungarici, vol. 13, pp. 301–330. (Bombyliidae pp. 318–330, 1 text fig.)
- 1915b. *Edmundiella*, novum genus Lomatiinarum. Wiener Ent. Zeitung, vol. 34, pp. 347–348, fig. 1 (wing).
- 1916. Beiträge zur Kenntnis einiger Gattungen der Bombyliiden. Ann. Mus. Nat. Hungarici, vol. 14, pp. 17–67, 2 figs.
- 1917. Nachtrag zu meinem Aufsatze. Beiträge zur Kenntnis einiger Gattungen der Bombyliiden. Ann. Mus. Nat. Hungarici, vol. 15, p. 382.
- 1920. *Conophorina*, novum genus Bombyliidarum (Dipt.). Ent. Mitt., vol. 9, pp. 181–184, 3 text-figs.
- 1923. Wissenschaftliche Ergebnisse der mit Unterstützung der Akademie der Wissenschaften in Wien aus der Erbschaft Treitl von F. Werner unternommenen Zoologischen Expedition nach dem Anglo-Agyptischen Sudan (Kordofan) 1914. Part VI. Diptera. Denkschr. Akad. Wiss. Wien, vol. 98, pp. 57–82, 6 figs. (Bombyliidae pp. 58–61, fig. 1.)
- 1926a. *In* Paramonov: Zur Kenntniss der Gattung *Anastoechus* O. S. (Bombyliidae, Diptera). Arch. Naturg., Berlin, vol. 91 (1925), Abt. A, pt. 1, pp. 46–55. (Becker p. 46;)
- 1926b. *In* Becker and Schnabl: Dipteren von W. W. Sowinsky an den Ufern der Baikal-Sees im Jahre 1902 gesammelt. Ent. Mitt., vol. 15, pp. 33-46. (Bombyliidae pp. 37-38.)

BEIER, MAX [see Hendel and Beier]

BERENDT, GEORG CARL
- 1830. Die Insekten im Bernstein, ein Beitrag zur Thiergeschichte der Vorwelt. Dansig, Verfasser. Erstes Heft, pp. 1–38. (Bombyliidae sp. l'Ambre du Samland, p. 30.)

BERG, CARLOS
- 1878. El género Streblota y los Notodontinas de la República Argentina. Ann. Soc. Sci. Argentina, vol. 5, pp. 177–188. (Bombyliidae p. 183. Records *Systropus nitidus* Wiedemann as *Systropus brasiliensis* Macquart.)

Berg, V. L.
1940. The external morphology of the immature stages of the bee fly, *Systoechus vulgaris* Loew (Diptera, Bombyliidae), a predator of grasshopper egg pods. Canadian Ent., vol. 72, no. 9, pp. 169–178, 6 figs.

Bergroth, E.
1894. Ueber einige Australische Dipteren. Stettiner Ent. Zeitung, vol. 55, pp. 71–75. (Bombyliidae p. 72.)

Berlese, Antonio
1909–1925. Gli Insetti. Milano.
    1909. Vol. 1, pp i–x, 1–1004, figs. 1–1292, 10 pls.
    1925. Vol. 2, pp. 1–9, 1–922. (Bombyliidae pp. 654, 711.)

Berthold, Arnold Adolph
1827. Natürliche Familien des Thierreichs. (German translation of Latreille, Familles naturelles du règne animal. . . . 1825.) 606 pp. Weimar. (Bombyliidae p. 506, *Cephenus*.)

Bertoloni, Giuseppe
1862. Illustrazioni dei Insetti Ditteri del Mozambico. Mem. Accad. Sci. Instit. Bologna, vol. 12. (Separate 1862, Bologna, pp. 1–4, 1–122, pl. 1, fig. 1.) (Bombyliidae p. 15.)

Bethune, C. J. S. (Rev.)
1881. Insects of the northern parts of British America. Canadian Ent., vol. 13, pp. 162–170. (Bombyliidae p. 166.)

Beutenmüller, William
1904. Types of Diptera in the collection of the American Museum of Natural History. Bull. Amer. Mus. Nat. His., vol. 22, art. 6, pp. 87–99. (Bombyliidae pp. 88–89.)

Bezrukov, J. G.
1922. A brief report on the work of the Omsk laboratory of the Siberian Entomological Bureau in 1919–1922. [In Russian.] IZV. Sibirsk. Ent. Byuro, vol. 1, pp. 26–30.

Bezzi, Mario
1892. Contribuzione alla fauna ditterologica della provincia Parvia. Boll. Soc. Ent. Italiana, vol. 24, pp. 97–149, 150–151. (Bombyliidae pp. 101–104, 136–137; 21 spp. listed.)
1894. I. Ditteri del Trentino. Atti Soc. Veneto-Trentino Sci. Nat. Padova, ser. 2, vol. 1, fasc. 2, pp. 209–272. (Bombyliidae pp. 248–250.)
1895–1900. Contribuzioni alla fauna ditterologica Italiana. Boll. Soc. Ent. Italiana.
    1895. Pt. I, Ditteri delle Calabria. Vol. 27, pp. 39–78. (Bombyliidae pp. 48–49.)
    1898. Pt. II, Ditteri delle Marche e degli Abruzzi. Vol. 30, pp. 19–50. (Bombyliidae pp. 35–36. Reprint pp. 19–20; 31 spp. listed.)
1901. Materiali per la conoscenza Fauna eritrea raccolti dal Dott. Paolo Magretti. Ditteri. Bull. Soc. Ent. Italiana, vol. 33, pp. 5–25. (Bombyliidae pp. 12–15.) Reprint pp. 3–23. (Bombyliidae pp. 10–13.)
1902. Neue Namen für einige Dipteren-Gattungen. Zeitschr. syst. Hymen. und Dipt. Teschendorf, vol. 2, pp. 190–192.
1903. *In* Becker Bezzi, Bischof, Kertész: Katalog der paläarktischen Dipteren. Budapest. Vol. 2, pp. 1–396. Orthorrhapha Brachycera by Bezzi. (Bombyliidae pp. 163–206.)
1905. Il genere *Systropus* Wied. nella fauna palearctica. Redia, vol. 2, pt. 2 (1904), pp. 262–279.
1906–1908. Ditteri Eritrei raccolti dal Dott. Andreini e dal Prof. Tellini. Boll. Soc. Ent. Italiana, Firenze.
    1906. Pt. 1, vol. 37 (1905), pp. 195–304. (Bombyliidae pp. 244–263: 381 spp. listed.)
    1908. Pt. 2, vol. 39, pp. 3–199. (Bombyliidae pp. 6–7, 10.)
1907–1908. Nomenklatorisches über Dipteren. Wiener Ent. Zeitung.
    1907. Vol. 26, pp. 51–56. (Bombyliidae p. 55.)

BEZZI, MARIO—Continued

1908a. Diagnoses d'espèces nouvelles de Diptères d'Afrique. Ann. Soc. Ent. Belgique, vol. 52, pp. 374–388. (Bombyliidae p. 375; 1 sp.)

1908b. *In* Schultze: Zool. und anthrop. Ergebnisse e Forschungsreise in Sudafrika. Bezzi: vol. 1, pt. 1: Simulidae, Bombyliidae, Empididae, Syrphidae. . . . Tachinidae, Drosophilidae, Geomyzidae, Agromyzidae, Conopidae. Denkschr. med. Ges., Jena, vol. 13, pp. 179–201. (Bombyliidae p. 180.)

1908c. Eine neue *Aphoebantus*-Art aus dem palaearktischen Faunengebiete (Dipt.). Zeitschr. syst. Hymen. Dipt. Teschendorf, vol. 8, pp. 26–36.

1909a. Diptera syriaca et aegyptia a cl. P. Beraud S. J. collecta. Brotéria, Lisbon, ser. Zool., vol. 8, pp. 37–67, pl. 9. (Bombyliidae pp. 45–51, pl. 9, figs. 1, 9, 26–31, 33–34, 37, 41.)

1909b. Diptères recueillis au Congo au cours du voyage de S. A. R. le Prince Albert de Belgique. Rev. Zool. Africaine, vol. 2, pp. 79–86. (Bombyliidae p. 79.)

1910. Zur synonymie und systematischen Stellung einiger Dipteren. Soc. ent. Stuttgart, vol. 25, pp. 65–72. (Bombyliidae p. 65.)

1912a. Dipteros do Brasil. Sobre tres interessantes Dipteros de S. Paulo. Brotèria, Lisbon, ser. Zool., vol. 10, pp. 76–84, figs. 1–2.)

1912b. Diptera Peninsulae Ibericae. Brotèria, ser. Zool., vol. 10, pp. 114–156, figs. 1–16. Bibliography with 44 entries. (Bombyliidae pp. 115, 116, 118, 119, 121, 122, 124, 126, 130, 132, 133, 137, 138, 139, 142, 143, 144, 147, 148, 149, fig. 9.)

1912c. Report on a collection of Bombyliidae (Diptera) from Central Africa, with description of new species. Trans. Ent. Soc. London (1911), pt. 4, pp. 605–656, pl. 50, 26 figs.

1913–1917. Studies in Philippine Diptera. I-II. Philippine Journ. Sci.
    1913. Pt. I, vol. 8, sect. D, pp. 305–332. (Bombyliidae pp. 309, 313.)
    1917. Pt. II, vol. 12, sect. D, pp. 107–159, 1 pl. (Bombyliidae pp. 126–128). Also published as Publ. 10, Dept. Inter. Bur. Sci., Manila, pp. 1–59, 1916.

1914a. Ditteri raccolti dal Prof. Silvestri durante il suo viaggio in Africa del 1912-13. Boll. Lab. zool. gen. Agric., Portici, vol. 8, pp. 279–308. (Bombyliidae pp. 279–280, 283–285.)

1914b. Ditteri raccolti da S. A. R. la Duchessa d'Aosta nella regione dei grandi laghi dell'Africa equatoriale. Ann. Mus. Zool. Roy. Univ. Napoli, n. ser., vol. 4, pp. 1–7. (Bombyliidae p. 5; 1 sp. listed.)

1914–1915. Contributo allo studio della fauna libica. Ann. Mus. Civ. Storia Nat. Genova.
    1914. Materiali raccolti nelle zone di Misurata e Homs (1912-13) Dal Dott. Alfredo Andreini, Capitano Medico. Ditteri. Ser. 3, vol. 6 (XLVI), pp. 165–181. Bibliography pp. 167–172. (Bombyliidae pp. 166, 176–177; 7 spp. listed.)
    1915. *Heterotropus trotteri*; nouva specie di dittero della Libia. Ser. 3, vol. 7 (XLVII), pp. 17–25.

1915. Ditteri raccolti nella Somalia Italiana meridionale. Missione Scientifica Stefanini-Paoli nella Somali Italiana Meridionale. Redia, Firenze, vol. 10 (1914), pp. 219–233. (Bombyliidae p. 224.)

1918. Studi sulla Ditterofauna nivale delle Alpi italiane. Mem. Soc. Italiana Sci. Nat. Mus. Civ. Storia Nat. Milano, vol. 9, fasc. 1, pp. 1–164, 2 pls. (Bombyliidae p. 50.)

1920. Ditteri raccolti da Leonarda Fea durante il suo viagio nell'Africa occidentale. Parte 2. Bombyliidae. Ann. Mus. Civ. Storia Genova, ser. 3, vol. 9 (XLIX), pp. 98–114.

1921a. On the bombyliid fauna of South Africa (Diptera) as represented in the South African Museum. Ann. South African Mus., vol. 18, pp. 1–180, pls. 1–2, 32 figs.

1921b. Additions to the bombyliid fauna of South Africa as represented in the South African Museum. Ann. South African Mus., vol. 18, pp. 469–478.

1921c. Ditteri di Cirenaica raccolti dal Prof. Alessandro Ghigi . . . Atti della Soc. Italiana Sci. Nat., vol. 60, pp. 432–443. (Bombyliidae pp. 434–435; 9 spp. listed.)

BEZZI, MARIO—Continued
- 1922a. Enumeratio Bombyliidarum (Dipt.) quae ex Africa meridionali Dr. H. Brauns misit. Brotèria Braga, vol. 20, pp. 64–86.
- 1922b. Contributo allo studio della fauna libica. Ditteri di Cirenaica. Mem. Soc. Ent. Italiana, Genoa, vol. 1, pp. 140–157, 1 fig. (Bombyliidae pp. 143–149, fig. 1; 20 spp. listed.)
- 1923a. Diptera, Bombyliidae and Myiodaria (Coenosiinae, Muscinae, Calliphorinae, Sarcophaginae, Dexiinae, Tachininae), from the Seychelles and neighboring islands. Parasitology, Cambridge, vol. 15, pp. 75–102, 9 figs. (Bombyliidae pp. 75–79, fig. 1.)
- 1923b. Voyage de Ch. Alluaud et R. Jeannel en Afrique Orientale (1911–1912). Résultats scientifiques. Insectes Diptères. Part 6. Bombyliidae et Syrphidae. Paris. Pp. 315–351, 1 photo. (Bombyliidae pp. 317–339, 350.)
- 1924. The Bombyliidae of the Ethiopian Region. British Mus. (Nat. Hist.), London, pp. i–viii, 1–390, 46 figs.
- 1925a. Missione del Dr. E. Festa in Cirenaica. XI. Ditteri di Cirenaica. Boll. Mus. Zool. Anat. Comp. Torino, n. ser. 18, vol. 39 (1924), pp. 1–26. (Bombyliidae pp. 4–7.)
- 1925b. Das noch nicht beschreibene Männchen der *Plesiocera flavifrons* Beck. (Dipt.). Konowia, Vienna, vol. 4, pp. 297–300.
- 1925c. Quelques notes sur les Bombyliides (Dipt.) d'Égypte, avec description d'espèces nouvelles. Bull. Soc. Roy. Ent. d'Égypte, vol. 8 (1924), pp. 159–242.
- 1926a. Materiali per lo studio della fauna Tunisina raccolti da G. e L. Doria. Ditteri. Ann. Mus. Civ. Storia Nat. Genoa, n. ser. 3, vol. 10 (L) (1925), pp. 97–139. (Bombyliidae pp. 103–117.)
- 1926b. Materiali per una fauna dell'arcipelago Toscano. XVII. Ditteri del Giglio. Ann. Mus. Civ. Storia Nat. Genoa, n. ser. 3, vol. 10 (L) (1925), pp. 291–354. (Bombyliidae pp. 314–316; 17 spp. listed.)
- 1926c. Notes additionelles sur les Bombyliides (Dipt.) d'Égypte. Bull. Soc. Roy. Ent. d'Égypte, vol. 9 (1925), pp. 244–273.
- 1926d. Nuove species de Ditteri della Cirenaica. Boll. Soc. Ent. Italiana, vol. 58, pp. 81–90. (Bombyliidae pp. 81–86.)

BEZZI, MARIO, and LAMB, C. G.
- 1926. Diptera (excluding Nematocera) from the Island of Rodriguez. Trans. Ent. Soc. London (1925), pp. 537–573. (Bombyliidae pp. 543–544.)

BEZZI, MARIO, and STEFANI-PEREZ, TEODOSIO DE
- 1897. Enumerazione dei Ditteri fino ad ora raccolti in Sicilia. Naturalista Siciliana, Palermo, vol. 2, n. ser. 1–3, pp. 25–72. Reprint pp. 1–48. (Bombyliidae pp. 24–26; 77 spp. listed.)

BIGOT, JACQUES MARIE FRANGILE
- 1852–1859. Essai d'une classification générale et synoptique de l'ordre des insectes Diptères. Ann. Soc. Ent. France.
  - 1858. Ser. 3, vol. 6, pp. 569–595. (Bombyliidae pp. 569–587.)
- 1856. Diptera in Ramon de la Sagra: Historia física, politica y natural de la Isla de Cuba. Secunda Parta: Historia natural. Paris. Vol. 7: Crustaceos, Aragnides e Insectos. Pp. i–xxxii, 1–371, 20 color pls. Diptera, pp. 328–349. (Bombyliidae pp. 332–333, pl. 20, fig. figs. 4, 4a-b.) Also in French, 1857, pp. 1–87, 1–868. Diptera, pp. 783–829, 1 pl. (Bombyliidae pp. 792–795; 1 n. sp; 6 spp. briefly described.)
- 1857a. Diptères nouveaux provenant du Chili. Ann. Soc. Ent. France, ser. 3, vol. 5, pp. 275–308, pl. 6, figs. 1–6, pl. 7, figs. 1–10. (Bombyliidae pp. 292–295, pl. 6, fig. 6, pl. 7, fig. 1, color.)
- 1857b. Note: Bombyliidae. Bull. Soc. Ent. France, ser. 3, vol. 5, p. xc.
- 1859. Diptères de Madagascar. Ann. Soc. Ent. France, ser. 3, vol. 7, pp. 115–135, 415–440, 533–558, 2 pls. (Bombyliidae pp. 429–430.)
- 1860. Diptères de Sicile recueillis par M. Bellier de la Chavignerie et description de onze espèces nouvelles. Ann. Soc. Ent. France, ser. 3, vol. 8, pp. 765–784. (Bombyliidae pp. 766, 771–776.)

BIGOT, JACQUES MARIE FRANGILE—Continued
- 1862a. Diptères nouveaux de la Corse. Ann. Soc. Ent. France, ser. 4, vol. 2, pp. 109–114, 1 pl. (Bombyliidae pp. 111–113.)
- 1862b. Diptères in Maillard: Notes sur l'ile de la Reunion. Ann. Soc. Ent. France, ser. 4, vol. 2, Annexe M, pp. 37–38. (Bombyliidae p. 38.)
- 1867–1892. Diptères nouveaux on peu connus. Ann. Soc. Ent. France.
  - 1880. Quelques Diptères de Perse et du Caucase. Part 20, ser. 5, vol. 10, pp. 139–154. (Bombyliidae pp. 140–141; 8 spp. listed.)
  - 1881. Ibid., part 24, ser. 6, vol. 1, pp. 13–24. (Bombyliidae pp. 22–23.)
  - 1892. Ibid., part 44, Bombylidi. 37 e partie (1), ser. 7, vol. 2, pp. 321–376.
- 1875a. Description d'une nouvelle espèce de Diptère. Bull. Soc. Ent. France, ser. 5, vol. 5, pp. clxxiv–clxxv.
- 1875b. *Thevenemyia* nov. gen. Bull. Soc. Ent. France, ser. 5, vol. 5, pp. clxxv–clxxvi.
- 1876. *Paranthrax* Bigot. Note: A new genus near *Anthrax* and *Argyramoeba*, for *Anthrax rufiventris* Blanchard. Bull. Soc. Ent. France, ser. 5, vol. 6, p. lxvi.
- 1886a. Genre *Rhabdopselaphus mus*, n. sp. Calif. Bull. Soc. Ent. France, ser. 6, vol. 6, pp. ciii–civ.
- 1886b. Genus *Peringueyimyia* (nov.). Bull. Soc. Ent. France, ser. 6, pp. cx–cxi.
- 1887a. *Stomylomyia*. Bull. Soc. Ent. France, ser. 6, vol. 7, p. xxxi.
- 1887b. Note: On *Eclimus, Epibates, Thevenemyia*. Bull. Soc. Ent. France, ser. 6, vol. 7, p. lx.
- 1888a. Enumération des Diptères recueillis en Tunisie dans la Mission de 1884 par M. Valéry Mayet. Exploration scientifique de la Tunisie. Zoologie. Diptères, pp. 1–11. Paris. (Bombyliidae pp. 5–6, 9.)
- 1888b. Description d'un nouveau genre de Diptère. Bull. Soc. Ent. France, ser. 6, vol. 8, p. cxl.
- 1892a. *Bombylius major* L. provenant du Japon. Bull. Soc. Ent. France, vol. 61, p. clxiii.
- 1892b. Catalogue of the Diptera of the Oriental Region by Mons. J. M. F. Bigot. Journ. Asiatic Soc. Bengal, vol. 61, pt. 1, pp. 250–282; pt. 2, pp. 133–177. (Bombyliidae pp. 159–163.)

BILIOTTI, E.
- 1958. Les parasites et predateurs de *Thaumetapoea pityocampa* Schiff (Lepidoptera). Entomophaga, vol. 3, pp. 23–24.

BILIOTTI, E., DEMOLIN, G., and MERLE, PAUL DU
- 1965. Parasitisme de la Processionnaire du Pin par *Villa quinquefasciata* Wied. Apud Meig. (Dipt., Bombyliidae). Ann. Epiphyties, 16, pp. 279–288, figs. 1–5.

BISCHOF, J.
- 1903. Neue Dipteren aus Afrika. Wiener Ent. Zeitung, vol. 22, pp. 41–42. (Bombyliidae p. 42.)

BLACKWELDER, R. E.
- 1947. The dates and editions of Curtis' British entomology. Smithsonian Misc. Coll., vol. 107, no. 5, 27 pp., 4 pls.

BLANCHARD, CHARLES EMILE
- 1840. Histoire naturelle des Insectes. *In* Castelnau: Histoire naturelle des Animaux articulés. III. Orthoptères, Néuroptères, Hémiptères, Hyménoptères, Lépidoptères, Diptères. Paris. Vol. 3, pp. 1–672, 72 pls. Diptera pp. 563–632, 5 pls. by Blanchard. (Anthracides.) (Bombyliidae pp. 584–587, pl. 3, figs. 6–7.)
- 1845. Histoire des Insectes traitant de leurs moeurs et de leurs métamorphoses en général. Paris. Pp. 1–524, pls. 11–20. (Bombyliidae pp. 466–470.)
- 1854. *In* Gay, Claudio: Histoire física y politica de Chile. Zoologia, vol. 7 (1852), pp. 1–471. Diptera pp. 327–468. (Bombyliidae [Antrosidas] pp. 377–385, pl. 3, figs. 9–10.)

BOEHM, RUDOLF
- 1913. Note complémentaire sur *Cleonus* Saint-Pierrei (Coléopt. curculion). Bull. Soc. Roy. d'Egypte (1912–1913), vol. 3, pp. 25–26. (Bombyliidae p. 26.)

BOHART, GEORGE E.
- 1952. Pollination by native insects. Yearbook, U.S. Dept. Agric. (1951), pp. 107–121, 10 text-figs. (Bombyliidae p. 118.)
- 1958. Helping helpful alkali bees. Agric. Research, vol. 7 (2), pp. 8–9.

BOHART, GEORGE E., and MACSWAIN, JOHN W.
- 1939. The life history of the sand wasp, *Bembix occidentalis beutenmuelleri* Fox and its parasites. Bull. Southern California Acad. Sci., vol. 38, pt. 2, pp. 84–95, pl. 16 with 18 figs. (Bombyliidae pp. 90–93.)

BOHART, GEORGE E., STEPHEN, W. P., and EPPLEY, R. K.
- 1960. The biology of *Heterostylum robustum* (Diptera: Bombyliidae), a parasite of the alkali bee. Ann. Ent. Soc. America, vol. 53, no. 3, pp. 425–435, figs. 1–16, 1 chart, 15 photographs.

BOHART, GEORGE E. [see Nye and Bohart]

BOISDUVAL, JEAN BAPTISTÉ ALPHONSE DECHAUFFOUR DE
- 1832–1835. Voyage de l'*Astrolabe*. Faune entomologique de l'Océan Pacifique. In 2 volumes and atlas with 12 color pls.
    - 1835. Vol. 2, pp. 1–7, 1–716, pls. 6–12. Coléoptères et outres ordres. Diptères pp. 660–669. (Bombyliidae pp. 664–665, pl. 12, figs. 12, 15.)

BONSDORFF, EVERHARD J.
- 1861. Finlands tvåvingade Insekter (Diptera). . . . I. Bidrag Finlands Naturk., Ethnogr. Stat., vol. 6, pp. 1–12, 37–301. (Bombyliidae pp. 129–134.)

BORKHAUSEN, MORITZ BALTHASAR
- 1797. Epitome Entomologiae Fabricianae sive Nomenclator entomologicus emendatus sistens Fabriciani systematis cum Linneano comparationem. Lipsiae. 240 pp.

BOROWSKI, GEORGE H. [see Herbst in Borowski]

BORROR, D. J., and DELONG, D.W.
- 1954. An introduction to the study of insects. 614 pp. New York: Rinehart and Co. (Bombyliidae pp. 614–615, figs. 27–57 a-b.) Edition 2, 1963, 819 pp. (Bombyliidae p. 486, figs. 27–58 a-b.)

BOWDEN, JOHN
- 1959–1962. Studies in African Bombyliidae, I-V. Journ. Ent. Soc. Southern Africa, Pretoria.
    - 1959a. I. Two new species of *Antonia* Loew. Vol. 22, pp. 13–17, 1 fig.
    - 1959b. II. New East African species of *Systoechus* Loew. Vol. 22, pp. 298–307, 7 figs.
    - 1960a. III. On *Gnumyia fuscipennis* Hesse. Vol. 23, pp. 209–211, 2 figs.
    - 1960b. IV. A new genus of the *Corsomyza-group* from East Africa. Vol. 23, pp. 212–217, 6 figs.
    - 1962. V. A new genus of Exoprosopinae from East Africa. Vol. 25, no. 1, pp. 116–120, figs. 1–6 on p. 119.
- 1962. Bombyliidae (Diptera Brachycera). Parc National de la Garamba, Mission H. de Saeger. Fasc. 32, no. 3, pp. 47–60, 6 text-figs.
- 1964. The Bombyliidae of Ghana. Mem. Ent. Soc. Southern Africa, Pretoria, no. 8, pp. 1–159, text-figs. 1–145.
- 1965. Diptera from Nepal (Bombyliidae). Bull. Brit. Mus. (Nat. Hist.), no. 17, pp. 203–208, 1 fig.
- 1967. Two new species of *Heterotropus* Loew (Diptera: Bombyliidae) from Turkey. Ent. Mon. Mag., vol. 103, pp. 36–39, 8 figs.
- 1968. Studies in African Bombyliidae: 6. A provisional classification of the Ethiopian Systropinae with descriptions of new and little known species. Journ. Ent. Soc. South Africa, vol. 30 (1967), pp. 126–173, 4 figs.
- 1971. Notes on the genus *Ligyra* Newman (Diptera: Bombyliidae) with descriptions of three new species from the New Guinea subregion. Journ. Australian Ent. Soc., vol 10, no. 1, pp. 5–12, illus.

Box, H. E.
- 1925. Porto-Rican cane-grubs and their natural enemies, with suggestions for the control of Lamellicorn larvae by means of wasp-parasites (Scoliidae). Journ. Porto Rico Dept. Agric., vol. 9, pp. 291–356. (Bombyliidae pp. 319–320.)

Brauer, Friedrich Moritz
- 1880–1883. Die Zwieflügler des Kaiserlichen Museums zu Wien. I-III. Denkschr. Akad. Wiss. Wien.
  - 1882. Vol. 46, pp. 59–108. Part II, in 3 sections (pp. 1–54.)
    - Section 2. Vergleichende Untersuchungen des Flugelgeäders der Dipteren nach Adolph's Theorie. Pp. 90–97 (pp. 34–41), pls. 1–2. (Bombyliidae pp. 93–94, 96, pl. 2, figs., wing of *Chalcochiton* and *Argyramoeba*.)
    - Section 3. Characteristik und mit *Scenopinus* verwandten Dipteren Familien und Gattungen. Pp. 98–108 (pp. 42–54). (Bombyliidae pp. 99–103, 107.)
  - 1883. Vol. 47, abt. 1, pp. 1–100, plates 1–5 with 110 figs. Part III. Systematische Studien auf Grundlage der Dipteren-Larven nebst einer Zusammenstellung von Beispielen aus der Literatur über dieselben und Beschreibung neuer Formen. (Bombyliidae pp. 5, 10–11, 15, 17–18, 27–28, 61–63, 94–96, 98–99, 100, pl. 4, figs. 64–71; pl. 5, figs. 104–105.)

Brauns, Adolph
- 1954a. Terricole Dipterenlarven. Untersuchunden zur Angewandten Bodenbiologie. Band I. Wissenschaftlichen Verlag. Gottingen. Pp. 1–179, 60 pls. with 96 figs., (3 in color), 3 tables. (Bombyliidae pp. 26, 86–88, fig. 49.)
- 1954b. Puppen terricoler Dipterenlarven. Band II. Gottingen. Pp. 1–156, figs. 1–75. (Bombyliidae pp. 35, 92–94, figs. 39 A-B, 40 A-E, 41 A-D, 42 A-C.)

Bree, William T.
- 1833. Some of the habits of *Bombylius major* L. Mag. Nat. Hist., vol. 6, pp. 73–74.

Bremi-Wolf, Johann Jacob
- 1846. Beitrag zur Kenntniss der Dipteren, insbesondere über das Vorkommen mehrerer Gattungen nach besonderen Localitaten und den Fang derselben, so wie auch über die Lebensweise einiger Larven. Isis, vol. 3, pp. 164–175.

Brèthes, Juan
- 1904. Insectos de Tucuman. Ann. Mus. Nac. Buenos Aires, ser. 3, vol. 4, pp. 17–24, 329–347. (Bombyliidae pp. 339–340.)
- 1908. Catálogo de los Dípteros de las Repúblicas del Plata. Ann. Mus. Nac. Buenos Aires, vol. 16 (1907), pp. 277–305. (Bombyliidae pp. 288–289.)
- 1909. Dípteros e Himenopteros de Mendoza. Ann. Mus. Nac. Buenos Aires, vol. 19, pp. 85–105. (Bombyliidae p. 89.)
- 1919–1920. Cueillette d'Insectes du Rio Blanco. Rev. Chileña Hist. Nat., Santiago.
  - 1919. Vol. 22, pp. 161–161.
  - 1920. Vol. 23, pp. 40–44. (Bombyliidae p. 40; *Mythicomyia hyalinipennis*, n. sp., as *Heterobybos*, n. g., ascribed to the Empididae by Melander, 1960.)
- 1925. Sur quelque Diptères chiliens. Rev. Chileña Hist. Nat., Santiago, vol. 28 (1924), pp. 104–111. (Bombyliidae p. 105.)

Brimley, Clement Samuel
- 1921. The Bee Flies of North Carolina (Bombyliidae, Dip.). Ent. News, vol. 32, pp. 170–172; 39 spp. listed.
- 1922. Additional data on North Carolina Tabanidae, Bombyliidae and Tachinidae (Diptera). Ent. News, vol. 33, pp. 230–232. (Bombyliidae pp. 230–231; 8 additional spp. listed.)
- 1938. The insects of North Carolina. North Carolina Dept. Agric., Div. Ent., pp. 1–560. Diptera pp. 313–390. (Bombyliidae pp. 340–343; 15 genera, 57 spp. listed.) See Wray, 1950, for second supplement.

Britton, W. E.
- 1920. Check list of the insects of Connecticut. Connecticut Geol. Nat. Hist. Survey Bull. 31, pp. 1–397. (Bombyliidae p. 176; 21 spp. listed.)

BRODIE, W., and WHITE, J. E.
    1883. Check list of insects of the Dominion of Canada. 67 pp. Toronto.

BROMLEY, STANLEY WILLARD
    1934. The robber flies of Texas (Diptera, Asilidae). Ann. Ent. Soc. America, vol. 27, pp. 74–113, 2 pls. (Bombyliidae as prey of Asilidae pp. 99, 101, 107, 108.)

BROOK, A. R.
    1952. Identification of bombyliid parasites and hyperparasites of Phalaenidae of the prairie provinces of Canada, with descriptions of six other bombyliid pupae (Diptera). Canadian Ent., vol. 84, no. 12, pp. 357–373, 46 text-figs. (Reviewed in Rev. Applied Ent., 1954, vol. 42, p. 137.)

BRUES, CHARLES THOMAS
    1939. The mimetic resemblance of flies of the genus *Systropus* to wasps. Psyche, vol. 46, no. 1, pp. 20–22.
    1946. Insect dietary. 476 pp. Cambridge, Massachusetts. (Bombyliidae p. 338; brief comment.)

BRUES, CHARLES THOMAS, and MELANDER, AXEL LEONARD
    1915. Key to the families of North American insects. 140 pp. Boston.
    1932. Classification of insects. Bull. Mus. Comp. Zool., vol. 73, pp. 1–672, figs. 1–1121. Third printing, 1945. (Bombyliidae pp. 294–297, 363, 367–377, figs. 544–548, 714, 758.)

BRUES, CHARLES THOMAS, MELANDER AXEL LEONARD, and CARPENTER, FRANK MORTON
    1954. Classification of insects. Revised. Bull. Mus. Comp. Zool., vol. 108, pp. 1–5, 1–917, figs. 1–1219. (Bombyliidae pp. 335–338, 400–401, 413, figs. 544–548, 714, 758.)

BRULLÉ, GASPARD AUGUSTE
    1832. Expédition scientifique de Morée. Paris. Insectes, vol. 3, pt. 2, pp. 1–29, 64–395, 22 pls. Diptera, pp. 289–325. (Bombyliidae pp. 301–302.)

BRUNETTI, ENRICO ADELELMO
    1909a. Revised and annotated catalogue of oriental Bombylidae with descriptions of new species. Rec. Indian Mus., Calcutta, vol. 2, pt. 5, pp. 437–492, pl. 12, figs. 1–28.
    1909b. New Indian Leptidae and Bombyliidae with a note on *Comastes* Osten Sacken, v. *Heterostylum* Macquart. Rec. Indian Mus., Calcutta, vol. 3, pt. 3, pp. 211–230, 2 figs. (Bombyliidae pp. 214–230, figs. 1–2.)
    1912. New oriental Diptera. I. Rec. Indian Mus., Calcutta, vol. 7, pp. 445–513, pl. 37, figs. 1–23 (Bombyliidae pp. 466–472, pl. 37, figs. 4–7).
    1917. Diptera of the Simla District. Rec. Indian Mus., Calcutta, vol. 13, pp. 59–101. (Bombyliidae pp. 75–78, fig. 1.)
    1920. Diptera brachycera, Fauna of British India . . . London. Vol. I, pp. i–ix, 1–401, 4 pls. (Bombyliidae pp. 173–295, text-figs. 14–21, pl. 3, figs. 1–16.)

BRYSON, H. R. [see Smith, Kelly, Dean, Bryson, Parker]

BUCKELL, E. R. [see Treherne and Buckell]

BUGNION, EDOUARD
    1886. Note: *Anthrax* (*Argyramoeba*) *sinuata* Fallen, a parasite on the nest of *Eumenes unguiculus* Villeneuve. Mittheil. Schweiz. Ent. Ges., vol. 7, p. 208.

BURGESS, EDWARD
    1878. Two interesting American Diptera: *Glutops singularis* and *Epibates osten-sackenii*. Proc. Boston Soc. Nat. Hist., vol. 19, pp. 320–324, 1 pl. (Bombyliidae pp. 323–324, pl. 9, figs. 1, 1a.)

BURKS, B. D.
    1951. *In* Muesebeck, Krombein, and Townes: Hymenoptera of America north of Mexico: Synoptic catalog. U.S. Dept. Agric. Monogr. 2, 1420 pp. (Tribe Bembecini, pp. 995–1000, refers to Bombyliidae.)

BURMEISTER, HERMANN CARL CONRAD
    1832–1855. Hanbuch der Entomologie. In 5 volumes. Berlin.
        1832. Vol. 1, 712 pp. (Fossil Bombyliidae p. 636.)
    1861. Reise durch die La Plata Staaten. In 2 volumes. Halle. Vol. 1, 510 pp., 1 pl., 1 map. (Bombyliidae p. 317.)

BUTLER, G. D., and RITCHIE, P. L., JR.
    1965. Additional biological notes on *Megachile concinna* Smith, Pan Pacific Entomologist, vol. 41, pp. 153–157.

BUYSSON, ROBERT DU
    1888. Note: On *Argyramoeba sinuata* Fall., a parasite of *Osmia*. Rev. Sci. Bourbonnais, vol. 1, p. 46.

CALVERT, FRANK
    1881. Letter quoted *in* Saunders, Waterhouse, and Fitch: Report to the Council of the Entomological Society of London about Bombyliidae destructive to locust eggs. Trans. Ent. Soc. London, July 6, vol. 29, pp. xiv–xix, pl. 14. Austen, 1937, pp. 52–54, again quotes Calvert's letter.

CAÑIZO, JOSÉ DEL
    1935. La plagas da langosta en España. Proc. Sixth Intern. Congress Ent., Madrid, pt. 2, pp. 845–865. (Bombyliidae pp. 859–860, figs. 11–12.)
    1944. Parásitos de la langosta en España. I. Dipteros Bombílidos. Bol. Pat. veg. Ent. Agric., Madrid, vol. 12 (1943), pp. 77–99, 3 pls., 12 figs.

CARILLO S, JOSE LUIS [see Gibson and Carillo S]

CARPENTER, FRANK MORTON [see Brues, Melander, and Carpenter]

CARRERA, MESSIAS, and D'ANDRETTA, MARIA A. V.
    1950. Sobre as espécies brasileiras de *Systropus* Wiedemann, 1830 (Diptera, Bombyliidae). Papéis Avulsos Dept. Zool. São Paulo, vol. 9, no. 20, pp. 295–319, 48 figs.

CARRERA, MESSIAS [see d'Andretta and Carerra]

CHAGNON, GUSTAVE
    1901. Preliminary list, no. 1, of Canadian Diptera. Ent. Student, vol. 2, pp. 5–16. (Bombyliidae p. 7; 12 spp. listed.)

CHAPMAN, JOHN A.
    1954. Studies on summit-frequenting insects in western Montana. Ecology, vol. 35, pp. 41–49, figs. 1–3. (Bombyliidae pp. 44, 48.)

CHAPMAN, T. ALGERNON
    1878. On the economy, etc. of *Bombylius*. Ent. Monthly Mag., vol. 14, pp. 196–200.
    1902. Exhibition cocoons of a limacodid moth from La Plata with empty pupa-cases of Dipterous parasite of the genus *Systropus* obtained from Herr Heyne. Proc. Ent. Soc. London, 1902, pp. viii–x.

CHENG, SHAN LIN [see Gillaspy, Evans and Cheng]

CHINKEVITSCH, V. (also written as Shimkevitch, or Schimkevic, or Schimitschek)
    1884. The problem of transformation of Dipterous parasites in the cocoon of locusts (title from Uvarov), or: Contributo alla questione dei Ditteri parassiti delle ooteche delle Cavalette (title from Paoli). Horae Ent. Rossia, vol. 18, pp. 11–16. (Russian title and Text.) [This paper concerns emergence of *Systoechus* (=*Anastoechus*) *nitidulus* Fabr., from egg pods of *Stauronotus* spp.]

CLAUSEN, CURTIS P.
    1928. *Hyperalonia oenomaus* Rond., a parasite of *Tiphia* larvae. (Dip., Bombyliidae). Ann. Ent. Soc. America, vol. 21, pp. 642–659, pl. 33 on p. 659, with figs. 1–6.
    1940. Entomophagous insects. New York, London: McGraw-Hill. 688 pp. 257 figs., 1064 references listed, all cited in text. (Bombyliidae pp. 19, 343–344, 346–350, 374–382, figs. 169–171.)

CLAUSEN, C. P., JAYNES, H. A., GARDNER, T. R.
    1933. Further investigations of the parasites of *Popillia japonica* in the Far East. U.S. Dept. Agric. Techn. Bull. no. 366, pp. 1–51. (Bombyliidae pp. 50–51.)

CLEMENTS, F. E., and LONG, F. L.
    1923. Experimental pollination an outline of the ecology of flowers and insects. Carnegie Inst. Washington, Publ. no. 336, pp. 1–274. (Bombyliidae p. 249; 4 genera, 13 species listed.)

COCKERELL, THEODORE DRU ALISON
    1895. *Phthiria sulphurea* Loew. Psyche, vol. 7, pp. 188–189. (A redescription.)

COCKERELL, THEODORE DRU ALISON—Continued

1896. New species of insects taken on a trip from Mesilla Valley to Sacramento Mountains, New Mexico. Journ. New York Ent. Soc., vol. 4, p. 206. (*Lordotus diversus* Coquillett, footnote.)

1902. Flowers and insects of New Mexico. Amer. Natur., vol. 36, pp. 809–817. (Bombyliidae p. 815, *Bombylius major* visiting *Iris missouriensis*.)

1905. The Diptera of Kansas and New Mexico. Trans. Kansas Acad. Sci., vol. 19, pp. 250–251. (Bombyliidae p. 250.)

1909a. Fossil Diptera from Florissant, Colorado. Bull. Amer. Mus. Nat. Hist., vol. 26, pp. 9–12, pl. 1, figs. 1–3. (Bombyliidae pp. 10–11, pl. 1, fig. 1.)

1909b. Fossil insects from Florissant, Colorado. Bull. Amer. Mus. Nat. Hist., vol. 26, pp. 67–76, pl. 16, figs. 1–5, 5a. (Bombyliidae pp. 70–72, pl. 16, figs. 4, 4a.)

1909c. Descriptions of Tertiary insects. Part 5. Some new Diptera. Amer. Journ. Sci., ser. 4, vol. 27, pp. 53–58, 1 text-fig. (Bombyliidae pp. 54–57, 1 fig.)

1910. Fossil insects and a crustacean from Florissant, Colorado. Bull. Amer. Mus., vol. 28, art. 25, pp. 275–288. (Bombyliidae pp. 284, 287–288.)

1911. Fossil insects from Florissant, Colorado. Bull. Amer. Mus. Nat. Hist., vol. 30, pp. 71–82, 3 text-figs., pl. 3, figs., 1–5. (Bombyliidae pp. 80–82, pl. 3, fig. 3.)

1913. Some fossil insects from Florissant, Colorado. Canadian Ent., vol. 45, pp. 229–233, fig. 9. Diptera pp. 230–231. (Bombyliidae pp. 230–231.)

1914a. Miocene fossil insects. Proc. Acad. Nat. Sci., Philadelphia, vol. 66, pp. 634–648. (Bombyliidae pp. 643–644.)

1914b. The fossil and recent Bombyliidae compared. Bull. Amer. Mus. Nat. Hist., vol. 33, pp. 229–236, figs. 1–20.

1914c. New and little-known insects from the Miocene of Florissant, Colorado. Journ. Geol., Univ. Chicago, vol. 22, pp. 714–724, 11 figs. (Bombyliidae pp. 720–724, figs. 5–11.)

1916. Some American fossil insects. Proc. U.S. Nat. Mus., vol. 51, no. 2146, pp. 89–106, 9 figs., pl. 2, figs. 1–7. (Bombyliidae pp. 93–94.)

1917a. New Tertiary insects. Proc. U.S. Nat. Mus., vol. 52, pp. 373–384, pl. 31, figs. 1–14. (Bombyliidae pp. 376–377.)

1917b. The fauna of Boulder County Colorado, III. Class Insecta, Order Diptera. Univ. Colorado Studies, vol. 12, no. 1, pp. 5–20. (Bombyliidae pp. 13–14; 21 spp. listed.)

1917c. A fossil tsetse fly and other Diptera from Florissant, Colorado. Proc. Biol. Soc. Washington, vol. 30, pp. 19–21. (Bombyliidae p. 20.)

1921. Fossil Arthropods in the British Museum. VI. Oligocene insects from Gurnet Bay, Isle of Wight. Ann. Mag. Nat. Hist., ser. 9, vol. 7, pp. 453–480, 50 figs. (Bombyliidae p. 469, fig. 32.)

COE, R. L.

1945. A list of British Diptera: Orthorrhapha. Reprinted from Kloet and Hincks, 1945: A check list of British Insects, pp. xxxiii, xlvii–li, 327–377. (Bombyliidae p. 363.)

COLE, FRANK RAYMOND

1912. Some Diptera of Laguna Beach. First Ann. Rept. Laguna Marine Lab., pp. 152–162. (Bombyliidae p. 152; 1 sp. listed.)

1916. A new species of *Exoprosopa* (Dip.). Ent. News, vol. 27, p. 463, 4 figs.

1917. Notes on Osten Sacken's group "*Poecilanthrax*," with descriptions of new species. Journ. New York Ent. Soc., vol. 25, pp. 67–80, pls. 3–7, 22 figs.

1923a. Expedition of the California Academy of Sciences to the Gulf of California in 1921. The Bombyliidae (Bee Flies). Proc. California Acad. Sci., ser. 4, vol. 12, no. 13, pp. 289–314, 38 figs.

1923b. Notes on California Bombyliidae with descriptions of new species. Journ. Ent. Zool., Pomona College, vol. 15, no. 2, pp. 21–26, 11 figs.

1923c. A revision of the North American two-winged flies of the family Therevidae. Proc. U.S. Nat. Mus., vol. 62, no. 2450, 138 pp., 13 pls. (Bombyliidae pp. 14–17, plate 1, figs. 1–4, *Caenotus*.)

Cole, Frank Raymond—Continued
- 1927. A study of the terminal abdominal structures of male Diptera (two-winged flies). California Acad. Sci., ser. 4, vol. 16, no. 14, pp. 397–499, figs. 1–287. (Bombyliidae p. 423, figs. 85, 88–89, 91–92, 94–95. Bibliography pp. 455–456, with 31 entries.)
- 1952. New bombyliid flies reared from anthophorid bees (Diptera: Brachycera). Pan-Pacific Ent., vol. 28, pp. 126–130, 2 figs.
- 1957. New bombyliid flies from Chiapas, Mexico. (Diptera). Pan-Pacific Ent., vol. 33, pp. 200–202, 1 fig.
- 1960. New names in Therevidae and Bombyliidae. Pan-Pacific Ent., vol. 36, p. 118.

Cole, Frank Raymond, and Lovett, Arthur Lester
- 1919. New Oregon Diptera. VII. Proc. California Acad. Sci., vol. 9, pp. 221–255, pls. 14–17. (Bombyliidae pp. 224–228, pl. 14, figs. 5, 5a–7; pl. 15, figs. 8–9, 9a.)
- 1921. An annotated list of the Diptera (flies) of Oregon. Proc. California Acad. Sci., ser. 4, vol. 11, pp. 197–344, 54 figs. (Bombyliidae pp. 243–252, fig. 21.)

Cole, Frank Raymond, Malloch, J. R., and McAtee, Waldo Lee
- 1924. District of Columbia Diptera: Tromoptera (Cyrtidae, Bombyliidae, Therevidae, Scenopinidae). Proc. Ent. Soc. Washington, vol. 26, no. 7, pp. 181–195. (Bombyliidae pp. 183–191.)

Cole, Frank R., and Schlinger, Evert
- 1967. The flies of the western United States of North America. University of California Press, 693 pages, frontispiece (color), 360 figs.

Collart, A.
- 1939. Résults scientifiques des Croisières du Navire-ecole Belge "Mercator." Vol. 2, pt. 16. Bombyliidae (Diptera) by A. Collart, p. 171. Mém. Mus. Hist. Nat. Belgique, ser. 2, fasc. 15, p. 171.

Colyer, Charles N., and Hammond, Cyril O.
- 1851. Flies of the British Isles. London. 383 pp, 103 pls., 17 figs. (Bombyliidae pp. 108–111, pl. 23 in color.)

Comstock, Anna Botsford [see Comstock and Comstock]

Comstock, John Henry
- 1918. The wings of insects. Ithaca, New York. Pp. i–xviii, 1–430, 10 pls. (Bombyliidae p. 355, fig. 368.)
- 1920. An introduction to Entomology. Ithaca, New York. Later editions: 1924, 1933, 1936, 1940, 1948. Revised by Herrick, 1940 edition. Ninth edition, 1948, pp. i–xix, 1–1064, figs. 1–1228. Revised. (Bombyliidae p. 838, figs. 1080–1081.)

Comstock, John Henry, and Comstock, Ann Botsford
- 1895. A manual for the study of insects. Ithaca, New York. 701 pp. Twelfth edition, 1914, pp. 1–17, 1–701. (Bombyliidae pp. 426, 463–464, figs. 563–564.)

Copello, Andres
- 1933. Biologia de *Hyperalonia morio* (Dip. Bomb.). Soc. Ent. Argentina, Buenos Aires, vol. 5, pp. 117–120, 2 text-figs.

Coquebert de Montbret, C. Antoine Jean
- 1798–1804. Illustratio inconographica insectorum quae in musaeis parisinis observavit et in lucem edidit John. Christ. Fabricius, praemissis ejusdem descriptionibus. Paris. I-III, pp. 1–142, 30 pls. (21–30 in color).
  - 1802. Pt. 2 (1801), pp. 47–90, 10 color pls. (Bombyliidae pp. 84–88, pl. 20, figs. 1, 4, 7–8.)
  - 1804. Pt. 3, pp. 91–142, 10 pls. (Bombyliidae p. 100, pl. 23, fig. 2 (*Anthrax*), pl. 26, figs. 1, 12.)

Coquillett, Daniel William
- 1886a. Monograph of the Lomatina of North America. Canadian Ent., vol. 18, pp. 81–87. (Reprint pp. 1–7.)
- 1886b. The North American genera of Anthracina. Canadian Ent., vol. 18, pp. 157–159.
- 1886c. The North American species of *Toxophora*. Ent. Americana, vol. 1, pp. 221–222.
- 1887a. Notes on the genus *Exoprosopa*. Canadian Ent., vol. 19, pp. 12–14.

Coquillett, Daniel William—Continued
- 1887b. Monograph of the species belonging to the genus *Anthrax* from America north of Mexico. Trans. American Ent. Soc., vol. 14, pp. 159–182.
- 1887c. Synopsis of the North American species of *Lordotus*. Ent. Americana, vol. 3, pp. 115–116.
- 1891a. New Bombyliidae from California. West American Scientist, vol. 7, pp. 197–200.
- 1891b. New Bombyliidae of the group *Paracosmus*. West American Scientist, vol. 7, pp. 219–222.
- 1891c. Revision of the bombylid genus *Aphoebantus*. West American Scientist, vol. 7, pp. 254–264.
- 1892a. Revision of the bombylid genus *Epacmus* (*Leptochilus*). Canadian Ent., vol. 24, pp. 9–11.
- 1892b. Notes and descriptions of Bombylidae. Canadian Ent., vol. 24, pp. 123–126.
- 1892c. Revision of the species of *Anthrax* from America north of Mexico. Trans. American Ent. Soc., Philadelphia, vol. 19, pp. 168–187.
- 1893. An anomalous empid. (*Mythicomyia rileyi*.) Ent. News, vol. 4, no. 6, pp. 208–210, 1 fig. wing.
- 1894a. Notes and descriptions of North American Bombylidae. Trans. American Ent. Soc., vol. 21, pp. 89–112.
- 1894b. A new *Anthrax* from California. Journ. New York Ent. Soc., vol. 2, pp. 101–102.
- 1895a. The bombylid genus *Acreotrichus* in America. Psyche, vol. 7, p. 273.
- 1895b. Revision of the North American Empidae—A family of two-winged insects. Proc. U.S. Nat. Mus., vol. 18, no. 1073, pp. 387–440. (Bombyliidae, as Empidae, p. 409, 2 spp. of *Mythicomyia*.)
- 1898. Report on a collection of Japanese Diptera, presented to the U.S. National Museum by the Imperial University of Tokyo. Proc. U.S. Nat. Mus., vol. 21, pp. 301–340. (Bombyliidae pp. 317–319.)
- 1900. Report on a collection of dipterous insects from Puerto Rico. Proc. U.S. Nat. Mus., vol. 22, pp. 249–270. (Bombyliidae p. 251; 2 spp.)
- 1901. Papers from the Hopkins-Stanford Galápagos expedition. Entomological results (2): Diptera. Proc. Washington Acad. Sci., vol. 3, pp. 371–379. (Bombyliidae pp. 373–374; 2 spp. listed.)
- 1902a. New Orthorrhaphous Diptera from Mexico and Texas. Journ. New York Ent. Soc., vol 10, pp. 136–141. (Bombyliidae pp. 138–139.)
- 1902b. New Diptera from North America. Proc. U.S. Nat. Mus., vol. 25, no. 1280, pp. 83–126. (Bombyliidae pp. 100–103.)
- 1903. *In* Baker: Reports on Californian and Nevadan Diptera. I. Invertebrata Pacifica, Diptera, vol. 1, pp. 17–39. (Bombyliidae p. 23.)
- 1904a. Diptera from southern Texas with descriptions of new species. Journ. New York Ent. Soc., vol. 12, pp. 31–35. (Bombyliidae pp. 32–33.)
- 1904b. New North American Diptera. Proc. Ent. Soc. Washington, vol. 6, pp. 166–192. (Bombyliidae pp. 172–177; 12 spp. *Phthiria*.)
- 1910a. The type-species of the North American genera of Diptera. Proc. U.S. Nat. Mus., vol. 37, no. 1719, pp. 499–647.
- 1910b. New species of North American Diptera. Canadian Ent., vol. 42, pp. 41–47. (Bombyliidae p. 41.)
- 1910c. Corrections to my paper on the type-species of the North American genera of Diptera. Canadian Ent., vol. 42, no. 11, pp. 375–378. (Bombyliidae pp. 376, 378.)

Corti, Emilio
- 1895a. Esplorazione del Guiba. VIII. Ditteri. Ann. Mus. Civ. Storia Nat. Genova, ser. 2, vol. 15 (35), pp. 129–148.
- 1895b. Aggiunte alla fauna ditterologica della Provincia di Pavia. Bull. Soc. Ent. Italiana, vol. 26 (1894), pp. 389–395. (Bombyliidae pp. 390–391; 10 spp.)

Costa, Achille
- 1857. Contribuzione alla fauna ditterologica italiana. Il Giambattista Vico, Napoli, vol. 2, pt. 3, pp. 438–460.

Costa, Achille—Continued
- 1863. Nuovi studie sulla entomologica della Calabria ulteriore. Atti Roy. Accad. Sci. Napoli, vol. 1, no. 2, pp. 1–80, 4 color pls. (Bombyliidae pp. 51–52, 77, pl. 4, fig. 13.)
- 1865. Descrizione di taluni insetti stranieri all'Europa. Annuar. Mus. Zool. Napoli, vol. 2, pp. 139–153. (Bombyliidae pp. 151–153.)
- 1882–1888. Notizie ed observazioni sulla Geo-Fauna Sarda. Atti Roy. Accad. Sci. Napoli.
  - 1882. Memoria prima, vol. 9, no. 11, pp. 1–41. (Bombyliidae p. 27.)
  - 1883. Memoria seconda, ser 2, vol. 1, no. 2, pp. 1–109. (Bombyliidae p. 103.)
  - 1884. Memoria terza, ser. 2, vol. 1, no. 9, pp. 1–64. (Bombyliidae p. 61.)
  - 1886. Memoria quinta, ser. 2, no. 7, pp. 1–24. (Bombyliidae p. 23.) (Also, Rend. Roy. Accad. Sci. Napoli, vol. 24, pp. 322–324.)
  - 1886. Memoria sesta, no. 8, pp. 1–40. (Bombyliidae p. 38.)
- 1884–1885. Diagnosi di nuovi Artropodi trovati in Sardegna. Ditteri. Boll. Soc. Ent. Italiana, vol. 15 (1883), pp. 332–341. (Bombyliidae p. 339.) Also, Mem. seconda, Atti Roy. Accad. Sci, Napoli, ser. 2, vol. 1, 1883.
- 1885. Diagnosi di nuovi Artropodi della Sardegna (1). Bull. Soc. Ent. Italiana, vol. 17, pp. 240–255. Ditteri, pp. 252–254. (Bombyliidae pp. 252–253.)
- 1893. Miscellanea entomologica. Mem. quarta. Rend. Roy. Accad. Sci. Napoli, ser. 2, vol. 7, pp. 99–102. (Bombyliidae p. 101.) Also, Atti Accad. Napoli, ser. 2, vol. 5, no. 14, pp. 1–30, pl. 4. Ditteri pp. 20–26. (Bombyliidae p. 23, pl. 4, fig. 9.)

Coucke, E., and Coucke, Louis
- 1894. Matériaux pour une étude des Diptères de Belgique. Bombylides. Ann. Soc. Ent. Belgique, vol. 38, pp. 281–292, 4 figs.

Coulson, Jack R., Sabrosky, Curtis W., and Muller, Irmgard
- 1965. Selected bibliography of North American Diptera in Stone and others: A catalog of the Diptera of America North of Mexico. Agric. Res. Serv., U.S. Dept. Agric., Washington, D.C., pp. 1117–1547.

Coulson, Jack R. [See also Stone, Sabrosky, Wirth, Foote, and Coulson.]

Crampton, Guy Chester
- 1914. On the misuse of the terms parapteron, hypopteron, tegula, squamula, patagium and scapula. Journ. New York Ent. Soc., vol. 22, pp. 248–261, 1 pl.
- 1923. The genitalia of male Diptera and Mecoptera compared with those of related insects from the standpoint of phylogeny. Trans. American Ent. Soc., vol. 48, pp. 207–225, pls. 8–10. (No special Bombyliidae comments.)
- 1926. A comparison of the neck and prothoracic sclerites throughout the orders of insects from the standpoint of phylogeny. Trans. American Ent. Soc., vol. 52, pp. 199–248, 8 pls. (Four Diptera considered and illustrated. No specific comments on Bombyliidae.)
- 1931. A phylogenetic study of the posterior metathoracic and basal abdominal structures of insects . . . Journ. New York Ent. Soc., vol. 39, pp. 323–357, 4 pls. (Comments on Diptera.)
- 1941. The terminal abdominal structures of male Diptera. Psyche, vol. 48, pp. 79–94, pls. 7–8. (No specific comments on Bombyliidae.)
- 1942. Guide to the insects of Connecticut. Part VI. The Diptera or true flies of Connecticut. First fascicle. The external morphology of the Diptera. State Geol. Nat. Hist. Surv. Connecticut, Bull. no. 64, pp. 10–174, 14 pls. with 145 figs. Bibliography with 343 entries. (Bombyliidae pp. 24, 66, 88, 94, 116, 117, 144, fig. 5L.)

Cresson, Ezra Townsend
- 1915a. A new genus and some new species belonging to the dipterous family Bombyliidae. Ent. News, vol. 26, pp. 200–207, figs. 1–3.
- 1915b. Note on the bombyliid genus *Rhabdopselaphus* Rondani (*Pseudogeron* Cress.) (Dip.). Ent. News, vol. 26, p. 305.
- 1915c. Some North American Diptera from the Southwest. Paper III. A revision of the species of the genus *Mythicomyia*. Ent. News, vol. 26, pp. 448–456, 1 fig.

CRESSON, EZRA TOWNSEND—Continued
- 1916. Dipterological notes. II. A study of the *lateralis*-group of the bombyliid genus *Villa* (*Anthrax* of authors, in part). Ent. News, vol. 27, pp. 439–444.
- 1919. Dipterological notes and descriptions. Proc. Acad. Nat. Sci. Philadelphia, vol. 71, pp. 171–194. (Bombyliidae pp. 177–188.)
- 1923. Records of some western Diptera, with descriptions of two new species of the family Bombyliidae. Proc. Acad. Nat. Sci. Philadelphia, vol. 75, pp. 365–367. (Bombyliidae pp. 365–367.)

CRIDDLE, NORMAN
- 1933. Notes on the habits of injurious grasshoppers in Manitoba. Canadian Ent., vol. 65, p. 98.

CROS, AUGUSTE
- 1929. Note sommaire sur les parasites des oothèques des sauterelles marocaines. Bull. Soc. Hist. Nat. Afrique du Nord, Algiers, vol. 20, pp. 141–142. (Bombyliidae p. 142; brief note.)
- 1932. *Toxophora maculata* Rossi (Diptera-Bombyliidae). Etude biologiques. Bull. Soc. Hist. Nat. Afrique du Nord, Algiers, vol. 23, pp. 67–73.

CURRAN, CHARLES HOWARD
- 1927. Descriptions of Nearctic Diptera. Canadian Ent., vol. 59, pp. 79–92, 5 figs., 2 keys. (Bombyliidae pp. 84–85.)
- 1927–1928. Diptera of the American Museum Congo Expedition, pts. I & III. Bull. American Mus. Nat. Hist.
    - 1927. Pt. I, vol. 57, pp. 33–89, figs. 1–2. (Bombyliidae pp. 34–42, figs. 1–2.)
- 1928. *In* Curran, Alexander, Twinn, and Van Duzee: Scientific survey of Porto Rico and the Virgin Islands. I. Insects of Porto Rico and the Virgin Islands. Diptera. New York Acad. Sci., vol. 11, pt. 1, pp. 1–118, 39 figs. (Bombyliidae pp. 18–22; 12 spp. in 6 genera listed.)
- 1929. New Diptera in the American Museum of Natural History. American Mus. Nov., no. 339, pp. 1–13, 2 figs. (Bombyliidae p. 6.)
- 1930a. New species of *Lepidanthrax* and *Parabombylius* (Bombyliidae, Diptera). American Mus. Nov., no. 404, pp. 1–7.
- 1930b. New Diptera from North and Central America. American Mus. Nov., no. 415, 16 pp., 1 fig. (Bombyliidae pp. 1–10.)
- 1930c. Report on the Diptera collected at the station for the study of insects, Harriman Interstate Park, N. Y., with appendix on the Tipulidae and Ptychopteridae by C. P. Alexander. Bull. American Mus. Nat. Hist., vol. 61, art. 2, pp. 21–115. (Bombyliidae pp. 42–43; 11 spp. in 5 genera listed.)
- 1931. First supplement to the Diptera of Porto Rico and the Virgin Islands. American Mus. Nov., no. 456, 23 pp., 4 figs. (Bombyliidae pp. 6–8, fig.)
- 1933. New North American Diptera. American Mus. Nov., no. 673, 11 pp., 1 fig. (Bombyliidae pp. 2–3.)
- 1934a. The families and genera of North American Diptera. New York, the author. 512 pp, 2 pls, many figs. (Bombyliidae pp. 191–200, 81 figs.)
- 1934b. The Templeton Crocker Expedition of the California Academy of Sciences, 1932. Diptera. Proc. California Acad. Sci., ser. 4, vol. 21, no. 13, pp. 147–172, 4 figs. (Bombyliidae p. 151; 2 spp. listed.)
- 1934c. *In* Curran and others: The Diptera of the Kartabo, Bartica District, British Guiana, with descriptions of new species from other British Guiana localities. Bull. American Mus. Nat. Hist., vol. 66, art. 3, pp. 287–532, including 7 pls. with 54 figs. (Bombyliidae pp. 361–364; 7 spp. listed with 2 n. sp. described.)
- 1935. New American Diptera. American Mus. Nov., no. 812, pp. 1–24. (Bombyliidae pp. 1–2.)
- 1936. *In* Curran, Alexander, and Cresson: The Templeton Crocker Expedition to Western Polynesian and Melanesian Islands. 1933. No. 30. Diptera. Proc. California Acad. Sci., ser. 4, vol. 22, pp. 1–67, 2 pls., 11 figs. (Bombyliidae p. 18.)

Curran, Charles Howard—Continued
- 1942a. American Diptera. Art. 3. Bull. American Mus. Nat. Hist., vol. 80, pp. 51–84. (Bombyliidae pp. 51–53.)
- 1942b. Key to families *in* Guide to the insects of Connecticut. Part VI. The Diptera or true flies of Connecticut. First fascicle. State Geol. Nat. Hist. Surv. Connecticut, Bull. no. 64, pp. 175–182. (Bombyliidae p. 178.)

Curtis, John
- 1823–1840. British entomology: Illustrations and descriptions of the genera of insects found in Great Britain and Ireland. First printing in 16 volumes, 769 plates in 192 parts, 1824–1839. Second printing bound in 8 volumes with Diptera-Omaloptera bound in volume 8. (See Blackwelder, 1947, and Stone and others, 1965, for further information.)
    - 1824. Bombyliidae, vol. 8, 9.2. and vol. 8, 9.3., 4 pls.
    - 1836. Bombyliidae, vol. 8, 613, 5, 7–8.
- 1829. A guide to the arrangement of British insects: A catalogue of all the named species hitherto discovered in Great Britain and Ireland. London. Pp. i–vi, 256 columns. Second edition, 1837, pp. i–vi, 294 columns. (Bombyliidae pls. 32–33.)

Custer, Clarence P.
- 1928. Parasites of some Anthidiine bees (Hym.: Megachilidae, Chrysididae; Dipt.: Bombyliidae). Ent. News, vol. 39, pp. 123–125, 1 fig. (Bombyliidae pp. 124–125, 1 fig.)

Custer, Clarence P., and Hicks, Charles H.
- 1927. Nesting habits of some Anthidiinebees. Biol. Bull. (Woods Hole), vol. 52, pp. 258–277, 1 fig. (Bombyliidae p. 276.)

Cuthbertson, Alexander
- 1937. Biological notes on some Diptera in Southern Rhodesia. Trans. Rhodesia Sci. Assoc., vol. 35, pp. 16–34.

Cuthbertson, Alexander [see Engel and Cuthbertson]

Cuvier, Georg Christian Leopold Dagobert
- 1817. Le règne animal. . . . I-IV. Paris. Vol. 3, Entomologie, by Latreille, pp. 1–29, 1–653, 2 pls. (See Latreille.)

Cyrillo, Dominico
- 1787. Entomologiae neapolitanae specimen primum. Napoli. Pp. 1–8, 12 color pls. (Bombyliidae pl. 9, figs, 4, 6; pl. 11, fig. 2.)

Czizěk, Karl
- 1907. Neue Beitrage zur Dipterenfauna Mährens. Brunn. Zeitschr. Mähr. Landesmus., vol. 7, pp. 157–177. (Bombyliidae p. 160.)

Dahlbom, Anders Gustav
- 1851. Anteckningar öfver Insekter; som bilifvit observerade pa Gottland och i en del af Calmar Län under sommaren. Kongl. Vetensk. Acad. Handl. (1850), pp. 157–229. (Bombyliidae p. 160.)

Dallas, William Sweetland
- 1866. *In* Gunther, A. C. L. G. (editor): The record of zoological literature, 1865. Vol. 2 (1865), 798 pp. Insecta, pp. 381–710. (Bombyliidae pp. 648–651. Designation type-species of *Hyperalonia*.)

Davidson, Anstruther, M. D.
- 1893. The nest and parasites of *Xylocopa orpifex*, Smith. Ent. News, vol. 4, no. 5, pp. 151–153. (Bombyliidae pp. 152–153.)
- 1900. A bee fly four years in the larval state. Is this a record? Ent. News, vol. 11, pp. 510–511. (*Chrysanthrax edititia*.)

Davis, Frank Marvin
- 1963. A biological study of two common mud-daubers, *Trypoxylon politum* Say and *Sceliphron caementarium* (Drury). Master of Science dissertation in the Department of Entomology, Mississippi State University, State College, Mississippi, May 1963. (Bombyliidae pp. 11–12, 31, 47, 53, 58.)

Davis, John June
    1916. A progress report on white grub investigations. Journ. Economic Ent., vol. 9, no. 2, pp. 261–281, 2 pls. (Bombyliidae p. 271; mentions *Sparnopolius fulvus* Wied., as a parasite.)
    1919. Contributions to a knowledge of the natural enemies of Phyllophaga. Bull. Illinois State Nat. Hist. Survey, vol. 13, pp. 53–138, pls. 3–15 with 61 figs., 2 charts, 1 table, text-figs. 1–46. (Bombyliidae pp. 57, 69–72, 76, 87–88; charts 1–2, pl. 3, fig. 7; pl. 4, fig. 12, text-figs. 7–10, 19.)

Dean, George A. [see Smith, Kelly, Dean, Bryson, and Parker]

DeGeer, Charles
    1752–1778. Mémoires pour servir à l'histoire des insectes. In 7 volumes. Stockholm.
        1776. Vol. 6, pp. i–viii, 1–523, 30 pls. (Bombyliidae pp. 189, 191–192, 206, 266–273; pl. 11, figs. 11–13; pl. 15, figs. 10–12; pl. 29, figs. 11–13; pl. 30, figs. 9–11.)

De Lepiney, J., and Mimeur, J. M.
    1930. Sur *Glossista infuscata* Meig. et *Anastoechus nitidulus* F., parasites marocains de *Dociostaurus maroccanus* Thunb. Rev. Pathol. Veg. Ent. agric., Paris, vol. 17, pp. 418–420.

Delgado de Torres, D.
    1929. Dipteros parásitos de la Langosta en España. Bol. de Patol. Veg. Ent. Agric, Madrid, An. IV, pp. 65–68, 2 figs.

DeLong, D. W. [see Borror and DeLong]

Demolin, G. [see Biliotti, E., Demolin, G., and du Merle, P.]

Dempster, J. P.
    1957. The population dynamics of the Moroccan Locust (*Dociostaurus maroccanus* Thunberg) in Cyprus. Anti-Locust Bull., no. 27, pp. 1–59.

De Serba, A. F.
    1901. As invasões de gafanhotos em Portugal. A proposito de um parasite notavel do *Stauronotus maroccanus* Thunberg. Archivo Rural, Lisboa, pp. 225–229, 246–251, 266–270, 286–295, 311–312.

Dietrich, Wilhelm
    1909. Die Facettenaugen der Dipteren. Zeitschr. Wiss. Zool., vol. 92, pp. 465–539, pls. 22–25.

Dimmock, George
    1881. The anatomy of the mouth-parts and of the sucking apparatus of some Diptera. Boston. 60 pp., 4 pls. (Bombyliidae pp. 22–26, 37–48, pl. 2, figs. 1–7.)

Disconzi, Don Francesco
    1865. [Note: Bombyliidae listed.] Ent. Vicentina. Padova. (Bombyliidae p. 218; 3 spp. listed.)

Doleschall, Carl Ludwig
    1856–1858. Bijdrage tot de Kennis der Dipterologische Fauna van Nederlandsch Indie. I-III. Natuurkundig Tijdschrift Nederlandsch Indie.
        1856. Pt. 1, new ser., vol. 7 (10), pp. 403–414, 12 pls. (Bombyliidae pp. 408–409, pl. 5, fig. 1.)
        1857. Pt. 2, ser. 3, vol. 4 (14), pp. 377æ418, 10 pls. (Bombyliidae pp. 399–402, pl. 9, figs. 1–2.)
        1858. Pt. 3, ser. 4, vol. 17, pp. 73–128. (Bombyliidae pp. 93–94.)

Donahue, Julian P.
    1968. *Geron calvus* (Diptera: Bombyliidae), a parasite of *Solenobia walshella* (Lepidoptera: Psychidae) in Michigan [description]. Michigan Ent., vol. 1, no. 8, p. 284.

Donovan, Edward
    1792–1813. The natural history of British insects. London. In 16 volumes, each with 36 color pls. Second edition, 1804, vol. 1 and 2 only.
        1793. Vol. 2, pp. 1–96, 1–6 index, pls. 37–72. (Bombyliidae pp. 78–80, pl. 66.)
        1796. Vol. 5, pp. 1–110, 1–6, pls. 145–180. (Bombyliidae pl. 146, fig. 1.)

Drury, Dru
    1770–1782. Illustrations of natural history, wherein are exhibited upwards of two hundred and forty figures of exotic insects. London. In 3 volumes. New edition by Westwood, 1837: *Illustrations of Foreign Entomology.*
        1773. Vol. 2, pp. 1–9, 1–90, 50 color pls. (Bombyliidae p. 74, pl. 39, fig. 3.)

Dufour, Léon
    1833. Description de quelques insectes diptères des genres *Astomella, Xestomyza, Ploas, Anthrax, Bombylius, Dasypogon, Laphria, Sepedon,* et *Myrmemorpha* observes en Espagne. Ann. Sci. Nat., vol 30, pp. 209–221, 1 pl. (Bombyliidae pp. 214–215.)
    1836. Beschreibung einiger zweyflügliger Insecten der Sippen *Astomella, Xestomyza, Ploas, Anthrax, Bombylius, Dasypogon, Laphria, Sepedon,* und *Myrmemorpha*—in Egypten beobachtet von Leon Dufour. S. 209–221. Isis (Oken) Leipzig, vol. 6, pp. 468–472. (Bombyliidae p. 470.)
    1850. Recherches anatomiques et physiologiques sur les les Diptères. Mém. Acad. Sci., Math. Phys., Paris, vol. 11, 1851, pp. 171–360, 11 pls. Separate published 1850. Paris. (Bombyliidae pp. 251–256, pl. 6, figs. 62–67.)
    1850–1852. Description et iconographie de quelques Diptères de l'Espagne. Ann. Soc. Ent. France.
        1850. Ser. 2, vol. 8, pp. 131–155. (Bombyliidae pp. 140–143, pl. 5, figs. 4–6, color.)
        1852. Ser. 2, vol. 10, pp. 5–10, pl 1. (Bombyliidae pp. 6–9, pl. 1 with 19 figs.; fine color plate.)
    1858. Histoire des métamorphoses de *Bombylius major.* Ann. Soc. Ent. France, ser. 3, vol. 6, pp. 503–511, pl. 13, figs. 1–9.

Duméril, André Marie Constant
    1823. Considérations générales sur la classe des insectes ... Strassbourg et Paris. 272 pp., 60 pls. (Bombyliidae pp. 226, 228, pl. 46, fig. 2; pl. 48, fig. 4.)

Du Merle, P.
    1964. Cycle biologique d'un Diptere Bombyliidae du genre *Villa.* Compt. Rend. Acad. Sci., Paris, vol. 259, pp. 1657–1659.
    1966a. Modalites de l'accouplement chez un Diptera Bombyliidae, *Villa quinquefasciata* Wied. apud Meig. Ann. Soc. Ent. France (n.s.), vol. 2, pp. 617–623.
    1966b. Modele de cage permettant d'obtenir la ponte d'un Diptere Bombyliidae, *Villa quinquefasciata* Wied. apud Meig. Entomophaga, vol. 11, pp. 325–330.
    1969a. Existence de deux diapauses facultative au cours du cycle biologique de *Villa brunnea* Beck. (Diptera: Bombyliidae). Compt. Rend. Acad. Sci., Paris, vol. 268, pp. 2433–2435.
    1969b. Le complexe parasitaire hypoge de *Thaumetapoea pityocampa* Schiff (Lepidoptera). Bol. Serv. Plagas Forestales, Madrid, vol. 12, no. 23, pp. 29–33, color plate.
    1971. Biologie de deux especes de genre *Usia* Latreille (Diptera: Bombyliidae). Ann. Soc. Ent. France, vol. 7, pp. 241–259.

Duncan, James
    1837–1838. Characters and descriptions of the dipterous insects indigenous to Britain. Mag. Zool. Bot.
        1838. Vol. 2, pp. 205–222. (Bombyliidae p. 210.)

Dusa, L. [also see Radu and Dusa]
    1964. Nouvelles contributs a l'etude des bombylides des Romanic IV. Studia Univ. Babes-Bolyai, ser. biol., no. 2, pp. 65–67.
    1966. New contribution to the knowledge of Bombyliidae (Diptera: Bombylides) from the Soviet Republic of Romania. Studia Univ. Babes-Bolyai, no. 1, pp. 67–71, 9 figs.

Dutt, Gobind Ram
    1912. Life-histories of Indian Insects. IV. (Hymenoptera). Mem. Dept. Agric. Indian Ent., Calcutta, ser. 4, no. 4, pp. 183–267, pls. 11–14, text-figs. 1–22. (Bombyliidae, pp. 195–196, fig. 41, pp. 210–212, text-figs. 8–9, *Hyperalonia sphinx,* pupa figured.)

Dyar, Harrison G.
- 1900. Life history of a South American slug-caterpillar *Sibine fusca* Stoll. Ent. News, vol. 11, pp. 517–526, pl. 13. (Bombyliidae p. 526.)

Edwards, Frederick Wallace
- 1915. Report on the Diptera collected by the British Ornithological Union Expedition and the Wollaston Expedition in Dutch New Guinea. Trans. Zool. Soc. London, vol. 20, pt. 13, pp. 391–422, pl. 38 (color), figs. 1–10. (Bombyliidae p. 406; 2 spp.)
- 1919. II. Diptera collected in Korinchi, West Sumatra, by Messrs. H. C. Robinson, and C. Boden Kloss. Journ. Federated Malay States, Singapore, vol. 8, pt. 3, pp. 7–59, pls. 3–6. (Bombyliidae pp. 36–37, fig. 21.)
- 1928. The family position of Pachyneres (Diptera). Entomologist, London, vol. 61, p. 208.
- 1930. Diptera of Patagonia and South Chile based mainly on material in the British Museum (Natural History). Part V. Fasc. 2. Bombyliidae, Nemestrinidae, and Cyrtidae. British Mus. (Nat. Hist.), pp. 162–197, 4 figs. (Bombyliidae pp. 162–179, fig. 10.)
- 1934. On the genus *Comptosia* and its allies (Bombyliidae). Diptera. Encycl. Ent., ser. B. II. Dipt., vol. 7, pp. 81–112, pls. 2–4, figs. 1–32.
- 1936. Bombyliidae from Chile and western Argentina. Rev. Chilena Hist. Nat., Santiago, vol. 40, pp. 31–41, 5 figs.

Efflatoun Bey, H. C.
- 1945. A monograph of Egyptian Diptera. Part 6, Family Bombyliidae. Section I: Subfamily Bombyliinae Homoeophthalmae. Bull. Soc. Fouad Ier d'Ent., pp. 1–483, pls. 1–38 with 551 figs. Life histories pp. 2, 6–14.

Egger, Johann von
- 1855. Bericht über eine Excursion am Neusiedlersee. (Diptera). Verh. zool.-bot. Ver. Wien, vol. 5, pp. 74–76. (Bombyliidae p. 76.)
- 1858–1863. Dipterologische Beiträge. Verh. zool.-bot. Ges. Wien.
    - 1859. Vol. 9, pp. 387–407. (Bombyliidae pp. 396–400.)

Emden, Fritz I. van
- 1953. The male genitalia of Diptera and their taxonomic value. Trans. 9th Int. Congress Ent., Amsterdam, 1951, vol. 2, pp. 22–26. (No Bombyliidae comment.)
- 1957. The taxonomic significance of the characters of immature insects. Ann. Rev. Ent., vol. 2, pp. 91–106. (Bombyliidae p. 97.) Bibliography pp. 102–106.

Emden, Fritz I. van, and Hennig, Willi
- 1956. Diptera *in* S. L. Tuxen: Taxonomist's glossary of genitalia in insects. Copenhagen, pp. 1–284, figs. 1–215. (No Bombyliidae comment.)

Enderlein, Gunther
- 1908. Biologisch-faunistiche Moor und Dünen-Studien ... Ber. bot.-zool. Ver. Danzig, vol. 30, pp. 54–238. (Bombyliidae pp. 179, 191.)
- 1926a. *In* Study, E.: Ueber einige mimetische Fliegen. Zool. Jahrb. Abt. allg. Zool., vol. 42, pp. 421–427, 2 pls. (Bombyliidae p. 426; Enderlein describes *Systropus studyi*.)
- 1926b. Zur Kenntnis der Bombyliiden-Subfamilie Systropodinae (Diptera). Wiener Ent. Zeitung, vol. 43, no. 2, pp. 69–92, 1 fig.
- 1930. Dipterologische Studien. XX. Deutsche Ent. Zeitschr., 1930, pp. 65–73. (Bombyliidae pp. 66–67.)
- 1934. Entomologische Ergebnisse der Deutsch-Russischen Alai-Pamir-Expedition, 1928. III. 1. Diptera. Deutsche Ent. Zeitschr. (1933), pp. 129–174, 3 figs. (Bombyliidae pp. 140–142, fig. 3.)
- 1936. Zweiflügler, Diptera *in* Brohmer, Ehrmann, Ulmer: Die Tierwelt Mitteleuropas, Leipzig, vol. 6, pt. 3, Abt. 16, pp. i–xvi, 1–259, figs. 1–317. (Bombyliidae pp. 89–91, figs. 172–173.)

Engel, Erich Otto [also see Werner and Engel]
- 1885. Uber von Herrn M. Quedenfeldt in Algier gesammelt Dipteren. Ent. Nachrichten, vol. 11, pp. 177–179, 1 text-fig. (Bombyliidae pp. 177–179, text-fig.)

Engel, Erich Otto—Continued
- 1887. Ueber Eigenthumlichkeiten im Bau der Flügelgeäders bei der Dipterenfamilie der Bombylarier. Ent. Nachrichten, vol. 13, pp. 46–47.
- 1932–1937. *In* Lindner: Die Fliegen der palaearktischen Region. Stuttgart. Bombyliidae 25. 619 pp., 239 figs., 15 pls., 1937.
  - 1932. Pp. 1–96, 41 figs., 2 pls.
  - 1933. Pp. 97–192, 62 figs., 2 pls.
  - 1934. Pp. 193–256, 10 figs.
  - 1935. Pp. 257–400, 47 figs., 3 pls.
  - 1936. Pp. 401–560, 64 figs., 6 pls.
  - 1937. Pp. 561–619, 15 figs., 2 pls.
- 1936. A new genus and new species of Bombyliidae (Dipt.) from Southern Africa. Occas. Papers, Rhodesia Mus., Bulawayo, no. 5, pp. 39–41, 2 figs.
- 1940. Über einige chinesische Bombyliiden und Asiliden (Dipt.). Mitt. Münchener ent Ges., vol. 30, pp. 72–84, 2 figs. (Bombyliidae pp. 72–75.)

Eppley, R. K. [see Bohart, Stephen, and Eppley]

Erichson, Wilhelm Ferdinand
- 1841–1849. *In* Wiegmenn's: *Archiv für Naturgeschichte* in Verbindung mit mehreren Gelehrten herausgegeben von Erichson. Volumes 7–15.
  - 1841. Vol. 7. (Bombyliidae p. 233; 2 spp.)
- 1848. Insecten *in* Schomburgk: Reisin in Britisch-Guiana, 1840–1844. Pt. 3, pp. 1–8, 1–1260. Insects pp. 553–617. (Bombyliidae (Bombyliarii) pp. 607–608.) [Also a reprint: *Versuch einer Fauna and Flora von British Guiana* with insects by Erichson.]

Esaki, T., Hori, H., Hozana, S., Shiraki, T., and others
- 1932. Iconographia Insectorum Japanicorum. Tokyo. Pp. 1–123, with plates and descriptions numbered in Japanese. (Bombyliidae includes 9 spp.)
- 1956. Iconographia Insectorum Japonicorum. Edition 2. Pp. 1–13, 1–1737, & index pp. 1–203. (In Japanese.) (Bombyliidae pp. 1593–1597, nos. 4577–4588.)

Essig, E. O.
- 1926. Insects of western North America. New York: Macmillan. xi + 1035 pp., 766 figs. (Bombyliidae pp. 560–562, fig. 450.)
- 1958. Insects and mites of Western North America. A revised edition of: *Insects of Western North America.* New York: Macmillan. 1050 pp.

Evans, A. M. [see Newstead, Evans, and Potts]

Evans, Howard E.
- 1957. Studies on the comparative ethology of digger wasps of the genus *Bembix*. Ithaca: Comstock Publ. Assoc. 258 pp., 52 figs. (Bombyliidae pp. 180–181, 211–213.)

Evans, Howard E. [see Gillaspy, Evans, and Cheng]

Eversmann, Eduardo von
- 1834. Diptera Wolgam fluvium inter et montes Uralenses observata. (Catalogue). Bull. Soc. Imp. Nat. Moscow, vol. 7, pp. 420–432. (Bombyliidae p. 423, 3 nomen nudum.)
- 1855. Beiträge zur Lepidopterologie Russlands, und Beschreibung einiger anderer Insecten aus den südlichen Kirgisteppen, den nördlichenfern des Aral-Sees und des Sir Darjás. Bull. Soc. Imp. Nat. Moscow, vol. 27 (1854), pp. 174–205, 1 color pl. (Bombyliidae pp. 201–204, pl. 1, 4 figs.)

Fabre, Jean Henri Casimir
- 1908. Souvenirs Entomologiques. 10 vols. (first edition, 1879). Seventh edition, Delegrave Paris. *Anthrax*, vol. 3, pp. 129–153, 189–223 (from Bohart and MacSwain, 1939). 1886, Souvenirs Entomologiques; études sur l'instinct et les moeurs des Insectes. (Troisième série). Paris. 435 pp. (Bombyliidae p. 129 and following, habits of *Anthrax trifasciata*; p. 205, primary larva; p. 131, secondary larva; p. 148, pupa. (Taken from Zoological Record.)
- 1913. The life of the fly, with which are interspersed some chapters of autobiography. Translated by Alexander Teixeira de Mattos. New York: Dodd, Mead & Co. 477 pp. (Bombyliidae pp. 28–62, 72, 78–81, 88–110.)

FABRE, JEAN HENRI CASIMIR—Continued
- 1914. The Mason Bees. Translated by Alexander Teixeira de Mattos. New York: Dodd, Mead & Co. 315 pp. (Bombyliidae pp. 196–199, 210, 212, 264, 274, 276, 279–281, 303.)
- 1915a. Bramble-bees and others. Translated by Alexander Teixeira de Mattos. New York: Dodd, Mead & Co. viii + 456 pp. (Bombyliidae pp. 84–85, 87–88, 126.)
- 1915b. The hunting wasps. Translated by Alexander Teixeira de Mattos. New York: Dodd, Mead & Co. 427 pp. (Bombyliidae pp. 281, 287, 291, 292–293, 297, 305–306.)
- 1939. Fabre's book of Insects. New York: Tudor Publ. Co., Dodd, Mead & Co. 271 pp. (Bombyliidae, Chapter 16, pp. 249–271, pl. 1.)

FABRE, JOSEPH LOUIS
- 1857. Mémoire sur l'hypermétamorphose et les moeurs des Méloides. Ann. Sci. Nat., ser. 4, vol. 7, pp. 299–365, 1 pl. (Bombyliidae p. 302, pl. 17, *Anthrax sinuata*.)

FABRICIUS, JOHANN CHRISTIAN
- 1775. Systema entomologiae, ... Flensburgi et Lipsiae. 862 pp. (Bombyliidae pp. 757–759, 797, 802–803.)
- 1781. Species Insectorum. ... In 2 volumes. Hamburgi et Kilonii. Vol. 2, 517 pp. (Bombyliidae pp. 414–416, 473–474.)
- 1787. Mantissa Insectorum. ... In 2 volumes. Hafniae: Proft. Vol. 2, 382 pp. (Bombyliidae pp. 329, 366–367.)
- 1792–1794. Entomologica Systematica. ... In 4 volumes. Hafniae.
  - 1794. Vol. 4, pp. 1–6, 1–472. (Bombyliidae pp. 256–262, 408–414.)
- 1796. Index alphabeticus in Entomologiam systematicam emendatam et actam, ordines, genera et species continens. Hafniae: Proft et Storch. Pp. 1–8, 1–175. (Bombyliidae p. 104.)
- 1798. Supplementum Entomologiae systematicae. Hafniae: Proft. et Storch. 572 pp. (Bombyliidae pp. 554, 569–570.)
- 1805. Systema Antliatorum. ... Brunsvigae. Pp. i–xiv, 1–372, 1–30. (Bombyliidae pp. 114–137.)

FALLEN, CARL FREDRICH
- 1810. Specimen entomologicum novam Diptera disponendi methodum exhibens. Lund. 26 pp., 1 pl. (Bombyliidae pp. 6–8.)
- 1814–1817. Diptera Sueciae.
  - 1814. Anthracides Sueciae. Lund. 16 pp.
  - 1815. Platypezinae et Bombylarii Sueciae. Lund. Pp. 1–12. (Bombyliidae pp. 8–12.)

FATTIG, P. W.
- 1933. Food of the robber fly, *Mallophora orcina* (Wied.) (Diptera). Canadian Ent., vol. 65, pp. 119–120. (Bombyliidae as prey of robber flies.)
- 1945. The Asilidae or robber flies of Georgia. Emory Univ. Mus. Bull., no. 3, pp. 1–33. (Bombyliidae: 3 spp. prey of 3 spp. of robber flies.)

FEUERBORN, H. J.
- 1922. Das Hypopygium "inversum" und "circumversum" der Dipteren. Zool. Anz., vol. 55, pp. 189–213, 13 figs. (Bombyliidae p. 189. An important article of general interest featuring psychodids and calliphorids.)

FIEDLER, O. G. H., and KLUGE, E. B.
- 1954a. The influence of the tsetse fly eradication campaign on the breeding activity of Glossinae and their parasites in Zululand. Onderstepoort Journ. Vet. Res., vol. 26, no. 3, pp. 389–393. (Bombyliidae pp. 390, 393.)
- 1954b. The parasites of tsetse flies in Zululand with special reference to the influence of the hosts upon them. Onderstepoort Journ. Vet. Res., vol. 26, no. 3, pp. 399–404. (Bombyliidae pp. 39–400, 403.)

FINLAYSON, THELMA [see also Finlayson and Finlayson]
- 1960. Taxonomy of cocoons and puparia, and their contents, of Canadian parasites of *Neodiprion sertifer* (Geoff.) (Hymenoptera: Diprionidae. Canadian Ent., vol. 92, pp. 20–47, figs. 1–21. (Bombyliidae pp. 40–41, figs. 19–20.)

FINLAYSON, L. R., and FINLAYSON, THELMA
    1958. Parasitism of the European pine sawfly, *Neodiprion sertifer* (Geoff.) (Hymenoptera: Diprionidae), in southwestern Ontario. Canadian Ent., vol. 90, pp. 223–225. (Bombyliidae pp. 223–224; *Hemipenthes sinuosa* Wiedemann parasite.)

FITCH, EDWARD A. [see Saunders, Waterhouse, and Fitch]

FLECK, EDUARD
    1904. Die Dipteren Rumäniens. Bul. Soc. St., Bucureşti, vol. 13, nos. 1–2, pp. 92–116.

FLETCHER, T. B.
    1916. One hundred notes on Indian insects. Bull. Agric. Res. Inst. Pusa, vol. 59, 17 pp. (*Systoechus socius* Walker parasitises *Colemania sphenarioides*.)

FLUKE, C. L., JR. [see Richter and Fluke]

FOOTE, RICHARD H. [see Stone, Sabrosky, Wirth, Foote, and Coulson]

FORBES, S. A.
    1907. On the life history, habits, and economic relations of the white-grubs and May beetles (Lachnosterna). Bull. Illinois Agric. Exper. Stat., no. 116, pp. 445–480. (Bombyliidae pp. 473–474.)
    1908. Twenty-fourth report of the state entomologist on the noxious and beneficial insects of the state of Illinois. Pp. 1–168, pls. 1–9. Life history, habits, etc., white grubs, pp. 135–168, 2 pls. (colored). (Bombyliidae pp. 160–161, pl. 9, fig. 1.)

FOURCROY, ANTOINE FRANCOIS DE
    1785. Entomologia parisiensis, sive catalogus insectorum, quae in agro parisiensi reperiuntur. Paris. 2 volumes. Vol. 2, vii + 544 pp.. The new species by Geoffroy. (Bombyliidae p. 459; see Goeffroy.)

FRANCOIS, F. J.
    1954a. Contribution à l'étude des Diptères de l'Urundi. V. Description d'un *Systropus* nouveau (Bombyliidae). Bull. Inst. Sci. Nat. Belgique, vol. 30, no. 18, pp. 1–4, 1 fig.
    1954b. Contribution à l'étude des Diptères de l'Urundi. VI. Synonymie du genre *Isotamia* Bezzi (Bombyliidae) et de son génotype. Bull. Ann. Soc. Ent. Belgique, vol. 90, pp. 108–182.
    1955a. Contribution à l'étude des Diptères de l'Urundi. VII. Bombyliidae du genre *Bombylius* (Espèces du groupe *analis*). Bull. Inst. Sci. Nat. Belgique, vol. 31, no. 7, pp. 1–7.
    1955b. Mission E. Janssens et R. Tollet en Grèce (juillet-août 1953). 4e note. Bull. Inst. Sci. Nat. Belgique, vol. 31, no. 28, pp. 1–6, 3 figs. (All Bombyliidae.)
    1960. Contribution à l'étude des Diptères de l'Urundi. VIII. Bombyliidae du Bugesera-Busoni. Bull. Ann. Soc. Ent. Belgique, vol. 96, pp. 284–290.
    1964a. Recoltes de M. A. Villiers dans les dunes cotieres de Senegal (1961). Diptères. Bombyliidae. Bull. Inst. franc. Afr. Noire, Dakar (A), vol. 26, pp. 924–943, 4 figs.
    1964b. Un *Exoprosopa* nouveau de Madagascar: *E. flammicoma* (Diptera: Bombyliidae). Bull. Ann. Soc. Ent. Belgique, vol. 100, pp. 309–313, 3 figs.
    1964c. Bombyliidae (Diptera) du Musee Royal de l'Afrique centrale. 1. *Palintonus*, un genre nouveau des Toxophorinae. Bull. Ann. Soc. Ent. Belgique, vol. 100, pp. 323–329, 3 figs.
    1967a. On the generic position of *Plesiocera flavifrons* Becker and of a few related species (Diptera, Bombyliidae). Bull. Inst. Sci. Nat. Belgique, vol. 43, pp. 1–8, 12 figs.
    1967b. Quelques Bombyliidae (Diptera) D'Israel. Bull. Ann. Soc. Ent. Belgique, vol. 103, pp. 276–282, 3 figs.
    1967c. Bombyliidae (Diptera) Meconnus: *Thyridanthrax fimbriatus* (Meig.), Bull. Ann. Soc. Ent. Belgique, vol. 103, pp. 289–293, 3 figs.
    1968a. Bombyliidae (Diptera) Meconnus: 2. Autres especes du groupe *Thyridanthrax afer*. Bull. Ann. Soc. Ent. Belgique, vol. 104, pp. 205–211, 6 figs.
    1968b. Contribution a l'etude de la faune de la basse Casamance (Senegal): 21. Diptera Bombyliidae (note complementaire). Bull. Inst. Fond. Afr. Noire, vol. 30, pp. 1477–1479, 4 figs.

FRANCOIS, C. J.—Continued
- 1968c. Contribution a la faune du Congo (Brazzaville). Mission A. Villiers et A. Descarpentries: 76. Dipteres Bombyliidae. Bull. Inst. Fond. Afr. Noire, vol. 30, pp. 787–789, 1 fig.
- 1968d. Expedition entomolicobatanique Mongole et Tchecoslovaque en Mongolie (1965-1966). 18. Diptera: Bombyliidae. Sub. Faun. Praci Ent. Odd. Nar. Mus. Praze, vol. 13, pp. 61–65, 10 figs.
- 1968e. Anthrax nouveaux du Sahara (Diptera: Bombyliidae). Bull. Ann. Soc. Ent. Belgique, vol. 104, pp. 91–96, 12 figs.
- 1969a. Bombyliidae (Diptera) from southern Spain, with descriptions of twelve new species. Ent. Meddel., vol. 37, pp. 107–160.
- 1969b. Bombyliidae (Diptera) Meconnus: 3. Essai de revision des *Villa* palearctiques du groupe cingulata-paniscus. Bull. Ann. Soc. Ent. Belgique, vol. 105.
- 1969c. Bombyliidae (Diptera) nouveaux D'Afghanistan. Bull. Ann. Soc. Ent. Belgique, vol. 105.
- 1969d. Notices sur des Bombyliidae palearctiques (Insecta: Diptera). Bull. Inst. Sci. Nat. Belgique, vol. 45, p. 28.
- 1969e. Le Parc National du Niokolo-Koba (Senegal). 25. Diptera: Bombyliidae. Inst. Roy. Sci. Nat. Belgique.
- 1970. Bombyliidae (Diptera) des Pyrenees (Provinces de Huesca et de Lerida). Pirineos, vol. 98, pp. 35–37.

FRAUENFELD, GEORG RITTER VON
- 1861. Beitrag zur Kenntniss der Insecten-Metamorphose. Verh. zool.-bot. Ges. Wien, vol. 11, pp. 163–174, pls. (Bombyliidae pp. 173–174, fig. pl. IID, figs. 14–15.)
- 1864. *Argyramoeba leucogaster* Mg. Verh. zool.-bot. Ges. Wien, vol. 14, pp. 688–689.

FREY, RICHARD
- 1911. Zur kenntnis der Dipterenfauna Finlands. Acta Soc. Fauna Flora Fenn., vol. 34, no. 6, pp. 1–59, 3 pls. (Bombyliidae pp. 44–53, pl. 2, figs. 7–12; pl. 3, figs. 13–17, scales off of legs of bee flies.)
- 1934. Diptera Brachycera von den Sunda-Inseln und Nord-Australien. Rev. Suisse Zool. Geneva, vol. 41, no. 15, pp. 299–339, 6 figs. (Bombyliidae pp. 318–320.)
- 1936. *In* Frey and others: Die Dipteren fauna der Kanarischen Inseln und ihre Probleme. Comment. Biol., Helsingfors, vol. 6, no. 1, pp. 1–237, 10 pls. with 87 figs., 4 graphs. (Bombyliidae pp. 46–51, pl. 1, figs. 2–3, 5; pl. 2, figs. 10–11.)
- 1938. Tiergeographische studien uber die Dipterenfauna der Azoren. I. Verzeichnis der bischer von den Azoren bekannten Dipteren. Comment. Biol. Soc. Sci. Fennica, vol. 8, no. 10, pp. 1–114, 4 pls. with 33 figs. (Bombyliidae pp. 35–36, 3 species listed.)
- 1952. Diptera *in* Lindberg and Saris: Insektfaunan i Pisavaara Naturpark (Finland, Prov. Ob). Acta Soc. Fauna Flora Fenn., vol. 69, no. 2, pp. 1–82, 4 pls., 1 map. (Bombyliidae p. 69, 1 species listed.)
- 1958. Entomologische Ergebnisse der finnländischen Kanaren-Expedition 1947–51. No. 15. Kanarische Diptera Brachycera p. p., von Hakan Lindberg gesammelt. Comment. Biol. Soc. Sci. Fenn., vol. 17, no. 4, pp. 1–63, 16 text-figs., 7 tables. (Bombyliidae pp. 14–18; 23 spp. listed, table 1, and pp. 4, 5, 6, 9.)

FREY, RICHARD UNTER MITWIRKUNG VON, HACKMAN, W., HERING, E. M., SABROSKY, C., and SPENCER, K.
- 1958. Ergebnisse der Zoologischen Forschungsreise von Prof. Dr. Hakan Lindberg nach den Kapverdischen Inseln im Winter 1953–54. No. 20. Zur Kenntnis der Diptera Brachycera p. p. der Kapverdischen Inseln. Comment. biol., Helsinki, vol. 18, no. 4, pp. 1–61, 20 figs., 1 map. (Bombyliidae pp. 8–11, fig. 4; 4 genera, 6 spp. [4 endemic].)

FRICK, KENNETH E.
- 1957. Biology and control of tiger beetles in alkali bee nesting sites. Journ. Economic Ent., vol. 50, pp. 503–505. (Bombyliidae p. 503.)
- 1962. Ecological studies on the alkali bee, *Nomia melanderi*, and its Bombyliid parasite, *Heterostylum robustum*, in Washington. Ann. Ent. Soc. America, vol. 55, pp. 5–15.

FRIEND, R. B.
    1942. Taxonomy. Wing venation. Connecticut State Geol. Nat. Hist. Surv. Bull., no. 64, pp. 166–174, figs. 15–17 with 29 wings pictured. (Bombyliidae p. 170, fig. 15, no. 10.)

FRIESE, H.
    1923. Die europäischen Bienen (Apidae). Das Leben und Wirken unserer Blumenwespen. Berlin and Leipsig. Pp. 1–6, 1–456, 33 pls. (color), 100 text-figs. (Bombyliidae p. 420; *Anthrax*, *Bombylius* living in the nests of *Anthophora*, *Chalicoderma* (pl. 20), *Andrena*, pl. 1 includes 1 Bombyliidae.)

FRISON, THEODORE H.
    1922. Notes on the life history, parasites and inquiline associates of *Anthophora abrupta* Say, with some comparisons with the habits of certain other Anthophorinae (Hymenoptera). Trans. Amer. Ent. Soc., vol. 48, pp. 137–156. (Bombyliidae p. 152.)

FRITSCH, KARL
    1875. Jährliche Periode de Insecterfauna von Österreich-Ungern. I. Die Fliegen (Diptera). Denkschr. Math.-Natur. Kaiser. Akad. wiss., vol. 34 (1874), pp. 33–114. (Bombyliidae pp. 40–41, 21 spp. listed.)

FRIVALDSZKY, JANOS
    1877. Adatok Temes és Krassó megyek faunájához. Magyar Tudományos Akad, Math. Természettud. Közlemények, vol. 13 (1875–1876), pp. 285–378. (Data and faunam Hungariae meridionalis comitatum Temes et Krasso.) The printer's note carries the date 1876. (Bombyliidae pp. 367–368, 9 spp. listed.)

FROST, S. W.
    1945. Spurious vein in the wings of *Exoprosopa fasciata* Macq. (Diptera). Ent. News, vol. 56, pp. 104–106.

FULLER, MARY E.
    1938a. Some flies associated with grasshoppers. Journ. Counc. Sci. Industr. Res. Australia, vol. 11, no. 2, pp. 202–203. (Bombyliidae p. 203.)
    1938b. Notes on *Trichopsidea oestracea* (*Nemestrinidae*) and *Cyrtomorpha flaviscutellaris* (Bombyliidae)—Two dipterous enemies of grasshoppers. Proc. Linnean Soc. New South Wales, vol. 63, pp. 95–104, text-figs. 1–19. (Bombyliidae pp. 100–104, text-figs. 10–19, pl. 6, figs. 6–7.)

GÄBLER, HELLMUTH
    1939. Beitrag zur Kenntnis der forstlich wichtigen Trauerschweber. Nachrichtenbl. Deutsch. Pflanzenschutzdienst (N.F.), vol. 3, pp. 55–57.
    1950. Larven- und Puppenformen der forstlich wichtigen Trauerschweber. Anz. Schädlingsk., Berlin, vol. 23, pp. 73–75, 6 figs. (Larvae and pupae of *Hemipenthes morio* and *Villa hottentotta*.)

GADEAU DE KERVILLE, HENRI
    1926. Liste.... des diptères recoltes en Syrie. Voyage zoologique d'Gadeau de Kerville in Syrie. Vol. 1, pp. 119–139. (Bombyliidae pp. 121–124.)

GARDNER, T. R. [see Clausen, Jaynes, and Gardner]

GEOFFROY, ETIENNE LOUIS
    1785. *In* Fourcroy: Entomologia parisiensis.... (see Fourcroy). (The new species are by Geoffroy. Bombyliidae p. 459.)

GERMAR, ERNST FRIEDRICH
    1814–1817. Reise durch Oesterreich, Tyrol nach Dalmatien und in das Gebiet von Ragusa. I-II. Leipzig. Vol. 2 also has title: Reise nach Dalmatien und in das Gebiet von Ragusa.
        1817. Vol. 2, 335 pp., 9 color pls, 2 maps.
    1817–1847. Fauna Insectorum Europae. Halle, Kümmel. Books 3–24, each book with 25 pages and 25 plates. Book 1–2 by Ahrens; book 3 with Kaulfuss, F.
        1817. Book 3. (Bombyliidae pp. 19–20, pls. 19–20.)
        1819. Book 6. (Bombyliidae p. 23, pl. 23.)
        1822. Book 7. (Bombyliidae p. 25, pl. 25. *Anthrax germani*, imperfect figure.)
        1823. Book 9. (Bombyliidae p. 22, pl. 22. *Bombylius nitidulus*.)

Germar, Ernst Friedrich—Continued
- 1831. Book 14. (Bombyliidae p. 24, pl. 24.)
- 1831. Book 15. (Bombyliidae pp. 22–23, pls. 22–23.)
- 1836. Book 19. (Bombyliidae p. 24, pl. 24. Fossil *Phthiria? dubia*, Rheni prope Bonnam. Not recognizable.)
- 1849. Ueber einige Insects aus Tertiarbildungen. Deutsche Geol. Zeitschr. Ges., vol. 1, pp. 52–66, pl. 2. (Bombyliidae, fossil, p. 64, pl. 2, figs. 7, 7a.)

Gerstfeldt, Georg
- 1853. Ueber die Mundtheile der saugenden Insecten. Dissert. inaug. Dorpat, Schuemann. 121 pp., 2 pls. (Also Mitau and Leipsig, G. A. Reyher.) (Bombyliidae pp. 14, 31–32, on mouthparts of *Bombylius*.)

Gibbs, Arthur E., and Barrand, Philip J.
- 1908. A preliminary list of Hertfordshire Diptera. Trans. Hertfordshire Nat. Hist. Soc., vol. 13, pp. 249–276. (Bombyliidae p. 257; 1 sp. listed.)

Gibson, William W., and Carillo S., Jose Luis
- 1959. Lista de insectos en la coleccion entomológica de la oficina de estudios especiales, S. A. G. Folleto Miscelaneo no. 9, Secr. de Agricultura y Ganaderia, Mexico City, pp. 1–18, 1–254. (Bombyliidae pp. 169–170; 27 determined spp. listed.)

Giebel, Christoph Gottfried Andreas
- 1862. Wirbelthier und Insektenreste im Bernstein. Zeitschr. Gessammten Naturwiss., vol. 20, pp. 311–321. (Fossil Bombyliidae pp. 318–319. *Lomatia gracilis*, n. sp.)

Gilbertson, G. I., and Horsfall, W. R.
- 1940. Blister beetles and their control. Bull. South Dakota Agric. Exper. Stat., no. 340, pp. 20–21.

Gil Collado, J.
- 1932. Dipteros de Ibiza recogidos por D. José Giner. Bol. Soc. Esp. Hist. Nat., Madrid, vol. 32, pp. 273–283, 7 figs. (Bombyliidae p. 273.)

Gillaspy, James E., Evans, Howard E., and Cheng, Shan Lin
- 1962. Observations on the behavior of digger wasps of the genus *Stictiella* (Hymenoptera: Sphecidae) with partition of the genus. Ann. Ent. Soc. America, vol. 55, pp. 559–566. (Bombyliidae pp. 563–564.)

Gimmerthal, Benjamin August
- 1842. [Note: *Exoprosopa pandora*.] Bull. Soc. Imp. Nat. Moscou. Vol. 15, p. 665.
- 1845–1847. Erster Beitrag zu einer künftig zu bearbeitenden Dipterologie Russlands. I-IV. Bull. Soc. Imp. Nat. Moscu.
    - 1847. Dritter Beitrag, vol. 20, pp. 175–223. (Bombyliidae p. 213.)
    - 1847. Vierter Beitrag, vol. 20, pp. 140–208. (Bombyliidae pp. 156–157.)

Giorna, Mich. Esprit
- 1791. Calendario entomologico, ossia osservazioni sulla stagioni proprie agl' Insetti nell' clima Piemontese e particolarmente ne' contorni di Torino. Torino. 154 pp. (From Horn and Schenkling, p. 423, 1928. See Lessona.)

Girschner, Ernst
- 1897. Uber die Postalar-Membran (Schüppchen, Squamulae) der Dipteren. Illustr. Wochenschr. Ent., 1897, pp. 1–32, 6 pls., 45 figs. (Bombyliidae p. 9.)

Gistel, Johnnes Nepomuk Franz Xaver (also written Gitl and Gistl)
- 1832–1837. Faunus. Zeitschr. Zool. und vergleichende Anatomie. 1832, vol. 1, pt. 1; 1834, pts. 2 and 3. 1835, vol. 2, pts. 1–3. (See Meigen.)

Gleason, Henry Allan [see Hart and Gleason]

Glover, Townsend
- 1867. *In* Report of the Commissioner of Agriculture, 1866. U.S. Dept. Agric., pp. 26–45. (Bombyliidae pp. 44–45.)
- 1871. Report of the Entomologist and Curator of the Museum, *in* Report of the Commissioner of Agriculture, 1870, U.S. Dept. Agric., pp. 65–91, figs. 1–60. (Bombyliidae p. 78, fig. 35.)
- 1878. *In* Report of the Commissioner of Agriculture, 1877, pp. 89–148, pls. 1–5 with 100 figs. (Bombyliidae p. 105.)

GMELIN, JOHANN FRIEDRICH
    1788–1793. Linné Systema Naturae. . . . Thirteenth edition. In 3 volumes, 10 parts. Edited by Gmelin.
        1790. Vol. 1, pt. 5, pp. 2225–3020. (Bombyliidae pp. 2830–2831, 2902, 2903.)

GOBERT, E.
    1887. Catalog des Diptères de France. Caen. 88 pp. (Bombyliidae pp. 20–22; 73 spp. listed.)

GOETGHEBUER, M.
    1931. Diptera of the plateau of Hautes-Fagnes. Bull. Ann. Soc. Ent. Belgique, vol. 71, pp. 171–182. (Bombyliidae p. 179.)

GOIN, FRANCOIS
    1950a. La maxille et son evolution chez les Diptères Brachyceres. Eighth Internat. Congress Ent., pp. 1–1030. Stockholm. Goin, pp. 551–553, 1 fig. (Bombyliidae 552.)
    1950b. Evolution de la region clypeo-cibariale chez les Diptères Brachyceres. Eighth Internat. Congress Ent., pp. 1–1030. Stockholm. Goin, pp. 554–556, 1 fig. (Bombyliidae p. 555.)

GOLDFUSS, GEORG AUGUST
    1831. [Note: *Hemipenthes*.] Acta Leop. Carol. Akad., Halle, vol. 7, no. 1, p. 118. (Bombyliidae p. 118, Fossil *Hemipenthes*.)

GONCALVES, CINCINNATO R.
    1946. *Systropus fumipennis* Westw. (Dipt. Bombyliidae). Parasita de *Miresa clarissa* (Stal.) (Lep. Eucleidae.) Livro de homenagem a R. F. d'Almeida, no. 19, pp. 119–204, pl. 9, figs. 1–5.

GONZALEZ-RINCONES, RAFAEL, and GUYON, LUISA
    1953. Classificacion general de los Dipteros. Universidad Central de Venezuela. Publ. Inst. Med. Exper. "Jose Gregorio Hernandez." Caracas. 239 pp. Braquiceros, pp. 50–166, figs. 1–102. (Bombyliidae p. 60, fig. 13.)

GOOD, RONALD
    1947. The geography of the flowering plants. London. 403 pp., 71 line drawings, 9 maps in color, 16 photogravure plates (refer to map of world regions).

GOWDEY, C. C.
    1926. Catalogus insectorum jamaicensis. Dept. Agric. Jamaica, Kingston, Ent. Bull., no. 4, pt. 1, pp. 1–114, xiv; pt. 2, pp. 1–10, ii. (Diptera pp. 72–90.)

GRAAF, HENRI WILLIAM DE
    1869. [Note on Bombyliidae.] Tijdschr. Ent., vol. 12, p. 192.

GRADL, HEINRICH
    1878. [Note: *Bombylius pictus* Panzer found on *Lamium purpureus*.] Biologisches und sonstiges. Ent. Nachricht, vol. 4, p. 238.

GRAENICHER, S.
    1909. Wisconsin flowers and their pollination. Compositae. Bull. Wisconsin Nat. Hist. Soc., vol. 7, pp. 19–77. (Bombyliidae p. 24.)
    1910a. Some new and rare Diptera from Wisconsin. Canadian Ent., vol. 42, pp. 26–29. (Bombyliidae pp. 26–28.)
    1910b. A preliminary list of the flies of Wisconsin belonging to the families Bombyliidae, Syrphidae and Conopidae. Bull. Wisconsin Nat. Hist. Soc., vol. 8, pp. 32–44. (Bombyliidae pp. 33–36; 32 spp. listed in 12 genera.)
    1910c. The bee-flies (Bombyliidae) in their relations to flowers. Bull. Wisconsin Nat. Hist. Soc., vol. 8, no. 2, pp. 91–101.
    1911. Wisconsin Diptera. A supplement to the preliminary list of Bombyliidae, Syrphidae and Conopidae. Bull. Wisconsin Nat. Hist. Soc., vol. 9, pp. 66–72. (Bombyliidae pp. 66–68; 18 spp.)
    1913. Records of Wisconsin Diptera. Bull. Wisconsin Nat. Hist. Soc., vol. 10 (1912), pp. 171–185. (Bombyliidae pp. 178–180.)

GRANDI, GUIDO
    1951. Introduzione allo studio dell'Entomologia Bologna. In 2 volumes. Vol. 1, pp. 1–950, figs. 1–780. (Bombyliidae p. 665.) Vol. 2, pp. 1–16, 1–332, figs. (Bombyliidae pp. 389, 392, 394–395, 396.)

GRASSE, P.
    1924. Les ennemis des Acridiens ravageurs français. Bordeaux. Rev. Zool. Agric. Appl., vol. 22, pp. 1–16, 45–53, 57–66.
    1964. Cycle biologique d'un Diptère Bombyliidae de genre *Villa*. Comp. Rend. Acad. Sci. Paris, vol. 31, pp. 1657–1659.

GRAVES, ROBERT C.
    1962. Predation on *Cicindela* by a dragon fly. Canadian Ent., vol. 94, p. 1231. (Comments upon a Bombyliid and Asilids.)

GRAY, GEORGE ROBERT
    1832. *In* Griffith: The animal kingdom arranged in conformity with its organization by the Baron Cuvier. London. New species of insects of all the orders, by Gray in vols. 14–15, 140 pls. Vol. 15, 769 pp. (Bombyliidae pp. 692–695, 753, 779, pl. 125, fig. 6; pl. 126, figs. 5–6; pls. 128, figs. 5–6. Eight genera notices.)

GREATHEAD, D. J. [see also Stower, Popov, and Greathead]
    1958a. A new species of *Systoechus* (Dipt., Bombyliidae) a predator on egg pods of the desert locust, *Schistocerca gregaria* (Forskal). Ent. Monthly Mag., vol. 94, pp. 22–23.
    1958b. Observations on two species of *Systoechus* (Diptera: Bombyliidae) preying on desert locust, *Schistocerca gregaria* (Forskal), in eastern Africa. Entomophaga, Paris, vol. 3, pp. 3–22, 29 figs., 1 photograph, 4 pls.
    1967. The Bombyliidae (Diptera) of northern Ethiopia. Journ. Nat. Hist., vol. 1, no. 2, pp. 195–284, illus., map.
    1969. Bombyliidae, and a first record of Nemestrinidae from Sokotra (Diptera). Bull. Brit. Mus. (Natur. Hist.) Ent., vol. 24, no. 3, pp. 67–82, illus.
    1970. Notes on Bombyliidae (Diptera) from the southern borderlands of the Sahara with descriptions of new species. Journ. Natur. Hist., vol 4, no. 1, pp. 89–118. Illus. Map.

GREENE, CHARLES TULL
    1921. A new genus of Bombyliidae (Diptera). Proc. Ent. Soc. Washington, vol. 23, no. 1, pp. 23–24, 1 fig.
    1924. New species of *Mythicomyia* and its relationship, with a new genus (Diptera). Proc. Ent. Soc. Washington, vol. 26, no. 3, pp. 60–64, 3 figs.

GRIFFINI, ACHILLE
    1896. Anthracidi del Piemonte. Studio monografico del Dott. Achille Griffini. Ann. Roy. Accad. Agric. Torino, vol. 39, pp. 3–50. [Also: Boll. Mus. Zool. Anat. Comp. Torino, vol. 11, no. 225, pp. 1–3.]

GRIFFITHS, K. J.
    1959. Observations on the European pine sawfly, *Neodiprion sertifer* (Geoff.), and its parasites in southern Ontario. Canadian Ent., vol. 91, pp. 501–512. (Bombyliidae p. 510.)

GRIMSHAW, PERCY H.
    1903–1904. Diptera Scotica III. The Fourth District. Ann. Scotland Nat. Hist.
        1903. Pp. 154–166, 213–266. (Bombyliidae p. 163.)
        1904. Pp. 27–33, 98–102.

GROTE, AUGUST RADCLIFFE
    1867. Description of two new species of North American Brachycerous Diptera. Proc. Ent. Soc. Philadelphia, vol. 6, p. 445 (Bombyliidae).

GUÉRIN-MÉNÉVILLE, FELIX EDOUARD
    1831–1839. Insectes. Diptères. *In* Duperrey, L. I.: Voyage autour du Monde La Coquille, 1822–1825. Zoologie.
        1838. Vol. 2, pt. 2, division 1, pp. i–vii, 1–319, 22 color pls. Insectes pp. 57–319. (Bombyliidae pp. 294–295, pl. 20, figs. 4, 9.)
    1831. Description du *Toxophora carcelii*. Guérin, Mag. Zool., vol. 1, no. 16, 1 pl. (color).

GUÉRIN-MÉNÉVILLE, FELIX EDOUARD—Continued
- 1829–1844 *In* Cuvier, G.: Iconographie du règne animal, du G. Cuvier. In 3 volumes commonly bound in 7 parts. Paris.
  - 1835. Atlas, 104 pls. (color). (Bombyliidae pl. 95, figs. 1, 4–6.)
  - 1844. Vol. 7, Insectes, 576 pp., 104 pls. Douzième ordre. Les Dipteres, pp. 531–559, 13 pls. (Bombyliidae pp. 537–539.)

GUNDLACH, JUAN (JOHANN)
- 1887. Fauna Puerto-Riqueña. Ann. Soc. Española Hist. Nat., vol. 16, pp. 174–200. (Bombyliidae pp. 180–181.)

GUYON, LUISA [see Gonzalez-Rincones and Guyon]

HAGEN, HERMANN AUGUST
- 1862–1863. Bibliotheca Entomologica. Leipzig.
  - 1862. Vol. 1, pp. 1–568.
  - 1863. Vol. 2, pp. 1–512.

HALIDAY, ALEXANDER HENRY
- 1833. Catalogue of Diptera occurring about Holywood in Downshire. Ent. Mag., London, vol. 1, pp. 147–180. (Bombyliidae p. 151; 1 sp.)
- 1851. Sendschreiben von Alexis H. Haliday an C. A. Dohrn über die Dipteren in London befindlichen Linneischen Sammlung. Stettiner Ent. Zeitung, vol. 12, pp. 131–145. (Bombyliidae pp. 136–138.)

HALL, JACK C. [see also Painter and Hall]
- 1952. A new species of *Lordotus* from southern California. (Bombyliidae: Diptera). Pan-Pacific Ent., vol. 28, no. 1, pp. 49–50.
- 1954a. Notes on the biologies of three species of Bombyliidae, with description of one new species. Ent. News, vol. 65, pp. 145–149.
- 1954b. A revision of the genus *Lordotus* Loew in North America (Diptera: Bombyliidae). Univ. California Publ. Ent., vol. 10, no. 1, pp. 1–33, 24 figs., 4 maps.
- 1956. A new species of *Anastoechus* Osten Sacken with notes on the congeners. Ent. News, vol. 67, pp. 199–203.
- 1957. Notes and descriptions of new California Bombyliidae (Diptera). Pan-Pacific Ent., vol. 33, pp. 141–148.
- 1958. A change of name in the bombyliid genus *Anastoechus* (Diptera). Ent. News, vol. 69, p. 195.
- 1967. A new spices of *Empidideicus* Becker from Texas (Diptera: Bombyliidae). Ent. News, vol. 78, pp. 85, 215–218.
- 1969. A review of the subfamily Cylleniinae with a world revision of the genus *Thevenemyia* Bigot (*Eclimus* auct.) (Diptera: Bombyliidae). Univ. California Publ. Ent., vol. 56, pp. 1–85, numerous illustr.
- 1972. New North American Heterotropinae (Diptera: Bombyliidae). Pan-Pacific Ent., vol. 48, pp. 37–50.

HALLOCK, H. C., and PARKER, L. B.
- 1926. Supplement to Smith's 1909 Diptera list. Circular 103, New Jersey Dept. Agric., pp. 1–20. (Bombyliidae p. 6.)

HAMMOND, CYRIL O. [see Colyer and Hammond]

HANDLIRSCH, ANTON
- 1906–1907. Die Fossilen Inseckten und die Phylogenie der rezenten Forman. Ein handbuch für Paläontologen and Zoologen. Leipzig: Englemann. In 9 parts, 1430 pp., 3 pls.
  - 1907. Pt. 7, pp. 961–1120, xxix–xl, pls. 37–51. (Bombyliidae pp. 1011–1012.)
- 1920. *In* Schroder: *Handbuch der Entomologie*. Jena. Paleontologie, vol. 3, chapter 7, pp. 117–306. (Bombyliidae p. 265; 5 subfamilies listed.) Systematisches Uebersicht, vol. 3, pp. 377–1143. (Bombyliidae pp. 989–993, fig. 899.)

HANSEN, HANS JACOB
- 1883. Fabrico oris dipterorum: Dipterernes Mund i anatomisk og systematisk Henseende. I. (Tabanidae, Bombyliidae, Asilidae, Thereva, Mydas, Apiocera.) Naturhist. Tidsskr., Kjøbenhavn, ser. 3, vol. 14, 220 pp., 15 pls. (Bombyliidae pp. 98–116, pl. 2, figs. 18–29; pl. 3, figs. 1–8.)

Hardy, D. Elmo
- 1960. Insects of Hawaii. Diptera: Nematocera-Brachycera. Univ. Hawaii Press, Honolulu. Vol. 10, pp. 1–7, 1–368, figs. 1–120. (Bombyliidae pp. 7, 319–322, fig. 113.)

Hardy, George Huddleston Hurlstone
- 1918. Notes on Tasmanian Diptera and description of new species. Papers Proc. Roy. Soc. Tasmania, vol. for 1917, pp. 60–66. (Bombyliidae p. 66.)
- 1920. Beeflies and their habits: The speediest creatures in the world. Scientific Australian, vol. 26, pp. 5–7, figs. 1–4.
- 1922a. The geographical distribution of genera belonging to the Diptera Brachycera of Australia. Australian Zool., vol. 2, pt. 4, pp. 143–147. (Bombyliidae p. 146.)
- 1922b. Australian Bombyliidae and Cyrtidae (Diptera). Papers Proc. Roy. Soc. Tasmania, 1921, pp. 41–83, pls. 16–17, figs. 1–19. (Bombyliidae pp. 41–75, pls. 16–17.)
- 1924. Notes on Australian Bombyliidae, mostly from the manuscript papers of the late Arthur White. Papers Proc. Roy. Soc. Tasmania, 1923, pp. 72–86.
- 1927a. On the phylogeny of some Diptera Brachycera. Proc. Linnean Soc. New South Wales, vol. 52, pt. 3, pp. 380–386, 2 figs. (Bombyliidae, scattered comments on pp. 380, 382–385.)
- 1927b. Critical remarks on *Pachyneres australis* Malloch, and its possible identity with Bibionidae. Australian Zool., vol. 4, pp. 337–338.
- 1928. Some further remarks on *Pachyneres australis* Mall. (appended to Malloch's paper of this title). Australian Zool., vol. 5, pp. 139–140.
- 1929. On the type locality of certain flies described by Macquart in "Diptères Exotiques," supplement four. Proc. Linnean Soc. New South Wales, vol. 54, pp. 61–64.
- 1933–1949. Miscellaneous notes on Australian Diptera. Proc. Linnean Soc. New South Wales.
    - 1933. (No subtitle.) Pt. I, vol. 58, pp. 408–420, figs. 1–4. (Bombyliidae pp. 114–116.)
    - 1941. Subfamily Lomatiinae. Pt. VIII, vol. 66, pts. 3–4, pp. 223–233, 4 figs.
    - 1942. Superfamily Asiloidea. Pt. IX, vol. 67, pp. 1970–204. (Bombyliidae pp. 197–202–204.)
    - 1944. Distribution, classification, and the *Tabanus posticus*-group. Pt. X, vol. 69, pp. 76–86, 1 fig. (Bombyliidae p. 84.)
    - 1945. Evolution of characters in the order; venation of the Nemestrinidae. Pt. XI, vol. 70, pp. 135–146, figs. 1–2. (Bombyliidae pp. 139, 140, 141.)
    - 1947. The origin of the *vena spuria*. Pt. XIII, vol. 72, pp. 229–232. (Bombyliidae p. 229.)
    - 1949. Venation and other notes. Pt. XIV, vol. 73, pp. 298–303, figs. 1–6. (Bombyliidae pp. 299, 300, figs. 3–5.)
- 1944. The copulation and the terminal segments of Diptera. Proc. Roy. Ent. Soc. London, ser. A, vol. 19, pts. 4–6, pp. 52–65. (Bombyliidae pp. 59, 61.)
- 1945. On flies that fold their wings. Ent. Monthly Mag., vol. 81, pp. 93–94. (Bombyliidae p. 94.)
- 1950. The twisting segments in Diptera. Ent. Monthly Mag., vol. 86, pp. 346–347.
- 1951a. Evolutionary trends in Diptera. Ent. Monthly Mag., vol. 87, pp. 56–59. (Bombyliidae pp. 56–57.)
- 1951b. Theories of the world distribution of Diptera. Ent. Monthly Mag., vol. 87, pp. 99–102. (Bombyliidae p. 100.)
- 1951c. The reticulation theory of wing venation in Diptera. Journ. Soc. British Ent., vol. 4, pt. 2, pp. 27–36, figs. 1–4. (Bombyliidae pp. 29, 30, 34, 35, fig. 4, pts. 3, 5.)
- 1951d. The phylogeny of Diptera. Ent. Monthly Mag., vol. 87, pp. 140–141.
- 1953. The evolution of antennae in the Diptera. Ent. Monthly Mag., vol. 89, pp. 79–80. (Mentions Brachycera.)

Hardy, George Huddleston Hurlstone—Continued
- 1954. Reduction of the median field in the wing venation of Diptera. Ent. Monthly Mag., vol. 90, pp. 2–3. (Bombyliidae mentioned p. 2.)
- 1955. The first median vein in the wings of Diptera. Ent. Monthly Mag., vol. 91, pp. 197–198. (Bombyliidae mentioned p. 198.)
- 1956a. The superfamily unit used in Diptera. Ent. Monthly Mag., vol. 92, pp. 213–215. (Bombylioidea discussed.)
- 1956b. The wing venation of Lomatiinae (Diptera-Bombyliidae). Proc. Linnean Soc. New South Wales, vol. 81, pp. 78–81, 1 fig.

Harned, Robey Wentworth
- 1921. Report of the Entomology Department, in Mississippi Agric. Exper. Stat., Ann. Report, no. 34, pp. 27–32. (Bombyliidae p. 30; *Poecilanthrax lucifer* listed as a parasite of *Laphygma frugiperda*, fall armyworm.)

Harrington, W. Nague
- 1900. Fauna Ottawaensis. Diptera. Ottawa Natur., vol. 14, pp. 127–135. (Bombyliidae, p. 130; 8 spp. listed.)

Harris, Moses
- 1782. An exposition of English insects. Third edition (earlier editions 1776 and 1781). London. Pp. i–viii, 1–166, 4 pp. index, 50 pls. (color). (Bombyliidae pl. 47 has 2 recognizable figs. of 2 spp. of *Bombylius*, without descriptions or comments.)

Harris, Thaddeus William
- 1841. A report on the insects of Massachusetts injurious to vegetation. Cambridge. Pp. 1–8, 1–459. 1842 (new title), A treatise on some of the insects of New England, which are injurious to vegetation. Cambridge. Same pages. (Bombyliidae p. 406.) Edition 2, 1852, pp. 1–8, 1–513. Boston. Edition 3, 1862, pp. 1–11, 1–640, 8 pls., 278 wood-cuts. (Bombyliidae p. 606, fig. 263.) New edition, 1880, same pages as third edition. (Bombyliidae pp. 603–604, fig. 263.)

Harrison, L.
- 1928. The composition and origin of the Australian fauna. 18th Meeting Australasian Assoc. Adv. Sci., vol. 18, pp. 332–396. (Has large bibliography.)

Hart, Charles A., and Gleason, Henry Allan
- 1907. On the biology of the sand areas of Illinois. Bull. Illinois State Lab. Nat. Hist., vol. 7, art. 7, pp. 137–272, 1 map, pls. 8–22. (Bombyliidae pp. 225, 250.)

Hayes, W. P.
- 1939. A bibliography of keys for the identification of immature insects. Part I, Diptera. Ent. News, vol. 49, pp. 246–251; vol. 50, pp. 5–10, 76–82.

Hegh, E. [See also Austen and Hegh]
- 1929. Les Tsé-Tsés. Imprimerie Industreille et Financière, Ministry des Colonies. Bruxelles. Pp. 1–4, 1–742, 15 color pls., 327 figs. (Bombyliidae pp. 712–714, figs. 320–321.)

Heikertinger, Franz
- 1919–1920. Exakte Begriffsfassung und terminologie in problem der Mimikry und verwandten erscheinungen. Zeitschr. Wiss. Insektenbiologie, Berlin. 1919, vol. 15, pp. 57–64; 1920, vol. 15, 162–174.
- 1921. Die morphologisch-analytische methode in der Kritik der Mimikryhypothese, dargelegt an der wespenmimikry (Sphekoidie) der Bockkäfer. Zool. Jahrb. Jena, Abt. f. syst., vol. 44, pp. 267–296, 1 pl.

Helm, Otto
- 1896. Beiträge zur Kenntniss der Insecten der Bernsteins. Schrift. Naturf. Ges. Danzig, vol. 9, pp. 220–231. (Of general interest; Bombyliidae not specifically mentioned.)

Hendel, Friedrich
- 1908. [Reprint of Meigen: Nouvelle classification des mouches à deux ailes. Notes by Hendel.] Verh. zool.-bot. Ges. Wien, vol. 58, pp. 43–69. (Bombyliidae pp. 57–58.)
- 1928. *In* Dahl, F., editor: Die Tierwelt Deutschlands und der angrenzenden Meerestiele. Jena: Gustav Fisher. Zweiflügler oder Diptera. II. Allgemeiner Teil. Vol. 11, pp. 1–135, figs. 1–224. (Bombyliidae p. 36.)

HENDEL, FRIEDRICH, and BEIER, MAX
    1936. *In* Kükenthal and Krumbach: Handbuch der Zoologie. Berlin. Vol. 4, in 2 pts., pp. 1–14, 1729–2756. Ordnung der Pterygogenea (Diptera). Pt. 2, (8–10), pp. 1729–1980, figs. 1–304. (Bombyliidae pp. 1925–1927, fig. 2052.)

HENNEGUY, L. FELIX
    1904. Les insectes morphologie, reproduction, embryogenie. Paris. 804 pp., 4 pls., 622 figs. (Bombyliidae p. 192, eyes; p. 454, brief comment on mouthparts.)

HENNIG, WILLI [see also Emden and Hennig]
    1936. Beziehungen zwischen geographischer Verbreitung und systematischer Gliederung bei einigen Dipterenfamilien: ein Beitrag zum Problem der Gliederung systematischer Kategorien höherer Ordnung. Zool. Anz., vol. 116, pp. 161–175, 3 figs.
    1941. Verzeichnis der Dipteren von Formosa. Ent. Beihefte Berlin-Dahlem, vol. 8, pp. 1–4, 1–239, figs. 1–35. (Bombyliidae pp. 67–70, figs. 1–10.)
    1948–1952. Die Larvenformen der Dipteren. In 3 parts. Berlin: Akademie-Verlag.
        1949. Pt. 1, 185 pp. 3 pls., 63 figs. (Bombyliidae p. 71; of general interest.)
        1952. Pt. 3, 628 pp., 21 pls., 338 figs. Bibliography pp. 517–616. (Bombyliidae pp. 82–83, 92–101, figs. 56–62, pl. 4, figs. 3–4; pl. 20, fig. 3.)
    1950. Grundzüge einer Theorie der phylogenetischen Systematik. Berlin. 370 pp.
    1954. Flügelgeäder und System der Dipteren unter Berücksichtigung der aus dem Mesozoikum beschreibenen Fossilien. Beiträge zur Ent. Deutsches Ent. Inst., vol. 4, nos. 3–4, pp. 245–388, 272 figs. (Bombyliidae pp. 350–354, 373–374, figs. 211–222, 266, 272; also pp. 336, 347, 357.)
    1960. Die Dipteren-Fauna von Neuseeland als systematisches und tiergeographisches Problem. Beiträge zur Ent., vol. 10, nos. 3–4, pp. 221–329. (Bombyliidae pp. 294–295.)
    1966. Bombyliidae im Kopal und im Baltischen Bernstein. Stuttgarter Beiträge zur Naturkunde, vol. 166, pp. 1–20, 27 figs.
    1967. Therevidae aus dem Baltischen Bernstein mit einigen Bemerkunge über Asilidae und Bombyliidae (Diptera Brachycera). [Therevidae from the Baltic amber with some remarks on Asilidae and Bombyliidae (Diptera: Brachycera.)] Stuttgarter Beiträge Naturkunde, vol. 176, pp. 1–14, illus., map.
    1968. Phylogenetic systematics. Translated by D. Dwight Davis and Rainer Zangerl. 263 pp., illus., Univ. Illinois Press.
    1969. Kritische Betrachtungen uber die phylogenetische bedeutung von Bernsteinfossilien: Die Gattungen *Proplatypygus* (Diptera, Bombyliidae) und *Palaeopsylla* (Siphonaptera). Estratto dalle Memorie della Società Entomologica Italiana. Vol. of the Centenario 48, pp. 57–67.

HERBST, JOHANN FRIEDRICH WILHELM
    1784–1787. Kurze Einleitung zur Kenntniss der Insecten für Ungeübte and Anfänger. In 3 volumes with 144 pls. Berlin. And *in* Borowski (as volumes 6–8): Gemeinnützige Naturgeschichte des Thierreichs.
        1787. Vol. 8, 200 pp., 48 pls. (Bombyliidae pp. 100–110, 120–122, pl. 347, fig. 1; pl. 349, fig. 1; also in other editions pl. 39, fig. 1; pl. 47, fig. 1; pl. 49, fig. 1.)

HERMANN, FREDRICH
    1907. Einige neue Bombyliden der palaearktischen Fauna. (Dipt.). Zeitschr. syst. Hymen. Dipt. Teschendorf, vol. 7, pp. 193–202, figs. 1–3.
    1909. *In* Kneucker: Zoologische Ergebnisse zweier in den Jahren 1902 und 1904 durch die Sinaihalbinsel unternommener botanischer Studienreise, nebst zoologischen Beobachtungen aus Aegypten, Palastina und Syria. Verh. Naturwiss. Ver. Karlsruhe, vol. 21 (1907–1908), pp. 147–160. (Bombyliidae pp. 148–151.)

HERRICK, G. W. [see Comstock]

HESSE, ALBERT J.
    1936. Scientific results of the Vernay-Lang Kalahari expedition, March to September, 1930. Bombyliidae (Diptera). Ann. Transvaal Mus., Pretoria, vol. 17, pp. 161–184, 5 figs.

Hesse, Albert J.—Continued
- 1938–1956. A revision of the Bombyliidae (Diptera) of Southern Africa. I-III. Ann. South African Mus., Cape Town.
  - 1938. Part I, vol. 34, 1053 pp., 332 figs.
  - 1956a. Part II, vol. 35 464 pp., 170 text-figs., pl. 1, 17 figs. (wings).
  - 1956b. Part III, vol. 35, pp. 465–947, index pp. 949–972, text-figs. 171–286, pl. 2, figs. 1–12 (wings); pl. 3, 6 figs. (wings).
- 1950. Some Bombyliidae in the Museu Dr. Alvaro de Castro, collected by Dr. A. Maria-Corinta Ferreira in Mozambique. Mem. Mus. Alvaro de Castro, Lourenço Marques, no. 1, pp. 21–34.
- 1952. Contribution à l'étude des Diptères de l'Urundi. II. Bull. Inst. Sci. nat. Belgique, vol. 28, no. 42, pp. 1–7, 4 figs. (Bombyliidae pp. 1–4, figs.)
- 1955. Diptera Bombyliidae *in* Hanström, Brinck, and Rudebeck: South African animal life, results of the Lund University Expedition in 1950–1951. Vol. 2, pp. 382–401, 2 figs.
- 1958. Exploration Parc nat. Upemda Miss. de Witte (1946–1949). Brussels. Bombyliidae (Diptera, Brachycera). Fasc. 50, pp. 69–79.
- 1960a. A new eastern province representative of the monotypic genus *Oniromyia* Bezzi (Diptera: Bombyliidae). Journ. Ent. Soc. South Africa, vol. 23, pp. 284–285.
- 1960b. Diptera, Bombyliidae. Mission zoologique de l'I. R. S. A. C. en Afrique orientale (P. Basilewsky et N. Leleup, 1957). Ann. Mus. Congo Belgique, Tervuren, ser. 8, Sci. Zool., vol. 88, pp. 315–317.
- 1960c. Supplementary contributions to the revision of the Bombyliidae (Diptera) of southern Africa. Rev. Fac. Cienc. Lisboa, ser. 2b, C, vol. 8, fasc. 1, pp. 51–95.
- 1962. *Apolysis lindneri* sp. nov., eine neue Bombyliida aus Sudafrika (Dipt.). Stuttgarter Beitrage zur Naturkunde, no. 80, pp. 1–2, 1 fig. (Jan. 15.)
- 1963a. New species of *Crocidium* and *Toxophora* (Bombyliidae). Ann. Natal Mus., Dorking, vol. 15, pp. 273–295. Diptera.
- 1963b. Supplementary contribution to the revision of the Bombyliidae (Diptera) of southern Africa: the genus *Systropus*. Ann. S. African Mus., Cape Town, vol. 46, pp. 393–405.
- 1965. Diptera (Brachycera) Bombyliidae, Cyrtosiinae, South African Animal Life, xi, pp. 482–484.
- 1967. Additions to the Cyrtosiinae (Bombyliidae) of South Africa. Ann. S. African Mus., 50 pp. 89–130, 8 text figs.

Hicks, Charles H. [see also Custer and Hicks]
- 1926. Nesting habits and parasites of certain bees of Boulder County, Colorado. Univ. Colorado Studies, vol. 15, pp. 217–252. (Bombyliidae p. 250.)

Hincks, Walter Douglas [see Kloet and Hincks]

Hine, James Stewart
- 1904. The Diptera of British Columbia. I. Canadian Ent., vol. 36, pp. 85–92. (Bombyliidae pp. 88–90, 7 genera, 14 spp., 1 n. sp.)
- 1908. Diptera of the 1905 University Museum Expedition to Isle Royal. Michigan Survey, pp. 308–316. (Bombyliidae p. 311; 2 spp. listed.)

Hoffmansegg, Graf Johann Centurius, von
- 1820. *In* Meigen: Systematische Beschriebung der bekannten europäischen zweiflügeligen Insecten. (Scattered references in volume 2.)

Holmes, S. J.
- 1913. Note on the orientation of *Bombylius* to light. Science, vol. 38, no. 972, p. 230.

Hood, Lewis E.
- 1892. The Leptidae and Bombyliidae of the White Mountains. Psyche, vol. 6, pp. 283–284. (Bombyliidae p. 284.)

Hori, H. [see Esaki, Hori, Hozana, Shiraki, others]

Horn, W., and Kahle, I.
 1935–1937. Ueber entomologische Sammlungen (ein Beiträge zur Geschichte der Entomo-Museologie). Ent. Beihefte Berlin.
 1935. Vol. 2, pp. 1–160, 16 pls.
 1936. Vol. 3, pp. 161–196, 10 pls.
 1937. Vol. 4, pp. 297–536, 12 pls.

Horn, Walther, and Schenkling, Sigm
 1928. Index literaturae entomologicae. Ser. I. Die Weltliteratur über die gesamte Entomologie bis inklusiv 1863. Berlin-Dahlem. Vol. I, pp. 1–352; vol. 2, pp. 353–704; vol. 3, pp. 705–1056; vol. 4, pp. i–xxi, 1057–1426.

Horsfall, W. R. [see Gilbertson and Horsfall]

Howard, Leland Ossian [see also Riley and Howard]
 1892. Note on the hibernation of carpenter bees. Proc. Ent. Soc. Washington, vol. 2, no. 3, p. 331.
 1901. The Insect Book. A popular account of the bees, wasps, ants, grasshoppers, flies, and other North American insects exclusive of the butterflies, moths and beetles. New York: Doubleday Page & Co. Second edition, 1904, New York. xxvii + 429 pp., 264 text-figs., 48 pls. (Bombyliidae pp. 137–138, text-figs. 76–78, pl. 17, figs. 1–9, 11; pl. 18, figs. 10, 13–22, 24–26, 29.)

Hoyt, C. P.
 1952. The evolution of the mouthparts of adult Diptera. Microentomology, vol. 17, pt. 3, pp. 61–125, figs. 38–72. (Bombyliidae pp. 78, figs. 68a-b.)

Hozana, S. [see Esaki, Hori, Hozana, Shiraki, others]

Hudson, G. V.
 1892. An elementary manual of New Zealand entomology. London. 128 pp., 29 pls. (color). (Bombyliidae pp. 54–55, pl. 6, fig. 3.)

Hull, Frank Montgomery
 1965. Notes and descriptions of Bombyliidae (Diptera). Ent. News, vol. 76, pp. 95–97.
 1966. Notes on the genus *Neodiplocampta* Curran and certain other Bombyliidae. Ent. News, vol. 77, pp. 225–227.
 1970. Some new genera and species of bee flies from South America (Diptera: Bombyliidae). Journ. Georgia Ent. Soc., vol. 5, pp. 163–166.
 1971a. A new genus and new species of bee flies. Proc. Ent. Soc. Washington, vol. 73, pp. 181–183.
 1971b. Some new genera and species of bee flies (Diptera: Bombyliidae) from South America and Australia. Journ. Georgia Ent. Soc., vol. 6, pp. 1–7.

Hungerford, H. B., and Williams, Francis X.
 1912. Biological notes on some Kansas Hymenoptera. Ent. News, vol. 23, pp. 241–260. (Bombyliidae p. 259; comment on protective value of clay arches.)

Hurd, Paul D., Jr.
 1947. Redescription of *Agenioideus humilis* (Cresson) with notes on its biology (Hymenoptera, Pompilidae). Pan-Pacific Ent., vol. 23, pp. 132–134. (Bombyliidae p. 132; unidentified sp. parasite reared.)
 1959. Beefly parasitism of the American carpenter bees belonging to the genus *Xylocopa* Latreille (Diptera; Hymenoptera). Journ. Kansas Ent. Soc., vol. 32, pp. 53–58, 1 fig.

Hurd, P. D., Jr., and Linsley, E. Gorton
 1950. Some insects associated with nests of *Dianthidium dubium dilectum* Timberlake, with a list of the recorded parasites and inquilines of *Dianthidium* in North America. Journ. New York Ent. Soc., vol. 58, pp. 247–250. (Bombyliidae pp. 247–249.)

Hutton, Frederic Wollaston
 1881. Catalogues of the New Zealand Diptera, Orthoptera, Hymenoptera, with description of species. Wellington. 132 pp., Diptera pp. 5–70. (Bombyliidae p. 24.)
 1901. I. Synopsis of the Diptera Brachycera of New Zealand. Trans. New Zealand Inst., vol. 33, pp. 1–95. (Bombyliidae pp. 23, 24–25.)
 1904. Index faunae Novae Zealandiae. London. 372 pp. (Bombyliidae p. 131.)

Hynes, H. B. N.
1947. Observations on *Systoechus somali* (Diptera, Bombyliidae) attacking the eggs of the desert locust (*Schistocerca gregaria* Forskål) in Somalia. Proc. Roy. Ent. Soc. London, ser. A., vol. 22, pp. 79–85.

Hyslop, J. A.
1912. Notes from the Pacific Northwest. (Hymenoptera.) Proc. Ent. Soc. Washington, vol. 14, pp. 100–101. (Bombyliidae p. 101.)
1915. Observations on the life history of *Meracantha contracta* (Beauv.). Psyche, vol. 22, pp. 44–48, 2 text-figs., 1 pl. (Bombyliidae p. 47, text-fig. 2.)

Imhoff, Ludwig
1834. Puppenhülle von *Bombylius major*, L. Isis, vol. 5, columns 536–537, pl. 12.

Ingenitsky, I. V.
1898. On the Acrididae of eastern Siberia and their parasites. [In Russian.] Horae Soc. Ent. Rossia, vol. 32, p. liii.

Ionescu, M. A., and Weinberg, Médée
1962. Contribution à l'étude des diptères da la R.P.R. (Fam. Asilidae; Fam. Bombyliidae). Trav. Mus. hist. Nat. "Gr. Antipa," vol. 3, pp. 183–203, figs. 1–21. (Bombyliidae pp. 192–203, figs. 16–21.)
1963. Ergebnisse der Albanien-Expedition 1961 des Deutschen Entomologischen Institutes. 17. Beitrag. Diptera: Bombyliidae. Beiträge zur Ent., vol. 13, no. 7/8, pp. 842–854.
1970. Diptere din colectia muzeului Bruckenthal-sibiu (Fam. Bombyliidae). Comun. Zool., vol. 4, pp. 37–46, illus.

Irwin-Smith, Vera
1923. Studies in life-histories of Australian Diptera Brachycera. Part 2. Proc. Linnean Soc. New South Wales, vol. 48, pp. 368–380. Bibliography pp. 370–374.

Iseley, Dwight
1913. The biology of some Kansas Eumenidae. Bull. Kansas Univ. Sci., vol. 8, no. 7 (whole ser. vol. 18, no. 7), pp. 235–309, pl. 34–37. (Bombyliidae p. 240.)

Iwata, K.
1933. Studies on the nesting habits and parasites of *Megachile sculpturalis* Smith (Hymenoptera, Megachilidae). Mushi, Fukuoka, vol. 6, pp. 6–24, 2 pls. 2 figs. (Bombyliidae pp. 17–19.)

Jacobi, A.
1913. Mimikry und verwandte Ercheinungen. Samml. Die Wissenschaft. Braunschweig, pp. i–iv, 1–215. (Reviewed by Prochnow, O.: Zeitschr. Wiss. Insektenbiologie, vol. 10, pp. 33–36.

Jacobs, J. C.
1906. Diptères de la Belgique. IV. Suite Mem. Soc. Ent. Belgique, vol. 12, pp. 24–75. (Bombyliidae pp. 26–27; 20 spp. listed.)

Jacquelin du Val, Pierre Nicolas Camille (Jacquelin-Duval)
1851. [Note: *Argyramoeba anthrax* (*sinuata*) Meigen.] Bull. Soc. Ent. France, ser. 2, vol. 9, p. lxxx.

Jaennicke, Johann Friedrich
1867a. Neue exotische Dipteren. . . . Abdruck. Abhandl. Senckenberg. Naturg. Gesellsch., Frankfurt, vol. 6, pp. 311–408, pls. 43–44. (Bombyliidae pp. 336–352, pl. 43, figs. 8–11; pl. 44, figs. 15–21.)
1867b. Beiträge zur Kenntniss der europäischen Bombyliden, Acroceriden, Scenopiniden, Thereviden und Asiliden. Berliner Ent. Zeitschr., vol. 11, pp. 63–94. (Bombyliidae pp. 63–76.)

Jannone, G.
1934. Osservazioni ecologiche e biologiche sul *Dociostaurus macroccanus* Thunb., *Calliptamus italicus* L. e loro parassiti in provincia di Napoli. Boll. Lab. Zool. Gen. Agric. Portici, vol. 28, pp. 75–151, 15 figs. (Spoleta, 1935.) (Bombyliidae pp. 144–154.)

Jansson, Anton
1922. Faunistika och biologiska studier över insektlivet vid Hornsjön på norra Öland. Ark. Zool. Stockholm, vol. 14, no. 23, 81 pp. (Bombyliidae p. 42.)

JAROSCHEFF, V. A.
 1877. Dopolnenie k spisku dvukrylykh nasiekomykh Khar'kova i ego okrestnosteĭ s ukazaniem rasprostraneniia ikh v predielakh Rossii. Trudy Kharkoff, vol. 11. (Bombyliidae pp. 345, 349.)

JAYNES, H. A. [see Clausen, Jaynes, and Gardner]

JAZYKOV, AL. A. [see Zakhvatkin]

JOHNSON, CHARLES WILLISON
 1894. List of the Diptera of Jamaica, with descriptions of new species. Proc. Acad. Nat. Sci. Philadelphia, 1894, pp. 271–281. (Bombyliidae pp. 274–275.)
 1895. Diptera of Florida, with additional descriptions of new genera and species by D. W. Coquillett. Proc. Acad. Nat. Sci. Philadelphia, vol. 47, pp. 303–304. (Bombyliidae pp. 325–326.)
 1898. Diptera collected by Dr. A. Donaldson Smith in Somaliland, Eastern Africa. Proc. Acad. Nat. Sci. Philadelphia, 1898, pp. 157–164, text-figs. 1–3. (Bombyliidae pp. 158–159; 3 spp. listed.) (Misspelled Bombycidae.)
 1900. *In* Smith: Insects of New Jersey. Twenty-seventh annual report, New Jersey State Board of Agriculture for 1899. Trenton. 755 pp. Insects in supplement. Diptera pp. 615–699. (Bombyliidae pp. 646–650, figs. 303–305.)
 1902. New North American Diptera. Canadian Ent., vol. 34, pp. 240–242. (Bombyliidae pp. 240–241.)
 1903. Descriptions of three new Diptera of the genus *Phthiria*. Psyche, vol. 10, pp. 184–185.
 1904. Some notes, and descriptions of four new Diptera. Psyche, vol. 11, pp. 15–20. (Bombyliidae p. 16.)
 1907. A review of the species of the genus *Bombylius* of the eastern United States. Psyche, vol. 14, pp. 95–100.
 1908a. Notes on New England Bombyliidae, with a description of a new species of *Anthrax*. Psyche, vol. 15, pp. 14–15.
 1908b. The Diptera of the Bahamas, with notes and description of one new species. Psyche, vol. 15, pp. 69–80. (Bombyliidae p. 72; 5 genera, 8 spp.)
 1910a. *In* Smith: Report of the insects of New Jersey. Report of New Jersey State Museum (1909). Trenton. Curator's report, pp. 1–14. Part 1, Insects, their classification and distribution, pp. 15–32. Part 2, Systematic list, pp. 33–888. Diptera pp. 703–814, figs. 293–340, by Johnson, C. W., Beutenmuller, William, Grossbeck, John A., Daecke, V. A. E., Greene, G. M., Greene, C. T., and Harbeck, Henry S. (Bombyliidae pp. 744–747, figs. 305–307; 14 genera, 43 spp. listed.)
 1910b. Some additions to the dipteran fauna of New England. Psyche, vol. 17, pp. 228–235. (Bombyliidae p. 229.)
 1913. Insects of Florida. I. Diptera. Bull. Amer. Mus. Nat. Hist., vol. 32, art. 3, pp. 37–90. (Bombyliidae pp. 54–58.)
 1914. The discovery of *Eclimus harrisi* in the White Mountains, N. H. Psyche, vol. 21, p. 123.
 1919a. New species of the genus *Villa* (*Anthrax*). Psyche, vol. 26, pp. 11–13.
 1919b. A revised list of the Diptera of Jamaica. Phoridae by C. T. Brues. Bull. Amer. Mus., vol. 41, art. 8, pp. 421–449. (Bombyliidae pp. 428–429.)
 1921. New species of Diptera. Occas. Papers Boston Soc. Nat. Hist., vol. 5, pp. 11–17, 2 figs. (Bombyliidae p. 12, fig.)
 1925a. Diptera of the Harris collection. Proc. Boston Soc. Nat. Hist., vol. 38, no. 2, pp. 57–99. (Bombyliidae pp. 72–73; 14 spp. listed.)
 1925b. Fauna of New England. Pt. 15. List of the Diptera of two-winged flies. Occas. Papers Boston Soc. Nat. Hist., vol. 7, pp. 1–326. (Bombyliidae pp. 108–111; 45 spp. listed.) Bibliography with 582 entries.
 1927. *In* Proctor: Biological survey of the Mount Desert Region. Pt. 1, The insect fauna, 247 pp. (Bombyliidae pp. 180–181.)

JOHNSON, D. ELMER, and JOHNSON, LUCILE MAUGHAN
 1957. New *Poecilanthrax*, with notes on described species (Diptera: Bombyliidae). Great Basin Nat., vol. 17, nos. 1–2, pp. 1–26, pl. 1 with 14 figs.

Johnson, D. Elmer, and Johnson, Lucile Maughan—Continued
- 1958. New and insufficiently known *Exoprosopa* from the Far West (Diptera: Bombyliidae). Great Basin Nat., vol. 18, pp. 69–84, 2 pls. with 28 figs.
- 1959. Notes on the genus *Lordotus* Loew, with descriptions of new species (Diptera: Bombyliidae). Great Basin Nat., vol. 19, pp. 9–26.
- 1960. Taxonomic notes on North American beeflies, with descriptions of new species (Diptera: Bombyliidae). Great Basin Nat., vol. 19 (1959), no. 4 pp. 67–74, 8 figs.

Johnson, D. Elmer, and Maughan, Lucile
- 1953. Studies in Great Basin Bombyliidae. Great Basin Nat., vol. 13, pp. 17–27, 9 figs.

Johnson, Lucile Maughan [see Johnson and Johnson]

Jones, C. G.
- 1940. Empididae: A. Hybotinae, Ocydromiinae, Clinocerinae and Hemerodromiinae. Ruwenzori Expedition 1934–5. British Mus. (Nat. Hist.), London, vol. 2, pp. 257–323, 2 pls., 17 figs. (Bombyliidae p. 323.)

Karl, O.
- 1930. Fliegen von der Insel Amrum. Ein Beitrag zur Fliegenfauna der nordfriesischen Inseln. (Dipt.). Diptera of Amrum I. Germany. Deutsche ent. Zeitschr., 1930, pp. 193–206. (Bombyliidae p. 194; 1 sp. listed.)
- 1935–1937. Die Fliegenfauna Pommerns. Diptera Brachycera. In 5 parts. Stettiner Ent. Zeitung.
    - 1935. Vol. 96, pp. 106–130, 242–261. (Bombyliidae pp. 115–116; 17 spp. listed.)

Karsch, Ferdinand Anton
- 1880. Die Spaltung der Dipterengattung *Systropus* Wiedemann. Zeitschr. Berliner Ent. Ges. Naturwiss., ser. 3, vol. 5 (LIII), pp. 654–658.
- 1886. Dipteren von Pungo-Andongo gesammelt von Herrn Major A. von Homeyer. Ent. Nachrichten, vol. 12, pp. 49–58, 257–264, 337–342. (Bombyliidae pp. 53–56.)
- 1887a. Bericht über die durch Herrn Lieutenant Dr. Carl Wilhelm Schmidt in Ost-Afrika gesammelten und von der Zoologischen Abtheilung des Königlichen Museums für Naturkunde in Berlin erworbenen Dipteren. Berliner Ent. Zeitschr., vol. 31, pp. 367–382, pl. 4. (Bombyliidae pp. 372–373, pl. 4, figs. 7–8.)
- 1887b. Dipterologisches von der Delagoabai. Ent. Nachrichten, vol. 13, pp. 22–25. (Bombyliidae pp. 24–25.)

Kaufman, Z. S. [see Zaitsev and Kaufman]

Keene, Eugene
- 1885. List of Diptera taken in the vicinity of Philadelphia from 1882–1884, inclusive. Canadian Ent., vol. 17, pp. 51–55. (Bombyliidae p. 53; 8 spp. listed.)

Keferstein, Ch.
- 1834. Die Naturgeschichte des Erdkörpers in ihren ersten Grundzügen dargestellt. Erster Theil. De Physiologie der Erde und Geognosie. Zweiter Theil. Die Geologie und Paläontologie, 896 pp. Die fossilen Insecten, pp. 325–347 (Chapter 6). E. Familie der fossilen Zweyflügler, Diptera, Flöhe, Fliegen, Bremsen, pp. 334–337. (Bombyliidae p. 337, "b. Sippf. Bombylico. Schwebfliegen. 2. *Bombilus*. (Schwebfliege) by Behrendt. Aus Bernstein. k. Sippf. Syrphoides. 38. Anthrax, (Hohrenfliege) der Art *semiatra* ähnlich. Burmstr. (Burmeister author). Aus Bernstein; auch in Schiefer von Oeningen.")

Keiser, Fred
- 1947. 18. Die Fliegen des Schweizerischen Nationalparks und seiner Umgebung. Pars I: Brachycera Orthorhapha. Ergebnisse wissenschaft. Untersuchung schweizerischen Nationalpark, Komm. Schweizerischen Naturforschenden Gesell. wissenschaft. Erforschung der Nationalparks. Vol. 2, 198 pp., 23 figs., 8 maps, 3 tables, 1 large map. Bibliography pp. 189–195. (Bombyliidae pp. 100–114, figs. 13–15.)

KELLY, E. G. [see Smith, Kelly, Dean, Bryson, and Parker]
KERTESZ, KOLOMAN
- 1901a. Neue und bekannte Dipteren in der Sammlung des Ungarischen National-Museums. Termés. Füzetek, vol. 24, pp. 403–432, pl. 20. (Bombyliidae pp. 406–407.)
- 1901b. Zoologische Ergebnisse der dritten Asiatischen Forschungsreise des Grafen Eugen Zichy. Ann. Mus. Nat. Hungarici, vol. 2, xli + 470 pp. Dipteren pp. 181–201. (Bombyliidae by Kertész p. 194.)
- 1908. Catalogus dipterorum hucusque descriptorum. Vol. 3. Stratiomyidae, Erinnidae, Coenomyidae, Tabanidae, Pantophthalmidae, Rhagionidae. 367 pp. Budapest. Zoological Record lists in 1907 as well as 1908. (Bombyliidae: contains *Mythicomyia* under Rhagionidae, p. 333.)
- 1909. Catalogus dipterorum hucusque descriptorum. V. Bombyliidae, Therevidae, Omphralidae. Mus. Nat. Hungaricum, G. Wesselenyi, Budapest. 199 pp. (Bombyliidae pp. 1–148.)
- 1912. A new *Glabellula* from Asia minor. Ann. Hist.-Nat. Mus. Hungarici, Budapest, vol. 10, pp. 638–639.

KHALAF, KAMEL T.
- 1960. Miscellaneous insects from Iraq. Bull. Iraq Nat. Hist. Mus. (Univ. Baghdad), vol. 1, pp. 1–7. (Bombyliidae p. 6 spp. listed.)

KIRBY, WILLIAM
- 1837. Insects *in* Richardson, J., Fauna Boreali-Americana; or the zoology of the northern parts of British America. Part IV. Insects. London, 364 pp., 8 color pls. (Bombyliidae p. 312, *Bombylius major*. Reproduced, 1881, Canadian Ent., vol. 13, p. 166.)

KIRBY, WILLIAM, and SPENCE, W.
- 1815–1826. An introduction to entomology, or elements of the natural history of insects. London. Edition 1.
  - 1815. Vol. 1, 535 pp., 3 pls.
  - 1817. Vol. 2, 529 pp., pls. 1–2.
  - 1826. Vol. 3, 732 pp., 20 pls.
  - 1826. Vol. 4, 634 pp., pls. 21–30.

KIRBY, WILLIAM FORSELL
- 1884. On the Diptera colllected during the recent expedition of H.M.S. *Challenger*. Ann. Mag. Nat. Hist., ser. 5, vol. 13, pp. 456–460. (Bombyliidae p. 458, 2 spp. listed.)

KITTEL, G., and KRIECHBAUMER, J.
- 1877. Systematische Uebersicht der Fliegen, welche in Bayern und in der nächsten Umgebung vorkommen. Abh. Naturh. Ges. Nürnberg, vol. 5, pp. 3–90. (Bombyliidae pp. 14–17; 35 spp. listed.)

KLOET, GEORGE SIDNEY, and HINCKS, WALTER DOUGLAS
- 1945. A check list of British insects. Stockport. 483 pp. (Bombyliidae p. 363; 5 genera, 12 spp.)

KLUG, JOHANN CHRISTOPH FRIEDRICH
- 1829–1845. *In* Klug and Ehrenberg: Symbolae Physicae, seu Icones et descriptiones Insectorum, quae ex itinere per Africam borealem et Asiam F. G. Hemprich et C. H. Ehrenberg studio novae aut illustratae redierunt. Berlin. In 5 parts. Text pages unnumbered.
  - 1832. Part 3, pp. a–l (unnumbered), 10 pls. (color). (Bombyliidae pp. 2, 8, 12 [unnumbered], pl. 30 [beautiful].)
- 1860. *In* Loew: Die Dipteren-Fauna Südafrika's. Abh. Naturwiss. Ver. Sacksen Thüringen in Halle, Berlin, vol. 2, pp. 73–402, 2 pls., figs. 1–52. (Bombyliidae p. 179, Klug, sp. sine descr.)

KLUGE, E. B. [see Fiedler and Kluge]

KNAB, FREDERICK
- 1913. A new *Heterostylum* from Mexico. (Diptera, Bombyliidae). Insecutor Insecitiae Menstruus, vol. 1, pp. 110–111.
- 1915. Some West Indian Diptera. Insecutor Insecitiae Menstruus, vol. 3, pp. 46–50. (Bombyliidae pp. 49–50.)

KNOWLTON, GEORGE F.
   1931. Notes on Utah Diptera. Canadian Ent., vol. 63, pp. 152–157. (Bombyliidae pp. 154–155; 22 spp. listed.)

KNUTH, PAUL
   1898–1899. Handbuch der Blütenkiologie. Wilhelm Engelmann, Leipzig, Germany. In 3 volumes. Translated by J. R. Ainsworth Davis, Oxford, England: 1906–1909.
      1906. Vol. 1, xix + 382 pp., 81 figs. Bibliography of 3748 titles. (Bombyliidae pp. 116, 139, 180–184.)
      1907. Vol. 2, viii + 703 pp., 210 figs. This volume gives visitors by flower families. (Bombyliidae interspersed.)
      1909. Vol. 3, iv + 644 pp., 208 figs. This volume also gives visitors by flower families. (Bombyliidae list p. 554.)

KOWARZ, FERDINAND
   1867. Beschreibung sechs neuer Dipteren-Arten. Verh. zool. bot. Ges. Wien. vol. 17, pp. 319–324. (Bombyliidae p. 324.)
   1872. Beitrag zur Dipteren-Fauna Ungarns. Verh. zool.-bot. Ges. Wien, vol. 23, pp. 453–464. (Bombyliidae p. 456; reprint p. 4; 12 spp. listed.)
   1883–1885. Beiträge zu einem Verzeichnisse der Dipteren Böhmens. I–V. Wiener Ent. Zeitung.
      1883. Vol. 2, pt. 1, pp. 108–110; pt. 2, pp. 168–170; pt. 3, pp. 241–243. (Bombyliidae pt. 2, pp. 168–170.)
   1894. Verzeichniss der Insekten Böhmens. Physiokratischen Gesellshaft, 1894, pp. 1–42. (Bombyliidae pp. 10–11; 28 spp. listed.)

KRIECHBAUMER, J. [see Kittel and Kriechbaumer]

KROMBEIN, KARL V.
   1936. Biological notes on some solitary wasps. (Hymenoptera: Sphecidae). Ent. News, vol. 47, pp. 93–99. (Bombyliidae p. 95; *Exoprosopa fascipennis* ovipositing in nest entrances of *Epibembix spinolae* Lep.)
   1967. Trap-nesting wasps and bees: Life histories, nests, and associates. Smithsonian Publication No. 4670, vi + 570 pp., 36 tables, 2 figs., 29 plates and frontispiece. (Bombyliidae pp. 53, 62, 69, 74, 89, 99, 101, 104, 108, 132, 139, 150, 155, 160, 187, 192, 209, 226, 251, 255, 262, 274, 286, 321, 327, 346, and summary 395 to 410. Plate 26, figs. 123 to 127.)

KÜNCKEL D'HERCULAIS, JULES PHILIPPE ALEXANDER
   1879. Recherches morphologiques et zoologiques sur le systeme nerveux des Insectes diptères. Comptes Rendus Acad. Sci., Paris, vol. 89, pp. 491–494. (Bombyliidae p. 494.)
   1889. [Note: Larvae of bombyliids destroying locust eggs.] Bull. Soc. Ent. France, ser. 6, vol. 9, p. vii.
   1893–1905. Invasions des Acridiens vulgo Sauterelles en Algérie. In 2 volumes. Imprimerie administrative Gouvernement général de l'Algérie. Paris, Challamel; Alger, Mustapha. Vol. 1, pp. 1–150, 1–1596, and corrections and additions pp. 1–40. Pt. 1, pp. 1–150, 1–592. Part 2, pp. 593–1576, includes: Causes naturelles de destruction des oefs, pp. 611–636. (Bombyliidae pp. 618, 631–635. Also Role des Insectes pp. 931–951, comments on work of Riley, Mulsant, Portschinsky.) Vol. 2, pp. 1–752, 10 maps, 34 pls. (Bombyliidae p. 633, figs. 1–4, showing *Thyridanthrax* (as *Anthrax*) *fenestrata* within egg sac and its emergence.) (In U.S. Dept. Agric. Library.)
   1894. Les Diptères parasites des Acridiens, les Bombylides. Hypnodie larvaire et métamorphose avec stade d'activité et métamorphose avec stade d'activité et stade repos. Comptes Rendus Acad. Sci., Paris, vol. 118, pp. 926–929, 1106, 1359. Reviewed, Ann. Mag. Nat. Hist., ser. 6, vol. 14, pp. 74–76, 1894.
   1904. Les Lépidoptères Limacodides et leurs Diptères parasites, Bombylides du genre *Systropus*. Adaptation parallèle de l'hôte et du parasite aux mêmes conditions d'existence. Comptes Rendus Acad. Sci., vol. 138, pp. 1623–1625. Also, 1905, Bull. Sci. France et Belgique, vol. 39, pp. 141–151, pls. 3–4. (All Bombyliidae.)

Künckel d'Herculais, Jules Philippe Alexander—Continued
 1905. Les Lépidoptères Limacodides et leurs Diptères parasites, Bombylides du genre *Systropus*-signification morphologique des pointes frontales de la chrysalide de l'hôte et de la nymphe du parasite. Comptes Rendus Assoc. Française.

Kuntze, A.
 1913. Dipterologische Sammelreise in Korsika des Herrn W. Schnuse in Dresden in Juni und Juli 1899. Deutsche Ent. Zeitschr., 1913, pp. 544–552. (Bombyliidae p. 548; 17 spp. listed.)

La Baume, W.
 1918. *In* Bücher and others: Die Heuschreckenplage und ihre Bekämpfung . . . auf Grund der in Anatolien und Syrien während der Jahre 1916 und 1917 gesammelten Erfahrungen dargestellt. Monographien zur angewandte Ent.-Beihefte zur vaterl. Naturdunde. Zeitschr. für angewand. Ent., Berlin. Pp. 1–13, 1–274, 20 pls. all on control measures, 11 maps, 33 figs. Chapter 4: Biologie der Marokkanischen Wanderheuschreke (*Stauronotus maroccanus* Thunb.), pp. 157–274 by La Baume. (Bombyliidae pp. 261–264, fig. 33.)

Laboulbène, Alexandre
 1857a. Note sur la nymphe de l'*Anthrax sinuata*. Ann. Soc. Ent. France, ser. 3, vol. 5, pp. 781–790, pl. 15, part II, with figs. 1–5.
 1857b. [Note: *Anthrax morio*.] Bull. Soc. Ent. France, ser. 3, vol. 5, p. xc.
 1858a. Note: L'*Anthrax sinuata* et la *Chrysis ignita*, parasites des Odyneres. Bull. Soc. Ent. France, ser. 3, vol. 6, pp. cxii–cxiii.
 1858b. Trouve dans Schaeffer les figures de la larve et de la nymphe d'un *Anthrax*. Bull. Soc. Ent. France, ser. 3, vol. 6, p. cxiii.
 1873. Note sur la nidification de l'*Heriades truncorum* et sur l'*Anthrax aethiops* parasite de cet hymènoptère. Ann. Soc. Ent. France, ser. 5, vol. 3, p. 57, pl. 5, pt. III, figs. 1–3.

Lacaze-Duthiers, Félix Josef Henri de
 1849–1853. Recherches sur l'armure génitale femelle des insectes. Ann. Sci. Nat. In 5 parts. Reprint, Paris, 1852.
  1850. Ser. 3, vol. 14, pp. 17–52, 3 pls.
  1853. Ser. 3, vol. 19, pp. 25–88, 203–237, 4 pls.

Lahille, F.
 1907. La langosta y sus Moscas parasitarias. Anal. Min. Agric. Argentina, Buenos Aires, vol. 3, no. 4, 136 pp., 7 pls. (5 color), 29 figs.

Lamarck, Jean Baptiste Pierre Antoine de Monet de
 1815–1822. Histoire naturelle des animaux sans vertèbres. Paris, Verdière. In 7 volumes.
  1816. Vol. 3, 586 pp. (Bombyliidae pp. 406–410.)

Lamb, C. G. [see Bezzi and Lamb]

Lamborn, W. A.
 1915. A preliminary report on the problem of controlling *Glossina* in Nyasaland. Bull. Ent. Res., London, vol. 6, pp. 59–65. (Bombyliidae pp. 53–64.)
 1915. Second report on *Glossina* investigations in Nyasaland. Bull. Ent. Res., vol. 6, pt. 3, pp. 249–266, pls. 4–5. (Bombyliidae p. 256.)

Lameere, August
 1906. Notes pour la classification des Diptères. Mem. Soc. Ent. Belgique, vol. 12, pp. 105–140. (Bombyliidae pp. 125–128, 140.)
 1936. Classification of Diptera, pp. 71–162, figs. 87–196, *in* Précis de Zoologique, vol. 5, 536 pp. Publ. l'Inst. Zool. Torley-Rosseau. Brussels, Liege. (Bombyliidae pp. 128–129, fig. 162.)

Lampert, K.
 1886. Die Mauerbiene und ihre Schmarotzer. Jahreshefte Ver. Württemberg, Stuttgart, vol. 42, pp. 89–101. (Bombyliidae pp. 97–99.) Summary: 1887, Journ. Roy. Microscopy, pp. 225–226. Summary: Naturforscher, vol. 20, pp. 15–16, 1887.

Langhoffer, August
- 1902. Einige Mitteilungen über den Blumenbesuch der Bombyliiden. Section 6: Arthropoda. Verh. Congresses Zool., pp. 848–851.
- 1910. Blütenbiologische Beobachtungen an Dipteren. Zeitschr. Wissensch. Insektenbiol., vol. 6. (Bombyliidae pp. 14–17, 57–61.)

Lassmann, R.
- 1912. *Anthrax morio* L. (Dipt.) als Schmarotzer 2. Grades. Mitt. ent. Ges. Halle, vols. 33–34, pp. 61–62.

Latreille, Pierre Andre
- 1796. Précis des caractères génériques des Insectes, disposés dans un ordre naturel par le Citoyen Latreille. Brive, Bordeaux. 229 pp. (Bombyliidae pp. 155–157.)
- 1802–1805. Histoire naturelle, générale et particulière des Crustacés et des Insectes. In 14 parts. Paris.
    - 1802. Part 3. Familles naturelles et genres, 479 pp. (Bombyliers, bombylarii pp. 427–430.)
    - 1804. Several articles in "Dictionnaire d'Histoire naturelle de Déterville." Paris. Also in vol. 24 "Ein Tableau méthodique des Insectes etc." (Taken from Horn and Schenkling, 1928.) (See *Ploas* Latreille.)
    - 1805. Part 14, pp. 1–432, pls. 104–112. (Bombyliidae p. 304, pl. 109, fig. 7.)
- 1806–1809. Genera Crustaceorum et Insectorum secundum ordinem naturalem in familias disposita, iconibus exemplisque plurimis explicata. In 4 vols. Paris.
    - 1809. Vol. 4, 399 pp. (Bombyliidae pp. 307, 311, 313.)
- 1810. Considérations générales sur l'ordre naturel des Animaux composant les classes des Crustaces, des Arachnides et des Insectes avec un tableau méthodique de leurs genres disposés en familles. Paris. 444 pp. (Bombyliidae p. 443; type designation for *Bombylius*.)
- 1817. *In* Cuvier: Le règne animal. I-IV. Paris. Entomologie by Latreille, vol. 3, 682 pp., 2 pls. (Bombyliidae pp. 610–611.)
- 1825. Familles naturelles du règne animal, exposées succinctement et dans un ordre analytique, avec l'indication de leurs genres. Paris. 570 pp. (Bombyliidae p. 496.) Republished, 1827. See Berthold for German translation of *Familles naturelles du règne animal*. . . . 1825.

Leach, William Elford
- 1817. On the genera and species of eproboscideous insects and on the arrangement of oestrideous insects. Edinburgh. Issued as a reprint, pp. 1–20, pls. 25–27. Also, 1818, Mém. Wernerian Nat. Hist. Soc., Edinburgh, vol. 2, pt. 2, pp. 547–566, pls. 25–27.
- 1819. Family 9. Anthracidae *in* Samouelle: The entomologist's useful compendium, or an introduction to the knowledge of British insects. London. 496 pp., 12 pls. (Bombyliidae pp. 295–296, pl. 9, fig. 10.)

LeConte, John Lawrence [see Thomas Say]

Lemmon, J. G.
- 1879. [Note: Bee-fly grubs.] Weekly Record-Union, Sacramento, California, Nov. 29, 1879.
- 1880. The locust scourge. San Francisco Weekly Bulletin, Sept. 15. 1880.

Leonard, Mortimer Demarest
- 1928. A list of the insects of New York with a list of the spiders and certain other allied groups. Mem. Cornell Univ. Agric. Exper. Stat., no. 101 (1926), pp. 1–1121. (Bombyliidae pp. 761–764; 45 spp. listed, in 11 genera.)

Lepeletier, Amédée Louis Michel, Comte de Saint Fargeau, and Serville, Jean Guillaume Audinet
- 1825. L'Encyclopédie méthodique. Paris. Dictionnaire des Insectes, vol. 10, 833 pp. (Bombyliidae pp. 161, 678, 679, 686.)

Lespes, L. [see Régnier, Lespes, and Rungs]

Lessona, M.
> 1873. Calendario Zoologico in Piemonte: *Calendario degli Insetti di Giorna il Figlio.* Ann. Roy. Accad. Agric. Torino, vol. 16. (Taken from Griffini, 1896, p. 3.)

Lichenstein, Anton August Heinrich
> 1796. Catalogus musei Zoologici ditissimi Hamburgi. Sect. 3, Insects, 237 pp. (From Horn and Schenkling, 1928, p. 730.) Kertesz, 1909, p. 126, gives: Catal. rerum natural. rariss. Hamburgi, sect. 3, Ins., 305. *Tabanus* ? (*charopus*) syn. of *Bombylius analis* Fabricius.)

Lindner, Erwin
> 1944. Dipterologisch-faunistiche Studien in Gebiet der Lunzer Seen. . . . Jber. Ver. Landesk. Hematpflege, Oberdonau, Linz, vol. 91, pp. 255–291.
> 1949. *In* Lindner, Erwin (editor): Die Fliegen der palaearktischen Region. Vol. 1, Handbuch. Stuttgart. Foreword pp. 1–10, pp. 1–422, 481 text-figs., 28 color pls. with 87 figs. (Bombyliidae pp. 146–148, text-figs. 224–228, pl. 10, fig. 25.)
> 1957. Ostafrikanische Bombyliidae (Diptera). Stuttgarter Beiträge Naturkunde, no. 3, pp. 1–7.
> 1962. Afrikanische Bombyliidae (Dipt.). Stuttgarter Beiträge Naturkunde, no. 81, Feb. 1, 1962, pp. 1–7.

Linné, Carl von
> 1735–1767. Systema naturae. . . . 12 editions by Linné, with slight changes in title. Edition 13 by Gmelin. (See Gmelin.)
>> 1758. Tenth edition. Systema naturae per regna tria naturae secundum classes, ordines, genera, species, cum characteribus, differentiis, synonymicis, locis. Holmiae. 824 pp. Pt. 6, Diptera, pp. 584–607. (Bombyliidae pp. 590, 606–607.)
>> 1766–1767. Twelfth edition. Holmiae. In 2 parts.
>>> 1767. Pt. 2, pp. 533–1327, and index 36 unnumbered pages. (Bombyliidae pp. 981, 983, 1009–1010.)
> 1746. Fauna svecica sistens animalia Sveciae regni: quädrupedia, aves, amphibia, pisces, insects, vermes, distributa per classes, et ordines, genera et species, cum differentiis specierum, synontmis autorum, nominibus incolarum, locus habitationum, descriptionibus insectorum. Stockholm. 411 pp. (Coulson, Sabrosky, Muller, 1965, state: "The first edition . . . is outside the scope of zoological nomenclature.)
>> 1761. Fauna svecica . . . Second edition. Stockholm. 46 pages without numbers, pp. 1–578, 2 pls. (Bombyliidae pp. 440–441, 470–471.)

Linsley, E. Gorton [also see Hurd and Linsley]
> 1958. The ecology of solitary bees. Hilgardia, vol. 27, pp. 543–599. (Bombyliidae pp. 578, 581, 582, 583.)
> 1960. Ethiology of some bee- and wasp-killing robber flies of southeastern Arizona and western New Mexico. (Diptera: Asilidae). Univ. California Publ. Ent., vol. 16, no. 7, pp. 357–392, pls. 48–55. (Bombyliidae pp. 363, 366, 371.)

Linsley, E. Gorton, and MacSwain, John W.
> 1942. The parasites, predators and inquiline associates of *Anthophora linsleyi.* Amer. Midl. Nat., vol. 27, pp. 402–417, figs. 1–11. (Bombyliidae pp. 403, 411, 412, 416; fig. 11.)
> 1952. Notes on some effects of parasitism upon a small population of *Diadosia bituberculata* (Cresson) (Hymenoptera: Anthophoridae). Pan-Pacific Ent., vol. 26, no. 3, pp. 131–135. (Bombyliidae pp. 132–134.)
> 1957. The nesting habits, flower relationships, and parasites of some North American species of *Diadasia* (Hymenoptera: Anthophoridae). The Wasmann Journ. Biol., vol. 15, no. 2, pp. 199–235. (Bombyliidae pp. 199, 211–212, 218, 229, 231.)

Linsley, E. Gorton, MacSwain, John W., and Smith, R. F.
> 1952a. Bionomics of *Diadasia consociata* Timberlake and some biological relationships of Emphorine and Anthophorine bees (Hymenoptera, Anthophoridae). Univ. California Publ. Ent., Univ. California Press, vol. 9, no. 3, pp. 267–290, pls. 1–6. (Bombyliidae pp. 275–276.)

LINSLEY, E. GORTON, MACSWAIN, JOHN W., and SMITH, R. F.—Continued
- 1952b. Outline for ecological life histories of solitary and semisocial bees. Ecology, vol. 33, pp. 558–567. (No Bombyliidae mentioned but valuable suggestions for rearing; 109 references.)

LIOY, PAOLO
- 1863–1865. Ditteri distribuiti secondo un nuovo methodo di classificazione naturale. . . . Atti Instuto Veneto, in 8 broken parts, ser. 3, vol. 9, pp. 187–236, 499–515, 569–604, 719–771, 879–910, 989–1027, 1087–1126, 1211–1352; ser. 3, vol. 10, pp. 59–84.
  - 1864. Ser. 3, vol. 9. (Bombyliidae pp. 727–733.)
- 1895. Manuali Hoepli: Entomologia. III. Ditteri Italiani, 356 pp., 227 figs.

LLOYD, L.
- 1916. Report on the investigations into the bionomics of *Glossina morsitans* in northern Rhodesia, 1915. Bull. Ent. Res. London, vol. 7, pp. 67–79. (Bombyliidae pp. 76–77, fig. 2.)

LOEW, HERMANN
- 1840. Ueber die im Grossherzogthum Posen aufgefundenen Zweyflügler; ein Beytrag zur genaueren Kritischen Bestimmung der europäischen Arten. Isis (Oken). Leipzig, vol. 7, columns 512–584, pl. 1, figs. 1–55. (Bombyliidae columns 513, 532–534, 581, fig. 34.) Reprint: Bemerkungen über die in der Posener Gegend einheimischen Arten mehrerer Zweiflüger Gattungen. Schulprogr. Posen, 1840, pp. i-iv, 1–40, 1 pl.
- 1844. Beschreibung einiger neuen Gattungen der europäischen Dipteren fauna. Stettiner Ent. Zeitung, vol. 5, pp. 114–130, 154–173, 2 pls. (Bombyliidae pp. 127–130, 154–162, pl. 1, figs. 12–18, pl. 2, figs. 6–16.)
- 1846. Fragmente zur Kenntniss der europäischen Arten einiger Dipterengattungen. Linn. Ent., vol. 1, pp. 319–530, pl. 3. (Bombyliidae pp. 365–422, pl. 3, figs. 20–30.)
- 1850. Ueber den Bernstein und die Bernstein fauna. Berlin. Progr. Realschule Meseritz, 1850, pp. 1–4, 1–44. (Bombyliidae p. 40.)
- 1852. Diagnosen der Dipt. von Peter's Reise in Mossambique. Monatsber. Akad. Wiss. Berlin, 1852, pp. 658–661. (Bombyliidae p. 659.)
- 1853–1862. Neue Beiträge zur Kenntniss der Dipteren. In 8 parts. Schulprogramm Meseritz. Berlin. Also published Berlin 1853–1861.
  - 1854. Zweiter Beitrag, pp. 1–24. (Bombyliidae pp. 2–4.)
  - 1855. Dritter Beitrag, pp. 1–52. (All Bombyliidae.)
  - 1856. Vierter Beitrag, pp. 1–57. (Bombyliidae pp. 9, 29–31.)
- 1855. Vier neue griechische Diptera. Stettiner Ent. Zeitung, vol. 16, pp. 39–41. (Bombyliidae p. 39.)
- 1856. *In* Rosenhauer: Die Thiere Andalusiens nach den Resultate einer Reise zusammengestellt, nebst den Beschreibungen von 249 neuen oder bis jetzt noch unbeschriebenen Gattungen und Arten. 429 pp., 3 pls., Insekten pp. 17–406, Dipteren pp. 376–389. (Bombyliidae by Loew pp. 377, 380–381; 22 spp. including 6 spp. *Usia*.)
- 1856–1863. Bidrag till kännendomen om Afrikas Dipteren. Ofvers. Kongl. Vet. Akad. Förhandl. In 5 parts. Stockholm.
  - 1861. Vol. 17 (1860), pp. 81–97. (Bombyliidae pp. 83–95.)
- 1875a. *Dischistus multisetosus* und *Saropogon aberrans*, zwei neue europäische Dipteren. Stettiner Ent. Zeitung, vol. 18, pp. 17–20. (Bombyliidae p. 17.)
- 1857b. Nachricht über syrische Dipteren. Verh. zool.-bot. Ver. Wien, vol. 7, pp. 79–86. (Bombyliidae pp. 79–84.)
- 1858a. Ueber die Schwinger der Dipteren. Berliner Ent. Zeitschr., vol. 2, pp. 225–230. (Bombyliidae pp. 228, 229, 230.)
- 1858b. Bericht über die neueren Erscheinungen auf dem Gebiete der Dipterologie. Berliner Ent. Zeitschr., vol. 2, pp. 325–349. (Bombyliidae p. 348.)
- 1860. Die Dipteren-Fauna Südafrika's. Abh. Naturwiss. Ver Sachsen Thüringen in Halle, Berlin, vol. 2, 330 pp. (pp. 57–402), 2 pls. (Bombyliidae pp. 173–245 (pp. 245–317), pl. 1, figs. 7–42.)

LOEW, HERMANN—Continued
- 1861a. Diptera aliquot in insula Cuba collecta. Wiener Ent. Monatsschr., vol. 5, pp. 33–43. (Bombyliidae pp. 34–35.)
- 1861b. Ueber die Dipteren-Fauna des Bersteins. Ber. 35. Vers. deutsch. Naturf., 1861 (1860), pp. 88–98.
- 1861–1972. Diptera Americae septentrionalis indigena. In 10 parts. Berliner Ent. Zeitschr. See also Stone et al, 1965, pp. 1324–1325.
    - 1863a. Centuria 3, vol. 7, pp. 1–55. (Bombyliidae 11–12.)
    - 1863b. Centuria 4, vol. 7, pp. 275–326. (Bombyliidae pp. 300–305.)
    - 1869. Centuria 8, vol. 13, pp. 1–52. (Bombyliidae pp. 12–31.)
    - 1872. Centuria 10, vol. 16, pp. 49–115. (Bombyliidae pp. 76–82.)
- 1862a. *In* Peters: Naturwissenschaftliche Reise nach Mossambique . . . in . . . 1842. Part 5, Insekten und Myriapoden. Diptera, pp. 1–34, 1 pl. (color). Berlin. (Bombyliidae pp. 10–14, pl. 1, figs. 8–9.)
- 1862b. Ueber einige bei Varna gefangene Dipt. Wiener Ent. Monatschr., vol. 6, pp. 161–175. (Bombyliidae pp. 161–162, 164–165.)
- 1862c. Ueber griechische Dipteren. Berliner Ent. Zeitschr., vol. 6, pp. 69–89. (Bombyliidae pp. 71, 77–83.)
- 1863a. Enumeratio Dipterorum quae C. Tollin ex Africa meridionali. Wiener Ent. Monatschr., vol. 7, pp. 9–16. (Bombyliidae pp. 12–15.)
- 1863b. Ueber bei Sliwno im Balkan gefangene Dipt. Wiener Ent. Monatschr., vol. 7, pp. 33–35. (Bombyliidae pp. 34–35.)
- 1865. Ueber einige bei Kutais in Imeretien gefangene Dipteren. Berliner Ent. Zeitschr., vol. 9, pp. 234–242. (Bombyliidae pp. 234, 237; 2 spp. listed.)
- 1868. Cilicische Dipteren und einige mit ihnen concurrirende Arten. Berliner Ent. Zeitschr., vol. 12, pp. 369–386. (Bombyliidae pp. 369, 378–384.)
- 1869. [Note: *Hemipenthes morio* and *seminigra*.] Berliner Ent. Zeitschr., vol. 13, p. 28.
- 1869–1873. Beschreibungen europäischer Dipteren. I-III. Halle.
    - 1869. Vol. 1, xvi + 310 pp. (Bombyliidae pp. 127–251.)
    - 1871. Vol. 2, 319 pp. (Bombyliidae pp. 204–221.)
    - 1873. Vol. 3, 320 pp. (Bombyliidae pp. 152–208.)
- 1870a. Turkestanskija dwukrilyia. (Diptera aus Turkestan, russisch.) Schriften kais. Ges. Freunde Natur. Moskau, 1870, pp. 52–59. (Bombyliidae pp. 56–57.)
- 1870b. Ueber die von Herrn Dr. G. Seidlitz in Spanien gesammelten Dipteren. Berliner Ent. Zeitschr., vol. 14, pp. 137–144. (Bombyliidae pp. 137–138, 142–143.)
- 1870c. Bermerkungen über die von Herrn v. d. Wulp in der Zeitschrift der Niederländischen entomologischen Gesellschaft für 1867 publizirten Nordamerikanischen Dipteren. Zeitschr. Ges. Naturwiss., vol. 36, pp. 113–120. (Bombyliidae pp. 113, 115.)
- 1873. Bemerkungen über die von Herrn F. Walker im 5. Bande des Entomologist beschriebenen ägyptischen und arabischen Dipteren. Zeitschr. Ges. Naturwiss., Berlin, new ser., vol. 8 (XLII), pp. 105–109. (Bombyliidae pp. 106–108.)
- 1874. Diptera nova a Hug. Theod. Christophe collecta. Zeitschr. Ges. Naturwiss., n. ser., vol. 9 (XLIII), pp. 413–420. (Bombyliidae p. 416.)
- 1876. *Eclimus* (generic characters amended, and the rendering *Eclimmus* by Agassiz repudiated) *hirtus*, sp. n. Deutsche Ent. Zeitschr., vol. 20, pp. 209–210.

LONG, F. L. [see Clements and Long]

LOVETT, A. L. [see Cole and Lovett]

LUBBOCK, J., SIR
- 1880. Larva (subsequently proved to belong to this family) destructive to eggs of locusts in the Troad. Proc. Ent. Soc. London, 1880, p. xxxiii.

LUCAS, PIERRE HIPPOLYTE
- 1849. Exploration scientifique de l'Algérie, Zoologie. Histoire naturelle des animaux articulés. In 3 volumes and Atlas, 89 color pls. Vol. 3, 527 pp. Diptères pp. 414–503, 6 pls. (Bombyliidae p. 456.)

Lucas, Pierre Hippolyte—Continued
- 1852. Note sur les transformations du *Bombylius Boghariensis*, nouvelle espece de Diptère qui habite les possessions Françaises du Nord de l'Afrique. Ann. Soc. Ent. France, ser. 2, vol. 10, pp. 11–18, 1 pl. (color) with 23 figs. (Bombyliidae part 2 of pl. 1.)

Lucia, Dusa
- 1968. Contributii La Studiul Ovipozitorului La Bombiliide. Biologia, II, pp. 77–86.

Lugger, Otto
- 1898. Third annual report of the Entomologist of State Experiment Station of the University of Minnesota to the Governor. 1898. The Orthoptera of Minnesota. 307 pp. (Bombyliidae pp. 42–44, figs. 15–16, after Riley of *Systoechus oreas* on grasshopper eggs.)
- 1899. Fourth annual report of the Entomologist . . . 1899. 279 pp., 237 figs., 24 pls. (Bombyliidae p. 97. *Systropus macer* parasitic on cherry slug caterpillar *Adoneta spinuloides* H. S.)

Lundbeck, William
- 1908. Diptera Danica. Part II. Asilidae, Bombyliidae, Therevidae, Scenopinidae. Copenhagen, London. 164 pp., 48 figs. (Bombyliidae pp. 87–132, figs. 33–44.)

Lynch Arribálzaga, Felix L.
- 1878. Notes dipterológicas sobre los Anthrácidos y Bombiliarios del Partido del Baradere (Provincia de Buenos Aires). Parte primera. Anthrácidos. Nat. Argentino, vol. 1, pp. 225–231, 263–275.

MacDonald, W. A.
- 1957. A calliphorid host of *Thyridanthrax abruptus* (Lw.) in Nigeria (Diptera, Bombyliidae). Bull. Ent. Res., vol. 48, p. 533.

Mackerras, I. M.
- 1950. The zoogeography of the Diptera. Australian Journ. Sci., vol. 12, no. 5, pp. 157–161. (Bombyliidae pp. 157, 159.)

MacLeay, William Sharp
- 1827. *In* King, Phillip P.: Narrative of a survey of the . . . coasts of Australia. London. Vol. 2, Annulosa, pp. 438–496. (Bombyliidae p. 468.)
- 1838. I. On some new Forms of Arachnida. Ann. Nat. Hist., vol. 2, no. 7, pp. 1–14, 2 pls. (Bombyliidae p. 12.)

Macquart, Pierre Justin Marie
- 1826–1834. Histoire naturelle. Insectes Diptères du nord de la France. Recueil Trav. Soc. Sci. Agric. et Arts, Lille.
  - 1826. Pt. 2, pp. 324–499, 3 pls. (Bombyliidae pp. 363–384, pl. 2, fig. 1.)
- 1834–1835. Histoire naturelle des insectes. Diptères. Suite à Buffon, edited by Roret. Paris. In 2 vols., 24 color pls.
  - 1834. Vol. 1, 578 pp., 12 pls. (Bombyliidae pp. 373–412, pl. 9, fig. 9; pl. 10, fig. 4.)
  - 1835. Vol. 2, 710 pp., pls. 13–24. (Bombyliidae p. 662.)
- 1835–1844. Diptères *in* Webb and Berthelot: Histoire naturelle des Îles Canaries. Paris. In 3 vols., atlas.
  - 1839. Entomologie, vol. 2, pt. 2, 119 pp., 8 pls. Diptera, pp. 97–119, 1 pl. (Bombyliidae pp. 105–106, pl. 4, fig. 8.)
- 1838–1855. Diptères exotiques nouveaux ou peu connus. Mem. Soc. Sci. Agric. et Arts, Lille. Also separates by Roret, Paris. Both paginations are given.
  - 1840. Vol. 2, pt. 1, pp. 5–135, 21 pls. (or pp. 283–413). (Bombyliidae pp. 32–118, pls. 6–21. (or pp. 310–396, 399–413.)
  - 1846. Suppl. 1, pp. 5–238, 20 pls. (or pp. 133–366). (Bombyliidae pp. 107–199, pl. 20, fig. 4.)
  - 1847. Suppl. 2, pp. 5–220 (or 21–236), 6 pls. (Bombyliidae pp. 66–71 (or pp. 60–54.)
  - 1848. Suppl. 3, pp. 1–77 (or 161–237), 7 pls. (Bombyliidae pp. 193–197 [or pp. 33–37, pl. 3, figs. 8, 10].)
  - 1849. Suppl. 4, pt. 1, pp. 5–175 (or 309–479), 14 pls. (Bombyliidae pp. 408–427, pls. 10–11 [or 104–122, pls. 10–11].)

Macquart, Pierre Justin Marie—Continued
- 1855. Suppl. 5, pp. 5–136 (or 25–156), 7 pls. (Bombyliidae pp. 88–104, pls. 3–4 [or pp. 68–84, pls.].)
- 1849. Diptères, *in* Lucas: Exploration scientifique de l'Algérie, Zoologie. Histoire naturelle des animaux articulés. Paris. In 3 vols. with atlas, 89 color pls.
  - 1849. Vol. 3, Insectes, 527 pp. Diptères, pp. 414–503, 6 pls. (Bombyliidae pp. 447–466, pl. 3, figs. 8–12; pl. 4, figs. 1–5, 7–8.)

MacSwain, J. W. [see Linsley, MacSwain, and Smith]

Malloch, J. R. [see also Cole, Malloch, and McAtee]
- 1915. Some additional records of Chironomidae for Illinois and notes on other Illinois Diptera. Bull. Illinois Lab. Nat. Hist., vol. 11, pp. 305–363, pls. 80–84. (Bombyliidae pp. 327–334, pl. 81.)
- 1916. The generic status of *Chrysanthrax* Osten Sacken. Proc. Biol. Soc. Washington, vol. 29, pp. 63–70, pl. 1.
- 1917. A preliminary classification of Diptera, exclusive of Pupipara, based upon larval and pupal characters, with keys to imagines in certain families. Part 1. Bull. Illinois State Lab. Nat. Hist., vol. 12, art. 3, pp. i–v, 161–409, pls. 28–57. (Bombyliidae pp. 389–396, pl. 52, figs. 1–7.)
- 1924. *Pachyneres australis* sp. n. Philippines. Australian Zool., Sydney, vol. 5, p. 205, fig.
- 1927–1931. Notes on Australian Diptera. Proc. Linnean Soc. New South Wales.
  - 1928. Vol. 53, pp. 295–309, 319–335, 343–366, 9 figs., 598–617, 651–662. (Bombyliidae p. 606, 607.)
- 1928. [Note: *Pachyneres australis* from Bibionidae, discussion.] Australian Zool., vol. 5, p. 138.
- 1932. Notes on exotic Diptera. Stylops, London, vol. 1, pt. 1, pp. 112–113; pt. 2, pp. 121–126, 6 figs. (Bombyliidae pp. 119–120, figs. 2–3.)

Marshall, August Friedrich
- 1873. Nomenclator zoologicus continens nomina systematica generum animalium tam viventium quam fossilium, secundum ordinem alphabeticum disposita. (Bombyliidae, *Bombylosoma* p. 323.)

Marshall, Guy Anstruther Knox
- 1902. XVII. Five years' observations and experiments (1896–1901) on the bionomics of South African insects. Trans. Ent. Soc. London, vol. 90, pp. 287–584. (Bombyliidae pp. 526–527, 529, pl. 20, fig. 22; pl. 22, fig. 17.)
- 1915. [Note: Bombyliidae.] Bull. Ent. Res., vol. 6, pt. 3, p. 256.

Marston, Norman
- 1963. A revision of the Nearctic species of the *Albofasciatus* group of the genus *Anthrax* Scopoli. (Diptera: Bombyliidae). Agric. Exper. Stat., Kansas State Univ., Tech. Bull. no. 127, pp. 1–79, 6 pls., 9 maps. Bibliography.
- 1964. The biology of *Anthrax limatulus fur* (Osten Sacken), with a key to and descriptions of pupae of some species in the *Anthrax albofasciatus* and *trimaculatus* groups (Diptera: Bombyliidae). Journ. Kansas Ent. Soc., vol. 37, no. 2, pp. 89–105.
- 1970. Revision of New World species of *Anthrax* (Diptera: Bombyliidae) other than the *Anthrax albofaciatus* group. Smithsonian Contrib. Zool., no. 43, 148 pp., 135 figs., 6 pls., 27 maps.
- 1971. Taxonomic study of the known pupae of the genus *Anthrax* (Diptera: Bombyliidae) in North and South America. Smithsonian Contrib. Zool., no. 100, 18 pp., 4 pls.

Martin, William Cline
- 1971. Insect parasites, hyperparasites, predators and scavengers of the common bagworm. Unpublished manuscript, University of Mississippi, 48 pages, 5 plates, 2 maps. (M.Sci. thesis: Library.)

Matsumura, Shonen
- 1905. Thousand insects of Japan. Tokyo. 1904–1907. Hemiptera, Diptera. 163 pp. (Bombyliidae p. 82, fig.)
- 1916. Thousand insects of Japan. [Text in Japanese.] Additamenta. vol 2. (Bombyliidae pp. 319–323, pl. 24, figs. 1, 4, 5, 6, 8.)

Matsumura, Shonen—Continued
    1931. 6000 illustrated insects of the Empire of Japan. [In Japanese.] Tokyo. Pp. ii–iii, 1–1497, 1–191, 10 pls., figs. (Bombyliidae 19 spp. on 5 pages.)

Maughan, Lucile [see Johnson and Maughan]
    1935. A systematical and morphological study of Utah Bombyliidae, with notes on species from intermountain states. Journ. Kansas Ent. Soc., vol. 8, pp. 27–80, 4 pls.

Maughan, Lucile, and Johnson, D. Elmer
    1936. Notes on Utah Bombyliidae (Diptera); No. I. Utah Acad. Sci., Arts and Letters, vol. 13, pp. 197–201.

Maxwell-Lefroy, H.
    1909. Indian insect life: A manual of the insects of plains (tropical India). Calcutta & Simla, India. xii + 786 pp., 536 text-figs. and 84 pls. (mostly color). (Bombyliidae pp. 597–600, pl. 63, figs. 1–4 (color).)

Mayr, Gustav L.
    1853–1855. Beiträge zur Insekten-Fauna von Siebenbürgen. Mitt. Siebenbürgischer Ver. Naturw. Hermannstadt, vols. 4–6. 1853, vol. 4, pp. 141–143. (Bombyliidae p. 143; 2 spp. listed.)

McArthur, Robert
    1960. On the relative abundance of species. Amer. Nat., vol. 94, pp. 25–36. (Demes, opportunistic species, equilibrium species discussed.)

McAtee, W. L. [also see Cole, Malloch, and McAtee]
    1932. Effectiveness in nature of the so-called protective adaptations in the animal kingdom, chiefly as illustrated by the feeding habits of nearctic birds. Smithsonian Misc. Coll., vol. 85, no. 7, publ. no. 3125, 201 pp. (Bombyliidae pp. 84, 86; 8 individual identifications out of 5934 determined Diptera.)

McDonald, W. A.
    1957. A calliphorid host of *Thyridanthrax abruptus* (Loew) in Nigeria (Diptera, Bombyliidae). Bull. Ent. Research, vol. 48, p. 533.

Meck
    1902. [Note: Bombyliidae.] Mitt. Zool. Mus. Berlin, vol. 2, p. 21.

Megerle von Mühlfeld, Johann Karl
    1820. *In* Meigen: Systematische Beschreibung der bekannten europäischen zweiflügeligen. Vol. 2. (Bombyliidae p. 140, and scattered ms. names, including *Mima*, ms. name for *Exoprosopa*.)

Meigen, Johann Wilhelm
    1800. Nouvelle classification des mouches à deux ailes. Paris. 40 pp. Reprinted 1908; see Hendel. (Bombyliidae pp. 28–29.)
    1803. Versuch einer neuen Gattungseintheilung der europäischen zweiflügeligen Insekten. VII. Mag. Insektenkunde, vol. 2, pp. 259–281. (Bombyliidae pp. 268, 270–271.)
    1804. Klassifikazion und Beschreibung der europäischen zweiflügeligen Insekten (Diptera). Braunschweig. One volume, 2 parts. Pt. 1, pp. i–xxviii, 1–152, 8 pls.; pt. 2, pp. i–vi, 153–314, 7 pls. (Bombyliidae pp. 199–211 and possibly others.)
    1818–1838. Systematische Beschreibung der bekannten europäischen zweiflügeligen Insekten. In 6 volumes and supplement. Achen (vols. 1–2) and Hamm (vols. 3–7). Reprinted 1851, vols. 1–2. Halle, Schmidt.
        1820. Vol. 2, pp. i–x, 1–365, pls. 12–21 (pls. 13–16, color). (Bombyliidae pp. 137–238, pl. 16, figs. 15–21; pl. 17, figs. 1–28; pl. 18, figs. 1–22; pl. 19, figs. 1–16). Reprint vol. 2, pp. 1–6, 1–276, Halle. (Bombyliidae pp. 103–141, 141–179, plates the same as 1820 edition.)
        1822. Vol. 3, pp. 1–10, 1–416, pls. 22–32. (Bombyliidae vortwort.)
        1830. Vol. 6, pp. i–iv, 1–401, pls. 55–66. Nachtrag, pp. 241–401. (Bombyliidae pp. 324–329.)
        1838. Vol. 7, pp. i–xii, 1–434, pls. 67–74. (Bombyliidae pp. 63–70.)
    1832–1837. *In* Gistel: Faunus. Zeitschr. Zool. und vergleichende Anatomie. München, Jaquet. Meigen: Neue Arten von Diptera aus der Umgegend von München benannt und beschreiben, pp. 66–72, in edition I, vol. 2, 1835, pts. 1–3, pp. 1–192. (Bombyliidae genus *Thlipsomyza*.) Edition II, 1837, vol. 2, pp. 65–128. (Bombyliidae p. 112, genus *Thlipsomyza*.)

MEIJERE, JOHANNES CORNELIS HENDRIK DE
- 1893–1904. Neue und bekannte Süd-Asiatische Dipteren. Bijdragen tot de Dierkunde, Amsterdam.
- 1906–1936. Uitkomsten der Nederlandsche Nieuw-Guinea-Expeditie in 1903, onder leiding von ... A. Wichmann (—in 1907 en. 1909 onder ... H. A. Lorentz—1912 en 1913 onder leiding van A. Fraussen Herderschee ...). Leiden. In 36 volumes.
    - 1913. Vol. 9, Dipteren I, pp. 305–386, pl. 10. (Bombyliidae pp. 322–323.); scattered references to bee flies in some other volumes, as 1906 (vol. 5), 1908 (p. 75).
- 1907–1924. Studien über südostasiatische Dipteren. 1–16. Gravenhage. Tijdschr. Ent.
    - 1907. Part 1, vol. 50, pp. 196–264, 2 pls. (Bombyliidae pp. 239–246.)
    - 1911. Part 6, vol. 54, pp. 258–432, 5 pls. (Bombyliidae pp. 295–300, pl. 19, figs. 17–19).
    - 1914. Part 8, vol. 56, suppl., pp. 1–99, pls. 1–3. (Bombyliidae pp. 33–34.)
    - 1914. Part 9, vol. 57, pp. 137–275. (Bombyliidae pp. 137–140, pl. 5, figs. 1–2.)
    - 1916. Part 10, vol. 58 (1915), pp. 64–97, 2 pls. (Bombyliidae pp. 73–75.)
    - 1924. Part 15. Dritter Beitrag zur Kenntnis der sumatranischen Dipteren. Vol. 67, suppl., pp. 1–64, 10 figs. (Bombyliidae p. 57.)
    - 1924. Part 16, vol. 67, pp. 197–224, 3 figs. (Bombyliidae pp. 209–210.)
- 1913. Dipteren from Ceram und Waigeu. I. Bijdragen tot de Dierkunde, vol. 19, pp. 45–67, 1 pl.
- 1915a. Fauna Simalurensis-Diptera. Tijdschr. Ent., vol. 58, pp. 1–63, pl. 1. (Bombyliidae p. 20.)
- 1915b. Diptera aus Nord-New-Guinea, gesammelt von P. N. van Kampen und K. Gjellerup in den Jahren 1910 und 1911. Tijdschr. Ent., vol. 58, pp. 98–139, 1 pl. (Bombyliidae p. 108.)
- 1916. Tweede Supplement op de Nieuwe Naamlijst van Nederlandsche Diptera. Tijdschr. Ent., vol. 59, pp. 293–320. (Bombyliidae p. 298.)
- 1917. Beiträge zur Kenntnis der Dipteren-Larven und -Puppen. Zool. Jahrb., Abth. Syst. Geogr. und Biol., vol. 40 (1916), pp. 177–322, pls. 4–14, 181 figs. (Bombyliidae pl. 13, fig. 169.)
- 1929. Fauna buruana. Syrphiden nebst einigen Brachyceren Orthorrhaphen. Treubia, Buitenzorg, vol. 7, pp. 378–387. (Bombyliidae p. 386–387; 6 spp. including 1 description.)

MELANDER, AXEL LEONARD [see also Brues and Melander; and Brues, Melander and Carpenter]
- 1902. A monograph of the North American Empididae. Part 1. Trans. Amer. Ent. Soc., vol. 28, pp. 195–367, pls. 5–9. (Bombyliidae pp. 337–338, under *Mythicomyia*.)
- 1906. Some new or little known genera of Empididae. Ent. News, vol. 17, pp. 370–379. (Bombyliidae pp. 372–373.)
- 1922. Collecting insects on Mount Rainier. Ann. Rept. Smithsonian Institution for 1921, pp. 415–422, 9 pls. (Bombyliidae p. 418, pl. 8, figs. 3, 6.)
- 1927. Genera Insectorum: Diptera. Fam. Empididae. Fasc. 185, 434 pp., 8 pls. (4 color) with 86 figs. (Bombyliidae, the genera *Mythicomyia* Coquillett, pp. 375–376, *Prorates* Melander, pp. 376–377, pl. 1, fig. 1.)
- 1946a. Some fossil Diptera from Florissant, Colorado. Psyche, vol. 53, nos. 3–4, pp. 43–49, pl. 2, figs. 1–3. (Bombyliidae pp. 47–48, pl. 2, fig. 3.)
- 1946b. Apolysis, Oligodranes, and Empidideicus in America (Diptera, Bombyliidae). Ann. Ent. Soc. America, vol. 39, no. 3, pp. 451–495, 1 pl., 13 figs.
- 1949. A report on some Miocene Diptera from Florissant, Colorado. Amer. Mus. Nov., no. 1407, pp. 1–63, 71 figs. (Bombyliidae pp. 31–35, figs. 37–42.)
- 1950a. *Aphoebantus* and its relatives *Epacmus* and *Eucessia* (Diptera: Bombyliidae). Ann. Ent. Soc. America, vol. 43, no. 1, pp. 1–45, 12 pls., 48 figs.

MELANDER, AXEL LEONARD—Continued
- 1950b. Taxonomic notes on some smaller Bombyliidae (Diptera). Pan-Pacific Ent., vol. 26, nos. 3–4, pp. 139–144; 145–156.
- 1961. The genus *Mythicomyia* (Diptera: Bombyliidae). Wasmann Journ. Biol., vol. 18 (1960), no. 2, pp. 161–261.

MELIS, A.
- 1934. Il Grillastro crociato (*Dociostaurus maroccanus* Thunb.) e le sue infestazioni in Sardegna. Atti Roy. Accad. Georgofili, ser, 5ª, vol. 30 (1933), pp. 399–504, 11 figs., 6 pls., 3 maps.

MERCIER, L.
- 1925. Diptères de la côte du Calvados (V^e liste). Bull. Ann. Soc. Ent. Belgique, vol. 65, pp. 173–182. (Bombyliidae p. 174, 2 spp. listed.)

MERLE, P. DU [see Biliotte, E., Demolin, G., and Merle, P. du]
- 1964. Cycle biologique d'un diptère *Bombyliidae* du genre *Villa*. Comptes Rendus Acad. Sci., Paris, vol. 259, pp. 1657–1659.
- 1966a. Modèle de cage permettant d'obtenir La Ponte d'un diptere Bombyliidae, *Villa quinquefasciata*. Entomophaga 3, pp. 325–330, figs. 1–3.
- 1966b. Modalites de l'accouplement chez un diptere Bombyliidae, *Villa quinquefasciata*. Ann. Soc. Ent. France, vol. 2, pp. 617–623.
- 1969a. Existence de deux diapauses facultatives au cours du cycle biologique de *Villa brunnea* Beck. (Diptera, Bombyliidae). Compt. Rend. Acad. Sci. Paris, t. 268, pp. 2433–2435.
- 1969b. Sur quelques facteurs qui régissent l'efficacite de *Villa brunnea* Beck. (Dipt., Bombyliidae) dans la regulation des populations de *Thaumetopoea pityocampa* Schiff. (Lep., *Thaumetopoeidae*). Institut National de la Recherche Agronomique, pp. 57–66.
- 1971. Biologie de deux especes du genre *Usia* Latreille (Dipt., Bombyliidae). Ann. Soc. Ent. Frace, vol. 7, pp. 241–259.

MERTON, L. F. H.
- 1959. Studies in the ecology of the Moroccan locust (*Dociastaurus maroccanus* Thunb.) in Cyprus. Anti-Locust Bull., London, no. 34, pp. 1–123, figs. 2–28, photos 29–44, 1 folding map. (Bombyliidae pp. 74, 75, 76, 78, 79, 80–83, 84–100, 102, 103–107; figs.)

METCALF, C. L.
- 1921. The genitalia of male Syrphidae: Their morphology, with especial reference to its taxonomic significance. Ann. Ent. Soc. America, vol. 14, pp. 169–226, pls. 9–19.

METZ, CHARLES W.
- 1916. Chromosome studies on the Diptera II. The paired association of chromosomes in Diptera and its significance. Journ. Exper. Zoology, vol. 21, pp. 213–280.

MEUNIER, FERNAND ANATOLE
- 1902. Etudes de quelques Dipteres de l'ambre. Ann. Soc. Nat. Zool. (Paris). Vol. 16, pp. 395–406.
- 1910. Un Bombylidae de l'ambre de la Baltique. (Dipt.) Bull. Soc. Ent. France, no. 19, pp. 349–350, figs. 1–2.
- 1914. Nouvelles Recherches sur quelques insectes du Sannoisien d'Aix en Provence. Bull. Soc. Geol. France, 4e série, vol. 14. (Bombyliidae pp. 195–196, figs. 8–9, pl. 7, fig. 5.)
- 1915. Nouvelles recherches sur quelques insectes des plâtrières d'Aix en Provence. Verh. Konink. Akad. Wetenschappen Amsterdam, sect. 2, vol. 18, no. 5, pp. 1–17, 5 pls. (Bombyliidae p. 13, pl. 4, fig. 13.)
- 1916. Sur quelques Diptères (Bombylidae, Leptidae, Dolichopodidae et Chironomidae) de l'ambre de la Baltique. Gravenhage. Tijdschr. Ent., vol. 59, pp. 274–286, 16 figs. (Bombyliidae pp. 274–286, 16 figs. (Bombyliidae pp. 274–277, figs. 1–3.)
- 1917. Sur quelques insectes de l'Aquitanien de Rott. Sept Montagnes (Prusse rhénane). Verh. Konink. Akad. Wetenschappen Amsterdam, sect. 2, vol. 20, no. 1, pp. 3–17, 4 pls., 22 figs. (Bombyliidae pp. 9–10, fig. 7.)

MICKEL, C. E.
    1928. The biotic factors in the environmental resistance of *Anthophora occidentalis* Cress. (Hym.: Apidae; Dipt., Coleop.). Ent. News, vol. 39, pp. 69–78.

MIK, JOSEF
    1864–1878. Dipterologische Beiträge. Verh. zool.-bot. Ges. Wien, vol. 14, pp. 785–798, pl. 21A. (Bombyliidae p. 786; 2 spp. mentioned.)
    1881–1882. Dipterologische Mittheilungen. Verh. zool.-bot. Ges. Wien.
        1881. I. Ueber einige Dipteren aus der Sammlung Dr. Emil Gobert's in Mont-de-Morzan. Vol. 30 (1880), pp. 587–610, pl. 17. (Bombyliidae p. 592; 4 spp. listed.)
    1885a. Dipterologische Winke. Ent. Nachrichten, vol. 11, pp. 341–343. (Bombyliidae p. 343.)
    1885b. Diptera des Gebietes *in* Becker: Hernstein in Nieder-Oesterreich. One volume and atlas, 711 pp. Wien. Dr. G. Beck was responsible for the part titled: Fauna von Hernstein in Nieder-Oesterreich und der wieteren Umgebung. Part II, pp. 43–77; part III, Fauna des Gebietes, pp. 467–711. (Bombyliidea pp. 55–56; 7 spp.)
    1886. Die Dipterengenera Paolo Lioy's. Ent. Nachrichten, Berlin, vol. 12, pp. 321–328. (Bombyliidae p. 323.)
    1886–1891. Dipterologische Miscellen. I–XIX. Wiener Ent. Zeitung.
        1887a. Part V, vol. 6, pp. 187–191. (Bombyliidae p. 187.)
        1887b. Part VI, vol. 6, pp. 238–242. (Bombyliidae p. 242.)
        1887c. Part VII, vol. 6, pp. 264–269. (Bombyliidae p. 269.)
        1888a. Part X, vol. 7, pp. 140–142. (Bombyliidae p. 140.)
        1888b. Part XIII, vol. 7, pp. 299–303. (Bombyliidae p. 299.)
    1890. Ueber die dipterologischen referate in den Jahrgängen 1882 bis inclusive 1890 der Wiener Entomologischen Zeitung. Wiener Ent. Zeitung, vol. 9, pp. 281–308. (Bombyliidae pp. 292–293.)
    1891. Ein Beitrag zur "Bibliotheca Entomologica." Wiener Ent. Zeitung, vol. 10, pp. 65–95. (Bombyliidae pp. 71, 83, 86.)
    1896. Dipterologische Miscellen. Ser. 2, part VII. Wiener Ent. Zeitung, vol. 15, pp. 106–114. (Bombyliidae p. 106.)

MIKAN, JOHANN CHRISTIAN
    1796. Monographia Bombyliorum Bohemiae. Prague. 71 pp., 4 pls. (color).
    1797. Entomologische Beobachtungen, Berichtigungen und Entdeckungen. Neue Abhandl. Bohm. Ges. Wissenschr., vol. 3, pp. 108–136. (Separate, Prague, 1797, pp. i–iv, 1–31.)

MILLER, DAVID
    1950. Catalogue of the Diptera of the New Zealand subregion. Dept. Sci. Industr. Res., Bull. 100, Ent Res. Stat. Publ., no. 5, pp. 1–194. (Bombyliidae p. 78.)

MIMEUR, J. M. [see De Lepiney and Mimeur]

MODEER, ADOLPH
    1776. Anmärkningar angående slägtet Gyrinus. Physiogr. Sallsk. Handl. Stockholm, vol. 1, pp. 155–162. (Bombyliidae, description of genus *Bombylius*.)

MORELET, P.
    1845. [Note: *Bombylius* obtained from a nest of *Halictus succinatus*.] Bull. Soc. Ent., France, p. xxiv.

MORITZ, L. D.
    1915. Biological observations on Acrididae of the Turgai Province. [In Russian.] Lyubitel Prirodui, St. Petersburg, 1915. Reprint 29 pp., 9 figs. 2 pls.

MUELLER, HERMANN
    1881. Alpenblumen, ihre Befruchtung durch Inseckten und ihre Anpassungen an dieselben. Leipzig. Vol. 4, 611 pp., 173 figs. (Bombyliidae p. 572, mostly spp. of *Bombylius* and *Systoechus* and 2 spp. of *Villa* as *Anthrax*.)

MUHLENBERG, M.
    1968. Zur morphologie der letzen abdominal-segmente bei weiblichen Wollschwebern (Diptera, Bombyliidae). Zool. Anz., vol. 181, nos. 3/4, pp. 277–279, illus.

MULLER, IRMGARD [see Coulson, Sabrosky, Muller]

MÜLLER, OTTO FRIEDRICH
- 1764. Fauna insectorum Fridrichsdalina, sila methodica descriptio insectorum agri fridrichsdalinensis. Hafniae et Lipsiae. xxiv + 96 pp. (Bombyliidae pp. 85, 88.)
- 1776. Zoologiae Danicae prodromus seu animalium Daniae et Norvegiae indigenarum. Hafniae. xxxii + 282 pp. Insects pp. 52–201. (Bombyliidae pp. 177, 182.)

MULSANT, ETIENNE
- 1852. Note pour servir à l'histoire des *Anthrax* (insectes Diptères), suivie de la description de trois espèces de ce genre nouvelles ou peu connus. Mem. Acad. Sci. Lyon, ser. 2, vol. 2, pp. 18–24. Also, opuscula Ent., vol. 1, pp. 178–184. Note: *Exoprosopa* (*interrupta*) *rutila*: Opuscula Ent., vol. 1, pp. 180–181.

NEDĚLKOV, N.
- 1912. Sesti prinos kăm entomologičnata fauna na Balgarija. Dvukrili. (6. Beitrag zur entomologischen Fauna Bulgariens. Zweiflügler.) Spis. Bulg. Acad. Nauk., kl. prirod.-matemat., vol. 2, pp. 177–218.

NEUHAUS, G. H.
- 1886. Diptera Marchica. Systematisches Verzeichniss der Zweiflügler der Mark Brandenburg, mit kurzer Beschreibung und analytischen Bestimmungs-Tabellen. Berlin. xvi + 371 pp., 6 pls. (Bombyliidae pp. 51–55.)

NEWMAN, EDWARD
- 1840–1842. Entomological notes. Entomologist, vol. 1, 424 pp., 16 pls. Art. 14, pp. 220–223, 1841. (Bombyliidae pp. 220–222.)

NEWSTEAD, R., EVANS, A. M., and POTTS, W. H.
- 1924. Guide to the study of Tsetse Flies. 332 pp. Liverpool School Tropical Medicine.

NICHOLS, G. E.
- 1933. The composition and biogeographical relations of the fauna of Western Australia. Report Australasian Assoc. Adv. Sci., vol. 21, pp. 93–138.

NICHOLSON, A. J.
- 1927. A new theory of mimicry in insects. Australian Zool., vol. 5, pt. 1, pp. 1–104, 3 text-figs., 14 pls. (2 color). (Bombyliidae pp. 33, 35, 57, text-figs. 1g, 3d, pl. 2, fig. 5; pl. 3, figs. 17–18.)

NIELSEN, E.
- 1925. Traek af Insekternes. Liv. II. Ent. Medd., Copenhagen, vol. 14, pp. 441–448, 4 figs. (Bombyliidae pp. 445–448, fig. 4.)

NIELSEN, J. C.
- 1903. Ueber die Entwicklung von *Bombylius pumilus* Meig., einer Fliege, welche bei *Colletes daviesana* Smith schmarotzt. Zool. Jahrb., Abth. Syst., Geogr. Biol. Thiere, Jena, vol. 18, pp. 647–658, pl. 28.

NININGER, H. H.
- 1916. Studies in the life histories of two carpenter bees of California, with notes on certain parasites. Journ. Ent. Zool., Claremont, vol. 8, pp. 158–165. (Bombyliidae pp. 162–163, pl. 1, figs. 1–6; pl. 2, figs. 7–12.)

NOWICKI, MAXIMILIAN SILA
- 1867. Beschreibung neuer Dipteren. Verh. zool.-bot. Ges. Wien, vol. 17, pp. 337–354, pl. 11. (Bombyliidae pp. 343–348.)

NURSE, C. G.
- 1922. New and little known Indian Bombyliidae. Journ. Bombay Nat. Hist. Soc., vol. 28, pp. 630–641, 1 pl.

NYE, W. P., and BOHART, GEORGE E.
- 1959. Photographing insects close up. Journ. Biol. Photographic Assoc., November 1959, pp. 139–145, figs. 1–10. (Beautiful photographs of *Heterostylum robustum* O. S., a serious parasite of the alkali bee, transforming from a larva to a pupa and drilling its way to the surface of the ground.)

OGILVIE, L.
- 1928. The insects of Bermuda. Dept. Agric. Bermuda, pp. 1–52.

OLDROYD, HAROLD
- 1938. Bombyliidae from Chile and Western Argentina. Rev. Chilena Hist. Nat., Santiago, vol. 41 (1937), pp. 83–93.

OLDROYD, HAROLD—Continued
- 1939. Rhagionidae, Tabanidae, Asilidae, Bombyliidae. Ruwenzori Expedition 1934–35. British Mus. (Nat. Hist.), vol. 2, pp. 13–47, 2 pls., 11 figs. (Bombyliidae pp. 46–47.)
- 1940. Entomological expedition to Abyssinia, 1926–27: Diptera-Brachycera, Tabanidae, Bombyliidae. Ann. Mag. Nat. Hist., ser. 11, vol. 5, pp. 192–203, 3 figs. (Bombyliidae pp. 201–203.)
- 1947a. Results of the Armstrong College expedition to Siwa Oasis (Libyan Desert), 1935. Bull. Soc. Fouad Ier Ent., vol. 31, pp. 113–120. (Bombyliidae p. 120; 3 determined spp. listed; 2 spp. undetermined.)
- 1947b. A new species of *Systoechus* (Diptera: Bombyliidae) bred from eggs of the desert locust. Proc. Roy. Ent. Soc. London, ser. B, vol. 16, pp. 105–107, 2 figs.
- 1951. A giant Bombyliid (Diptera) bred from the pupa of a Cossid moth. Proc. Roy. Ent. Soc. London, ser. B, vol. 20, pp. 49–50, 6 figs.
- 1961. Ergebnisse der Deutscher Afghanistan-Expedition 1956 der Landessammlungen fur Naturkunde Karlsruhe. Bombyliidae, Therevidae (Diptera). Beitr. Naturk.-forsch. Südwestdeutsch., Karlsruhe, vol. 19, 1961, pp. 301–303, 2 figs.
- 1964. The natural history of Flies. London: Wiedenfeld & Nicolson, 324 pp., 40 figs. (Bombyliidae pp. 8, 119, 131–133, 136, 145, 157, fig. 16.)

OLIVIER, ANTOINE GUILLAUME
- 1789–1825. *In*: Encyclopédie méthodique. Dictionnaire des Insectes. Paris, Pankouke. In 10 vols., 398 pls.
    - 1789. Vol. 4, 331 pp. (Bombyliidae pp. 323–329; 27 spp. listed in general terms on pp. 323–325; 14 Linnaean and Fabrician spp. listed, and 9 new genera described pp. 326–329.)

OLIVIER, ERNEST
- 1877. [Transmitted note: *Bombylii* in large numbers flying in company with *Anthophora*.] Proc. Ent. Soc. London, 1877, p. ii.

OSTEN SACKEN, CARL ROBERT BARON VON
- 1858. Catalogue of the described Diptera of North America. . . . Smithsonian Misc. Coll., vol. 3, vii + 95 pp. (Bombyliidae pp. 38–44 (Bombyliarii), including *Hirmoneura*.)
- 1862. Entomologischer Notizen. XIII. Ueber einige Fälle von Parasitismus unter Hymenoptern und Diptern (*Toxophora, Eumenes; Trypoxylon, Pelopaeus; Somula, Vespa*). Stettiner Ent. Zeitung, vol. 23, pp. 411–412.
- 1876. Report upon the collection of Diptera made in portions of Colorado and Arizona during the year 1873. Rept. Geogr. Geol. Survey West 100th Meridian, vol. 5, pp. 305–807. (Bombyliidae pp. 806–807.)
- 1877. Art. XIII. Western Diptera: Descriptions of new genera and species of Diptera from the region west of the Mississippi and especially from California. Bull. U.S. Geol. Geogr. Survey Territories, vol. 3, no. 2, pp. 189–354. (Bombyliidae pp. 225–274.)
- 1878. Catalogue of the described Diptera of North America. Second edition. Smithsonian Misc. Coll., no. 270, vol. 16, xlvi + 276 pp. (Bombyliidae pp. ix, 85–95; notes on 47 genera and spp. on pp. 237–239.)
- 1880. [Quoting Lemmon: Habits of *Bombylius*.] Ent. Monthly Mag., vol. 17, p. 161.
- 1881. Habits of *Bombylius*. Ent. Monthly Mag., vol. 17, pp. 206–207.
- 1882a. Enumeration of the Diptera of the Malay Archipelago collected by Prof. Odoardo Beccari, Mr. L. M. D'Albertis, and others. Ann. Mus. Civ. Storia Nat. Genova, vol. 16 (1881), pp. 393–492. (Bombyliidae pp. 432–433; 7 spp., 2 genera listed. Reprint pp. 44–45.)
- 1882b. Diptera from the Philippine Islands, brought home by Dr. Carl Semper. Berliner Ent. Zeitschr., vol. 26, pp. 83–120, 187–252. (Bombyliidae p. 112.)
- 1882c. [Note: Diptera Eremochaeta generally contiguous eyes, Diptera Chaetophora generally noncontiguous eyes.] Wiener Ent. Zeitung, vol. 1, pp. 91–92. (Bombyliidae p. 92.)
- 1885. Bericht über eine russischer Sprache erschienene dipterologische Arbeit. Wiener Ent. Zeitung, vol. 4, pp. 9–10. (Bombyliidae p. 9.)

OSTEN SACKEN, CARL ROBERT BARON VON—Continued
- 1886–1887. *In*: Biologia Centrali-Americana. Insecta, Diptera. London. 43 Vols. (Diptera, vol. 1).
  - 1886. Diptera, vol. 1, pp. i–viii, 1–128, pls. 1–3. (Bombyliidae, pp. 75–128, pl. 1, figs. 12–18; pl. 2, figs. 1–19; pl. 3, figs. 1–5, 15.)
  - 1887. Diptera, vol. 1, pp. 129–216, pls. 4–6. (Bombyliidae pp. 129–162, pl. 5, figs. 1–11.)
- 1890. Suggestions towards a better grouping of certain families of the order Diptera. Ent. Monthly Mag., ser. 2, vol. 2, pp. 35–39. (Bombyliidae p. 38.)
- 1896. Preliminary notice of a subdivision of the suborder Orthorrhapha Brachycera (Dipt.) on chaetotactic principles. Berliner Ent. Zeitschr., vol. 41, no. 4, pp. 365–373. (Bombyliidae pp. 367–370.)

PACKARD, ALPHEUS SPRING
- 1868. The home of the bees. Amer. Natur., vol. 1 (1867), pp. 364–378, pl. 10, figs. 1–15. (Bombyliidae pp. 368–369, pl. 10, fig. 5, tunnel containing pollen and young; fig. 6, the larva; fig. 7, the pupa of *Anthrax sinuosa*.)
- 1869. Guide to the study of insects, and a treatise on those injurious and beneficial to crops: for the use of colleges, farm-schools, and agriculturists. Salem: Essex Institute Press, and London. 702 pp., 651 text-figs., 11 pls. (Bombyliidae pp. 132, pl. 4, figs. 6–7, pp. 363, 396–397.)
- 1897. Notes on the transformations of higher Hymenoptera. III. Journ. New York Ent. Soc., vol. 5, no. 3, pp. 109–120. (Bombyliidae pp. 113–114.)

PAINTER, ELIZABETH M. [see Painter and Painter]

PAINTER, REGINALD H.
- 1925. A review of the genus *Lepidophora* (Diptera, Bombyliidae). Trans. Amer. Ent. Soc., vol. 51, pp. 119–127, no. 870, figs, 1a, 1b.
- 1926a. Notes on the genus *Parabombylius* (Diptera). Ent. News, vol. 37, pp. 73–78.
- 1926b. The *lateralis* group of the bombylid genus *Villa*. Ohio Journ. Sci., vol. 26, no. 4, pp. 205–212.
- 1930a. A review of the bombyliid genus *Heterostylum* (Diptera). Journ. Kansas Ent. Soc., vol. 3, no. 1, pp. 1–7.
- 1930b. Notes on some Bombyliidae (Diptera) from the Republic of Honduras. Ann. Ent. Soc. America, vol. 23, no. 4, pp. 793–807, 1 pl., 7 figs.
- 1932a. A monographic study of the genus *Geron* Meigen as it occurs in the United States (Diptera: Bombyliidae). Trans. Amer. Ent. Soc., vol. 58, pp. 139–167, pls. 10–11.
- 1932b. The Bombyliidae of China and near-by regions. Lingnan Sci. Journ., vol. 11, no. 3, pp. 341–374, pls. 4–5, 16 figs.
- 1933a. New subgenera and species of Bombyliidae (Diptera). Journ. Kansas Ent. Soc., vol. 6, no. 1, pp. 5–18.
- 1933b. Notes on some Bombyliidae (Diptera) from Panama. Amer. Mus. Nov., no. 642, pp. 1–10, figs. 1–11, 12a-d.
- 1934. Two new species of North American *Exoprosopa* (Bombyliidae; Diptera). Journ. Kansas Ent. Soc., vol. 7, pp. 68–70.
- 1939. Two new species of South American Bombyliidae. Arb. morph. taxon. Ent. Berlin-Dahlem, vol. 6, pp. 42–45, figs. 1–8.
- 1940. Notes on type specimens and descriptions of new North American Bombyliidae. Trans. Kansas Acad. Sci. (1939), vol. 42, pp. 267–301, 2 pls.
- 1946. Bombyliidae *in* Stuardo: Catalogo de los Dipteros de Chile. Ministry Agric. Santiago de Chile, 253 pp. (Bombyliidae by Painter, pp. 88–97; 109 spp. listed in 27 genera.)
- 1959. A new genus of Bombyliidae (Diptera). Journ. Kansas Ent. Soc., vol. 32, pp. 73–75, 1 fig.
- 1962. The taxonomy and biology of *Systoechus* and *Anastoechus* bombyliid (Diptera) predators in grasshopper egg pods. Journ. Kansas Ent. Soc., vol. 35, no. 2, pp. 255–269.
- 1968. Review of the genus *Desmatomyia* Williston (Diptera: Bombyliidae). Journ. Kansas Ent. Soc., vol. 41, pp. 408–412, 1 pl., 12 figs.

PAINTER, REGINALD H., and HALL, JACK C.
    1960. A monograph of the genus *Poecilanthrax* (Diptera: Bombyliidae). Tech. Bull. no. 106, Agric. Exper. Stat., Kansas State Univ., pp. 1–132, pls. 1–8, maps 1–31, photographs 1–34.

PAINTER, REGINALD H., and PAINTER, ELIZABETH M.
    1962. Notes on the redescriptions of types of North American Bombyliidae (Diptera) in European museums. Journ. Kansas Ent. Soc., vol. 35, no. 1, pp. 1–164, 20 figs.
    1963. A review of the subfamily Systropinae (Diptera: Bombyliidae) in North America. Journ. Kansas Ent. Soc., vol. 36, no. 4, pp. 278–348, photographs 1–11, pls. 1–16, (pls. 1–2 with drawings, pls. 3–6 with 1–39 photographs.)
    1965. Family Bombyliidae *in* Stone and others: A catalog of the Diptera of America North of Mexico. Agric. Res. Serv., U.S. Dept. Agric., Washington, D.C., pp. 407–446.
    1968a. A review of the genus *Hyperalonia* Rondani (Bombyliidae, Diptera) from South America. Papeis Avulsos de Zoologia São Paulo, vol. 22, pp. 107–121.
    1968b. Review of the genus *Desmatomyia* Williston (Diptera: Bombyliidae). Journ. Kansas Ent. Soc., vol. 41, no. 3, pp. 408–412, illus.
    1969. New Exoprosopinae from Mexico and Central America (Diptera: Bombyliidae). Journ. Kans. Ent. Soc., vol 42, pp. 5–34, 2 plates with 35 figures.

PALLAS, PETER SIMON
    1818. Aus Pallas Dipterologischem Nachlasse. Zool. Mag., Wiedemann, Kiel, vol. 1, pt. 2, pp. 1–40. (Bombyliidae pp. 8–24.)

PALM, JOSEF
    1876. Beitrag zur Dipteren-Fauna Oesterreichs. Verh. zool.-bot. Ges. Wien, vol. 25 (1875), pp. 411–422. (Bombyliidae pp. 412–414.)

PANTEL, J.
    1910–1913. Récherches du les Diptères à larves entomobies I. Charactères parasitiques aux points de vue biologique, éthologique et histologique.
        1910. Cellule Louvain, vol. 26, pp. 25–216, text-figs. 1–26, pls. 1–4.
        1913. Cellule Louvain, vol. 29, pp. 1–289, text-figs. 1–26, pls. 1–7.

PANZER, GEORG WOLFGANG FRANZ
    1793–1813. Faunae insectorum germanicae initiae oder Deutschlands Insecten. Nürnberg. In 109 parts (hefts) by Panzer (190 in all), each part with 24 pages and with 24 color plates. (Bombyliidae, 1794, pt. 24, pl. 24; 1796, pt. 32, pls. 18, 19; 1797, pt. 45, pls. 16, 17.)

PAOLI, GUIDO
    1919. Notizie sulla lotta contro le Cavallette nella provincia di Foggia nel 1919 e su proposte di nuovi metodi. La Propaganda Agricila e l'Agric. Pugliese, Bari, ser. 2ª, vol. 11, 5 pp.
    1920. Considerazioni sui rapporti biologici fra le cavallette e i loro parassiti oofagi. Riv. Biol., Roma, vol. 2, pp. 387–397.
    1932. Osservazioni sulla biologia del *Dociostaurus marroccanus* Thnb. In Italia nelle fasi gregaria e solitaria e sull'azione di alcumi insetti parassiti. Vᵉ Congrès Intern. d'Entom., vol. 2 (Travaux), pp. 633–643, Paris. Also, Nuovi Ann. Agric., Roma, vol. 12, pp. 627–639, 2 figs.
    1937a. Osservazioni sulla morfologia dell'estremo addome della femmina dei Ditteri Bombiliidi. Redia, Firenze, vol. 23, pp. 1–4, figs. 1–3 (several parts each fig.)
    1937b. Ricerche sulla morfologia e anatomia del capo delle larve dei Ditteri Bombiliidi. Redia, Firenze, vol. 23, pp. 5–16, figs. 1–12.
    1937c. Studi sulle Cavallette di Foggia (*Dociostaurus marroccanus* Thunb.). E sui loro oofagi (Ditteri Bombiliidi e Coleotteri Meloidi). Ed acari Ectofagi (Eritreidi e Trombidiidi). Redia, Firenze, vol. 23, pp. 27–206, 99 figs., 3 pls. (Bombyliidae pp. 103–129, figs. 34–61.)

PARAMONOV, SERGEI J.
    1924a. Zwei neue Bombyliiden-Arten aus dem palaearktisch Gebiet. Mém. Acad. Sci. d'Ukraine, vol. 1, livr. 2, pp. 59–62.
    1924b. Zwei neue Bombyliiden-Arten (Diptera) aus Transkaspien. Konowia, vol. 3, pp. 136–139, 2 figs.

PARAMONOV, SERGEI J.—Continued
- 1924–1926. Zur Kenntnis der Gattung *Lomatia* (Bombyliidae, Diptera). I-III. Neue Beitr. syst. Insektenk., Berlin.
  - 1924. Pt. I, vol. 3, no. 6, pp. 41–46.
  - 1925. Pt. II, vol. 3, nos. 8–9, pp. 78–84, no. 10, pp. 95–100, 112–116.
  - 1926. Pt. III, vol. 3, nos. 17–18, pp. 176–183.
- 1925a. Zur Kenntnis der Gattung *Aphoebantus* (Bombyliidae, Diptera). Mém. Acad. Sci. d'Ukraine, vol. 1, livr. 3, pp. 26–29.
- 1925b. Zur Kenntnis der Gattung *Toxophora* (Bombyliidae, Diptera). Mém. Acad. Sci. d'Ukraine, vol. 1, livr. 3, pp. 43–48.
- 1925c. Zwei neue *Exoprosopa*-Arten (Bombyliidae, Diptera) aus dem palaearktischen Gebiet. Konowia, vol. 4, pp. 43–47.
- 1925d. Zur Kenntnis der Gattung *Heterotropus* (Diptera, Bombyliidae). Konowia, vol. 4, pp. 110–114.
- 1925e. Drei neue Bombyliiden-Arten aus dem palaearktischen Gebiet (Bombyliidae, Diptera). Zool. Anz., Leipzig, vol. 64, nos. 3–4, pp. 91–95.
- 1925f. Zwei neue *Villa*-Arten (Bombyliidae, Diptera) aus Turkestan (nebst einingen Bermerkungen über andere turkestanische Bombyliiden). Zool. Anz., Leipzig, nos. 5–6, pp. 144–148.
- 1925g. Zur Kenntnis der Insektenfauna (hauptsächlich Diptera) von Bessarabien und der Ukraine. Soc. Ent., Stuttgart, vol. 40, no. 6, pp. 21–23. (Bombyliidae pp. 21–22.)
- 1925h. Zwei neue *Bombylius*-Arten (Bombyliidae, Diptera) aus dem paläarktischen Gebiet. Soc. Ent., Stuttgart, vol. 40, no. 9, pp. 33–34.
- 1926a. Beiträge zur Monographie der Gattung *Bombylius* L. (Fam. Bombyliidae, Diptera). Mém. Acad. Sci. d'Ukraine, vol. 3, livr. 5, pp. 77–184.
- 1926b. Zur Kenntnis der Gattung *Platypygus* Löw (Bombyliidae, Diptera). Konowia, vol. 5, pp. 85–92.
- 1926c. Zur Kenntnis der Gattung *Anastoechus* O. S. (Bombyliidae, Diptera). Arch. Naturg. Berlin, vol. 91 (1925), abt. A, pt. 1, pp. 46–55. (Becker p. 46.)
- 1926d. Zur Kenntnis der Gattung *Anastoechus* (Dipt. Bombyliidae). (Beschreibung neuer Arten und eine Bestimmungstabelle.) Part II. Neue Beitr. syst. Insektenk., Berlin, vol. 3, pp. 127–137.
- 1926e. Zur Kenntnis der Gattung *Dischistus* Lw. (Dipt., Bombyl.) nebst einer Bestimmungstabelle. Neuer Beitr. syst. Insektenk., Berlin, vol. 3, nos. 15–16, pp. 155–161.
- 1926f. Zur Kenntnis der Gattung *Hemipenthes* Lw. Encycl. Ent., ser. B., II, vol. 3, pp. 150–192.
- 1926g. Generis *Prorachthes* Lw. (Diptera, Bombyliidae) species quattour novae palaearcticae. (In Russian with Latin descriptions.) Ann. Mus. Zool., Leningrad, vol. 27, pp. 76–87, pl. 5, 12 figs., 8 of antennae and 4 of wings.
- 1926–1947. Dipterologische Fragmente. I-XXXVIII.
  - 1926. I. Mém. Acad. Sci. d'Ukraine, vol. 4, livr. 2, pp. 95–104. Trav. Mus. Zool. Keiv, no. 1, pp. 77–86.
  - 1927a. VII. Über *Sparnopolius asiaticus* Becker. Mém. Acad. Sci. d'Ukraine, vol. 4, livr. 4, p. 325. Trav. Mus. Zool. Keiv, no. 2, p. 81.
  - 1927b. IX. Synonymische Bermukungen. Mém. Acad. Sci. d'Ukraine, vol. 7, livr. 1, p. 168. Trav. Mus. Zool. Kiev, no. 3, p. 168.
  - 1927c. X. Über die *Anthrax nomas* Eversm. Mém. Acad. Sci. d'Ukraine, vol. 7, livr. 1, p. 169. Trav. Mus. Zool. Kiev, no. 3, p. 169.
  - 1928a. XIII. Über die Gattung *Lomatia*. Mém. Acad. Sci. d'Ukraine, vol. 6, livr. 3, pp. 507–509. Trav. Mus. Zool. Kiev, no. 5, pp. 203–205.
  - 1928b. XIV. Über die gattung *Exoprosopa*. Mém. Acad. Sci. d'Ukraine, vol. 6, livr. 3, pp. 509–510. Trav. Mus. Zool. Kiev, no. 5, pp. 205–206.
  - 1928c. XV. Über die gattung *Bombylius*. Mém. Acad. Sci. d'Ukraine, vol. 6, livr. 3, pp. 510–511. Trav. Mus. Zool. Kiev, no. 5, pp. 206–207.
  - 1929a. XX. Über die gruppe von *Hyperalonia venus* Karsch. Mém. Acad. Sci. d'Ukraine, vol. 13, livr. 1, pp. 186–187. Trav. Mus. Zool., Kiev, no. 7, pp. 188–189.

Paramonov, Sergei J.—Continued

    1929b. XXI. Über neue *Bombylius*-Formen. Mém. Acad. Sci. d'Ukraine, vol. 13, livr. 1, pp. 188–190. Trav. Mus. Zool., Kiev, no. 7, pp. 190–192.

    1929c. XXII. Nachtrag zu der Gattung *Exoprosopa*. Mém. Acad. Sci. d'Ukraine, vol. 13, livr. 1, pp. 190–192. Trav. Mus. Zool., Kiev, no. 7, pp. 192–194.

    1931a. XXVI. Über einige im Berliner Museum befindliche Typen und wenig bekannte arten der Bombyliiden. Mém. Acad. Sci. d'Ukraine, no. 5, pp. 228–237.

    1931b. XXVII. Zwei neue Bombyliiden aus Afrika. Mém. Acad. Sci. d'Ukraine, no. 5, pp. 237–239. (*Systoechus marshalli* bred from egg pod of Acridid *Acrotylus deustus*.)

    1933a. XXIII. Über neue und alte *Antonia*-Arten. Inst. de Recherches Zool. et Biol. a'La Academie Sci. Ukraine. Trav. Mus. Zool. Kiev, no. 12, pp. 47–49.

    1933b. XXIX. Bombyliidologische Zusätzes. Inst. de Recherches Zool. et Biol. a'La Academie Sci. Ukraine. Trav. Mus. Zool. Kiev, no. 12, pp. 50–52.

    1933c. XXX. Über einige Bombyliiden-Typen des Leningrader Museums. Inst. de Recherches Zool. et Biol. a'La Academic Sci. Ukraine. Trav. Mus. Zool. Kiev, no. 12, pp. 53–56.

    1947. XXXVIII. Bombyliiden-Notizen. Eos, Rev. Española Ent., vol. 23, pp. 79–101.

1928. Beiträge zur Monographie der Gattung *Exoprosopa* Macq. (Bombyliidae, Diptera). Mém. Acad. Sci. d'Ukraine, vol. 6, livr. 2, pp. 181–303, 3 text-figs., 6 pls., 60 figs. (Trav. Mus. Zool. Kiev, no. 4, pp. 1–125.)

1929. Beiträge zur Monographie einiger Bombyliiden-Gattungen (Diptera). (Russian summary.) Mém. Acad. Sci. d'Ukraine, vol. 11, livr. 2, pp. 65–224. (Trav. Mus. Zool., Kiev, no. 6, pp. 1–161.)

1930. Beiträge zur Monographie der Bombyliiden-Gattungen *Cytherea*, *Anastoechus*, etc. Bombyliiden (Diptera). Mém. Acad. Sci. d'Ukraine, Classe Sci. Phys. Math., vol. 15, livr. 3, pp. 355–481. (Trav. Mus. Zool. Kiev, no. 9, pp. 1–128.)

1931a. Beiträge zur Monographie der Bombyliiden-Gattungen *Amictus*, *Lyophlaeba*, etc. (Diptera). Mém. Acad. Sci. d'Ukraine, vol. 9, pp. 1–218. (Trav. Mus. Zool., Kiev, no. 11.)

1931b. Die Verbreitung der Gattung *Usia* Latr. (Bombyliidae, Diptera) und die Probleme der Krimschen Fauna Zool. Anz., Leipzig, vol. 96, pp. 282–284.

1932. Die Verbreitung der Gattung *Usia* Latr. (Bombyliide, Diptera) und die Probleme der Krimschen Fauna. [In Russian with German summary.] Journ. Cycle Bio-zool., Kiev, no. 3, pp. 65–66.

1933a. Beiträge zur Monographie der palaearktischen Arten der Gattungen *Toxophora* (Bombyliidae, Diptera). Inst. Recherches Zool. Biol. Acad. Sci. D'Ukraine, Kiev, no. 12, pp. 33–46. (Trav. Mus. Zool. Kiev, no. 12, pp. 33–46.)

1933b. Bombyliidae in the New Zealand fauna. Entomologist, London, vol. 66, p. 178.

1934a. Einige neue Ausgaben Eiher die Artbildung als Folge der Inselisolation. Mém. Acad. Sci. d'Ukraine, vol. 4. (Trav. Mus. Zool., Kiev, no. 13, pp. 172–174.)

1934b. Unentbehrliche kritische Bemerkungen zu der Arbeit von Dr. Engel über Bombyliiden *in* Lindner: Die Fliegen der paläarktischen Region. Konowia, Vienna, vol. 13, pp. 10–21.

1934c. Ueber einige exotische (hauptsächlich südamerikanische) Bombyliiden (Dipteren). Konowia, vol. 13, pp. 22–34.

1934d. Schwedisch-chinesische wissenschaftliche Expedition nach den nordwestlichen Provinzen Chinas, unter Leitung von Dr. Sven Hedin und Prof. Su Ping-Chang. (Insekten gesammelt vom schwedischen Arzt der Expedition Dr. David Hummel 1927–1930.) 9. Diptera. 1. Bombyliidae. Arkiv Zool., Stockholm, vol. 26A, no. 4, pp. 1–7.

1934e. Ueber einige interessante Dipterenfinde in Armenien. Journ. Cycle bio-zool., Kiev, no. 4, pt. 8, pp. 31–39.

Paramonov, Sergei J.—Continued

1934f. Neue und alte Bombyliiden (Diptera). Stylops, vol. 3, pt. 5, pp. 107–111.

1935a. Ist das südliche Untergebiet der paläarktis (nach Bartenew) aetiopischer Herkunft? [In Ukrainian with summaries in Russian and German.] Acad. Sci. d'Ukraine, Trav. Instit. Zool. Biol., vol. 5. (Trav. Mus. Zool. Kiev, no. 14, pp. 43–57.)

1935b. Ist das südliche Untergebiet der paläarktis (nach Bartenew) aetiopischer Herkunft? [In Russian with German summary.] Rev. Zool. russe, Moscow, vol. 14, pp. 397–409.

1935c. In Visser: Wissenshaftliche Ergebnisse der Niederlandischen Expedition in den Karakorum und die Angrenzenden Gebiet, 1922, 1925 und 1929–1930. Leipzig. Vol. 1, Zoologie, 517 pp., 8 pls., 230 figs.; Insects pp. 205–405. (Bombyliidae p. 399.)

1935d. In Sjöstedt: Entomologische Ergebnisse der schwedischen Kamtchatka-Expedition 1920–1922. 37. Abschluss und Zusammenfassung. Ark. Zool., vol. 28A, no. 7, pp. 1–19, 1 fig. (Bombyliidae by Paramonov, p. 11; 2 spp. listed.)

1936a. Beiträge zur Monographie der Gattung *Anthrax* (Bombyliidae). Acad. Sci. d'Ukraine, vol. 11, pp. 3–31. Trav. l'Inst. Zool. Biol., Kiev, no. 16, pp. 3–31. [German summary in form of a key to species pp. 17–31; pp. 3–16 in Russian.]

1936b. Ueber einige interessante Dipteren funde im Mariupolschen Gebiet. Acad. Sci. d'Ukraine, vol. 11; Trav. l'Inst. Zool. Biol., Kiev, no. 16, pp. 113–123, 6 figs. (Bombyliidae, 19 spp. listed.) [Russian pp. 113–121; German summary pp. 122–123.]

1936c. Materialien zur Monographie der Gattung *Anthrax* (Bombyliidae, Diptera). Acad. Sci. R.S.S.D. Ukraine, Trav. Inst. Zool. Biol., Kiev, no. 13. (Trav. Mus. Zool., Kiev, no. 18, pp. 69–159.)

1936d. Uber neue und alte *Antonia*-Arten (Bombyl. Dipt.). Mitt. deutsche. ent. Ges., Berlin, vol. 7, pp. 27–31.

1936e. Schwedisch-chinesische Wissenschaftliche Expedition nach den nordwestlichen Provinzen Chinas. 45. Diptera. 13. Bombyliidae (bis). Ark. Zool., vol. 27A, no. 26, pp. 1–7.

1937. Ueber einige Bombyliiden-Typus (Diptera) des Berliner Museums. Mitt. zool. Mus., Berlin, vol. 22, pp. 286–303.

1939a. Kritische Ubersicht der gegenwärtigen und fossilen Bombyliiden-Gattungen (Diptera) der ganzen Welt. Rep. Inst. Zool. Biol., Kiev, no. 23, pp. 23–50 (of Acad. Sci. d'Ukraine, U.S.S.R.). [In Ukrainian with Russian and German summaries]

1939b. Ein Grundriss der Biologie, der Verbreitung und der oekonomischen Bedeutung der Bombyliiden (Diptera). Rep. Inst. Zool. Biol., Kiev, no. 23, pp. 51–88 (of Acad. Sci. d'Ukraine, U.S.S.R.). [In Ukrainian, with Russian and German summaries.]

1940a. Ueber einige aussereuropäische (hauptsächlich amerikanische) Bombyliiden-Gattungen. Eos, Rev. Española Ent., Madrid, vol. 13 (1937), pp. 13–43.

1940b. Faune de l'U.R.S.S. Insectes Dipteres, vol. 9, no. 2, Fam. Bombyliidae (sous-fam. Bombyliinae). [In Russian with German descriptions.] Inst. Zool. Acad. Sci. U.R.S.S., Leningrad, new ser., no 25, 423 pp., 328 figs. [Pp. 1–324 in Russian; pp. 325–413 in German with key, descriptions, and index.]

1944. *Exoprosopa rhymnica* Eversm. (Bombyliidae)—eine vergessene Art der paläarktischen Dipterenfauna. Mitt. Deutsche. Ent. Ges., Berlin, vol. 12, pp. 41–43.

1946. Unentbehrliche kritische Bemerkungen zu der Arbeit von Dr. E. O. Engel ueber Bombyliiden *in* Lindner: Die Fliegen der paläarktischen region. Encycl. Ent., ser. B., II. Diptera., vol. 10, pp. 15–22.

1947a. Zur Kenntnis der Amerikanischen Bombyliiden-Gattung *Triploechus* Edw. (Diptera). Rev. Ent., Rio do Janeiro, vol. 18, pts. 1–2, pp. 183–192.

1947b. Kurze Uebersicht der *Sericosoma*-Arten (Bombyliidae, Diptera). Rev. Ent., Rio de Janeiro, vol. 18, pt. 3, pp. 361–369.

Paramonov, Sergei J.—Continued
- 1947c. Uebersicht der mit der Gattung *Usia* Latr. (Bombyliidae, Diptera) Naechstverwandten Gattungen. Eos, Rev. Española Ent., Madrid, vol. 23, pp. 207–220.
- 1948. Uebersicht der Bombyliiden-Gattung *Lyophlaeba* Rond. (Diptera), nebst einer Bestimmungstabelle. Rev. Ent., Rio de Janeiro, vol. 19, pts. 1–2, pp. 115–148.
- 1949. Revision of the species of *Lepidophora* Westw. (Bombyliidae, Diptera). Rev. Ent., Rio de Janeiro, vol. 20, pts. 1–3, pp. 631–643.
- 1950. Bestimmungstabelle der *Usia*-Arten der Welt (Bombyliidae, Diptera). Eos, Rev. Española Ent., Madrid, vol. 26, pp. 341–378.
- 1950a. Notes on Australian Diptera: I. The localities referred to by Macquart as "Cap des Aiguilles," "Isle Sydney," and "Oceanie." Ann. Mag. Nat. Hist., ser. 12, vol. 3, pp. 515–519. (Bombyliidae pp. 516–518.)
- 1950b. Notes on Australian Diptera: IV. A review of Australian species of the genus *Antonia* (Bombyliidae). Ann. Mag. Nat. Hist., ser. 12, vol. 3, pp. 529–533.
- 1951a. Notes on Australian Diptera: VI. A review of the genus *Docidomyia* White (Bombyliidae). Ann. Mag. Nat. Hist., ser. 12, vol. 4, pp. 745–752.
- 1951b. On two South American species of *Walkeromyia* Param. (Bombyliidae, Diptera). Rev. Ent., Rio de Janeiro, vol. 22, pts. 1–3, pp. 353–356.
- 1953a. Notes on Australian Diptera: XI. A new species of *Antonia* (Bombyliidae). Ann. Mag. Nat. Hist., ser. 12, vol. 6, pp. 204–205.
- 1953b. A note on the group *Ligyra* (*Hyperalonia olim*) *venus* Karsch (Diptera: Bombyliidae). Proc. Roy. Ent. Soc. London, ser. B., vol. 22, pts. 11–12, pp. 220–222.
- 1953c. Uebersicht der palaearktischen *Toxophora-Arten* (Bombyliidae). Encycl. Ent., Paris, ser. B. II. Diptera, vol. 11, p. 93–117.
- 1954a. A note on some African species of *Toxophora* Meigen (Diptera: Bombyliidae). Proc. Roy. Ent. Soc. London, ser. B, vol. 23, pts. 11–12, pp. 213–214.
- 1954b. Two new genera of Bombyliidae (Diptera) from the Belgian Congo. Proc. Roy. Ent. Soc. London, ser. B, vol. 23, pp. 26–28.
- 1955a. Notes on some African species of *Ligyra* and *Exoprosopa* (Diptera: Bombyliidae). Proc. Roy. Ent. Soc. London, ser. B, vol. 24, pts. 3–4, pp. 58–61.
- 1955b. African species of the *Bombylius discoideus* Fabricius group (Diptera: Bombyliidae). Proc. Roy. Ent. Soc. London, ser. B, vol. 24, pts. 9–10, pp. 159–164.
- 1957. Zur Kenntnis der Gattung *Spongostylum* (Bombyliidae, Diptera). Eos, Rev. Española Ent., Madrid, vol. 33, pts. 1–4, pp. 123–155.
- 1959. Zoogeographical aspects of the Australian Dipterofauna. Biogeography and ecology in Australia. Monogr. Biol., vol. 8, pp. 164–191. (Bombyliidae pp. 165, 166, 167, 168, 172, 183, 184, 186, 187, 188, 189.)
- 1960a. Notes on African species of *Eurycarenus* Loew and *Sisyrophanus* Karsh (Diptera: Bombyliidae). Proc. Roy. Ent. Soc. London, ser. B, vol. 29, pp. 75–76.
- 1960b. Some notes on African species of *Bombylius*, *Systoechus* and *Dischistus* (Diptera: Bombyliidae). Proc. Roy. Ent. Soc. London, ser. B, vol. 29, pp. 99–102.
- 1961a. Notes on Diptera, XXXIV: additions and corrections to "A monograph of Egyptian Diptera, pt. VI. Bombyliidae," by Efflatoun Bey, 1945. Eos, vol. 37, contained on pp. 71–90.
- 1961b. Notes on Diptera, XL: On some bombyliid types in British Museum. Eos, vol. 37, contained on pp. 71–90.
- 1967. A review of the Australian species of the genus *Ligyra* Newman (Hyperalonia Olim) (Bombyliidae: Diptera). Australian Journ. Zool., vol. 15, pp. 123–144.

Parker, John Bernard
- 1917. A revision of the bembicine wasps of America north of Mexico. Proc. U.S. Nat. Mus., vol. 52, art. 2173, 155 pp. (Bombyliidae p. 141.)

Parker, J. R., and Wakeland, Claude
- 1957. Grasshopper egg pods destroyed by larvae of bee flies, blister beetles, and ground beetles. U.S. Dept. Agric., Tech. Bull. no. 1165, pp. 1–29, 3 figs. (Bombyliidae pp. 15–19.)

Parker, L. B. [see Hallock and Parker]
Parker, R. L. [see Smith, Kelly, Dean, Bryson, and Parker]
Pawlik, E.
 1930. Von heimische Schwebern (Bombyliidae). Ent. Zeitschr., vol. 44, pp. 102–104.
Pearce, E. K.
 1915. Typical flies, a photographic altas, including *Aphaniptera*. Cambridge Univ. Press, 47 pp., 45 pls. (Bombyliidae pp. 22–23, figs. 73–76.)
 1921. Typical flies, a photographic atlas. Second series. Cambridge Univ Press, 38 pp., 36 pls. (Bombyliidae pp. 15–16, figs. 46–50.)
 1928. Typical flies, a photographic atlas. Third series. Cambridge Univ. Press, 64 pp. (Bombyliidae pp. 23, 25–26, 59, figs. 47, 52–53, 153.)
Percheron, Achille Remy
 1835–1838. Genera des insectes ou expedition detaillée de tous les caractères propres a chacun des genres de cette classe animaux (with Guerin). Paris. 6 lief., 10 color pls. (Bombyliidae pl. 1, fig. of pupa of *Anthrax sinuata* and cell of earth.)
Perris, Edouard
 1839. Notice sur quelques Diptères nouveaux. Ann. Soc. Ent. France, vol. 8, pp. 47–57, pl. 7. (Bombyliidae pp. 54–56, pl. 7.)
Petagna, Vincenz (Vincentii)
 1787. Specimen insectorum ulterioris Calabriae. Francofurti et Moguntiae. Pp. i–vi, 1–46, 1 color pl. (Bombyliidae p. 45; 2 spp. listed.) Republished, Lipsiae, 1820, pp. i–vi, 1–46, 1 color pl.
Peters, Wilhelm Carl Hartwig
 1862. Naturwissenshaftliche Reise nach Mosambique. Berlin. Pp. i–iv, 1–21, and 1–566, 35 pls., mostly color. The insects (part 5) by Gerstaecher, Hagen, Hopfer, Klug, Loew, Schaum. (See Loew, 1862a.)
Peterson, Alvah
 1916. The head capsule and mouth parts of Diptera. Illinois Biol. Monogr., vol. 3, no. 2, 112 pp., 25 pls., with 606 figs. (Bombyliidae pp. 11, 16, 19, 24, 28, 30, 33, 40, 48, 49; figs. 29, 98, 162, 216, 285, 361, 426–429, 482, 549–550.)
 1939. Keys to the orders of immature stages (exclusive of eggs and pronymphs) of North American insects. Ann. Ent. Soc. America, vol. 32, pp. 267–278.
 1951. Larvae of insects. An introduction to Nearctic species. Part 2. Coleoptera, Diptera, Neuroptera, Siphonaptera, Mecoptera, Tricoptera. Vol. 2, 416 pp., 104 figs. Diptera pp. 219–348. Columbus, Ohio. (Bombyliidae pp. 292–293, fig. D15.)
Philippi, Rudolph Amandus
 1865. Aufzahlung der chilenischen Dipteren. Verh. zool-bot. Ges. Wien, vol. 15, pp. 595–782, pls. 23–29. (Bombyliidae pp. 649–654, 663–680, pl. 28, fig. 53.)
 1873. Chilenischen Insekten. Stettiner Ent. Zeitung, vol. 34, nos. 7–9, pp. 296–316. (Bombyliidae pp. 307–308.)
Plateau, F.
 1877. L'instinct des insectes peut-il être mis en défaut par des fleurs artificielles. Assoc. française pour l'Avancement des Sciences. 6 pp.
Poda von Neuhaus, Nicolaus
 1761. Insecta Musei Graecensis, quae in ordines, genera et species joxta Systema naturae Linnaei digessit. Graecii. Pp. i–v, 1–127, 1–8, 2 pls. (Bombyliidae pp. 114, 119.)
Pokorny, Emanuel
 1887–1893. Beitrag zur dipterenfauna Tirols. Verh. zool.-bot. Ges. Wien. (Abhandl.)
   1887. III. Vol. 37, pp. 381–420, pl. 7. (Bombyliidae p. 392; 6 spp.)
   1889. IV. Vol. 39, pp. 543–574. (Bombyliidae p. 548; 3 spp.)
   1893. V. Vol. 43, pp. 1–19. (Bombyliidae p. 2; 1 sp.)
Popov, G. B. [also see Stower, Popov, and Greathead]
 1958. Ecological studies on oviposition by swarms of the desert locust (*Schistocerca gregaria* Forskål.) in eastern Africa. Anti-Locust Bull., no. 31, pp. 1–70, 51 figs.

PORTER, J. C.
- 1951. Notes on the digger-bee *Anthophora occidentalis*, and its inquilines. Iowa State College Journ. Sci., Ames, vol. 26, pp. 23–30, 19 figs. (Bombyliidae p. 23.)

PORTSCHINSKY, JOSIFA ALOIZIOVICH
- 1881–1892. Diptera europaea et asiatica nova aut minus cognita . . . Horae Soc. Ent. Rossicae.
    - 1881. Pt. 1, vol. 16, pp. 136–145. (Bombyliidae p. 136.)
    - 1887. Pt. 6, vol. 21, pp. 176–200, pls. 6–7. (Bombyliidae pp. 182–187, pl. 6, figs. 1–3.)
    - 1892. Pt. 7, vol. 26 (1891), pp. 201–227, pl. 1 (color). (Bombyliidae pp. 207–209, pl. 1, figs. 4–5.)
- 1894. Über die den Saaten und Gräsern in den Gouvernements Perm, Tobolsk und Orenburg schädlichen Heuschrecken. Trudy Bur. Ent., vol. 1, no. 1, St. Petersburg, Dept. Agric., pp. 1–131. [In Russian.] Summary, 1895, p. 285, Zool. Centralbl., vol. 2, in German.
- 1895. Die Parasiten der schädlichen Feldheuschrecken Russlands. Dept. Agric. St. Petersburg. [In Russian.] Pp. 1–32. (Bombyliidae pp. 9, 14, fig.) Summary, 1895, p. 285, Zool. Centralbl., vol. 2, in German.
- 1915. [Note: *Villa* spp. of *ixion* group parasitic on Tenebrionia larvae.] Samovod (The Horticulturist), p. 244, 1915. (From Bezzi, p. 11, 1924.)

POTGIETER, J. T.
- 1929. A contribution to the biology of the brown swarm locust *Locustana paradalina* (Wlk.) and its natural enemies. Union South Africa Dept. Agric. Forestry, Pan-African Agric. Veterinary Conference, Pretoria, Aug. 1–17, 1929, p. 292, figs. 6a-d. Also in Sci. Bull. Dept. Agric. South Africa, no. 82, pp. 1–48. (Bombyliidae pp. 32–33, 35, fig. 6 (2a-c).)

POTTS, W. H. [see Newstead, Evans, and Potts]

PRIDDY, RALPH B.
- 1939a. Preliminary report on the Bombyliidae of Southern California. Journ. Ent. Zool., Pomona College, Calif., vol. 31, no. 3, pp. 1–2.
- 1939b. List of Bombyliidae collected in Southern California and Yuma County, Arizona. Journ. Ent. Zool., Pomona College, Calif., vol. 31, no. 3, pp. 45–53.
- 1954. Three new species of Nearctic *Conophorus* (Diptera, Bombyliidae). Journ. Kansas Ent. Soc., vol. 27, pp. 53–56.
- 1958. The genus *Conophorus* in North America (Diptera, Bombyliidae). Journ. Kansas Ent. Soc., vol. 31, no. 1, pp. 1–33, 6 figs.

PROCTER, WILLIAM
- 1946. Biological survey of the Mount Desert region. Part 7. The insect fauna (a revision of part 1, 1927, and part 6, with the addition of 1100 species). Wistar Inst. Anat. Biol., Philadelphia. 566 pp. Diptera pp. 328–437. (Bombyliidae p. 370; 6 genera, 12 spp.)

RADDATZ, A.
- 1873. Uebersicht der in Meckelburg bis jetzt beobachteten Fliegen (Diptera). Archiv. Ver. Freunde Naturg. Mecklenburg, Rostock, vol. 27, pp. 22–131. (Bombyliidae pp. 29–30.)

RADU, V. G., and DUSA, L.
- 1963. Contributii la cunoasterea bombiliidelor (diptera, brachicere) din tara noastra (II). Stud. Univ. Babes-Bolyai, ser. Biol. 8 (fasc. 2), pp. 60–68.
- 1965. Contributii la studiul aparatului genital mascul la Bombiliide. Stud. Univ. Babes-Bolyai, ser. Biol. 10 (fasc. 2), pp. 73–81.

RATZEBURG, JULIUS THEODOR CHRISTIAN
- 1837–1844. Die Forst-Insekten oder Abbildung und Beschreibung der in der Wäldern Preussens und der Nachbarstaaten als schädlich oder nützlich bekannt gewordenen Insekten. . . . Berlin. In 3 volumes.
    - 1844. Vol. 3. Die Ader-, Zwei-, Halb-, Netz-, und Geradflügler. 314 pp., 16 pls. (Bombyliidae p. 154, footnote.)

RAU, NELLIE [see Rau and Rau]

Rau, Phil
- 1916. The biology of mud-daubing wasps. Journ. Animal Behavior, vol. 6, pp. 27–63.
- 1926. The ecology of a sheltered clay bank, a study in insect ecology. Trans. Acad. Sci. St. Louis, vol. 25, pp. 158–260, pls. 14–21. (Bombyliidae p. 162.)
- 1940. Some mud-daubing wasps of Mexico and their parasites. Ann. Ent. Soc. America, vol. 33, pp. 590–595. (Bombyliidae pp. 594–595.)
- 1946. Notes on a few dipterous and hymenopterous parasites of mud-daubing wasps (Diptera, Hymenoptera). Ent. News, vol. 57, pp. 195–196.

Rau, Phil, and Rau, Nellie
- 1916. The sleep of insects: an ecological study. Ann. Ent. Soc. America, vol. 9, no. 3, pp. 227–274. (Bombyliidae p. 249.)
- 1918. Wasp studies afield. Princeton Univ. Press, New Jersey, pp. 1–5, 1–372. (Bombyliidae pp. 24, 32, 38, 94.)

Réaumur, Réné Antoine Ferchault, de
- 1734–1742. Mémoires pour servir à l'histoire des Insectes. Paris. In 6 volumes. [Also published in Amsterdam, 1737–1748.]
    - 1740. Vol. 5, 774 pp., 38 pls. (Bombyliidae p. 196, pl. 8, figs. 11–13, 18.)
    - 1742. Vol. 6, 690 pp., 48 pls. (Bombyliidae pp. 272, 276, 290, pl. 27, figs. 1, 13.)

Reed, Edwyn C.
- 1888. Catálogo de los insectos dipteres de Chile. Anal. Univ. Chile, Santiago, vol. 73, pp. 271–316. (Bombyliidae pp. 294–297.)

Régnier, P. R.
- 1931. Les invasions d'Acridiens au Maroc de 1927 à 1931. Dir. Gén. Agric. Comm. et Colonisation-Défense des Cultures. Rabat, no. 3, pp. 1–139, 9 maps.

Régnier, P., Lespes, L., and Rungs, C.
- 1931. Sur l'habitat de *Schistocerca gregaria* Forsk. et la succession des générations chez cette espècie. Comptes Rendus Acad. Sci. France, vol. 192, pp. 1485–1487.

Remaudière, G.
- 1947. Sur les principaux parasites du criquet migrateur (*Locusta migratoria* L.) dans ses foyers des Landes de Gascogne. I. Ennemis des oeufs et des oothèques. Bull. Soc. Ent. France, vol. 52, pp. 53–64. (Bombyliidae p. 64.)

Ricardo, Gertrude
- 1901. Notes on Diptera from South Africa. Ann. Mag. Nat. Hist., vol. 7, pp. 89–110. (Bombyliidae pp. 89–104.)
- 1903. *In* Forbes, editor: Natural history of Socotra and Abd-el-kuri. Liverpool. Pp. i–xiv, 1–598. Diptera pp. 359–378, pl. 22 (color). (Bombyliidae pp. 365, 366, 367, pl. 22, figs 1, 1a, 2, 2a, 3, 3a, 4, 4a.)

Richter, P. O., and Fluke, Jr., C. L.
- 1935. *Exoprosopa fasciata* Macq., white grub pupal parasite. Journ. Econ. Ent., vol. 28, p. 248.

Richter, Willi, and Schüz, Ernst
- 1959. Zoologische Arbeiten des Stuttgarter Museums über Iran (Bibliographie). Stuttgarter Beiträge zur Naturkunde Statt. Mus. Naturkunde in Stuttgart, no. 22, pp. 1–8, 1 map. (Bombyliidae p. 2.)

Riley, Charles Valentine
- 1877. Ninth annual report on the noxious, beneficial and other insects of the State of Missouri, made to the State Board of Agriculture. Jefferson City, Missouri. Pp. 1–7, 1–129, 1–3, figs. 1–33. (Bombyliidae p. 96, fig. 24, an undetermined sp. of egg parasite of the Rocky Mountain Locust.)
- 1878. First annual report U.S. Entomological Commission for the year 1877, relating to the Rocky Mountain locust. Dept. Interior, U.S. Geol. Surv., vol. 16, pp. 1–477, 105 figs., 5 pls., 1 map. Chapter 11. Invertebrate enemies, pp. 284–334. (Bombyliidae pp. 304–305, fig. 37, unidentified larvae.)
- 1880a. Second report U.S. Entomological Commission for the years 1878, 1879, relating to the Rocky Mountain locust, etc. Dept. Interior. Chapter 13. Further facts about the natural enemies of the locusts, pp. 259–271. (Bombyliidae pp. 262–270, pl. 16 with 7 figs.)

RILEY, CHARLES VALENTINE—Continued
- 1880b. On the natural history of certain beeflies (Bombyliidae). Amer. Ent., vol. 3 (new ser., vol. 1), pp. 279–283, figs. 147–151.
- 1881a. Larval habits of beeflies (Bombyliidae). Amer. Natur., vol. 15, pp. 143–145, 3 text-figs.
- 1881b. Larval habits of beeflies. Amer. Natur., vol. 15, pp. 438–447, pl. 6 (color), 7 figs.
- 1881c. Report on Bombyliidae destructive to locust eggs. Proc. Ent. Soc. London, 1881, pp. xxxviii–xl.
- 1881d. *Systoechus oreas* and *Triodites mus* Ost.-Sack. Larvae destructive to locust eggs. Proc. Amer. Assoc. Adv. Sci., vol. 29 (1880), p. 649; p. 33 of reprint.

RILEY, CHARLES VALENTINE, and HOWARD, LELAND OSSIAN
- 1890. *Anthrax* parasitic on cut-worms. Insect Life, vol. 2, pp. 353–354, fig. (woodcuts).

RITSEMA, C.
- 1868. [Note: *Anthrax hottentotta* L. parasite of *Noctua porphyrea*.] Tijdschft. Ent., ser. 2, vol. 4 (XII). (Bombyliidae pl. 7, fig. 2, Verslag p. 192.)

ROBERTS, FREDERICK H. S.
- 1928–1929. A revision of the Australian Bombyliidae (Diptera). I–III. Proc. Linnean Soc. New South Wales.
  - 1928a. Pt. 1, vol. 53, pp. 90–144, 4 figs.
  - 1928b. Pt. 2, vol. 53, pt. 4, pp. 413–455.
  - 1929. Pt. 3, vol. 54, pt. 5, pp. 553–583.
- 1929. A list of the Australian Bombyliidae of the subfamilies Exoprosopinae, Anthracinae and Bombylinae in the German Entomological Museum, Berlin. Proc. Linnean. Soc. New South Wales, Sydney, vol. 54, pp. 517–518.

ROBERTSON, CHARLES
- 1895. The philosophy of flower seasons, etc. Amer. Natur., vol. 29, pp. 97–117. (Gives a flight curve for Bombyliidae.)
- 1928. Flowers and insects. Lists of visitors of four hundred and fifty-three flowers. Science Press, Lancaster, Pa. 221 pp. (Bombyliidae: Lists all bee fly species observed visiting 173 species of flowers.)

ROBINEAU-DESVOIDY, J. B.
- 1863. Histoire naturelle des diptères des environs de Paris. 2 volumes. Vol. 1, 1143 pp.; Vol. 2, 920 pp.

ROEDER, VICTOR VON
- 1882. Zur synonymie einiger Chilenischer Dipteren. Stettiner Ent. Zeitung, vol. 43, pp. 510–511. (Bombyliidae pl. 510.)
- 1883. Dipteren von den Canarischen Inseln. Wiener Ent. Zeitschr., vol. 2, pp. 93–95. (Bombyliidae p. 93; 1 sp. *Anthrax*.)
- 1884. Dipteren von der Insel Sardienien. Wiener Ent. Zeitung, vol. 3, pp. 40–42. (Bombyliidae pp. 40–41.)
- 1885a. Bemerkungen über 2 Dipteren. Berliner Ent. Zeitschr., vol. 29, p. 137.
- 1885b. Dipteren von der Insel Portorico. Stettiner Ent. Zeitung, vol. 46, pp. 337–349. (Bombyliidae p. 339.)
- 1886. Ueber die Nordamerikanischen Lomatina von Mr. Coquillett in den "Canadian Entomologist." Wiener Ent. Zeitung, vol. 5, pp. 263–265.
- 1887a. Uebersicht der beim Dorf Elos bei Kisamos auf der Insel Creta von Herrn E. v. Oertzen gesammelten Dipteren. Berliner Ent. Zeitschr., vol. 31, pp. 73–75. (Bombyliidae p. 73.)
- 1887b. Eine neue *Exoprosopa* aus Syrien. Berliner Ent. Zeitschr., vol. 31, pp. 75–76.
- 1888. Bemerkungen zur Dipteren-Gattung *Exoprosopa*. Wiener Ent. Zeitung, vol. 7, pp. 97–98.
- 1889. Ueber die Dipteren-Gattung *Clitodoca* Lw. Ent. Nachrichten, vol. 15, p. 291.
- 1896. *Spongostylum flavipes*, nov. spec. Dipt. Wiener Ent. Zeitung, vol. 15, p. 273.

ROEHRICH, R.
- 1951. Parasites et prédators du criquet migrateur (*Locusta migratoria gallica* Rem.) dans les Landes de Gascogne de 1945 à 1950. Ann. Epiphyties, vol. 2, nos. 3–4, pp. 479–495, 7 figs., 18 references. (Bombyliidae pp. 480–482, figs. 1–3.)

ROGENHOFER, ALOIS FRIEDRICH (Custodian Hofmuseum in Wien)
> [Credited by Brauer, 1883, pp. 61–62, with observing *Anthrax flava* emerge from *Agrotis segetum* and *forcipula*; and *Argyromoeba subnotata* emerge from *Chalcodoma muraria*.]

ROHDENDORF, B. B.
> 1964. Istoritcheskoje razvitie dvukrylych nasjekomych-Trudy paleout. Inst. Akad. Nauk. SSSR, vol. 100, pp. 1–311, Moscow.

RONDANI, A. CAMILLO
> 1848. Esame di varie specie d'insetti ditteri brasiliani. *In* Truqui, Studi Ent., vol. 1, fasc. 3, pp. 63–112. (Bombyliidae pp. 97–98.) Reprint pp. 1–50, 1 pl.
> 1856–1880. Dipterologiae Italicae prodromus.
>> 1856. Genera Italica ordinis dipterorum ordinatim disposita et distincta et in familias et stirpes aggregata. Vol. 1, pp. 1–228. Parmae. (Bombyliidae pp. 162–165.)
>> 1861. Species Italicae.... Pars 3, vol. 4, pp. 1–174. (Bombyliidae p. 8.)
> 1863. Diptera exotica revisa et annotata. 99 pp., 1 pl. Modena. [Also published under the title "Dipterorum species et genera aliqua exotica," in Arch. Zool. l'Anat. Fisiol., vol. 3, no. 1, (1863), pp. 1–99, pl. 5, 1864. Modena.]
> 1868. Diptera aliqua in America meridionali lecta a Prof. A. Strobel annis 1866 et 1867. Ann. Soc. Nat. Modena, vol. 3, pp. 24–40, pl. 4. (Bombyliidae pp. 34–36, pl. 4, figs. 7–9.)
> 1872. Degli insetti parassiti e delle loro vittime. Bull. Soc. Ent. Italiana, vol. 4, pp. 321–342. Diptera pp. 321–336. (Bombyliidae p. 321; on the finding of *Hemipenthes hottentota* L. [as *Anthrax flavus* Meigen] in chrysalids of *Hadema brassicae* L., a species of Noctuida.)
> 1873–1878. Muscaria exotica musei civici januensis observata et distincta. Ann. Mus. Civ. Storia Nat. Genova.
>> 1873. Fragmentum 2, vol. 4, pp. 295–300. (Bombyliidae pp. 296, 299.)
>> 1875. Fragmentum 3. Species in insula Bonae Fortunae (Borneo), Provincia Sarawak, annis 1865–1868 lectae a March J. Doria et Doct. O. Beccari. Vol. 7, pp. 421–464, 5 figs. (Bombyliidae p. 453.)
> 1877. Repertorio degli insetti parassiti e delle loro Vittime. Supplemento Alla Parte Prima Parassiti Muscarii-Diptera. Bull. Ent. Italiana, Firenze, vol. 9, pp. 55–66. (Bombyliidae p. 55; note on *Anthrax palumbii* Rondani, no description, "trovato allo stato di ninifa entro en bozzolo di *Bombyx quercus* Linné speditomi dalla Sicilia. La ninfa e distintissima e maggiore di quelle delle *Anthrax flava* ed *ottentota* conosciura brucivore.")

ROSER, C. L. F., VON
> 1834–1840. Erster Nachtrag zu dem in Jahre 1834 bekant gemachten Verzeichnisse in Württemberg vorkommender zweiflügliger Insekten. K. Württemberg Landwirthsch. Ver., Stuttgart, Correspondenzbl., vol. 37 (n. s. 17).
>> 1840. Nachtrag 1, pt. 1, pp. 49–64. (Bombyliidae p. 52.)

ROSS, EDWARD S.
> 1953. Insects close up. A pictorial guide for the photographer and collector featuring 125 photographs and drawings. California Acad. Sci., Berkeley, pp. 1–81, 126 illustrations (7 in color). (Bombyliidae pp. 23, 43, with 2 photographs.)

ROSSI, PETER
> 1790. Fauna Etrusca, sistens Insecta, quae in provinciis Florentina et Pisana praesertim collegit. Ligurni, Masi. In 2 vols., 272 pp. and 348 pp., 10 color pls., 2 frontispieces. (Bombyliidae vol. 1, pp. 255–265; vol. 2, pp. 255, 265, 275, 328, pl. 4, figs. 11, 14.) Second edition, 1807, vi + 511 pp. (Bombyliidae pp. 425–429, 498.)
> 1792–1794. Mantissa Insectorum, exhibens species nuper in Etruria collectas, adjectis faunae Etruscae illustrationibus ac emendationibus. Pisa.
>> 1792. Vol. 1, 148 pp. (Bombyliidae p. 59, pl. 1, fig. 11.)
>> 1794. Vol. 2, 154 pp. (Bombyliidae p. 78.)

Rubio, Jose M. Andreu
- 1959. Bombilidos marroquies del Instituto Español de Entomologia (Diptera). Eos, Rev. Español Ent., Madrid, vol. 35, pp. 7–19.
- 1961. Los dipteros bombilidos españoles y su distribucion geográfica. (Murcia, Instituto de Orientacion y Assistencia tecnica del Sureste, p. 42.)

Ruiz Pereira, Hermano Flaminio
- 1929. Breves notas biológicas sobre *Exoprosopa erythrocephala* (Fabr.). Rev. Chilena Hist. Nat., Santiago, vol. 32 (1928), pp. 57–60, 1 fig. (Dipt.).
- 1930. Nuevas observaciones sobre la biologia de *Exoprosopa erythrocephala* Fabricius. Rev. Chilena Hist. Nat., Santiago, vol. 34, pp. 155–158.
- 1939. El genero *Comptosia* Macquart en Chile (Dipt. Bombyliidae). Rev. Univ. Santiago, vol. 24, pp. 111–126.

Ruiz Pereira, Flaminio, and Stuardo Ortiz, Carlos
- 1936. Insectos de las Termas de Chillan. Rev. Chilena Hist. Nat., Santiago, vol. 39 (1935), pp. 313–322, figs. 49–50. (Bombyliidae pp. 315–316; 10 spp. listed.)

Rungs, C. [see Regnier, Lespes, and Rungs]

Russell, Harold
- 1922. On Indian parasitic flies. Journ. Bombay Nat. Hist. Soc., vol. 28, I., pp. 370–380 (continued). (Bombyliidae pp. 372, 374–376). II, III, pp. 703–718 (continued).

Ruthe, Johann Friedrich
- 1831. Einige Bemerkungen und Nachträge zu Meigen's "Systematischer Beschreibung der europäischen zweiflügeligen Insecten." Isis (Oken's), Leipzig, 1831, pp. 1203–1222. (Bombyliidae p. 1216.)

Sabrosky, Curtis W. [also see Stone, Sabrosky, Wirth, Foote, and Coulson]
- 1971. Additional corrections to a Catalog of the Diptera of America North of Mexico. Bull. Ent. Soc. America, vol. 13, pp. 115–125.

Sack, Pius
- 1899a. Uberliegen von Dipteren-Puppen. Illustr. Zeitschr. Ent., vol. 4, p. 8.
- 1899b. Summary *in* Xambeu: Moeurs et Metamorphoses de l'*Usia atrata* (Fabricius) *in*: Le Naturaliste, no. 275, Seite 189 und 190. Illustr. Zeitschr. Ent., vol. 4, p. 48.
- 1906. Diptera *in* Graeffe: Beiträge zur Insektenfaune von Tunis. Verh. zool.-bot. Ges. Wein, vol. 56, pp. 446–471. Diptera pp. 468–471. (Bombyliidae pp. 469–470.)
- 1909. Die Palaearktischen Spogostylinen. Abhand. Senckenberg. Naturf. Ges., vol. 30, pp. 501–548, pls. 19–22.

Samouelle, George
- 1819. The entomologist's useful compendium, or an introduction to the knowledge of British insects. London. 496 pp., 12 pls. Second edition, 1824, with new title page. (See Leach for Bombyliidae.)

Sartor, Clyde Flake, Jr.
- 1966. Life history studies of the parasites of the mining bee *Anthophora abrupta* (Anthophoridae). Unpublished manuscript, University of Mississippi, 35 pages, 12 figures, 4 maps. (M.Sci. thesis: Library.)

Sasscer, E. R.
- 1947. A tabulation of insects recovered from aircraft entering the United States at Miami, Fla., during the period July 1943 through December 1944. U.S. Dept. Agric., Foreign Plant Quarantine Memorandum no. 474, pp. 1–35. (Bombyliidae p. 26; 4 spp. *Anthrax*.)

Saunders, Sidney Smith, Waterhouse, Charles, O., and Fitch, Edward Asa
- 1881a. Report to the Council of the Entomological Society of London about Bombyliidae destructive to locust eggs. Trans. Ent. Soc. London, July 6, vol. 29, pp. xiv–xix, pl. 14, figs. 1, 1a, 1b, 2, 2a, 2b, 3, 3a, 4.
- 1881b. Report to the Council of the Entomological Society of London about an insect destructive to locust eggs. Trans. Ent. Soc. London, December 7, 1881, vol. 29, pp. xxxviii–xl.

SAUNDERS, WILLIAM WILSON
- 1841. Descriptions of four new dipterous insects from central and northern India. Trans. Ent. Soc. London, vol. 3, pp. 59–61, 1 pl. (color). (Bombyliidae p. 59, pl. 5, fig. 5.)

SAY, THOMAS
- 1823. Descriptions of dipterous insects of the United States. Journ. Acad. Nat. Sci. Philadelphia, vol. 3, pp. 9–54, 73–104. (Bombyliidae pp. 41–66.) (Bombyliidae in Complete Writings, vol. 2, pp. 58–62.)
- 1825. *In* Keating: Narrative of an expedition to the source of St. Peter's River . . . under command of Stephen H. Long. In 2 volumes. Philadelphia. Insects by Say, vol. 2, appendix, pp. 268–378; Diptera, pp. 357–378. (Bombyliidae pp. 371–373.) (Bombyliidae in Complete Writings, vol. 1, pp. 252–255.)
- 1829–1830. Descriptions of North American dipterous insects. Journ. Acad. Nat. Sci. Philadelphia.
    - 1829. Vol. 6, no. 1, pp. 149–178. (Bombyliidae pp. 156–157.) (Bombyliidae in Complete Writings, vol. 2, pp. 353–354.)
- 1859. The complete writings of Thomas Say on the entomology of North America, with a memoir of the author by George Ord. Edited by Le Conte. In 2 volumes. New York: Ballière. Vol. 1, 433 pp., 54 pls. (color), pls. 1–36 original. (Bombyliidae pp. 252–255.) Vol. 2, 818 pp., 1 pl. (Bombyliidae pp. 58–62, 353–354.) Again published 1883, Boston, same pages.

SCHÄFFER, JACOB CHRISTIAN
- 1764–1779. Abhandlungen von Insecten. Regensburg. In 3 volumes, 48 pls. (color).
    - 1764. Vol. 2, 344 pp., 18 pls.; in 8 sections. Section I. Die Mauerbiene. Pp. 1–38, pls. 1–5. (Bombyliidae pp. 22–23; explanation of pls. pp. 37–38; pl. 5, figs. 11–14. Figs. 11 and 12 are larvae and pupae; figs. 13 and 14 are male Schimmerfliege.) (Book at U.S. Dept. Agric. Library, book no. 422, SchlA, vol. 2.)
- 1766–1779. Icones Insectorum circa Ratisbonam indigenorum coloribus naturam referentibus expressae. Natürlich ausgemahlte Abbildungen Regensburgscher Insecten. Regensburg. In 3 volumes, with 280 pls. (color).
    - Vol. 1, part 2 (without year), pp. 1–6, 1–50, index pp. 1–2, pls. 51–100. (Bombyliidae pl. 78, fig. 3; pl. 79, fig. 5.)

SCHAFFNER, J. V., JR.
- 1959. Microlepidoptera and their parasites reared from field collections in the northeastern United States. Miscellaneous Publ. 767. Forest Service, U.S. Dept. Agric., pp. 1–4, 1–97. (Bombyliidae pp. 47–48.)

SCHELLENBERG, JOHANN RUDOLF
- 1803. Genres des mouches Diptères. . . . Gattungen der Fliegen in 42 Kupfertafeln, entworfen . . . durch zwei Liebhaber der Insecten kunde. Zürich. (French and German on opposite pages.) Pp. 1–95, 42 pls. (color). (Bombyliidae pp. 30, 31, pp. 90–91, pl. 32, figs. 1–3, pl. 34, fig. 2.)

SCHENKLING, SIGMUND [see Horn and Schenkling]

SCHIMITSCHEK, W. [see Chinkevitsch, V.]

SCHINER, JGNAZ RUDOLF (Ignaz)
- 1860. Vorläufiger Commentar zum dipterologischen Theile der "Fauna Austriaca," mit einer näheren Begründung der in derselben aufgenommenen Dipterengattungen. Part 1. Wiener Ent. Monatschr., vol. 4, pp. 47–55. (Bombyliidae pp. 50–52.)
- 1860–1864. Fauna Austriaca. Die Fliegen, Diptera. Nach der analytischen Methode bearbeitet. In 2 parts. Vienna.
    - 1860–1862. Part 1, pp. 1–80, 1–674, pls. 1–2. (Bombyliidae pp. 46–71; uses Family Bombylides. Key to 19 genera pp. 49–50.)
- 1864. Catalogus systematicus dipterorum europae. Wien. 127 pp.
- 1867. Zweiter Bericht über die von der Weltumseglungsreise der K. Fregatte *Novara* mitgebrachten Dipteren. Section 1. Diptera orthorhapha: Division: 2. Brachycera. Verh. zool.-bot. Ges. Wien, vol. 17, pp. 303–314. (Bombyliidae pp. 312–314; *Diplocampta, Tritoneura, Callynthrophora*.)

SCHINER, JGNAZ RUDOLF (Ignaz)—Continued
1868. Diptera *in* Wüllerstorf-Urbair, von: Reise der österreichischen Fregatte *Novara* 1857–1859. Zool. Theil, Band 2, Abt. 1, pt. 1, pp. i–vi, 1–388, 4 pls. (Bombyliidae pp. 113–140, pl. 2, fig. 9.)

SCHLOTHEIM, E. F. VON
1820. Petrefaktenkunde. (Bombyliidae p. 43.)

SCHMIDT-GOEBEL, HERMANN MAX
1876. Coleopterologische Kleinigkeiten. Stettiner Ent. Zeitung, vol. 37, pp. 388–401. (Bombyliidae pp. 392–393. *Bombylius* parasitic on pupae of *Colletes fodiens*.)

SCHNABL, J. [see Becker and Schnabl]

SCHOLTZ, HEINRICH
1850–1851. Beiträge zur Kunde der schlesischen Zweiflügler. In 2 parts. Ent. Zeitschr. Breslau.
    1851. Pt. 2, vol. 5, no. 17, pp. 41–60. (Bombyliidae pp. 41–43.)

SCHOMBURGK, MORITZ RICHARD
1847–1848. Reisen in Britisch-Guiana inden Jahren 1840–1844. In 3 volumes. Leipzig.
    1848. Vol. 3. Versuch einer Fauna und Flora von Britisch-Guiana sind die Insekten von Erichson bearbeitet.

SCHRANK, FRANZ VON PAULA
1781. Enumeratio insectorum austriae indigenorum. Agustae Vindelicorum, Klett. 24 (unnumbered) + 548 pp., 4 pls. (Bombyliidae pp. 439, 490–493.)
1798–1804. Fauna Boica. Beyträge zur Beobachtungskunst in der Naturgeschichte. Durchgedachte Gesichte der in Baiern einheimischen und Zahmen Thiere, von Franz von Paula Schrank. Nürnberg. In 3 volumes, 6 parts.
    1803. Vol. 3, pt. 1, viii + 272 pp. (Bombyliidae pp. 90, 173–174.)

SCHRODER, CHRISTOPH WILHELM MARCUS
1925–1929. Handbuch der Entomologie. Jena.
    1928. Vol. 1, 1442 pp. (Bombyliidae pp. 142, 149, 152, 207, 261, 1228, 1273.)
    1929. Vol. 2, 1426 pp. (Bombyliidae pp. 41, 45, 130, 151, 330, 406, 531, 1251.)
    1925. Vol. 3, 1209 pp. (Bombyliidae pp. 265, 922, 989–993.)

SCHULER, VON
[Brauer, 1883, p. 61, credits von Schuler with observing the emergence of *Anthrax modesta* from *Agrotis signifera*.]

SCHUMMEL, THEODORE EMIL
1838. Fortsetzung der Zusätze zur schlesischen Fauna aus der Ordnung der Dipteren in Meigen's Tom. II und VI. beschreiben. Arbeit Schlesische Gesell. vaterländische Kultur. Breslau, 1838, pp. 57–59. (Bombyliidae pp. 58–59.)

SCHÜZ, ERNST [see Richter and Schüz]

SCINCHIEVIC, V.
1883–1884. Contributo alla questione dei Ditteri parassiti delle ooteche delle Cavallette. [In Russian.] Horae Ent. Ross. Pietroburgo, vol. 18, pp. 11–16.

SCOPOLI, JOHANN ANTON
1763. Entomologia Carniolica exhibens insects Carnioliae indigena et distributa in ordines, genera, species, varietates, methodo Linneana. Vindobonae. xxxvi + 421 pp., 43 pls. (Bombyliidae pp. 358, 375, 376.)
1772. Annus historico naturalis. Lipsiae. Vol. 5, 128 pp. Note: *Bombylius ?minimus* described. (Bombyliidae p. 123.)

SCUDDER, SAMUEL HUBBARD
1882. Nomenclator zoologicus. U.S. Nat. Mus. Bull. 19, preface, xix pp. Part 1. Supplemental list of genera in zoology. 367 pp. Part 2. Universal index to genera in zoology, 340 pp. (Bombyliidae pt. 1, p. 108; pt. 2, p. 99.)
1890. A classed and annotated bibliography of fossil insects. Bull. U.S. Geol. Surv., Dept. Interior, no. 69, pp. 1–101.
1891. Index to the known fossil insects of the world including Myriapods and Arachnids. Bull. U.S. Geol. Surv., no. 71.

SEABRA, A. F. DE
- 1901. As invasões de gafanhotos em Portugal. Archivo Rural, Lisbon, 1901. Reprint pp. 1–25, 1 pl. [Uvarov, p. 111, 1928, states *Anastoechus nitidulus* F. parasitises *Dociostaurus maroccanus*.]

SÉGUY, EUGÈNE
- 1926. *In* Lechevalier: Faune de France. 13. Diptères (Brachycères), (Stratiomyiidae, Erinnidae, Coenomyiidae, Rhagionidae, Tabanidae, Oncodidae, Nemestrinidae, Mydaidae, Bombyliidae, Therevidae, Omphralidae). Paris. 305 pp., 685 figs. (Bombyliidae pp. 178–254, 186 figs.)
- 1929. Étude systématique d'une collection de diptères d'Espagne formée par le R. P. Longin Navás, S. J. Mem. Soc. Ent. Zaragoza, vol. 3a, pp. 1–30, 6 figs. (Bombyliidae pp. 14–15; 7 spp. listed.
- 1930a. Description d'un Africain *Heterotropus*. Encycl. Ent., Paris, ser. B, II. Dipt., vol. 5 (1929), p. 62.
- 1930b. Contribution à l'étude des diptères du Maroc. Mém. Soc. sci. nat. Maroc., Rabat, no. 24, pp. 1–206, 115 figs. (Bombyliidae pp. 90–107, figs. 74–82.)
- 1930c. Note sur quatre Toxophorines de l'Amérique centrale et méridionale. Rev. Chilena Hist. nat., Santiago de Chile, vol. 33 (1929), pp. 532–536, 4 figs. (Dipt.).
- 1930d. Risultati zoologici della missione inviata dalla R. Società Geografica Italiana per l'esplorazione dell' oasi di Giarabub (1926–1927). Insectes Diptères. Ann. Mus. Storia Nat. Genova, vol. 55, pp. 75–93, 5 figs. (Bombyliidae pp. 78–82, 1 fig.)
- 1931a. Spedizione del Barone Raimondo Franchetti in Dancalia. Insectes Diptères. Ann. Mus. Civ. Storia Nat. Genova, vol. 55, pp. 234–247, 5 figs. (Bombyliidae pp. 237–239, fig. 2.)
- 1931b. Un nouvel *Heterotropus* de Tunisia. Ann. Soc. Ent. France, vol. 100, p. 106.
- 1932a. Étude sur les Diptères parasites ou prédateurs des sauterelles. Encycl. Ent., Paris, ser. B, II. Dipt., vol. 6, pp. 11–40, 34 figs. (Bombyliidae pp. 14–18, figs. 2–6.)
- 1932b. Trois Diptères nouveaux de Madagascar. Bull. Soc. Ent. France, vol. 37, pp. 160–163, 3 figs. (Bombyliidae pp. 160–161, fig. 1.)
- 1932c. Contribution à l'étude de fauna du Mozambique. Voyage de M. P. Lesne 1928–29. 3e Note. Diptères. Part I. Bull. Mus. Hist. Nat. Paris, ser. 2, vol. 3 (1931), pp. 113–124, 1 fig. (Bombyliidae pp. 115–118.)
- 1932d. Diptères nouveaux ou peu connus. Encycl. Ent., Paris, ser. B., II. Dipt., vol. 6, pp. 125–132, 2 figs. (Bombyliidae pp. 126–130, fig.)
- 1932e. Spedizione scientifica all'oasi di Cufra (Marzo-Luglio 1931). Insectes diptères. Ann. Mus. Storia Nat. Genova, vol. 55 (1930–1931), pp. 490–511, figs. 1–3.
- 1933a. Contribution à l'étude de la faune du Mozambique. Voyage de M. P. Lesne 1928–1929. 3e Note. Diptères (2e partie). Mem. Mus. Zool. Univ. Coimbra, ser. 1, no. 67, pp. 5–80, 19 figs. (Bombyliidae pp. 13–15, figs. 2–3.)
- 1933b. Notes Scientifiques. Une nouvelle espèce de *Toxophora* de Madagascar. Terre et la Vie, Paris, vol. 4, pp. 366–367, fig. of *Toxophora seyrigi*, female.
- 1934a. Diptères d'Afrique. Encycl. Ent., Paris, sér. B., II. Dipt., vol. 7, pp. 63–80, 12 figs. (Bombyliidae pp. 72–74.)
- 1934b. Diptères d'Espagne. Etude systematique basée principalement sur les collections formees par le R. P. Longin Navas, S. J. Mem. Acad. Cienc. Zaragoza, vol. 3, pp. 1–54, 7 figs. (Bombyliidae pp. 32–37.)
- 1935. Étude sur quelques diptères nouveaux de la Chine orientale. Mus. Heude, Notes Ent. chinoise, Shanghai, vol. 2, pp. 175–184. (Bombyliidae pp. 176–178.)
- 1936. Diptères des Açores. Ann. Soc. Ent. France, vol. 105, pp. 11–26, 2 figs. (Bombyliidae p. 15.)
- 1938a. Étude sur les diptères recueillis par M. H. Lhote dans le Tassili des Ayyer. (Sahara Touareg.) Encycl. Ent., Paris, ser. B., II. Dipt., vol. 9, pp. 37–45, 3 figs. (Bombyliidae pp. 39–42.)
- 1938b. Mission scientifique de l'Omo. Diptera I. Nematocera et Brachycera. Mém. Mus. Hist. nat. Paris, new ser., vol. 8, pp. 319–380, 55 figs. (Bombyliidae pp. 332–333, text-fig. 15.)

Séguy, Eugène—Continued
- 1939. Diptera. Missione biologica paese Borana. Reale Accad. d'Italia, Rome, vol. 3, no. 2, 1466 pp., 4 figs. Diptera pp. 123–148. (Bombyliidae pp. 130–131.)
- 1940. Sur l'*Exoprosopa oculata* Macquart. Rev. française Ent., vol. 7, pp. 142–143.
- 1941a. Récoltes der R. Paulian et A. Villiers dans le haut Atlas Marocain, 1938 (xvii note). Rev. française Ent., vol. 8, pp. 25–33. (Bombyliidae p. 29.)
- 1941b. Diptères recueillis par M. L. Chopard d'Alger à la Côte d'Ivoire. Ann. Soc. Ent. France, Paris, vol. 109 (1940), pp. 109–130, 6 figs. (Bombyliidae pp. 111–112.)
- 1941c. Diptères recueillis par M. L. Berland dans le Sud Marocain. Ann. Soc. Ent. France, vol. 110, pp. 1–23, 21 figs. (Bombyliidae pp. 7–10, fig. 4.)
- 1949a. Un *Cyrtosia* nouveau et synopsis des espèces méditerranéennes. (Dipt. Bombyliidae.) Rev. Française Ent., Paris, vol. 16, pp. 83–85, figs. 1–12.
- 1949b. Diptères du Sud-Moracain (Vallée du Draa) recueillis par M. L. Berland en 1947. Rev. Française Ent., Paris, vol. 16, fasc. 3, pp. 152–161, 2 figs. (Bombyliidae pp. 154–155; 8 spp. listed (3 new sp.), key to 6 spp. *Bombylius*.)
- 1950a. Contribution a l'étude de l'Aïr. Diptères. Mem. Inst. française Afrique noire, Paris, no. 10, pp. 1–562. (Bombyliidae p. 274.)
- 1950b. La biologie des diptères. Encycl. Ent., Paris, ser. A, vol. 26, pp. 1–609, 10 pls. (7 color), 225 figs. Bibliography with 309 entries on pp. 528–543. (Bombyliidae pp. 172, 206, 497; figs. 93, 102.)
- 1951a. Les Diptères de France, Belgique Suisse. Nouvel atlas d'entomologie. Paris. In 2 volumes. Sous directeur au Muséum National d'Histoire naturelle.
  - 1951. Vol. I. Introduction et caractères généraux nématocères-brachycères, pp. 5–175, text-figs. 1–79, pls. 1–12 (color) with 146 figs. (Bombyliidae pp. 145–148, pl. 8, figs. 87–91, text-fig. 71.)
- 1951b. Contribution a l'étude du peoplement de la Mauritanie. Bull. Inst. Française Afrique noire, vol. 13, pp. 317–318. (Bombyliidae p. 317; 3 spp. listed.)
- 1953. Diptères de Maroc. Encycl. Ent., Paris, ser. B, II. Dipt., vol. 11, pp. 77–92. (Bombyliidae p. 83; 13 spp. listed.)
- 1955. Introduction à l'étude biologique et morphologique des insectes Diptères. Publicações Avulas Mus. Nac., no. 17, pp. 1–260. (Bombyliidae paragraphs 170, 260, 261, 321, 325, 332, 336, 337, 341. *Bombylius* 260–262; *Systropus* 146; *Toxophora* 263; Anthracines 263; *Anthrax* 260, 263, 332.)
- 1963. Microbombyliides de la Chine palearticque (Insectes Dipteres). Bull. Mus. Hist. nat., Paris (2), vol. 35, pp. 253–256, 2 figs.

Séguy, Eugène, and Baudot, E.
- 1922. Note sur les premiers états du *Bombylius fugax* Wied. (Dipt. Bombyliidae). Bull. Soc. Ent. France, Paris, 1922, pp. 139–141, text-figs. 1–8.

Senior-White, R. A.
- 1921–1925. New Ceylon Diptera. Parts I-IV. Spolia Zeylanica, Colombo.
  - 1922. Pt. II, vol. 12, pp. 195–206. (Bombyliidae pp. 203–205.)
  - 1924. Pt. III, vol. 12, pp. 375–406, 5 text-figs. (Bombyliidae pp. 392–394, text-fig. 3.)
- 1923. Catalog of Indian Insects. Part 3. Bombyliidae. Government of India, Central Publ. Bureau, Calcutta.

Serville, Jean Guillaume Audinet [see Lepeletier and Serville]

Seydel, Ch.
- 1934. [Note: Biologique sur *Parasa urda* et *Systropus marshalli*.] Rev. Zool. Bot. africaine, Brussels, vol. 26, p. 26, figs. 1–2 on unnumbered pl.

Shannon, Raymond C.
- 1916. Two new North American Diptera. Insecutor Inscitiae Menstruus, vol. 4, pp. 69–72. (Bombyliidae pp. 71–72.)
- 1923. The pleural sclerites of Diptera. Canadian Ent., vol. 55, pp. 219–220.
- 1924. Some special features of the wings of Diptera. Insecutor Inscitiae Menstruus, vol. 12, pp. 34–36, 1 pl.

SHANNON, RAYMOND C., and BROMLEY, STANLEY WILLARD
  1924. Radial venation in the Brachycera (Diptera). Insecutor Inscitiae Menstruus, vol. 12, nos. 7–9, pp. 137–140, pl. 5, figs. 1–9. (Bombyliidae pp. 137, 139, 140, pl. 5, figs. 7–9.)

SHARP, D.
  1909. *In* Verrall: British Flies, vol. 5. On the metamorphoses of the Diptera Brachycera included in this volume. . . . by D. Sharp, pp. 31–39, figs. 54–69. (Bombyliidae p. 37, fig. 64.)

SHELFORD, VICTOR E.
  1913a. The life-history of a bee-fly (*Spogostylum anale* Say), parasite of the larva of a tiger beetle (*Cicindela scutellaris* Say var. *lecontei* Hald.). Ann. Ent. Soc. America, vol. 6, no. 2, pp. 213–225, figs. 1–17.
  1913b. Animal communities in temperate America as illustrated in the Chicago region. A study in animal ecology. The Geographic Society of Chicago, Bull. no. 5, Univ. Chicago Press, xiii + 368 pp., 9 diagrams, map I, frontispiece, and large map of Chicago region, 68 tables, 306 figs. Edition 2, 1937, with corrections and additions to bibliography. (Bombyliidae pp. 223–224, 229–230, 232, 252, 285, figs. 188, 204–205, 210.)

SHERBORN, CAROLO DAVIES
  1902. Index Animalium. . . . Cantabrigiae. xviii + 1195 pp.

SHIRAKI, T. [see Esaki, Hori, Hozana, Shiraki, others] also [see Adachi, Aoki, Shiraki and others]

SHOTWELL, R. L.
  1939. The species and distribution of grasshoppers in the 1939 outbreaks. U.S. Bureau Ent. and Plant Quar., Insect Pest Survey Bull. No. 19 (supp. 5), pp. 190–191.

SHIMKEVITSCH, V. (Chimkévitsch)
  1884. Metamorphose de *Systoechus nitidulus* parasite des Orthoptères du genre *Stauronotus*. [In Russian.] Horae Soc. Ent. Rossia, vol. 18, pp. 11–16.

SIEBKE, H.
  1874. Enumeratio insectorum norvegicorum. . . . 255 pp. (Bombyliidae pp. 11–13.)

SKAIFE, S. H.
  1954. African insect life. London, Cape Town. iv + 387 pp., 75 pls. (5 color), 190 figs. (Bombyliidae pp. 282–284, fig. 137.)

SKINNER, HENRY
  1903. Diptera of Beulah, New Mexico. Trans. Amer. Ent. Soc., vol. 29, pp. 104–106. (Bombyliidae; mentions *Anastoechus nitidulus* (*barbatus*).)

SMART, JOHN
  1958. The tergal depressor of the trochanter muscle in the Diptera. Proc. Tenth Internat. Congress Ent., Montreal, 1956, vol. 1, pp. 551–555.
  1959. Notes on the Mesothoracic Musculature of Diptera *in* Studies in Invertebrate Morphology. Smithsonian Misc. Coll., vol. 137, pp. 331–364, text-figs. 1–4, pl. 1, figs. a-f. (Bombyliidae pp. 354, 356.)

SMITH, C. W.
  1940. An exchange of grasshopper parasites between Argentina and Canada with notes on parasitism of native grasshoppers. Report Ent. Soc. Ontario, vol. 70 (1939), pp. 57–62. (Bombyliidae pp. 57, 58, 60.)

SMITH, JOHN B.
  1890. A contribution to a knowledge of the mouth parts of the Diptera. Trans. Amer. Ent. Soc., vol. 17, pp. 319–339, 22 figs. (Bombyliidae pp. 330–331, 336, figs. 11–12.)
  1900. Insects of New Jersey. Twenty-seventh annual report, New Jersey Board of Agriculture for 1899. Trenton. 755 pp. Insects in supplement. (See Johnson, C. W. for Bombyliidae.)
  1910. Report of the insects of New Jersey. Report of New Jersey State Museum (1909). Trenton. Curator's Report, pp. 1–14. Part 1, Insects, their classification and distribution, pp. 15–32. Part 2, Systematic List, pp. 33–888. (See Johnson, C. W. for Bombyliidae.)

SMITH, ROGER C.
  1934. Notes on the Neuroptera and Mecoptera of Kansas with keys for the identification of species. Journ. Kansas Ent. Soc., vol. 7, pp. 120–145, 1 pl. (Bombyliidae pp. 136–137.)
  1955. Guide to the literature of the zoological sciences. Revised edition. Burgess Publication Co.
SMITH, ROGER C., KELLY, E. G., DEAN, GEORGE A., BRYSON, H. R., and PARKER, R. L.
  1943. Common insects of Kansas. Report Kansas State Board Agric., 440 pp., 464 text-figs., 6 pls. (color). (Bombyliidae pp. 348–349, text-figs. 368–369.)
SNODGRASS, R. E.
  1957. A revised interpretation of the external reproductive organs of male insects. Smithsonian Misc. Coll., vol. 135, no. 6, 60 pp., 15 figs.
SNOW, STERLING J.
  1925. Observations on the cutworm, *Euxoa auxiliaris* Grote, and its principal parasites. Journ. Econ. Ent., vol. 18, pp. 602–609. (Bombyliidae pp. 605, 608–609.)
SÖRENSEN, WILLIAM
  1884. [Note: *Anthrax erythrocephalus* Fabr.] Ent. Tidscrift, Stockholm, vol. 5, p. 18.
SPEARS, SYLVIA ALLEN
  1965. A study of the parasites of mud-dauber wasps. Unpublished manuscript, University of Mississippi.
SPEISER, PAUL
  1910. *In* Sjöstedt: Wissenschaftliche Ergebnisse der Schwedischen zoologischen Expedition nach dem Kilimandjaro, dem Maru und dem umgebenden Massaisteppen Deutsch-Ostafrikas, 1905–1906 . . . Stockholm. Diptera, band 2, hefte 8, 4 (Orthorrhapha), pp. 65–112. (Bombyliidae pp. 75–81.)
  1914. Beiträge zur Dipterenfauna von Kamerun. Deutsche Ent. Zeitschr. Pt. 2, pp. 1–16. (Bombyliidae pp. 5–7; 3 spp., 2 of them n. spp.)
  1920. Zur Kenntnis der Diptera Orthorrhapha Brachycera. Zool. Jahrb., vol. 43, pp. 195–220, 7 text-figs. (Bombyliidae p. 213.)
SPENCE, W. [see Kirby and Spence]
SPENCER, G. J.
  1958. The natural control complex affecting grasshoppers in the dry belt of British Columbia. Proc. Tenth Internat. Congress Ent., vol. 4, pp. 497–502. (Bombyliidae p. 497.)
STACKELBERG, A.
  1933. Opredeliteli muh evropeiskoi ciasti S.S.S.R. Leningrad.
STATZ, G.
  1940. Neue Dipteren (Brachycera et Cyclorhapha) aus dem Oberoligozan von Rott. Palaeontographica, Stuttgart (A), vol. 91, pp. 120–174, 9 pls., 1 fig. (fossil). (Bombyliidae pp. 131–132, pl. 19, figs. 10–11; pl. 23, figs. 61–62.)
STEFANI-PÉREZ, TEODOSIO DE [see also Bezzi and Stefani-Pérez]
  1913. Cavalette, loro invasioni e lotta contro di esse in Sicilia. Giorn. Sci. Nat. Econ. Palermo, vol. 30, pp. 117–199, 21 figs.
STEIN, PAUL [see Becker and Stein]
STEIN, J. P. E. F.
  1881. Die Löw'sche Dipteren-Sammlung. Stettiner Ent. Zeitung, vol. 42, pp. 489–491. (Bombyliidae pp. 489, 491.)
STEIN, RICHARD RITTER VON
  1885. Biologische Mittheilungen. Zur Naturgeschichte von *Argyramoeba sinuata* Fallen. Ent. Nachrichten, vol. 11, pp. 306–309.
STEPANOV, P. J.
  1881. Die Parasiten der Heuschrecken. Trudy obschtsch. ispyt. prirod. Imp. Charkovsk. Univ., vol. 13 (1879), pp. 101–114. [In Russian.] (Bombyliidae p. 109.)
  1882a. Die Metamorphose der Dipterenfamilie Bombyliidae. Trudy obschtsch, ispyt. prirod. Imp. Charkovsk. Univ., vol. 15 (1881), pp. 1–9, 1 pl. [In Russian.]
  1882b. Bemerkungen über die Parasiten von *Stauronotum vastator*. Trudy obschtsch. ispyt. prirod. Imp. Charkovsk. Univ., vol. 16, pp. 1–3. (Bombyliidae: *Systoechus leucophaeus* from *Stauronotus*.) [In Russian.]

Stephen, W. P. [see Bohart, Stephen, and Eppley]
Stephens, J. F.
   1829. A systematic catalogue of British insects: Being an attempt to arrange all the hitherto discovered indigenous insects in accordance with their natural affinites. London. Vol. 2, 388 pp. (Bombyliidae; Anthracidae pp. 273–274. *Bombylius* p. 273.)
Steyskal, George C.
   1953. A suggested classification of the lower brachycerous Diptera. Ann. Ent. Soc. America, vol. 46, no. 2, pp. 237–242. (Bombyliidae not mentioned, but a useful paper.)
Stiles, C. W.
   1914. International Commission of Zoological Nomenclature, Opinion 65. Case of a genus based upon erroneously determined species. Smithsonian Institution Publ. no. 2256, pp. 152–153.
Stone, Alan, Sabrosky, Curtis W., Wirth, Willis W., Foote, Richard H., and Coulson, Jack R.
   1965. A catalog of the Diptera of America North of Mexico. (Prepared cooperatively by specialists on the various groups of Diptera under the direction of the above.) Agric. Res. Serv., U.S. Dept. Agric., Washington, D. C. Pp. i–iv, 1–1116, index pp. 1549–1696. (Bombyliidae see Painter and Painter.)
Stower, W. J., Popov, G. B., and Greathead, D. J.
   1958. Oviposition behaviour and egg mortality of the desert locust *Schistocerca gregaria* (Forskål) on the coast of Eritrea. Anti-Locust Bull. 30, pp. 1–33, 17 figs. (Bombyliidae p. 27.)
Strand, Embrik
   1928. Miscellanea nomenclatorica zoologica et palaeontologica. Arch. Naturg., Berlin, vol. 92 (1926), Abt. A., part. 8, pp. 30–75. (Bombyliidae p. 48.)
Streich, Ivo von
   1910. Zum Begattungsakt des *Bombylius venosus* Mikn. (Dipt.). Deutsche Ent. Zeitschr., Berlin (1910), p. 314.
Strickland, E. H.
   1938. An annotated list of the Diptera (flies) of Alberta. Canadian Journ. Res. D., vol. 16, pp. 175–219. (Bombyliidae pp. 195–196.)
Strobl, P. Gabriel
   1893. Beiträge zur Dipterenfauna des österreichischen Littorale. Wiener Ent. Zeitung, vol. 12, pp. 29–42, 74–80, 89–108, 121–136, 160–170. (Bombyliidae pp. 31–33.)
   1893–1910. Die Dipteren von Steiermark. Mitt. naturw. Ver. Steiermark.
      1893. Vol. 29 (1892), pp. 1–298. (Bombyliidae pp. 36–39.)
      1898. Part IV. Nachträge. Vol. 34, pp. 192–298. (Bombyliidae p. 197.)
      1910. II. Nachtrag. Granz Mitt. naturw. Ver. Steiermark, vol. 46 (1909), pp. 45–293. (Bombyliidae pp. 45, 55.)
   1896. Siebenburgische Zweiflüger. Verh. Mitt. Siebenbürgischer Ver. Hermannstadt, vol. 46, pp. 11–48.
   1898a. Spanische Dipteren. I, part 1. Wiener Ent. Zeitung, vol. 17, pp. 294–304. (Bombyliidae pp. 300–302.)
   1898b. Dipterous fauna of Bosnia, Herzegovina and Dalmatia. [In Serbian and Latin.] Glasnik zem. Mus. Bosni Hercegovini, Sarajevo, vol. 10, pp. 387–466, 561–616. (Bombyliidae p. 396.)
   1900. Dipteren fauna von Bosnien, Hercogovina und Dalmatia. Wiss. Mitt. Bosnien, vol. 7, pp. 552–670. (Bombyliidae p. 559.)
   1901. Dipteren *in* Kertész: Zoologische Ergebnisse der dritten asiatischen Forschungareise des Grafen Eugen Zichy. Ann. Mus. Nat. hungarici, vol. 2, xli + 470 pp. Dipteren pp. 181–201. (See Kertész.)
   1902. Contribution to the dipterous fauna of the Balkan peninsula. [In Serbian and Latin.] Glasnik zem. Mus. Bosnien Hercegovini, Sarajevo, vol. 14, pp. 461–517. (Bombyliidae pp. 465–467.)
   1905. Neue Beiträge zur Dipterenfauna der Balkanhalbinsel. Wiss. Mitt. Bosnien, Herzegovina, vol. 9 (1904), pp. 519–581. (Bombyliidae pp. 524–525.)

STROBEL, P. GABRIEL—Continued
- 1906. Spanische Dipteren. II. Beitrag. Madrid. Mem. Soc. españa Hist. Nat., vol. 3 (1905), pp. 271–422. (Bombyliidae pp. 281–288.)
- 1909. *In* Czerny and Strobl: Spanische Dipteren. III. Verh. zool.-bot. Ges. Wien, vol. 59, pp. 121–301. (Bombyliidae pp. 145–153.)

STUARDO ORTIZ, CARLOS [see also Ruiz Pereira, and Stuardo]
- 1946. Catálogo de los Dipteros de Chile. Ministry Agric. Santiago de Chile, pp. 1–253. (Bombyliidae by Painter.)

STUDY, E.
- 1926. Uber einige mimetische Fliegen. Zool. Jahrb., vol. 42, Abt. f. allg. Zool. u. Physiol., pp. 421–427, pls. 13–14. (Bombyliidae by Enderlein, pp. 425–426, pl. 14, fig. 4.)

STURTEVANT, A. H.
- 1925. The seminal receptacle and accessory glands of Diptera, with special reference to the Acalypterae. Journ. New York Ent. Soc., vol. 33, no. 4, pp. 195–215; vol. 34, pp. 1–21, pls. 1–3. (Bombyliidae pp. 204–205.)

SULZER, JOHANN HEINRICH
- 1776. Abgekürzte Geschichte der Insekten nach dem Linnaeischen System. Winterthur. I-II. Pt. 1, pp. 1–27, 1–274; pt. 2, pp. 1–71, 32 color pls. (Bombyliidae pp. 174, 225, pl. 28, figs. 22–23.)

SWEETMAN, HARVEY L.
- 1958. The principles of biological control. Interrelation of hosts and pests and utilization in regulation of animal and plant populations. xii + 560 pp., 328 figs. Dubuque, Iowa: Wm. C. Brown. (Bombyliidae pp. 99, 108, 110, 111, 209, 255, 258–259, 282, 296, 297, 321, 322, 323, 325, 467, 482, figs. 77–78, 257.)

SWEZEY, OTTO H.
- 1915. Some hyperparasites of white grubs. Proc. Hawaiian Ent. Soc., vol. 3, no. 2, pp. 71–72. (*Chrysanthrax fulvohirta* reared from *Elis*.)

SWYNNERTON, C. F. M.
- 1936. The Tsetse flies of East Africa. Trans. Roy. Ent. Soc. London, vol. 84, 579 pp. (Bombyliidae pp. 230, 236. Lists 3 spp. of *Thyridanthrax—abruptus, lineus, argentifrons*.)

SZILADY, Z.
- 1921. Thermokopische Farben im Tierreich. Festschrift 2. II. Ferienhochschulkurs, Nagyszeben, pp. 65–84. (Bombyliidae p. 79.)
- 1940. Dipteren. Explorationes zoologicae ab E. Csiki in Albania peractae XVIII. A Magyar Tudományos Akadémia Balkán—Kutatásainak Tudományos Eredményei, vol. 1, pt. 2, pp. 316–328. Budapest.
- 1942. Ostafrikanische Syrphiden und Bombyliden (Dipt.) im Ungarischen nationalmuseum. Ann. Hist. Nat. Mus. hungarici, Budapest, vol. 35 (Zool.), pp. 91–101, 2 figs. (Bombyliidae pp. 98–101; 3 spp. *Exoprosopa*, 1 sp. *Dischistus* described; 34 spp. listed.)

TAKEUCHI, KICHIZO
- 1955. Coloured illustrations of the insects of Japan. 190 pp., 68 color pls. Hoikusha, Osaka, Japan. (Republished 1962.) (Bombyliidae p. 147, pl. 65, figs. 1057–1063.)

TASCHENBERG, ERNST LUDWIG
- 1884. Insects *in* Brehm: Tierleben. Vol. 9, 741 pp. Leipzig. (Bombyliidae pp. 462–463, fig. of *Anthrax semiatra* and emerging pupa, also *Bombylius venosus*.)

THALHAMMER, JÁNOS
- 1911. *Empidideicus hungaricus*, Dipteren novum ex Hungaria. Ann. Hist.-Nat. Mus. Nat. hungarici, Budapest, vol. 9, pp. 388–389.

THÉOBALD, NICHOLAS
- 1937a. Les insectes fossiles des terrains oligocènes de France. Nancy: Imprimerie Georges Thomas. 473 pp., 29 pls., 7 maps. (Bombyliidae pp. 287, 344, 349–351, 378.)

THEOBALD, NICHOLAS—Continued
  1937b. Note complémentaire sur les insectes fossiles oligocènes des gypses d'Aix-en-Provence. Bull. Soc. des Sci. de Nancy, n. ser., 1937, pp. 157–178, 7 text-figs., pls. 1–2 with 10 figs. (Bombyliidae pp. 168–171, text-figs. 5, pl. 1, fig.6; pl. 2, fig. 2.)

THOMPSON, W. R.
  1923. Recherches sur la biologie des Diptères parasites. Bull. Biol. France Belgique, vol. 57, pp. 174–237.
  1944. A catalogue of the parasites and predators of insect pests. Imper. Agric. Bureau, Insect Entomology. Canadian Parasite Service.
  1951. A catalogue of the parasites and predators of insect pests. Commonwealth Agric. Bureau, Commonweath Institute Biol. Control, Ottawa, Canada, sect. 2, pt. 1, pp. 1–147.

THOMSON, CARL GUSTAV
  1869. Diptera in Kongliga Svenska Fregatten *Eugenies* Resa . . . 1851–53. Svenska Vetenskaps-Akademien. Vol. 2, Zool., pt. 1, Insekt. Haft 12. Diptera, pp. 443–614, pl. 9. (Bombyliidae pp. 479–488.)

THUNBERG, CARL PETER
  1827. Tanyglossae septendecim novae species descriptae. Nova Acta Upsala, vol. 9, pp. 63–75, 1 pl. (Bombyliidae: *Exoprosopa* (*Tanyglossa*) ?*maura*, p. 73, pl. 1, 1, fig. 11; also 4 synonyms.)

TIEGS, O. W.
  1955. The flight muscles of insects—their anatomy and histology. . . . Philos. Trans. Roy. Soc. London, ser. B. vol. 238, no. 656, pp. 221–359.

TILLYARD, R. J.
  1926. The insects of Australia and New Zealand. Sydney. 571 pp., 465 text-figs., 44 pls. (8 color). (Bombyliidae pp. 343, 361, 363–364, pl. 25, figs. 21–22, text-figs. W50–W55.)

TIMON-DAVID, JEAN
  1937. Recherches sur le peuplement des hautes montagnes. Diptères de la vallée de Chamonix et du massif du Mont Blanc. Ann. Facultes Sci. Marseille, vol. 10, fasc. 1, pp. 7–54, 14 text-figs., 34 references. (Bombyliidae pp. 11, 21–22; 2 spp. listed.)
  1944. Insects fossiles de l'Oligocene inférieur des Camoins (Bassin de Marseille). In 2 parts. Bull. Soc. Ent. France. Pt. 1, Diptères, brachycères. Vol. 48 (1943), pp. 128–134, 7 text-figs., 1 pl. (Bombyliidae pp. 132–133, fig. 5.)
  1949. Observations sur le peuplement entomologique (Diptères) des Pyrénées Ariégeoises. Thirteenth Congress Internat. Zool. Paris, 1948, pp. 475–476. (Bombyliidae p. 476; 1 sp. listed.)
  1950. Diptères des Pyrénées Ariégeoises notes écologiques et biogéographiques. Bull. Soc. Hist. Nat. Toulouse, vol. 85, pp. 11–25, 5 figs. (Bombyliidae pp. 16–17, 19–21, 23.)
  1952. Contribution à la connaissance de la faune entomologique du Maroc, Diptera: Asilidae, Bombyliidae, Nemestrinidae et Syrphidae. Bull. Soc. Sci. Nat. Maroc, vol. 31, pp. 131–148, 3 pls., 8 figs. (Bombyliidae pp. 139–144, pl. 1, figs. 5–7.)

TONNOIR, A. L.
  1927. Descriptions of new and remarkable New Zealand Diptera. Rec. Canterbury (New Zealand) Mus., Christchurch, vol. 3, pt. 3, pp. 101–112, 4 figs. (Bombyliidae pp. 101–104, fig. 1.)

TORRES, DEMETRIO DELGADO DE
  1929. Dipteros parasitos de la langosta en españa. Bol. Patol. Veg. Ent. Agric., vol. 4, pp. 65–68, 2 figs.

TOTH, S.
  1964. Angaben uber die Dipteren der Tari-Tales. I. Bombyliidae und Tabanidae. [In Hungarian with German summary.] Folio ent. hung., Budapest (S.N.), vol. 17, pp. 67–73.

TOWNSEND, C. H. TYLER
- 1893a. Description of the pupa of *Toxophora virgata* O. S. Psyche, vol. 6, pp. 455–457.
- 1893b. The pupa of *Argyramoeba oedipus*. Amer. Natur., vol. 27, pp. 60–63.
- 1901. New and little known Diptera from the Organ mountains and vicinity in New Mexico. Trans. Amer. Ent. Soc., vol. 27, pp. 159–164. (Bombyliidae pp. 159–160.)

TRAUTMANN, W.
- 1916a. *Argyromoeba sinuata* Fall. Biologisches, Intern. ent. Zeitschr. Guben, vol. 10, p. 56.
- 1916b. Beitrag zur Goldwespenfauna Brankens. Biologisches, Intern. ent. Zeitschr. Guben, vol. 10, pp. 58–59. (Bombyliidae p. 58; mentioned under Trauerfliegen.)

TREHERNE, R. C., and BUCKELL, E. R.
- 1924. Grasshoppers of British Columbia. Bull. Dept. Agric. Canada, no. 39 (Ent. Bull. no. 26), pp. 1–40, 18 figs., 3 pls.

TROITSKY, D.
- 1914. Acridid pests in the Semipalatinsk province in 1912. [In Russian.] Nuzhdui Zapad.-Sibirsk. Selsk. Khoz., vol. 2, no. 1, pp. 23–49.

TROJAN, P.
- 1967. Muchowski, Diptera. Zeszyt: 24. Bujanki-Bombyliidae. Klucze. Oz Nauz, Owadpol., vol. 28, 84 pages, 71 figs.

TSACAS, L.
- 1962. Contribution à la connaissance des Diptères de Grèce. III. Bombyliidae de Macédonie. Rev. franç. Ent., vol. 29, pp. 287–305, 19 figs.

TUCKER, ELBERT S.
- 1907. Some results of desultory collecting of insects in Kansas and Colorado. Kansas Univ. Sci. Bull., vol. 4, no. 2, pp. 51–112. (Bombyliidae pp. 89–91.)

UHLER, P. R.
- 1877. Article 31. Report upon the insects collected by P. R. Uhler during the explorations of 1875, etc. Bull. U.S. Geol. Geogr. Surv. Terr., vol. 3, pp. 765–796. (Bombyliidae pp. 780–781; 13 spp.)

UNDERHILL, H. M. J. [see Allen and Underhill]

UVAROV, B. P.
- 1928. Locusts and grasshoppers. Imperial Bur. Ent., London. 365 pp., 118 text-figs., 9 pls., 1 map; 445 references on pp. 331–346. (Bombyliidae pp. 109–111, text-figs. 50–51.)
- 1931. Insects and climate. Trans. Ent. Soc. London, vol. 79, pp. 1–247.

VAN DUZEE, E. P.
- 1931. Swarming of two species of Diptera. Pan-Pacific Ent., vol. 7, no. 3, p. 104.

VASSILIEW, IVAN VON
- 1905. Beitrag zur Biologie der Gattung *Anthrax* Scop. (Fam. Bombyliidae). Zeitschr. Wiss. Insektenbiologie, vol. 1, heft 4, pp. 174–175.

VENTURI, F.
- 1948. Notulae dipterologicae II. Sulla distribuzione geografica e cronologica delle *Usia* (Dipt. Bombyliidae) in Italia. Redia, Florence, vol. 33, pp. 127–142, 2 figs., 2 maps, 4 tables.
- 1958. Su alcune *Villa* Lioy (Dipt. Bombyliidae) della collezione di M. le Dr. M. Bequaert di Gand. Bull. Inst. Sci. Nat. Beligique, vol. 34, no. 33, pp. 1–4.
- 1962. Some Bombyliidae taken by Giordani Soika in N. Africa. Boll. Mus. Stor. Nat. Venezia, vol. 14, pp. 37–41.

VERHOEFF, C
- 1891. Biologische Aphorismen über einige Hymenopteren, Dipteren, und Coleopteren. Verh. Naturhist. Ver. preuss. Rheinl. Westphal., vol. 48, pp. 1–80, pls. 1–3. (Bombyliidae pp. 40–57.)

VERRALL, G. H.
- 1882–1884. *In* Scudder: Nomenclator Zoologicus. U.S. Nat. Mus. Bull, preface xix pp. Part I. Supplement list, 376 pp. Part II. Universal Index, 340 pp. (Bombyliidae p. 355; *Xystropus* Wiedemann (=*Systropus* Wiedemann), Diptera Bombyliidae, Verrall.)

VERRALL, G. H.—Continued
- 1901. List of British Diptera. Second edition, 47 pp. London.
- 1909. British flies. London: Gurney & Jackson. Vol. 5, 814 pp., 407 figs. (Bombyliidae relations to other families pp. 470–473; keys and descriptions pp. 474–536, 755, 760–763, figs. 269–309; immature stages pp. 502–503.)

VILLENEUVE, JOSEPH
- 1901. [Note: *Anthrax flava* victims.] Bull. Soc. Ent. France, 1901, p. 279.
- 1903a. Etude sur quelques Diptères. Bull. Soc. Ent. France, 1903, pp. 125–127. (Bombyliidae pp. 125–126.)
- 1903b. Les Bombyles de Meigen an Museum de Paris. (Dipt.). Bull. Soc. Ent. France, 1903, pp. 237–239.
- 1904. Contribution au catalogue des diptères de France. Feuille Natur., vol. 34, pp. 69–73, 166–173, 225–229. (Bombyliidae pp. 69–73.)
- 1905. Les types de Meigen au Museum de Paris. Ann. Soc. Ent. France, vol. 74, pp. 304–310. (Bombyliidae pp. 304–306.)
- 1908. Travaux diptèrologiques. Wiener Ent. Zeitung, vol. 27, pp. 281–288. (Bombyliidae p. 283.)
- 1910. Diptères nouveaux du Nord de l'Afrique. Wiener Ent. Zeitung, vol. 29, pp. 301–304. (Bombyliidae pp. 301–302.)
- 1912. Diptères nouveaux recueillis en Syrie par M. Henri Gadeau de Kerville et descrits. Bull. Soc. Sci. Nat. Rouen, vol. 47, pp. 40–55. (Bombyliidae p. 43.)
- 1912–1914. Notes synonymiques. Wiener Ent. Zeitung.
  - 1912. Vol. 31, pp. 96–97. (Bombyliidae p. 97.)
  - 1914. Vol. 33, pp. 207–208. (Bombyliidae p. 207.)
- 1913. Description d'une espèce nouvelle de Diptère du genre *Lomatia* Meig. recueillie par M. H. Gadeau de Kerville en Syrie. Bull. Soc. Sci. Nat. Rouen, vol. 47, pp. 73–77. (Bombyliidae p. 75.)
- 1920. Diptères paléarctiques nouveaux ou peu connus. Ann. Soc. Ent. Belgiques, Bruxelles vol. 60, pp. 114–120. (Bombyliidae pp. 115–116.)
- 1930. Diptères inédits. Bull. Ann. Soc. Ent. Belgique, Bruxelles, vol. 70, pp. 98–104. (Bombyliidae pp. 100–101.)
- 1932. Descriptions de Diptères nouveaux du Nord Africain. Bull. Soc. Ent. France, Paris, vol. 37, pp. 32–34. (Bombyliidae pp. 32–33.)
- 1934. Notes diptèrologiques. Rev. franç. Ent., Paris, vol. 1, pp. 180–183. (Bombyliidae pp. 182–183.)

VILLENEUVE, JOSEPH, and SURCOUF, J.
- 1908. Ditteri, pp. 85–89, *in* Gadeau de Kerville: Voyage zoologique en Khroumirie (Tunisie). . . . 1906. Paris.

VILLERS, CHARLES JOSEPH DE
- 1789. Caroli Linnaei entomologia, Faunae Suecicae descriptionibus aucta. . . . In 4 volumes. Vol. 3, pp. 1–657, 4 pls. (Bombyliidae pp. 408–410, 421, 427, 603–609; pl. 9, fig. 11 [*Musca*].)

VIMMER, ANTON
- 1909. O kukláh několika Bombylidů. (Ueber das Gehäuse einiger Bombyliidae.) Prag Čas. Čecké Spol. Ent., 1909, pp. 24–27.

VOLLENHOVEN, SNELLEN SAMUEL CONSTANTINUS VAN
- 1863. Beschrijving van eenige neiuwe soorten van Diptera. Versl. Medded. Akad. Wetensch. Amsterdam Afd. Natuurk., vol. 15, pp. 8–18, 1 pl. (Bombyliidae pp. 8–9, pl. 1, fig. 4.) Also in Nederl. Tijdschr. Dierkunde, vol. 1, pp. 349–355.

VORONTSOVSKY, D. (Vorontzovskiĭ)
- 1926. New cases of parasitism in Acrididae. [In Russian.] La Défense des Plantes, Leningrad, vol. 3, p. 280. (*Systoechus autumnalis* Pallas parasitises *Arcyptera microptera*, *Dociostaurus kraussi*, *D. albicornis*, Western Siberia.)

WAHLBERG, PETER FREDRIK
- 1838. Bidrag till Svenska Dipternas kännendom. Kongl. Vetensk. Akad. Handl., 1838, pp. 1–23. (Bombyliidae pp. 9–10.)
- 1854. Bidrag till kännendomen om de nordiska Diptera. Öfvers. Svenska Vet.-Akad. Forhandl., vol. 11, pp. 211–216. (Bombyliidae p. 213.)

WAHLGREN, EINAR
- 1907. Svensk Insekfauna. 11. Uppsala. Tvåvingar. Diptera Första unterordningen Orthorhapha. Andra Gruppen Flugor Brachycera. Fam. 14–23. Ent. Tidskr., Stockholm, vol. 28, pp. 129–192. (Bombyliidae pp. 180–186, figs. 23–25.)
- 1915. Det öländska alvarets djurvärld. Ark. Zool., Stockholm, vol. 9, no. 19, pp. 1–135, 4 pls., 8 pictures. (Bombyliidae p. 41; 4 spp. listed.)

WAKELAND, CLAUDE [see Parker and Wakeland]

WALKDEN, H. H.
- 1950. Cutworms, armyworms and related species attacking cereal and forage crops in the Central Great Plains. Circular no. 849, U.S. Dept. Agric., pp. 1–52, 7 text-figs., 5 tables. (Bombyliidae pp. 12, 14, 21, 32, 36.)

WALKER, FRANCIS
- 1835. Characters of some undescribed New Holland Diptera. Ent. Mag., vol. 2, pp. 468–473. (Bombyliidae p. 473.)
- 1837. Descriptions, etc., of the insects collected by Capt. P. P. King, R.N., F.R.S., in the Survey of the Straits of Magellan. Diptera. Trans. Linnean Soc. London, vol. 17, pp. 331–359. (Bombyliidae pp. 339–340.)
- 1848–1855. List of the specimens of dipterous insects in the collections of the British Museum. In 4 parts and 3 supplements. London.
    - 1849a. Pt. 2, pp. 231–484. (Bombyliidae pp. 235–299.)
    - 1849b. Pt. 4, pp. 689–1172. (Bombyliidae pp. 1154–1155.)
- 1850. Characters of undescribed Diptera in the British Museum. Zoologist (Newman's), vol. 8, Appendix pp. lxv, xcv–xcix, cxxi. (Bombyliidae p. xcvii.)
- 1850–1856. *In* Insecta saundersiana (W. W. Saunders, ed.): or characters of undescribed insects in the collection of W. W. Saunders. In 5 parts, pp. 1–474, 8 pls. London.
    - 1852. Part 3, pp. 157–252, 2 pls. (Bombyliidae pp. 165–202, pl. 5, figs. 5, 8.)
- 1851–1856. Insecta Britannica, Diptera. In 3 volumes, with 30 pls. London.
    - 1851. Vol. 1, 319 pp., pls. 1–10. (Bombyliidae pp. 72–73, pl. 2.)
- 1857a. Catalogue of the dipterous insects collected at Singapore and Malacca by Mr. A. R. Wallace, with descriptions of some new species. Journ. Proc. Linnean Soc. London, vol. 1, pp. 4–39, pls. 1–2. (Bombyliidae p. 15.)
- 1857b. Catalogue of the dipterous insects collected at Sarawak, Borneo, by Mr. A. R. Wallace with descriptions of new species. Journ. Proc. Linnean Soc. London, vol. 1, pp. 105–136. (Bombyliidae pp. 118–119.)
- 1857–1861. Characters of undescribed Diptera in the collection of W. W. Saunders. Trans. Ent. Soc. London. In 2 parts.
    - 1857. Part 1, ser. 2, vol. 4, pp. 119–158; 1858, pp. 190–235. (Bombyliidae pp. 135–147.)
    - 1860. Part 2, ser. 2, vol. 5, pp. 268–296. (Bombyliidae pp. 285–286.)
- 1858. Catalogue of the dipterous insects collected in the Aru Islands by Mr. A. R. Wallace, with descriptions of new species. Journ. Proc. Linnean Soc. London, vol. 3, pp. 77–110. (Bombyliidae pp. 90–91; 4 spp.)
- 1859. Catalogue of the dipterous insects collected at Makassar in Celebes, by Mr. A. R. Wallace, with descriptions of new species. Journ. Proc. Linnean Soc. London, vol. 4, pp. 90–144. (Bombyliidae pp. 111–113.)
- 1860. Catalogue of the dipterous insects collected in Amboyna by Mr. A. R. Wallace, with descriptions of new species. Journ. Proc. Linnean Soc. London, vol. 5, pp. 144–168. (Bombyliidae pp. 148–149.)
- 1861a. Catalogue of the dipterous insects collected at Dorey, New Guinea, by A. R. Wallace, with descriptions of new species. Journ. Proc. Linnean Soc. London, vol. 5, pp. 229–254. (Bombyliidae p. 237.)
- 1861b. Catalogue of the dipterous insects collected at Manado in Celebes, and in Tond, by Mr. A. R. Wallace, with descriptions of new species. Journ. Proc. Linnean Soc. London, vol. 5, pp. 258–270. (Bombyliidae pp. 260, 266.)

WALKER, FRANCIS—Continued
- 1861c. Catalogue of the dipterous insects collected in the Batchian, Kaisan and Makian, and at Tidon in Celebes by Mr. A. R. Wallace, with descriptions of new species. Journ. Proc. Linnean Soc. London, vol. 5, pp. 270–303. (Bombyliidae pp. 282–283, 301–302.)
- 1862. Catalogue of the dipterous insects collected at Gilolo, Ternate and Ceram by Mr. R. Wallace. Journ. Proc. Linnean Soc. London, vol. 6, pp. 4–23. (Bombyliidae pp. 8–9.)
- 1864. Catalogue of the dipterous insects collected in Waigiou, Mysol, and North Ceram by Mr. A. R. Wallace, Esq., with descriptions of new species. Journ. Proc. Linnean Soc. London, vol. 7, pp. 202–238. (Bombyliidae pp. 209, 224, 233.)
- 1865. Descriptions of new species of the dipterous insects of New Guinea. Journ. Proc. Linnean Soc. London, vol. 8, pp. 102–108. (Bombyliidae p. 111.)
- 1866. Synopsis of the Diptera of the Eastern Archipelago discovered by Mr. Wallace, and noticed in the "Journal of the Linnean Society." Journ. Proc. Linnean Soc. London, vol. 9, pp. 1–30. (Bombyliidae pp. 4, 14–15.)
- 1871. List of Diptera collected in Egypt and Arabia by J. K. Lord, Esq., with descriptions of the species new to science. Entomologist, vol. 5, pp. 255–263, 271–275, 339–346. (Bombyliidae pp. 260–263, 271–272.)

WALKER, FRED H., JR.
- 1936. Observations on sunflower insects in Kansas. Journ. Kansas Ent. Soc., vol. 9, pp. 16–22. (Bombyliidae p. 22; 3 spp. on blooms.)

WALSH, BENJAMIN DANN
- 1864. On certain remarkable or exceptional larvae, coleopterous, lepidopterous and dipterous. Proc. Boston Soc. Nat. History, Boston, vol. 9, pp. 286–318. (Bombyliidae p. 300.)

WALTON, W. R.
- 1914. Report on some parasitic and predaceous Diptera from northeastern New Mexico. Proc. U.S. Nat. Mus., vol. 48, art. 2070, pp. 171–186, pls. 6–7. (Bombyliidae pp. 173–174; 14 spp. listed.)

WANDOLLECK, BENNO
- 1897. Die Dipterengattungen *Systropus* Wiedem. und *Cephenus* Latr. Ent. Nachrighten, vol. 23, pp. 198–199.

WASHBURN, F. L.
- 1906. The Diptera of Minnesota. Two-winged flies affecting the farm, garden, stock and household. Bull. Minnesota Agric. Exper. Stat., no. 93 (1905), pp. 18–169, index, 2 pls. (color), and 163 text-figs. (Bombyliidae pp. 89–91, figs. 82–85, pl. 1, fig. 15, pl. 2, fig. 17.)

WATERHOUSE, C. O. [see Saunders, Fitch, and Waterhouse]

WEBER, FRIEDRICH
- 1801. Observationes entomologicae, continentes novorum quae condidit generum characteres, et nuper detectarum specierum descriptiones. Kiliae. 128 pp. (Bombyliidae p. 115.)

WEBSTER, F. M.
- 1890. Report of observations upon insects affecting grains. *In* Reports of observations and experiments in the practical work of the division made under the direction of the Entomologist. U.S. Dept. Agric., Division of Ent. (old ser.), Bull. no. 22, pp. 1–110. Webster, pp. 42–72. (Bombyliidae p. 44; *Anthrax* sp. parasitic on *Agrotis herilis*.)

WEINBERG, MEDEEA [see Ionescu and Weinberg]

WEINBERG, MEDEEA, and LUCIA DUSA
- 1968. Ord. Diptera—Fam. Bombyliidae. Trav. Mus. Hist. Natur. Grigore Antipa, no. 9, pp. 293–295.

WELLES, S. PAUL
- 1967. A new host record for *Anthrax limatulus valicola* Marston (Diptera: Bombyliidae). Pan-Pacific Ent., vol. 44, no. 4.

WERNER, FRANZ, and ENGEL, ERICH OTTO
- 1934. Ergebniss einer zoologischen Studien- und Sammelreise nach den Inseln des Aegäischen Meeres. V. Arthropoden. Sitzb. Akad. Wiss. Wien, vol. 143, Abt. 1, pp. 159–168, 4 figs. Insecta pp. 159–162. (Bombyliidae pp. 161–162.)

WESCHÉ, WALTER
- 1907. The genitalia of both the sexes in Diptera, and their relation to the armature of the mouth. Trans. Linnean Soc. London, ser. 2, Zool., vol. 9 (1906), pt. 10, pp. 339–386, pls. 23–30. (Bombyliidae p. 365; comment only on number of receptacula.)

WESTMAAS, A. DE ROO VAN
- 1872. *In* Snellen van Vollenhoven: Nederlandsch Insecten. Sep. 2, vol. 2. (Bombyliidae p. 195, pl. 42, figs. a, b.)

WESTWOOD, JOHN OBADIAH
- 1835. Insectorum novarum exoticorum descriptiones. London and Edinburgh Philos. Mag., ser. 3, vol. 6, pp. 280–281, 447–449. (Bombyliidae p. 447.)
- 1839–1840. An introduction to the modern classification of insects, founded on the natural habits and corresponding organization of the different families. London: Longman. In 2 volumes.
    - 1840. Vol. 2, pp. i–xi, 1–587, figs. 57–133. (Family Bombyliidae pp. 538, 542–543, figs. 128–11, 12, 14. Family Anthracidae pp. 543–545, figs. 128–15, 129–1.)
- 1840. Summary: An introduction to the modern classification of Insects. Isis (Oken), vols. 7–8, columns 584–588. Leipzig. (Bombyliidae column 588); vol. 10, columns 781–795. (Bombyliidae column 794.)
- 1842. Generis dipterorum Monographia Systropi. *In* Guérin-Méneville, Mag. Zool., ser. 2, vol. 4 (XII), 4 pp., pl. 90.
- 1850. Diptera nonnulla exotica descripta. Trans. Ent. Soc. London, vol. 5, pp. 231–236, 1 pl. (Bombyliidae p. 233, pl. 23, fig. 6.)
- 1876a. Nota Dipterologicae. No. 1. Bombylii at Pompeii. Trans. Ent. Soc. London, 1876, pp. 497–499.
- 1876b. Notae Dipterologicae. No. 4. Monograph of the genus *Systropus*, with notes on the economy of a new species of that genus. Trans. Ent. Soc. London, 1876, pp. 571–579, pl. 10, figs. 1–15.

WHITE, ARTHUR
- 1917. The Diptera-Brachycera of Tasmania. Part 3. Families Asilidae, Bombyliidae, Empididae, Dolichopodidae and Phoridae. Papers Proc. Roy. Soc. Tasmania (1916), pp. 148–266. (Bombyliidae pp. 183–214.)

WHITE, GILBERT
- 1795. A naturalist's calendar with observations in various branches of natural history. ... London: White. 170 pp. Insects pp. 97–118. Later editions: 1813, 1822, 1836–38, 1840, 1843, 1851. Paoli, 1937, gives 1837. The natural history and antiquities of Selborn, with the naturalist's calendar and miscellaneous observations with notes by E. T. Bennet and others. London. 640 pp.

WHITE, J. E. [see Brodie and White]

WHITE, M. J. D.
- 1949. Cytological evidence on the phylogeny and classification of the Diptera. Evolution, vol. 3, pp. 252–261.

WIEDEMANN, CHRISTIAN RUDOLPH WILHELM
- 1817–1825. Zoologisches Magazin, Kiel. In 2 volumes.
    - 1817. Ueber einige neue Fliegengattungen. Vol. 1, pt. 1, pp. 57–61. (Bombyliidae pp. 58–59.)
    - 1818a. Aus Pallas dipterologischem Nachlasse. Vol. 1, pt. 2, pp. 1–40. (Bombyliidae pp. 8–24.)
    - 1818b. Neue Insecten vom Vorgebirge der Guten Hoffnung. Vol. 1, pt. 2, pp. 40–48. (Bombyliidae pp. 40–42.)
    - 1819a. Beschreibung neuer Zweiflugler aus Ostindien und Afrika. Vol. 1, pt. 3, pp. 3–39. (Bombyliidae pp. 8–12.)
    - 1819b. Brasilianische Zweifluger (beschrieben vom herausgeber.). Vol. 1, pt. 3, pp. 40–56. (Bombyliidae pp. 46–47.)

WIEDEMANN, CHRISTAIN RUDOLPH WILHELM—Continued
- 1820a. Munus rectoris in Academia Christiano-Albertina iterum aditurus nova dipterorum genera offert iconibusque illustrat. Kiliae, Mohr, pp. 1–23, 1 pl. (Bombyliidae pp. 9–18.)
- 1820b. *In* Meigen, J. W.: Systematische Beschreibung der bekannten europaischen zweiflugeligen Insekten. Achen. (Contains many descriptions attributed to Wiedemann.)
- 1821. Diptera exotica. Kiliae. Part 1, xix + 244 pp., 2 pls. [In Latin.] (Bombyliidae pp. 118–154, 158–179, 242–243, pl. 1, figs. 1, 3–5, 7; pl. 2, figs. 4, 6.)
- 1824. Analecta entomologica ex Museo Regio Hafniensi maximae congesta. Kiliae. 60 pp., 1 pl. (Bombyliidae pp. 22–24, 60.)
- 1828–1830. Aussereuropäische zweiflügelige Insekten, als Fortsetzung des Meigenschen Werkes. Hamm. In 2 volumes.
  - 1828. Vol. 1, xxxii + 608 pp., 7 pls. (Bombyliidae pp. 252–363, pl. 3, figs. 1–8; pl. 4, figs. 1–8; pl. 5, figs. 1–4, 6.)
  - 1830. Vol. 2, xii + 684 pp., 5 pls. (Bombyliidae pp. 632–644, pl. 10B, fig. 12.)

WILLIAMS, FRANCIS X. [see Hungerford and Williams]

WILLISTON, SAMUEL WENDELL
- 1879. An anomolous bombyliid. Canadian Ent., vol. 11, pp. 215–216.
- 1885. On the classification of North American Diptera. Ent. Americana, vol. 1, pp. 10–13, 114–120, 152–155. (Bombyliidae p. 10; discussion of Tanystoma.)
- 1887. Table of families of Diptera. Trans. Kansas Acad. Sci., vol. 10 (1885–1886), pp. 122–128. (Bombyliidae pp. 126, 128.)
- 1888a. Synopsis of the families and genera of North American Diptera, exclusive of the Nematocera and Muscidae. 88 pp. (Bombyliidae pp. 34–38.) (Page 8 gives Brauer system. New Haven, Connecticut.
- 1888b. Table of the families of Diptera. Trans. Kansas Acad. Sci., vol. 10, pp. 122–128.
- 1893a. New or little-known Diptera. Kansas Univ. Quart., vol. 2, pp. 59–78. (Bombyliidae pp. 64–66.)
- 1893b. *In* Riley: Diptera of the Death Valley Expedition. U.S. Dept. Agric., North American Fauna, no. 7, pp. 253–259. (Bombyliidae pp. 254–255.)
- 1893c. A list of species of Diptera from San Domingo. Canadian Ent., vol. 25, pp. 170–171.
- 1894. On the genus *Dolichomyia* with the description of a new species from Colorado. Kansas Univ. Quart., vol. 3, no. 1, pp, 41–43.
- 1895. New Bombyliidae. Kansas Univ. Quart., vol. 3, no. 4 (1894), pp. 267–269.
- 1896a. On the Diptera of St. Vincent (West Indies). Trans. Ent. Soc. London, 1896, pp. 253–308, 346–446, pls. 8–14, 170 figs. (Dolichopodidae by Aldrich, pp. 309–345.) (Bombyliidae p. 306, pl. 11, fig. 81; 1 sp.)
- 1896b. Manual of the families and genera of North American Diptera. New Haven, Connecticut. Second edition, liv + 167 pp. (Bombyliidae pp. 63–68.)
- 1896c. Bibliography of North American dipterology 1878–1895. Kansas Univ. Quart., vol. 4, pp. 129–144, 199–204.
- 1899. On the genus *Thlipsogaster* Rond. Psyche, vol. 8, pp. 331–332.
- 1900–1901. *In*: Biologia Centrali-Americana. Insecta, Diptera. London. Pl. 1–6.
  - 1900. Vol. 1, supplement, pp. 217–248.
  - 1901. Vol. 1, supplement, pp. 249–332, 377–378, pls. 4–5, part of pl. 6. (Bombyliidae, supplement, pp. 269–296, pl. 5, figs. 1–11a.)
- 1907a. Dipterological notes. Journ. New York Ent. Soc., vol. 15, pp. 1–2. (Bombyliidae p. 1.)
- 1907b. The antennae of Diptera; a study of phylogeny. Biol. Bull. Woods Hole, vol. 13, pp. 324–332.
- 1908. Manual North American Diptera. Third edition. New Haven, Connecticut. 405 pp., 163 figs. (Bombyliidae pp. 210–217, figs. 82–86; fig. 86 with 26 parts.)

WILSON, C.C.
- 1936. Notes on the warrior grasshopper, *Camnula pellucida* (Scudder), and its egg parasite, *Aphoebantus hirsutus* Coquillett in northern California, 1928–29. Journ. Econ. Ent., vol. 29, pp. 413–416.

WIRTH, WILLIS W. [see Stone, Sabrosky, Wirth, Foote, and Coulson]

WNUKOWSKY, W. VON
- 1930. Zur Fauna der Dipteren des Bezirks Kamenj (sudwestliches Sibierien, fruheres Gouvernement Tomsk). Deutsche Ent. Zeitschr., 1930, pp. 71–73. (Bombyliidae p. 72; 2 spp. listed.)

WOLCOTT, GEORGE N.
- 1923–1924. Insectae Portoricensis. A preliminary annotated checklist of the insects of Porto Rico, with descriptions of some new species. Insular Exper. Stat., Rio Piedras, P. R., vol. 7, no. 1, Journ. Dept. Agric. Porto Rico, 313 pp. Diptera pp. 209–235. (Bombyliidae pp. 214–215; 11 spp. listed.)
- 1936. Insectae Borinquenses. A revised annotated checklist of the insects of Puerto Rico. Journ. Agric. Univ. Puerto Rico (continuation of Journ. Dept. Agric. Puerto Rico), vol. 20, 600 pp., illustrations not numbered, with a host-plant index by José I. Otero, pp. 601–627. Diptera pp. 320–392, 13 figs. (Bombyliidae pp. 338–340, 1 fig., 8 genera, 18 spp.)
- 1948. The insects of Porto Rico. Journ. Agric. Univ. Porto Rico, vol. 32, no. 3, pp. 417–748. (Bombyliidae pp. 450–452; 16 spp. listed.)

WOOD, JOHN GEORGE
- 1874. Insects abroad. 780 pp. (Bombyliidae pp. 761–762, figs. 505–506.) Also, new edition, 1883, London. 90 pp.

WOOD, WILLIAM
- 1821. Illustrations of the Linnean genera of insects. London: Taylor. In 2 volumes, pp. 1–118, 1–161, 86 pls. (color). (Bombyliidae p. 106, pl. 71.)

WRAY, DAVID L.
- 1950. Insects of North Carolina. Second supplement. North Carolina Dept. Agric., Div. Ent., pp. 1–59. (Bombyliidae p. 29.)

WU, CHENFU, F.
- 1940. Catalogus insectorum sinensium. (Catalogue of Chinese Insects.) Fan Mem. Inst. Biol., Peiping, vol. 5 (1939), 530 pp. Order Diptera, including Siphonaptera. (Bombyliidae pp. 219–236; 15 genera and 50 spp.)

WULP, FREDERICK MAURITS VAN DER
- 1868a. Eenige Noord-Amerikaansch Diptera. Tijdschr. Ent., ser. 2, vol. 2 (X) (1867), pp. 125–164, pls. 3–5. (Bombyliidae pp. 141–142, pl. 4, figs. 1–4.)
- 1868b. Diptera uit den Oost-Indischen Archipel. Tijdschr. Ent., ser. 2, vol. 3 (XI), pp. 97–119, pls. 3–4 (color). (Bombyliidae pp. 105–111, pl. 3, figs. 10–12; pl. 4, figs. 1–3.)
- 1875. Observations on a collection, chiefly Bombyliidae, received from Weyenbergh, with indications of new species. Tijdschr. Ent., vol. 18, Verslag pp. xv–xviii. (Bombyliidae pp. xv–xviii.)
- 1878. On *Diplocampta paradoxa* Jaennicke and allied species. Tijdschr. Ent., vol. 21, pp. 189–193.
- 1880. Eenige Diptera van Nederlandsche Indie. Tijdschr. Ent., vol. 23, pp. 155–194, pls. 10–11. (Bombyliidae pp. 164–166, pl. 10, fig. 8, in color.)
- 1881–1884. Amerikaansche Diptera. Tijdschr. Ent. In 3 parts, and note.
    - 1881. Vol. 24, pp. 141–168, pl. 15, figs. 10–14. (Bombyliidae pp. 162–168, pl. 15, figs. 10–14.)
    - 1882. Vol. 25, pp. 77–136, pls. 9–10 (color). (Bombyliidae pp. 77–88, pl. 9, figs. 1–12.)
- 1882. Remarks on certain American Diptera in the Leyden Museum and descriptions of nine new species. Notes Leyden Mus., vol. 4, note 5, pp. 73–92. (Bombyliidae pp. 73–76.)
- 1885. On exotic Diptera. Notes Leyden Mus., vol. 7, pp. 57–86, pl. 5. (Bombyliidae pp. 81–86, pl. 5, fig. 8.)
- 1887–1888. Nieuwe Argentijnsche Diptera. Tijdschr. Ent.
    - 1888. Vol. 31, pp. 359–376, pls. 9–10 (color). (Bombyliidae pp. 366–368, pl. 9, figs. 7–8; pl. 10, fig. 1.)
- 1896. Catalogue of the described Diptera from South Asia. Hague. Dutch Ent. Soc. 219 pp. (Bombyliidae pp. 69–75.)

WULP, FREDERICK MAURITS VAN DER—Continued
   1898. Dipteren aus Neu-Guinea in de Sammlung des ungarischen National-Museums. Természetrajzi Füzetek, Budapest, vol. 21, pp. 409–426, pl. 20, figs. 1–8b. (Bombyliidae p. 419.)

XAMBEU, VICTOR
   1898. Moeurs et métomorphoses de l'*Usia atrata* Fabricius, Diptère du groupe des Asilides. Naturaliste, vol. 20, pp. 189–190. (Summary by Sack, 1899b.)
   1902. Moeurs et métamorphoses des insectes. Ann. Soc. Linnéenne Lyon, vol. 48 (1901), pp. 1–40. (Bombyliidae pp. 35–37.)
   1904. Moeurs et métamorphoses des insectes. Ann. Soc. Linnéenne Lyon, vol. 50 (1903), pp. 79–129, 167–221.

YERBURY, JOHN WILLIAM
   1900. [Note: *Anthrax paniscus*, pupal habits.] Proc. Ent. Soc. London, 1900, pp. xxii–xxiii.

YOUNG, BENJAMIN PERCY
   1921. Attachment of the abdomen to the thorax in Diptera. Mem. Cornell Univ. Agric. Exper. Stat., no. 44, pp. 255–282, pls. 9–32. (Bombyliidae pp. 259, 267, 272, pl. 16, fig. 24.)

ZAITSEV, V. F.
   1958. Notes on the bombyliid fauna (Diptera, Bombyliidae) of Transcaspia. [In Russian with English summary.] Ent. Obozrenie, Moscow, vol. 37, pp. 200–205.
   1961. New and rare Palearctic species of the genus *Spongostylum* Macq. (Diptera, Bombyliidae). [In Russian.] Ent. Obozrenie, Moscow, vol. 40, pp. 413–428, 38 figs.
   1964. New species of Bombyliidae (Diptera) from Kazakhstan. [In Russian.] Trudy Zool. Inst. Leningrad, vol. 34, pp. 283–285, 4 figs.
   1966a. Parasitic flies of the family Bombyliidae in the fauna of Transcaucasia. [In Russian.] Contributions to the Fauna of the U.S.S.R. no. 92, Academy of Sciences Moscow–Leningrad, pp. 196–200, 230–231.
   1966b. Reviziya paraziticheskikh dvukrylykh roda *Hemipenthes* Lw. (Diptera, Bombyliidae) Palarktichesckoi oblasti. [Revision of parasitic flies of the genus *Hemipenthes* (Diptera: Bombyliidae) from the palearctic area.] Tr. Vses. Entomol. Obshchest, no. 51, pp. 157–205. [From: Ref. Zh. Biol., 1967, No. 2E172.]
   1967. Novye vidy mukh semeistva Bombyliidae (Diptera) iz Palearktiki. [New beeflies of the family Bombyliidae (Diptera) from the Palaearctic region (*Geron asiaticus*, n. sp., *Antonia rufa*, n. sp., *Villoestrus stackelbergi*, n. sp., *Prorachthes stackelbergi*).] Ent. Obozrenie, vol. 46, no. 2, pp. 409–414, illus. [Translation: Ent. Rev., Washington, vol. 46, p. 242.]
   1969. Novye vidy zhuzhzall roda *Conophorus* Meigen (Diptera, Bombyliidae) iz Srednei Azii i Mongolii. [New species of beeflies of the genus *Conophorus* Meigen (Diptera: Bombyliidae) from Central Asia and Mongolia.] Ent. Obozrenie, vol. 48, no. 3, pp. 663–668, illus.

ZAITSEV, V. F., and KAUFMAN, Z. S.
   1962. K morfologii gipopigija samtsov mukh-zhuzhzhal (Diptera, Bombyliidae). [Morphology of the male hypopygium in the Bombyliidae.] Ent. Obozr., vol. 41, no. 3, pp. 579–582 (Ent. Review, vol. 41, no. 3, pp. 355–357).

ZAJONC, I.
   1959. Weitere Kenntnisse über die Familie der Bombyliidae (Diptera) in Gebiete des Kreises Nitra. [In Czech with German and Russian summaries.] Biológia, vol. 14, pp. 785–790, 3 figs.

ZAKHAVATKIN, AL. A. (Zachvatkine) (Jazykov)
   1931. Parasites and hyperparasites of the egg-pods of injurious locusts (Acridoidea) of Turkestan. Bull. Ent. Res., vol. 22, pp. 385–391. (Bombyliidae pp. 385–386, 389–391.)
   1934. Les parasites du criquet marocain in Azerbaidjan. Lenin Akad. Agric. Sci. U.S.S.R.; Inst. for Plant Protection; Bull. Plant Protection, Leningrad, ser. 1, Entomology no. 9, pp. 52–71, 5 figs. (Zasch. Rast. ot Vred.)

Zakhavatkin, Al. A. (Zachvatkine) (Jazykov)—Continued
- 1954. Parasites of Acrididae near the river Angara. [In Russian.] Trudy vsesoyuzn ent. Obshch., Moscow, vol. 44, pp. 240–300, 30 figs. (Bombyliidae p. 289, figs.)

Zanden, G. van der
- 1955. Faunistische notities betreffende Diptera I. Levende Natuur, vol. 58, no. 5, pp. 100–101. (Bombyliidae, 1 sp. mentioned.)

Zavrel, J.
- 1930. Praemandibeln einiger Dipterenlarven. XI. Congress Intern. Zool. Padova II, pp. 1000–1004.

Zeller, Philipp Christoph
- 1840–1842. Beytrag zur Kenntniss der Dipteren aus den Familien: Bombylier, Anthracier und Asiliden. In 2 parts. Isis (Oken), Leipzig, vol. 1, columns 10–77. (Bombyliidae columns 14–34, 76–77. Bombylier columns 14–23; Anthracier columns 23–34.) 1842. Columns 807–848.
- 1841. Nachricht über die Seefelder bei Reinerz in entomologischer Beziehung. Stettiner Ent. Zeitung, vol. 2, pp. 171–176, 178–182. (Bombyliidae p. 179.)

Zetterstedt, Johan Wilhelm
- 1838–1840. Insecta Lapponica descripta. Lipsiae. vi + 1140 pp.
  - 1838. Sect. 3, Diptera pp. 477–868. (Bombyliidae pp. 490, 509–510, 520–522, 574.)
- 1842–1860. Diptera Scandinaviae. Disposita et descripta. Lundae. In 14 volumes, xvi + 6609 pp.
  - 1842. Vol. 1, pp. 1–440. (Bombyliidae pp. 17–19, 188–202; 18 spp., 4 genera.)
  - 1849. Vol. 8, pp. 2935–3366. (Bombyliidae pp. 2977–2981.)
  - 1852. Vol. 11, pp. v–xii, 4091–4546. (Bombyliidae pp. 4264–4265.)
  - 1855. Vol. 12, pp. v–xx, 4547–4942. (Bombyliidae pp. 4582–4586.)
  - 1859. Vol. 13, pp. v–xvi, 4943–6190. (Bombyliidae p. 4968.)

Zumpt, F.
- 1936. Die Tsetsefliegen. 144 pp. Jena: Fischer.

Zimsen, Ella
- 1964. The type material of J. C. Fabricius. 656 pp. Copenhagen, Munksgaard. (Diptera pp. 449–503.)

Zwaluwenburg, H. H. van
- 1914. Preliminary check-list of Porto Rican insects. 62 pp.
- 1915. Additions to Porto Rican check list.

Indexes

Page number of principal entries are in **boldface**. Synonyms are in *italics*.

Index to Genera and Subgenera

A

Acanthogeron Bezzi, 76, **88**; figs. 2, 237, 467, 784
Acoecus, new subgenus, 265, **276**
Acreotrichus Macquart, 195, **199**; figs. 98, 315, 534, 536
Acrodisca Bezzi, 369, **413**
Acrophthalmyda Bigot, 74, **174**; figs. 18, 251, 437, 450, 797, 798, 799
Actherosia, new subgenus, 280, **296**
Adelidea Macquart, 74, **119**; figs. 34, 256, 266, 478, 516, 739, 1006, 1007, 1008
Adelogenys Hesse, 75, **190**; figs. 223, 301, 564, 1003, 1004, 1005
Aetheoptilus Hesse, 265, **271**; fig. 344
Agenosia, new subgenus, 195
Agitonia Hull, 367, **389**, 390
Aldrichia Coquillett, 71, **160**
Aleucosia Edwards, 306
Allocotus Loew, 296
Alloxytropus Bezzi, 227
Alonipola Rondani, 287
Alyosia Rondani, 306, **349**
Amictogeron Hesse, 203, **207**; figs. 106, 310, 530, 537, 776
Amictus Wiedemann, 280, **284**; figs. 109, 357, 611, 641, 741, 752, 779, 890, 891, 892
Amphicosmus Coquillett, 280, **293**; figs. 110, 114, 361, 608, 618
Anastoechus Osten Sacken, 75, **84**; figs. 8, 239, 240, 447, 465, 778, 780, 791, 792
Anisotamia Macquart, 306, **311**
Anomaloptilus Hesse, 265, **275**; figs. 353, 586, 595
Anthocolon, new subgenus, 307
Anthrax Scopoli, 341, **436**; figs. 159, 160A, 163, 337, 409, 712, 734, 966, 967, 968
Antonia Loew, 305, **319**; figs. 136, 137, 138, 384, 624, 637, 977, 979, 981
Antoniaustralia Becker, 319
Apatomyza Wiedemann, 75, **190**; figs. 97B, 349, 544
Aphochantus Hyslop, 322
Aphoebantes Bigot, 322
Aphoebantus Loew, 305, **322**; figs. 126, 205, 213, 227, 369, 405, 706, 719, 906, 907
Apolysis Loew, 213, **217**; figs. 81, 329, 332, 549, 560, 855, 856
Apystomyia Melander, 223, **229**; figs. 85, 324, 599
Argyramoeba, E. and L. Coucke, 436
Argyromaeba Schiner, 436
Argyrospila Verrall, 413
Argyrospyla Rondani, 413, **416**
Arthroneura Hull, 76, **180**
Aspiloptera Kunckel, 369
Astrophanes Osten Sacken, 367, **374**; figs. 172, 395, 660, 666, 927, 928
Atrichochira Hesse, 367, **421**; figs. 206, 431A, 711, 721

B

Bombyliosoma Verrall, 107
Bombylisoma Rondani, 74, **107**; figs. 27, 273, 473, 513
Bombylius Linné, **76**; figs. 1, 4, 7, 244, 245, 247, 432, 458, 487, 520, 786, 787, 788
Bombylodes Paramonov, 74, **91**; figs. 5, 262, 438
Bombylosoma Marshall, 107
Brachydemia, new subgenus, 306, **314**
Bromoglycis Hull, 74, **128**; figs. 38, 271
Brychosoma, new genus, 76, **102**; fig. 14
Bryodemina Hull, 306, **314**; figs. 207, 375, 695, 898, 899

C

Cacoplox Hull, 73, **130**; figs. 44, 276, 489, 515
Caenotus Cole, 76, 223, **226**; figs. 108A, 574, 697, 867, 868
Calledax, new subgenus, 335, 336, **338**
Callistoma Kertész, 149
Callostoma Macquart, 71, **149**; figs. 74, 76, 306, 511, 531
Callynthrophora Schiner, 72, **166**; figs. 292, 480, 493
Calopelta Greene, 156
Canaria Becker, 307
Canariellum Strand, 306, **307**
Cephalodromia Becker, 254
Cephene Latreille, 243
Cephenius Enderlein, 243
Cephenus Karsch, 243
Cephenus Berthold, 243
Ceratolaemus Hesse, 254, **256**; figs. 334, 585, 592
Chalcamoeba Sack, 436
Chalcochiton Loew, 71, **147**
Chasmoneura Hesse, 74, **108**; figs. 41, 42, 280, 562
Cheilohadrus Hesse, 161
Chelina, new subgenus, 306
Chiasmella Bezzi, 304, **348**; figs. 378, 644, 654
Chionamoeba Sack, 304, **342**; figs. 228, 387, 645, 651
Choristus Walker, 76
Chrysamoeba Sack, 436
Chrysanthrax Osten Sacken, 368, **376**; figs. 170, 179, 663, 936, 937
Cladella, new subgenus, 273
Cladodisca Bezzi, 369, 413, **416**
Codionus Rondani, 71, **156**
Collosoptera new genus, 368, **430**; figs. 201, 418, 714, 969, 970
Comastes Osten Sacken, 138
Compsella, new subgenus, 307, **354**
Comptosia Macquart, 306, **349**; figs. 150, 156, 158, 209, 366, 424, 693, 735, 921, 922, 923, 924, 925, 926
Coniomastix Enderlein, **445**; fig. 412
Conogaster Hermann, 322
Conomyza Hesse, 304, **338**
Cononedys Hermann, 305, **322**
Conophorina Becker, 72, **158**; figs. 33, 269, 481
Conophorus Meigen, 71, 76, **156**; figs. 66, 69, 70, 71, 72, 284, 484, 490, 820, 821
Coptodicrus Enderlein, 243
Coptopelma Enderlein, **243**
Coquillettia Williston, 442
Corsomyza Wiedemann, 73, **164**; figs. 63, 64, 68, 287, 462, 497, 729, 817, 818, 819
Corycetta, new subgenus, 413
Coryprosopa Hesse, 304, **330**; figs. 132, 386, 632, 647
Crocidium Loew, 75, **183**; figs. 57, 58, 59, 327, 526, 528, 840, 841, 842
Cryomyia, new genus, 74, **131**; figs. 39, 265
Cyananthrax Painter, 367, **378**; figs. 175, 373, 675, 938, 939, 440, 941
Cyclorhynchus Macquart, 195
Cyclorrhynchus Bezzi, 195
Cyllenia Latreille, 280, **281**; figs. 116, 117, 358, 616, 756, 771, 777
Cyrthosia Rondani, 254
Cyrtisiopsis Séguy, 254, **262**; figs. 330, 600, 997, 998, 999
Cyrtoides Engel, 265, **275**
Cyrtomorpha White, 254, **259**; figs. 95, 222, 320, 581, 587, 875, 876, 877
Cyrtomyia Bigot, 232, **238**; figs. 88, 319, 584, 736, 745, 749, 872, 873, 874
Cyrtophorus Bigot, 238
Cyrtosia Perris, **254**; figs. 321, 593
Cytherea Fabricius, 71, **145**; figs. 73, 75, 296, 298, 459, 503, 521, 569, 825, 826, 827

D

Dagestania Paramonov, **221**
Dasypalpus Macquart, 170
Defilippia Lioy, 369, **416**
Denamyza, new subgenus, **164**
Desmatomyia Williston, 73, **186**; figs. 60, 61, 343, 576, 691, 846, 847, 848
Desmatoneura Williston, 304, **346**; figs. 133, 379, 636, 649, 908, 909, 910
Deusopora Hull, 368, **379**; figs. 173, 277, 664
Diaerops Enderlein, 243
Diatropomma Bowden, 368, **423**; figs. 194A, 199, 404
Dicranoclista Bezzi, 436, **442**; figs. 160, 403, 679, 753, 1018, 1019, 1020
Dimelopelma Enderlein, 243
Dimorpaphus (Walker) Loew in Scudder, 319
Dimorphaphus Walker, 319
Dimorphophora Walker, 319
Dipalta Osten Sacken, 367, **381**; figs. 186, 370, 662, 665, 944, 945
Diplocampta Schiner, 367, **382**; figs. 195, 391, 671, 674
Dischistus Loew, 75, **105**; figs. 25, 29, 275, 453, 980, 982, 984
Docidomyia White, 280, 306, **318**; figs. 142, 372, 623, 633
Doddosia Edwards, 307, **358**; figs. 149, 155, 431B, 698, 722, 931, 932
Dolichomyia Wiedemann, 243, **248**; figs. 93, 94, 335, 579, 589, 881, 882, 883
Doliogethes Hesse, 74, 75, **110**; figs. 37, 260, 468, 471, 744, 772
Doliopteryx Hesse, 265, **272**; fig. 323
Dumontiella Séguy, 161

E

Eclimnus Verrall, 178
Eclimus Loew, 76, **178**; figs. 15, 258, 454, 572
Edmundiella Becker, 306, **316**
Efflatounia Bezzi, 71, **143**; figs. 300, 529, 557
Empidideicus Becker, 265, **273**; figs. 107, 108, 328, 348, 591, 594
Empidigeron Painter, **203**
Enica Macquart, 287
Eniconeura Macquart, 232
Epacmoides Hesse, 304, **332**; figs. 129, 380, 646, 650
Epacmus Osten Sacken, 305, **324**; figs. 141, 143, 628, 631, 913, 914, 915, 916, 917
Epibates Osten Sacken, 180
Epidosina, new subgenus, 306, **349**
Euanthobates Hesse, 265, **276**; figs. 350, 758
Eucessia Coquillett, 305, **339**; figs. 124, 374, 635, 658
Eucharimyia Bigot, **102**
Euchariomyia Wulp, 102
Euprepina Hull, 73, **133**; figs. 49, 419, 455, 815, 816
Eurycarenus Loew, 71, **141**; figs. 51, 52, 299, 525, 533, 831, 832, 833
Eusurbus Roberts, 73, **100**; figs. 22, 257, 439, 451
Exepacmus, Coquillett, 305, **339**; fig. 293

Exhyalanthrax Becker, 403, **404**
Exoprosopa Macquart, 368, **413**; figs. 197, 198, 208, 210, 376, 680, 747, 971, 972
Exoptata Coquillett, 368, **413**

F

Francoisia Hesse, 425

G

Geminaria Coquillett, 72, **115**; figs. 35, 281, 461, 571
Geron Meigen, **203**; figs. 104, 105, 309, 523, 543, 754, 755, 859, 860, 861
Glabella Loew, 269
Glabellula Bezzi, 265, **269**; figs. 85A, 351, 597, 606
Glossista Rondani, 145
Glycophorba, new subgenus, **295**
Gnumyia Bezzi, **72**; figs. 289, 483, 506
Gonarthrus Bezzi, 74, **111**; figs. 24, 263, 561, 1015, 1016, 1017
Gyrocraspedum Becker, 71, **150**; figs. 56, 303, 501, 510

H

Hallidia Hull, 74, **134**; figs. 36, 278, 519, 812, 813, 814
Hemipenthes Loew, 367, **384**; figs. 177, 178, 183, 185, 187, 413, 672, 678, 942, 943
Henica Macquart, 280, **287**; figs. 226, 406, 620, 973, 978, 1030
Heniconeura Verrall, 232
Heteralonia Rondani, 368, 415, **424**; fig. 194
Heterostylum Macquart, 71, **138**; figs. 54, 55, 283, 504, 559, 834, 835, 836
Heterotropus Loew, 76, **223**; figs. 96, 527, 547, 774, 782, 785, 837, 838, 839, 976
Hyalanthrax Osten Sacken, 369
Hyperalonia Rondani, 431
Hyperusia Bezzi, 73, **169**; figs. 62, 295, 492, 500

I

Isocnemus Bezzi, 75, **92**; figs. 246, 441, 445
Isopenthes Osten Sacken, 384
Isotamia Bezzi, 304, 368, **425**; figs. 193, 402, 684, 688

K

Karakumia Paramonov, 70, **144**; fig. 307

L

Lagochilus Loew, 287
Lasioprosopa Macquart, 164
Laurella Hull, 75, **135**; figs. 23, 274, 514
Legnotomyia Bezzi, 73, **162**; figs. 31, 254, 255, 469, 474, 800, 801, 802
Legnotus, Loew, 162
Legonotus Bischof, 162
Lepidanthrax Osten Sacken, 367, **386**; figs. 211, 567, 702, 946, 947

Lepidochlanus Hesse, 73, **118**; figs. 43, 270, 464, 476, 773
Lepidophora Westwood, 232, **236**; figs. 90, 318, 457, 540, 869, 870, 871
Leptochilus Loew, 324
Leucamoeba Sack, 436
Leylaiya Efflatoun, 265, **277**; fig. 346
Ligira Mik, 431
Ligyra Newman, 368, **431**; figs. 192, 202, 425, 704, 731, 962, 963
Lissomerus Austen, 76, **90**; figs. 13, 238, 435, 448, 781
Litorrhynchus Bezzi, 426
Litorrhynchus Macquart, 368, **426**; figs. 191, 408, 687, 689
Logcocerius Rondani, 145
Lomatia Meigen, 306, **307**; figs. 130A, 139, 144, 145, 146, 216, 354, 701, 733, 903, 904, 905
Lonchocerius Bezzi, 145
Loncocerius Kertész, 145
Lordotus Loew, 72, **113**; figs. 45, 46, 47, 48, 290, 575, 690, 807, 808, 809
Lyophlaeba Rondani, **307**; figs. 147, 151, 154, 157, 161, 162, 368, 656, 673, 988-993
Lyophleba Bezzi, 354

M

Macrocondyla Rondani, 307, **354**
Mallophthiria Edwards, 75, **185**; figs. 341, 522, 532
Malthacotricha Becker, 223
Mancia Coquillett, 367, **388**; figs. 168, 396, 661, 738
Mariobezzia Becker, 72, **173**; figs. 291, 488, 496
Marleyimyia Hesse, 366, **412**; fig. 174
Marmasoma White, 232, **239**; figs. 89, 313, 542, 578, 878, 879, 880
Meganedys, new subgenus, **322**
Megapalpus Macquart, 73, **170**; figs. 67, 288, 485, 505, 742
Mesoclis Bezzi, 369, **413**
Metacosmus Coquillett, 280, **295**; figs. 118, 355, 615
Metapenta Bezzi, 369, **413**
Micomitra Bowden, 368, **428**; figs. 203, 421, 694
Mima Megerle, in Meigen, 413
Molybdamoeba Sack, 436
Mulio Latreille, 145
Myonema Roberts, 303, **317**; figs. 135, 140, 219, 388, 626, 629, 757
Mythicomyia Coquillett, 265, **266**; figs. 101, 102, 342, 347, 596, 605, 884, 885, 886

N

Neacreotrichus, Cockerell, 195
Nectaropota Philippi, 75, **98**; figs. 20, 253, 1027, 1028, 1029
Neodiplocampta Curran, 367, **389**; figs. 167, 189, 392, 657, 658, 669, 743
Neodischistus Painter, 75, **127**; figs. 26, 261, 472, 518
Neosardus Roberts, 279, **300**; figs. 121, 398, 621, 639, 900, 901, 902
Neuria Newman, 349
Nomalonia Rondani, 280, **289**; figs. 224, 225, 363, 609, 612, 696, 728, 994, 995, 996

O

Oestranthrax Bezzi, 366, **409**; figs. 166, 230, 233, 394, 683, 686, 740, 775, 783, 955, 956, 957
Oestrimyza, new genus, **411**; figs. 415, 703
Ogcodocera Macquart, 306, **312**; figs. 122, 713, 724, 911, 912
Ogilviella Paramonov, 425
Oligodranes Loew, 213, **219**; figs. 77, 231, 338, 550, 552, 553, 554, 737, 857, 858
Onchopelma Hesse, 254, **256**; figs. 352, 538, 580, 582, 590
Oncodocera Osten Sacken, 312
Oncodosia Edwards, 306, **361**; figs. 152, 153, 427, 699, 723, 933, 934, 935
Oniromyia Bezzi, 71, **154**; figs. 80, 297, 498, 509, 822, 823, 824
Opsonia, new subgenus, 306
Oriellus, new subgenus, 403, **404**
Ostentator Jaennicke, 174
Othniomyia Hesse, 72, **117**; figs. 264, 470, 475

P

Pachyneres Greene, 269
Palintonus François, **241**
Pantarbes Osten Sacken, 71, **151**; figs. 82, 83, 305, 499, 507, 828, 829, 830
Pantostomus Bezzi, 281, **292**; figs 111, 356, 602, 614
Parabombylius Williston, 75, **95**; figs. 9, 272, 456, 732, 793, 794
Paracosmus Osten Sacken, 280, **296**; figs. 112, 119, 360, 619, 896, 897
Paradosia, new subgenus, 306
Parageron Paramonov, 213
Paranthracina Paramonov, 408
Paranthrax Paramonov, 408
Paratoxophora Engel, 73, **176**; figs. 12, 268, 460, 803, 804
Paravilla Painter, 368, **391**; figs. 176, 184, 407, 707, 717, 951, 952
Parisus Walker, 77
Peringueyimyia Bigot, 280, **298**; figs. 120, 359, 607, 627, 642, 893, 894, 895
Petrorossia Bezzi, 304, **344**; figs. 134, 234, 383, 634, 652, 918, 919, 920
Phthiria Meigen, 194, **195**; figs. 99, 100, 215, 308, 566, 570, 849, 850, 851
Phtyria Rondani, 195
Pilosia new genus, 73, **137**; figs. 28, 282, 512
Pioperna Enderlein, 244
Pipunculopsis Bezzi, 304, **344**
Platamodes Loew, 125
Platamomyia Brethes, 76, **125**; figs. 21, 286, 486, 1009, 1010, 1011
Platygaster Zetterstedt, 269
Platypygus Loew, 254, **261**; fig. 331
Plesiocera Macquart, 304, **335**; figs. 125, 127, 381, 643, 648
Ploas Latreille, 156
Podolepida, new subgenus, 386
Poecilanthrax Osten Sacken, 368, **392**; figs. 180, 190, 198A, 417, 710, 720, 948, 949, 950
Poecilognathus Jaennicke, **195**
Popsia Costa, 261

Prorachthes Loew, 72, **161**; figs. 115A, 259, 479, 494, 1012, 1013, 1014
Prorates Melander, 223, **227**; figs. 97A, 322, 326, 573, 700
Prorostoma Hesse, 304, **329**; figs. 130, 389, 622, 640
Prothaplocnemis Bezzi, **398**; fig. 214
Psamatamoeba Sack, 436
Pseudempis Bezzi, 209
Pseudoamictus Bigot, 203, **209**; figs. 103, 311, 548, 555, 862, 863, 864
Pseudogeron Cresson, 219
Pseudoglabellula Hesse, 265, **273**; fig. 345
Pseudopenthes Roberts, 367, **399**; figs. 196, 393, 668, 676
Psiatholasius Becker, 162
Psiloderoides Hesse, 254, **257**; fig. 340
Pteraulacodes Hesse, **328**
Pteraulax Bezzi, 305, **326**; figs. 123, 382, 625, 630
Pterobates Bezzi, 369, 413, **416**
Pusilla Paramonov, 164
Pygocona, new subgenus, 195

R

Rhabdopselaphus Bigot, 219
Rhynchanthrax Painter, 367, **406**; figs. 232, 401, 716, 727, 964, 965

S

Satyramoeba Sack, 436
Scinax Loew, 174
Semiramis Becker, 75, **192**; figs. 325, 535
Sericosoma Macquart, 71, **153**; figs. 40, 302, 502, 508, 1000, 1001, 1002
Sericusia Edwards, 74, **124**; figs. 30, 50, 267, 491, 517
Sinaia Becker, 280, **286**; figs. 212, 365, 604, 617, 638, 759, 973, 974, 975
Sisyromyia White, 74, **97**; figs. 11, 252, 436, 443
Sisyrophanus Karsch, 76, **86**; figs. 241, 243, 434, 446, 466, 477
Sobarus Loew, 119
Sosiomyia Bezzi, 74, **121**; figs. 32, 279, 452, 810, 811
Sparnopolius Loew, 75, **123**; figs. 204, 420, 692, 805, 806
Sphaerogaster Zetterstedt, 269
Sphenoidoptera Williston, 280, **283**; figs. 235, 364, 601, 610, 1024, 1025, 1026
Spogostylum Macquart, 436
Spongostylum Williston, 436
Staurostichus Hull, 76, **104**
Stomylomyia Bigot, 304, **334**; figs. 131, 385
Stonyx Osten Sacken, 367, **400**; figs. 200, 429, 708, 718, 960, 961
Stygia Meigen, 307
Stymphalina, n.n., 445
Symballa Enderlein, **243**
Synthesia Bezzi, 368, **402**; figs. 220, 715, 726
Systoechus Loew, 76, **82**; figs. 6, 10, 236, 442, 444, 789, 790
Systrophopus Karsch, 243
Systrophus Latreille, 243
Systropus Wiedemann, **243**; figs. 86, 87, 336, 565, 588, 887, 888, 889

T

Tamerlania Paramonov, 75, **192**
Thevenemya Bigot, 180
Thevenemyia Bigot, 76, **180**; figs. 16, 19, 248, 249, 422, 563, 795, 796
Thevenetimyia Bigot, 180
Thlipsogaster Coquillett, 284
Thlipsomyza Wiedemann, 284
Thlyposogaster Rondani, 284
Thyridanthrax Osten Sacken, 368, **403**; figs. 169, 171, 188, 217, 218, 220, 221, 705, 725, 958, 959
Tillyardomyia Tonnoir, 76, **182**; figs. 17, 250, 440
Tomomyza Wiedemann, 280, 281, **290**; figs. 113, 115, 362, 603, 613, 1021, 1022, 1023
Toxomyia, new subgenus, **232**
Toxophora Meigen, **232**; figs. 91, 92, 316, 317, 558, 577, 865, 866
Trinaria Mulsant, 369, 413, **416**
Triodina, new subgenus, 306, **322**
Triodites Osten Sacken, **322**
Triodytus Riley, 322
Triplasius Loew, 76
Triploechus Edwards, 71, **140**; figs. 53, 304, 524, 551, 843, 844, 845
Tritoneura Schiner, 354
Truquia Rondani, 284
Tumulus, new subgenus, **403**

U

Ulosometa, new genus, 307, **359**; figs. 431c, 929, 930
Usia Latreille, 213, **214**; figs. 78, 79, 312, 545, 556, 852, 853, 854

V

Velocia Coquillett, 431
Villa Lioy, **367**; figs. 181, 182, 390, 667, 681, 953, 954
Villoestrus Paramonov, 366, **408**; figs. 410, 682, 685
Voluccella Fabricius, 214
Volucella Meigen, 214

W

Walkeromyia Paramonov, 436, **444**; figs. 164, 165, 411, 426, 677, 709, 730, 746, 748, 750, 751

X

Xenoprosopa Hesse, **251**; figs. 84, 339, 539, 760
Xeramoeba Hesse, 304, **340**; figs. 128, 655
Xystropus Verrall, 244

Y

Ylasoia Speiser, 307, **356**; figs. 148, 367, 659, 670, 985, 986, 987

Z

Zaclava, new genus, 243, **250**; figs. 97, 333, 583
Zinnomyia Hesse, 75, **94**; figs. 3, 242, 449, 468
Zygodipla Bezzi, 368, **413**
Zyxmyia Bowden, **172**; figs. 294, 546, 770

Index to Species

A

abbadon Fabricius, Villa, 373
abbadon Meigen, Villa, 373
abbreviata Paramonov, Exoprosopa, 419
abbreviatus Wiedemann, Villa, 372
abdominalis Wiedemann, Bombylius, 80
abdominalis Johnson and Johnson, Lordotus, 114
abdominalis Bezzi, Sisyrophanus, 88; figs. 243, 434, 446
abdominalis Abreu, Villa, 373
aberrans Paramonov, Anastoechus, 86
aberrans Walker, Cytherea, 148
aberrans Paramonov, Exoprosopa, 419
aberrans Hesse, Systoechus, 84
aberrans Hesse, Thyridanthrax, 406
abjecta Nurse, Exoprosopa, 421
abnormis Coquillett, Aphoebantus, 324
abragi Efflatoun, Cyrtosia, 256
abreviata Villeneuve, Lomatia, 311
abrogata Nurse, Exoprosopa, 421
abruptoides Hesse, Thyridanthrax, 406
abruptus Loew, Thyridanthrax, 29, 406
absalon Wiedemann, Villa, 374
accola Becker, Usia, 216
acourti Cockerell, Systropus, fossil, 59
acridophagus Hesse, Systoechus, 9, 84
acrodiscoides Bezzi, Exoprosopa, 420
acroleuca Wiedemann, Anthrax, 441
acroleuca Painter, Lepidophora, 238
acroleucus Bezzi, Bombylius, 81
acrospila Bezzi, Exoprosopa, 420
acrostichalis Melander, Oligodranes, 220
acrostichalis matutinus Melander, Oligodranes, 220
actites Melander, Mythicomyia, 269
actuosus Paramonov, Anthrax, 441
acuminatus Enderlein, Systropus, 248
acuta Melander, Mythicomyia, 269
acutangula Loew, Lomatia, 311
acutangula transvaalensis Hesse, Lomatia, 311
acuticornis Macquart, Anastoechus, 86
acutus Bezzi, Bombylius, 81
acutus Painter, Systropus, 248
adelaidica Macquart, Exoprosopa, 421
adelphia Becker, Exoprosopa, 419
adonis Osten Sacken, Astrophanes, 376; figs. 172, 395, 660, 666, 927, 928
adumbrata Coquillett, Chrysanthrax, 378
adumbrata Paramonov, Cytherea, 148
aduncus Loew, Conophorus, 158
aduncus Loew, Conophorus, 158
adustus Loew, Villa, 371
aeacus Meigen, Exoprosopa, 419
aegeriiformis Westwood, Lepidophora, 236
aegiale Walker, Thevenemyia, 182
aegina Wiedemann, Exoprosopa, 419
aegyptiaca Bezzi, Efflatounia, 144
aegyptiaca Paramonov, Exoprosopa, 419
aegyptiaca Engel, Mariobezzia, 174
aegyptiaca Bezzi, Stomylomyia, 335
aegyptiaca Efflatoun, Toxophora, 236
aegyptiacum Bezzi, Crocidium, 185
aegyptiacus Paramonov, Amictus, 286
aegyptiacus Paramonov, Anastoechus, 86

aegyptiacus Bezzi, Conophorus, 158
aegyptiacus Paramonov, Heterotropus, 226; figs. 96, 527, 547
aegyptiacus Macquart, Villa, 373
aegypticola Paramonov, Anthrax, 441
aemulus Hesse, Bombylius, 81
aenea Rossi, Usia, 17, 216; figs. 79, 556
aenea Coquillett, Villa, 371
aeneoides Paramonov, Usia, 216
aequa Walker, Anthrax, 441
aequalis Fabricius, Bombylius, 80
aequalis Painter, Geron, 206
aequisexus Paramonov, Bombylius, 81
aerata Hesse, Micomitra, 430
aernophilus Paramonov, Anthrax, 441
aetheocoma Hesse, Anthrax, 442
aethiops Fabricius, Anthrax, 441; fig. 163
aethiops Laboulbene, Anthrax, 441
aethiops bezzii Paramonov, Anthrax, 441
afer Wiedemann, Henica, 289
afer Fabricius, Thyridanthrax, 406; fig. 169
affinis Hesse, Systoechus, 84
affinis discrepans Hesse, Systoechus, 84
affinissima Senior-White, Exoprosopa, 421
afra Wiedemann, Henica, 288
afra Macquart, Nomalonia, 290; figs. 224, 225, 363, 609, 612, 696, 728
afra Wiedemann, Nomalonia (Cyllenia), 290
africanus Paramonov, Paranthracina, 408
agassizii Loew, Exoprosopa, 419
agilis Olivier, Bombylius, 80
agilis Melander, Mythicomyia, 269
aglota Séguy, Cyrtosia, 256
agnitionalis Austen, Thyridanthrax, 406
agrestis Coquillett, Lepidanthrax, 388
agrippina Osten Sacken, Villa, 371; figs. 953, 954
aharonii Engel, Sinaia, 287
alagoezicus Paramonov, Anthrax, 441
alashanicum Paramonov, Anthrax, 441
alatus Becker, Hemipenthes, 386
alba Cole, Geron, 206
albaminis Séguy, Bombylius, 80
albarius Painter, Geron, 206
albata Bezzi, Exoprosopa, 420
albata novaeformis Bezzi, Exoprosopa, 420
albata Hesse, Lomatia, 311
albata Roberts, Villa, 374
albatus Séguy, Acanthogeron, 90
albavitta Macquart, Bombylius, 82
albescens Brunetti, Geron, 207
albescens Loew, Villa, 374
albibarbis Zetterstedt, Bombylius, 80
albibarbis Engel, Systoechus, 84
albicans Paramonov, Anastoechus, 86
albicans Bezzi, Eurycarenus, 143
albicans Macquart, Systoechus, 84
albicapillus Loew, Bombylius, 79
albicapillus diegoensis Painter, Bombylius, 79
albiceps Macquart, Bombylius, 82
albicerus Hesse, Anastoechus, 86
albicincta Macquart, Exoprosopa, 419
albicincta Macquart, Ligyra, 434
albicincta Hesse, Lomatia, 311
albicincta Cole, Villa, 371, 372

albicinctus Paramonov, Aphoebantus, 324
albicinctus Macquart, Bombylius, 82
albicingula Austen, Thyridanthrax, 406
albicollaris Painter, Exoprosopa, 419
albicollaris Cole, Villa, 372
albicoma Hesse, Lomatia, 311
albida Walker, Exoprosopa, 421
albida Wiedemann, Phthiria, 199
albida Becker, Villa, 373
albidipennis Loew, Geron, 206
albidipennis Loew, Heterotropus, 226
albidipennis sudanensis Becker, Heterotropus, 226
albidus Walker, Geron, 206
albidus Hall, Lordotus, 114
albidus Loew, Systoechus, 84
albidus auripilus Hesse, Systoechus, 84
albifacies Macquart, Bombylius, 82
albifacies Paramonov, Exoprosopa, 419
albifacies Macquart, Petrorossia, 346
albifacies Rondani, Villa, 372
albifrons Gmelin, Bombylius, 82
albifrons Loew, Cytherea, 148
albifrons Hesse, Epacmoides, 334; figs. 380, 646, 650
albifrons pallidulum Hesse, Epacmoides, 334
albifrons Curran, Exoprosopa, 419
albifrons Roberts, Lepidanthrax, 388
albina Becker, Thyridanthrax, 406
albipectus Macquart, Bombylius, 80
albipectus Walker, Petrorossia, 346
albipectus Hesse, Systoechus, 84
albipectus Macquart, Villa, 371
albirufus Walker, Villa, 374
albissima Oldroyd, Villa, 372
albissimus Zaitsev, Bombylius, 80
albiventris Macquart, Bombylius, 81
albiventris Macquart, Exoprosopa, 420
albiventris Macquart, Ligyra, 434
albiventris Thomson, Ligyra, 434
albiventris Frey, Villa, 373
albivilla Pallas, Cytherea, 148
albivittata Bowden, Toxophora, 236
albizonata Hesse, Lomatia, 311
albobasalis Roberts, Villa, 374
albocapitis Roberts, Phthiria, 199
albofasciata Thomson, Comptosia, 353
albofasciatus Macquart, Anthrax, 19, 440
albofasciatus cascadensis Marston, Anthrax, 440
albofasciatus picea Marston, Anthrax, 440
albofimbriata Bezzi, Exoprosopa, 420
albofulvus Walker, Villa, 374
albogilva Séguy, Phthiria, 199
albohirtus Roberts, Systoechus, 84
albolineata Bezzi, Cytherea, 148
albomicans Loew, Bombylius, 80
albonigra Bezzi, Exoprosopa, 420
albopectinatus Becker, Anastoechus, 86
albopenicillatus Bigot, Parabombylius, 97
albopilosus Cole, Oligodranes, 220
albosegmentatus Engel, Thyridanthrax, 406
albosparsus Bigot, Anthrax, 440
albosparsus Bigot, Bombylius, 82
albovittata Macquart, Villa, 371
albula Loew, Villa, 373

INDEX

albulata Hesse, Lomatia, 311
albulus Paramonov, Anastoechus, 86
albus Cole, Geron, 206
alcyon Say, Poecilanthrax, 27, 398
aldrichi Johnson, Phthiria, 198
alecto Loew, Lomatia, 311
alexandri Paramonov, Bombylius, 80
alexandrina Becker, Cytherea, 148
alexon Walker, Exoprosopa, 421
alfierii Paramonov, Oesthanthrax, 411
algeciras Strobl, Villa, 373
algericus Villeneuve, Bombylius, 81
algira Villeneuve, Anthrax, 441
algira Fabricius, Exoprosopa, 419
algira Macquart, Plesiocera, 338; figs. 127, 381, 643, 648
algirus Macquart, Dischistus, 107
algirus Paramonov, Platypygus, 262
alienus Hardy, Bombylius, 82
alliopterus Hesse, Thyridanthrax, 29, 406
allothyris Bezzi, Litorryhychus, 428
alpha Osten Sacken, Poecilanthrax, 398
alpicola Villeneuve, Conophorus, 158
alta Tucker, Chrysanthrax, 378
altaica Paramonov, Exoprosopa, 420
altaicus Paramonov, Bombylius, 80
alterans Williston, Phthiria, 199
altercinctus Melander, Aphoebantus, 324
alternans Macquart, Exoprosopa, 421
alternans Macquart, Villa, 374
alternata Say, Villa, 26, 371; figs. 182, 390
alternata nigropectus Cresson, Villa, 371
alternus Walker, Villa, 374
altivolans Hesse, Systoechus, 84
altus Walker, Bombylius, 82
alula Bezzi, Ligyra, 434
alveolus Becker, Bombylius, 80
amabilis Osten Sacken, Conophorus, 158
amabillis Wulp, Dischistus, 107
amasia Wiedemann, Villa, 372
ambigua Lynch Arribalzaga, Villa, 372
ambustus Pallas in Wiedemann, Bombylius, 80
americana Wulp, Exoprosopa, 419
americana Coquillett, Phthiria, 198
americana Guerin-Meneville, Toxophora, 235
americanus Melander, Platypygus, 262
amicula Séguy, Toxophora, 236
ammophila Paramonov, Exoprosopa, 419
ammophiloides Townsend, Systropus, 248
ammophilus Hesse, Bombylius, 81
ammophilus Paramonov, Heterotropus, 226
amnicola Bowden, Cyrtosia, 256
amoenus Austen, Thyridanthrax, 406
amphitea Walker, Toxophora, 16, 42, 235; figs. 6, 91, 92, 316, 865, 866
amplicella Coquillett, Phthiria, 198
ampla Walker, Oncodosia, 363; figs. 933, 934, 935
analis Macquart, Anthrax, 440
analis Say, Anthrax, 19, 440; fig. 159A
analis Fabricius, Bombylius, 80
analis Olivier, Bombylius, 80
analis Thunberg, Bombylius, 80
analis Williston, Ogcodocera, 314
analis Melander, Oligodranes, 220
anastoechoides Hesse, Bombylius, 81
anceps Bezzi, Corsomyza, 166
anceps Hesse, Geron, 207
ancepsoides Hesse, Corsomyza, 166
ancilla Bezzi, Micomitra, 430

andalusiaca Strobl, Apolysis, 219
andalusiacus Paramonov, Anastoechus, 86
andinus Brethes, Astrophanes, 376
androgynus Loew, Bombylius, 80
angularis Thomson, Villa, 374
angulata Loew, Exoprosopa, 420
angulata Brethes, Villa, 372
angulatus Macquart, Bombylius, 80
angulatus Karsch, Systropus, 248; figs. 87, 336, 565, 588, 887, 888, 889
angulosus Bezzi, Bombylius, 81
angulus Osten Sacken, Lepidanthrax, 388
angusta Edwards, Comptosia, 353
angusta Paramonov, Cytherea, 148
angusta Bezzi, Exoprosopa, 420
angusta Melander, Mythicomyia, 269
angusta Paramonov, Toxophora, 236
angustatus Doleschall, Villa, 374
angusteoculatus Becker, Thyridanthrax, 406
angustibasalis Hesse, Petrorossia, 346
angustibasalis buziana Hesse, Petrorossia, 346
angustifrons Paramonov, Anastoechus, 86
angustifrons Becker, Usia, 216
angustipennis Macquart, Anthrax, 441
angustipennis Edwards, Triploechus, 141
anisospilus Hesse, Thyridanthrax, 406
anna Coquillett, Villa, 371
annandalei Brunetti, Exoprosopa, 421
annexus Roberts, Anastoechus, 86
annulata Melander, Mythicomyia, 269
annulatus Engel, Systropus, 248
annuliventris Hesse, Bombylius, 81
anomala Wiedemann, Adelidea, 121; figs. 34, 256, 739
anomala fuligineipennis Hesse, Adelidea, 121
anomala Williston, Desmatomyia, 187; figs. 61, 343, 576, 691, 846, 847, 848
anomala Painter, Exoprosopa, 419
anomala Melander, Mythicomyia, 269
anomala Hesse, Tomomyza, 292
anomalus Hesse, Amictogeron, 208
anomalus Paramonov, Anastoechus, 86
anomalus Hesse, Bombylius, 81
anomalus Bezzi, Prorates, 228; fig. 326
anomalus Painter, Sparnopolius, 124
antecedens Walker, Anthrax, 440
antecedens Walker, Bombylius, 82
antecedens Walker, Villa, 374
antecessor Melander, Mythicomyia, 269
antennatus Paramonov, Conophorus, 158
antennipilosus Paramonov, Bombylius, 80
antenoreus Lioy, Bombylius, 80
anthonomus Melander, Oligodranes, 220
anthophilus Hesse, Systoechus, 84
anthophoroides Hesse, Villa, 374
anthracina Thomson, Lomatia, 311
anthracina Bowden, Micomitra, 430
anthracina Becker, Prothaplocnemis, 399; fig. 214
anthracoides Jaennicke, Exoprosopa, 419
anthracoides Wiedemann, Tomomyza, 292; fig. 113
anthrax Schrank, Anthrax, 19, 441; fig. 337
antica Walker, Exoprosopa, 419
anticus Walker, Villa, 372
antiopa Bezzi, Anthrax, 441
antipodesia, new species, Comptosia, 349; fig. 366

antiqua Cockerell, Protolomatia, fossil, 59
anus Wiedemann, Exoprosopa, 419
anus Wiedemann, Thyridanthrax, 406
anus Becker, Usia, 216
apache Painter and Hall, Poecilanthrax, 398
aperta Melander, Apolysis, 219
apertus Macquart, Sparnopolius, 124
apertus Walker, Villa, 374
aphelesticta Hesse, Exoprosopa, 420
apicalis Meigen, Bombylius, 80
apicalis Macquart, Comptosia, 353
apicalis Klug, in Loew, Exoprosopa, 419; fig. 210
apicalis Wiedemann, Exoprosopa, 31, 420
apicalis Hesse, Lomatia, 311
apicifera Walker, Villa, 374
apicola Cole, Paravilla, 28, 392
apiculus Coquillett, Lordotus, 114
apiformis Hesse, Exoprosopa, 420
apiformis Hesse, Villa, 374
apparens Walker, Villa, 374
appendiculata Bigot, Anthrax, 442
appendiculata Macquart, Anthrax, 441
appendiculata Macquart, Lepidophora, 236
appendiculata Macquart, Petrorossia, 346
appendiculatus Bezzi, Bombylius, 80
aprica Roberts, Villa, 374
apricaria Hesse, Xeramoeba, 342
apricata Melander, Mythicomyia, 269
apulus Cyrillo, Bombylius, 80
arabicus Paramonov, Anastoechus, 86
arabicus Paramonov, Oesthanthrax, 411
arabicus Macquart, Villa, 373
araxana Paramonov, Cytherea, 148
araxana Paramonov, Exoprosopa, 419
araxana Paramonov, Plesiocera, 338
araxis Paramonov, Anastoechus, 86
araxis nigrisetosa Paramonov, Anastoechus, 86
archimedea Bigot, Exoprosopa, 419
arctica Zetterstedt, Glabellula, 271
arctica Strobl, Systoechus, 84
arcticus, Strobl, Systoechus, 84
arcuata Macquart, Exoprosopa, 420
arcus de Meijere, Anthrax, 442
ardens Walker, Bombylius, 82
arenacea Paramonov, Antonia, 321
arenacea Becker, Exoprosopa, 419
arenaria Hesse, Lomatia, 311
arenicola Melander, Aphoebantus, 324
arenicola Paramonov, Cytherea, 148
arenicola Johnson and Johnson, Exoprosopa, 419
arenicola Painter, Geron, 206
arenicola Cole, Villa, 372
arenicola, Johnson, Villa, 372
arenicolus Hesse, Thyridanthrax, 406
arenivagum Austen, Anthrax, 441
arenivagus Paramonov, Heterotropus, 226; figs. 774, 782, 785
arenivagus chivaensis Paramonov, Heterotropus, 226
arenivagus flavoscutellatus Paramonov, Heterotropus, 226
arenivagus normalipes Paramonov, Heterotropus, 226
arenivagus repeteki Paramonov, Heterotropus, 226
arenophilus Paramonov, Anthrax, 441
arenosa Coquillett, Chrysanthrax, 378

arenosus Paramonov, Bombylius, 80
areolata Abreu, Anthrax, 442
areolatus Walker, Bombylius, 82
arethusa Osten Sacken, Poecilanthrax, 398
argentarius Séguy, Bombylius, 80
argentatus Cole, Anthrax, 440
argentatus Becker, Bombylius, 81
argentatus Fabricius, Bombylius, 81
argentiapicalis Brunetti, Anthrax, 442
argenticincta Bigot, Ligyra, 434
argentifacies Austen, Bombylius, 80
argentifasciata Macquart, Exoprosopa, 419
argentifer Walker, Bombylius, 81
argentifer Becker, Thyridanthrax, 406
argentifera Walker, Exoprosopa, 421
argentifluus Philippi, Villa, 372
argentifrons Cole, Aphoebantus, 324
argentifrons Loew, Bombylius, 80
argentifrons Macquart, Cytherea, 148
argentifrons Williston, Desmatoneura, 348; figs. 133, 379, 636, 649, 908, 909, 910
argentifrons Bowden, Eurycarenus, 143
argentifrons Macquart, Exoprosopa, 420
argentifrons scalaria Bezzi, Exoprosopa, 420
argentifrons Brunetti, Geron, 18, 207
argentifrons Becker, Hemipenthes, 386
argentifrons Austen, Thyridanthrax, 29, 406
argentilatus Walker, Villa, 374
argentina Bezzi, Villa, 374
argentinae Paramonov, Lyophlaeba, 356
argentinae Paramonov, Sericosoma, 154
argentipennis White, Villa, 374
argentosa Painter, Villa, 372
argillocosmia Hesse, Exoprosopa, 420
argutus Painter, Geron, 206
argyrocephala Macquart, Cytherea, 148
argyrocephala Macquart, Exoprosopa, 420
argyrocomus Hesse, Anastoechus, 86
argyrolepis Bezzi, Litorryhynchus, 428
argyroleucus Hesse, Systoechus, 84
argyrolomus Bowden, Bombylius, 81
argyrolophus Hesse, Thyridanthrax, 406
argyrophora Bezzi, Exoprosopa, 420
argyropila, Cryomyia, n. sp., 133; figs. 39, 40, 265
argyropogonus Hesse, Systoechus, 84
argyropus Loew, Chasmoneura, 110
argyropyga Doleschall, Anthrax, 442
argyropygus Wiedemann, Anthrax, 19, 440; fig. 159
argyropygus Wiedemann, Chasmoneura, 110; figs. 41, 280, 414, 562
argyropygus Macquart, Dischistus, 107
argyura de Meijere, Ligyra, 434
aridicola Hesse, Anthrax, 442
aridicola Hesse, Exoprosopa, 420
aridicolus Hesse, Doliogethes, 111; figs. 37, 270, 468, 471, 772
ariditata Cole, Villa, 372
aridus Painter, Geron, 206
arizonensis Johnson and Johnson, Amphicosmus, 295
arizonensis Coquillett, Chrysanthrax, 378
arizonensis Johnson and Johnson, Lordotus, 114
arizonicus Banks, Systropus, 248
armada Cresson, Mythicomyia, 269; fig. 102

armeniaca Paramonov, Antonia, 321
armeniaca Paramonov, Cytherea, 148
armeniaca Paramonov, Lomatia, 311
armeniaca Paramonov, Anthrax, 441
armeniacus Paramonov, Aphoebantus, 324
armeniacus Paramonov, Bombylius, 80
armenicus Paramonov, Aphoebantus, 324
armipes Cresson, Mythicomyia, 269
arnaudi Johnson and Johnson, Lordotus, 114
arnoldi Hesse, Bombylius, 81
artemesia Marston, Anthrax, 440
asaphoedesma Hesse, Lomatia, 311
asiaticus Becker, Anastoechus, 86
asiaticus albulus Paramonov, Anastoechus, 86
asiaticus Becker, Bombylodes, 92
asiaticus Paramonov, Conophorus, 158
asiaticus Becker, Sparnopolius, 124
aspiculata Hesse, Villa, 374
aspilota Melander, Mythicomyia, 269
assimilis Macquart, Hemipenthes, 386
astarte Wiedemann, Chrysanthrax, 378
ater Roberts, Anthrax, 442
ater Scopoli, Bombylius, 81
ater Lamarck, Conophorus, 158
ater Cresson, Oligodranes, 220
ater Coquillett, Parabombylius, 97
aterrima Hesse, Lomatia, 311
aterrimus Bigot, Anthrax, 20, 440; figs. 966, 967, 968
atherix Newman, Comptosia, 353
atlanticus Abreu, Bombylius, 80
atlanticus Séguy, Heterotropus, 226
atomus Frey, Anomaloptilus, 276
atra Cresson, Mythicomyia, 269
atra Cresson, Oligodranes, 220
atra Melander, Protophthiria, fossil, 59
atra Painter, Villa, 372
atrata Hesse, Exoprosopa, 420
atrata Coquillett, Thyridanthrax, 29, 406
atrata Fabricius, Usia, 17, 216; figs. 312, 545
atratulus Loew, Conophorus, 158
atratus Meigen, Conophorus, 158
atratus Coquillett, Phthiria, 199
atratus Macquart, Systropus, 248
atrella Hesse, Exoprosopa, 420
atrella Hesse, Lomatia, 311
atricapillus Wulp, Bombylius, 79
atricapillus Hesse, Litorrhynchus, 428
atricauda Austen, Villa, 373
atriceps Loew, Bombylius, 79
atriceps fulvibasoides Painter, Bombylius, 79
atriceps Loew, Phthiria, 199
atriceps Bowden, Systoechus, 84
atricosta Bezzi, Ligyra, 434
atrinasis Speiser, Exoprosopa, 420
atripes Cole, Exoprosopa, 419
atriplex Marston, Anthrax, 20, 22
atrisquama Hesse, Exoprosopa, 420
atrisquama Bezzi, Villa, 374
atrita Melander, Mythicomyia, 269
atriventris Hesse, Thyridanthrax, 406
atronotatus Hesse, Bombylius, 81
atropos Egger, Lomatia, 311
attenuatus Macquart, Systropus, 248
atterimus Doleschall, Villa, 374
audouinii Macquart, Ligyra, 434

aurantiaca Guerin, Exoprosopa, 421
aurantiacus Macquart, Bombylius, 81
aurata Bowden, Antonia, 321
aurata Macquart, Lomatia, 311
aurata Fabricius, Usia, 216
aurata loewii Becker, Usia, 216
auratus Priddy, Conophorus, 158
auratus Walker, Sisyromyia, 98; fig. 252
auratus Williston, Thevenemyia, 182
aurea Fabricius, Cytherea, 148
aurea Macquart, Toxophora, 236
aureohirta Brunetti, Villa, 374
aureola Melander, Mythicomyia, 269
aureosquamosus Marston, Anthrax, 440
aureosquamosus chaparralus Marston, Anthrax, 440
aureus Hesse, Systoechus, 84
auricomatus Bowden, Systoechus, 84
auricomus Bezzi, Bombylius, 81
aurifacies Greathead, Systoechus, 9, 84
aurifer Osten Sacken, Bombylius, 79
aurifer pendens Cole, Bombylius, 79
aurifera Melander, Mythicomyia, 269
aurifera Rondani, Toxophora, 236
auriferus Hesse, Bombylius, 81
auriferus melanus Hesse, Bombylius, 81
auriferus nigripes Hesse, Bombylius, 81
aurifluus Bezzi, Dischistus, 107
aurifrons Efflatoun, Anastoechus, 86
aurifrons Macquart, Comptosia, 353
aurimystax Hesse, Bombylius, 81
auripila Hull, Laurella, 137; figs. 23, 274, 514
auripilus Efflatoun, Acanthogeron, 90
auripilus Séguy, Acanthogeron, 90
auripilus Hesse, Systoechus, 84
auripilus Bigot, Thevenemyia, 182
auripilus Osten Sacken, Thevenemyia, 182
auriplena Walker, Exoprosopa, 421
auripuncta Painter, Aldrichia, 161
aurocinctus Bigot, Villa, 374
aurovittatus Macquart, Bombylius, 80
aurulans Bezzi, Exoprosopa, 420
aurulentus Paramonov, Bombylius, 80
aurulentus Wiedemann, Systoechus, 84
ausana Hesse, Pteraulax, 328
austeni Brunetti, Anthrax, 442
austini Painter, Bombylius, 79
austeni Paramonov, Palintonus, 241
austeni Bezzi, Systoechus, 84
australensis Schiner, Lomatia, 311
australia Hesse, Toxophora, 236
australianus Bigot, Bombylius, 82
australis Guerin-Meneville, Bombylius, 82
australis Loew, Bombylius, 80
australis Hesse, Geron, 207
australis Macquart, Geron, 207
australis Malloch, Glabellula, 271
australis Walker, Villa, 374
austrandina Edwards, Phthiria, 199
autumnalis Cole, Poecilanthrax, 398
autumnalis Pallas, Systoechus, 9, 84
autumnalis albibarbis Engel, Systoechus, 84
autumnalis Becker, Thyridanthrax, 406
axillaris Meigen, Bombylius, 80
aygula Fabricius, Villa, 374
aygulus Fabricius, Anthrax, 442
azaleae Shannon, Bombylius, 80
aztec Painter, Phthiria, 199

B

baccha Loew, Exoprosopa, 419
badia Coquillett, Phthiria, 198
badipennis Hesse, Systoechus, 84
badius Hesse, Anthrax, 442
badius Hesse, Systoechus, 84
bagdadensis Becker, Exoprosopa, 420
bagdadensis Becker, Exoprosopa, 420
bagdadensis Macquart, Exoprosopa, 419
bahirae Becker, Anastoechus, 86
bahirae pyramidum Paramonov, Anastoechus, 86
baigakumensis Paramonov, Anastoechus, 12, 86
balioptera Loew, Exoprosopa, 420
balteatus Melander, Aphoebantus, 324
balteatus Philippi, Villa, 372
baluchiana Brunetti, Villa, 373
bancrofti Edwards, Comptosia, 353
banksi Johnson, Dipalta, 382
barbara Sack, Cytherea, 148
barbata Philippi, Phthiria, 199
barbata Rondani, Phthiria, 199
barbatula Bezzi, Tomomyza, 292
barbatus Bezzi, Amictogeron, 208
barbatus Osten Sacken, Anastoechus, 12, 39, 86; figs. 3, 791, 792
barbatus Melander, Aphoebantus, 324
barbiellinii Bezzi, Systropus, 248
barbiventris Rondani, Villa, 372
barbula Pallas, Bombylius, 80
barbula Loew, Dischistus, 107
barnardi Hesse, Exoprosopa, 420
barnardi Hesse, Systropus, 18, 248
basalis Walker, Comptosia, 353
basalis Ricardo, Exoprosopa, 420
basalis Ricardo, Litorrhynchus, 428
basalis Macquart, Villa, 374
basifascia Walker, Exoprosopa, 420
basifumatus Speiser, Bombylius, 81
basilare Wiedemann, Heterostylum, 140
basilaris Painter, Systropus, 248
basilinea Loew, Bombylius, 80
bastardi Macquart, Villa, 371
basutoensis Hesse, Amictogeron, 208
basutoensis Hesse, Anomaloptilus, 276
basutoensis Hesse, Lomatia, 311
batrachoides Bezzi, Exoprosopa, 420
bechuanus Hesse, Bombylius, 81
bechuanus Hesse, Geron, 207
bechuanus Hesse, Litorrhynchus, 428
bechuanus Hesse, Systoechus, 84
bechuanus Hesse, Thyridanthrax, 406
beckeri Paramonov, Cytherea, 148
beckeri Bezzi, Empidideicus, 275
beckeri Austen, Exoprosopa, 419
beckeri Paramonov, Prorachthes, 162
beckeri Paramonov, Thyridanthrax, 406
beckerianus Bezzi, Thyridanthrax, 406
bedfordi Hesse, Systoechus, 84
bedouinus Efflatoun, Bombylius, 80
bella Curran, Antonia, 321
bella Loew, Lomatia, 311; fig. 216
bella tibialis Loew, Lomatia, 311
bella Austen, Micomitra, 430
bella Melander, Mythicomyia, 269
bellulus Philippi, Villa, 372
bellus Philippi, Bombylius, 80
bellus Roberts, Bombylius, 82
bellus Becker, Conophorus, 158
bellus Loew, Platypygus, 262
belzebul Fabricius, Lomatia, 311; figs. 144, 145, 733
belzebul loewi Paramonov, Lomatia, 311
belzebul Panzer, Lomatia, 311
bembesiana Hesse, Lomatia, 311
beneficus Austen, Thyridanthrax, 29, 406
bengalensis Macquart, Exoprosopa, 421
bergi Paramonov, Bombylius, 81
berzeliaphila Hesse, Lomatia, 311
bevisii Hesse, Lomatia, 311
bezzi Paramonov, Exoprosopa, 419
bezzi Paramonov, Toxophora, 236
bezzianus Paramonov, Anthrax, 441
bezzii Hesse, Amictogeron, 208
bezzii Paramonov, Anthrax, 441
bezzii Hesse, Bombylius, 81
bezzii Paramonov, Exoprosopa, 419
bezzii Paramonov, Prorates, 228
bezzii Paramonov, Pseudoamictus, 211
biappendiculatus Macquart, Villa, 374
bibosa Melander, Mythicomyia, 269
bicellaris Becker, Conophorina, 159; figs. 33, 269, 481
bicellata Macquart, Exoprosopa, 421
bicinctus Wiedemann, Hemipenthes, 386
bicinctus Wiedemann, Sparnopolius, 124
bicingulatus Macquart, Villa, 373
bicolor Loew, Bombylius, 80
bicolor Macquart, Comptosia, 353
bicolor Bezzi, Corsomyza, 166
bicolor Abreu, Cyrtosia, 256
bicolor Wulp, Heterostylum, 140
bicolor Melander, Oligodranes, 220
bicolor Bezzi, Phthiria, 199
bicolor Coquillett, Phthiria, 199
bicolor Efflatoun, Usia, 216
bicolor Macquart, Usia, 216
bicoloratus Bezzi, Bombylius, 81
bicoloricornis Macquart, Bombylius, 82
bicoloripennis Hesse, Systropus, 248
bicornis Painter, Systropus, 248
bicuspis Bezzi, Systropus, 18, 248
bifarius Hesse, Anthrax, 442
bifarius Melander, Oligodranes, 220; figs. 857, 858
bifasciata Blanchard, Lyophlaeba, 356
bifasciata Macquart, Lyophlaeba, 356; figs. 988, 989, 990
bifasciata Philippi, Lyophlaeba, 356
bifasciatus Meigen, Hemipenthes, 386
bifenestrata Bigot, Thyridanthrax, 406
bifidus Bezzi, Bombylius, 81
biflexus Loew, Anthrax, 442
biflexus Loew, Villa, 374
bifrons Walker, Bombylius, 82; figs. 432, 487
bifurca Loew, Exoprosopa, 419
bigoti Edwards, Lyophlaeba, 356
bigotianum Edwards, Sericosoma, 154
bigotii Macquart, Corsomyza, 166
bigradata Loew, Hemipenthes, 385
biguttata Edwards, Comptosia, 353
biguttatus Macquart, Villa, 374
bilineata Engel, Anthrax, 441
bilineatus Hesse, Ceratolaemus, 256
bilineatus Melander, Oligodranes, 220
bilobatus Bezzi, Aphoebantus, 324
bilychnis Melander, Mythicomyia, 269
bimacula Walker, Villa, 373
bimaculatus Macquart, Villa, 373
binotata Wiedemann, Anthrax, 20, 441
binotata Painter, Desmatomyia, 187; fig. 60
binotata Macquart, Exoprosopa, 421
binotatus Macquart, Villa, 372
bipartita Bigot, Exoprosopa, 419
bipartitus Painter, Lordotus, 114; fig. 47
bipartitus Jaennicke, Villa, 372
bipenicillatus Bigot, Villa, 372
bipunctata Fabricius, Anthrax, 442
bipunctata Pallas, Cytherea, 148
bipustulata Bezzi, Corsomyza, 166
biroi Becker, Acanthogeron, 90
bisalbifrons Bezzi, Cytherea, 148
bisglaucus Bezzi, Heterotropus, 226
bisniphas Bezzi, Anthrax, 441
bistella Walker, Villa, 372
bisulcus Osten Sacken, Aphoebantus, 324
bitinctus Becker, Anastoechus, 86
bituberculatus Becker, Aphoebantus, 324
biumbonatum Bezzi, Epacmoides, 334; fig. 129
bivirgata Austen, Villa, 373
bivittata Cresson, Oligodranes, 220
bivittata Bezzi, Petrorossia, 346
bivittatus Loew, Bombylius, 81; figs. 247, 433, 520
bivittatus Loew, Conophorus, 158
bivulneris Melander, Mythicomyia, 269
bizona Walker, Exoprosopa, 419
bizonata Becker, Villa, 373
blanchardi Paramonov, Lyophlaeba, 356
blanchardiana Jaennicke, Exoprosopa, 419
blanchariana Jaennicke, Hemipenthes, 386
blanchardii Philippi, Thyridanthrax, 406
blanchei Efflatoun, Acanthogeron, 90
blanda Loew, Villa, 373
blumei Vollenhoven, Systropus, 248
boghariensis Lucas, Bombylius, 8, 80
bolbocera Hesse, Exoprosopa, 420
bolbocerus Hesse, Thyridanthrax, 406
boliviana Paramonov, Lyophlaeba, 356
bombiformis Bezzi, Bombylius, 81
bombiformis Becker, Villa, 373
bombyciformis Dufour, Exoprosopa, 419
bombycinus Hesse, Systoechus, 84
bombycinus bedfordi Hesse, Systoechus, 84
bombycinus pallidispinis Hesse, Systoechus, 84
bombyliformis Macleay, Ligyra, 434
bombyliiformis Loew, Conophorus, 158
bombyliiformis Becker, Legnotomyia, 163
boninensis Matsumura, Anthrax, 441
borealis Cole, Aphoebantus, 324
borealis Johnson, Phthiria, 199
bouillonae Séguy, Antonia, 321
bourdariei Séguy, Heterotropus, 226
bovei Macquart, Exoprosopa, 419
bovis Becker, Exoprosopa, 419
brachialis Thomson, Villa, 372
brachipleuralis Hesse, Exoprosopa, 420
brachycera Bezzi, Micomitra, 430
brachyrhynchus Bezzi, Bombylius, 81
brahma Schiner, Exoprosopa, 421
bramah Wulp, Exoprosopa, 421
brasiliensis Macquart, Systropus, 248
braunsi Bezzi, Adelidea, 121
braunsi Bezzi, Bombylius, 81
braunsi Bezzi, Pteraulax, 328
braunsii Hesse, Adelogenys, 192
braunsii Hesse, Corsomyza, 166

bravae Bezzi, Villa, 374
brevicornis Hesse, Corsomyza, 166; fig. 63
brevicornis Loew, Sparnopolius, 124
brevifacies Hesse, Thyridanthrax, 29, 406
brevinasis Bezzi, Exoprosopa, 419, 420
brevipennis Bezzi, Chiasmella, 349; figs. 378, 644, 654
brevirostratus Austen, Bombylius, 80
brevirostris Hesse, Apolysis, 219
brevirostris Macquart, Bombylius, 82
brevirostris Meigen, Bombylius, 80
brevirostris Olivier, Bombylius, 80
brevirostris Olivier, Cytherea, 148
brevirostris Williston, Exoprosopa, 419
brevirostris Bezzi, Gnumyia, 169; figs. 289, 483, 506
brevirostris Macquart, Sisyromyia, 98; figs. 436, 443
brevirostris Macquart, Sparnopolius, 124
brevis Becker, Anthrax, 441
brevistilus Hesse, Anomaloptilus, 276
brevistylata Williston, Exoprosopa, 419
brevistylus Coquillett, Aphoebantus, 324
breviusculus Loew, Dischistus, 107
breyeri Hesse, Geron, 207
brinchki Hesse, Zinnomyia, 95
brunea Roberts, Villa, 374
brunetti Senior-White, Bombylius, 82
brunipennis Macquart, Villa, 374
brunnea Edwards, Comptosia, 353
brunnea Becker, Villa, 373
brunnescens Loew, Oesthanthrax, 411; fig. 233
brunnibasis Hesse, Systoechus, 84
brunnicosa Becker, Anthrax, 441
brunnipennis Loew, Bombylius, 81
brunnipennis Macquart, Lomatia, 311; fig. 130A
brunnipennis Wulp, Lomatia, 311
brunnitincta Hesse, Lomatia, 311
bucerus Coquillett, Lordotus, 114
bucharense Paramonov, Anthrax, 441
bucharensis Paramonov, Cytherea, 148
bucharensis Paramonov, Exoprosopa, 419
bucinator Melander, Mythicomyia, 269
bulawayoensis Hesse, Lomatia, 311
bullulatus Hesse, Pantostomus, 293
burnsi Paramonov, Ligyra, 434
burtii Hesse, Thyridanthrax, 29, 406
busiris Jaennicke, Exoprosopa, 419
busonicus François, Anthrax, 442
butleri Johnson and Johnson, Exoprosopa, 419
butleri Johnson and Johnson, Poecilanthrax, 398
buttneri Enderlein, Systropus, 248
buziana Hesse, Petrorossia, 346

C

cachinnans Osten Sacken, Bombylius, 79
cadericina Bezzi, Exoprosopa, 420
cadicerina Bezzi, Exoprosopa, 420
caffer Hesse, Anthrax, 20, 442
caffra Wiedemann, Exoprosopa, 420
caffrariae Hesse, Oniromyia, 156
caffrariae Hesse, Thyridanthrax, 406
caffrariana Hesse, Exoprosopa, 420
cairensis Paramonov, Anthrax, 441
cala Melander, Mythicomyia, 269
calabyana Paramonov, Ligyra, 434
californiae Walker, Exoprosopa, 419
californica Greene, Mythicomyia, 269
californica Bigot, Thevenemyia, 182
californicus Cole, Poecilanthrax, 398; figs. 190, 198A, 710, 720, 948, 949, 950
caligula Melander, Mythicomyia, 269
caliptera Say, Exoprosopa, 31, 419
callima Melander, Mythicomyia, 269
callopterus Loew, Bombylius, 80
callopterus umbripennis Paramonov, Bombylius, 80
callynthrophorus Schiner, Systoechus, 84
calochromatus Bezzi, Thyridanthrax, 406
calogaster Philippi, Villa, 372
calophthalma Thomson, Comptosia, 353
caloptera Schiner, Anthrax, 441
caloptera Pallas, Exoprosopa, 419
caloptera Macquart, Lomatia, 311
calopterus Philippi, Villa, 372
calopus Bigot, Systropus, 248
calva Melander, Mythicomyia, 269
calva Loew, Usia, 216
calvescens Philippi, Villa, 372
calviniensis Hesse, Bombylius, 81
calvus Loew, Geron, 18, 206
caminarius Wiedemann, Sparnopolius, 124
campbelli Paramonov, Ligyra, 434
campestris Hesse, Exoprosopa, 420
campestris Coquillett, Lepidanthrax, 388
campestris Fallen, Phthiria, 199
campicola Hesse, Corsomyza, 166
campicola Eversman, Exoprosopa, 419
camptocladius Bezzi, Anthrax, 442
cana Philippi, Phthiria, 199
cana Meigen, Villa, 373
canadensis Curran, Bombylius, 79
canalis Coquillett, Geminaria, 117; figs. 35, 281, 461, 571
canariensis Engel, Cyrtosia, 256
canariensis tenuis Frey, Cyrtosia, 256
canariensis Frey, Glabellula, 271
cancescens Loew, Phthiria, 199
candida Sack, Anthrax, 441
candidata Bezzi, Villa, 374
candidifrons Austen, Bombylius, 80
candidopex Austen, Anthrax, 441
candidulus Hesse, Anthrax, 441, 442
candidulus Loew, Systoechus, 84
candidus Loew, Bombylius, 80
candidus Hesse, Systoechus, 84
canescens Mikan, Bombylius, 8, 80
canescens Hesse, Lomatia, 311
canescens Philippi, Lyophlaeba, 356
canescens Loew, Phthiria, 199
canescens Hesse, Systoechus, 84
canicapillis Bowden, Systoechus, 84
canipectus Hesse, Systoechus, 84
cantonensis Enderlein, Systropus, 248
canus Melander, Caenotus, 227
canus Philippi, Geron, 206
canus Macquart, Systoechus, 84
capax Coquillett, Aphoebantus, 324
capax Coquillett, Oligodranes, 220
capensis Linné, Bombylius, 81
capensis Schiner, Callynthrophora, 168; figs. 292, 480, 493
capensis Hesse, Corsomyza, 166
capensis Wiedemann, Exoprosopa, 420
capensis Walker, Geron, 206
capensis Wiedemann, Megapalpus, 172; figs. 67, 288, 485, 505, 742, 762, 764
capensis Hesse, Pantostomus, 293
capensis Bigot, Peringueyimyia, 300; figs. 120, 359, 607, 627, 642, 893, 894, 895
capensis Wiedemann, Phthiria, 199
capensis Philippi, Systropus, 248
capicolus Hesse, Amictogeron, 208
capillatus J. Palm, Bombylius, 80
capitalata Mulsant, Anthrax, 442
capito Loew, Dischistus, 107
capito longirostris Hesse, Dischistus, 107
capito Osten Sacken, Pantarbes, 153; fig. 82
capnoptera Bezzi, Exoprosopa, 420
caprea Coquillett, Villa, 371
capucina Fabricius, Anthrax, 440
capucina Fabricius, Exoprosopa, 419
carbo Rondani, Anthrax, 442
carbonarius Osten Sacken, Aphoebantus, 324
carbonarius Walker, Villa, 374
carcassoni Bowden, Diatropomma, 424 figs. 194A, 199, 404
carcelii Guerin, Toxophora, 236
carmelitensis Becker, Cytherea, 148
carmelitensis Becker, Usia, 216
carnata Bezzi, Sosiomyia, 123; figs. 32, 279, 810, 811
carptura Melander, Mythicomyia, 269
carthaginiensis Becker, Empidideicus, 275; fig. 348
carus Cresson, Lordotus, 115
cascadensis Marston, Anthrax, 440
caspica Paramonov, Lomatia, 311
castanea Jaennicke, Paravilla, 392; fig. 176
castaneus Macquart, Amictus, 286
castaneus Jaennicke, Villa, 372
castanipes Bigot, Hemipenthes, 385
castilla Painter, Exoprosopa, 419
catenarius Melander, Aphoebantus, 324
catherinae Efflatoun, Mariobezzia, 174
catheriniensis Efflatoun, Bombylius, 80
catulina Coquillett, Hemipenthes, 30, 385
catulus Coquillett, Aphoebantus, 324
caucasicum Zaitsev, Anthrax, 441
caucasicus Paramonov, Anastoechus, 86
caucasicus Paramonov, Bombylius, 80
caucasicus Zaitsev, Conophorus, 158
caudatus Meigen, Anastoechus, 86
cautor Coquillett, Villa, 371
ceballosi Rubio, Villa, 373
cedens Walker, Anthrax, 440
celebensis Enderlein, Systropus, 248
celebesi Paramonov, Ligyra, 434
celer Cole, Thevenemyia, 182; figs. 795, 796
celeris Cole, Exoprosopa, 419
celeris Wiedemann, Hemipenthes, 385
cellularis Bowden, Systoechus, 84
cellulifer Hesse, Anomaloptilus, 276; figs. 353, 586, 595
centralis Macquart, Lomatia, 311
cephalotes Walker, Anastoechus, 86
cephalotes Walker, Bombylius, 82
cephus Fabricius, Anthrax, 441
cerberus Fabricius, Ligyra, 434
cerberus Macquart, Ligyra, 434
cercoplecta Hesse, Antonia, 321
cerdo Osten Sacken, Systropus, 248
ceria Williston, Villa, 372
cervina Bezzi, Exoprosopa, 420
cervinus Loew, Aphoebantus, 324
cervinus Loew, Systoechus, 84
ceuthodonta Hesse, Exoprosopa, 420
ceylonica Brunetti, Anthrax, 442

ceylonicus Brunetti, Aphoebantus, 324
ceyx Loew, Poecilanthrax, 398
chalciurus Hesse, Anthrax, 442
chalcoides Pallas, Micomitra, 430
chalybaea Roeder, Exoprosopa, 419
chalybaea chrysogaster Bezzi, Exoprosopa, 419
chalybeus, Melander, Oligodranes, 220
chan Paramonov, Exoprosopa, 419
chaparralus Marston, Anthrax, 440
chapini Curran, Petrorossia, 346
charopus Lichtenstein, Bombylius, 80
chellicterus Hesse, Amictogeron, 208
chiastoneura, Staurostichus, new species, 105
chilena Rondani, Phthiria, 199
chilensis Paramonov, Cyrtomyia, 239; figs. 88, 319, 541, 584, 736, 745, 749, 872, 873, 874
chilensis Paramonov, Eurycarenus, 143
chilensis Rondani, Ligyra, 434
chilensis Paramonov, Lyophlaeba, 356
chilensis Philippi, Systropus, 248
chilensis Philippi, Villa, 372
chimaera Osten Sacken, Hemipenthes, 385
chimaera Osten Sacken, Villa, 372
chinensis Paramonov, Anastoechus, 86
chinensis Paramonov, Bombylius, 80
chinensis Paramonov, Conophorus, 158
chinensis Bezzi, Systropus, 248
chinnicki Paramonov, Ligyra, 434
chinooki Priddy, Conophorus, 158
chinophorus Bezzi, Thyridanthrax, 406
chioleucus Hesse, Gonarthrus, 112
chionalis Hesse, Villa, 374
chionanthrax Bezzi, Anthrax, 441
chionea Bezzi, Micomitra, 430
chioneus Bezzi, Gonarthrus, 112
chionoleucus Hesse, Doliogethes, 111
chionostigma Tsacas, Anthrax, 441
chitona Melander, Mythicomyia, 269
chivaensis Paramonov, Anthrax, 441
chivaensis Paramonov, Heterotropus, 226
chlamydicterus Hesse, Systoechus, 84
chlorizans Rondani, Conophorus, 158
chloroxanthus Hesse, Gonarthrus, 112
chopardi Séguy, Toxophora, 236
chorassani Becker, Hemipenthes, 386
choreutes Bowden, Chionamoeba, 344; fig. 228
chraecoptera Hesse, Lomatia, 311
chromolepida Cole, Villa, 371
chrysanthemi Loew, Platypygus, 262
chrysendetus White, Bombylius, 82
chrysogaster Bezzi, Exoprosopa, 419
chrysolampis Jaennicke, Ligyra, 434
chrysonotum Hesse, Crocidium, 185
chrysothrix Bezzi, Villa, 374
chrystallina Bezzi, Micomitra, 430
chrystallinus Bezzi, Systoechus, 84
cidarellus Hesse, Thyridanthrax, 406
cincinnatus Becker, Bombylius, 81
cincta Edwards, Comptosia, 353
cinctalis Hesse, Pteraulax, 328
cinctiventris Roberts, Systoechus, 84
cincturus Williston, Amphicosmus, 295; figs. 110, 361, 608, 618
cincturus Coquillett, Oligodranes, 220; figs. 77, 332, 552, 554, 737, 765
cinefacta Coquillett, Chrysanthrax, 378
cineracea Austen, Legnotomyia, 163; fig. 254

cinerarius Pallas, Bombylius, 80
cinerarius eversmanni Paramonov, Bombylius, 80
cinerarius karelini Paramonov, Bombylius, 80
cinerarius pallasi Paramonov, Bombylius, 80
cinerascens Bigot, Amictus, 286
cinerascens Mikan, Bombylius, 80
cinerascens Loew, Cyrtosia, 256
cinerea Perris, Apolysis, 219
cinerea Fabricius, Cytherea, 148
cinerea Wiedemann, Cytherea, 148
cinerea Cole, Paravilla, 392
cinerea Cole, Villa, 372
cinereitincta Hesse, Chasmoneura, 110
cinereola Hesse, Lomatia, 311
cinereus Zaitsev, Anthrax, 441
cinereus Bigot, Bombylius, 80
cinereus Meigen, Bombylius, 80, 81, 84
cinereus Lynch Arribalzaga, Cytherea, 148
cinereus Melander, Oligodranes, 220
cinereus Meigen, Systoechus, 84
cinerivus Painter, Bombylius, 80
cingularis Hesse, Exoprosopa, 420
cingulata Hesse, Apolysis, 219
cingulata Johnson and Johnson, Exoprosopa, 419
cingulata Wulp, Ligyra, 434
cingulata Loew, Phthiria, 199
cingulata Meigen, Villa, 373
cingulatus Hesse, Eurycarenus, 143
cingulatus Johnson and Johnson, Lordotus, 114
cingulatus lineatus Johnson and Johnson, Lordotus, 114
cingulatus Meigen, Villa, 373
cingulatus Walker, Villa, 373
cingulatus Zetterstedt, Villa, 373
cingulum Wiedemann, Villa, 373; fig. 181
circe Klug, Thyridanthrax, 406
circeoides Paramonov, Exoprosopa, 419
circeoides nigrofasciata Paramonov, Exoprosopa, 419
circumdata Becker, Hemipenthes, 386
circumdata quinquefasciata Becker, Hemipenthes, 386
circumdatus Roberts, Neosardus, 301
circumdatus Meigen, Villa, 373
circumdatus Meigen, Villa, 373
circumdatus algeciras Strobl, Villa, 373
circumdatus fulvimaculatus Abreu, Villa, 373
cirratus Melander, Epacmus, 326
cirrhata Bezzi, Antonia, 321
citraria Hesse, Lomatia, 311
citrinus Loew, Bombylius, 80
citrinus Hesse, Gonarthrus, 112
cladurus Cockerell, Verrallites, fossil, 59
claripennis Brunetti, Anthrax, 442
claripennis Becker, Aphoebantus, 324; fig. 227
claripennis Macquart, Bombylius, 81
claripennis Becker, Cytherea, 148
claripennis Hesse, Exoprosopa, 420
claripennis Melander, Prorates, 228; figs. 97A, 322, 573, 700
claripennis Macquart, Usia, 216
claripennis Becker, Villa, 373
claripennis Kowarz, Villa, 373
clarissimus Loew, Villa, 373

clarki Curran, Exoprosopa, 419
clarus Walker, Villa, 374
clathrata Bezzi, Exoprosopa, 420
claudia François, Diatropomma, 424
clausa Brunetti, Anthrax, 442
clausa Becker, Exoprosopa, 419
clauseni Aldrich, Aphoebantus, 324
clausina Bezzi, Exoprosopa, 420
clavatus Karsch, Systropus, 248
clavicornis Wiedemann, Corsomyza, 166
clavicornis Hesse, Nomalonia, 290; figs. 994, 995, 996
clavifemorata Hardy, Zaclava, 250
clavirostris Hesse, Gonarthrus, 112
clelia Osten Sacken, Stonyx, 402
cleomene Egger, Exoprosopa, 419
clio Williston, Bombylius, 80
clotho Wiedemann, Stonyx, 402; figs. 429, 708, 718, 960, 961
clunalis Melander, Epacmus, 326
cockerelli Melander, Alepidophora, fossil, 59
cockerelli Hesse, Bombylius, 81
cockerelli Melander, Oligodranes, 220
coeruleiventris Macquart, Ligyra, 434
coeruleiventris Karsch, Toxophora, 236
coeruleopennis Doleschall, Ligyra, 434
cognata Walker, Comptosia, 353
cognata Engel, Cyrtosia, 256
cognata trisignata Engel, Cyrtosia, 256
cognata Walker, Lomatia, 311
cognata Hesse, Phthiria, 199
colei, nomen nudum, Geron, 206
colei Melander, Oligodranes, 220
colei Johnson and Johnson, Poecilanthrax, 398
coleoptrata Bezzi, Ligyra, 434
collaris Becker, Bombylius, 80
collaris Wiedemann, Exoprosopa, 421
collaris Macquart, Litorryhynchus, 428
collaris Painter, Neodischistus, 128; figs. 26, 261, 472, 518
collina Melander, Mythicomyia, 269
collini Priddy, Conophorus, 158
coloradensis Grote, Sparnopolius, 124
coloratus Hesse, Antoechus, 86
columbianus Karsch, Systropus, 248
columbiensis Marston, Anthrax, 440
columbiensis Priddy, Conophorus, 158
comanche Painter, Bombylius, 79
comanche Painter, Hemipenthes, 385
comastes Brunetti, Bombylius, 82
comata Austen, Lomatia, 311
comata Bezzi, Sosiomyia, 123
comatum Austen, Anthrax, 441
combinatus Walker, Villa, 374
comma Melander, Mythicomyia, 269
commiles Walker, Villa, 374
commistus Macquart, Villa, 374
commoni Paramonov, Ligyra, 434
comosus Melander, Oligodranes, 220
compacta Brunetti, Lepidanthrax, 388
compar Bezzi, Exoprosopa, 420
comparata Tucker, Chrysanthrax, 378
comparata Melander, Mythicomyia, 269
completa Loew, Exoprosopa, 419
completa Paramonov, Lomatia, 311
completa Paramonov, Toxophora, 236
compressa Painter, Villa, 372
compressus Fabricius, Amictus, 286
compressus Wiedemann, Amictus, 286
compsocoma Hesse, Lomatia, 311

compta Melander, Mythicomyia, 269
compta Roberts, Toxophora, 236
comptus Paramonov, Anthrax, 441
concinna Melander, Mythicomyia, 269
concinnus Coquillett, Aphoebantus, 324
concinnus Meigen, Villa, 373
concisa Macquart, Hemipenthes, 386
concisa Macquart, Villa, 374
concisus Macquart, Anthrax, 442
conclusus Walker, Villa, 372
concolor Mikan, Bombylius, 80
concolor Zeller, Bombylius, 80
concrescens Melander, Mythicomyia, 269
condemnatus Cockerell, Pachysystropus, fossil, 59
confirmatus Walker, Villa, 374
confluens Macquart, Villa, 374
confluensis Roberts, Anthrax, 442
confrater Loew, Bombylius, 80
confusemaculata Macquart, Anthrax, 442
confusus Wiedemann, Sparnopolius, 124
confusus Becker, Thyridanthrax, 406
congruens Hesse, Anastoechus, 86
congruus Walker, Villa, 374
coniceps Macquart, Exoprosopa, 419
conicera Hesse, Lomatia, 311
conifacies Macquart, Chrysanthrax, 378
connectens Melander, Epacmus, 326
connexa Macquart, Villa, 372
connexus Bigot, Villa, 372
connivens Bezzi, Exoprosopa, 420
conocephala Macquart, Lomatia, 311
conochila Bezzi, Exoprosopa, 420
conopas Philippi, Villa, 372
conopoides Kunckel, Systropus, 18, 248
conostoma Hesse, Lomatia, 311
consanguinea Macquart, Exoprosopa, 419
consanguineus Macquart, Bombylius, 80
consanguineus Macquart, Villa, 371
consessor Coquillett, Villa, 372
consimilis Thomson, Villa, 374
consobrina Philippi, Lyophlaeba, 356
consobrina Bowden, Petrorossia, 346
consobrinus Hesse, Anthrax, 442
consobrinus suffusipunctis Hesse, Anthrax, 442
consobrinus Macquart, Bombylius, 82
consobrinus Hesse, Doliogethes, 111
consors Hesse, Amictogeron, 208
consors Hesse, Lomatia, 311
consors Osten Sacken, Phthiria, 199
conspersipennis Hesse, Prorachthes, 162; figs. 1012, 1013, 1014
conspersipennis xerophilus Hesse, Prorachthes, 162
conspicabilis Austen, Lomatia, 311
conspicienda Austen, Exoprosopa, 419
conspicua Loew, Phthiria, 199
conspurcata Wiedemann, Anthrax, 442
constitutus Walker, Villa, 372
consul Osten Sacken, Paravilla, 392
contigua Loew, Anthrax, 440
contiguus Melander, Aphoebantus, 324
contiguus separatus Melander, Aphoebantus, 324
contorta Bezzi, Exoprosopa, 420
contrarius Becker, Thyridanthrax, 406
contrasta Paramonov, Ligyra, 434
conturbata Loew, Exoprosopa, 420
conurus Osten Sacken, Aphoebantus, 324
convergens Loew, Phthiria, 199
convergens Loew, Systoechus, 84

convexus Walker, Villa, 372
coquilletti Cockerell, Lithocosmus, fossil, 59
coquilletti Williston, Parabombylius, 97; figs. 9, 272, 456, 732, 793, 794
coquilletti Painter, Paravilla, 392
coquilletti Johnson, Phthiria, 199
coracinus Loew, Dischistus, 107
corculum Newman, Comptosia, 353
corcyreus Frey, Geron, 206
corrigiolata Rondani, Villa, 372
corsikana Paramonov, Anastoechus, 86
corticalis Bezzi, Litorrhynchus, 428
corticeus Bezzi, Litorrhynchus, 428
corticeus corticalis Bezzi, Litorrhynchus, 428
corvina Loew, Exoprosopa, 420
corvinoides Hesse, Exoprosopa, 420
cosmoptera Bezzi, Heteralonia, 425
costalis Paramonov, Aphoebantus, 324
costalis subcostalis Paramonov, Aphoebantus, 324
costalis Edwards, Comptosia, 353
costalis Macquart, Exoprosopa, 420
costalis Wiedemann, Villa, 372
costata Bigot, Cytherea, 148
costatus Say, Villa, 372
costilabre Hesse, Crocidium, 185
cothurnatus Bigot, Geron, 207
crassicornis Greene, Glabellula, 271
crassilabris Macquart, Eusurbus, 102; figs. 439, 451
crassipalpis Villeneuve, Prorachthes, 162
crassirostris Macquart, Bombylius, 82
crassirostris Hesse, Cyrtisiopsis, 263; figs. 330, 600, 997, 998, 999
crassirostris Loew, Paracorsomyza, fossil, 59
crassitarsis Paramonov, Bombylius, 80
crassus Walker, Bombylius, 82
crepuscularis Lynch Arribalzaga, Villa, 372
cressoni Melander, Mythicomyia, 269
crinipes Becker, Usia, 216
crinita Bigot, Anthrax, 441
cristata Melander, Mythicomyia, 269
cristatus Painter, Conophorus, 158; fig. 66
croaticus Kertész, Dischistus, 107
crocea Séguy, Cyrtosia, 256
crocea Hesse, Doliopteryx, 273; fig. 323
croceum Painter, Heterostylum, 140; fig. 54
crocina Coquillett, Chrysanthrax, 378
crocina Melander, Mythicomyia, 269
crocisops Hesse, Toxophora, 236
crocogramma Hesse, Phthiria, 199
crosi Villeneuve, Anthrax, 441
crossodesma Hesse, Lomatia, 311
cruciatus Fabricius, Bombylius, 80
cruciatus leucopygus Macquart, Bombylius, 80
crudelis Westwood, Systropus, 18, 248
cryptochaunum Hesse, Epacmoides, 334
ctenopterus Walker, Bombylius, 80
ctenopterus Mikan, Systoechus, 84; fig. 10
cubana Loew, Exoprosopa, 419
culiciformis Hesse, Gonarthrus, 112
culiciformis Walker, Lepidophora, 238
culiciformis Hull, Thevenemyia, 182
culicoides Hesse, Adelogenys, 192; figs. 223, 301, 564, 568, 1003, 1004, 1005
cumatillis Grote, Sparnopolius, 124
cunctator Hesse, Anthrax, 442

cuneata Edwards, Comptosia, 353
cuneata Painter, Lepidophora, 238
cunicula Osten Sacken, Paravilla, 392
cuniculus Osten Sacken, Villa, 372
cupido Bezzi, Ligyra, 434
cuprea Fabricius, Toxophora, 236
cuprea Macquart, Usia, 216
currani Painter, Neodischistus, 128
currani d'Andretta and Carrera, Systropus, 248
curta Loew, Hemipenthes, 385
curvata Becker, Cyrtosia, 256
curvata de Meijere, Ligyra, 434
curvicornis Bezzi, Exoprosopa, 420
curvirostris Thomson, Villa, 372
curvistoma Hesse, Plesiocera, 338
cuspidicauda Enderlein, Systropus, 248
cuthbertsoni Hesse, Antrax, 442
cuthbertsoni Engel, Paratoxophora, 178; figs. 12, 268, 460, 803, 804
cyanoceps Johnson, Phthiria, 199
cyanolepida Hesse, Toxophora, 236
cyanopterus Wiedemann, Cyananthrax, 379; figs. 175, 373, 675, 940, 941
cybele Coquillett, Anthrax, 440
cyclops Osten Sacken, Aphoebantus, 324
cygnus Bigot, Gonarthrus, 112
cylindricus Bezzi, Gonarthrus, 112; figs. 24, 263, 561
cylla Melander, Mythicomyia, 269
cyllenia Lamarck, Cyllenia, 283
cyprignus Rondani, Villa, 373
cypris Meigen, Chrysanthrax, 28, 378; figs. 170, 663, 936, 937
cyrenaica Bezzi, Cytherea, 148
cytherea frontalis Wiedemann, Chionamoeba, 344

D

daemon Panzer, Hemipenthes, 386
dagniensis Paramonov, Thyridanthrax, 406
dagomba Bowden, Anthrax, 442
dalamatinus Strobl, Bombylius, 80
dalamatinus Loew, Systoechus, 84
damarensis Hesse, Bombylius, 81
damarensis Hesse, Exoprosopa, 420
damerensis Hesse, Litorrhynchus, 428
damarensis Hesse, Systoechus, 84
daphne Osten Sacken, Anthrax, 440
darlingi Hesse, Bombylius, 81
dasycerus Hesse, Amictogeron, 208
dasypolium Hesse, Crocidium, 185
daubei Guerin, Exoprosopa, 420
daveyi Bezzi, Isotamia, 426; figs. 193, 402, 684, 688
dayas Séguy, Acanthogeron, 90
deani Painter, Heterostylum, 140
debilis Loew, Bombylius, 80
decastroi Hesse, Exoprosopa, 420
decedens Walker, Comptosia, 353
decemmacula Walker, Villa, 372
deceptus Hesse, Systoechus, 84
decipiens Bezzi, Anthrax, 442
decipiens Engel, Conophorus, 158
decipiens Loew, Conophorus, 158
decipiens Bezzi, Exoprosopa, 420
decipula Austen, Villa, 373
decolor Bezzi, Exoprosopa, 420
decora Loew, Exoprosopa, 419
decorata Rondani, Acrophthalmyda, 176

INDEX

decorata Paramonov, Antonia, 321
decorata Walker, Sisyromyis, 98; fig. 11
decrepita Wiedemann, Exoprosopa, 419
decta Schiner, Dolichomyia, 250
decusys Bezzi, Anthrax, 442
dedecor Hermann, Anthrax, 441
dedecor Loew, Exoprosopa, 419
dedecor minoides Paramonov, Exoprosopa, 419
dedecoroides Paramonov, Exoprosopa, 419
deducta Bowden, Petrorossia, 346
degenera Walker, Anthrax, 442
delicata Becker, Cytherea, 148
delicatulus Melander, Amphicosmus, fossil, 59
delicatulus Walker, Villa, 372
delicatus Wiedemann, Bombylius, 81
delicatus Hesse, Geron, 207
delila Loew, Anthrax, 440
delineata Becker, Exoprosopa, 420
demogorgon Walker, Poecilanthrax, 398
demonstrans Walker, Villa, 374
denigratus Linné, Hemipenthes, 386
dentata Becker, Anthrax, 442
dentata Roberts, Ligyra, 434
dentei d'Andretta and Carrera, Aphoebantus, 324
dentiferus Bezzi, Litorrhynchus, 428
denudatus Melander, Aphoebantus, 324
depereti Meunier, Bombylius, fossil, 59
depressifrons Hesse, Corsomyza, 166
depressifrons Hesse, Crocidium, 185
depressus Loew, Platamomyia, 127; figs. 21, 286, 486, 1009, 1010, 1011
depressus Loew, Platypygus, 262
deserta Paramonov, Exoprosopa, 419
deserta Paramonov, Toxophora, 236
deserta tadzhikorum Paramonov, Toxophora, 236
deserticola Hall, Anastoechus, 86
deserticola Paramonov, Bombylius, 80
deserticola Paramonov, Cytherea, 148
deserticola albifrons Paramonov, Cytherea, 148
deserticola Paramonov, Exoprosopa, 419
deserticola Efflatoun, Usia, 216
deserticolus Hesse, Anastoechus, 86
deserticolus coloratus Hesse, Anastoechus, 86
desertivagus Paramonov, Bombylius, 80
desertorum Paramonov, Anthrax, 441
desertorum Paramonov, Bombylius, 81
desertorum Loew, Callostoma, 14, 150
desertus Coquillett, Aphoebantus, 324
deses Meigen, Acanthogeron, 90
desmophora Hesse, Lomatia, 311
detectus Walker, Villa, 372
deustum Thunberg, Heterostylum, 140
devectus Walker, Villa, 374
dia Wiedemann, Villa, 374
diadela Melander, Mythicomyia, 269
diadema Meigen, Anastoechus, 86
diademata Bezzi, Chasmoneura, 110
diademus Lindner, Chasmoneura, 110
diagonalis Wiedemann, Bombylius, 81
diagonalis Loew, Paravilla, 392
diana Walker, Villa, 372, 374
diana Williston, Villa, 372
dichopticum Hesse, Crocidium, 185
dichopticus Bezzi, Eurycarenus, 143; figs. 52, 831, 832, 833
dichroma Paramonov, Cytherea, 148

dichromatopus Bezzi, Aphoebantus, 324
dichromus Bigot, Geron, 207
didesma Hesse, Exoprosopa, 420
dido Osten Sacken, Ligyra, 434
diegoensis Painter, Bombylius, 79
difficulis Wiedemann, Anthrax, 442
diffusa Wiedemann, Anthrax, 20, 442
diffusa decusys Bezzi, Anthrax, 442
diffusa fuscopurpuratus Hesse, Anthrax, 442
diffusa hybridus Hesse, Anthrax, 442
diffusa majusculus Bezzi, Anthrax, 442
diffusa pallidulus Hesse, Anthrax, 442
diffusa suffusipennis Hesse, Anthrax, 442
digitaria Cresson, Geron, 206
dilatata Bezzi, Exoprosopa, 420
dilatatus Bezzi, Litorrhynchus, 32, 428
diloga Melander, Mythicomyia, 269
diluta Bezzi, Exoprosopa, 420
dilutus Wiedemann, Bombylius, 80
dimidiata Wiedemann, Anthrax, 441
dimidiata Macquart, Exoprosopa, 420
dimidiata Roberts, Exoprosopa, 421
dimidiata Hesse, Lomatia, 311
dimidiata Macquart, Ogcodocera, 314
dimidiatipennis Hesse, Anthrax, 442
dimidiatus Macquart, Bombylius, 82
dimidiatus Wiedemann, Bombylius, 80
dimidiatus Bezzi, Dischistus, 107
dimidiatus Curran, Systropus, 248
diminutivus Macquart, Villa, 372
dimorphus Hesse, Villoestrus, 409
dionysii Bigot, Exoprosopa, 420
diplasus Hall, Lordotus, 114
diploptera Speiser, Toxophora, 236
dipura Melander, Mythicomyia, 269
diremptus Enderlein, Systropus, 248
diropeda Melander, Mythicomyia, 269
discepes Becker, Cytherea, 148
discoideus Fabricius, Bombylius, 80
discoideus waterbergensis Hesse, Bombylius, 80
discolor Meigen, Bombylius, 80
discolor Mikan, Bombylius, 80
discolor shelkovnikovi Paramonov, Bombylius, 80
discrepans Hesse, Systoechus, 84
discriminata Bezzi, Exoprosopa, 420
disjuncta Melander, Apolysis, 219
disjuncta Becker, Villa, 373
disjunctus Bezzi, Bombylius, 81
disjunctus Wiedemann, Lepidanthrax, 388
dispar Meigen, Bombylius, 80
dispar Coquillett, Chrysanthrax, 378
dispar Loew, Cytherea, 148
dispar Loew, Exoprosopa, 419
dispar interrupta Paramonov, Exoprosopa, 419
dispar Macquart, Geron, 207
disparilis Hesse, Amictogeron, 208
disparilis Hesse, Oestranthrax, 411
disparoides Paramonov, Cytherea, 148
disrupta Walker, Exoprosopa, 419
dissimilis Hesse, Corsomyza, 166
dissimilis Melander, Oligodranes, 220
dissimilis Hesse, Villa, 374
dissoluta Nurse, Exoprosopa, 421
dissors Hesse, Geron, 207
distigma Wiedemann, Anthrax, 20, 441, 442
distinctus Walker, Bombylius, 82
distinctus Melander, Oligodranes, 220
distinctus Meigen, Villa, 373

ditaenia Wiedemann, Villa, 372
diversa Coquillett, Phthiria, 199
diversus Coquillett, Lordotus, 114; fig. 48
diversus diplasus Hall, Lordotus, 114
diversus Williston, Sparnopolius, 124
dives Bigot, Bombylius, 82, 102
dives Bigot, Eucharimyia, 102
dives Walker, Exoprosopa, 419
divisa Coquillett, Exoprosopa, 419
divisus Cresson, Lordotus, 114
divisus Melander, Oligodranes, 221
divisus Walker, Villa, 372
divulsus Séguy, Systropus, 248
dizona Loew, Villa, 374
doctum Paramonov, Anthrax, 441
doddi Roberts, Systropus, 248
dodrans Osten Sacken, Exoprosopa, 419
dodrina Curran, Exoprosopa, 419
dolichoptera Bezzi, Exoprosopa, 420
doliops Hesse, Anthrax, 442
doliops fulviventris Hesse, Anthrax, 442
doliops gamka Hesse, Anthrax, 442
dolomatus Bowden, Anthrax, 442
dolorosa Williston, Phthiria, 199
dolorosus Melander, Oligodranes, 221
dolorosus Williston, Parabombylius, 97
dolorosus Williston, Systropus, 248
dolosus Hesse, Anastoechus, 86
dolosus Jaennicke, Villa, 373
dorcadion Osten Sacken, Exoprosopa, 419
dorcadion Osten Sacken, Exoprosopa, 419
doriae Bezzi, Villa, 373
doris Osten Sacken, Exoprosopa, 419
dorsalis Loew, Bombylius, 80
dorsalis Olivier, Bombylius, 80
dorsalis Walker, Comptosia, 353
dorsalis Walker, Villa, 372
doryca Boisduvall, Ligyra, 434
druias Melander, Apolysis, 219; figs. 81, 855, 856
dryitis Séguy, Toxophora, 236
dubia Macquart, Cytherea, 148
dubia Ricardo, Exoprosopa, 420
dubiosus Hesse, Geron, 207
dubium Zaitsev, Anthrax, 441
dubius Macquart, Anthrax, 442
ducens Walker, Comptosia, 353
dulcis Roberts, Bombylius, 82
dulcis Austen, Exoprosopa, 419
duncani Painter, Bombylius, 79
duodecimpunctatus Philippi, Villa, 372
durvillei Macquart, Villa, 372
duvaucelii Macquart, Anthrax, 442
dux Wiedemann, Exoprosopa, 420

E

ebneri Becker, Mariobezzia, 174
eboreus Painter, Bombylius, 79
ectophaeus Hesse, Litorrhynchus, 428
edititia Say, Chrysanthrax, 28, 378; fig. 179
edititoides Painter, Paravilla, 392; figs. 951, 952
edwardsi Hardy, Comptosia, 353
edwardsi Paramonov, Lyophlaeba, 356
edwardsi Brunetti, Systropus, 248
edwardsi Oldroyd, Villa, 372
edwardsii Coquillett, Hemipenthes, 385
adwardsii Loew, Paracosmus, 298
efflatounbeyi François, Acanthogeron, 90
efflatoun beyi Paramonov, Exoprosopa, 419

efflatouni Paramonov, Anthrax, 441
efflatouni Bezzi, Aphoebantus, 324
efflatouni Engel, Empidideicus, 275; fig. 107
efflatouni Bezzi, Exoprosopa, 419
efflatouni Venturi, Usia, 216
effrenus Coquillett, Poecilanthrax, 398; fig. 180
egerminans Loew, Phthiria, 199
ehrmanni Coquillett, Aldrichia, 161; figs. 65, 285, 482, 495
elbae Efflatoun, Usia, 216
electrica Hennig, Proglabellula, fossil, 59
elegans Coquillett, Amphicosmus, 295; fig. 114
elegans Paramonov, Anastoechus, 86
elegans Wiedemann, Bombylius, 81
elegans Paramonov, Cytherea, 148
elegans Hesse, Oligodranes, 221; figs. 338, 550, 553
elegans Wiedemann, Thyridanthrax, 406
elegantula Bigot, Acrophthalmyda, 176
elephantinus Séguy, Heterotropus, 226
elinguis Melander, Apystomyia, 230; figs. 85, 324, 599
elipsis Bezzi, Exoprosopa, 420
elongata Ricardo, Exoprosopa, 420
elongatus Rossi, Bombylius, 80
eluta Loew, Exoprosopa, 420
elutus Bezzi, Anthrax, 442
emarginata Macquart, Exoprosopa, 419
emarginatus Macquart, Villa, 374
emiliae Zaitsev, Geron, 206
emissus Walker, Villa, 374
emittens Walker, Villa, 374
emulata Painter, Paravilla, 392
enderleini Paramonov, Ligyra, 434
enderleini Paramonov, Ligyra, 434
engelhardti Painter, Heterostylum, 140
engeli Paramonov, Conophorus, 158
engeli Paramonov, Usia, 216
engyoptera Hesse, Exoprosopa, 420
enoria Melander, Mythicomyia, 269
epargyra Hermann, Toxophora, 236
epargyroides Hesse, Toxophora, 236
epheba Osten Sacken, Paravilla, 392
ephebus Osten Sacken, Villa, 372
epilais Wiedemann, Villa, 372
epolceus Séguy, Bombylius, 80
erebus Walker, Exoprosopa, 420
erectus Brunetti, Bombylius, 82
eremia Hesse, Lomatia, 311
eremica Melander, Mythicomyia, 269
eremicola Melander, Aphoebantus, 324
eremicus Painter and Hall, Poecilanthrax, 398
eremita Osten Sacken, Exoprosopa, 31, 419
eremitis Melander, Oligodranes, 221
eremobia Hesse, Corsomyza, 166
eremobia braunsii Hesse, Corsomyza, 166
eremobius Hesse, Anthrax, 442
eremochara Hesse, Exoprosopa, 420
eremophila Loew, Apolysis, 219
eremophilia Hesse, Pteraulax, 328
eremophilus Hesse, Systoechus, 84
erinaceus Bezzi, Anastoechus, 86
erinnis Loew, Lomatia, 311
erinnis erinnoides Paramonov, Lomatia, 311
erinnis obscuripennis Loew, Lomatia, 311
erinnoides Paramonov, Lomatia, 311
erivanensis Paramonov, Legnotomyia, 163
ermae Hall, Lordotus, 114

erronea Bezzi, Micomitra, 430
erythraeus Bezzi, Litorrhynchus, 428
erythraeus allothyris Bezzi, Litorrhynchus, 428
erythraspis Hermann, Aphoebantus, 324; fig. 213
erythrocephala Fabricius, Ligyra, 434
erythrocerus Bezzi, Bombylius, 81
erythrogaster Paramonov, Anthrax, 441
erythropus Bezzi, Geron, 206
erythrostoma Rondani, Chionamoeba, 344
escheri Bezzi, Aphoebantus, 324
etrusca Fabricius, Anthrax, 441; fig. 160A
eucnemis François, Villa, 374
eudora Coquillett, Chrysanthrax, 378
eulabiatus Bigot, Sisyromyia, 98
eumenes Osten Sacken, Hemipenthes, 385; figs. 942, 943
eumenoides Westwood, Systropus, 248
euplanes Loew, Anthrax, 441
eupogonatus Bigot, Systoechus, 84
eurhinatus Bezzi, Bombylius, 81
eurhinatus bechuanus Hesse, Bombylius, 81
europaea Loew, Stomylomyia, 335; figs. 131, 385
europaea aegyptiaca Bezzi, Stomylomyia, 335
europaea nigrirostris Bezzi, Stomylomyia, 335
eurrhinatus Bigot, Villa, 372
eurymetopa Hesse, Pteraulax, 328
eurypterus Hesse, Anthrax, 442
eurystephus Hesse, Anastoechus, 86
euzona Loew, Villa, 373
evanescens Becker, Exoprosopa, 419
evanida Bezzi, Ligyra, 434
eversmanni Paramonov, Bombylius, 80
exalbidus Wiedemann, Anastoechus, 86
exalbidus Meigen, Systoechus, 84
excisus Enderlein, Systropus, 248
excisus Walker, Villa, 372
exigua Macquart, Exoprosopa, 420
exiguus Walker, Bombylius, 80
exiguus Hesse, Systoechus, 84
exilipes Bezzi, Systoechus, 84
exilis Coquillett, Metacosmus, 296
exilis Philippi, Phthiria, 199
eximia Macquart, Bryodemina, 316
eximius Becker, Bombylodes, 92; figs. 5, 263, 438
exoprosopoides Paramonov, Exoprosopa, 419
expletus Loew, Bombylius, 80
exsuccus Séguy, Systropus, 248
extensa Walker, Comptosia, 353
extensa Wulp, Exoprosopa, 419
extraneus Hesse, Bombylius, 81
extremitis Coquillett, Paravilla, 392
extremitis Coquillett, Villa, 372
eyreana Paramonov, Ligyra, 434

F

facialis Cresson, Bombylius, 79
falcata, Ulosometa, new species, 361; figs. 431c, 929, 930
fallaciosa Loew, Exoprosopa, 419
fallax de Meijere, Anthrax, 442
fallax Austen, Bombylius, 80
fallax Green, Conophorus, 158

fallax Hesse, Phthiria, 199
fallax Austen, Villa, 373
falsus Hesse, Systoechus, 84
famula Bezzi, Micomitra, 430
farinosa Loew, Cytherea, 148
farinosa Becker, Exoprosopa, 419
farinosus Bezzi, Dischistus, 107
farinosus Johnson and Maughan, Oestranthrax, 411
fasciata Williston, Bryodemina, 316
fasciata Fabricius, Comptosia, 353
fasciata Johnson and Johnson, Dicranoclista, 444; figs. 1018, 1019, 1020
fasciata Dufour, Exoprosopa, 419
fasciata Macquart, Exoprosopa, 31, 419
fasciata Melander, Glabellula, 271
fasciata Paramonov, Ligyra, 434
fasciatus Johnson and Johnson, Poecilanthrax, 398
fasciatus Meigen, Villa, 373
fasciculata Loew, Lomatia, 311
fasciculata Villeneuve, Toxophora, 236
fasciculata Becker, Villa, 373
fasciculatus Macquart, Systoechus, 84
fascifrons Macquart, Sericosoma, 154
fasciolaris Walker, Lomatia, 311
fasciolata Melander, Mythicomyia, 269
fasciolus Coquillett, Oligodranes, 221
fascipenne Macquart, Callostoma, 14, 150; figs. 74, 76, 306
fascipenne palaestinae Paramonov, Callostoma, 150; figs. 511, 531
fascipennis Becker, Anastoechus, 86
fascipennis Macquart, Comptosia, 353; figs. 156, 209, 424, 735, 921, 923, 926
fascipennis Say, Exoprosopa, 31, 419
fascipennis albicollaris Painter, Exoprosopa, 419
fascipennis noctula Wiedemann, Exoprosopa, 419
fascipennis Bezzi, Legnotomyia, 31, 163
fascipennis Williston, Thevenemyia, 182
fasciventris Curran, in Twinn, Phthiria, 199
fasciventris Macquart, Villa, 373
fastidiosa Bezzi, Exoprosopa, 420
faustinus Osten Sacken, Villa, 372
faustus Hesse, Systoechus, 84
favillaceus Meigen, Bombylius, 80, 81
fedtschenkoi Loew, Antonia, 321
femoralis Bezzi, Bombylius, 81
femorata Loew, Glabellula, 271
femoratus Karsch, Systropus, 248
fenestralis Hesse, Bombylius, 81
fenestralis Wahlgren, Thyridanthrax, 406
fenestralis Macquart, Villa, 374
fenestralis Wiedemann, Villa, 372
fenestrata Loew, Cytherea, 148
fenestrata barbara Sack, Cytherea, 148
fenestrata Roberts, Pseudopenthes, 440; figs. 196, 393, 668, 676
fenestratoides Coquillett, Thyridanthrax, 406
fenestratus Osten Sacken, Conophorus, 158; figs. 70, 71
fenestratus Hull, Oestrimyza, n.sp., 412; figs. 415, 703
fenestratus Fallen, Thyridanthrax, 29, 406; figs. 377, 725, 958, 959
fenestratus Meigen, Thyridanthrax, 406
fenestratus montana Paramonov, Thyridanthrax, 406

fenestrellus Wiedemann, Villa, 372
fenestrulata Loew, Cytherea, 148
ferrea Walker, Ligyra, 434
ferreirae Hesse, Exoprosopa, 420
ferruginea Klug, Exoprosopa, 419
ferrugineum Fabricius, Heterostylum, 140
ferrugineus Macquart, Systoechus, 84
fervidus Walker, Villa, 374
festae Bezzi, Eclimus, 180
festivus Philippi, Villa, 372
filia Osten Sacken, Exoprosopa, 419
filicornis Hesse, Villa, 374
fimbriatella Bezzi, Exoprosopa, 420
fimbriatella furvalis Hesse, Exoprosopa, 420
fimbriatus Meigen, Bombylius, 80
fimbriatus expletus Loew, Bombylius, 80
fimbriatus ventralis Loew, Bombylius, 80
fimbriatus Hesse, Lepidochlanus, 119; figs. 43, 260, 464, 476, 761, 763
fimbriatus Meigen, Thyridanthrax, 406
firjuzanus Paramonov, Amictus, 286
firjuzanus Paramonov, Anastoechus, 86
firjuzanus Paramonov, Bombylius, 80
fissa Bigot, Villa, 372
flagrans Bezzi, Bombylius, 81
flammea Brunetti, Exoprosopa, 421
flammiger Walker, Villa, 374
flava Sack, Anthrax, 441
flava Paramonov, Exoprosopa, 419
flava Hardy, Phthiria, 199
flavalis Hesse, Villa, 374
flaveola Coquillett, Phthiria, 199
flaveolus Becker, Anastoechus, 86
flaveolus Macquart, Villa, 374
flavescens Sack, Anthrax, 441
flavescens J. Palm, Bombylius, 80
flavescens Meigen, Conophorus, 158
flavescens Loew, Villa, 374
flavibarbus Loew, Dischistus, 107
flavicans Bowden, Eurycarenus, 143
flavicans Bezzi, Exoprosopa, 420
flavicans Bowden, Petrorossia, 346
flavicapillis Bowden, Systoechus, 84
flaviceps Macquart, Bombylius, 81
flaviceps Loew, Poecilanthrax, 27, 398
flaviceps fuliginosus Loew, Poecilanthrax, 27, 398
flavicincta, Phthiria, 199
flavicincta Cole, Villa, 372
flavicornis Enderlein, Systropus, 248
flavicoxa Enderlein, Systropus, 248
flavida Melander, Mythicomyia, 269
flavifrons Melander, Empidideicus, 275
flavifrons Hesse, Lomatia, 311
flavifrons Becker, Plesiocera, 338
flavigenualis Hesse, Phthiria, 199
flavilabris Hesse, Plesiocera, 338
flavinervis Macquart, Exoprosopa, 419
flavipectus Enderlein, Systropus, 248
flavipennis Sack, Anthrax, 441
flavipennis Brunetti, Exoprosopa, 421
flavipes Roeder, Anthrax, 441
flavipes Wiedemann, Bombylius, 80
flavipes Hesse, Chasmoneura, 110
flavipes Cresson, Mythicomyia, 269
flavipes Efflatoun, Usia, 216
flavipes Loew, Villa, 374
flavipilosa Cole, Bombylius, 80
flavipilosa Cole, Paravilla, 392
flavipilosa Cole, Villa, 372

flavipleurus Bowden, Oligodranes, 221
flaviscutellaris Roberts, Cyrtomorpha, 17, 261
flavisetis Paramonov, Dischistus, 107
flaviventris Doleschall, Exoprosopa, 421
flaviventris Melander, Mythicomyia, 269
flavocostalis Painter, Villa, 372
flavofasciata Macquart, Ligyra, 434
flavo-ornatus Roberts, Systropus, 248
flavopilosa Hull, Pilosia, 138; figs. 28, 282, 512
flavorufa Strobl, Cyrtosia, 256
flavoscutellatus Paramonov, Heterotropus, 226
flavosericatus Hesse, Anastoechus, 86
flavosparsa Bigot, Ligyra, 434
flavospinosus Brunetti, Systoechus, 84
flavotomentosa Paramonov, Hemipenthes, 386
flavotomentosa de Meijere, Ligyra, 434
flavovillosus Roberts, Systoechus, 84
flavum Macquart, Heterostylum, 140
flavus Meigen, Bombylisoma, 108
flavus Macquart, Bombylius, 81
flavus Jaennicke, Lordotus, 114
flavus Paramonov, Oligodranes, 221; fig. 231
flavus Curtis, Villa, 373
flavus Meigen, Villa, 373
fletcheri Brunetti, Anthrax, 442
flexicornis Bezzi, Pteraulax, 328; figs. 625, 630
floccosus Loew, Bombylius, 80
flora Frey, Ligyra, 434
floralis Meigen, Bombylius, 80
floralis Coquillett, Phthiria, 199
florea Fabricius, Usia, 216; figs. 78, 769A
florea Loew, Usia, 216
florea Meigen, Usia, 216
florea Schiner, Usia, 216
floridana Macquart, Hemipenthes, 385
floridensis Macquart, Hemipenthes, 385
fodillus Séguy, Systropus, 248
foenoides Westwood, Systropus, 248
forcipata Brulle, Usia, 216
formosa Dufour, Anthrax, 441
formosa Melander, Mythicomyia, 269
formosana Paramonov, Ligyra, 434
formosanus Enderlein, Systropus, 248
formosula Bezzi, Exoprosopa, 420
formosus Roberts, Dischistus, 107
formosus Cresson, Oligodranes, 221
fornicata Loew, Plesiocera, 338
fracida Bowden, Anthrax, 442
fratellus Wiedemann, Bombylius, 80
fratellus Becker, Cytherea, 148
fratercula Bowden, Anthrax, 442
fraudulentus Johnson, Bombylius, 79
froggatti Paramonov, Docidomyia, 319
frontalia Melander, Mythicomyia, 269
frontalis Wiedemann, Chionamoeba, 344
frontalis Loew, Dischistus, 107
frontatus Philippi, Bombylius, 80
fruticicolus Hesse, Pantostomus, 293
fucatipennis Hesse, Lomatia, 311
fucatus Coquillett, Aphoebantus, 324
fucatus Bezzi, Bombylius, 81
fucoides Bezzi, Synthesia, 402; figs. 229, 430, 715, 726
fugax Wiedemann, Bombylius, 81
fulgida Melander, Mythicomyia, 269

fuligineipennis Hesse, Adelidea, 121
fuligineus Loew, Systoechus, 84
fuliginosus Wiedemann, Bombylius, 80
fuliginosus polypogon Loew, Bombylius, 80
fuliginosus rhodius Loew, Bombylius, 80
fuliginosus tavrizi Paramonov, Bombylius, 80
fuliginosus Wiedemann, Conophorus, 158; figs. 284, 484, 490
fuliginosus Loew, Poecilanthrax, 398
fulva Hesse, Lomatia, 311
fulva Meigen, Phthiria, 199
fulva Gray, Toxophora, 235
fulvago Philippi, Villa, 372
fulvastra Bowden, Anthrax, 442
fulvescens Becker, Anastoechus, 86
fulvescens Wiedemann, Bombylius, 80
fulvescens candidus Loew, Bombylius, 80
fulvianus Say, Villa, 26, 372
fulvianus nigricauda Loew, Villa, 372
fulvibasis Macquart, Bombylius, 80
fulvibasoides Painter, Bombylius, 79
fulviceps Macquart, Bombylius, 82
fulviceps Bezzi, Megapalpus, 172
fulvicoma Coquillett, Paravilla, 392
fulvida Coquillett, Phthiria, 199
fulvifacies Austen, Thyridanthrax, 406
fulvimaculatus Abreu, Villa, 373
fulviops Szilady, Exoprosopa, 420
fulvipeda Rondani, Villa, 372
fulvipennis Tucker, Anastoechus, 86
fulvipes Loew, Anthrax, 442
fulvipes Bigot, Bombylius, 82
fulvipes Villers, Bombylius, 80
fulvipes Bigot, Comptosia, 353
fulvipes Bigot, Comptosia, 353
fulvipes Séguy, Heterotropus, 226
fulvipes Loew, Petrorossia, 346
fulvipes Coquillett, Villa, 372
fulvipleura Hesse, Lomatia, 311
fulvipleura Hesse, Villa, 374
fulviventris Hesse, Anthrax, 442
fulvohirta Wiedemann, Chrysanthrax, 378
fulvonotatus Wiedemann, Bombylius, 81
fulvosetosus Hesse, Bombylius, 81
fulvula Wiedemann, Anthrax, 442
fulvus Wiedemann, Sparnopolius, 124
fulvus Meigen, Systoechus, 84
fumalis Hesse, Apolysis, 219
fumea Roberts, Villa, 374
fumicosta Painter, Villa, 372
fumida Coquillett, Paravilla, 392
fumida Coquillett, Villa, 372
fumidus Coquillett, Aphoebantus, 324
fumiflamma Walker, Poecilanthrax, 398
fuminervis Dufour, Conophorus, 158
fumipennis Schiner, Conophorus, 158
fumipennis Loew, Oligodranes, 221
fumipennis Hesse, Petrorossia, 346
fumipennis Painter, Systoechus, 84
fumipennis Westwood, Systropus, 18, 248
fumipennis Wiedemann, Villa, 374
fumitinctus Hesse, Systoechus, 84
fumosa Hesse, Anthrax, 442
fumosa Cresson, Exoprosopa, 419
fumosus Coquillett, Aphoebantus, 324
fumosus Dufour, Bombylius, 80
fumosus Hesse, Systropus, 248
funebris Paramonov, Amictus, 286
funebris Macquart, Villa, 372
funereus Costa, A., Systropus, 248

funesta Walker, Ligyra, 434
funestus Osten Sacken, Thevenemyia, 182
fur Osten Sacken, Anthrax, 20, 440
furcatus Enderlein, Systropus, 248
furcifer Hesse, Geron, 207
furiosus Walker, Bombylius, 81
furvalis Hesse, Exoprosopa, 420
furvicostatus Roberts, Thevenemyia, 182
furvipennis Hesse, Exoprosopa, 420
furvum Edwards, Sericosoma, 154
furvus Hesse, Anthrax, 442
fusca Cockerell, Alomatia, fossil, 59
fusca Séguy, Cyrtosia, 256
fuscanus Macquart, Bombylius, 82
fuscianulatus Hesse, Anastoechus, 86
fuscicosta Bezzi, Petrorossia, 346
fuscicostatus Macquart, Villa, 374
fuscilobus Bezzi, Bombylius, 81
fuscipennis Macquart, Adelidea, 121
fuscipennis Becker, Anastoechus, 86
fuscipennis Ricardo, Anthrax, 442
fuscipennis Macquart, Conophorus, 158
fuscipennis Macquart, Corsomyza, 166
fuscipennis Macquart, Exoprosopa, 421
fuscipennis Hesse, Gnumyia, 169
fuscipennis Portschinsky, Lomatia, 311
fuscipennis Macquart, Poecilanthrax, 398
fuscipennis Macquart, Toxophora, 236
fuscipes Hesse, Amictogeron, 208
fuscipes Bigot, Sparnopolius, 124
fusciventris Hesse, Systoechus, 84
fusciventris Loew, Villa, 373
fusconotata Becker, Exoprosopa, 419
fuscopurpuratus Hesse, Anthrax, 442
fuscula Bezzi, Exoprosopa, 420
fuscus Paramonov, Anastoechus, 86
fuscus Fabricius, Bombylius, 80
fuscus Thunberg, Bombylius, 82
fuscus Fabricius, Villa, 372
fusicornis Macquart, Acreotrichus, 201
fusicornis Hesse, Corsomyza, 166

G

gabbroensis Handlirsch, Hemipenthes, fossil, 59
gaedei Wiedemann in Meigen, Phthiria, 199; fig. 100
gagathea Bigot, Usia, 216
galathea Osten Sacken, Villa, 372
galbea Melander, Mythicomyia, 269
gallus Loew, Thyridanthrax, 406
gamka Hesse, Anthrax, 442
garagniae Efflatoun, Geron, 206
gargantua Knab, Ligyra, 434
gariepina Hesse, Villa, 374
gariepinus Hesse, Geron, 207
gaudanicus Paramonov, Hemipenthes, 386
gausa Melander, Mythicomyia, 269
gayi Macquart, Villa, 372
gazophylax Loew, Ligyra, 434
gebeli Efflatoun, Amictus, 286
gebleri Loew, Ligyra, 434
geijskesi Curran, Systropus, 248
gemella Coquillett, Villa, 372
gemina Hardy, Comptosia, 353
gemmeus Bezzi, Dischistus, 107
gentilis Brunetti, Anthrax, 442
gentilis Bezzi, Exoprosopa, 420
geometrica Macquart, Comptosia, 353
georgicus Macquart, Anthrax, 440
germari Wiedemann, Exoprosopa, 420

gestroi Brunetti, Anthrax, 442
gibba Melander, Mythicomyia, 269
gibbera Melander, Mythicomyia, 269
gibbicornis Macquart, Acroetrichus, 201; figs. 98, 315, 534, 536
gibbicornis Bezzi, Dischistus, 107
gibbiventris Bezzi, Pantostomus, 293
gibbosa Walker, Phthiria, 199
gibbosus Meigen, Geron, 206
gibbosus Olivier, Geron, 18, 206; fig. 105
gibbosus erythropus Bezzi, Geron, 206
gibbosus halteralis Wiedemann, Geron, 206
gibbosus subflavofemoratus Rubio, Geron, 206
gibbus Loew, Lordotus, 114; figs. 45, 290, 575, 690, 807, 808, 809
gibbus striatus Painter, Lordotus, 114
gideon Fabricius, Anthrax, 441
gideon Macquart, Anthrax, 441
gideon, authors, not Fabricius, Hemipenthes, 386
gigantea Bezzi, Lomatia, 311
giganteus Villeneuve, Dischistus, 107
glabellula Strobl, Systoechus, 84
glauca Melander, Apolysis, 219
glaucescens Hesse, Anthrax, 442
glaucescens Loew, Conophorus, 158
glauciella Hesse, Lomatia, 311
glaucus Becker, Heterotropus, 226
globiceps Loew, Cyllenia, 283
globulus Bezzi, Bombylius, 81
gloriosa Walker, Ligyra, 434
glossalis Cockerell, Melanderella, fossil, 59
gluteatus Melander, Aphoebantus, 324
gnatus Walker, Villa, 374
goliath Bezzi, Exoprosopa, 420
goliath Oldroyd, Oestranthrax, 29, 441
goliath Bezzi, Systoechus, 84
gomez menori Rubio, Systoechus, 84
gonioneura Hesse, Exoprosopa, 420
gonucera Hesse, Corsomyza, 166
gorgon Fabricius, Villa, 372
goyaz Macquart, Bombylius, 80
gracilipes Becker, Bombylius, 80
gracilis Williston, Dolichomyia, 250; figs. 31, 93, 335, 579, 589, 881, 882, 888
gracilis Loew, Eclimus, 180; figs. 258, 454, 572
gracilis festae Bezzi, Eclimus, 180
gracilis Melander, Mythicomyia, 269
gracilis Walker, Phthiria, 199
gracilis Enderlein, Systropus, 248
gracilis Hesse, Systropus, 248
gracilis Tonnoir, Tillyardomyia, 183; figs. 17, 250, 440, 763
gracilis Macquart, Villa, 372
gracilis Bowden, Xeramoeba, 342; figs. 128, 655
gracilis Giebel, Xeramoeba, fossil, 59
gradatus Wiedemann, Systoechus, 9, 84
gradatus gallicus Villeneuve, Systoechus, 84
gradatus lucidus, Loew, Systoechus, 84
gradatus tesquorum Becker, Systoechus, 84
gradatus validus Bezzi, Systoechus, 84
gradatus Macquart, Villa, 372
graeca Paramonov, Lomatia, 311
graeca erinnoides Paramonov, Lomatia, 311
graecus Paramonov, Thyridanthrax, 406
grahami Hesse, Lomatia, 311
grajugena Loew, Lomatia, 311

grandis Efflatoun, Acanthogeron, 90
grandis Schiner, Comptosia, 353
grandis Wiedemann, Exoprosopa, 419
grandis Painter, Geron, 206
grandis Paramonov, Systoechus, 84
grata Coquillett, Exoprosopa, 419
grata Loew, Usia, 216
gratiosa Loew, Lomatia, 311
gratiosa Bezzi, Petrorossia, 346
greeni Austen, Conophorus, 158
griqua Hesse, Exoprosopa, 420
grisea Paramonov, Tamerlania, 193
grisea Paramonov, Usia, 217
grisea Efflatoun, Usia, 216
grisealis Hesse, Lomatia, 311
griseata Hull, Cacoplox, 131; figs. 44, 276, 489, 515
griseifrons Hesse, Thyridanthrax, 406
griseipennis Macquart, Exoprosopa, 419
griseohirta Nurse, Mariobezzia, 174; figs. 291, 488, 496
griseohirta aegyptiaca Engel, Mariobezzia, 174
griseolus Klug, Thyridanthrax, 406
grisescens Paramonov, Anthrax, 441
griseus Fabricius, Conophorus, 158
griseus Paramonov, Usia, 217
grossbecki Johnson, Anthrax, 440
guerini Macquart, Ligyra, 434
guianica Curran, Anthrax, 441
guillarmodi Hesse, Exoprosopa, 420
guillarmodi Hesse, Tomomyza, 292
gujaratica Nurse, Exoprosopa, 421
gulperii Efflatoun, Cyrtosia, 256
gussakovskiji Paramonov, Heterotropus, 226
guttatipennis Brunetti, Anthrax, 442
guttatipennis Macquart, Bombylius, 82
guttipennis Bezzi, Exoprosopa, 419

H

habra Melander, Mythicomyia, 269
haemorhoidalis Bezzi, Bombylius, 81
haemorhoidalis waterbergenisis Hesse, Bombylius, 81
haemorrhoicum Loew, Heterostylum, 140
haesitans Becker, Villa, 373
hafniensis Hennig, Glaesamictus, fossil, 59
halli Hull, Thevenemyia, 182
halteralis Weidemann, Geron, 206
halteralis Hesse, Lomatia, 311
halteralis Kowarz, Villa, 373
halteratus Melander, Aphoebantus, 324
halycon Wiedemann, Poecilanthrax, 398
hamata Macquart, Exoprosopa, 420
hamata Austen, Lomatia, 311
hamata Melander, Mythicomyia, 269
hamifera Loew, Hemipenthes, 386
hamifera Becker, Lomatia, 19, 311
hamilkar Paramonov, Conophorus, 158
hamula Hesse, Exoprosopa, 420
handfordi Curran Villa, 26, 372; fig. 3
hannibal Paramonov, Conophorus, 158
harpyia Wiedemann, Exoprosopa, 419
harrisi Johnson, Thevenemyia, 182
harrisi Osten Sacken, Thevenemyia, 182
harroyi François, Villa, 374
harveyi Hine, Villa, 372
hassani Paramonov, Anthrax, 441
hastaticornis Hesse, Callynthrophora, 168

haustellata Bezzi, Exoprosopa, 420
haywardi Edwards, Bombylius, 80
haywardi Edwards, Lyophlaeba, 356
hecate Loew, Lomatia, 311
hecate Meigen, Lomatia, 311
hedickei Paramonov, Bryodemina, 316
hela Erichson, Ligyra, 434
helena Loew, Ligyra, 434
helenae Paramonov, Anthrax, 441
helvus Wiedemann, Bombylius, 80
hemichroa Hesse, Lomatia, 311
hemifusca Roberts, Ligyra, 434
hemiphaea Hesse, Exoprosopa, 420
hemiteles Schiner, Comptosia, 353
henicoides Hesse, Nomalonia, 290
hermanni Becker, Antonia, 321
heros Wiedemann, Exoprosopa, 420
heros litoralis Hesse, Exoprosopa, 420
heros melanthia Hesse, Exoprosopa, 420
heros protuberans Bezzi, Exoprosopa, 420
herzi Portschinsky, Exoprosopa, 419
hesperidum Frey, Geron, 206
hesperus Rossi, Petrorossia, 346; fig. 134
hesperus tropicalis Bezzi, Petrorossia, 346
hessei Hall, Anastoechus, 86
hessei François, Systropus, 248
hessii Wiedemann, Anthrax, 442
heterocera Bezzi, Exoprosopa, 420
heterocerus Macquart, Dischistus, 107
heterocoma Hesse, Lomatia, 311
heteroneurus Macquart, Triploechus, 141; figs. 53, 304, 524, 551, 843, 844, 845
heteropilosus Timon-David, Conophorus, 158
heteropogon Bowden, Systoechus, 84
heteropterus Macquart, Amictus, 286
heteropterus Macquart, Dischistus, 107
heteropterus Bezzi, Pseudoamictus, 211
heteropterus Wiedemann, Pseudoamictus, 211; figs. 103, 311, 548, 555, 862, 863, 864
heteropterus Paramonov, Thyridanthrax, 406
heteropyga Sack, Anthrax, 441
hetrusca Fabricius, Anthrax, 441
hians Melander, Aphoebantus, 324
hiata Melander, Mythicomyia, 269
hilarii Macquart, Villa, 372
hilaris Walker, Bombylius, 82
hilaris Eversman, Exoprosopa, 420
hilaris White, Geron, 207
hilaris Walker, Phthiria, 199
hiltoni Priddy, Conophorus, 158
himalayensis Brunetti, Anthrax, 442
hindlei Paramonov, Conophorus, 158
hippolyta Wiedemann, Anthrax, 441
hircanus Wiedemann, Anastoechus, 86
hircina Coquillett, Chrysanthrax, 378
hircinus Coquillett, Villa, 372
hirsutus Coquillett, Aphoebantus, 19, 324
hirsutus Villers, Hemipenthes, 386
hirsutus Williston, Villa, 372
hirta Loew, Thevenemyia, 182
hirticeps Bezzi, Acanthogeron, 90
hirticeps Bezzi, Dischistus, 107
hirticeps karooensis Hesse, Dischistus, 107
hirticornis Latreille, Conophorus, 158
hirtipes Macquart, Corsomyza, 166
hirtipes Loew, Exoprosopa, 420
hirtus Loew, Bombylius, 81
hirtus Bezzi, Dischistus, 107

hispanus Loew, Thyridanthrax, 406
histrio Walker, Heterostylum, 140
hohlbecki Paramonov, Heterotropus, 226
holaspis Speiser, Systropus, 248
hololeucus Loew, Bombylius, 81
holosericea Fabricius, Cytherea, 15, 148
holosericea Wiedemann, Cytherea, 148
holosericeus Wiedemann, Bombylius, 81
holosericeus Walker, Geron, 206
holosericeus Meigen, Hemipenthes, 386
homeyeri Karsch, Sisyrophanus, 88
homogeneus Bezzi, Anthrax, 442
hoplites Melander, Mythicomyia, 269
hoppo Matsumura, Systropus, 248
hormatha Melander, Mythicomyia, 269
horni Paramonov, Bombylius, 81
horni Hesse, Chasmoneura, 110
hospes Melander, Caenotus, 227; figs. 108A, 574, 697, 867, 868
hottentotta Jaennicke, Villa, 373
hottentotta Linné, Villa, 26, 373
hottentotta modestus Meigen, Villa, 373
hottentottus Schellenberg, Villa, 373
hottentottus Walker, Villa, 373
hottentotus Hesse, Bombylius, 81
hulli Painter, Exoprosopa, 419
humeralis Abreu, Cyrtosia, 256
humile Roberts, Myonema, 318; figs. 135, 140, 388, 629, 757
humilis Loew, Apolysis, 219
humilis Melander, Empidideicus, 275; fig. 108
humilis Osten Sacken, Phthiria, 199
humilis Ruthe, Villa, 373
hummeli Paramonov, Anastoechus, 86
hungaricus Thalhammer, Empidideicus, 275
hurdi Hall, Lordotus, 114
hyalacrus Wiedemann, Villa, 372, 374
hyalinipennis Painter and Hall, Poecilanthrax, 398
hyalinipennis Blanchard, Villa, 372
hyalinus Fabricius, Bombylius, 80
hyalinus Wiedemann, Villa, 374
hyalipennis Macquart, Bombylius, 80
hyalipennis Cole, Exoprosopa, 419
hyalipennis Paramonov, Exoprosopa, 419
hyalipennis Cole, Lepidanthrax, 388
hyalipennis Séguy, Oligodranes, 221
hyalipennis Macquart, Usia, 216
hyalodisca Bezzi, Exoprosopa, 420
hyaloptera Hesse, Exoprosopa, 420
hybos Melander, Mythicomyia, 269
hybrida Hesse, Villa, 374
hybridus Hesse, Anthrax, 442
hybridus Bezzi, Geron, 207
hybridus Meigen, Geron, 206
hybus Coquillett, Geron, 206
hylesia Hesse, Lomatia, 311
hypargira Bezzi, Exoprosopa, 420
hypargyroides Hesse, Exoprosopa, 420
hypoleuca Wiedemann, Phthiria, 199
hypoleucus Wiedemann, Bombylius, 81
hypomelaena Bezzi, Exoprosopa, 420
hypomelaena Macquart, Villa, 372
hypomelas Macquart, Villa, 372
hypoxantha Macquart, Thyridanthrax, 406
hypoxanthus Loew, Bombylius, 81
hyrcanus Pallas, Anastoechus, 86
hyrcanus aralicus Paramonov, Anastoechus, 86

I

ichneumoniformis Hesse, Systropus, 248
icteroglaenus Hesse, Bombylius, 81
idolus Hesse, Thyridanthrax, 406
ignava Loew, Exoprosopa, 420
ignea Macquart, Villa, 372
ignifera Walker, Villa, 374
ignorata Becker, Usia, 216
illatus Walker, Villa, 374
illustris Melander, Mythicomyia, 269
imbecillus Karsch, Systropus, 248
imbellis Melander, Mythicomyia, 269
imbuta Walker, Ligyra, 434
imbutata Hesse, Petrorossia, 346
imbutatus Hesse, Doliogethes, 111
imitans Schiner, Anthrax, 441
imitans Schiner, Villa, 372
imitata Hesse, Nomalonia, 290
imitator Hesse, Bombylius, 81
imitator Loew, Dischistus, 107
immaculata Bezzi, Adelidea, 121
immaculata Bowden, Anthrax, 442
immaculatum Bezzi, Crocidium, 185
immutatus Walker, Bombylius, 82
imperator Nurse, Callostoma, 150
impiger Coquillett, Chrysanthrax, 378
impurus Loew, Bombylius, 81
inaequalipes Loew, Exoprosopa, 420
inaequalis Becker, Plesiocera, 338
inappendiculatum Bigot, Anthrax, 441
inappendiculatus Bigot, Acreotrichus, 201
inaurata Klug, Thyridanthrax, 406
inauratus Coquillett, Lepidanthrax, 388
incanus Johnson, Bombylius, 79
incanus Klug, Thyridanthrax, 406
incipiens Bezzi, Thyridanthrax, 406
incisa Walker, Hemipenthes, 385
incisa Becker, Phthiria, 199
incisa Wiedemann, Usia, 216, 217
incisiva Painter, Hemipenthes, 385
incisuralis Macquart, Anthrax, 442
incisuralis ardicola Hesse, Anthrax, 442
incisuralis fumosa Hesse, Anthrax, 442
incisuralis glaucescens Hesse, Anthrax, 442
incisus Macquart, Villa, 374
incitus Paramonov, Anthrax, 441
inclusa Walker, Comptosia, 353
incognita Paramonov, Usia, 216
incomptus Walker, Villa, 374
inconspicua Becker, Phthiria, 199
inconspicuus, Loew, Thyridanthrax, 406
inconstans Becker, Villa, 373
incultus Coquillett, Villa, 372
indecisa Walker, Exoprosopa, 420
indecisa Curran, Lepidanthrax, 388
indecora Wulp, Comptosia, 354
indicata Nurse, Anthrax, 442
indicata Melander, Mythicomyia, 269
indicus Nurse, Heterotropus, 226; figs. 397, 837, 838, 839, 976
indigena Becker, Anthrax, 441
indigenus Becker, Thyridanthrax, 406
indogatus Séguy, Systropus, 248
inermis Coquillett, Aphoebantus, 324
inermis Bezzi, Atrichochira, 423
inermis Hesse, Bombylius, 81
inexacta Walker, Villa, 372
infernalis Loew, Lomatia, 311
infernalis Schiner, Lomatia, 311
infumata Bezzi, Exoprosopa, 420

infumata Philippi, Lyophlaeba, 356
infuscata Meigen, Cytherea, 15, 148
infuscata Bezzi, Lomatia, 311
infuscatus Bezzi, Litorrhynchus, 428
infuscatus Karsch, Systropus, 248
ingens Cresson, Exoprosopa, 419; figs. 971, 972
ingens Johnson and Johnson, Poecilanthrax, 398
inglorius Philippi, Villa, 372
injii Efflatoun, Cyrtosia, 256
innocens Austen, Thyridanthrax, 406
innocuus Bezzi, Anastoechus, 86
innominata Austen, Lomatia, 311
inops Coquillett, Hemipenthes, 385
inordinatus Hesse, Systoechus, 84
inordinatus Rondani, Villa, 372
inornata Loew, Exoprosopa, 420
inornata Loew, Lomatia, 311
inornata Coquillett, Phthiria, 199
inornata Engel, Usia, 216
inornatus Walker, Bombylius, 82
inornatus Cole, Caenotus, 227
inquieta Becker, Villa, 373
inquinita Roberts, Ligyra, 434
insignis Loew, Amictus, 286
insignis Walker, Comptosia, 354
insignis Macquart, Ligyra, 434
insignis Melander, Mythicomyia, 269
insignis Austen, Villa, 373
insolens Coquillett, Paracosmus, 298; fig. 15
instabilis Melander, Oligodranes, 221
institutus Walker, Villa, 374
insularis Rondani, Amictus, 286
insularis Frey, Empidideicus, 275
insularis Ricardo, Exoprosopa, 419
insularis Bigot, Geron, 206
insularis Cole, Geron, 206
insularis Walker, Villa, 374
insulata Walker, Exoprosopa, 421
integrum Bezzi, Prorostoma, 330; figs. 130, 389, 622, 640
interlitus Séguy, Systropus, 248
intermedia Brunetti, Anthrax, 442
intermedia Melander, Mythicomyia, 269
intermedius Becker, Anastoechus, 86
intermedius Hesse, Anthrax, 442
intermedius Walker, Bombylius, 80
intermedius Paramonov, Hemipenthes, 386
intermedius Paramonov, Thyridanthrax, 406
interrupta Mulsant, Exoprosopa, 420
interrupta Paramonov, Exoprosopa, 419
interrupta Wiedemann, Exoprosopa, 419
interruptus Coquillett, Aphoebantus, 324
interruptus Painter, Poecilanthrax, 398
interstitialis Nurse, Exoprosopa, 421
intonsus Bezzi, Geron, 206
introrsa Melander, Mythicomyia, 269
inversus Melander, Aphoebantus, 324
io Williston, Bombylius, 80
iopterus Wiedemann, Villa, 372
iota Osten Sacken, Exoprosopa, 419
iranicus Paramonov, Bombylius, 80
iranicus Paramonov, Conophorus, 158
iridipennis Hesse, Exoprosopa, 421
iridipennis Nurse, Micomitra, 430
iris Szilady, Dischistus, 107
iris Loew, Micomitra, 430
irrorata Macquart, Anthrax, 441
irroratus Say, Anthrax, 20, 440
irrorellus Klug, Anthrax, 441
irrorellus Klug, Thyridanthrax, 406
irrupta Melander, Mythicomyia, 269
irvingi Hesse, Gonarthrus, 112
isis Meigen, Anthrax, 20, 441; fig. 399
italica Rossi, Exoprosopa, 419
italica Wiedemann, Exoprosopa, 419
italica megaera Wiedemann, Exoprosopa, 419
ixion Megerle, Anthrax, 441
ixion Fabricius, Villa, 373

J

jacchoides Bezzi, Exoprosopa, 420
jacchoides Bezzi, Exoprosopa, 421
jacchus Fabricius, Exoprosopa, 419; figs. 198, 680, 747
jacchus baccha Loew, Exoprosopa, 419
jacchus maenas Loew, Exoprosopa, 419
jacchus quadripunctata Paramonov, Exoprosopa, 419
jaennickeana Osten Sacken, Hemipenthes, 386
jansei Hesse, Lomatia, 311
javana Macquart, Exoprosopa, 421
javana Wiedemann, Toxophora, 236
jazykovi Paramonov, Anthrax, 20, 441, jeanneli Séguy, Cyrtosia, 256
jezoensis Matsumara, Anthrax, 441
jocchus Fabricius, Exoprosopa, 419
johnsoni Coquillett, Exepacmus, 339; figs. 7, 293
johnsoni Painter, Geron, 206
johnsoni Painter, Villa, 372
johnsonorum Painter and Hall, Poecilanthrax, 398
jonesi Cresson, Exoprosopa, 419
jubatipes Hesse, Exoprosopa, 420
junceus Coquillett, Lordotus, 114
junctura Coquillett, Chrysanthrax, 378
junta Curran, Exoprosopa, 419

K

kalaharica Hesse, Lomatia, 311
kalaharicus Hesse, Gonarthrus, 112
kalaharicus venustus Hesse, Gonarthrus, 112
kalaharicus Hesse, Systoechus, 84
kaokoana Hesse, Lomatia, 311
kaokoensis Hesse, Anthrax, 442
kaokoensis Hesse, Bombylius, 81
kaokoensis Hesse, Chasmoneura, 110
kaokoensis Hesse, Exoprosopa, 420
kaokoensis Hesse, Heteralonia, 425; fig. 194
kaokoensis Hesse, Litorrhynchus, 428
kaokoensis Hesse, Villa, 374
karasana Hesse, Villa, 374
karasanus Hesse, Bombylius, 81
karavaievi Paramonov, Anthrax, 441
karavajevi Paramonov, Oestranthrax, 411; fig. 230
karelini Paramonov, Bombylius, 80
karooana Hesse, Corsomyza, 166
karooana Hesse, Exoprosopa, 420
karooana Hesse, Petrorossia, 346
karooana Hesse, Tomomyza, 292
karooanum Hesse, Crocidium, 185
karooanum Hesse, Onchopelma, 257
karooanus Hesse, Amictogeron, 209
karooensis Hesse, Bombylius, 81
karooensis Hesse, Dischistus, 107
karooensis Hesse, Pteraulacodes, 329
karooensis Hesee, Villa, 374
karooensis Hesse, Zinnomyia, 95; figs. 3, 242, 449, 463
karoonana Hesse, Tomomyza, 292
kassanovskiji Paramonov, Platypygus, 262
kaupii Jaennicke, Ligyra, 434
kazanovskyi Paramonov, Heterotropus, 226
keenii Coquillett, Stonyx, 402
killimandjaricus Speiser, Bombylius, 81
kirgizorum Paramonov, Exoprosopa, 419, 420
kiritshenkoi Paramonov, Anthrax, 441
klugii Wiedemann, Ligyra, 434
knabi Cresson, Oligodranes, 221
kneuckeri Becker, Sinaia, 287; figs. 212, 604, 617, 638, 759, 973, 974, 975
koreanus Paramonov, Bombylius, 80
koshunensis Matsumura, Anthrax, 441
koslowskyi Edwards, Ligyra, 434
koslowskyi Edwards, Lyophlaeba, 356
kozlovi Paramonov, Anthrax, 441
kozlovi Paramonov, Bombylius, 80
kozlovi Paramonov, Conophorus, 158
krugeri Bezzi, Exoprosopa, 420
krymensis Paramonov, Geron, 206
kurdorum Paramonov, Platypygus, 262
kurdorum persicus Paramonov, Platypygus, 262
kushkaensis Paramonov, Thyridanthrax, 406
kutshurganicus Paramonov, Bombylius, 80

L

labiosus Melander, Epacmus, 326
labiosus Hesse, Gonarthrus, 112
lacerata Engel, Exoprosopa, 419
lacerus Wiedemann, Stonyx, 402
lachesis Egger, Lomatia, 311
lachesis graeca Paramonov, Lomatia, 311
lactea Paramonov, Exoprosopa, 420
lacteipennis Strobl, Phthiria, 199
lactipenne Hesse, Crocidium, 185
lactipennis Hesse, Geron, 207
lacunaris Coquillett, Paravilla, 392
lacus Bowden, Systoechus, 84
laeta Loew, Exoprosopa, 420
laeta Bezzi, Phthiria, 199
laeta xerophiles Hesse, Phthiria, 199
laetus Paramonov, Anthrax, 441
laevifrons Loew, Systoechus, 84
laevipennis Paramonov, Anthrax, 441
laevis Bigot, Cyllenia, 283
laevis Becker, Villa, 373
lanata Edwards, Mallophthiria, 186; figs. 341, 522, 532
lanata Edwards, Sericusia, 124; figs. 30, 267, 491, 517, 767
lanatus Bowden, Eurycarenus, 143
lancifer Osten Sacken, Bombylius, 79
landbecki Philippi, Bombylius, 80
landbecki Philippi, Lyophlaeba, 356
lanei d'Andretta and Carrera, Systropus, 248
laniger Cresson, Thevenemyia, 182
lanigera Bezzi, Phthiria, 199
lanigerus Fourcroy, Bombylius, 80

laqueatus Enderlein, Systropus, 248
lar Fabricius, Litorrhynchus, 428; figs. 191, 408, 687, 689
larrea Marston, Anthrax, 440
lasiocornis Hesse, Amictogeron, 209; figs. 530, 537
lasius Melander, Oligodranes, 221
lasius Wiedemann, Villa, 374
lassenensis Johnson and Johnson, Bombylius, 79
lata Loew, Usia, 216
latelimbata Bigot, Exoprosopa, 421
latelimbatus Bigot, Anthrax, 440
lateralis Fabricius, Bombylius, 81
lateralis Newman, Comptosia, 353
lateralis Rondani, Cytherea, 148
lateralis Brunetti, Exoprosopa, 421
lateralis Meigen, Lomatia, 311
lateralis atropos Egger, Lomatia, 311
lateralis Say, Villa, 372
lateralis Thomson, Villa, 372
lateralis arenicola Johnson, Villa, 372
lateralis atra Painter, Villa, 372
lateralis johnsoni Painter, Villa, 372
lateralis nigra Cresson, Villa, 372
lateralis semifulvipes Painter, Villa, 372
laticeps Loew, Eurycarenus, 143; figs. 51, 299, 525, 533
laticeps Bigot, Heterostylum, 140
laticlavia Melander, Mythicomyia, 269
latifacies Hesse, Pteraulax, 328
latifascia Walker, Villa, 373
latifimbria Walker, Villa, 372
latifrons Loew, Amictus, 286
latifrons Macquart, Anastoechus, 86
latifrons Paramonov, Aphoebantus, 324
latifrons Paramonov, Cytherea, 148
latifrons Loew, Eurycarenus, 143
latifrons Bezzi, Exoprosopa, 420
latifrons Hesse, Geron, 207
latifrons Hesse, Lomatia, 311
latifrons Bezzi, Petrorossia, 346
latipectus Hesse, Bombylius, 81
latipennis Brunetti, Colossoptera, 431; figs. 201, 418, 714, 969, 970
latipennis Brunetti, Exoprosopa, 421
latipennis Paramonov, Hemipenthes, 386
latipennis Paramonov, Thyridanthrax, 406
latisoma, new species, Bryodemina, 316
latissima Bezzi, Micomitra, 430
latiuscula Loew, Exoprosopa, 419
latiuscula Loew, Lomatia, 311
lativentris Loew, Platypygus, 262
latona Wiedemann, Thyridanthrax, 406
latreillii Wiedemann, Ligyra, 434
latus Dufour in Verrall, Conophorus, 158
lautus Coquillett, Lepidanthrax, 388
lawrencei Hesse, Lomatia, 311
lebedevi Paramonov, Toxophora, 236
ledereri Loew, Proracthes, 162
leechi Hall, Thevenemyia, 182
lejostomus Loew, Bombylius, 80
lelia Williston, Villa, 372
lemimelas Speiser, Anthrax, 442
lemniscatus Philippi, Villa, 372
lenticularis Melander, Mythicomyia, 269
leonina Bigot, Stomylomyia, 335
leoninus Bowden, Systoechus, 84
lepida Hermann, Chionamoeba, 344
lepida Austen, Lomatia, 311
lepidiota Roberts, Anthrax, 442

lepidocera Wiedemann, Lepidophora, 16, 238; figs. 90, 318, 457, 540, 869, 870, 871
lepidocera d'Andretta and Carrera, Toxophora, 236
lepidogastra Bezzi, Exoprosopa, 420
lepidota Osten Sacken, Hemipenthes, 386; fig. 183
lepidota (not Osten Sacken), Chrysanthrax, 378
lepidotoides Johnson, Chrysanthrax, 378
lepidotus Osten Sacken, Villa, 372
lepidulus Austen, Thyridanthrax, 406
lepidus Loew, Dischistus, 107
lepidus Bowden, Geron, 207
lepidus Roberts, Neosardus, 301
leptocerus Bezzi, Amictogeron, 209
leptocerus Bezzi, Sisyrophanus, 88
leptogaster Loew, Systropus, 248
leptopus Thomson, Villa, 374
letho Wiedemann, Petrorossia, 346; figs. 383, 634, 652, 918, 919, 920
letho liliputiana Bezzi, Petrorossia, 346
leuchochroicus Hesse, Anastoechus, 86
leucocephala Wulp, Anthrax, 441
leucochila Bezzi, Villa, 374
leucochlaena Hesse, Lomatia, 311
leucochroicus Hesse, Anastoechus, 86
leucogaster Wiedemann, Anthrax, 442
leucolasiua Hesse, Bombylius, 81
leucomallus Philippi, Villa, 372
leucon Séguy, Toxophora, 236
leuconoe Jaennicke, Ligyra, 434
leucophaeus Wiedemann, Systoechus, 84
leucophasia Hesse, Lomatia, 311
leucophys Bigot, Gonarthrus, 112
leucopogon Bezzi, Anthrax, 442
leucopogon Meigen, Bombylius, 80
leucoprocta Wiedemann, Ogcodocera, 314; figs. 122, 428, 713, 724, 911, 912
leucoproctus Loew, Thyridanthrax, 29
leucoproctus Loew, Villa, 374
leucopsis Hesse, Lomatia, 311
leucopyga Macquart, Anthrax, 441
leucopyga Wiedemann, Toxophora, 16, 235; figs. 558, 577
leucopygus Macquart, Anthrax, 442
leucopygus Macquart, Bombylius, 347
leucopygus Wulp, Systoechus, 84
leucopygus Macquart, Villa, 374
leucosoma Bezzi, Anastoechus, 86
leucostictus Hesse, Systoechus, 84
leucostigma Wulp, Villa, 374
leucostomum Hesse, Crocidium, 185
leucostomus Meigen, Villa, 373
leucostomus Wiedemann, Villa, 374
leucotaeniatus Engel, Thyridanthrax, 406
leucotelus Walker, Ogcodocera, 314
leucotelus Walker, Villa, 374
leucothoa Wiedemann, Villa, 372
leucothyrida Hesse, Exoprosopa, 420
leucurus Hesse, Anthrax, 442
leviculus Coquillett, Aphoebantus, 324
leviculus Coquillett, Villa, 372
levigata Melander, Mythicomyia, 269
leyladea Efflatoun, Legnotomyia, 163
leyladea Efflatoun, Toxophora, 236
lherminierii Macquart, Sparnopolius, 12, 124; figs. 204, 420, 692, 805, 806
lichtwardti Becker, Mariobezzia, 174
lightfooti Hesse, Systoechus, 84
liliputiana Bezzi, Petrorossia, 346

limacodidarum Enderlein, Systropus, 248
limatulus Say, Anthrax, 20, 37, 440; figs. 1, 4
limatulus artemesia Marston, Anthrax, 20, 440
limatulus columbiensis Marston, Anthrax, 20, 440
limatulus larrea Marston, Anthrax, 20, 440
limatulus vallicola Marston, Anthrax, 20, 440
limatus Séguy, Platypygus, 262
limbata Megerle, Lomatia, 311
limbatus Loew, Conophorus, 158
limbatus Bigot, Sparnopolius, 124
limbatus Enderlein, Systropus, 248
limbatus Coquillett, Villa, 373
limbipennis Macquart, Bombylius, 82
limbipennis Macquart, Exoprosopa, 419
limitarsis Brunetti, Anthrax, 442
limpidus Walker, Villa, 374
lindneri Hesse, Apolysis, 219
linearis Bezzi, Exoprosopa, 420
lineata Hesse, Coryprosopa, 332; figs. 132, 386, 632, 647
lineatus Johnson and Johnson, Lordotus, 114
lineatus Walker, Villa, 372
lineifera Walker, Phthiria, 199
lineus Loew, Thyridanthrax, 406
linguata Roberts, Lepidanthrax, 388
lioyi Griffin, Chionamoeba, 344
lipposa Bigot, Cytherea, 148
literalis Hesse, Exoprosopa, 420
liticen Melander, Mythicomyia, 269
litoralis Hesse, Exoprosopa, 420
litoralis Painter, Geron, 206
litoralis Painter and Hall, Poecilanthrax, 398
litoralis Bowden, Systoechus, 84
liturata Loew, Lomatia, 311
litus Coquillett, Epacmus, 326; figs. 143, 628, 631
livia Osten Sacken, Villa, 372
livida Pallas, Exoprosopa, 419
lloydi Austen, Thyridanthrax, 29
lloydi Austen, Villa, 374
lobalis Thomson, Bombylius, 82
loewi Hesse, Chasmoneura, 110
loewi Paramonov, Conophorus, 158
loewi Hesse, Eurycarenus, 143
loewi Paramonov, Exoprosopa, 420
loewi Bezzi, Lomatia, 311
loewi Painter, Phthiria, 199
loewiana Bezzi, Exoprosopa, 420
loewii Jaennicke, Bombylius, 82
loewii Hesse, Villa, 374
longibarbus Efflatoun, Geron, 206
longimana Melander, Mythicomyia, 269
longipalpis Hesse, Corsomyza, 166
longipalpis Hardy, Thevenemyia, 182
longipennis Loew, Exoprosopa, 420
longipennis Loew, Litorrhynchus, 428
longipennis Macquart, Villa, 373
longirostris Wiedemann, Adelidea, 121
longirostris Wulp, Anastoechus, 86
longirostris Meigen, Bombylius, 80
longirostris Hesse, Ceratolaemus, 256
longirostris Paramonov, Corsomyza, 166
longirostris Paramonov, Dagestania, 221
longirostris Hesse, Dischistus, 107
longirostris Macquart, Exoprosopa, 419
longirostris Wiedemann, Henica, 288; figs. 226, 406, 620, 766, 978, 983, 1030

longirostris Melander, Oligodranes, 221
longirostris Bezzi, Prorachthes, 162; figs. 115A, 259, 479, 494
longirostris Becker, Systoechus, 84
longitarsis Becker, Petrorossia, 346
longitarsis Becker, Villa, 373
longitarsus Séguy, Heterotropus, 226
longitudinalis Loew, Lomatia, 311
longiventris Efflatoun, Geron, 206
loricatus Melander, Oligodranes, 221
lotus Williston, Thevenemyia, 182
lotus, Loew, Thyridanthrax, 406
loxospila Hesse, Exoprosopa, 420
lucens Melander, Mythicomyia, 269
lucens Walker, Villa, 374
lucida Becker, Anthrax, 441
lucidifrons Becker, Exoprosopa, 419
lucidus Loew, Systoechus, 87
lucidus Walker, Villa, 374
lucifer Fabricius, Poecilanthrax, 27, 398
lucifer Aldrich, Thevenemyia, 182
luctifer Osten Sacken, Thevenemyia, 182
luctifera Bezzi, Exoprosopa, 420
luctuosa Macquart, Anthrax, 441
luctuosus Loew, Conophorus, 158
luctuosus Bezzi, Pseudoamictus, 211
luculenta Séguy, Villa, 373
lugens Bezzi, Bombylius, 81
lugens Paramonov, Exoprosopa, 420
lugens Melander, Oligodranes, 221
lugens Loew, Thyridanthrax, 29, 406
lugubris Loew, Bombylius, 80
lugubris Loew, Cytherea, 148
lugubris Macquart, Exoprosopa, 419
lugubris Philippi, Lyophlaeba, 356
lugubris Wiedemann, Lyophlaeba, fig. 368
lugubris Rondani, Lyophlaeba, 356; figs. 157, 991, 992, 993
lugubris Osten Sacken, Systropus, 248
lurida Walker, Phthiria, 199
luridus Wiedemann, Conophorus, 158
luridus Hesse, Doliogethes, 111
luridus Walker, Walkeromyia, 26, 445; figs. 164, 411, 677, 709, 730, 746, 748, 750, 751
lusitanicus Meigen, Bombylius, 80
lusitanicus Guerin-Meneville, Conophorus, 158
lutea Macquart, Exoprosopa, 419
lutea Painter, Lepidophora, 238
luteicincta Hesse, Exoprosopa, 420
luteicosta Bezzi, Exoprosopa, 420
luteicosta metapleuralis Hesse, Exoprosopa, 420
luteifacies Bezzi, Hyperusia, 170; figs. 295, 492, 500
luteipennis Bezzi, Bombylius, 81
luteiventris Bezzi, Cyrtosia, 256
luteiventris minima Efflatoun, Cyrtosia, 256
luteocera Hesse, Exoprosopa, 420
luteolus Hall, Lordotus, 115
luteolus Bezzi, Thyridanthrax, 406
lutescens Loew, Dischistus, 107
lutescens Johnson and Johnson, Lordotus, 114
lutescens Bezzi, Usia, 217
lutescens minor Efflatoun, Usia, 217
lutulentus Bezzi, Thyridanthrax, 406
lutzi Curran, Lepidanthrax, 388

M

macassarensis Paramonov, Ligyra, 434
maccus Enderlein, Systropus, 248
macer Loew, Systropus, 18, 41, 248; fig. 5
macilentus Macquart, Systropus, 248
macilentus Wiedemann, Systropus, 248
macquarti d'Andretta and Carrera, Anthrax, 441
macquarti Bezzi, Thyridanthrax, 406
macraspis Thomson, Ligyra, 434
macrocerus Paramonov, Anthrax, 441
macroglossa Dufour, Conophorus, 158
macrophthalmus Bezzi, Anastoechus, 86
macrops Portschinsky, Thyridanthrax, 406
macroptera Loew, Exoprosopa, 420
macroptera Loew, Litorrhynchus, 428
macropterus Loew, Geron, 206
macropterus Ricardo, Litorrhynchus, 428
macrorrhynchus Bezzi, Anastoechus, 86
macula Cole, Thyridanthrax, 406
maculata Hesse, Adelidea, 121; figs. 1006, 1007, 1008
maculata Latreille, Cyllenia, 283
maculata Wiedemann, Phthiria, 199
maculata Meigen, Toxophora, 236
maculata Rossi, Toxophora, 16, 236
maculata completa Paramonov, Toxophora, 236
maculatissimus Villeneuve, Heterotropus, 226
maculatus Melander, Aphoebantus, 324
maculatus Fabricius, Bombylius, 80
maculatus Melander, Oligodranes, 221
maculatus Macquart, Villa, 374
maculifacies Hesse, Geron, 207
maculifer Walker, Bombylius, 80, 82
maculifera Bezzi, Exoprosopa, 420
maculipennis Macquart, Anthrax, 442
maculipennis Eversmann, Bombylius, 82
maculipennis Macquart, Bombylius, 80
maculipennis melanopus Timon-David, Bombylius, 81
maculipennis Macquart, Comptosia, 353
maculipennis Cole, Phthiria, 199
maculipennis Hull, Thevenemyia, 182
maculipennis Karsch, Toxophora, 236
maculithorax Paramonov, Bombylius, 80
maculithorax Engel, Cyrtosia, 256
maculiventris Brunetti, Exoprosopa, 421
maculiventris Bezzi, Heterotropus, 226
maculiventris Loew, Platypygus, 262
maculosa Sack, Anthrax, 441
maculosa Newman, Comptosia, 353
maculosa Wiedemann, Exoprosopa, 420
maculosus Painter, Parabombylius, 97
madagascariensis Macquart, Exoprosopa, 420
madagascariensis Macquart, Villa, 374
maenas Loew, Exoprosopa, 419
magister Melander, Apolysis, fossil, 59
magna Osten Sacken, Thevenemyia, 182; figs. 16, 563
magnifrons Bezzi, Callynthrophora, 168
magnipennis Bezzi, Exoprosopa, 420
magnirostris Bezzi, Heterotropus, 226
maherniaphila Hesse, Apolysis, 219; figs. 329, 549, 560
maimon Fabricius, Villa, 372
major Linné, Bombylius, 8, 80; figs. 1, 245, 453, 786, 787, 788

major australis Loew, Bombylius, 80
major basilinea Loew, Bombylius, 80
major consanguineus Macquart, Bombylius, 80
major Samouelle, Bombylius, 80
major Ricardo, Exoprosopa, 420
major Strobl, Phthiria, 199
major Macquart, Usia, 216, 217
majuscula Hesse, Exoprosopa, 420
majusculus Bezzi, Anthrax, 442
maldonadensis Macquart, Exoprosopa, 419
mallochi Hesse, Pantostomus, 293
manca Edwards, Lyophlaeba, 356
manca Loew, Usia, 216
mancipennis Coquillett, Metacosmus, 296
manifestus Walker, Villa, 374
mansfeldi Hesse, Psiloderoides, 17, 259; fig. 340
mara Walker, Exoprosopa, 420
marcidus Coquillett, Aphoebantus, 324
margaritifer Dufour, Villa, 374
marginalis Rondani, Cytherea, 148
marginalis Walker, Exoprosopa, 420
marginalis Cresson, Oligodranes, 221
marginalis Loew, Thyridanthrax, 406
marginalis Wiedemann, Villa, 373
marginata Loew, Cyllenia, 283; figs. 116, 117, 358, 616, 756, 771, 777
marginata Perris, Cyrtosia, 256
marginata Melander, Mythicomyia, 269
marginata Coquillett, Phthiria, 199
marginata Osten Sacken, Thevenemyia, 182
marginata Brunetti, Usia, 216
marginatus Cole, Aphoebantus, 324
marginatus Cyrillo, Bombylius, 80
marginatus Johnson and Johnson, Poecilanthrax, 398
marginatus Walker, Villa, 374
marginellus Bezzi, Bombylius, 81
marginicollis Gray, Exoprosopa, 421
marginifrons Bezzi, Callythrophora, 168
maria Williston, Villa, 372
mariouti Efflatoun, Empidideicus, 275
marleyi Hesse, Exoprosopa, 420
marleyi Hesse, Lomatia, 311
marmoreus Johnson and Johnson, Poecilanthrax, 398
maroccana Becker, Cytherea, 148
maroccanus Séguy, Acanthogeron, 90
mars Bezzi, Ligyra, 434
mars vulcanus Bezzi, Ligyra, 434
mars Curran, Systropus, 248
marshalli Hesse, Amictogeron, 209
marshalli Paramonov, Systoechus, 9, 84
marshalli Bezzi, Systropus, 18, 248
masienensis Hesse, Exoprosopa, 420
masienensis Hesse, Petrorossia, 346
massauensis Jaennicke, Anthrax, 442
massinissa Wiedemann, Anthrax, 441
matabeleensis Hesse, Lomatia, 311
matutinus Walker, Bombylius, 82
matutinus Melander, Oligodranes, 220
maura Wiedemann, Phthiria, 199
mauritanicus Bigot, Conophorus, 158
mauritanus Olivier, Bombylius, 81
maurus Olivier, Bombylius, 80
maurus Mikan, Conophorus, 158
maurus Linné, Hemipenthes, 30, 386; fig. 178
maurus flavotomentosa Paramonov, Hemipenthes, 386

INDEX

maurus Bezzi, Litorrhynchus, 428
maurus Thunberg, Litorrhynchus, 428
maxima Coquillett, Toxophora, 236
mayeti Bigot, Exoprosopa, 419
media Séguy, Petrorossia, 346
medii Strobl, Bombylius, 80
medius Linné, Bombylius, 80
medius albomicans Loew, Bombylius, 80
medius caucasicus Paramonov, Bombylius, 80
medius dalmatinus Strobl, Bombylius, 80
medius pallipes Loew, Bombylius, 80
medius pictipennis Loew, Bombylius, 80
medius punctipennis Loew, Bombylius, 80
medius seminiger Becker, Bombylius, 80
medius Scopoli, Bombylius, 80
medorae Painter, Bombylius, 80
megacephalus Portschinsky, Bombylius, 80
megachile Bowden, Zyxmyia, 173; figs. 294, 546, 770
megaera Wiedemann, Exoprosopa, 419
megaspilus Bezzi, Bombylius, 81
megerlei Meigen, Exoprosopa, 419
megerlei consanguinea Macquart, Exoprosopa, 419
megerlei deserticola Paramonov, Exoprosopa, 419
megerlei vesperugo A. Costa, Exoprosopa, 419
meigenii Wiedemann, Exoprosopa, 419
melaena Loew, Exoprosopa, 419
melaena Bowden, Hemipenthes, 386
melaena abbreviata Paramonov, Exoprosopa, 419
melaleuca Loew, Cytherea, 148
melaleucus Wiedemann, Villa, 372
melalophus Hesse, Gonarthrus, 112
melampogon Philippi, Bombylius, 80
melampogon Loew, Lomatia, 311
melampogon Bezzi, Systoechus, 84
melanaspis Bezzi, Exoprosopa, 420
melania Wulp, Anthrax, 442
melanista Bezzi, Anthrax, 441
melanocephalus Fabricius, Dischistus, 107
melanoceratus Bigot, Conophorus, 158
melanochlaenus Loew, Thyridanthrax, 406
melanogaster Bigot, Villa, 372
melanohalteralis Tucker, Anastoechus, 12, 86
melanohalteralis fulvipennis Tucker, Anastoechus, 86
melanoleuca Becker, Cytherea, 148
melanoloma Hesse, Lomatia, 311
melanolomus Hesse, Bombylius, 81
melanopalis Hesse, Crocidium, 185; figs. 327, 526, 528, 840, 841, 842
melanopleurus Bezzi, Thyridanthrax, 406
melanopogon Bigot, Anthrax, 440
melanopogon Bigot, Thevenemyia, 182
melanops Hesse, Doliogethes, 111
melanoptera Pallas, Exoprosopa, 419
melanoptera Bowden, Ligyra, 434
melanopus Bezzi, Bombylius, 81
melanopygus Bigot, Bombylius, 80
melanosa Williston, Thevenemyia, 182
melanoscuta Coquillett, Phthiria, 199
melanostola Hesse, Exoprosopa, 420
melanosus Johnson and Johnson, Lordotus, 115
melanotidus Hesse, Pantostomus, 293
melanozona Hesse, Exoprosopa, 420
melanthia Hesse, Exoprosopa, 420

melanthia Hesse, Lomatia, 311
melanura Bigot, Exoprosopa, 419
melanura Bigot, Exoprosopa, 419
melanurus Loew, Bombylius, 81
melanurus Bigot, Dischistus, 107
melanurus Bezzi, Eurycarenus, 143
melanurus Loew, Villa, 373
melanus Hesse, Bombylius, 81
melanus Bowden, Hemipenthes, 386
melasoma Wulp, Thyridanthrax, 406
melasoma Wulp, Villa, 374
melia Williston, Stonyx, 402
melinoproctus Loew, Platypygus, 262
mellea Bezzi, Glabellula, 271
melleus Bezzi, Platypygus, fig. 331
melleus Loew, Platypygus, 262
melli Enderlein, Systropus, 248
mellivorus Hesse, Euanthobates, 277; fig. 350
meltoni Hesse, Bombylius, 81
mendax Austen, Anthrax, 441
mendax Austen, Bombylius, 80
mendozana Brethes, Villa, 372
mentiens Bezzi, Systoechus, 84
mercedis Coquillett, Paravilla, 392
meridiana Paramonov, Lomatia, 311
meridionalis Bezzi, Anastoechus, 86
meridionalis Hesse, Chionamoeba, 344
meridionalis Rondani, Cyrtosia, 256
meridionalis Hesse, Doliogethes, 111
meridionalis Paramonov, Exoprosopa, 419
meridionalis François, Glabellula, 271
meridionalis Hesse, Pseudoglabellula, 273; fig. 345
meridionalis Cole, Villa, 372
meromelanus Hesse, Amictogeron, 209; fig. 106
merope Wiedemann, Exoprosopa, 420
mervensis Paramonov, Cytherea, 148
mervensis Paramonov, Thyridanthrax, 406
mesoleuca Hesse, Lomatia, 311
mesomelas Wiedemann, Bombylius, 80
mesopleuralis Bezzi, Exoprosopa, 420
metapleuralis Hesse, Exoprosopa, 420
metapleuralis Bezzi, Litorrhynchus, 428
metatarsalis Melander, Glabellula, 271
metopargyra Hesse, Exoprosopa, 420
metopium Osten Sacken, Bombylius, 80
mexicana Cole, Anthrax, 441
mexicanus, Wiedemann, Bombylius, 80
micans Fabricius, Bombylius, 81
micans Meigen, Bombylius, 81
micrargyra Walker, Villa, 373
microcephalus Loew, Systoechus, 84
micromelaena Bigot, Villa, 372
micropyga Melander, Aphoebantus, 324
microstictum Hesse, Crocidium, 185
midas Fabricius, Villa, 372
miegi Dufour, Exoprosopa, 420
mimetes Hesse, Anthrax, 442
mimetica Hesse, Exoprosopa, 420
mimnermia Efflatoun, Leylaiya, 277; fig. 346
mimus Hesse, Gonarthrus, 112
minas Macquart, Hemipenthes, 386
minerva Paramonov, Ligyra, 434
minerva Wiedemann, Villa, 372
miniata Oldroyd, Villa, 372
minima Efflatoun, Cyrtosia, 256
minima Melander, Mythicomyia, 269
minima Roberts, Zaclava, 250
minimaculatus Oldroyd, Anthrax, 441

minimus Schrank, Bombylisoma, 108; figs. 27, 273, 473, 513
minimus Macquart, Bombylius, 80, 82
minimus Scopoli, Bombylius, 80
minimus Bezzi, Eurycarenus, 143
minimus Fabricius, Systoechus, 84
ministra Melander, Mythicomyia, 269
minoana Paramonov, Exoprosopa, 419
minoides Paramonov, Exoprosopa, 419
minois Loew, Exoprosopa, 419
minor Melander, Alepidophora, fossil, 59
minor Austen, Amictus, 286
minor Curtis, Bombylius, 80
minor Linné, Bombylius, 8, 80
minor Meigen, Bombylius, 80, 81
minor ochraceus Paramonov, Bombylius, 80
minor Bezzi, Hyperusia, 170
minor Melander, Mythicomyia, 269
minor Bezzi, Sisyrophanus, 88
minor Edwards, Triploechus, 141
minor Efflatoun, Usia, 217
minor Macquart, Villa, 374
minos Meigen, Exoprosopa, 419; figs. 108, 208, 376
minos pharaonis Paramonov, Exoprosopa, 420
minuscula Hesse, Corsomyza, 166
minuscula Melander, Mythicomyia, 269
minuscula Efflatoun, Usia, 216
minusculus Efflatoun, Bombylius, 80
minusculus Hesse, Bombylius, 81
minusculus pallidiventris Hesse, Bombylius, 81
minuta Paramonov, Lyophlaeba, 356
minuta Greene, Mythicomyia, 269
minuta Fabricius, Phthiria, 199
minutissima Melander, Apolysis, 219
minutus Cole, Caenotus, 227
minutus Macquart, Thyridanthrax, 406
mira Coquillett, Chrysanthrax, 378
mira Hesse, Exoprosopa, 420
mira Melander, Mythicomyia, 269
mirandus Cockerell, Megacosmus, fossil, 59
mirifica Melander, Mythicomyia, 269
miscellus Coquillett, Lordotus, 115
miscellus melanosus Johnson and Johnson, Lordotus, 115
miscellus Coquillett, Villa, 372
miscens Walker, Bombylius, 81
mischanensis Paramonov, Hemipenthes, 386
misellus Loew, Thyridanthrax, 406; fig. 217
mitis Loew, Lomatia, 311
mitis Cresson, Oligodranes, 221
mitrata Melander, Mythicomyia, 269
mittrei Séguy, Acanthogeron, 90
mixta Loew, Anthrax, 442
mixteca Painter, Phthiria, 199
mixtus Coquillett, Aphoebantus, 324
mixtus Wiedemann, Systoechus, 84
mobile Coquillett, Hemipenthes, 386
mobilis Loew, Bombylius, 80
mobilis obscuripennis Paramonov, Bombylius, 80
modesta Melander, Mythicomyia, 269
modestus Loew, Bombylius, 80
modestus alexandri Paramonov, Bombylius, 80
modestus phaeopterus Bezzi, Bombylius, 80
modestus Loew, Epacmus, 326

modestus Loew, Oligodranes, 221
modestus Meigen, Villa, 373
moerens Philippi, Villa, 372
moffitti Painter and Hall, Poecilanthrax, 398
mohavea Melander, Apolysis, 219
moldavanicus Paramonov, Bombylius, 80
molitor Wiedemann, Bombylius, 81
molitor Loew, Villa, 20, 372
mollihirtus Hesse, Bombylius, 81
mollis Bezzi, Bombylius, 81
mollivestis Hesse, Lomatia, 311
monacha Sack, Anthrax, 20, 441
monacha Klug, Ligyra, 434
monacha Melander, Mythicomyia, 269; figs. 101, 342, 884, 885, 886
moneta Osten Sacken, Villa, 372
mongolica Paramonov, Exoprosopa, 420
mongolicus Paramonov, Anastoechus, 86
mongolicus Paramonov, Anthrax, 441
monogolicus Paramonov, Heterotropus, 226
montana Enderlein, Coniomastix, 445; fig. 412
montana Hesse, Corsomyza, 166
montana Paramonov, Lomatia, 311
montana Edwards, Lyophlaeba, 356
montana Philippi, Lyophlaeba, 356; figs. 151, 154, 656, 673
montana Paramonov, Thyridanthrax, 406
montanorum Austen, Thyridanthrax, 406
montanum Paramonov, Anthrax, 441
montanus Hesse, Amictogeron, 209
montanus Hesse, Ceratolarmus, 256
montanus Melander, Oligodranes, 221
montanus Painter and Hall, Poecilanthrax, 398
montanus Hesse, Systoechus, 84
monticola Paramonov, Anastoechus, 86
monticola Paramonov, Anthrax, 441
monticola Paramonov, Bombylius, 81
monticola Paramonov, Conophorus, 158
monticola Hesse, Exoprosopa, 421
monticola Paramonov, Heterotropus, 226
monticola Hesse, Lomatia, 311
monticola Johnson and Johnson, Poecilanthrax, 398
monticolanus Hesse, Systoechus, 84
monticolus François, Bombylius, 81
monticolus Hesse, Gonarthrus, 112
monticolus Hesse, Thyridanthrax, 406
montium Becker, Anastoechus, 86
montium François, Bombylius, 81
montivagus Hesse, Bombylius, 81
montivagus Séguy, Systropus, 248
moretonii Macquart, Comptosia, 354
morio Linné, Anthrax, 441
morio Olivier, Bombylius, 80
morio Linné, Hemipenthes, 30, 386; fig. 177
morio Fabricius, Hyperalonia, 431; fig. 192
morio Fabricius, Ligyra, 28, 434; fig. 192
morioides Say, Hemipenthes, 386
mormon Melander, Aphoebantus, 324
morna Curran, Lepidanthrax, 388
morosa Loew, Exoprosopa, 421
morosus de Meijere, Bombylius, 82
morrisoni Osten Sacken, Paracosmus, 298; figs. 112, 360, 619, 896, 897
morsicans Melander, Epacmus, 326
moussayensis Efflatoun, Bombylius, 80
mozambica Hesse, Exoprosopa, 421
mozambica Hesse, Lomatia, 311

mucidus Zeller, Villa, 373
mucorea Klug, Exoprosopa, 420
mucoreus Loew, Villa, 372
mucronatus Enderlein, Systropus, 248
mulsea Melander, Mythicomyia, 269
multibalteata Austen, Villa, 373
multicolor Bigot, Cytherea, 148
multifasciata Austen, Lomatia, 311
multisetosus Loew, Bombylodes, 92
munda Loew, Exoprosopa, 420
munda rivularis Griffin, Exoprosopa, 420
mundus Loew, Bombylius, 81
munroi Hesse, Anthrax, 442
munroi willowmorensis Hesse, Anthrax, 442
munroi Hesse, Geron, 207
munroi Bezzi, Heterotropus, 226
munroi Hesse, Systropus, 248
muricata Osten Sacken, Thevenemyia, 182
murina Newman, Comptosia, 354
murina Melander, Mythicomyia, 269
murinus Philippi, Villa, 372
mus Osten Sacken, Aphoebantus, 19, 324
mus barbatus Melander, Aphoebantus, 324
mus Bigot, Bombylius, 80
mus Curran, Exoprosopa, 419
mus Bigot, Oligodranes, 221
mus Becker, Villa, 373
muscaria Klug, Anthrax, 442
muscarius Coquillett, Villa, 372
muscarius Pallas, Villa, 373
muscoides Hesse, Bombylius, 81
muscoides Hesse, Hyperusia, 170
mutabilis Bowden, Gonarthrus, 112
mutabilis Melander, Mythicomyia, 269
muticum Bezzi, Anthrax, 442
mutillatus Bezzi, Bombylius, 81
mutilus Loew, Thyridanthrax, 406
mutuus Walker, Villa, 372
mydasiformis Bezzi, Exoprosopa, 421
mystaceus Macquart, Anthrax, 441
mystacinus Bezzi, Geron, 207
mystax Wiedemann, Dischistus, 107; figs. 29, 275, 453, 980, 982, 984

N

namaensis Hesse, Amictogeron, 209
namaensis Hesse, Anthrax, 442
namaensis Hesse, Gonarthrus, 112
namaensis Hesse, Oligodranes, 221
namana Hesse, Corsomyza, 166
namaqua Hesse, Lomatia, 311
namaquense Hesse, Crocidium, 185; fig. 58
namaquensis Hesse, Adelogenys, 192
namaquensis Hesse, Bombylius, 81
namaquensis Hesse, Cyrtosia, 256
namaquensis Hesse, Systoechus, 84
namaquensis Hesse, Systropus, 248
nana Coquillett, Mancia, 389; figs. 168, 396, 661, 738
nana Becker, Villa, 373
nanella Melander, Glabellula, 271
nanellus Paramonov, Anthrax, 441
nanus Hesse, Anthrax, 442
nanus Paramonov, Anthrax, 441
nanus Meigen, Bombylius, 80
nanus Walker, Bombylius, 82
napaea Melander, Mythicomyia, 269
nasalis Melander, Exepacmus, 339
nasutus Bezzi, Geron, 207

natalensis Hesse, Glabellula, 271
natalensis Hesse, Gonarthrus, 112
natalensis Hesse, Marleyimyia, 413; fig. 174
natalensis Hesse, Platypygus, 262
natalicola Hesse, Lomatia, 311
neavei Bezzi, Lomatia, 311
neavei Bezzi, Sisyrophanus, 88
nebritus Coquillett, Epacmus, 326
nebulo Coquillett, Villa, 372
nebulosa Coquillett, Chrysanthrax, 378
nebulosa Hesse, Exoprosopa, 421
nebulosus Dufour, Thyridanthrax, 406
neglectus Hesse, Systoechus, 84
neithokris Jaennicke, Bombylius, 81
nemakagonensis Graenicker, Paravilla, 392
nemesis Fabricius, Exoprosopa, 421
nemestrinus Bezzi, Isocnemus, 94; figs. 246, 441, 445
nephoneura Hesse, Exoprosopa, 421
nero Fabricius, Villa, 372
neurospila Bezzi, Exoprosopa, 421
neuter Melander, Oligodranes, 221
nidicola Cole, Anthrax, 20, 440
nieuwveldensis Hesse, Bombylius, 81
niger Cresson, Lordotus, 114
niger Walker, Systropus, 248
niger Macquart, Thevenemyia, 82
nigerrima Bezzi, Exoprosopa, 421
nigerrimus Bezzi, Anthrax, 442
nigerrimus ocellatus Bezzi, Anthrax, 442
nigerrimus Hesse, Geron, 207
nigra Wiedemann, Dolichomyia, 250
nigra Paramonov, Karakumia, 145; fig. 307
nigra Meigen, Phthiria, 199
nigra Macquart, Thevenemyia, 182
nigra Cresson, Villa, 372
nigralis Roberts, Geron, 207
nigrapicalis Roberts, Thevenemyia, 182
nigrata Wiedemann, Exoprosopa, 419
nigratus Roberts, Neosardus, 301
nigrescens Newman, Comptosia, 353
nigrescens Hesse, Doliopteryx, 273
nigrescens Ricardo, Lomatia, 311
nigrescens aterrima Hesse, Lomatia, 311
nigrescens bulawayoensis Hesse, Lomatia, 311
nigribarba Hesse, Phthiria, 199
nigribarbus Loew, Systoechus, 84
nigribarbus falsus Hesse, Systoechus, 84
nigricans Melander, Mythicomyia, 269
nigricauda Loew, Villa, 372
nigricaudus Brunetti, Systropus, 248
nigricephalus Séguy, Dischistus, 107
nigriceps Loew, Dischistus, 107
nigriceps Macquart, Villa, 373
nigriceps abdominalis Abreu, Villa, 373
nigricinctus Roberts, Eusurbus, 102; figs. 22, 257
nigricirratus Becker, Anastoechus, 86
nigricornis Philippi, Bombylius, 80
nigricosta Macquart, Villa, 374
nigricosta Schiner, Villa, 372
nigricostalis Guerin-Meneville, Villa, 374
nigrifacies Bezzi, Crocidium, 185
nigrifacies Hesse, Geron, 207
nigrifemorata Paramonov, Anastoechus, 86
nigrifemoratus Brunetti, Anthrax, 442
nigrifemoris Hesse, Adelidea, 121
nigrifemoris Hesse, Amictogeron, 209
nigrifera Walker, Exoprosopa, 420
nigrifimbriata Hesse, Exoprosopa, 421

nigrifrons Bezzi, Antonia, 321
nigrifrons Becker, Bombylisoma, 108
nigrifrons Becker, Bombylius, 80
nigrifrons Macquart, Villa, 373
nigrilobus Bezzi, Bombylius, 81
nigrilobus Collart, Bombylius, 81
nigrimanus Séguy, Heterotropus, 226
nigrina Bezzi, Exoprosopa, 421
nigrina Hardy, Phthiria, 199
nigripecten Bezzi, Bombylius, 81
nigripecten cinctutus Hesse, Bombylius, 81
nigripennis Loew, Conophorus, 158; figs. 820, 821
nigripennis Loew, Exoprosopa, 421
nigripennis Loew, Ligyra, 434
nigripennis Cole, Poecilanthrax, 398
nigripes Hesse, Bombylius, 81
nigripes Macquart, Bombylius, 81
nigripes Strobl, Bombylius, 81
nigripes Wiedemann, Corsomyza, 166; figs. 68, 287, 497, 729
nigripes turneri Hesse, Corsomyza, 166
nigripes Painter, Geron, 206
nigripes Loew, Systoechus, 84
nigripes nomteleensis Hesse, Systoechus, 84
nigripes plebeius Hesse, Systoechus, 84
nigripes Painter, Systropus, 248
nigrirostris Bezzi, Stomylomyia, 335
nigrisetosa Paramonov, Anastoechus, 86
nigrispina Bezzi, Exoprosopa, 421
nigrita Cyrillo, Bombylius, 80
nigrita Fabricius, Exoprosopa, 419
nigritarsis Engel, Heterotropus, 226
nigritarsis Enderlein, Systropus, 248
nigritella Bezzi, Exoprosopa, 421
nigritus Fabricius, Thyridanthrax, 406
nigritus Fabricius, Villa, 372
nigriventris Johnson and Johnson, Lordotus, 115
nigriventris Philippi, Sparnopolius, 124
nigrofasciata Paramonov, Exoprosopa, 419
nigrofemorata Brunetti, Anthrax, 442
nigrofemoratus Painter, Parabombylius, 97
nigrofimbriatus Williston, Villa, 372
nigronasica Painter, Paravilla, 392
nigropectus Cresson, Villa, 371
nigropenicillatus Bigot, Bombylius, 81
nigrovenosa Bezzi, Exoprosopa, 421
nigrum Austen, Anthrax, 441
niloticus Jaennicke, Villa, 374
niphas Hermann, Anthrax, 441
niphas bilineata Engel, Anthrax, 441
niphobleta Loew, Villa, 373
niphobletoides Hesse, Villa, 374
nitens Hesse, Anastoechus, 86
nitens Loew, Cyrtosia, 256
nitida Macquart, Corsomyza, 166
nitida Melander, Mythicomyia, 269
nitida Cole, Villa, 372
nitidapex Bezzi, Cytherea, 148
nitidifrons Austen, Thyridanthrax, 406
nitidifrons Hesse, Thyridanthrax, 406
nitidilabre Hesse, Crocidium, 185
nitidissima Engel, Cyrtosia, 256
nitidofasciatus Portschinsky, Hemipenthes, 386
nitidula Melander, Mythicomyia, 269
nitidula Hesse, Tomomyza, 292
nitidulus Fabricius, Anastoechus, 12, 86; fig. 240

nitidulus aberrans Paramonov, Anastoechus, 86
nitidulus hummeli Paramonov, Anastoechus, 86
nitidum Austen, Anthrax, 441
nitidus Cole, Epacmus, 326
nitidus Macquart, Megapalpus, 172
nitidus Cole, Metacosmus, 296; figs. 118, 355
nitidus Wiedemann, Systropus, 18, 248
nitobei Matsumura, Systropus, 248
nivalis Hesse, Gonarthrus, 112
nivalis Brunetti, Systoechus, 84
nivea Rossi, Chionamoeba, 344
nivea lioyi Griffin, Chionamoeba, 344
nivea Hesse, Hyperusia, 170
nivea Cole, Villa, 372
nivearia Hesse, Villa, 374
niveicauda Brunetti, Anthrax, 442
niveicollis Enderlein, Anastoechus, 86
niveicomatus Austen, Lissomerus, 91; figs. 13, 238, 435, 448, 781
niveicomatus Bowden, Systoechus, 84
niveifrons Walker, Bombylius, 81
niveifrons Becker, Edmundiella, 316
niveifrons Bezzi, Ligyra, 434
niveifrons Hesse, Thyridanthrax, 406
niveisquamis Brunetti, Anthrax, 442
niveiventris Brunetti, Exoprosopa, 421
niveoides Cole, Geron, 206
niveus Hermann, Anastoechus, 86
niveus Macquart, Bombylius, 81, 82
niveus Meigen, Bombylius, 81
niveus hololeucus Loew, Bombylius, 81
niveus Macquart, Dischistus, 107
niveus Cresson, Geron, 206
niveus Hesse, Geron, 207
niveus Bezzi, Lepidochlanus, 119
nivifrons Engel, Bombylius, 81
nivifrons Walker, Bombylius, 81
nivosa Hesse, Lomatia, 311
nobilis Loew, Amictus, 286
nobilis Loew, Conophorus, 158
nobilis iranicus Paramonov, Conophorus, 158
nobilis Kertész, Glabellula, 271; fig. 351
nobilis palaestinensis Engel, Glabellula, 271
noctilio Klug, Exoprosopa, 420
noctilunus Walker, Villa, 374
noctula Wiedemann, Exoprosopa, 419
nomadicus Hesse, Geron, 207
nomadicus breyeri Hesse, Geron, 207
nomas Paramonov, Anastoechus, 86
nomas Eversmann, Villa, 373
nomteleensis Hesse, Systoechus, 84
nonna Becker, Exoprosopa, 420
normalipes Paramonov, Heterotropus, 226
normalis Loew, Exoprosopa, 420
norrisi Paramonov, Docidomyia, 319
noscibilis Austen, Thyridanthrax, 406
notabilis Macquart, Exoprosopa, 421
notabilis Macquart, Villa, 374
notata Bigot, Phthiria, 199
notata Loew, Phthiria, 199
notata Loew, Usia, 199, 216
notatipennis Macquart, Bombylius, 82
notatus Hesse, Anomaloptilus, 276
notatus Engel, Dischistus, 107
nova Ricardo, Exoprosopa, 421
no aeformis Bezzi, Exoprosopa, 420
novakii Strobl, Usia, 216

novus Williston, Triploechus, 141
nox Walker, Exoprosopa, 421
nubeculosa Loew, Exoprosopa, 420
nubeculosa Coquillett, Phthiria, 199
nubeculosus Hesse, Anthrax, 442
nubifera Loew, Exoprosopa, 419
nubila Bezzi, Cyrtosia, 256
nubilus Mikan, Bombylius, 81
nubilus algericus Villeneuve, Bombylius, 81
nubilus monticola Paramonov, Bombylius, 81
nubilus Bezzi, Empidideicus, 275
nubilis Meigen, Systoechus, 84
nubilus Loew, Thyridanthrax, 406
nucalis Bezzi, Dischistus, 107
nucleorum Becker, Cytherea, 148
nuda, Hull, Euprepina, 134; figs. 49, 419, 455, 456, 815, 816
nudiusculus Thomson, Villa, 372
nudum Efflatoun, Crocidium, 185
nudus Villers, Bombylius, 81
nudus Painter, Geron, 206
nugator Coquillett, Thyridanthrax, 406
numeratus de Meijere, Systropus, 248
numida Macquart, Bombylius, 81
nuragasana Hesse, Exoprosopa, 421
nyasae Hesse, Litorrhynchus, 428
nyasae Ricardo, Litorrhynchus, 428
nycthemerus Macquart, Hemipenthes, 386
nycthemerus Wiedemann, Hemipenthes, 386

O

obesulus Loew, Conophorus, 158
obesus Bezzi, Bombylius, 81
obesus Loew, Oestranthrax, 411; figs. 166, 394, 683, 686, 740, 775, 783, 955, 956, 957
obesus olfierii Paramonov, Oestranthrax, 411
obesus pallifrons Bezzi, Oestranthrax, 411
obliqua Macquart, Exoprosopa, 421
obliquebifasciata Macquart, Exoprosopa, 421
obliquenotatus Austen, Amictus, 286
obliquisquamosa Hesse, Toxophora, 236
obliquus Brulle, Bombylius, 81
obliquus Macquart, Villa, 372
obliteratus Loew, Thyridanthrax, 406
oblongus Fabricius, Amictus, 286
obscura Sack, Anthrax, 441
obscura Walker, Comptosia, 354
obscura Fabricius, Cyrtosia, 256
obscura Fabricius, Cytherea, 15, 148; figs. 73, 296, 459, 569, 825, 826, 827
obscura beckeri Paramonov, Cytherea, 148
obscura Cresson, Oligodranes, 221
obscura Coquillett, Paravilla, 392
obscurifrons Brunetti, Anthrax, 442
obscurinotata Hesse, Exoprosopa, 421
obscurior Bowden, Petrorossia, 346
obscuripennis Paramonov, Bombylius, 311
obscuripennis Hesse, Exoprosopa, 421
obscuripennis, Loew, Lomatia, 311
obscuripennis Loew, Oligodranes, 221
obscuripes Loew, Cyrtosia, 256
obscuripes maculithorax Engel, Cyrtosia, 256
obscuripes serena Becker, Cyrtosia, 256
obscuripes Bigot, Villa, 372

obscurus Macquart, Villa, 374
obscurus Macquart, Villa, 374
obscurus Weber, Villa, 373
obsoleta Loew, Anthrax, 440
obsoleta Loew, Cyllenia, 283
obtectus Melander, Aphoebantus, 324
obtiosus Coquillett, Anthrax, 441
obtusa Bezzi, Exoprosopa, 421
obtusus Bezzi, Bombylius, 81
obuchovae Paramonov, Anthrax, 441
obumbratus Bezzi, Litorrhynchus, 428
occidentalis Johnson, Anthrax, 440
occidentalis Bowden, Antonia, 321
occidentalis Hesse, Bombylius, 81
occidentalis Rondani, Cyrtosia, 256
occidentis Roberts, Zaclava, 250
occiduus Hesse, Thyridanthrax, 406
occipitalis Loew, Thyridanthrax, 406
occlusa Loew, Exoprosopa, 420
occlusa nubeculosa Loew, Exoprosopa, 420
occlusiodes Paramonov, Exoprosopa, 420
occlusiodes lactea Paramonov, Exoprosopa, 420
occultus Wiedemann, Villa, 373
oceanus Becker, Bombylius, 81
ocellata Newman, Comptosia, 354; fig. 158
ocellatus Bezzi, Anthrax, 442
ochraceus Bigot, Bombylius, 80
ochraceus Paramonov, Bombylius, 80
ochrostoma Hesse, Corsomyza, 166
ocreata Melander, Mythicomyia, 269
oculata Macquart, Heteralonia, 425
ocyale Wiedemann, Anthrax, 441
oedipus Fabricius, Anthrax, 441
oenomaus Rondani, Ligyra, 32, 434
oenomaus flora Frey, Ligyra, 434
offuscata Bezzi, Exoprosopa, 421
ogasawarensis Matsumura, Anthrax, 441
ogilviei Hesse, Exoprosopa, 421
ogilviei Hesse, Lomatia, 311
ogilviei Hesse, Sisyrophanus, 88
okahandjanus Hesse, Bombylius, 81
oldroydi d'Andretta and Carrera, Systropus, 248
oligocenica Timon-David, Phthiria, fossil, 59
olivaceus Paramonov, Anastoechus, 86
olivaceus corsikana Paramonov, Anastoechus, 86
olivierii Macquart, Bombylius, 81
olivierii Macquart, Exoprosopa, 420
olivierii Macquart, Geron, 206
olmeca Painter, Phthiria, 199
olsufjevi Paramonov, Bombylius, 81
oneilii Hesse, Corsomyza, 166
onusta Walker, Exoprosopa, 420
oophagus Paramonov, Anthrax, 20, 441
oophagus parva Paramonov, Anthrax, 441
opaca Loew, Cyrtosia, 256
opalina Hesse, Micomitra, 430; figs. 203, 421, 694
ophioneus Westwood, Systropus, 248
oporina Melander, Mythicomyia, 269
optata Melander, Mythicomyia, 269
orbitalis Williston, Hemipenthes, 386
orbitalis Williston, Villa, 373
orchestes Melander, Mythicomyia, 269
orcus Walker, Exoprosopa, 419
oreas Osten Sacken, Systoechus, 9, 84; fig. 6
oreoica Hesse, Lomatia, 311
oreophila Hesse, Lomatia, 311

orest Paramonov, Ligyra, 434
orientalis Macquart, Bombylius, 82
orientalis Bezzi, Exoprosopa, 411
orientalis Paramonov, Ligyra, 434
orientalis Zakavatkin, Systoechus, 84
orientalis Paramonov, Usia, 217
ornata Melander, Mythicomyia, 269
ornata Engel, Usia, 216
ornatifrons Séguy, Cyrtosia, 256
ornatula Melander, Mythicomyia, 269
ornatus Wiedemann, Bombylius, 81
ornatus pleuralis Bezzi, Bombylius, 81
ornatus Curtis, Thyridanthrax, 406
ornatus Rondani, Triploechus, 141
orthoperus Hesse, Geron, 207
osten sackenii Burgess, Thevenemyia, 182
ostenta Melander, Mythicomyia, 269
othello Szilady, Exoprosopa, 421
otiosa Coquillett, Thyridanthrax, 406
ovamboana Hesse, Exoprosopa, 421
ovamboensis Hesse, Lomatia, 311
ovata Loew, Villa, 373
ovatus Bezzi, Dischistus, 107
ovatus Bezzi, Doliogethes, 111

P

pachyceratus Bigot, Oniromyia, 156; figs. 80, 297, 498, 509, 822, 823, 824
pachycerum Hesse, Crocidium, 185
pachystyla Hesse, Villa, 374
paganica White, Cyrtomorpha, 261; figs. 95, 320, 581, 587, 875, 876, 877
painteri Priddy, Conophorus, 158
painteri Maughan, Poecilanthrax, 398
painterorum Johnson and Johnson, Exoprosopa, 419
palaestinae Paramonov, Callostoma, 150; fig. 511
palaestinensis Engel, Glabellula, 271; figs. 597, 606
palaestinus Paramonov, Bombylius, 81
palestinae Paramonov, Legnotomyia, 163
pallasii Paramonov, Bombylius, 80
pallasii Loew, Cytherea, 148; figs. 75, 503, 521
pallasii Nowicki, Cytherea, 148
pallasii Wiedemann, Cytherea, fig. 298
pallasii Wiedemann, Exoprosopa, 420
pallens Wiedemann, Bombylius, 81
pallens Bigot, Exoprosopa, 419
pallens Nurse, Heterotropus, 226
pallescens Hesse, Bombylius, 81
pallescens Johnson and Maughan, Bombylius, 80
pallescens Engel, Phthiria, 199
pallescens Becker, Usia, 216
palliata Loew, Paravilla, 392
pallida Bezzi, Exoprosopa, 421
pallida Tsacas, Lomatia, 311
pallida Coquillett, Thyridanthrax, 406
pallida d'Andretta and Carrera, Toxophora, 236
pallidicris Brulle, Bombylius, 81
pallidifacies Hesse, Exoprosopa, 421
pallidipes Hesse, Corsomyza, 166
pailidipes Hesse, Exoprosopa, 421
pallidipilosus Austen, Systoechus, 84
pallidisetigera Austen, Exoprosopa, 420
pallidispinis Hesse, Systoechus, 84
pallidiventris Hesse, Bombylius, 81
pallidoventer Roberts, Dischistus, 107

pallidula Coquillett, Chrysanthrax, 378
pallidulum Hesse, Epacmoides, 334
pallidulus Hesse, Anthrax, 442
pallidulus Walker, Bombylius, 81
pallidulus Hesse, Doliogethes, 111
pallidulus Greathead, Systoechus, 84
pallidus Cresson, Epacmus, 326
pallidus Roberts, Systoechus, 84
pallifrons Bezzi, Oestranthrax, 411
palliolatus White, Bombylius, 82
pallipennis Paramonov, Lyophlaeba, 356
pallipes Loew, Anthrax, 441
pallipes Loew, Bombylius, 80
pallipes A. Costa, Cyrtosia, 256
pallipes Bigot, Heterostylum, 140
pallipes Bigot, Phthiria, 199
pallipes Bezzi, Tomomyza, 292
pallipes Edwards, Triploechus, 141
pallipes Bigot, Villa, 373
paloides Painter, Systropus, 248
palpalis Melander, Oligodranes, 221
palpalis Cockerell, Protophthiria, fossil, 59
paludosa de Meijere, Ligyra, 434
palumbii Rondani, Villa, 373
palustris Bezzi, Exoprosopa, 421
pamirense Paramonov, Anthrax, 441
pamirensis Paramonov, Cytherea, 148
panamensis Curran, Exoprosopa, 419
panamensis Curran, Lepidanthrax, 388
pandora Fabricius, Exoprosopa, 420
pandora Macquart, Exoprosopa, 419
pandora Meigen, Exoprosopa, 419
paniscoides Bezzi, Villa, 374
paniscus Rossi, Villa, 26, 373
panneus Melander, Oligodranes, 221
pantostomoides Hesse, Tomomyza, 292
paradoxa Jaennicke, Neodiplocampta, 390; figs. 167, 392, 658, 669, 743
paradoxa Hesse, Xenoprosopa, 252; figs. 84, 339, 539, 760
paraduncus Paramonov, Conophorus, 158
parallelus Bezzi, Bombylius, 81
paramonovi Hull, Neosardus, 301
parbifasciata Paramonov, Lyophlaeba, 356
pardus Osten Sacken, Exoprosopa, 419
pargrisescens Paramonov, Anthrax, 441
paris Bezzi, Ligyra, 434
parkeri Melander, Aphoebantus, 324
parkeri Melander, Oligodranes, 221
parma Melander, Mythicomyia, 269
parobscurum Paramonov, Anthrax, 441
partita Newman, Comptosia, 354
parva Loew, Exoprosopa, 419
parva Ricardo, Exoprosopa, 421
parva Paramonov, Thyridanthrax, 406
parvicellula Bezzi, Micomitra, 430
parvicornis Loew, Rhynchanthrax, 28, 408; figs. 232, 401, 716, 727, 964, 965
parvidus Painter, Geron, 206
parvissima Roberts, Cyrtosia, 256
parva Ricardo, Exoprosopa, 421
parvula Bezzi, Exoprosopa, 421
parvula Efflatoun, Usia, 216
parvus Bowden, Bombylius, 81
parvus Hesse, Geron, 207
paterculus Walker, Bombylius, 81
patula Walker, Oncodosia, 363
paucispina Hesse, Exoprosopa, 421
paulseni Philippi, Acrophthalmyda, 176
pauper Loew, Anthrax, 440
pauper Walker, Exoprosopa, 420

pauper Becker, Thyridanthrax, 406
pausarius jaennicke, Systoechus, 84
pavida Williston, Exoprosopa, 419
pavidus Coquillett, Aphoebantus, 324
pealei Cockerell, Alepidophora, fossil, 59
pectinigulus Hesse, Euanthobates, 277; figs. 598, 758
pectoralis Loew, Chasmoneura, 108; fig. 42
pectoralis Loew, Exoprosopa, 420
pedemontana Griffin, Anthrax, 442
pediformis Bezzi, Atrichochira, 423; figs. 206, 431A, 711, 721
pedunculata Hesse, Lomatia, 311
pegasus Wiedemann, Ylasoia, 358; figs. 148, 367, 659, 670, 985, 986, 987
pellucida Coquillett, Geminaria, 117
pellucida Paramonov, Mariobezzia, 174
pellucida Coquillett, Toxophora, 16, 236
pellucidus Coquillett, Aphoebantus, 324
pellucidus Walker, Villa, 374
pelops Walker, Ligyra, 434
pendens Cole, Bombylius, 79
penicillatus Macquart, Bombylius, 82
penicula Melander, Mythicomyia, 269
peninsularis Cole, Villa, 373
pennata Nurse, Exoprosopa, 421
pennipes Wiedemann, Corsomyza, 166
pennipes Wiedemann, Exoprosopa, 421
pentala Macquart, Exoprosopa, 421
pentaspilus Bezzi, Bombylius, 81
penthoptera Bezzi, Exoprosopa, 421
peociloptera Bezzi, Exoprosopa, 420
perfecta Becker, Anthrax, 441
perfecta Becker, Cyrtosia, 256
perfecta Becker, Villa, 373
perfecta Becker, Villa, 373
perfectus Becker, Empidideicus, 275
pericaustus Loew, Bombylius, 81
perimele Wiedemann, Villa, 373
peringueyi Hesse, Amictogeron, 209
peringueyi Bezzi, Bombylius, 81
peringueyi Hesse, Geron, 207
perlucida Hesse, Micomitra, 430
permixtus Hesse, Bombylius, 81
perniveus Bezzi, Acanthogeron, 90
pernotata Hesse, Plesiocera, 338
perparvus Roberts, Dischistus, 107
perplexa Coquillett, Paravilla, 28, 392
perplexus Bezzi, Litorrhynchus, 428
perplexus Johnson and Johnson, Lordotus, 115
perpulchra Bezzi, Exoprosopa, 421
perpusillum Austen, Anthrax, 441
perpusillus Austen, Thyridanthrax, 406
persica Paramonov, Legnotomyia, 163
persica Paramonov, Lomatia, 311
persica Paramonov, Prothaplocnemis, 399
persicana Paramonov, Antonia, 321
persicana Becker, Cyrtosia, 256
persicana Becker, Cytherea, 148
persicum Paramonov, Callostoma, 150
persicus Paramonov, Anthrax, 441
persicus Becker, Aphoebantus, 324
persicus Paramonov, Bombylius, 81
persicus Paramonov, Platypygus, 262
persicus Macquart, Villa, 373
personata Bezzi, Exoprosopa, 421
perspicillaris Loew, Eclimus, 180
perspicillaris Loew, Thyridanthrax, 29, 406; fig. 171
perspicuus Roberts, Anastoechus, 86
pertusa Loew, Thyridanthrax, 406; fig. 188

petena Melander, Mythicomyia, 269
petes Melander, Mythicomyia, 269
petiolata Melander, Apolysis, 219
petiolata Melander, Mythicomyia, 269
phacodes Melander, Mythicomyia, 269
phaenochilum Hesse, Crocidium, 185
phaenostigma Hesse, Lomatia, 311
phaeoptera Meigen, Exoprosopa, 420
phaeopteralis Hesse, Anthrax, 442
phaeopteralis Hesse, Crocidium, 185; fig. 59
phaeopteris Hesse, Amictogeron, 209
phaeopterus Bezzi, Bombylius, 84
phaeopterus Bezzi, Systoechus, 84
phaeotaenia Bezzi, Villa, 374
phalerata Melander, Mythicomyia, 269
phaleratus Hesse, Anastoechus, 86
phaleratus albicerus Hesse, Anastoechus, 86
phallophorus Bezzi, Geron, 206
pharao Paramonov, Micomitra, 430
pharaonis Paramonov, Anthrax, 441
pharaonis Paramonov, Exoprosopa, 420
pharetra Melander, Mythicomyia, 269
philadephica Macquart, Exoprosopa, 419
philadelphicus Johnson, Bombylius, 79
philadelphicus Macquart, Bombylius, 80
philerema Hesse, Plesiocera, 338
phileremus Hesse, Gonarthrus, 112
phileremus Hesse, Thyridanthrax, 406
philippi Paramonov, Lyophlaeba, 356
philippiana Rondani, Phthiria, 199
philippii Rondani, Villa, 373
philoxera Hesse, Tomomyza, 292; fig. 115
phlegethon Walker, Exoprosopa, 419
phloeochromus Bezzi, Litorrhynchus, 428
phthinoxantha, Bowden, Petrorossia, 346; fig. 234
picea Marston, Anthrax, 440
picta Edwards, Doddosia, 359; figs. 149, 155, 431B, 698, 722, 931, 932
picta Melander, Mythicomyia, 269
picta Philippi, Phthiria, 199
picta Wiedemann, Exoprosopa, 419
pictilipennis Austen, Exoprosopa, 420
pictinervis Rondani, Lyophlaeba, 356
pictipennis Loew, Bombylius, 81
pictipennis Macquart, Bombylius, 82
pictipennis Macquart, Conophorus, 158
pictipennis Bigot, Cyrtomyia, 239
pictipennis Wiedemann, Lomatia, 311
pictipennis Bezzi, Tomomyza, 292; figs. 362, 603, 613, 1021, 1022, 1023
pictipes Coquillett, Mythicomyia, 269
picturata Coquillett, Phthiria, 199
pictus Loew, Amictus, 286
pictus Panzer, Bombylius, 81
pilad Paramonov, Ligyra, 434
pilatei Macquart, Ligyra, 434
pilirostris Loew, Bombylius, 81
pilirostris Hesse, Phthiria, 199
pilosula Strobl, Anthrax, 442
pilosulus Strobl, Anthrax, 441
pilosulus tadzhikorum Paramonov, Anthrax, 441
pilosulus Hesse, Pantostomus, 293
pilosus Cole, Poecilanthrax, 398
pima Painter, Hemipenthes, 386
pinguis Walker, Bombylius, 82
pirioni Edwards, Phthiria, 199
pithecius Fabricius, Anthrax, 442
plagiatus Bezzi, Bombylius, 81
plagiatus Walker, Villa, 373

plagosa Coquillett, Rhynchanthrax, 408
plana Walker, Oncodosia, 363; figs. 152, 153, 427, 699, 723
planicornis Fabricius, Bombylius, 81
planus Osten Sacken, Lordotus, 115
platycheira Melander, Mythicomyia, 269
platyrus Walker, Bombylius, 82
platysoma Cockerell, Geron, fossil, 59
plena Walker, Comptosia, 354
plerophaia Hesse, Petrorossia, 346
plerosticta Hesse, Exoprosopa, 421
pleroxantha Hesse, Exoprosopa, 421
plesius Curran, Anthrax, 440
pleskei Paramonov, Exoprosopa, 420
pleskei Becker, Gyrocraspedum, 151; figs. 56, 303, 501, 510
pleskei Paramonov, Prorachthes, 162
pleuralis Bezzi, Bombylius, 81
pleuralis Williston, Villa, 373
plocamoleuca Hesse, Lomatia, 311
plorans Bezzi, Bombylius, 81
plumipalpis Bezzi, Dischistus, 107
plumipalpis Hesse, Dischistus, fig. 25
plumipes Hesse, Anthrax, 442
plumipes Drury, Bombylius, 80
plumipes Philippi, Walkeromyia, 445
plumipilosa Hull, Hallidia, 135; figs. 36, 278, 519, 812, 813, 814
pluricella Williston, Villa, 373
pluricellata Macquart, Henica, 288
pluricellus Williston, Anthrax, 440
plurinota Bigot, Anthrax, 441
pluto Wiedemann, Anthrax, 440
podagricus Paramonov, Bombylius, 81
poecilogaster Osten Sacken, Poecilanthrax, 398
poecilogaster interruptus Painter, Poecilanthrax, 398
poecilophora Schiner, Anthrax, 441
poecilopterum Loew, Crocidium, 185
poedes Osten Sacken, Aphoebantus, 324
polioleucus Hesse, Systoechus, 84
polistoides Westwood, Systropus, 248
polius Melander, Oligodranes, 221
polygena Melander, Mythicomyia, 269
polygona Bezzi, Lomatia, 311
polyphemus Wiedemann, Thyridanthrax, 406
polyphemus pumilio Austen, Thyridanthrax, 406
polyphemus Wiedemann, Villa, 373
polypogon Loew, Bombylius, 80
polyspila Bezzi, Exoprosopa, 421
polysticta Hesse, Exoprosopa, 421
polystigma Sack, Anthrax, 441
polyzona Loew, Lomatia, 311
polyzona meridiana Paramonov, Lomatia, 311
ponderosus Melander, Epacmus, 326
porectella Hesse, Exoprosopa, 421
poricella Hesse, Exoprosopa, 421
porteri Oldroyd, Villa, 373
portschinskyi Paramonov, Prorachthes, 162
portshinskiji Paramonov, Exoprosopa, 420
posticus Fabricius, Bombylius, 80
posticus Meigen, Bombylius, 81
potrix Melander, Mythicomyia, 269
poweri Hesse, Systoechus, 84
praeargentata Macleay, Comptosia, 354
praecisus Loew, Hemipenthes, 386
praedicans Walker, Villa, 374
praefica Loew, Exoprosopa, 421

praeterita Oldroyd, Villa, 373
pravipes Melander, Mythicomyia, 269
pretendens Walker, Villa, 374
pretiosa Loew, Villa, 373; figs. 667, 681
pretiosus Coquillett, Villa, 372
priapeus Bezzi, Geron, 206
primitivus Walker, Villa, 373
primogenitus Walker, Bombylius, 82
princeps Bezzi, Anthrax, 442
principius Roberts, Neosardus, 301; figs. 121, 398, 621, 639, 900, 901, 902
probellus Hardy, Bombylius, 82
proboscideus Loew, Lepidanthrax, 388
procedens Walker, Villa, 373
procne Osten Sacken, Exoprosopa, 419
proconcisa Hardy, Anthrax, 442
proconcisa Hardy, Villa, 374
productus Bezzi, Litorrhynchus, 428
proferens Walker, Villa, 374
prometheus Macquart, Exoprosopa, 419
propinqua Schiner, Anthrax, 441
propinqua Schiner, Villa, 373
propinquus Brunetti, Bombylius, 82
propinquus Hesse, Eurycarenus, 143
propleuralis Melander, Empidideicus, 275
proprius Roberts, Bombylius, 82
proserpina Wiedemann, Ligyra, 434
prosimplex Hardy, Anthrax, 442
protuberans Bezzi, Exoprosopa, 420
provincialis Handlirsch, Hemipenthes, fossil, 59
pruinosa Melander, Mythicomyia, 269
pruinosulus Hesse, Bombylius, 81
pruinosus Hesse, Anastoechus, 86
psammina Cole, Villa, 372
psammobates Hesse, Geron, 207
psammocharus Hesse, Doliogethes, 111
psammophila Hesse, Plesiocera, 338; fig. 125
psammophila Paramonov, Toxophora, 236
psammophilus Hesse, Pantostomus, 293
pseudaduncus Paramonov, Conophorus, 158
pseudoargentatus Paramonov, Bombylius, 81
pseudocollaris Bezzi, Litorrhynchus, 428
pseudofasciata Hesse, Lomatia, 311
pseudoflammiger Bezzi, Thyridanthrax, 406
pseudopsis Hesse, Bombylius, 81
pseudoterminalis Senior-White, Bombylius, 82
psi Cresson, Phthiria, 199
pterosticha Hesse, Exoprosopa, 421
pterosticta Hesse, Adelidea, 121; figs. 266, 478, 516
pterosticta Hesse, Lomatia, 311
pterostictum Hesse, Crocidium, 185
pubera Loew, Usia, 216
pubescens Bezzi, Phthiria, 199
pubipes Edwards, Sericosoma, 154
pueblensis Jaennicke, Exoprosopa, 31, 419
puellaris White, Docidomyia, 319; figs. 142, 400, 623, 633
puellus Williston, Lordotus, 115
puerula Brunetti, Exoprosopa, 421
pulchella Wulp, Eucharimyia, 102
pulchella Williston, Phthiria, 199
pulchellum Hesse, Onchopelma, 257; figs. 538, 582
pulchellus Macquart, Amictus, 286
pulchellus Eversmann, Bombylius, 82
pulchellus Loew, Bombylius, 80

pulchellus Roberts, Bombylius, 82
pulchellus Wulp, Bombylius, 82
pulchellus Austen, Dischistus, 107
pulchellus Wulp, Eucharimyia, 102
pulcher Paramonov, Anastoechus, 86
pulcher Melander, Oligodranes, 221
pulcher Painter, Parabombylius, 97
pulcher Williston, Systropus, 248
pulcherrima Paramonov, Exoprosopa, 420
pulcherrimus Aldrich, Lordotus, 115
pulchra Brychosoma, new species, 102; fig. 14
pulchra Walker, Exoprosopa, 421
pulchriceps Loew, Lomatia, 311
pulchriceps linganaui Hesse, Lomatia, 311
pulchriceps ogilviei Hesse, Lomatia, 311
pulchriceps tinctella Hesse, Lomatia, 311
pulchripes Austen, Phthiria, 199
pulchrissimus Williston, Lordotus, 115
pulchrissimus luteolus Hall, Lordotus, 115
pulchrum, Brychosoma, new species, 104
pulicaria Mikan, Phthiria, 199; fig. 215
pulicaria major Strobl, Phthiria, 199
pulicaria Zetterstedt, Phthiria, 199
pulla Melander, Mythicomyia, 269
pulla Bezzi, Phthiria, 199
pullata Coquillett, Hemipenthes, 386
pullatus Hesse, Doliogethes, 111
pullatus Melander, Oligodranes, 221
pulvereus Melander, Epacmus, 326
pumila Melander, Glabellula, 271
pumilio Loew, Platypygus, 262
pumilio Becker, Systoechus, 84
pumilio Austen, Thyridanthrax, 406
pumilus Meigen, Bombylius, 38, 80, 81; fig. 2
pumilus Zetterstedt, Bombylius, 80
punctata Meigen, Anthrax, 441
punctata Macquart, Exoprosopa, 419
punctata Meigen, Phthiria, 199
punctatelloides Hesse, Bombylius, 81
punctatillus Bezzi, Bombylius, 81
punctatus De Geer, Bombylius, 80
punctatus Fabricius, Bombylius, 81; fig. 7
punctatus Macquart, Villa, 373
punctifer Bezzi, Bombylius, 81
punctifrons Bezzi, Exoprosopa, 421
punctifrons Bezzi, Lomatia, 311
punctinervis Becker, Exoprosopa, 420
punctipennis Zaitsev, Anthrax, 441
punctipennis Jaennicke, Acrophthalmyda, 176
punctipennis Wiedemann, Anthrax, 441, 442; fig. 409
punctipennis Wiedemann, Apatomyza, 190; figs. 349, 544
punctipennis Loew, Bombylius, 80
punctipennis Thomson, Bombylius, 82
punctipennis Macquart, Cytherea, 148
punctipennis Ricardo, Exoprosopa, 419
punctipennis Macquart, Ligyra, 434
punctipennis Walker, Phthiria, 199
punctipennis Becker, Semiramis, 192; figs. 325, 365, 535
punctipennis Bezzi, Toxophora, 236
punctipennis Loew, Usia, 216
punctulata Macquart, Exoprosopa, 421
punctulatus Macquart, Villa, 374
punctulatus Macquart, Villa, 373
punctum Walker, Anthrax, 441
punctum Loew, Thyridanthrax, 406
punctum Loew, Villa, 373

puncturellus Hesse, Anthrax, 442
puniceus Hesse, Doliogethes, 111
punicisetosa Hesse, Anthrax, 442
punjabensis Nurse, Exoprosopa, 421
purpuraria Walker, Exoprosopa, 421
purpuratus Wiedemann, Villa, 374
purpureus Bezzi, Bombylius, 81
purpuripennis Hesse, Lomatia, 311
pusilla Loew, Cyrtosia, 256
pusilla Paramonov, Dagestania, 221
pusilla Macquart, Exoprosopa, 421
pusilla Melander, Mythicomyia, 269
pusilla Bezzi, Plesiocera, 338
pusilla Macquart, Usia, 216
pusilla Meigen, Usia, 216
pusillima Edwards, Mythicomyia, 269; figs. 347, 596, 605
pusillus Paramonov, Aphoebantus, 324
pusillus Meigen, Bombylius, 80, 81
pusillus Loew, Conophorus, 158
pusillus Wiedemann, Villa, 374
pusio Meigen, Bombylius, 81
pusio Wiedemann, Dischistus, 107
pusio Osten Sacken, Pantarbes, 153; figs. 305, 499, 507
pusio Macquart, Villa, 373
pusio Philippi, Villa, 373
putealis Matsumura, Anthrax, 441
putilla Becker, Usia, 216
pycnopeltis Hesse, Anthrax, 442
pycnorrhynchus Thomson, Bombylius, 82
pygarga Loew, Villa, 26, 373
pygmaea Fabricius, Phthiria, 199
pygmaeus Fabricius, Bombylius, 80
pygmaeus Macquart, Bombylius, 80
pygmaeus canadensis Curran, Bombylius, 80
pygmaeus Cole, Oligodranes, 221
pygmalion Fabricius, Exoprosopa, 419, 420
pygmalion Macquart, Exoprosopa, 420
pyramidum Paramonov, Anastoechus, 86
pyrrhocerus Bezzi, Sisyrophanus, 88; figs. 241, 466, 477

Q

quadrata Williston, Thevenemyia, 182
quadratus Loew, Systoechus, 84
quadricellulata Hesse, Toxophora, 236
quadricinctus Philippi, Villa, 373
quadricinctus Rondani, Villa, 373
quadrifarius Loew, Bombylius, 81
quadrinotata Loew, Phthiria, 199
quadripennis Walker, Comptosia, 354
quadripunctata Paramonov, Exoprosopa, 419
quadripunctata Cole, Villa, 373
quadripunctatus Séguy, Systropus, 248; fig. 86
quadripunctatus Williston, Systropus, 248
quedenfeldti Engel, Thevenemyia, 182
quinquefasciata Becker, Hemipenthes, 386
quinquefasciatus Wiedemann, Villa, 26, 373
quinqueguttata Roberts, Villa, 374
quinquemaculatum Bezzi, Anthrax, 442
quinquenotatus Johnson, Oligodranes, 221
quinquepunctatus Thomson, Villa, 373
quinta Becker, Villa, 373
quivera Painter, Rhynchanthrax, 408

R

ramsesi Paramonov, Anthrax, 441
rasa Loew, Exoprosopa, 421
rattus Osten Sacken, Aphoebantus, 324
rava Roberts, Villa, 373, 374
ravus Loew, Bombylius, 80
ravus Painter, Villa, 373
recedens Walker, Bombylius, 82
rectifascia Bezzi, Exoprosopa, 421
rectus Walker, Villa, 373
recurrens Loew, Exoprosopa, 421
recurrens Cockerell, Protolomatia, fossil, 59
recurvus Coquillett, Triploechus, 141
reductus Walker, Villa, 374
referta Bezzi, Exoprosopa, 421
regiomontana Hennig, Amictites, fossil, 59
repertus d'Andretta and Carrera, Systropus, 248
repertus Walker, Villa, 373
repeteki Paramonov, Bombylius, 81
repeteki submodestus Paramonov, Bombylius, 81
repeteki Paramonov, Heterotropus, 226
repetekianus Paramonov, Anthrax, 441
repletus Bezzi, Litorrhychus, 428
resplendens Brunetti, Dischistus, 107
restitutus Walker, Villa, 373
resurgens Walker, Villa, 374
retardatus Becker, Anastoechus, 86
reticulata Loew, Exoprosopa, 421
retracta Bezzi, Exoprosopa, 421
retrogradus Becker, Anastoechus, 86
retrorsa Brunetti, Exoprosopa, 421
retrorsus Melander, Oligodranes, 221
rex Osten Sacken, Rhynchanthrax, 408
rex Curran, Systropus, 248
rhadamanthus Meigen, Micomitra, 430
rhaeba Melander, Mythicomyia, 269
rhea Osten Sacken, Exoprosopa, 419
rhodesiana Hesse, Chasmoneura, 110
rhodesiana Hesse, Systropus, 248
rhodesianus Hesse, Systoechus, 84
rhodesiensis Hesse, Anthrax, 442
rhodesiensis Hesse, Exoprosopa, 421
rhodesiensis Hesse, Gonarthrus, 112
rhodesiensis Hesse, Plesiocera, 338
rhodius Loew, Bombylius, 80
rhomboidalis Hesse, Bombylius, 81
rhomphaes Séguy, Phthiria, 199
rhymnica Eversmann, Exoprosopa, 420
ricardoi Bezzi, Litorrhynchus, 428
ridibundus A. Costa, Platypygus, 262
rieki Paramonov, Antonia, 321; fig. 769
rileyi Coquillett, Mythicomyia, 269
rivularis Griffin, Exoprosopa, 419, 420
rivularis Meigen, Exoprosopa, 420
rivularis munda Loew, Exoprosopa, 420
rivulosa Becker, Exoprosopa, 420
rjabovi Paramonov, Conophorus, 158
robertii Macquart, Exoprosopa, 421
robertsi Paramonov, Bombylius, 82
robertsi Paramonov, Ligyra, 434
robiginosa Melander, Mythicomyia, 269
robustalis Hesse, Anthrax, 442
robustum Osten Sacken, Heterostylum, 13, 140; figs. 55, 283, 504, 559, 834, 835, 836
robustus Hull, Bromoglycis, 130; figs. 38, 271
robustus Cresson, Geron, 206
robustus Johnson and Johnson, Poecilanthrax, 398
robustus Bezzi, Systoechus, 84
roddi Paramonov, Antonia, 321
roederi Curran, Neodiplocampta, 390
roepkei de Meijere, Systropus, 248
rogenhoferi Nowicki, Lomatia, 311
rogenhoferi caspica Paramonov, Lomatia, 311
rogersi Osten Sacken, Systropus, 248
rohdendorfi Paramonov, Thyridanthrax, 406
rohweri Cockerell, Pachysystropus, fossil, 59
rossicus Paramonov, Conophorus, 158
rostrata Loew, Litorrhynchus, 428
rostratus Loew, Litorrhynchus, 428
rostrifera Jaennicke, Exoprosopa, 419
rottensis Meunier, Systropus, fossil, 59
rotundifacies Paramonov, Thyridanthrax, 406
rotundipennis Melander, Glabellula, 271
rubella Bezzi, Exoprosopa, 421
rubens Coquillett, Eucessia, 340; figs. 374, 635, 653
rubescens Bezzi, Exoprosopa, 421
rubicunda Bezzi, Chionamoeba, 344
rubicunda Hesse, Exoprosopa, 421
rubicundus Bezzi, Anastoechus, 86
rubicundus Bezzi, Dischistus, 107
rubicundus Bezzi, Doliogethes, 111
rubicundus Melander, Paracosmus, 298; fig. 119
rubidus Roberts, Systoechus, 84
rubiginipennis Macquart, Anthrax, 441
rubiginosa Macquart, Exoprosopa, 419
rubricosus Wiedemann, Systoechus, 84
rubrifera Bigot, Comptosia, 354
rubriventris Paramonov, Anastoechus, 86
rubriventris Bigot, Bombylius, 82
rubriventris Paramonov, Oestranthrax, 411
rubriventris Paramonov, Oligodranes, 221
rubriventris Paramonov, Thyridanthrax, 406
rudebecki Hesse, Systoechus, 84
rufa Wiedemann, Lomatia, 311
rufescens Hesse, Bombylius, 81
rufescens Walker, Ligyra, 434
rufiarticularis Hesse, Systoechus, 84
ruficeps Macquart, Bombylius, 81
ruficeps Macquart, Villa, 373
ruficollis Bigot, Villa, 373
ruficollis Saunders, Exoprosopa, 421
ruficornis Bezzi, Adelidea, 121
ruficornis Macquart, Anisotamia, 312
ruficornis, Bezzi, Corsomyza, 166
rufidulus Bowden, Systropus, 248
rufifemur Enderlein, Systropus, 248
rufilabris Macquart, Bombylius, 82
rufipes Macquart, Geron, 206
rufipes Becker, Hemipenthes, 386
rufipes Macquart, Petrorossia, 346
rufirostris Bezzi, Dischistus, 107
rufiventris Macquart, Bombylius, 81
rufiventris Blanchard, Exoprosopa, 419
rufiventris Hesse, Plesiocera, 338
rufiventris Osten Sacken, Systropus, 248
rufoanalis Macquart, Bombylius, 80
rufoantennatus Becker, Bombylius, 81
rufolimbatus Bigot, Lepidanthrax, 388
rufoscutellata Jaennicke, Comptosia, 353
rufotibialis Johnson and Johnson, Lordotus, 115
rufulum Zaitsev, Anthrax, 441
rufulus Osten Sacken, Conophorus, 158
rufum Olivier, Heterostylum, 140
rufus Macquart, Bombylius, 81
rugosus Bezzi, Systropus, 248
ruizi Edwards, Bombylius, 80
rungsi Timon-David, Cytherea, 148
rustica Rossi, Cyllenia, 283
rustica Loew, Phthiria, 199
rutila Pallas, Exoprosopa, 420
rutiloides Bezzi, Exoprosopa, 421
rutilus Walker, Bombylius, 82

S

sabaea Meigen, Exoprosopa, 420
sabaea Fabricius, Lomatia, 311; figs. 354, 701, 903, 904, 905
sabinus Osten Sacken, Villa, 372
sabulina Becker, Exoprosopa, 421
sabulonis Becker, Chionamoeba, 344; figs. 387, 645, 651
sabulonis rubicunda Bezzi, Chionamoeba, 344
sabulosa Coquillett, Chrysanthrax, 378
sabulosus Paramonov, Heterotropus, 226
sabulosus nigritarsis Engel, Heterotropus, 226
sackeni Williston, Exoprosopa, 419
sackeni Williston, Heterostylum, 140
sackenianus Williston, Villa, 373
sackenii Johnson and Maughan, Conophorus, 158
sackenii Coquillett, Poecilanthrax, 398
sackenii monticola Johnson and Johnson, Poecilanthrax, 27, 398
sacki Paramonov, Anthrax, 441
sagata Loew, Hemipenthes, 386
salebrosus Painter, Villa, 372
sallei A. Costa, Systropus, 248
salmayensis Efflatoun, Phthiria, 199
salpinx Melander, Mythicomyia, 269
salticola Hesse, Lomatia, 311
salticolus Hesse, Systoechus, 84
salutaris Austen, Thyridanthrax, 29, 406
samarkandicus Paramonov, Thyridanthrax, 406
sanctipauli Macquart, Exoprosopa, 419
sanguineus Bezzi, Systropus, 248
sapphirina Hull, Deusopora, 381; figs. 173, 277, 664
sardii Theobald, Praecytherea, fossil, 59
sardous Macquart, Villa, 373
saskae Szilady, Exoprosopa, 421
satanas Becker, Villa, 373
satellitia Walker, Villa, 374
saturatum Bezzi, Anthrax, 442
satyrus Rossi, Anthrax, 441
satyrus Fabricius, Ligyra, 434
sauteri Enderlein, Systropus, 248
sauvipennis Macquart, Exoprosopa, 420
scabrirostris Bezzi, Systoechus, 84
scalaris Melander, Aphoebantus, 324
scalaris Bezzi, Exoprosopa, 420
scaligera Bezzi, Exoprosopa, 421
scapularis Melander, Oligodranes, 221
scapulata Melander, Mythicomyia, 269
scapulatus Melander, Oligodranes, 221
sceliphronina Séguy, Petrorossia, 346
schineri Nowicki, Cytherea, 148

schineri Enderlein, Systropus, 248
schlingeri Hall, Aphoebantus, 324
schmidti Karsch, Exoprosopa, 421
scintillans Brunetti, Bombylius, 82
scitula Coquillett, Chrysanthrax, 378
scolopax Osten Sacken, Phthiria, 199
scopulicornis Cockerell, Acreotrichites, fossil, 59
scriptus Say, Anthrax, 441
scriptus Coquillett, Aphoebantus, 324
scrobiculata Loew, Villa, 372
scrutatus Wiedemann, Villa, 373
scutellaris Loew, Amictus, 286
scutellaris Melander, Empidideicus, 275
scutellaris Wiedemann, Phthiria, 199
scutellaris Thomson, Sisyromyia, 98
scutellaris Wiedemann, Systoechus, 84
scutellata Coquillett, Mythicomyia, 269
scutellatus Meigen, Aphoebantus, 324
scutellatus Macquart, Systoechus, 84
scylla Osten Sacken, Hemipenthes, 386
scylla Osten Sacken, Villa, 373
secunda Paramonov, Diplocampta, 384; figs. 195, 391, 671, 674
secundus Cockerell, Megacosmus, fossil, 59
secutor Walker, Lepidophora, 238
sedophila Brunetti, Usia, 216
segetus Bowden, Systoechus, 84
seia Séguy, Cyrtosia, 256
sejungendus Rondani, Villa, 373
selene Osten Sacken, Anthrax, 441
selene Osten Sacken, Thyridanthrax, 406
semialba Wiedemann, Exoprosopa, 420
semialbus Painter, Systropus, 248
semiargentea Macquart, Cytherea, 148
semiargyrea Strobl, Cytherea, 148
semiater Meigen, Hemipenthes, 386
semicinctus Bigot, Villa, 373
semiclara Hesse, Lomatia, 311
semiflavida Becker, Exoprosopa, 420
semifulvipes Painter, Villa, 372
semifura Sack, Anthrax, 441
semifuscata Brunetti, Ligyra, 434
semifuscata Brunetti, Villa, 374
semifuscus Meigen, Bombylius, 81
semifuscus Séguy, Geron, 207
semifuscus Engel, Thyridanthrax, 406; fig. 218
semihyalinus Meijere, Villa, 374
semiiscita Walker, Anthrax, 442
semilautus Hesse, Thyridanthrax, 406
semilimpidus Wiedemann, Villa, 374
semilucida Walker, Exoprosopa, 421
semilugens Philippi, Thyridanthrax, 406
semimacula Walker, Villa, 374
seminiger Becker, Bombylius, 80
seminiger Loew, Hemipenthes, 386
seminigra Loew, Hemipenthes, 386
semirufa Sack, Anthrax, 441
semirufa Sack, Chionamoeba, 344
semirufella Hesse, Conomyza, 339
semirufella karoonana Hesse, Conomyza, 339
semirufus Loew, Bombylius, 80
semiscitus Walker, Anthrax, 442
semitincta Schiner, Villa, 373
semitristis Philippi, Thyridanthrax, 406
senecio Loew, Villa, 373
senegalensis Macquart, Bombylius, 81
senegalensis Macquart, Exoprosopa, 421
senex Wiedemann, Acanthogeron, 90; figs. 2, 237, 467, 762, 776, 784

senex deses Meigen, Acanthogeron, 90
senex violaceipes Strobl, Acanthogeron, 90
senex Rondani, Bombylius, 81
senex violaceipes Strobl, Bombylius, 81
senex Melander, Heterotropus, 226
seniculus Wiedemann, Exoprosopa, 420
seniculus Philippi, Systoechus, 84
senilis Jaennicke, Bombylius, 81
senilis Klug, Exoprosopa, 420
senilis Fabricius, Geron, 206
senilis Weidemann, Sparnopolius, 124
separata Efflatoun, Cyrtosia, 256
separata Walker, Paravilla, 392
separatus Becker, Acanthogeron, 90
separatus Melander, Aphoebantus, 324
sepia Hull, Neodiplocampta, 390; figs. 189, 657
septentrionis Roberts, Ligyra, 434
septoptera Hesse, Lomatia, 311
seratus Aldrich, Aphoebantus, 324
serena Becker, Cyrtosia, 256
serenus Becker, Empidideicus, 275
seriatus Wiedemann, Doliogethes, 111
seriatus pullatus Hesse, Doliogethes, 111
seriatus puniceus Hesse, Doliogethes, 111
seriatus vagens Hesse, Doliogethes, 111
sericans Macquart, Bombylius, 82
sericatus Hesse, Anastoechus, 86
sericeus Meigen, Systoechus, 84
sericeus Bezzi, Systropus, 248
sericophorus Hesse, Anastoechus, 86
sericophorus congruens Hesse, Anastoechus, 86
sericosoma Hesse, Lomatia, 311
seriepunctatus Osten Sacken, Anthrax, 441
serpentata Loew, Exoprosopa, 420
serpentiger Walker, Comptosia, 354
serpentiger Walker, Villa, 374
serpentina Osten Sacken, Dipalta, 29, 382; figs. 186, 370, 662, 665, 944, 945
serratus Coquillett, Conophorus, 158
serva Bezzi, Micomitra, 430
servillei Macquart, Bombylius, 81
servillei Macquart, Ligyra, 434
sessilis Bezzi, Eurycarenus, 143
setaria Hesse, Pteraulax, 328; figs. 123, 382
setigera Philippi, Nectaropota, 100; figs. 20, 253, 464A, 1027, 1028, 1029
setosa Paramonov, Cytherea, 148
setosa Paramonov, Lyophlaeba, 356
setosus Loew, Amictus, 286
setosus Loew, Anastoechus, 86
setosus Cresson, Oligodranes, 221
setosus Cockerell, Protepacmus, fossil, 59
sexfasciatus Wiedemann, Villa, 26, 374
sexnotatus Macquart, Villa, 374
seyrigi Séguy, Toxophora, 236
shafiki Efflatoun, Amictus, 286
shah Paramonov, Bombylius, 81
sharonae Johnson and Johnson, Exoprosopa, 419
shawii Johnson, Villa, 372
shelkovnikovi Paramonov, Anthrax, 441
shelkovnikovi Paramonov, Bombylius, 80
shelkovnikovi Paramonov, Exoprosopa, 420
shelkovnikovi Paramonov, Lomatia, 311
shelkovnikovi Paramonov, Toxophora, 236
sheppardi Hesse, Systropus, 248
shibakawae Matsumura, Bombylius, 81
shirakii Paramonov, Ligyra, 434
sibiricus Becker, Anastoechus, 86
sibiricus Becker, Systoechus, 84

siccifolius Bezzi, Litorrhynchus, 428
sicula Macquart, Exoprosopa, 419
sicula Egger, Usia, 216
siderata Pallas, Exoprosopa, 420
sigma Coquillett, Oligodranes, 221
sigmoidea Bezzi, Exoprosopa, 421
signatipennis Cole, Poecilanthrax, 398
signifer Walker, Bombylius, 82
sikkimensis Enderlein, Systropus, 248
silvaticus Hesse, Systoechus, 84
silvaticus turneri Hesse, Systoechus, 84
silvestrii Bezzi, Systropus, 248
silvus Cole, Bombylius, 80
sima Osten Sacken, Exoprosopa, 419
similis Paramonov, Amictus, 286
similis Loew, Bombylius, 80
similis Kertész lapsus, Conophorus, 158
similis Coquillett, Exoprosopa, 420
similis Coquillett, Ligyra, 434
similis Hall, Paracosmus, 298
similis Coquillett, Phthiria, 199
similis Williston, Systropus, 248
similis Paramonov, Usia, 216
simillima Hesse, Exoprosopa, 421
simmillimus Hesse, Anthrax, 442
simmondsi Hesse, Phthiria, 199
simmondsi Hesse, Thyridanthrax, 406
simonyi Becker, Phthiria, 199
simplex Macquart, Anthrax, 442
simplex Loew, Conophorus, 158
simplex Wiedemann, Corsomyza, 166; figs. 64, 817, 818, 819
simplex Walker, Geron, 207
simplex Wiedemann, Lomatia, 311
simplex Loew, Systoechus, 84
simplicipennis Bezzi, Bombylius, 81
simpsoni Bezzi, Dicranoclista, 444; figs. 403, 679, 753
simson Fabricius, Anthrax, 441
simulans Austen, Bombylius, 81
simulans Hesse, Bombylius, 81
simulator Loew, Dischistus, 107
sinaiticus Efflatoun, Dischistus, 107
sinaiticus Efflatoun, Systoechus, 84
singularis Séguy, Cyrtisiopsis, 263
singularis Schiner, Diplocampta, 384
singularis Macquart, Dischistus, 107
singularis Macquart, Exoprosopa, 419
sini Cole, Villa, 372
sinuata Meigen, Anthrax, 441
sinuatifascia Macquart, Exoprosopa, 421
sinuatifascia Hardy, Ligyra, 434
sinuatifascia Macquart, Ligyra, 434
sinuatus Mikan, Bombylius, 80
sinuosa Wiedemann, Hemipenthes, 30, 386; figs. 413, 672, 678
sinuosa blanchardiana Jaennicke, Hemipenthes, 386
sinuosa jaennickeana Osten Sacken, Hemipenthes, 386
sinuosa Hesse, Lomatia, 311
sipho Paramonov, Exoprosopa, 421
sipho Melander, Oligodranes, 221
sirius Hoffmannsegg, Thyridanthrax, 406
sisyphus Fabricius, Exoprosopa, 421
sisyphus Fabricius, Ligyra, 434
siva Nurse, Exoprosopa, 421
slossonae Johnson, Anthrax, 440
smirnovi Paramonov, Anastoechus, 86
smirnovi nigrifemorata Paramonov, Anastoechus, 86

snowi Painter, Geron, 206
snowi Adams, Systropus, 248
sobria Walker, Comptosia, 356
sobricula Walker, Lomatia, erratum, 311
sobricula Walker, Villa, 374
socia Osten Sacken, Exoprosopa, 419
socius Walker, Bombylius, 82
socius Walker, Systoechus, 9, 84
sodalis Williston, Thevenemyia, 182
sodom Williston, Villa, 372
sokotrae Ricardo, Villa, 374
sola Painter, Exoprosopa, 419
solitus Walker, Systoechus, 84
solitus Walker, Villa, 373
somali Greathead, Systoechus, 9, 84
somali Oldroyd, Systoechus, 84
sonorensis Cole, Villa, 373
sorbens Melander, Mythicomyia, 269
sorbicula Walker, Lomatia, 311
sordida Loew, Exoprosopa, 419
sordidus Sack, Anthrax, 441
soror Loew, Callostoma, 150
soror Bezzi, Hyperusia, 170; fig. 62
sororculus Williston, Lordotus, 115
sororculus nigriventries Johnson and Johnson, Lordotus, 115
sororia Williston, Phthiria, 199
spadix Painter, Exoprosopa, 419
spaldingi Painter, Paravilla, 392
spathistylus Hesse, Anthrax, 442
speciosa Loew, Cytherea, 148
speciosus Hesse, Thyridanthrax, 406
spectabilis Loew, Anthrax, 442
spectrum Speiser, Exoprosopa, 421
speculifer Melander, Oligodranes, 221
speiserianus Bezzi, Oestranthrax, 411
sperryorum Melander, Aphoebantus, 324
sphegoides Walker, Systropus, 248
sphenoptera Loew, Acrophthalmyda, 176; figs. 18, 251, 437, 450, 768, 797, 798, 799
sphinx Fabricius, Ligyra, 434
spiloneura Bezzi, Exoprosopa, 420
spiloptera Bezzi, Lomatia, 311
spilopterus Wiedemann, Villa, 373
spinibarbus Bezzi, Bombylius, 81
spinipes Thomson, Bombylius, 82
spinithorax Bezzi, Systoechus, 84
spinosus Meunier, Palaeoamictris, fossil, 59
spoliata Bezzi, Exoprosopa, 420
spoliata Hesse, Exoprosopa, misid., 421
sporanthera Hesse, Nomalonia, 290
squalida Philippi, Anthrax, 441
squamea Mulsant, Exoprosopa, 420
squamifer Jaennicke, Villa, 373
squamigera Coquillett, Villa, 372
squamiventre Edwards, Sericosoma, 154; figs. 40, 302, 502, 508, 1000, 1001, 1002
squamosus Coquillett, Aphoebantus, 324
squamosus Coquillett, Aphoebantus, 324
stackelbergi Paramonov, Anastoechus, 86
stackelbergi Paramonov, Anthrax, 441
stackelbergi Paramonov, Prorachthes, 162
stannusi Bezzi, Exoprosopa, 421
stellans Loew, Anthrax, 441
stellifer Walker, Exoprosopa, 421
stenogaster A. Costa, Petrorossia, 346
stenolopha Hesse, Tomomyza, 292
stenometaena Bezzi, Exoprosopa, 421
stenometopa Hesse, Lomatia, 311
stenoptera Hesse, Lomatia, 311
stenozona Loew, Villa, 372, 373

stepensis Paramonov, Anthrax, 441
sternurus Loew, Aphoebantus, 324
stevensoni Hesse, Exoprosopa, 421
stevensoni Hesse, Systoechus, 84
stheno Wiedemann, Villa, 373
sticticalis Hesse, Anthrax, 442
sticticus Boisduval, Bombylius, 81
stigmalis Cockerell, Geronites, fossil, 59
stigmatias Knab, Heterostylum, 140
stigmaticus Bezzi, Heterotropus, 226
stigmulus Klug, Thyridanthrax, 406; fig. 221
stoechades Jaennicke, Villa, 373
stramineus Becker, Anastoechus, 86
stramineus Wiedemann, Anastoechus, 86
strenua Loew, Exoprosopa, 421
stria Walker, Comptosia, 354
striata Pallas, Cytherea, 148
striata Bischof, Legnotomyia, 163
striatifrons, Becker, Bombylius, 81
striatus Becker, Bombylius, 81
striatus Painter, Lordotus, 114
stricticus Klug, Anthrax, 441
striligatus Loew, Amictus, 286
stuckenbergi Hesse, Cyrtosia, 256
studyi Enderlein, Systropus, 248
stupida Rossi, Micomitra, 430
stupida pharao Paramonov, Micomitra, 430
stylata Brunetti, Exoprosopa, 421
stylicornis Macquart, Systoechus, 84
stylicornis Hesse, Thyridanthrax, 406
stymphalis Wiedemann, Exoprosopa, 419
suavipennis Macquart, Exoprosopa, 420
suavissima Loew, Antonia, 321; figs. 136, 137, 384, 624, 637
subacutus Hesse, Bombylius, 81
subaequalis Lynch Arribalzaga, Villa, 373
subandina Philippi, Anthrax, 441
subannalus Walker, Villa, 374
subanthrax, Bezzi, Anthrax, 442
subarcuatus Loew, Hemipenthes, 386
subater Paramonov, Eurycarenus, 143
subauratus Loew, Geron, 206
subcaliga Hesse, Lomatia, 311
subcinctus Wiedemann, Bombylius, 80
subcingulatus Enderlein, Systropus, 248
subcontiguus Hesse, Systoechus, 84
subcostalis Paramonov, Aphoebantus, 324
suberosus Bezzi, Litorrhynchus, 428
subfascia Walker, Exoprosopa, 419
subfasciata Engel, Exoprosopa, 420; fig. 197
subflavofemoratus Rubio, Geron, 206
subflavus Painter, Parabombylius, 97
subluna Walker, Bombylius, 81
submacrocerus Paramonov, Anthrax, 441
submacula Walker, Villa, 374
submixtus Séguy, Systropus, 248
submodestus Paramonov, Bombylius, 81
subnitens Loew, Phthiria, 199
subnotata Walker, Anthrax, 441
subnotata Wiedemann, Anthrax, 441
subobscura Hardy, Villa, 374
subperspicillaris Bezzi, Thyridanthrax, 406
subsenex Walker, Comptosia, 354
subsenex Walker, Villa, 374
subsinuata Oldroyd, Diplocampta, 384
subtropicalis Hesse, Gonarthrus, 112
subulinus Bowden, Systoechus, 84
subvarius Johnson, Bombylius, 80
succandidus Roberts, Bombylius, 82
succedens Walker, Villa, 374
succincta Coquillett, Hemipenthes, 386

sudanensis Becker, Heterotropus, 226
suffusa Klug, Exoprosopa, 420
suffusipennis Hesse, Anthrax, 442
suffusipennis Bezzi, Exoprosopa, 421
suffusipennis Brunetti, Ligyra, 434
suffusipunctis Hesse, Anthrax, 442
suffusus Walker, Bombylius, 80
suffusus Walker, Villa, 373
sugens Melander, Mythicomyia, 269
sulcifacies Hesse, Isotamia, 426
sulphurea Loew, Phthiria, 18, 42, 199; fig. 6
sulphureus Fabricius, Bombylisoma, 108
sulphureus Paramonov, Heterotropus, 226
sulphureus Mikan, Systoechus, 9, 84
sulphureus aurulentus Wiedemann, Systoechus, 84
sulphureus convergens Loew, Systoechus, 84
sulphureus dalmatinus Loew, Systoechus, 84
sulphureus orientalis Zakavatkin, Systoechus, 84
sumatrensis de Meijere, Ligyra, 434
sumptuosum White, Marmasoma, 241; figs. 89, 313, 542, 578, 878, 879, 880
superba Loew, Lomatia, 311
superbus Engel, Oligodranes, 221
supina Coquillett, Villa, 372
suprema Becker, Villa, 373
surinamensis Rondani, Ligyra, 434
suspensus Bezzi, Litorrhynchus, 428
suzukii Matsumura, Anastoechus, 86
suzukii Matsumura, Systropus, 248
svenhedini Paramonov, Villa, 373
swani Paramonov, Docidomyia, 319
sylvana Fabricius, Comptosia, 354
syndesmus Coquillett, Parabombylius, 97
syphax Fabricius, Villa, 373
syrdarpensis Paramonov, Anastoechus, 86
syriaca Loew, Cytherea, 148
syriaca Paramonov, Usia, 216
syriacus Villeneuve, Acanthogeron, 90
syriacus Paramonov, Conophorus, 158
syrphoides Walker, Antonia, 321
syrticola Hesse, Nomalonia, 290
syrtis Coquillett, Paravilla, 392
sytshuana Paramonov, Exoprosopa, 420
sytshuanensis Paramonov, Bombylius, 81

T

tabaninus Bezzi, Thyridanthrax, 406
tabanoides Bezzi, Exoprosopa, 421
tadzhikorum Paramonov, Anthrax, 441
tadzhikorum Paramonov, Toxophora, 236
taeniolata A. Costa, Usia, 216
tagax Melander, Mythicomyia, 269
talboti Séguy, Acanthogeron, 90
talyshensis Zaitsev, Conophorus, 158
tamerlan Portschinsky, Exoprosopa, 420
tamerlan bezzi Paramonov, Exoprosopa, 420
tanbarkeris Painter and Hall, Poecilanthrax, 398
tankerinus Bigot, Villa, 373
tantalus Fabricius, Ligyra, 434
tantilla Coquillett, Chrysanthrax, 378
tardus Coquillett, Aphoebantus, 324
tashkentica Paramonov, Villa, 373
taurica Becker, Cytherea, 148
taurica Paramonov, Lomatia, 311

taurica completa Paramonov, Lomatia, 311
tauricus Paramonov, Platypygus, 262
tavrizi Paramonov, Bombylius, 80
tegminipennis Say, Poecilanthrax, 27, 398
telamon Loew, Exoprosopa, 420
telluris Coquillett, Villa, 372
tendens Walker, Comptosia, 354; fig. 150
tenebrosa Paramonov, Ligyra, 434; figs. 202, 425, 704, 731, 962, 963
tenebrosum Paramonov, Anthrax, 441
tenella Loew, Plesiocera, 338
tener Becker, Amictus, 286
tenera Loew, Lomatia, 311
tenthes Melander, Mythicomyia, 269
tenuicornis Macquart, Bombylius, 82
tenuirostris Roberts, Bombylius, 82
tenuirostris Hesse, Gonarthrus, 112
tenuirostris Macquart, Villa, 373
tenuis Frey, Cyrtosia, 256
tenuis Macquart, Exoprosopa, 420
tenuis Walker, Geron, 207
tenuis Enderlein, Systropus, 248
tenuis Walker, Villa, 373
tephroleuca Loew, Exoprosopa, 420
tephroleucus Loew, Bombylius, 81
terminalis Brunetti, Bombylius, 82
terminalis Wiedemann, Ogcodocera, 314
terminalis Wulp, Villa, 374
terminatus Becker, Bombylius, 81
terminus Walker, Villa, 374
ternarius Bezzi, Thyridanthrax, 406
ternarius speciosus Hesse, Thyridanthrax, 406
terrenus Coquillett, Villa, 372
tertiaria Cockerell, Dolichomyia, fossil, 59
tertiarius Handlirsch, Hemipenthes, fossil, 59
tesquorum Becker, Systoechus, 84
tessellatus Vollenhoven, Systropus, 248
tessmanni Enderlein, Systropus, 248
testacea Macquart, Phthiria, 199
testaceiventris Paramonov, Bombylius, 81
testaceiventris bergi Paramonov, Bombylius, 81
testaceus Macquart, Thyridanthrax, 406
testea Melander, Melanderella, fossil, 59
tetragramma Bezzi, Cyrtosia, 256
tetragramma canariensis Engel, Cyrtosia, 256
tetraspilus Hesse, Anthrax, 442
tewfiki Paramonov, Exoprosopa, 421
tewfiki Paramonov, Heterotropus, 226
tewfiki Efflatoun, Usia, 216
texana Curran, Exoprosopa, 419
texana Painter, Rhynchanthrax, 408
texanus Painter, Bombylius, 80
thlipsomyzoides Jaennicke, Phthiria, 199; fig. 314
thomae Fabricius, Exoprosopa, 421
thoracica Bezzi, Exoprosopa, 421
thoracicus Fabricius, Bombylius, 80
thornei Hesse, Apolysis, 219
thornei Hesse, Bombylius, 81
thyridophora Bezzi, Cytherea, 148
thyridophora Bezzi, Ligyra, 434
thyridus Hesse, Thyridanthrax, 406
thysanomela Hesse, Lomatia, 311
tibialis Walker, Bombylius, 82
tibialis Loew, Lomatia, 311
tibialis Coquillett, Mythicomyia, 269
tiburonensis Cole, Exoprosopa, 419
tigrinus Austen, Amictus, 286

tigrinus De Geer, Anthrax, 20, 25, 42, 441; figs. 6, 712, 734
timberlakei Melander, Aphoebantus, 324
timberlakei Melander, Apolysis, 219
timurensis Paramonov, Thyridanthrax, 406
tinctella Hesse, Lomatia, 311
tinctellipennis Hesse, Pantostomus, 293; figs. 111, 356, 602, 614
tinctellipennis Hesse, Phthiria, 199
tinctipenne Hesse, Crocidium, 185
tinctipennis Hesse, Bombylius, 81
tinctipennis thornei Hesse, Bombylius, 81
tinctus Walker, Bombylius, 82
tinctus Thomson, Villa, 373
tipuloides Westwood, Systropus, 248
tisiphone Loew, Lomatia, 311
titanus Megerle, Exoprosopa, 420
titubans Osten Sacken, Exoprosopa, 419
togatus Melander, Oligodranes, 221
tollini Loew, Exoprosopa, 421
tollini Bezzi, Litorrhynchus, 428
tollini Loew, Litorrhynchus, 428
tolteca Painter, Phthiria, 199
tomentosa Engel, Usia, 216
tomentosa Becker, Villa, 373
tomentosus Melander, Epacmus, 326
torquatus Loew, Bombylius, 81
trabalis Loew, Exoprosopa, 419
transatlanticus Philippi, Dischistus, 107
transandina Edwards, Lyophlaeba, 356
transcaspia Becker, Cytherea, 148
transcaspica Paramonov, Usia, 216
transcaspicus Paramonov, Anthrax, 441
transcaspicus Paramonov, Aphoebantus, 324
transcaspicus Paramonov, Dischistus, 107
transcaspicus Paramonov, Hemipenthes, 386
transcaspicus Paramonov, Thyridanthrax, 406
transcaucasica Paramonov, Exoprosopa, 420
transiens Bezzi, Ligyra, 434
transiens Bezzi, Tyridanthrax, 29, 406
transitus Coquillett, Aphoebantus, 324
transitus Hesse, Bombylius, 81
translatus Walker, Villa, 373
transvaalensis Hesse, Geron, 207
transvaalensis Hesse, Hyperusia, 170
transvaalensis Hesse, Lomatia, 311
transvaalensis Hesse, Systoechus, 84
transversa Brunetti, Lepidanthrax, 388
travassosi d'Andretta and Carrera, Toxophora, 236
triangularis Bezzi, Thyridanthrax, 406
triatomus Hesse, Anthrax, 442
tricellata Macquart, Comptosia, 354
tricellula Cole, Paravilla, 28, 392
tricellulata Hesse, Corsomyza, 166
trichorhoea Loew, Legnotomyia, 163; figs. 255, 469, 474, 800, 801, 802, 803
trichurus Pallas, Bombylius, 81
tricolor Guerin-Meneville, Bombylius, 82
tricolor Macquart, Exoprosopa, 421
tricolor orientalis Bezzi, Exoprosopa, 421
tricolor Bezzi, Phthiria, 199
tricuspis Enderlein, Systropus, 248
tridentata Paramonov, Isotamia, 426
tridentata Hull, Thevenemyia, 182; figs. 19, 248
trifaria Becker, Cytherea, 148
trifaria Melander, Mythicomyia, 269

trifasciata Meigen, Anthrax, 20, 441
trifasciata leucogaster Wiedemann, Anthrax, 20, 442
trifidus Melander, Oligodranes, 221
trifiguratus Walker, Villa, 373
triformis Melander, Mythicomyia, 269
trigonalis Bezzi, Systropus, 248
trigonus Bezzi, Dischistus, 107
trigradata Hesse, Exoprosopa, 421
triguttellus Hesse, Anthrax, 442
trilineatum Hesse, Onchopelma, 257; figs. 352, 580, 590
triloculina Hesse, Exoprosopa, 421
trimacula Walker, Villa, 373
trimaculata Becker, Anthrax, 442
trimaculata Engel, Anthrax, 442
trimaculata Macquart, Anthrax, 20, 442
trimaculata Sack, Anthrax, 442
trimaculatus Bezzi, Anthrax, 442
trimaculatus Macquart, Villa, 373
trimaculatus Wulp, Villa, 374
trinotata Dufour, Anthrax, 442
trinotata areolata Abreu, Anthrax, 442
tripartita Hesse, Exoprosopa, 421
triplex Bezzi, Exoprosopa, 421
tripolitanus Paramonov, Acanthogeron, 90
tripudians Bezzi, Bombylius, 81
tripunctata Wiedemann, Anthrax, 442
tripunctata Wulp, Anthrax, 441, 442
tripunctata Edwards, Comptosia, 354
tripunctatus Macquart, Bombylius, 82
tripunctatus Macquart, Doliogethes, 111
trisignata Engel, Cyrtosia, 256
trisignatus Portschinsky, Anastoechus, 86; figs. 8, 239, 447, 465, 778, 780
trisignatus retrogradus Becker, Anastoechus, 86
trisignatus werneri Paramonov, Anastoechus, 86
trisinatus Hesse, Anthrax, 442
tristis Wulp, Exoprosopa, 421
tristis Melander, Mythicomyia, 269
tristis Bigot, Phthiria, 199
tristis Séguy, Toxophora, 236
trivergatus Hesse, Doliogethes, 111
trivincula Roberts, Villa, 374
trivittata Bezzi, Toxophora, 236
trivittatus Macquart, Villa, 374
trixus Bowden, Anthrax, 442
trochilides Williston, Geron, 206
trochilus Coquillett, Geron, 206
trochilus Coquillett, Oligodranes, 221
troglodyta Fabricius, Villa, 374
tropicalis Bezzi, Petrorossia, 346
trotteri Bezzi, Heterotropus, 226
truquii Rondani, Exoprosopa, 420
tubicen Melander, Mythicomyia, 269
tuckeri Hesse, Bombylius, 81
tuckeri Bezzi, Exoprosopa, 421
tumescens Melander, Mythicomyia, 269
tumidifrons Bezzi, Systoechus, 84
turanica Paramonov, Cytherea, 148
turanica Paramonov, Exoprosopa, 419
turanicus Paramonov, Anastoechus, 86
turanicus Paramonov, Anthrax, 442
turanicus Paramonov, Bombylius, 81
turbata Coquillett, Chrysanthrax, 378
turbidus Loew, Villa, 374
turcmenicum Paramonov, Anthrax, 442
turcmenicus Paramonov, Bombylius, 81
turcomana Portschinsky, Exoprosopa, 419

turcomanus Paramonov, Thyridanthrax, 406
turkestanica Paramonov, Cyllenia, 283
turkestanica Paramonov, Cytherea, 148
turkestanica Paramonov, Exoprosopa, 420
turkestanica mongolica Paramonov, Exoprosopa, 420
turkestanica Paramonov, Plesiocera, 338
turkestanica Paramonov, Toxophora, 236
turkestanica angusta Paramonov, Toxophora, 236
turkestanicus Paramonov, Anastoechus, 86
turkestanicus Paramonov, Conophorus, 158
turkestanicus Paramonov, Emipideicus, 275
turkmenica Paramonov, Cytherea, 148
turkmenica Paramonov, Usia, 217
turkmenicola Paramonov, Anthrax, 442
turkmenicus Paramonov, Aphoebantus, 324
turkmenicus Paramonov, Dischistus, 107
turkmenicus flavisetis Paramonov, Dischistus, 107
turkmenorum Paramonov, Anastoechus, 86
turkmenorum Paramonov, Anthrax, 442
turkmenorum Paramonov, Platypygus, 262
turneri Hesse, Bombylius, 81
turneri Hesse, Corsomyza, 166
turneri Hesse, Empidideicus, 275; figs. 328, 591, 594
turneri Hesse, Geron, 207
turneri Hesse, Gonarthrus, 112
turneri melalophus Hesse, Gonarthrus, 112
turneri Hesse, Systoechus, 84
turneri Hesse, Villa, 374
tylopelta Hesse, Othniomyia, 118; figs. 264, 470, 475
tysiphone Loew, Lomatia, 311

U

udei Enderlein, Systropus, 248
ultraareolata Abreu, Anthrax, 442
umbra Walker, Exoprosopa, 421
umbrifer Walker, Exoprosopa, 421
umbripennis Paramonov, Bombylius, 80
umbripennis Bezzi, Geron, 207
umbripennis Loew, Phthiria, 199
umbrosa Loew, Exoprosopa, 421
una Oldroyd, Villa, 373
uncata Melander, Mythicomyia, 269
uncinus Loew, Hemipenthes, 386
unctus Loew, Thyridanthrax, 406
undans Walker, Exoprosopa, 421
undatus Mikan, Bombylius, 81
undatus diagonalis Wiedemann, Bombylius, 81
undatus Wiedemann, Bombylius, 81
unicinctus Bigot, Thyridanthrax, 406
unicinctus Guerin, Thyridanthrax, 406
unicolor Jaennicke, Cyllenia, 283
unicolor Loew, Dischistus, 107
unicolor Strobl, Glabellula, 271
unicolor Bezzi, Phthiria, 199
unicolor Becker, Systoechus, 84
unicolor Becker, Thyridanthrax, 406
unicolor Loew, Usia, 216
unifasciata Ricardo, Exoprosopa, 421
unifasciatus Macquart, Villa, 374
uniformis Paramonov, Bombylius, 81
unimaculata Coquillett, Phthiria, 199
uniplaga Hesse, Lomatia, 311

unistriata Engel, Thyridanthrax, 406
uralensis Paramonov, Anthrax, 442
uroganus Hesse, Thyridanthrax, 406
ursula Melander, Aphoebantus, 324
uruguayi Macquart, Exoprosopa, 419
ushinskii Paramonov, Bombylius, 81
ussuriensis Paramonov, Conophorus, 158
ussuriensis Paramonov, Ligyra, 434
utahensis Johnson and Johnson, Exoprosopa, 419
utahensis Maughan, Thyridanthrax, 406
uvarovi Paramonov, Villoestrus, 409; figs. 410, 682, 685
uzbekorum Paramonov, Bombylius, 81
uzbekorum Paramonov, Exoprosopa, 419, 420

V

vagabundus Meigen, Bombylius, 82
vacans Coquillett, Villa, 372
vagans Meigen, Bombylius, 81
vagans Loew, Phthiria, 199
vagans pallescens Engel, Phthiria, 199
vagans Loew, Thyridanthrax, 406; fig. 220
vagans beckeri Paramonov, Thyridanthrax, 406
vagans contrarius Becker, Thyridanthrax, 406
vagans parvus Paramonov, Thyridanthrax, 406
vagans unistriata Engel, Thyridanthrax, 406
vagans Becker, Usia, 216
vagens Hesse, Doliogethes, 111
valdezi Bezzi, Systropus, 248
valdivianus Philippi, Bombylius, 80
valdivianus Rondani, Bombylius, 80
valida Wiedemann, Bryodemina, 316; figs. 207, 375, 695, 898, 899
validicornis Bezzi, Villa, 374
validus Loew, Amictus, 286; figs. 109, 357, 611, 691, 752, 779, 890, 891, 892
validus Loew, Bombylius, 80
validus Loew, Conophorus, 158
validus Bezzi, Systoechus, 84
vallicola Marston, Anthrax, 440
vana Coquillett, Chrysanthrax, 378
vanduzeei Cole, Amphicosmus, 295
vanduzeei Cole, Villa, 372
vandykei Coquillett, Dicranoclista, 444; fig. 160
vansoni Hesse, Bombylius, 81
varia Fabricius, Anthrax, 442
variabilis Loew, Bombylius, 80
variata Coquillett, Chrysanthrax, 378
varicolor Bigot, Anthrax, 441
varicolor Macquart, Exoprosopa, 419
variegata Paramonov, Lomatia, 311
variegata Austen, Phthiria, 199
variegatus Meigen, Amictus, 286
variegatus De Geer, Bombylius, 80
variegatus Macquart, Dischistus, 107
variegatus Jaennicke, Thyridanthrax, 406
variegatus Pallas, Thyridanthrax, 406
varinervis Macquart, Exoprosopa, 420
varipecten Bezzi, Anastoechus, 86
varipennis Williston, Sphenoidoptera, 284; figs. 235, 364, 601, 610, 1024, 1025, 1026
varipennis, Williston, Toxophora, 236
varipennis Macquart, Villa, 373

varipennis Roberts, Villa, 374
varipes Austen, Phthiria, 199
varipes Edwards, Systropus, 248
varius Coquillett, Aphoebantus, 324
varius Fabricius, Bombylius, 80
varius Painter and Hall, Poecilanthrax, 398
vasatus Melander, Aphoebantus, 324
vassiljevi Paramonov, Exoprosopa, 420
vasta Coquillett, Toxophora, 236
vastitatis Cole, Villa, 373
vastus Coquillett, Paravilla, 392
velox Loew, Anthrax, 442
velox White, Anthrax, 442
velutinus Meigen, Hemipenthes, 29, 30, 386; fig. 187
venosa Wiedemann, Exoprosopa, 421
venosa Bigot, Thevenemyia, 182
venosus Meigen, Bombylius, 80
venosus Mikan, Bombylius, 81
ventrale Bezzi, Anthrax, 442
ventralis Loew, Bombylius, 80
ventricosus Bezzi, Systoechus, 84
ventrimacula Doleschall, Ligyra, 434
ventruosa Loew, Villa, 373
venus Karsch, Ligyra, 434
venustulus Austen, Thyridanthrax, 406
venustus Hesse, Gonarthrus, 112
venustus Meigen, Villa, 373
verdensis Oldroyd, Villa, 373
vernayi Hesse, Litorrhynchus, 428
verona Curran, Toxophora, 236
versfeldi Hesse, Gonarthrus, 112
versicolor Macquart, Acanthogeron, 90
versicolor Fabricius, Bombylius, 81
versicolor Painter, Geron, 206
versicolor Loew, Ligyra, 434
versicolor Fabricius, Usia, 216
vertebralis Dufour, Bombylius, 81
vespertilio Wiedemann, Exoprosopa, 419
vespertilio Séguy, Prorachthes, 162
vesperuga A. Costa, Exoprosopa, 419
vespiformis Enderlein, Systropus, 248
vestis Melander, Mythicomyia, 269
vestita Macquart, Usia, 216
vestita Walker, Villa, 372
vetula Wiedemann, Thyridanthrax, 406
vetusta Walker, Lepidophora, 238
vetustus Walker, Bombylius, 82
vetustus Meunier, Palaeogeron, fossil, 59
vexativus Painter and Hall, Poecilanthrax, 398
vicina Macquart, Usia, 216
vicinalis Hesse, Lomatia, 311
vicinalis Hesse, Thyridanthrax, 406
vicinus Brunetti, Bombylius, 82
vicinus Macquart, Bombylius, 80
vicinus Painter, Systropus, 248
vicinus Blanchard, Villa, 373
vicinus Macquart, Villa, 373
vicinus Macquart, Villa, 372
victoriana Paramonov, Docidomyia, 319
viduata Loew, Villa, 373
viduus Becker, Anastoechus, 86
viduus Walker, Bombylius, 82
viereicki Cresson, Anthrax, 441
viereicki Cresson, Triploechus, 141
vigilans Coquillett, Paravilla, 392
vilis Melander, Mythicomyia, 269
villaeformis Bezzi, Exoprosopa, 421
villicus Philippi, Villa, 373
villosa Klug, Anthrax, 441

villosa Bezzi, Exoprosopa, 421
villosus Paramonov, Anastoechus, 86
vinula Bezzi, Petrorossia, 346
violaceipes Strobl, Acanthogeron, 90
violaceipes Strobl, Bombylius, 81
violaceo Bezzi, Exoprosopa, 421
violacescens Enderlein, Systropus, 248
virescens Fabricius, Conophorus, 158; figs. 69, 72
virgata Melander, Mythicomyia, 269
virgata Austen, Phthiria, 199
virgata Osten Sacken, Toxophora, 16, 236
virgatus Austen, Amictus, 286
virgo Egger, Anthrax, 442
virgo pedemontana Griffin, Anthrax, 442
virgo pilosula Strobl, Anthrax, 442
virgo Bezzi, Ligyra, 434
virgo Melander, Mythicomyia, 269
vitre Walker, Villa, 374
vitrea Bigot, Exoprosopa, 421
vitreicosta Walker, Micomitra, 430
vitripennis Loew, Dischistus, 107
vitripennis Brunetti, Exoprosopa, 421
vitripennis Loew, Geron, 206
vitripennis Loew, Geron, 206
vitripennis Bezzi, Toxophora, 236
vitripennis Loew, Villa, 26, 374
vitripennis Philippi, Villa, 373
vittata Edwards, Comptosia, 354
vittata Ricardo, Ligyra, 434
vittata, Phthiria, 199
vittatus Coquillett, Aphoebantus, 324
vittatus Curran, Parabombylius, 97
vittatus Painter, Parabombylius, 97
vittipes Bezzi, Dichistus, 107
vittipes Bezzi, Doliogethes, 111
vittiventris Coquillett, Phthiria, 199
vlasovi Paramonov, Anastoechus, 86
vlasovi albicans Paramonov, Anastoechus, 86
vlasovi Paramonov, Bombylius, 81
vlasovi Paramonov, Exoprosopa, 420
volitans Wiedemann, Anthrax, 442
volitans Wiedemann, Exoprosopa, 420
volucer Hesse, Bombylius, 81
vulcanus Bezzi, Ligyra, 434
vulgaris Philippi, Lyophlaeba, 356
vulgaris Philippi, Phthiria, 199
vulgaris Loew, Systoechus, 10, 39, 84; figs. 3, 236, 442, 444, 789, 790
vulnerata Melander, Mythicomyia, 269

vulpecula Coquillett, Aphoebantus, 324
vulpecula Philippi, Villa, 373
vulpina Sack, Anthrax, 442
vulpina Coquillett, Chrysanthrax, 378
vulpinus Wiedemann, Bombylius, 8, 38, 81; fig. 2
vulpinus desertorum Paramonov, Bombylius, 81
vulpinus palaestinus Paramonov, Bombylius, 81
vulpinus Becker, Systoechus, 84
vumbuensis Hesse, Exoprosopa, 421
vumbuensis Hesse, Gonarthrus, 112

W

wadensis Bezzi, Aphoebantus, 324
wadensis Efflatoun, Bombylius, 81
wadensis Efflatoun, Cytherea, 148
walkeri Edwards, Comptosia, 354
waltoni Hesse, Amictogeron, 209
waltoni Hesse, Systoechus, 84
watanabei Matsumura, Bombylius, 82
webberi Johnson, Hemipenthes, 386
werneri Paramonov, Anastoechus, 86
wilcoxi Painter, Hemipenthes, 386
wilkinsi Edwards, Comptosia, 354
willistoni Osten Sacken, Pantarbes, 153; figs. 83, 828, 829, 830
willistoni Curran, Systropus, 248
willistonii Coquillett, Poecilanthrax, 27, 398
willowmorensis Hesse, Anthrax, 442
willowmorensis Hesse, Gonarthrus, 112
winburni Painter, Geron, 206

X

xanthaspis Bezzi, Phthiria, 199
xanthina Painter, Exoprosopa, 419
xanthina Painter, Paravilla, 392; fig. 184
xanthinus Bezzi, Gonarthrus, 112
xanthinus Painter, Systropus, 248
xanthobase Curran, Heterostylum, 140
xanthocerus Bezzi, Bombylius, 81
xanthogaster Hesse, Apolysis, 219
xanthogramma Bezzi, Antonia, 321; figs. 977, 979, 981
xanthogrammus Hesse, Ceratolaemus, 256; figs. 334, 585, 592

xantholeucus Bowden, Gonarthrus, 112
xanthoplocamus François, Systoechus, 84
xanthothorax Efflatoun, Heterotropus, 226
xerophiles Hesse, Phthiria, 199
xerophilum Hesse, Epacmoides, 334
xerophilus Hesse, Prorachthes, 162
xerophilus Hesse, Systoechus, 10, 84
xerozous Hesse, Anthrax, 442
xylotona Germar, Anthracida, fossil, 59

Y

yamashiroensis Matsumara, Anthrax, 442
yaqui Painter, Hemipenthes, 386
yellowstonei Cole, Poecilanthrax, 398
yosemite Cresson, Thevenemyia, 182

Z

zambesiana Hesse, Exoprosopa, 421
zanoni Bezzi, Exoprosopa, 420
zarudnyi Paramonov, Bombylius, 81
zarudnyi Paramonov, Heterotropus, 81, 226
zarudnyi Becker, Mariobezzia, 174
zarudnyji Paramonov, Exoprosopa, 420
zikani d'Andretta and Carrera, Toxophora, 236
zilpa Walker, Toxophora, 236
zimini Paramonov, Anastoechus, 86
zimini Paramonov, Exoprosopa, 420
zimini Paramonov, Heterotropus, 226
zimini monticola Paramonov, Heterotropus, 226
zimini Paramonov, Oestranthrax, 411
zimini Paramonov, Usia, 217
zimmermanni Nowicki, Phthiria, 199
zinamominus Becker, Amictus, 286
zinnii Hesse, Thyridanthrax, 406
zionensis Johnson and Johnson, Poecilanthrax, 398
zona Bigot, Exoprosopa, 420
zonabriphaga Portschinsky, Anthrax, 20, 442
zonata Hesse, Exoprosopa, 421
zonus Coquillett, Lordotus, 115; fig. 46
zoutpansbergianus Hesse, Bombylius, 81
zoutpansbergianus occidentalis Hesse, Bombylius, 81
zuluensis Hesse, Aethioptilus, 272; fig. 344
zuluensis Hesse, Systropus, 248

Index to Genera and Species of Fossil Bee Flies

A

Acreotrichites scopulicornis Cockerell, 59
Alepidophora cockerelli Melander, 59
Alepidophora minor Melander, 59
Alepidophora pealei Cockerell, 59
Alomatia fusca Cockerell, 59
Amictites regiomontana Hennig, 59
Amphicosmus delicatulus Melander, 59
Anthracida xylotona Germar, 59
Apolysis magister Melander, 59

B

Bombylius sp. Berendt, 59
Bombylius depereti Meunier, 59
Bombylius sp. Schlotheim, 59

D

Dolichomyia tertiaria Cockerell, 59

G

Geron? platysoma Cockerell, 59
Geronites stigmalis Cockerell, 59
Glaesamictus hafniensis Hennig, 59

H

Hemipenthes s.l.sp. Burmeister, 59
Hemipenthes s.l.sp. Goldfuss, 59

Hemipenthes s.l.sp. Keferstein, 59
Hemipenthes s.l. gabbroensis Handlirsch, 59
Hemipenthes s.l. provincialis Handlirsch, 59
Hemipenthes s.l. tertiarius Handlirsch, 59

L

Lithocosmus coquilletti Cockerell, 59

M

Megacosmus mirandus Cockerell, 59
Megacosmus secundus Cockerell, 59
Melanderella glossalis Cockerell, 59
Melanderella testea Melander, 59

P

Pachysystropus condemnatus Cockerell, 59
Pachysystropus rohweri Cockerell, 59
Palaeoamictus spinosus Meunier, 59
Palaeogeron vetustus Meunier, 59
Paracorsomyza crassirostris Loew, 59
Phthiria oligocenica Timon-David, 59
Praecythera sardii Theobald, 59
Proglabellula electrica Hennig, 59
Protepacmus setosus Cockerell, 59
Protolomatia antiqua Cockerell, 59
Protolomatia recurrens Cockerell, 59
Protophthiria atra Melander, 59
Protophthiria palpalis Cockerell, 59

S

Systropus acourti Cockerell, 59
Systropus rottensis Meunier, 59

U

Usia atra Statz, 59

V

Verrallites cladurus Cockerell, 59

X

Xeramoeba gracilis Giebel, 59

THE LIBRARY
ST. MARY'S COLLEGE OF MARYLAND
ST. MARY'S CITY, MARYLAND 20686

72842

FOR REFERENCE

NOT TO BE TAKEN FROM THE ROOM